Structural Analysis

STRUCTURAL ANALYSIS

FIRST EDITION

UNDERSTANDING BEHAVIOR

Bryant G. Nielson
Clemson University

and

Jack C. McCormac
Clemson University

PUBLISHER	Laurie Rosatone
EDITORIAL DIRECTOR	Don Fowley
DEVELOPMENTAL EDITOR	Chris Nelson
EXECUTIVE MARKETING MANAGER	Dan Sayre
MARKET SOLUTIONS ASSISTANT	Courtney Jordan/Adriana Alecci
SENIOR CONTENT MANAGER	Valerie Zaborski
SENIOR PRODUCTION EDITOR	Laura Abrams
DESIGN DIRECTOR	Harry Nolan
SENIOR DESIGNER	Thomas Nery
SENIOR PHOTO EDITOR	Billy Ray
COVER PHOTO	©baspa/iStockphoto

This book was set in 11/14 Kepler Std Regular at Cenveo®, Inc. and printed and bound by Lightning Source, Inc.

Founded in 1807, John Wiley & Sons, Inc. has been a valued source of knowledge and understanding for more than 200 years, helping people around the world meet their needs and fulfill their aspirations. Our company is built on a foundation of principles that include responsibility to the communities we serve and where we live and work. In 2008, we launched a Corporate Citizenship Initiative, a global eff ort to address the environmental, social, economic, and ethical challenges we face in our business. Among the issues we are addressing are carbon impact, paper specifications and procurement, ethical conduct within our business and among our vendors, and community and charitable support. For more information, please visit our website: www.wiley.com/go/citizenship.

ISBN: 978-1-119-32957-2 (PBK)
ISBN: 978-1-119-32157-6 (EXALC)

Library of Congress Cataloging-in-Publication Data on File

The inside back cover will contain printing identification and country of origin if omitted from this page. In addition, if the ISBN on the back cover differs from the ISBN on this page, the one on the back cover is correct.

This book is dedicated to our spouses Kimberly Nielson and Mary McCormac. And also to our children Andrew, Heidi, Tyler, Alex, Josh, Mary Christine, and Becky.

Brief Contents

CONTENTS

PREFACE

Structural Analysis in Today's World

When it comes to learning or teaching structural engineering concepts, there exist two main schools of thought with respect to courses on structural analysis. Some claim that today's students need to have a broad exposure to many of the classical techniques of analysis such as conjugate beam, moment-area, slope deflection, virtual work and moment distribution. Others will argue that learners today will benefit more from computer-based analysis experiences that involve parametric studies. The authors of this text find themselves somewhere in the middle ground with the belief that, while our students no longer need to know every classical technique available, they still need a fundamental knowledge of the concepts that come from studying a subset of the existing classical techniques. This foundation of knowledge is then strengthened by the use of structural analysis software in activities designed to promote self-discovery of structural concepts and behaviors. It is with this goal in mind that this text was formulated.

Most chapters within this textbook have a section devoted to the use of computer software, including the commercially available structural analysis software SAP2000. In some cases, the use of spreadsheets for analysis tasks is also presented. Parametric exercises are provided as part of most problem sets that are designed to let students discover the influence that various modeling parameters have upon the response of structures. The authors believe that a well-rounded problem set includes work both by hand and by computer.

The foundation for the current textbook is the 4th Edition of Jack C. McCormac's notable text titled: *Structural Analysis: Using Classical and Matrix Methods.* However, over a five-year period, the authors have undertaken to rework the content, format, layout and emphasis of this text. In many cases, traditional classical techniques, such as slope-deflection, that are not widely used today have been removed altogether from the text. Other classical techniques like moment distribution, which still has some tie-in with today's design codes, have been left in. Indeed, anecdotal evidence from our interaction with today's practicing engineers has helped to shape which topics were retained and which were removed. By dropping some classical techniques, room was made for a new chapter dealing with the lateral load path in a building and its relevant components—a topic for which many graduating students find themselves ill prepared.

Key Components of this Text

- The chapters have been organized in what is believed to be a natural sequence of material presentation. Of note, a major deviation from most structural analysis textbooks is that the topic of influence lines is not presented early in the textbook. Rather, experience has shown that the traditional way of presenting influence lines shortly after the discussion of shear and moment diagrams is a major source of confusion for many students. Rather, in this text, the students are given plenty of time to become familiar with shear and moment diagrams prior to the discussion of influence lines.

- Margin notes have been added throughout the text to highlight key concepts, point out trouble spots, and to provide clarifying remarks. This addresses the major issue that many "pearls of wisdom" contained in textbooks are often buried within a lot of text—thus they are never recognized by the reader for what they truly are.

- Video examples for each chapter have been developed to demonstrate the processes presented in the text. Experience has shown that, if the students have access to such examples, they can often answer their own questions during the homework exercises. While every effort has been made to keep the videos short, the authors feel that it is important that students are able to see a problem worked out from start to finish with appropriate explanations.

- A tutorial video on how to use SAP2000 has been provided. This video is intended to be a basic introduction to the tools and modeling process required to complete the modeling exercises contained in this text. Some of the early chapters will add to this knowledge by highlighting various functions or modeling strategies. The content related to SAP2000 is located in the separate chapters and cannot stand alone. The intent is that the video tutorial is viewed initially; then the tutorial slides found in Appendix B can be referred to as a reference.
- Spreadsheets are introduced for handling some analysis calculations. Some functional spreadsheets are provided to the reader to help in learning about the implementation process.
- Any graphics or photos provided within the text have been placed there for a purpose. While descriptions are not always provided, the reader should ask themselves: "Why was this photo placed here?"
- Problem sets for most chapters are quite extensive and constitute varying degrees of difficulty.
- The chapters on the matrix-based stiffness approach cover a wide range of topics yet are streamlined to ease the introduction and implementation of the subject matter. Not only do they introduce the basic method, but they also address special topics that are suitable for a class in matrix structural analysis. These chapters demonstrate the use of spreadsheets in the implementation of this approach.

Acknowledgments

As mentioned previously, this textbook represents a five-year endeavor to refine the included material and to ensure a positive and effective learning experience for the reader. While it has not been easy, the authors would like to express their gratitude to the many individuals who have supported or encouraged the development of this text, including intrepid family members who were willing to help generate some of the graphics—even though they had other things they would like to do with their time. We would like to express thanks to Dr. Scott D. Schiff who provided valuable consultation in the early stages of the process and helped to shape the format of the final project and Professor Stephen F. Csernak for his early implementation and feedback on the text.

We hope you enjoy the experience!

Bryant G. Nielson, SE
Jack C. McCormac, PE

Clemson, SC, USA
December, 2016

Determinate Structures

Part 1

1 Introduction

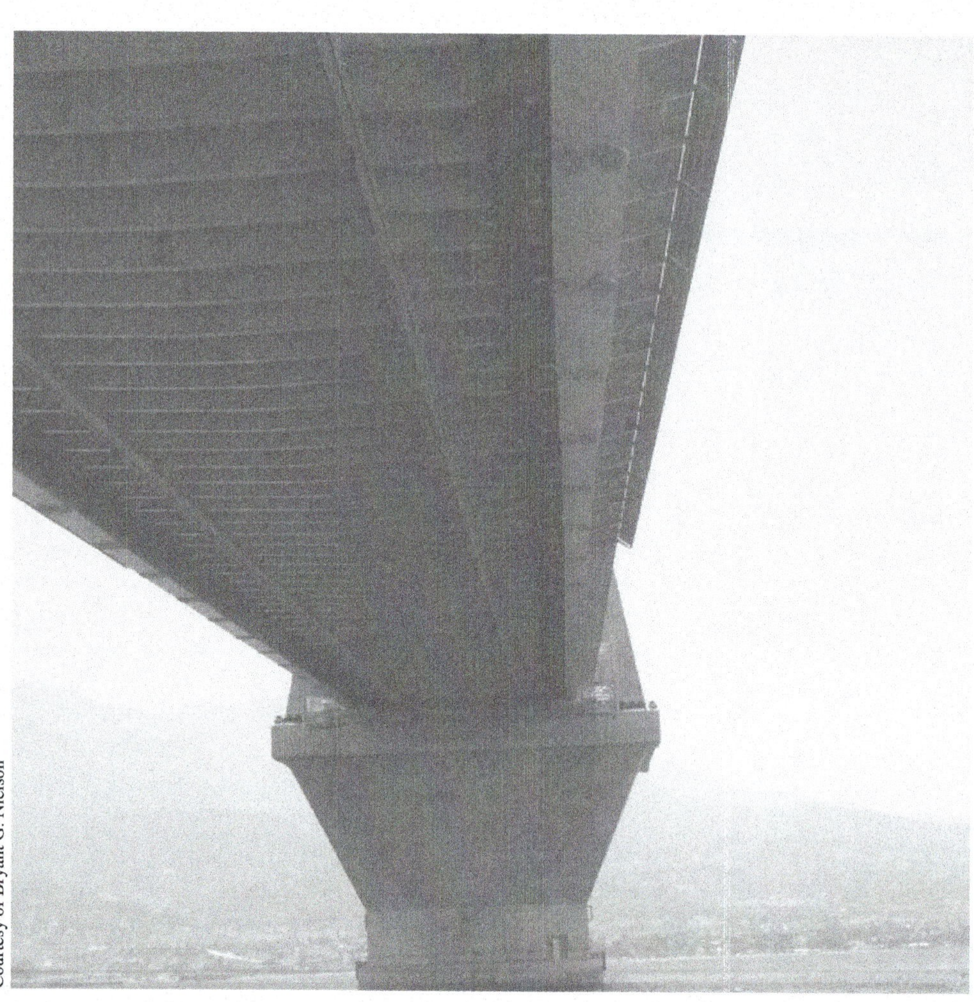

Courtesy of Bryant G. Nielson

1.1 Structural Analysis and Design

The application of loads to a structure causes the structure to deform. Due to the deformation, various forces are produced in the components that comprise the structure. Calculating the magnitude of these forces and the deformations that caused them is referred to as *structural analysis*, which is an extremely important topic to society. Indeed, almost every branch of technology becomes involved at some time or another with questions concerning the strength and deformation of structural systems.

Courtesy of Bryant G. Nielson

FIGURE 1.1 Ancient Coliseum in Rome, Italy.

Structural design includes the arrangement and proportioning of structures and their parts so they will satisfactorily support the loads that they may be subjected to. More specifically, structural design involves the following:

1. General layout of the structural system
2. Studies of alternative structural configurations that may provide feasible solutions
3. Consideration of loading conditions
4. Preliminary structural analyses and design of the possible solutions
5. Selection of a solution
6. Final structural analysis and design of the structure
7. Preparation of design drawings

QUICK NOTE

The design of structures need not be restricted to the design of new structures. Rather, many structural engineers find themselves upgrading and retrofitting existing buildings and/or bridges. Such activities generally require a thorough analysis of the existing structure before any changes are formulated.

This book is devoted to structural analysis with only occasional remarks concerning the other phases of structural design. Structural analysis can be so interesting to engineers that they become completely attached to it and have the feeling that they want to become 100% involved in the subject. Although analyzing and predicting the behavior of structures and their parts is an extremely important part of structural design, it is only one of several important and interrelated steps. Consequently, it is rather unusual for an engineer to be employed solely as a structural analyst. An engineer, in almost all probability, will be involved in several or all phases of structural design.

It is said that Robert Louis Stevenson studied structural engineering for a time, but he apparently found the "science of stresses and strains" too dull for his lively imagination. He went on to study law for a while before devoting the rest of his life to writing prose and poetry.[1] Most of us who have read *Treasure Island*, *Kidnapped*, or his other works would agree that the world is a better place because of his decision. Nevertheless, there are a great number of us who regard structural analysis and design to be extremely interesting topics. In fact, some of us have found it so interesting that we have gone on to practice in the field of structural engineering. The authors hope that this book will inspire more engineers to do the same.

1.2 History of Structural Analysis

Structural analysis as we know it today evolved over several thousand years. During this time, many types of structures such as beams, arches, trusses, and frames were used in construction for hundreds or even thousands of years before satisfactory methods of analysis were developed for them. Though ancient engineers showed some understanding of structural behavior (as evidenced

[1]Proceedings of the First United States Conference on Prestressed Concrete (Cambridge, Mass.: Massachusetts Institute of Technology, 1951), 1.

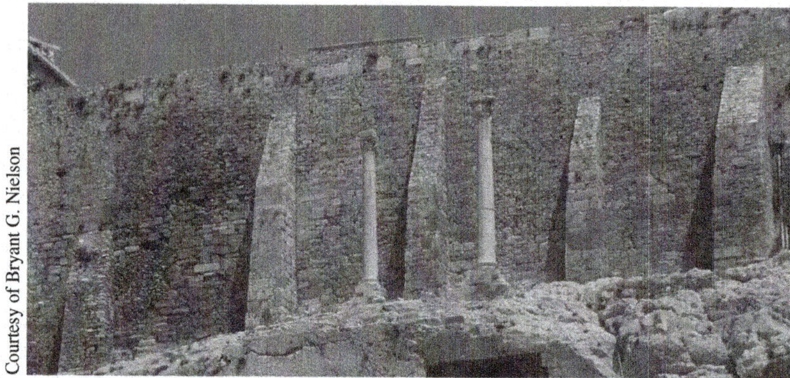
Courtesy of Bryant G. Nielson

FIGURE 1.2 Buttressed retaining wall around Acropolis in Athens, Greece.

by their successful construction of great bridges, cathedrals, sailing vessels, and so on), real progress with the theory of structural analysis occurred only in the last 175 years.

The Egyptians and other ancient builders surely had some kinds of empirical rules drawn from previous experiences for determining the sizes of structural members. There is, however, no evidence that they had developed any theory of structural analysis.

OF INTEREST
The Egyptian Imhotep who built the great step pyramid of Sakkara in about 3000 B.C.E. sometimes is referred to as the world's first structural engineer.

Although the Greeks built some magnificent structures, their contributions to structural theory were few and far between. Pythagoras (about 582–500 B.C.E.), who is said to have originated the word *mathematics*, is famous for the right angle theorem that bears his name. This theorem actually was known by the Sumerians in about 2000 B.C.E. Further, Archimedes (287–212 B.C.E.) developed some fundamental principles of statics and introduced the term *center of gravity*.

The Romans were excellent builders and very competent in using certain structural forms such as semicircular masonry arches. But, as did the Greeks, they too had little knowledge of structural analysis and made even less scientific progress in structural theory. They probably designed most of their beautiful buildings from an artistic viewpoint. Perhaps their great bridges and aqueducts were proportioned with some rules of thumb; however if these methods of design resulted in proportions that were insufficient, the structures collapsed and no historical records were kept. *Only their successes endured.*

One of the greatest and most noteworthy contributions to structural analysis, as well as to all other scientific fields, was the development of the Hindu–Arabic system of numbers. Unknown Hindu mathematicians in the 2nd or 3rd century B.C.E. originated a numbering system of one to nine. In about 600 C.E., the Hindus invented the symbol *sunya* (meaning empty), which we call zero. The Mayan Indians of Central America, however, had apparently developed the concept of zero about 300 years earlier.[2]

In the 8th century C.E., the Arabs learned this numbering system from the scientific writings of the Hindus. In the following century, a Persian mathematician wrote a book that included the system. His book later was translated into Latin and brought to Europe.[3] In around 1000 C.E., Pope Sylvester II decreed that the Hindu–Arabic numbers were to be used by Christians.

C. NAVIER
Published a textbook in 1826 in which the strengths and deflections of beams, columns, arches, suspension bridges, and other structures was discussed.

Before real advances could be made with structural analysis, it was necessary for the science of *mechanics of materials* to be developed. By the middle of the 19th century, much progress had been made in this area. French physicist Charles Augustin de Coloumb (1736–1806) and French engineer-mathematician Claude Louis Marie Henri Navier (1785–1836), building upon the work of numerous other investigations over hundreds of years, are said to have founded the science of mechanics of materials.

DID YOU KNOW?
Squire Whipple's analysis of trusses often is said to have signaled the beginning of modern structural analysis.

Andrea Palladio (1508–1580), an Italian architect, is thought to have been the first person to use modern trusses. He may have revived some ancient types of Roman structures and their empirical rules for proportioning them. It was actually 1847, however, before the first rational method of analyzing jointed trusses was introduced by Squire Whipple (1804–1888). His was the first significant American contribution to structural theory. Since that time, there has been an almost continuous series of important developments in the subject.

[2] *The World Book Encyclopedia* (Chicago, IL, 1993, Book N–O), 617.
[3] Ibid.

Courtesy of Angelina Stasulis

FIGURE 1.3 Roller coaster at Cedar Point in Sandusky, OH.

Several excellent methods for calculating deflections were published in the 1860s and 1870s, which further accelerated the development of structural analysis. Among the important investigators and their accomplishments were James Clerk Maxwell (1831–1879) of Scotland, for the reciprocal deflection theorem in 1864; Otto Mohr (1835–1918) of Germany, for the method of elastic weights presented in 1870; Carlo Alberto Castigliano (1847–1884) of Italy, for the least-work theorem in 1873; and Charles E. Greene (1842–1903) of the United States, for the moment-area theorems in 1873.

The advent of railroads gave a great deal of impetus to the development of structural analysis. It was suddenly necessary to build long-span bridges capable of carrying very heavy moving loads. As a result, the computation of stresses and strains became increasingly important—as did the need to analyze statically indeterminate structures.

One method for analyzing continuous statically indeterminate beams—the three-moment theorem—was introduced in 1857 by the Frenchman B. P. E. Clapeyron (1799–1864) and was used for analyzing many railroad bridges. In the decades that followed, many other advances were made in indeterminate structural analysis that were based upon the recently developed deflection methods.

OTTO MOHR

A German engineer who worked with railroads and is said to have reworked into practical, usable forms many of the theoretical developments up to his time. Particularly notable in this regard was his 1874 publication of the method of consistent distortions for analyzing statically indeterminate structures.

In the United States, two great developments in statically indeterminate structure analysis were made by G. A. Maney (1888–1947) and Hardy Cross (1885–1959). In 1915, Maney presented the slope deflection method, whereas Cross introduced moment distribution in 1924, which is discussed in Chapters 13 and 14 of this text.

In the first half of the 20th century, many complex structural problems were expressed in mathematical form, but sufficient computing power was not available for practically solving the resulting equations. This situation continued into the 1940s, when much work was done with matrices for analyzing aircraft structures. Fortunately, the development of digital computers made the use of matrices practical for these and many other types of complex structures, including high-rise buildings.

Some particularly important historical references on the development of structural analysis include those by Kinney,[4] Timoshenko,[5] and Westergaard.[6] They document the slow but steady development of the fundamental principles involved. It seems ironic that the college student of

[4] J. S. Kinney, *Indeterminate Structural Analysis* (Reading, Mass.: Addison-Wesley, 1957), 1–16.

[5] S. P. Timoshenko, *History of Strength of Materials* (New York: McGraw-Hill, 1953), 1–439.

[6] H. M. Westergaard, "One Hundred Fifty Years Advance in Structural Analysis," (*ASCE-94*, 1930), 226–240.

today can learn in a few months the theories and principles of structural analysis that took many scholars several thousand years to develop.

1.3 Basic Principles of Structural Analysis

DID YOU KNOW?
Plans have long been considered to build a bridge connecting Sicily to mainland Italy over the Strait of Messina. Its suspended main span is projected to be an almost unbelievable 2.05 miles.

Structural engineering embraces an extensive variety of structural systems. When speaking of structures, people typically think of buildings and bridges. There are, however, many other types of systems with which structural engineers deal, including sports and entertainment stadiums, radio and television towers, arches, storage tanks, aircraft and space structures, concrete pavements, and fabric air-filled structures. These structures can vary in size from a single member—as is the case of a light pole—to buildings or bridges of tremendous size. The Willis Tower in Chicago is over 1450 ft tall, while the Taipei 101 building in Taiwan has a height of 1670 ft. Among the world's great bridges are the Humber Estuary Bridge in England, which has a suspended span of over 4626 ft, and the Akashi-Kaikyo bridge in Japan with its main suspended clear span of 6530 ft.

To be able to analyze this wide range of sizes and types of structures, a structural engineer must have a solid understanding of the basic principles that apply to all structural systems. It is unwise to learn how to analyze a particular structure or even a few different types of structures. Rather, it is more important to learn the fundamental principles that apply to all structural systems, regardless of their type or use. One never knows what types of problems the future holds or what type of structural system may be conceived for a particular application, but a firm understanding of basic principles helps us to analyze new structures with confidence.

The fundamental principles used in structural analysis are Sir Isaac Newton's laws of inertia and motion, which are

1. A body will exist in a state of rest or in a state of uniform motion in a straight line unless it is forced to change that state by forces imposed on it.
2. The rate of change of momentum of a body is equal to the net applied force.
3. For every action there is an equal and opposite reaction.

These laws of motion can be expressed by the equation

$$\sum F = ma$$

In this equation, $\sum F$ is the summation of all the forces that are acting on the body, m is the mass of the body, and a is its acceleration.

In this textbook, we will be dealing with a particular type of equilibrium called *static equilibrium*, in which the system is not accelerating. The equation of equilibrium thus becomes

$$\sum F = 0$$

STRUCTURAL RESPONSE
This refers to the displacement of the system and the forces that occur in each component of the structural system.

These structures either are not moving, as is the case for most civil engineering structures, or are moving with constant velocity, such as space vehicles in orbit. Using the principle of static equilibrium, we will study the forces that act on structures and methods to determine the response of structures to these forces. This emphasis should provide readers with a solid foundation for advanced study, and hopefully convince them that structural theory is not difficult and that it is not necessary to memorize special cases.

1.4 Structural Components and Systems

All structural systems are composed of components. The following are considered to be the primary components in a structure:

Ties Those members that are subjected to axial tension forces only. Load is applied to ties only at the ends. Ties cannot resist flexural forces.

Struts Those members that are subjected to axial compression forces only. Like ties, struts can be loaded only at their ends and cannot resist flexural forces.

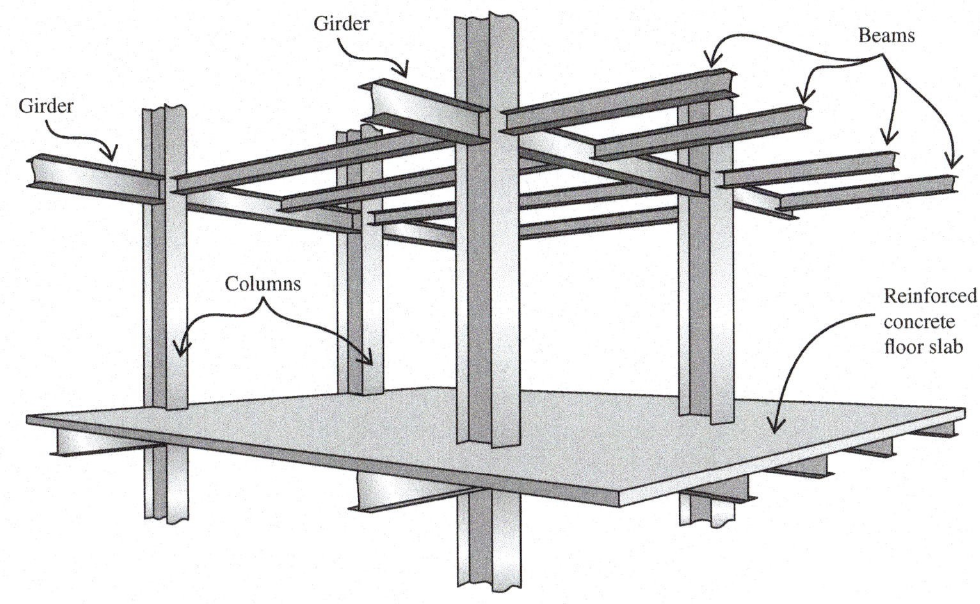

FIGURE 1.4 A typical building frame.

Beams and Girders Those members that are primarily subjected to flexural forces. They usually are thought of as being horizontal members that are primarily subjected to gravity forces; but there are frequent exceptions (e.g., inclined rafters).

Columns Those members that are primarily subjected to axial compression forces. A column may be subjected to flexural forces also. Columns usually are thought of as being vertical members, but they may be inclined.

Diaphragms Structural components that are flat plates. Diaphragms generally have very high in-plane stiffness. They are commonly used for floors and shear-resisting walls. Diaphragms usually span between beams or columns. They may be stiffened with ribs to better resist out-of-plane forces.

Structural components are assembled to form structural systems. In this textbook, we will be dealing with typical framed structures. A building frame is shown in Figure 1.4. In this figure, a girder is considered to be a large beam that supports smaller beams.

A truss is a special type of structural frame. It is composed entirely of struts and ties. That is to say, all of its components are connected in such a manner that they are subjected only to axial forces. All of the external loads acting on trusses are assumed to act at the joints and not directly on the components, where they might cause bending in the truss members. An older type of bridge structure consisting of two trusses is shown in Figure 1.5. In this figure, the top and bottom chords and the diagonals are the primary load carrying components of trusses. Floor beams are used to support the roadway. They are placed under the roadway and perpendicular to the trusses.

QUICK NOTE

There are other types of structural systems. These include fabric structures (e.g., tents and outdoor arenas) and curved shell structures (e.g., dams or sports arenas). The analysis of these types of structures requires advanced principles of structural mechanics and is beyond the scope of this book.

FIGURE 1.5 Some components of a railroad bridge truss.

1.5 Structural Forces

A structural system is acted upon by forces. Under the influence of these forces, the entire structure is assumed to be in a state of static equilibrium, and as a consequence, each component of the structure also is in a state of static equilibrium. The forces that act on a structure include the applied loads and the resulting reaction forces.

QUICK NOTE
The loads that are applied to structures come from many sources. Such sources include gravity, wind, earthquake, soil, impact, etc. A more complete introduction to such loads is presented in Chapter 2.

The applied loads are the known loads that act on a structure. They can be the result of the structure's own weight, occupancy loads, environmental loads, and so on. The *reactions* are the forces that the supports exert on a structure. They are considered to be part of the external forces applied and are in equilibrium with the other external loads on the structure.

To introduce loads and reactions, three simple structures are shown in Figure 1.7. The beam shown in part (a) of the figure is supporting a uniformly distributed gravity load and is itself supported by upward reactions at its ends. The barge in part (b) of the figure is carrying a group of containers on its deck. It is in turn supported by a uniformly distributed hydrostatic pressure provided by the water beneath. Part (c) shows a building frame subjected to a lateral wind load. This load tends to overturn the structure, thus requiring an upward reaction at the right-hand support and a downward one at the left-hand support. These forces create a couple that offsets the effect of the wind force. A detailed discussion of reactions and their computation is presented in Chapter 5.

1.6 Calculation Accuracy

DID YOU KNOW?
The loads that a structural engineer uses are only an estimate. The analysis that the engineer performs is approximate at best. Poor engineering judgment results from an engineer not understanding these two truths. This does not eliminate the need for the engineer to be exact and methodical in their approach to the analysis, but it does negate the need for a lot of significant digits in the answers.

A most important point that many students with their superb pocket calculators and personal computers have difficulty understanding is that structural analysis is not an exact science for which answers can be confidently calculated to eight or more significant digits. Computations to only three places probably are far more accurate than the estimates of material strengths and magnitudes of loads used for structural analysis and design. The common materials dealt with in structures (wood, steel, concrete, and a few others) have ultimate strengths that only can be estimated. The loads applied to structures may be known within a few hundred pounds or no better than a few thousand pounds. It therefore seems inconsistent to require force computations to more than three or four significant figures.

Several partly true assumptions can be made about the construction of trusses, such as: Truss members are connected with frictionless pins, the deformation of truss members under load is so slight as to cause no effect on member forces, and so on. These deviations from actual conditions emphasize that it is of little advantage to carry the results of structural analysis to many significant

Courtesy of Angelina Stasulis

FIGURE 1.6 Industrial digester under construction.

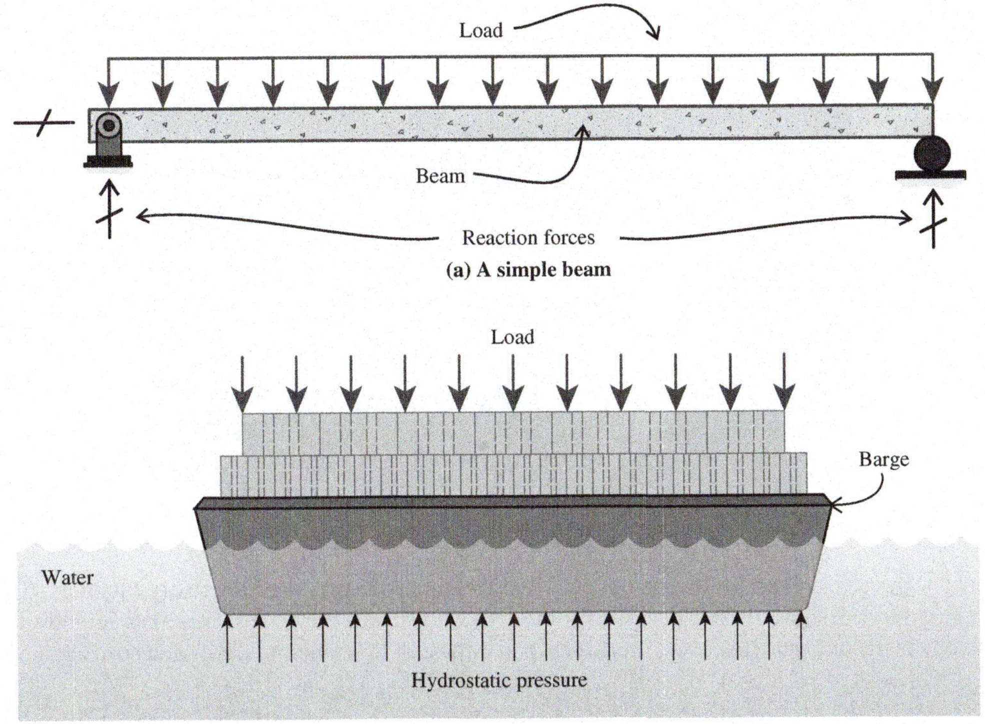

(a) A simple beam

(b) Forces on a barge

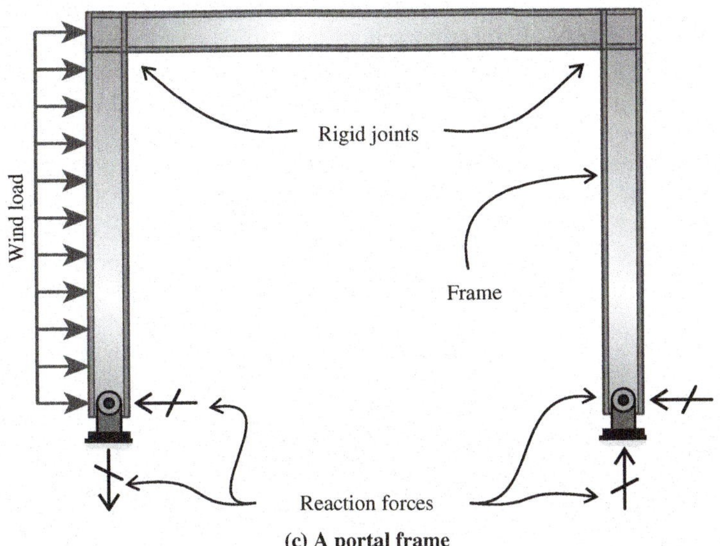

(c) A portal frame

FIGURE 1.7 Loads and reactions for three simple structures.

figures. Furthermore, calculations to more than three or four significant figures may be misleading in that they may give you a false sense of precision.

1.7 Checks on Problems

A definite advantage of structural analysis is the possibility of making either mathematical checks on the analysis by some method other than the one initially used or by the same method from some other position on the structure. You should be able, in nearly every situation, to determine if your work has been done correctly.

All of us, unfortunately, have the weakness of making exasperating mistakes, and the best that can be done is to keep them to an absolute minimum. The application of simple arithmetic

Courtesy of Parsons in Denver, CO

FIGURE 1.8 Dubai Metro transit rail in Dubai, United Arab Emirates.

checks suggested in the following chapters will eliminate many of these costly blunders. The best structural designer is not necessarily the one who makes the fewest mistakes initially but probably is the one who discovers the largest percentage of his or her mistakes and corrects them.

1.8 Using Computers for Structural Analysis

The availability of personal computers has drastically changed the way structures are analyzed and designed. In nearly every engineering school and office, computers are used to address structural engineering problems. Within an engineering office, computers are used to increase efficiency. While in school, computers are used to teach concepts and aid in the understanding of structural behavior.

Many structural engineering professors feel that students should first learn the theories involved in structural analysis and the solution of problems with their pocket calculators before computer applications are introduced. As a result, the authors have placed computer applications at the end of many chapters so that they can either be used at that time, skipped, or temporarily bypassed until a later date. However, the authors believe that computer applications can help students to

Courtesy of Bryant G. Nielson

FIGURE 1.9 Movable roof for City Center in Salt Lake City, Utah.

more readily gain an understanding of structural behavior and may be a used in a supplementary fashion as the theories are taught.

Use of the structural analysis computer program, SAP2000, has been incorporated into this textbook. This is a widely used, commercially available analysis software. A student version of this software is available on the textbook website at www.wiley.com/college/nielson. Also available on this website is a Powerpoint presentation and accompanying video that provide an essential introduction to this software package. Many chapters within the text contain a section specific to modeling structures within the SAP2000 software. Within these sections, some guidance is provided, but the material within the text *cannot stand alone*. Instead, these sections are written with the expectation that the software introduction video is viewed first.

The sections titled *SAP2000 Exercises* located at the end of many chapters are designed to explore the behavior of structures while learning to utilize the software. The use of the computer allows a student to explore the impact and influence of many structural parameters without the tedium. Many of the classical techniques taught in a structural analysis course can be time consuming and as such make it difficult for a student to explore the influence a certain structural parameter has on the behavior of a structure.

QUICK NOTE

Most commercial structural analysis programs are very similar in structure and operation. While this textbook uses one such program, the lessons learned using it can be readily extended to alternate software packages.

BEWARE

Most commercial structural analysis packages provide nice renderings of the structure being modeled. However, a "pretty picture" does not guarantee that the structure was modeled correctly. The modeler must be aware of details associated with the behavior they desire to capture.

1.9 Overview of this Textbook

The composition of this textbook is intended to serve a couple of key purposes to the structural engineering student. The first purpose is to teach the fundamental theories behind the analysis of structures—including the calculations of reaction forces, internal forces, and deflections. The second purpose is to help the student understand the behavior of structures that are subjected to different loads and are of different configurations.

To help with the understanding of theory and behavior, the use of margin notes is used extensively throughout the text. These notes (e.g., "Quick Note", "Did You Know?", etc.) are intended to highlight key ideas. Of particular interest are the margin notes titled "Common Mistake." The wise reader will pay particular attention to these and will try to avoid these common pitfalls.

Another resource made available with this text is a set of video-based examples associated with most chapters. The intent is to allow the reader the opportunity to follow along as a relevant example is worked and discussed similar to what would be done in a classroom. These videos may be found on the textbook website.

Finally, the content presented in this text first focuses on basic background knowledge of loads and load paths presented in Chapters 2 through 4. The next part deals with the analysis of determinate structures followed by a part dealing with the analysis of indeterminate structures. Both of these parts omit some of the classical techniques presented in traditional textbooks (e.g., moment-area, conjugate beam, and slope-deflection). This is done to allow the student to focus on fewer methods, learn those well, and still allow space and time for exploring structural behavior using SAP2000.

The next part deals specifically with the notion of influence lines. This is placed towards the back of the text to avoid the common confusion students experience when dealing with influence lines immediately after learning about shear and moment diagrams. The last part of the text introduces matrix methods, which are the basis for the computerization of structural analysis.

The authors hope you enjoy learning structural analysis through the use of this text and its accompanying resources. Remember that it is always a *beautiful day for studying structures.*

DID YOU KNOW?

Within the practice of structural engineering, one quickly finds out that not every design or analysis situation encountered is prescribed by building codes or textbooks. At times like this, the engineer must apply judgment based upon sound principles. **Good engineering judgment** comes from being able to describe general structural behavior without ever performing a calculation. During the learning phase it is wise to always ask "Why does it do that?" and "What happens if I modify that parameter?"

2 Structural Loads

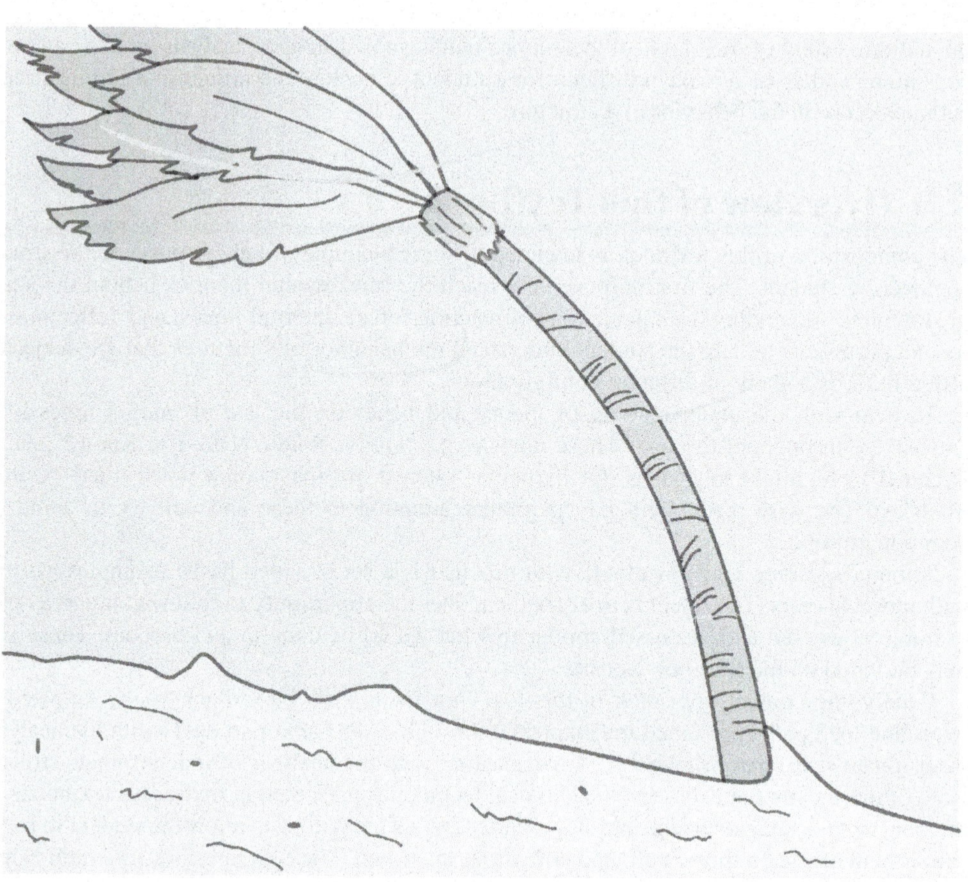

2.1 Introduction

DID YOU KNOW?
Structural load calculations are not an exact representation of actual loads that a structure will experience. Instead, they generally are conservative estimates based on a current understanding of the physical world around us.

On rather frequent occasions, professors of structural engineering ask their former students if they feel they were adequately prepared by their structures courses for their initial jobs. So often the answer is "Yes" in most areas, but probably not in the area of estimating design loads. This is a rather disturbing answer, because the appropriate estimation of the magnitude and character of the loads that a structure will have to support during its life is probably the designer's most important task.

A very large number of different types of structures (beams, frames, trusses, and so forth) subjected to all sorts of loads are introduced in this text. The student may as a result wonder "Where in the world did the author get all of these loads?" This very important question is addressed in this chapter and also in Chapters 3 and 4.

Structural engineers today generally use computer software in their work. Although the typical software enables them to quickly analyze and design structures after the loads are established, it provides little help in selecting the loads in many cases.

In this chapter, various types of loads are introduced and standards are presented from which the individual magnitudes of the loads may be estimated. Our objective is to be able to answer questions such as "How heavy could the snow load be on a school in Minneapolis? What maximum wind force might be expected on a hotel in Miami? How large of a rain load is probable for a flat roof in Houston?"

The methods used for estimating loads are constantly being refined and may involve some very complicated formulas. Do not be concerned about committing such expressions to memory. Rather, learn the types of loads that can be applied to a structure and where to obtain information for estimating the magnitude of these loads.

The authors are quite concerned that the enclosed sections on wind and seismic loads may be a little overwhelming for students just beginning the study of structures. They have included the material primarily to be used as an introduction and as a reference for future study and not as an essential part of an elementary course in structural analysis. The estimation of the magnitudes of wind and seismic loads are so involved that they each are often the subject of entire college courses. It is to be realized that the procedures for estimating wind and seismic loads are constantly changing through the years as a result of continuing research in these topics.

Once the authors opened the door to these loads, they found it very difficult to find a reasonable stopping place. It is probable that instructors of elementary structural analysis classes, for which this book was prepared, will not require students to learn the information presented in detail. The authors' purpose here is to give the student an idea of the items involved in estimating the magnitude of wind and earthquake loads and to serve as a starting point and reference for further study when it becomes necessary in later work.

2.2 Structural Safety

A structure must be adequate to support all of the loads it may foreseeably be subjected to during its lifetime. Not only must it safely support these loads, but it must do so in such a manner that deflections and vibrations are not so great as to frighten the occupants or to cause unsightly cracks.

QUICK NOTE

A structural engineer's responsibility is to design structures that will transfer *all applied loads* **safely** to the foundation. To do this, they must have reasonable estimates of current and future loads.

The reader might think that all a structural designer has to do is to look at some structures similar to the one he or she is preparing to design, estimate the loads those structures are supporting, and then design his or her structure to be sufficiently strong to support those loads—and a little more for safety's sake. However, it's not quite that simple, because there are so manys uncertainties in design. Some of these uncertainties follow:

1. Material strengths may vary appreciably from their assumed values, and they will vary more with time due to creep, corrosion, and fatigue.
2. The methods of analysis of structures are often subject to appreciable errors.
3. Environmental occurrences such as earthquakes, hurricanes, rain and snow storms cause loads that are extremely difficult to predict.
4. There are technological changes that cause increased loads to occur, such as larger trucks, trains, or army tanks crossing our bridges.
5. Loads occurring during construction operations can be severe, and their magnitudes are difficult to predict.
6. Among other uncertainties that structures face are variations in member sizes, residual stresses, and stress concentrations.

Many years of design experience, both favorable and unfavorable, have led to detailed specifications and building codes. This chapter is devoted to some of the load requirements of those specifications. Ultimately, the safety of the public is the major issue in this topic of selecting the magnitudes of loads for design.

2.3 Codes, Standards, and Specifications

The design of most structures is governed by relevant building codes or specifications. Even if they are not so controlled, the designer probably will refer to them as a guide. No matter how many structures a person has designed, it is impossible for them to have encountered every situation. By referring to standards and specifications, an engineer is making use of the best available material on the subject. Engineering specifications that are developed by various organizations present the best opinion of those organizations as to what represents good practice.

Municipal and state governments concerned with the safety of the public have established or adopted building codes that they use to control the construction of various structures within their jurisdiction. These codes, which actually are laws or ordinances, provide direction for the calculation of design loads, design stresses, construction types, material quality, and other factors. They can vary considerably from city to city, which can cause some confusion among architects and engineers.

The International Code Council has developed the *International Building Code*.[1] This code was developed to meet the need for a modern building code for building systems that emphasize performance. The *International Building Code* (IBC-2015) is primarily used in the United States and is intended to serve as a set of model code regulations to safeguard the public in all communities across the nation. The IBC accomplishes this by requiring engineers to use appropriate standards and specifications (see definitions in the side note).

Readers should note that logical and clearly written codes are quite helpful to structural engineers. Furthermore, far fewer structural failures occur in areas with good codes that are strictly enforced. The specifications published by various organizations are frequently used to estimate the maximum loads to which buildings, bridges, and some other structures may be subjected during their estimated lifetimes.

Standards and specifications on many occasions clearly prescribe the minimum loads for which structures are to be designed. Despite the availability of this information, however, the designer's ingenuity and knowledge of the situation are often needed to predict the loads a particular

DEFINITIONS

Model Building Code: A consensus document that is a collection of rules and regulations related to the building of safe and functional structures. These are adopted by governmental agencies as legal documents governing the design, construction, and operation of various structures.

Standards: A set of rules and requirements concerned with the definition of terms, outlining procedures, identifying allowable materials, etc. All standards must be adopted by the model building code to become enforceable.

Specifications: *Recommendations* for the practice of structural design developed by technical trade associations. Usually focused on the use of a particular building material (e.g., concrete, steel, masonry, wood) or structure type (e.g., bridges). Specifications also must be adopted by the model building code to become enforceable.

FIGURE 2.1 A few of the codes, standards, and specifications currently used for buildings.

[1] 2015 *International Building Code*, (Falls Church, Virginia, International Code Council Inc., 2015).

structure will have to support in years to come. For example, over the past several decades, insufficient estimates of future traffic loads by bridge designers have resulted in many bridges being replaced with wider and stronger structures.

This chapter introduces the basic types of loads with which the structural engineer needs to be familiar. Its purpose is to help the reader develop an understanding of structural loads and their behavior and to provide a foundation for estimating their magnitudes. It should not be regarded, however, as a complete essay on the subject of the loads that might be applied to any and every type of structure the engineer may design.

Since building loads are the most common type encountered by designers, they are the loads most frequently referred to in this text. The standards required by the International Building Code for estimating the loads to be applied to buildings is the ASCE 7 minimum loads standard produced by the American Society of Civil Engineers.[2] Much information in this text is based on the 2016 edition of this standard that is called ASCE 7-16.

An engineer should always view *minimum* design standards with some skepticism. The design standards are excellent and well prepared for most situations. However, there may be a building configuration, or a building use, for which the specified design loads are not adequate. A structural engineer should evaluate the minimum specified design loads to determine whether they are adequate for the structural system being designed.

QUICK NOTE

When studying the information provided in this chapter or reviewing any standard or specification providing design loads, the reader is reminded that minimum design load standards are presented.

2.4 Types of Structural Loads

Structural loads usually are categorized by means of their character and duration. Common loads that are applied to buildings are categorized as follows.

Dead loads Those loads of constant magnitude that remain in one position throughout the life of the structure. They include the weight of the structure under consideration as well as any fixtures that are permanently attached to it.

Live loads Those loads that can change in magnitude and position. They include occupancy loads, warehouse materials, construction loads, overhead service cranes, and equipment operating loads. In general, live loads are caused by gravity.

Environmental loads Those loads caused by the environment in which the structure is located. For buildings, the environmental loads are caused by rain, snow, ice, wind, temperature, and earthquakes. Strictly speaking, these are also live loads, but they are the result of the environment in which the structure is located.

DID YOU KNOW?

Codes, standards, and specifications cannot address every possible situation. Therefore, the responsibility for the design of a safe structure ultimately lies with the structural engineer.

The common loads encountered are more specifically identified here by ASCE 7.

D = dead loads

E = seismic or earthquake load effects

F = loads due to the weight and pressure of fluids

W = wind load

L_r = roof live loads

R = rain loads

T = total effects of temperature, creep, shrinkage, and differential settlement

L = floor live load

S = snow loads

H = loads due to weight and lateral earth pressure of soils, groundwater, or bulk materials

DID YOU KNOW?

The earthquake effect (E) is split into the effects due to horizontal shaking (E_h) and vertical shaking (E_v). Thus, $E = E_h + E_v$.

The determination of the magnitude of these loads is only a part of dealing with the structural loads. The structural engineer must be able to determine which loads can be reasonably expected

QUICK NOTE

Load combinations account for the fact that not all loads act on a structure at the same time. Neither is it reasonable to assume that all loads be maximum at the same time.

[2]American Society of Civil Engineers, *ASCE 7-16, Minimum Design Loads and Associated Criteria for Buildings and Other Structures*, Reston, Virginia, 2016.

to act concurrently on a structure. For example, would a highway bridge completely covered with ice and snow be simultaneously subjected to fast moving lines of heavily loaded trucks in every lane and a 110-mile-per-hour lateral wind? Instead, is some lesser combination of these loads more reasonable and realistic? The proper combination of loads is presented and discussed in the next two sections.

2.5 Loading Conditions for Allowable Stress Design

There are two general methods used for the design of structures. These are the allowable stress and the strength design procedures. This section is concerned with the allowable stress method while the next section is concerned with the strength method.

With the allowable stress design (ASD) method, the most severe loading conditions are estimated, and elastic stresses in the members are computed. These stresses are limited to certain maximums that are appreciably below the ultimate stresses that the materials can withstand.

To determine the most severe loadings that the structure must be able to safely support, it is necessary to consider which of the loads (dead, live, and environmental) can occur simultaneously—or rather which simultaneous loads are reasonable to consider. In accordance with Chapter 2 of ASCE 7-16, the following possible simultaneous load situations may occur and should be considered for determining the *most severe yet reasonable situations*. These equations are called *load combinations*.

QUICK NOTE

The loads H, F, and T do not presently show up in the load combinations given here. However, ASCE 7-16 does provide guidance on how to include these loads if they exist.

1. D
2. $D + L$
3. $D + (L_r \text{ or } S \text{ or } R)$
4. $D + 0.75L + 0.75(L_r \text{ or } S \text{ or } R)$
5. $D + 0.6W$
6. $D + 0.75L + 0.75(0.6W) + 0.75(L_r \text{ or } S \text{ or } R)$
7. $0.6D + 0.6W$
8. $D + 0.7E_v + 0.7E_h$
9. $D + 0.525E_v + 0.525E_h + 0.75L + 0.75S$
10. $0.6D - 0.7E_v + 0.7E_h$

Note that all of these loads, other than dead loads, vary appreciably with time—there is not always snow on the structure and the wind is not always blowing, for example. Observe that in the seventh and tenth loading combinations that the full dead load is not being considered. The two variable loads in these combinations—the wind and earthquake loads—generally have a lateral component. As such, they tend to cause the structure to overturn. A dead load, on the other hand, is a gravity load, which tends to keep the structure from overturning. Consequently, a more severe condition can occur if for some reason the full dead load is not acting or was simply overestimated.

QUICK NOTE

Remember that the standard is listing the minimum conditions that must be considered. The design engineer retains the right to increase the loads beyond this level.

Most likely when two or more loads are acting on a structure in addition to dead load, the loads other than dead load are not likely to achieve their absolute maximum values simultaneously. Load surveys seem to bear out this assumption. The ASCE loads standard permits the load effects in these loading conditions, except dead load, to be multiplied by 0.75, provided the result is not less than that produced by dead load and the load causing the greatest effect.

These load combinations are only the recommended minimum load combinations that need to be considered. As with the determination of the loads themselves, the engineer must evaluate the structure being analyzed and determine whether these load combinations comprise all of the possible combinations for a particular structure. Under some conditions, other loads and load combinations may be appropriate.

EXAMPLE 2.1

A girder in the observation deck at an airport is subject to the following uniformly distributed design loads. Using the ASD load combinations from ASCE 7-16, what combined loads can be reasonably expected on this girder?

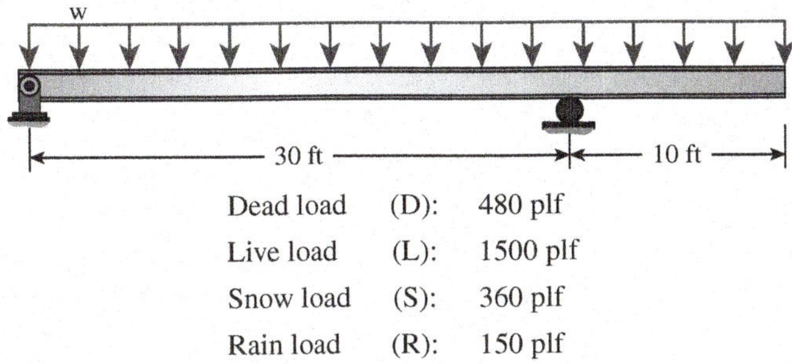

Dead load	(D):	480 plf
Live load	(L):	1500 plf
Snow load	(S):	360 plf
Rain load	(R):	150 plf

SOLUTION STRATEGY

Evaluate each of the combination equations. Make sure that you also evaluate the conditions when the word "or" appears in the equation. This is simply done by comparing the values of L_r, S, and R and by choosing the one which controls and then using it.

DID YOU KNOW?
The units of lb/ft can be written as plf and is read as *pounds per linear foot.*

The units of k/ft can be written as klf and is read as *kips per linear foot.*

SOLUTION

Since $L_r = 0$ plf, S = 360 plf, and R = 150 plf, S will control any related equations.

1. w = 480 plf
2. w = 480 plf + 1500 plf = 1980 plf ⇐ **controls**
3. w = 480 plf + 360 plf = 840 plf
4. w = 480 plf + 0.75(1500 plf) + 0.75(360 plf) = 1875 plf
5. w = 480 plf + 0.7(0) = 480 plf
6. w = 480 plf + 0.75(1500 plf) + 0.75(0) + 0.75(360 plf) = 1875 plf
7. w = 0.6(480 plf) + 0.6(0) = 288 plf
8. w = 480 plf + 0.7(0) + 0.7(0) = 480 plf
9. w = 480 plf + 0.525(0) + 0.525(0) + 0.75(1500 plf) + 0.75(360 plf) = 1875 plf
10. w = 0.6(480 plf) − 0.7(0) + 0.7(0) = 288 plf

QUICK NOTE
The loads calculated using the ASD combinations are also known as *service level loads.*

SUMMARY

The girders must be designed to support a maximum load of w = 1980 plf or 1.98 klf.

2.6 Loading Conditions for Strength Design

A design philosophy that has become more common in recent decades is the strength-design procedure. With this method, the estimated loads are multiplied by certain load factors that are almost always larger than 1.0, and the resulting ultimate (U) or "factored" loads are used for designing the structure for strength. The structure is proportioned to have a design ultimate strength sufficient to support the ultimate loads.

The purpose of load factors is to increase the loads to account for the uncertainties involved in estimating the magnitudes of the various loads. For instance, how close could you estimate the largest wind or snow loads that ever will be applied to the building that you are now occupying?

QUICK NOTE
Dead load estimates are less variable and hence have less uncertainty than other types of loads. Thus, dead loads have smaller load factors.

The recommended load combinations and load factors for strength design usually are presented in ASCE 7 but may also appear in the specifications of the different technical trade associations, like the American Institute of Steel Construction (AISC) or the American Concrete Institute (ACI). The load factors for use with strength design are determined statistically, and consideration is given to the type of structure upon which the loads are acting. As such, the load combinations and factors for a building will be different from those used for a bridge. Both of these will likely be different from those used for an offshore oil production platform. A structural analyst always should refer to the design guide or recommendations that are appropriate for the system being analyzed.

The following are the load combinations for building structures as recommended by ASCE 7-16.

QUICK NOTE

In the third, fourth, and sixth equations, the load factors for live loads may be reduced to 0.5, except for garages, for areas used for public assembly, and for all areas where the live loads are larger than 100 psf.

1. $U = 1.4D$
2. $U = 1.2D + 1.6L + 0.5(L_r \text{ or } S \text{ or } R)$
3. $U = 1.2D + 1.6(L_r \text{ or } S \text{ or } R) + (1.0L \text{ or } 0.5W)$
4. $U = 1.2D + 1.0W + 1.0L + 0.5(L_r \text{ or } S \text{ or } R)$
5. $U = 0.9D + 1.0W$
6. $U = 1.2D + 1.0E_v + 1.0E_h + 1.0L + 0.2S$
7. $U = 0.9D - 1.0E_v + 1.0E_h$

The load combinations 5 and 7 contain a 0.9D value. This 0.9 factor accounts for cases where larger dead loads tend to reduce the effects of other loads. One obvious example of such a situation occurs in tall buildings that are subject to lateral wind and seismic forces where overturning may be a possibility. As a result, the dead loads are reduced by 10% to take into account situations where they may have been overestimated.

ASCE 7-16 provides seismic and wind loads at strength-level values (that is, they have already been multiplied by a load factor). This is the situation assumed in the fourth through seventh load combination equations.

The reader must realize that the sizes of the load factors do not vary in relation to the seriousness of failure. You may think that larger load factors should be used for hospitals or high-rise buildings than for cattle barns, but such is not the case. The load factors were developed on the assumption that designers would consider the seriousness of possible failure in specifying the magnitude of their service loads. Furthermore, the ASCE load factors are minimum values, and designers are perfectly free to use larger factors as they desire.

Example 2.2 presents the calculation of factored loads for the girder of Example 2.1 using the strength load combinations. The largest value obtained is referred to as the *critical* or *governing load combination* and is the value to be used in design.

EXAMPLE 2.2

Consider the girder and unscaled loads that are presented in Example 2.1. Using the strength load combinations from ASCE 7-16, what are the combined loads which can reasonably be expected on this girder? (w_U)

SOLUTION STRATEGY

Repeat the procedure outlined in Example 2.1 using the strength-based equations.

SOLUTION

Since $L_r = 0$ plf, $S = 360$ plf, and $R = 150$ plf, S will control any related equations.

1. $w = (1.4)480 \text{ plf} = 672 \text{ plf}$
2. $w = (1.2)480 \text{ plf} + (1.6)1500 \text{ plf} + (0.5)360 \text{ plf} = \underline{\underline{3156 \text{ plf}}}$ ⇐ **controls**
3. $w = (1.2)480 \text{ plf} + (1.6)360 \text{ plf} + (1.0)\ 1500 \text{ plf} = 2652 \text{ plf}$
4. $w = (1.2)480 \text{ plf} + (1.0)0 + (1.0)1500 \text{ plf} + (0.5)360 \text{ plf} = 2256 \text{ plf}$
5. $w = (0.9)480 \text{ plf} + (1.0)0 = 432 \text{ plf}$
6. $w = (1.2)480 \text{ plf} + (1.0)0 + (1.0)0 + (1.0)1500 \text{ plf} + (0.2)360 \text{ plf} = 2148 \text{ plf}$
7. $w = (0.9)480 \text{ plf} - (1.0)0 + (1.0)0 = 432 \text{ plf}$

SUMMARY

The girders must be designed to support a maximum ultimate load of $w_U = 3156$ plf or 3.16 klf.

Example 2.3 presents the calculation of factored loads for a column using the strength combinations. The largest value obtained is referred to as the ultimate or governing load combination and is the value to be used in design. *Notice that the values of the wind and seismic loads can have two values depending on the direction of those forces, and it may be possible for the sign of those loads to be different (that is, compression or tension).* As a result, we may have to apply the applicable equations two times each to take into account the different values. This same situation can occur with the load combinations required for allowable stress design that were described in the last section. The substitution into all of these load combinations is a little bit tedious but can be easily handled with computer programs, like spreadsheets.

EXAMPLE 2.3

The axial loads for a building column have been estimated with the following results: D = 150 k, live load from roof L_r = 60 k, L = 300 k, compression wind W = 112 k, tensile wind W = 96 k, seismic compression load $(E_v + E_h)$= 50 k, and tensile seismic load $(-E_v + E_h) = 40$ k. Determine the critical design load using ASCE 7-16's strength-based load combinations, P_U.

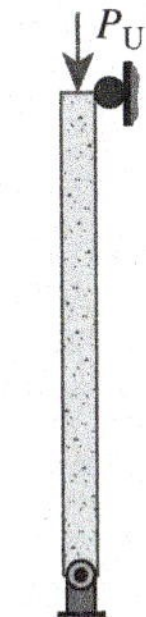

SOLUTION STRATEGY

Using the strength-based load combination equations, make sure that you consider both tension and compression cases for each relevant equation.

SOLUTION

1. $P_U = (1.4)150 \text{ k} = 210 \text{ k}$
2. $P_U = (1.2)150 \text{ k} + (1.6)300 \text{ k} + (0.5)60 \text{ k} = \underline{\underline{690 \text{ k}}} \Leftarrow$ **controls**
3. a. $P_U = (1.2)150 \text{ k} + (1.6)60 \text{ k} + (1.0)300 \text{ k} = 576 \text{ k}$
   b. $P_U = (1.2)150 \text{ k} + (1.6)60 \text{ k} + (0.5)112 \text{ k} = 332 \text{ k}$
   c. $P_U = (1.2)150 \text{ k} + (1.6)60 \text{ k} + (0.5)(-96 \text{ k}) = 228 \text{ k}$
4. a. $P_U = (1.2)150 \text{ k} + (1.0)112 \text{ k} + (1.0)300 \text{ k} + (0.5)60 \text{ k} = 622 \text{ k}$
   b. $P_U = (1.2)150 \text{ k} + (1.0)(-96) + (1.0)300 \text{ k} + (0.5)60 \text{ k} = 414 \text{ k}$
5. a. $P_U = (0.9)150 \text{ k} + (1.0)112 \text{ k} = 247 \text{ k}$
   b. $P_U = (0.9)150 \text{ k} + (1.0)(-96 \text{ k}) = 39 \text{ k}$
6. $P_U = (1.2)150 \text{ k} + (1.0)50 \text{ k} + (1.0)300 \text{ k} + (0.2)0 = 530 \text{ k}$
7. $P_U = (0.9)150 \text{ k} + (1.0)(-40 \text{ k}) = 95 \text{ k}$

QUICK NOTE

When dealing with wind and seismic loads, it is possible that the wind or seismic uplift can be more than the 0.9D in Equations 5 and 7. If this happens, the designer must design the column for the maximum compressive force and the maximum tensile force.

SUMMARY

The column must be designed to carry an ultimate axial compressive load of P_U = 690 k.

2.7 Dead Loads

The *dead loads* that must be supported by a particular structure include all of the loads that are permanently attached to that structure. They include the weight of the structural frame and also the weight of the walls, roofs, ceilings, stairways, and so on. Permanently attached equipment, described as "fixed service equipment" in ASCE 7-16, is also included in the dead load applied to the building. This equipment includes ventilating and air-conditioning systems, plumbing fixtures, electrical cables, support racks, and so forth. Depending upon the use of the structure, kitchen equipment such as ovens and dishwashers, laundry equipment such as washers and dryers, or suspended walkways also are included in the dead load.

The dead loads acting on the structure are determined by reviewing the architectural, mechanical, and electrical drawings for the building. From these drawings, the structural engineer can estimate the size of the frame necessary for the building layout and the equipment and finish details can be indicated. Standard handbooks and manufacturers' specifications can be used to determine the weight of floor and ceiling finishes, equipment, and fixtures. The

Table 2.1 **Weight of Some Common Building Materials**

Building material	Unit weight	Building material	Unit weight
Reinforced concrete	150 pcf	2 × 12 @ 16-in. double wood floor	7 psf
Acoustical ceiling tile	1 psf	Linoleum or asphalt tile	1 psf
Suspended ceiling	2 psf	Hardwood flooring (7/8-in.)	4 psf
Plaster on concrete	5 psf	1-in. cement on stone-concrete fill	32 psf
Asphalt shingles	2 psf	Movable steel partitions	4 psf
3-ply ready roofing	1 psf	Wood studs w/ 1/2-in. gypsum	8 psf
Mechanical duct allowance	4 psf	Clay brick, 4-in. wythe	39 psf

approximate weights of some common materials used for walls, floors, and ceilings are shown in Table 2.1.

The estimates of building weight or other structural dead loads may have to be revised one or more times during the analysis–design process. Before a structure can be designed, it must be analyzed. Among the loads used for the first analysis are the estimates of the weights of the components of the frame, which then is designed using those values. The component weights may be recomputed with the sizes just calculated and compared with the initially estimated values. If there are significant differences, the structure should be reanalyzed using the revised weight estimates. This cycle is repeated as many times as necessary.

EXAMPLE 2.4

A set of architectural drawings gives a detailed wall section for a two-story building, as shown in the accompanying figure. Calculate the dead loads for both the roof and the wall. Include mechanical duct allowance in the roof. The following weights will be helpful.

Rigid insulation (per inch thickness) = 0.5 psf
Steel decking = 2.5 psf
Roof framing estimate = 5.0 psf
Steel stud wall = 2.0 psf
5/8-in. gypsum board = 2.5 psf
4-in. batt insulation = 1.0 psf

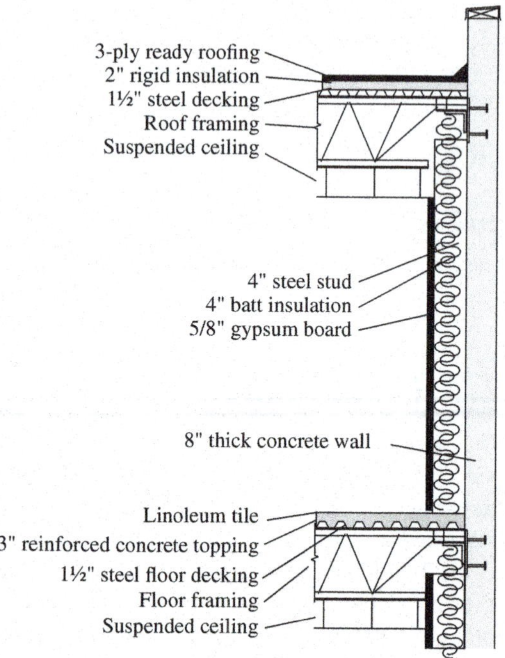

SOLUTION STRATEGY

For both the roof and the wall, create a table of all materials contained in them, including the calculated self-weights. Where appropriate, pull needed values from Table 2.1.

SOLUTION

The weight of the rigid insulation is (0.5 lb/ft^2/in)(2 in.) = 1.0 lb/ft^2 = 1.0 psf. To calculate the weight of the concrete wall, the weight density of concrete as found in Table 2.1 is 150 pcf (pounds per cubic foot) and the wall thickness is 8 in. So (150 lb/ft^3)(8 in.)(1 ft/12 in.) = 100 lb/ ft^2 = 100 psf.

Source	Roof weight (psf)	Source	Wall weight (psf)
3-ply ready roofing	1	5/8-in. gypsum board	2.5
Rigid insulation (2 in. thick)	1	Batt insulation	1
1-1/2-in. steel decking	2.5	Steel studs	2
Roof framing (self-weight)	5	8-in. concrete wall	100
Suspended ceiling	2	**Total Wall Dead Load (D)**	**105.5**
Mechanical duct allowance	4		
Total Roof Dead Load (D)	**15.5**		

COMMON MISTAKE

Many forget to make sure they have consistent units in their calculations.

SUMMARY

The primary focus of dead load calculations is to account for all sources of weight within the structure. To do this, one must make reasonable estimates—not too little and not too much—of the weight of each component. ASCE 7-16 is a great resource, but don't forget that many manufacturers provide meaningful information.

2.8 Live Loads

Live loads are those loads that can vary in magnitude and position with time. They are caused by the building being occupied, used, and maintained. Most of the loads applied to a building that are not dead loads are live loads (see Figure 2.2). Environmental loads, which are actually live loads by our usual definition, are listed separately in ASCE 7-16. Although environmental loads do vary with time, they are not all caused by gravity or operating conditions, as is typical with other live loads.

Some typical live loads that act on building structures are presented in Table 2.2. The loads shown in the table are taken from Table 4-1 in ASCE 7-16. They are acting downward and are distributed uniformly over the entire floor or roof area.

In addition to uniformly applied live loads, Section 4.4 of ASCE 7-16 also requires the consideration of concentrated loads in design. This standard states that the designer must consider the effect of certain concentrated loads as an alternative to the previously discussed uniform loads.

QUICK NOTE

Floor live loads which act over large areas may be reduced from the values in Table 2.2 when certain conditions are met. A full discussion of the rationale and procedure is presented in Chapter 3.

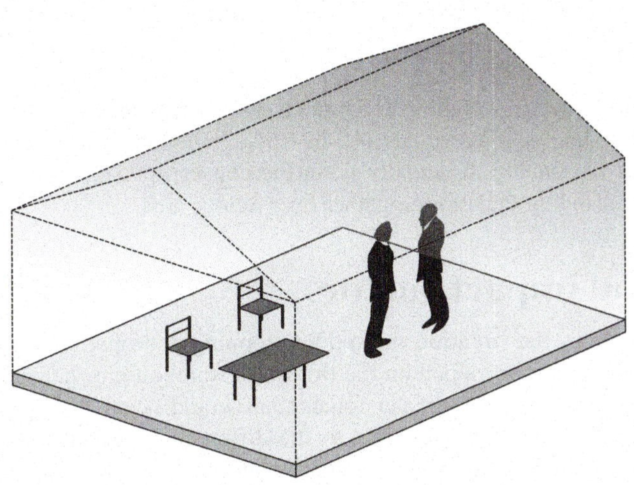

FIGURE 2.2 Depiction of typical live load sources.

Table 2.2 Some Typical Uniformly Distributed Live Loads

Area utilization	Live load	Area utilization	Live load
Lobbies of assembly areas	100 psf	Classrooms in schools	40 psf
Dance halls and ballrooms	100 psf	Upper floor corridors in schools	80 psf
Library reading rooms	60 psf	Stairs and exitways	100 psf
Library stack rooms	150 psf	Heavy storage warehouses	250 psf
Light manufacturing buildings	125 psf	Retail stores - first floor	100 psf
Offices in office buildings	50 psf	Retail stores - upper floors	75 psf
Residential dwelling areas	40 psf	Walkways and elevated platforms	60 psf

Table 2.3 Typical Concentrated Live Loads

Area or structural component	Concentrated live load
Elevator machine room grating on 4-in^2	300 lbs
Office floors	2000 lbs
Center of stair tread on 4-in^2	300 lbs
Sidewalks	8000 lbs
Accessible ceilings	200 lbs

DEFINITION

Live loads are transitory in nature and can have a relatively short duration with respect to the life of the structure. Examples include:

- Occupants
- Furniture
- Vehicles
- Mobile equipment

The intent, of course, is that the loading used for design be the one that causes the most severe internal loads and stresses.

Presented in Table 4-1 of ASCE 7-16 are the minimum concentrated loads to be considered and some typical values from this ASCE table are shown in Table 2.3. The appropriate loads are to be positioned on a particular floor or roof so as to cause the greatest stresses (a topic to be discussed in more detail in Chapters 3, 16, and 17 of this text). Unless otherwise specified, each of the concentrated loads is assumed to be uniformly distributed over a square area of 2.5 ft × 2.5 ft (6.25 ft^2).

When estimating the magnitudes of the live loads that may be applied to a particular structure during its lifetime, engineers need to consider the future utilization of that structure. For example, modern office buildings are often constructed with large open spaces that may later be divided into offices and other work areas by means of partitions. These partitions may be moved, removed, or added to during the life of the structure. Section 4.3.2 of ASCE 7-16 requires that a partition load not less than 15 psf be considered if the floor live load is less than 80 psf—even if partitions are not shown on the drawings.

QUICK NOTE

The partitions referred to in ASCE 7-16 are not restricted to the portable partitions seen in some offices. It also includes the floor to ceiling walls constructed of metal studs and gypsum board—even though you may think they look permanent.

To convince the reader that the specified loads are reasonable, a brief examination of one of the specified values is considered. The example used here is the 100-psf live load specified by ASCE 7-16 for the lobbies of theaters and for assembly areas. Determine if such a load is reasonable for a crowd of people standing quite close together. Assume that the area in question is full of average adult males each weighing 165 pounds and each occupying an area 20 in. × 12 in. (1.67 ft^2). The average load applied equals 165 lb/1.67 ft^2 = 98.8 psf. As such, the 100-psf live load specified seems reasonable. It actually is on the conservative side, as it would be rather difficult to have men standing that close together over a floor area.

2.9 Live Load Impact Factors

DID YOU KNOW?

The magnitude of the specified live loads represents a maximum live load that would occur on average every 50 years. The life of an average building is assumed to be 50 years.

Impact loads are caused by the vibration and sudden stopping or dropping of moving or movable loads. It is obvious that a crate dropped on the floor of a warehouse or a truck bouncing on the uneven pavement of a bridge cause greater forces than would occur if the loads were applied gently and gradually. Impact loads are equal to the difference between the magnitude of the loads actually caused and the magnitude of the loads had they been purely static. In other words,

Table 2.4 Live Load Impact Factors

Equipment or component	Impact factor
Motor-driven machinery	20%
Reciprocating machinery	50%

impact loads result from the dynamic effects of a load as it is applied to a structure. For static loads, these effects are short lived and do not necessitate a dynamic structural analysis. They do, however, cause an increase in stress in the structure that must be considered. Impact loads usually are specified as percentage increases of the basic static live load. Table 2.4 shows the impact percentages for buildings given in Section 4.6 of ASCE 7-16 with the presence of machinery.

QUICK NOTE

The live load table given in ASCE 7-16 is assumed to already include enough allowance for ordinary impact conditions.

2.10 Live Loads on Roofs

The live loads that act on roofs are handled in ASCE 7-16 in a little different manner than are the other building live loads. The pitch of the roof (see Figure 2.3) affects the amount of load that realistically can be placed upon it. As the pitch increases, the amount of load that can be placed on the roof before it begins to slide off decreases. Furthermore, as the area of the roof that contributes to the load acting on a supporting component increases, it is less likely that the entire area will be loaded at any one time to its maximum live-load value.

The largest roof live loads usually are caused by repair and maintenance operations that probably do not occur simultaneously over the entire roof. This is not true of the environmental rain and snow loads, however, which are considered in Sections 2.11 and 2.12.

In the equations presented in this section, the term *tributary area* is used. This term is discussed in detail in Chapter 3 but is defined in the side margin for current use. When a building is being analyzed, even though it may not be exact, it is customary for the analyst to assume that the load supported by a member is the load that is applied to its tributary area. The tributary area for a column is shown in Figure 2.4 where it is defined by drawing lines halfway between each line of columns.

The basic minimum roof live load to be used in design is 20 psf. This value is specified in Table 4.1 of ASCE 7-16. Depending on the size of the tributary area and the pitch of the roof, this value may be reduced. The actual value to be used is determined with

$$L_r = (20 \text{ psf})R_1R_2$$

$$12 \text{ psf} \leq L_r \leq 20 \text{ psf}$$

The term L_r represents the roof live load—in psf—of the horizontal projection, whereas R_1 and R_2 are reduction factors. R_1 is used to account for the size of the tributary area A_t, and R_2 is

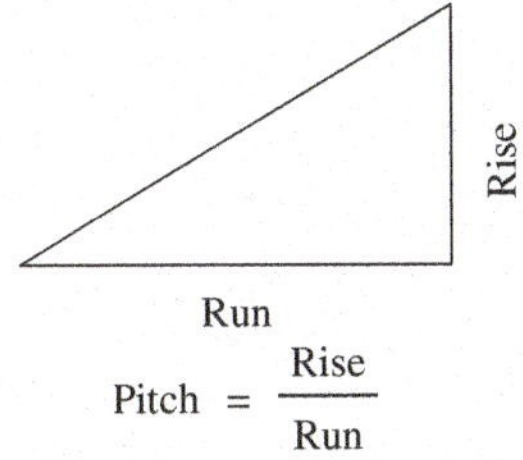

FIGURE 2.3 Illustration of the meaning of roof pitch, which is also shown as rise:run.

DEFINITION

Tributary area (A_t): The loaded area of a structure that directly contributes to the load applied to a particular member. The tributary boundary is often defined by drawing dividing lines midway between adjacent supporting members.

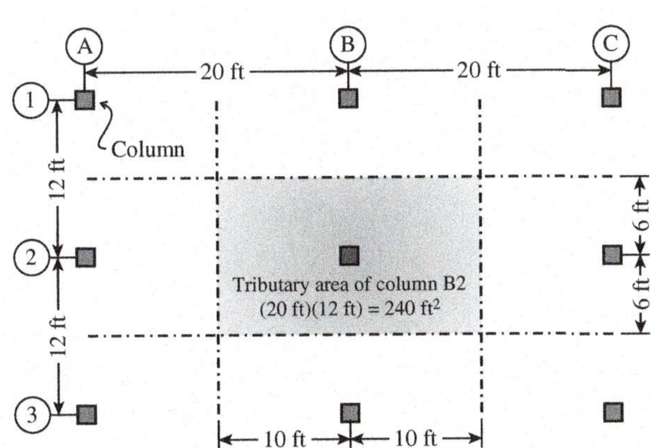

FIGURE 2.4 Tributary area for a column.

included to estimate the reduction effect of the pitch (rise) of the roof. The greater the tributary area (or the greater the rise of the roof), the larger the applicable reduction factor, and the smaller the roof live load. The maximum roof live load is 20 psf and the minimum is 12 psf. Expressions for computing R_1 and R_2 are

$$R_1 = \begin{cases} 1.0 & \text{if } A_t \leq 200 \text{ ft}^2 \\ 1.2 - 0.001A_t & \text{if } 200 \text{ ft}^2 < A_t < 600 \text{ ft}^2 \\ 0.6 & \text{if } A_t \geq 600 \text{ ft}^2 \end{cases}$$

$$R_2 = \begin{cases} 1.0 & \text{if } F \leq 4 \\ 1.2 - 0.05F & \text{if } 4 < F < 12 \\ 0.6 & \text{if } F \geq 12 \end{cases}$$

Where the term F represents the number of inches of rise of the roof per 12 inches (1 foot) of run.

EXAMPLE 2.5

For the column shown in Figure 2.4 (column B2) find the appropriate roof live load (L_r) that should be used if allowed reductions are applied. The roof has a pitch of 6:12 ⇒ rise:run.

SOLUTION STRATEGY

Find the reduction factors R_1 and R_2 for the tributary and the pitch, respectively. Compute the reduced live load, but ensure that it does not fall below the allowable minimum of 12 psf.

SOLUTION

Reduction Factor for R_1:

From Figure 2.4, the tributary area is calculated to be $A_t = 240$ ft^2. Since

$$200 \text{ ft}^2 \leq 240 \text{ ft}^2 \leq 600 \text{ ft}^2$$

Then

$$R_1 = 1.2 - 0.001A_t = 1.2 - 0.001(240) = 0.96$$

QUICK NOTE

For standard roofs, L_o is equal to 20 psf, which then may be reduced as shown. However, a roof garden should use a base live load of 100 psf with a different type of reduction scheme (explained in Chapter 3).

Reduction Factor R_2:

The pitch of the roof is 6:12, which means the rise $(F) = 6$. Since

$$4 \leq 6 \leq 12$$

Then

$$R_2 = 1.2 - 0.05F = 1.2 - 0.05(6) = 0.9$$

The reduced roof live load L_r is found from the following equation, where L_o is 20 psf for standard roofs.

$$L_r = L_o R_1 R_2 = (20 \text{ psf})(0.96)(0.9) = 17.3 \text{ psf}$$

Since

$$L_r > 12 \text{ psf}$$

we may use 17.3 psf.

SUMMARY

A reduction in the applied roof live load is appropriate where large tributary areas and large roof slopes exist. It is not reasonable to anticipate the full 20 psf developing everywhere under such conditions.

2.11 Rain Loads

Ponding is a problem that historically has been experienced by many flat roofs. If water accumulates more rapidly on a roof than it runs off, ponding results. The increased load causes the roof to deflect into a dish shape that can hold more water, which causes greater deflections, and so on—see Figure 2.5. This process continues until equilibrium is reached or collapse occurs. Through the proper selection of loads and creating a good design that provides adequate roof stiffness, the designer tries to avoid the latter situation.

During a rainstorm, water accumulates on a roof for two reasons. First, when rain falls, time is required for the rain to run off the roof. Therefore, some water will accumulate. Second, roof drains may not be even with the roof surface (Figure 2.6) and they may become clogged. Generally, roofs with slopes of 0.25 in. per ft or greater are not susceptible to ponding unless the roof drains become clogged, thus enabling deep ponds to form.

In addition to ponding, another problem may occur for very large, flat roofs (with perhaps an acre or more of surface area). During heavy rainstorms, strong winds frequently occur. If there is a great deal of water on the roof, a strong wind may very well push a large quantity of the water toward one end. The result can be a dangerous water depth influencing the load on that end of the roof. For such situations, *scuppers* are sometimes used. Scuppers are large holes or tubes in walls or parapets that enable water above a certain depth to quickly drain off the roof—see Figure 2.7.

Generally, two different drainage systems are provided for roofs. These normally are referred to as the primary and secondary drains. Usually, the primary system will collect the rainwater through surface drains on the roof and direct it to storm sewers. The secondary system consists of scuppers, pipes, or other openings through the walls or roof that permits the rainwater to drain if necessary—see Figure 2.8. The inlets of the secondary drains are normally located at elevations (d_s) above the inlets to the primary drains to prevent them from becoming clogged when the primary drain is clogged.

The secondary drainage system is used to provide adequate drainage of the roof in the event that the primary system becomes clogged or disabled in some manner. The rainwater design load therefore is based on the amount of water that can accumulate before the secondary drainage system becomes effective.

The determination of the water that can accumulate on a roof during a rainstorm depends upon local conditions, the elevation, and the type of secondary drains used. Chapter 8 of ASCE 7-16 specifies that the rain load (in psf) on an undeflected roof can be computed from

$$R = 5.2(d_s + d_h)$$

The term d_s is the depth of water (in inches) on the undeflected roof up to the inlet of the secondary drainage system when the primary drainage system is blocked. This is the *static head*, which can be determined from the drawings of the roof system. The term d_h is the additional depth of water on the undeflected roof above the inlet of the secondary drainage system at its design flow. This is the *hydraulic head*. It is dependent upon the capacity of the drains installed and the rate at which rain falls.

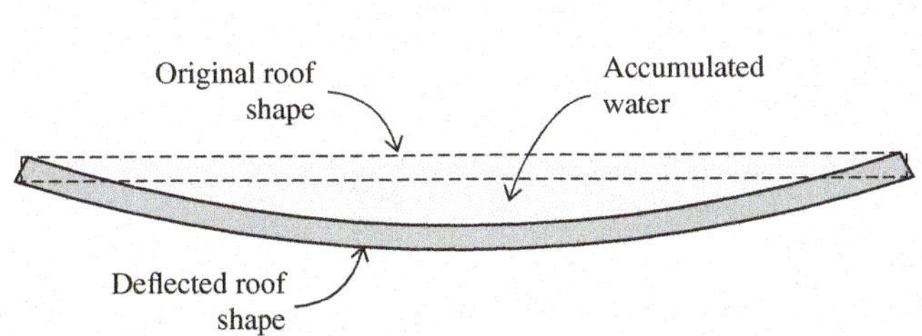

FIGURE 2.5 Illustration of the roof ponding effect.

Courtesy of Bryant G. Nielson

FIGURE 2.6 Standing water on a flat roof.

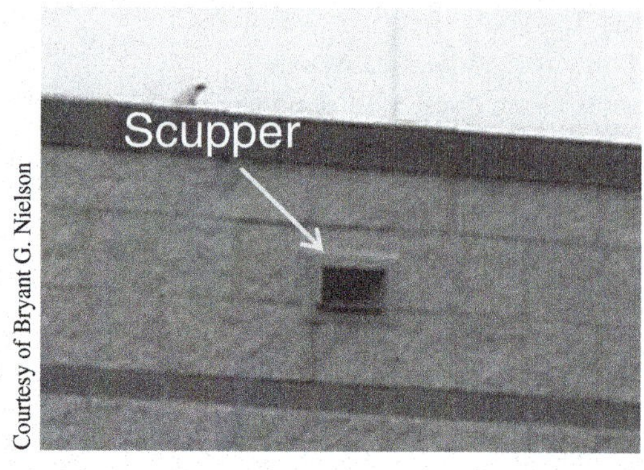

FIGURE 2.7 Scupper in the side of a parapet wall used to drain water from roof.

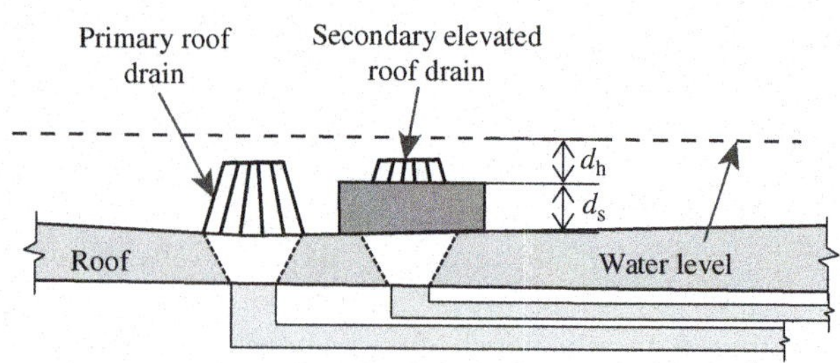

FIGURE 2.8 Roof drainage system with both primary and secondary drains. Notice the protective ring around the secondary drain to prevent clogging.

From Section 8.3 of the ASCE 7-16 Commentary, the flow rate (in gallons per minute) that a particular drain must accommodate can be computed from

$$Q = 0.0104Ai$$

The term A is the area of the roof (in square feet) that is served by a particular drain, and i is the rainfall intensity (in inches per hour). The rainfall intensity is specified by the code that has jurisdiction in a particular area. After the flow quantity is determined, the hydraulic head can be determined from Table 2.5 (from ASCE 7-16, Table C8.3-3) for the type of drainage system being used. If the secondary drainage system is simply runoff over the edge of the roof, the hydraulic head will equal zero.

Table 2.5 Flowrate Q (in Gallons per Minute) of Various Drainage Systems and Hydraulic Heads

	Hydraulic head d_h (in.)									
Drainage system	**1**	**2**	**2.5**	**3**	**3.5**	**4**	**4.5**	**5**	**7**	**8**
4-in. diameter drain	80	170	180							
6-in. diameter drain	100	190	270	380	540					
8-in. diameter drain	125	230	340	560	850	1100	1170			
6-in.-wide-channel scupper	18	50		90		140		194	321	393
24-in.-wide-channel scupper	72	200		360		560		776	1284	1572
6-in. wide, 4-in. high, closed scupper	18	50		90		140		177	231	253
24-in. wide, 4-in. high, closed scupper	72	200		360		560		708	924	1012
6-in. wide, 6-in. high, closed scupper	18	50		90		140		194	303	343
24-in. wide, 6-in. high, closed scupper	72	200		360		560		776	1212	1372

Note: Interpolation is appropriate, including between scupper widths. Closed scuppers are four–sided and channel scuppers are open–topped.

EXAMPLE 2.6

A roof measuring 240 feet × 160 feet has 6-in.-wide, channel-shaped scuppers serving as secondary drains. The scuppers are 4 inches above the roof surface and are spaced 20-feet apart along the two long sides of the building. The design rainfall for this location is 3 inches per hour. What is the design roof rain load?

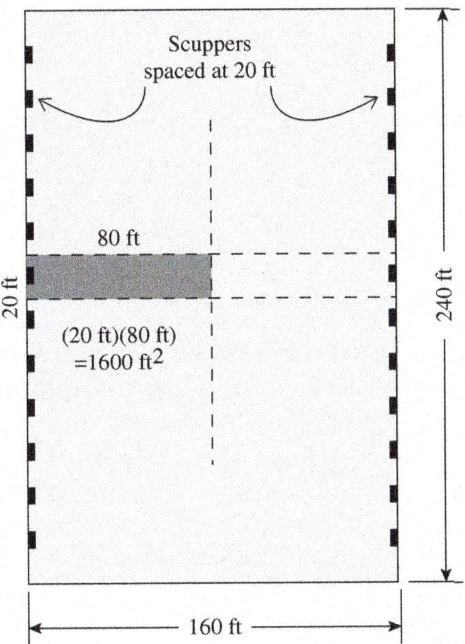

SOLUTION STRATEGY

Finding the area of the roof where each of the scuppers service is the biggest challenge. Find area A by calculating the spacing of the scuppers in each direction. Once Q is found, the hydraulic depth is found from Table 2.5.

SOLUTION

Sketch a schematic of the roof as shown. Each scupper will drain water from a region that is 20-ft wide (i.e., spacing in the long direction) and 80-ft wide.

$$A = (20\ \text{ft})(80\ \text{ft}) = 1600\ \text{ft}^2$$

Use the given equation to calculate the peak flow Q that is based on the area and intensity of the rainfall.

$$Q = 0.0104A_i = 0.0104(1600\ \text{ft}^2)(3\ \text{in./hr}) = 49.96\ \text{gpm}$$

Referring to Table 2.5, observe that the hydraulic head at this flow rate for the scupper is 2 in. ($d_h = 2$ in.). The static depth is taken as the vertical dimension from the primary to secondary drainage systems ($d_s = 4$ in.). Then

$$R = 5.2(d_s + d_h) = 5.2(4\ \text{in.} + 2\ \text{in.}) = \underline{\underline{31.2\ \text{psf}}}$$

QUICK NOTE

It is common in codes, standards and specifications to give equations that appear to be using inconsistent units. However, the constant in the equation (e.g., 0.0104) is designed to account for all appropriate unit conversions. **Make sure you know what your input and output units should be for a given equation.**

$$\text{ft}^3\left(\frac{\text{in}}{\text{hr}}\right)\left(\frac{1\text{hr}}{60\ \text{min}}\right)\left(\frac{7.48\text{gal}}{\text{ft}^3}\right) = 0.0104\frac{\text{gal}}{\text{min}}$$

SUMMARY

The rain load R is really the load applied to the roof at the deepest portion of the water at the location of the primary drain (see Figure 2.8). It may be appropriate to reduce this load in other portions of the roof where the water is not as deep due to the roof slope.

2.12 Snow Loads

In the colder states, snow and ice loads are often quite important. One inch of snow is equivalent to approximately 0.5 psf, but it may be higher at lower elevations where snow is denser. For roof design, snow loads are often in the range of 10 to 40 psf. The magnitude depends primarily on the geographic location, environmental conditions, slope of the roof, and—to a lesser degree—on the character of the roof's surface. The larger values are used for flat roofs and the smaller values for sloped roofs. Snow tends to slide off sloped roofs, particularly those with metal or slate surfaces. Studies of snowfall records in areas with severe winters indicate the occurrence of snow loads much greater than 40 psf with values as high as 150 psf in ski resort areas and the like.

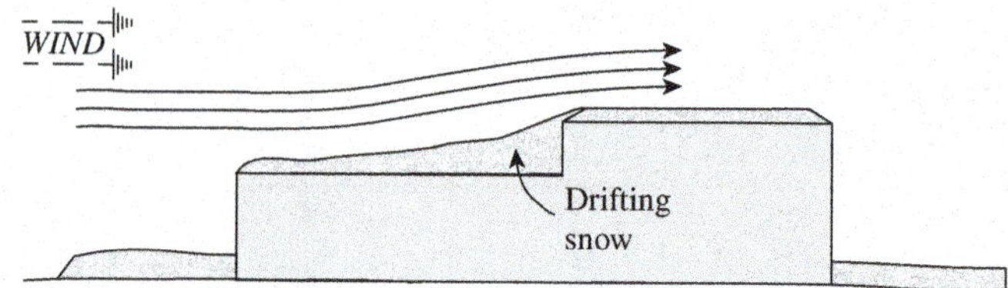

FIGURE 2.9 Wind can cause snow to accumulate (drift) against any projections present on the roof or ground.

Snow is a variable load that may cover an entire roof or only part of it. There can be drifts against walls or buildup in valleys or between parapets as seen in Figure 2.9. Snow may slide off one roof onto a lower one. The snow may blow off one side of a sloping roof, or it may crust over and remain in position—even during very heavy winds.

The snow loads that are applied to a structure are dependent upon many factors, including geographic location, the pitch of the roof, amount of shelter, and the shape of the roof. The discussion that follows is intended to provide only a cursory introduction to the determination of snow loads on buildings. When estimating these loads, consult ASCE 7-16 for information that is more complete and deals explicitly with issues related to drifting and sliding snow.

According to Section 7.3 of ASCE 7-16, the basic snow load to be applied to structures with flat roofs in the contiguous United States can be obtained from

$$p_f = 0.7C_eC_tI_Sp_g \tag{2.1}$$

QUICK NOTE

p_f stands for flat roof snow loads. ASCE 7-16, Section 7.4 shows how to account for snow on roofs having slopes (p_S) above a 1:12 pitch.

This expression is for unobstructed flat roofs with slopes equal to or less than 5° (a 1 in./ft slope is equal to 4.76°). In Equation 2.1, C_e is the exposure index. It is intended to account for the snow that can be blown from the roof because of the surrounding locality. The exposure coefficient is lowest for highly exposed areas and is highest when there is considerable sheltering. In essence, if a building is surrounded by many tall trees then the wind is going to have a more difficult time blowing the snow off of the roof—thus the structure experiences a higher snow load. Values of C_e are presented in Table 2.6 (from Table 7.3-1 in ASCE 7-16).

The terrain category and roof exposure condition chosen must be representative of the anticipated conditions during the life of the structure. In this table, the following definitions are used for the exposure of the roof:

Partially Exposed All roofs except as described as fully exposed or sheltered.

Fully Exposed Roofs exposed on all sides with no shelter afforded by terrain, higher structures, or trees. Roofs that contain several large pieces of mechanical equipment or other obstructions are not in this category.

Sheltered Roofs located tight in among conifers that qualify as obstructions.

The term C_t is the thermal index. Values of this coefficient are shown in Table 2.7 (from Table 7.3-2 in ASCE 7-16). As shown in the table, the coefficient is equal to 1.0 for heated structures, 1.1 for structures that are minimally heated to keep them from freezing, and 1.2 for unheated structures.

Table 2.6 Exposure Coefficients for Snow Loads

Terrain category (See ASCE 7-16 section 26.7)	Exposure of the roof		
	Fully exposed	Partially exposed	Sheltered
B: Urban and suburban areas	0.9	1	1.2
C: Open terrain with scattered obstructions	0.9	1	1.1
D: Unobstructed areas with wind over open water	0.8	0.9	1
Above the tree line in windswept mountainous areas	0.7	0.8	N/A
Alaska in areas with trees not within 2 miles of the site	0.7	0.8	N/A

Table 2.7 Thermal Factor for Snow

Representative anticipated winter thermal conditions	Ct
All structures except as indicated	1.0
Structures kept just above freezing with cold ventilated roofs in which the thermal resistance between ventilated and heated space exceeds 25° Fhft^2/BTU	1.1
Unheated and open air structures	1.2
Freezer building	1.3
Continuously heated greenhouses with a roof having a thermal resistance of less than 2.0° Fhft^2/BTU.	0.85

QUICK NOTE

As more heat escapes the building, it will tend to melt some of the snow and thus does not allow as large of a load to build up on the roof.

Risk Category and Importance Factor

The loads calculated for buildings in ASCE 7-16 are often selected to represent a maximum loading which can be expected to occur—on average—within a certain number of years. This recurrence interval is known as the *return period*. The shorter the return period (i.e., more frequent occurrences) that is considered, the smaller the computed design load is. An adjustment from the basic return period is given in the terms of an importance factor. The *importance factor* scales the base value either up or down depending on the level of importance that is assigned to the building. This importance assignment is based on risk categories that the structural engineer must assign the building based on its use and risk to life and property. Section 1.5 of ASCE 7-16 gives the criteria for making this assignment. A general interpretation of these risk categories is presented in Table 2.8.

The values of the importance factors, I_S, for snow loads are shown in Table 2.9 (from Table 1.5-2 in ASCE 7-16).

DID YOU KNOW?

Risk is more than just the potential of a building to fail. It includes the losses (both human life and economic) which could occur if the building does fail. Hence, the design loads for high-risk categories are higher than typical structures. This is done to lessen the likelihood that it fails.

Ground and Flat Roof Snow Loads

The last term in Equation (2.1) used for calculating roof snow is the ground snow load, p_g, in pounds per square foot (psf). Typical ground snow loads for the United States are shown in Appendix A. These values are dependent upon the climatic conditions at each site. If data are available that show local conditions are more severe than the values given in the ground snow figure in Appendix A, the local conditions should always be used and should be able to be obtained from local building officials. The minimum value of p_f is $(p_g)(I_S)$ in areas where the ground snow load is less than or equal to 20 psf. In other areas, the minimum value of p_f is $(20\text{ psf})(I_S)$.

QUICK NOTE

Other snow load conditions considered in ASCE 7-16 include:

- Snow drift
- Unbalanced snow
- Sliding snow
- Rain on snow
- Partial loading
- Ice dams

Table 2.8 Risk Categories

Category	Types of structures
I	Low risk to life—agriculture and other low populated buildings
II	All structures not fitting into categories I, III, or IV
III	Substantial risk to life and economics—high occupancy areas such as arenas, auditoriums, and also storage facilities for toxic substances.
IV	Essential facilities—hospitals, fire stations, etc.

Table 2.9 Importance Factor for Snow

Building risk category	I_S
I	0.8
II	1.0
III	1.1
IV	1.2

EXAMPLE 2.7

A shopping center is being designed for a location in Chicago. The building will be located in a residential area with minimal obstructions from surrounding buildings and the terrain. It will contain large department stores and enclosed public areas in which more than 300 people can congregate. The roof will be flat, but to provide for proper drainage, it will have a slope equal to 0.5 in./ft. What is the roof snow load that should be used for design?

SOLUTION STRATEGY

Because the slope of the roof is less than 5°, it can be designed as a flat roof. Look up appropriate modification factors based on local conditions along with the ground snow load. Be sure to also check the minimum required design loads.

SOLUTION

From the snow map in Appendix A, the ground snow load p_g for Chicago is 25 psf. The exposure factor C_e can be taken to equal 0.9 because there are minimal obstructions—though not necessarily an absence of all obstructions. Furthermore, because the building will be located in a residential area, it is unlikely there will be any obstructions to wind blowing across the roof. The thermal factor C_t is 1.0, because this will have to be a heated structure. Lastly, the risk category and associated importance factor I_S for the building are III and 1.10, respectively. This is because more than 300 people can congregate in one area. The design snow load to be used is then

$$p_f = 0.7C_eC_tI_Sp_g = 0.7(0.9)(1.0)(1.1)(25\text{ psf}) = 17.3\text{ psf}$$

Check the minimum requirement for the case where $p_g > 20$ psf.

$$p_f = (20\text{ psf})\,(1.1) = \underline{\underline{22\text{ psf}}} \Leftarrow \textbf{minimum controls}$$

SUMMARY

Snow load is based on many environmental and geometric conditions around the structure, but reasonable estimates may be made if this procedure is followed.

2.13 Wind Loads

A survey of engineering literature for the past 150 years reveals many references to structural failures caused by wind. Perhaps the most infamous of these have been bridge failures, such as those of the Tay Bridge in Scotland in 1879 (which caused the deaths of 75 persons) and the Tacoma Narrows Bridge (Tacoma, Washington) in 1940. However, there were some disastrous building failures due to wind during the same period, such as the Union Carbide Building in Toronto in 1958. It is important to realize that a large percentage of building failures due to wind have occurred during their construction.[3]

Considerable research has been conducted in recent decades on the subject of wind loads. Nevertheless, a great deal more study is needed, as the estimation of wind forces can by no means be classified as an exact science. Indeed, each subsequent edition of ASCE 7 continues to adjust and refine these calculations and procedures.

The average structural designer would love to have a simple rule with which he or she could compute the magnitude of the design wind loads, such as when the wind pressure is to be 20 psf for all parts of structures 100 ft or less above the ground and 30 psf for parts that are more than 100 ft above the ground. However, a simple specification such as this one—though of the type used for many years—has never been satisfactory. If we are to prevent future, perhaps catastrophic, mishaps, we must do better.

[3] "Wind Forces on Structures, Task Committee on Wind Forces". Committee on Loads and Stresses, Structural Division, ASCE, Final Report, *Transactions ASCE* 126, Part II (1961): 1124–1125.

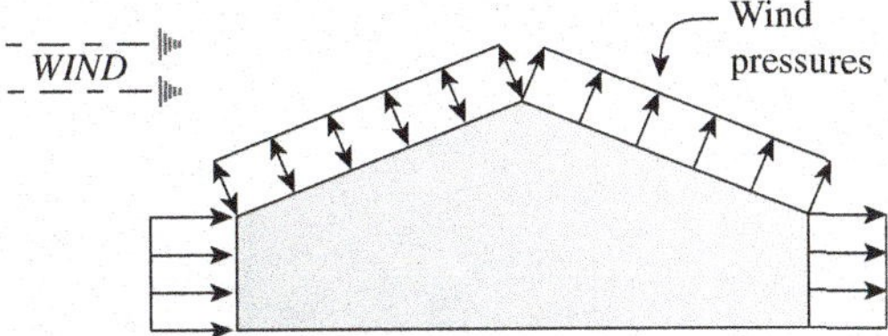

FIGURE 2.10 Pressures caused on a structure due to wind blowing. Arrows indicate the possible directions of the pressures for the given wind load direction.

Wind forces act as pressures on vertical windward surfaces, pressures or suctions on sloping windward surfaces (depending on slope), and suction or uplift on flat surfaces and on leeward vertical and sloping surfaces (due to the creation of negative pressures or vacuums) see Figure 2.10. The student may have noticed this definite suction effect where shingles or other roof coverings have been lifted from the roof surfaces of buildings. Suction or uplift easily can be demonstrated by holding a piece of paper horizontally at two of its adjacent corners and blowing above it. You will see that the far end of the paper moves upwards. For some common structures, uplift loads may be as large as 20 or 30 psf or even more.

The ASCE 7-16 standard provides equations for wind pressures that can be estimated for various parts of buildings. Though the use of these equations is complicated, the work is somewhat simplified with the tables and charts presented in the specification, and one is shown in Appendix A of this book with the permission of the ASCE. The reader should particularly note that the information provided is for buildings of regular shape. Should domes, A-frames, buildings with roofs sloped at angles greater than 45 degrees, or buildings with unusual floor plans such as H or Y shapes or others be encountered, it will be desirable to conduct wind tunnel studies. Guidelines for making such studies are presented in Chapter 31 of ASCE 7-16.

QUICK NOTE
Wind loads calculated using ASCE 7-16 do not consider tornadic winds.

The introductory discussion of wind forces presented in this section provides only a brief introduction to the topic. Furthermore, only wind forces applied to the main wind force resisting systems are discussed and then only for low-rise buildings with roof slopes of less than 45 degrees.

There are many factors which affect wind pressures. Among them are wind speed, the exposures of structures, characteristics of surrounding terrain, presence of nearby structures, and so on. The factors mentioned here are briefly discussed in the following paragraphs.

Design Wind Speed, V

The basic wind speed to be used in design for the locality involved may be estimated from wind speed maps provided in ASCE 7-16. (See the side bar note for practical determination of wind speeds.) The values provided in these maps are not applicable to mountainous areas, gorges, and other regions where unusual wind conditions may exist. For such areas, special studies will have to be made. The velocities obtained are the estimated worst 3-second gust speeds in miles per hour (mph) that would occur at 33 feet above the ground.

USEFUL TOOL
The wind speed maps provided in ASCE 7-16 can be difficult to read with any level of accuracy. An alternate tool to look up design wind speeds based on latitude/longitude or zip code is found at http://www.atcouncil.org/windspeed/

Risk Category

As with the snow load, the risk category to which a building belongs determines in part the magnitude of the wind load. Those structures which pose a higher risk to human life should be designed out of necessity for wind speeds and subsequent pressures that are higher. Unlike the snow load, importance factors are not used for this adjustment. Rather, there are four different wind speed maps provided in ASCE 7-16. One map is for each individual risk category. Wind speeds for all risk categories can be found using the tool identified in the side bar.

Surface Roughness Categories

In ASCE 7-16 Section 26.7, ground surfaces around structures are classified by their roughness as being B, C, or D. These classifications, alluded to in Table 2.6, run from urban and suburban areas with numerous closely spaced obstructions around a building to smooth mud or salt flats or unbroken ice. The roughness of the ground does impact the speed of the wind. Figure 2.11 illustrates wind velocity profiles for various surface roughness cases. The more rough the ground

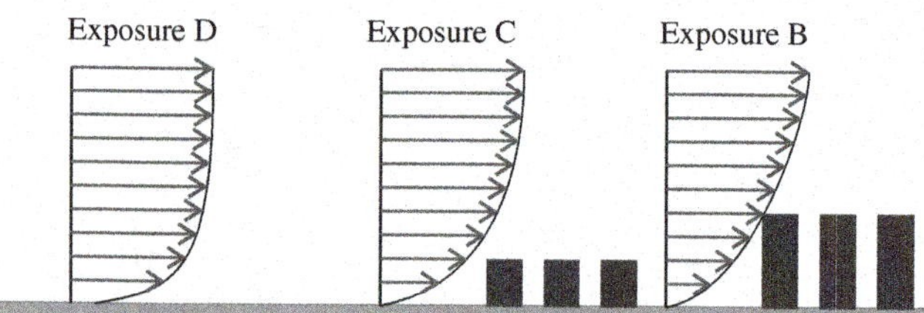

FIGURE 2.11 Effects of surface roughness on wind velocity.

surface is, the more the wind gets slowed down. An engineer is responsible for projecting the possible roughness conditions that will be present throughout the life of the structure.

A number of methods are presented by ASCE 7-16 for estimating wind loads. Among these methods is one which is called the *envelope procedure* (sometimes still called the simplified procedure). This procedure is outlined in Sections 28.4 and 28.5 of ASCE 7-16. Because it is quite simple to apply, it is restricted to buildings that meet a specific set of conditions. The other methods are more versatile but can be quite complex in their application and as such are beyond the scope of this text.

2.14 ASCE Envelope Procedure for Estimating Wind Loads

Section 28.4 of the ASCE Standard presents a simplified method for estimating wind loads. However, the procedure is only satisfactory for buildings that meet certain conditions. These conditions, listed in detail in Section 28.5.2 of the ASCE Standard, are summarized as follows:

1. The building must be low-rise, enclosed, rigid, nearly symmetrical, have a regular shape, and have a simple diaphragm (for flat roof or gable/hip roofs with slope $\leq 45°$).
2. Its roof height must not exceed 60 ft.
3. There may not be any expansion joints or separations in the building structure.
4. There are also some requirements concerning wind-borne debris and response characteristics (no topographical effects included).

QUICK NOTES

Wind loads for main structural members and systems, denoted as MWFRS, are different than those for wall and roof coverings, which belong to a group called components and cladding (C&C).

Chapter 28 of ASCE 7-16 gives loads for the MWFRS, while Chapter 31 gives the loads for the C&C. This text only demonstrates the MWFRS loads calculations.

The wind pressures on the *main wind force resisting system* (MWFRS) for such buildings may be estimated with

$$p_s = \lambda K_{zt} p_{S30}$$

To use this expression, the following values need to be determined:

V = design wind velocity based on location and risk category of building
p_{S30} = design wind pressure for exposure B at 30-ft height (see Appendix A)
λ = factor to adjust p_{S30} for actual building height and exposure (see Appendix A)
K_{zt} = topographic factor (assumed to be 1.0 for typical terrain)

The basic wind pressure, p_{S30}, is a function of where it is acting on the building. In zones close to the edge of building walls and roofs, the wind pressure tends to be higher than those acting on the inner regions of the walls and roofs (see Appendix A). The figure in Appendix A shows two cases—A and B. When the roof is sloped, both cases should be shown. However, if the building has a flat roof (less than 5°), case A should be used for both orthogonal loading directions.

Example 2.8 presents sample calculations for estimating wind pressures on the MWFRS using the simplified envelope procedure.

EXAMPLE 2.8

The building shown here is to be constructed in a suburban area with numerous closely spaced small buildings in Tampa on the Florida Gulf Coast (Latitude: 27.959°; Longitude: −82.413°). Its primary use will be for hotel rooms, and there will be no areas in which more than 300 people can congregate. Compute the estimated wind pressures acting on the various areas of this enclosed, rigid-frame, simple diaphragm structure using the ASCE 7-16 simplified envelope procedure.

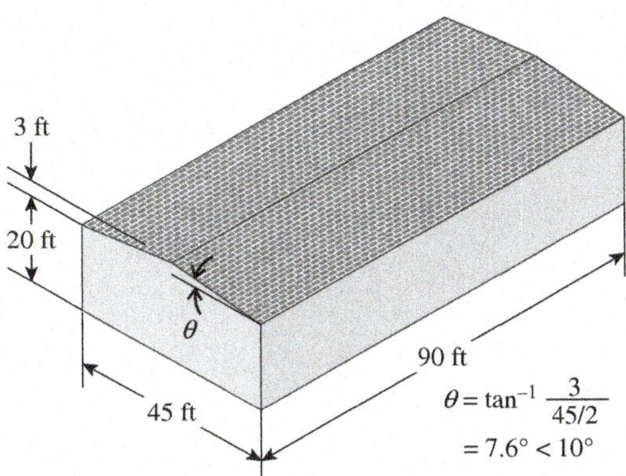

SOLUTION STRATEGY

We must classify the building into a risk category to get the design wind speed. This speed will then allow us to look up the pressures for each of the zones on the building, as shown in the figure in Appendix A. Make adjustments to these pressures based on the exposure and height of the building.

SOLUTION

Because the building is not considered to have high congregation areas nor is it considered to be an essential facility, it is classified as risk category II.

- Wind speed = 140 mph for Tampa (From http://windspeed.atccouncil.org for the given latitude/longitude)
- Surface roughness category will be C (from Table 2.6)
- $\lambda = 1.29$ (by interpolation from Appendix A for 20 ft)
- $K_{zt} = 1.0$ (no terrain irregularities)

Table 2.10 Estimated Wind Pressures for Building

	Horizontal pressure, p_{S30} (psf)			Adjusted pressure (psf),
Zone	If $\theta = 5°$	If $\theta = 10°$	If $\theta = 7.6°$	$p_s = \lambda K_{zt} p_{S30}$
A	31.1	35.1	33.2	42.8
B	−16.1	−14.5	−15.3	−19.7
C	20.6	23.3	22.0	28.4
D	−9.6	−8.5	−9.0	−11.6
E	−37.3	−37.3	−37.3	−48.1
F	−21.2	−22.8	−22.0	−28.4
G	−26.0	−26.0	−26.0	−33.5
H	−16.4	−17.5	−17.0	−21.9

Estimated pressures or suctions are represented by p_{S30} in the table in Appendix A. The applicable values are selected for the building in question and are recorded in Table 2.10. Then these values are multiplied by (λ) to adjust them for our particular conditions where $\lambda = 1.29$. The values obtained are shown in Table 2.10 for the various zones A to H in the associated drawing shown here. Note that negative signs represent suctions.

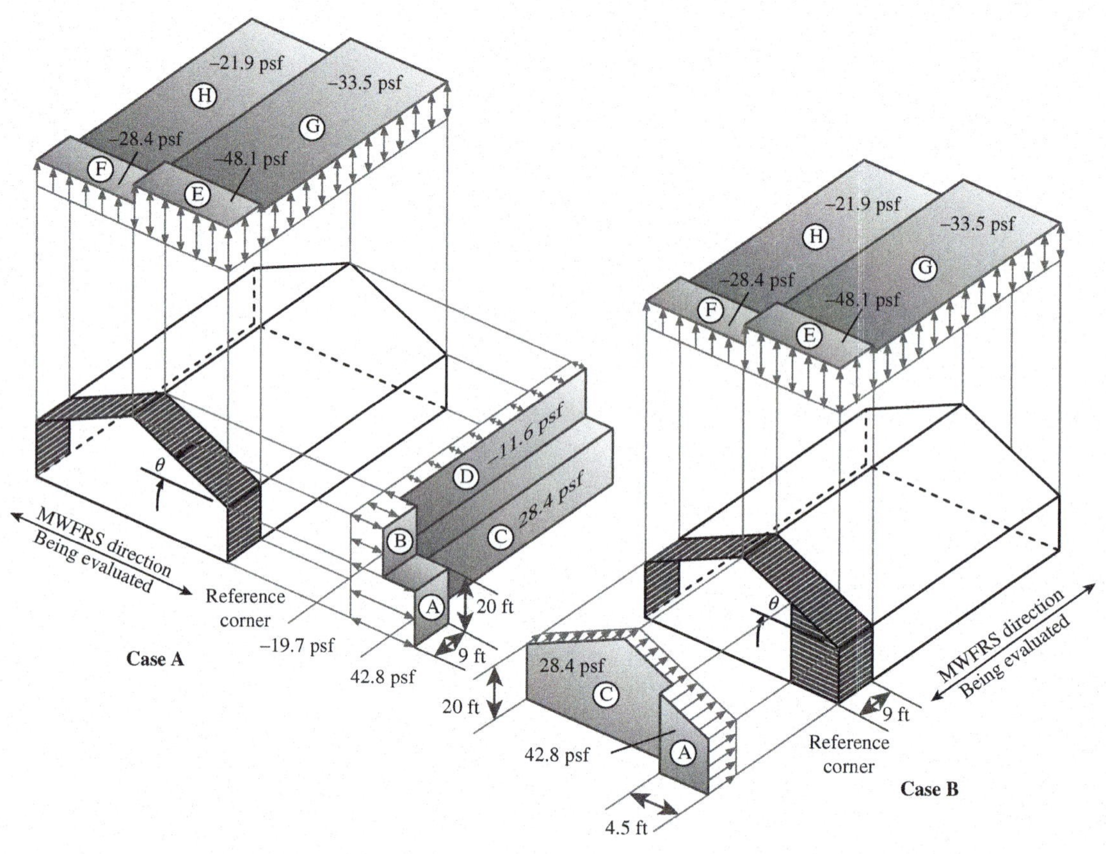

2.15 Seismic Loads

DID YOU KNOW?
The Interagency Committee on Seismic Safety in Construction has prepared an order (signed by the president) stating that all federally owned or leased buildings as well as all federally regulated and assisted construction should be constructed to mitigate seismic hazards.

Many areas of the world fall into "earthquake territory," and in those areas, it is necessary to consider seismic forces in the design of all types of structures. Through the centuries, there have been catastrophic failures of buildings, bridges, and other structures during earthquakes. For example, the 1989 Loma Prieta and 1994 Northridge earthquakes in California caused many billions of dollars of property damage as well as considerable loss of life.

The earth's outer crust is composed of hard plates as large or larger than entire continents. These plates float on the soft molten rock beneath. Their movement is very slow—perhaps only a few inches per year. It has been noted that this is slower than your fingernails grow. When the plates contact each other, significant horizontal and vertical motion of the ground surface may be caused. These motions may cause very large inertial forces in structures. The distribution and characteristics of the geologic makeup in a particular area will greatly affect the magnitudes of the ground motions caused.

Recent earthquakes have clearly shown that the average building or bridge that has not been designed for earthquake forces can be destroyed by a relatively moderate earthquake. Most structures can be economically designed and constructed to withstand the forces caused during most earthquakes. On the other hand, the cost of providing seismic resistance to existing structures (called retrofitting) can be extremely high.

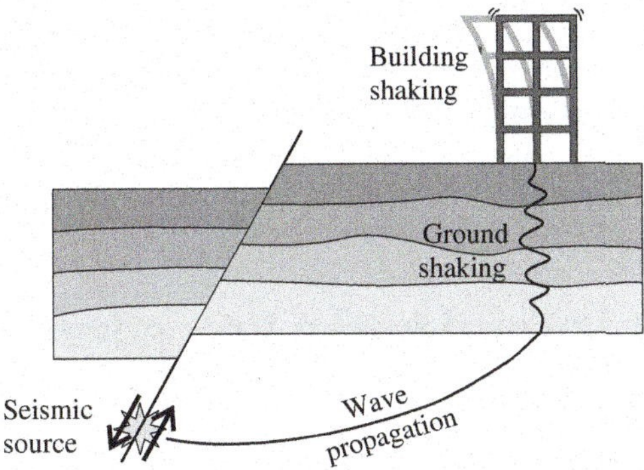

FIGURE 2.12 Schematic of earthquake-induced building response.

Seismic loads are different in their action and are not proportional to the exposed area of the building as are wind loads. Rather, they are proportional to the distribution of the mass (weight) of the building above the particular level being considered.

Building codes around the United States where earthquakes are most likely to occur require that some type of seismic design be used. Even in areas less prone to earthquakes, seismic loading should be seriously considered for high-rise buildings, hospitals, nuclear power plants, and other important structures.

To assess the importance of designing for earthquake forces, a seismic-induced structure acceleration map, such as the one shown on the USGS website,[4] should be examined. This particular map presents what are thought to be the estimated structural accelerations that might very well occur in various parts of the United States. High-risk areas for damaging earthquakes, such as coastal California, are quite obvious on this map, while low-risk areas are also clearly shown in Florida and parts of Texas.

Earthquakes apply loads to structures in an indirect fashion. The ground is displaced, and since the structures are connected to the ground, they are subject to sudden movements. These movements generate accelerations in the building leading to differential movement of the building levels (see Figure 2.12). These deformations cause horizontal shears to be produced. From this information, it is clear that no external forces are actually applied to buildings above ground by earthquakes.

The usual procedure for analyzing structures for seismic forces is to represent them with sets of equivalent static loads. The magnitudes of the loads selected are based upon the distribution of mass in the various structures and the accelerations of the structure that account for the dynamic characteristics of the systems. Another factor to be considered in seismic design is the soil condition. Almost all of the structural damage and loss of life in the 1989 Loma Prieta earthquake located in the bay area of California occurred in areas having soft clay soils. Apparently, these soils amplified the motions of the underlying rock. Figure 2.12 shows this amplification as the seismic waves travel vertically through the soil.

Chapters 12 and 16 of ASCE 7-16 present five different seismic analytical procedures for estimating seismic forces to be used in structural analysis and design. The first of the methods is a 'simplified analysis.' As such, it is quite restrictive on where and how it may be applied. Indeed, ASCE 7-16 Section 12.14.1.1 lists 12 conditions that must be met to allow the use of this method.

The *equivalent lateral force* (ELF) procedure is the method today's designer needs to be the most familiar with, as it is the most commonly used method in practice in the United States. Consequently, it is the only one of the five methods illustrated in the next section of this text.

The third method is called the *modal analysis* procedure and is generally used instead of the ELF method when some floors of a building are quite different from the other floors with respect to stiffness, weight, and so forth. The final two methods are called the *linear response history* analysis and *nonlinear response history* analysis and are used in design practice today on a limited basis. This is because of the length of time required to do such an analysis. However, these two methods are used extensively in various research studies.

QUICK NOTE

The acceleration terms given in ASCE 7-16's earthquake maps are called spectral accelerations (S_S and S_1). The spectral accelerations give some idea of the accelerations that occur in a structure. They are presented in terms of %g, which is a % of gravity.

RECALL

Newton's second law states that the force on an object is a function of its mass and the acceleration to which it is subjected. Thus,

$$F = ma$$

[4] http://earthquake.usgs.gov/hazards/designmaps/pdfs/?code=ASCE+7&edition=2010.

QUICK NOTE

Table 12.6-1 of ASCE 7-16 provides guidance as to what seismic analysis methods may be used under various conditions.

As discussed in Sections 2.5 and 2.6, the calculated loadings on structures and their components due to seismic effects are combined with other loads. In making the combinations, the reader should note that the methods just mentioned are based on a strength limit state beyond the first yielding of the structures.

2.16 Equivalent Lateral Force Procedure for Seismic Loads

DID YOU KNOW?

The seismic loads calculated for the design of buildings assume that the structure is going to yield (sustain damage) under load. Generally speaking, an engineer's objective in seismic design is not to eliminate damage but rather control where and how much of it occurs.

In this section, the authors briefly introduce the equivalent lateral force procedure for estimating seismic forces. As the student studies this method, the authors want them to give a great deal of thought to the accuracies obtained. With this procedure, equivalent static loads are computed to estimate the effect of dynamic seismic forces. In addition, the structures considered are assumed to resist the design loads in an elastic manner—but their resistance is actually inelastic during a design event.

With the equivalent lateral force procedure, a total base shear is estimated based on properties of the structure and the ground motion expected at the building site. Empirical equations are presented to estimate the total lateral shear applied to the building and to apportion that shear to the various floor levels. These equations in the following steps contain several terms that are listed and defined here in the order they will be used in the calculations.

PERIOD CALCULATION VARIABLES

Steel moment frame
$C_t = 0.028$ and $x = 0.8$

Concrete moment frame
$C_t = 0.016$ and $x = 0.9$

Eccentrically braced frame
$C_t = 0.03$ and $x = 0.75$

Steel buckling restrained frame
$C_t = 0.02$ and $x = 0.75$

A description of structure types can be found in Section 4.3 of this text.

1. The *fundamental natural period of a building*, which is represented by the letter T and given in units of seconds, is the time required for the building to go through one complete cycle of motion. Its magnitude is dependent upon both the mass of the structure and its stiffness and can be estimated with

$$T_a = C_t h_n^x \tag{2.2}$$

The application of Equation 2.2 usually provides periods that are somewhat smaller than the real periods of the structures involved. Such a situation causes the calculated shears to be a little high and hopefully puts us on the conservative side. In Equation 2.2 the term h_n is the height of the highest level of the building. Typical values for the other terms as given in ASCE 7-16 are shown in the side margin.

2. The *design spectral accelerations*, represented by the terms S_{D1} and S_{DS}, may be determined from seismic maps or software put out by the United States Geological Survey (USGS). Such tools provide the estimated intensities of design earthquakes with $T = 1$ second (S_{D1}) and with $T = 0.2$ seconds (S_{DS}) and are dependent on the location and soil type under the structure. The numbers obtained represent proportions of g, which is the gravitational acceleration. For example, for a site in Salt Lake City, Utah (latitude = 40.79° and longitude = −111.98°) located on rock (soil type B), the design values are $S_{DS} = 1.07g$ and $S_{D1} = 0.37g$.

USEFUL TOOL

The spectral maps provided in ASCE 7-16 can be difficult to read with any level of accuracy. An alternate tool is to look up design spectral accelerations based on latitude/longitude or zip code and soil type at **http://geohazards.usgs.gov/designmaps/us/**

3. The *response modification factor*, R, is used to estimate the ability of a structure to resist seismic forces. Its value varies from 1.0 up to 8.0 with the high values applicable to ductile structures and the lower values applicable to brittle structures. For structures with reinforced concrete shear walls, R values of about 4 are used. For structural steel frames and for reinforced concrete frames that are ductile with rigid joints, R will equal approximately 8. Other values are provided in ASCE 7-16. The larger the R value, the smaller are the computed seismic design forces.

4. The *importance factor*, I_e, of a structure provides a measure of the consequences of failure. The higher the number, the more important is the structure. For instance, ASCE 7-16 (see Table 2.11) provides a value of 1.5 for hospitals, police stations, and other public buildings but only 1.0 for office buildings.

Table 2.11 Importance Factor for Earthquake Loads

Building risk category	I_e
I	1.0
II	1.0
III	1.25
IV	1.5

5. The *effective seismic weight* of a building, W, includes the total dead load of the structure plus applicable portions of other loads. For instance, a minimum of 25% of floor live loads must also be included if the space is used for storage. It must also include a 10 psf allowance for partitions if they are present. Furthermore, W must include the total weight of permanent equipment. Where flat roof snow loads exceed 30 psf, 20% of the snow load is included in the seismic weight.

6. Section 12.8 of ASCE 7-16 provides the following guidance for estimating the total static lateral base shear in a given direction for a building:

$$V = \frac{S_{D1}}{T_a\left(R/I_e\right)}W \tag{2.3}$$

When very stiff structures with small T_a values are involved, Equation 2.3 yields unnecessarily high values. Therefore, the calculated shear, V, need not be greater than the value obtained with.

$$V_{max} = \frac{S_{DS}}{\left(R/I_e\right)}W$$

A practical minimum value of shear is given with

$$V_{min} = 0.044 S_{DS} I_e W$$

7. The portion of the base shear V to be distributed to a particular floor is determined for the seismic force as

$$F_x = \frac{w_x h_x^k}{\sum_{i=1}^{n} w_i h_i^k} V$$

where

F_x = lateral seismic force to be applied to level x

w_i and w_x = the weights assigned to levels i and x.

h_i and h_x = height of levels i and x above the ground

k = a distribution exponent related to the fundamental natural period of the structure in question. If T_a is 0.5 seconds or less, k = 1.0. Should T_a be > 0.5 seconds and < 2.5 seconds, k can be determined as

$$k = 1 + \frac{T - 0.5}{2}$$

if $T_a > 2.5$ seconds then $k = 2.0$

QUICK NOTE

Ductility is a measure of how far past yield a structure can be loaded without breaking. Steel is typically a very ductile material, whereas glass is a very brittle material. Ductile structures are preferred for earthquake-prone regions due to their improved safety.

QUICK NOTE

The subscripts in the equation for F_x can tend to be confusing.

x = the building level being considered

n = the total number of building levels

i = an index which can take values from 1 to n

EXAMPLE 2.9

Using the ASCE 7-16 equivalent lateral force procedure, compute the lateral force to be applied to the third floor of the proposed office building shown. The building, which is to be located in Salt Lake City, is a steel frame having special moment-resisting connections (R = 8). Values of $S_{DS} = 1.07g$ and $S_{D1} = 0.37g$ have been identified. The estimated weight, w, of each level is 500 k—except the roof, which is 200 k.

SOLUTION STRATEGY

It is important to identify all of the information given for a building and the site it is located on. This includes the structure type. Once this is found, the seven steps listed can be performed.

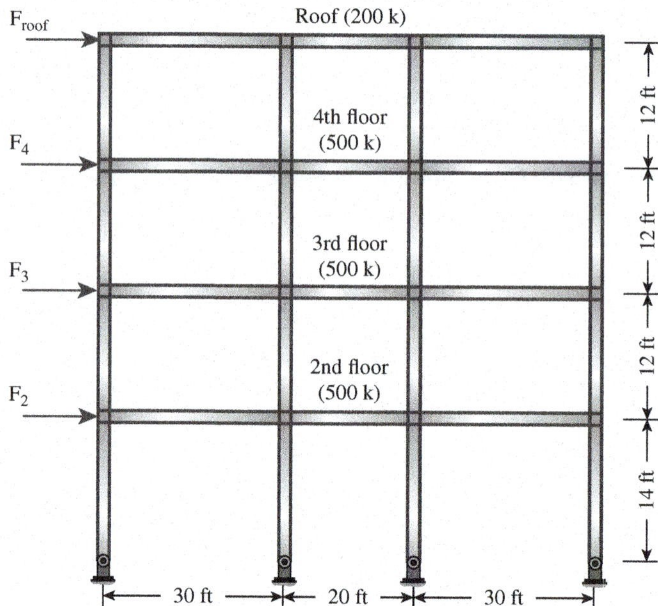

SOLUTION

1. $T_a = C_t h_n^x = 0.028(50\text{ ft})^{0.8} = 0.64s$ (C_t and x values are for a steel moment frame building. h_n is the total height of the building).
2. $S_{DS} = 1.07g$ and $S_{D1} = 0.37g$ (given).
3. R = 8 (special moment resisting steel frame).
4. Since the building is an office building it belongs to a risk category II (see Table 2.8). Looking at Table 2.11, we find the seismic importance factor $I_e = 1.00$.
5. Add up the weight of the entire building.

$$W = 3(500\ k) + (200\ k) = 1700\text{ k}$$

6. Calculate the total base shear V in the building.

$$V = \frac{S_{D1}}{T_a\left(R/I_e\right)} W = \frac{0.37}{0.64\left(8/1.0\right)}(1700\text{ k}) = 122.9\text{ k}$$

Not more than

$$V_{max} = \frac{S_{DS}}{(R/I_e)}W = \frac{1.07}{(8/1.0)}(1700\text{ k}) = 227.4\text{ k}$$

but not less than

$$V_{min} = 0.044S_{DS}I_eW = 0.044(1.07)(1.0)(1700\text{ k}) = 80.0\text{ k}$$

So use

$$V = 122.9\text{ k}$$

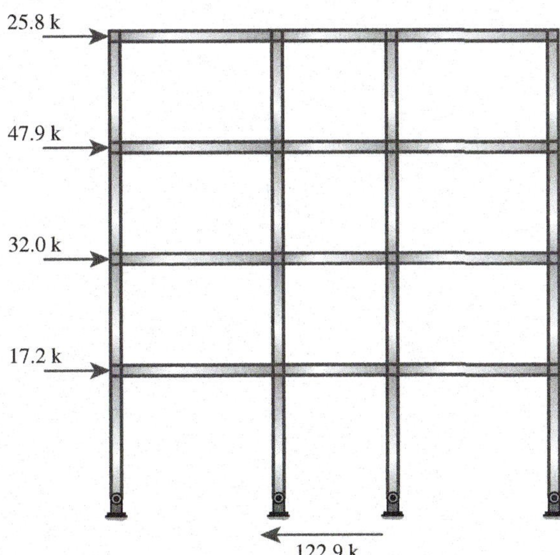

7. Determine the lateral force applied to the third floor (F_3) by first solving for k, which is based on the value for T_a. So

$$k = 1 + \frac{T_a - 0.5}{2} = 1 + \frac{0.64 - 0.5}{2} = 1.07$$

Now the following equation may be used to find F_3

$$F_3 = \frac{w_3h_3^k}{\sum_{i=2}^{5} w_ih_i^k}$$

$$= \frac{(500\text{ k})(26\text{ ft})^{1.07}}{(500\text{ k})(14\text{ ft})^{1.07} + (500\text{ k})(26\text{ ft})^{1.07} + (500\text{ k})(38\text{ ft})^{1.07} + (500\text{ k})(50\text{ ft})^{1.07}}(122.9\text{ k})$$

$$= (0.26)(122.9\text{ k}) = \underline{\underline{32.0\text{ k}}}$$

SUMMARY

In a like fashion, the equivalent seismic forces that are applied at the other levels can be found. All level forces are shown in the second figure. Note that the sum of the forces is equal to the base shear of 122.9 k.

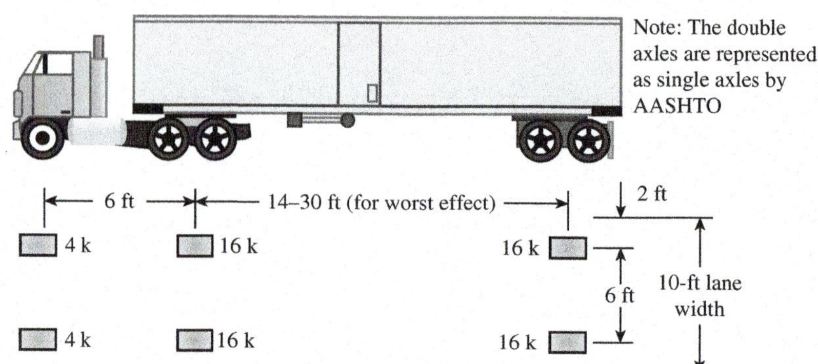

FIGURE 2.13 ASSHTO design truck.

2.17 Highway Bridge Loads

DID YOU KNOW?

The design tandem accounts for high loadings caused by military vehicles.

DESIGN LANE LOAD

The design load for a 10-ft-wide lane is to be taken as

$$640\text{lb/ft} = 0.64\text{ klf}$$

Until about 1900, bridges in the United States were 'proof loaded' before they were considered acceptable for use. Highway bridges were loaded with carts filled with stone or pig iron, and railway bridges were loaded with two locomotives in tandem. Such procedures were probably very useful in identifying poor designs and/or poor workmanship, but they were no guarantee against overloads and fatigue stress situations.[5]

As discussed in *America's Highways 1776–1976*, during much of the 19th-century highway bridges were designed to support live loads of approximately 80 to 100 psf applied to the bridge decks. These loads supposedly represented large, closely spaced crowds of people walking across the bridges. In 1875, the American Society of Civil Engineers (ASCE) recommended that highway bridges should be designed to support live loads varying from 40 to 100 psf—the smaller values were to be used for very long spans. The Office of Public Roads published a circular in 1913 recommending that highway bridges be designed for a live loading of (a) a series of electric cars or (b) a 15-ton road roller and a uniform live load on the rest of the bridge deck.

Although highway bridges must support several different types of vehicles, the heaviest possible loads are caused by a series of trucks. In 1931, the American Association of State Highway and Transportation Officials (AASHTO) Bridge Committee issued its first printed edition of the *AASHTO Standard Specification for Highway Bridges*.

Today the *AASHTO LRFD Bridge Design Specifications* state that highway bridges should be designed for motor trucks occupying 10-ft-wide lanes. Specifically, it uses what is called the 1993 highway loading (HL-93). This loading consists of a design truck, a design tandem, and a design lane load. The designer must place either a design truck or a design tandem in each design lane of the bridge concurrently with the design lane load. The design truck is represented in Figure 2.13, the design tandem in Figure 2.14, and the two-lane loading conditions are represented in Figure 2.15.

QUICK NOTE

Be aware that the actual loading which is applied to a single bridge girder is a function of the percentage of a single lane of traffic that the girder takes. This depends on the bridge type and girder spacing and is accounted for in what is called a *distribution factor*.

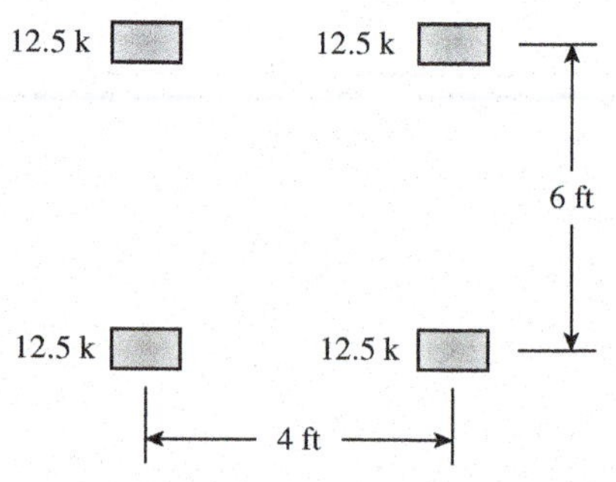

FIGURE 2.14 AASHTO design tandem.

[5]U.S. Department of Transportation, Federal Highway Administration, *America's Highways 1776–1976* (Superintendent of Documents, U.S. Government Printing Office, 1976), 429–432.

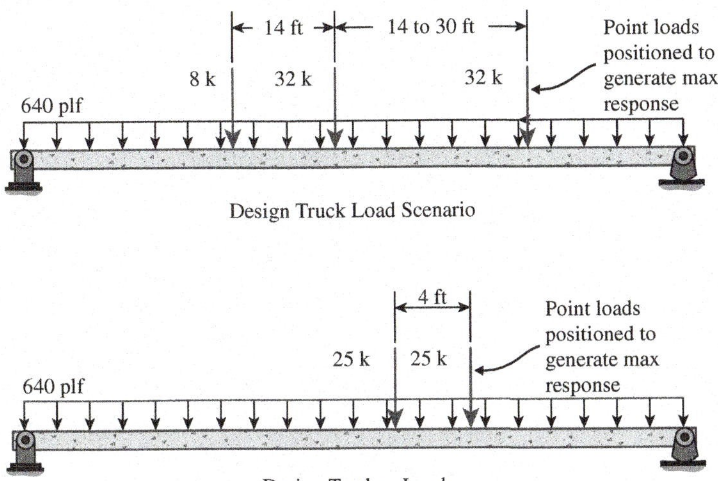

QUICK NOTE

The point loads in Figure 2.15 represent the total load for an entire axle load. For example, the single patch load in a design tandem is 12.5 k. This creates an axle load of $(12.5\text{ k})(2) = 25\text{ k}$.

FIGURE 2.15 AASHTO lane load scenarios.

2.18 Railway Bridge Loads

Railway bridges are commonly analyzed for a series of loads devised by Theodore Cooper in 1894. His loads, which are referred to as E loads, represent two locomotives with their tenders followed by a line of freight cars, as shown in Figure 2.16(a). A series of concentrated loads is used for the locomotives, and a uniform load represents the freight cars, as pictured in Figure 2.16(c). The E-40 train is assumed to have a 40-kip load on the driving axle of the engine. Today bridges are designed based on an E-80 loading or larger.

If information is available for one E loading, the information for any other E loading can be obtained by direct proportion. The axle loads of an E-75 are 75/40 of those for an E-40; those for an E-60 are 60/72 of those for an E-72; and so on. As such, the axle loads for an E-80 load are twice those shown in Figure 2.16.

The American Railway Engineering Association also specifies an alternate loading. That loading is shown in Figure 2.17. This load or the E-80 load—whichever causes the greatest stress in the components—is to be used.

As we can see from the modern locomotives in Figure 2.16(b), Cooper's loads do not accurately picture today's trains. Nevertheless, they are still in general use despite the availability of several more modern and more realistic loads.

QUICK NOTE

Today's railway loads are significantly larger than the E-40 load. For example, E-72 and E-80 are not uncommon.

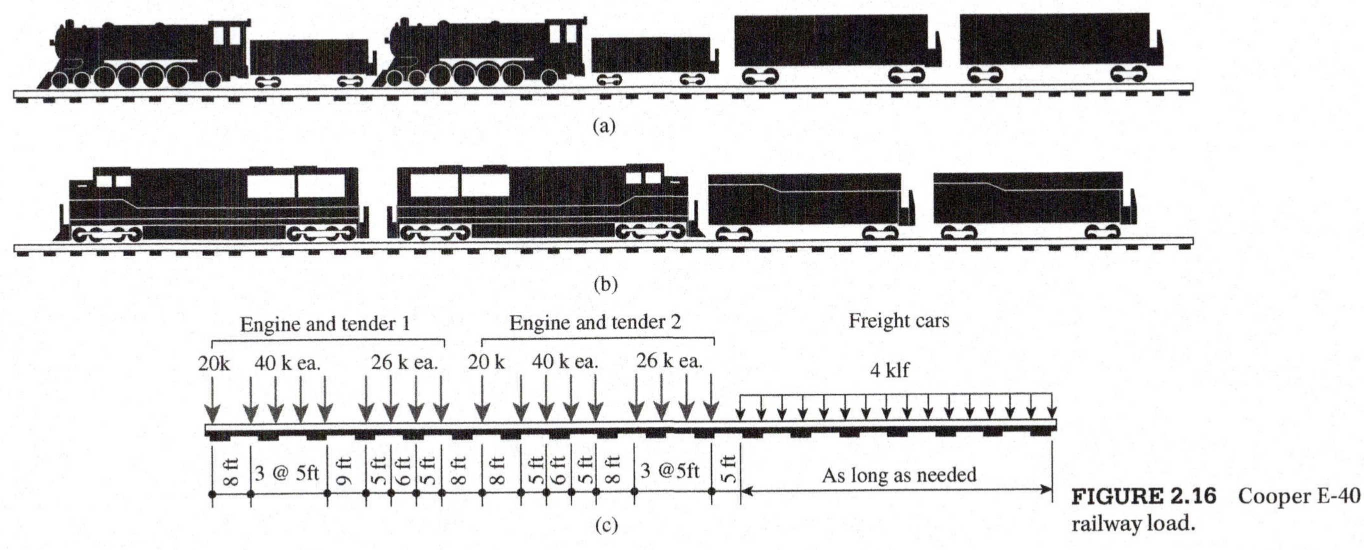

FIGURE 2.16 Cooper E-40 railway load.

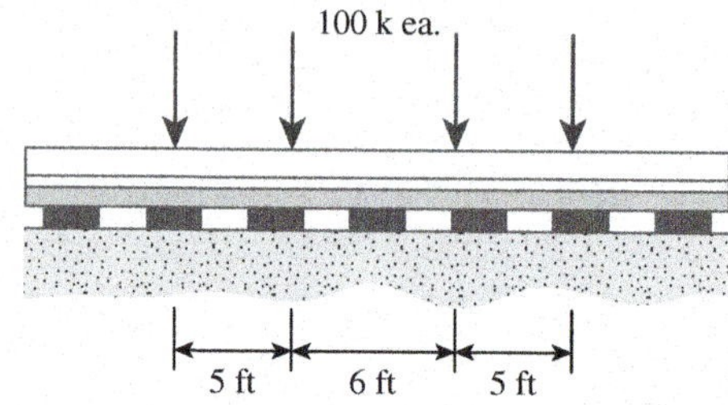

FIGURE 2.17 Alternate railroad loading.

2.19 Other Loads

There are quite a few other kinds of loads that the designer may occasionally face. These include the following:

Ice Loads

Ice has the potential of causing the application of extraordinarily large forces to structural members. Ice can emanate from two sources: (1) surface ice on frozen lakes, rivers, and other bodies of water and (2) atmospheric ice (freezing rain and sleet). The latter can form even in warmer climates.

In colder climates, ice loads often will greatly affect the design of marine structures. One such situation occurs when ice loads are applied to bridge piers. For such situations, it is necessary to consider dynamic pressures caused by moving sheets of ice, pressures caused by ice jams, and uplift or vertical loads in water of varying levels causing the adherence of ice.

The breaking up of ice and its movement during spring floods can significantly affect the design of bridge piers. During the breakup, tremendous chunks of ice may be heaved upward, and when the jam breaks up, the chunks may rush downstream, striking and grinding against the piers. Furthermore, the wedging of pieces of ice between two piers can be extremely serious. It is thus necessary to either keep the piers out of the dangerous areas of the stream or to protect them in some way. Section 3 of the AASHTO specifications provides formulas for estimating the dynamic forces caused by moving ice.

Bridges and towers—any structure for that matter—are sometimes covered with layers of ice from 1 to 2 in. thick. The weight of the ice runs up to about 10 psf. A factor that influences wind loads is the increased surface area of the ice-coated members.

Atmospheric icing is discussed in Chapter 10 of ASCE 7-16. Detailed information for estimating the thickness and weight of ice accumulations is provided. This ice typically accumulates on structural members. The thickness of the ice accumulation must be determined from historic data for the site. It can also be determined from a meteorological investigation of the conditions at the site. The extent of ice accumulation is such a localized phenomenon that general tables cannot be reasonably prepared.

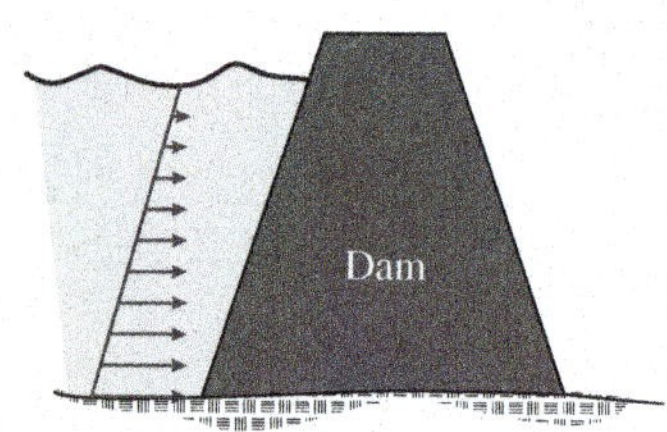

FIGURE 2.18 Hydrostatic load.

Miscellaneous Loads

Here are some of the many other loads with which the structural designer will have to contend.

Soil pressures The exertion of lateral earth pressures on walls or upward pressures on foundations

Hydrostatic pressures Water pressure such as on dams (Figure 2.18), inertial forces of large bodies of water during earthquakes, and uplift pressures on tanks and basement structures

Flooding Caused by heavy rain or melting snow and ice

Blast loads Caused by explosions, sonic booms, and military weapons

Thermal forces Due to changes in temperature resulting in structural deformations and internal structural forces

Centrifugal forces Those caused on curved bridges by trucks and trains, or similar effects on roller coasters, etc.

Longitudinal loads Caused by stopping trucks or trains on bridges, ships running into docks, and the movement of traveling cranes that are supported by building frames

2.20 Examples with Video Solutions

VE2.1 A roof framing plan for a given building is shown in Figure VE2.1. Calculate the roof live load to be applied to the column located on grid point A4. Take reductions if appropriate. (Assume that the roof pitch is 0.5 inches in 1 foot.)

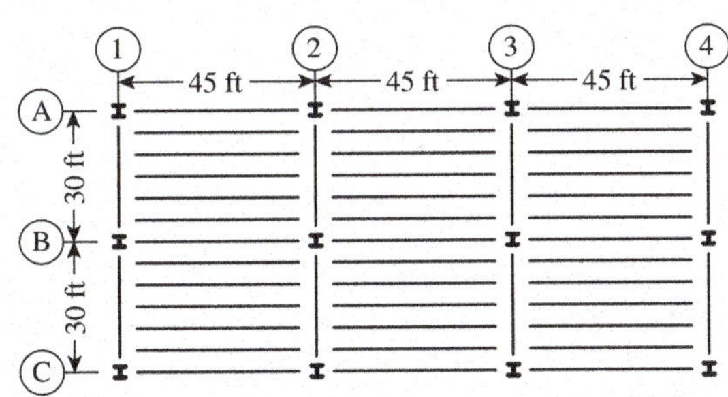

Figure VE2.1

VE2.2 For the intermediate concrete moment frame (R = 5.0) structure shown in Figure VE2.2, calculate the equivalent seismic loads to be applied at all levels (F_2, F_3, F_4, F_{roof}). The structure will be used as a hospital building located in Cincinnati, Ohio (latitude = 39.16° longitude = −84.46°), where the soil is classified as a type B.

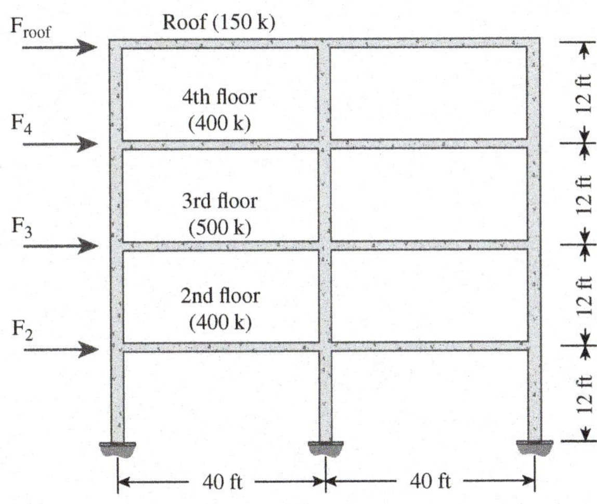

Figure VE2.2

2.21 Problems for Solution

Section 2.5

For Problems 2.1 through 2.6, given the loads specified, compute the maximum combined load using the ASCE 7-16 load combinations for allowable stress design.

P2.1 D = 50 psf, L_r = 75 psf, R = 8 psf, and S = 20 psf. (Ans: case 3 = 125 psf)

P2.2 D = 45 psf and L = 60 psf.

P2.3 D = 2750 lb, L = 4500 lb, L_r = 1500 lb, R = 1250 lbs, and S = 1000 lbs. (Ans: case 6 = 7250 lb)

P2.4 D = 87 psf and L = 150 psf.

P2.5 D = 75 psf, L_r = 35 psf, and R = 12 psf. (Ans: case 3 = 110 psf)

P2.6 D = 13 k, L = 32 k, W = −22 k, $E_v + E_h$ = 16 k, and $-E_v + E_h$ = −16 k.

Section 2.6

For Problems 2.7 through 2.12, given the loads specified, compute the maximum combined load using the ASCE 7-16 load combinations for strength design.

P2.7 Repeat Problem 2.1. (Ans: case 3 = 180 psf)

P2.8 Repeat Problem 2.2.

P2.9 Repeat Problem 2.3. Assume that L comes from a live load less than 100 psf. (Ans: case 2 = 11,250 lb)

P2.10 Repeat Problem 2.4.

P2.11 Repeat Problem 2.5. (Ans: case 3 = 146 psf)

P2.12 Repeat Problem 2.6. Assume that L comes from a live load less than 100 psf.

Section 2.7–2.10

Use the basic building layout shown in the figure for the solution of Problems 2.13 through 2.14. Determine the requested loads using ASCE 7-16. You do not need to determine the weight of the beams, girders, and columns for these problems. You are not expected to have a copy of this standard; sufficient information is provided herein for solving these problems.

P2.13 The roof of the building is flat. It is composed of 3-ply ready roofing on 3 in. of reinforced concrete. The ceiling beneath the roof is a suspended ceiling. Determine

a. The roof dead load (Ans: 40.5 psf)

b. The roof live load in psf to be applied to Column B2 (Ans: 12 psf)

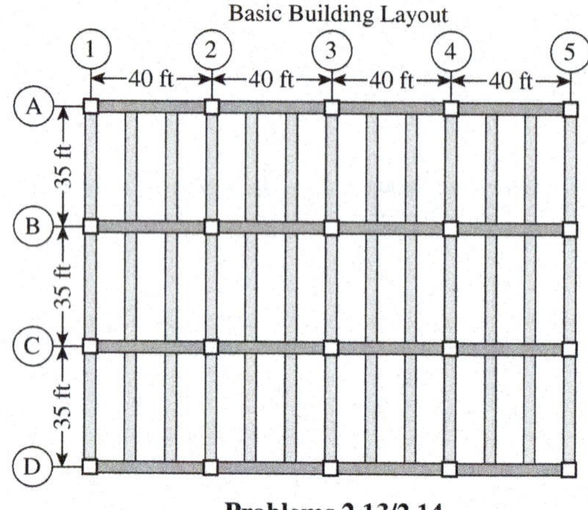

Problems 2.13/2.14

P2.14 The roof of the building is flat. It is composed of 2 in. of reinforced concrete on 18-gauge metal decking that weighs 3 psf. A single-ply waterproof sheet of 0.7 psf will be used. The ceiling beneath the roof is unfinished, but allowance for mechanical ducts should be provided. Determine

a. The roof dead load

b. The roof live load in psf to be applied to column A1

P2.15 Determine the dead load and live load for the second floor of a library in which any area can be used for stacks. Assume that there will be a steel channel ceiling system (2 psf) and asphalt tile on the floors. The floors are 6-in. reinforced concrete. Allowance should be provided for mechanical ducts. (Ans: D = 82 psf, L = 150 psf)

P2.16 Determine the dead load and live load for a floor in a light manufacturing warehouse/office complex in which any area can be used for storage. Assume that there will be no ceiling or floor finish and that the floors are 4-in. reinforced concrete. Allowance for mechanical ducts should be provided.

P2.17 Determine the dead load and live load for a typical upper floor in an office building with movable steel partition walls. The ceiling below is a suspended steel channel system (2 psf) and the floors have a linoleum finish. The floors are 3-in. reinforced concrete. Allowance should be provided for mechanical ducts. (Ans: D = 48.5 psf, L = 50 psf)

P2.18 Determine the loads on an upper floor in a school with steel stud walls and a 3-in. reinforced concrete floor with asphalt tile covering. Allowance should be made for mechanical ducts.

Section 2.11

P2.19 Determine the design rain load on the roof of a building that is 250-ft wide and 500-ft long. The architect has decided to use 6-in. diameter drains spaced uniformly along the long sides of the building at 40-ft intervals for the secondary drainage system. The secondary drains are located 1.5 in. above the roof surface (primary drain). The rainfall intensity at the location of this building is 2.5 in. per hour. (Ans: R = 14.7 psf)

Section 2.12

P2.20 A new auditorium is being built in a region where the ground snow load is 75 psf. The surroundings of the building site can be classified as an exposure C with partial shelter. It is adequately insulated and kept well above freezing. What is the flat roof design snow load, in psf, that should be used for this building?

P2.21 A new fire station is being built in a region where the ground snow load is 15 psf. The surroundings of the building site can be classified as an exposure B, and the roof is sheltered. It is adequately insulated and kept well above freezing. What is the flat roof design snow load, in psf, that should be used for this building? (Ans: 18 psf)

P2.22 A pole-barn is being constructed to store hay and other farm equipment in a region where the ground snow load is 20 psf. The surroundings of the building site can be classified as an exposure D and the roof is unsheltered. It is an unheated building. What is the flat roof design snow load, in psf, which should be used for this building?

Section 2.13–2.14

P2.23 A convenience store is to be built in New Orleans, Louisiana (latitude = 29.97°, longitude = −90.06°) in a location that could be classified as an exposure C. Calculate the design wind pressures for the MWFRS of the building for the wind direction shown in the accompanying figure. Provide a sketch of the wind pressure zones similar to what is presented in Example 2.8. (Ans: zone A = 40.9 psf, zone E = −49.1 psf)

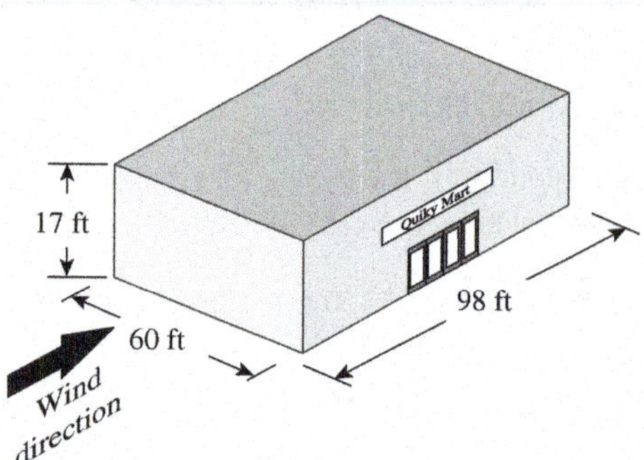

Problems 2.23/2.24

P2.24 Rework Problem 2.23 if the convenience store is located in Lincoln, Nebraska (latitude = 40.80°, longitude = −96.67°) with an exposure B category.

P2.25 For the structure shown in Example 2.8, find the design wind pressures of the MWFRS for both wind directions. Assume that the design wind speed for the structure is 120 mph in an exposure B category. Provide a sketch of the wind pressures on the building similar to Example 2.8. (Ans: zone A = 24.4 psf, zone E = −27.4 psf)

Section 2.15—2.16

P2.26 For the structure shown in Example 2.9, calculate the equivalent seismic loads that should be applied at all levels (F_2, F_3, F_4, and F_{roof}). This structure will be used as a hospital located in Buffalo, New York (latitude = 42.89°, longitude = −78.88°). It will be located on a soil type C and will be constructed of a reinforced concrete moment frame (R = 5).

P2.27 For the ordinary steel moment frame (R = 3.5) structure shown, calculate the equivalent seismic loads that should be applied at all levels (F_2, F_3, and F_{roof}). The structure will be used as an office building and will be located in Atlanta, Georgia (latitude = 33.75°, longitude = −84.39°) where the soil type is classified as a type D. (Ans: F_2 = 12.4 k, F_3 = 13.76 k, F_{roof} = 4.84 k)

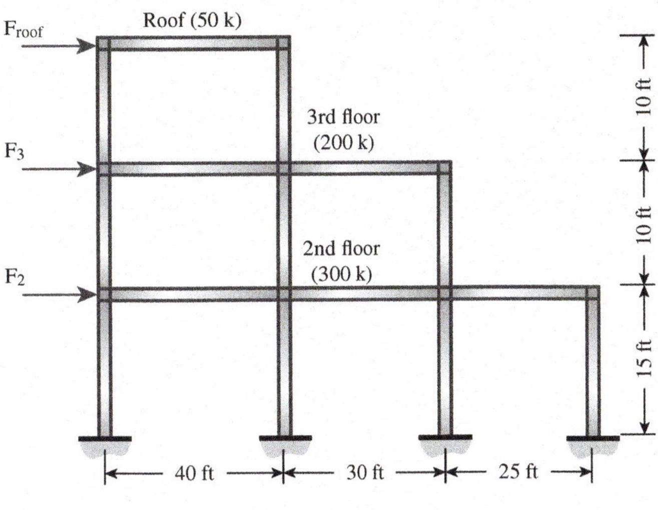

Problem 2.27

3 Vertical System Loading and Behavior

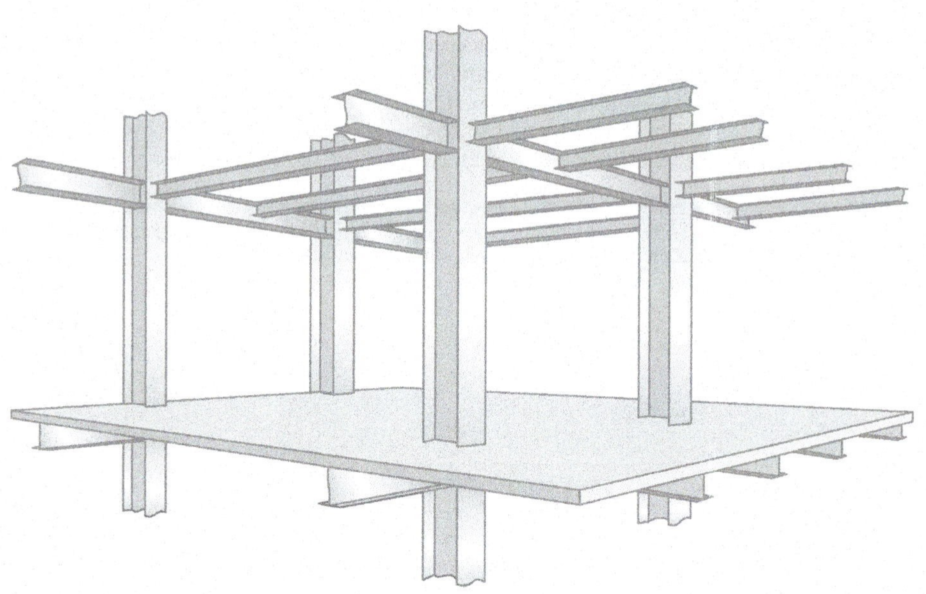

3.1 Introduction

STRUCTURAL LOAD PATH
The path that all applied loads follow to get from the point of application all the way to the foundation. For example, the next time you are on an upper floor of a multistory building, ask yourself how your own weight is transferred to the ground. What structural components carry your weight (e.g., slab, beams, columns, connections, etc.)?

Identifying this load path is the responsibility of the structural engineer.

In Chapter 2, different types of loads that might be applied to structural systems were discussed. Methods for estimating the individual magnitudes of the loads were presented. In that discussion, however, we did not address *how* and *where* to place them on the structure to cause the maximum system response.

System response is a catch-all phrase that really refers to a particular quantity of structural behavior. The response could be the negative bending moment in a floor beam, the displacement at a particular location in the structure, or the force at one of the structural supports. The student probably knows very little about how to calculate these aspects of response at this time.

After the magnitudes of the loads have been computed, the next step in the analysis of a particular structure includes the placing of the loads on the structure and the calculation of its response to those loads. When placing the loads on a structure, two distinct tasks must be performed.

1. We must decide which loads can be reasonably expected to act concurrently in time. Because different loads act on the structure at different times, several different loading conditions must be evaluated. Each of these loading conditions cause the structural system to respond in a different manner. This issue is addressed by the load combination equations presented in Sections 2.5 through 2.6.

2. We also need to determine where and how to place those loads on the structure. This includes the essential task of tracking the loads through the system; in other words, a load path must be identified within the structure. After loads are placed on the structure and tracked through the system, the response of the structure is computed. If the same loads are placed on the structure in different positions, the response of the system will be different. We need to determine where to place the loads to obtain the maximum response. For example, would the *bending moments* in the floor beams be greater if we placed the floor live loads on every span or on every other span?

Placing the live loads to cause the worst effects on any member of a structure is the responsibility of the structural engineer. Theoretically, their calculations are subject to the review of the appropriate building officials, but seldom do such individuals have the time and/or the ability to make significant reviews. Consequently, these calculations remain the responsibility of the engineer.

3.2 Structural Idealization

Before a further discussion of load path is presented, the notion of structural idealization needs to be discussed. To calculate the forces in the various parts of a structure with reasonable simplicity and accuracy, it is necessary to represent the structure in a simple manner that is conducive to analysis. Structural components (e.g., beams, columns, etc.) have width and thickness. Concentrated forces rarely act at a single point; rather, they are distributed over small areas. If these characteristics are taken into consideration in detail, however, an analysis of the structure will be very difficult—if not impossible—to perform.

The process of replacing an actual structure with a simple system conducive to analysis is called *structural idealization*. Most often, lines that are located along the centerlines of the components represent the structural components. The sketch of a structure idealized in this manner usually is called a *line diagram*.

The preparation of line diagrams is shown in Figure 3.1. In part (a) of the figure, the girder shown supports several floor beams and in turn is supported by three concrete-block walls. The actual distribution of the forces acting on the beam is shown in Figure 3.1(b). For purposes of analysis, though, we can conservatively represent the beam and its loads and reactions with the line diagram of part (c). The loaded spans are longer with the result that shears and moments are higher than actually occur.

Another line diagram is presented in Figure 3.2 for the floor system of a steel frame building. Various other line diagrams are presented throughout the text as needed.

Sometimes the idealization of a structure involves assumptions about the behavior of the structure. As an example, the bolted steel roof truss of Figure 3.3(a) is considered. The joints in

DID YOU KNOW?
If the representation of a structure and its loads were not simplified to some degree, we as structural engineers would not be able to analyze them with any degree of efficiency or effectiveness.

DID YOU KNOW?
Often in structural drawings, when floor framing members are connected with simple connections (i.e., pinned), the lines representing the beams/girders are often drawn with a visible gap as shown in Figure 3.2(b). (If no gap exists, you should still assume a simple connection if it is not stated otherwise.) If there are moment connections, a triangle is often placed on the sketch as shown.

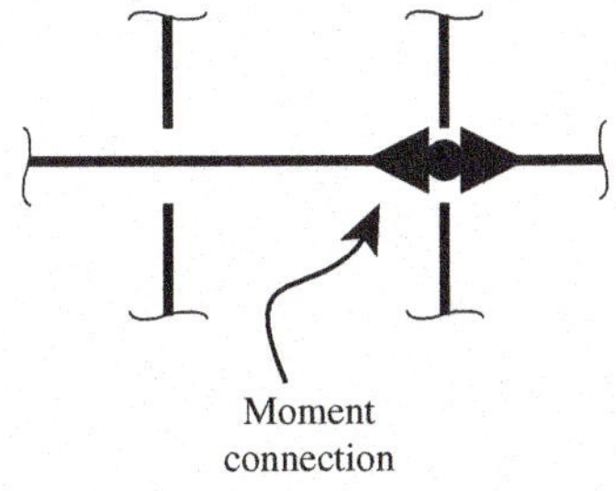

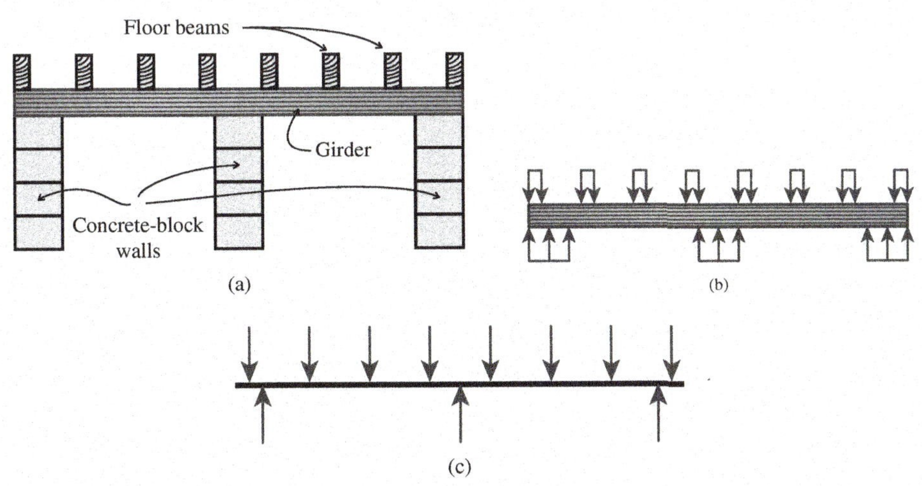

FIGURE 3.1 Replacing a structure and its forces with a line diagram.

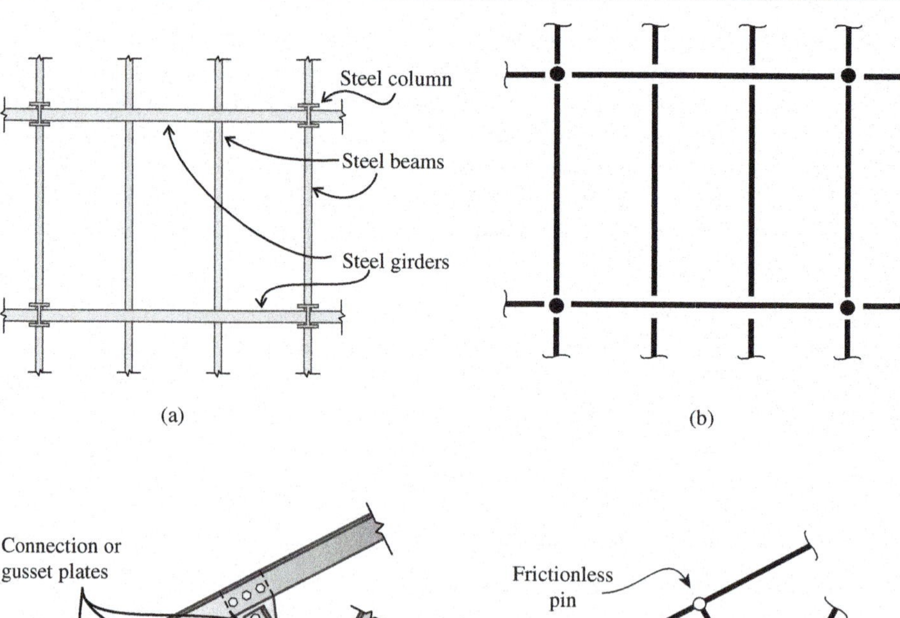

FIGURE 3.2 Line diagram for part of the floor system of a steel frame building.

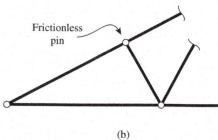

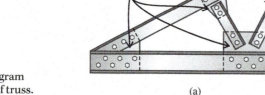

FIGURE 3.3 A line diagram for a portion of a steel roof truss.

trusses are often made with large connection or gusset plates and, as such, can transfer moments to the ends of the members. However, experience has shown that the stresses caused by the axial forces in the members greatly exceed the stresses caused by flexural forces. As a result, for purposes of analysis, we can assume that the truss consists of a set of pin-connected lines, as shown in Figure 3.3(b).

QUICK NOTE
The lines in a line diagram are located at the centroid of the member it represents.

Beam section with actual depth.

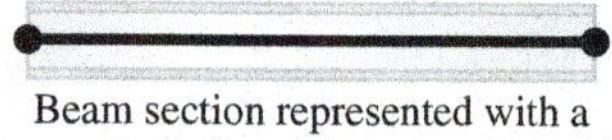

Beam section represented with a single line at its centroid.

Although the use of simple line diagrams for analyzing structures will not result in perfect analyses, the results usually are quite acceptable. Sometimes though, there may be some doubt in the mind of the analyst as to the exact line diagram or model to be used for analyzing a particular structure. For instance, should beam lengths be clear spans between supports, or should they equal the distances from center to center of those supports? Should it be assumed that the supports are free to rotate under loads, are fixed against rotation, or do they fall somewhere in between? Because of many questions such as these, it may be necessary to consider different models and perform the analysis for each one to determine the worst cases. A treatment of support idealization is considered extensively in Chapter 5.

3.3 Vertical Load Path

To identify the load path is to track all loads from their point of application on the structure to the foundation/supports of the structure. A *lateral load path* tracks a lateral load being applied to a building or building component through the structural system all the way to the supports or foundation of the building.

To adequately identify the load path, one must consider the source and "point of application" of the load. *Gravity loads* such as live load, snow load, and rain load are applied vertically from an outside source to the structure element(s). For example, the snow on a roof applies a direct pressure in the vertical direction regardless of the orientation of the structural member (Figure 3.4). The snow load shows up on the exterior surface of the roof. Live load and rain loads behave in a similar manner.

FIGURE 3.4 Snow load depiction on a roof frame.

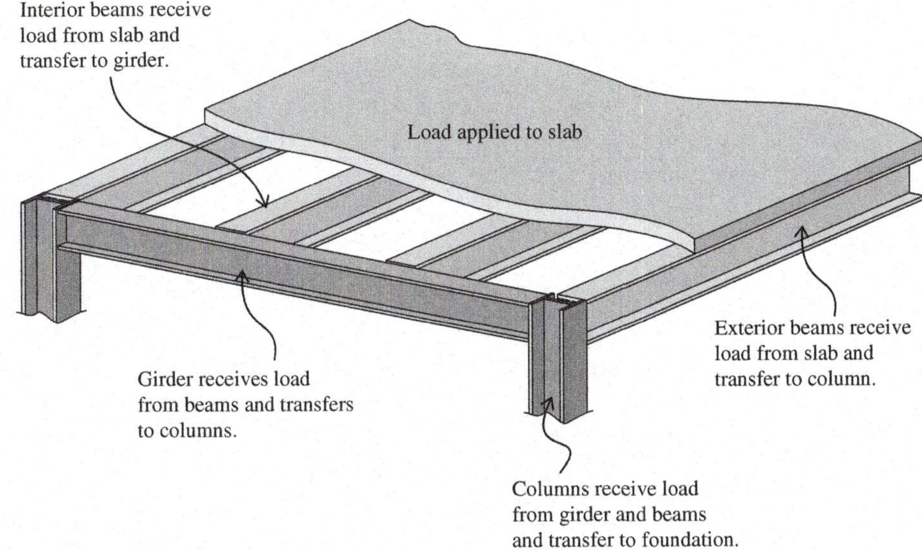
Interior beams receive
load from slab and
transfer to girder.
Load applied to slab
Exterior beams receive
load from slab and
transfer to column.
Girder receives load
from beams and transfers
to columns.
Columns receive load
from girder and beams
and transfer to foundation.

FIGURE 3.5 Typical floor framing.

Some loads can occur from "within" the structural member, such as a load due to self-weight. The acceleration due to gravity causes a force to develop within the structural members due to their own mass. However, for convenience, we can sketch such loads as if they are externally applied to the member. This action simplifies the analysis of the structural response without compromising the integrity of the results significantly.

VERTICAL LOAD PATH HIERARCHY

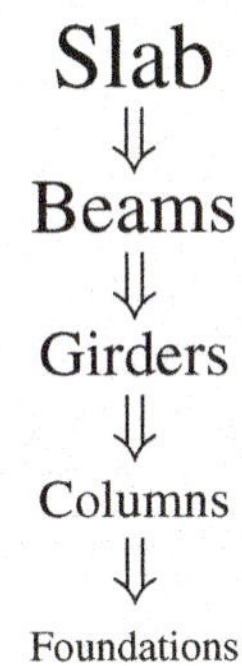
Slab
Beams
Girders
Columns
Foundations

Hierarchy of Vertical Load Path

Identifying the *vertical load path* begins with identifying the point of load application. It then becomes an exercise in identifying what provides support to the structural element and then transferring the load to those supports. A typical floor framing system is shown in Figure 3.5, where the load is usually applied directly to the slab. The slab is supported by the beams. The beams are supported by the girder or columns, and the columns are supported by the foundation.

The general hierarchy is presented here to familiarize the student with how loads flow through a floor system. However, there is no substitution for the student looking at the details of the framing. This point is illustrated in Figure 3.6, where the only change in the framing is the elevation at which the beams are set. This does nothing to change the load path of the two interior beams, but it does change the load path of the exterior beams. The load in them must first transfer

COMMON MISTAKE

Students will often take the load that is in the girder and dump it to the beams.

One easy way to avoid this mistake is to try asking the following questions. *"If I remove the beam, will the girder fall down? If I remove the girder, will the beam fall down?"* In this case, if we remove the girder, we clearly have no support for the beams. But the reverse argument cannot be made. This tells us that the girder is supporting the beam and not the beam supporting the girder.

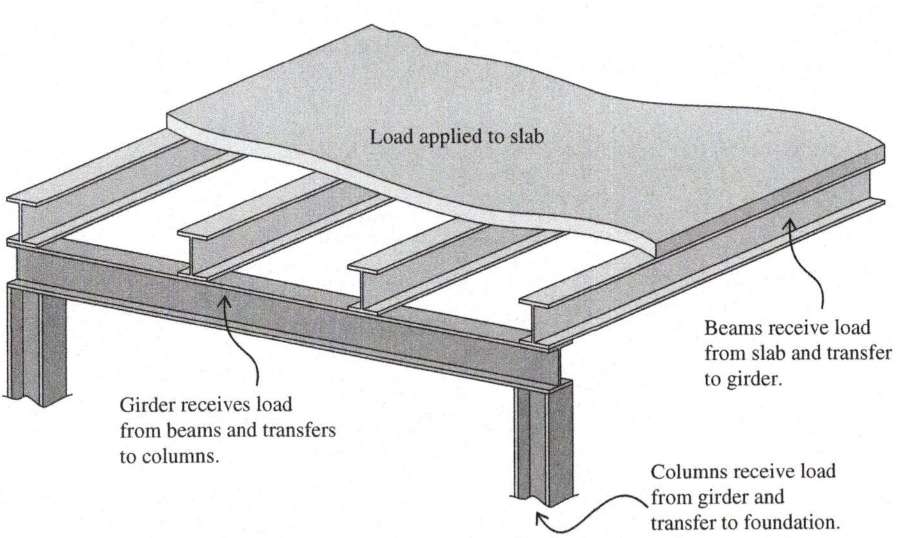
Load applied to slab
Beams receive load
from slab and transfer
to girder.
Girder receives load
from beams and transfers
to columns.
Columns receive load
from girder and
transfer to foundation.

FIGURE 3.6 Alternate floor framing.

QUICK NOTE
All loads that are applied to the structure must be accounted for and **tracked to the foundation**. Failure to do so has resulted in numerous structural failures in the past.

Courtesy of Bryant G. Nielson

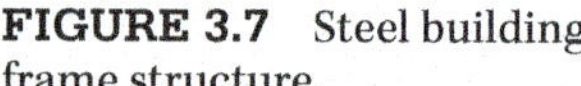
FIGURE 3.7 Steel building frame structure.

through the girder *before* it goes into the column, whereas in Figure 3.5 this is not the case. This configuration also affects how the load transfers from member to member, which influences the type of connection that will be used.

Customized Vertical Load Path

QUICK NOTE
The connections between structural elements are critical components in the vertical load path and should not be ignored.

Generally speaking, the vertical load path can be described as given in the previous section. However, there is no replacement for the engineer looking closely at the framing details of the building and explicitly defining the load path that will be followed. Example 3.1 is presented to illustrate this point.

EXAMPLE 3.1

On the structure shown, a roof load is applied to the roof slab which in turn applies a distributed load to the roof beam. Explicitly describe the vertical load path that the roof load follows through the system and make a sketch.

SOLUTION STRATEGY

Examine and identify all of the building components noting what supports each element.

COMMON MISTAKE
Students are often tempted to let the load transfer from the columns into the third-floor beams.

Once a load makes it into the columns, it will remain in the columns all the way to the foundation unless the column is interrupted, such as columns B and C shown in this example.

SOLUTION

The elements that exist in the building are the roof beams, third-floor beams, second-floor transfer girder, columns A through D, and the foundation supports.

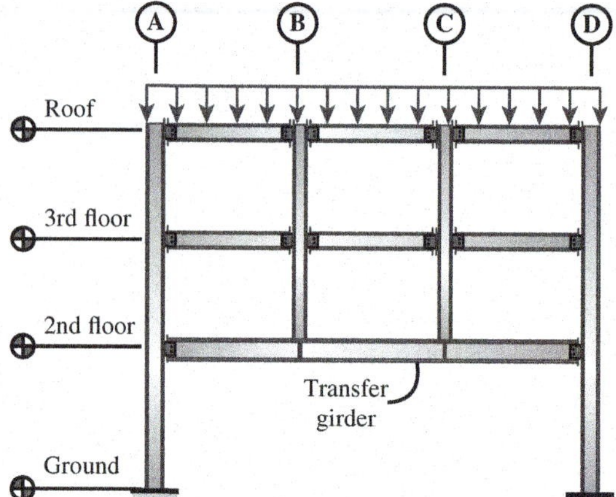

- All four columns support the roof beams.
- All four columns support the third-floor beams.
- The transfer girder supports columns B and C.
- Columns A and D support the transfer girder.
- The foundations support columns A and D.

The roof load travels from the roof beam to all four columns. It then travels down all four columns until it gets into the transfer girder. The load in the transfer girder is then transmitted to columns A and D and then down to the foundation (see accompanying figure).

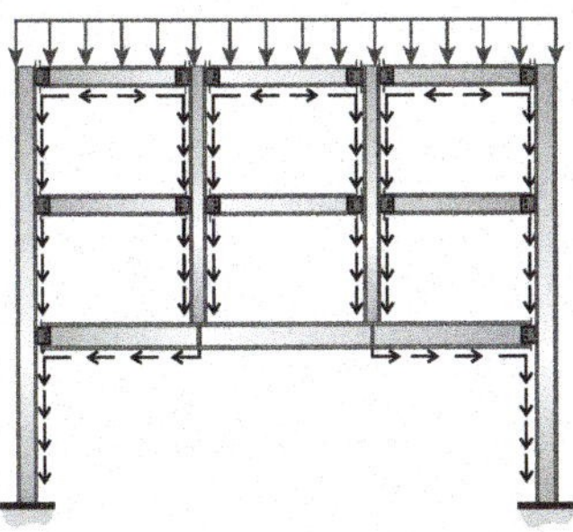

SUMMARY

A generalized vertical load path for a building system can be given as slab ⇒ beam ⇒ girder ⇒ columns ⇒ foundations. However, there is no substitute for looking at the configuration of the structural elements. Note how the third-floor beams never take any of the roof load.

3.4 Tributary Areas

Once the vertical load path has been established, the magnitude and nature of the load going into each member must be established. This task is made easier using the notion of a *tributary area* (A_t) which was briefly defined in Section 2.10. In this section, this term is discussed at greater length. In the next section, a related term, the influence area (A_I), is introduced.

The tributary area is the loaded area of a particular structure that directly contributes to the load applied to a particular member in the structure. It is best defined as the area that is bounded by lines located halfway to the next beam or to the next column. Tributary areas are shown for several beams and columns by the shaded areas in Figure 3.8 for a structure with one-way bending between the beams.

QUICK NOTE

To delineate the tributary area for a certain structural component, draw a line halfway between the closest supporting components (see the dashed lines in Figure 3.8). The component that the tributary area serves is darkly shaded.

Tributary Width

When dealing with beams and girders in a framing system, the notion of tributary width (b_t) is a useful concept to understand. The slab in a building must be supported by beams and girders. The slab often is loaded with a uniform load over its area, as discussed in Section 2.8 of this text dealing with floor live loads. The slab then takes this surface load—units of force per length squared (e.g., lb/ft^2)—and distributes it to the supporting beams. This load shows up on beams as a line load with units of force per length (e.g., lb/ft). The calculation of the magnitude of this line load is found by taking the area load (p) and multiplying by the width of slab that is tributary to the supporting beam (b_t).

$$w_{\text{beam}} = (p)\,(b_t)$$

NOMENCLATURE

Beams are members that support transverse loads. They are usually thought of as being used in horizontal positions, but there are many exceptions.

The term *girder* is often used rather loosely to refer to a large beam. However, many reserve the term to refer to beams that support other smaller beams.

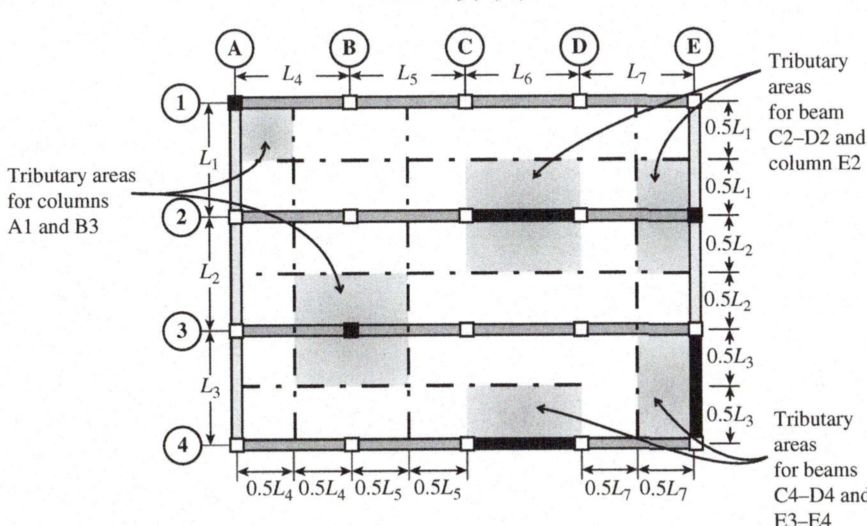

FIGURE 3.8 Tributary areas for selected columns and beams.

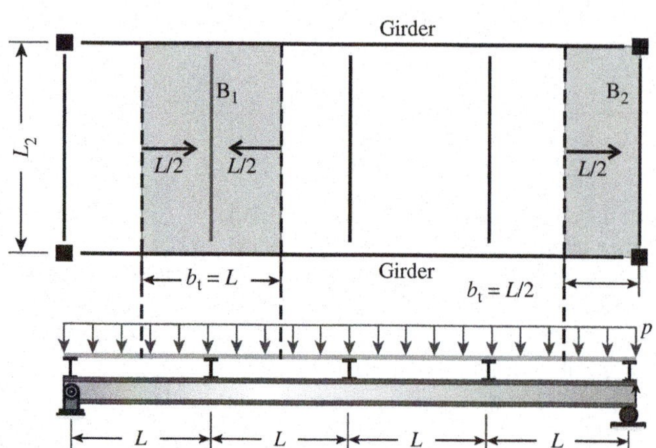

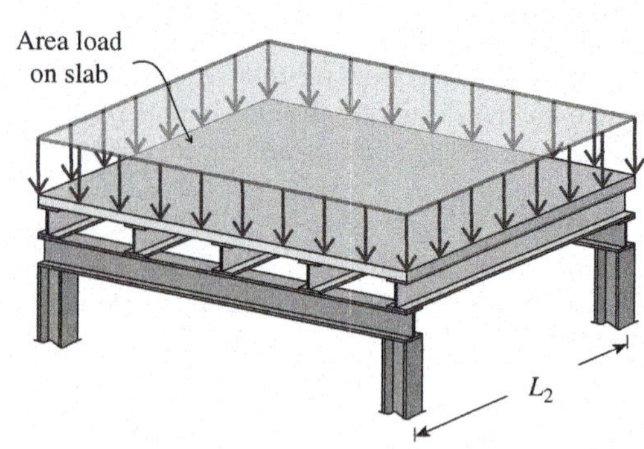

FIGURE 3.9 Illustration of beam tributary width.

The calculation of the tributary width is found by drawing dividing lines halfway between the beams supporting the slab. This concept is illustrated in Figure 3.9, where the beam labeled B_1 receives load from both sides of the beam while B_2 receives load only from the left side of the beam. The tributary width is $b_t = L$ for B_1 and $b_t = L/2$ for B_2.

EXAMPLE 3.2

Consider the floor framing system shown in Figure 3.9. Assume that the area load applied to the slab is $p = 60$ psf, $L = 6$ ft, and $L_2 = 15$ ft. Calculate and sketch the load which should be applied to B1, B2, and the girder.

SOLUTION STRATEGY

Identify the tributary width for each of the beams and multiply these widths by 60 psf. Apply a uniformly distributed load to each beam with the calculated magnitude. Calculate the reactions for the beams and apply these loads to the girder.

SOLUTION

The following tributary widths are calculated by referring to Figure 3.9.

$$B_1 \Rightarrow b_t = (2)(6\text{ ft})/2 = 6\text{ ft}$$

$$B_2 \Rightarrow b_t = (6\text{ ft})/2 = 3\text{ ft}$$

The applied load to each of the beams is calculated as

$$B_1 \Rightarrow w_1 = (6\text{ ft})(60\text{ lb/ft}^2) = 360\text{ lb/ft} = 360\text{ plf}$$

$$B_2 \Rightarrow w_2 = (3\text{ ft})(60\text{ lb/ft}^2) = 180\text{ lb/ft} = 180\text{ plf}$$

Sketch the loaded beam diagrams where the length of each beam is $L_2 = 15$ ft. The reactions for each beam are shown here for completeness, but the method for doing so is discussed at length in Chapter 5.

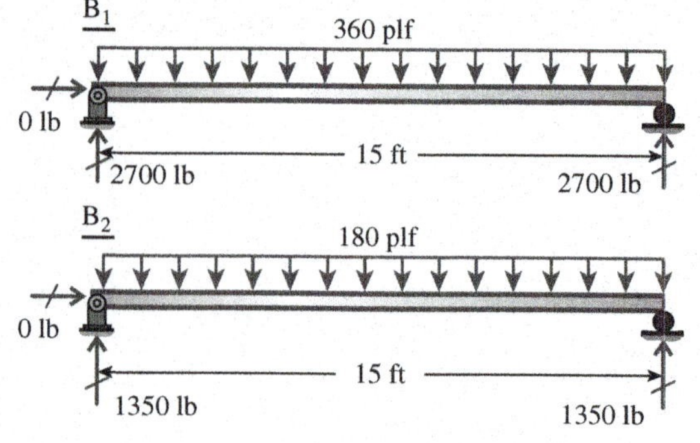

QUICK NOTE

The term slab, as it is used throughout this text, refers to any planar structural element that serves to distribute a vertical load to the supporting beams, girders, and walls. Examples include:

- Reinforced concrete slab
- Steel decking
- Plywood sheathing
- Solid wood sheathing

Courtesy of Bryant G. Nielson

Since the girder supports the beams, the reactions of the beams show up on the girder as point loads acting in the opposite direction as the reactions.

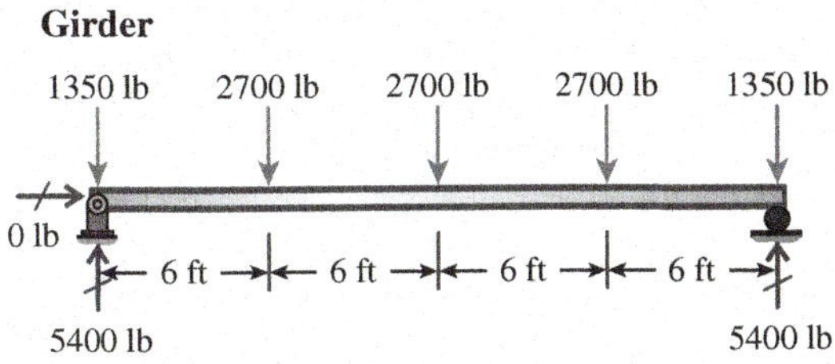

DID YOU KNOW?

The reason the distributed load on the beam is rectangular is because the tributary width is constant along the length of the beam. If the tributary width is variable along the length, the load intensity will be variable also.

Slab Behavior

The tributary areas shown for the beams in Figure 3.8 have traditionally been the tributary areas used in common practice for one-way or two-way bending. For two-way bending, this can produce some relatively conservative estimates for the loads. If a slab is supported on all four edges, it theoretically behaves as a two-way slab. This tributary area is defined by drawing a 45° line projecting out from the intersection of two of the slab-supporting elements and also drawing dividing lines halfway between the supporting elements, as shown in Figure 3.10. The regions delineated by the projected lines define the tributary areas.

The theoretical tributary area for typical interior and edge beams are shown in Figure 3.10 for a structure with a two-way floor system spanning between the beams. We see that for the middle beam the tributary area extends halfway to the next beam in each direction. At the ends of the beam, however, load is supported partly by the beams in the perpendicular direction. Therefore, the theoretical boundary of the tributary area falls halfway between the two, that is, at a 45° angle. We thus see that the tributary width along the length of a beam varies. This can produce either a trapezoidal or triangular load on the beam.

QUICK NOTE

A one-way slab distributes its load along a single direction (usually the shorter) perpendicular to the supporting beams. A two-way slab distributes some of the load in each of the two orthogonal directions.

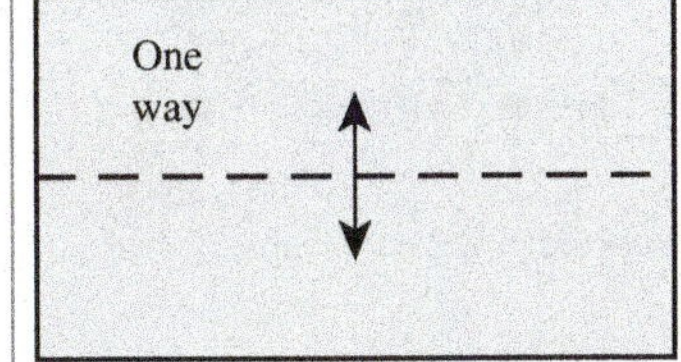

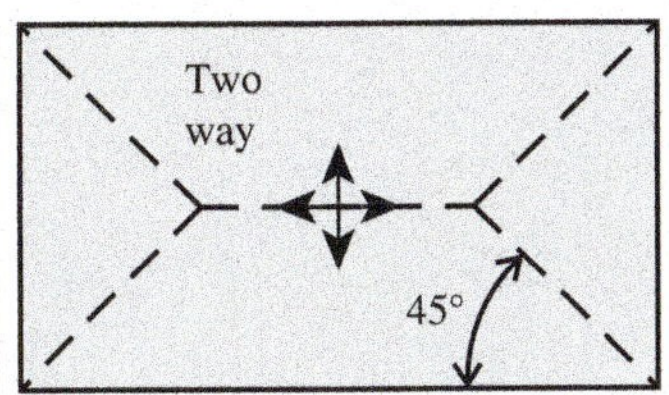

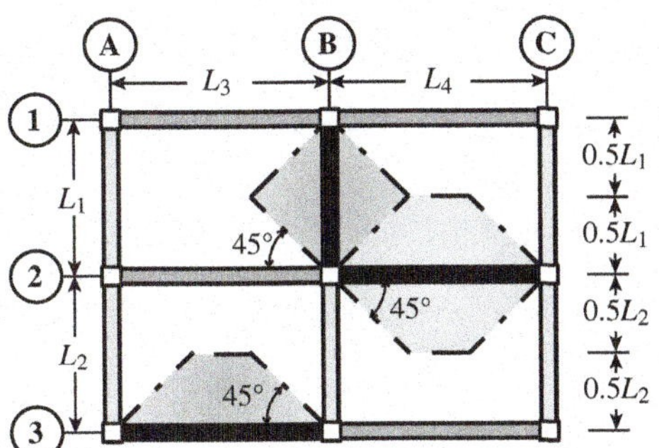

FIGURE 3.10 Theoretical tributary areas for beams supporting two-way slabs.

EXAMPLE 3.3

Consider the floor framing system shown here. Assume that the area load applied to the slab is 50 psf. Assuming that the slab behaves like a two-way slab, calculate and sketch the load that should be applied to beam B_1 and girder G_2.

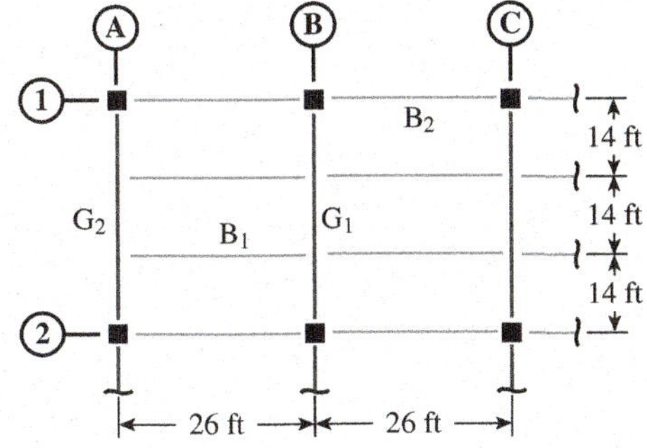

SOLUTION STRATEGY

Sketch the dividing lines for the tributary areas remembering that lines projecting from the intersection of two supporting members must be drawn as a 45° line. Identify the tributary width for each element and multiply these widths by 50 psf. Apply a distributed load to each beam that reflects the shape of the tributary area. Calculate the reactions for the beams and apply these loads to the girder. Do not forget to also apply the load that is transferred directly from the slab to the girder.

SOLUTION

The dividing lines are sketched, and the tributary areas are shaded in the following sketch. This results in the following tributary widths.

COMMON MISTAKES

Students will not draw the dividing lines at a 45° angle.

Students will only place the 45° lines at the columns and come up with tributary areas shown.

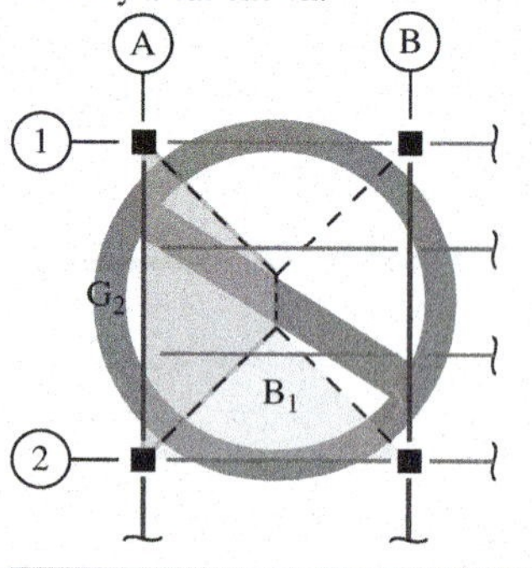

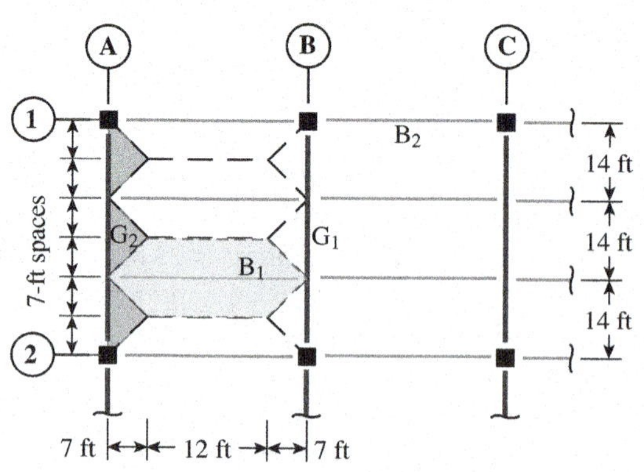

$$B_1 \Rightarrow b_t = (2)(14\text{ ft})/2 = 14\text{ ft} \quad (0\text{ ft at the ends})$$

$$G_2 \Rightarrow b_t = (14\text{ ft})/2 = 7\text{ ft} \quad (0\text{ ft at the ends})$$

The applied load to each of the components is calculated as

$$B_1 \Rightarrow w_1 = (14\text{ ft})(50\text{ lb/ft}^2) = 700\text{ lb/ft} = 700\text{ plf}$$

$$G_2 \Rightarrow w_2 = (7\text{ ft})(50\text{ lb/ft}^2) = 350\text{ lb/ft} = 350\text{ plf}$$

QUICK NOTE

The analysis of the reactions is not given in this chapter, rather it is presented in Chapter 5.

Sketch the loaded beam diagram where the distributed load is 0 plf at the beam ends and reaches 700 plf at a distance of 7 ft in from the ends.

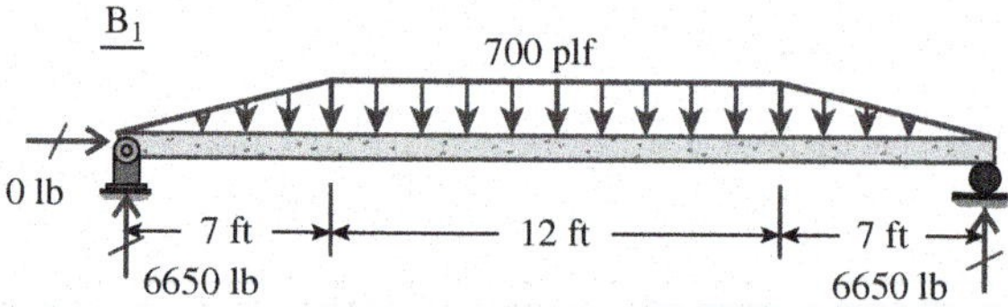

We can see that there is a series of triangular loads that apply directly to the girder G_2. We must also recall that G_2 is supporting two beams like B_1. Thus, G_1 sees both the triangular loads and the point loads from the reactions of B_1.

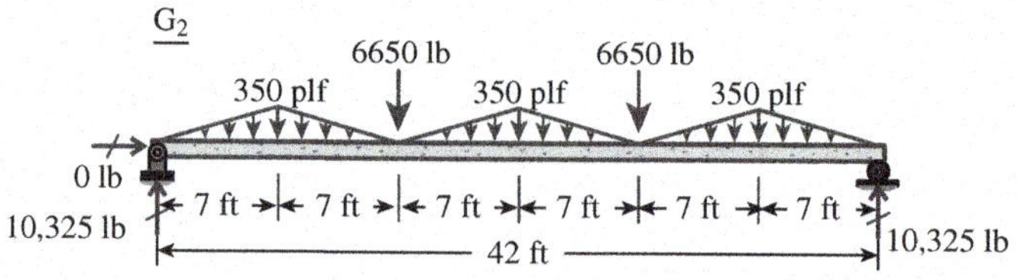

We need to contrast the results we get from assuming a two-way slab behavior with those we get from assuming a one-way slab behavior. Thus, Example 3.4 looks at the same floor framing layout as did Example 3.3.

EXAMPLE 3.4

Consider the floor framing system shown in Example 3.3. Assume that the area load applied to the slab is 50 psf. Assuming that the slab behaves like a one-way slab, calculate and sketch the load to be applied to beam B_1 and girder G_2.

SOLUTION STRATEGY

Sketch the dividing lines for the tributary areas located halfway between the beams. Identify the tributary width for each element and multiply these widths by 50 psf. Apply a distributed load to each beam that reflects the shape of the tributary area. Calculate the reactions for the beams and apply these loads to the girder.

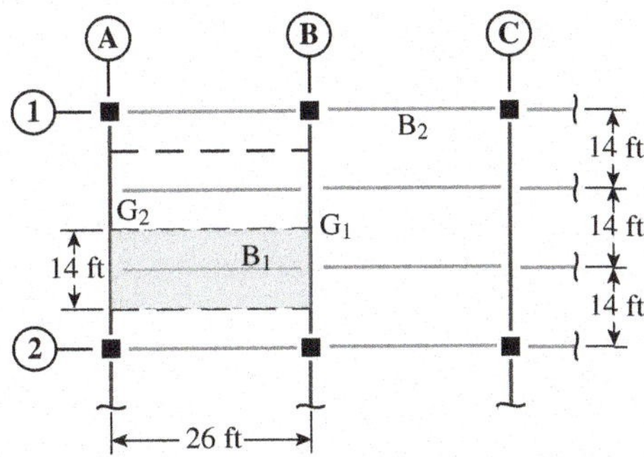

SOLUTION

The dividing lines are sketched, and the tributary area for B_1 is shaded in the previous figure. So

$$B_1 \Rightarrow b_t = (2)(7 \text{ ft}) = 14 \text{ ft}$$

The applied load to B_1 is calculated as

$$B_1 \Rightarrow w_1 = (14 \text{ ft})(50 \text{ lb/ft}^2) = 700 \text{ lb/ft} = 700 \text{ plf}$$

Sketch the loaded beam diagram where the load is uniform across the length because the tributary width is constant along the length.

In the floor framing figure, we see that there is no load from the slab that transmits directly into the girder. Rather, the girder only supports beam B_1 and will thus only have point loads present.

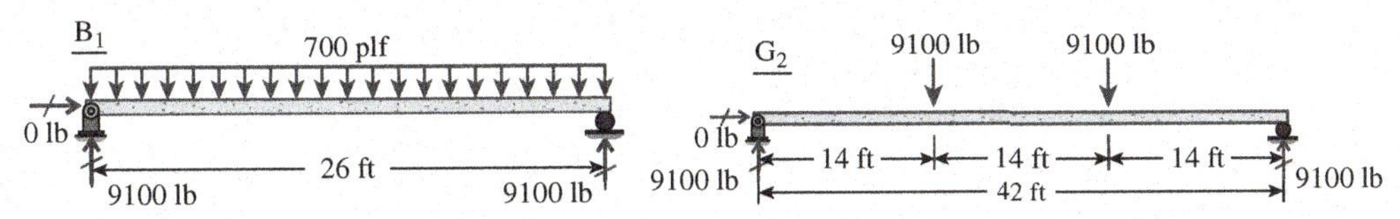

QUICK NOTE

The total load assumed to go into the girder is dependent upon the behavior assumed for the slab. One can see this difference by examining the reactions of G_2 in Examples 3.3 and 3.4. Essentially, the one-way or two-way assumption is a statement regarding the perceived load path.

Naturally, the question arises regarding when a slab should be treated as a one-way slab and when it should be treated as a two-way slab. This question is not as easily answered as we would like, but there are a few guiding principles that can be used to make an appropriate assignment. These principles include the following.

- *A slab can only distribute a load directly to an element with which it makes physical contact.* For example, the position of the girders portrayed in the floor systems of Figures 3.5 and 3.6 help make the decision of slab behavior. For the system in Figure 3.5, the slab bears directly on both the beams and the girders. This configuration makes a two-way slab assumption to

DID YOU KNOW?

The majority of floor and roof systems employed into today's structures are one-way slabs. However, there are some buildings, such as "flat-plate" concrete buildings, which will consistently exhibit two-way behavior. Thus, it is imperative that the structural engineer is comfortable with the slab behavior assumption.

be physically possible. However, the configuration of Figure 3.6 out of necessity requires a one-way slab behavior because the slab is only supported directly by the beams.

SLAB ASPECT RATIO

The aspect ratio is defined by the length of the slab segment L and the width of the slab segment B. L should always be taken as the long dimension, and B should be taken as the short dimension.

$L/B \geq 2.0 \Rightarrow$ one-way slab

$L/B < 2.0 \Rightarrow$ two-way slab

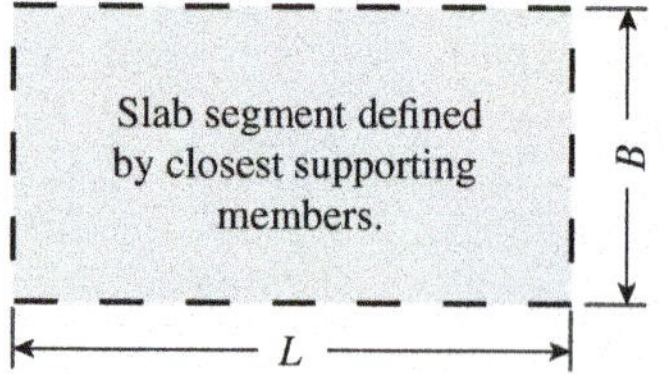

- *The relative stiffness/strength of a slab in the two orthogonal directions can limit the slab behavior to be one-way.* For example, the slab (i.e., steel decking) shown in the picture in Example 3.2 has a very high stiffness and strength in the axis perpendicular to the beams compared to the strength and stiffness it has in the axis parallel to the beams. This type of floor/roof system is designed to distribute load in only one direction (i.e., perpendicular to the beams.)
- *The aspect ratio of the slab segments can influence the perceived behavior of a slab that is supported on all edges.* The larger the aspect ratio is, the closer its behavior mimics a one-way behavior. In contrast, the closer the aspect ratio of the supported slab segment gets to one, the more a two-way slab behavior becomes significant. There is no hard-fast rule regarding the aspect ratio value where it becomes justified to assume a one-way behavior instead of the more precise two-way behavior. However, a well-accepted value for this transition point can be taken as 2.0 for the aspect ratio (see note in sidebar).

EXAMPLE 3.5

If the floor system presented in Examples 3.3 and 3.4 is framed up as shown in Figure 3.5 (i.e., beams and girders are the same level) and the slab has similar strength and stiffness in both orthogonal axes, make a decision on whether it should be treated as a one-way or a two-way slab.

SOLUTION STRATEGY

Identify the boundary of the slab segment by selecting any point on the slab. Then identify the slab supports (e.g., beams) located nearest to that point. Use the aspect ratio of the segment to determine slab behavior.

SOLUTION

A point of interest is randomly selected on the framing plan as shown. The nearest supporting elements to that point are indicated with a dashed line. The slab segment enclosed by the dashed line is shaded where $L = 26$ ft and $B = 14$ ft. Thus, $L/B = 26$ ft/14 ft $= 1.86 < 2.0$. Thus this segment is classified as a two-way slab and should be analyzed as shown in Example 3.3.

COMMON MISTAKE

Often one is tempted to define the slab dimensions as 26 ft by (3)(14ft) = 42 ft. This is done because students often think that the slab boundary is defined by the column lines and not the nearest supporting members.

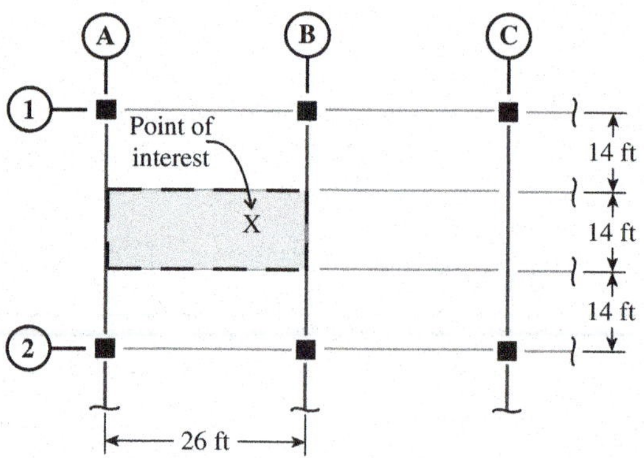

Loads to Columns

Column tributary areas are not dependent on the slab behavior assumed. This phenomenon can be demonstrated by looking at the results from Examples 3.3 and 3.4. Column A2 shown in Example 3.3 supports two girders (both labeled G_2) and one beam (B_1). Thus, the total load to the column can be calculated by summing up the reactions of the three members it supports. For the two-way slab assumption, the total column load is calculated as

$$P_{A2} = (2)(10,325 \text{ lb}) + (6650 \text{ lb}) = 27,300 \text{ lb} = 27.3 \text{ k}$$

For the one-way slab assumption of Example 3.4, the total load to the column located on grid A2 is calculated as

$$P_{A2} = (2)(9100 \text{ lb}) + (9100 \text{ lb}) = 27,300 \text{ lb} = 27.3 \text{ k}$$

Since the slab behavior doesn't influence the load distribution to the columns, the easiest way to define the tributary area to a column and the resulting load is to draw dividing lines half-way between the column lines. Calculate the area enclosed by the dashed lines that is shared by the column of interest as illustrated in Figure 3.8. The load is calculated by multiplying the area load by the tributary area.

EXAMPLE 3.6

For the floor system presented in Example 3.3, find the total load going to the column located on gridline A2. Recall the distributed area load is 50 psf.

SOLUTION STRATEGY

Draw dividing lines halfway between column lines A and B, between 1 and 2, and between 2 and 3. Dimension the distance between the dividing lines and use these to calculate the tributary area. Multiply the tributary area by 50 psf.

SOLUTION

The dividing lines are not dependent on the beam configuration or the slab behavior. The dimensions of the area tributary to column A2 is 42 ft by 13 ft.

$$A_t = (42 \text{ ft})(13 \text{ ft}) = 546 \text{ ft}^2$$

The total load to the column is calculated as

$$P_{A2} = (546 \text{ ft}^2)(50 \text{ lb/ft}^2) = 27,300 \text{ lb} = \underline{\underline{27.3 \text{ k}}}$$

SUMMARY

The tributary approach to a column load calculation produces the same result as when we track the loads through the reactions of the beams and girders. This approach proves to be useful when loads in columns of multistory buildings are being evaluated.

3.5 Influence Area

Influence areas (A_I) are those areas that affect the forces in a particular member of a structure when loaded. These areas are different from the tributary areas previously described. Referring to the drawing of the building floor shown in Figure 3.11, it can be seen that a load placed anywhere in the upper-left rectangle of the building floor directly affects the force applied to the upper-left column A1. This entire rectangle is referred to as the *influence area* for that particular column. In other words, if a point load is placed somewhere in the upper-left rectangle, column A1 will know about it. The tributary area for this same column is only the upper-left quarter of the same rectangle defined by dividing lines drawn halfway between the column lines. We can see that the influence area for this column is four times its tributary area.

The influence areas for several different beams and columns in this same floor system are shown in Figure 3.11. In each case, the member in question is shown darkened in the figure.

DID YOU KNOW?

Surveys have shown that the average magnitude of the live load that acts on a floor system tends to decrease as the influence area for a structural member increases.

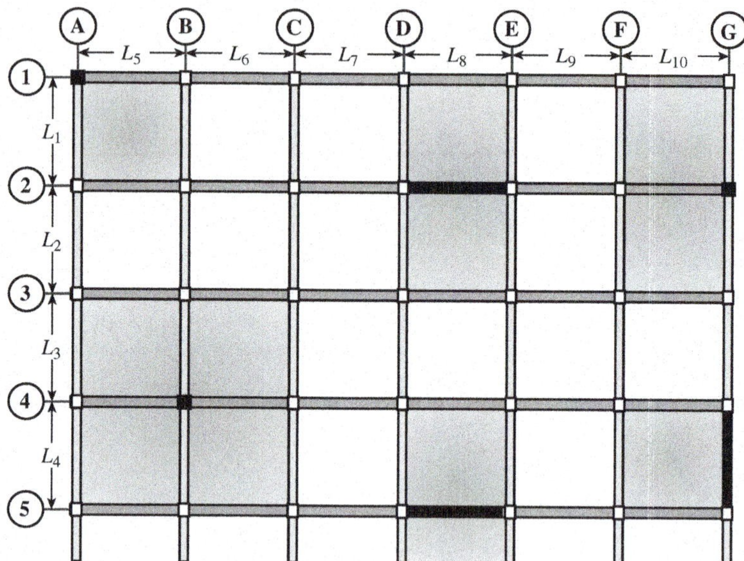

FIGURE 3.11 Influence areas of selected beams and columns.

QUICK NOTE

The concept of the live load element factor K_{LL} is introduced to make it easier on the structural engineers to find the approximate influence area for a structural element. All they need to do is calculate the tributary area (A_t) and look up the appropriate K_{LL} factor.

In ASCE 7-16, the influence area is defined as

$$A_I = K_{LL} A_t$$

The relationship between the tributary area and the influence area for any element can be captured with the live load element factor, K_{LL} the live load element factor is calculated from building geometry or taken from Table 3.1.

From Figure 3.11 and Table 3.1, you can see that the influence area for an interior column is four times as large as its tributary area, whereas that for an interior beam is twice as large as its tributary area.

3.6 Floor Live Load Reductions

Under some circumstances, the code-specified live loads for a building can be reduced. For this discussion, it is assumed that the maximum specified live load for a particular building floor is 100 psf. If the influence area for a particular member is 1000 ft^2, the likelihood of having the maximum live load of 100 psf applied to every single square foot of that area seems much less likely than if the area was 200 ft^2. Thus, we say that, as the influence area contributing to the load applied to a particular member increases, the possibility of having the full design live load applied to every square foot decreases. Consequently, building codes usually permit some reduction in

Table 3.1 Live load element factor, K_{LL}*

Type of element	K_{LL}
Interior columns	4
Exterior columns without cantilever slabs	4
Edge columns with cantilever slabs	3
Corner columns with cantilever slabs	2
Edge beams without cantilever slabs	2
Interior beams	2
All other beams not identified including: edge beams with cantilever slabs cantilever beams, two-way slabs, and members without provisions for continuous shear transfer normal to their span.	1

*Adapted from ASCE 7-16 Table 4-2

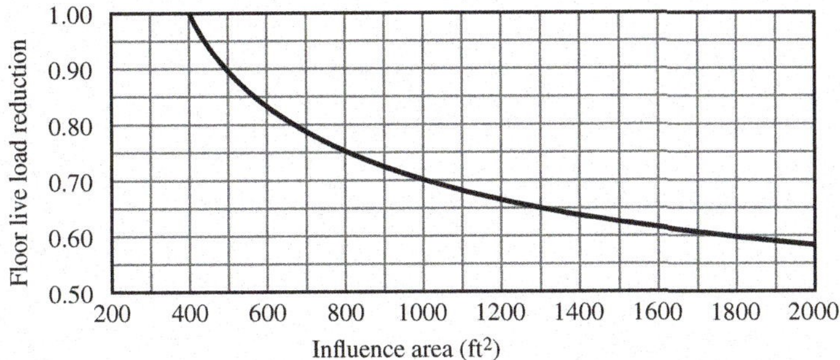

FIGURE 3.12 Floor live load reduction factor versus influence area.

the specified live loads when large areas are involved. In Section 4.7 of ASCE 7-16, the allowable reduction is

$$L = L_0\left(0.25 + \frac{15}{\sqrt{K_{LL}A_t}}\right) \tag{3.1}$$

In this equation, L is the reduced live load, L_0 is the code-specified live load, and the term in parentheses is the reduction factor. The terms K_{LL} and A_t were defined in the last section and show that as the influence area ($A_I = K_{LL}A_t$) increases the design floor live load decreases.

From this equation, we can show that live loads are reduced only when the influence area is greater than 400 ft^2. There are limits, though, as to how much the live load can be reduced. If the structural member supports load from one floor only, the live load cannot be reduced by more than 50%. For members supporting load from more than one floor, the live load cannot be reduced by more than 60%. Usually, individual beams are involved in the support of only one floor. On the other hand, columns are often involved in the support of more than one floor; in fact, they support all of the floors above them. A plot of the floor live-load reduction factor as obtained from the quantity inside the parenthesis of Equation 3.1 is presented in Figure 3.12 as a function of the influence area.

However, live-load reduction is not permitted in all cases by ASCE 7-16. If the unit live load is greater than 100 psf or if the loaded area is used for a place of public assembly, the reduction cannot be taken. When the unit live load exceeds 100 psf, a maximum reduction of 20% can be taken for structural members supporting more than one floor, which would be the case for the transfer girder shown in Example 3.1. The basis for this 20% reduction is that higher unit live loads tend to occur in buildings used for storage and warehousing. In this type of building, several adjacent spans may be loaded concurrently, but studies have indicated that rarely is an entire floor loaded to more than 80% of its rated design load.

The provision for live load reduction and the related limitations have two significant implications on structural analysis. First, the loads used to obtain column design forces and those used to obtain floor beam design forces may be different. This situation occurs because the live load reduction factors for each are likely to be different. Also, because roofs and floors are treated differently, it does not appear that the live-load reduction is always permitted when columns support a floor and the roof. Typical of such columns are those supporting the top story in a building. When a column supports a floor and the roof, that column should be considered to support a single floor for purposes of determining the permissible live-load reduction. The treatment of columns in multistory buildings is discussed in the next section.

> **QUICK NOTE**
> Reductions for roof live loads are handled in a different fashion than for floor live loads. The reduction procedure for roof live loads was presented in Section 2.10 of this text. The two procedures (i.e., for roof and for floor) should not be confused.

> **QUICK NOTE**
> The reduction limitations are mathematically expressed as follows.
> Supporting only one floor
> $$L \geq 0.5L_0$$
> Supporting more than one floor
> $$L \geq 0.4L_0$$
> Supporting more than one floor and $L_0 > 100$ psf
> $$L \geq 0.8L_0$$

> **QUICK NOTE**
> Floor live load reductions are generally not permitted for the following conditions:
> - $L_0 > 100$ psf
> - Assembly area
> - Parking garage
> - $A_I < 400$ ft^2
>
> There are some exceptions to these limitations which should be understood.

EXAMPLE 3.7

For the floor system presented in Example 3.4, find the reduced live load L to be used for the design of beam B_1, girder G_2, and column A2. Take the unreduced floor live load as 50 psf and assume that each of these elements supports only one floor.

SOLUTION STRATEGY

For each of the three elements, the tributary areas must be calculated and the appropriate K_{LL} factors identified from Table 3.1. Reduced loads can be calculated but must be ensured to not fall below the 50% reduction limit for elements supporting only one floor.

> **QUICK NOTE**
> The tributary area to a girder is found by asking "What slab loads eventually make it to the girder, and what slab loads go through a different path?"

QUICK NOTE

It is circumstantial that the tributary areas to both the beam and the girder are the same in this example. Usually these elements will have differing areas and thus differing reduced loads.

SOLUTION

The tributary areas for beam B_1 and girder G_2 are shown in parts (a) and (b), respectively, and both calculate out to be 364 ft². The tributary area for column A2 is shown from Example 3.6 to be 546 ft². From Table 3.1, we have

B_1 is an interior beam with $K_{LL} = 2.0$

G_2 is an exterior beam with $K_{LL} = 2.0$

Column A2 is an exterior column where $K_{LL} = 4.0$

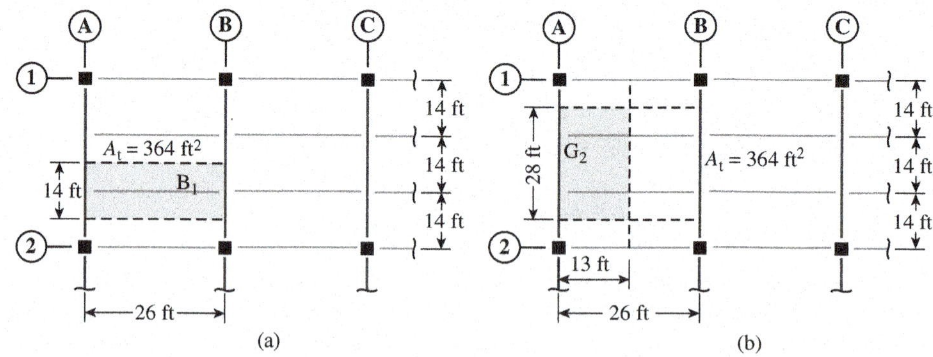

(a) (b)

Reduced Live Loads for B_1 and G_2:

$$K_{LL}A_t = (2.0)(364\ \text{ft}^2) = 728\ \text{ft}^2 > 400\ \text{ft}^2 \Leftarrow \textbf{reduction permitted}$$

$$L = 50\ \text{psf}\left(0.25 + \frac{15}{\sqrt{728\ \text{ft}^2}}\right) = (50\ \text{psf})(0.806)$$
$$= \underline{\underline{40.3\ \text{psf}}} > (50\ \text{psf})(0.5)$$

Column A2:

$$K_{LL}A_t = (4.0)(546\ \text{ft}^2) = 2184\ \text{ft}^2 > 400\ \text{ft}^2 \Leftarrow \textbf{reduction permitted}$$

$$L = 50\ \text{psf}\left(0.25 + \frac{15}{\sqrt{2184\ \text{ft}^2}}\right) = (50\ \text{psf})(0.571)$$
$$= \underline{\underline{28.5\ \text{psf}}} > (50\ \text{psf})(0.5)$$

DID YOU KNOW?

When it actually comes down to sketching the load diagrams for the structural elements, you can develop them using 40.3 and 28.5 psf instead of 50 psf.

An alternative is to generate load diagrams just like we did in Examples 3.4 and 3.6 and then multiply them by their respective reduction factors (0.806 or 0.571). Each approach has its pros and cons.

3.7 Columns in Multistory Buildings

Columns in multistory buildings must be able to support the loads from all of the stories above the column. Indeed, this is where the floor live load reduction becomes more important. Imagine a first story column in a 10-story building that must carry all of the load from the roof and the eight floors above it. Ask yourself the question "*How probable is it that each of the eight floors above experience their full design load simultaneously?*" The answer to this question is that it is highly improbable. Therefore, we don't world's really want to design the column for that specific scenario. Rather, we will allow for a reduction in this design load. The challenge in all of this is to identify the proper tributary area to each column at each level of the building.

DID YOU KNOW?

Many structural design firms will not bother with taking floor live load reductions for any element that supports only one floor. However, they do recognize the need to reduce the loads on columns in multistory buildings so that they are not wasting money on designing for a highly improbable load condition.

EXAMPLE 3.8

The elevation view for a multistory building is given here, while the partial floor plan and roof plan for this building is given in Example 3.3. The roof live load is 20 psf (flat roof), the floor live load for the second, third, and fifth floors is 50 psf. The floor live load for the fourth floor is 125 psf because it is used for light storage. If the roof dead load is 25 psf and the floor dead load is 45 psf, find the total reduced live load and also the total dead load present on the column located between the ground and the second floor and also on grid point A2.

SOLUTION STRATEGY

Find the tributary area of column A2 for each level. The total live load will be the sum of the contributions of four floors and one roof where appropriate reductions must be taken for each level. Recall that a different reduction factor must be used for the roof and also for the fourth floor. The total dead load should sum up the dead loads for all of the floors and the roof.

SOLUTION

The tributary area to column A2 of each floor/roof level is found in Example 3.6 to be 546 ft^2.

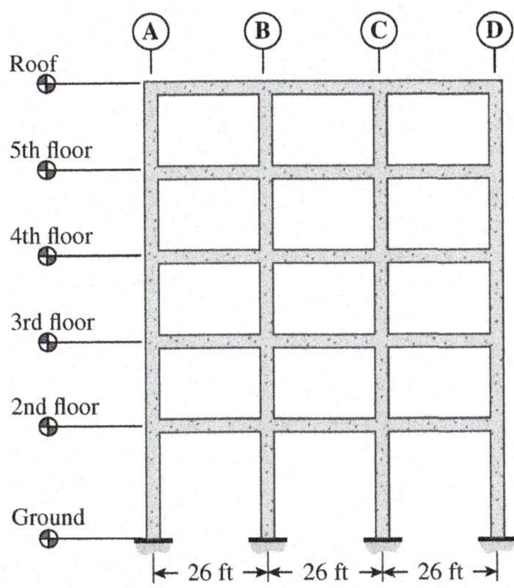

Find the reduced loads for each load type:

Roof: $R_1 = 1.2 - 0.001(546\ \text{ft}^2) = 0.654$ (see Section 2.10)
$R_2 = 1.0$ (flat roof)
$L_r = (20\ \text{psf})(0.654)(1.0) = 13.1\ \text{psf}$

Fourth floor: For multistory columns, floor live loads may be reduced by 20% if the floor live load is over 100 psf.
$L = (0.8)(125\ \text{psf}) = 100\ \text{psf}$

Floors: Find A_t for bottom story column considering only floors that have an unreduced live load of 50 psf.
$A_t = (546\ \text{ft}^2)(3\ \text{floors}) = 1638\ \text{ft}^2$
$K_{LL} = 4.0$ for an exterior column

$$L = 50\ \text{psf}\left(0.25 + \frac{15}{\sqrt{4.0(1638\ \text{ft}^2)}}\right) = (50\ \text{psf})(0.435)$$
$$= \underline{\underline{21.8\ \text{psf}}} > (50\ \text{psf})(0.4)$$

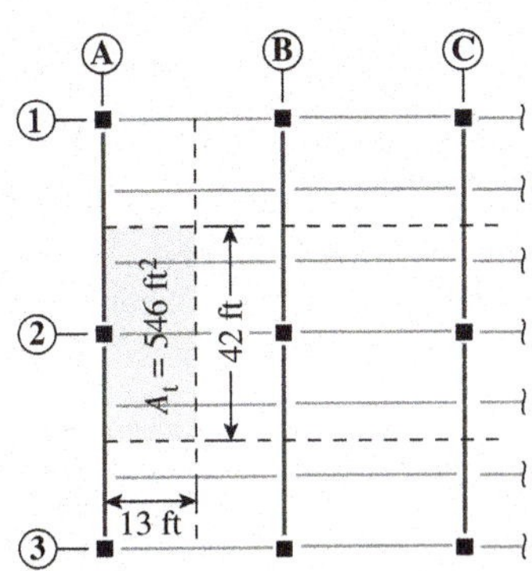

The total reduced live load must account for the contribution from all stories.

$$P_L = (13.1\ \text{psf})(546\ \text{ft}^2) + (100\ \text{psf})(546\ \text{ft}^2) + (21.8\ \text{psf})(1638\ \text{ft}^2)$$
$$= 97,461\ \text{lb} = \underline{\underline{97.5\ \text{k}}}$$

The total dead load is found using a similar approach, but no reductions are allowed.

$$P_D = (25\ \text{psf})(546\ \text{ft}^2) + (45\ \text{psf})(546\ \text{ft}^2)(4\ \text{floors}) = 111,930\ \text{lb} = \underline{\underline{111.9\ \text{k}}}$$

QUICK NOTE

The fourth-floor reduction factor is

$$0.25 + \frac{15}{\sqrt{4(546\ \text{ft}^2)}} = 0.57$$

So use 0.8, since $L_0 > 100$ psf

3.8 Examples with Video Solutions

VE3.1 A roof framing plan for a given building is shown in Figure VE3.1. Assume that all girders and beams come in contact with the slab and that the slab has equal strength and stiffness in both orthogonal directions. Classify the structure behavior as either a one-way or a two-way slab.

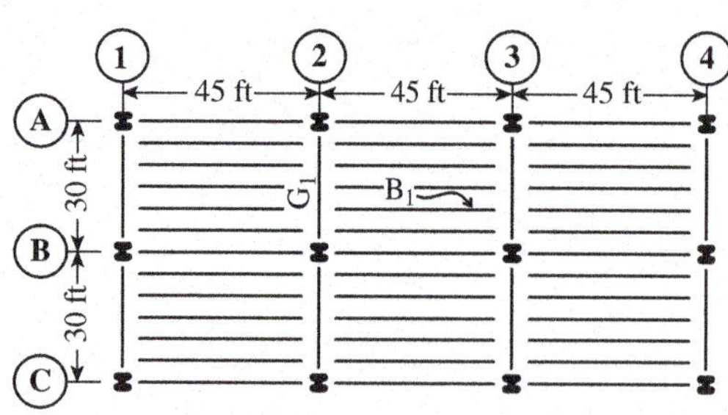

Figure VE3.1

VE3.2 Assume that the slab behavior for the system shown in Figure VE3.1 is one-way. If the dead load over the roof system is 40 psf, sketch the dead load diagrams for B_1, G_1, and column B3 (assume only one story on the column).

VE3.3 A floor system has a framing layout as shown in Figure VE3.1. If the live load over the floor system is 40 psf, sketch the reduced live load diagrams for B_1, G_1, and column B3 (assume only one story on the column).

VE3.4 A floor framing plan for a given building is shown in Figure VE3.4. Assume that all girders and beams come in contact with the slab and that the slab has equal strength and stiffness in both orthogonal directions. Classify the structure behavior as either a one-way or a two-way slab.

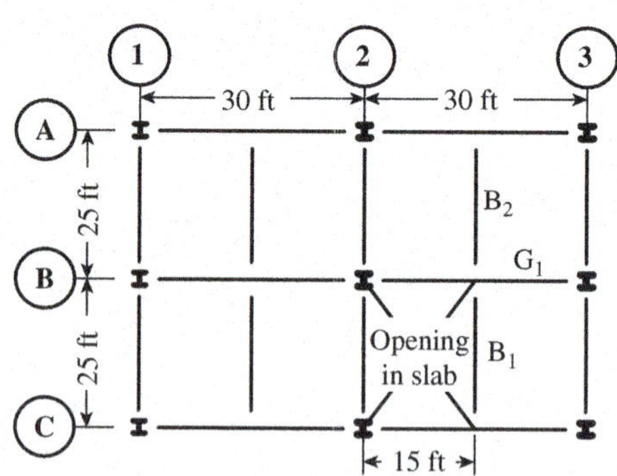

Figure VE3.4

VE3.5 Assume that the slab behavior for the roof system shown in Figure VE3.4 is two-way. If the dead load over the roof system is 65 psf, sketch the dead load diagrams for B_1, B_2, and G_1.

3.9 Problems for Solution

Sections 3.2–3.4

P3.1 For the floor framing system and loads shown in Example 3.3, sketch the loading diagrams for beam B_2 and girder G_1 and the load on column B1. Assume the column supports only the single floor shown and no reductions are taken. (Ans: $P_{B1} = 27.3$ k, girder point loads = 13.3 k)

P3.2 For the floor framing system and loads described in Example 3.4, sketch the loading diagrams for beam B_2 and girder G_1 and the load on column B1. Assume the column supports only the single floor shown and no reductions are taken. Assume one-way slab behavior.

P3.3 For the floor framing system and loads described in Example 3.3, compute the tributary area, the influence area, and the load assigned to column B2. Assume the column supports only the single floor shown and no reductions are taken. (Ans: $A_t = 1092$ ft^2, $A_I = 4368$ ft^2)

For Problems 3.4 through 3.8, compute the values requested using the basic floor framing plan shown in the accompanying figure. Consider floor live-load reductions following the provisions of ASCE 7-16 as appropriate. For purposes of these problems, assume that this is the uppermost floor in a multistory office building. Also assume that it has a one-way slab behavior.

Basic Building Layout

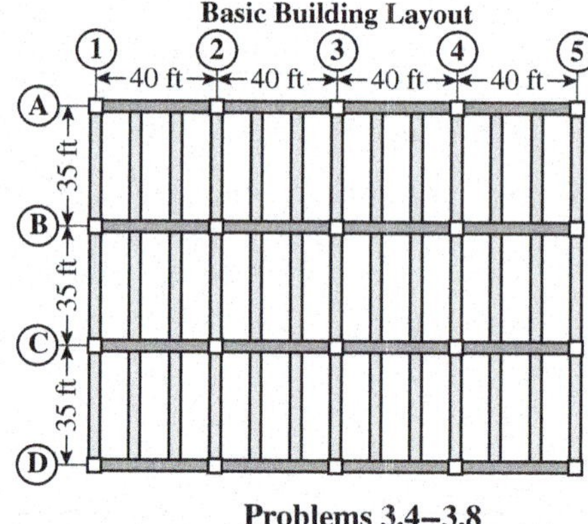

Problems 3.4–3.8

P3.4 Load on column B3 contributed by this floor if the floor live load is 150 psf.

P3.5 Load on column A3 contributed by this floor if the floor live load is 75 psf. (Ans: $P_{A3} = 27.8$ k)

P3.6 Load on column A1 contributed by this floor if the floor live load is 100 psf.

P3.7 Load on an interior floor beam if the floor live load is 75 psf. (Ans: w = 740 plf)

P3.8 Load on Girder B2–B3 if the floor live load is 50 psf.

Section 3.4–3.7

For Problems 3.9 through 3.27 compute the values requested using the basic floor/roof framing plan shown in the figure part (a). The building consists of five levels as shown in the figure part (b). If live load reductions are considered, use the appropriate provisions given in ASCE 7-16.

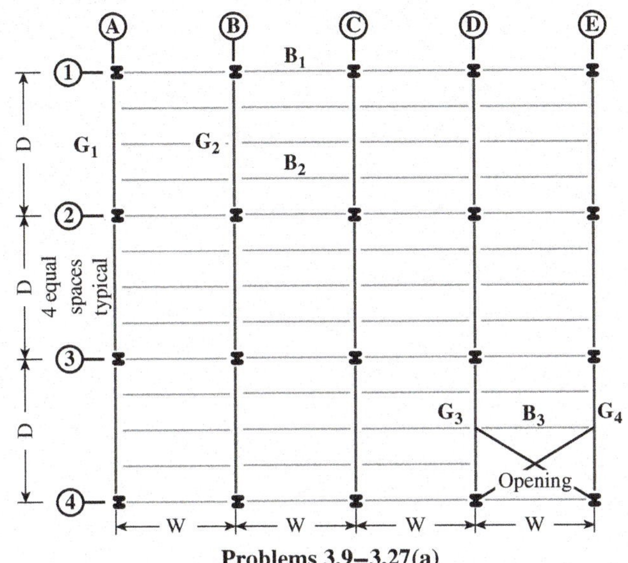

Problems 3.9–3.27(a)

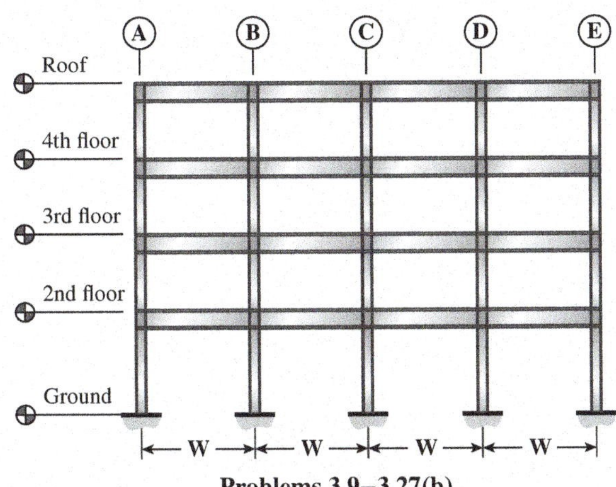

Problems 3.9–3.27(b)

P3.9 The roof is built using a metal deck that has a dominant strength and stiffness in the direction perpendicular to the roof beams. The tops of all roof girders and roof beams are located at the same elevation (see Figure 3.5). If W = 30 ft and D = 30 ft, classify the appropriate behavior of the slab. (Ans: one-way slab)

P3.10 The roof is built using a metal deck that has a dominant strength and stiffness in the direction perpendicular to the roof beams. The tops of all roof girders and roof beams are located at the same elevation (see Figure 3.5). If W = 25 ft and D = 60 ft, classify the appropriate behavior of the slab.

P3.11 The floor is built using a reinforced concrete slab that has similar strength and stiffness in both orthogonal directions. The tops of all floor girders and floor beams are located at the same elevation (see Figure 3.5). If W = 30 ft and D = 30 ft, classify the appropriate behavior of the slab. (Ans: one-way slab)

P3.12 The floor is built using a reinforced concrete slab that has similar strength and stiffness in both orthogonal directions. The tops of all floor girders and floor beams are located at the same elevation (see Figure 3.5). If W = 25 ft and D = 60 ft, classify the appropriate behavior of the slab.

P3.13 Assume W = 20 ft and D = 28 ft and that the slab behavior can be classified as a one-way slab. If the dead load for the roof is found to be 22 psf, compute and sketch the load diagrams for B_1, B_3, G_3, and G_4 located in the roof. (Be careful of the opening.) (Ans: for B_1, $B_3 \Rightarrow W = 77$ plf)

P3.14 Assume W = 20 ft and D = 28 ft and that the slab behavior can be classified as a one-way slab. If the dead load for the roof is found to be 22 psf, compute and sketch the load diagrams for B_1, B_2, G_1, and G_2 located in the roof.

P3.15 Assume W = 25 ft and D = 30 ft and that the slab behavior can be classified as a one-way slab. If the snow load for the roof is found to be 45 psf and the wind load is an uplift of 25 psf, compute and sketch the load diagrams for B_3 and G_3 located in the roof. (Be careful of the opening.) (Ans: wind point loads 2344 lb, 3047 lb, 4688 lb)

P3.16 Assume W = 25 ft and D = 30 ft and that the slab behavior can be classified as a one-way slab. If the snow load for the roof is found to be 45 psf and the wind load is an uplift of 25 psf, compute and sketch the load diagrams for B_2 and G_2 located in the roof.

P3.17 Assume W = 20 ft and D = 28 ft and that the slab behavior can be classified as a one-way slab. If the dead load for the roof is found to be 22 psf and the dead load of each floor is 65 psf, compute the total dead load in the following columns: A1 between third and fourth floor, D2 between second and third floor, and E2 between ground and second floor. (Ans: $P_{A1} = 12.2$ k, $P_{D2} = 85.1$ k, $P_{E2} = 60.8$ k)

P3.18 Assume W = 20 ft and D = 48 ft and that the slab behavior can be classified as a two-way slab. If the dead load for the roof is found to be 22 psf, compute and sketch the load diagrams for B_1, B_3, G_3, and G_4 located in the roof. (Be careful of the opening.)

P3.19 Assume W = 20 ft and D = 48 ft and that the slab behavior can be classified as a two-way slab. If the dead load for the roof is found to be 22 psf, compute and sketch the load diagrams for B_1, B_2, G_1, and G_2 located in the roof. (Ans: point loads on G_1 = 1848 lb, point loads on G_2 = 3696 lb)

P3.20 Assume W = 20 ft and D = 48 ft and that the slab behavior can be classified as a two-way slab. If the dead load for the roof is found to be 22 psf and the dead load of each floor is 65 psf, compute the total dead load in the following columns; A1 between third and fourth floor, D2 between second and third floor, and E2 between ground and second floor.

P3.21 Assume W = 20 ft and D = 30 ft and that the slab behavior can be classified as a one-way slab. If the unreduced roof live load is 20 psf, compute and sketch the load diagrams for B_1, B_2, G_1, and G_2 located in the roof. Take appropriate reductions considering the roof slope to be flat. (Ans: point loads on G_1 = 1463 lb, point loads on G_2 = 2250 lb)

P3.22 Assume W = 20 ft and D = 30 ft and that the slab behavior can be classified as a one-way slab. If the unreduced roof live load is 20 psf, compute and sketch the load diagrams for B_1, B_3, G_3, and G_4 located in the roof. Take appropriate reductions considering the roof slope to be flat.

P3.23 Assume W = 20 ft and D = 30 ft and that the slab behavior can be classified as a one-way slab. If the unreduced floor live load is 50 psf, compute and sketch the load diagrams for B_1, B_2, G_1, and G_2 located in the fourth floor. Take appropriate reductions. (Ans: point loads on G_1 = 3593 lb, point loads on G_2 = 5625 lb)

P3.24 Assume W = 20 ft and D = 30 ft and that the slab behavior can be classified as a one-way slab. If the unreduced floor live load is 50 psf, compute and sketch the load diagrams for B_1, B_3, G_3, and G_4 located in the fourth floor. Take appropriate reductions.

P3.25 Assume W = 40 ft and D = 40 ft. The unreduced roof live load is 20 psf. The live load on the second and third floors is 65 psf, and the live load on the fourth floor is 40 psf. Compute the total live load in the following columns; A1 between third and fourth floor, D2 between second and third floor, and E2 between ground and second floor. (Ans: P_{A1} = 16.4 k, P_{D2} = 86.4 k, P_{E2} = 64.4 k)

P3.26 Assume W = 30 ft and D = 30 ft. The unreduced roof live load is 20 psf. The live load on the second and third floors is 65 psf, and the live load on the fourth floor is 40 psf. Compute the total live load in the following columns; A1 between third and fourth floor, D2 between second and third floor, and E2 between ground and second floor.

P3.27 Assume W = 45 ft and D = 30 ft. The unreduced roof live load is 20 psf. The live load on the second and third floors is 65 psf, and the live load on the fourth floor is 80 psf. Compute the total live load in the following columns; A1 between third and fourth floor, D2 between second and third floor, and E2 between ground and second floor. (Ans: P_{A1} = 23.6 k, P_{D2} = 94.5 k, P_{E2} = 67.2 k)

P3.28 The loading for a single frame-line of a three story building is given in the figure. The roof is subjected to a snow load and a dead load, while the two upper floors are subjected to floor live load and dead load. Using the tributary approach and without taking any reductions, find the following:

a. The total dead load, live load and snow load present in column B between the second and third floors.
b. Sketch the loading diagram including all relevant loads for the transfer girder located at the second floor.
c. The total dead load, floor live load, and snow load present in column A between the ground and the second floor.
d. Using the strength-based load combinations given in Chapter 2, find the total design axial load that should be used for the column described in part c. (Assume the floor live load results from a load less than 100 psf.)
e. Using the ASD-based load combinations given in Chapter 2, find the total design axial load that should be used for the column described in part c.

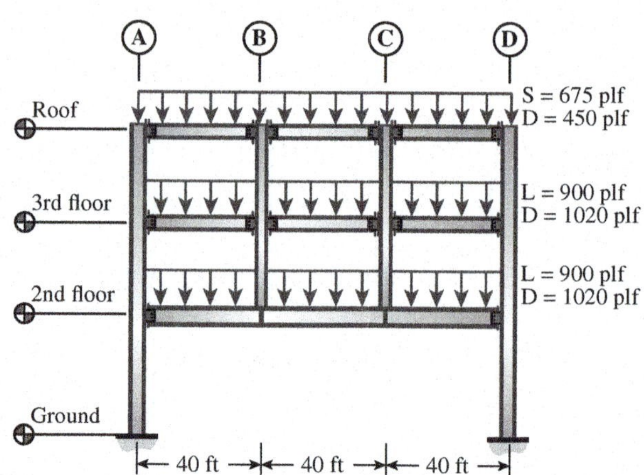

Problem 3.28

Lateral System Loading and Behavior

4

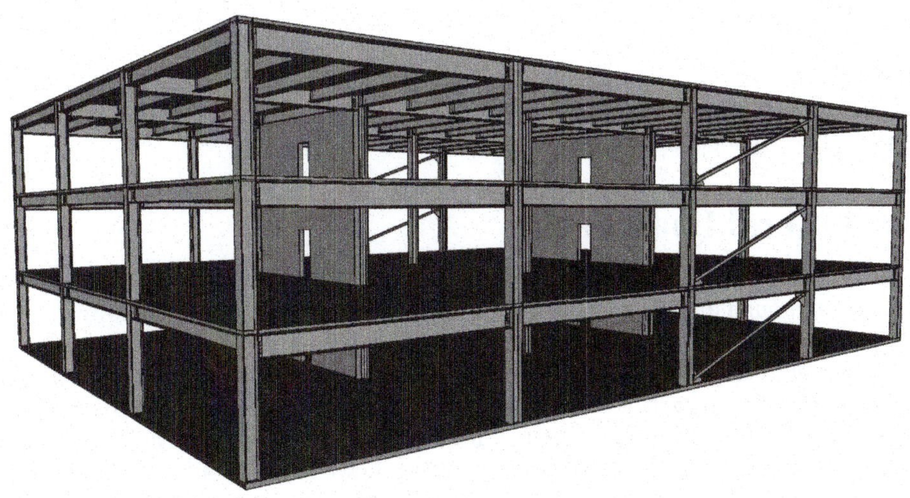

4.1 Introduction

In the previous chapter, we discussed the need to track vertical loads through a structural system and ensure that all of those loads reach the foundation. As seen in Chapter 2, which is focused on the calculation of structural loads, many structural systems are subjected to lateral loads (e.g., wind, earthquake, soil etc.) in addition to the more obvious vertical or gravity type loads. Dealing with these lateral loads appropriately should be every bit as important to the structural engineer as are the vertical loads. Indeed, the lateral system behavior is key to ensuring overall stability of the structural system. For example, Figure 4.1(a) shows a simple portal frame that is subject to both a vertical and a lateral load. The frame is constructed of three members that are all connected using frictionless pins. Using statics, one can resolve the vertical loads and show that equilibrium can be satisfied. However, when the lateral load is applied, equilibrium cannot be satisfied, and the structure undergoes rigid body motion and suffers collapse. In contrast, Figure 4.1(b) shows a system that is designed to handle the lateral load P and distribute it to the supports. This lateral system includes but is not limited to the cross-braces.

The purpose of this chapter is to introduce how lateral loads are applied to and distributed through building systems. When wind hits the side of a building, it is not going to show up as a point load on the side of a building. Rather, it is some kind of a distributed load that is applied to the exterior walls. If this is the case, how does a point load like shown in Figure 4.1 show up on the frame? Once it does show up on the frame how will it be transferred to the foundation/supports? Tracking these loads through the system is really a fundamental and critical exercise for a structural engineer.

This chapter also introduces the basic structural components that make up the lateral load path through a structural system. This includes the out-of-plane walls, diaphragms and **vertical lateral**

QUICK NOTE

Common sources of lateral load on buildings include:

- Wind
- Earthquake
- Soil
- Fluid

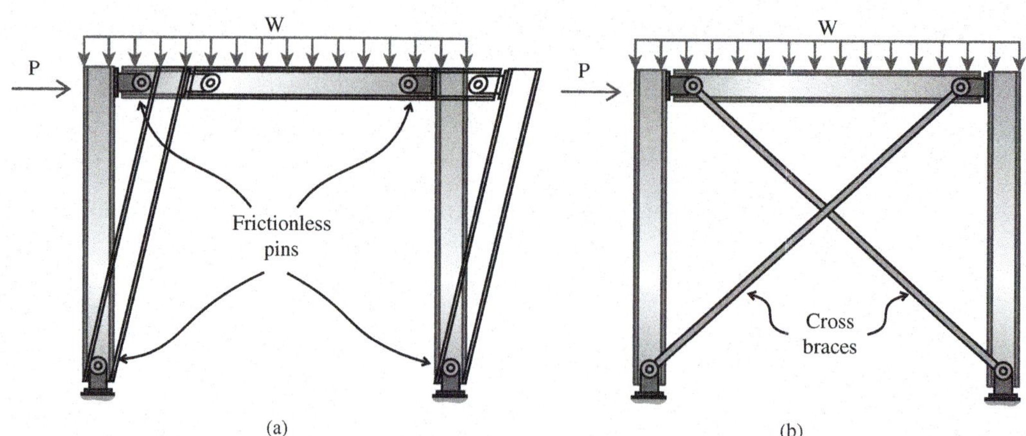

FIGURE 4.1 Portal frame subjected to a vertical and lateral load (a) without a lateral system and (b) with a lateral system.

force resisting systems (VLFRS). Focus is given in this chapter to calculating the loads that are applied to these elements and sketching appropriate load diagrams. Analysis of these elements are treated throughout the remainder of the text.

4.2 Lateral Load Path

To identify the lateral load path is to track a lateral load being applied to a building or building component through the structural system. The loads must be tracked all the way to the supports or foundation of the building.

To adequately identify the load path, one must consider the source and "point of application" of the load. Lateral loads such as wind, fluid, and earth pressure are applied from an outside source to the structure element(s). For example, the soil backfilled against a basement wall applies a direct pressure perpendicular to the wall surface as shown in Figure 4.2. The soil pressure shows up on the exterior surface of the wall. Wind and fluid pressures behave similarly.

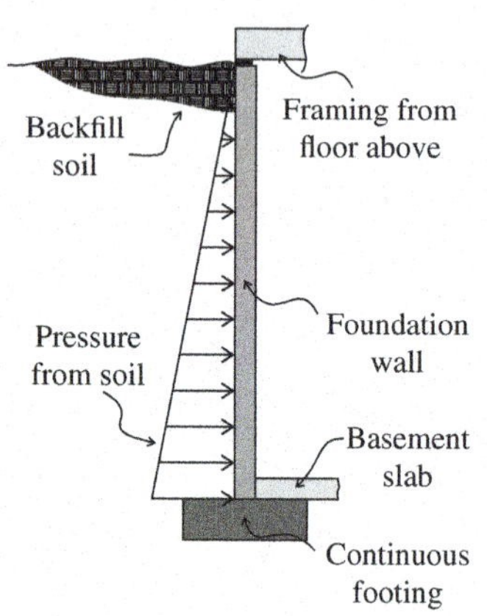

FIGURE 4.2 Side view of basement wall subjected to lateral soil pressures.

Lateral loads due to earthquakes become a little more challenging to picture, because they are due to the shaking of the ground. The shaking of the ground causes the structure to vibrate, which produces inertial loads caused from the acceleration of the mass of the structure and any attached elements. The load is then "applied" to the structure wherever the masses are located. For instance, in typical buildings, much of the mass of the structure is located at the level of the floors and will thus result in earthquake loads being applied at those levels. One must remember that the walls of a building also have mass that will result in earthquake forces being applied directly to them as well.

Hierarchy of Lateral Load Path

Identifying the lateral load path of course begins with identifying the point of load application. It then becomes an exercise in identifying what provides support to the structural element and transferring the load to those supports. Take a look at Figure 4.2 and ask a couple of questions regarding the diagram. If the soil pressure is being applied directly to the foundation wall, what is preventing the basement wall from sliding to the right and/or rotation clockwise about the footing? For this case, the sliding or translation is prevented by the footing (i.e., the wall is tied into the footing) and also the basement slab (i.e., the wall would push against the slab.) So what is preventing wall from rotating clockwise? It is the floor system that is tied into the top of the wall. Hence, one can state that the soil load is applied to the wall and then distributes into the footing, the slab, and the floor system. This is the first step in the load path. But the process of identifying the load path cannot be complete until the engineer can answer how the entire load is transferred back into the soil. The following questions signify the type of questions that should be asked. "If the load is distributed into the floor system—also known as a *diaphragm*—what prevents the floor from freely translating to the right? If the footing and slab prevent the base of the wall from sliding, then what is preventing the footing and slab from sliding?" Such questions must be asked until all loads have reached the ground.

Lateral Load Path Hierarchy

Out-of-plane wall
⇓
Diaphragms and/or footing
⇓
Vertical lateral force resisting system (in-plane)
⇓
Foundations

FIGURE 4.3 Typical hierarchy for lateral load path.

In most building structures, there is a general hierarchy (Figure 4.3) that can be identified. This hierarchy shows that the load is generally applied to the out-of-plane surface of a wall. The wall then distributes the load to the diaphragm(s) and/or footing. The diaphragm then distributes the load to the VLFRS, which can be viewed as an in-plane wall or system.

This hierarchy can be traced out on the simple building system depicted in Figure 4.4 with its exploded view shown in Figure 4.5. The building is a simple rectangular layout with four walls attached to footings and a roof attached to the walls. Envision that wind is blowing on one of the walls and is creating a pressure on the face of the wall. Since the load is being applied perpendicular to the plane of the wall, the loading is seen as an out-of-plane load. The label of "out-of-plane wall" really refers to the notion that the wall is loaded out of its plane. Some of the load travels down the wall to the footing, while the remainder of the load travels up the wall to the roof that serves as a diaphragm [Figure 4.5(a)]. The load that went into the footing is transferred into the soil. However, the load that goes into the diaphragm is transferred into the VLFRS, which is shown as a shear wall in this depiction [Figure 4.5(b)]. The load that is transferred to the VLFRS shown in Figure 4.5(c) is applied at the top of the wall and applied in the plane of the wall, resulting in the term "in-plane wall."

Since seismic loads originate at the location of mass in the structure, does this cause the load path to be different than what was previously described? Whatever load originates in the out-of-plane wall will follow the hierarchy presented earlier. Loads that originate elsewhere simply cause the load path to be abbreviated. For example, whatever load originates in the diaphragm due to its own mass will travel through the diaphragm, into the VLFRS, and into the footings. Whatever load originates in the VLFRS travels through the VLFRS and into the footings. In other words, the hierarchy is not changed—it is only truncated based on where the load originates.

QUICK NOTE

The connections between structural elements are critical components in the lateral load path and should not be ignored.

COMMON MISTAKE

Students often assign the load from the diaphragm to be resisted by the out-of-plane wall. One must remember that the diaphragm supports the out-of-plane wall, but the out-of-plane wall doesn't support the lateral loads of diaphragm.

NOMENCLATURE

Out-of-plane wall: The lateral load is applied perpendicular to the plane of the wall.

In-plane wall: The lateral load is applied parallel to the plane of the wall.

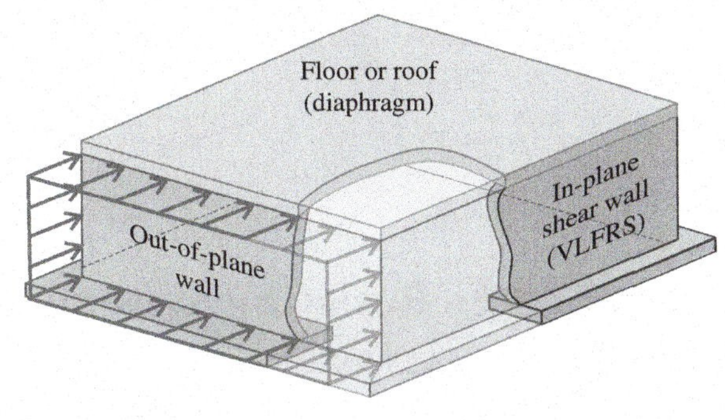

FIGURE 4.4 Typical lateral load resisting system for buildings.

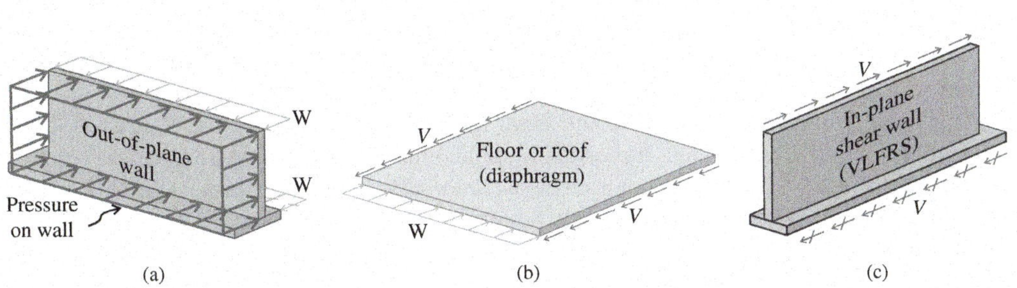

FIGURE 4.5 Free-body diagrams for the building components under lateral loads.

Customized Lateral Load Path

Generally speaking the lateral load path can be described as given in the previous section. However, there is no substitute for the engineer looking closely at the framing details of the building and explicitly defining the load path that will be followed. Example 4.1 is presented to illustrate this point.

EXAMPLE 4.1

NOMENCLATURE

Cladding: External covering of a building which could include siding, glazing, paneling, roofing, etc.

Girt: Horizontal beam-type element that spans between vertical columns. This usually takes load directly from the cladding and distributes it to the columns.

On this structure, wind applies pressure on one face of the building. Explicitly describe the lateral load path that the wind follows through the system.

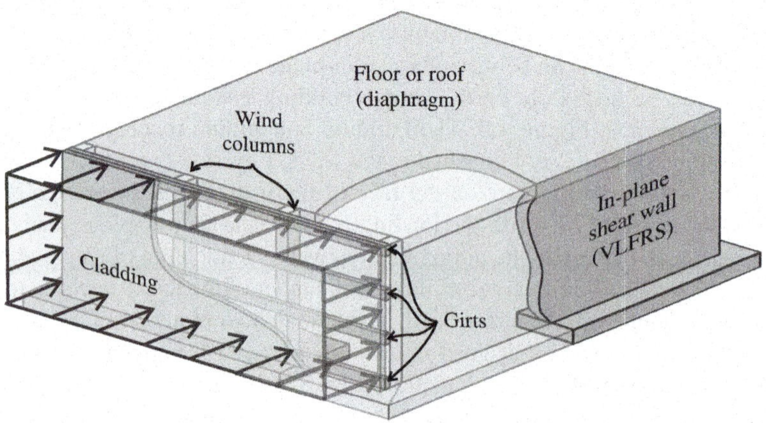

SOLUTION STRATEGY

Examine and identify all of the building components, noting what supports each element.

SOLUTION

The elements that exist in the building are

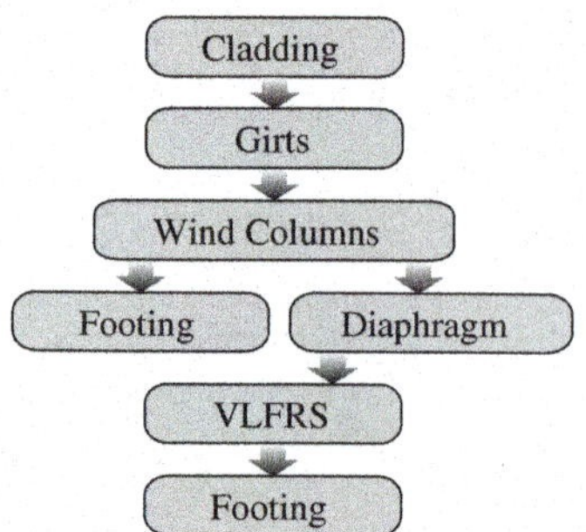

4.3 Vertical Lateral Force Resisting Systems

The **vertical lateral force resisting system (VLFRS)** is defined as the element or combination of elements that are used to resist the in-plane lateral loads and are oriented vertically. For instance, the roof shown in Figure 4.4 is loaded in-plane, but its plane is oriented horizontally. The plane of the shear wall shown in the same figure is oriented vertically. Since the shear wall is oriented vertically and receives in-plane loads, as shown in Figure 4.5(c), it is classified as a component in the VLFRS.

Whether an element or group of elements is part of the VLFRS is highly dependent on the direction that the load is being applied. If the load is applied in a particular direction, certain elements may be part of the VLFRS while other elements are not. If the load is reoriented, the elements that were originally part of the VLFRS may no longer be. This is most easily seen by looking at Figure 4.6, where the VLFRS is shown for two different loading scenarios for the same building. Note that the out-of-plane wall in one scenario becomes the in-plane wall in the other scenario

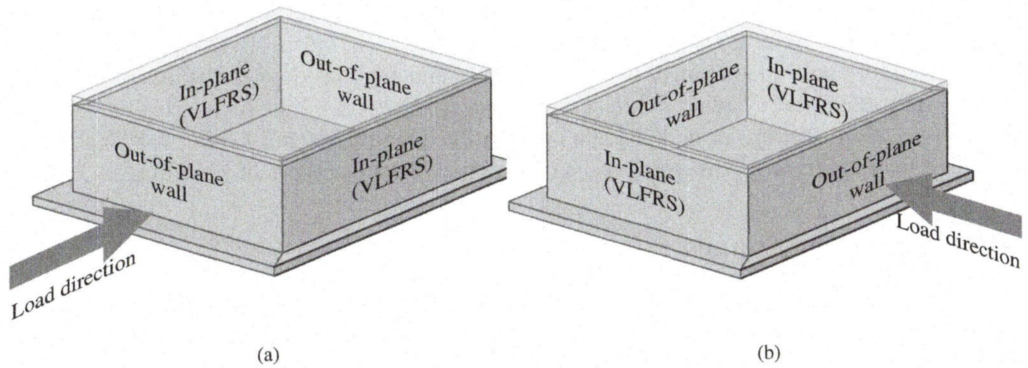

FIGURE 4.6 Lateral system designations change depending on which direction the load is applied.

and vice versa. Indeed, it is not only possible but is highly probable that a wall can be loaded both in-plane and out-of-plane simultaneously due to wind blowing on the surface of both walls.

While the examples presented thus far have demonstrated the use of shear walls as the VLFRS, there are many options that are available to the engineer to achieve the required lateral resistance. These VLFRS systems can be classified into four major categories as outlined in the sidebar and shown in Figure 4.7. Each class of VLFRS has its advantages and disadvantages related to issues such as building height limitation, stiffness, and architectural versatility. However, they do have certain characteristics in common. They are all loaded in-plane with a shear load—either distributed as shown in the figure or as a point load. They all must resist shear but must also resist overturning due to the moment caused by the location of load application.

CLASSES OF VLFRS

- Shear wall systems
- Moment resisting frames
- Braced frames
- Dual systems using a combination of these

Shear Wall Systems

Shear wall systems are characterized by wall segments that are oriented to be in-plane with the direction of the applied load. They are commonly constructed using concrete, masonry block, or wood panels (sheathing). However, the use of steel plates for creating shear walls is becoming

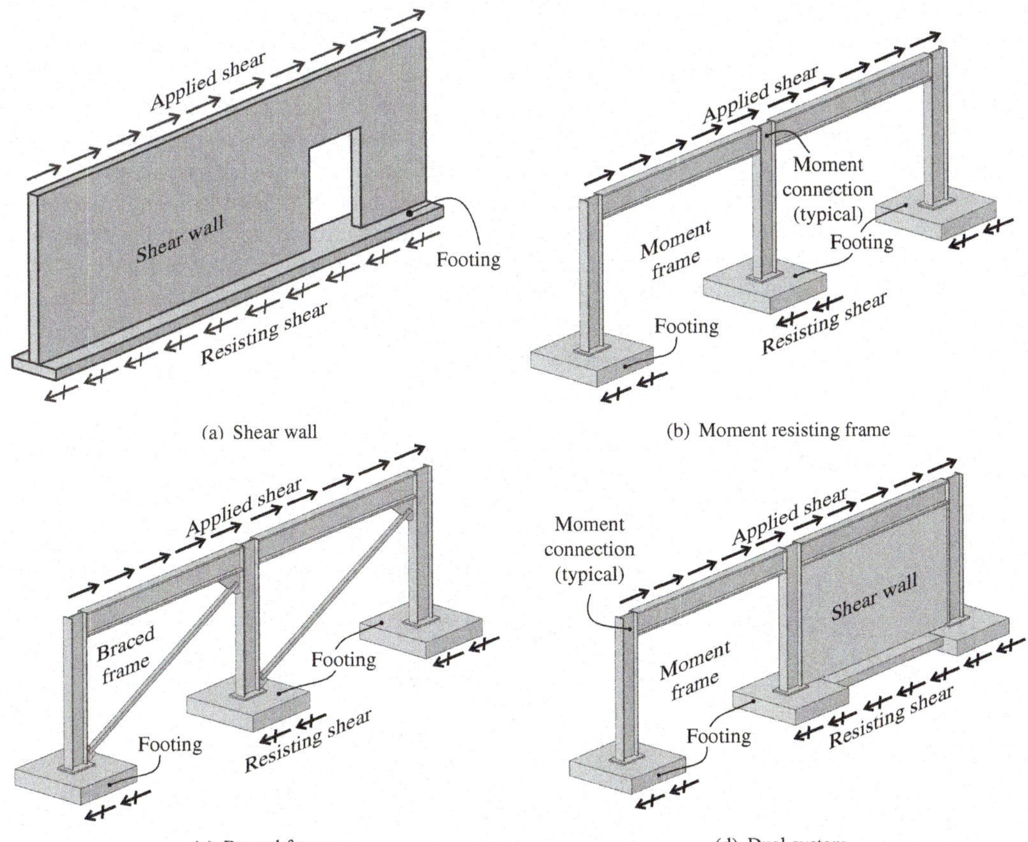

FIGURE 4.7 Major classes of vertical lateral force resisting systems.

more widely used and is thus noteworthy. The shear resistance is achieved through in-plane shear capacity through the entire height of the wall.

Shear walls are quite stiff relative to other VLFRS's and are considered to be one of the most architecturally restrictive systems. While a shear wall can have openings, such as is illustrated in Figure 4.7(a), there still has to be adequate uninterrupted lengths of full-height wall to achieve the desired load resistance. This restriction can limit the visual openness of a building.

Moment Resisting Frames

Moment resisting frames are characterized by an assemblage of beams and columns that are fastened together using connections that can transfer moment. The connections into the foundation/footing either can be pinned or fixed. The resistance from this assemblage is through the shear and bending resistance of the individual elements. The use of beams and columns in the configuration shown in Figure 4.7(b) is present in many of today's structures. However, a moment resisting frame is not limited to beams and columns that are oriented perpendicular to one another. Figure 4.8 illustrates a moment resisting frame configuration that can be found in many pre-fabricated metal buildings.

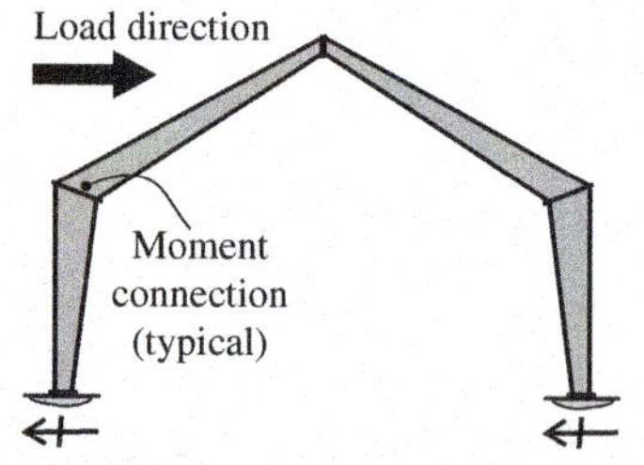

FIGURE 4.8 Moment resisting gable frame. (The term gable describes the sloped shape of the roof.)

The architectural versatility of moment resisting frames tends to be a highly desirable characteristic. This is because each bay (section between columns) does not have any elements that will obstruct the view or use of the space. This comes at a bit of a price though in that moment resisting frames are considerably more flexible than a shear wall of comparable length. That is to say, the deflections in moment resisting frames can get relatively large. Common building materials for these frames are steel and concrete.

Braced Frames

Braced frames usually involve an assemblage of beams and columns that are then braced by at least one element connected at an incline to the beams, as seen in Figure 4.7(c). While the beams, columns, and braces work together to provide the lateral resistance of this system, the primary resistance comes from the axial resistance of the brace. The beams and columns may be connected with either pin or moment connections as demonstrated in Figure 4.1(b). The braces are then attached in any number of configurations to achieve the desired resistance. The configuration of the bracing is classified into two main types: *concentric* or *eccentric* bracing. Within each class, there are many different configurations. However, a few of the more common configurations within the concentric and eccentric classes are demonstrated in Figures 4.9 and 4.10, respectively.

Relative to moment resisting frames, braced frames tend to be less flexible and consequently experience smaller deflections. However, they become a little bit more architecturally restrictive because of the obstruction of the braces. There still exists free space in the braced bay used to install windows and doorways. This becomes even more versatile when eccentrically braced frames are used—a point seen when comparing the systems in Figure 4.9 with the systems in Figure 4.10. The most common material for these kinds of frames is steel. Occasionally, the beams and columns are made of wood or concrete.

NOMENCLATURE

Concentrically Braced Frame: A frame where the centerline of all members at a connection pass through the same point.

Eccentrically Braced Frame: A frame where the centerline of not every member at a connection passes through the same point.

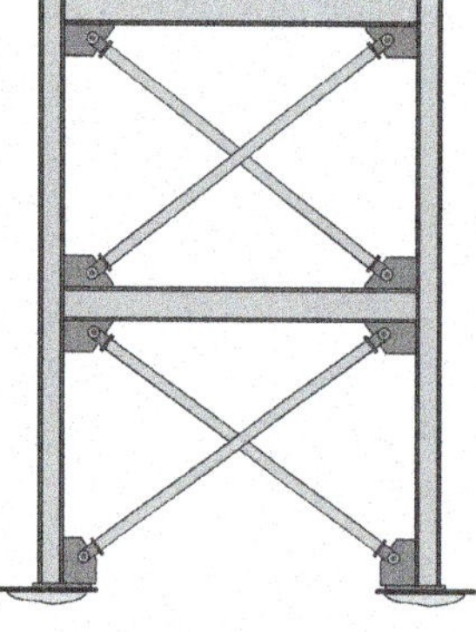

(a) Cross bracing

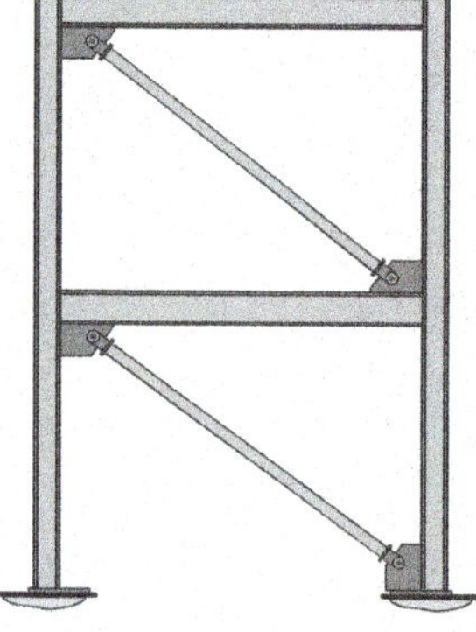

(b) Diagonal bracing

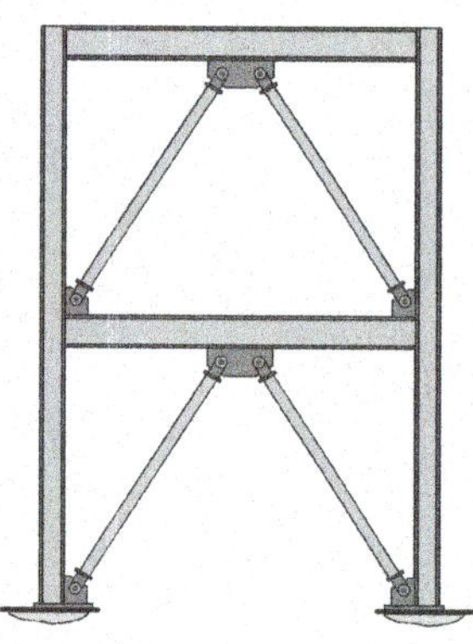

(c) Chevron bracing

FIGURE 4.9 Two story concentrically braced frames.

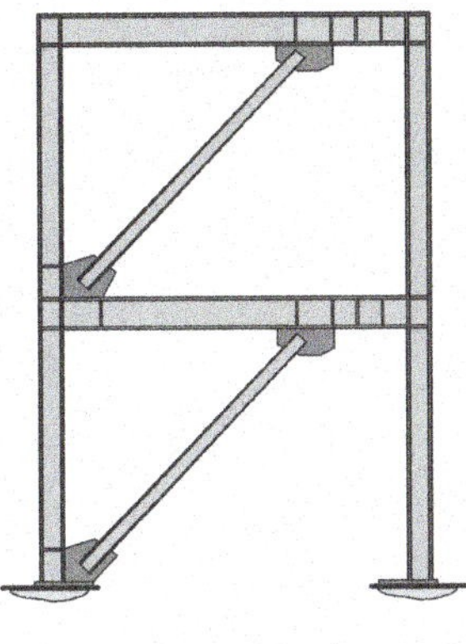

(a) Diagonal bracing

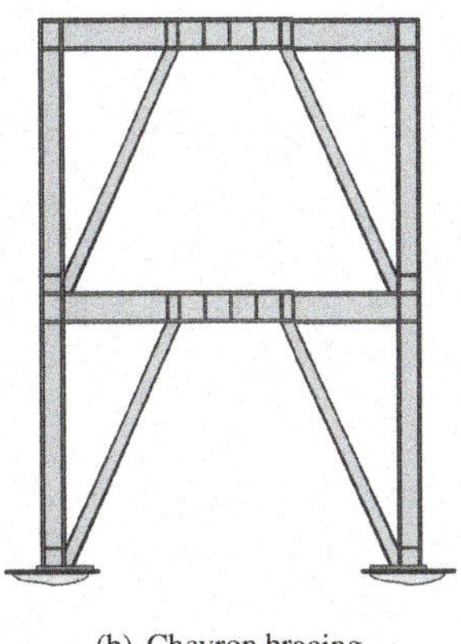

(b) Chevron bracing

FIGURE 4.10 Two story eccentrically braced frames.

Dual Systems

Sometimes the advantages of two systems can be achieved by using a combination of two different systems. A system that uses two of the previously discussed systems is referred to as a dual system. Figure 4.7(d) shows a moment resisting frame where one bay contains a shear wall to help increase the lateral resistance and stiffen the system. Such systems are common in taller buildings where the structural engineer wants to maintain a sense of openness around the perimeter of the building (e.g., moment resisting frames) and still take advantage of the resistance of concrete or masonry walls surrounding stair wells and elevator shafts (e.g., shear walls).

4.4 Diaphragms

The purpose of discussing diaphragms in this chapter is to clarify the role that a diaphragm plays in the lateral load path and to calculate the loads that act on the diaphragm. The analysis of diaphragms takes some consideration and cannot be treated until the student has a clear grasp of structural analysis tools and techniques.

The hierarchy of the lateral load path indicates that the load applied to the out-of-plane wall or cladding must be distributed to a diaphragm in the building. The diaphragm then distributes the load to the VLFRS. Indeed, the main role of the diaphragm is to tie the structure together (i.e., the out-of-plane walls with the VLFRS). It provides stability to the overall building structure. Without the diaphragm, the out-of-plane walls do not have support and fall over. Furthermore, the load is never distributed to the VLFRS.

QUICK NOTE

Diaphragms provide *stability* to the structure by providing a *continuous load path* for lateral loads.

In reality, the structural components that make up the diaphragm in a building will most likely need to serve another purpose as well. Floor and roof systems can serve as diaphragms that are characterized by in-plane loads [see Figure 4.5(b)], but they also must be able to carry the gravity loads (e.g., dead, live, snow, etc.) that are applied to them. Indeed, when an individual thinks of a floor in a building, they inherently acknowledge its purpose of carrying vertical load which is applied out-of-plane to the floor (see Chapter 3). One is less likely to see the role that the same floor system plays in providing the requisite load path and stability for lateral loads.

Floors are always going to be horizontal elements, and as such, this tends to make the student think that all diaphragms must be horizontal. The International Building Code (IBC), which governs the design of buildings throughout much of the United States, states that a diaphragm is

> *A horizontal or sloped system acting to transfer lateral forces to the vertical-resisting elements. When the word diaphragm is used, it shall include horizontal bracing systems.*[1]

If the floors and roof of a building serve as diaphragms, one must recognize that not all roofs are horizontal (i.e., flat) as seen in Figures 4.11 and 4.12.

Courtesy of Bryant G. Nielson

FIGURE 4.11 Building having an inclined diaphragm in the roof.

Courtesy of Bryant G. Nielson

FIGURE 4.12 Building having a slightly arched diaphragm in the roof.

Diaphragm Construction Classifications

Diaphragms can be classified into two main groups based upon their construction type—namely *panelized* and *braced* types (Figure 4.13). Panelized diaphragms are characterized by in-plane

[1] 2012 *International Building Code*, (Falls Church, Virginia, International Code Council Inc., 2012).

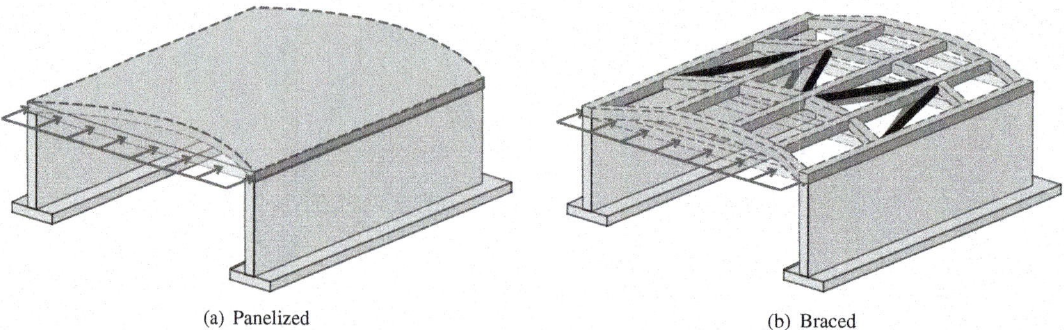
(a) Panelized (b) Braced

FIGURE 4.13 Two classes of diaphragms.

sheets, panels, or slabs typically made of wood, steel, or concrete. These types of diaphragms resist shear by developing in-plane shear in the panels or slabs and transferring it out to the VLFRS. Some examples of a panelized diaphragm include:

1. Plywood sheathing nailed to wood or steel framing members
2. Steel decking welded or screwed to steel joists
3. Steel decking with a concrete topping
4. Cast-in-place concrete slab
5. Precast concrete panels with concrete topping

This list of examples is not exhaustive but gives a good sampling of a large majority of existing diaphragms.

Panelized diaphragms can be viewed as a beam that is loaded horizontally and is supported by the elements of the VLFRS. Figure 4.14(a) illustrates this concept in relation to Figure 4.13(a). The beam analogy is very helpful for the analysis of diaphragms.

Under certain conditions, the cladding or paneling attached to a roof is not sufficiently strong to withstand and transfer in-plane shear. A braced diaphragm may be used in a case like this to transfer the requisite loads. Pre-fabricated metal buildings are a common place where such bracing is found (see Figure 4.15). This is because light-gauge metal sheathing is used as the exterior panels and is insufficient to transfer the loads.

A braced diaphragm can be seen as a truss, such as is shown in Figure 4.14(b). It is clear that this type of diaphragm resists the in-plane shear through the development of axial loads in the diaphragm members. A structural engineer who is comfortable with the analysis of trusses can adequately handle the analysis of a braced diaphragm.

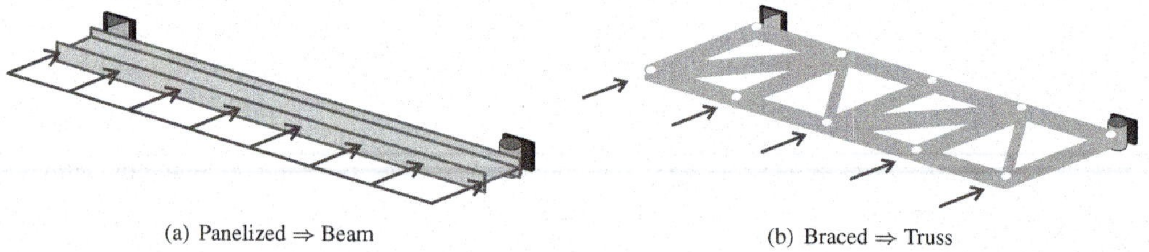

(a) Panelized ⇒ Beam (b) Braced ⇒ Truss

FIGURE 4.14 Analysis analogy for both diaphragm types.

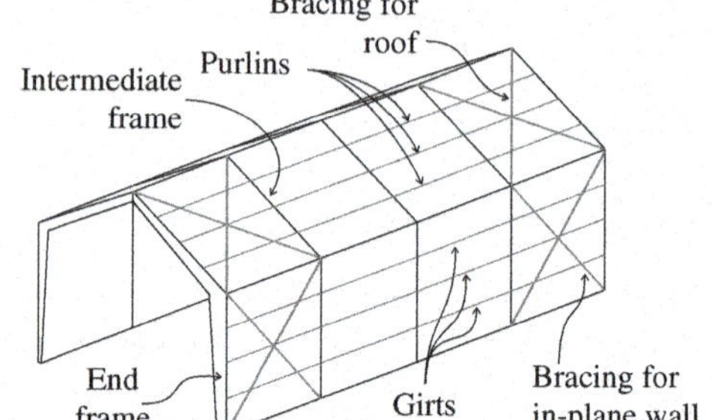

FIGURE 4.15 Framing of pre-fabricated metal building. Note the bracing in the roof and walls.

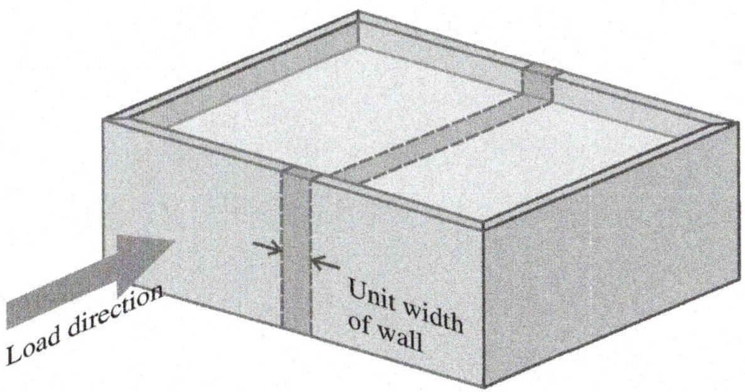

FIGURE 4.16 Tributary width of wall.

4.5 Tributary Approach

We have presented the basic components present in the lateral load path of a building, namely the out-of-plane walls, diaphragm, and VLFRS—not forgetting the connections of these components. The development of load diagrams for these components is the final objective of this chapter. Specifically, developing appropriate load diagrams for the out-of-plane walls and the diaphragms will be presented. The development of loads to the VLFRS is a basic concept of tracking the loads from the diaphragm into the VLFRS. However, the actual application of this can be complex at times, as it can require the analyst to understand indeterminate structural analysis.

In practice, the tributary approach, which was previously discussed in Chapter 3 for vertical loads, is generally used for tracking loads through the lateral load path. The basic tributary approach is to distribute half of the load acting on each component (e.g., out-of-plane wall) to each of its respective supports (e.g., diaphragms, footings). However, the tributary width for an out-of-plane wall is usually taken as a unit width (see Figure 4.16). For example, one would calculate the load on the wall for a strip that is 1.0-ft wide.

EXAMPLE 4.2

Sketch the loading diagram for both the front and the back out-of-plane walls of Figure 4.16 if the load on the building cross-section is as shown.

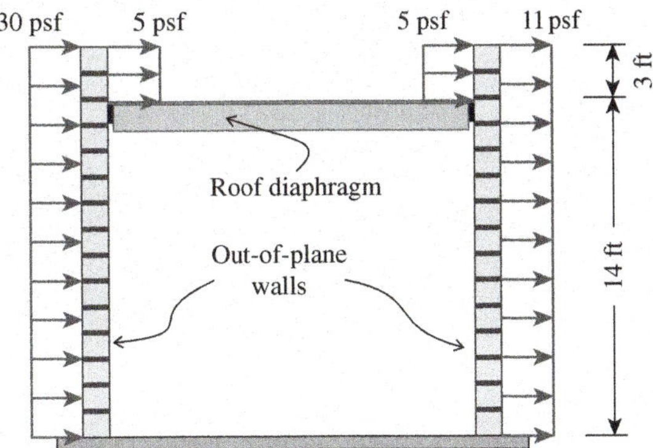

SOLUTION STRATEGY

Make a sketch of a 1.0-ft wide strip of wall and identify its idealized supports. Determine how much load is applied to that 1.0-ft width. One must recognize that the footings and the diaphragm provide the supports to both the front and the back walls.

SOLUTION

The base of the wall is assumed to be pinned, which is a reasonable assumption for most wall systems. The support provided by the diaphragm prevents the wall from translating to the right. As such, a roller support is appropriate. The idealized wall is shown here. This representation can be used for both the front and the back walls because of our understanding that rollers can provide reactions acting to the left or the right (see Chapter 5).

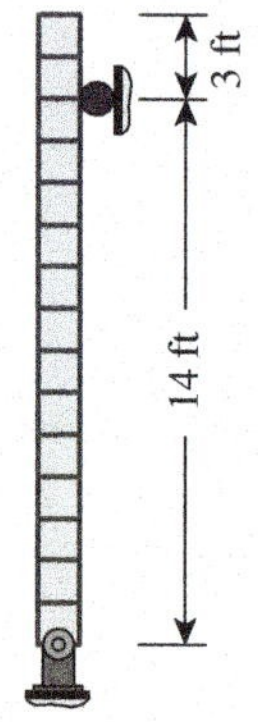

The tributary width to a 1.0-ft width segment of wall is 1.0-ft. Hence, the distributed loads on the front wall can be calculated as follows.

Front Wall:

$$w_1 = \left(30\,\tfrac{\text{lb}}{\text{ft}^2}\right)(1\text{ ft}) = 30\,\tfrac{\text{lb}}{\text{ft}} = 30\text{ plf}$$

$$w_2 = \left(5\,\tfrac{\text{lb}}{\text{ft}^2}\right)(1\text{ ft}) = 5\,\tfrac{\text{lb}}{\text{ft}} = 5\text{ plf}$$

Back Wall:

$$w_1 = \left(11\,\tfrac{\text{lb}}{\text{ft}^2}\right)(1\text{ ft}) = 11\,\tfrac{\text{lb}}{\text{ft}} = 11\text{ plf}$$

$$w_2 = \left(5\,\tfrac{\text{lb}}{\text{ft}^2}\right)(1\text{ ft}) = 5\,\tfrac{\text{lb}}{\text{ft}} = 5\text{ plf}$$

We can now complete the loading diagrams for both the front and back walls.

QUICK NOTE

The diaphragm is receiving load from both the front and the back walls.

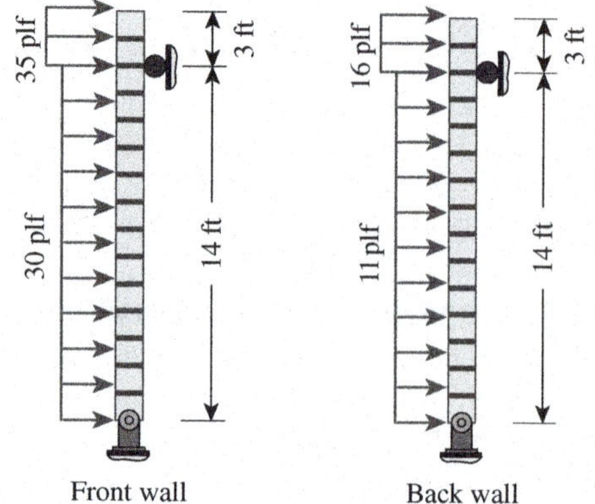

EXAMPLE 4.3

The front wall of the building shown is constructed of cladding, girts, and wind columns. If the load applied to the cladding is 24 psf uniformly distributed over the entire wall, sketch the loading diagrams for the cladding, the girts (g1 and g2), and the wind columns (wc1 and wc2). Treat the cladding as continuous over the girts and treat the girts as simple spans between the columns.

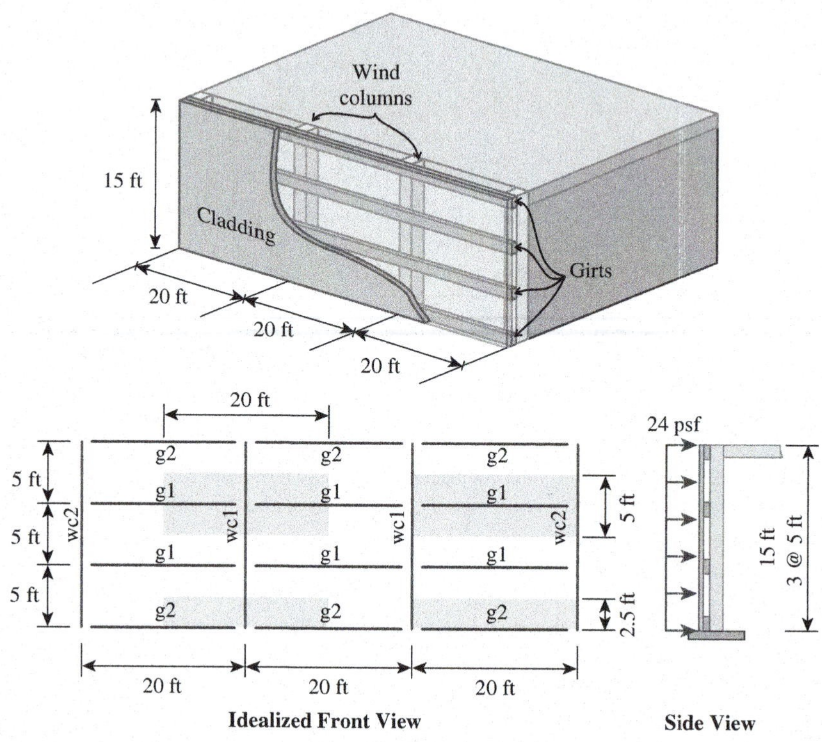

SOLUTION STRATEGY

The framing of the wall is very similar to the framing of a floor system. The load diagrams of the framing members (e.g., girts and columns) can be developed using the same principles as discussed in Chapter 3, which deals with vertical loads. Label all unique framing members. Identify the sequence of the load path, as demonstrated in Example 4.1, then identify tributary widths and areas for each of the elements. Multiply these widths and areas by the applied load of 24 psf.

SOLUTION

The loading diagram for the cladding is constructed for a 1.0-ft wide strip similar to what is shown in Figure 4.16. The cladding is supported by four girts that create a three-span condition.

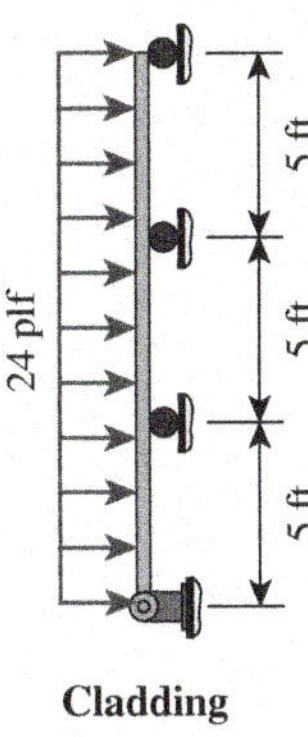

Cladding

Cladding:

$$w = \left(24 \tfrac{\text{lb}}{\text{ft}^2}\right)(1 \text{ ft}) = 24 \tfrac{\text{lb}}{\text{ft}} = 24 \text{ plf}$$

The load travels from the cladding directly onto the girts. There are two unique girts in the wall namely those that receive load from the cladding from two sides (g1) and those that receive load from one side only (g2). The cladding in this building spans one direction (i.e., one-way), because the support conditions require it. Note that the cladding only attaches to the girts and not to the columns. Thus, it must span from girt to girt. The front view identifies the tributary widths for girts g1 and g2.

Girts:

$$g1_{\text{trib}} = 5 \text{ ft}$$
$$g2_{\text{trib}} = 2.5 \text{ ft}$$

The distributed loads on each of the two girts are calculated as follows and shown with appropriate loading diagrams.

$$w_{g1} = \left(24 \tfrac{\text{lb}}{\text{ft}^2}\right)(5 \text{ ft}) = 120 \tfrac{\text{lb}}{\text{ft}} = 120 \text{ plf}$$
$$w_{g2} = \left(24 \tfrac{\text{lb}}{\text{ft}^2}\right)(2.5 \text{ ft}) = 60 \tfrac{\text{lb}}{\text{ft}} = 60 \text{ plf}$$

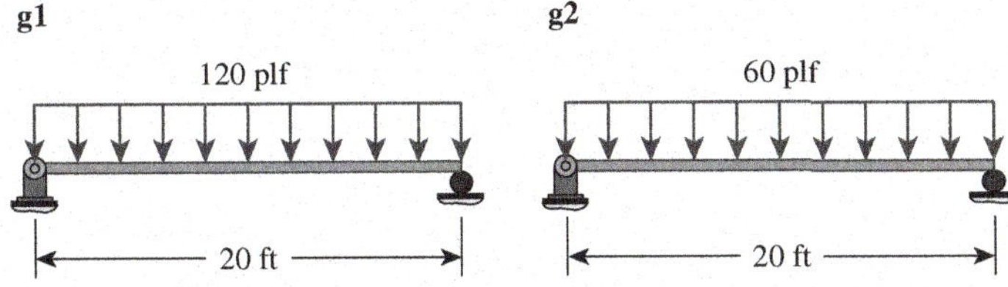

None of the load transfers directly from the cladding to the wind columns. Rather, it passes through the girts and the girts pass it to the wind columns. The load shows up on the wind columns in the form of *point loads*. One can calculate the magnitude of the point load by multiplying the tributary area by the distributed load 24 psf. Tributary areas to each point load for wc1 are shown in the figure.

Wind Columns Tributary Areas:
For wc1,

$$A_{t-\text{end}} = (2.5 \text{ ft})(20 \text{ ft}) = 50 \text{ ft}^2$$
$$A_{t-\text{mid}} = (5.0 \text{ ft})(20 \text{ ft}) = 100 \text{ ft}^2$$

The tributary areas to wc2 are half of those to wc1.
For wc2,

$$A_{t-\text{end}} = (2.5 \text{ ft})(10 \text{ ft}) = 25 \text{ ft}^2$$
$$A_{t-\text{mid}} = (5.0 \text{ ft})(10 \text{ ft}) = 50 \text{ ft}^2$$

Finally, the magnitude of the point loads may be calculated by multiplying the distributed load by the respective tributary areas. The diaphragm provides support to the wind columns.

For wc1,

$$P_{end} = \left(50\ \text{ft}^2\right)\left(24\tfrac{\text{lb}}{\text{ft}^2}\right) = 1200\ \text{lb} = 1.2\ \text{k}$$

$$P_{mid} = \left(100\ \text{ft}^2\right)\left(24\tfrac{\text{lb}}{\text{ft}^2}\right) = 2400\ \text{lb} = 2.4\ \text{k}$$

For wc2,

$$P_{end} = \left(25\ \text{ft}^2\right)\left(24\tfrac{\text{lb}}{\text{ft}^2}\right) = 600\ \text{lb} = 0.6\ \text{k}$$

$$P_{mid} = \left(50\ \text{ft}^2\right)\left(24\tfrac{\text{lb}}{\text{ft}^2}\right) = 1200\ \text{lb} = 1.2\ \text{k}$$

1.2 k
2.4 k
2.4 k
1.2 k
wc1
0.6 k
1.2 k
1.2 k
0.6 k
wc2
5 ft
5 ft
5 ft

Tributary Load to Diaphragms

QUICK NOTE

The tributary approach for diaphragm loads assumes that the wall spans are simply supported between diaphragms.

When the out-of-plane walls span vertically, such as from floor-to-floor or foundation-to-roof, the tributary width/height to a diaphragm can be calculated by taking half of the wall height below and half the wall height above the diaphragm. The one exception to this rule is that all of the wall/roof height above the roof is attributed to the roof diaphragm. Figure 4.17 demonstrates this concept by showing a three-story building that has two floor diaphragms and one roof diaphragm. Envision that a uniformly distributed load is applied in the direction indicated over the entire front wall and roof. The load that is attributed to the first floor diaphragm is the one in the lower shaded region of Figure 4.17(a). The tributary height is then taken as half the height of the first

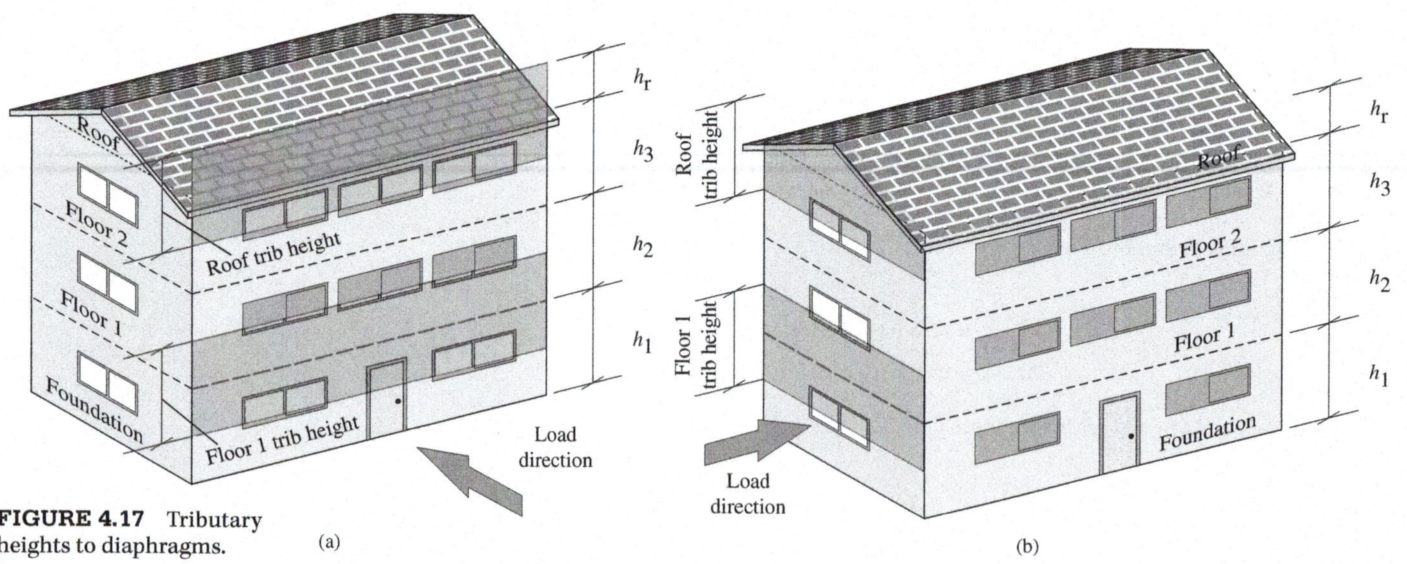

FIGURE 4.17 Tributary heights to diaphragms.

story and half the height of the second story and can be expressed as

$$\text{tributary}_{\text{floor}-1} = \frac{h_1 + h_2}{2} \quad (4.1)$$

The tributary height to the roof diaphragm can be expressed as

$$\text{tributary}_{\text{roof}} = \frac{h_3}{2} + h_r \quad (4.2)$$

where the entire roof height is included in the tributary height. This can be deduced rationally by recognizing that any load that is applied above the uppermost diaphragm has nowhere to go except to the nearest diaphragm. In Figure 4.17(b), the tributary area to the roof diaphragm is in the shape of the gable end, whereas the tributary area to the roof in Figure 4.17(a) is rectangular. This is because the load acts on the vertical projection of the structure.

EXAMPLE 4.4

Sketch the loading diagrams for the roof and first floor diaphragms of the structure shown. Cross-sections of the structure are provided with the load which is applied in each direction. View each diaphragm as a simply supported beam which has supports wherever the VLFRS is present. In this case, assume that the VLFRS is only on the perimeter of the building.

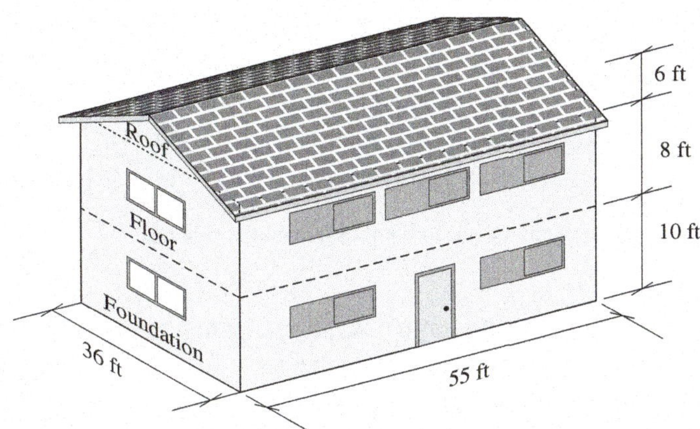

SOLUTION STRATEGY

Identify the diaphragm configurations by noting where the supports (i.e., VLFRS) are located. Sketch the diaphragm as if it is a beam [see Figures 4.13(a) and 4.14(a)], noting the span length in each case. Sketch the tributary shape to each diaphragm in each direction and note the tributary height of each one. Multiply the value of the distributed load by the tributary height in each region. You should end up with units of force/length.

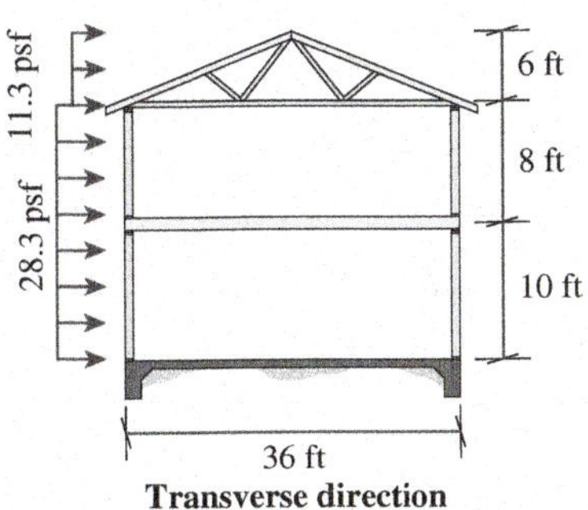

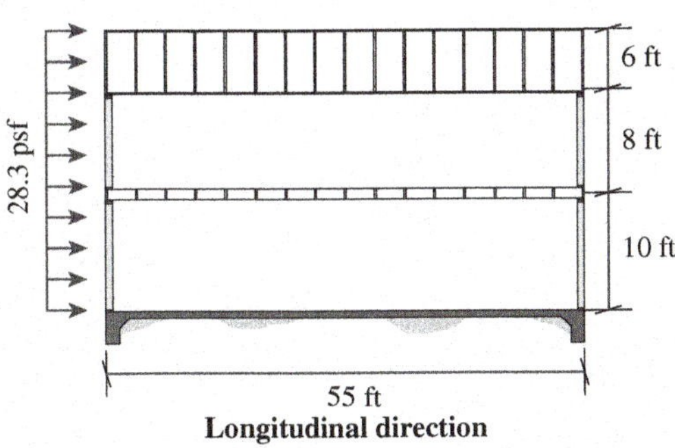

SOLUTION

Tributary areas for each diaphragm (roof and floor) are found by taking half of the height of each segment and assigning it to its respective level.

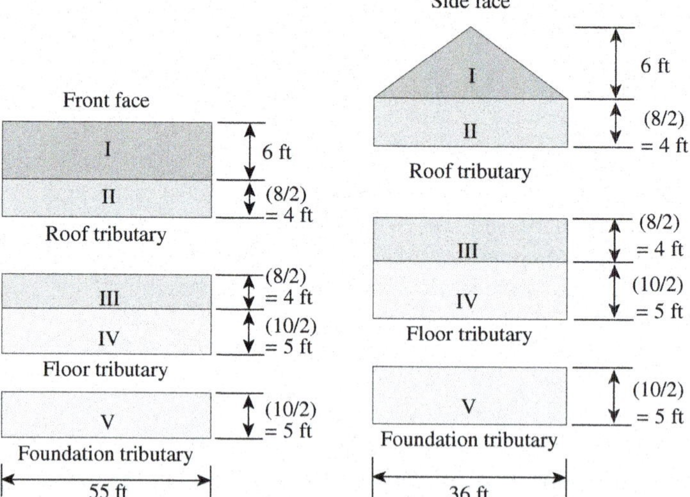

> **QUICK NOTE**
> There is a different pressure in region I and in region II on the front face. On the side face, the pressure in these regions is the same.

The load assigned to the roof, based on the pressures applied to the front face (see the transverse detail on previous page), is found by multiplying each tributary height by the applied load in that region. Do the same for the floor and foundation.

Front Face:

$$w_{roof} = (11.3\text{ psf})(6\text{ ft}) + (28.3\text{ psf})(4\text{ ft}) = 181.0\text{ plf}$$

$$w_{floor} = (28.3\text{ psf})(4\text{ ft}) + (28.3\text{ psf})(5\text{ ft}) = 254.7\text{ plf}$$

$$w_{foundation} = (28.3\text{ psf})(5\text{ ft}) = 141.5\text{ plf}$$

Now find the load on the diaphragms due to the pressure on the side face.

Side Face:

$$w_{roof-peak} = (28.3\text{ psf})(6\text{ ft}) + (28.3\text{ psf})(4\text{ ft}) = 283.0\text{ plf}$$

$$w_{roof-low} = (28.3\text{ psf})(4\text{ ft}) = 113.2\text{ plf}$$

$$w_{floor} = (28.3\text{ psf})(4\text{ ft}) + (28.3\text{ psf})(5\text{ ft}) = 254.7\text{ plf}$$

$$w_{foundation} = (28.3\text{ psf})(5\text{ ft}) = 141.5\text{ plf}$$

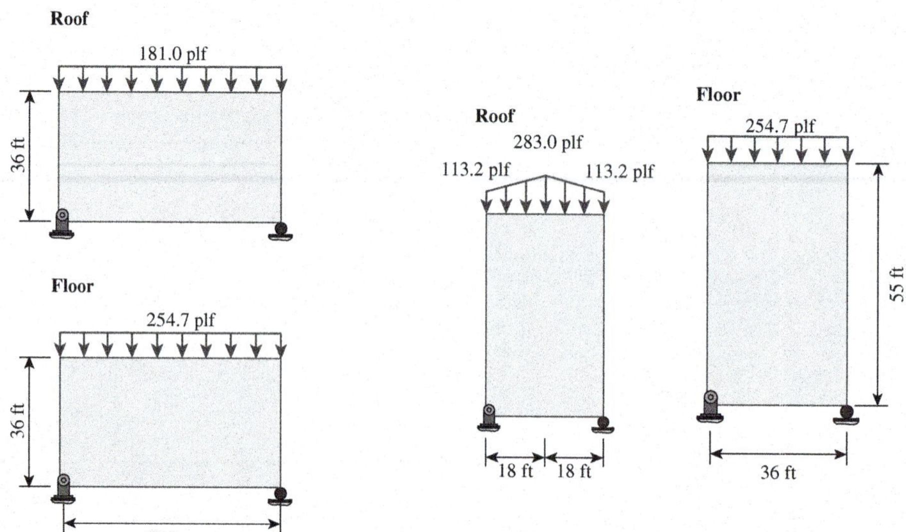

EXAMPLE 4.5

For the building shown, the wind loading acts on both the windward (front) wall and the leeward (back) wall. The loads are as shown. Sketch the resulting load diagram for the diaphragm.

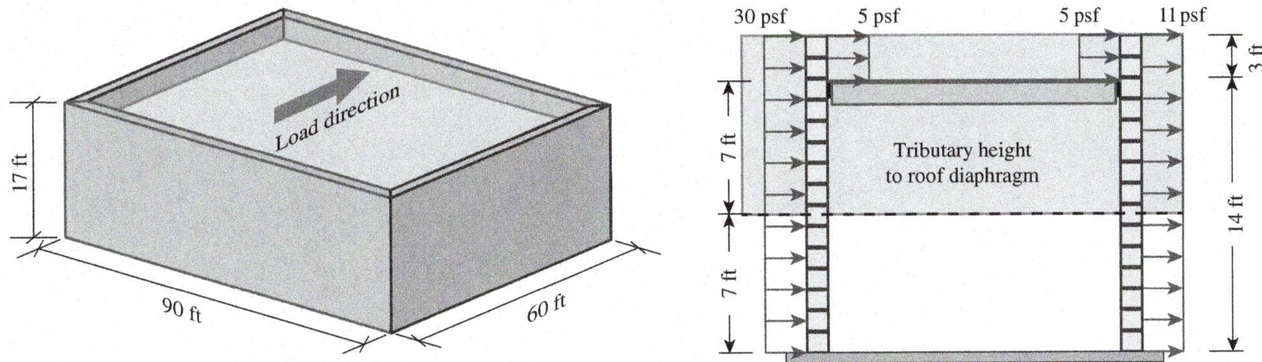

SOLUTION STRATEGY

Determine the tributary height to the roof diaphragm by using the one-half rule. Note that the diaphragm is responsible for supporting the front wall and also the back wall. Both loads must be accounted for in the loading diagram.

SOLUTION

The roof receives 7 ft of tributary wall height and 3 ft of parapet height. So

$$w_{roof-front} = (30\text{ psf})(7\text{ ft}) + (30\text{ psf} + 5\text{ psf})(3\text{ ft}) = 315\text{ plf}$$

$$w_{roof-back} = (11\text{ psf})(7\text{ ft}) + (11\text{ psf} + 5\text{ psf})(3\text{ ft}) = 125\text{ plf}$$

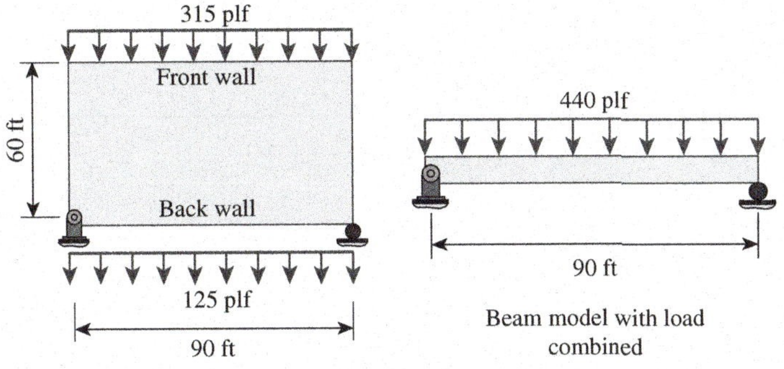

EXAMPLE 4.6

For the building and loading discussed in Example 4.3, sketch the diaphragm load diagram. The span of the diaphragm is 60 ft. Take the depth of the diaphragm to be 40 ft. The load is only applied on the front wall.

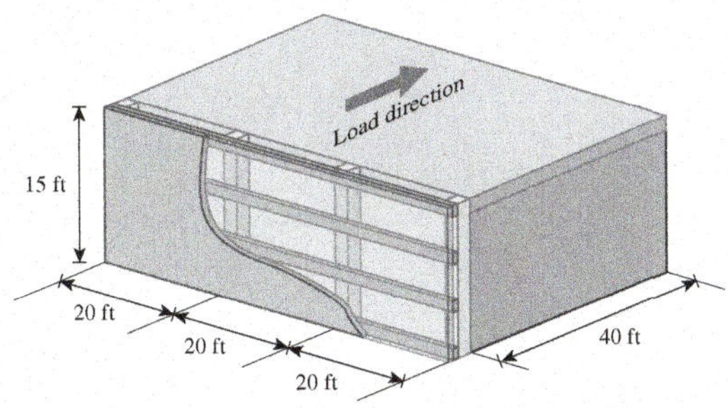

SOLUTION STRATEGY

One must always be careful of the load path in the building. Simply taking half of the wall height will not be sufficient for this building. While the load that is applied to the upper half of the wall will eventually make it to the diaphragm, the load on the diaphragm will not be uniformly distributed as in previous examples. Rather, the load makes it to the diaphragm through contact with the wind columns—this produces point loads.

SOLUTION

Calculate the reactions of the wind columns shown in Example 4.3 and then apply these loads to the diaphragm.

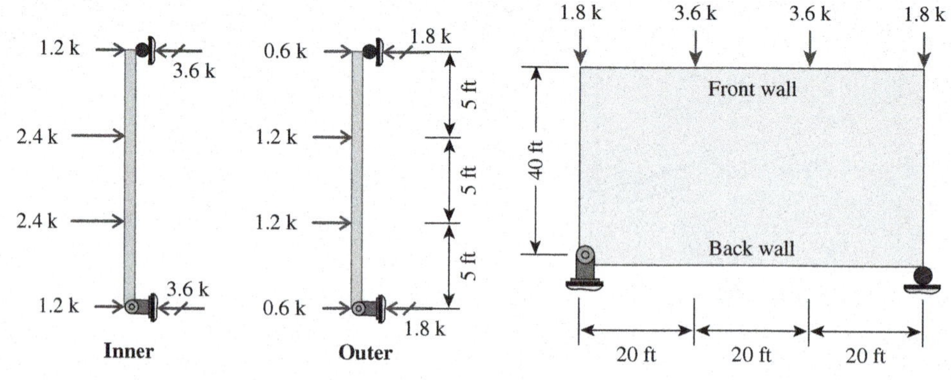

4.6 Examples with Video Solutions

VE4.1 A 12-story building has a floor plan that is 120-ft square and has 12 stories that are 15-ft high in Figure VE4.1(a). The wind load on the windward (front) face gets larger at the higher elevation stories but is a constant value on the leeward (back) wall [Figure VE4.1(b)]. Sketch the diaphragm load diagrams for the roof diaphragm and the sixth-floor diaphragm. Assume that the VLFRS is around the perimeter of the building and also down the centerline of the building (i.e., 60 ft in from the outside).

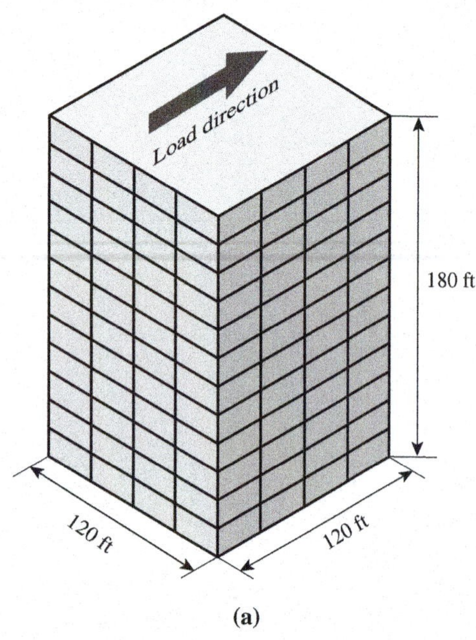

(a)

Figure VE4.1

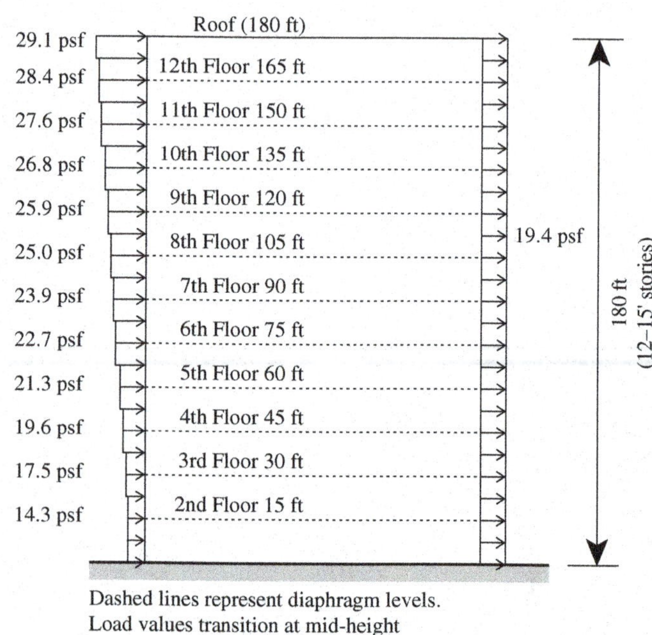

(b)

Figure VE4.1 (continued)

VE4.2 Wind loads are calculated for a two-story low-rise hotel building [Figure VE4.2(a)]. The loads for the MWFRS for one direction are shown in Figure VE4.2(b). Sketch the loading diagram for the roof diaphragm. Assume the VLFRS is around the perimeter of the building.

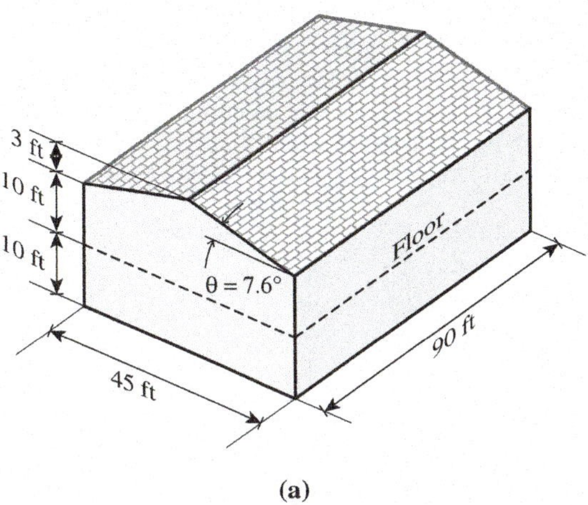

(a)

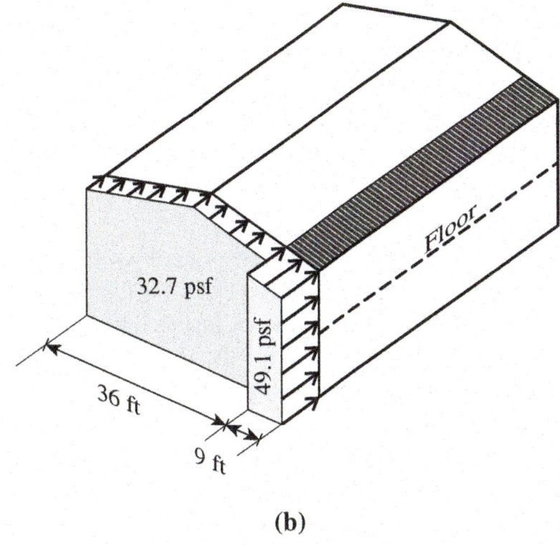

(b)

Figure VE4.2

4.7 Problems for Solution

Section 4.6

P4.1 The end wall of a gymnasium is constructed of vertical wind columns that span between the foundation and the roof diaphragm with lengths varying between 22 ft and 28 ft. Concrete masonry block spans 10-ft side-to-side between columns in the figure. Wind pressure is applied to both the inside and outside faces of the wall where the inside pressure can either act into or out of the surface of the wall. Sketch the two loading diagrams for a 1.0-ft wide strip of the masonry wall. (Ans: w = 35 plf or 22 plf)

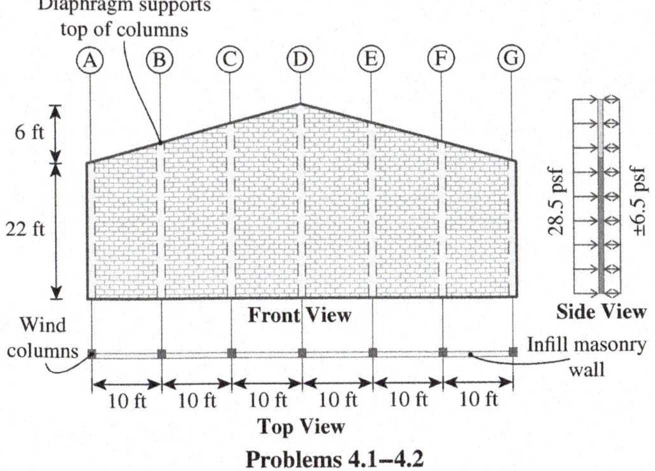

Problems 4.1–4.2

P4.2 For the wall described in Problem P4.1 and shown in the figure, sketch the two loading diagrams for the wind columns located on lines A, B, C, and D.

P4.3 A two-story building has pressures acting on its front and back walls as shown in the figure. The building will also experience an internal pressure that can act either into or out of the face of the wall. Sketch loading diagrams for a 1.0-ft wide strip of the front and back walls if a positive internal pressure of 5.3 psf exists. This means that the pressure on the inside of the building will act into the wall. (Ans: on front wall w = 31.1 plf, w = 21.3 plf, w = 16.7 plf)

P4.4 Assume that the internal pressure in the building described in Problem P4.3 is a negative pressure of 5.3 psf. This means that the pressure on the inside of the building will act pointing out from the wall surface. Sketch the loading diagrams for a 1.0-ft wide strip of both the front and back walls.

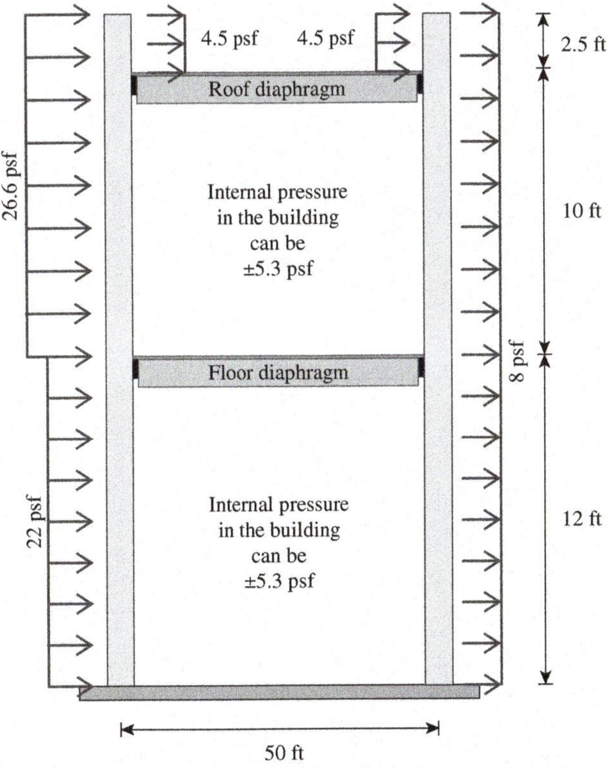

Problems 4.3–4.4

P4.5 A 12-story building has a floor plan that is 120 ft square and has 12 stories that are 15-ft high in the figure VE4.1 part(a). The wind load on the windward (front) face gets larger at the higher elevation stories but is a constant value on the leeward (back) wall in the figure VE4.1 part(b). Sketch the diaphragm load diagrams for the tenth- and second-floor diaphragms. Assume that the VLFRS is around the perimeter of the building and also down the centerline of the building (i.e., 60 ft in from the outside). (Ans: level 2: w = 505.5 plf, level 10: w = 693 plf)

P4.6 Wind loads are calculated for a two story low-rise hotel building in the figure VE4.2 part(a). The loads for the MWFRS for the longitudinal direction are shown in the figure VE4.2 part(b). Sketch the loading diagram for the floor diaphragm. Assume the VLFRS is around the perimeter of the building.

P4.7 Wind loads are calculated for a two-story low-rise hotel building. The loads for the MWFRS for the transverse direction are shown in the figure. Sketch the loading diagram for the floor diaphragm. Assume the VLFRS is around the perimeter of the building and that the dimensions of the building are 50 ft × 100 ft. (Ans: w = 353 plf)

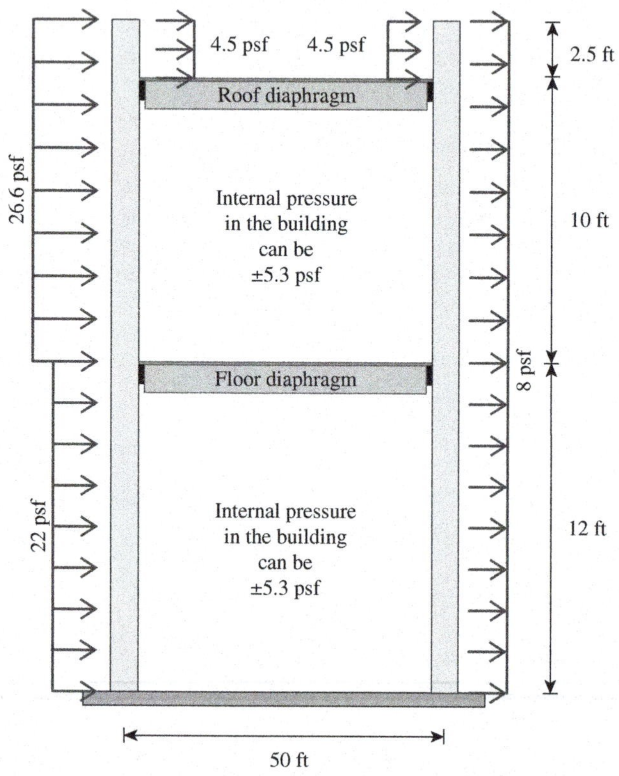

Problems 4.7–4.8

P4.8 For the structure described in Problem P4.7, sketch the loading diagram for the roof diaphragm. Assume the VLFRS is around the perimeter of the building.

P4.9 A foundation wall for a basement is shown in the figure. The basement slab provides sliding resistance at the base of the wall, and the floor diaphragm provides lateral resistance to the top of the wall. Using the tributary approach, what is the per foot load distributed to the diaphragm and the per foot load provided to the slab by the foundation wall? (Ans: at floor w = 375 plf, at slab w = 1125 plf)

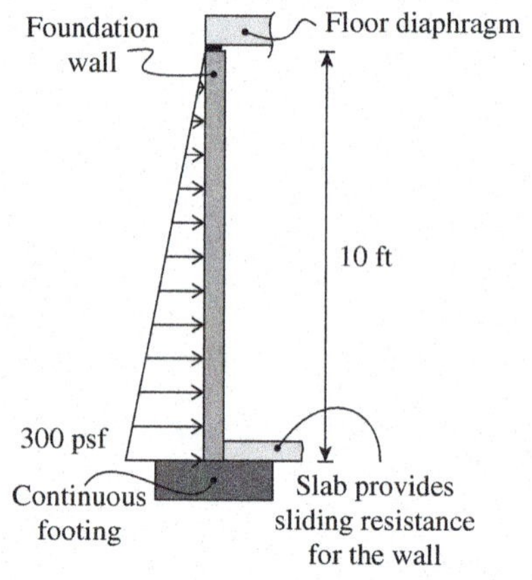

Problems 4.9–4.10

P4.10 A foundation wall for a basement is shown in the figure. The basement slab provides sliding resistance at the base of the wall, and the floor diaphragm provides lateral resistance to the top of the wall. Using a statics approach, what is the per foot load distributed to the diaphragm and the per foot load provided to the slab by the foundation wall?

Reactions

5

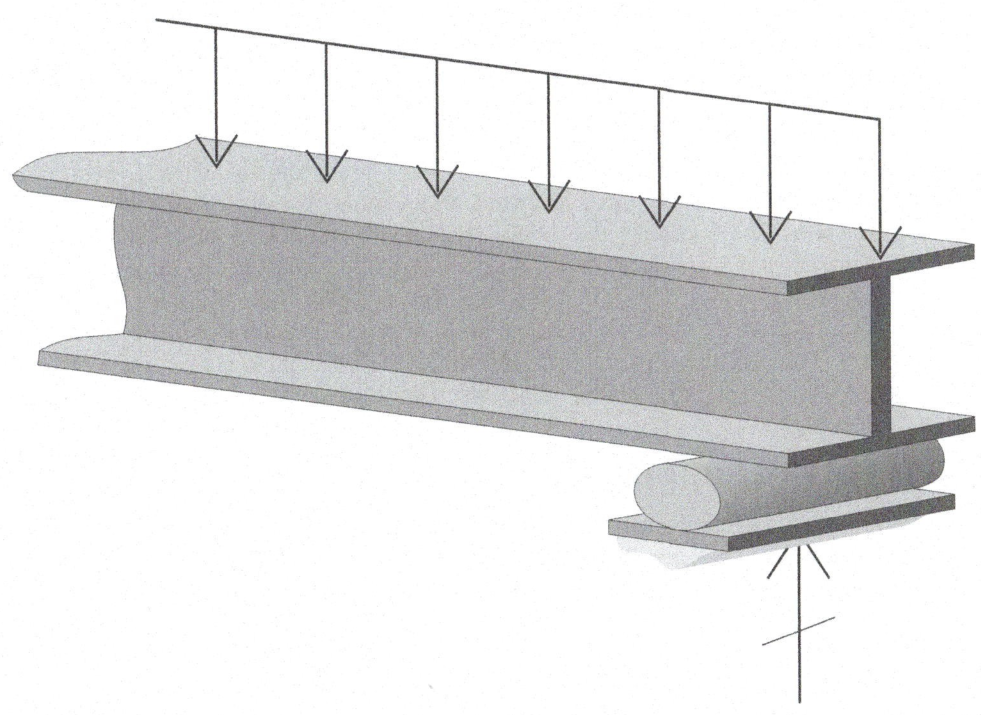

5.1 Equilibrium

A body at rest is said to be in *static equilibrium*. The resultant of the external forces acting on the body—including the supporting forces, which are called reactions—is zero. Not only must the sum of all forces (or their components) acting in any possible direction be zero, but the sum of the moments of all forces about any axis also must be equal to zero—otherwise acceleration of the object will ensue.

If a structure, or part of a structure, is to be in equilibrium under the action of a system of loads, it must satisfy the six equations of static equilibrium. Using the Cartesian x, y, and z coordinate system, the equations of static equilibrium can be written as

$$\sum F_x = 0 \qquad \sum F_y = 0 \qquad \sum F_z = 0$$
$$\sum M_x = 0 \qquad \sum M_y = 0 \qquad \sum M_z = 0$$

For purposes of analysis and design, the large majority of structures are considered to be planar structures without loss of accuracy. For these structures, which are usually assumed to be in the x–y plane, the sum of the forces in the x and y directions and the sum of the moments about an axis perpendicular to the plane must be zero. The equations of equilibrium reduce to

$$\sum F_x = 0 \qquad \sum F_y = 0 \qquad \sum M_z = 0$$

The third equation is commonly written as

$$\sum M_A = 0$$

QUICK NOTE

Summing forces in the horizontal and vertical directions is the most common because of the way structures are configured. However, it is sometimes convenient to sum forces in a rotated set of orthogonal axes.

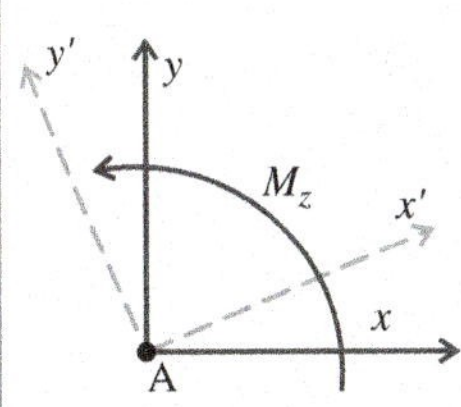

where the subscript, in this case A, represents the point where moments about a z-axis are taken. These equations cannot be proved algebraically; they are merely statements of Sir Isaac Newton's observation that for every action on a body at rest there is an equal and opposite reaction. Whether the structure under consideration is a beam, a truss, a rigid frame, or some other type of assembly supported by various reactions, the equations of static equilibrium must apply if the body is to remain at rest.

5.2 Calculation of Unknowns

QUICK NOTE

Reactions in this textbook are presented as an arrow with a hatch mark across it. This is to distinguish reactions from other applied or internal forces sketched on a free-body diagram.

The direction shown for each reaction arrow is assumed to be the positive direction (see previous Quick Note).

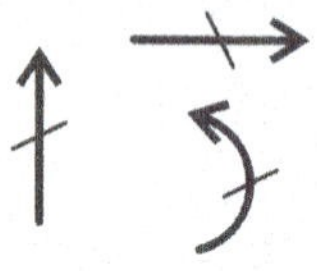

To completely define a force, three properties of that force must be defined. These properties are its *magnitude*, its *line of action*, and the *direction* in which it acts along the line of action. All of these properties are generally known for each of the externally applied loads. However, when dealing with structural reactions, only the points at which the reaction forces act, and perhaps their directions, are known. The magnitude of the reaction forces, and sometimes the directions in which they act, are unknown quantities that must be determined.

The number of unknowns that can be determined using the equations of static equilibrium is limited by the number of independent equations of static equilibrium available. For any structure lying in a plane, there are only three independent equations of static equilibrium. These three equations can be used to determine at most three unknown quantities for the structure. When there are more than three unknowns to evaluate, additional equations must be used in conjunction with the equations of static equilibrium. In some instances, because of special construction features, equations of condition are available in addition to the usual equations of static equilibrium. In later chapters, we will learn to write some other equations for use in analysis. These equations will pertain to the compatibility of displacements in the structure.

5.3 Types of Supports

Courtesy of Bryant G. Nielson

Rocker supporting a bridge girder.

QUICK NOTE

Rollers/rockers may be installed in such a manner that they can resist movement either toward or away from the supporting surface.

Supports for structures can come in many different shapes and sizes, including bolted, welded, nailed, and bearing types. Regardless of the many and various configurations, all supports can be classified into several basic categories based upon their dominant behavior. These categories are hinges/pins, rollers/rockers, fixed ends, or links. A description of each of these general classes is given in the following paragraphs. As they are discussed, keep in mind that these classifications are idealized and the structural engineer must assign an actual support to one of these classes based upon which one most closely approximates the real support behavior—not based on appearance.

A *hinge* or pin-type support is assumed to be connected to the structure with a frictionless pin. This type of support prevents movement in the horizontal and vertical directions, but it does not prevent slight rotation about the hinge. There are two unknown forces at a hinge and are generally taken as the magnitude of the force required to prevent horizontal movement and the magnitude of the force required to prevent vertical movement. In reality, a pin can be seen as providing restraint against translation in any direction while still permitting rotations. This results in two orthogonal reaction forces but no developed moment. Figure 5.1 illustrates the graphical symbols used to illustrate pin behavior in this text and their resulting reactions.

A *roller* type or *rocker* type of support is assumed to offer resistance to movement only in a direction perpendicular to the supporting surface beneath the roller. There is no resistance to slight rotation about the roller or to movement parallel to the supporting surface. The magnitude of the force required to prevent movement perpendicular to the supporting surface is the one unknown. Should the temperature of the beam be raised, the beam will elongate. However, since there is no longitudinal restraint supplied at the roller or expansion support, no stress will be developed

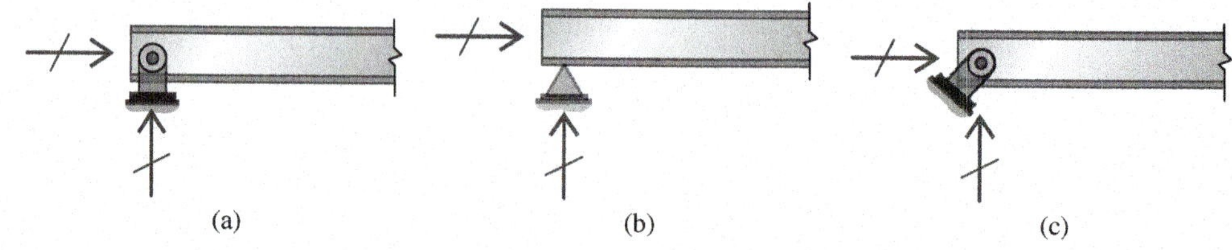

FIGURE 5.1 Hinge/pin support diagrams and the resulting reactions (all produce two orthogonal reactions).

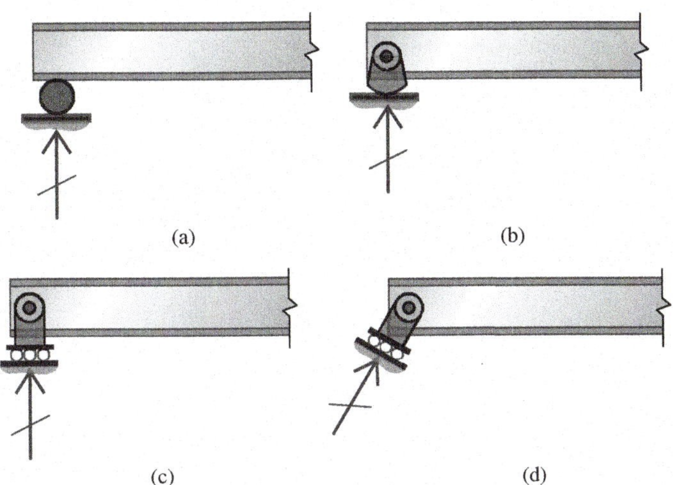

FIGURE 5.2 Common symbols used to represent rollers/rockers.

in the beam or in the supporting walls or other structural members. Figure 5.2 demonstrates the symbols used to represent a roller in this text.

A *fixed-end* support (see Figure 5.3) is assumed to offer resistance to both rotation about the support and translation vertically and horizontally. There are three unknowns: the magnitude of the force required to prevent horizontal movement, the magnitude of the force required to prevent vertical movement, and the magnitude of the force required to prevent rotation.

A *link* type of support (Figure 5.4) is similar to the roller in its action because the pins at each end are assumed to be frictionless. The line of action of the supporting force must be in the direction of the link and through the two pins, which is a characteristic of a two-force member. One unknown is present: the magnitude of the force in the direction of the link.

Figure 5.5 shows *hinge* and *expansion* (or roller) type connections as they might be used for a steel beam. A hinge or simple connection theoretically should be free for some rotation when loads are applied to the member. The ends of the structural steel beam shown in Figure 5.5 are fairly free to rotate, since the connection is only made to the web near the neutral axis of the section. Two other types of end connections for steel members that have a similar rotation capacity are shown in Figure 5.6.

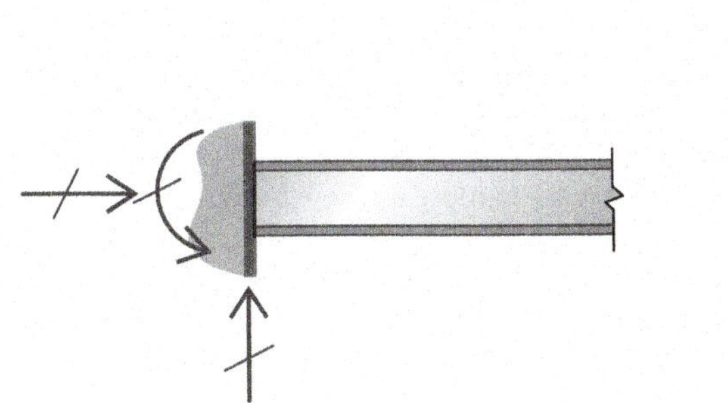

FIGURE 5.3 Fixed support.

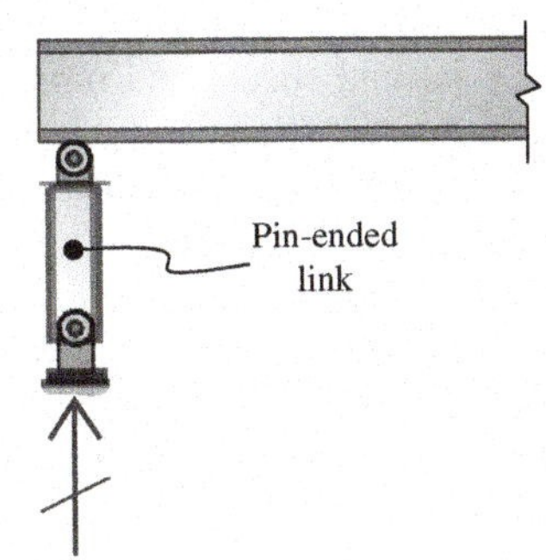

FIGURE 5.4 Link support.

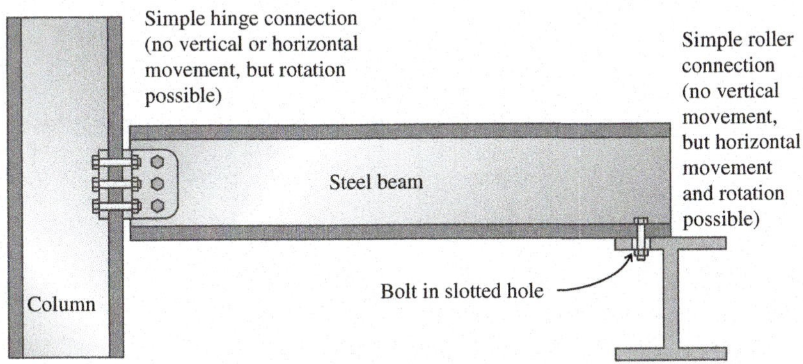

FIGURE 5.5 Simple connections for a steel beam (i.e., pin and roller).

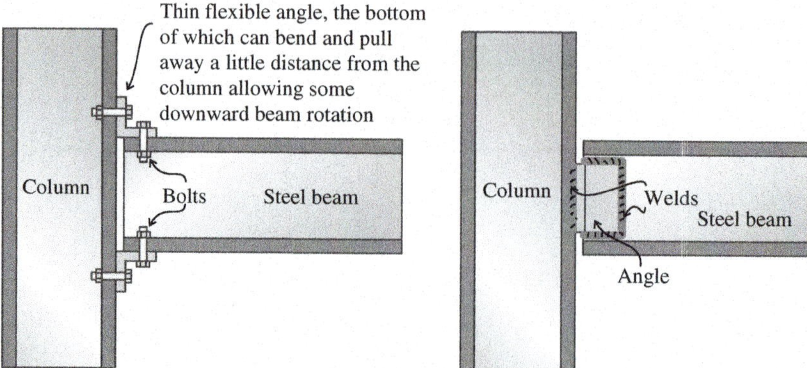

FIGURE 5.6 More simple hinge connections.

DID YOU KNOW?
Supports never provide perfect restraint against translation or rotation. However, for most cases, each support has a dominant behavior that can be adequately captured by one of the idealized supports.

COMMON MISTAKE
Students often want to have a translational spring take some moment or have a rotational spring take a translational force. Translational springs can only take a force along its axis and a rotational spring can only take a moment.

Two connections that provide considerable moment resistance, and thus approach fixed ends, are shown in Figure 5.7. Notice in each case that some type of connection is provided at the top and bottom of the ends of the beams to prevent downward rotation. Other connections will be discussed as needed at various places in the text.

5.4 Role and Analysis of Springs

The connections/supports in Figures 5.4 through 5.7 can never really provide perfect restraint against translation and rotation. When more refinement of the actual behavior is needed, the supports can be modeled using springs—both translational and rotational. Springs provide resisting forces or moments but also allow some translation or rotation. If properly understood, they are also useful tools to use in structural analysis. Spring behavior is governed by its stiffness—translational (k_T) or rotational (k_R)—depending on the spring type. Units for these stiffnesses are ($k_T \rightarrow$ force/length) and ($k_R \rightarrow$ (force $\cdot$ length)/rad). Figure 5.8 shows the resulting reactions when either a translational spring or rotational spring is used to support a structure.

5.5 Internal Releases

Connections between members in planar structures by default transfer two orthogonal forces (i.e., axial and shear) and moment about the axis perpendicular to the plane. However, some connections are specifically designed to not transfer one of these internal forces between adjacent members. In other words, the connection releases one of these forces. Additionally, some analysis techniques presented later in this textbook use the concept of internal releases to our advantage. For planar structures, there are three possible releases—namely a *shear release/slider*, a *moment release/internal hinge*, and an *axial release*.

In many statics and mechanics textbooks, the *shear release* is often called a slider, and the *moment release* is often called a hinge. Within this textbook, we use the terminology interchangeably with the intent to get the reader more comfortable with the concept of releases. This is because commercial analysis software uses the term *release*, and this term is also very descriptive of the internal force/moment that is not present in the connection. Figure 5.9 demonstrates the

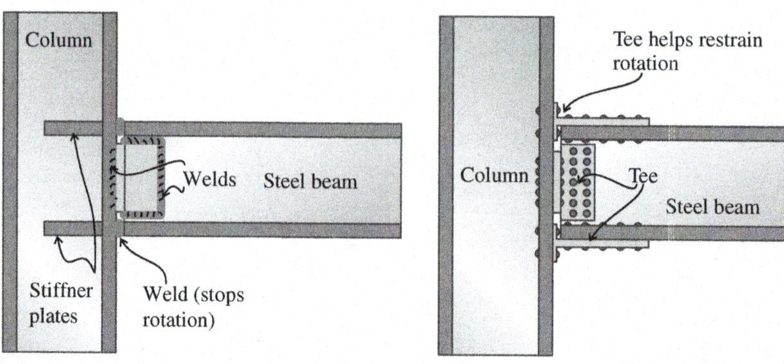

FIGURE 5.7 Moment resisting connections.

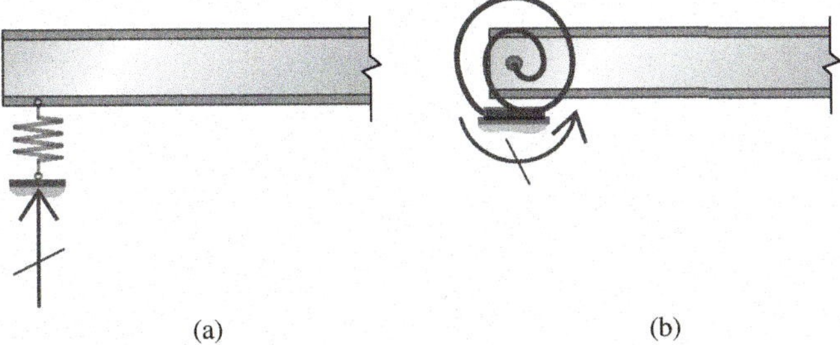

FIGURE 5.8 (a) Translational spring and (b) rotational spring.

internal forces that are present in each connection. A shear release has an internal moment and axial force but no shear. A moment release has an internal shear and axial but no moment. An axial release has an internal shear and moment but no axial force. Once the engineer understands this, they will be able to use this knowledge of internal forces to deal with fairly complex structures.

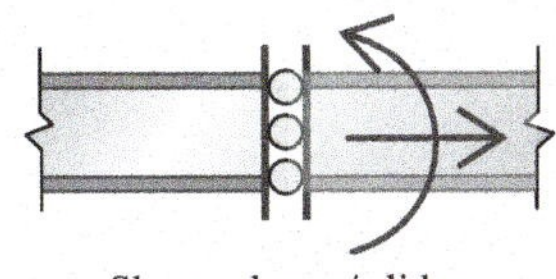

Shear release / slider

Moment release / internal hinge

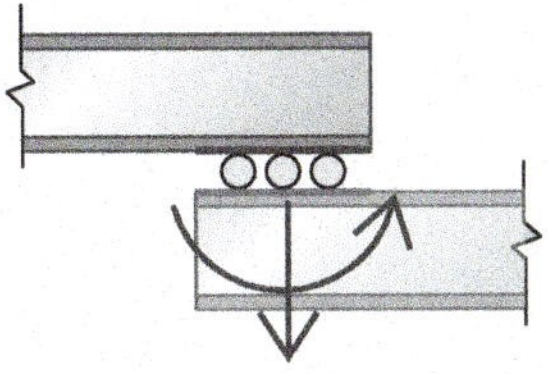

Axial release

FIGURE 5.9 Typical releases in structures and their resulting internal forces.

5.6 Stability and Statical Determinacy

The discussion of supports shows there are three unknown reaction components at a fixed end, two at a hinge, and one at a roller or link. If the total number of forces of reaction for a particular structure is equal to the number of equations of static equilibrium available, all of the unknowns may be determined, and the structure is said to be externally *statically determinate*. Should the number of unknowns be greater than the number of available equilibrium equations, the structure is externally *statically indeterminate*. If the number of unknowns is less than the number of available equations of equilibrium it is *unstable* externally. From this discussion, note that stability, determinacy, and indeterminacy are dependent upon the configuration of the structure; they are not dependent upon the loads applied to the structure. The following examples demonstrate the application of these concepts to structural systems. Pay close attention to each example because they each illustrate key points and the process of calculation.

EXAMPLE 5.1

Determine the statical classification of the simply supported beam shown here.

QUICK NOTE

Checking determinacy, indeterminacy, and external stability is a simple process of comparing the number of unknowns with the number of available equilibrium equations.

SOLUTION STRATEGY

Sketch the free-body diagram of the structure with all possible reactions based upon support conditions. Count up the number of unknown forces (F) and compare with the number of available equilibrium equations (N). Classify the structure based on the conditions identified in the margin.

DETERMINACY CONDITIONS

$F < N \Rightarrow$ unstable
$F = N \Rightarrow$ statically determinate
$F > N \Rightarrow$ statically indeterminate

SOLUTION

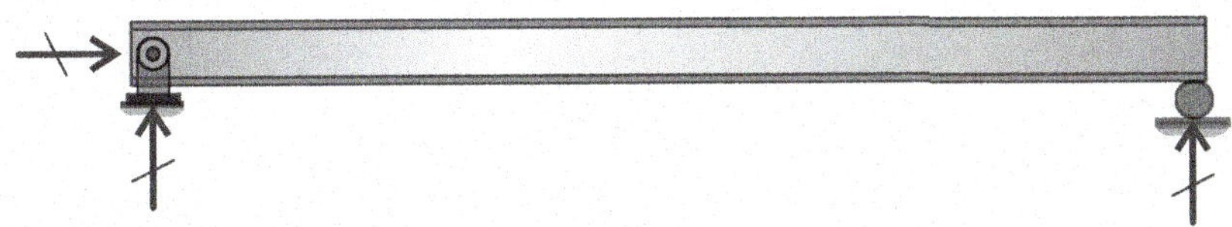

There are three unknown forces on the free-body diagram, so $F = 3$.

There are three applicable equations of static equilibrium: summation of forces vertically, summation of forces horizontally, and summation of moments. So $N = 3$.

$3 = 3 \Rightarrow$ Beam is stable and statically determinate externally.

SUMMARY

The number of unknowns is equal to the number of equations of static equilibrium.

EXAMPLE 5.2

Determine the statical classification of the simply supported beam shown here.

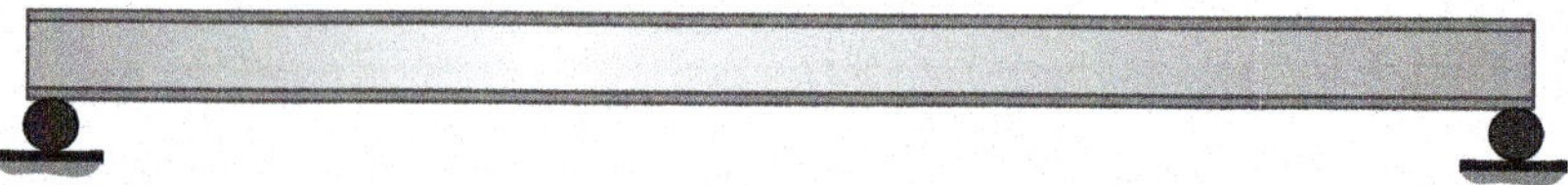

SOLUTION STRATEGY

Sketch the free-body diagram of the structure with all possible reactions based upon support conditions. Count up the number of unknown forces (F) and compare with the number of available equilibrium equations (N). Classify the structure.

SOLUTION

QUICK NOTE
A structure may be stable under one arrangement of loads, but if it is not stable under any other conceivable set of loads, it is considered unstable.

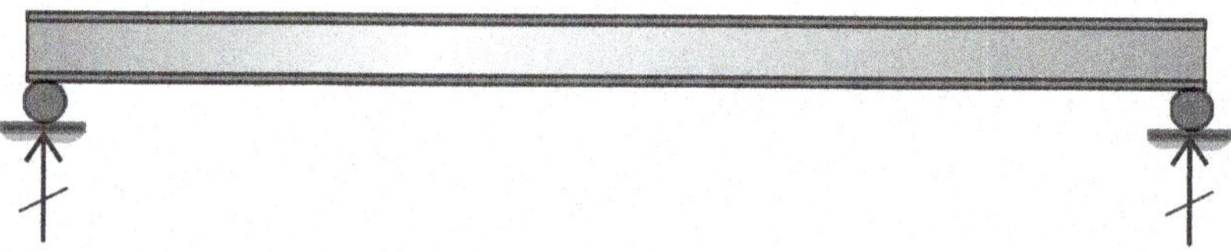

There are two unknown forces on the free-body diagram, so $F = 2$.
There are three applicable equations of static equilibrium, so $N = 3$.
$2 < 3 \Rightarrow$ Beam is externally unstable.

SUMMARY

This beam cannot hold its position when subjected to a horizontal load, therefore it is unstable.

EXAMPLE 5.3

Determine the statical classification of the multi-supported beam shown here.

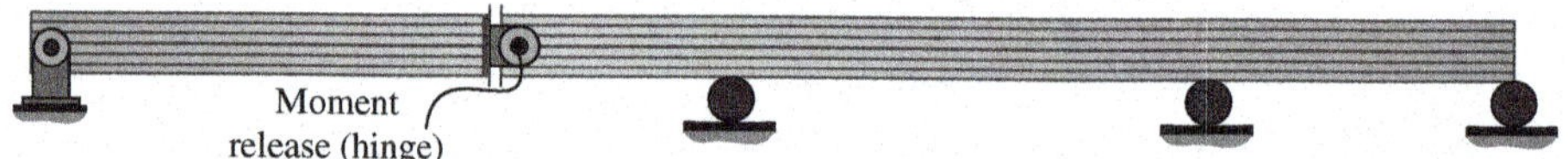

SOLUTION STRATEGY

Cut the structure at any internal releases and sketch the free-body diagram of the structure with all possible reactions and internal forces. Count up the number of unknown forces (F) and compare with the number of available equilibrium equations (N), so there are three for each section you have following each cut. Classify the structure.

QUICK NOTE
Always cut at any internal release (e.g., hinge, slider, or axial release) before sketching free-body diagrams.

SOLUTION

DID YOU KNOW?
The degree of statical indeterminacy (SI) can be calculated by

$$SI^\circ = F - N$$

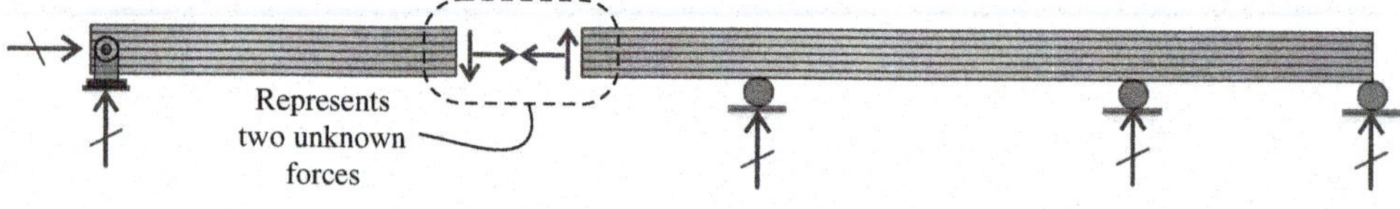

There are five unknown reaction forces and two unknown internal forces on the free-body diagram, so $F = 7$.

There are three applicable equations of static equilibrium for each section, so $N = (2)(3) = 6$.
$7 > 6 \Rightarrow$ Beam is statically indeterminate to the first degree (SI 1°).

SUMMARY

If the beam had not been cut at the internal hinge, we would have come to the incorrect conclusion that this beam is statically indeterminate to the second degree (SI 2°).

The internal arrangement of some structures is such that one or more equations of condition are available. This situation can occur when there are hinges or sliders in the structure. A special condition exists because the internal moment at the hinge must be zero, regardless of the loading. Likewise, the internal shear at a slider is zero. A similar statement cannot be made for any continuous section of the beam. As shown in Example 5.3, the surest way to account for these equations of condition is to make a cut at all internal releases. Students may avoid making the cut if they recognize that an internal release simply provides one more equation to solve for all the unknowns, but this can confuse many if they don't first make the cut.

EXAMPLE 5.4

Determine the statical classification of the spring supported beam shown here.

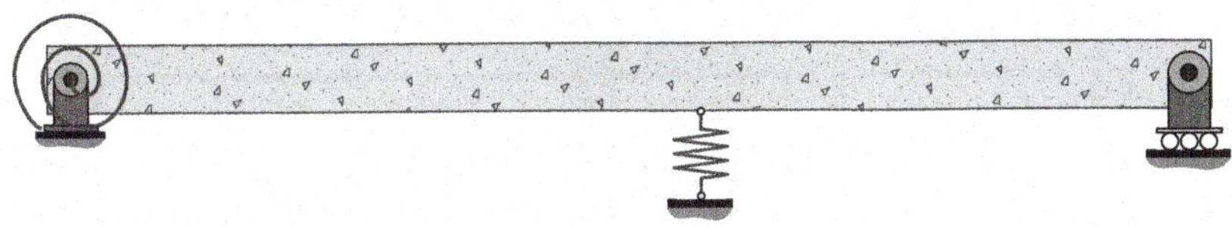

SOLUTION STRATEGY

Sketch the free-body diagram of the structure with all possible reactions, ensuring that the reactions located at the springs are correct according to spring behavior. Count up the number of unknown forces (F) and compare with the number of available equilibrium equations (N). Classify the structure.

SOLUTION

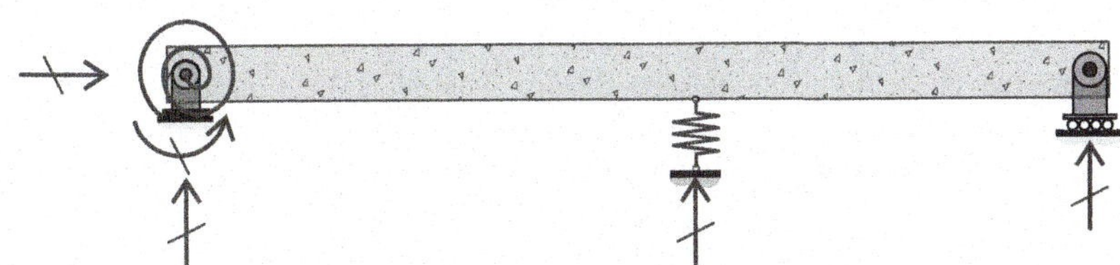

QUICK NOTE

The vertical and horizontal reactions at the left-most support are a result of the pin support. The moment reaction is a consequence of the rotational spring.

There are no internal releases, so no cuts need to be made. Thus, only five unknown reaction forces exist on the free-body diagram, so $F = 5$.

There are three applicable equations of static equilibrium, so $N = 3$.

$5 > 3 \Rightarrow$ Beam is stable and statically indeterminate to the second degree (SI 2°).

EXAMPLE 5.5

Determine the statical classification of the two-bay two-story frame shown here.

SOLUTION STRATEGY

Identify any closed loops that exist in the structure and cut them open. Sketch the free-body diagram of each section with all possible reactions and internal forces. Count up the number of unknown forces (F) and compare with the number of available equilibrium equations (N). Classify the structure.

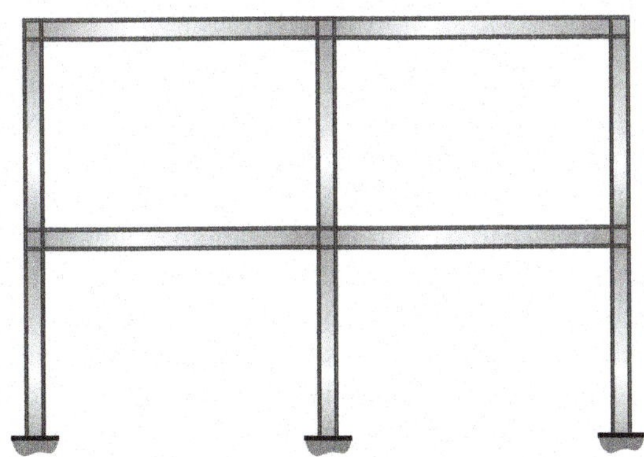

SOLUTION

DID YOU KNOW?
Any time you have members within a structure creating a closed loop, you will inappropriately identify the determinacy unless you break open the loops (i.e., cut them open). Remember the rule:

NO LOOPS!!!

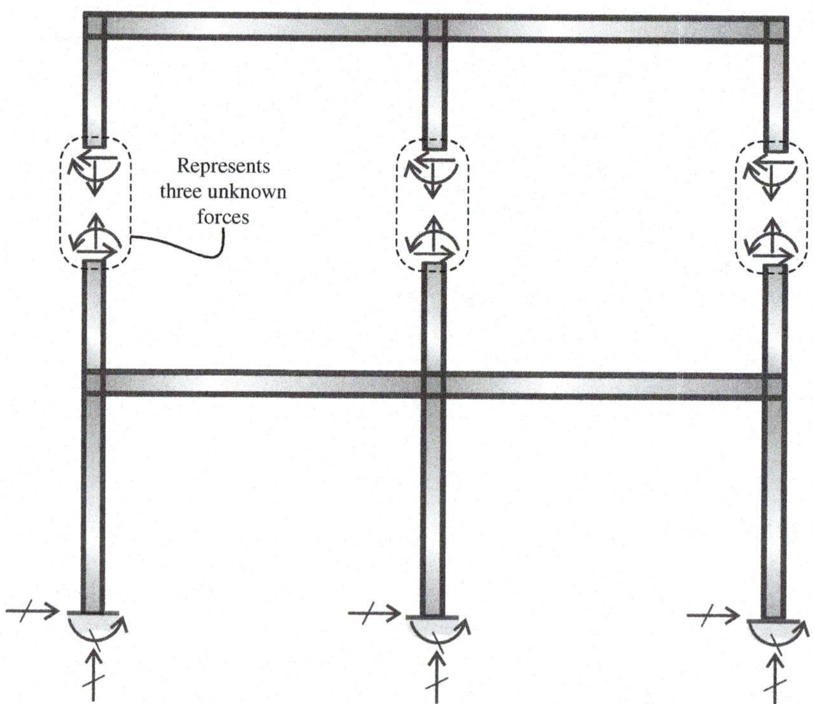

DID YOU KNOW?
An alternate way of cutting would result in the same conclusion of SI 12°.

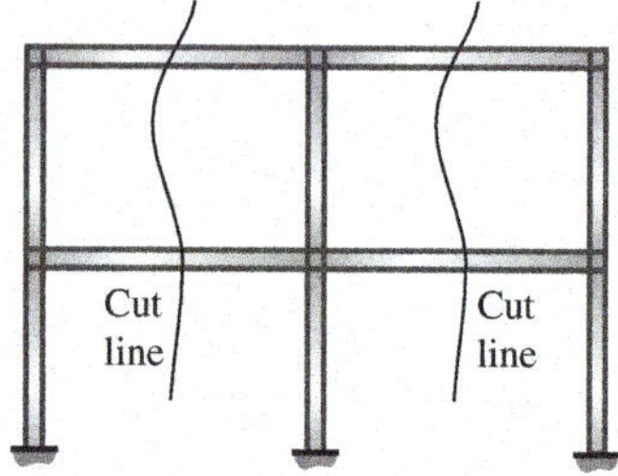

This structure has two closed loops in the second story. A horizontal cut through this story will break these loops open. The location of the cut in each member represents three unknown internal forces. There are nine unknown reaction forces and nine unknown internal forces that exist on the free-body diagram, so $F = 9 + 9 = 18$.

There are three applicable equations of static equilibrium for each of the two sections, so $N = (2)(3) = 6$. $18 > 6 \Rightarrow$ Frame is stable and statically indeterminate to the twelfth degree (SI 12°).

SUMMARY

If the cut had not been made, we would have made an inappropriate classification of SI 6°. You can never make too many cuts—but you can make too few.

5.7 Unstable Equilibrium and Geometric Instability

The ability of a structure to adequately support the loads applied to it is dependent not only on the number of reaction components but also on the arrangement of those components. A structure can be unstable yet be stable under a certain set of loads. The beam previously shown in Example 5.2 is an example of such a structure. This beam is supported on its ends with rollers only and is unstable. The beam will slide laterally if any horizontal force is applied. However, the beam can support vertical loads and is stable if only vertical loads are applied. This condition is sometimes referred to as *unstable equilibrium.* Note that the chair in Figure 5.10 is considered unstable because all reactions are parallel, so when an individual sits in it, they have a case of unstable equilibrium. This is why it is not recommended to use a chair like this to stand on.

FIGURE 5.10 Structure with all five reactions parallel.

It is also possible for a structure to have as many or more reaction components than there are equations available yet still be unstable. This condition is referred to as geometric instability. The frame of Figure 5.11 has three reaction components and three equations available for their solution. However, a study of the moment at B shows that the structure is unstable, since equilibrium cannot be satisfied when load P is applied. The line of action of the reaction at A passes through the reaction at B. Unless the line of action of the force P passes through the same point, the sum of

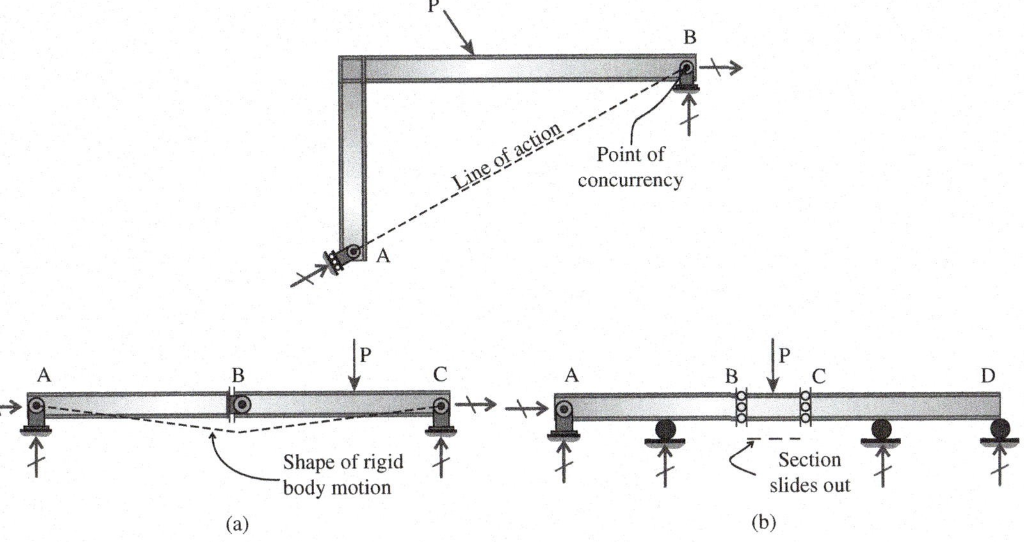

FIGURE 5.11 Frame with geometric instability due to concurrent reactions.

FIGURE 5.12 Structures with geometric instability due to rigid-body motion.

the moments about B cannot be equal to zero. There would be no resistance to rotation about B, and the frame would immediately begin to rotate. It might not collapse, but it would rotate until a stable situation was developed when the line of action of the reaction at A did not pass through B. Of prime importance to the engineer is for a structure to hold its position under load (even though it may deform) and not permit rigid body motion. One that does permit such motion is unstable.

Another geometrically unstable structure is shown in Figure 5.12(a). Four equations are available to compute the four unknown forces of reaction—three equations of static equilibrium and one equation of condition. Nevertheless, rotation will instantaneously occur about the hinge at B. After a slight vertical deflection at B, the structure will probably become stable.

GEOMETRIC INSTABILITY CHECK

Three conditions to look for when identifying a geometric instability condition.

1. Are all reactions parallel to one another?
2. Are the action lines of all reactions coincident?
3. Is there a collapse mechanism that permits rigid-body motion?

Any one of these conditions indicates instability.

5.8 Free-Body Diagrams

For a structure to be in equilibrium, every part of the structure also must be in equilibrium. The equations of static equilibrium are as applicable to each piece of a structure as they are to an entire structure. It is therefore possible to draw a diagram of any part of a structure, including all of the forces that are acting on that part of the structure, and apply the equations of static equilibrium to that part. Such a diagram is called a *free-body diagram* (FBD). The forces acting on the free body are the external forces acting on that piece of the structure, including structural reactions, and the internal forces applied from the adjoining parts of the structure.

A simple beam is shown in Figure 5.13(a). This beam has two supports and is acted upon by two loads. A free-body diagram of the entire beam in Figure 5.13(b) shows all of the reaction forces.

QUICK NOTE

You must discover over and over that free-body diagrams (FBDs) **open the way to the solution** of structural problems.

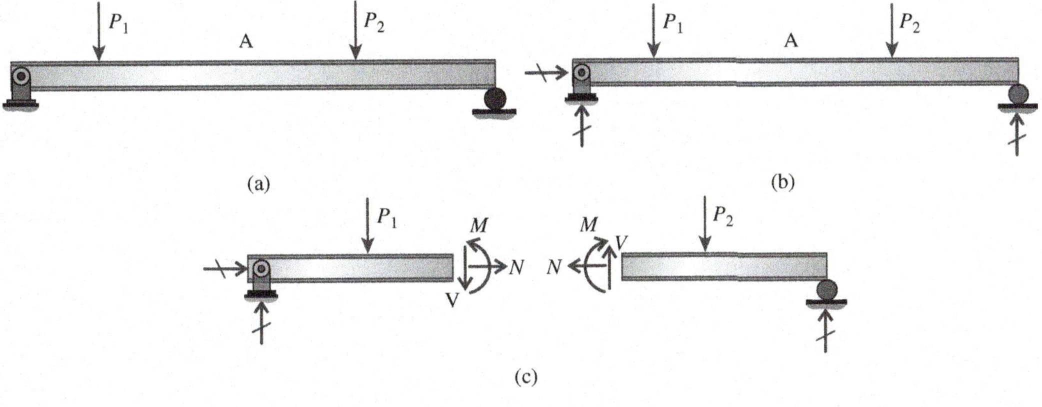

COMMON MISTAKE

Students often fail to sketch all of the forces—both known and unknown—on the free-body diagram (FBD).

FIGURE 5.13 A beam and two free-body diagrams.

We also can cut the beam at point A and draw a free-body diagram for each of the two pieces, as shown in Figure 5.13(c). Observe that we now have included the internal forces at the location of the cut on the diagrams. The internal forces are the same on the two pieces, but the direction in which they act is reversed. In essence, the effects of the right side of the beam on the left side are shown on the left free body and vice versa. For instance, the right-hand part of the beam tends to push the left-hand free body down, while the left-hand part is trying to push the right-hand free body up.

Isolating certain sections of structures and considering the forces applied to those sections is the basis of all structural analysis. It is doubtful that this procedure can be overemphasized to the reader.

5.9 Reactions for Single Rigid-Body Systems

Reactions by Proportions

DID YOU KNOW?
k or kip is the abbreviation for kilopound or in other words 1000 pounds.

The calculation of reactions is fundamentally a matter of proportions. To illustrate this point, reference is made to Figure 5.14(a). The load P is three-fourths of the distance L as measured from support A. By proportions, support B will carry three-fourths of the load, and support A will carry the remaining one-fourth of the load. In other words, the load carried by a support is inversely proportional to its distance from the applied load.

Similarly, for the beam of Figure 5.14(b), the 10-kip load is one-half of the distance from A to B, and each support will carry half of it or 5 kips. The 20-kip load is three-fourths of the distance from A to B. The B support will carry three-fourths of it or 15 kips, and the A support will carry one-fourth or 5 kips. In this manner, the total reaction at the A support is found to be 10 kips and the total reaction at the B support 20 kips.

Reactions Calculated by Equations of Statics

QUICK NOTE
Solving reaction problems using proportions is very handy for simple structures and loads. This helps with top-of-the-head calculations.

Reaction calculations by the equations of statics can be applied to all types of structures—not just simple ones. Examples 5.6 to 5.9 illustrate the application of these equations. In applying the $\sum M = 0$ equation, a point usually may be selected as the center of moments so that the lines of action of all but one of the unknowns pass through the point. The unknown is determined from the moment equation, and the other reaction components are found by applying the $\sum F_x = 0$ and $\sum F_y = 0$ equations. The solutions of reaction problems may be checked by taking moments about the other support, as illustrated in Example 5.6. For future examples, space is not taken to

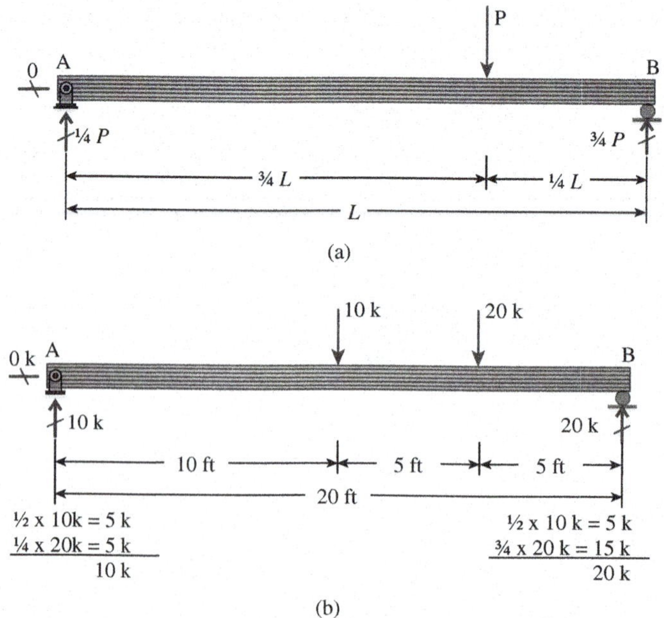

FIGURE 5.14 Reactions of simply supported beam by proportions.

show the checking of calculations. *A problem, however, should be considered incomplete until a mathematical check of this nature is made.*

EXAMPLE 5.6

Compute the reaction components for the beam shown here.

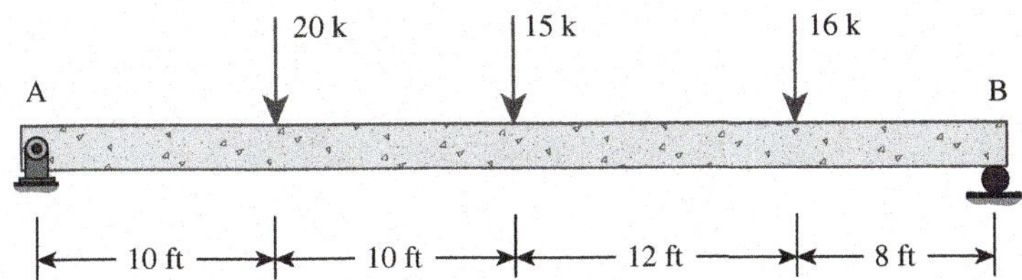

SOLUTION STRATEGY

Sketch the FBD of the beam, including all properly labeled reactions. Use equations of equilibrium. Try to formulate equations that have at most one unknown force in each one.

SOLUTION

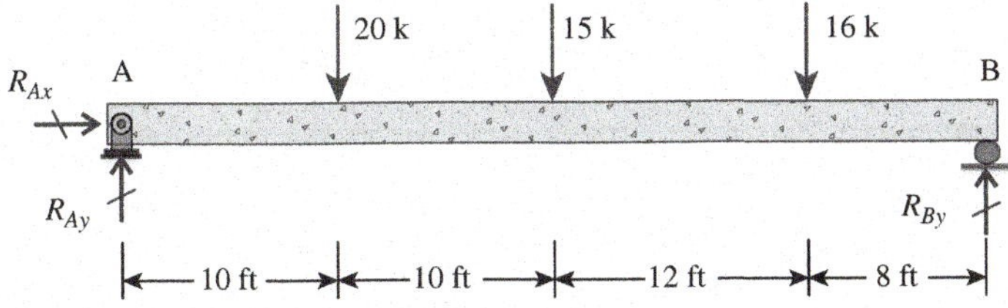

The reaction forces and their assumed direction are shown on the FBD. Begin the solution by summing forces horizontally to determine R_{Ax}:

$$\xrightarrow{+} \sum F_x = R_{Ax} = 0$$

$$\therefore R_{Ax} = \underline{\underline{0 \text{ kips}}}$$

Next, sum moments clockwise about the left support. By doing so, we obtain the equation

$$\circlearrowright^{+} \sum M_A = -(20\text{ k})(10\text{ ft}) - (15\text{ k})(20\text{ ft}) - (16\text{ k})(32\text{ ft}) + R_{By}(40\text{ ft}) = 0$$

$$\therefore R_{By} = \underline{\underline{25.3 \text{ kips} \uparrow}}$$

The result for R_{By} is positive, so the assumed direction is correct; the reaction at B acts upward. Lastly, forces are summed vertically to compute the remaining reaction:

$$^{+}\uparrow \sum F_y = R_{Ay} - 20\text{ k} - 15\text{ k} - 16\text{ k} + R_{By} = 0$$

$$R_{Ay} - 20\text{ k} - 15\text{ k} - 16\text{ k} + 25.3\text{ k} = 0$$

$$\therefore R_{Ay} = \underline{\underline{25.7 \text{ kips} \uparrow}}$$

Again, the computed reaction at A is positive, so the assumed direction is correct. We can sum moments about B to check our calculations:

$$\circlearrowright^{+} \sum M_B = -(25.7\text{ k})(40\text{ ft}) + (20\text{ k})(30\text{ ft}) + (15\text{ k})(20\text{ ft}) + (16\text{ k})(8\text{ ft})$$

$$\sum M_B = 0$$

SUMMARY

Because the summation of moments is equal to zero, the computed reactions are correct.

QUICK NOTE

The symbols $\circlearrowright^{+}$, $\rightarrow^{+}$, and $^{+}\uparrow$ are indicators of the sign convention used to write the equation. They are used to help you correctly formulate the equation and allows others to follow your work.

EXAMPLE 5.7

Compute the reaction components for the structure shown here.

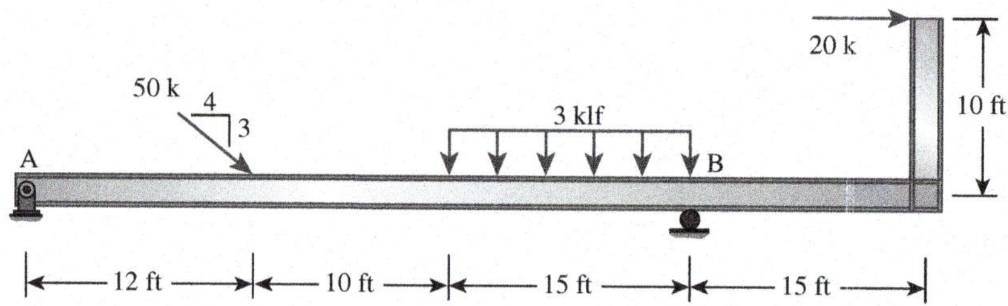

SOLUTION STRATEGY

Sketch the FBD of the structure, including all properly labeled reactions. Resolve any inclined forces into x and y components. Calculate and locate any distributed loads into an equivalent resultant. Use equations of equilibrium.

QUICK NOTE

It is good practice to resolve any inclined forces into x and y components. This will often make the calculation of moment arms much simpler.

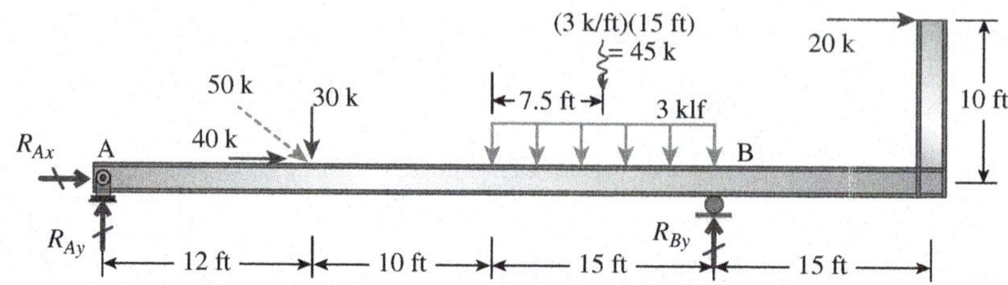

QUICK NOTE

Resultants of distributed loads in this textbook are presented as a squiggly arrow. This is to distinguish resultants from other applied or internal forces sketched on a free-body diagram.

Observe that the inclined force has been replaced with its horizontal and vertical components, and the distributed load has been replaced with an equivalent concentrated force acting at its centroid.

Begin by summing forces horizontally to compute the horizontal reaction at A:

$$\xrightarrow{+} \sum F_x = 40 \text{ k} + 20 \text{ k} + R_{Ax} = 0$$

$$\therefore R_{Ax} = -60 \text{ kips} = \underline{\underline{60 \text{ kips} \leftarrow}}$$

Observe that the computed value is negative, so the actual direction of the reaction is opposite that assumed. Next, moments are taken counterclockwise about A to determine the vertical reaction at B.

$$\circlearrowleft^{+} \sum M_A = -(30 \text{ k})(12 \text{ ft}) - (45 \text{ k})(29.5 \text{ ft}) + R_{By}(37 \text{ ft}) - (20 \text{ k})(10 \text{ ft}) = 0$$

$$\therefore R_{By} = \underline{\underline{51 \text{ kips} \uparrow}}$$

Because the computed sign on R_{By} is positive, the assumed direction is correct. Forces are summed vertically to compute the vertical reaction at A.

$$+\uparrow \sum F_y = R_{Ay} - 30 \text{ k} - 45 \text{ k} + R_{By} = 0$$

$$R_{Ay} - 30 \text{ k} - 45 \text{ k} + 51 \text{ k} = 0$$

$$\therefore R_{Ay} = \underline{\underline{24 \text{ kips} \uparrow}}$$

Again, the assumed direction of R_{Ay} is correct.

The next example includes a roller support which is bearing on an inclined surface. Recall that the reaction resultant is perpendicular to the supporting surface and represents only one unknown force.

EXAMPLE 5.8

Compute the reaction components for the frame shown here.

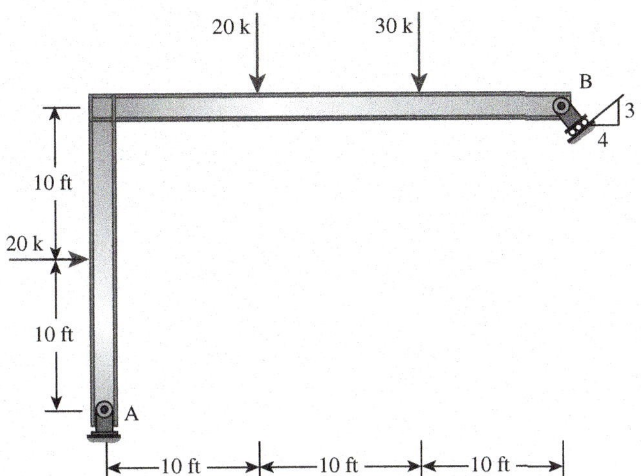

DID YOU KNOW?

When the slope of a line is given in the rise:run (y:x) format, the slope of a perpendicular line is simply the reciprocal run:rise.

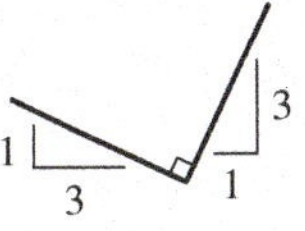

SOLUTION STRATEGY

Sketch the FBD of the structure, including all properly labeled reactions. Resolve any inclined reactions into x and y components. Use equations of equilibrium.

SOLUTION

The free-body diagram for this frame is shown next. Observe that the inclined force of reaction at B has been replaced with its x and y components. These components act at the same point as the reaction.

We begin by summing moments about A to determine the reaction at B. In the calculation, we use the components of the reaction.

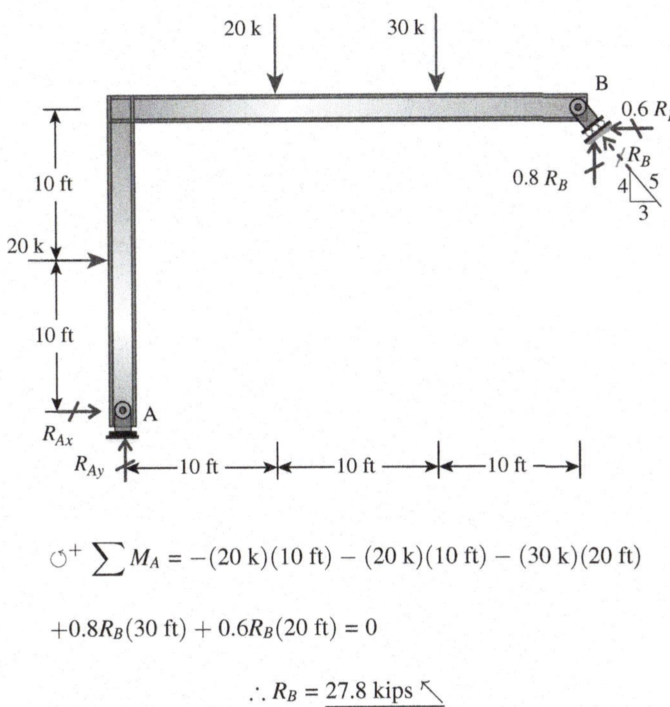

QUICK NOTE

When sketching the x and y components of the inclined reaction you need to draw them in the direction which is consistent with the line of action of the resultant.

$$\circlearrowleft^{+} \sum M_A = -(20\text{ k})(10\text{ ft}) - (20\text{ k})(10\text{ ft}) - (30\text{ k})(20\text{ ft})$$

$$+0.8R_B(30\text{ ft}) + 0.6R_B(20\text{ ft}) = 0$$

$$\therefore R_B = \underline{\underline{27.8\text{ kips}}} \nwarrow$$

The sign on the computed reaction at B is positive, so the assumed direction is correct. Next, forces are summed vertically to obtain the vertical reaction at A:

$$+\uparrow \sum F_y = R_{Ay} - 20\text{ k} - 30\text{ k} + 0.8R_B = 0$$

$$R_{Ay} - 20\text{ k} - 30\text{ k} + (0.8)(27.8\text{ k}) = 0$$

$$\therefore R_{Ay} = \underline{\underline{27.8\text{ kips}\uparrow}}$$

Again, the sign on the computed reaction is positive, so the assumed direction is correct. By summing the horizontal forces, the horizontal reaction at A can be evaluated.

$$\overset{+}{\rightarrow} \sum F_x = R_{Ax} + 20\text{ k} - 0.6R_B = 0$$

$$R_{Ax} + 20\text{ k} - (0.6)(27.8\text{ k}) = 0$$

$$\therefore R_{Ax} = -3.3\text{ kips} = \underline{\underline{3.3\text{ kips}\leftarrow}}$$

The sign on the computed reaction is negative, so the force of reaction is actually acting to the left—opposite the direction indicated on the FBD.

EXAMPLE 5.9

COMMON MISTAKE

When rotational springs are used, it is common for students to sketch either vertical or horizontal reactions at the spring location.

The rotational spring will only provide a moment reaction.

Compute the reaction components for the structure shown here.

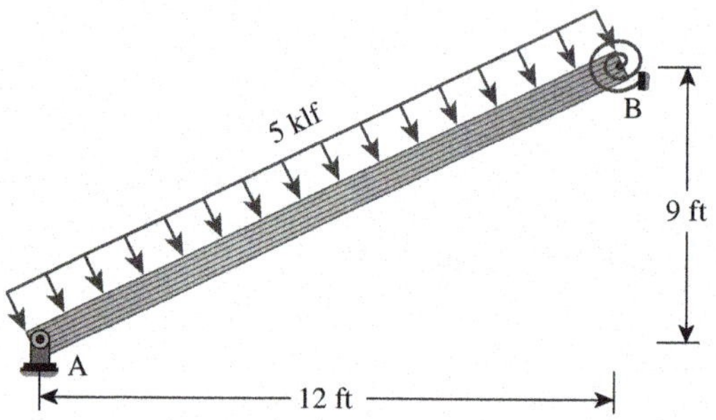

5 klf
B
9 ft
A
12 ft

SOLUTION STRATEGY

ALTERNATE APPROACH

The horizontal and vertical components of the distributed load may also be calculated by taking the load intensity and multiplying it by the projected length of the member that is perpendicular to the direction of the computed component.

Vertical component:

$$(5\text{ k/ft})(12\text{ ft}) = \underline{\underline{60\text{ k}}}$$

Horizontal component:

$$(5\text{ k/ft})(9\text{ ft}) = \underline{\underline{45\text{ k}}}$$

Sketch the FBD of the structure, including all properly labeled reactions. Find the resultant of the distributed load and then resolve into x and y components. Use equilibrium equations.

SOLUTION

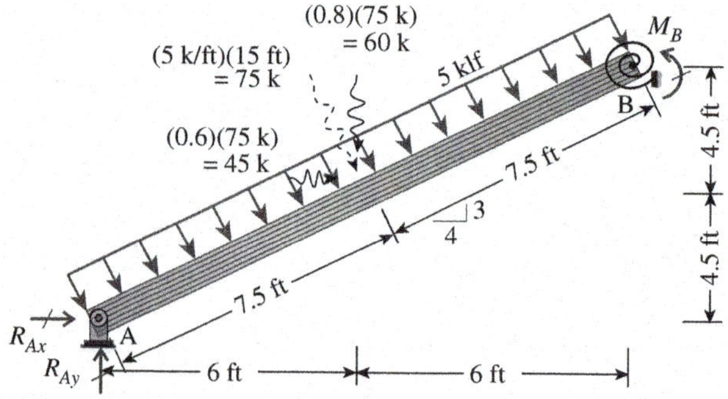

(0.8)(75 k) = 60 k
(5 k/ft)(15 ft) = 75 k
(0.6)(75 k) = 45 k
5 klf
M_B
B
7.5 ft
7.5 ft
3
4
4.5 ft
4.5 ft
R_{Ax}
R_{Ay}
A
6 ft
6 ft

In the free-body diagram for this frame, only a moment reaction occurs at the location of the rotational spring.

We begin by summing moments about A to determine the moment reaction at B. For convenience, we use the resultant and its perpendicular distance from point A:

$$\circlearrowleft^{+} \sum M_A = -(75\text{ k})(7.5\text{ ft}) + M_B = 0$$
$$\therefore M_B = \underline{\underline{562.5\text{ k}\cdot\text{ft} \circlearrowleft}}$$

The sign on the computed moment reaction at B is positive, so the assumed direction is correct. Next, forces are summed vertically to obtain the vertical reaction at A:

$$+\uparrow \sum F_y = R_{Ay} - 60\text{ k} = 0$$
$$\therefore R_{Ay} = \underline{\underline{60\text{ kips} \uparrow}}$$

Again, the sign on the computed reaction is positive, so the assumed direction is correct. By summing the horizontal forces, the horizontal reaction at A can be evaluated as

$$\rightarrow^{+} \sum F_x = R_{Ax} + 45\text{ k} = 0$$
$$\therefore R_{Ax} = -45\text{ kips} = \underline{\underline{45\text{ kips} \leftarrow}}$$

The sign on the computed reaction is negative, so the force of reaction is actually acting to the left—opposite the direction indicated on the FBD.

5.10 Reactions for Multiple Connected Rigid-Body Systems

Cantilevered Structures

Moments in structures that are simply supported increase rapidly as their spans become longer. We will see that bending increases approximately in proportion to the square of the span length. Stronger and more expensive structures are required to resist the greater moments. For very long spans, moments are so large that it becomes economical to introduce special types of structures that will reduce the moments. One of these types is cantilevered construction, as illustrated in Figure 5.15(b).

A cantilevered type of structure is substituted for the three simple beams of Figure 5.15(a) by making the beam continuous over the interior supports B and C and introducing hinges or moment releases in the center span as indicated in part (b). The reader can verify that, with the presence of the hinges, the structure is statically determinate.

The moment advantage of cantilevered construction is illustrated in Figure 5.16. The diagrams shown give the variation of bending moment in each of the structures of Figure 5.15 due to a uniform load of 3 klf for the entire spans. The maximum moment for the cantilevered type is seen to be considerably less than that for the simple spans, and this permits lighter and more economical construction. The plotting of moment diagrams is fully explained in Chapter 6. For the cantilevered structure of Figure 5.15(b), it is possible to better balance the positive and negative moments by moving the unsupported hinges closer to supports B and C.

QUICK NOTE

A multiple connected structure is one that consists of multiple members that are connected with internal moment releases or shear releases.

Cantilever structures are common examples of such structures.

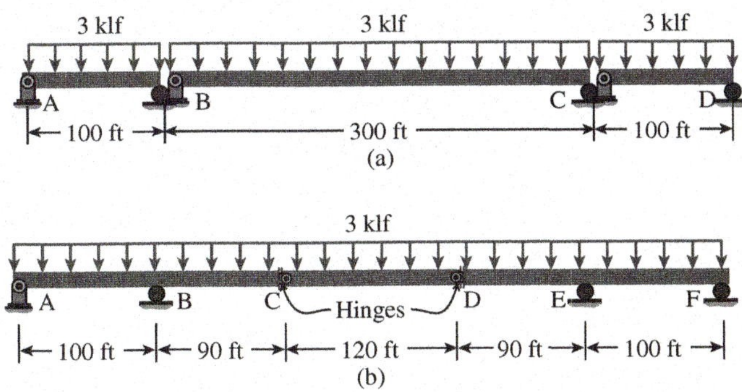

FIGURE 5.15 Simple span and cantilevered structure.

QUICK NOTE

The key to solving multi-connected structures is to **cut** it apart at the internal releases whether they are hinges, sliders or a combination thereof then sketch the FBDs.

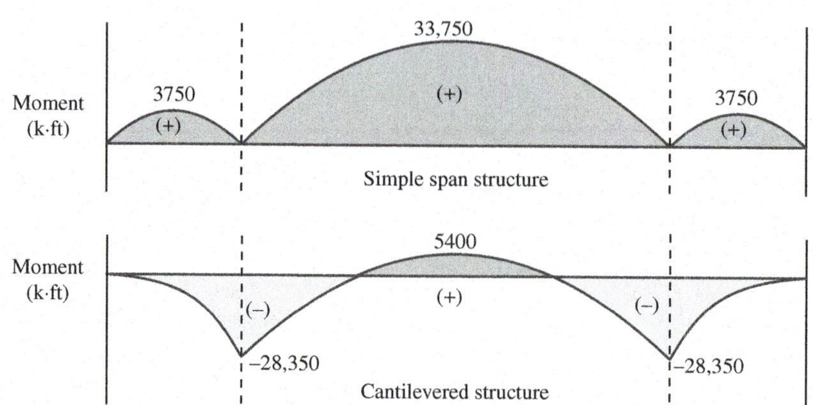

FIGURE 5.16 Moment diagrams for beams in Figure 5.15.

Reaction Calculation Strategy

DID YOU KNOW?

If you make a single cut in a structure, you gain three more equations of equilibrium, but you also gain three more unknowns—no real benefit is realized.

If you make a cut at a release, you gain three more equations of equilibrium, but you only gain two more unknowns.

The internal releases in cantilever and other types of structures provide the structural analyst with a knowledge of the internal shear or moment at the location of the release(s). For example, the moment release that is located 90 ft to the right of support B in Figure 5.15 indicates that we know the internal moment at that location is zero. To solve structures of this type, we must be able to use the information about the moments being zero at the hinges to get the reaction forces. We do this by cutting the structure apart at the releases. This gives us access to the internal forces at the cut.

The beam in Figure 5.15 has five external reactions. If we leave the structure fully assembled, we have only three equations of equilibrium, which clearly will not work out. However, if we cut the structure at the hinges, we end up with nine unknowns and nine equations.

EXAMPLE 5.10

Compute all reactions for the cantilevered structure illustrated here.

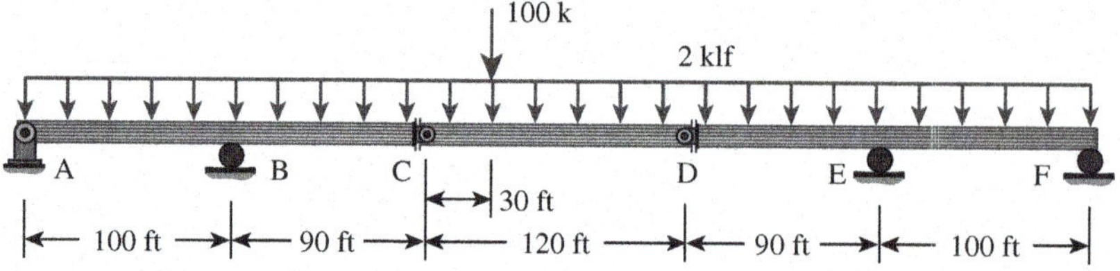

SOLUTION STRATEGY

Cut the structure apart at the internal hinges. Sketch the FBD of each section of the structure, including all properly labeled unknowns. Find the resultant of the distributed load. Starting with the section having the fewest unknowns, use equilibrium equations to solve for both internal forces and reactions. Once you find internal forces on one section, you should transfer them to the adjacent section.

SOLUTION

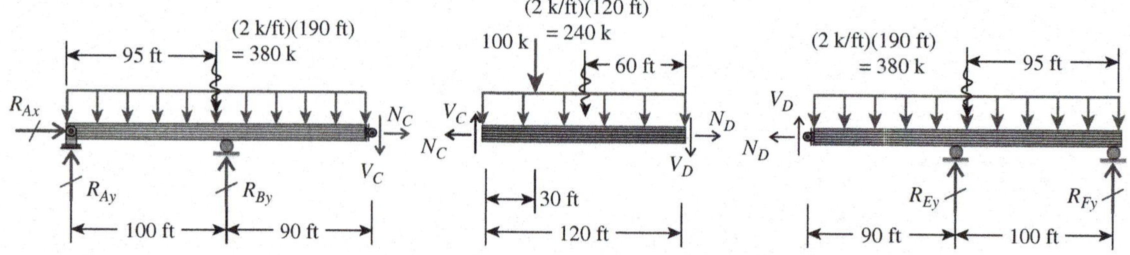

When we cut the structure at the internal hinges at C and D, we only have an internal shear (V) and an internal axial force (N). The left and right sections each have three unknown vertical forces, while the middle section has only two. Thus, we use the free body of the center section to compute the forces V_C and V_D as

$$\circlearrowleft^{+} \sum M_C = -(100\text{ k})(30\text{ ft}) - (240\text{ k})(60\text{ ft}) - V_D(120\text{ ft}) = 0$$
$$\therefore V_D = \underline{\underline{-145\text{ kips}}}$$

Then by summing forces on the center section in the y-direction, we have

$$+\uparrow \sum F_y = V_C - 100\text{ k} - 240\text{ k} - V_D = 0$$
$$V_C - 100\text{ k} - 240\text{ k} - (-145\text{ k}) = 0$$
$$\therefore V_C = \underline{\underline{195\text{ kips}}}$$

Finally, by summing forces horizontally on the center section, a relationship is obtained between the horizontal forces:

$$\rightarrow^{+} \sum F_x = -N_C + N_D = 0$$
$$\therefore N_C = N_D$$

Now that we have computed the forces acting on the center section, we turn to the right section. First, sum moments about the right end:

$$\circlearrowleft^{+} \sum M_F = -(V_D)(190\text{ ft}) + (380\text{ k})(95\text{ ft}) - R_{Ey}(100\text{ ft}) = 0$$
$$-(-145\text{ k})(190\text{ ft}) + (380\text{ k})(95\text{ ft}) - R_{Ey}(100\text{ ft}) = 0$$
$$\therefore R_{Ey} = \underline{\underline{636.5\text{ kips}\uparrow}}$$

We then sum forces vertically and horizontally to obtain the other components of reaction on the right section.

$$+\uparrow \sum F_y = V_D - 380\text{ k} + R_{Ey} + R_{Fy} = 0$$
$$-145\text{ k} - 380\text{ k} + 636.5\text{ k} + R_{Fy} = 0$$
$$\therefore R_{Fy} = -111.5\text{ kips} = \underline{\underline{111.5\text{ kips}\downarrow}}$$

$$\rightarrow^{+} \sum F_x = -N_D = 0$$
$$\therefore N_D = \underline{\underline{0\text{ kips} = N_C}}$$

We now work with the left section to compute the remaining reaction forces. Begin by summing moments counterclockwise about A:

$$^{+}\circlearrowleft \sum M_A = -(V_C)(190\text{ ft}) - (380\text{ k})(95\text{ ft}) + R_{By}(100\text{ ft}) = 0$$
$$-(195\text{ k})(190\text{ ft}) - (380\text{ k})(95\text{ ft}) + R_{By}(100\text{ ft}) = 0$$
$$\therefore R_{By} = \underline{\underline{731.5\text{ kips}\uparrow}}$$

The remaining vertical reaction is found by summing forces in the y-direction:

$$+\uparrow \sum F_y = R_{Ay} - 380\text{ k} + R_{By} - V_C = 0$$
$$R_{Ay} - 380\text{ k} + 731.5\text{ k} - 195\text{ k} = 0$$
$$\therefore R_{Ay} = -156.5\text{ kips} = \underline{\underline{156.5\text{ kips}\downarrow}}$$

Finally, the horizontal reaction is computed as

$$\xrightarrow{+} \sum F_x = R_{Ax} + N_C = 0$$
$$R_{Ax} + 0 = 0$$
$$\therefore R_{Ax} = \underline{\underline{0\text{ kips}}}$$

QUICK NOTE

If you select the section with the fewest number of unknown forces, you can keep the number of unknowns in any equation to a minimum.

The center section is the best place to start for solving the vertical forces.

The right section is the best place to start for the horizontal forces.

EXAMPLE 5.11

Compute all reactions for the multi-connected beam illustrated here.

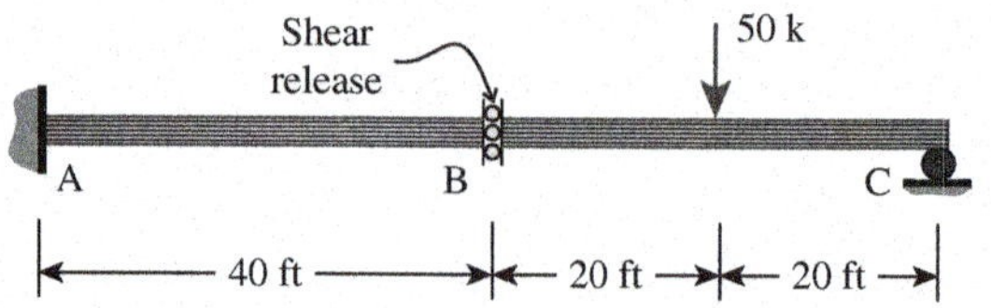

SOLUTION STRATEGY

Cut the structure apart at the internal shear release. Sketch the FBD of each section of the structure, including all properly labeled unknowns. Starting with the section having the fewest unknowns, use equilibrium equations to solve for both internal forces and reactions. Once you find internal forces on one section, you should transfer them to the adjacent section.

RECALL
The internal forces at a shear release are an axial force and a moment (see Figure 5.9).

SOLUTION

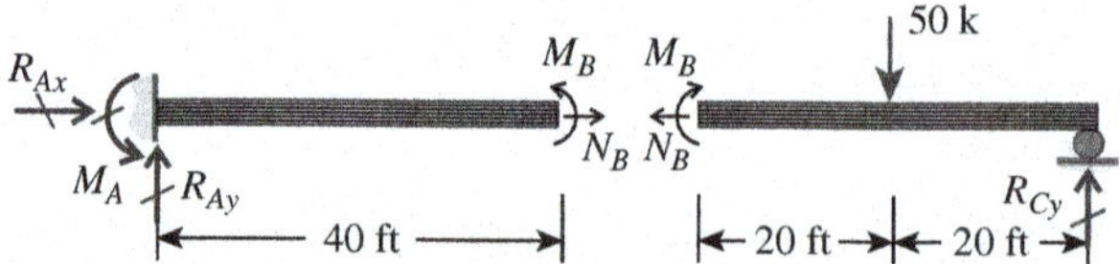

The left section of the FBD has five unknowns, and the right section has three unknowns. We start with the right section and sum forces in the y-direction:

$$+\uparrow \sum F_y = R_{Cy} - 50\text{ k} = 0$$
$$\therefore R_{Cy} = \underline{\underline{50\text{ kips}\uparrow}}$$

Then summing moments about point B, we have

$$+\circlearrowleft \sum M_B = -M_B - (50\text{ k})(20\text{ ft}) + R_{Cy}(40\text{ ft}) = 0$$
$$-M_B - (50\text{ k})(20\text{ ft}) + (50\text{ k})(40\text{ ft}) = 0$$
$$\therefore M_B = \underline{\underline{1000\text{ k}\cdot\text{ft}}}$$

and summing forces in the x-direction gives

$$\overset{+}{\rightarrow} \sum F_x = -N_B = 0$$
$$\therefore N_B = \underline{\underline{0\text{ kips}}}$$

QUICK NOTE
When giving values for reactions we always provide arrows indicating the direction of the reaction.

When giving values for internal forces arrows are not used. This is because the direction of the arrow would depend on which section you were looking at. Rather we will use signs (+) or (−).

Now we can transfer the known forces from the right section to the left section and sum forces in the x-direction:

$$\overset{+}{\rightarrow} \sum F_x = R_{Ax} + N_B = 0$$
$$R_{Ax} + 0 = 0$$
$$\therefore R_{Ax} = \underline{\underline{0\text{ kips}}}$$

Next, summing the moments about point A will find M_A.

$$+\circlearrowleft \sum M_A = M_A + M_B = 0$$
$$M_A + 1000\text{ k}\cdot\text{ft} = 0$$
$$\therefore M_A = -1000\text{ k}\cdot\text{ft} = \underline{\underline{1000\text{ k}\cdot\text{ft}\circlearrowright}}$$

Finally, a sum of forces in the y-direction finishes out the problem, so

$$+\uparrow \sum F_y = R_{Ay} = 0$$
$$\therefore R_{Ay} = \underline{\underline{0\text{ kips}}}$$

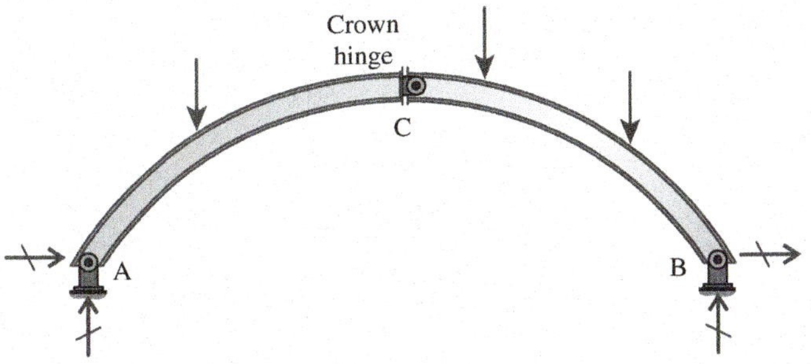

FIGURE 5.17 Three-hinged arch.

Three-Hinged Frames/Arches

Simple frames and arches are sometimes constructed using three hinges as shown in Figure 5.17. The analysis of such structures really follows the same approach as for other multi-connected structures. It requires one to cut the structure at the center hinge and utilize the knowledge of zero moment at that location to come up with the final solution to the reactions.

EXAMPLE 5.12

Compute all reactions for the three-hinged arch shown here. The vertical dimension is measured from the center of the bottom hinge to the center of the top hinge.

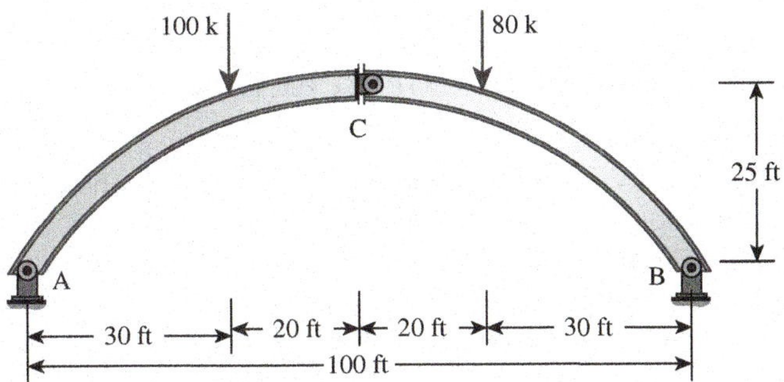

SOLUTION STRATEGY

Find vertical reactions by looking at the FBD of the entire arch. Horizontal reactions only can be found by cutting the structure at the crown hinge and looking at either the left or right section. Use equilibrium equations to solve for unknowns.

SOLUTION

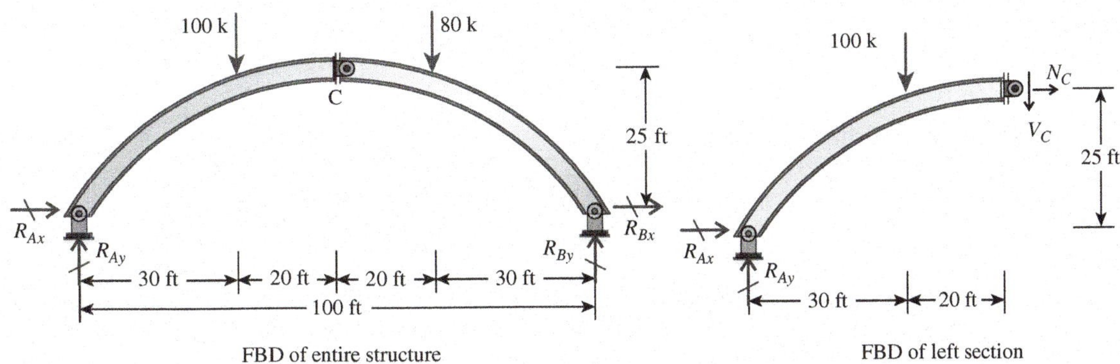

Summing moments about support A of the entire structure gives

$$\stackrel{+}{\circlearrowleft} \sum M_A = -(100\text{ k})(30\text{ ft}) - (80\text{ k})(70\text{ ft}) + R_{By}(100\text{ ft}) = 0$$

$$\therefore R_{By} = \underline{\underline{86\text{ kips}\uparrow}}$$

Sum the forces in the *y*-direction to calculate R_{Ay}:

$$+\uparrow \sum F_y = R_{Ay} - 100\text{ k} - 80\text{ k} + R_{By} = 0$$

$$R_{Ay} - 100\text{ k} - 80\text{ k} + 86\text{ k} = 0$$

$$\therefore R_{Ay} = \underline{\underline{94\text{ kips}\uparrow}}$$

AVOID EXTRA WORK

Notice that we did not need to find the internal shear (V_C) and axial (N_C) at C to be able to find the reactions. Every time you calculate another equation, you increase the possibility of making a mistake.

To compute the magnitude of the horizontal reactions R_{Ax} and R_{Bx}, we may use the FBD of the left section. Sum moments about point C:

$$\overset{+}{\circlearrowleft} \sum M_C = R_{Ax}(25\text{ ft}) - R_{Ay}(50\text{ ft}) + (100\text{ k})(20\text{ ft}) = 0$$

$$R_{Ax}(25\text{ ft}) - (94\text{ k})(50\text{ ft}) + (100\text{ k})(20\text{ ft}) = 0$$

$$\therefore R_{Ax} = \underline{\underline{108\text{ kips}\rightarrow}}$$

Now sum forces in the *x*-direction on the entire structure:

$$\overset{+}{\rightarrow} \sum F_x = 108\text{ k} + R_{Bx} = 0$$

$$\therefore R_{Bx} = -108\text{ kips} = \underline{\underline{108\text{ kips}\leftarrow}}$$

5.11 Matrix Formulation for Reactions

In the previous example, a three-hinged arch was analyzed for its reactions. The process for solving for its reactions seems fairly straightforward. However, this is because both pinned supports are at the same position vertically. This makes it easy to solve for the vertical reactions in the arch. Once the pins have a vertical offset, writing equations of equilibrium having only one unknown is no longer possible. Rather, the analyst is required to solve a set of simultaneous equations. The solution of these equations can be readily handled if a matrix formulation is used.

Assume that two equations of equilibrium are written for R_{Ax} and R_{Ay}. The equations are written in terms of constants C_1 through C_6 as

$$C_1 R_{Ax} + C_2 R_{Ay} - C_3 = 0$$

$$C_4 R_{Ax} + C_5 R_{Ay} - C_6 = 0$$

These equations can be rewritten as

$$C_1 R_{Ax} + C_2 R_{Ay} = C_3$$

$$C_4 R_{Ax} + C_5 R_{Ay} = C_6$$

which can be written in matrix form as

$$\begin{bmatrix} C_1 & C_2 \\ C_4 & C_5 \end{bmatrix} \begin{Bmatrix} R_{Ax} \\ R_{Ay} \end{Bmatrix} = \begin{Bmatrix} C_3 \\ C_6 \end{Bmatrix}$$

The reader can refer to Appendix C to refresh on basic matrix math. The solution to the two unknowns is found by first multiplying the vector of constants with the inverse of the square coefficient matrix. The inverse of a matrix is noted as $[C]^{-1}$. Most calculators and spreadsheets can perform this calculation.

$$\begin{Bmatrix} R_{Ax} \\ R_{Ay} \end{Bmatrix} = \begin{bmatrix} C_1 & C_2 \\ C_4 & C_5 \end{bmatrix}^{-1} \begin{Bmatrix} C_3 \\ C_6 \end{Bmatrix}$$

EXAMPLE 5.13

Compute all reactions for the three-hinged frame shown. Use the matrix formulation to solve for all four reactions simultaneously.

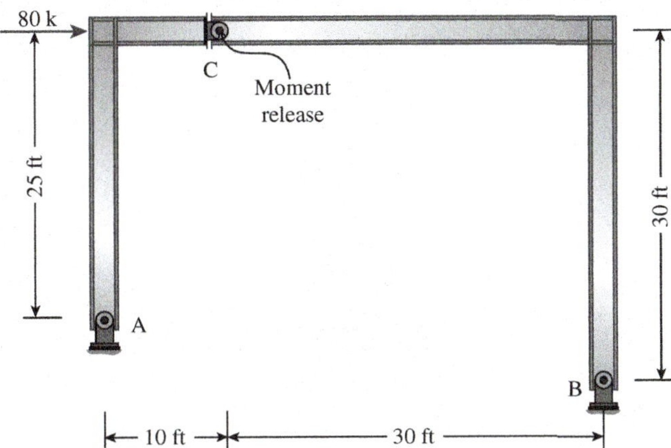

SOLUTION STRATEGY

Write a set of four equilibrium equations using the FBD for the full structure and also for one of the sections after cutting at the moment release. Write the equations in terms of the four unknown reactions. Assemble in matrix form and solve for the unknowns simultaneously.

SOLUTION

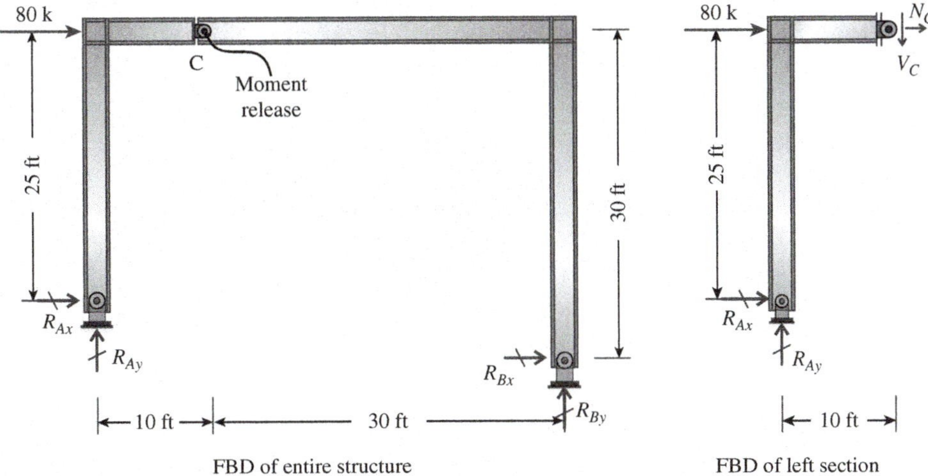

We can write up to three equations using the FBD for the entire structure and up to three for the left section. Since the entire structure contains all four unknown reactions, we can write three of the four required equations of equilibrium using the full FBD. The fourth equation can come from the FBD of the left section.

Sum moments at A on the entire structure:

$$^{+}\circlearrowleft \sum M_A = -(80\text{ k})(25\text{ ft}) + R_{By}(40\text{ ft}) + R_{Bx}(5\text{ ft}) = 0$$

$$R_{By}(40\text{ ft}) + R_{Bx}(5\text{ ft}) = 2000\text{ k}\cdot\text{ft}$$

Rewrite the equation in terms of all four unknown reactions for

$$R_{Ax}(0\text{ ft}) + R_{Ay}(0\text{ ft}) + R_{Bx}(5\text{ ft}) + R_{By}(40\text{ ft}) = 2000\text{ k}\cdot\text{ft}$$

QUICK NOTE

To use the matrix formulation, one must have a coefficient (C) for each unknown reaction in each equation of equilibrium. If an unknown does not exist in an equation of equilibrium, then you must add it in—but you will do so with a coefficient of zero.

Look at the equation for $\sum M_A$; it was originally written in terms of R_{By} and R_{Bx} only. However, it needed to be in terms of all four unknown reactions. Therefore, we added in R_{Ay} and R_{Ax} and then multiplied them by zero.

Sum the forces in the y-direction:

$$+\uparrow \sum F_y = R_{Ay} + R_{By} = 0$$
$$R_{Ax}(0) + R_{Ay}(1) + R_{Bx}(0) + R_{By}(1) = 0$$

Now sum forces in the x-direction on the entire structure:

$$\xrightarrow{+} \sum F_x = R_{Ax} + R_{Bx} + 80 = 0$$
$$R_{Ax}(1) + R_{Ay}(0) + R_{Bx}(1) + R_{By}(0) = -80\text{ k}$$

Finally, sum moments about point C on the left section FBD:

$$+\circlearrowleft \sum M_A = R_{Ax}(25\text{ ft}) - R_{Ay}(10\text{ ft}) = 0$$
$$R_{Ax}(25\text{ ft}) - R_{Ay}(10\text{ ft}) + R_{Bx}(0\text{ ft}) + R_{By}(0\text{ ft}) = 0$$

Write the four equations into matrix form. The factors in front of the unknowns go into the coefficient matrix—which should be square. The constants on the right-hand side of the equations go into a vector as shown.

$$\begin{bmatrix} 0 & 0 & 5 & 40 \\ 0 & 1 & 0 & 1 \\ 1 & 0 & 1 & 0 \\ 25 & -10 & 0 & 0 \end{bmatrix} \begin{Bmatrix} R_{Ax} \\ R_{Ay} \\ R_{Bx} \\ R_{By} \end{Bmatrix} = \begin{Bmatrix} 2000 \\ 0 \\ -80 \\ 0 \end{Bmatrix}$$

Now solve for the unknowns by pre-multiplying the constant vector by the inverse of the coefficient matrix.

$$\begin{Bmatrix} R_{Ax} \\ R_{Ay} \\ R_{Bx} \\ R_{By} \end{Bmatrix} = \begin{bmatrix} 0 & 0 & 5 & 40 \\ 0 & 1 & 0 & 1 \\ 1 & 0 & 1 & 0 \\ 25 & -10 & 0 & 0 \end{bmatrix}^{-1} \begin{Bmatrix} 2000 \\ 0 \\ -80 \\ 0 \end{Bmatrix} = \begin{Bmatrix} -22.9\text{ k} \\ -57.1\text{ k} \\ -57.1\text{ k} \\ 57.1\text{ k} \end{Bmatrix}$$

Remember that the negative signs indicate that we assumed the wrong direction to begin with.

$$\therefore R_{Ax} = -22.9\text{ kips} = \underline{\underline{22.9\text{ kips} \leftarrow}}$$

$$\therefore R_{Ay} = -57.1\text{ kips} = \underline{\underline{57.1\text{ kips} \downarrow}}$$

$$\therefore R_{Bx} = -57.1\text{ kips} = \underline{\underline{57.1\text{ kips} \leftarrow}}$$

$$\therefore R_{By} = \underline{\underline{57.1\text{ kips} \uparrow}}$$

5.12 SAP2000 Computer Applications

HOW TO USE SAP2000

For an introduction on how to use SAP2000, refer to Appendix B and view the tutorial video on this textbook's website.

This chapter focuses on the how to assess the determinacy of a structure—whether it is stable or not—and how to compute the magnitude of external reactions. The analysis technique used in SAP2000, which will be explained in the latter chapters of this textbook, can be used for both determinate and indeterminate structures. However, it does require that the structure be stable. If SAP2000 detects that a structure is unstable, it will attempt to "fix" the problem by providing some additional stiffness, like the spring in Figure 5.18, with k_T being very small. While this

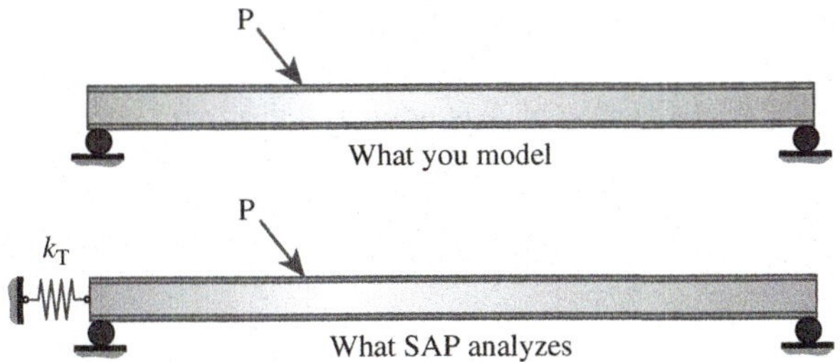

FIGURE 5.18 Unstable beam and the SAP2000 "fix".

sounds like a good idea, the analyst needs to understand that there is potentially an inherent physical problem in the structural system that needs to be addressed. When reviewing the results of an analysis, the analyst is cautioned to look for displacements that are excessively large in nature. Do you really want to have a structure that needs to translate over a million inches to reach equilibrium? This is a clear indication that the structure is really unstable.

As was discussed earlier in the chapter, you are quite comfortable sitting on a chair that is supported by a number of casters and you are not concerned about stability as long as your body is only applying a downward load to the chair. Imagine analyzing a horizontal beam with roller supports at each end resting on a flat horizontal surface. Clearly, this beam is unstable because it also cannot resist a force that has a horizontal component. SAP2000 would solve this problem by placing a very flexible horizontal spring support at one of the roller supports to provide some minor restraint. If the beam only supports vertical loads, the force in the added spring support would be zero and thus would not have an elongation and no calculated rigid body motion. Therefore, the remaining results (e.g., the vertical reactions) are correct. However, if this same beam was subjected to a horizontal force, SAP2000 would calculate a horizontal reaction at the end of the added spring support. This is to provide the necessary force for equilibrium. The results would indicate the beam had an excessive horizontal translation. As an analyst, you would need to question the validity of these results, since the original support system provided no means to resist horizontal forces and the out-of-balance horizontal force would actually lead to horizontal acceleration, velocity, and displacement of the beam.

NOMENCLATURE

To be **stable** means that no rigid-body motion of either the entire structure or a portion of the structure occurs.

QUICK NOTE

Some of the homework problems will explicitly ask you to create models using SAP2000. Keep in mind, however, that you can also use a computer model to check the answers for other homework problems that you are asked to complete.

SAP2000 EXERCISE 5.1

The SAP2000 input file *C5EX6.sdb* is a functional model of Example 5.6 in this chapter. The model has the correct dimensions, support conditions, and loadings of the beam. Note that pinned support A is located at joint 1 and that roller support B is located at joint 2. After starting SAP2000, load the *C5EX6.sdb* input file and execute the analysis. Verify through the graphical output of the reactions that SAP2000 calculated the correct results for the reactions at A and B.

DID YOU KNOW?

See Appendix B for additional information on loading an input file into SAP2000, executing the analysis, and viewing graphical results.

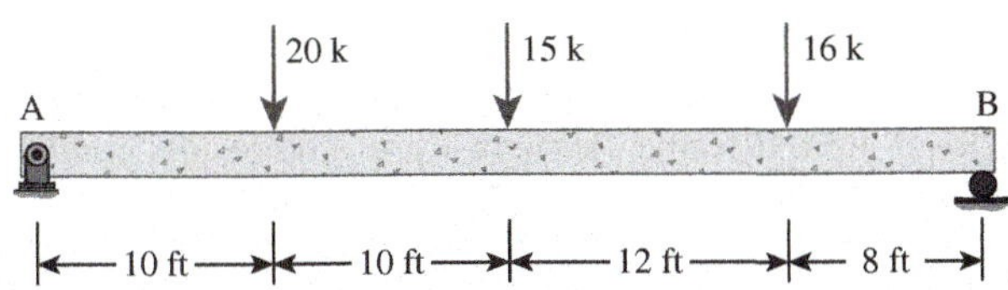

One of the keys in using analysis programs is to anticipate the results so that you have the ability to do a quick sanity check on those results before accepting them as being correct. For example, what would happen to the vertical reaction at A in example C5EX6 if all of the loads would move to the left by 4 feet? It would be reasonable to assume that the vertical reaction would increase, since the loads are moving closer to this support. We know that the vertical reaction at A cannot exceed 51 kips as this is the total applied loading to this simply supported beam.

Edit the loading on the beam in C5EX6 by replacing the existing loading with a new loading where each of the original loads are moved 4 feet to the left. Rerun the analysis and verify that the new vertical reaction at A is now 30.8 kips acting upwards.

SAP2000 EXERCISE 5.2

COMMON MISTAKE

Improper development of a numerical model of a physical system will yield results, but not the results that you were looking for from your analysis. The model needs to represent the proper geometry of the structure, elements connections, support conditions, and loadings.

The SAP2000 input file *C5EX7.sdb* is a functional model of Example 5.7 in this chapter. This model is composed of three frame elements—a horizontal element from A to B, a horizontal element from B to the corner, and a vertical element representing the vertical leg of the frame. Thus there are four joints in the structural system. Also note the 50-kip force is indirectly applied to the element in SAP2000 by using the vertical and horizontal components, since SAP2000 has limitations on how forces are described in the input. The uniform distributed load is applied to the right 15 feet of the element between A and B. The 20-kip force is applied as a joint force, since it acts at a joint, but it could have been applied as a member force acting at the top of the vertical leg. Typical convention is to apply forces acting at joints as joint forces and to apply all other forces as member forces. Run the analysis using *C5EX7.sdb* as provided.

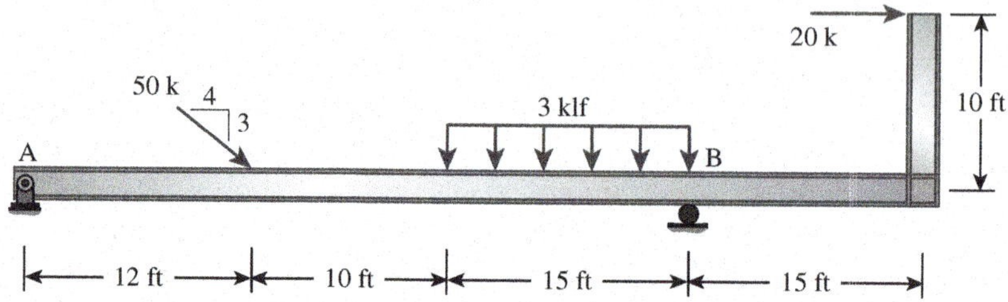

Now remove the joint load from the top of the vertical leg and replace it with an equivalent member force applied to the end of the vertical leg. Be sure that the force is applied to the correct end and in the horizontal direction acting to the right. Rerun the results and verify that you get the same external reactions.

SAP2000 EXERCISE 5.3

The SAP2000 input file *C5EX11.sdb* is a functional model of Example 5.11 in this chapter. This model is composed of two frame elements—a horizontal element from A to B and a horizontal element from B to C. In SAP2000, the shear release is applied to the end of one of the members framing into the joint at B. The model given in C5EX11 has the shear release applied to the right end of the left frame element. Run the analysis using C5EX11 as provided and then move the shear release to the left end of the right frame element. Be sure to remove the shear release from the left frame element. Rerun the results and verify that you get the same external reactions.

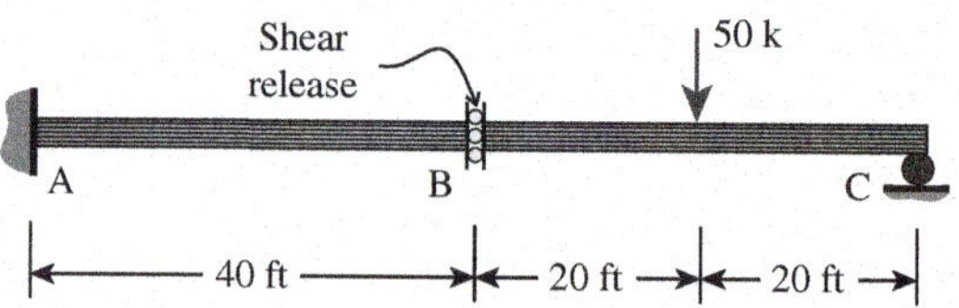

5.13 Examples with Video Solutions

VE5.1 For the following four structures, determine the statical classification.

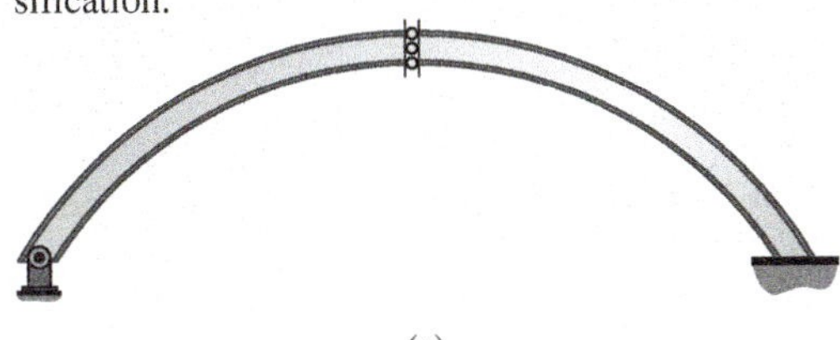

(a)

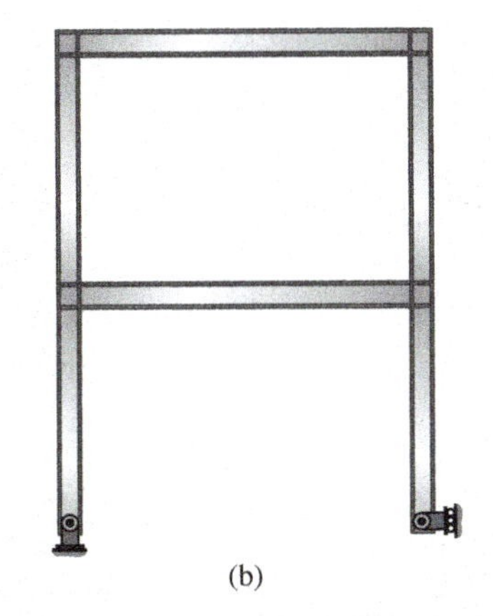

(b)

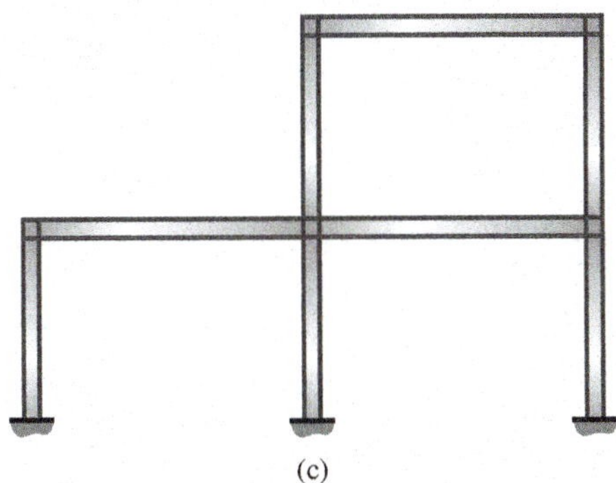

(c)

Figure VE5.1

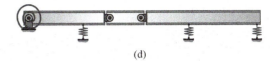

(d)

Figure VE5.1(continued)

VE5.2 For the frame in Figure VE5.2, calculate the reactions resulting from the applied loads.

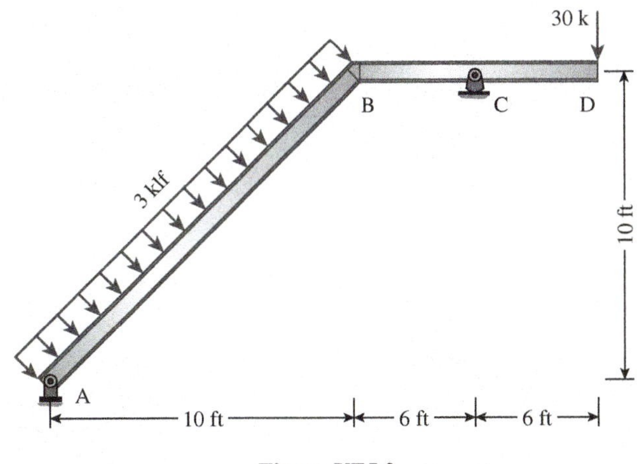

Figure VE5.2

VE5.3 For the beam shown in Figure VE5.3, calculate the reactions resulting from the applied loads.

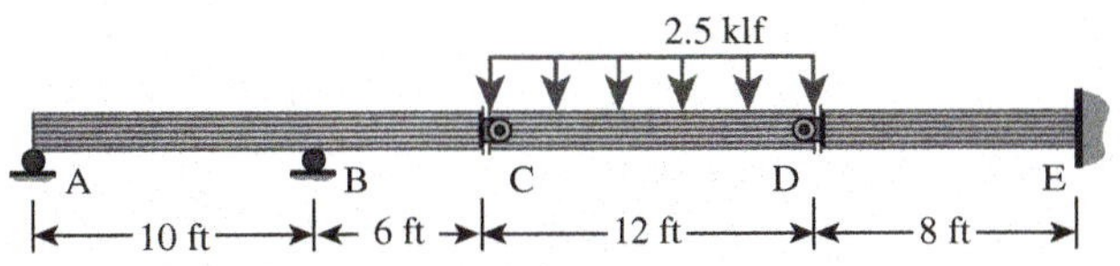

Figure VE5.3

5.14 Problems for Solution

Section 5.6–5.7

P5.1 Determine which of the structures shown in the accompanying illustration are statically determinate, statically indeterminate (including the degree of indeterminacy), and unstable in regards to outer forces. [Ans: (a) SI 2°, (b) unstable, and (e) determinate]

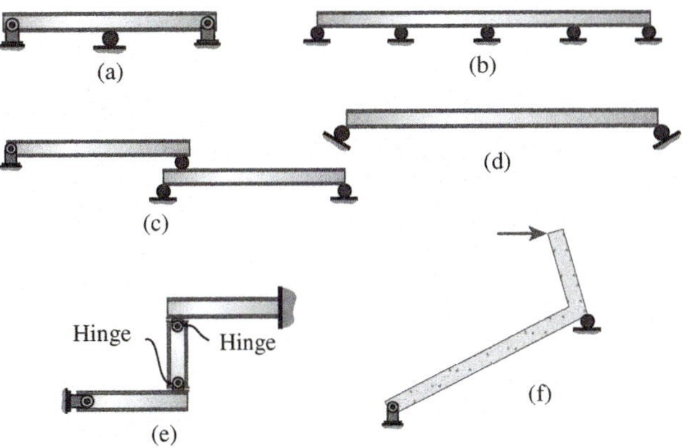

Problem 5.1

P5.2 Determine which of the structures shown in the accompanying illustration are statically determinate, statically indeterminate (including the degree of indeterminacy), and unstable in regards to outer forces.

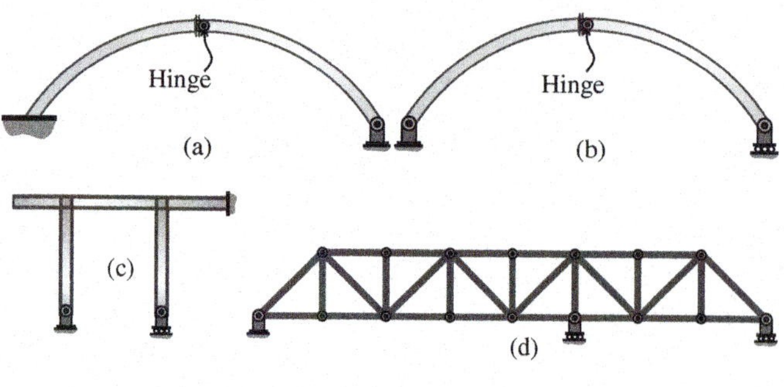

Problem 5.2

P5.3 Determine which of the structures shown in the accompanying illustration are statically determinate, statically indeterminate (including the degree of indeterminacy), and unstable in regards to outer forces. [Ans: (a) SI 1°, (b) SI 2°, and (d) unstable]

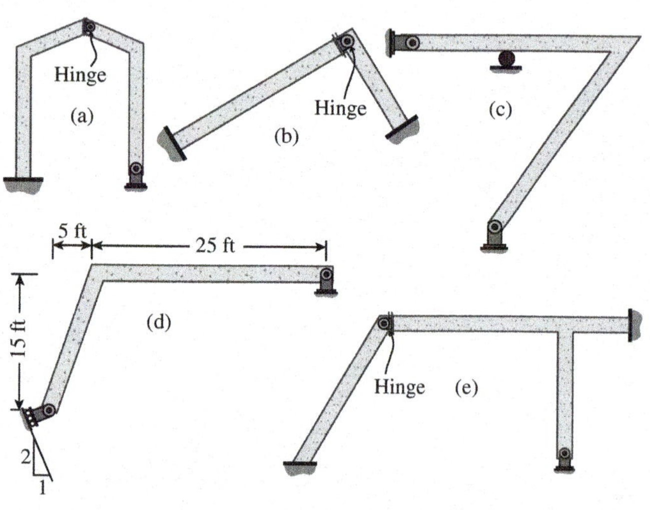

Problem 5.3

P5.4 Determine which of the structures shown in the accompanying illustration are statically determinate, statically indeterminate (including the degree of indeterminacy), and unstable in regards to outer forces.

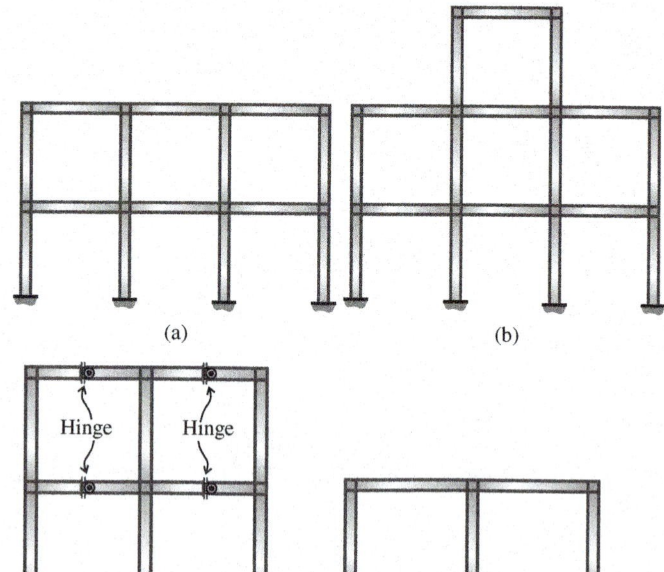

Problem 5.4

P5.5 Determine which of the structures shown in the accompanying illustration are statically determinate, statically indeterminate (including the degree of indeterminacy), and unstable in regards to outer forces. [Ans: (a) SI 5°, (b) unstable, and (d) SI 4°]

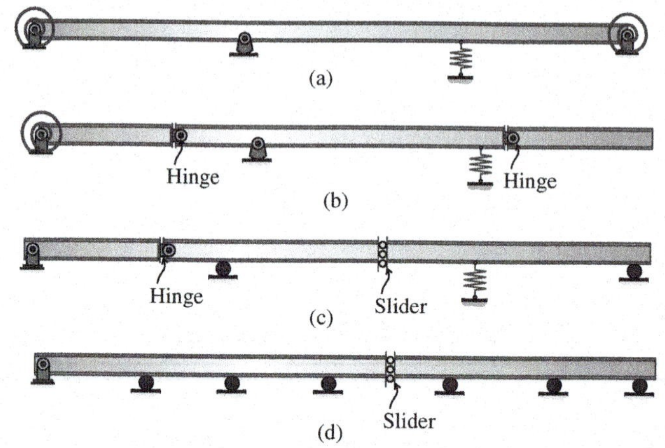

Problem 5.5

Section 5.9

For Problems 5.6 to 5.35, compute the reactions for the structures.

P5.6

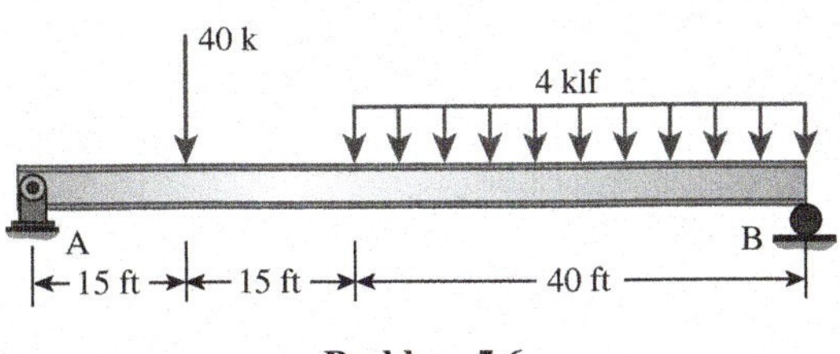

Problem 5.6

P5.7 (Ans: $R_{Ay} = 62.9$ k, $R_{By} = 57.1$ k)

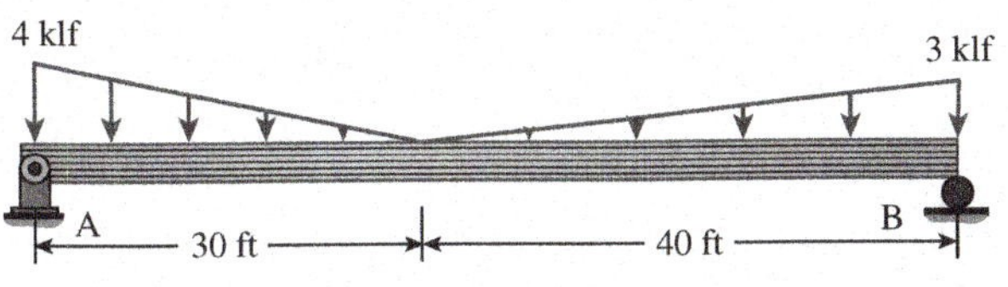

Problem 5.7

P5.8

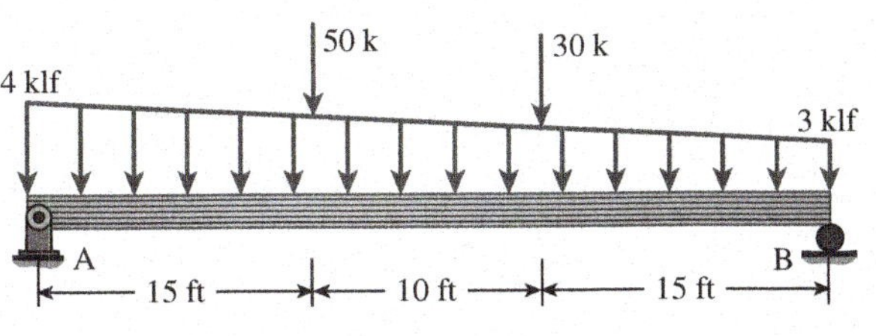

Problem 5.8

P5.9 (Ans: $R_{Ay} = 115$ k, $R_{By} = 95$ k)

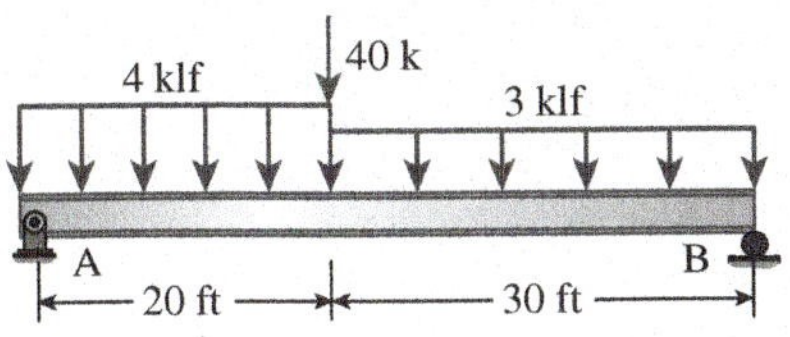

Problem 5.9

P5.10

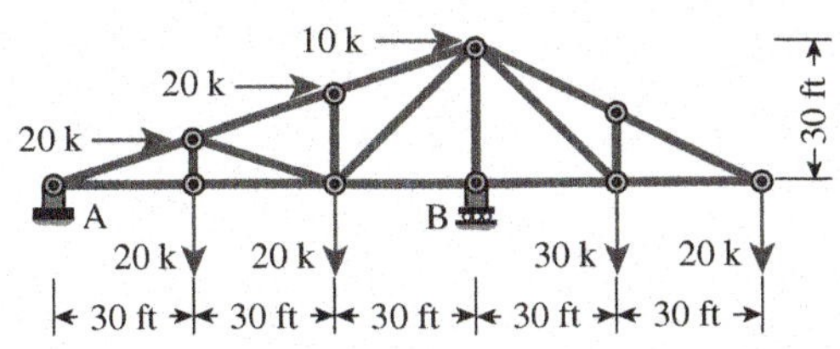

Problem 5.10

P5.11 (Ans: $R_{Ax} = 235$ k, $R_{Bx} = 235$ k, $R_{By} = 170$ k)

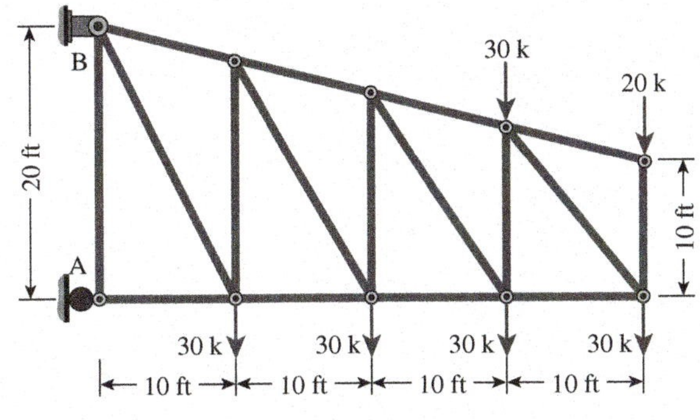

Problem 5.11

P5.12

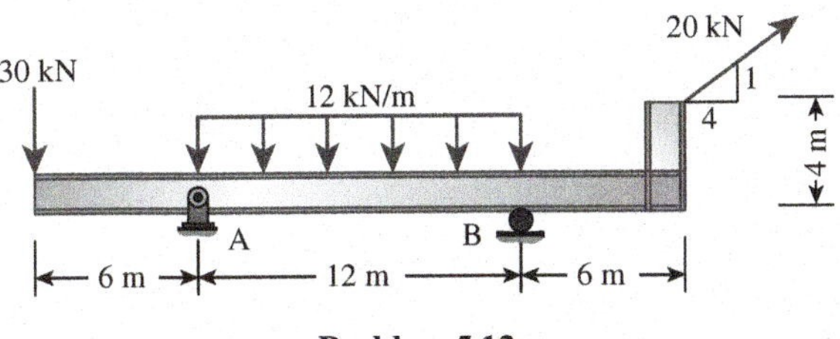

Problem 5.12

P5.13 (Ans: $R_{Cy} = 72.1$ k, $R_{By} = 77.9$ k)

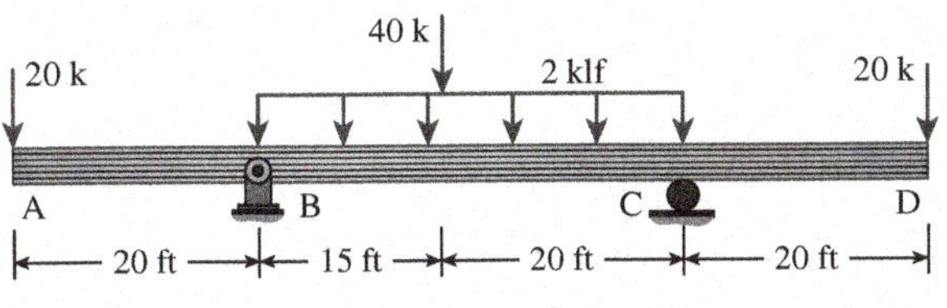

Problem 5.13

P5.14

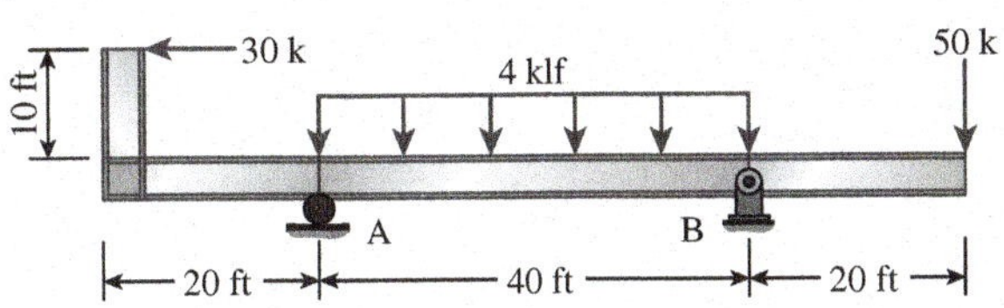

Problem 5.14

P5.15 (Ans: $R_{Ay} = 135.6$ k, $R_{By} = 7.6$ k, $R_{Bx} = 77.7$ k)

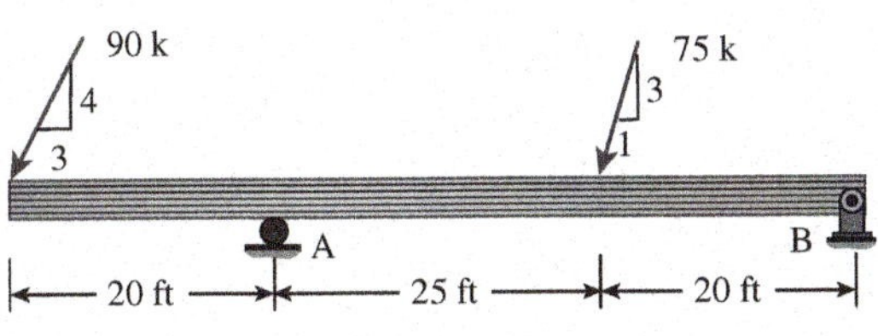

Problem 5.15

P5.16

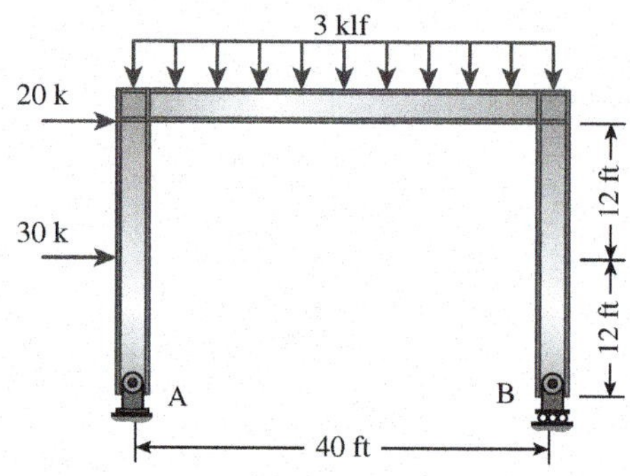

Problem 5.16

P5.17 (Ans: $R_{Ax} = 90$ k, $R_{By} = 82.5$ k, $R_{Ay} = 137.5$ k)

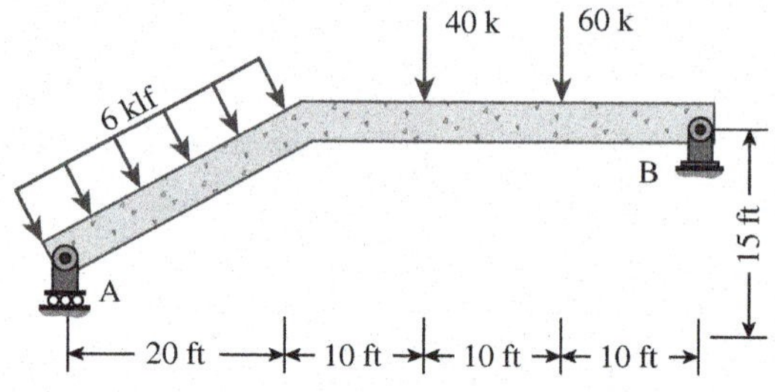

Problem 5.17

P5.18

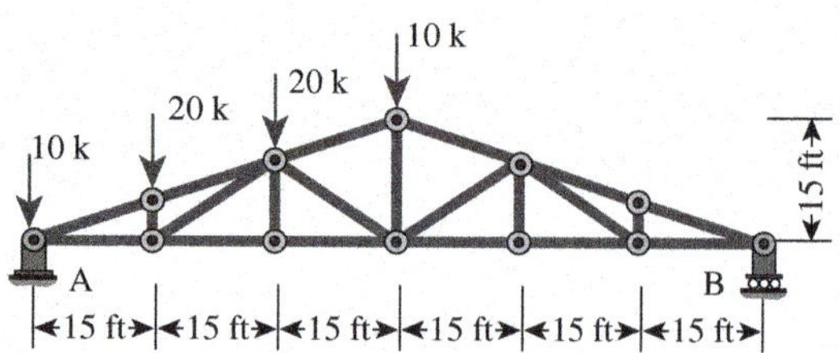

Problem 5.18

P5.19 (Ans: $R_{Ay} = R_{By} = 5120$ lb)

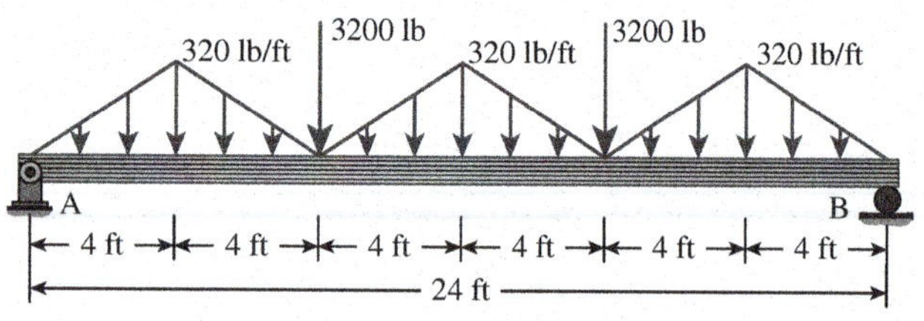

Problem 5.19

P5.20

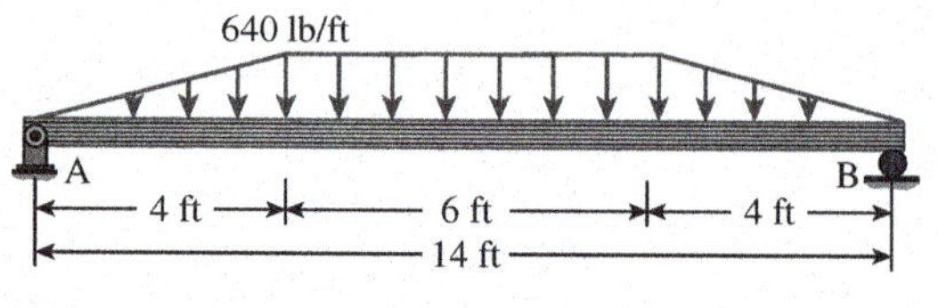

Problem 5.20

P5.21 (Ans: $R_{Ay} = 60$ k, $R_{By} = 50$ k, $R_{Bx} = 30$ k)

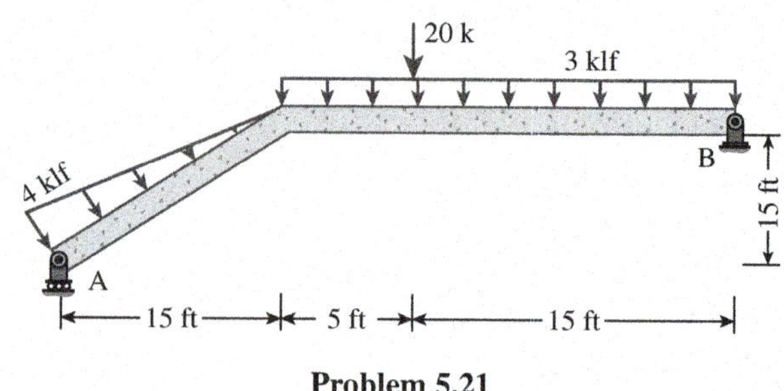

Problem 5.21

P5.22

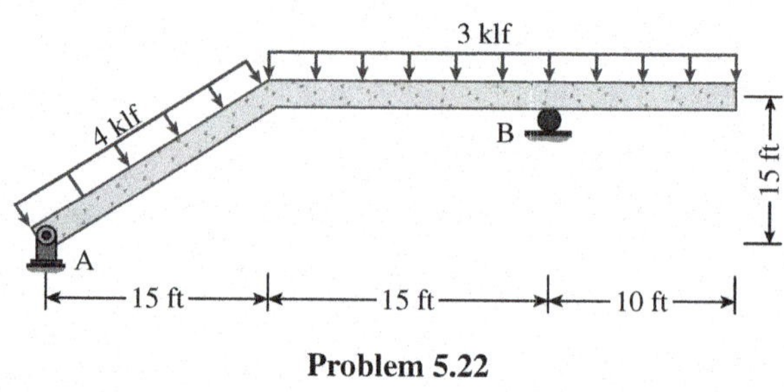

Problem 5.22

P5.23 (Ans: $R_{Ay} = 65.4$ k, $R_{Ax} = 18.4$ k, $R_B = 30.7$ k)

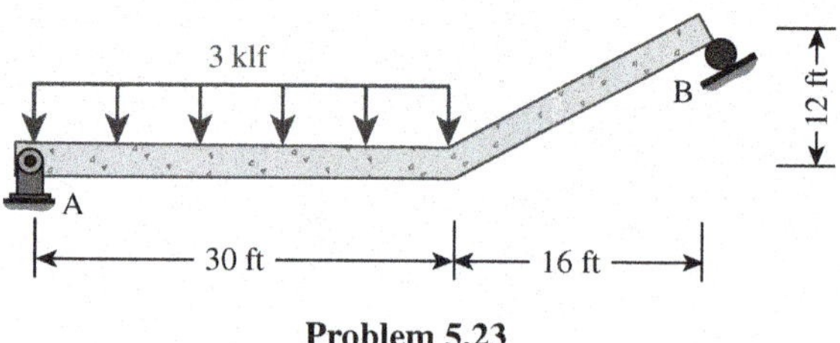

Problem 5.23

P5.24

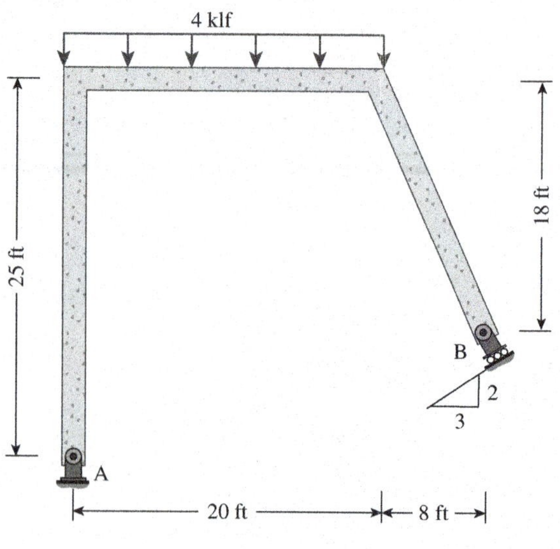

Problem 5.24

P5.25 (Ans: $R_{Ay} = 66.5$ k, $R_{Ax} = 33.3$ k, $R_B = 59.7$ k)

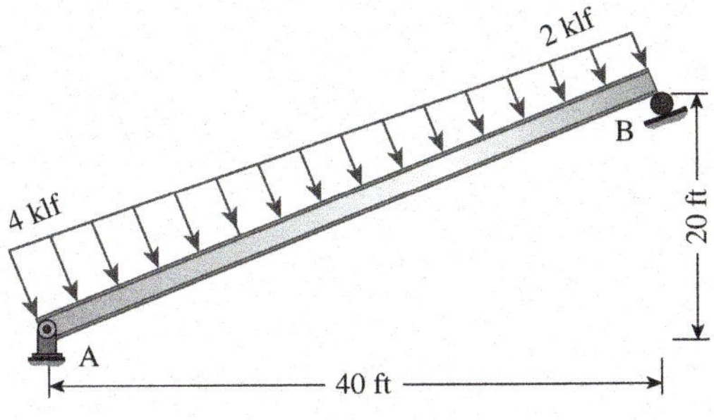

Problem 5.25

P5.26

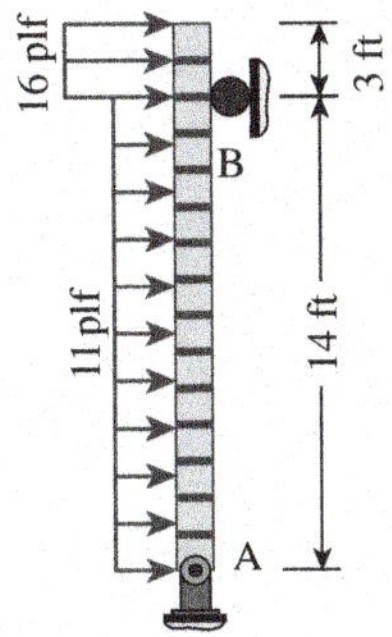

Problem 5.26

P5.27 (Ans: $R_{Ay} = 4808.5$ lb, $R_{By} = 5313.7$ lb)

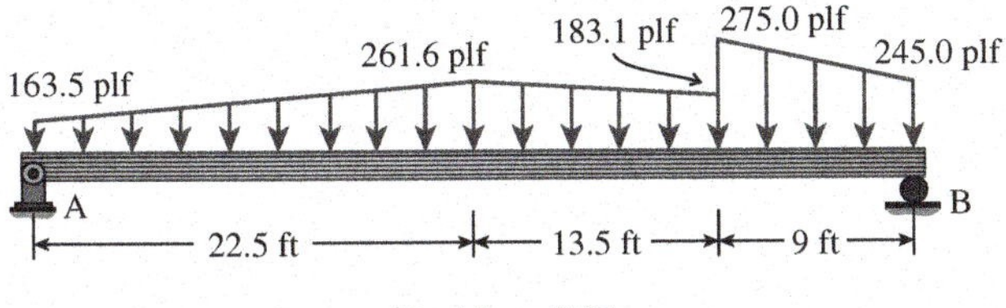

Problem 5.27

P5.28

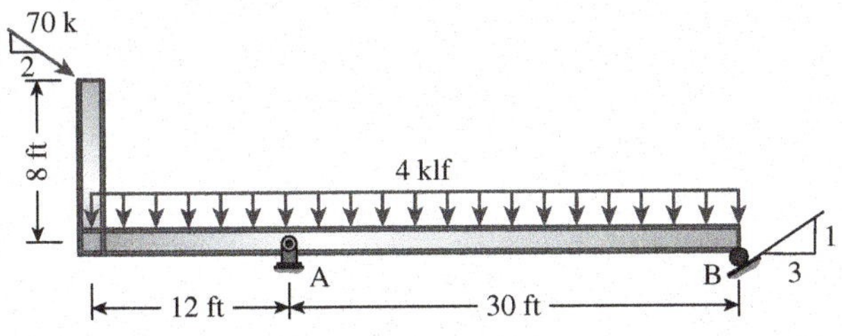

Problem 5.28

P5.29 (Ans: $R_{Ay} = 106.4$ k, $R_{By} = 45.6$ k)

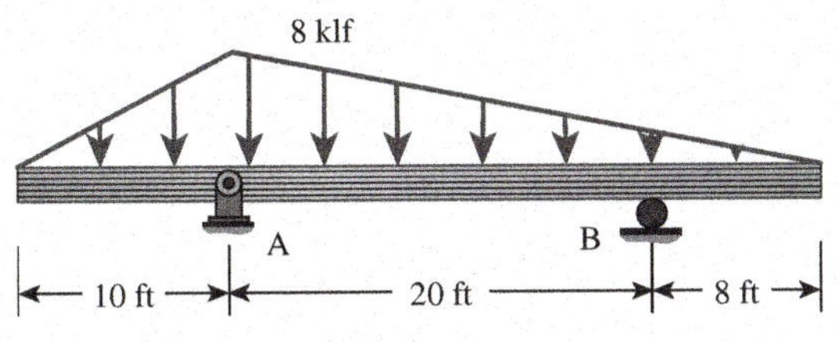

Problem 5.29

P5.30

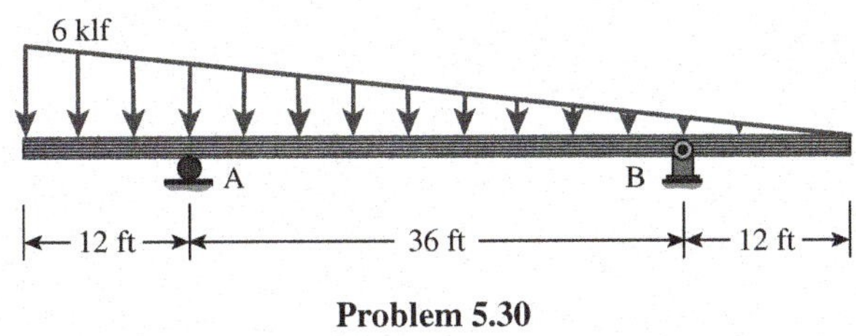

Problem 5.30

P5.31 (Ans: $R_{Ay} = 147.2$ k, $R_{By} = 192.8$ k)

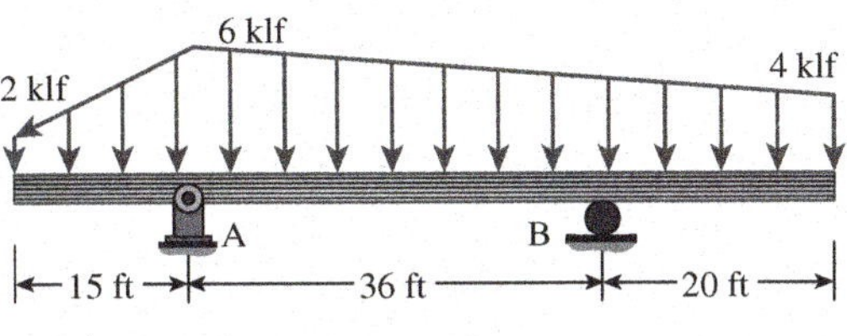

Problem 5.31

P5.32

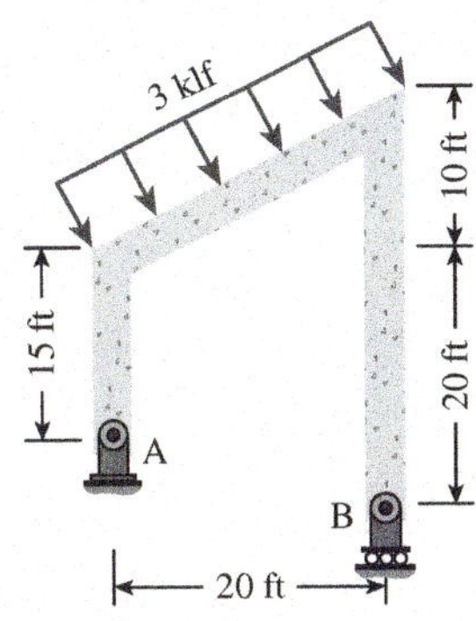

Problem 5.32

P5.33 (Ans: $R_{Ay} = 130$ k, $R_{Ax} = 22.5$ k, $R_{Bx} = 22.5$ k)

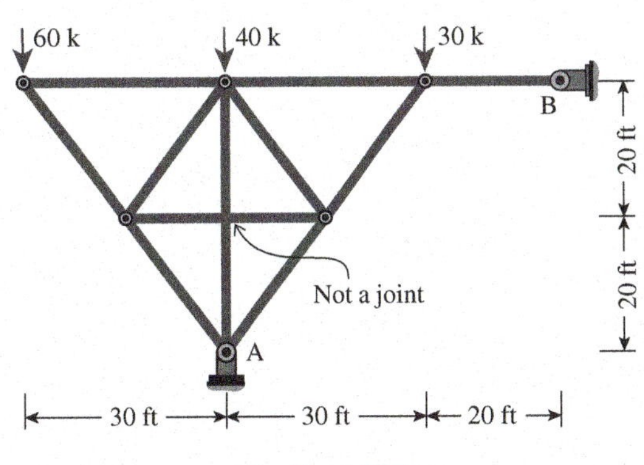

Problem 5.33

P5.34

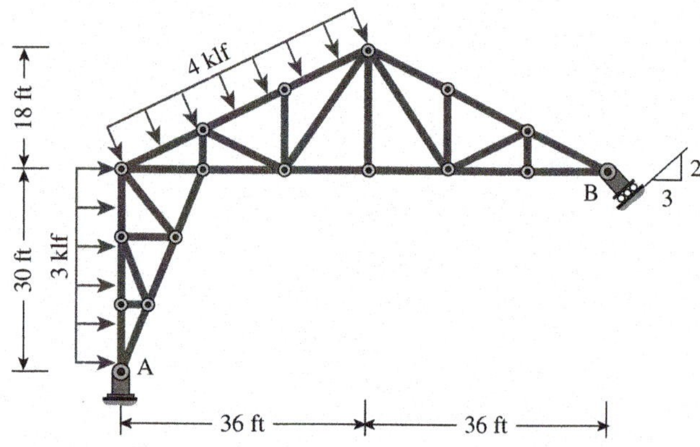

Problem 5.34

P5.35 (Ans: R_{Ay} = 13.1 k, R_{Ax} = 18.8 k, R_B = 28.6 k)

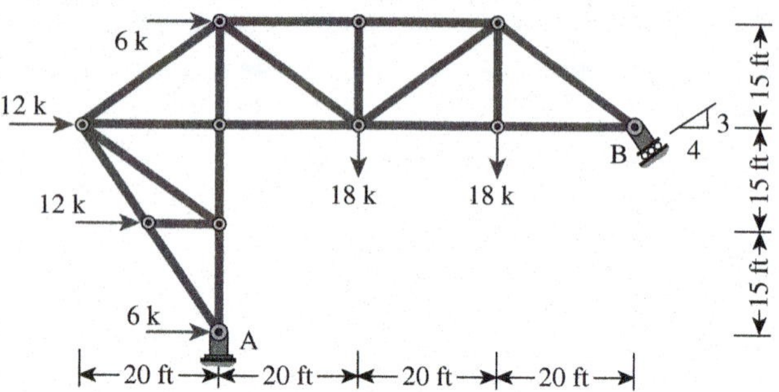

Problem 5.35

Sections 5.10–5.11

For Problems 5.36 to 5.53, compute the reactions for the multi-connected structures.

P5.36

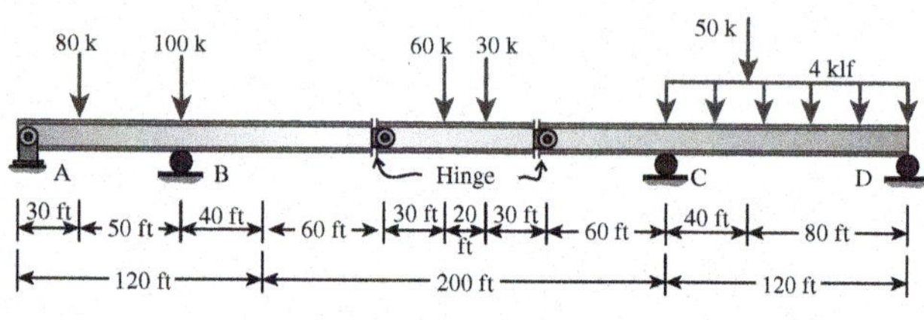

Problem 5.36

P5.37 (Ans: R_{Ay} = 186.0 k, R_{By} = 714.4 k, R_{Cy} = 20.4 k)

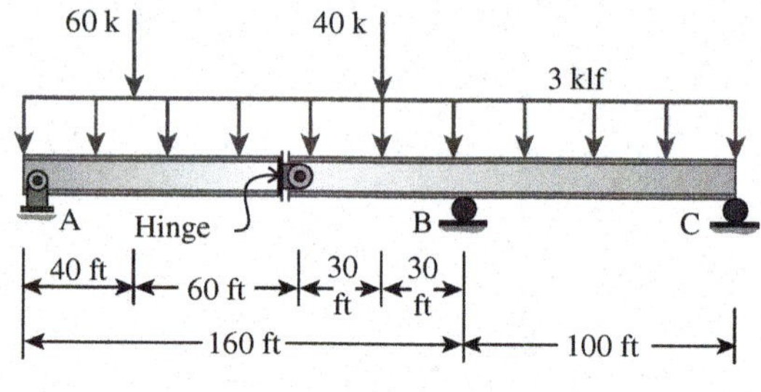

Problem 5.37

P5.38

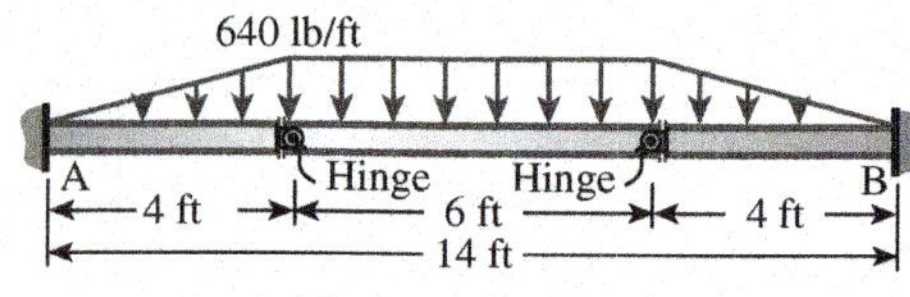

Problem 5.38

P5.39 (Ans: R_{Ay} = 7 kN, R_{Cy} = 4 kN, M_A = 18 kN·m)

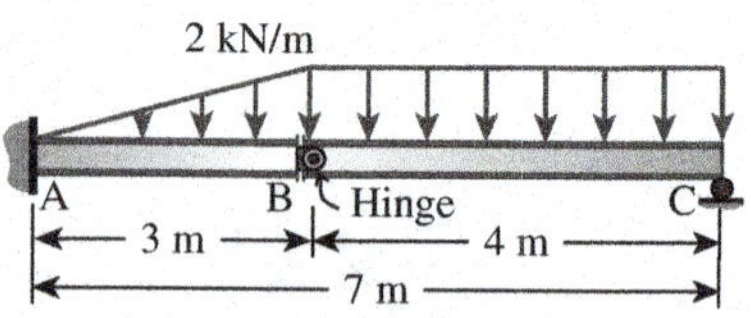

Problem 5.39

P5.40

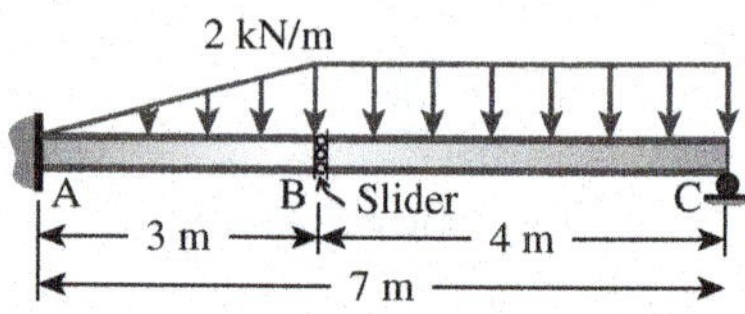

Problem 5.40

P5.41 (Ans: R_{Ay} = 8 k, R_{By} = 24 k, M_A = 128 k·ft)

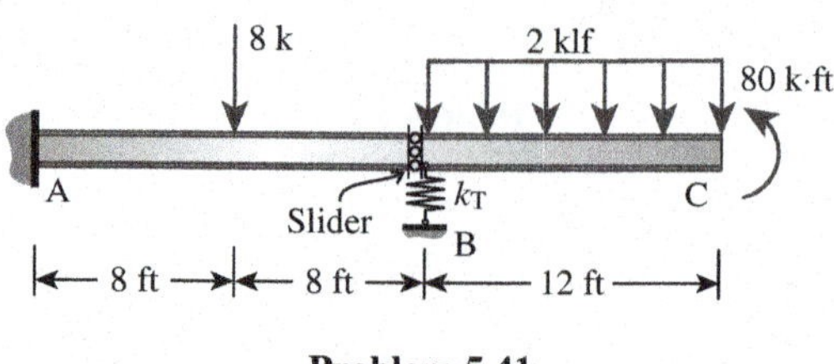

Problem 5.41

P5.42

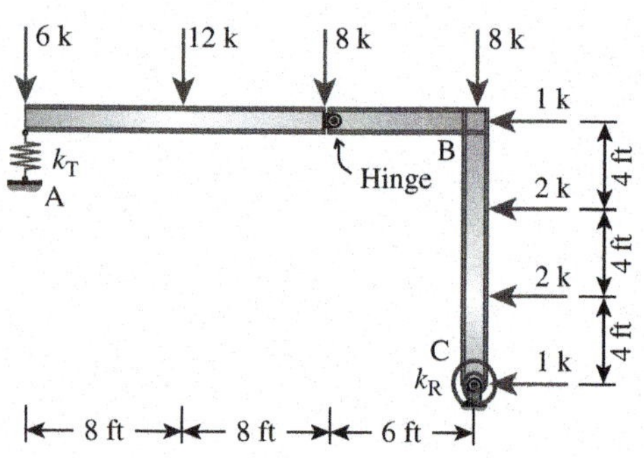

Problem 5.42

P5.43 (Ans: $R_{Cy} = 270$ k, $R_{Dy} = 80$ k)

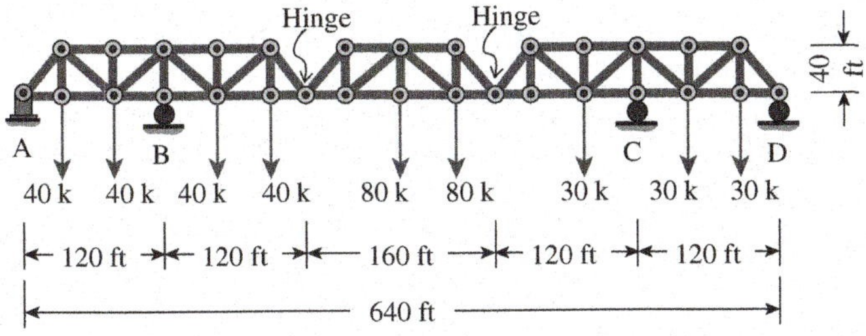

Problem 5.43

P5.44

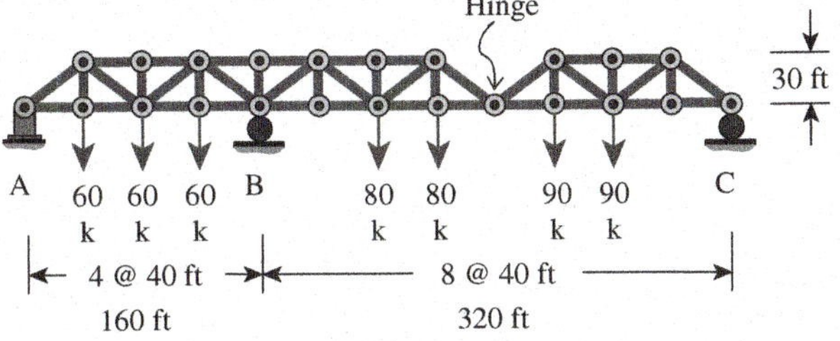

Problem 5.44

P5.45 (Ans: $R_{Ay} = 85$ k, $R_{Ax} = 63.3$ k)

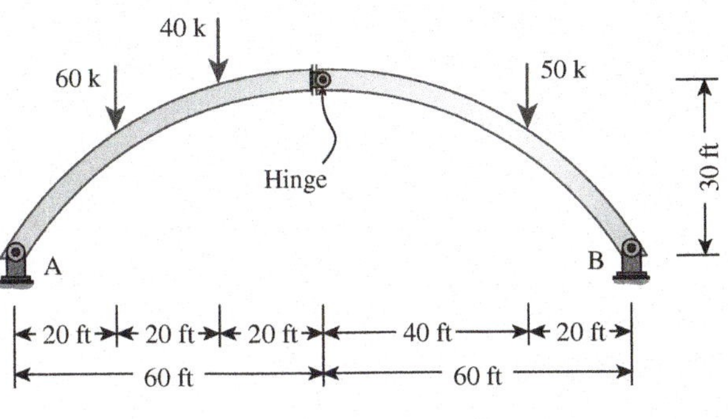

Problem 5.45

P5.46

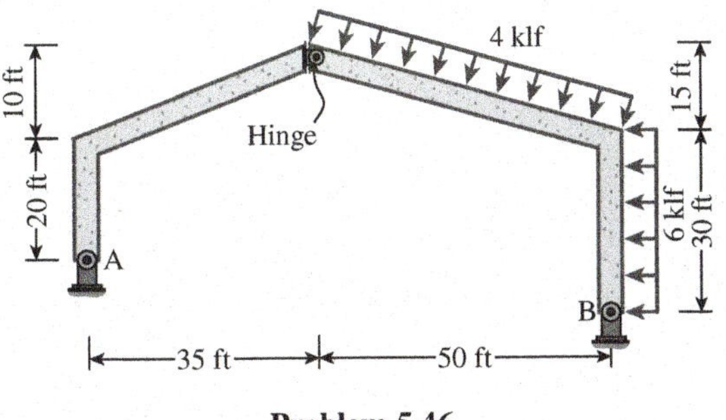

Problem 5.46

P5.47 Solve for all reactions using the matrix formulation. (Ans: $R_{Ay} = 55.5$ kN, $R_{Ax} = 67.5$ kN)

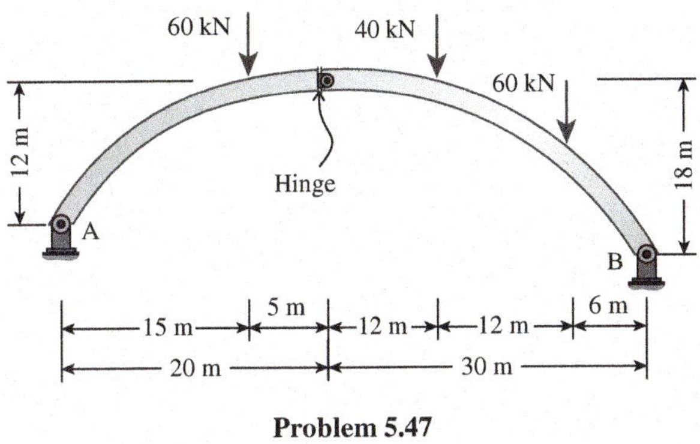

Problem 5.47

P5.48 Solve for all reactions using the matrix formulation.

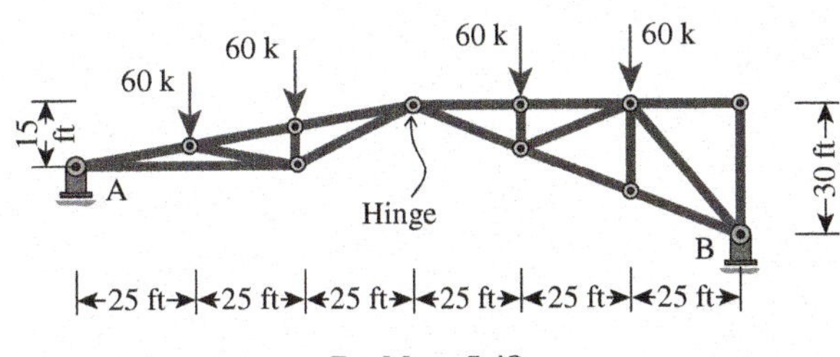

Problem 5.48

P5.49 Solve for all reactions using the matrix formulation. (Ans: $R_{Ay} = 0.0$ k, $R_{Ax} = 125.0$ k)

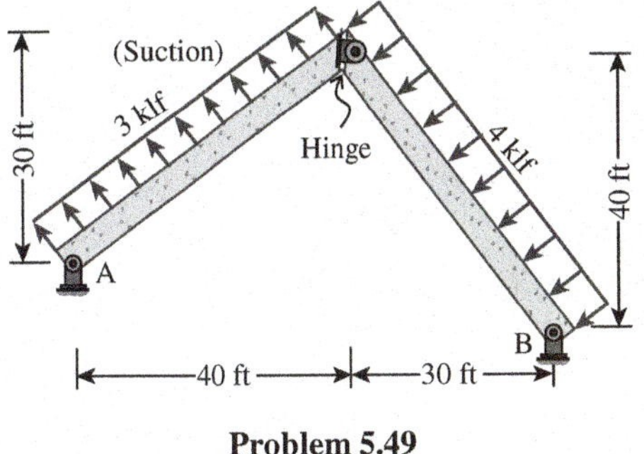

Problem 5.49

P5.50 Repeat Problem 5.49 if the 3 klf load is increased to 4 klf and the 4 klf load is increased to 6 klf.

P5.51 Repeat Problem 5.49 if the 4 klf load is removed. (Ans: $R_{Ay} = 60$ k, $R_{Ax} = 45$ k)

P5.52

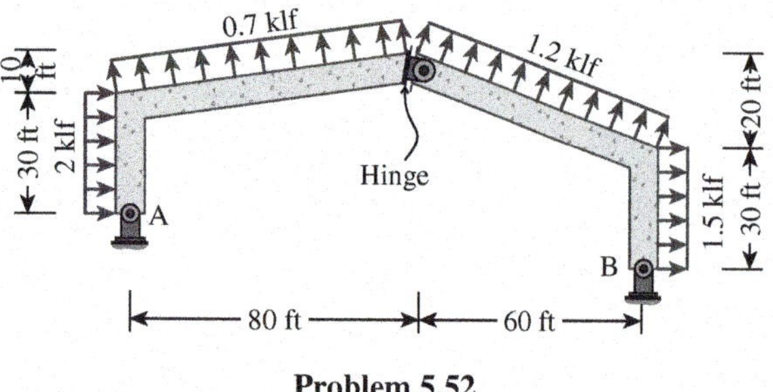

Problem 5.52

P5.53 (Ans: $R_{Ay} = 98.6$ k, $R_{Ax} = 115.7$ k)

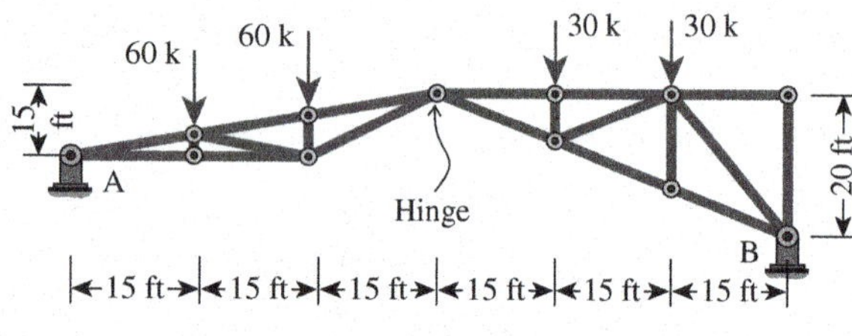

Problem 5.53

Section 5.12 SAP2000

P5.54 Using the structure provided in Problem 5.28, conduct a parametric study using SAP2000 to investigate the influence of the orientation of the roller support on the external reactions. Assume the roller support is along a rise/run ratio of (a) 0, (b) 0.125, (c) 0.25, (d) 0.5, (e) 1.0, (f) 2, (g) 4, (h) 8, and (i) a vertical surface. Plot each of the reactions against roller support orientation. Make an observation on the trends.

P5.55 Analyze the structure given in Problem 5.38 by assuming that there is a moment release at each end of the interior supported span. Reanalyze the structure but now assume the moment releases are located in the end spans. Verify equilibrium of the structure by summing forces and moments using the results from the SAP2000 analysis. Compare the reactions from the two modeling conditions.

P5.56 Analyze the structure given in Problem 5.38 by assuming that there is a moment release at each end of the interior supported span. Reanalyze the structure but now assume that the left moment release is shear release. Verify equilibrium of the structure by summing forces and moments using the results from the SAP2000 analysis. Compare the reactions from the two modeling conditions.

Axial Force, Shear Force, and Bending Moment

6

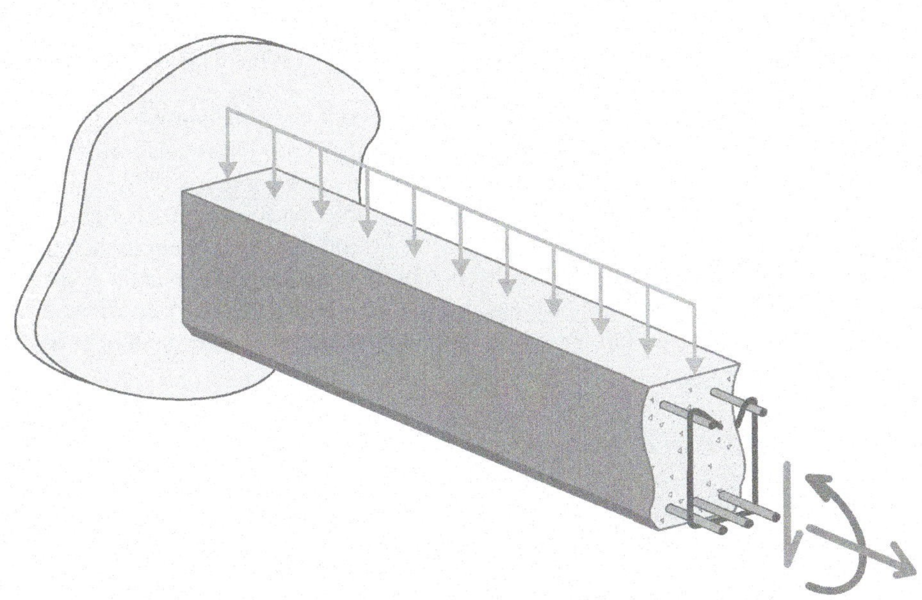

6.1 Introduction

An important part of structural engineering—and indeed, the understanding of structural behavior—is the understanding of the axial forces, shearing forces, and bending moments that exist within a structural system. Equations for these forces are sometimes needed to compute structural deflections. Very often the shearing force and bending moment are represented on diagrams to provide a visualization of structural response. Although less common—but no less useful—is a diagram that represents the internal axial force in a structural member. All of these diagrams, from which the values of axial force, shearing force, and bending moment at any point in a beam are immediately available, are very convenient in design since they visually provide the magnitude and location of maximum design forces. In this chapter, we examine methods for developing the equations for shearing force and bending moment in structural systems, as well as methods for constructing axial force, shearing force, and bending moment diagrams. It is doubtful that there is any other topic for which careful study will give more reward in structural engineering knowledge.

NOTE OF CAUTION

Aside from basic statics errors, the proper generation of axial, shear, and moment diagrams appears to be persistently troublesome to the structural engineering student. Devoting adequate attention to learning the process will pay great dividends in later topics and classes.

6.2 Member Internal Forces

To examine the internal forces at any location in a structure, a cut must be made at the point of interest. A complete free-body diagram is generated and studied to see what internal forces (axial, shear, and moment for a planar structure) must be present for the body to remain in a state of equilibrium.

Axial force is defined as the algebraic summation of the forces to the left or to the right of a section that are parallel to the axis of the member. *Shear force* is defined as the algebraic

QUICK NOTE

This text focuses on planar structures where all of the loads are acting in the same plane. Thus, the internal forces of interest are axial, shear, and moment. Torsion also develops under certain conditions, and a discussion of such can be found in any mechanics of materials text.

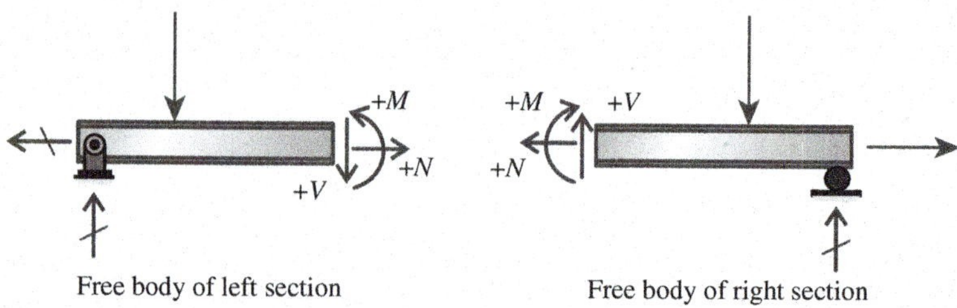

FIGURE 6.1 Commonly used sign convention for internal shear, moment, and axial forces.

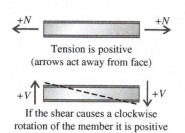

summation of the forces to the left or to the right of a section that are perpendicular to the axis of the member. *Bending moment* is the algebraic summation of the moments generated by all the forces to the left or to the right of a particular section—the moments being taken about an axis through the centroid of the section.

The selection of a sign convention for shear (V), moment (M), and normal or axial force (N) is essential to avoid calculation errors. It is actually a matter of personal choice, but it is convenient to be consistent with others. The convention shown in Figure 6.1 is commonly used in many texts and will be used throughout this one when looking at internal forces. In this figure, the authors have passed a section through a beam and shown free bodies of that beam to the left and to the right of the section cut. For each of these free bodies, internal values for plus shear, plus moment, and plus axial or normal force are given using this convention. It is often convenient to remember this sign convention by following a few simple guidelines. These guidelines are shown in Figure 6.2.

+N +N

Tension is positive (arrows act away from face)

+V +V

If the shear causes a clockwise rotation of the member it is positive

If the moment causes the member to "smile" then it is positive

FIGURE 6.2

Finding Internal Forces—Procedure

1. Solve for reactions.
2. Make a single cut through the structure at the desired location.
3. Draw a COMPLETE free-body diagram for either the left or right section.
   (a) Assume all unknown internal forces to be positive according to the sign convention given in Figure 6.1.
   (b) When dealing with distributed loads, make sure that cuts are made prior to the calculation of the resultant of the distributed load.
4. Use equilibrium to solve for the unknown internal forces. Answers for internal forces should not include arrows—unlike as done for reporting reactions. Rather, the sign associated with the internal forces is appropriate and sufficient. The use of arrows is inappropriate because their direction depends on whether you are looking at the left or the right section.

EXAMPLE 6.1

Find the internal shear and moment at sections a–a and b–b for the beam shown. Reactions were calculated in Example 5.6.

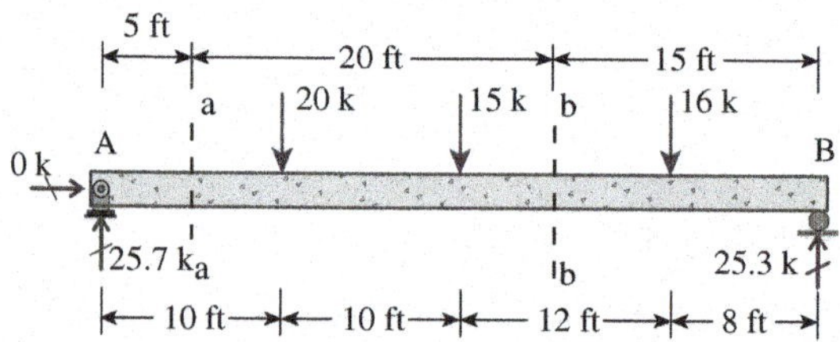

DID YOU KNOW?

When cutting a structure, you may use either section to solve for your unknown forces. Either one will give you the correct results. Choosing the section with the fewest number of forces will minimize the length of the equilibrium equations and thus minimize potential for mistakes.

SOLUTION STRATEGY

Whenever the internal force is desired at a particular location, a cut must be made. Cut the beam at a–a and draw the complete FBD. Choose either the left or right section and use equilibrium to solve for the unknowns (V, M).

SOLUTION

Cut the beam at section *a–a* and use the left section because it will have fewer forces on it than the right section. Assume all unknown internal forces to be positive.

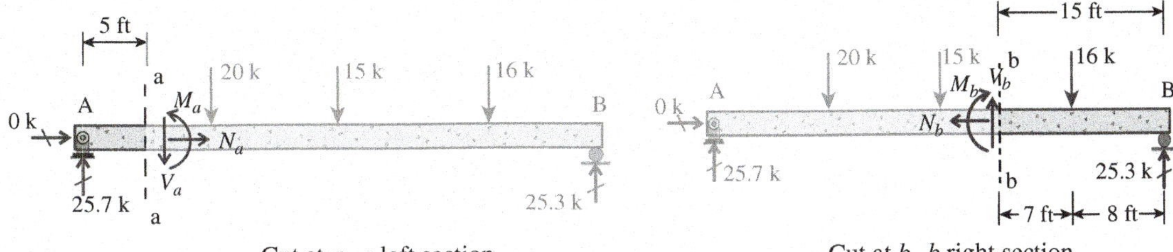

Cut at *a–a* left section

Cut at *b–b* right section

Sum forces in the *y*-direction to solve for V_a:

$$+\uparrow \sum F_y = 25.7\text{ k} - V_a = 0$$

$$\therefore V_a = \underline{\underline{25.7\text{ kips}}}$$

Sum moments about the cut *a–a* to solve for M_a:

$$+\circlearrowleft \sum M_{a-a} = M_a - (25.7\text{ k})(5\text{ ft}) = 0$$

$$\therefore M_a = \underline{\underline{128.5\text{ k·ft}}}$$

To find the internal forces at *b–b*, we must make a cut at *b–b*. We use the right section because it contains the fewest forces.

$$+\uparrow \sum F_y = V_b - 16\text{ k} + 25.3\text{ k} = 0$$

$$V_b = \underline{\underline{-9.3\text{ kips}}}$$

$$+\circlearrowleft \sum M_{b-b} = -M_b - (16\text{ k})(7\text{ ft}) + (25.3\text{ k})(15\text{ ft}) = 0$$

$$\therefore M_a = \underline{\underline{267.5\text{ k·ft}}}$$

COMMON MISTAKE

Students are very tempted to include forces in the equation of equilibrium that don't appear on the FBD of the section. If a force doesn't appear on the FBD, then don't include it in your equations.

EXAMPLE 6.2

Find the internal forces, *V*, *M*, and *N* at section *a–a* for the structure shown. Reactions were calculated in Example 5.7.

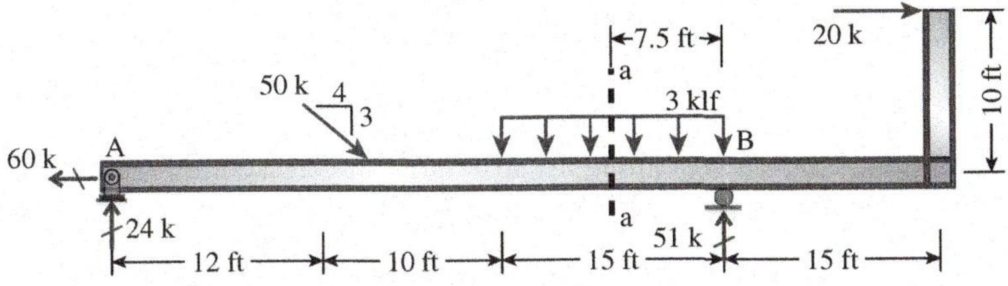

SOLUTION STRATEGY

Cut the structure at *a–a* and draw the complete FBD for one of the sections. Choose either the left or right section and use equilibrium to solve for the unknowns (*V*, *M*, *N*).

COMMON MISTAKE

Students sometimes calculate the resultant of the distributed load to first solve for the reactions. This is correct as seen in Example 5.7, however, they often make the mistake of treating the distributed load as a point load for every other step of the analysis. Let the resultant revert back to the distributed load before making the cut. Then a new resultant can be calculated based on the cut.

SOLUTION

Cut the beam at section *a–a* and use the right section. There is no real advantage to using the right over the left except that we won't need to resolve the 50-k load into its components. Be sure to locate the resultant of the distributed load. The magnitude of this resultant should only account for the distributed load located on the right section.

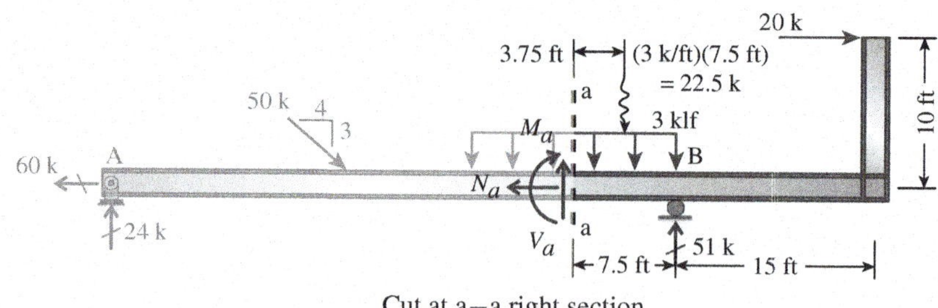

Cut at a—a right section

Sum forces in the *y*-direction to solve for V_a:

$$+\uparrow \sum F_y = V_a - 22.5\text{ k} + 51\text{ k} = 0$$

$$\therefore V_a = \underline{\underline{-28.5\text{ kips}}}$$

Sum moments about the cut *a–a* to solve for M_a:

$$+\circlearrowleft \sum M_{a-a} = -M_a - (22.5\text{ k})(3.75\text{ ft}) + (51\text{ k})(7.5\text{ ft}) - (20\text{ k})(10\text{ ft}) = 0$$

$$\therefore M_a = \underline{\underline{98.1\text{ k·ft}}}$$

Sum forces in the *x*-direction to solve for N_a:

$$\overset{+}{\rightarrow} \sum F_x = -N_a + 20\text{ k} = 0 \Rightarrow \therefore N_a = \underline{\underline{20\text{ kips}}}$$

6.3 Axial, Shear, and Bending Moment Equations

An important part of structural engineering is the writing or derivation of equations for the axial forces, shears, and bending moments in different parts of a structure. The ability to write these equations is extremely important for the understanding of much of the information discussed later in this text. For instance, a procedure used for the determination of slopes and deflections of various points in structures requires us to be able to write these types of equations. In this section, axial, shear, and moment expressions are prepared for beams. Using the same procedure, the student learns how to prepare similar equations for frames.

The basic notion of developing equations follows the same process as finding the internal forces at a point in a structure: MAKE A CUT. The only difference is that, instead of cutting at a known location, we make the cut at a variable distance *x*. All other steps are the same.

QUICK NOTE

Load discontinuities can be found by screening the structure for

- Point loads
- Applied moments
- Structure supports
- Beam ends
- Abrupt changes in load intensity

Finding Equation for Internal Forces—Procedure

1. Solve for reactions.
2. Identify any load discontinuities along the structure. These discontinuities define the boundaries of segments in the structure. Each segment will need a unique set of equations.
3. Define the coordinate system for each structure segment. Common practice is to set the origin of the coordinate system to the far left of the structure. This is okay, but sometimes there is an advantage to selecting different origins.
4. Extend some arbitrary distance into the segment for which you want to find the equations. Make a single cut through the structure at this point.

5. Draw a COMPLETE free-body diagram for either the left or right section.
   (a) Assume all unknown internal forces to be positive according to the sign convention given in Figure 6.1.
   (b) When dealing with distributed loads, make sure that cuts are made prior to the calculation of the resultant of the distributed load.

6. Use equilibrium to solve for the unknown internal forces. These forces should result in equations that are a function of x. To be complete, the equation must contain the domain over which x is valid.

EXAMPLE 6.3

Find the equations for internal shear and moment as a function of x for the entire length of the beam. Reactions are already given, and x for all segments should be measured from the left end of the beam.

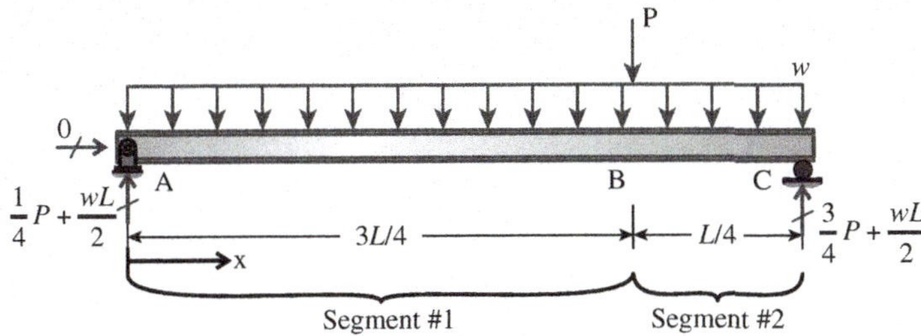

SOLUTION STRATEGY

Identify and clearly label all appropriate segments of the beam based on the presence of load discontinuities. Find the equations for each segment by measuring a distance of x as measured from the left end and make a cut. A complete FBD of either the left or right section is a must. Use $\sum F_y$ to get the shear equation and use $\sum M$ to get the moment equation.

QUICK NOTE

Each segment needs its own unique set of equations. Thus, the overall internal forces in the beam are described by piecewise equations.

SOLUTION

1. Reactions are given.
2. Load discontinuities exist at the beam ends and also at point B where the point load is applied. This results in two distinct segments of the beam, as labeled in the figure.
3. The coordinate for both segments is chosen to occur from the left end of the beam.
4. Cut segment #1 a distance x from the left end.
5. Draw the complete FBD of the left segment because it has fewer loads on it.

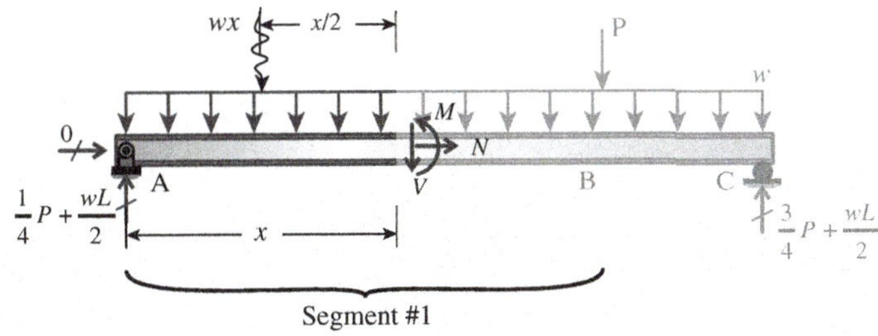

6. Use equilibrium. Sum forces in the y-direction to solve for V:

$$+\uparrow \sum F_y = \frac{1}{4}P + \frac{wL}{2} - wx - V = 0$$

$$\therefore V = \frac{1}{4}P + w\left(\frac{L}{2} - x\right) \qquad 0 \le x < \frac{3L}{4}$$

QUICK NOTE

It doesn't matter which section you use to write your equilibrium equations. Both the left and the right section result in the same equation.

The equations look different if you selected another origin for x (e.g., not at the left end of the beam).

Sum moments about the cut to solve for M.

$$+\circlearrowleft \sum M_{cut} = -\left(\frac{1}{4}P + \frac{wL}{2}\right)x + wx\left(\frac{x}{2}\right) + M = 0$$

$$\therefore M = \frac{1}{4}Px + \frac{w}{2}(Lx - x^2) \qquad 0 \leq x < \frac{3L}{4}$$

COMMON DIFFICULTY

Students seem to have difficulty placing correct dimensions in terms of x after the cut is made. For example, the dimension between point B and the cut should have a length of $\left(x - \frac{3L}{4}\right)$.

Students are encouraged to add up all of the lengths once they have identified them to ensure they give the original length.

$$\left(\frac{3L}{4}\right) + \left(x - \frac{3L}{4}\right) + (L - x) = L$$

OKAY

Repeat steps 4 through 6 for segment #2.

4. Cut segment #2 a distance x from the left end.

5. Draw the complete FBD of the right segment because it has fewer loads on it.

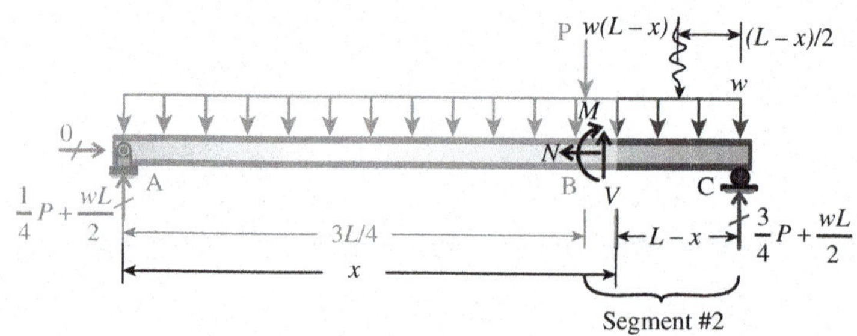

Cut at x right section

6. Use equilibrium. Sum forces in the y-direction to solve for V:

$$+\uparrow \sum F_y = \frac{3}{4}P + \frac{wL}{2} - w(L - x) + V = 0$$

$$\therefore V = -\frac{3}{4}P + w\left(\frac{L}{2} - x\right) \qquad \frac{3L}{4} < x \leq L$$

EQUATION FOR AXIAL FORCE

Equations for axial force are composed in the same way they are for shear. The only difference is that you must sum the forces in the x-direction (i.e., along the axis of the member) instead of the y-direction (i.e., perpendicular to the member).

Sum moments about the cut to solve for M:

$$+\circlearrowleft \sum M_{cut} = -M - w(L - x)\left(\frac{L - x}{2}\right) + \left(\frac{3}{4}P + \frac{wL}{2}\right)(L - x) = 0$$

$$M = w\left(\frac{Lx}{2} - \frac{x^2}{2}\right) + \frac{3}{4}P(L - x) \qquad \frac{3L}{4} < x \leq L$$

Plotting the shear and moment equations often can be useful, as they give a quick visual reference for how the internal forces are distributed in the structure. Example 6.4 demonstrates what such a set of plots looks like.

EXAMPLE 6.4

For the beam given in Example 6.3, set the variables to be

$$w = 1.5 \text{ klf} \qquad P = 18 \text{ kips} \qquad L = 12 \text{ ft}$$

Plot the shear and the moment equations developed in the same example.

SOLUTION STRATEGY

Substitute in the given variables, set up an appropriate set of axes, and plot the equations. Note the location of the load discontinuity with a vertical dashed line.

SOLUTION

For $0 \le x < 9$ ft,

$$V = 13.5 - 1.5(x) \qquad \text{units of kips}$$
$$M = 13.5x - 0.75x^2 \qquad \text{units of k·ft}$$

For $9 < x \le 12$ ft,

$$V = -1.5(x) - 4.5 \qquad \text{units of kips}$$
$$M = -0.75x^2 - 4.5x + 162 \qquad \text{units of k·ft}$$

QUICK NOTE

Since the distributed load is in units of kips and feet, the units of x should be feet in order for the units to be consistent.

Moment (k·ft)

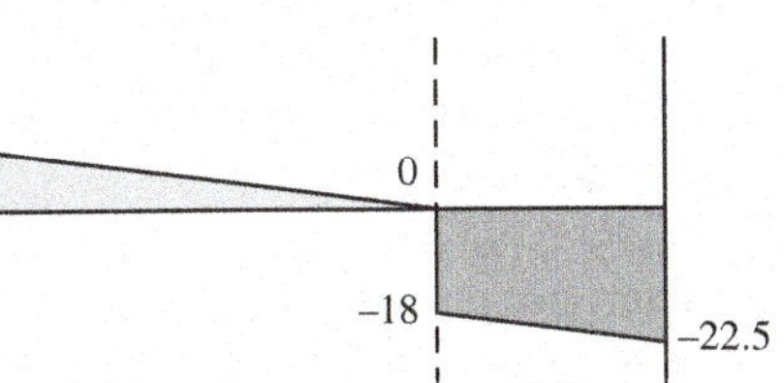

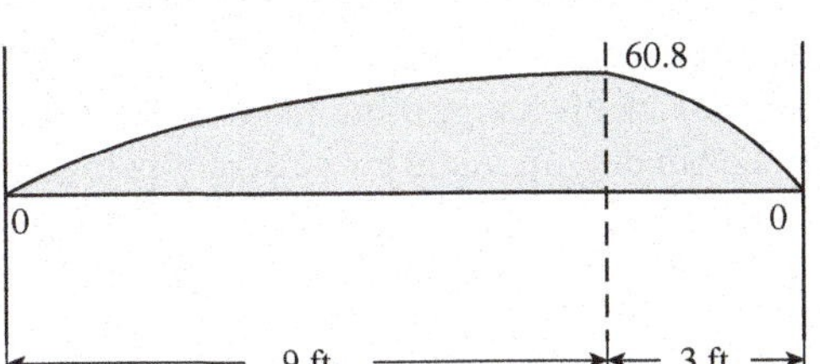

6.4 Relation Between Load, Shear, and Moment

There are significant mathematical relations between the loads (w), shears (V), and moments (M) in a beam. These relations are discussed in the following paragraphs with reference to Figure 6.3.

For this discussion, an element of a beam of length dx shown in Figure 6.3(b) is considered. This particular element is loaded with a uniform load of magnitude w (it doesn't have to be uniform). The shear and moment at the left end of this element at section 1–1 may be written as follows.

$$V_{1-1} = R_{Ay} - P - wa$$
$$M_{1-1} = R_{Ay}x - P(a+b) - \frac{wa^2}{2}$$

If we move a distance dx from section 1–1 to section 2–2 at the right end of the element, the new values of shear and moment can be written as

$$V_{2-2} = V_{1-1} + dV = R_{Ay} - P - wa - w(dx)$$
$$M_{2-2} = M_{1-1} + dM = R_{Ay}x - P(a+b) - \frac{wa^2}{2} + V_{1-1}(dx) - \frac{w(dx)^2}{2}$$

The changes in these values may be expressed as dV and dM, respectively. From these expressions and with reference to Figure 6.3(b), the changes in shear and moment in a dx distance are

$$\frac{dV}{dx} = -w$$
$$\frac{dM}{dx} = V \text{ (this is neglecting the infinitesimal higher order term } (dx)^2$$

RELATIONSHIP SUMMARY

Looking in terms of integrals, the following relationships exist.

load = w

shear = $V = -\int wdx$

moment = $M = \int Vdx = -\iint wdxdx$

Looking in terms of derivatives, the following relationships exist.

load = $w = -\frac{dV}{dx} = -\frac{d^2M}{dx^2}$

shear = $V = \frac{dM}{dx}$

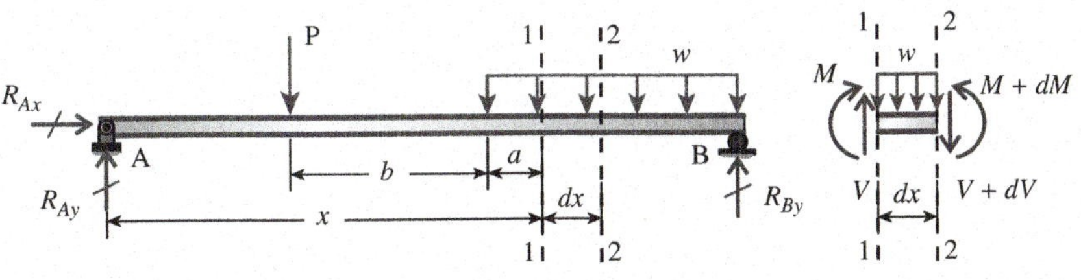

FIGURE 6.3

From the preceding expressions, the change in shear from one section to another can be written as

$$\Delta V = \int_{1-1}^{2-2} -w dx \tag{6.1}$$

and the change in moment in the same distance is

$$\Delta M = \int_{1-1}^{2-2} V dx \tag{6.2}$$

QUICK NOTE

For future reference it is helpful to remember the following.

- The derivative of an equation will provide the slope of that equation.
- The integration of an equation between two points gives the area under that equation.

These two relationships are very useful to the structural designer. Equation 6.1 indicates that the rate of change of shear at any point equals the load per unit length at the point, meaning that the slope of the shear diagram at any point is equal to the load at that point. Equation 6.2 indicates that the rate of change of moment at any point equals the shear. This relation means that the slope of the bending-moment curve at any point equals the shear.

The procedure for drawing shear and moment diagrams, to be described in Section 6.5, is based on these equations and is applicable to all structures regardless of loads or spans. Before the process is described, it is recommended to examine the equations more carefully. A particular value of dV/dx or dM/dx is good only for the portion of the structure at which the function is continuous.

EXAMPLE 6.5

QUICK NOTE

Note that the shear equation derived from the moment equation is identical to the shear equation developed through equilibrium.

The load equation found through double derivation is constant at 1.5 klf. This agrees with the definition of Example 6.4.

For the beam given in Example 6.3 and the V and M equations given in Example 6.4, show that the following conditions are true:

$$V = \frac{dM}{dx} \qquad w = -\frac{dV}{dx}$$

SOLUTION STRATEGY

From Example 6.4, we know that w = 1.5 k/ft. For each piecewise equation, take the first and second derivatives of the moment to show that they equal the equation for shear and load, respectively.

SOLUTION

For $0 \leq x < 9$ ft,

$$V = \frac{dM}{dx} = \frac{d}{dx}\left(13.5x - 0.75x^2\right) = 13.5 - 1.5x$$

$$w = -\frac{d^2M}{dx^2} = -\frac{d^2}{dx^2}\left(13.5x - 0.75x^2\right) = 1.5$$

DID YOU KNOW?

The value on the shear diagram is the same as the slope on the moment diagram at that point.

The value of the load is the same as the negative of the slope on the shear diagram.

For $9 < x \leq 12$ ft,

$$V = \frac{dM}{dx} = \frac{d}{dx}\left(-0.75x^2 - 4.5x + 162\right) = -1.5x - 4.5$$

$$w = -\frac{d^2M}{dx^2} = -\frac{d^2}{dx^2}\left(-0.75x^2 - 4.5x + 162\right) = 1.5$$

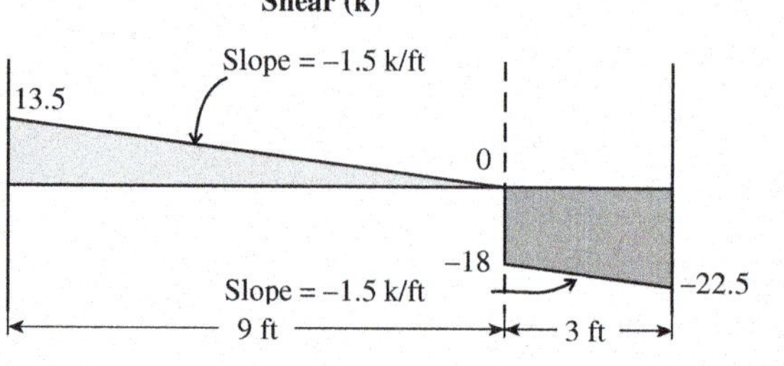

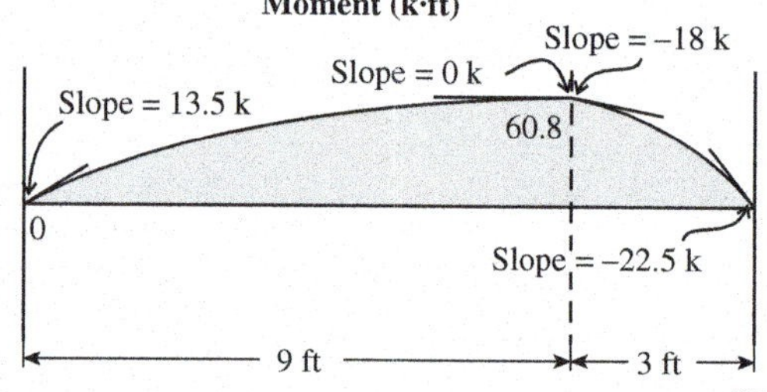

6.5 Shear and Bending Moment Diagrams for Beams

From a design perspective, structural engineers are often only interested in knowing what minimum and maximum internal forces exist in a given structure and their corresponding locations. We are also interested in gaining a general sense of how these forces vary across the structure. As such, it is very useful to have accessibility to shear and moment diagrams and not just the equations for these internal forces. A quick glance at these diagrams, as shown in Example 6.3, gives engineers all of the information they need to know for most design-oriented tasks.

Since these diagrams have a practical use in everyday practice, being able to efficiently and accurately create them is of significant value to the engineer. This section demonstrates how the generation of these diagrams may be achieved without needing to first develop the equations for shear and moment. Understanding the relationship between the load, shear, and moment is necessary to successfully achieving this task.

QUICK NOTE

The most critical step in creating accurate shear and moment diagrams is to make sure that complete and accurate FBDs are available for use. This includes the proper magnitude and sense of reactions.

Shear Diagrams

There are a couple of basic rules using previously learned concepts that aid in the generation of a shear diagram. These rules are described here and are graphically depicted in Figure 6.4.

1. **A point load on a structure will cause a jump in the shear diagram.** The magnitude of the jump in the diagram is equal to the magnitude of the point load. If the load points up, the jump is positive, and if it points down, the jump is negative.
2. **The change in shear between two points (ΔV) is equal to the negative area under the load diagram.** Section 6.4 shows that the change in shear between any two points can be found by integrating the load equation between those two points.
3. **The slope at any point on the shear diagram is equal to the negative intensity of the distributed load, w, at that point.** This means that if there is no distributed load, the shear will be constant. If the distributed load is constant, the shear will be a first-order line. If the distributed load is a first-order line, the shear will be a second-order line and so forth.

DID YOU KNOW?

A point load will cause a jump in a shear diagram, but a concentrated moment will NOT cause a jump in the shear diagram.

FRIENDLY ADVICE

Learn the rules now!! A little bit of effort at this point can save a significant amount of trouble as your study of structural analysis continues.

Shear Diagram—Procedure

1. Calculate reactions and draw a complete FBD of the structure element (e.g., beam).
2. Draw a set of x versus shear (V) axes and be sure to list the units that will be used. Anywhere there is a load discontinuity (see Section 6.3), draw a vertical dashed line on the axes.
3. Plot the value of shear at each end of the beam/segment. Remember to use the sign convention for shear. Only a point load or reaction at the end of a beam will cause the shear at the end of the beam to be nonzero. If only a distributed load is present at the end of the beam, the shear will be zero at the end.
4. Start from the left and work to the right. Follow the rules as outlined previously to identify the value of shear at each of the load discontinuities. If you have done this correctly, you will get the same values by the time you get to the end of the beam as you have already plotted. This serves as a check.

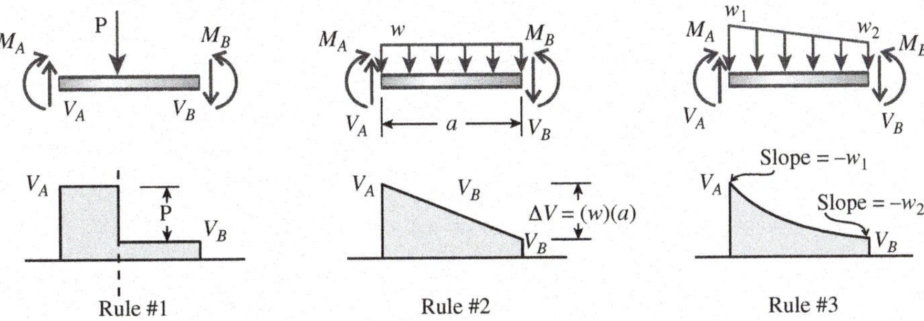

FIGURE 6.4 Graphical depiction of the rules for generating shear diagrams.

5. Connect the points on the diagram using Rule #3. Remember that a shear curve is one order higher than the load curve because of the integration relationship between the two. Locate and dimension all points of zero shear.

Moment Diagrams

As with the shear diagrams, there are a couple of basic rules using previously learned concepts that aid in the generation of a moment diagram. These rules are described here and are graphically depicted in Figure 6.5.

DID YOU KNOW?
A concentrated moment will cause a jump in a moment diagram, but a point load will NOT cause a jump in the moment diagram.

4. **A concentrated moment on a structure will cause a jump in the moment diagram**. The magnitude of the jump in the diagram is equal to the magnitude of the concentrated moment. If the moment is clockwise (CW), the jump is positive, and if it is counterclockwise (CCW), the jump is negative.
5. **The change in moment between two points (ΔM) is equal to the area under the shear diagram**. Section 6.4 shows that the change in moment between any two points can be found by integrating the shear equation between those two points.
6. **The slope at any point on the moment diagram is equal to the value of the shear, V, at that point**. This means that if the shear is constant, the moment diagram will be a first-order line. If the shear is a first-order line, the moment diagram will be a second-order line and so forth.

Moment Diagram—Procedure

1. Construct the shear diagram. Make sure all points of zero shear are identified and dimensioned.
2. Draw a set of x versus moment (M) axes and be sure to list the units that will be used. Anywhere there is a load discontinuity (see Section 6.3) or a point of zero shear, draw a vertical dashed line on the axes.
3. Plot the value of moment at each end of the beam/segment. Remember to use the sign convention for moment. Only a concentrated moment or moment reaction at the end of a beam will cause the moment at the end of the beam to be nonzero.

QUICK NOTE
The points of zero shear constitute the locations of either a local minimum or local maximum value in the moment diagram.

4. Start from the left and work to the right. Follow the rules as outlined previously to identify the value of moment at each of the load discontinuities and points of zero shear. If you have done this correctly, you will get the same values by the time you get to the end of the beam as you have already plotted. This serves as a check.
5. Connect the points on the diagram using Rule #6 for moment diagrams. Remember that a moment curve is one order higher than the shear curve because of the integration relationship between the two.

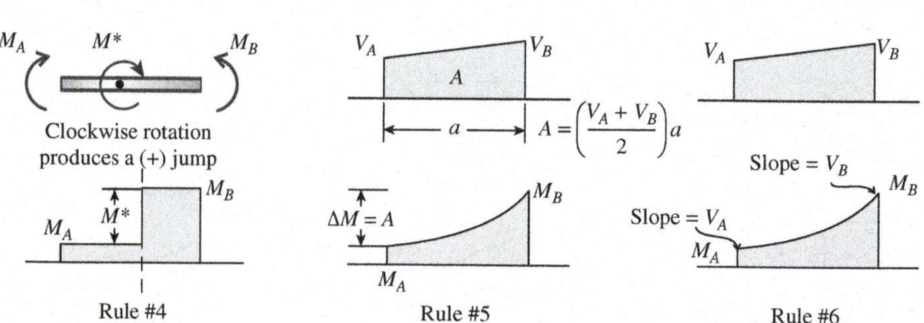

FIGURE 6.5 Graphical depiction of rules the for generating moment diagrams.

EXAMPLE 6.6

Draw the shear and moment diagrams for the beam shown.

SOLUTION STRATEGY

Follow the steps and rules as outlined in the procedures presented previously.

1. Reactions are given.
2. Axes are drawn. Points of load discontinuity occur at the point loads.
3. Shear at the left end of the beam is 70 k based on sign convention. Shear at the right end of the beam is −50 k. Remember, if the shear causes a CCW rotation, it is (−). The moment at both ends of the beam is zero (i.e., no concentrated moments at the ends).

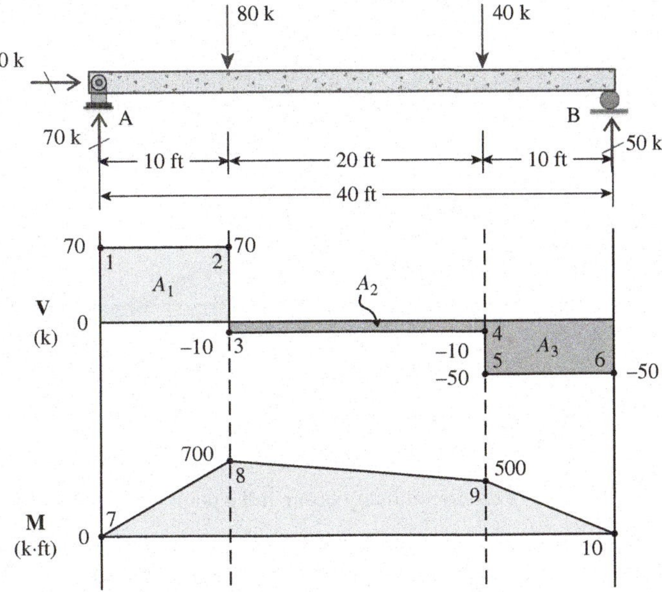

4. Shear diagram. The following rules are used to go from point to point.
   (a) 1 to 2 use Rule #2 (no area under load)
   (b) 2 to 3 use Rule #1 (negative jump)
   (c) 3 to 4 use Rule #2 (no area under load)
   (d) 4 to 5 use Rule #1 (negative jump)
   (e) 5 to 6 use Rule #2 (no area under load)

   In going from points 5 to 6, we find $V = -50$ k at the end. This agrees with the value we already have there, so we have done it correctly.
5. Moment diagram. The following rules are used to go from point to point.
   (a) 7 to 8 use Rule #5 ($A_1 = 700$ k·ft) $A_1 = (10\text{ ft})(70\text{ k}) = 700$ k·ft
   (b) 8 to 9 use Rule #5 ($A_2 = -200$ k·ft) $A_2 = (20\text{ ft})(-10\text{ k}) = -200$ k·ft
   (c) 9 to 10 use Rule #5 ($A_3 = -500$ k·ft) $A_3 = (10\text{ ft})(-50\text{ k}) = -500$ k·ft

   In going from 9 to 10, we find $M = 0$ k·ft, which is what we found in step 3. This indicates we have done it correctly.
6. Connect the dots in the diagrams using Rules #3 and #6 for the V and M diagrams, respectively. Note that the distributed load is zero, the shear is constant, and the moment is first order.

EXAMPLE 6.7

Draw the shear and moment diagrams for the beam shown.

SOLUTION STRATEGY

Follow the steps and rules as outlined in the procedures presented previously.

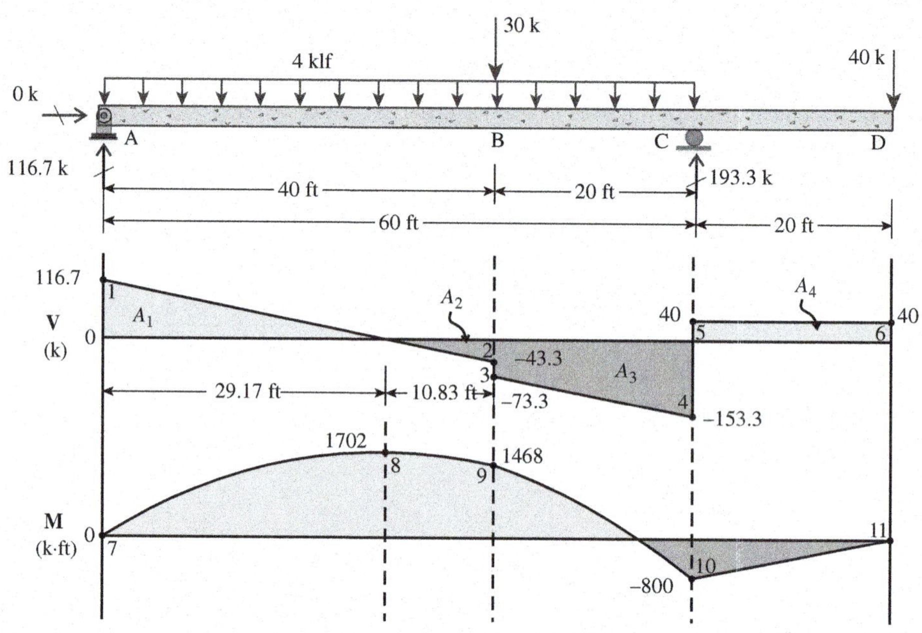

1. Reactions are given.
2. Axes are drawn. Points of load discontinuity occur at the point loads and reactions.
3. Shear at the left end of the beam is 116.7 k. Shear at the right end of the beam is 40 k. The moment at both ends of the beam is zero (i.e., no concentrated moments at the ends).
4. Shear diagram. The following rules are used to go from point to point.
   (a) 1 to 2 use Rule #2 [area under load = (−4 klf)(40 ft) = −160 k]
   (b) 2 to 3 use Rule #1 (negative jump)
   (c) 3 to 4 use Rule #2 [area under load = (−4 klf)(20 ft) = −80 k]
   (d) 4 to 5 use Rule #1 (positive jump)
   (e) 5 to 6 use Rule #2 (no area under load)

   In going from 5 to 6, we find $V = 40$ k at the end. This agrees with the value we already have there, so we have done it correctly.
5. Moment diagram. The following rules are used to go from point to point.
   (a) 7 to 8 use Rule #5 ($A_1 = 1702$ k·ft)
   Note that moment is max where shear is zero.
   (b) 8 to 9 use Rule #5 ($A_2 = -234$ k·ft)
   (c) 9 to 10 use Rule #5 ($A_3 = -2266$ k·ft)
   (d) 10 to 11 use Rule #5 ($A_4 = 800$ k·ft)

   In going from 10 to 11, we find $M = 0$ k·ft, which is what we found in step 3. This indicates we have done it correctly.
6. Connect the dots in the diagrams using Rules #3 and #6 for the V and M diagrams, respectively. Note that when the distributed load is constant, the shear is first order (i.e., straight line) and the moment is second order. The concavity of the second-order line is found by looking at the shear diagram at relative slopes. For example, point 7 on the moment diagram should have a positive slope because the

DID YOU KNOW?
The location of the point of zero shear can be found for a first-order line by using the slope and intercept of the shear.

$$x = \frac{\text{intercept}}{\text{slope}} = \frac{V_O}{w}$$

$$x = \frac{116.7\ \text{k}}{4\ \text{klf}} = 29.17\ \text{ft}$$

value of shear at that point is positive. The slope at point 8 should be zero because the value of shear at that point is zero and so forth.

The remainder of the examples present various challenges to the generation of correct shear and moment diagrams. For the sake of brevity, these examples do not present a step-by-step or point-by-point description of the process. Rather, the lessons that each example is presenting are highlighted.

EXAMPLE 6.8

Draw the shear and moment diagrams for the cantilever-type structure shown. This is the same beam used in Example 5.1.

SOLUTION STRATEGY

Follow the steps and rules as outlined in the procedures presented. Nothing special is needed for treating the hinges. The hinges were used to calculate the reactions. Follow the procedure as normal.

$$A_1 = \left(\frac{-156.5\text{ k} - 356.5\text{ k}}{2}\right)(100\text{ ft}) = -25650\text{ k·ft}$$

$$A_2 = \left(\frac{375\text{ k} + 135\text{ k}}{2}\right)(120\text{ ft}) = 30600\text{ k·ft}$$

$$A_3 = \frac{1}{2}(35\text{ k})(17.5\text{ ft}) = 306.3\text{ k·ft}$$

$$A_4 = \frac{1}{2}(-325\text{ k})(162.5\text{ ft}) = -262406.3\text{ k·ft}$$

$$A_5 = \left(\frac{311.5\text{ k} - 111.5\text{ k}}{2}\right)(100\text{ ft}) = 21150\text{ k·ft}$$

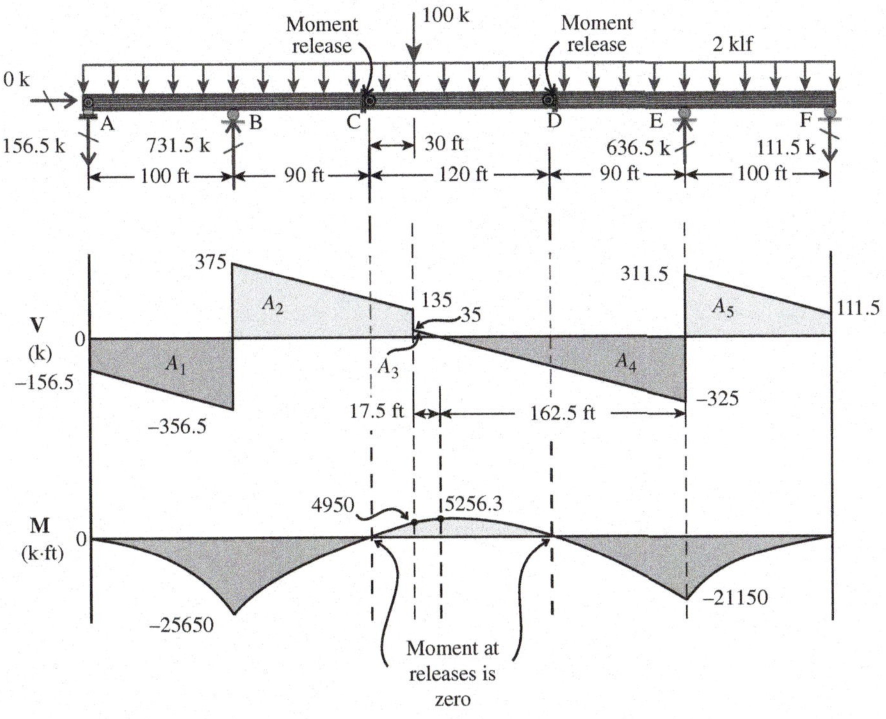

QUICK NOTE

The distance to point of zero shear is found as

$$x = \frac{35\text{ k}}{2\text{ klf}} = 17.5\text{ ft}$$

The slope everywhere on the shear diagram is the intensity of the load diagram.

$$\text{slope} = -2\text{ klf}$$

The concavity of the moment diagram is found by looking at the values of the shear diagram. For example, the slope at point A is −156.5 k, and the slope just to the left of B is −356.5 k. As such, the slope of the moment diagram increases traveling from left to right.

SUMMARY

No special approach needs to be followed when generating the shear and moment diagrams for beams that have internal moment releases (hinges). However, if these diagrams have been generated correctly, the moment diagram will pass through zero at the location of the releases. One could always find the value of moment at these locations using Rule #5 and verify that the moment is indeed zero at this location.

EXAMPLE 6.9

Draw the shear and moment diagrams for the beam shown.

SOLUTION STRATEGY

Follow the steps and rules as outlined in the procedures presented. Remember to use Rule #5 appropriately.

$A_1 = 468$ k·ft $\quad A_2 = 140$ k·ft

$A_3 = -144$ k·ft $\quad A_4 = -360$ k·ft

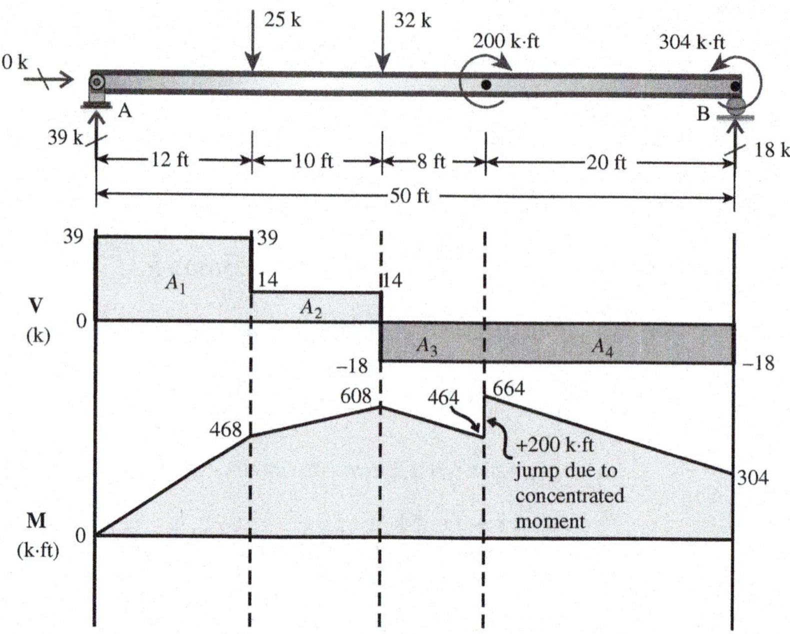

QUICK NOTE

The concentrated moment located 20 ft from the right end causes:

- A +200 k·ft jump in the moment diagram because it is a clockwise moment as per Rule #5.
- No jump in the shear diagram.

The concentrated moment at B causes a positive moment to be at B according to the sign convention given in Figure 6.2.

Students at first may have a little difficulty in drawing shear and moment diagrams for structures that are subjected to triangular loads. Example 6.10 is presented to demonstrate how to deal with them.

EXAMPLE 6.10

Draw the shear and moment diagrams for the beam shown.

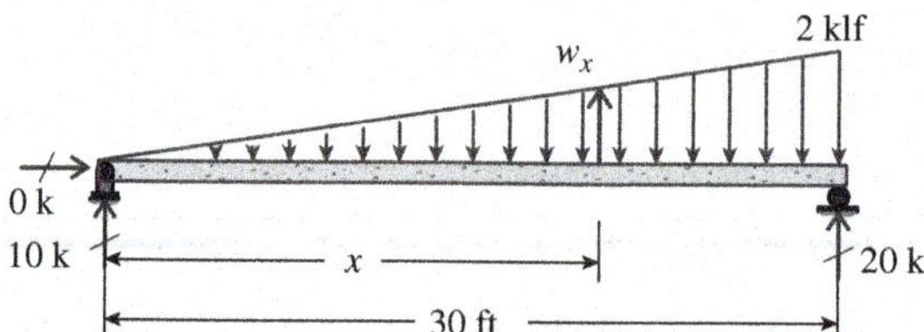

QUICK NOTE

The basic form of a linear equation is given as

$$y = mx + b$$

where m is the slope and b is the intercept. In this example, the intercept of the load equation at the left end is zero, so only the slope needs to be found. This is done by placing the rise over the run.

$$m = \frac{2 \text{ klf}}{30 \text{ ft}} = \frac{1k}{15 \text{ ft}^2}$$

SOLUTION STRATEGY

Follow the steps and rules as outlined in the procedures presented. It is necessary to develop an equation for the load diagram to be able to find the point of zero shear. Find the max moment by making a cut at the point of zero shear.

SOLUTION

Plot the shear and the moment at the two ends as usual. No load discontinuities exist in the beam, so we should be able to connect the dots on the shear diagram. At the left end of the beam, the load intensity is zero, so this means the slope of the shear diagram at that location is zero. At the right end of the beam, the load intensity is −2 klf, so the slope of the shear diagram at that location is negative. This tells us that the shear diagram is concave down.

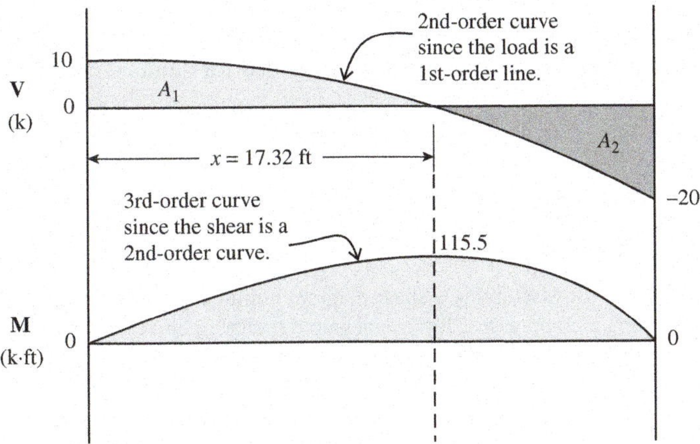

Finding the point of zero shear is a bit more challenging and requires that an equation of the load diagram be developed. The load diagram is linear and can be written as (see side bar)

$$w = \frac{1}{15}x;\ 0\text{ ft} \leq x \leq 30\text{ ft}$$

Since the shear at the left end is 10 k and we want to know the distance x to the point of zero shear, we really want to know at what distance of x will cause the area under the load diagram to be 10 k between 0 and x. We have

$$A_{\text{load}} = \frac{1}{2}(x)\left(-\frac{x}{15}\frac{\text{k}}{\text{ft}^2}\right) = -10\text{ k} \Rightarrow x = \underline{\underline{17.32\text{ ft}}}$$

Finding the area under a second-order curve (A_1 and A_2) can be a bit difficult without an equation for the shear. One of the easiest ways to handle this is to avoid calculating the areas. Rather, simply make a cut at $x = 17.32$ ft and solve for the internal moment at that point.

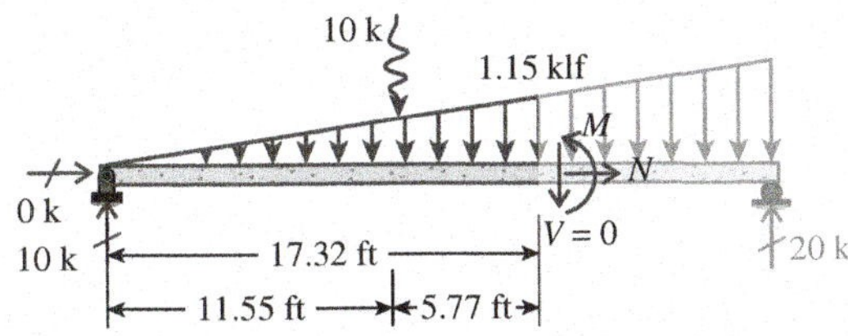

$$+\circlearrowleft \sum M_{\text{cut}} = -(10\text{ k})(17.32\text{ ft}) + (10\text{ k})(5.77\text{ ft}) + M = 0$$

$$M = \underline{\underline{115.5\text{ k·ft}}} \Rightarrow \text{place on moment diagram}$$

Alternative Since the left end of the shear diagram has a zero slope, the simplified area equation may be used to find A_1.

$$\Delta M = A_1 = \frac{2}{3}(17.32\text{ ft})(10\text{ k}) = 115.5\text{ k·ft}$$

DID YOU KNOW?

The area under a second-order curve can be found easily if certain conditions are met. Consider a rectangle that has a width of b and a height of h. The area of this rectangle is bh. Let it be subdivided by a second-order curve as shown.

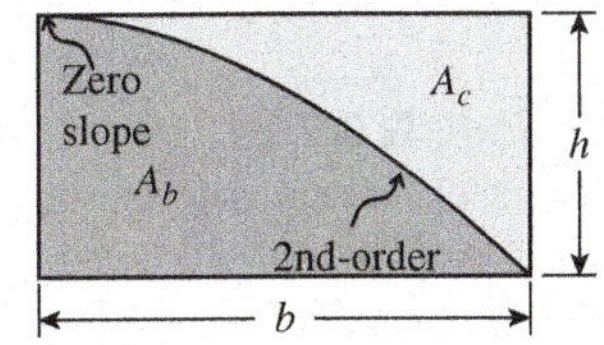

$$A_b = \frac{2}{3}bh \quad A_c = \frac{1}{3}bh$$

These are only valid if one end of the shape has a zero slope.

6.6 Axial Diagrams

Axial diagrams for structural members are not used as often as shear and moment diagrams. The reason for this is that the axial force in a member is often constant along its length. In cases like this, it is just easier to report the axial force for the member than to provide a diagram. However, there are times when the axial force along the length of a member varies. It is particularly useful at times like these to sketch the axial force diagram. The rules for this will not be laid out specifically, rather we will simply indicate that the rules are very similar to the first two rules used for creating the shear diagram. Therefore, a point load causes a jump, and a distributed load causes a change in axial force. Two examples are provided for demonstrating this process.

QUICK NOTE
An axial force pointing to the right causes a negative jump, while an axial force pointing to the left causes a positive jump

DID YOU KNOW?
Members loaded with uniform loads along their length are commonly found in diaphragms (see Chapter 4).

EXAMPLE 6.11

Draw the axial force diagram for the member shown.

SOLUTION STRATEGY

Follow the same basic steps for assembling a shear diagram noting any load discontinuities. Plot the axial force at the two ends noting a compressive force is negative (pointing into the member) and a tensile force is positive (pointing out of the member).

SOLUTION

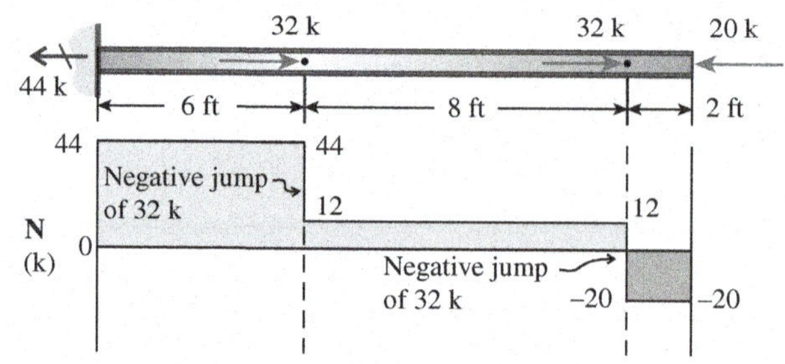

QUICK NOTE
Since no point loads/reactions exist at the end of the member, the axial force at the ends are zero.

CALCULATIONS
From A to B,

$$\Delta N_{AB} = (-4 \text{ klf} + 2 \text{ klf})(12 \text{ ft}) = -24 \text{ k}$$

From B to C,

$$\Delta N_{BC} = (+2 \text{ klf})(16 \text{ ft}) = 32 \text{ k}$$

From C to D,

$$\Delta N_{CD} = (-4 \text{ klf} + 2 \text{ klf})(4 \text{ ft}) = -8 \text{ k}$$

EXAMPLE 6.12

Draw the axial force diagram for the member shown.

SOLUTION STRATEGY

Follow the same basic steps for assembling a shear diagram noting any load discontinuities (e.g., B and C). Find the change in axial force between two points by calculating the area under the load.

SOLUTION

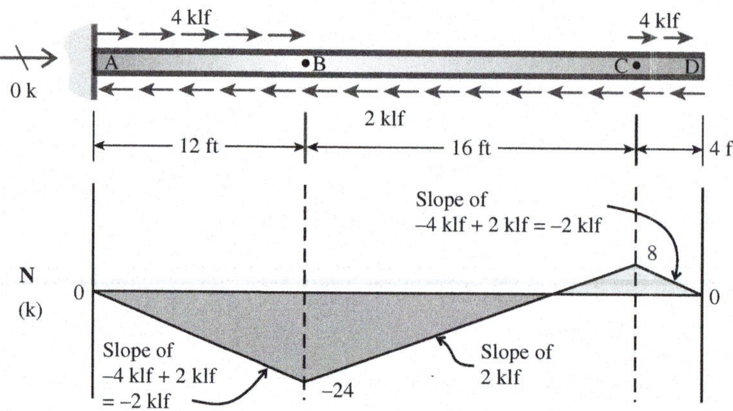

6.7 Shear and Bending Moment Diagrams for Frames

Shear, moment, and axial diagrams are very useful for rigid frames as well as for individual members. The members of such frames cannot rotate or translate with respect to each other at their connections. As a result, axial forces, shear forces, and bending moments are transferred between the members at the joints. These values must be accounted for in the preparation of such diagrams.

The process for generating diagrams for frame members is very similar to the processes used for single members. However, there are a couple of additional considerations that should be acknowledged to successfully generate the diagrams.

- **Complete FBDs must be made for each member of the frame individually**. Any loads or end forces on the member must be resolved into components that are either parallel or perpendicular to the axis of the member. This is because axial and shear forces are defined as forces that are parallel and perpendicular to the member, respectively.
- **Coordinate system for each member must be defined**. Since members may not necessarily be oriented in the horizontal direction, the sign convention used for moment can become confusing. To aid in this, the student needs to define a starting end for each member and then interpret the sign convention (both shear and moment) working left to right (see note in side margin).

QUICK NOTE

The sign given to a moment in a frame is dependent upon the starting end you choose. In the following figure, if you choose x_1 as the origin (i.e., placed to your left), then the moment in the figure would be called positive. If x_2 is used (i.e., placed to your left), the moment in the figure would be called negative. This may change the sign, but it doesn't change the look of the resulting diagram.

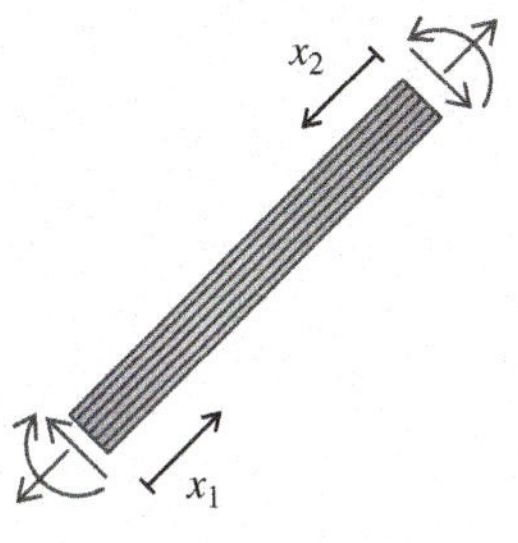

EXAMPLE 6.13

Draw the shear and moment diagrams for the frame shown.

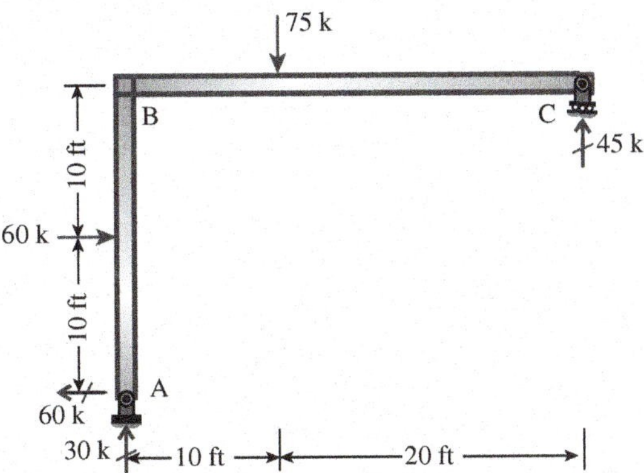

SOLUTION STRATEGY

Cut the frame at joint B and isolate the joint. Use basic equilibrium for each section to solve for the unknown internal forces. Identify a coordinate for each member, then follow the process for generating diagrams for each member as was done for beams.

QUICK NOTE

The FBD for joint B doesn't really need to be sketched for this frame. However, whenever a point load, reaction, or concentrated moment exists at a joint, it is wise to draw the FBD of that joint so that you can ensure that equilibrium does work out.

SOLUTION

Make a cut just below B and also a cut just to the right of B. The following are the resulting FBDs.

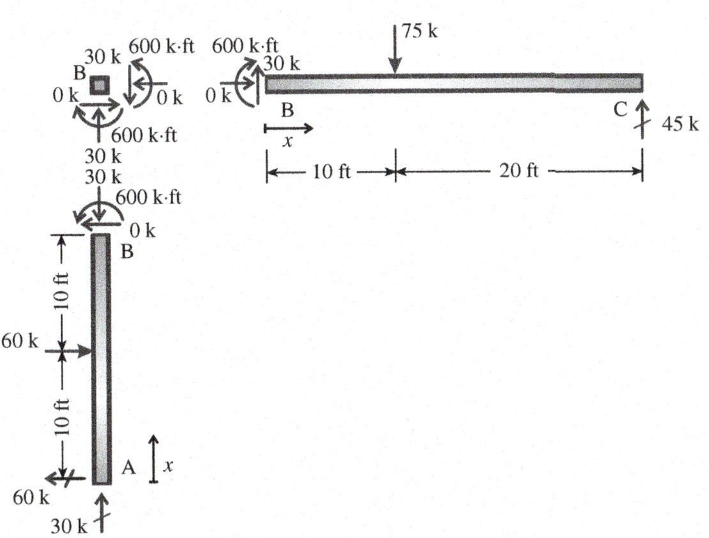

QUICK NOTE

For simple frames, it is often convenient to define the origin for each section using a clockwise rotation in the structure. For example, looking at element AB, we place the origin at A pointing to B because it is a clockwise direction in the structure. The same is true for BC.

The shear and moment diagrams are generated for each segment by considering each one to be an individual beam. For example, to generate the diagrams for AB, rotate the book clockwise 90° so that the AB is horizontal with A being on the left. Then follow the same rules (1 through 6) as before. There is no need to rotate BC, since it is already horizontal and the origin is at the left end.

$$A_1 = (10\text{ ft})(60\text{ k}) = 600\text{ k}\cdot\text{ft}$$

$$A_2 = (10\text{ ft})(30\text{ k}) = 300\text{ k}\cdot\text{ft}$$

$$A_3 = (20\text{ ft})(45\text{ k}) = -900\text{ k}\cdot\text{ft}$$

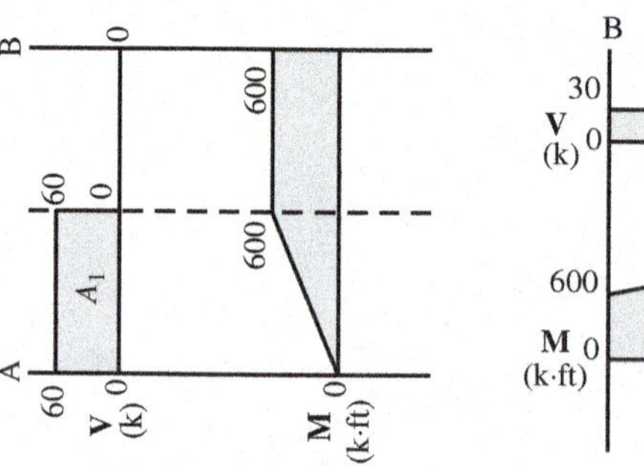

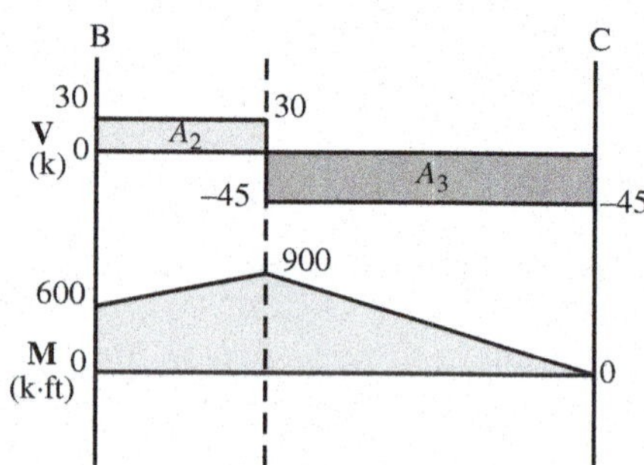

EXAMPLE 6.14

Draw the shear and moment diagrams for the frame shown here.

COMMON MISTAKE

Students often feel that, when making a cut in a frame, the shear at a point should be equal and opposite. For example, at joint C on member BC, the shear is −20 k. But the shear at joint C in member CD is −14.14 k. The difference comes because the members are not in the same orientation, and shear is defined as the internal force perpendicular to the axis of the member.

DID YOU KNOW?

A member end moment on a FBD points to the side of the member on which the moment diagrams should be sketched. For example, the moment at C is pointing to the upside of the member. The resulting moment diagram also ended up on the upside of the member. **This is a handy little piece of information when double checking your moment diagrams.**

SOLUTION STRATEGY

Cut the frame at joints B and C. Isolate joint B because it has a point load applied to it. Generate the FBD for each member making sure that all forces are resolved into components that are either parallel or perpendicular to the member. Be particularly careful with CD. Identify a coordinate for each member, then follow the process for generating diagrams for each member as was done for beams.

SOLUTION

Make a cut just below B and also a cut just to the right of B. Also make a cut at C. The following are the resulting FBDs. Joint C is exploded to help work out the transformation of forces.

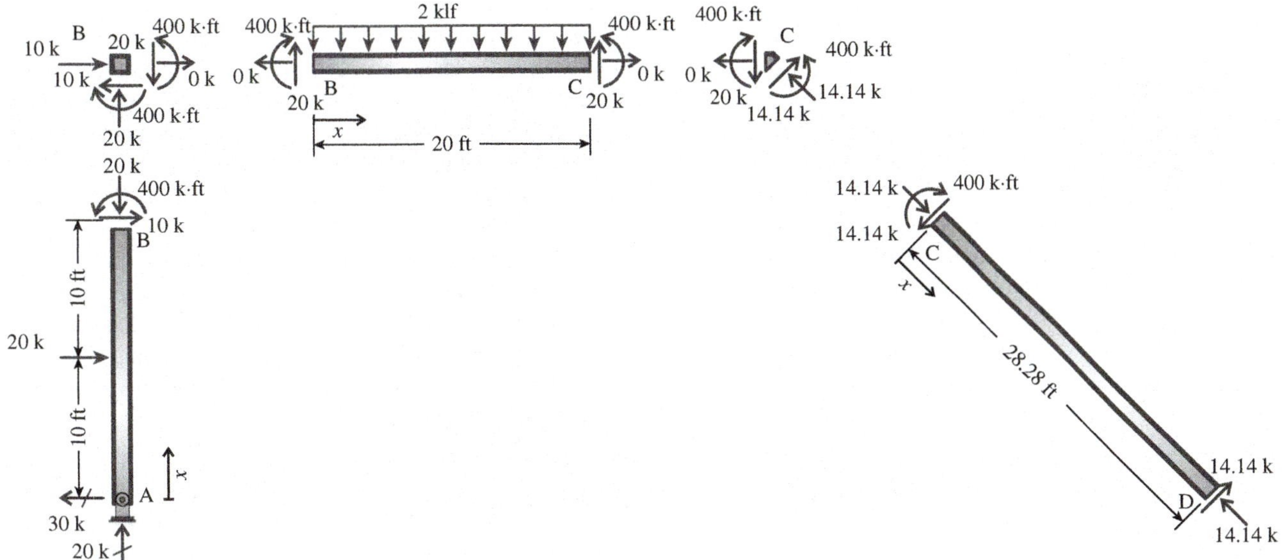

The shear and moment diagrams are generated for each segment by considering each one to be an individual beam. Use the coordinate for each segment as shown. Follow the same rules (1 through 6) as before. Segment AB is the only one with a load discontinuity.

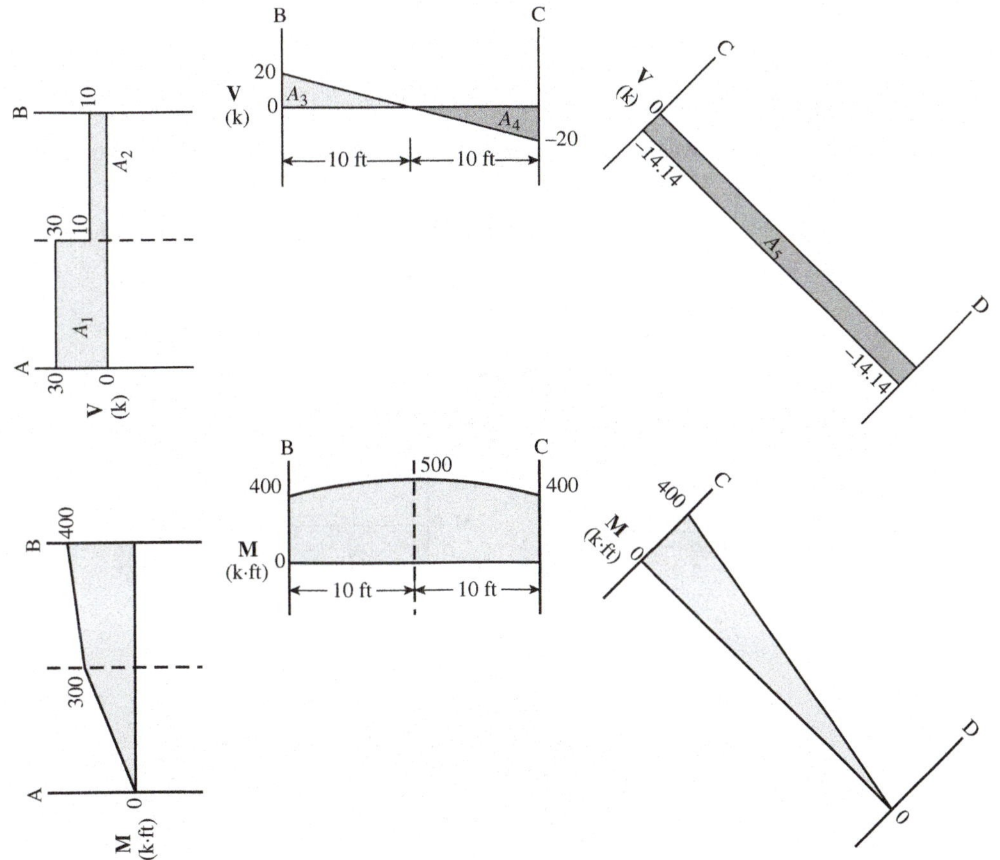

SUMMARY

Geometry and trigonometry tend to be the most troublesome part of the analysis of frames. Take your time and ensure that the FBDs are correct and that all forces have been resolved into parallel and perpendicular components.

6.8 Moment Diagrams Using Superposition

QUICK NOTE
These moment diagrams are very easily derived using the processes previously presented. However, it will be to your advantage to commit these to memory.

Although not obvious at this point, there are circumstances where we would like to know some basic geometric properties of the moment diagrams. These properties include the calculation of area under the moment diagram and also the location of the centroid of that area. For the moment diagrams shown in Example 6.13, the calculation of these properties is very straightforward. However, this task can become much more complicated and difficult for moment diagrams like shown in Example 6.7. For this reason, it is useful to generate moment diagrams using the method of superposition.

Superposition allows for the treatment of each applied load/reaction separately and then adding their effects up. As will be shown, the moment diagrams resulting from the individual loads are generally quite simple and direct. Indeed, if a student learns how to sketch the moment diagram for a few simple cantilever beams without first generating the shear diagram, they find that they can generate the superimposed moment diagrams very rapidly and correctly. Figure 6.6 illustrates the basic cantilever beams and their resulting moment diagrams.

Cantilever Beam Method of Superposition

QUICK NOTE
We don't use the superposition method to develop moment diagrams used for design. That process has been well documented in previous sections. When creating superimposed moment diagrams, it is not intended that we actually add the moment diagrams back together. This is difficult to do. Rather, we leave them separated for future use (see Chapter 9).

The cantilever beam method of superposition is quite flexible and consists of constructing a series of cantilever beams that when added together are equal to the original beam or element under consideration. The procedure is as follows.

Cantilever Beam Method—Procedure

1. Solve for reactions.
2. Draw complete FBDs of each element in the structure.
3. For each element/beam, select a point on the element and fix that point. This requires us to provide a fixed support at this point.
4. For each individual load—including reactions—on the element, construct a separate moment diagram.

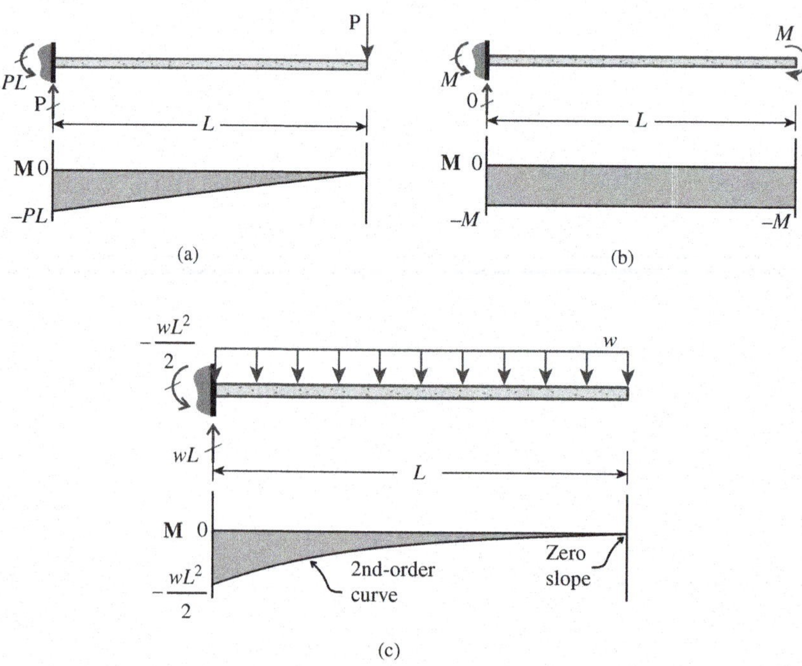

FIGURE 6.6 Moment diagrams for basic cantilever beams.

EXAMPLE 6.15

For the beam given in Example 6.7, create a series of load and moment diagrams. Select point C to be the point of cantilever.

SOLUTION STRATEGY

The reactions and FBD are given in Example 6.7. Fix point C and then consider each load/reaction to be independent.

SOLUTION

Look at the basic beam diagrams shown in Figure 6.6 and use these to handle each of the loads. Generate the individual moment diagrams as shown. If you sum up the moment reactions for the left cantilever beams, they equal −800 k·ft. If you sum up the moment reactions for the right cantilever beam, it also equals −800 k·ft. If the cantilever beam method is correctly done, the reactions on both sides will be equal. Look back at Example 6.7 and note that the internal moment at point C is equal to −800 k·ft. This is not coincidence.

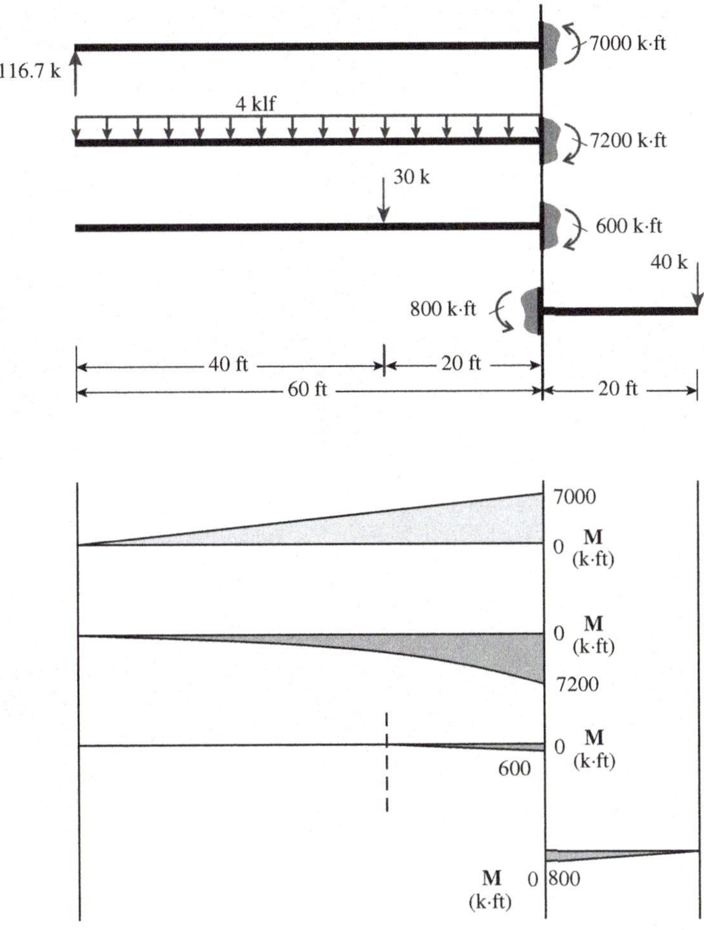

QUICK NOTE

This method can be used for members of frames as well. For example, if member BC of Example 6.14 is fixed at point B, then three separate cantilever beams will be needed: One for the 2 klf distributed load, one for the 20 k end shear, and one for the 400 k·ft end moment.

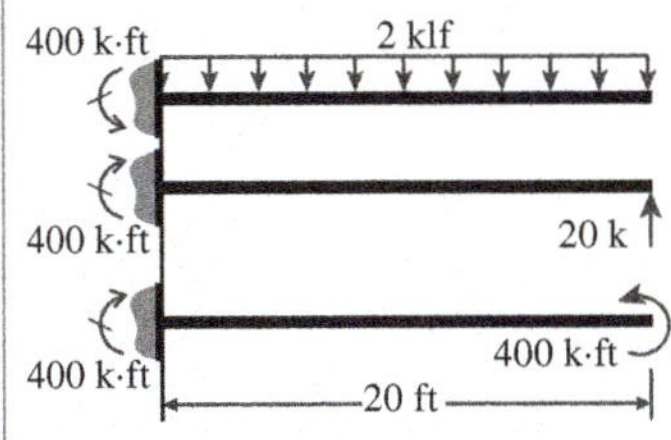

QUICK NOTE

We can easily calculate areas and centroids for all of these shapes.

6.9 Structural System Consideration

In Chapter 3, we presented the notion of a vertical load path. Generally speaking, the load is applied directly to the slab, travels to the beams, and then into the girders. From the girders, the load travels into the columns and finally into the foundation. For a floor system with a one-way slab, the load on the beams results in a uniformly distributed load, while the load on the girders results in a series of point loads (see Example 3.2). This is a very good representation of the way the loads are distributed throughout the floor system.

QUICK NOTE

Simply supported beams having a distributed load are extremely common in building roof and floor systems. It is helpful for an engineer to know the basic shear and moment diagram from memory—especially the maximum values.

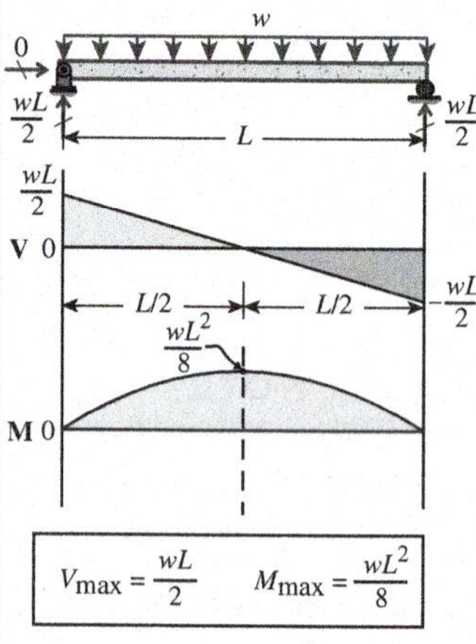

Girder Simplification for Design Purposes

In the design of beams and girders, the structural engineer is often interested in simply knowing what the maximum shear and moment are in a beam. As seen in the side margin, the equations for maximum shear and moment in a beam/girder are very simple if it is subjected to a uniformly distributed load. The calculation of maximum internal forces can be considerably more tedious when the beam/girder is subjected to a series of point loads. This is especially true when there are a large number of point loads present on the beam.

Experience has shown that for design purposes the load on girders may be treated conservatively as a uniformly distributed load—even though the actual load is a series of point loads. This simplification is used very commonly by professional engineers when there are at least three equally spaced point loads of equal magnitude on a girder. This will produce a conservative estimate for the actual shear in the girder and will give a very close approximation for the maximum moment. We should note, however, that a series of point loads does not produce the same shape of shear and moment diagrams as does a distributed load. Example 6.16 is presented to demonstrate the process of implementing this simplification and the resulting error from doing so.

EXAMPLE 6.16

For the two framing floor plans shown, assume that the distributed load applied to the slab is 60 psf. Sketch the load diagram for the girder in each floor and generate the associated shear and moment diagrams. Next, assume that the girder is subjected to a uniformly distributed load that is calculated using the tributary approach. Sketch the shear and moment diagrams for this loading and compare maximum values calculated using both methods.

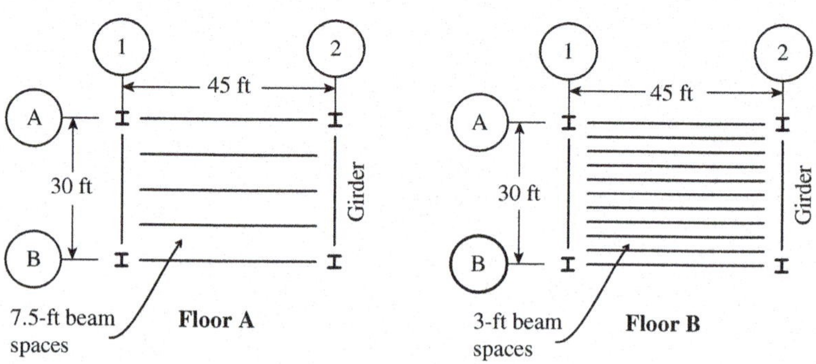

SOLUTION STRATEGY

Refer back to Example 3.2 to refresh on how to generate load diagrams for a one-way slab. Generate shear and moment diagrams following the basic rules set forth previously. To calculate the distributed load for the girder, consider half the span of the beam as the tributary width to the girder.

SOLUTION

For floors A and B, the tributary width for each beam is 7.5 ft and 3 ft, respectively. The reactions of each beam (P_A and P_B) are calculated as follows, and results in the load and shear and moment diagrams as shown for the girders.

$$P_A = (60 \text{ psf})\,(7.5 \text{ ft})\left(\frac{45 \text{ ft}}{2}\right) = 10125 \text{ lbs} = 10.125 \text{ k}$$

$$P_B = (60 \text{ psf})\,(3.0 \text{ ft})\left(\frac{45 \text{ ft}}{2}\right) = 4050 \text{ lbs} = 4.05 \text{ k}$$

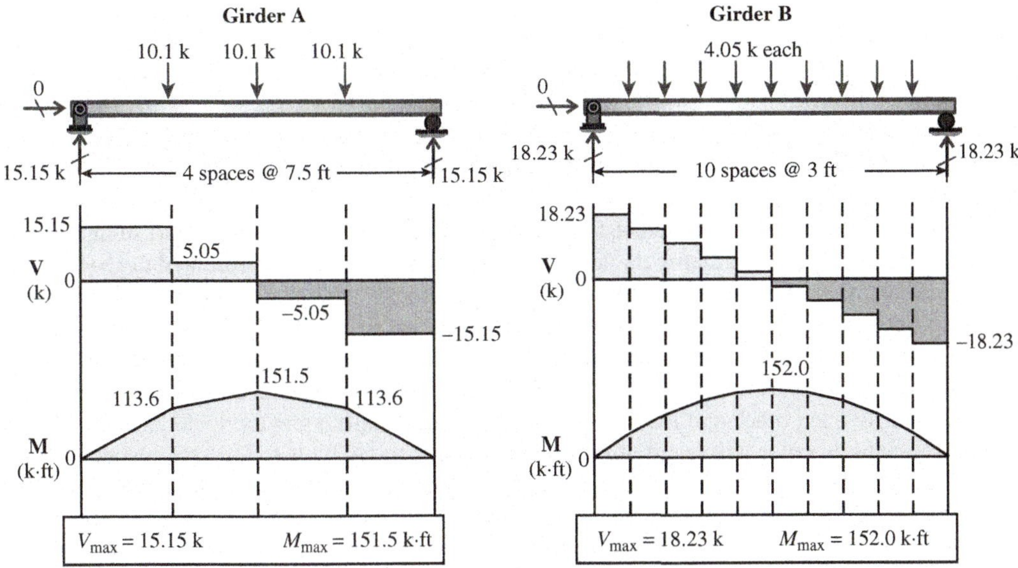

The preceding shear and moment diagrams result from a proper treatment of the way loads are actually placed on the girder. Next consider a simplification of assigning the load to the girder as a uniformly distributed load. The load is calculated using a tributary approach as

$$w_{A \text{ or } B} = (60 \text{ psf})\,(22.5 \text{ ft}) = 1350 \text{ plf} = 1.35 \text{ klf}$$

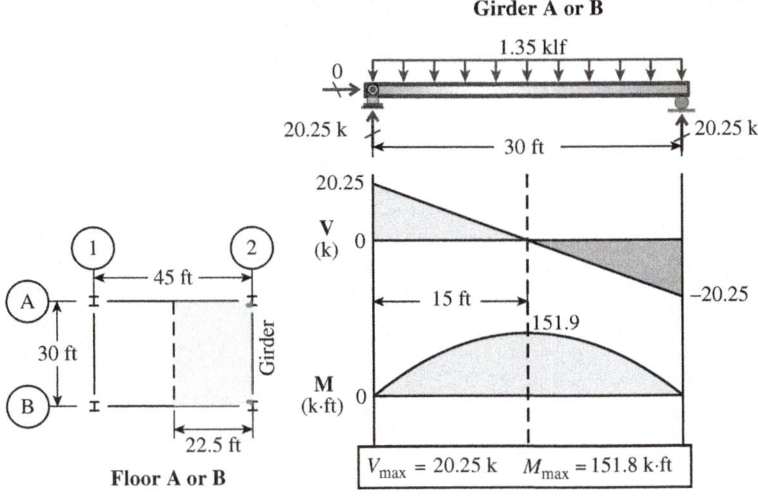

SUMMARY

In all three cases, the maximum moment estimates are practically the same. The shear produces a rather significant difference. However, you should notice that the shear for the uniform load simplification is a conservative estimate of the actual shear. Thus, for practical use, many engineers prefer to use the uniform load estimate.

6.10 SAP2000 Computer Applications

The displaying and reporting of results is important in any structural analysis program. There are some specifics you must know about SAP2000 so that you can access and understand the output.

Graphical Display

Once an analysis is run, SAP2000 has all of the results available for display. A graphical display is a convenient way to represent internal forces such as shear, moment, or axial. The display tools are found in the pull-down menu: *DISPLAY → SHOW FORCES/STRESSES → FRAMES/CABLES/TENDONS*. This will open the *internal force* display window shown in Figure 6.7. You will have the opportunity to display axial force, shear force in two different directions, torsion, or moment about one of two axes.

To understand which axis to select for plotting internal shear or moment, the student must understand something about member axes. The structure as a whole is represented in global axes X, Y, and Z. However, this set of axis descriptions often does not coincide with the member axes. For this purpose it becomes necessary to define a separate set of axes for each individual member. These member (or local) axes are then used to describe the internal forces. They are labeled as 1, 2, or 3 where the 1-axis is always along the length of the member and the 2 and 3 axes are perpendicular to each other and the 1-axis. One may graphically view the local axes which are then represented by color where red, green, and blue correspond with 1, 2, and 3, respectively.

If shear is present in the 2 axis (noted as 2–2) then it will produce a moment about the 3–3 axis. If shear is present in the 3–3 direction then it will cause a moment about the 2–2 direction. It is most common for planar (2-D) structures for the shear to develop in the 2–2 direction and the moment in the 3–3 direction.

A few key points about viewing shear and moment diagrams in SAP2000 are made below.

- It is usually helpful to show moment values on the diagrams. Select the radio button shown in the lower-left hand corner of Figure 6.7.
- Select the relevant load case in the upper-right hand corner of Figure 6.7. Most often there will only be one case to choose: *Dead*.
- Shear diagrams are plotted just opposite from the way they are constructed and presented in this text. This variation should simply be understood and accepted. It does not change the value of the shear though.

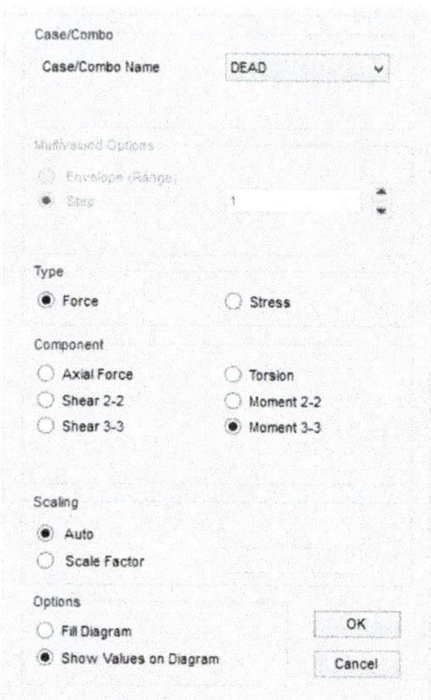

FIGURE 6.7

VIEWING MEMBER LOCAL AXES

Clicking the check box symbol found in the upper ribbon in the SAP window will open a dialog box. This box will have many display options available including the display of local axes for Frames/Cables/Tendons.

DID YOU KNOW?

Constructing 2-D structures in the X–Z plane will prevent the student from being confused about the local axes. This will mean that shear will occur in the 2–2 direction and moment will occur in the 3–3 direction.

QUICK NOTE

The global units you have set at the time of table generation will dictate the units which are reported in the tables. Units can be changed through the pull-down box located in the lower-right hand corner of the SAP2000 window.

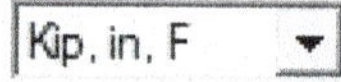

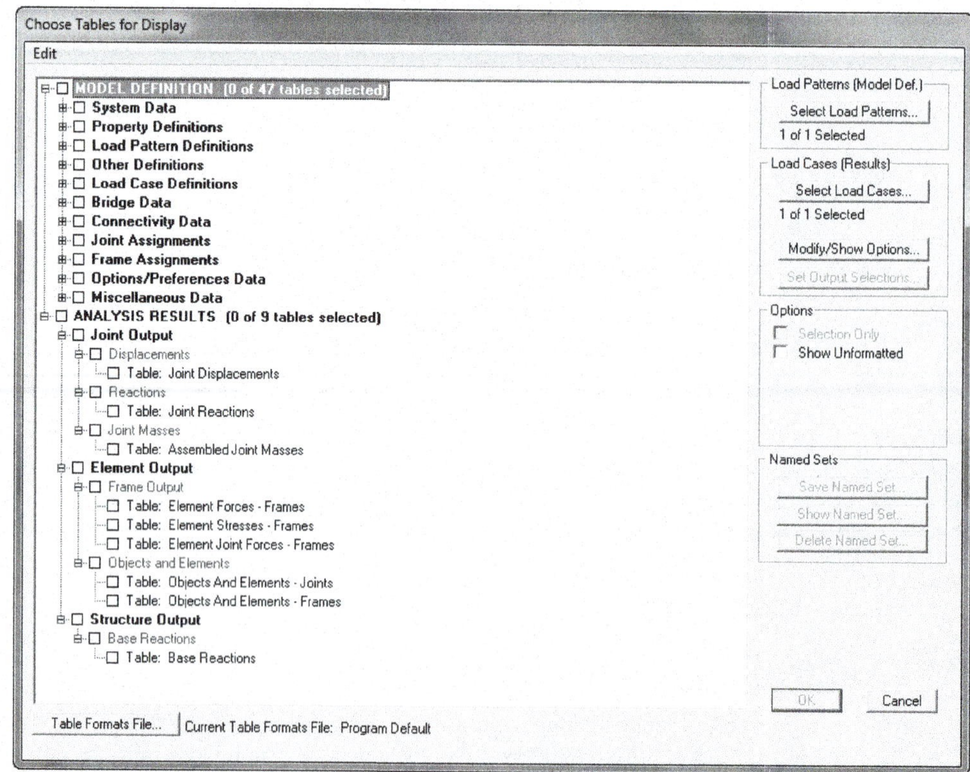

FIGURE 6.8 Table output dialog box for SAP2000.

- Moment diagrams are plotted on the tension side of the member by default. This is opposite of what is generated in this text, which produces moment diagrams on the compression side of the member. One does have the option of changing this setting by going to the following pull-down menu selection. *OPTIONS/MOMENT DIAGRAMS ON TENSION SIDE*. Doing this is recommended.

Textual Output

Text based output for internal forces and reactions can be obtained by the following pull-down menu. *DISPLAY/SHOW TABLES*. This will bring up the dialog box shown in Figure 6.8.

Many options are available through this dialog but the student will find that they most often will look for *ELEMENT FORCES → FRAMES*, *JOINT REACTIONS* and *JOINT DISPLACEMENTS*. The tables then become available for viewing or export as needed. One challenge students often face is making sure they know what information they are viewing in the table. There is simply no replacement for taking the time to familiarize oneself with the data presented.

QUICK NOTE

The vertical lines that appear in both the shear and moment diagrams are stations along the beam for which internal forces are plotted and also reported in the tables. If more or fewer stations are desired for reporting, you can change this by the following measures.

- Ensure the model is unlocked.
- Right-click on the member for which you want to change the station locations.
- Select the *ASSIGNMENTS* tab.
- Double-click on the *max station spacing cell*.

SAP2000 EXERCISE 6.1

The SAP2000 input file *C6EX1.sdb* is a functional model of Example 6.1 in this chapter. The model has the correct dimensions, support conditions, and loading of the beam. Note that pinned support A is located at joint 1 and that roller support B is located at joint 2. After starting SAP2000, load the *C6EX1.sdb* input file and execute the analysis. Provide the shear and moment diagrams and report maximum values by showing the relevant tables.

Shear Diagram (k)

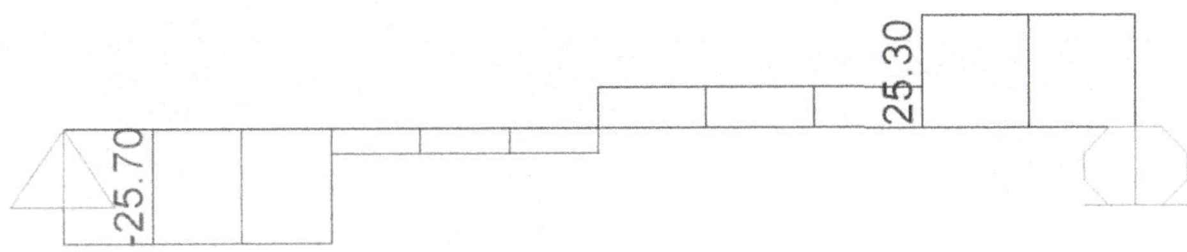

Moment Diagram (k·ft)

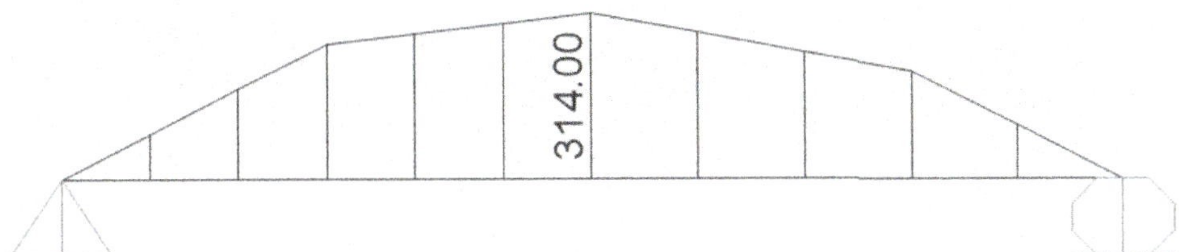

Frame Text	Station ft	OutputCase Text	CaseType Text	P Kip	V2 Kip	V3 Kip	T Kip-ft	M2 Kip-ft	M3 Kip-ft
1	0	DEAD	LinStatic	0	-25.7	0	0	0	0
1	3.3333	DEAD	LinStatic	0	-25.7	0	0	0	85.6667
1	6.6667	DEAD	LinStatic	0	-25.7	0	0	0	171.3333
1	10	DEAD	LinStatic	0	-25.7	0	0	0	257
1	10	DEAD	LinStatic	0	-5.7	0	0	0	257
1	13.3333	DEAD	LinStatic	0	-5.7	0	0	0	276
1	16.6667	DEAD	LinStatic	0	-5.7	0	0	0	295
1	20	DEAD	LinStatic	0	-5.7	0	0	0	314
1	20	DEAD	LinStatic	0	9.3	0	0	0	314
1	24	DEAD	LinStatic	0	9.3	0	0	0	276.8
1	28	DEAD	LinStatic	0	9.3	0	0	0	239.6
1	32	DEAD	LinStatic	0	9.3	0	0	0	202.4
1	32	DEAD	LinStatic	0	25.3	0	0	0	202.4
1	36	DEAD	LinStatic	0	25.3	0	0	0	101.2
1	40	DEAD	LinStatic	0	25.3	0	0	0	-4.547E-13

Max Shear

Max Moment

Edit the loading on the beam in C6EX1 by replacing the existing loading with a new loading where each of the original loads is moved 4 feet to the left. Rerun the analysis and verify the new maximum shear is 30.8 k and the new maximum moment is 292.8 k·ft.

SAP2000 EXERCISE 6.2

The SAP2000 input file *C6EX2.sdb* is a functional model of Example 6.2 in this chapter. Generate the shear, moment, and axial diagrams for the structure. What is the maximum axial force that exists? What is the maximum moment that exists?

Shear Diagram (k)—max = 50.01 k

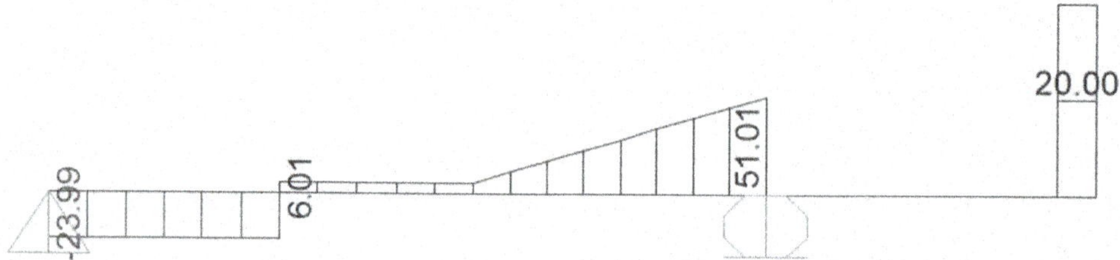

Moment Diagram (k·ft)—max = 287.84 k·ft

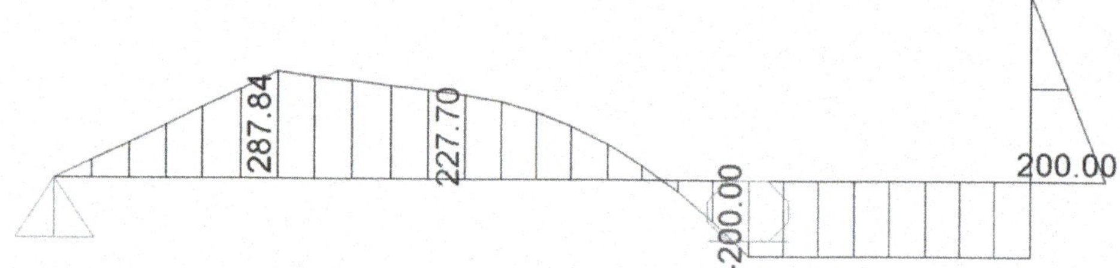

Axial Diagram (k)—max = 60 k

QUICK NOTE

The vertical leg has no axial force in it—a fact that can be verified through a quick equilibrium check.

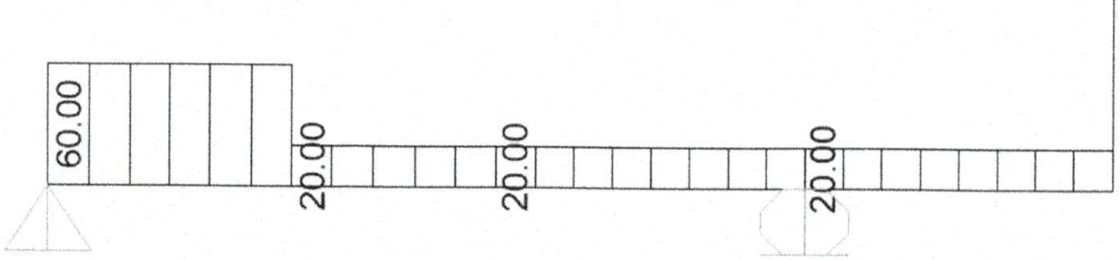

Now rotate the 20-kip joint load to act downward vertically at the top of the vertical leg. Rerun the analysis; plot the shear, moment, and axial diagrams; and verify that you get a maximum axial force of 40 k and a maximum moment of −300 k·ft.

SAP2000 EXERCISE 6.3

The SAP2000 input file *C6EX3.sdb* is a functional model of the beam shown with a spring stiffness $k_T = 1$ k/in. This beam is supported on the left with a pin, on the right with a roller, and in the middle with a translational spring. Plot the moment diagram for the beam when the spring stiffness is $k_T = 1.0$ k/in., when it is 25 k/in., and when it is 83.3 k/in. What is the maximum moment that exists in each case?

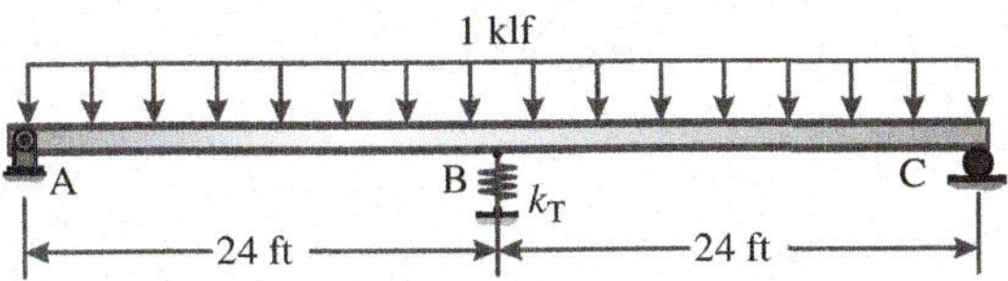

Moment Diagram (k·ft) → k_T = 1 k/in.

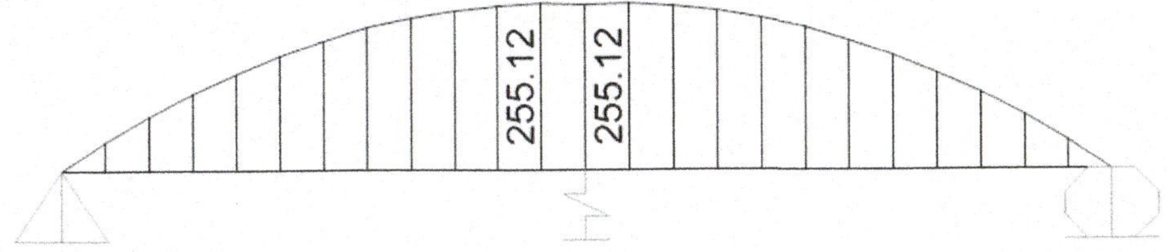

Moment Diagram (k·ft) → k_T = 25 k/in.

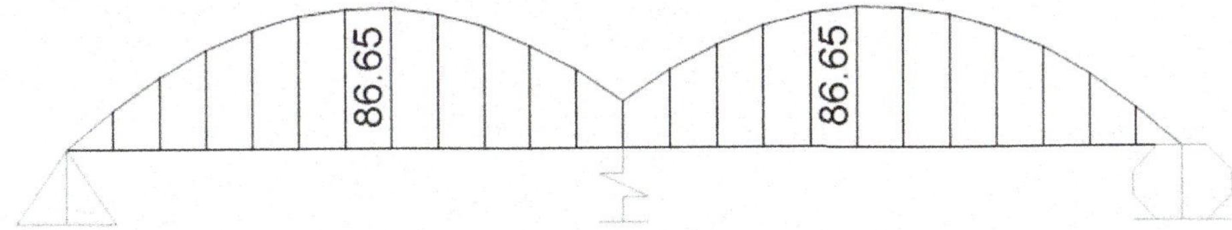

Moment Diagram (k·ft) → k_T = 83.3 k/in.

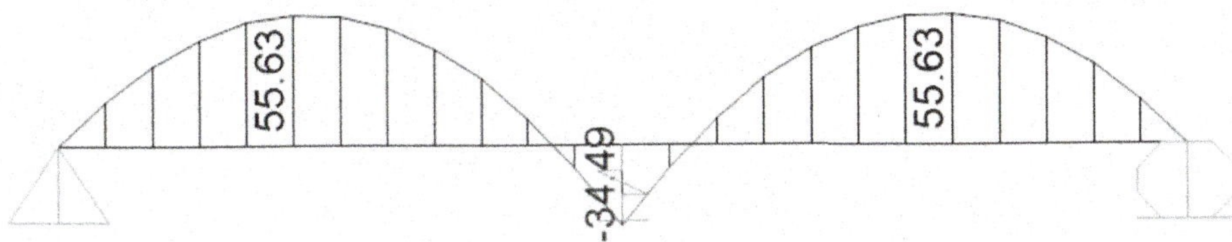

Now generate results tables to report the reactions in each case. Note that the reaction at the spring is almost zero when the stiffness is low. However, the spring takes more of the load when the spring stiffness gets larger. Plot the stiffness of the spring versus the reaction at the spring. Is it a linear trend?

QUICK NOTE

Springs are defined under the pull-down menu: *ASSIGN/JOINT/SPRINGS.* Remember to select the joint first or the previous menu option will be grayed out.

Spring stiffensses are defined in terms of the local coordinate system for the joint. The default is that the local coordinate system 1,2, and 3 corresponds with the structure coordinate system X, Y, and Z, respectively. In this example, the spring was added in the 3 direction.

DID YOU KNOW?

As the stiffness of the spring approaches zero, the behavior of the beam is as though there is no support at B. As the stiffness approaches infinity, the behavior of the beam is as though there is a roller at B.

6.11 Examples with Video Solutions

VE6.1 For the following beam developed in Example 3.3, develop the shear and moment equations for each of the three segments using the coordinates indicated.

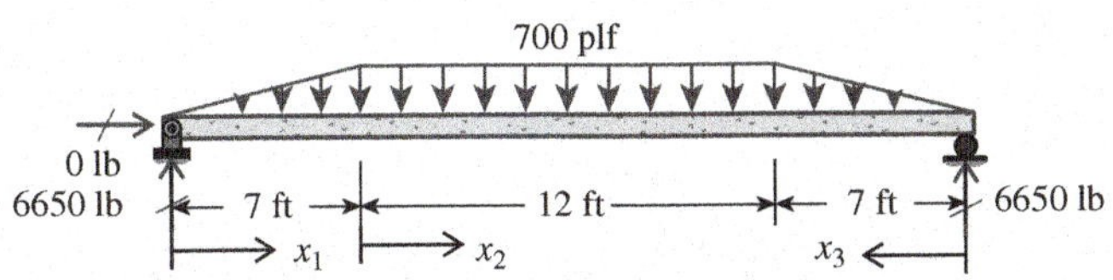

Figure VE6.1

VE6.2 For the beam shown in Figure VE6.2, generate the shear and moment diagrams.

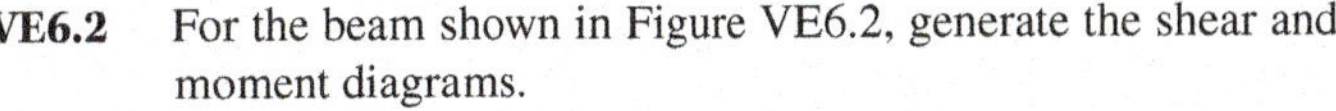

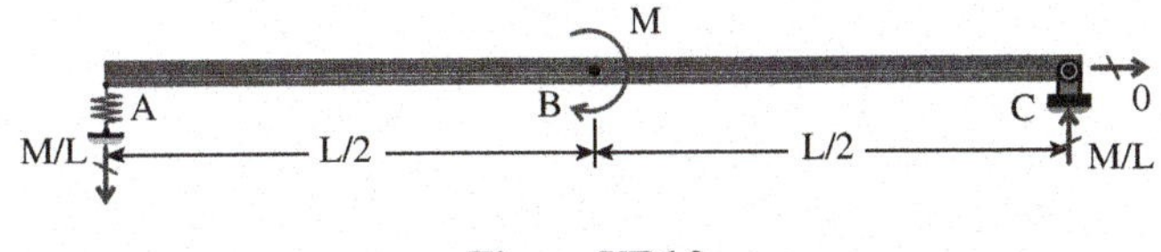

Figure VE6.2

VE6.3 For the beam shown in Figure VE6.3, generate the shear and moment diagrams.

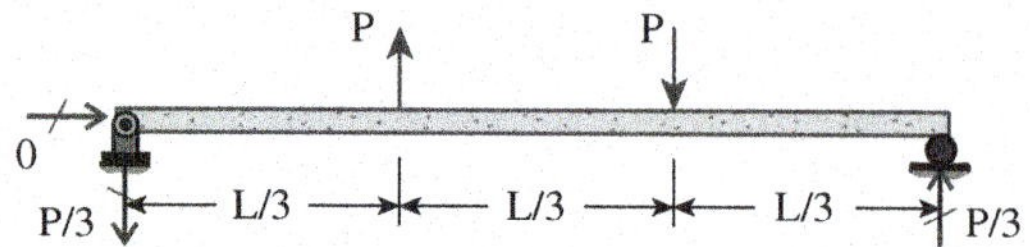

Figure VE6.3

VE6.4 For the beam shown in Figure VE6.4, generate the shear and moment diagrams.

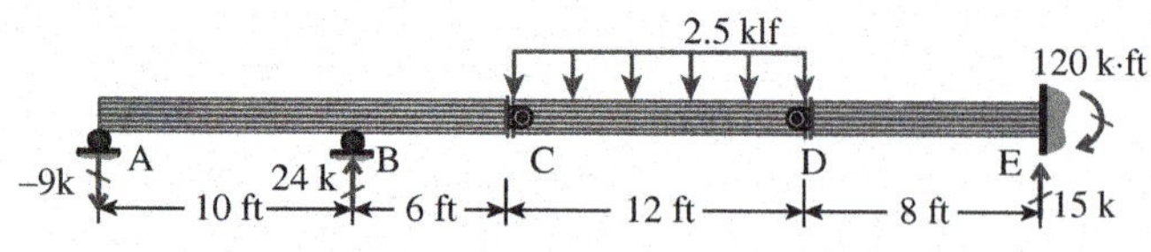

Figure VE6.4

VE6.5 For the frame shown in Figure VE6.5, generate the axial, shear, and moment diagrams.

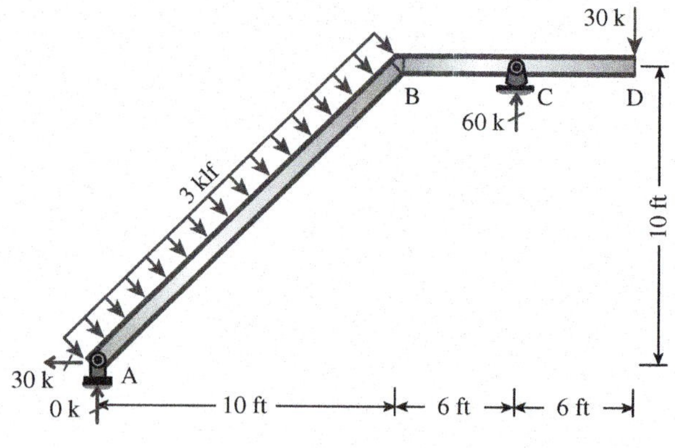

Figure VE6.5

VE6.6 For the beam shown in Figure VE6.4, generate the moment diagrams using the principle of superposition. Select point C to be the cantilever point. Once the diagrams have been created, check to be sure that the moments at point C sum up to be equal to zero on each side of the cantilever, since point C is an internal moment release.

6.12 Problems for Solution

Section 6.3

For Problems 6.1 through 6.10, write the equations for shear and bending moments throughout the structures shown using the x coordinate(s) given.

P6.1 (Ans: $M_3(x) = -64x + 3200$; $40 \text{ ft} < x \leq 50 \text{ ft}$)

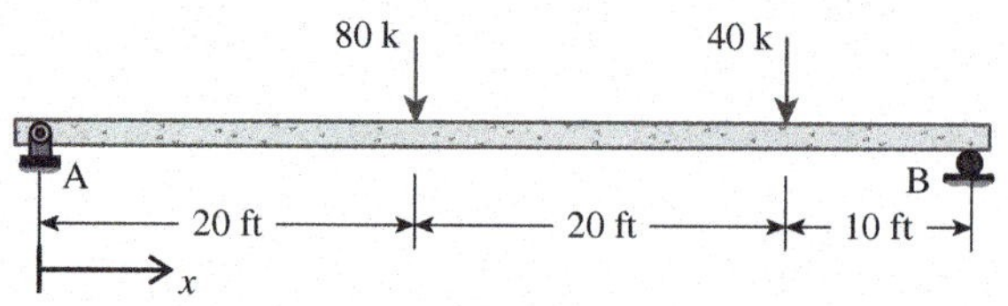

Problem 6.1

P6.2

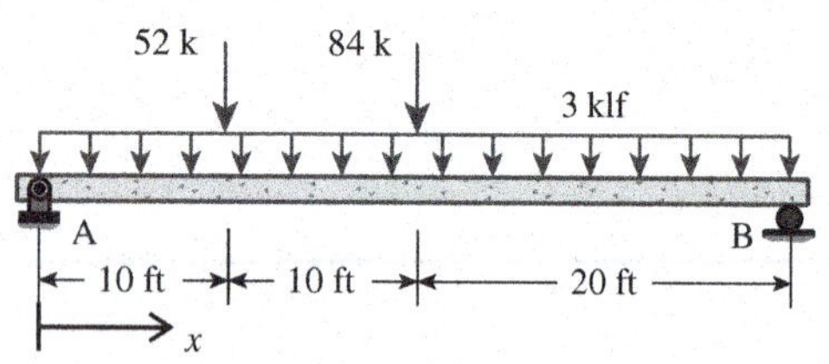

Problem 6.2

P6.3 (Ans: $M_3(x) = -1.5x^2 - 21x + 3240$; $30 \text{ ft} < x \leq 40 \text{ ft}$)

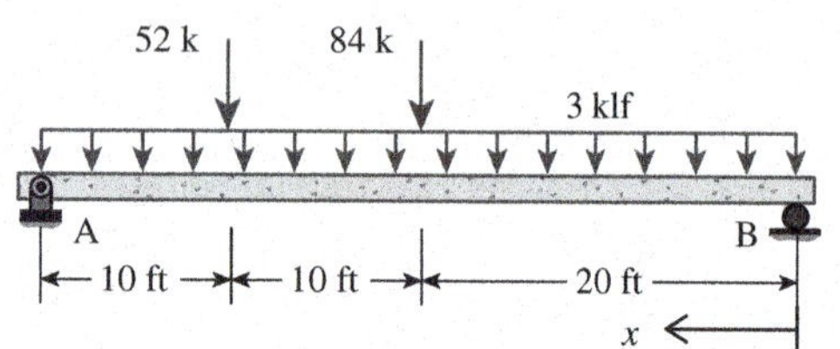

Problem 6.3

P6.4

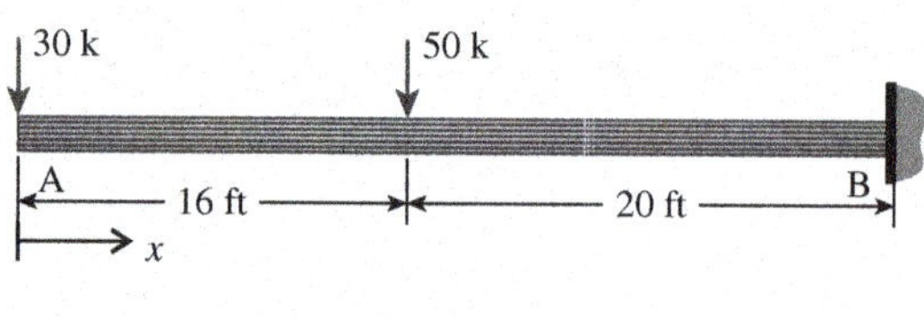

Problem 6.4

P6.5 (Ans: $M(x_2) = -80x_2 - 480$; $0 \text{ ft} < x_2 \leq 20 \text{ ft}$)

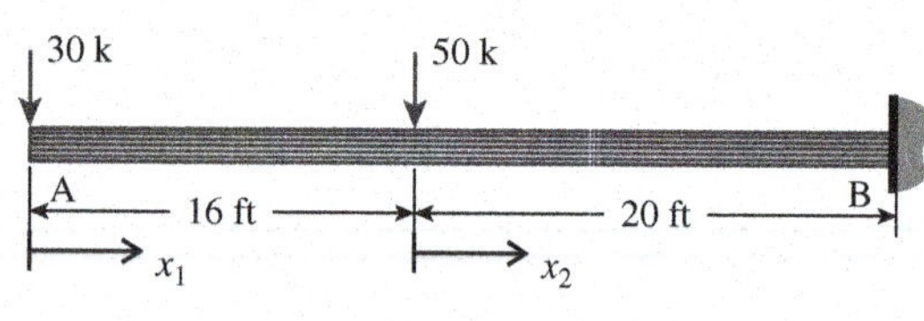

Problem 6.5

P6.6

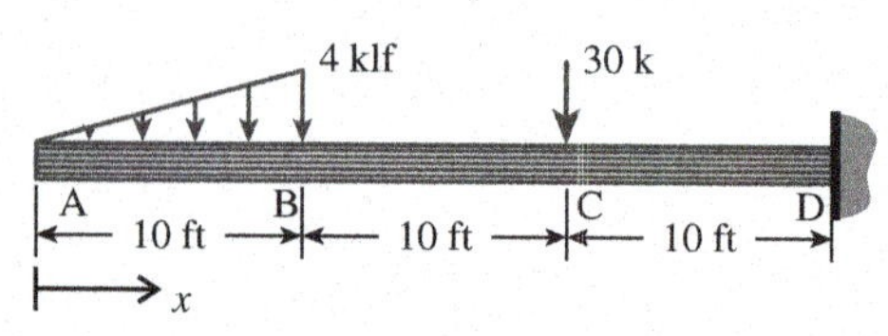

Problem 6.6

P6.7 (Ans: $M_2(x) = -x^2 + 96x - 2304;\quad 36 \text{ ft} < x \leq 48 \text{ ft}$)

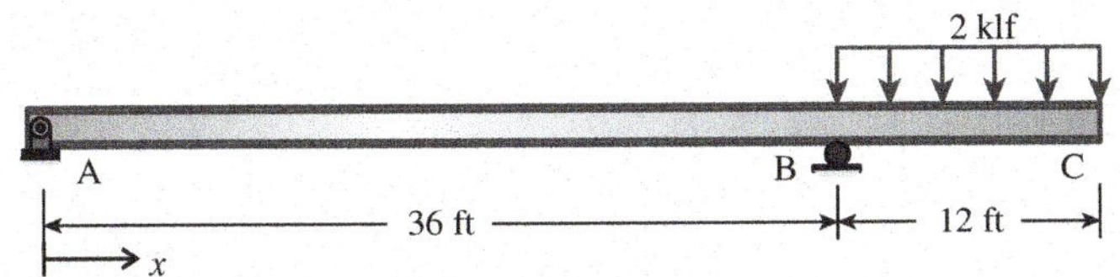

Problem 6.7

P6.8

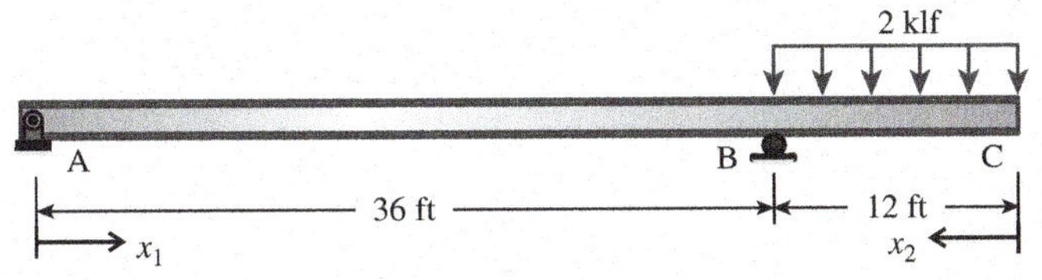

Problem 6.8

P6.9 (Ans: $M(x_2) = 800;\quad 0 \text{ ft} < x_2 \leq 10 \text{ ft}$)

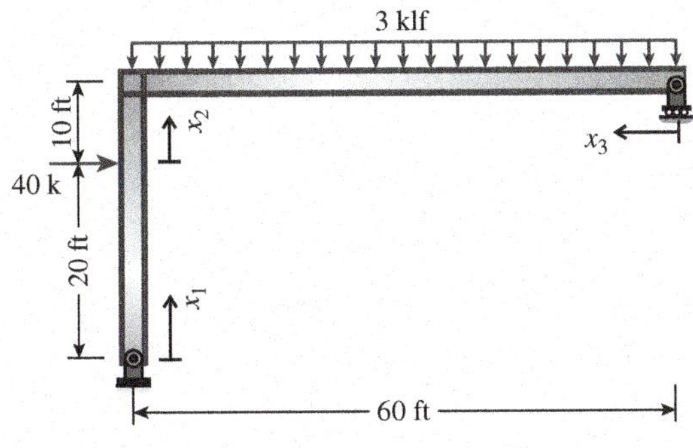

Problem 6.9

P6.10

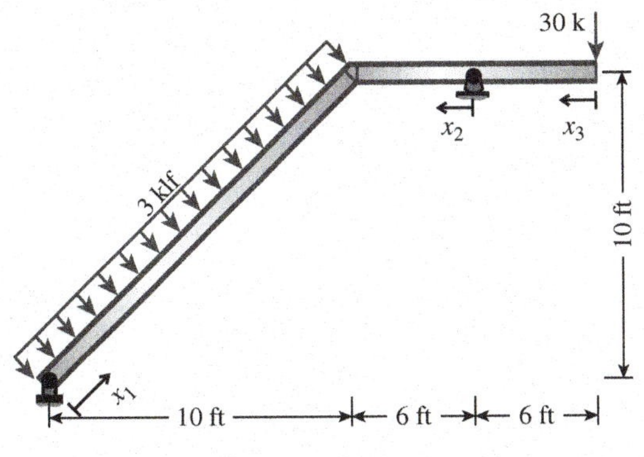

Problem 6.10

Section 6.4

For Problems 6.11 through 6.14, the piecewise moment equations are given for a beam where the origin of x is located at the left end of the beam. Derive the shear and load equations for the beam. Then plot the load diagram, shear diagram, and moment diagram. Assume upward forces are supports and that units in all cases are kips and feet.

P6.11

$$M(x) = \begin{vmatrix} -x^2 + 5x & 0 \leq x \leq 10 \text{ ft} \\ -2x^2 + 60x - 450 & 10 < x \leq 15 \text{ ft} \end{vmatrix}$$

(Ans: for the first section, $V(x) = -2x + 5$, $w(x) = 2$ klf)

P6.12

$$M(x) = \begin{vmatrix} -13x + 8 & 0 \leq x \leq 2 \text{ ft} \\ 6x - 30 & 2 < x \leq 4 \text{ ft} \\ -1.5x^2 + 18x - 54 & 4 < x \leq 6 \text{ ft} \end{vmatrix}$$

P6.13

$$M(x) = \begin{vmatrix} -10x & 0 \leq x \leq 2.5 \text{ ft} \\ 19x - 72.5 & 2.5 < x \leq 5.5 \text{ ft} \\ -16x + 120 & 5.5 < x \leq 7.5 \text{ ft} \end{vmatrix}$$

(Ans: for middle section: $V(x) = 19$ k, $w(x) = 0$ k)

P6.14

$$M(x) = -\frac{x^3}{12} - x^2 + 9x \qquad 0 \leq x \leq 6 \text{ ft}$$

Section 6.5

For Problems 6.15 through 6.55, draw the shearing force and bending moment diagrams for the structures shown.

P6.15 Use the beam given in Problem 6.1 (Ans: max $V = -64$ k, max $M = 1120$ k·ft)

P6.16 Repeat Problem 6.15 if the 80-k load is changed to 120 k.

P6.17 (Ans: max $V = -70$ k, max $M = 312.5$ k·ft)

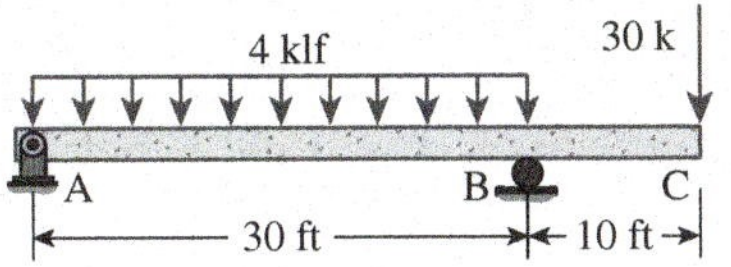

Problem 6.17

P6.18 Repeat Problem 6.17 if the 4-klf load is changed to 2.5 klf.

P6.19 (Ans: max $V = -65$ k, max $M = -337.5$ k·ft)

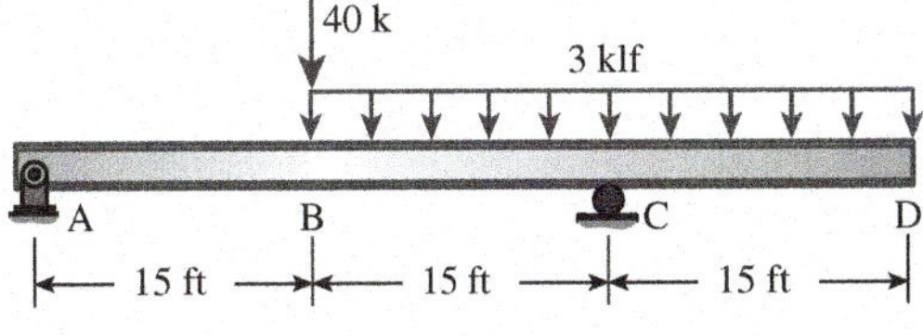

Problem 6.19

P6.20

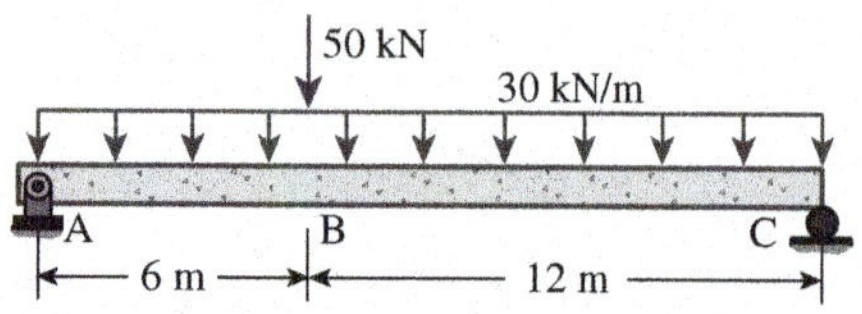

Problem 6.20

P6.21 (Ans: max $V = -64$ k, max $M = -1706$ k·ft)

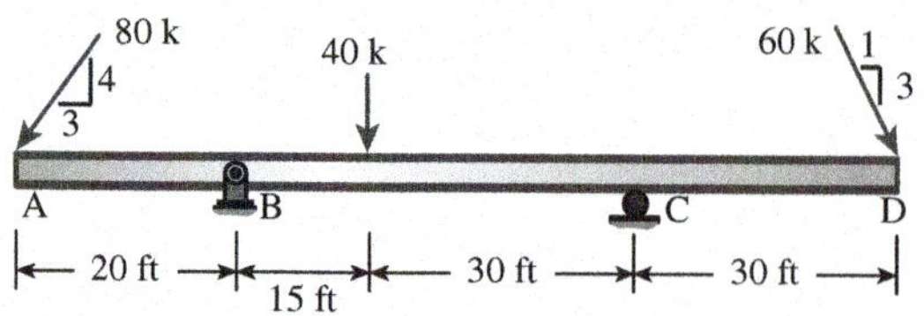

Problem 6.21

P6.22 Use the beam given in Problem 6.2.

P6.23 (Ans: max $V = 120$ k, max $M = -1200$ k·ft)

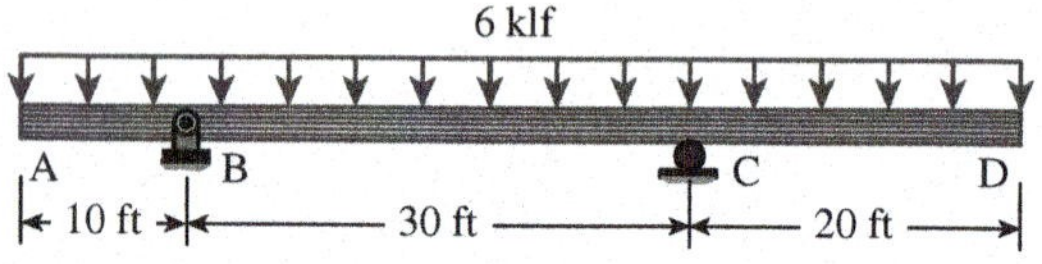

Problem 6.23

P6.24

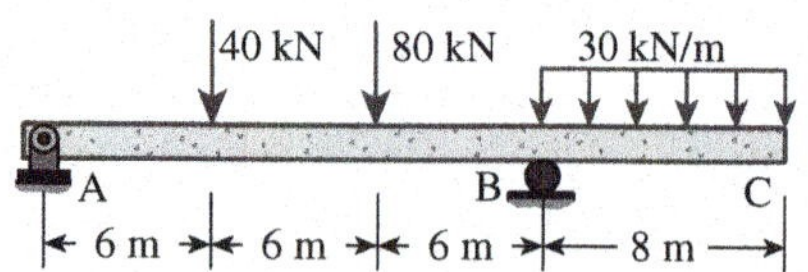

Problem 6.24

P6.25 (Ans: max $V = -145.6$ k, max $M = -1458$ k·ft)

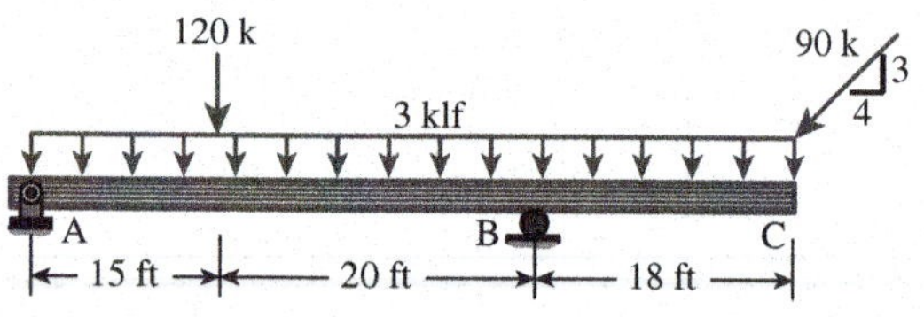

Problem 6.25

P6.26 Use the beam given in Problem 6.4.

P6.27 (Ans: max $V = 204$ k, max $M = -4032$ k·ft)

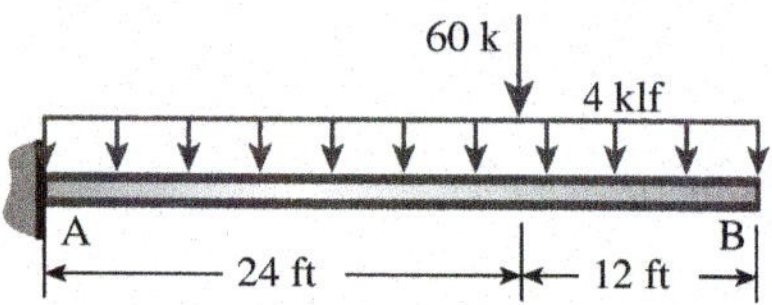

Problem 6.27

P6.28

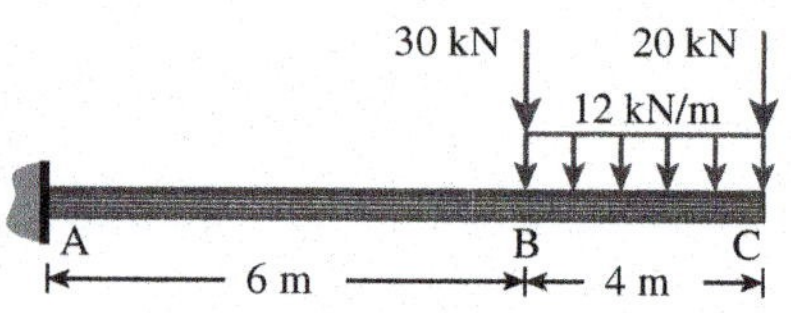

Problem 6.28

P6.29 (Ans: max $V = -90$ k, max $M = -900$ k·ft)

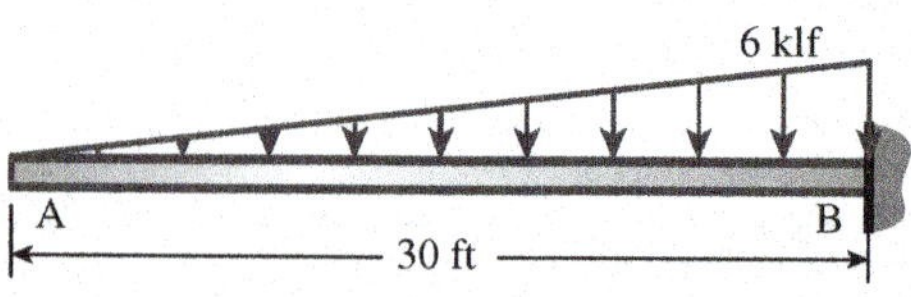

Problem 6.29

P6.30 Use the beam given in Problem 6.6.

P6.31 Use the beam given in Problem 6.7. (Ans: max $V = 24$ k, max $M = -144$ k·ft)

P6.32

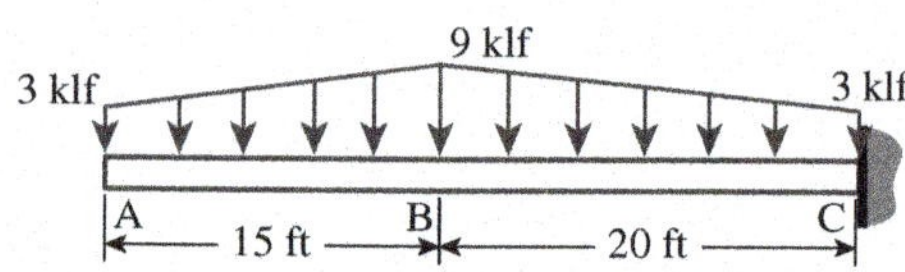

Problem 6.32

P6.33 (Ans: max $V = -117.5$ k, max $M = -2762.5$ k·ft)

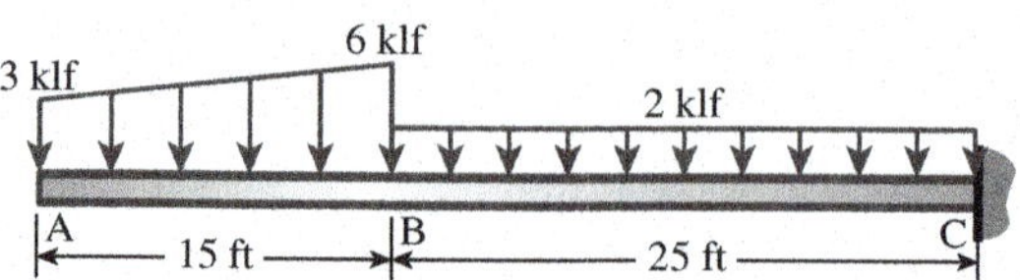

Problem 6.33

P6.34

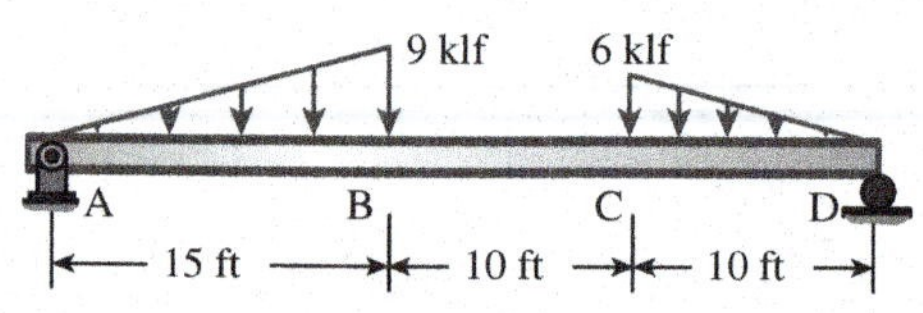

Problem 6.34

P6.35 (Ans: max $V = -95.4$ k, max $M = 453.5$ k·ft)

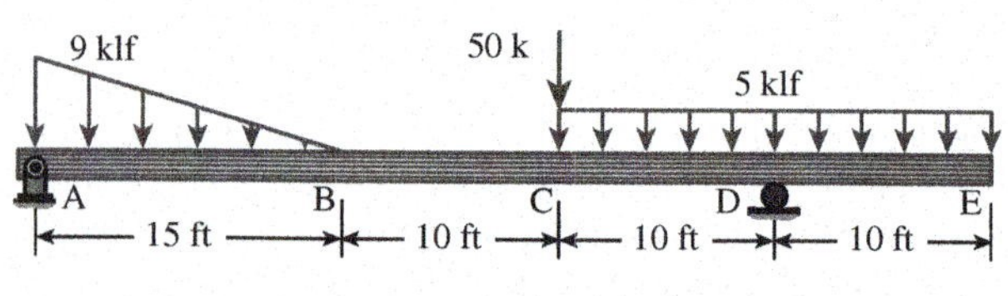

Problem 6.35

P6.36

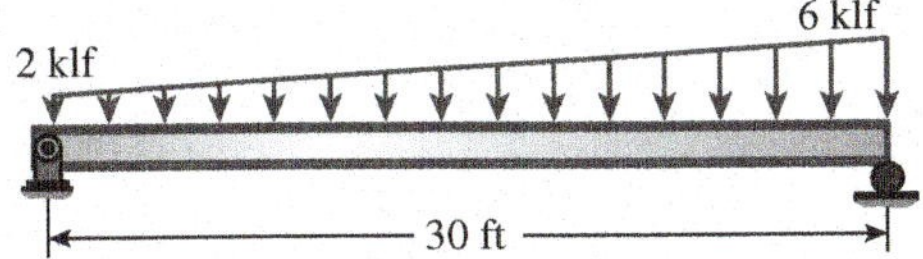

Problem 6.36

P6.37 (Ans: max $V = -73$ k, max $M = -524$ k·ft)

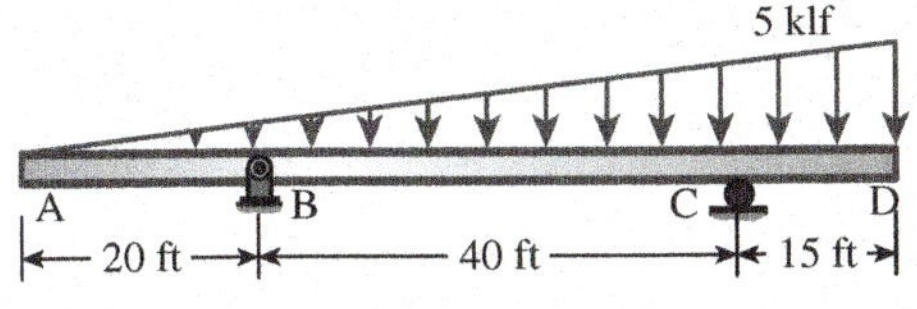

Problem 6.37

P6.38

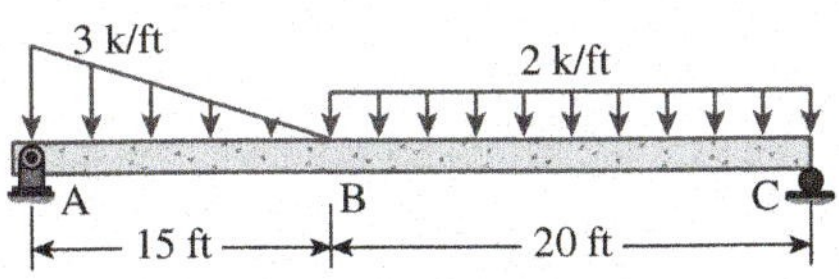

Problem 6.38

P6.39 (Ans: max $V = 44.7$ k, max $M = -533$ k·ft)

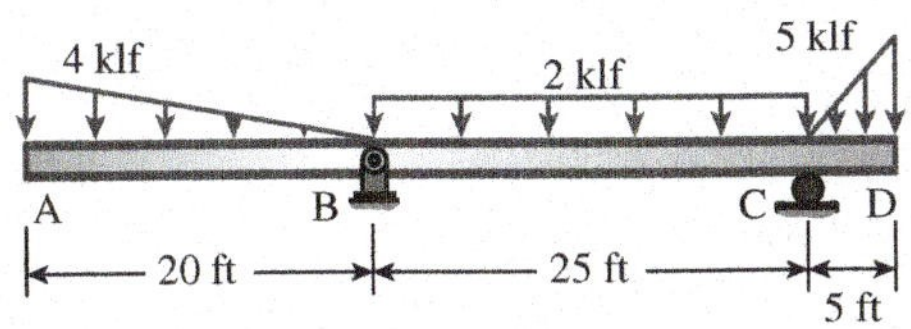

Problem 6.39

P6.40

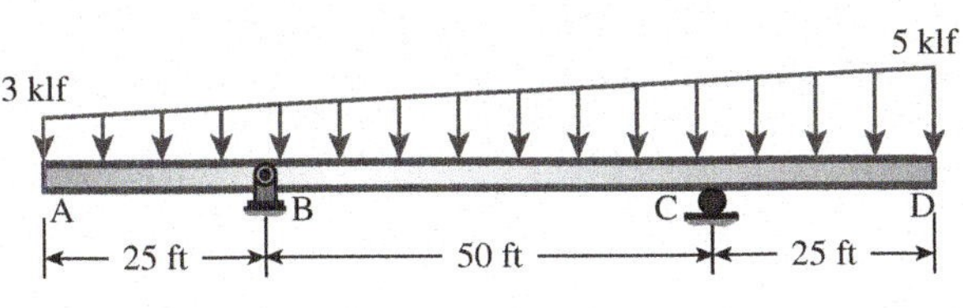

Problem 6.40

P6.41 (Ans: max $V = 5120$ lb, max $M = 37547$ lb·ft)

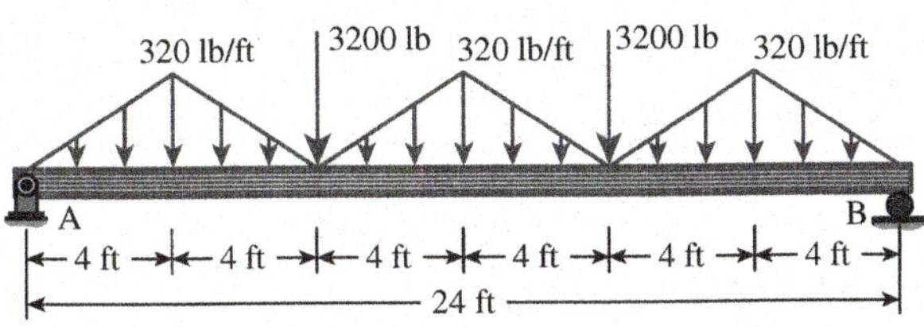

Problem 6.41

P6.42

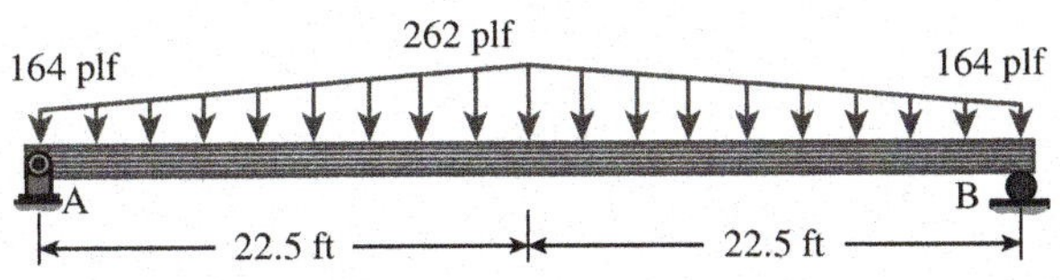

Problem 6.42

P6.43 (Ans: max $V = -50$ k, max $M = -920$ k·ft)

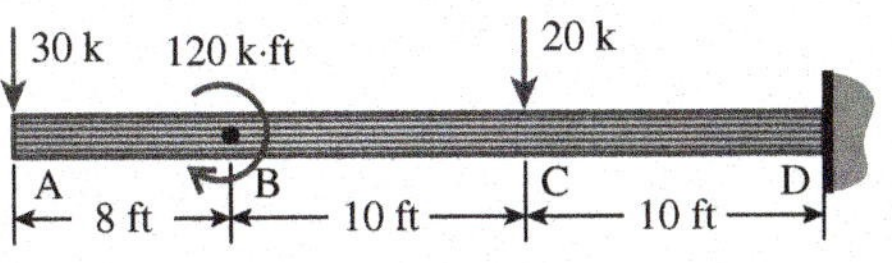

Problem 6.43

P6.44

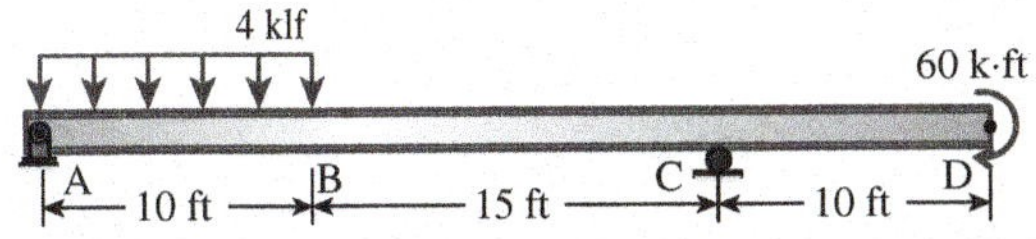

Problem 6.44

P6.45 (Ans: max $V = 137.9$ k, max $M = -1920$ k·ft)

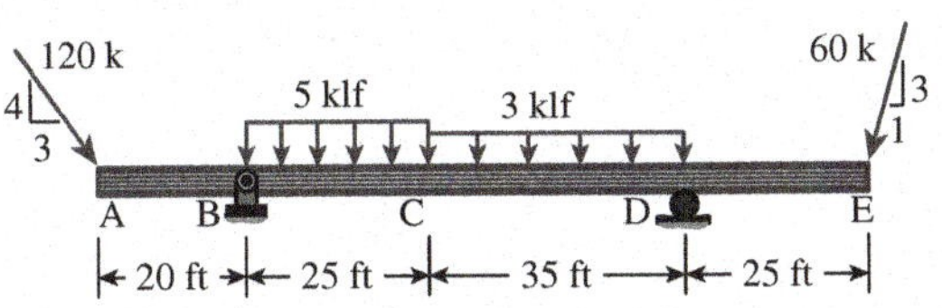

Problem 6.45

P6.46

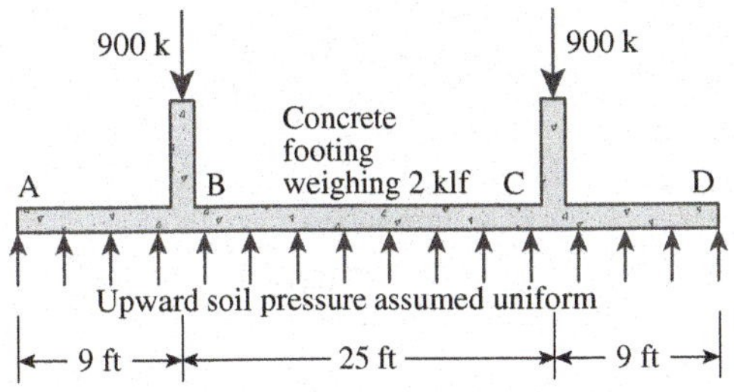

Problem 6.46

P6.47 (Ans: max $V = 96.7$ k, max $M = 2933$ k·ft)

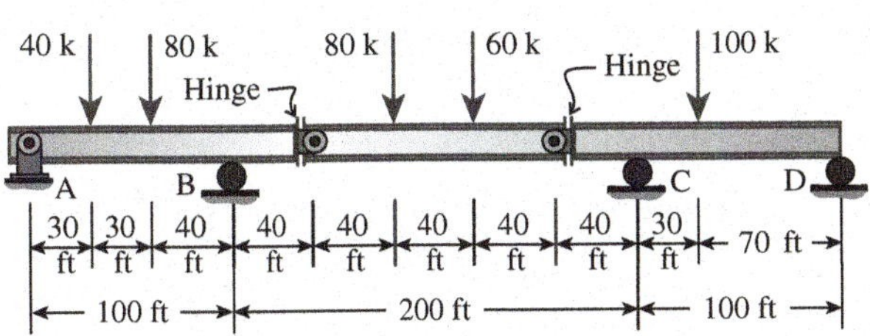

Problem 6.47

P6.48

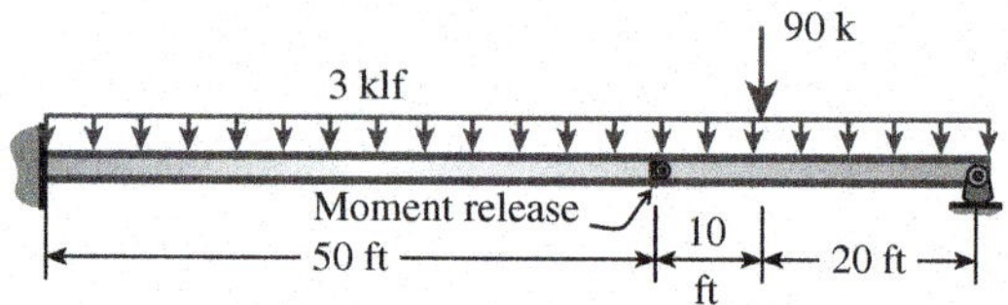

Problem 6.48

P6.49 (Ans: max $V = 105$ k, max $M = -3400$ k·ft)

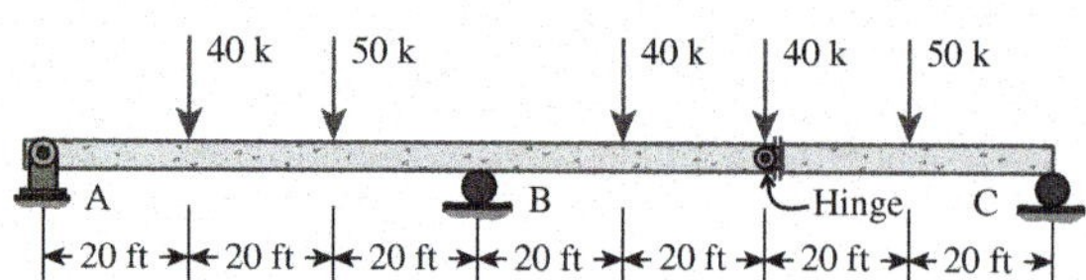

Problem 6.49

P6.50

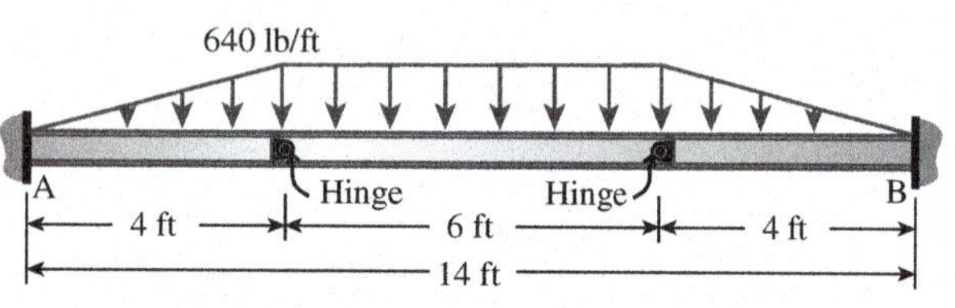

Problem 6.50

P6.51 (Ans: max $V = 7$ kN, max $M = -18$ kN·m)

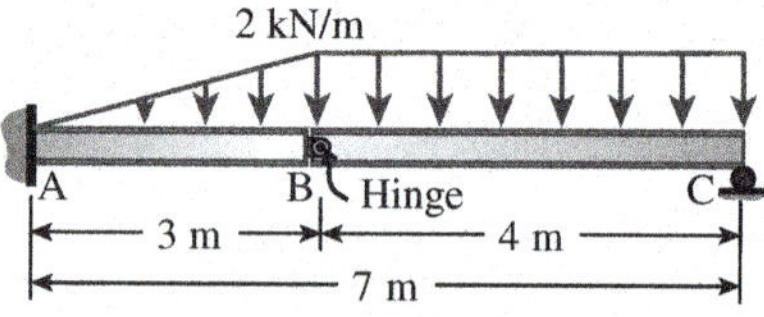

Problem 6.51

P6.52

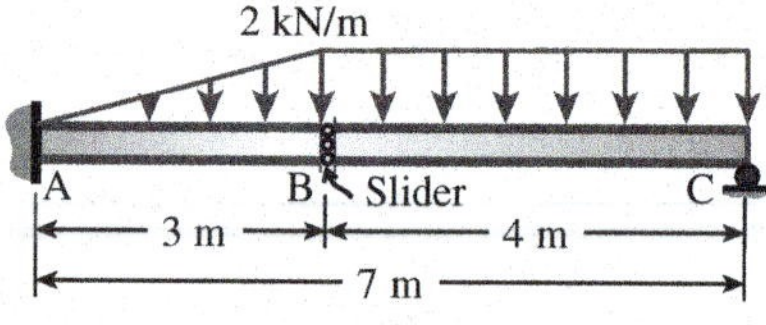

Problem 6.52

P6.53 (Ans: max $V = 24$ k, max $M = -128$ k·ft)

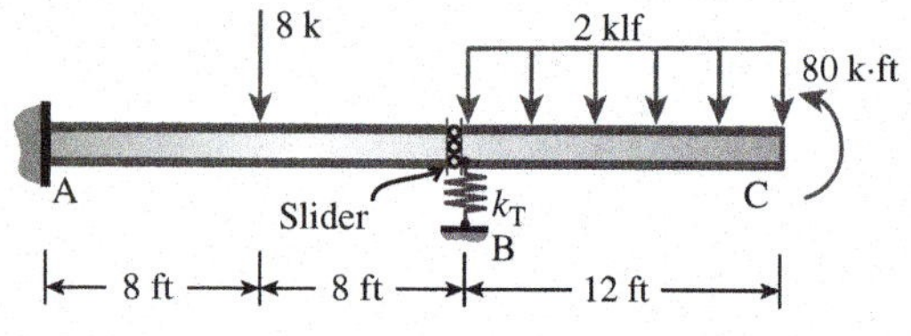

Problem 6.53

P6.54

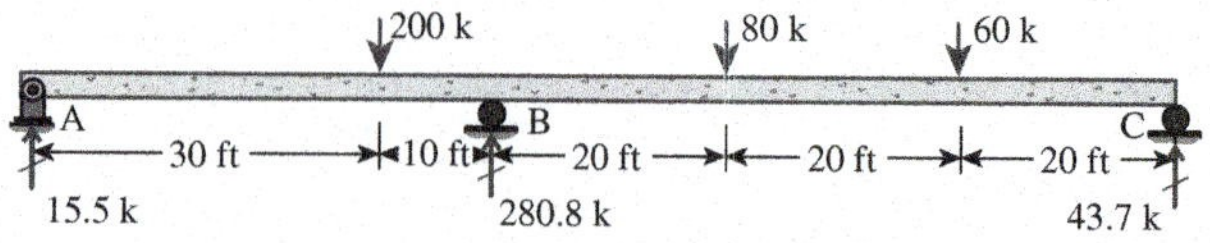

Problem 6.54

P6.55 (Ans: max $V = 25.5$ k, max $M = -164$ k·ft)

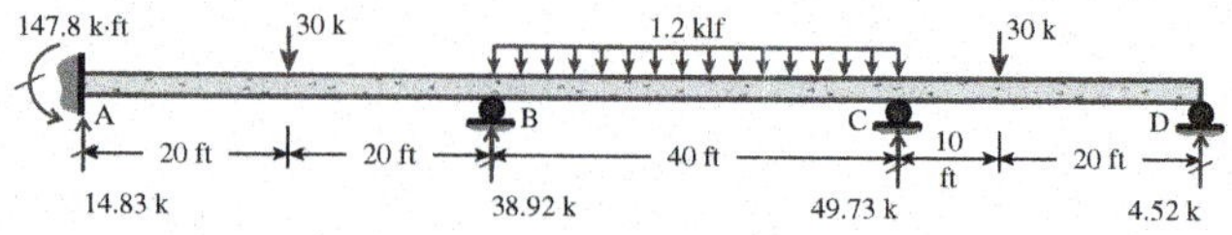

Problem 6.55

For Problems 6.56 through 6.58, draw the shear diagrams and load diagrams for the moment diagrams and dimensions given. Assume that upward forces are reactions.

P6.56

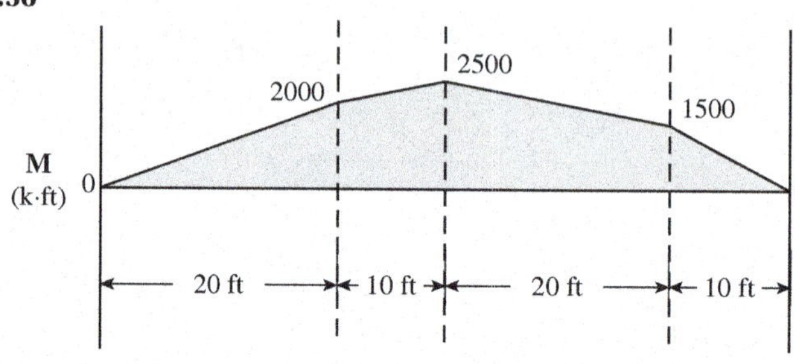

Problem 6.56

P6.57 (Ans: forces from left to right: 108 k, −93.3 k, −56.7 k, −20 k, and 62 k)

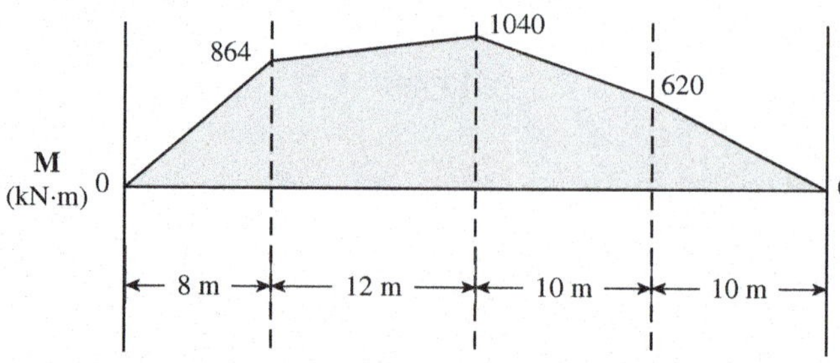

Problem 6.57

P6.58 Assume all lines in the moment diagram are second order.

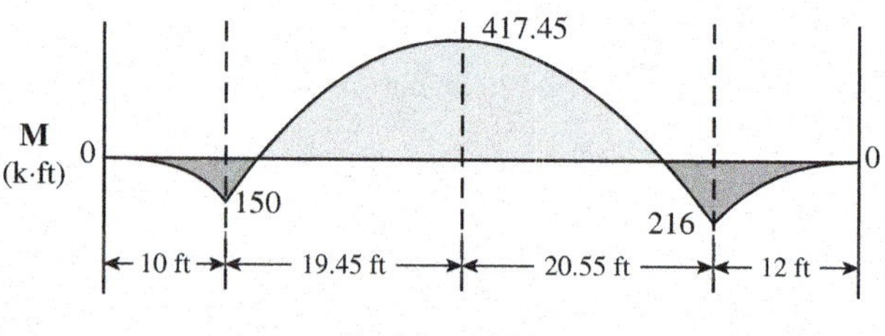

Problem 6.58

Section 6.6

For Problems 6.59 through 6.63, draw the axial force diagrams for the structures shown.

P6.59 (Ans: max $N = -75.9$ k)

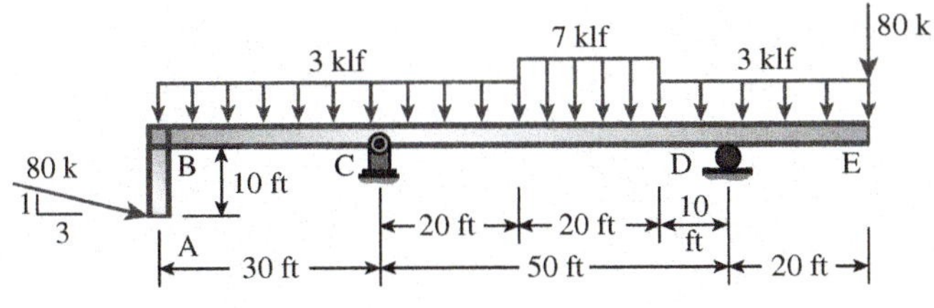

Problem 6.59

P6.60 The structure in Problem 6.21.

P6.61 The structure in Problem 6.45. (Ans: max $N = -72$ k)

P6.62

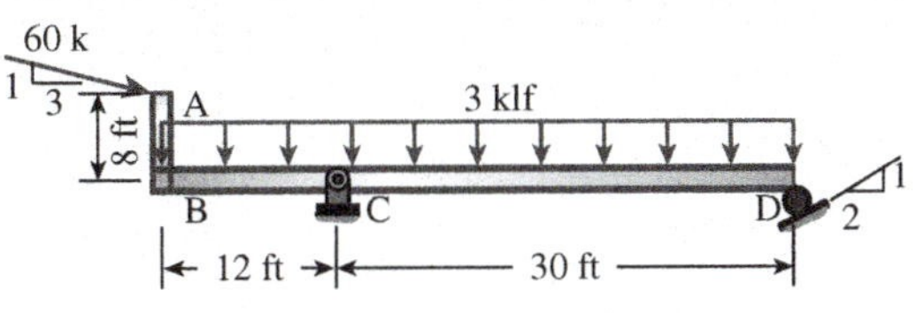

Problem 6.62

P6.63 (Ans: max $N = -42.4$ k)

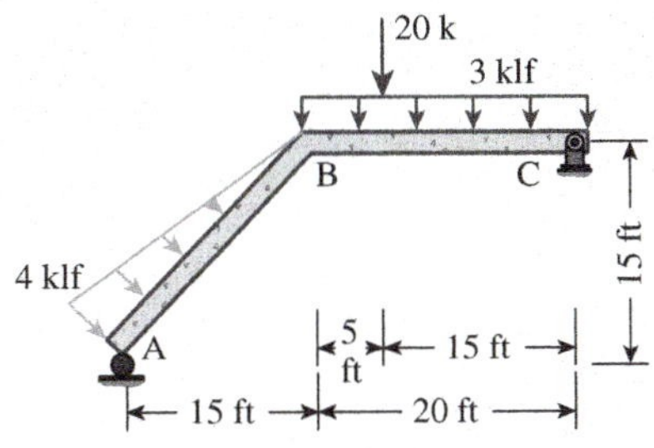

Problem 6.63

For Problems 6.64 through 6.65, draw the axial force diagrams for a bar that is subjected to distributed axial loads over portions of its length. These diagrams are typically encountered when analyzing diaphragms like those presented in Chapter 4.

P6.64

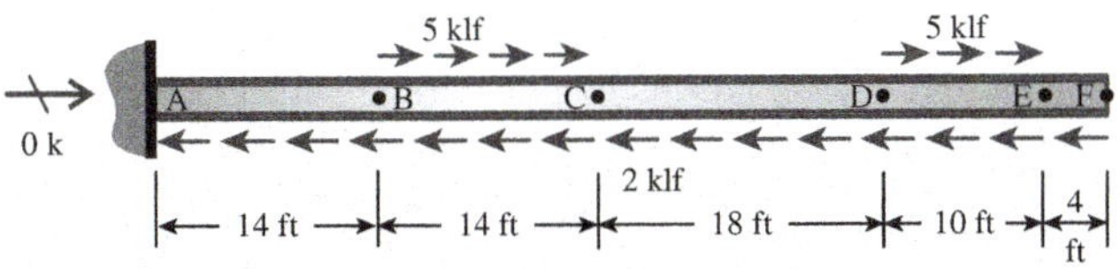

Problem 6.64

P6.65 (Ans: max $N = 10$ k)

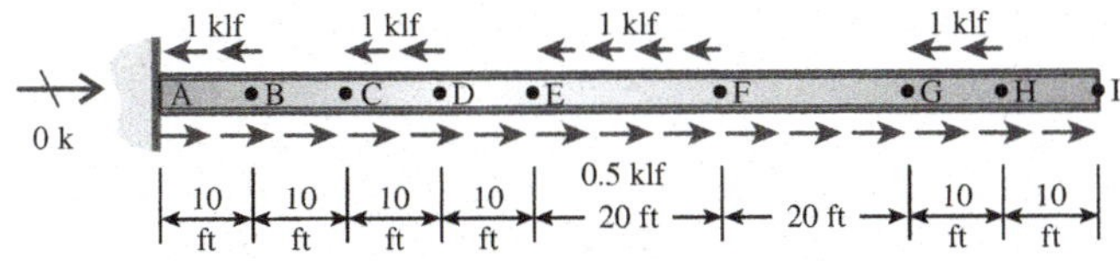

Problem 6.65

Section 6.7

For Problems 6.66 through 6.79, draw the shear and moment diagrams for the frames shown.

P6.66 Use the structure in Problem 6.59.

P6.67 Use the structure in Problem 6.62. (Ans: max $V = 56.9$ k, max $M = 455$ k·ft)

P6.68 Use the structure in Problem 6.63.

P6.69 (Ans: max $V = -66.7$ k, max $M = 1240$ k·ft)

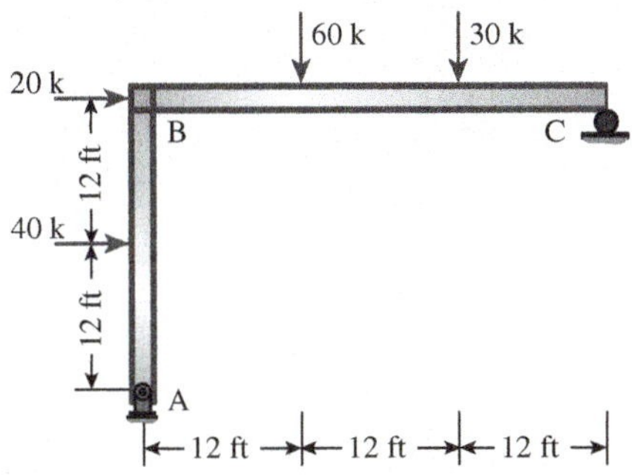

Problem 6.69

P6.70 Repeat Problem 6.69 but swap the pin and roller supports.

P6.71 (Ans: max $V = -78.5$ k, max $M = 1339$ k·ft)

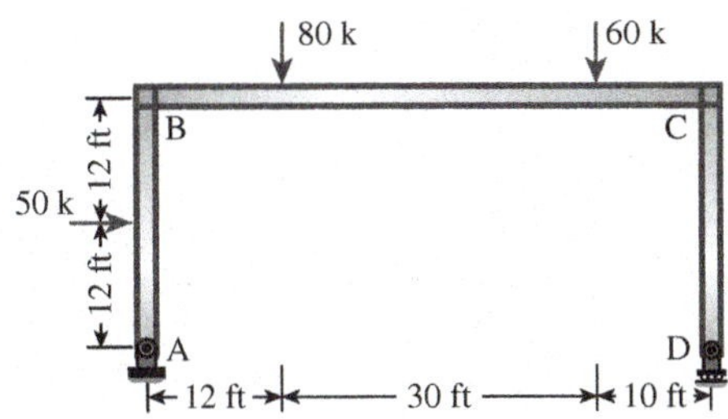

Problem 6.71

P6.72

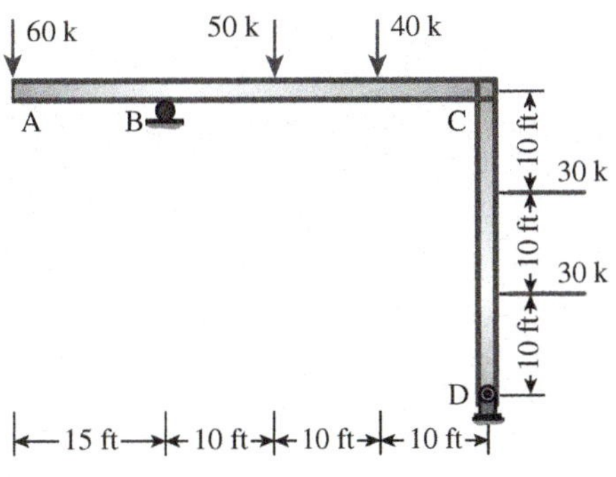

Problem 6.72

P6.73 (Ans: max $V = -261$ k, max $M = 5682$ k·ft)

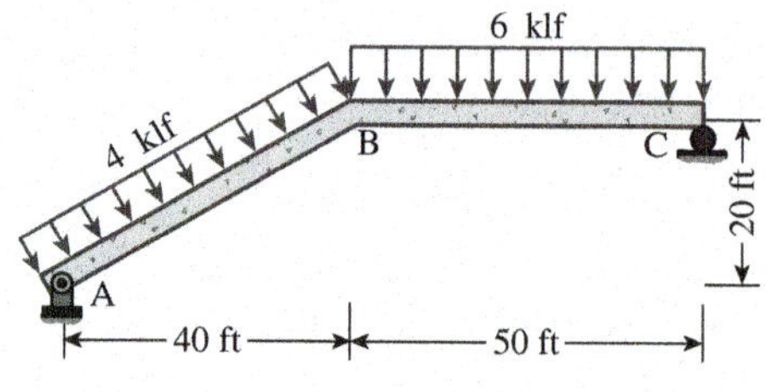

Problem 6.73

P6.74

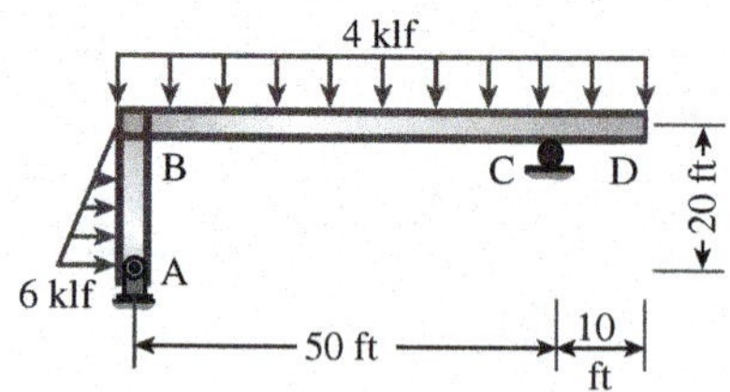

Problem 6.74

P6.75 (Ans: max $V = 126.5$ k, max $M = -3300$ k·ft)

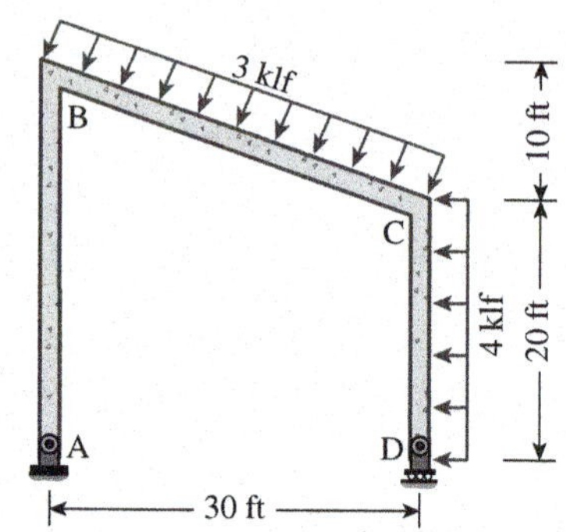

Problem 6.75

P6.76

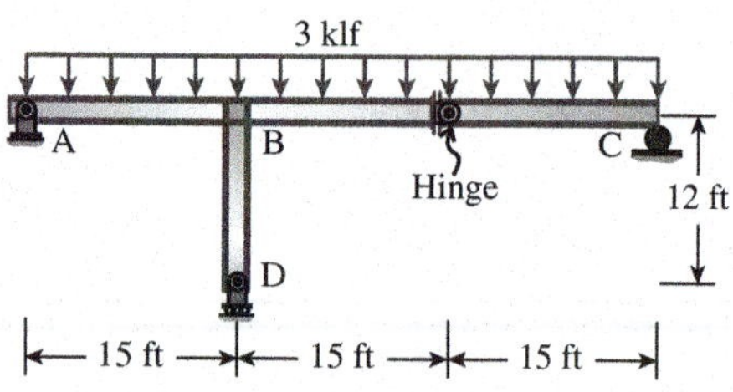

Problem 6.76

P6.77 (Ans: max $V = 110$ k, max $M = 1050$ k·ft)

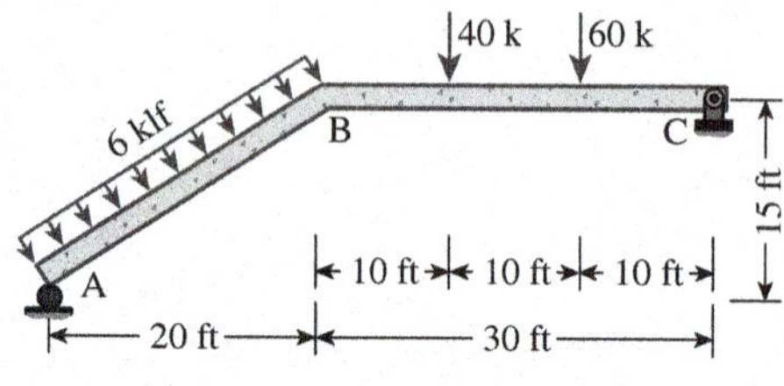

Problem 6.77

P6.78

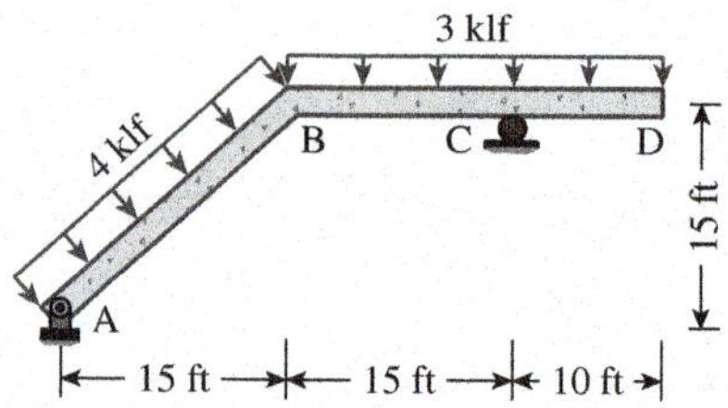

Problem 6.78

P6.79 (Ans: max $V = -14$ k, max $M = -120$ k·ft)

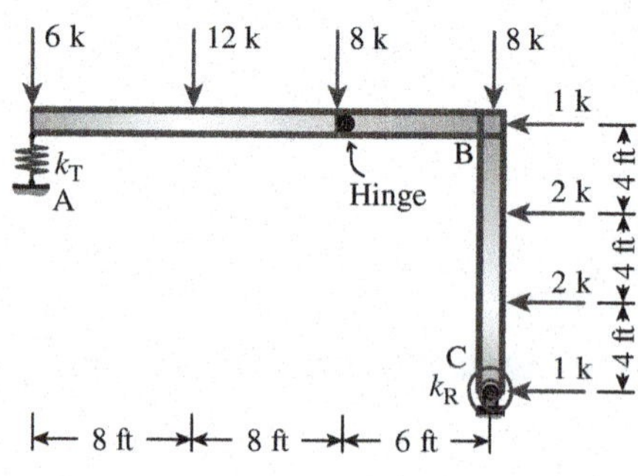

Problem 6.79

Section 6.8

For Problems 6.80 through 6.89, draw moment diagrams using superposition for the structure identified. You do not need to add all the moment diagrams together.

P6.80 Use the structure in the Problem 6.1 cantilever at left support.

P6.81 Use the structure in the Problem 6.1 cantilever at 80 k load. (Ans: total M at point of cantilever 1120 k·ft)

P6.82 Use the structure in the Problem 6.2 cantilever at 84 k load.

P6.83 Use the structure in the Problem 6.4 cantilever at right support. (Ans: total M at point of cantilever −2080 k·ft)

P6.84 Use the structure in the Problem 6.23 cantilever at roller support.

P6.85 Use the structure in the Problem 6.43 cantilever at right support. (Ans: total M at point of cantilever −920 k·ft)

P6.86 Use the structure in the Problem 6.49 cantilever at internal hinge.

P6.87 Use the structure in the Problem 6.54 cantilever at middle roller. (Ans: total M at point of cantilever −1380 k·ft)

P6.88 Use the structure in the Problem 6.73 cantilever at joint between sloped and horizontal member.

P6.89 Use the structure in the Problem 6.72 cantilever at joint between vertical and horizontal member. (Ans: total M at point of cantilever 900 k·ft)

Section 6.10 SAP2000

P6.90 Using the structure provided in Problem 6.17, conduct a parametric study using SAP2000 to investigate the influence of the position of the 30-k load in the maximum positive internal moment. First position the 30-k load just over the left support. Then move the load in 5-ft increments to the right. The structure should be analyzed for the 30-k load in the following locations as measured from the left: (a) 0 ft, (b) 5 ft, (c) 10 ft, (d) 15 ft, (e) 20 ft, (f) 25 ft, (g) 30 ft, (h) 35 ft, and (i) 40 ft. Plot the location versus the maximum positive moment in the beam. What do you notice happens to the internal moment once the point load is placed to the right of the roller?

P6.91 Analyze the structure given in Problem 6.41 and plot its shear and moment diagrams. Now calculate by hand the resultant of the loads applied, apply it as a point load in the middle of the beam, and get the shear and moment diagrams. Finally, take the load resultant, divide it by the length of the beam, apply as a uniformly distributed load over the length of the beam, and then get the shear and moment diagrams. Make comment on your observations relative to the magnitude and shape of the three sets of diagrams.

P6.92 Analyze the structure given in Problem 6.52 where the slider is modeled as a shear release in SAP2000. Plot the shear and moment diagrams. Now change the shear release to a moment release and then remove the release altogether. Compare all three sets of shear and moment diagrams and make observations.

P6.93 Analyze the structure given in Problem 6.75. Plot the shear, moment, and axial diagrams. Now model the roller as a pin and re-plot the shear, moment, and axial diagrams. Make an observation regarding the influence of the added support on the values and shapes of the diagrams. Let all members have the same material and cross-sectional properties: $E = 29{,}000$ ksi, $I = 158$ in.^{4}, and $A = 6.6$ in.^{2}

7 Plane Trusses

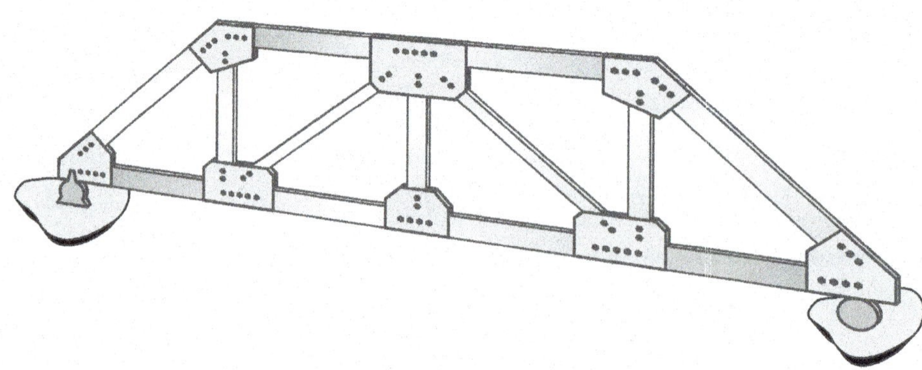

7.1 Introduction

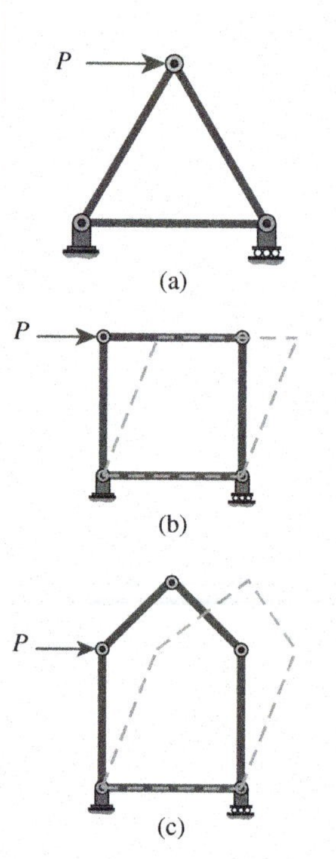

FIGURE 7.1 Stable versus unstable pin-connected structures.

The Italian architect Andrea Palladio (1508–1580) is thought to have first used modern trusses, although his design basis is not known. He may have revived some old Roman designs and probably sized the truss components by some rules of thumb (perhaps old Roman rules). Palladio's extensive writings in architecture included detailed descriptions and drawings of wooden trusses quite similar to those used today. After his time, trusses were forgotten for 200 years, until they were reintroduced by Swiss designer Ulric Grubermann.

A *truss* is a structure formed by a group of members arranged in the shape of one or more triangles. Because the members are assumed to be connected with frictionless pins, the triangle is the only stable shape. Study of the truss of Figure 7.1(a) shows that it is impossible for the triangle to change shape under load—except through deformation of the members—unless one or more of the sides is bent or broken. Pin-connected structures of four or more sides are not stable and may collapse under load, as seen in Figures 7.1(b) and (c). These structures may be deformed even if none of the members change length. We will see, however, that many stable trusses can include one or more shapes that are not triangles. Careful study reveals that trusses consist of separate groups of triangles that are connected together according to definite rules, forming non-triangular but stable elements in between.

Design engineers often are concerned with selecting either a truss or a beam to span a given opening. Should no other factors be present, the decision probably would be based on consideration of economy. The smallest amount of material will nearly always be used if a truss is selected for spanning a certain opening; however, the cost of fabrication and erection of trusses probably will be appreciably higher than that required for beams. For shorter spans, the overall cost of beams (material cost plus fabrication and erection cost) will definitely be less, but as the spans become greater, the higher fabrication and erection costs of trusses will be more than offset by the reduction in the total weight of material used. A further advantage of a truss is that for the same amount of material it can have greater stiffness than a beam with the same span.

A lower limit for the economical span of trusses is impossible to give. Trusses may be used for spans as small as 30 to 40 ft and as large as 300 to 400 ft. Beams may be economical for some spans much greater than the lower limits mentioned for trusses.

7.2 Assumptions for Truss Analysis

The following assumptions are made to simplify the analysis of trusses.

1. ***Truss members are connected with frictionless pins.*** In reality, pin connections are used for very few trusses erected today, and no pins are frictionless. A heavy bolted or welded joint is very different from a frictionless pin. If frictionless pins are assumed, then a truss bar is nothing more than a two-force member discussed in any statics class.
2. ***Truss members are straight.*** If they were not straight, the axial forces would cause them to have bending moments.
3. ***The displacement of the truss is small.*** The applied loads cause the members to change length, which then causes the truss to deform. The deformations of a truss are not of sufficient magnitude to cause appreciable changes in the overall shape and dimensions of the truss. If the dimensions did change significantly, the dimension changes would need to be included in the equilibrium equations.
4. ***Loads are applied only at the joints.*** Members are arranged so that the loads and reactions are only applied at the truss joints.

QUICK NOTE

A frictionless pin in this textbook is denoted by the following symbol.

Examination of roof and bridge trusses prove this last statement to be generally true. Beams, columns, and bracing members frame directly into the truss joints of buildings with roof trusses. Roof loads are transferred to trusses by horizontal beams, called *purlins*, that span between the trusses. The roof is supported directly by the purlins. The roof may also be supported by rafters, or sub-purlins, that run parallel to trusses and are supported by the purlins. The purlins are placed at the truss joints unless the top-chord panel lengths become exceptionally long; in this case, it is sometimes economical to place purlins between the joints, although some bending will develop in the top chords. Some types of roofing, such as corrugated steel and gypsum slabs, may be laid directly on the purlins. In this case, the purlins have to be spaced at intermediate points along the top chord to provide a proper span for the supported roof. Similarly, the loads supported by a highway bridge are transferred to the trusses at the joints by floor beams and girders underneath the roadway.

The effect of the four assumptions is to produce an ideal truss whose members have only axial forces. A member with only axial force is subjected to axial tension or compression; there is no bending moment or shear present. Be aware, however, that even if all the assumptions about trusses were completely true, there still would be some bending in a member because of its own weight. The weight of the member is distributed along its length rather than being concentrated at the ends. Compared to the forces caused by the applied loads, the forces caused by self-weight are small and generally can be neglected when calculating the forces in the components.

Component forces obtained using some or all of these simplifying assumptions are very satisfactory in most cases and are referred to as *primary forces*. Forces caused by conditions not considered in the primary force analysis are said to be *secondary forces*.

DID YOU KNOW?

Most truss joints are not connected with pins. Rather they could be connected as shown in Figure 7.2. This appears to be a very rigid connection and is far removed from the frictionless pin connection. However, if the centerline of the members coincide, the truss members are still dominated by axial loads.

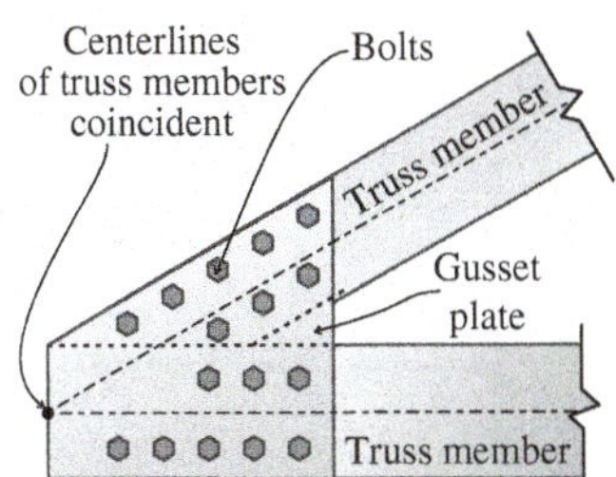

FIGURE 7.2 Typical joint in a steel truss.

Truss Terminology

The notation used for trusses in this textbook is to identify each joint with a letter in the alphabet, as shown in Figure 7.3. This allows the truss member to be identified using two letters. There are other notation systems frequently used for trusses. For instance, for computer-programming purposes, it is convenient to assign a number to each joint and each member of a truss as seen in the output of SAP2000. However, for hand calculations, it becomes less confusing to use the alphabet as joint identifiers when there are fewer than 26 joints in the truss.

Defining a few basic terms will aid the student in communicating with others.

Chords Those members forming the outline of the truss, such as members AB and GH.

Verticals These are named on the basis of their direction in the truss, such as members AG and BH.

QUICK NOTE

Not every truss will have

- Verticals
- End posts
- Clearly definable panels

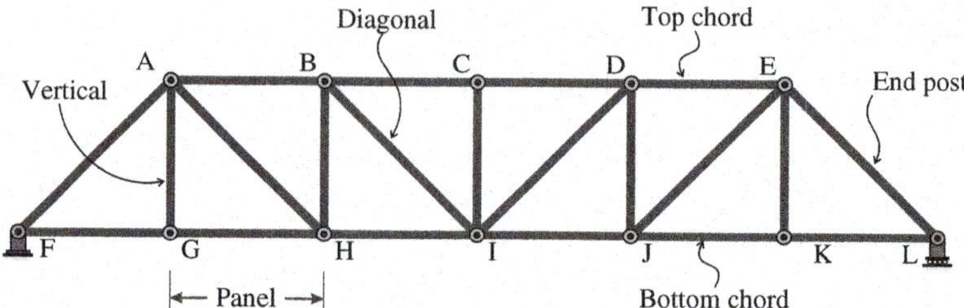

FIGURE 7.3 Labeling of truss joints.

Diagonals These are named on the basis of their direction in the truss, such as members AH and BI.

End Posts The members at the ends of the truss, such as members AF and EL.

Web Members These include the verticals and diagonals of a truss. Most engineers consider the web members to include the end posts. This is the most generic and common term used for the members located between the top and bottom chords.

Panels The spaces delineated by the vertical web members as seen between G and H. There are six panels in Figure 7.3.

Panel Point Another term used for truss joints. G and H are panel points.

7.3 Roof Trusses

The purposes of roof trusses are to support the roofs that keep out the elements (rain, snow, wind), the loads connected underneath (ducts, piping, ceiling), and their own weight.

Roof trusses can be flat or peaked. In the past, peaked roof trusses probably have been used more for short-span buildings and flatter trusses for longer spans. The trend today for both long and short spans, however, seems to be away from the peaked trusses and toward the flatter trusses—except for in structures like houses. The change is predominantly due to the desired appearance of the building and perhaps a more economical construction of roof decks. Figure 7.4 illustrates several types of roof trusses that have been used in the past.

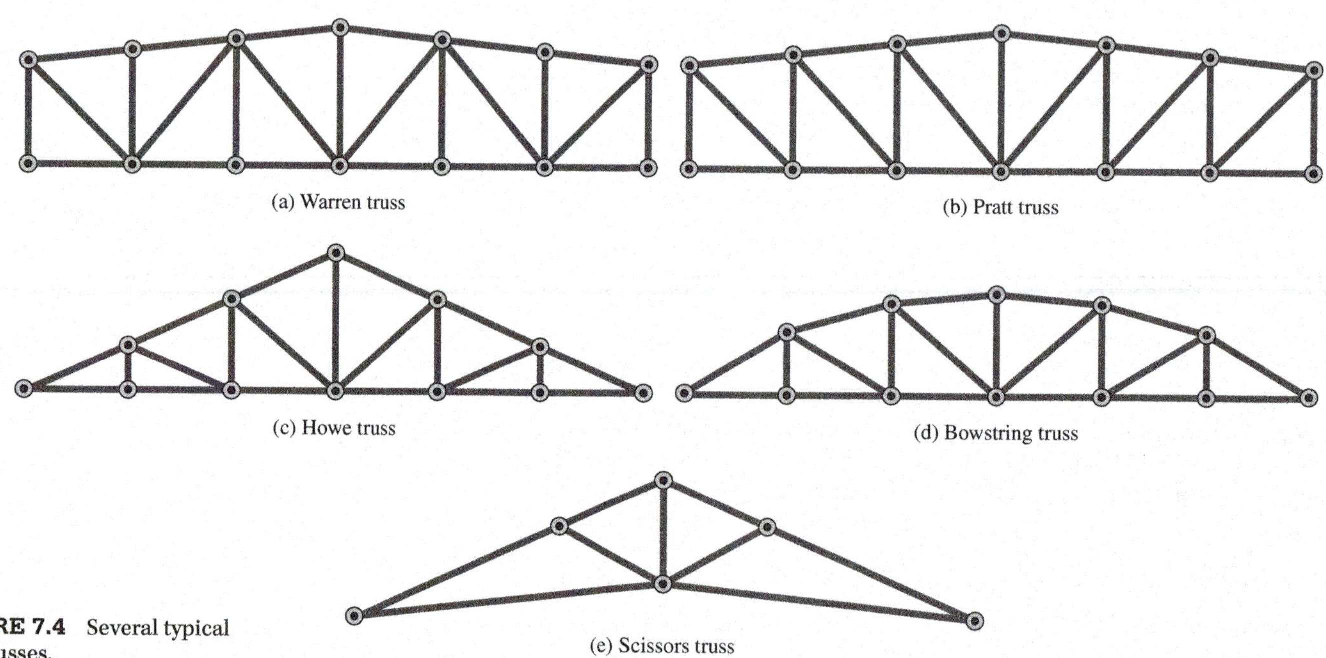

FIGURE 7.4 Several typical roof trusses.

7.4 Bridge Trusses

As bridge spans become longer and loads heavier, trusses begin to be economically competitive with beams. Early bridge trusses were constructed with wood, and they had several disadvantages. First, they were subject to deterioration from wind and water. As a result, covered bridges were introduced, and such structures often would last for quite a few decades. Nevertheless, wooden truss bridges, particularly railroad bridges, were subject to destruction by fire. In addition, the movement of traffic loads back and forth across the bridge spans could cause a gradual loosening of the fasteners.

Because of the several preceding disadvantages, wooden truss bridges faded from use toward the end of the 19th century. Although there were some earlier iron truss bridges, structural steel bridges became predominant. Steel bridges did not require extensive protection from the elements, and their joints had higher fatigue resistance.

Today, existing steel truss bridges are being steadily replaced with steel, precast-concrete, or prestressed-concrete beam bridges. The age of steel truss bridges appears to be over, except for bridges with spans of more than several hundred feet, which represent a very small percentage of the total. Even for these longer spans, there is much competition from other types of structures, such as cable-stayed bridges and prestressed-concrete box girder bridges.

Some highway bridges have trusses on their sides and overhead lateral bracing between the trusses. This type of bridge is called a *through bridge*. The floor system is supported by floor beams that run under the roadway and between the bottom-chord joints of the trusses. A through truss bridge is shown in Figure 7.5.

Shown in Figure 7.6 are several types of bridge trusses. Parallel-chord trusses are shown in parts (a) through (d) of the figure. The Baltimore truss is said to be a subdivided truss because the unsupported lengths of some of the members have been reduced by short members called sub-diagonals and sub-verticals.

In a *deck bridge*, the roadway is placed on top of the trusses or girders. Deck bridges have many advantages over through bridges: There is unlimited overhead horizontal and vertical clearance, future expansion is more feasible, and supporting trusses or girders can be placed close together, which reduces lateral moments in the floor system. Other advantages of the deck truss are simplified floor systems and possible reduction in the sizes of piers and abutments due to reductions in their heights. Finally, the very pleasing appearance of deck structures is another reason for their popularity. The only real disadvantage of a deck bridge is clearance beneath the bridge. The bridge may need to be set high to allow adequate clearance for ships and vehicles to pass underneath.

DID YOU KNOW?
If two parallel chord trusses have the same size of chords but have different overall depths, the truss that is deeper will have a larger moment capacity. This is because the space between chords is the moment arm for calculating capacity.

DID YOU KNOW?
Parallel-chord trusses are typically less efficient for long span bridges than are curved-chord trusses like the Parker truss shown in Figure 7.6(e).

FIGURE 7.5 A through truss bridge over the Ohio river.

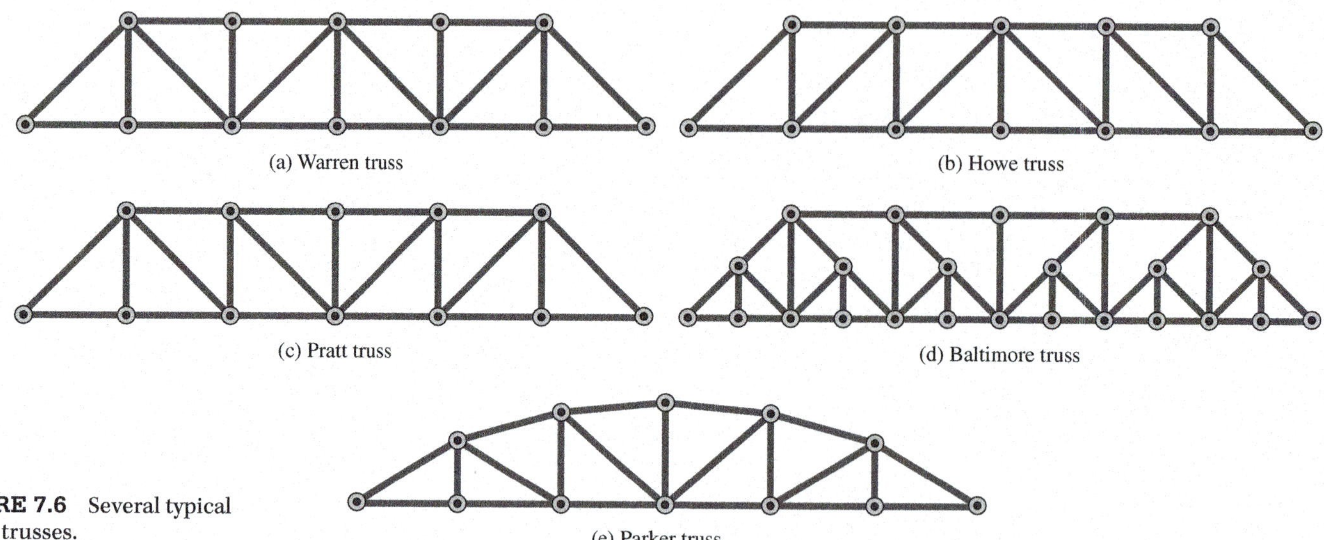

FIGURE 7.6 Several typical bridge trusses.

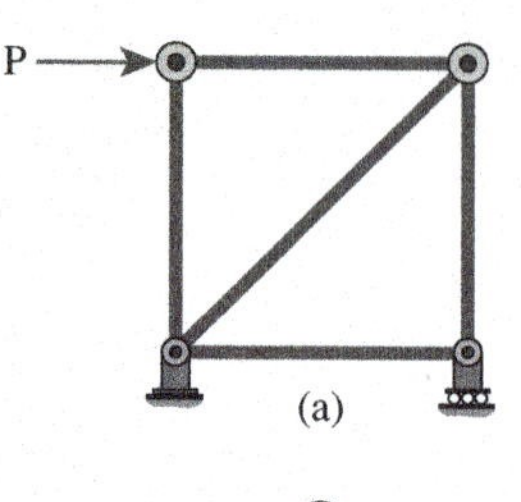

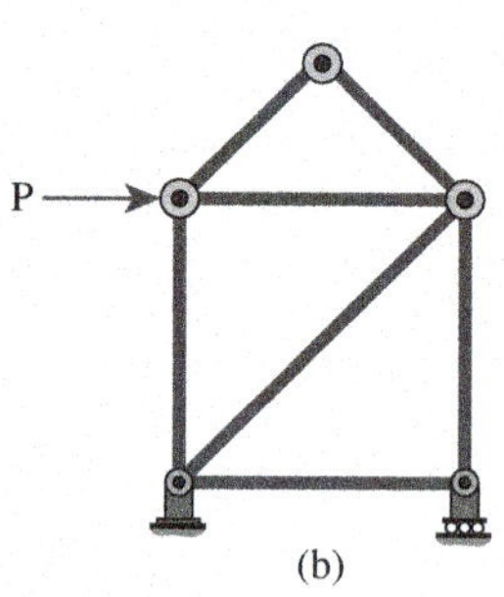

FIGURE 7.7 Stabilized trusses.

7.5 Arrangement of Truss Members

The triangle has been shown to be the basic shape from which trusses are developed because it is the only stable shape. For the following discussion, remember that the members of trusses are assumed to be connected at their joints with frictionless pins.

Other shapes, such as those shown in Figure 7.1(b) and (c), are obviously unstable and may possibly collapse under load. Structures such as these, however, can be made stable by one of the following methods.

- Add members so that the unstable shapes are altered to consist of triangles. The structures in Figures 7.1(b) and (c) are stabilized in this manner in Figures 7.7(a) and (b), respectively.
- Use a member to tie the unstable structure to a stable support. Member AB performs this function in Figure 7.8.
- Make some or all of the joints of an unstable structure rigid, so they become moment resisting. A structure with moment-resisting joints, however, does not coincide with the definition of a truss; the members are no longer connected with frictionless pins.

7.6 Stability and Statical Determinacy of Trusses

The simplest form of truss, a single triangle, is illustrated in Figure 7.9(a). To determine the unknown forces and reaction components for this truss, it is possible to isolate the joints and write two equations of equilibrium for each. These equations of equilibrium involve the summation of vertical and horizontal forces (see side bar).

QUICK NOTE

For each joint in a truss, there are only two equations of equilibrium available:

$$\sum F_x = 0$$

$$\sum F_y = 0$$

The moment in a pin joint is zero, so the $\sum M = 0$ is inherently satisfied and of no use when doing joint equilibrium.

Determinacy

The single-triangle truss may be expanded into a two-triangle truss by the addition of two new members and one new joint. In Figure 7.9(b), triangle ABD is added by installing new

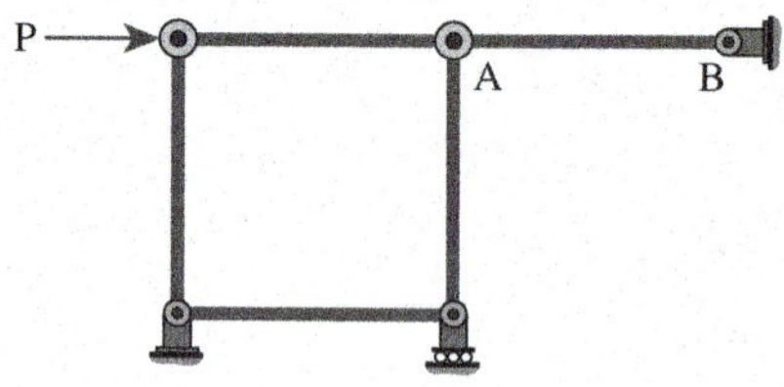

FIGURE 7.8 Stable arrangement of truss members.

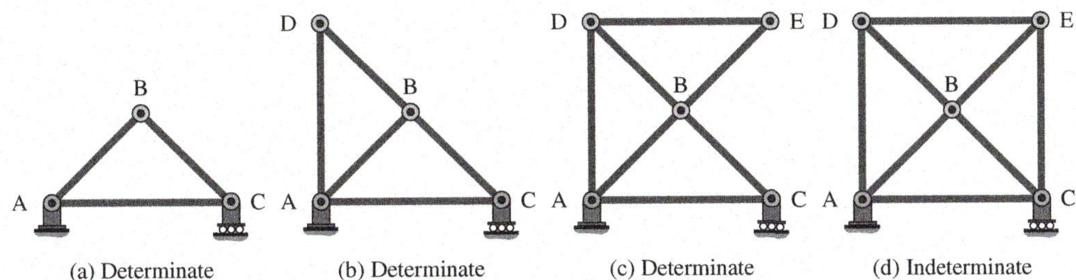

FIGURE 7.9 Expanding a simple truss.

members AD and BD and the new joint D. A further expansion with a third triangle is made in Figure 7.9(c) by the addition of members BE and DE and joint E. For each of the new joints, D and E, a new pair of equations of equilibrium is available for calculating the two new-member forces. As long as this procedure of expanding the truss is followed, the truss will be statically determinate internally. Should new members be installed without adding new joints, such as member CE in Figure 7.9(d), the truss will become statically indeterminate because no new equations of equilibrium are available to find the new member forces.

Using this information, an expression can be written for the relationship that must exist between the number of joints, the number of members, and the number of reactions for a particular truss if it is to be statically determinate internally. The identification of externally determinate structures has been discussed in Chapter 5. In the following discussion, m is the number of members, j is the number of joints, and r is the number of reaction components.

If the number of available equations of static equilibrium, which is $2j$ (two per joint), is sufficient to compute the unknown forces $(m + r)$, the structure is statically determinate. Thus, for a determinate truss, the following equation is satisfied.

$$2j = m + r \Rightarrow \textbf{determinate}$$

It is possible to build trusses that have more members and/or more reactions than can be analyzed using only the equations of static equilibrium. Such trusses are statically indeterminate and are identified by the condition:

$$2j < m + r \Rightarrow \textbf{indeterminate}$$

If there are more members or reactions present than are necessary for maintaining overall stability, then out of necessity they produce forces that cannot be solved using just the equations of equilibrium. These "extra" forces—either bar forces or reactions—are known as *redundant forces*.

If the following condition is met, the truss is considered to have too many joints and too few forces to keep it stable.

$$2j > m + r \Rightarrow \textbf{unstable}$$

Issues surrounding the stability of trusses are discussed more in-depth later. Figure 7.10 gives numerous examples of the application of the previously discussed conditions.

Little explanation is necessary for most of the structures shown, but some remarks may be helpful for a few. The truss of Figure 7.10(e) has five reaction components and is statically indeterminate to the second degree. If we remove two of the vertical reactions, the structure becomes determinate. Compare this to the truss in Figure 7.10(g), which is also SI 2°. If we remove two reactions, from this truss, it would not be able to support itself. This means that the redundant forces are not as a result of too many reactions but rather from having too many members. If we removed one member from each of the two middle panels, the structure will become determinate. If any other members are removed than the two indicated, the structure will become unstable, as discussed in the next section.

RECALL

Statical indeterminacy says that there are not enough equations of equilibrium to solve for all of the unknown forces. It is possible that you may be able to solve for some—but not ALL. Unknown forces are the bar forces and the external reactions.

QUICK NOTE

The equations to identify determinacy of trusses are only valid if external stability of a truss is ensured. Review this concept presented in Section 5.7.

DID YOU KNOW?

The determinacy and stability of a truss is not dependent on the applied loads.

QUICK NOTE

The degree of statical indeterminacy is calculated by the following equation. It indicates the number of redundant forces in the truss.

$$\text{SI}^\circ = m + r - 2j$$

QUICK NOTE

Remember that a **minimum of three** non-coincident reactions must be present to maintain overall stability of the structure.

QUICK NOTE

The values of the redundant forces may be obtained by applying certain simultaneous equations known as compatibility equations. This procedure is presented in Chapter 12.

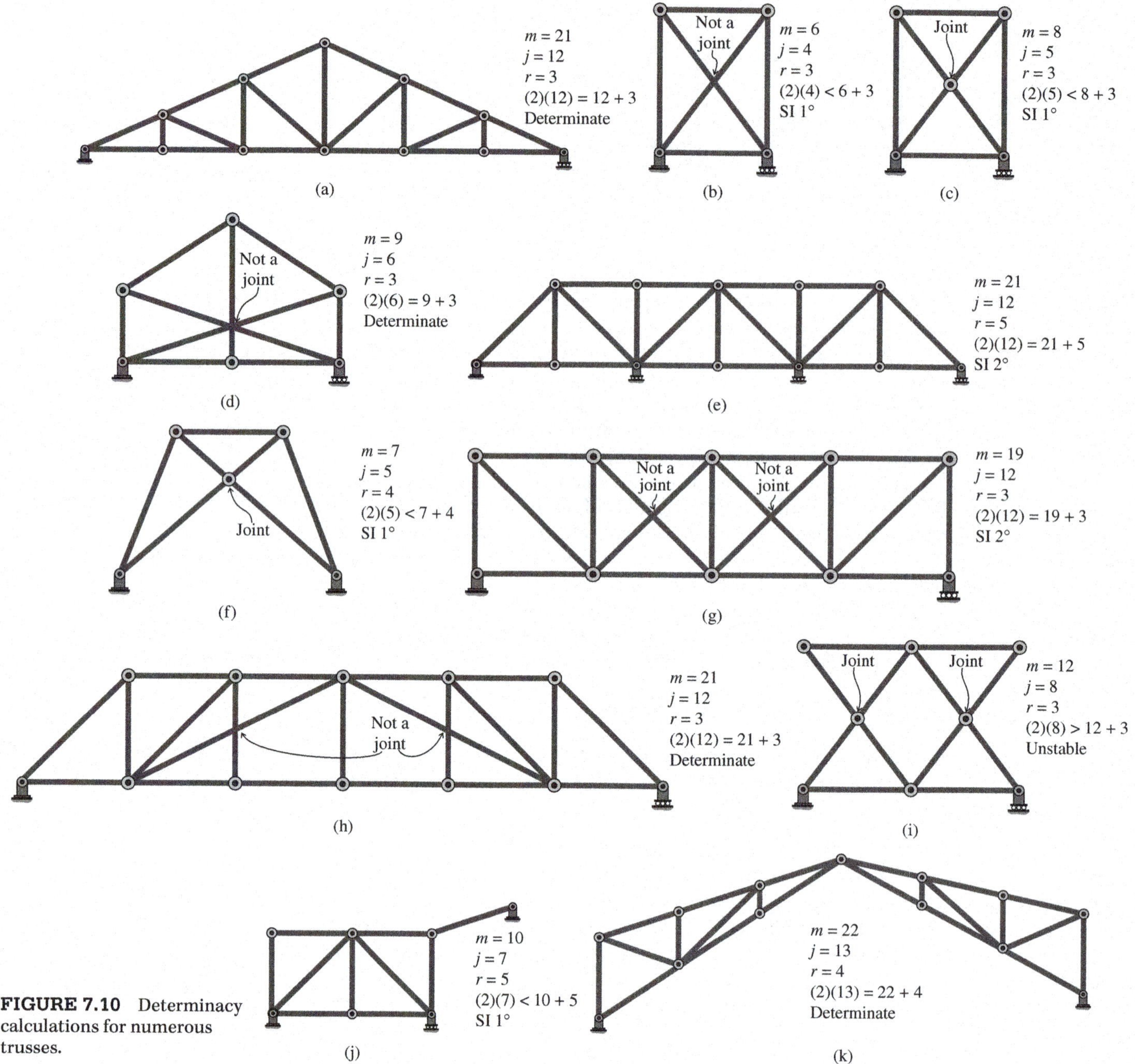

FIGURE 7.10 Determinacy calculations for numerous trusses.

Stability

The stability of a truss always can be determined through structural analysis. The members of a truss must be arranged to support the external loads. What will support the external loads satisfactorily is a rather difficult question to answer with only a glance at the truss under consideration, but an analysis of the structure will always provide the answer. If the structure is stable, the analysis will yield reasonable results and equilibrium will be satisfied at all of the joints in the truss. On the other hand, if a truss is unstable, equilibrium cannot be concurrently satisfied at all of the joints.

Various means for quickly identifying unstable trusses are very valuable to the analyst. The careful analysis and checking of the work for a truss can be so time consuming and frustrating if it is finally discovered that the truss is unstable and all of that time was wasted. Several methods

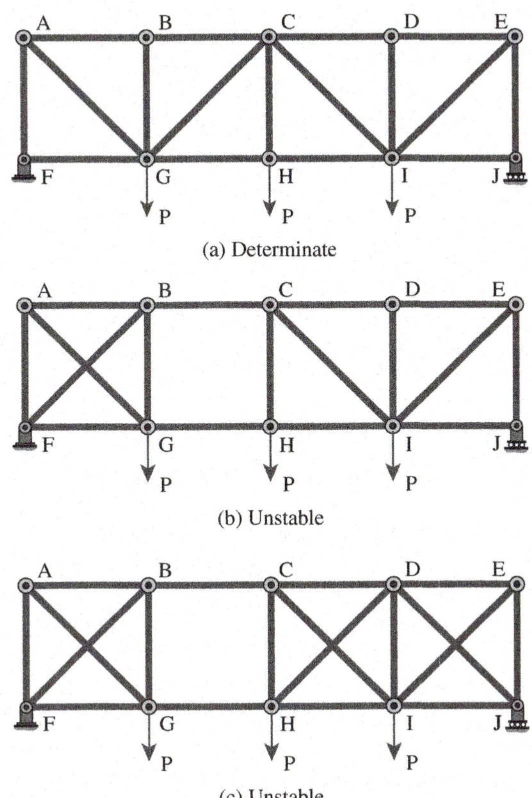

FIGURE 7.11 Stable (a) and unstable truss geometry (b) and (c).

for quickly identifying such structures, without the necessity of trying to analyze the structure first, are identified in the next few paragraphs.

$2j > m + r \Rightarrow$ unstable

A truss that has too many joints is unstable. It will not have enough members to restrain the joints in all directions. This characteristic is easily identified by the determinacy equation $2j > m + r$. Nevertheless, a truss can have what seems to be the correct balance of joints, reactions, and members and yet can still be unstable. That is, it can satisfy the equation of $2j \leq m + r$ and still be unstable. For example, the truss of Figure 7.11(a) satisfies the $2j \leq m + r$ relationship. It is statically determinate and stable.

If the diagonal in the second panel is removed and added to the first panel, as shown in Figure 7.11(b), the truss is unstable even though $2j = m + r$. The part of the truss to the left of the second panel can move with respect to the part of the truss to the right of the second panel because the second panel is a rectangle. As previously indicated, a rectangular shape is unstable unless restrained in some way.

Similarly, the addition of diagonals to the third and fourth panels, as in Figure 7.11(c), will not prevent the truss from being unstable. For this case, $2j < m + r$, and it is seemingly statically indeterminate to the second degree; but it is actually unstable because the second panel is unstable.

CAUTION

- When the condition ($2j > m + r$) exists, the truss is necessarily unstable.
- When the condition ($2j \leq m + r$) exists, the truss is not necessarily stable.

QUICK NOTE

Whenever you have non-triangular shapes in a truss, you should take a closer look. If you can spot a collapse mechanism, then it is an unstable truss. A non-triangular section does not guarantee an unstable truss.

A **collapse mechanism** is where you can experience displacement of the joints without any of the members changing length. This is also known as *rigid-body motion*.

Trusses Consisting of Shapes That Are Not All Triangles

As you become more familiar with trusses, in most cases, you will be able to tell with a brief glance if a truss is stable or unstable. For the present, though, it may be a good idea to study trusses in detail if you think there is a possibility of instability. When a truss has some non-triangular shapes in its geometry, we should be aware that instability is indeed a possibility. The trusses shown in Figures 7.11(b) and (c) fall into this category.

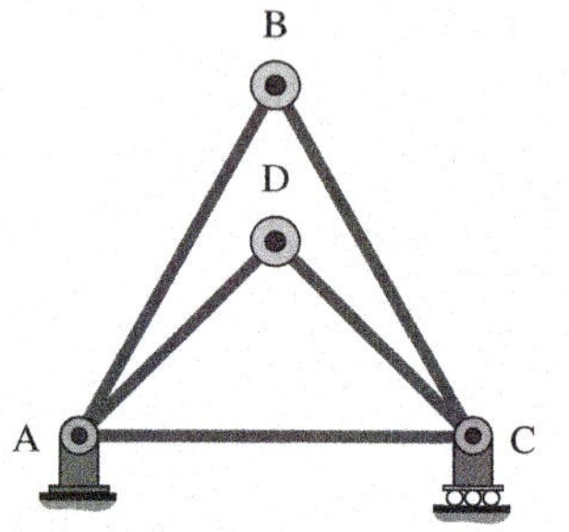

FIGURE 7.12 A stable truss that is not consisting entirely of triangles.

The fallacy of this statement about triangular shapes is that there is an endless number of perfectly stable trusses that can be assembled and do not consist entirely of triangles. As an example, consider the truss shown in Figure 7.12. The basic triangle ABC has been extended by the addition of the joint D and the members AD and CD. A stable truss is maintained, even though the shape ABCD is not a triangle. Joint D cannot move without changing the length of one or more members. Therefore, the truss is stable.

Unstable Supports

A structure cannot be stable if its supports are unstable. To be stable, the truss must be supported by at least three nonparallel, non-concurrent forces. This subject was discussed in detail in Section 5.7.

7.7 Methods of Analysis and Conventions

An indispensable and essential tool in truss analysis is the ability to divide the truss into pieces, construct a free-body diagram for each piece, and then from these free-body diagrams determine the forces in the components. The free-body diagrams can be single joints in the truss or a large part of the truss. As an example, the truss of Figure 7.13 is considered. First a section is drawn completely around joint D. Then a vertical section **1–1** is passed through the second panel of the truss and the free body to its left is considered. These sections are shown in Figure 7.13(a) and the free bodies developed are shown in parts (b) and (c) of the same figure. When working with free-body diagrams such as the one shown in Figure 7.13(b), we are said to be evaluating the member forces using the **method of joints**. On the other hand, if we evaluate the member forces using free-body diagrams such as that shown in Figure 7.13(c), we are said to be using the **method of sections**. Both methods are presented in this chapter.

Applying the equations of static equilibrium to isolated free bodies enables us to determine the forces in the cut members. The free bodies must be carefully selected so that the section cuts do not pass through too many members whose forces are unknown. When using the method of joints, there are only two relevant equations of equilibrium for each free body: $\sum F_x$ and $\sum F_y$. There are three applicable equations of equilibrium for each free body when using the method of sections: $\sum F_x$, $\sum F_y$, and $\sum M_z$.

After you have analyzed a few trusses, you will have little difficulty (in most cases) in selecting satisfactory locations for the sections. You are not encouraged to remember specific sections for specific trusses, although you will probably fall into such a habit unconsciously as time passes. At this stage, you need to consider each case individually without reference to other similar trusses.

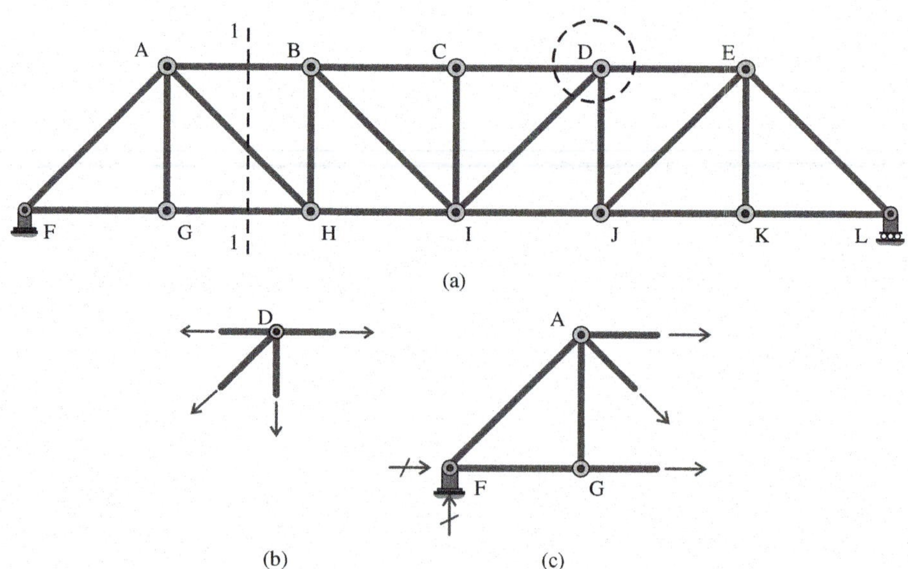

FIGURE 7.13 Free-body diagrams of a truss joint and a truss section.

FIGURE 7.14 Tension arrows for both joints and members.

Sign Convention

Good practice dictates that a sign convention be adopted when analyzing trusses and that this convention be used consistently for all trusses. By doing so, many errors inherent with changing sign conventions are eliminated. The authors use the sign convention that all **unknown forces** in a truss are assumed to be tensile forces. A depiction of this convention, as it shows up on joints and truss members, is seen in Figure 7.14. If all unknown forces are assumed to be tensile, then any value that is numerically positive is understood to be in tension, while a negative number is understood to be in compression. This convention is demonstrated through the examples.

After some practice analyzing trusses, the sense of the forces in many of the members can be determined by examination. Try to picture whether a member is in tension or compression before making actual calculations; by doing so, you will achieve a better understanding of the action of trusses under load. The next section demonstrates that it is possible to determine entirely by mathematical means the sense as well as the numerical value of the forces.

CAUTION

On any free body we use, there cannot be more unknown forces than there are relevant equations of static equilibrium.

FOLLOW THIS SIGN CONVENTION

- A positive force is a tensile force (T).
- A negative force is a compression force (C).

When performing all of the calculations, the analyst should keep track of (+) and (−) signs of the bar forces. However, it is often convenient when providing a summary of the bar forces to report them as positive values with either a (T) or a (C) denoting the sense of the force.

7.8 Method of Joints

An imaginary section may be passed around a joint in a truss, regardless of its location, completely isolating it from the remainder of the truss. The joint has become a free body in equilibrium under the forces applied to it. The applicable equations of equilibrium, $\sum F_x = 0$ and $\sum F_y = 0$, may be applied to the joint to determine the unknown forces in the members meeting there. It should be evident that no more than two unknowns can be determined at a joint with these two equations.

When writing equations of equilibrium for a joint, it becomes necessary to resolve all forces into two orthogonal directions. Often these two orthogonal directions are chosen to be in the x-horizontal and y-vertical directions. However, there are practical reasons why a coordinate system other than horizontal and vertical may be selected—an illustration that is presented in Example 7.2. When breaking forces into orthogonal components, one may use either trigonometric functions or geometric relationships. The primary approach in this text is to use geometry. For example, using the information presented in Figure 7.15, one may calculate the vertical component of a force vector (N) by multiplying it by the ratio $\frac{\text{rise}}{\text{hyp}}$ or $\frac{V}{L}$—whichever form is preferred. In a like fashion, the horizontal component of the force is found by multiplying the force resultant (N) by $\frac{\text{run}}{\text{hyp}}$ or $\frac{H}{L}$.

A student learning the method of joints may initially find it necessary to draw a free body sketch for every joint in a truss he or she is analyzing. After you have computed the forces in two or three trusses, it is necessary to draw the diagrams for only a few joints, because you can easily visualize the free bodies involved. Drawing large sketches helps visualization. The most important thing for you to remember is that you are interested in only one joint at a time. Keep your attention away from the loads and forces at other joints. Your concern is only with the forces at the one joint on which you are working. The method of joints is illustrated through the examples that follow.

DID YOU KNOW?

When using the method of joints, the $\sum M_Z = 0$ equation is inherently satisfied. Therefore, it is a non-informative equation to use, leaving us with just two equations of equilibrium per joint.

QUICK NOTE

The rise and run of a truss member is most often represented by dividing V and H by the greatest common multiple. For the example shown in Figure 7.15, H and V are both divided by 20 ft to obtain rise = 1 and run = 1—a finding used in Example 7.1.

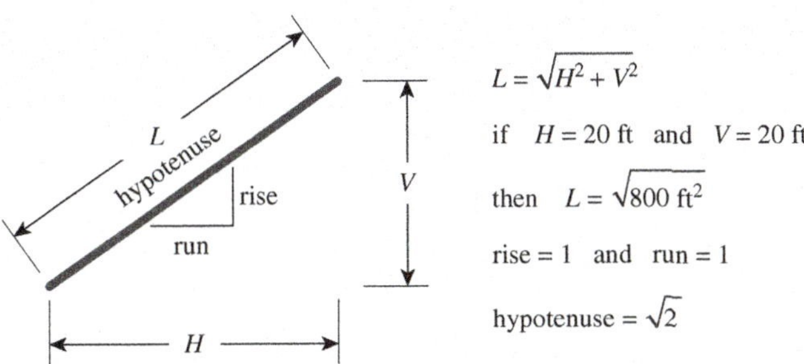

FIGURE 7.15 Geometric illustration of orthogonal components.

EXAMPLE 7.1

Using the method of joints, find all forces in the truss shown.

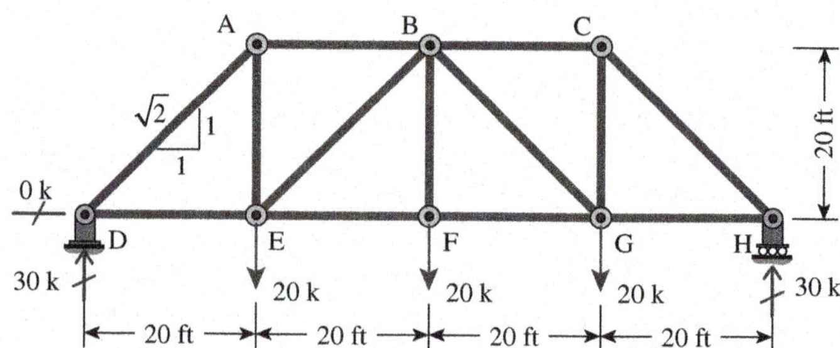

QUICK NOTE

The evaluation of trusses can be tedious at times. Being very methodical throughout the analysis process reduces many errors—including sign errors.

QUICK NOTE

When analyzing a joint, you may want to start with the equilibrium equation that has only one unknown in it if possible. This keeps the mathematics simpler and causes the analyst less work. For example, starting with $\sum F_y$ for joint D ensures that only one unknown is in the resulting equation of equilibrium. The same could not be said about $\sum F_x$.

SOLUTION STRATEGY

The reactions for this example are already given, but this is usually the first step of the process. Start with a joint that has at least one known force and no more than two unknown forces. Draw a FBD of that joint and use equilibrium equations to solve for the unknown forces. Once those forces are known, move to another joint that meets the same set of criteria. Proceed until all unknown bar forces are found. Remember to report whether the bar is in tension (T) or compression (C) when all of the forces are summarized.

SOLUTION

Start with joint D. Draw a FBD of this joint and sketch all unknown forces (N_{DA} and N_{DE}) acting out from the joint. Any external forces (i.e., reactions and applied forces) may be sketched in the direction they act. For example, the vertical reaction at D acts into the joint, so it is drawn as such. Resolve all forces into x and y components. Then use equilibrium equations to solve for the two unknown forces.

Joint D

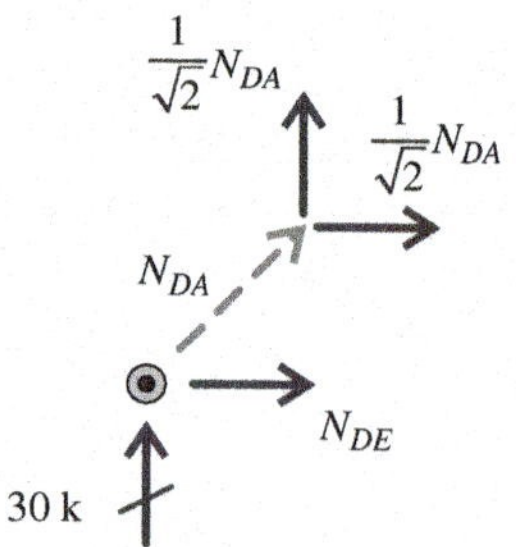

$$+\uparrow \sum F_y = 30\text{ k} + \tfrac{1}{\sqrt{2}}N_{DA} = 0$$
$$N_{DA} = -42.43\text{ k} = \underline{\underline{42.43\text{ k}\quad(\text{C})}}$$

Once N_{DA} is found, we sum forces in the x direction to find the force in bar DE (N_{DE}).

$$\xrightarrow{+} \sum F_x = \tfrac{1}{\sqrt{2}}N_{DA} + N_{DE} = 0$$
$$N_{DE} = 30\text{ k} = \underline{\underline{30\text{ k}\quad(\text{T})}}$$

With these two internal bar forces found, the analyst can move to the next joint. Moving to joint E results in three unknown forces (N_{AE}, N_{EB}, and N_{EF}), whereas moving to joint A results in only two unknown forces that need to be dealt with (N_{AB}, N_{AE}). Therefore, it is necessary to consider joint A next.

Joint A

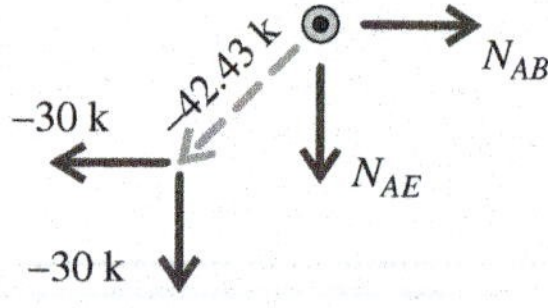

$$+\uparrow \sum F_y = -(-30\text{ k}) - N_{AE} = 0$$
$$N_{AE} = 30\text{ k} = \underline{\underline{30\text{ k}\quad(\text{T})}}$$

$$\xrightarrow{+} \sum F_x = -(-30\text{ k}) + N_{AB} = 0$$
$$N_{AB} = -30\text{ k} = \underline{\underline{30\text{ k}\quad(\text{C})}}$$

It is now okay to move to joint E. The internal forces in two of the bars are currently known (N_{AE}, N_{DE}), leaving only two bar forces that are unknown (N_{EB}, N_{EF}).

Joint E

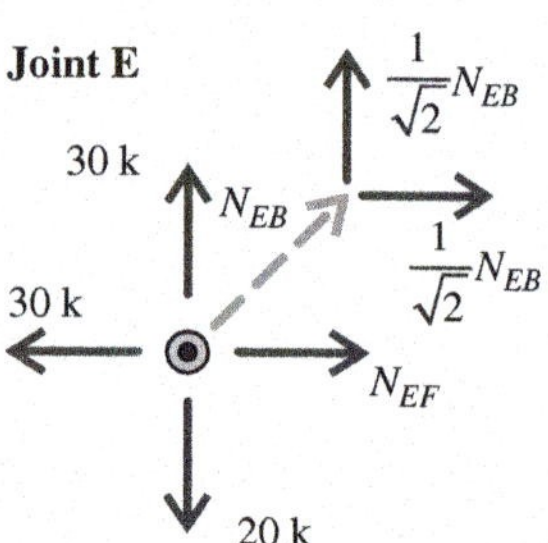

$$+\uparrow \sum F_y = \tfrac{1}{\sqrt{2}}N_{EB} + 30\text{ k} - 20\text{ k} = 0$$
$$N_{EB} = -14.14\text{ k} = \underline{\underline{14.14\text{ k}\quad(\text{C})}}$$

$$\xrightarrow{+} \sum F_x = \tfrac{1}{\sqrt{2}}N_{EB} + N_{EF} - 30\text{ k} = 0$$
$$N_{EF} = 40\text{ k} = \underline{\underline{40\text{ k}\quad(\text{T})}}$$

This particular truss and its applied loading are symmetric. From an efficiency perspective, the entire truss need not be analyzed—rather just half of it. In this particular truss, an odd number of members (13) is present, which means just over half of the bar forces (7) need to be evaluated. Six bar forces are already known, so this will require that one more bar force (N_{BF}) be evaluated. This also requires that a FBD of either joint B or joint F be constructed. Joint B would not be a good choice because it would have three unknown bar forces. Joint F becomes the obvious choice because of the number of unknowns and also its simplicity.

We evaluate $\sum F_y$ to find N_{BF}, but we will not need to evaluation $\sum F_x$ because it will only give us redundant information.

$$+\uparrow \sum F_y = N_{BF} - 20\text{ k} = 0$$

$$N_{BF} = 20\text{ k} = \underline{\underline{20\text{ k} \quad (\text{T})}}$$

A summary of all of the forces in the truss are shown here.

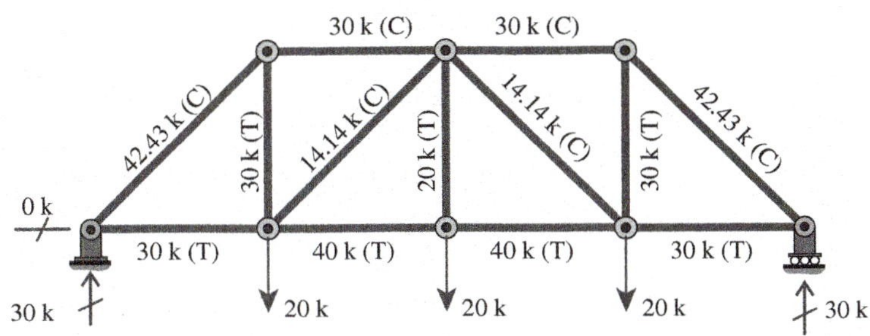

Joint F

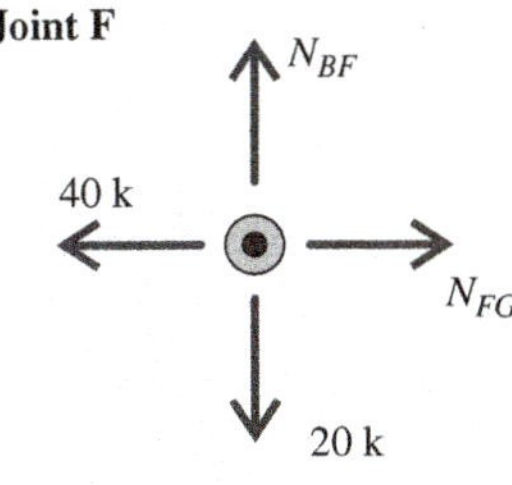

EXAMPLE 7.2

Use the method of joints to find all the forces in the truss shown in the figure.

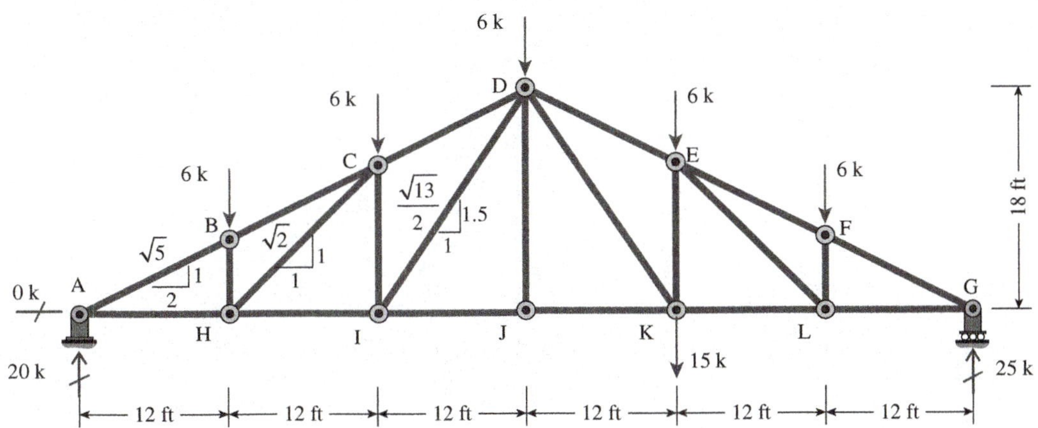

SOLUTION STRATEGY

The reactions for this example are already given. Start with a joint that has at least one known force and no more than two unknown forces. Draw a FBD of that joint then use equilibrium equations to solve for the unknown forces. Once those forces are known, move to another joint which meets the same set of criteria. Proceed until all unknown bar forces are found.

SOLUTION

Start with joint G. Draw a FBD of this joint and sketch all unknown forces (N_{FG} and N_{LG}) acting out from the joint. Any external forces (i.e., reactions and applied forces) may be sketched in the direction they act.

$$+\uparrow \sum F_y = \tfrac{1}{\sqrt{5}} N_{FG} + 25\text{ k} = 0$$

$$N_{FG} = -55.90\text{ k} = \underline{\underline{55.90\text{ k} \quad (\text{C})}}$$

$$\overset{+}{\rightarrow} \sum F_x = -\tfrac{2}{\sqrt{5}} N_{FG} - N_{LG} = 0$$

$$N_{LG} = 50\text{ k} = \underline{\underline{50\text{ k} \quad (\text{T})}}$$

Joint G

$\frac{1}{\sqrt{5}} N_{FG}$

$\frac{2}{\sqrt{5}} N_{FG}$

N_{FG}

N_{LG}

25 k

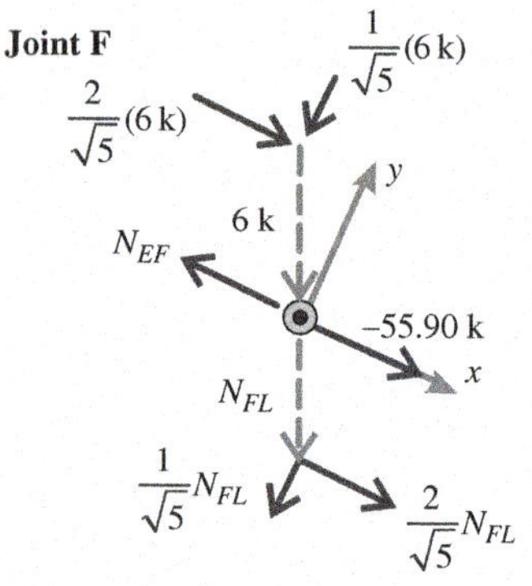

The next joint to be evaluated is joint F because it only has two unknown bar forces at this stage in the analysis. One may analyze this joint the way all previous joints have been by resolving all forces into vertical and horizontal components. This will produce two equations and two unknowns that need to be solved simultaneously. An alternative is to use a rotated coordinate system for joint F, as shown in the FBD. The *x-axis* is aligned with member FG. This allows for the writing of an independent equation, as shown hereafter, and avoids the need for simultaneous equations.

$$+\nearrow \sum F_y = -\tfrac{1}{\sqrt{5}}(6\text{ k}) - \tfrac{1}{\sqrt{5}}N_{FL} = 0$$

$$N_{FL} = -6\text{ k} = \underline{\underline{6\text{ k} \quad (\text{C})}}$$

$$+\searrow \sum F_x = \tfrac{2}{\sqrt{5}}(6\text{ k}) + \tfrac{2}{\sqrt{5}}N_{FL} - N_{EF} - 55.90\text{ k} = 0$$

$$N_{EF} = -55.90\text{ k} = \underline{\underline{55.90\text{ k} \quad (\text{C})}}$$

It is reasonable to ask at this stage if the rotated axes for joint F are worth the effort. The answer is a conditional yes. If the student is proficient with the geometry, they will find this approach to be useful. For those still struggling with the concepts of geometry, working all joints in an unrotated orientation is likely the least confusing approach.

Following the same process for all remaining joints, all bar forces may be determined. Like in Example 7.1, this truss is symmetric. However, the loading is not symmetric, so an analysis of all joints is required. The following figure is the solution for all truss bars. The reader is encouraged to check one or two more joints and see if they can get the correct answers.

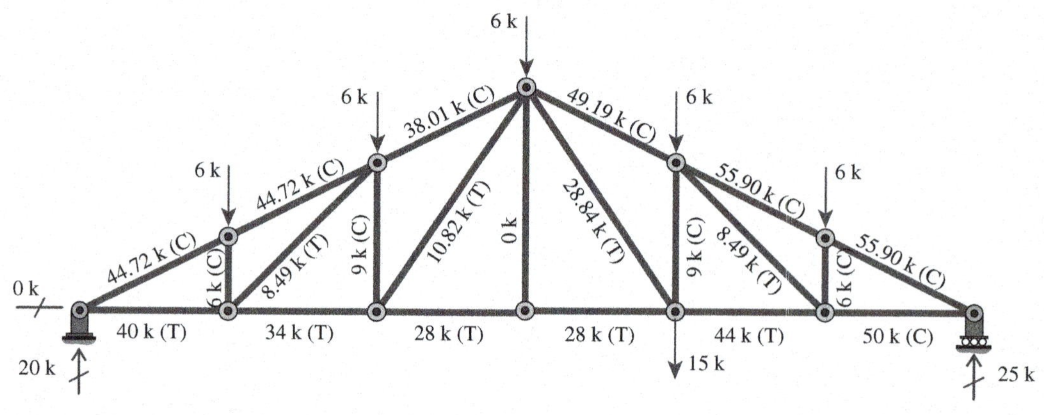

QUICK NOTE

The matrix formulation can become cumbersome as the number of joints increases. However, it still may be a preferable approach when dealing with complex truss configurations. Chapter 19 shows an alternate approach to assembling these matrices.

7.9 Matrix Formulation for Reactions and Bar Forces

In the previous section, the method of joints was introduced where one joint is analyzed at a time and at most two unknown bar forces are determined. These bar forces are then carried forward to the next joint. An alternative to working on one joint at a time is to write a system of equations for all of the joints within the truss and then solve the equations simultaneously. This approach will not only solve for the unknown member forces but will also provide values for the unknown reactions. The equations then can be written in matrix form and solved simultaneously—similar to what was done in Section 5.11 for reactions. Example 7.3 shows how this is done for a three member—three joint truss.

EXAMPLE 7.3

Find all of the forces in the truss shown by simultaneously solving all equilibrium equations.

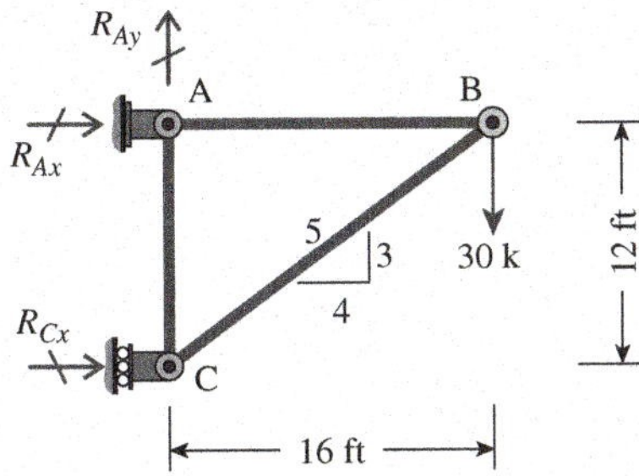

SOLUTION STRATEGY

The reactions for this truss are not known, but equilibrium equations for each joint still can be written. There are three unknown reactions and three unknown bar forces that produce a total of six unknowns. There are three joints for which two equations of equilibrium may be written for each. Place the coefficients of each unknown force into a matrix and the known forces into a column vector. Then solve the matrix using a calculator or some math program.

SOLUTION

For the sake of brevity, the FBD for each joint is not shown. However, proper sketching of joint FBDs should be followed by the students in their work.

DID YOU KNOW?
When using this approach, there is no benefit or restriction as to which joint you start with.

Joint A:

$$\overset{+}{\rightarrow} \sum F_x = R_{Ax} + N_{AB} = 0$$

$$+\uparrow \sum F_y = R_{Ay} - N_{AC} = 0$$

Joint B:

$$\overset{+}{\rightarrow} \sum F_x = -N_{AB} - \tfrac{4}{5}N_{BC} = 0$$

$$+\uparrow \sum F_y = -30\text{ k} - \tfrac{3}{5}N_{BC} = 0$$

Joint C:

$$\overset{+}{\rightarrow} \sum F_x = R_{Cx} + \tfrac{4}{5}N_{BC} = 0$$

$$+\uparrow \sum F_y = N_{AC} + \tfrac{3}{5}N_{BC} = 0$$

RECALL
The coefficient matrix contains the coefficients for the unknown forces (N, R). If an equation doesn't contain a particular unknown force, the coefficient for that force should be taken as zero. See Example 5.13 for a full explanation.

Rewrite these six equations in matrix form.

$$\begin{bmatrix} 1 & 0 & 0 & 1 & 0 & 0 \\ 0 & 1 & 0 & 0 & 0 & -1 \\ 0 & 0 & 0 & -1 & -0.8 & 0 \\ 0 & 0 & 0 & 0 & -0.6 & 0 \\ 0 & 0 & 1 & 0 & 0.8 & 0 \\ 0 & 0 & 0 & 0 & 0.6 & 1 \end{bmatrix} \begin{Bmatrix} R_{Ax} \\ R_{Ay} \\ R_{Cx} \\ N_{AB} \\ N_{BC} \\ N_{AC} \end{Bmatrix} = \begin{Bmatrix} 0 \\ 0 \\ 0 \\ 30\text{ k} \\ 0 \\ 0 \end{Bmatrix}$$

Now solve for the unknowns by pre-multiplying the constant vector by the inverse of the coefficient matrix.

$$\begin{Bmatrix} R_{Ax} \\ R_{Ay} \\ R_{Cx} \\ N_{AB} \\ N_{BC} \\ N_{AC} \end{Bmatrix} = \begin{bmatrix} 1 & 0 & 0 & 1 & 0 & 0 \\ 0 & 1 & 0 & 0 & 0 & -1 \\ 0 & 0 & 0 & -1 & -0.8 & 0 \\ 0 & 0 & 0 & 0 & -0.6 & 0 \\ 0 & 0 & 1 & 0 & 0.8 & 0 \\ 0 & 0 & 0 & 0 & 0.6 & 1 \end{bmatrix}^{-1} \begin{Bmatrix} 0 \\ 0 \\ 0 \\ 30\text{ k} \\ 0 \\ 0 \end{Bmatrix} = \begin{Bmatrix} -40\text{ k} \\ 30\text{ k} \\ 40\text{ k} \\ 40\text{ k} \\ -50\text{ k} \\ 30\text{ k} \end{Bmatrix}$$

An interpretation of these results is given as

$$\therefore R_{Ax} = \underline{\underline{40\text{ k} \leftarrow}} \qquad \therefore R_{Ay} = \underline{\underline{30\text{ k} \uparrow}} \qquad \therefore R_{Cx} = \underline{\underline{40\text{ k} \rightarrow}}$$

$$\therefore N_{AB} = \underline{\underline{40\text{ k}\quad (T)}} \qquad \therefore N_{BC} = \underline{\underline{50\text{ k}\quad (C)}} \qquad \therefore N_{AC} = \underline{\underline{30\text{ k}\quad (T)}}$$

7.10 Zero-Force Members

QUICK NOTE

A question often arises when a student encounters zero-force members within a truss. *Why have the truss bar present if it has zero force in it?* This is a legitimate question and really has two answers.

- They are present to provide stability to the truss as a whole. Remember that members in compression are prone to buckle (i.e., instability).
- You may be analyzing the truss for a set of loads that result in zero-force members. However, the bar may be there for handling a different load scenario as would be common in the analysis of highway bridges.

Frequently, some readily identifiable truss members have zero forces (assuming secondary forces due to member weights, load eccentricities, and so forth, are neglected). The ability to spot these members will, on occasion, appreciably expedite truss analysis. Zero-force members usually can be identified by making brief examinations of the truss joints. Zero-force members can be identified visually by looking to see if one of two conditions exists at a truss joint.

1. If three members meet at a joint, two of the members are colinear, and there is no external force at the joint (neither an applied force or reaction force), then the third member, out of necessity, must have a zero force. Figure 7.16(a) illustrates this condition. Notice that if we $\sum F_y$, then N_3 is found to be zero. If $N_3 = 0$, then $\sum F_x$ reveals that $N_1 = N_2$.
2. If two members meet at a joint, they are not colinear, and there is no external force at the joint, then out of necessity both bar forces must be zero. Figure 7.16(b) illustrates this condition. If we $\sum F_y$, then it is easily seen that $N_1 = 0$. Then if we $\sum F_x$, we find that $N_2 = 0$.

Procedure

The following procedure makes the process of identifying zero-force members quite simple. A student must be very familiar with the previously presented joint conditions before embarking on this process.

1. Scan the truss for any joint that only has two or three members.
2. See if any of these identified joints meet either of the two conditions for being zero force.
3. If a member or members are identified as being zero force, mentally erase those members from the truss.
4. With the known zero-force members erased, re-evaluate the truss looking for two and three member joints and see if they meet either of the two conditions.
5. Repeat steps 3 and 4 until no other zero-force members are identified.

(a) Three-member condition

(b) Two-member condition

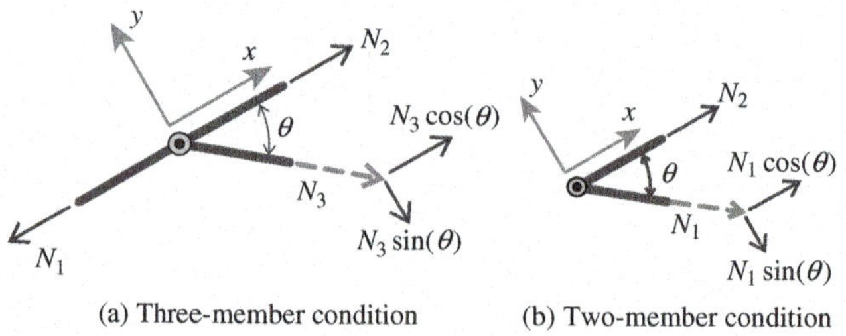

FIGURE 7.16 Conditions which identify the presence of zero-force members.

EXAMPLE 7.4

The zero-force members have been identified for the truss shown. Verify these findings.

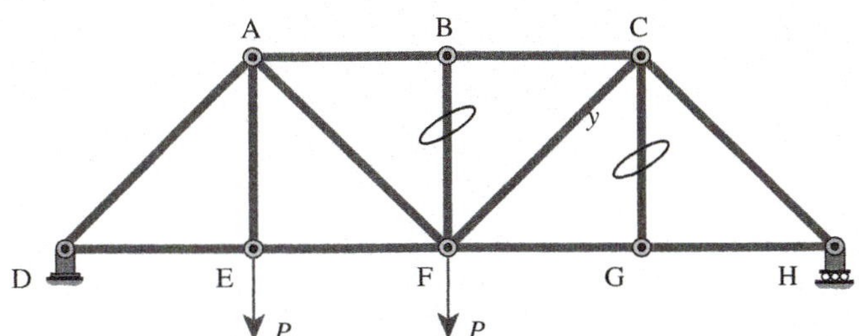

SOLUTION STRATEGY

Make sure both conditions are well understood. Scan all eight joints of the given truss following the procedure outlined in this text.

SOLUTION

The following table is presented with the findings of each joint condition.

Joint	Condition	Finding
A	Four members	None
B	Condition #1	$N_{BF} = 0$
C	Four members	None
D	External force	None
E	External force	None
F	Condition #1	$N_{CG} = 0$
G	External force	None

Mentally remove members BF and CG. Re-scan the truss. With the two members removed, there are no joints that meet either condition 1 or 2.

SUMMARY

Two zero-force members exist in this truss: BF and CG.

COMMON MISTAKE

Students commonly say that member BF cannot be zero force because when they look at joint F it has five members, and they think that none of the members have a zero force.

Because joint F has five members, we cannot use the joint to identify any zero-force members. This does not mean that none of the five members are zero force. It is clear when we look at joint B that member BF is indeed zero force.

EXAMPLE 7.5

The zero-force members have been identified for the truss shown. Verify these findings.

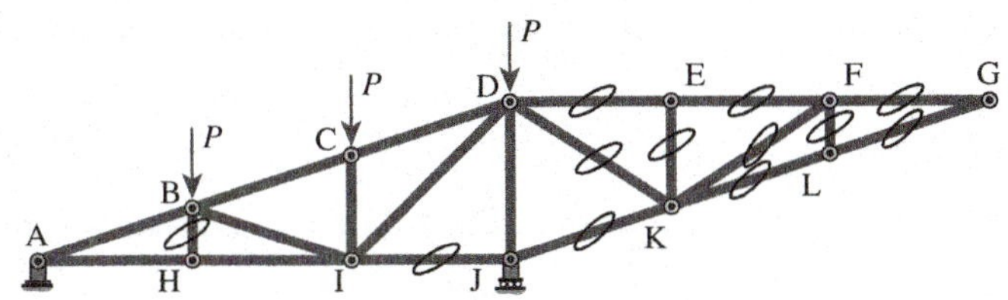

SOLUTION STRATEGY

Make sure both conditions are well understood. Scan all 12 joints of the given truss following the procedure outlined in this text.

QUICK NOTE

Member IJ is a zero-force member, but joint J will never meet either of the two defined conditions. Rather, identification of this zero-force member relies on you understanding the principle behind all conditions. This means that a member is zero force if there is no other member or force that has a component in the direction of the member in question. For example, once we find that JK is zero force, joint J has no *X* component force. This means that IJ (only has a horizontal component) must also be zero. A quick equilibrium equation at the joint verifies this.

SOLUTION

The following table is presented with the findings of each joint condition.

Joint	Condition	Finding
A	External force	None
B	External force	None
C	External force	None
D	External force	None
E	Condition #1	$N_{EK} = 0$
F	Four members	None
G	Condition #2	$N_{FG} = N_{LG} = 0$
H	Condition #1	$N_{BH} = 0$
I	Five members	None
J	External force	None
K	Five members	None
L	Condition #1	$N_{FL} = 0$

The first scan through the truss revealed the existence of five zero-force members. After removing the zero-force members, a sequence of subsequent scans reveals that all members to the right of joints D and J are zero-force members.

7.11 Method of Sections

QUICK NOTE

When using the *method of sections* there are three equations of equilibrium that are available for each section ($\sum F_X$, $\sum F_Y$, and $\sum M_Z$). As such, you must ensure that you do NOT end up with more than three unknown bar forces when making a cut. Sometimes this requires creative cutting.

Applying the equations of equilibrium to free-body diagrams of sections of a truss is the basis of force computation by the method of sections, just as it was when using the method of joints discussed previously. When using the method of sections to determine the force in a particular member, an imaginary plane is passed completely through the truss, which cuts it into two sections as shown in Figure 7.17(a). The resulting free-body diagrams are shown in Figures 7.17(b) and 7.17(c). The location at which the sections are cut is selected to identify the desired bar forces and to ensure that there are at least as many equations of equilibrium available as there are unknown forces.

The equations of static equilibrium may be applied to either of the free bodies to determine the magnitude of the unknown forces. An advantage of the method of sections is that the force in one member of a truss can be computed in most cases without having to compute the forces in other members of the truss. If the *method of joints* were used, calculating the forces in other members, joint by joint from the end of the truss up to the member in question, would be necessary.

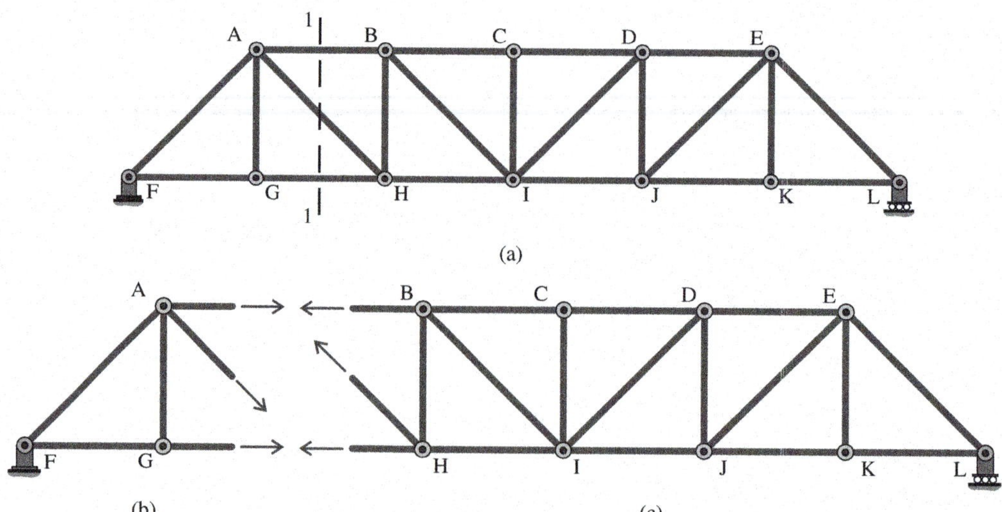

FIGURE 7.17 A truss and the internal forces at a section.

Nevertheless, both methods are used in the analysis of trusses. In fact, both methods are often used at the same time. Depending upon the geometry of the truss, some member forces are more easily calculated with the method of sections while other members may be more easily calculated using the method of joints—a nuance more readily distinguished with experience.

Particular attention should be paid to the point about which moments are summed when applying the equations. Often moments of the forces can be taken about a point so that only one unknown force appears in the equation. As such, the value of that force can be obtained directly. This objective usually is attained by selecting a point along the line of action of one or more of the other members' forces and making it the point about which moments are summed. This point does not necessarily have to be on the cut section. Some familiar trusses have special locations for placing sections that greatly simplify the work involved. Some of these are discussed in the pages to follow.

FRIENDLY ADVICE

Writing equations of equilibrium that contain only one unknown should really be a common consideration. It saves time and is less prone to making calculation mistakes.

Application of The Method of Sections

When using the method of sections, we need to establish a sign convention for the sense of the forces in the cut members just as we did when using the method of joints. As with the method of joints, assume that all unknown member forces are acting in tension. This sign convention was illustrated in Figure 7.14. Upon analysis, positive results indicate that members are in tension, and negative results indicate that members are in compression. By using this sign convention, we obtain consistent results with all of the methods of truss analysis. The errors that result from confusing the sense of the member forces is greatly reduced.

Examples 7.6 and 7.7 illustrate the computation of member forces using the method of sections. To demonstrate the principles of this method, only the forces in selected members are computed. The forces in the other members could be computed by cutting additional sections or with the method of joints.

EXAMPLE 7.6

Find the forces in members EB, EF, and BC for the truss shown.

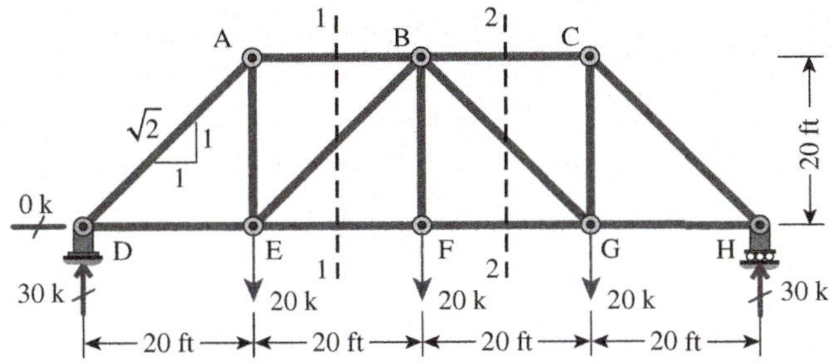

SOLUTION STRATEGY

Reactions should always be the first quantities found. They are given in this example, so the next step is to examine the truss and determine which section(s) need to be cut to access the internal forces for the bars indicated. Sketch the FBD of either the left or right section of the truss. Look to identify equilibrium equations that only contain one unknown force.

SOLUTION

Making a cut at section 1–1 will give access to two of the desired bar forces. Sketch the FBD of the left section because it contains the fewest number of total forces.

DID YOU KNOW?

The final answer does not depend on whether the left or the right section is selected. Working with either section will give the same internal bar force. By choosing the section with the fewest number of total forces, we may reduce the number of terms in any given equilibrium equation and thus reduce the number of buttons that need to be pushed on the calculator.

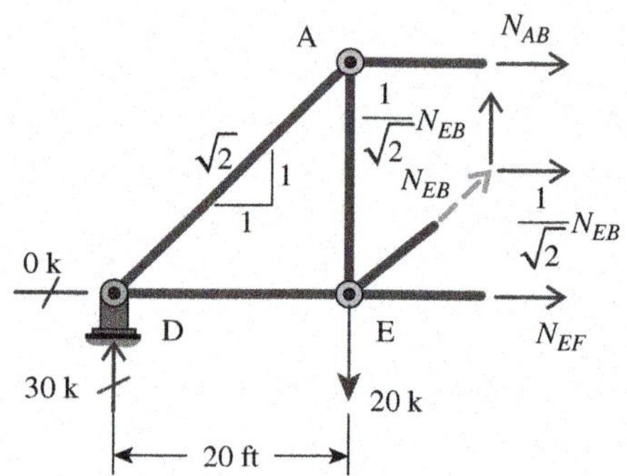

QUICK NOTE

The selection for the point about which to sum moments was determined based on the line of action of two of the three unknown forces. Their point of coincidence then becomes the selection point (B). Points about which moments are summed DO NOT need to be located on the section itself.

The first equation that will be used is $\sum F_y$. This will allow an equation to be written where the only unknown is N_{EB}.

$$+\uparrow \sum F_y = 30\text{ k} - 20\text{ k} + \tfrac{1}{\sqrt{2}}N_{EB} = 0$$

$$N_{EB} = -14.14\text{ k} = \underline{\underline{14.14\text{ k} \quad (\text{C})}}$$

Recognizing that the other required bar force is N_{EF}, it is to our advantage to sum moments around point B. The reason for this selection is that the resultants of both N_{EB} and N_{AB} coincide at that point. Thus, they don't show up in the equilibrium equation.

$$^{+}\circlearrowleft \sum M_B = (-30\text{ k})(40\text{ ft}) + (20\text{ k})(20\text{ ft}) + N_{EF}(20\text{ ft}) = 0$$

$$N_{EF} = 40\text{ k} = \underline{\underline{40\text{ k} \quad (\text{T})}}$$

The bar force for BC is found by making a cut at section 2–2. Once again the selection of which section to use is based on the one with the fewest number of forces. Thus, the right section is selected. To find N_{BC}, point G becomes an optimal location about which to sum moments.

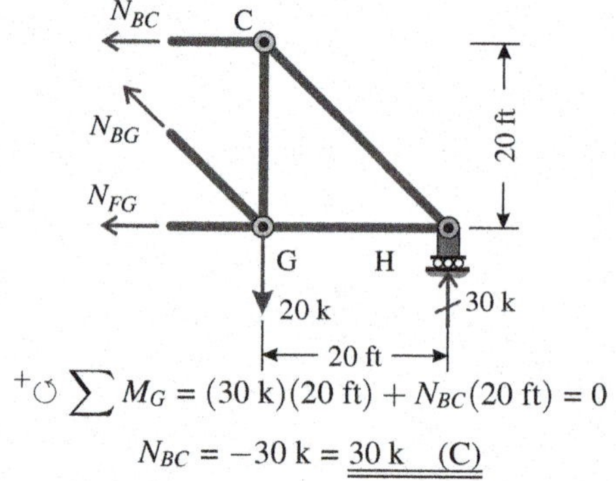

$$^{+}\circlearrowleft \sum M_G = (30\text{ k})(20\text{ ft}) + N_{BC}(20\text{ ft}) = 0$$

$$N_{BC} = -30\text{ k} = \underline{\underline{30\text{ k} \quad (\text{C})}}$$

EXAMPLE 7.7

For the truss shown, use the method of sections to find the internal forces for bars BC, BG, and FG.

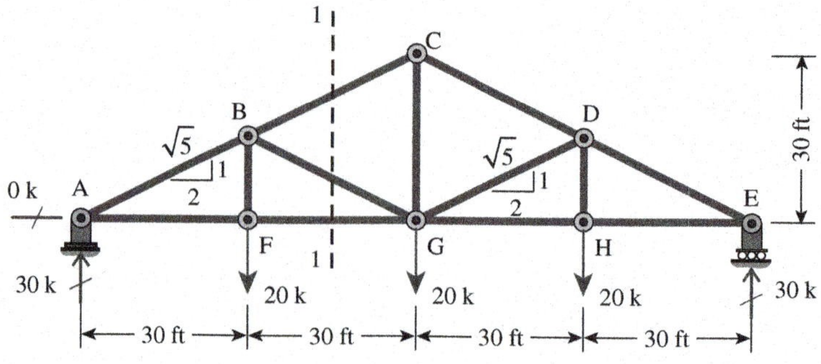

SOLUTION STRATEGY

Since only a few bar forces need to be found, selecting an appropriate cut is really the critical task at hand. In this example, a single cut at 1–1 will cut through the three bars of interest. Look for equations of equilibrium that minimize the number of unknowns.

SOLUTION

Make at cut at 1–1 and draw the FBD of the left section. The first equation of equilibrium can be $\sum M_A$, which will then solve directly for N_{BG}.

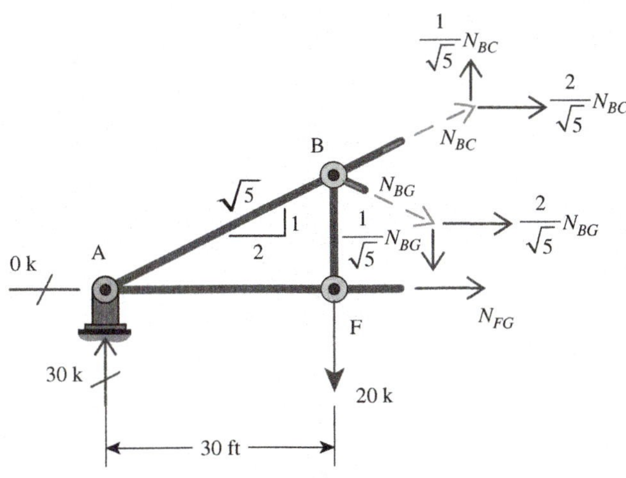

$$\circlearrowleft^{+} \sum M_A = -(20\text{ k})(30\text{ ft}) - \tfrac{1}{\sqrt{5}}N_{BG}(60\text{ ft}) = 0$$

$$N_{BG} = -22.36\text{ k} = \underline{\underline{22.36\text{ k} \quad (\text{C})}}$$

Now sum moments about point B to solve for N_{FG}.

$$\circlearrowleft^{+} \sum M_B = -(30\text{ k})(30\text{ ft}) + N_{FG}(15\text{ ft}) = 0$$

$$N_{FG} = 60\text{ k} = \underline{\underline{60\text{ k} \quad (\text{T})}}$$

Finally a sum of moments about point G will find the N_{BC}. You will want to translate the vertical and horizontal components of N_{BC} up to point C to make the equation of equilibrium easier to compose.

$$\circlearrowleft^{+} \sum M_G = -(30\text{ k})(60\text{ ft}) + (20\text{ k})(30\text{ ft}) - \tfrac{2}{\sqrt{5}}N_{BC}(30\text{ ft}) = 0$$

$$N_{BC} = -44.72\text{ k} = \underline{\underline{44.72\text{ k} \quad (\text{C})}}$$

DID YOU KNOW?

The components of the force in a truss bar may be positioned anywhere up and down the length of the member itself. When summing moments about A, it will be to your advantage to translate the horizontal and vertical components of BG down to point G. Thus, only the vertical component will show up in the equation of equilibrium.

SUMMARY

It is interesting to note that a judicious selection of the points about which moments were summed led to three independent equilibrium equations. Indeed, none of the equations relied on the correctness of the force calculations found in previous equations. This helps to minimize the propagation of mistakes.

Examples 7.8 and 7.9 are presented in a different fashion than the previous examples. In each example, a truss with a given loading will be presented and all the resulting bar forces will be given. The "solution" that is provided in each case describes the thought process followed to analyze the entire truss.

EXAMPLE 7.8

Determine the forces in all of the bars of the truss shown using a combination of the method of joints and the method of sections.

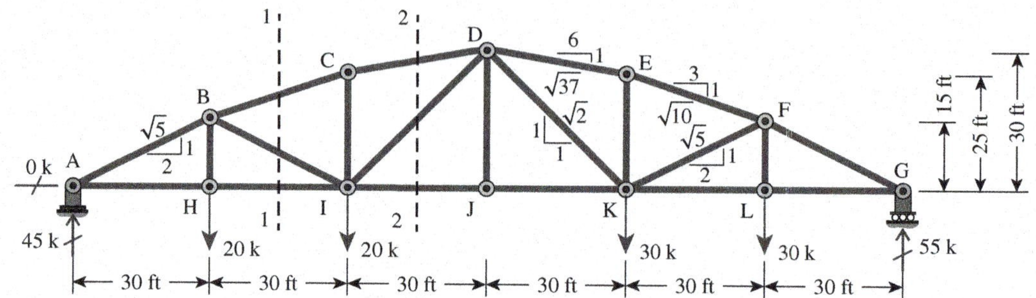

SOLUTION STRATEGY

Throughout the process, consider what steps may be taken to keep the equations of equilibrium as simple as possible.

SOLUTION

The following describes the steps followed to find the bar forces starting from the left side of the truss.

1. Scan all joints looking for zero force members. Joint J reveals that DJ is a zero force member.
2. The forces in bars meeting at joints A and H are quickly determined using the method of joints (AB, AH, BH, and HI).
3. A section cut 1–1 is made through the truss to solve for two more forces. Use $\sum M_I \Rightarrow N_{BC}$ and $\sum M_O \Rightarrow N_{BI}$.

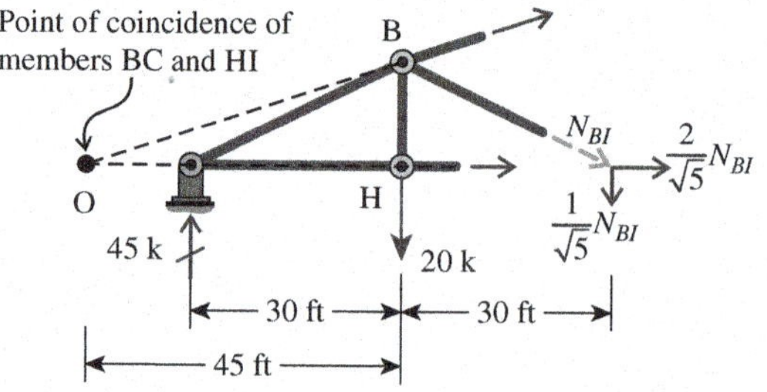

4. Joint C can be evaluated by method of joints (CD and CI).
5. Cut a section through 2–2 and solve for the forces in ID and IJ. Use $\sum M_D \Rightarrow N_{IJ}$ and $\sum F_x \Rightarrow N_{ID}$.

Follow a similar procedure to solve for the remaining bar forces.

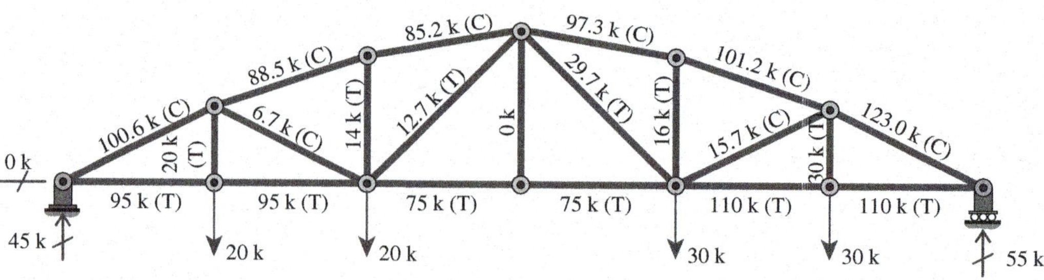

DID YOU KNOW?

One may sum moments about any point. The point DOES NOT need to lie on the structure. The intersection point of the line of actions of two members is often a convenient point about which to sum.

EXAMPLE 7.9

Determine the force in member CG of the truss shown.

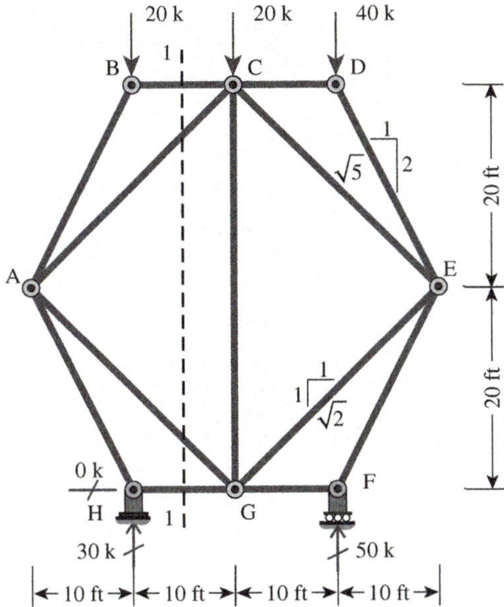

SOLUTION STRATEGY

Look for a joint or a cut that can be made to directly get the bar force of interest. The force in the member in question cannot be determined immediately by joints or by sections. Use a combination of the method of joints and method of sections to find enough bar forces to allow the solution of CG.

SOLUTION

1. The forces in bars BA, BC, DC, and DE are found by evaluating joints B and D.
2. Take a cut through section 1–1 and sum moments about G to solve for the force in bar AC.
3. The forces in bars CE and CG are found by evaluating joint C.

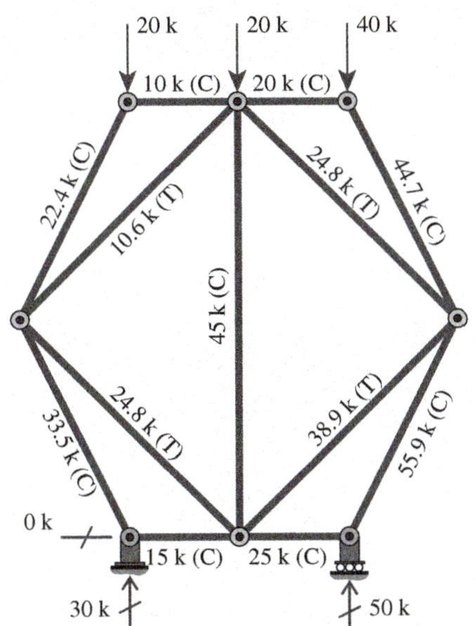

7.12 Simple, Compound, and Complex Trusses

Occasionally, you hear a truss referred to as a *simple truss* or a *compound truss*. These are references to the geometric form of the truss or the "building blocks" that form the truss. They are not references to the complexity of the analysis to determine member forces. For completeness, we briefly discuss the categories of trusses from which these references arise. We also discuss some issues regarding analysis of trusses in each category.

Simple Trusses

As we have seen, the first step in forming a truss is connecting three members at their ends to form a triangle. Subsequent segments are incorporated by adding two members and one joint; the new members meet at the new joint, and each is pinned at its opposite ends into one of the existing joints. Trusses formed in this way are said to be simple trusses. The authors are not suggesting, however, that all trusses formed in this manner are "simple" to analyze.

Compound Trusses

A *compound truss* is a truss made by connecting two or more simple trusses. The simple trusses may be connected by three nonparallel non-concurrent links, by one joint and one link, by a connecting truss, by two or more joints, and so on. An almost unlimited number of trusses may be formed in this way. The Fink truss shown in Figure 7.18(a) consists of the two shaded simple trusses that are connected with a joint and a link. All of the methods for checking stability and for analysis can be used on compound trusses with equal success. See Example 7.10.

Complex Trusses

There are a few trusses that are statically determinate that do not meet the requirements necessary to fall within the classification of either simple or compound trusses. Such a truss is shown in Figure 7.18(b). These are referred to as *complex trusses*.

The members of simple and compound trusses usually are arranged so that sections may be passed through three members at a time; moments are taken about the intersection of two of them, and the force in the third member is found. Complex trusses may not be analyzed in this manner. Not only does the method of sections fail to simplify the analysis, the method of joints is also of no avail. The difficulty lies in the fact that there are three or more members meeting at almost every joint. Consequently, there are too many unknowns at every location in the truss to pass a section and obtain the force in any member directly using the equations of static equilibrium.

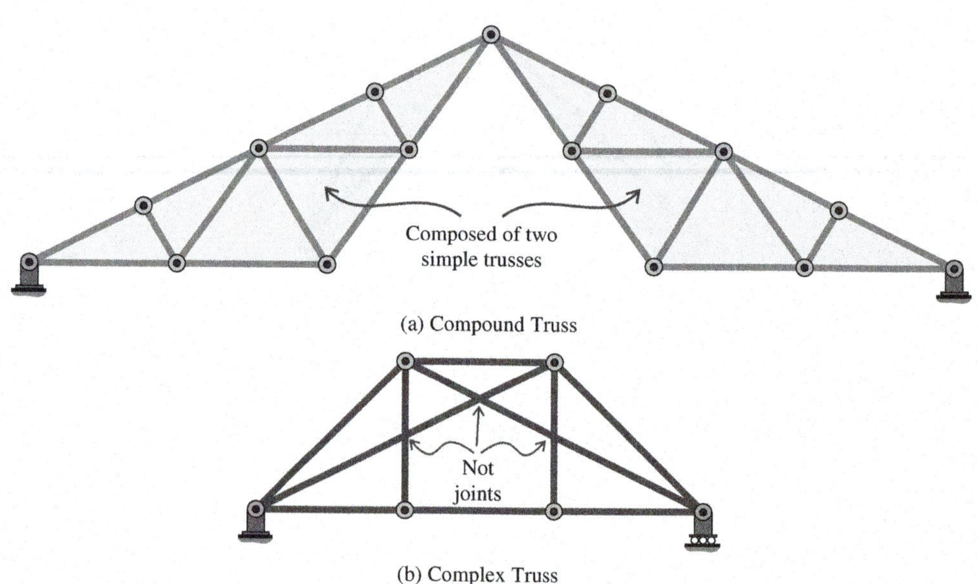

FIGURE 7.18 Non-simple truss examples.

One method of computing the forces in complex trusses is to write the equations of static equilibrium at each joint, which yields $(2)(j)$ equations. These equations may be solved simultaneously for the member forces and external reactions as discussed in Section 7.9. It is often possible to calculate the external reactions initially, and their values may be used as a check against the results obtained from the solution of the simultaneous equations. This method will work for any statically determinate complex truss, but the solution of the equations is very tedious unless a computer is used.

Generally, there is little need for complex trusses, because it is possible to select simple or compound trusses that will serve the desired purpose equally well. Nevertheless, for a more comprehensive discussion of complex trusses, you may refer to the *method of substitute members* described in *Theory of Structures* by S. P. Timoshenko and D. H. Young.[1]

EXAMPLE 7.10

For the compound truss shown, explain the process one would follow to find all of the bar forces.

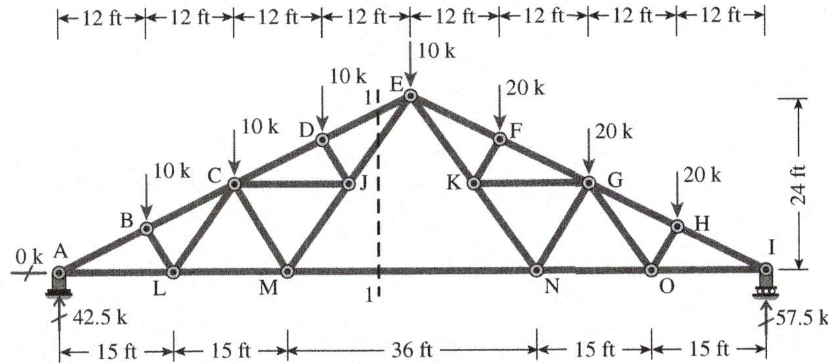

SOLUTION STRATEGY

Scan the truss looking for any joints that have no more than two unknowns. Also look for any places where section cuts might be made where no more than three members are intersected.

SOLUTION

1. Start with joints in this sequence: A, B, and L.
2. Make a section cut at 1–1. This will solve for forces in bars DE, JE and MN.
3. Work the joints in this order: D, C, and M.
4. Work the right side of the truss in a similar fashion with joints in this order: I, H, O, E, F, G, and N.

QUICK NOTE

The first three joints working from the left of the truss can be solved using the method of joints. However, neither joint C or M could be solved because they would still have three unknowns. We must look for a section cut.

7.13 Structural System Consideration

There are two basic underlying assumptions for use in the analysis of trusses in this text.

1. All loads are applied at the joints.
2. All truss members are connected using frictionless pins.

There are many *truss-like* structural elements that do not meet these conditions and are used to build everyday structures. A classic example is a pre-fabricated wood truss in the roof system of a residential structure or the like. Figure 7.19 shows a typical connection and loading scenario that is faced in residential wood construction. Note that the top and bottom chords are continuous

QUICK NOTE

Very few, if any, trusses would ever meet this set of criteria exactly. As such, one should recognize that the results of the analysis are at best a close approximation of the actual forces.

[1] S. P. Timoshenko and D. H. Young, *Theory of Structures.* 2nd Ed., (New York: McGraw-Hill, 1965), 92–103.

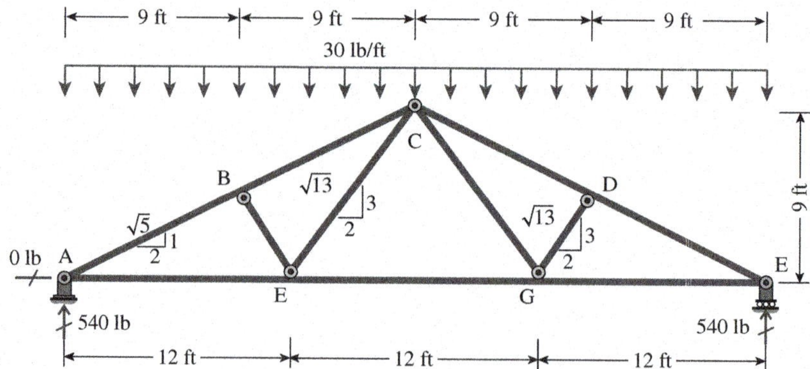

FIGURE 7.19 Typical Fink truss found in residential construction.

> **QUICK NOTE**
> Such an approach for analysis can cause significant errors in member force approximations depending on the relative size (stiffness) of the individual members. Nevertheless, it is a useful approach when hand calculations are necessary.

from the heel of the truss up to the peak. The bottom chord is also continuous from support to support, while the webs are connected with pins. A structural component like this still benefits from truss-like action even though it does not meet the assumption criteria directly.

In practice, a structural component like that shown in Figure 7.19 is sometimes analyzed in a simplistic fashion.[2] The following steps are taken.

1. The intersection of all members are considered to be pins. Take for instance joint B. Assumption is that ALL three members are pin connected at that joint. This is usually a conservative approximation but not always.
2. The the resultant load applied to each joint is based on the tributary width and the distributed load.
3. The axial forces in the members are evaluated by doing a basic truss analysis resulting in Figure 7.20(a).
4. The maximum moment in each of the top chords is estimated by treating each member (e.g., AB) as a simply supported beam and then analyzing it accordingly. The results of this are shown in Figure 7.20(b).
5. The truss members are then designed for both the axial forces and bending moment determined from this analysis.

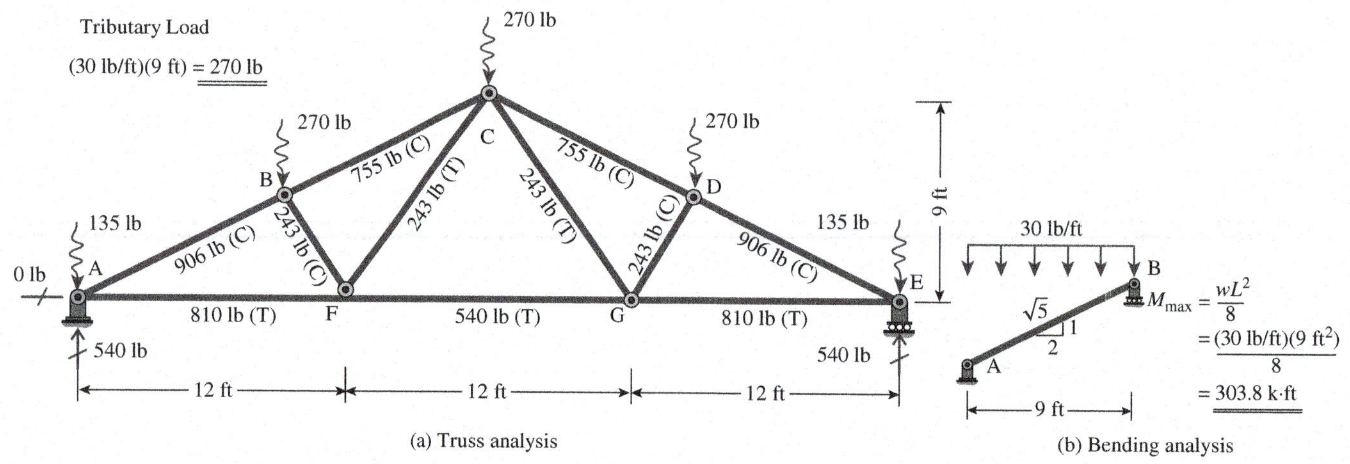

FIGURE 7.20 Simplified analysis of a residential wood truss.

[2] D.E. Breyer, *Design of Wood Structures* 3rd Ed (New York: McGraw-Hill, 1993), 321–322.

7.14 SAP2000 Computer Applications

The modeling of trusses in SAP2000 is very similar to the analysis of frames. The one significant difference between a frame element and a truss element is that frame elements are connected with moment-resisting connections, while the truss elements are connected with pins (i.e., moment releases).

Trusses can be drawn using the "New Model" truss wizard. While convenient for assembling some trusses, the wizard can be quite restrictive as to what kind of truss you will be able to model. A more generic truss may be modeled by first setting up a grid and drawing the members in the desired configuration. Remember that a truss element is pinned at both ends. It is best to set these conditions at the time of drawing the members. Figure 7.21(a) shows the default settings that are used when drawing structure elements. Notice that next to the term *Moment Releases* the value is *continuous*. This will produce moment-resisting connections between the members. Click on the word *continuous* and select *pinned* from the pull-down menu. This is shown in Figure 7.21(b). Now proceed to draw all of the elements needed for the truss.

To look at results either graphically or tabular, you need to review the material presented in Section 6.10. Remember that a true truss should not have any internal forces other than an axial force. If other forces exist, then something is wrong in the modeling. Two common mistakes are that not all members were modeled as pin-ended members or the self-weight of the structure was not turned off.

QUICK NOTE

It will be worth your while to play around with the "New Model" wizard available in SAP2000. It can save you a lot of time in the basic modeling process of many structures.

DID YOU KNOW?

If you forget to draw all of the elements with pin-ended connections, you may modify this through the *ASSIGN* menu. First select all members you need to modify. Then choose the following menu: *ASSIGN* ⇒ *FRAME* ⇒ *RELEASES/PARTIAL FIXITY....* Then place moment releases at both ends of the member about the 3–3 axis as shown in Figure 7.22.

SAP2000 EXERCISE 7.1

The SAP2000 input file *C7EX1.sdb* is a functional model of the truss structure shown. Run SAP and report the axial forces in the diagonal members.

Properties of Object

Line Object Type	Straight Frame
Section	truss
Moment Releases	Continuous
XY Plane Offset Normal	0.
Drawing Control Type	None <space bar>

(a) Draw frame element

Properties of Object

Line Object Type	Straight Frame
Section	truss
Moment Releases	Pinned
XY Plane Offset Normal	0.
Drawing Control Type	None <space bar>

(b) Draw truss element

FIGURE 7.21 Simplified analysis of a residential wood truss.

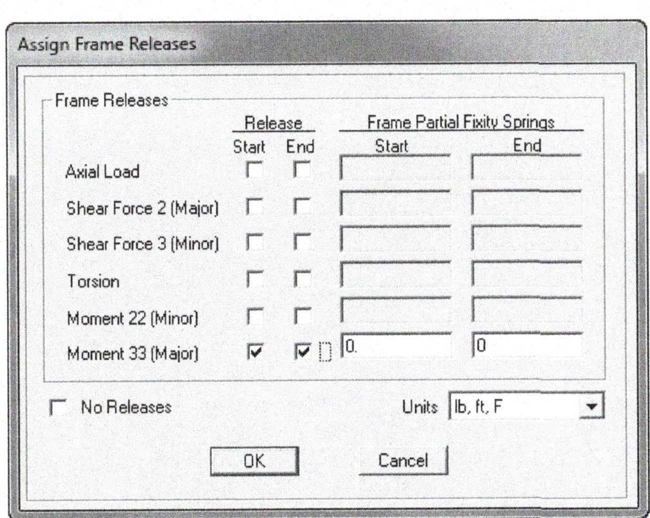

FIGURE 7.22 Assigning pin-ended members (i.e., truss members).

QUICK NOTE

You should note that SAP2000 uses a number to identify each truss member rather than the alphabet based on the joints of that member. SAP will number the members in the order you draw them. If you want to reassign the number (label) given to a member, you need to right-click on that member.

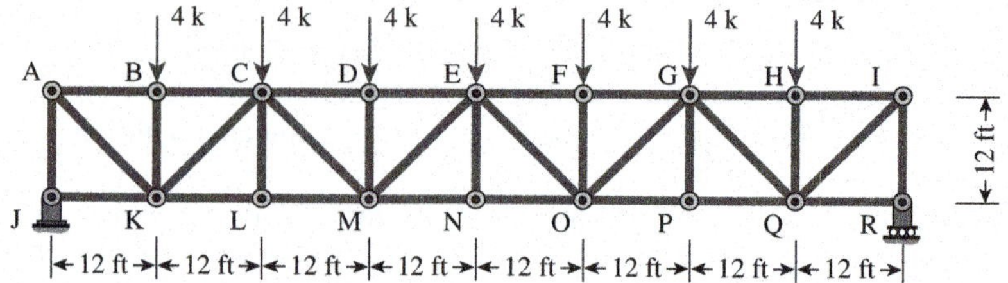

This is the SAP model with the numbering adopted for each member.

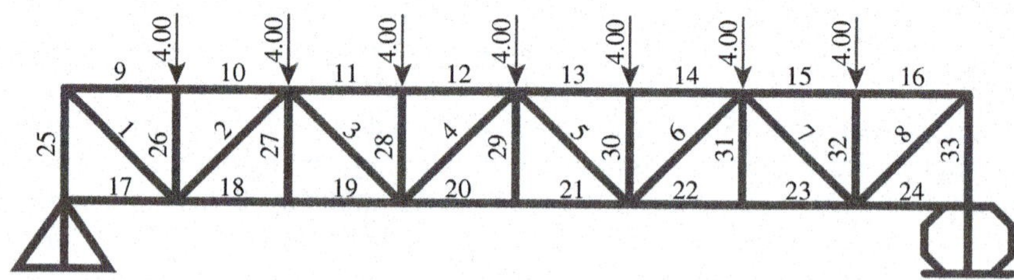

Member	Force (k)	T or C
1	19.799	T
2	14.142	C
3	8.485	T
4	2.828	C
5	2.828	C
6	8.485	T
7	14.142	C
8	19.799	T

Now place 28-k at joint E and no load at the other joints. Find the forces in the diagonal bars and compare with the previous findings.

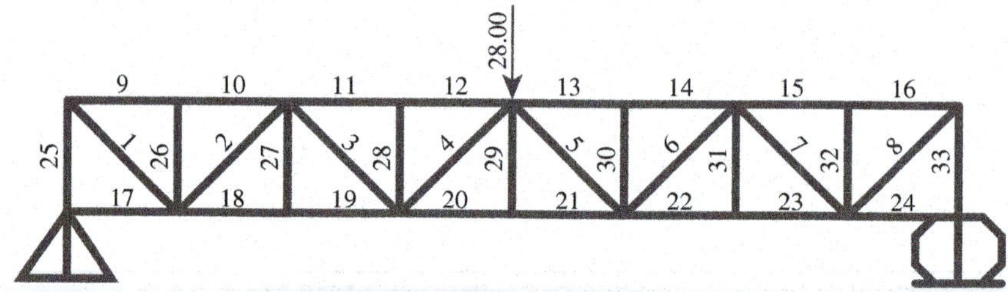

Member	Force (k)	T or C
1	19.799	T
2	19.799	C
3	19.799	T
4	19.799	C
5	19.799	C
6	19.799	T
7	19.799	C
8	19.799	T

SUMMARY

The force in the diagonal bar is a direct result of the shear within each panel of the truss. In the first truss, the shear at the left end started at 14 k and then reduced by 4 k in each panel until mid-span. After that, the shear in each panel increased by 4 k until it reached 14 k at the support. For the case when all of the load is at the middle, the shear in all the panels left of mid-span are the same and the shear in all the panels to the right of mid-span are the same. The diagonals must resist all of the shear $N = \sqrt{2}(14\text{ k}) = 19.8\text{ k}$.

SAP2000 EXERCISE 7.2

For the parallel chord truss presented in SAP Exercise 7.1 (*C7EX1.sdb*), remove all of the 4-k loads, place 28 k at joint A, and analyze the truss. Record the axial force in diagonal bars AK (bar 1) and EO (bar 5). Then move the load to point B and do the same thing. Repeat for all joints A through I. Create a plot of the load location versus the axial force in the two bars. Comment on the observed trends.

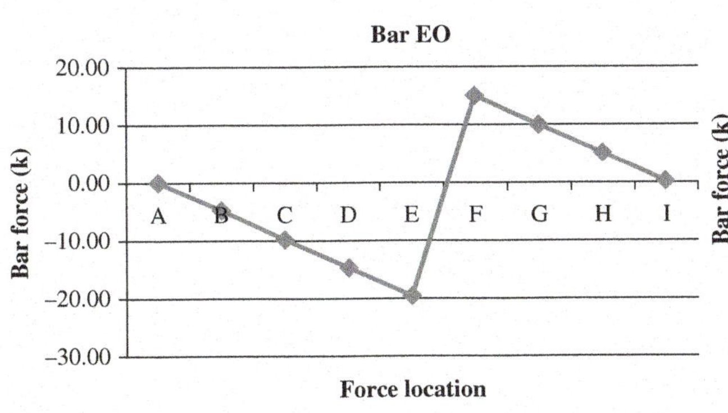

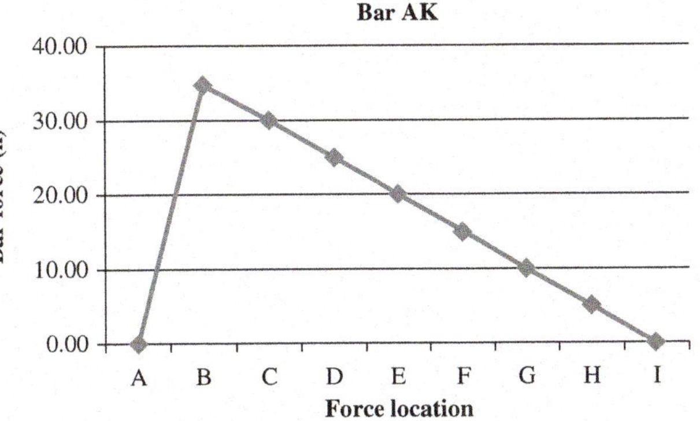

Load Location Joint	N_{AK} (k)	N_{EO} (k)
A	0.00	0.00
B	34.65	4.95
C	29.70	9.90
D	24.75	14.85
E	19.80	−19.80
F	14.85	14.85
G	9.90	−9.90
H	4.95	−4.95
I	0.00	0.00

SUMMARY

When the loads are positioned right over the supports (A and I), all of the load goes into the supports and none of it goes into the span. The trend for bar AK really is of no surprise. As the applied force is positioned closer to this bar, the bar sees more internal force. The internal force in bar EO was in compression until the load passed over the panel EFON, then it jumped to a tension force. This is a direct consequence of whether the shear in panel EFON is positive or negative.

7.15 Examples with Video Solutions

VE7.1 For the two trusses in Figure VE7.1, indicate whether the truss is determinate, indeterminate, or unstable. If the truss is indeterminate, indicate the degree of indeterminacy.

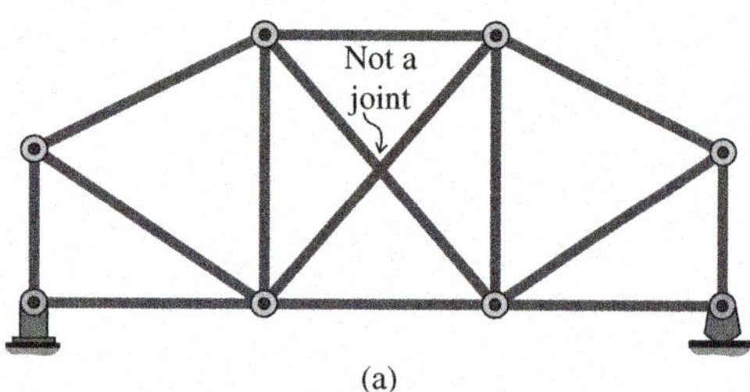

(a)

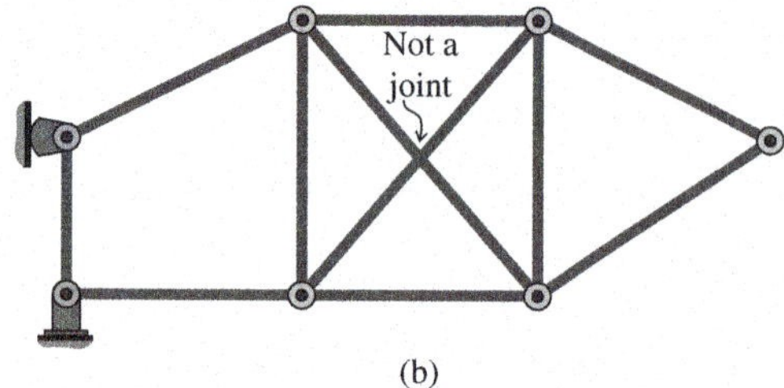

(b)

Figure VE7.1

VE7.2 For the truss in Figure VE7.2, use the method of joints to solve for all bar forces. Note the symmetry of the truss.

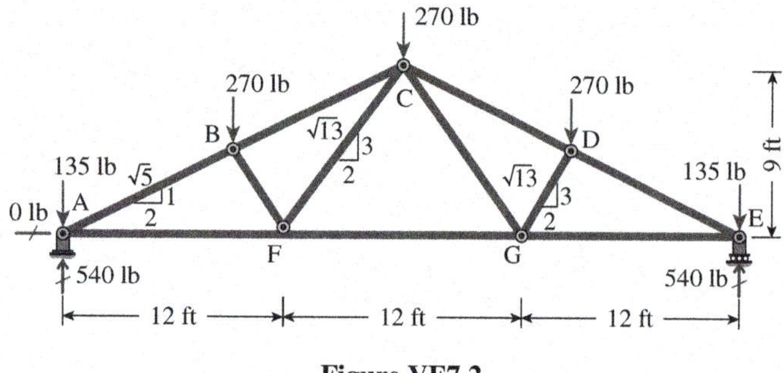

Figure VE7.2

VE7.3 For the truss in Figure VE7.3, identify all zero-force members.

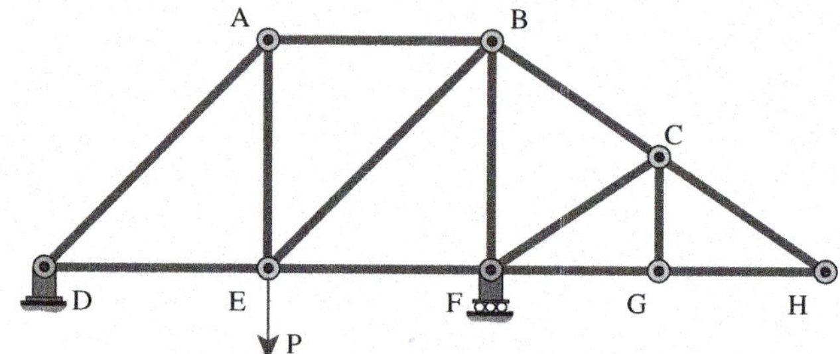

Figure VE7.3

VE7.4 For the truss in Figure VE7.4, use the method of sections to find the internal forces in members CD, CM, and LM. Use equilibrium equations that are written only in terms of one unknown force.

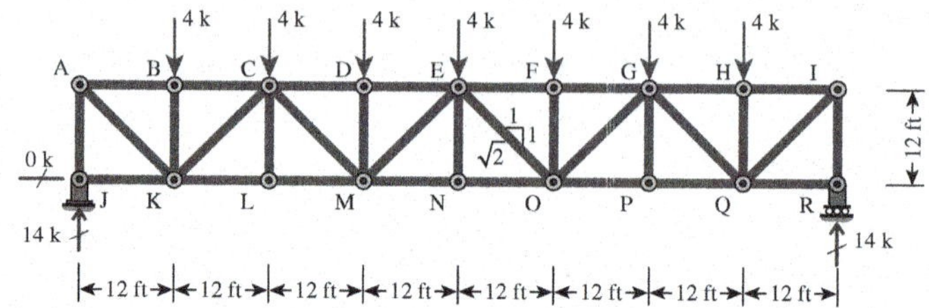

Figure VE7.4

VE7.5 For the truss in Figure VE7.5, indicate the equilibrium equation (e.g., $\sum F_x$, $\sum F_y$, and $\sum M$) that could be used to find the forces in the bars BC, FC, and FG if the method of sections is to be used. Choose the equilibrium equation such that each equation would have only one unknown force.

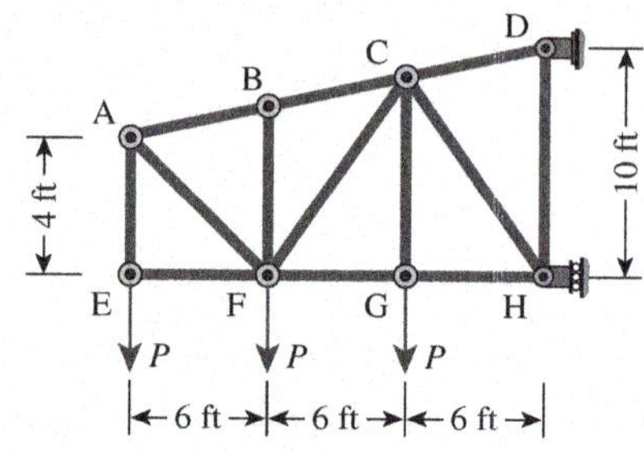

Figure VE7.5

7.16 Problems for Solution

Section 7.6

For Problems 7.1 through 7.22, classify the structures as to their internal and external stability and determinacy. For statically indeterminate structures, include the degree of indeterminacy. (The small circles on the trusses indicate the joints.)

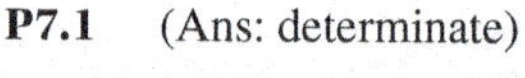

P7.1 (Ans: determinate)

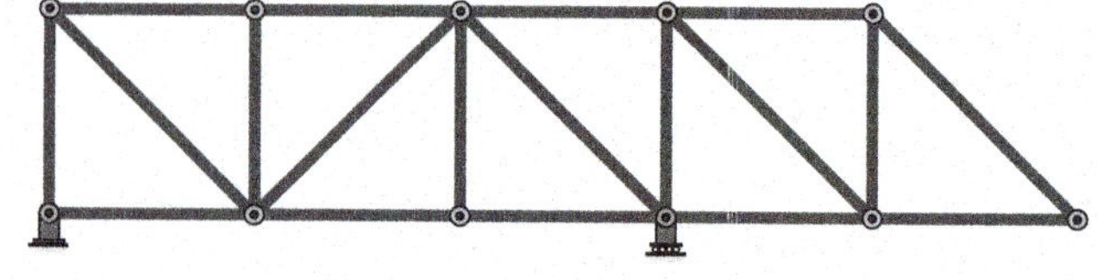

Problem 7.1

P7.2

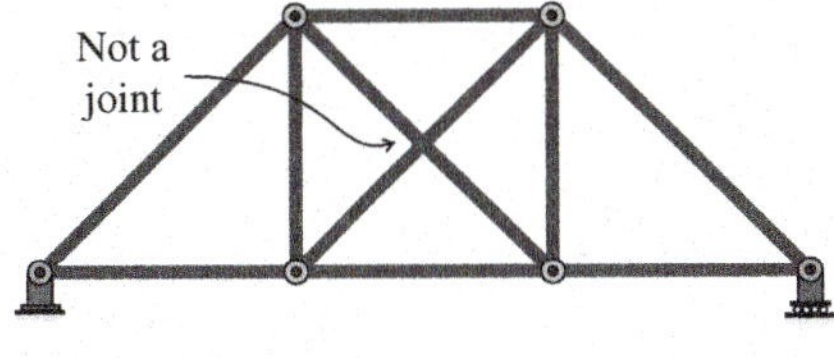

Problem 7.2

P7.3 (Ans: SI 1°)

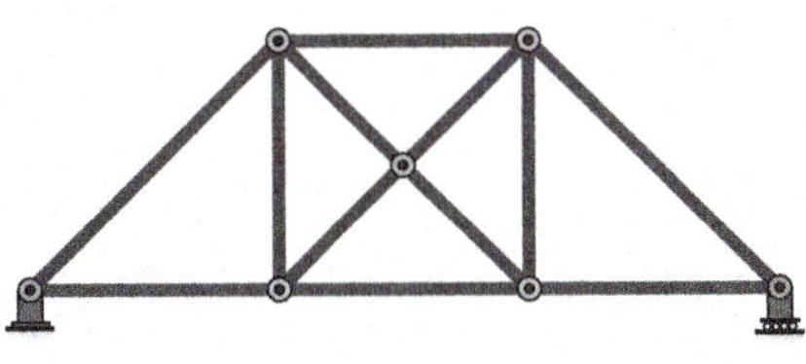

Problem 7.3

P7.4

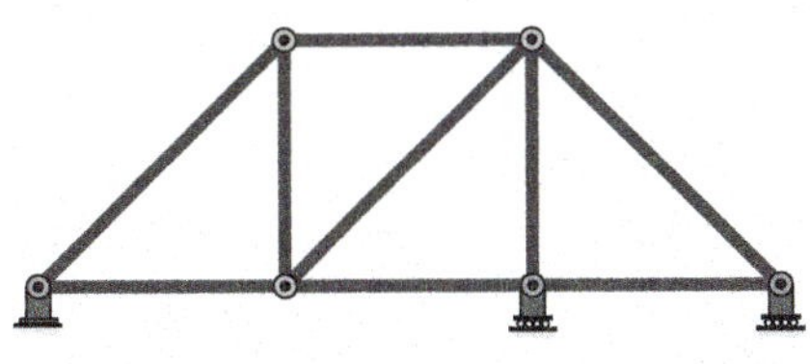

Problem 7.4

P7.5 (Ans: SI 1°)

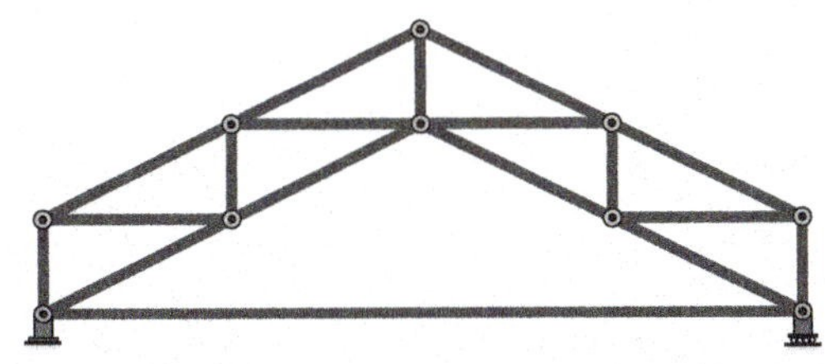

Problem 7.5

P7.6

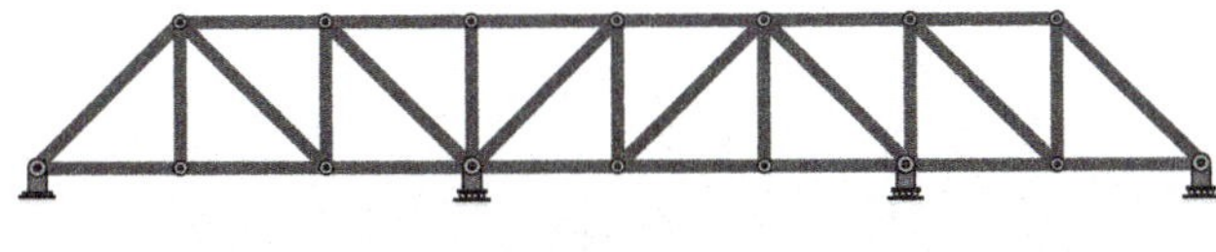

Problem 7.6

P7.7 (Ans: SI 1°)

Problem 7.7

P7.8

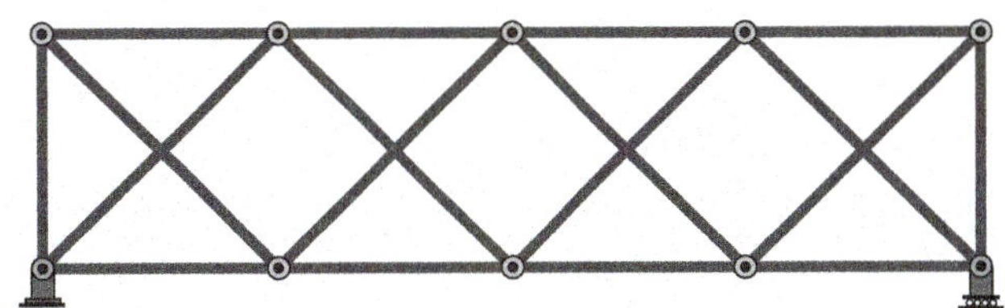

Problem 7.8

P7.9 (Ans: determinate)

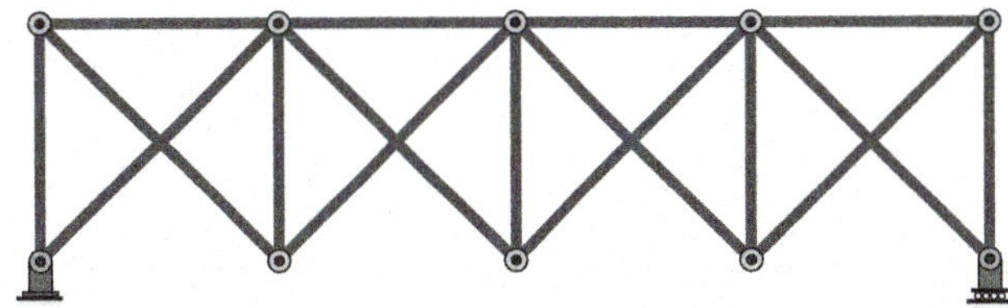

Problem 7.9

P7.10

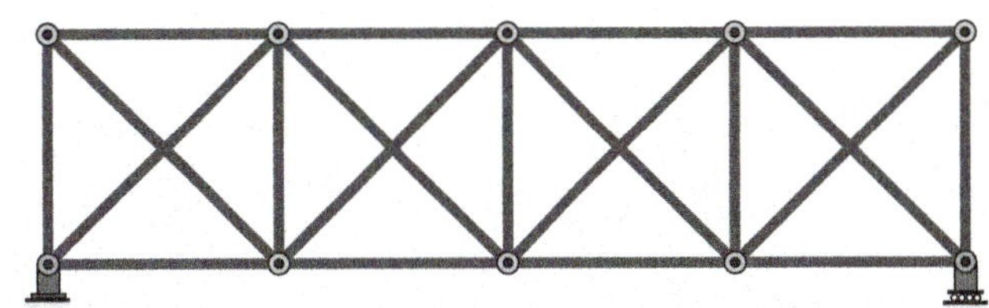

Problem 7.10

P7.11 (Ans: internally unstable)

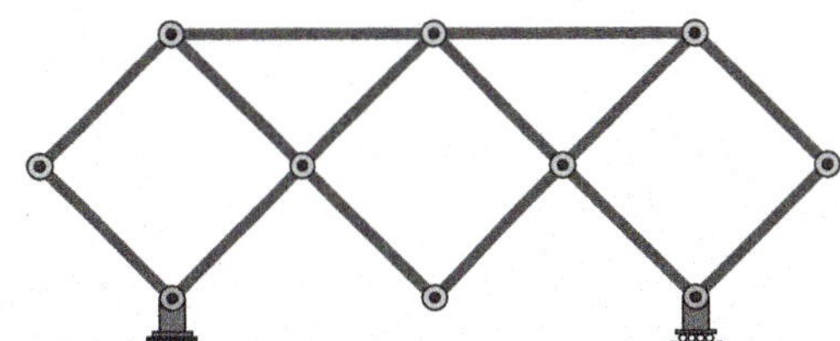

Problem 7.11

P7.12

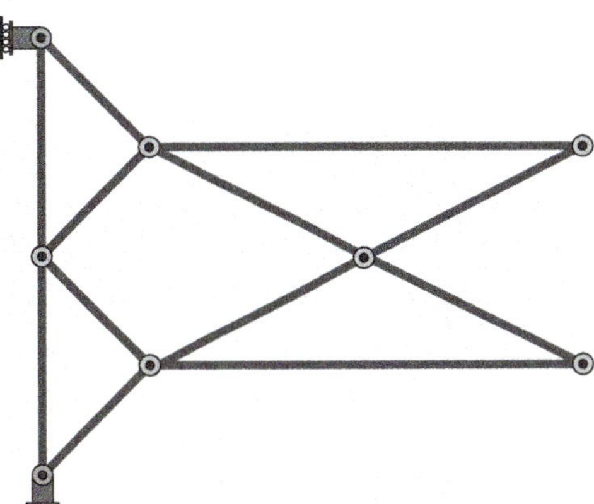

Problem 7.12

P7.13 (Ans: determinate)

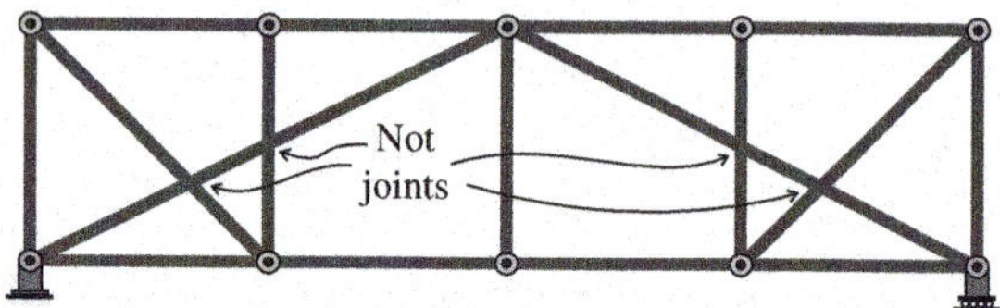

Problem 7.13

P7.14

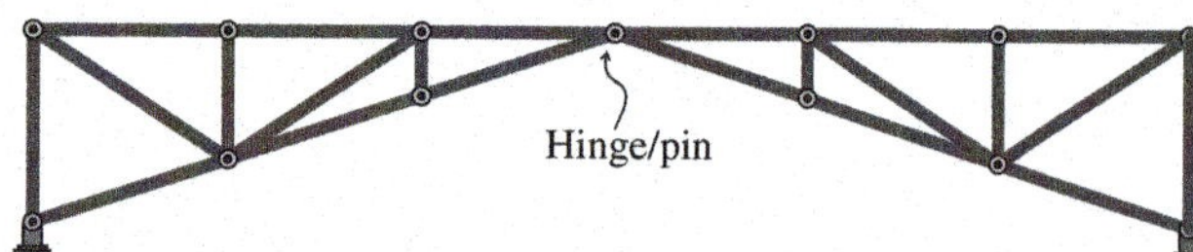

Problem 7.14

P7.15 (Ans: SI 1°)

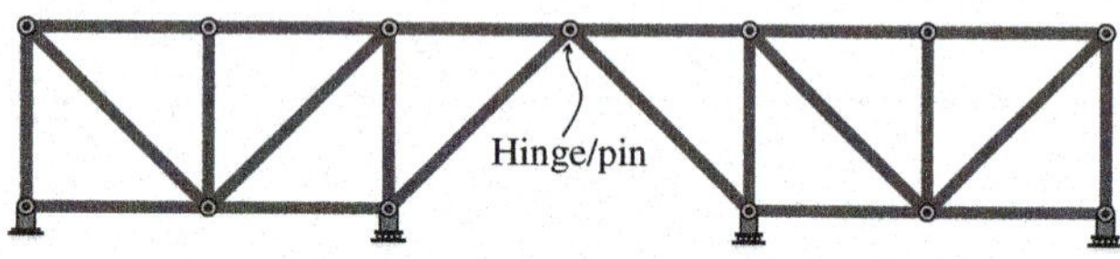

Problem 7.15

P7.16

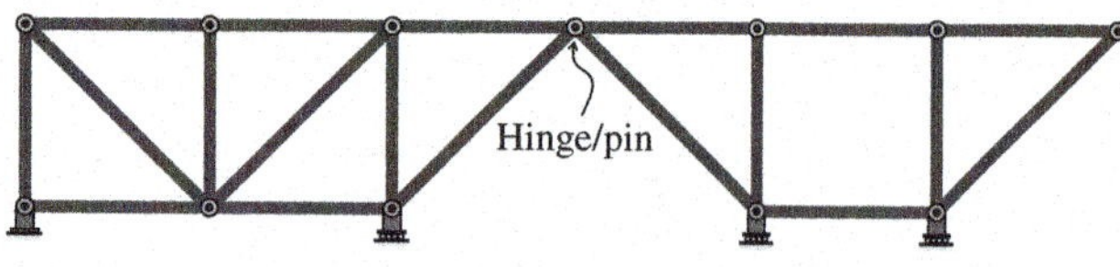

Problem 7.16

P7.17 (Ans: unstable)

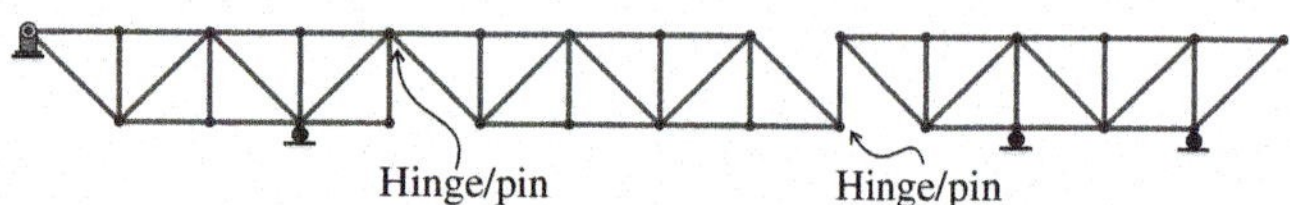

Problem 7.17

P7.18

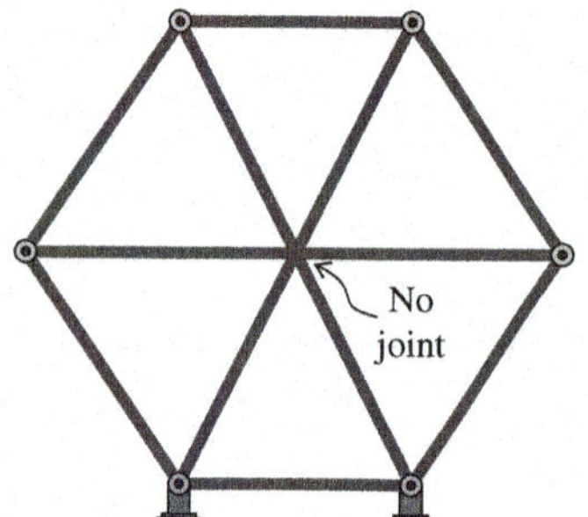

Problem 7.18

P7.19 (Ans: SI 1°)

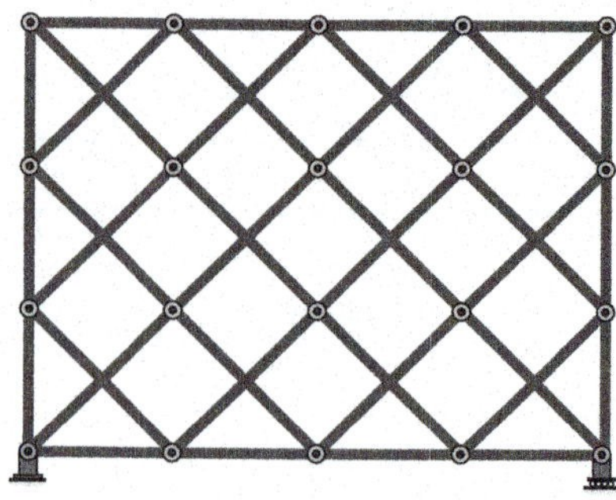

Problem 7.19

P7.20

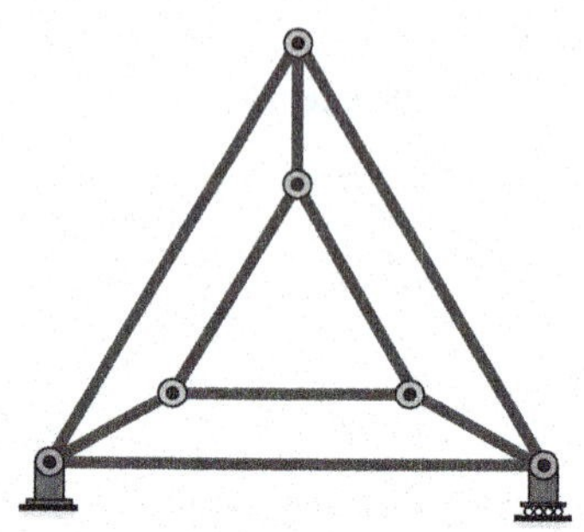

Problem 7.20

P7.21 (Ans: SI 2°)

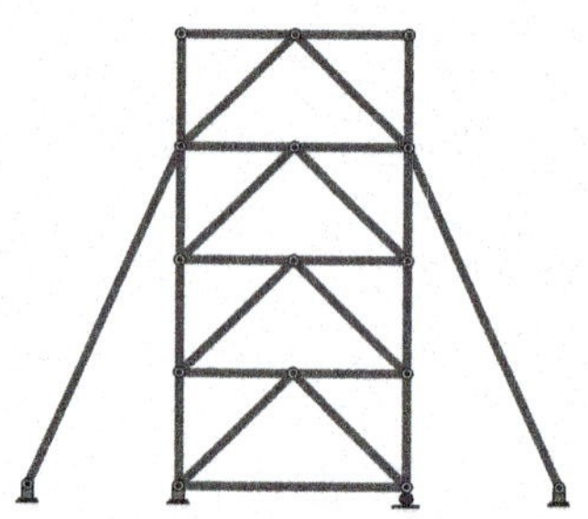

Problem 7.21

P7.22

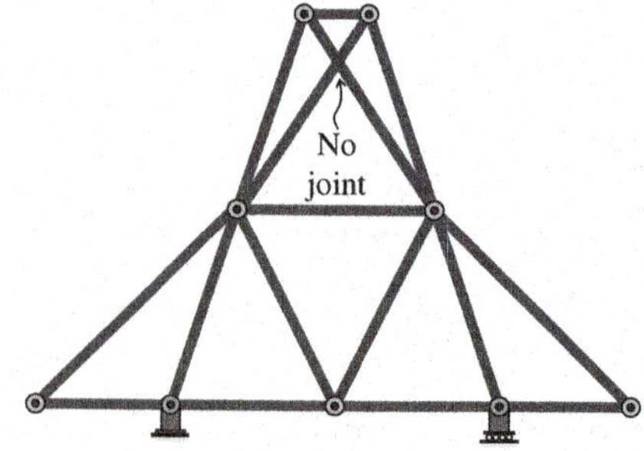

Problem 7.22

Section 7.8

For Problems 7.23 through 7.40, use the method of joints to compute the internal forces in all bars of each truss.

P7.23 (Ans: $N_{EB} = 4.7$ k, $N_{CH} = 42.4$ k)

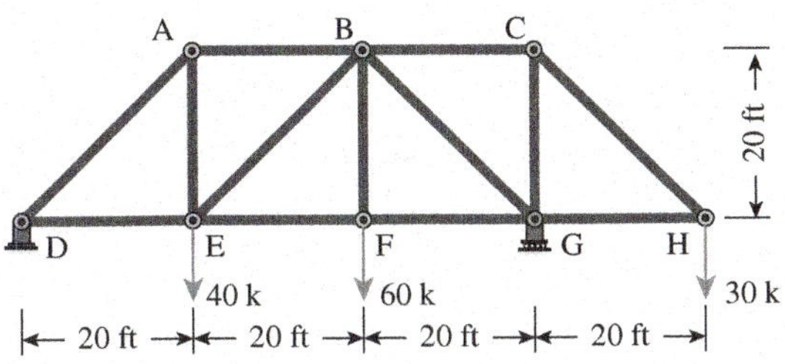

Problem 7.23

P7.24

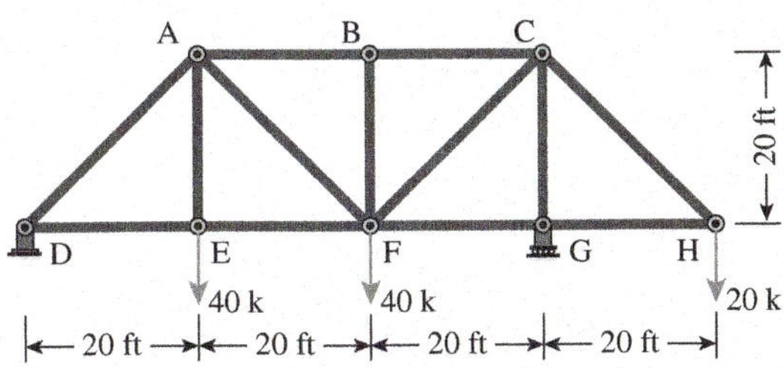

Problem 7.24

P7.25 (Ans: $N_{CG} = -50.0$ k, $N_{FC} = 58.3$ k)

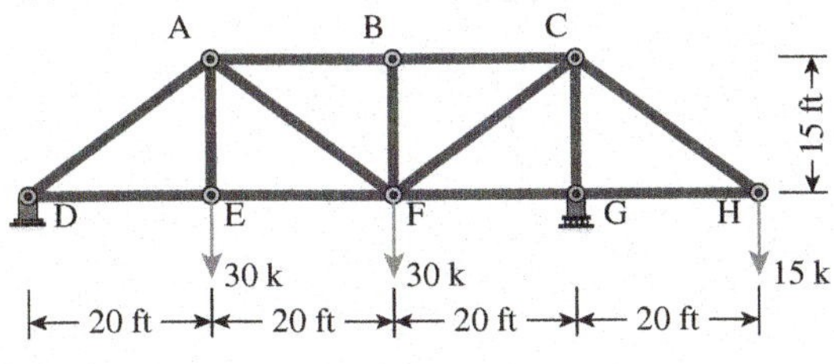

Problem 7.25

P7.26

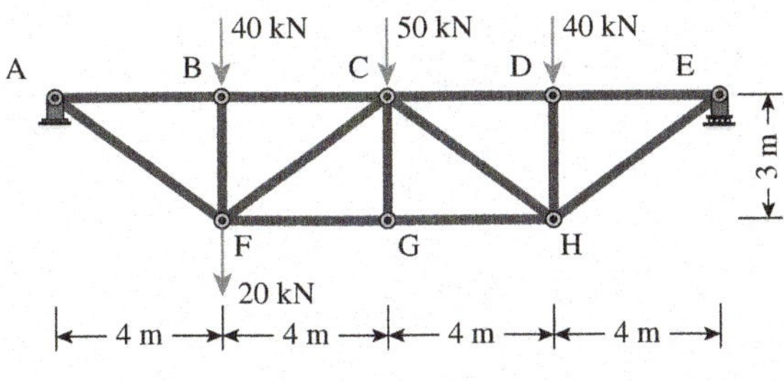

Problem 7.26

P7.27 (Ans: $N_{FC} = 70.7$ k, $N_{CD} = -167.7$ k)

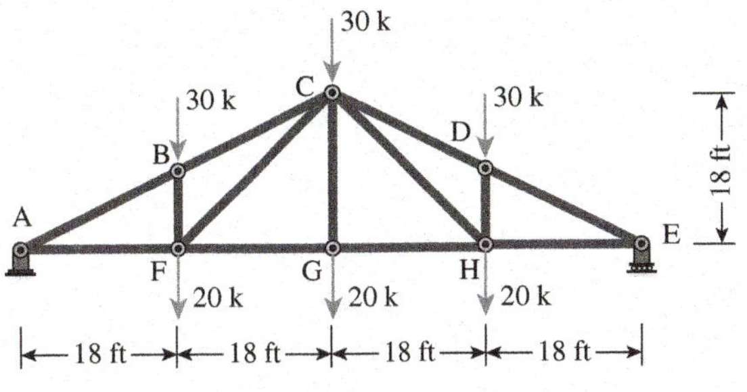

Problem 7.27

P7.28

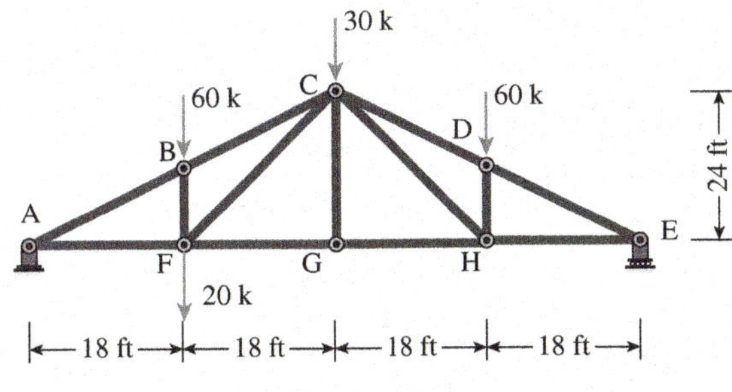

Problem 7.28

P7.29 (Ans: $N_{CF} = -134.2$ k, $N_{DF} = -140.0$ k)

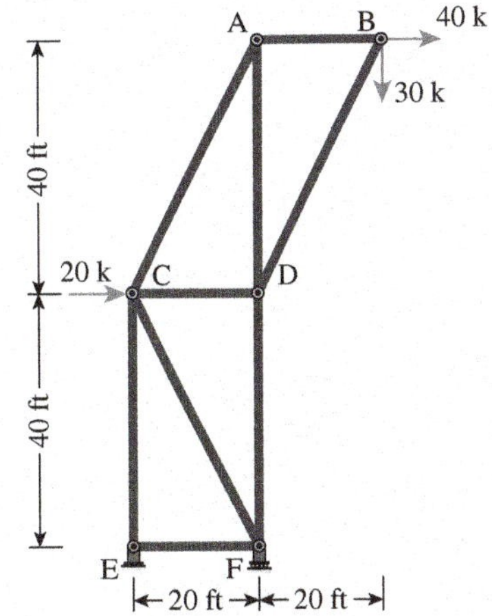

Problem 7.29

P7.30

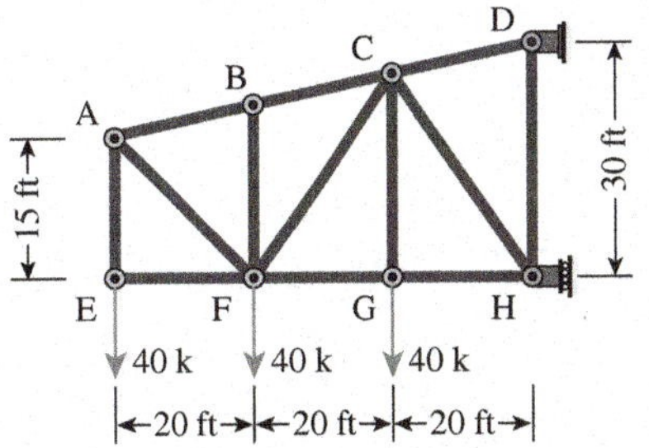

Problem 7.30

P7.31 (Ans: $N_{EF} = -134.2$ k, $N_{BE} = -89.4$ k)

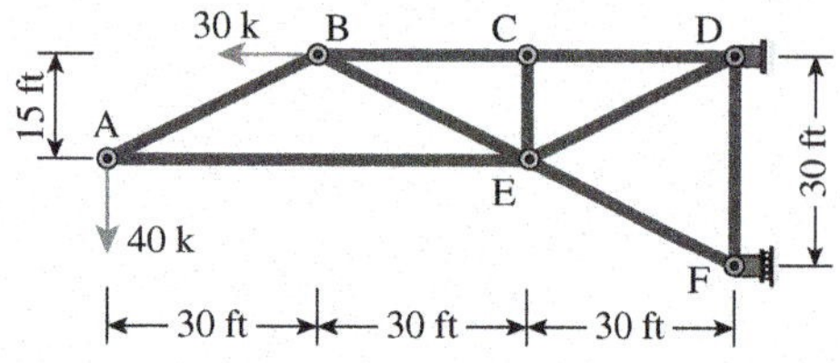

Problem 7.31

P7.32

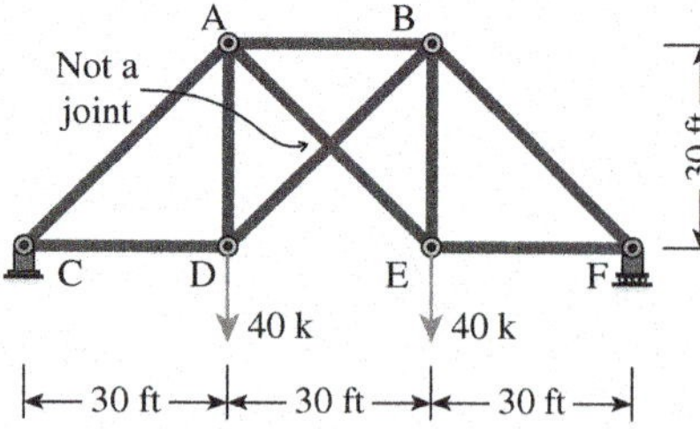

Problem 7.32

P7.33 (Ans: $N_{BC} = 53.3$ k, $N_{EF} = -113.9$ k)

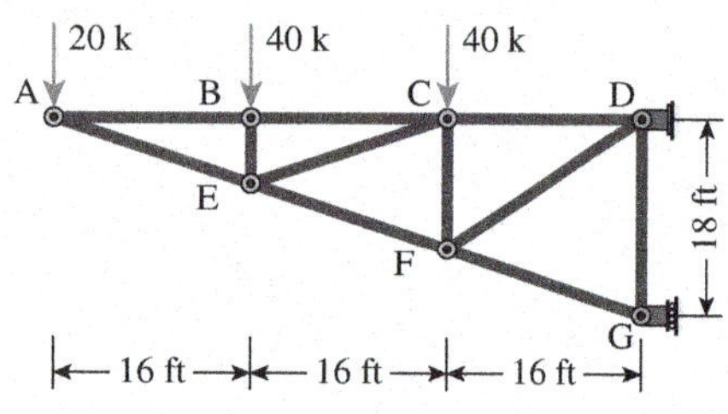

Problem 7.33

P7.34

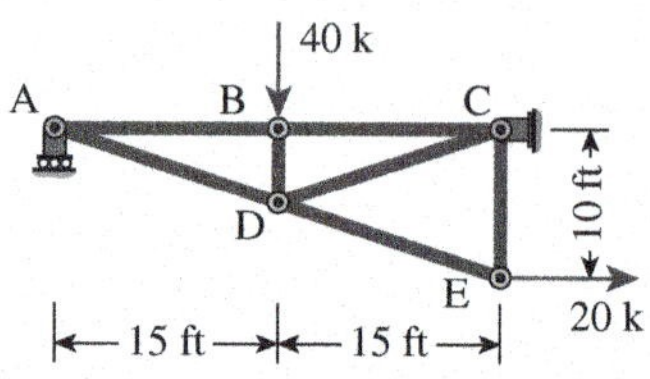

Problem 7.34

P7.35 Repeat Problem 7.34 if the roller support (due to friction, corrosion, etc.) is assumed to supply one-third of the total horizontal force resistance needed, with the other two-thirds supplied by the pin support. (Ans: $N_{DE} = 21.1$ k, $N_{AB} = -73.3$ k)

P7.36

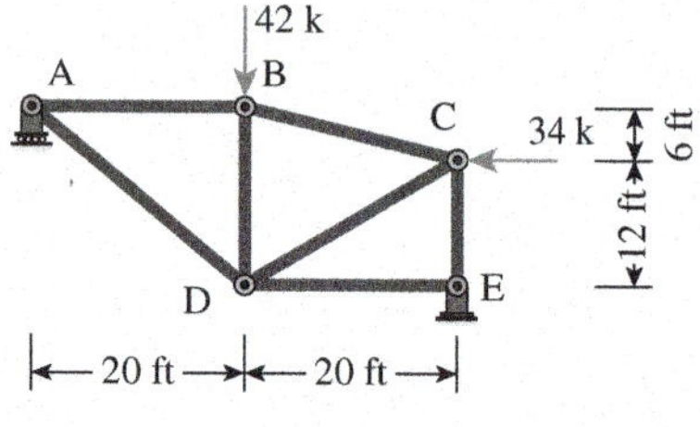

Problem 7.36

P7.37 (Ans: $N_{BC} = -17.1$ k, $N_{DB} = -8.8$ k)

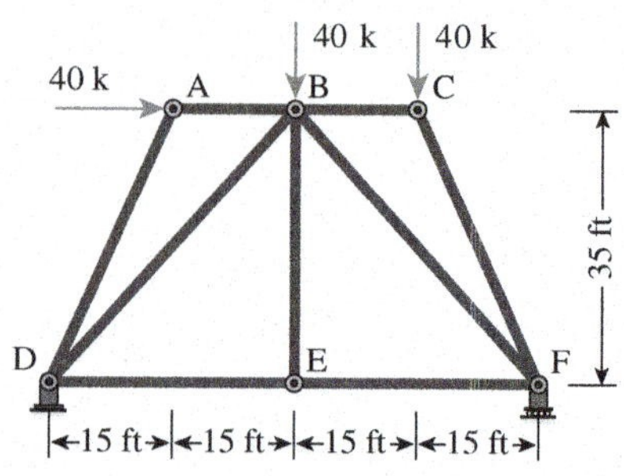

Problem 7.37

P7.38 Rework Problem 7.37 with the support at F replaced with the following inclined roller.

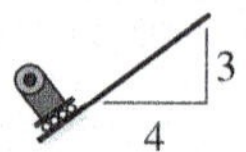

Problem 7.38

P7.39 (Ans: $N_{CF} = -62.5$ k, $N_{DF} = -45.0$ k)

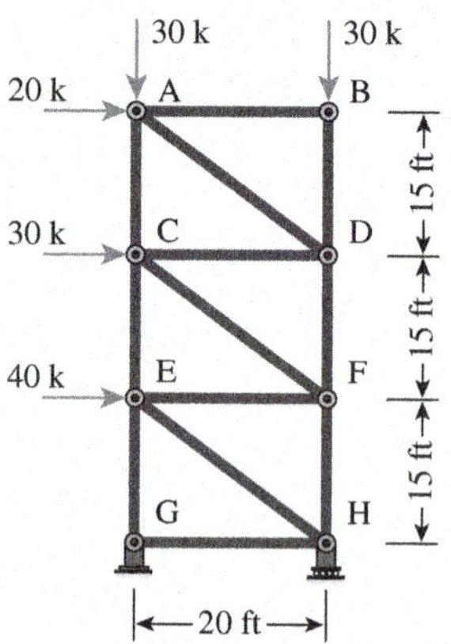

Problem 7.39

P7.40

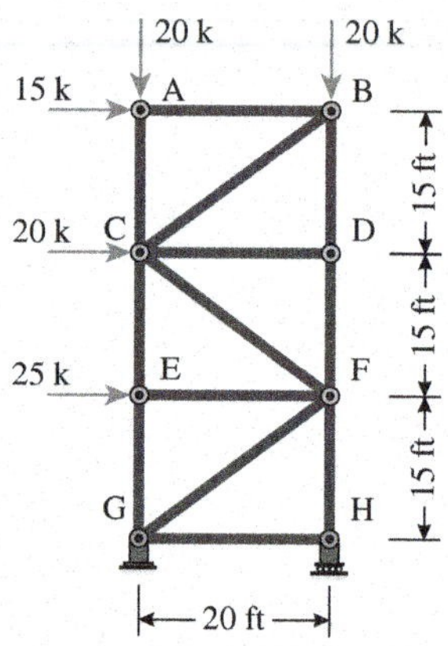

Problem 7.40

Section 7.9

P7.41 Using the matrix formulation, solve for all bar forces and reactions for the truss shown in Problem 7.34. (Ans: $N_{DC} = 63.2$ k, $N_{DE} = 21.1$ k)

P7.42 Using the matrix formulation, solve for all bar forces and reactions for the truss shown in Problem 7.36.

P7.43 Using the matrix formulation, solve for all bar forces and reactions for the truss shown. (Ans: $N_{BA} = 18.8$ k, $N_{BC} = 0.0$ k)

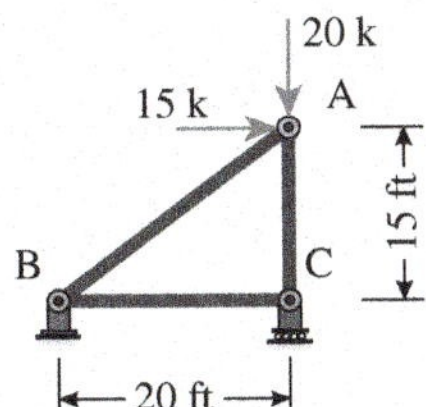

Problem 7.43

P7.44 Using the matrix formulation, solve for all bar forces and reactions for the truss shown.

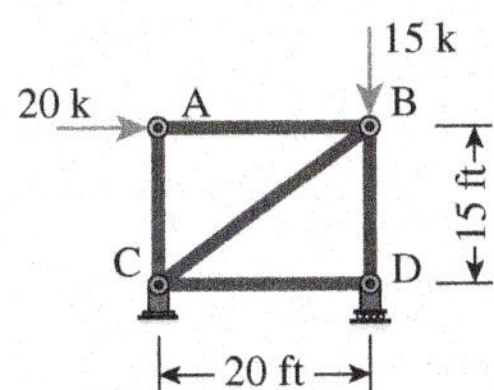

Problem 7.44

P7.45 Using the matrix formulation, solve for all bar forces and reactions for the truss shown in Problem 7.37. (Ans: $N_{BD} = -8.8$ k, $N_{DE} = 45.7$ k)

Section 7.10

P7.46 Using the principles outlined in Section 7.10, identify all zero-force members in the truss given in Problem 7.24.

P7.47 Using the principles outlined in Section 7.10, identify all zero-force members in the truss given in Problem 7.26. (Ans: member CG)

P7.48 Using the principles outlined in Section 7.10, identify all zero-force members in the truss given in Problem 7.31.

P7.49 Using the principles outlined in Section 7.10, identify all zero-force members in the truss given in Problem 7.37. (Ans: members BE, AD)

P7.50 Using the principles outlined in Section 7.10, identify all zero-force members in the truss given in Problem 7.40.

P7.51 Using the principles outlined in Section 7.10, identify all zero-force members in the truss given. (Ans: members BG, CH, CD, DI, DJ, DE, IJ, EJ, EK, JK)

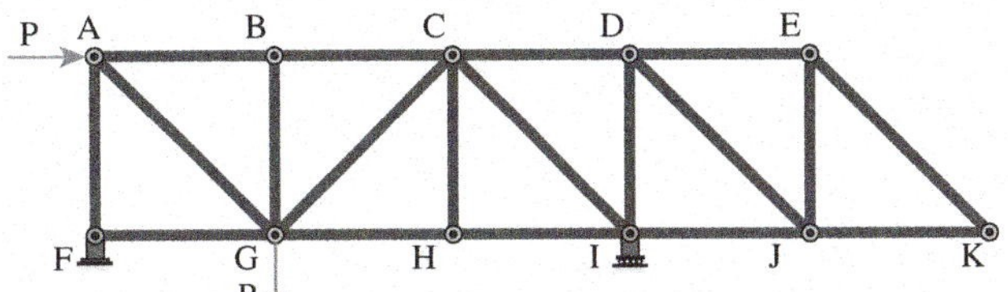

Problem 7.51

P7.52 Using the principles outlined in Section 7.10, identify all zero-force members in the truss given in Problem 7.60.

Section 7.11 through 7.12

For Problems 7.53 through 7.69, use the method of sections to determine the forces for the members indicated. The reactions are given for all trusses. Try to solve for each requested bar force using a single independent equilibrium equation. Indicate whether the members are in tension or compression.

P7.53 Determine for members BC, HC, JE, and JK. (Ans: $N_{BC} = -160.0$ k, $N_{JK} = 100.0$ k)

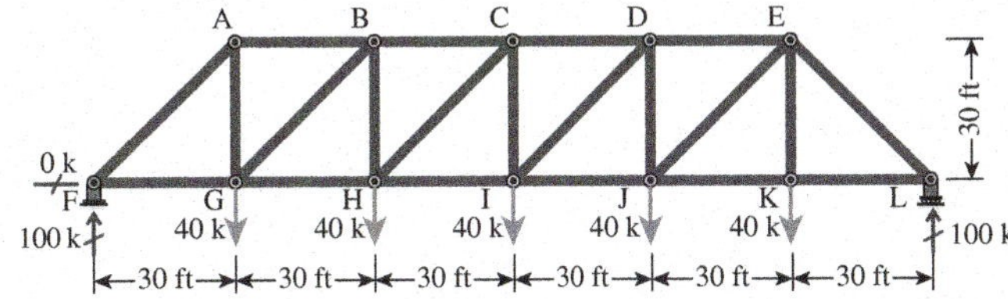

Problem 7.53

P7.54 Determine for members CD, GD, and GH.

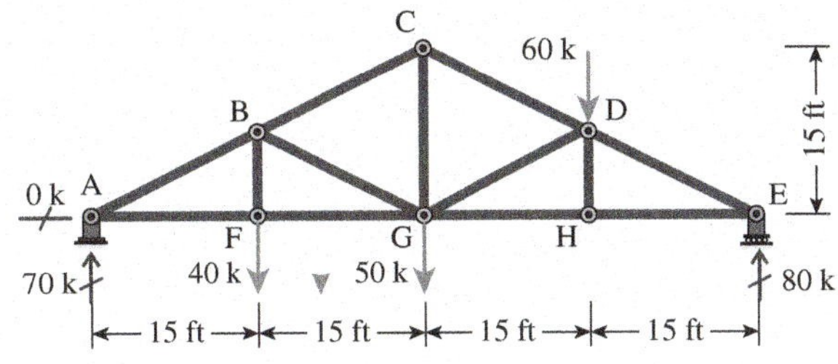

Problem 7.54

P7.55 Determine for members CH, GH, and BC. (Ans: $N_{CH} = 10.5$ k, $N_{BC} = -116.3$ k)

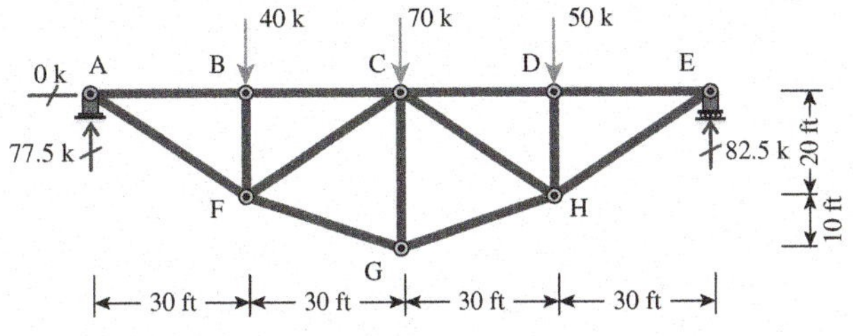

Problem 7.55

P7.56 Determine for members AB, CD, and AD.

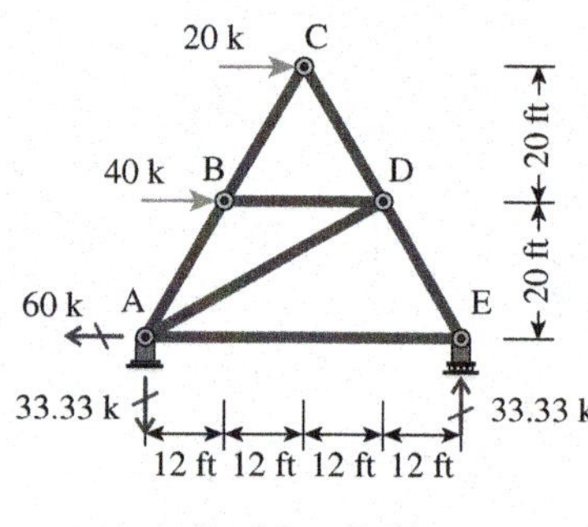

Problem 7.56

P7.57 Determine for members BC, EC, and EF. (Ans: $N_{EF} = 46.1$ k, $N_{BC} = -76.7$ k)

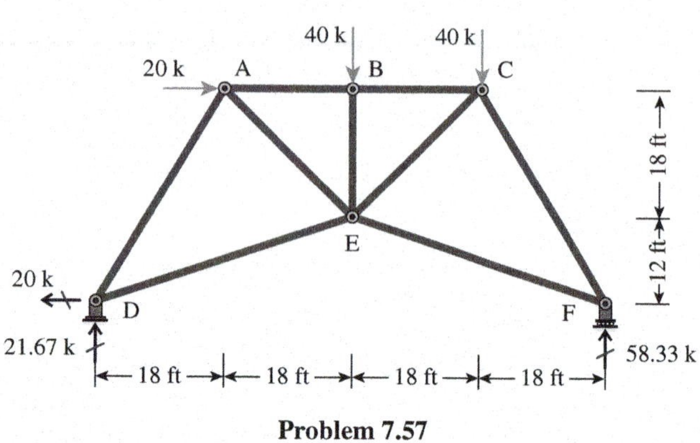

Problem 7.57

P7.58 Determine the force in member BC using section 1–1. Having that value, calculate the force in member FC making use of section 2–2.

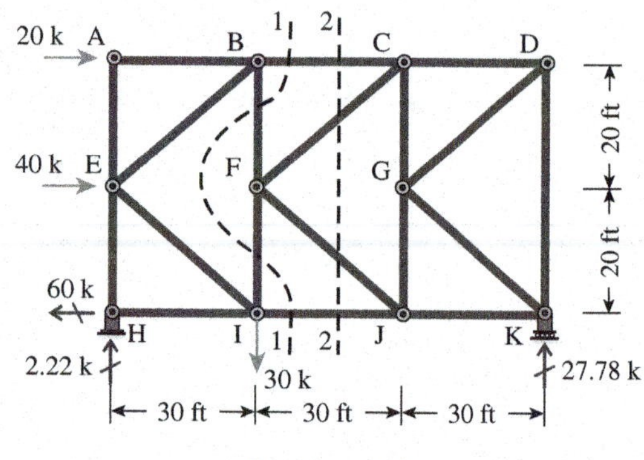

Problem 7.58

P7.59 Determine for members CD, CF, and EF. (Ans: $N_{CD} = -13.4$ k, $N_{EF} = 12.0$ k)

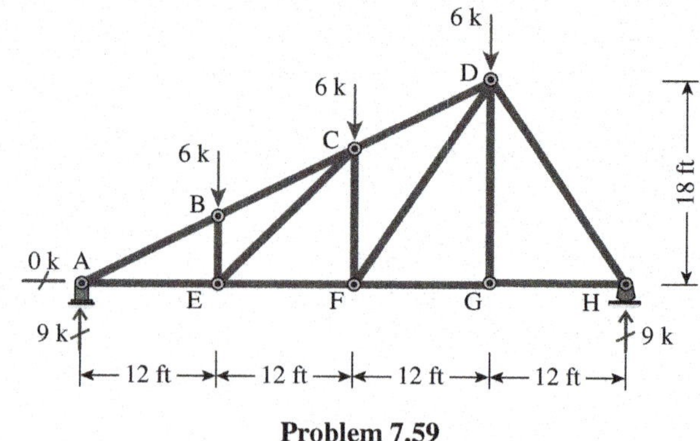

Problem 7.59

P7.60 Determine for members BC, CI, and IJ.

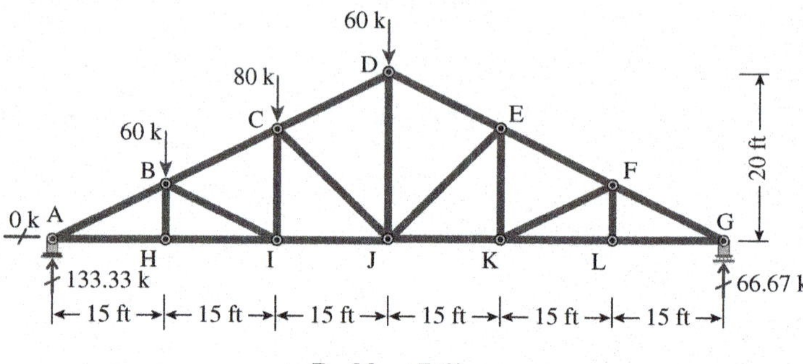

Problem 7.60

P7.61 Consider the truss from Problem 7.60 for members EF, KF, and KL. (Ans: $N_{EF} = -164.2$ k, $N_{KL} = 150.0$ k)

P7.62 Determine for members CD, GD, and CG.

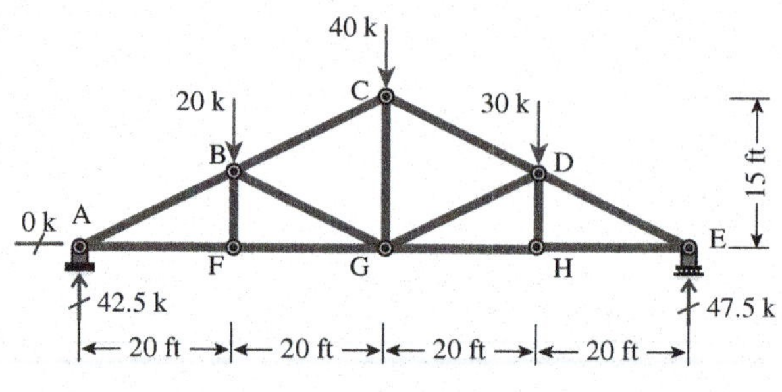

Problem 7.62

P7.63 Rework Problem 7.62 if the panels are changed from 20-ft width to 15-ft width. The truss height should be changed from 15 ft to 20 ft and the 40-k load is doubled. (Ans: $N_{CD} = -27.0$ k, $N_{GD} = -94.7$ k)

P7.64 Determine for members BC, CG, GD, and GH.

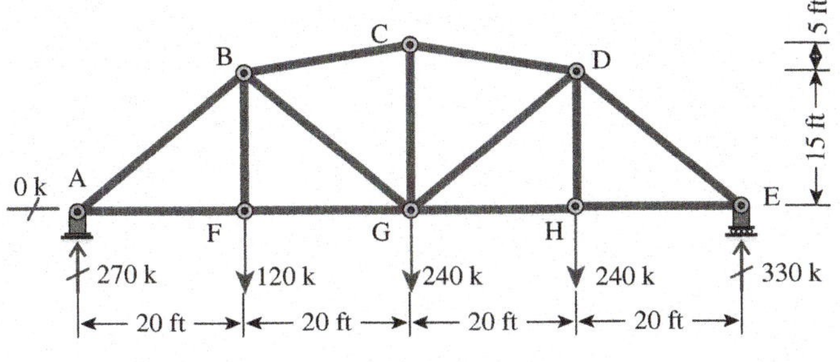

Problem 7.64

P7.65 Determine for members MN, CD, CJ, and MJ. (Ans: N_{CJ} = 10.0 k, N_{CD} = −97.3 k)

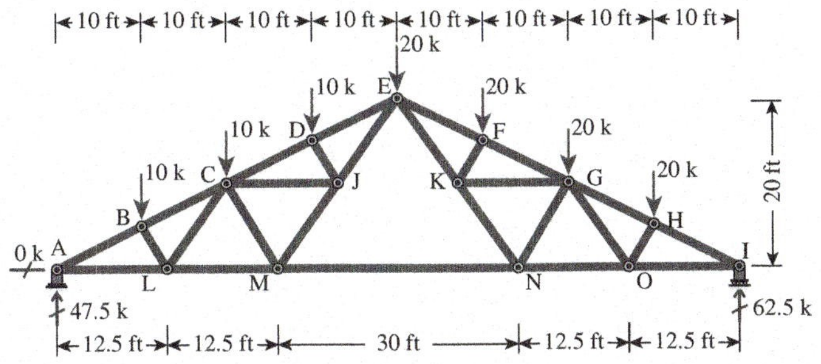

Problem 7.65

P7.66 Determine for members CD, RD, RL, and KL. (*Hint*: Look to Problem 7.58 for a possible approach.)

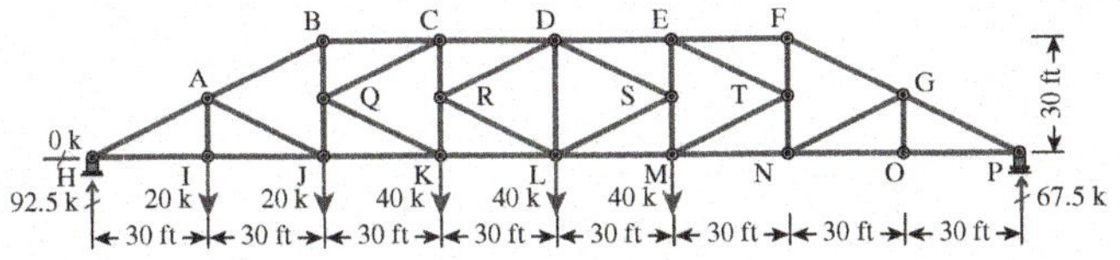

Problem 7.66

P7.67 Looking at truss in Problem 7.66 find forces in members DE, ES, MT, and MN. (*Hint*: Look to Problem 7.58 for a possible approach.) (Ans: N_{DE} = −202.5 k, N_{MN} = 135.0 k)

P7.68 Determine for members BC, EC, EF, and EG.

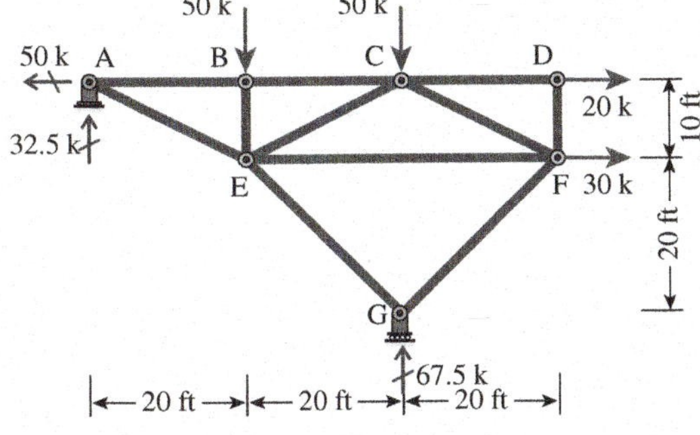

Problem 7.68

P7.69 Determine for members BE, FC, and AD. (*Hint*: You might draw a section around triangle DEF then take moments at the intersection of two of the cut diagonals to determine the force in the third diagonal.) (Ans: N_{BE} = 25.8 k, N_{AD} = −17.7 k)

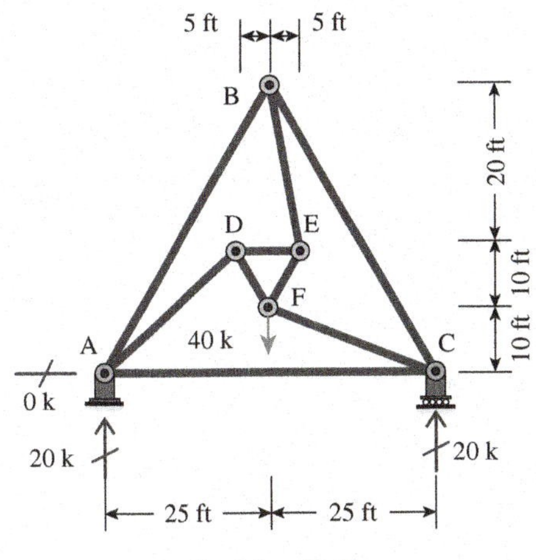

Problem 7.69

Section 7.13

For Problems 7.70 through 7.73, use the simplified approach presented in Section 7.13 to find the approximate internal axial force and maximum bending moment.

P7.70 For the truss in Problem 7.62, add a 3-klf uniform load to the bottom chord of the truss. Remove the existing point loads. Find the internal forces in member FG.

P7.71 Determine for members BC and AF. (Ans: N_{BC} = −3488 lb, $M_{max\text{-}BC}$ = 1238 lb·ft)

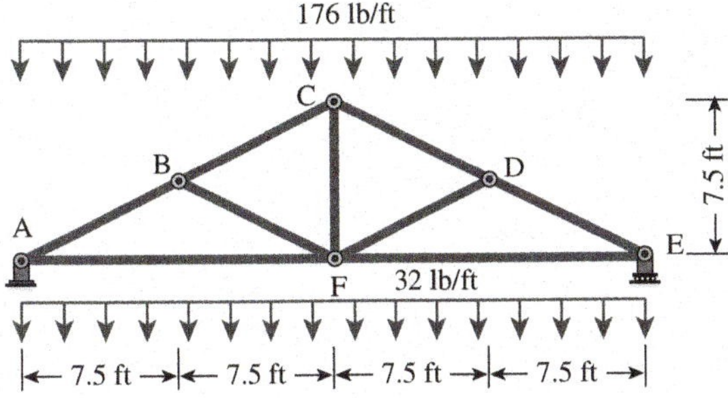

Problem 7.71

P7.72 Determine for members AB and BC.

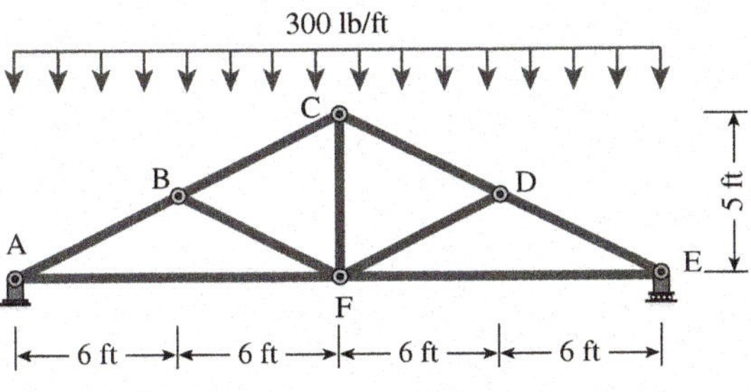

Problem 7.72

P7.73 Determine for members AB and BC. (Ans: $N_{AB} = -1594$ lb, $M_{max\text{-}AB} = 441$ lb·ft)

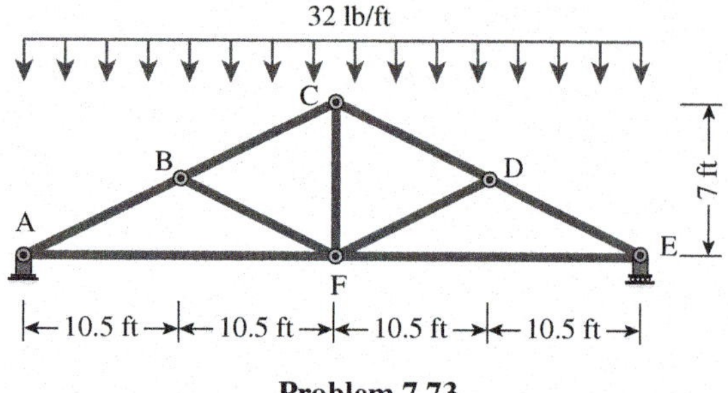

Problem 7.73

Section 7.14 SAP2000

P7.74 Model the truss provided in Problem 7.28 in SAP2000. Give all members a cross-sectional area of 2.0 in^2 and I = 35 in^4. Use the steel material (E = 29,000 ksi). Find all of the bar forces and report them in a table. Now change the cross-sectional area of all bars to A = 3.0 in^2. What change, if any, occurs in the bar forces of this statically determinate truss?

P7.75 Model the truss provided in Problem 7.39 in SAP2000. Give all members a cross-sectional area of 2.0 in^2 and I = 35 in^4. Use the steel material (E = 29,000 ksi). Find all of the bar forces and report them in a table. Now take the horizontal 30-k and 40-k loads and point them to the left instead of right. Indicate the change in the bar forces due to the change in the applied forces.

P7.76 Model the truss provided in Problem 7.58 in SAP2000. Give all members a cross-sectional area of 2.0 in^2 and I = 35 in^4. Use the steel material (E = 29,000 ksi). Find all of the bar forces and report them in a table.

P7.77 Model the truss provided in Problem 7.65 in SAP2000. Give all members a cross-sectional area of 2.0 in^2 and I = 35 in^4. Use the steel material (E = 29,000 ksi). Find all of the bar forces and report them in a table. Now remove all point loads and then apply a downward vertical point load of 110 k at point E. Find all bar forces. Does this loading scenario produce any peculiar behavior in the truss?

P7.78 Model the truss provided in Problem 7.69 in SAP2000. Give all members a cross-sectional area of 2.0 in^2 and I = 35 in^4. Use the steel material (E = 29,000 ksi). Find forces in bars FE and FC. Now rotate the 40-k load counterclockwise 10° and reanalyze the truss. Continue to do so in increments of 10° until the load is horizontal pointing to the right. Create a plot of orientation (0° to 90°) versus the bar forces of FE and FC.

Deflections and Angle Changes in Structures

8

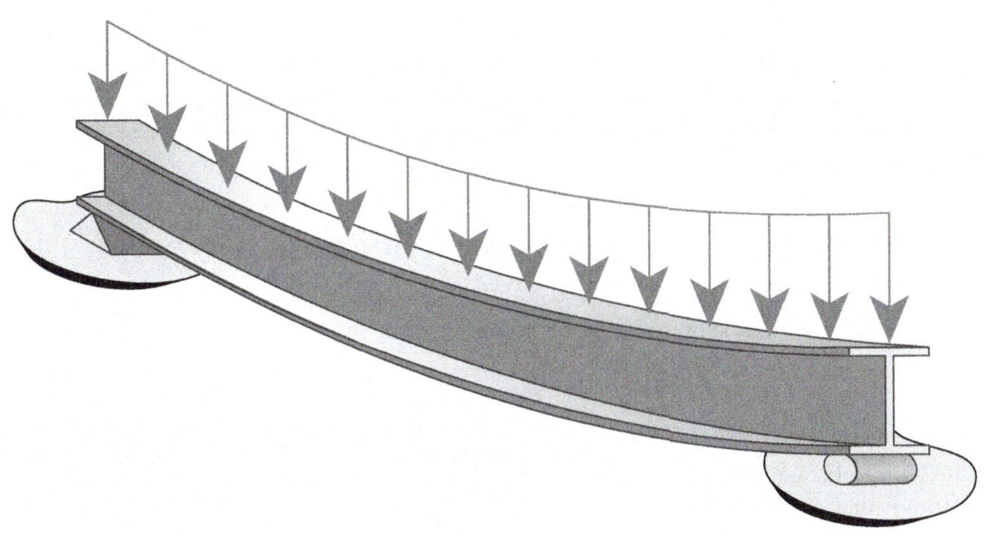

8.1 Introduction

This chapter and the next are concerned with the elastic deformations of structures. Both the linear displacements of points (deflections) and the rotational displacements of lines (slopes) are considered. The word *elastic* is used to mean:

1. Stresses are proportional to strains.
2. There is a linear variation of stress from the neutral axis of a beam to its extreme fibers.
3. The members will return to their original geometry after loads are removed.

The deformations of structures are caused by bending moments, by axial forces, and by shearing forces. For beams and frames, the largest contributions to overall deflections are caused by bending moments, whereas for trusses the largest contribution to overall deflections are caused by axial forces. Deflections caused by shearing forces, as a percentage of total beam deflections, increase as the ratio of beam depth to span increases. For the usual depth/span ratio of 1/12 to 1/6, the percentages of shear deflections to bending deflections vary from about 1 to 8%. For a depth/span ratio of one-quarter, the percentages can be as high as 15 to 18%. An approach for including deflections due to shear deformations is presented in Chapter 9.

There are many classical methods available for calculating deflections within a structure.

Moment-area	Conjugate beam
Elastic weights	Newmark's
Double integration	Castigliano's 2nd theorem
Virtual work	Real work

QUICK NOTE
The philosophy of this text is that it is better to learn a couple of versatile methods well rather than learn a lot of restrictive methods casually.

A traditional textbook on structural analysis would present many of these methods for the student to learn. Admittedly, each of these methods has a reason for having been developed, and they all have their strengths and limitations. However, as computing technology continues to improve, the broad exposure to so many methods loses its appeal and effectiveness. To this end, only two of the previously mentioned deflection calculation methods are addressed in this text.

- **Double integration** This allows for a broad description of deflections in a beam. It gives a clear picture of how beam charts are derived. It also lays the groundwork for the introduction and use of shape functions for computer analysis (see Chapter 20).
- **Virtual work** This method is one of the most versatile methods available. It can calculate displacements and rotations in beams, frames, grids and trusses. It can also account for deflections due to shearing forces.

In this chapter, displacements are computed using the double-integration method, which includes the development of equations for deflection. In Chapter 9, displacements are determined with an energy method known as *virtual work* that is based on the conservation of energy principle. Both of the methods presented provide identical results for deflections due to bending moments.

8.2 Reasons for Computing Deflections

QUICK NOTE

When the 3-D truss was first installed, it experienced deflections so large that the glass curtain wall could not be installed. The truss was removed, modified and re-installed.

The members of all structures are made up of materials that deform when loaded. If deflections within a structure are not controlled, there can be many undesirable consequences. The following lists some examples of these consequences.

- May visually detract from the appearance of the structures.
- The materials and finishes attached to the members may be damaged. For example, a floor beam that deflects too much may cause cracks in the ceiling below or a frame that has excessive lateral deflection may cause windows to leak or even break.
- Pieces of the structure may not fit together as planned. See the note in the side bar for an example.
- The use of a floor supported by beams that "give" appreciably does not inspire confidence, although the beams may be perfectly safe. Excessive vibration may occur in a floor of this type, particularly if it supports machinery. These movements can cause discomfort and fear even though collapse is very unlikely.

DID YOU KNOW?

Serviceability is the ability of a structure to serve its intended function. For example, a tall building that sways too much in the wind can cause occupants to have motion sickness and not be able to work. Thus, it does not serve its function of being a work place. *Deflection checks during structural design are often called serviceability checks.*

Standard American practice is to limit deflections caused by live load to L/360, where L is the length of the span. This figure probably originated for beams supporting plaster ceilings and was thought to be sufficient to prevent plaster cracks. The reader should realize that a large part of the deflections in a building are due to dead load, most of which will have occurred before plaster is applied.

The L/360 deflection criterion is only one of many maximum deflection values in use because of different loading situations, different designers, and different specifications. For situations in which precise and delicate machinery is supported, maximum deflections may be limited to L/1500 or L/2000. The 2010 AASHTO specifications limit deflections in steel beams and girders due to live load and impact to L/800. This value, which is applicable to both simple and continuous spans, is preferably reduced to L/1000 for bridges in urban areas that are used in part by pedestrians. Corresponding AASHTO values for cantilevered arms are L/300 and L/375.

Structural members subject to large downward deflections often are unsightly and may even cause users of the structure to be frightened. Such members may be cambered so their displacements do not appear to be so large. The members are constructed of such a shape that they will become straight under some loading condition (usually dead load). A simple beam would be constructed with a slight convex bend so that, under gravity loads, it would become straight as assumed in the calculations. Some designers take into account for both dead and live loads when figuring the amount of camber to be installed.

Despite the importance of deflection calculations, rarely will structural deformations be computed—even for statically indeterminate structures—for the purpose of modifying the original dimensions on which computations are based. The deformations of the materials used in ordinary work are quite small when compared to the overall dimensions of the structure. For example, the strain (ϵ) that occurs in a steel section that has a modulus of elasticity (E) of 29×10^6 psi when the stress (σ) is 20,000 psi is only

$$\epsilon = \frac{\sigma}{E} = \frac{20 \cdot 10^3}{29 \cdot 10^6} = 0.000690$$

This value is only 0.0690 percent of the member length. So modifying the geometry of the structure and reworking the statics because of this slight change will not produce any marked difference.

NOMENCLATURE

Camber is a built-in slight arch. It is often measured in units of length in terms of the total rise at the center of the span. For example, a 1.0 inch camber means the center of the beam is 1.0 inch higher than the beam ends.

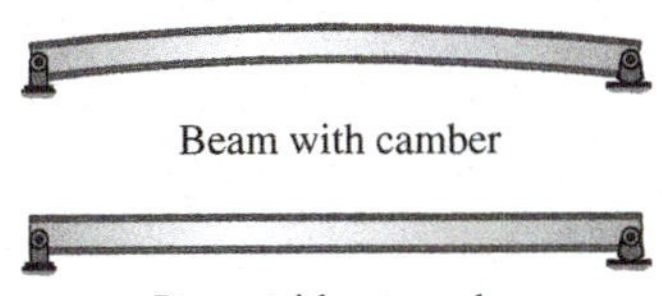

Beam with camber

Beam without camber

8.3 Long Term Deflections

Under sustained loads, concrete will continue to deform for long periods of time. This additional deformation is called creep or plastic flow. If a compressive load is applied to a concrete member, an immediate or elastic shortening occurs. If the load is left in place for a long time, the member will continue to shorten over a period of several years, and the final deformation may be as much as two or three (or more) times the initial deformation. Creep is dependent on such items as humidity, temperature, curing conditions, age of concrete at time of loading, ratio of stress to strength, and other items.

When sustained loads are applied to reinforced-concrete beams, their compression sides will become shorter and shorter over time, and the result will be larger deflections. The American Concrete Institute states that the total long-term deflection in a particular member should be estimated by the following method:

1. Calculating the instantaneous deflection caused by all the loads.
2. Computing the part of the instantaneous deflection that is caused by the sustained loads.
3. Multiplying the value from step 2 by an empirical factor from the ACI Code, which is dependent on the time elapsed.
4. Adding this value to the overall instantaneous deflection of step 1.

The sustained loads for a building include the dead load plus some percentage of the live load. For an apartment house or for an office building, perhaps only 20 to 25% of the live load should be considered as being sustained, while as much as 70 to 80% of the live load of a warehouse might fall into this category.

DID YOU KNOW?

Wood is also highly susceptible to creep. Seasoned timber members subjected to long-term loads develop a permanent deformation or sag approximately equal to twice the deflection computed for short-term loads of the same magnitude.

8.4 Sketching Deformed Shapes of Structures

Before learning methods for calculating structural displacements, there is considerable benefit in learning to qualitatively sketch the expected deformed shapes of structures under load. Understanding the displacement behavior of structural systems is a very important part of understanding how structures perform. A structural analyst should sketch the anticipated deformed shape of structures under load before making actual calculations. Such a practice provides an appreciation of the behavior of the structure and provides a qualitative check of the magnitudes and directions of the computed displacements.

To sketch the anticipated deformed shape of a structural system, there are only a few general rules to follow. Some of these rules apply to beams and columns and others apply to the joints between the components. By applying these simple rules, we can obtain reasonable qualitative indications of the deflection response of beams and frames. Only by applying the

QUICK NOTE

The proper sketching of a deformed structure can also provide a sense of the direction in which the reactions act. Such an understanding can serve as a check of both hand and computer calculations without performing any additional calculations.

quantitative methods discussed in this and the next chapter can we obtain the actual numerical deflections.

Procedure

The quantitative approach to sketching deflected shapes is not an exact process. However, the following procedure provides a good foundation for generating reasonably accurate sketches.

1. Identify all support conditions of the structure (e.g., pin, roller, fixed, spring, etc.) and note what each of those support conditions tell you about possible motion at those supports.
2. Identify what motion each of the joints undergo (i.e., rotation and/or translation) and mark this movement on the sketch. Make sure that all rules listed hereafter are followed.
3. Connect the joints (i.e., sketch the members) following the rules for members provided hereafter. You must ensure that you have not violated any of the conditions prescribed by the joints.

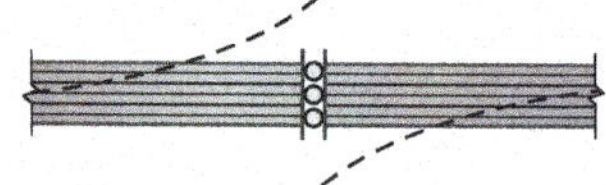

Slope on either side of a shear release must be the same—displacements can be different.

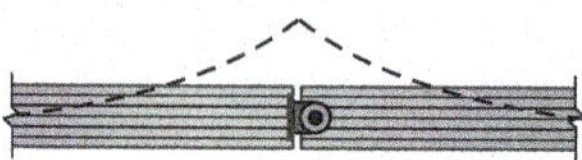

Displacements on either side of a moment release must be the same—slopes can be different.

FIGURE 8.1 Permissible motion for internal releases.

Rules for Joints

1. A joint in a structure is assumed to be rigid. A rigid joint can displace but it cannot deform—the joint does not change size or shape as it displaces. For example, if two members are 90° relative one to another before the deformation occurs, they are still 90° apart after deformation occurs.
2. A joint that is pin-connected (i.e., has a moment release) does permit relative rotation between connected members (see Figure 8.1).
3. A joint can only displace in accordance with the external supports acting on it.
   - A joint at a pin support is free to *rotate* but cannot *translate* in any direction.
   - A joint at a roller support is free to *rotate* and is only permitted to *translate* in the direction parallel to the support surface.
   - A joint at a fixed support is not able to *rotate* or *translate* in any direction.
4. The distance between two joints does not change. This is to say that the length of a member doesn't change. (See Rule #5 for members.)

Rules for Members

1. A member deforms in the direction of the load applied to it.
2. Deflections of loaded members are sketched first. Deflections of unloaded members are sketched last.
3. Unless there is a hinge between a member and a joint, the end of the member and the joint displace in the same manner.
4. Members with smaller stiffnesses (EI/L) tend to deform more than do members with larger stiffnesses. That is to say, long slender members deform more than short stocky ones.
5. When sketching the qualitative deformed shape of a structure, the beams and columns are assumed to retain their original lengths.

In Figure 8.2, the approximate deflected shapes of several loaded structures are sketched by applying these rules. In each case, the member weight is neglected. You should note that the corners of the frame of part (e) of the figure are free to rotate, but the angles between the members are assumed to be constant.

Several examples for sketching the qualitative deformed shapes of structures follow. In these examples, the thought process for preparing each sketch is presented.

DID YOU KNOW?
If the moment diagrams have been prepared previously, they can be helpful in making the sketches where we have both positive and negative moments. Direction for this is given in Section 8.6.

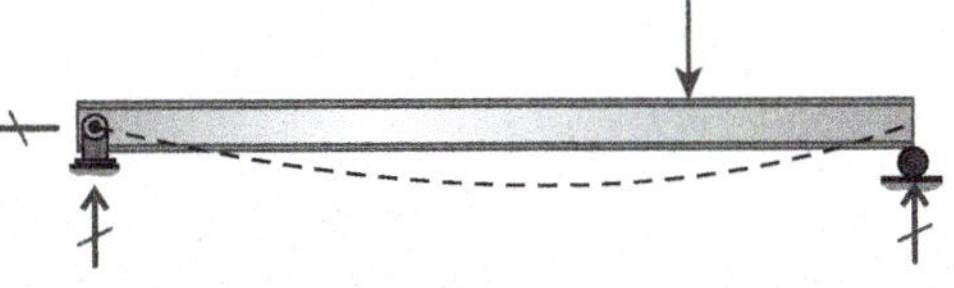
(a) Point of maximum deflection is somewhere to the left of this off-center load.

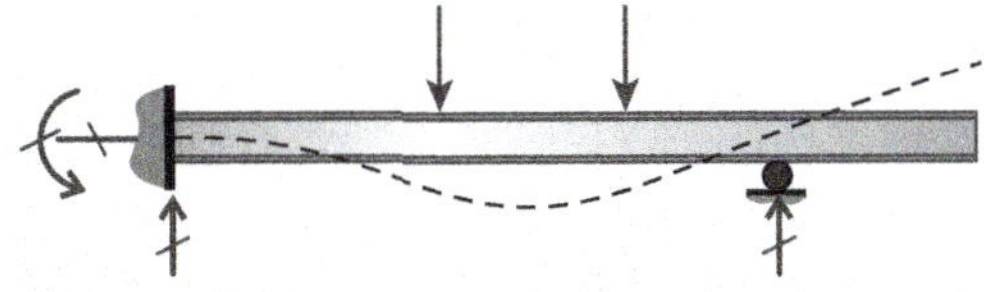
(b) Tangent at the fixed end is horizontal and the right end deflects upward.

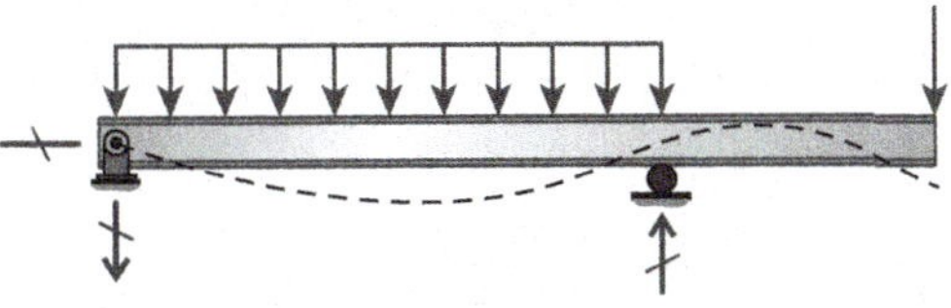
(c) Without calculations, we do not know whether the deflection at the right end is up or down. The concentrated load tends to push the right end down while the uniform load tends to push it up.

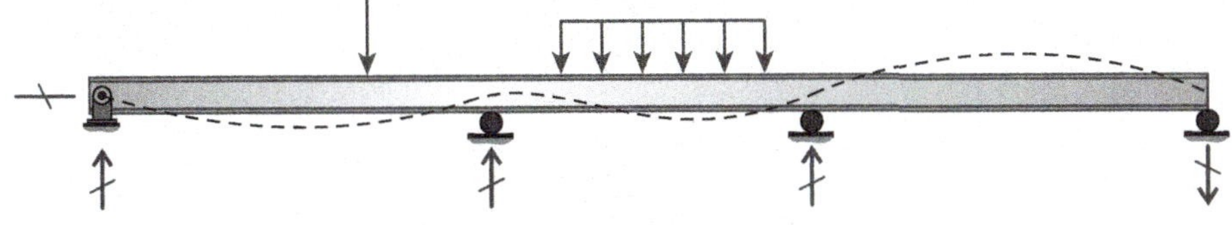
(d) Note the upward deflection in the third span.

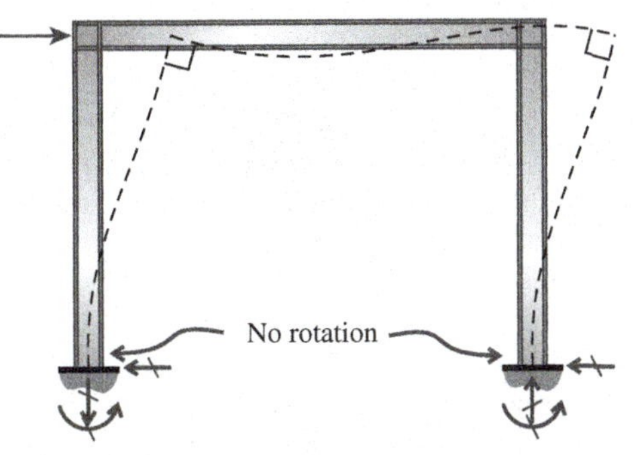

(e) The frame sways to the right. Both joints rotate clockwise and stay the same distance apart.

FIGURE 8.2 Qualitative deformed shapes of some structures under load.

EXAMPLE 8.1

Consider the three-span continuous beam shown. Sketch the qualitative deflected shape for the beam.

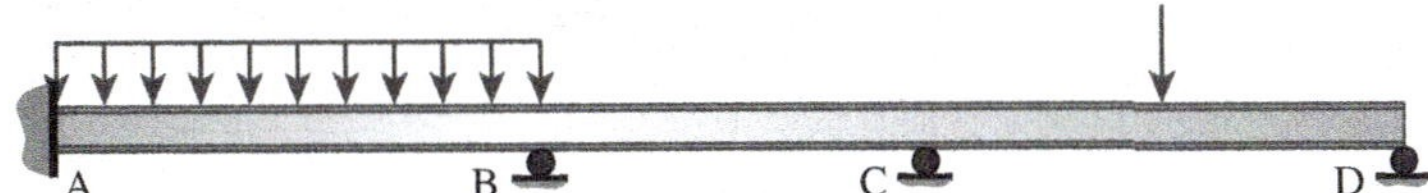

SOLUTION STRATEGY

Identify the motion of the joints first. Pay attention to the types of supports. Only once the joint movement is identified should the member displacements be sketched. Sketch the loaded members before the unloaded members.

SOLUTION

None of the joints can translate up and down in this structure. All joints are allowed to rotate except joint A. Joints B and D both rotate counterclockwise due to the applied load, and joint C rotates clockwise. This should result in the sketch shown.

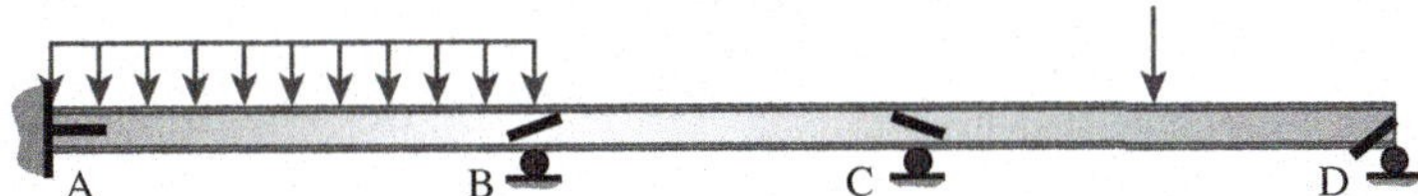

DID YOU KNOW?
The sketched deflected shapes actually represent an exaggeration of the true magnitude of deflection. For example, a 30-ft beam may deflect an inch at mid-span. If we were to sketch the deflection to true scale, it would be very difficult to see the deflected shape.

Since members AB and CD have loads applied to them, sketch these members first, making sure they deflect in the direction of the applied load and also matching up the motion of the joints that had been identified previously.

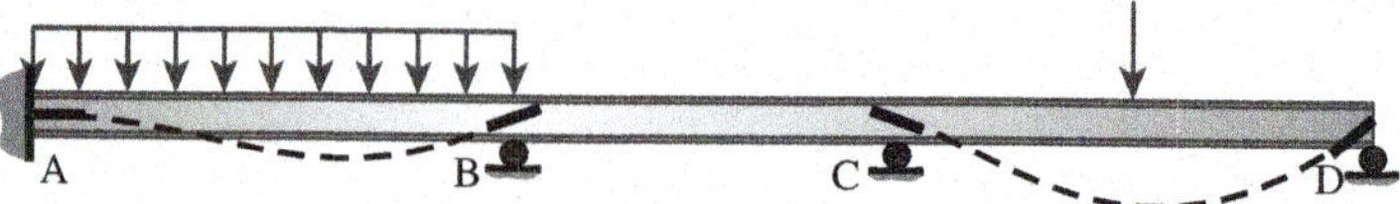

Now member BC can be sketched connecting the two joints while paying attention to the joint rotations.

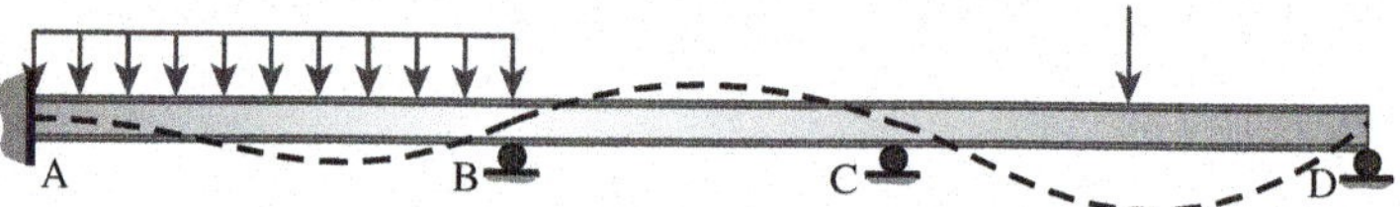

EXAMPLE 8.2

Sketch the qualitative deflected shape of the overhanging beam shown.

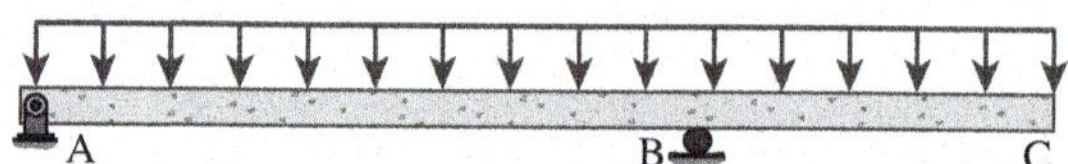

SOLUTION STRATEGY

Identify the motion of the joints first. Pay attention to the types of supports. Only once the joint movement is identified should the member displacements be sketched.

SOLUTION

Joint A will rotate clockwise due to the distributed load on member AB. Joint C will translate downwards and rotate clockwise due to the distributed load on member BC. The rotation of joint B is not as easy to address. The load on member AB causes it to rotate counterclockwise, while the load on member BC causes the joint to rotate clockwise. These counter-balancing motions cause the rotation of the joint at B to be smaller than if only one of the members was loaded. We assume that joint B slightly rotates counterclockwise (see side note). This should result in the sketch shown.

QUICK NOTE
The actual motion of joint B is dependent on the magnitude of the distributed load and the relative span lengths. Even then, one would need to perform a calculation to be able to answer this question. Therefore, when such ambiguities are encountered during the sketching process, an assumption of *clockwise*, *counterclockwise*, or *no rotation* must be made. None of them would be wrong, and all three sketches would look similar.

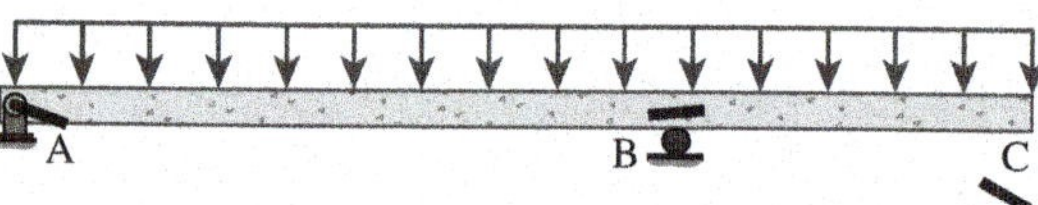

Since both members are loaded, it doesn't matter which one you start with. In this example, we sketch member AB first. Make sure the joint rotations, as previously sketched, are honored.

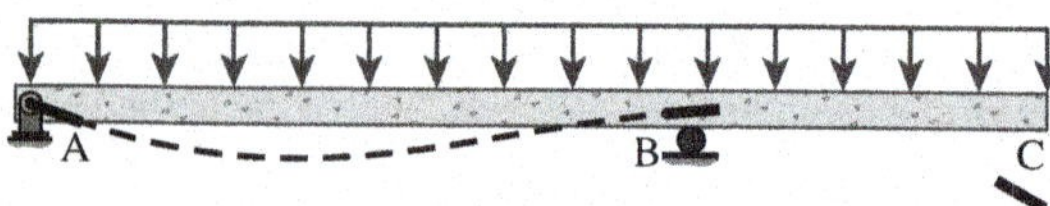

Now sketch member BC.

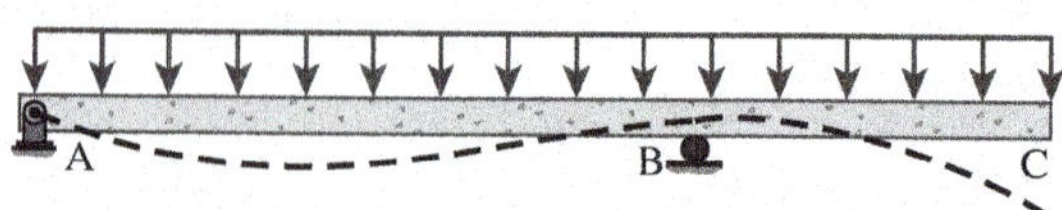

EXAMPLE 8.3

Sketch the qualitative deflected shape of the beam shown.

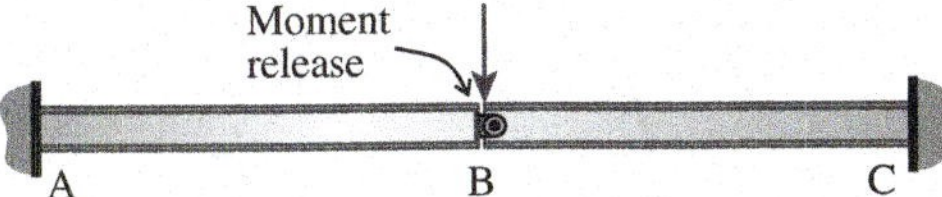

SOLUTION STRATEGY

Identify the motion of the joints first. Pay attention to the types of supports. Only once the joint moment is identified should the member displacements be sketched.

SOLUTION

Joints A and C will not rotate or translate. Joint B has the load applied directly to it. The moment release at B will allow different rotations to occur on either side of the release (see Figure 8.2). This should result in the sketch shown.

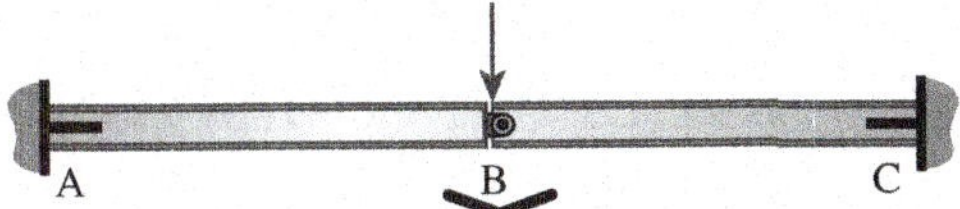

Since no load is applied to the members directly, they can all be sketched simultaneously by connecting the joints while honoring the joint conditions.

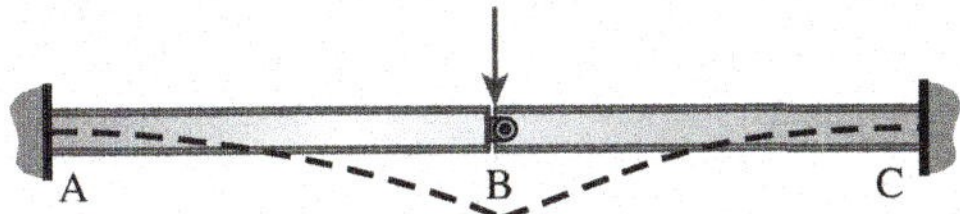

EXAMPLE 8.4

Sketch the qualitative deformed shape of the moment frame shown.

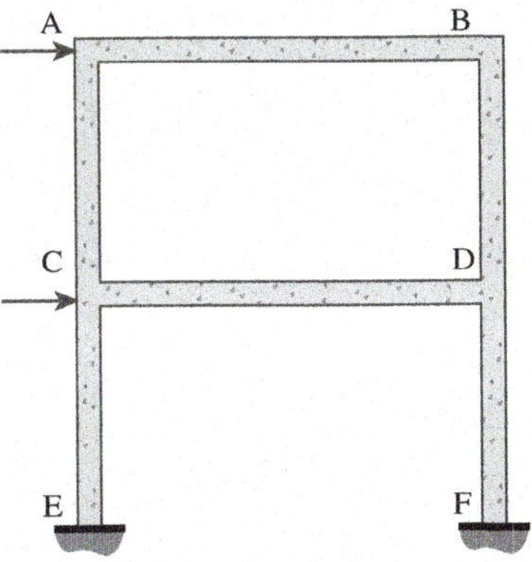

SOLUTION STRATEGY

Identify the motion of the joints first. Pay attention to the types of supports. Only once the joint movement is identified should the member displacements be sketched.

SOLUTION

Joints E and F will not rotate or translate. Joints C and D will translate to the right and rotate clockwise. Joints A and B will translate further to the right than C and D and will also rotate clockwise. Then we need to connect the joints while honoring the position and rotation of each of the joints.

QUICK NOTE

The distance between all of the joints did not change in the deflected shape. This is because we assume the length of the members doesn't change. Students can become confused by this because we exaggerate the deflected shape so significantly. Just follow the rules, and you won't be fooled by this.

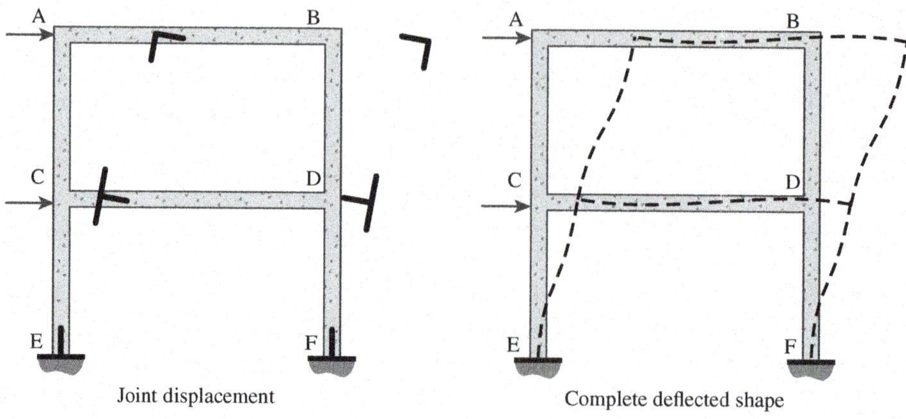

Joint displacement

Complete deflected shape

8.5 Determining Sense of Reactions from Deformed Shape

One of the advantages of being able to quickly sketch the qualitative deflected shape of a structure is that it helps the structural engineer to gain a sense of the direction of the reactions. Such an exercise can provide a sanity check for many computer analysis results. It also helps in the development of *engineering judgment* relative to structural behavior.

A simple process may be followed to determine the direction of the reactions from a qualitative deflected shape.

QUICK NOTE

The formalized process presented here for identifying the direction of reactions is really intended to help a student develop the thought process. With a bit of experience, one should be able to determine reaction directions without having to make physical sketches. Rather, a little thought can reveal such things.

1. Sketch the deflected shape for the structure.
2. Select the reaction for which the direction is to be determined.
3. Remove the support that is causing the selected reaction.
4. Sketch the deflected shape for the structure with that support removed.
5. Now ask yourself the question: "What direction do I need to apply a force/moment to push the structure back into the shape sketched in step 1?"
6. Follow steps 2 through 5 for all desired reactions.

EXAMPLE 8.5

Consider the beam discussed in Example 8.1. Determine the direction of all of the reactions—neglecting the horizontal reaction that is known to be zero.

SOLUTION STRATEGY

Work from left to right on the beam, selecting one reaction at a time, and remove the support restraint associated with that reaction. Follow the procedure previously outlined.

SOLUTION

Select the moment at A (M_A) as the first reaction to be considered. Remove the rotational restraint at point A and sketch the deflected shape as shown.

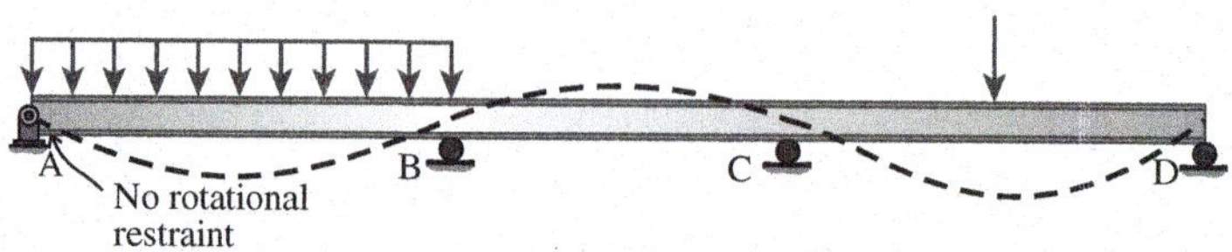

Compare this deflected shape with the one sketched in Example 8.1, and it is clear to see that a counter-clockwise moment needs to be added to this structure to get the deflected shape to match the one shown in Example 8.1.

Now replace the rotational restraint at A and remove the vertical restraint at A to determine the direction of the vertical reaction at A. It is clear that an upward force must be applied to A to push the structure back into place.

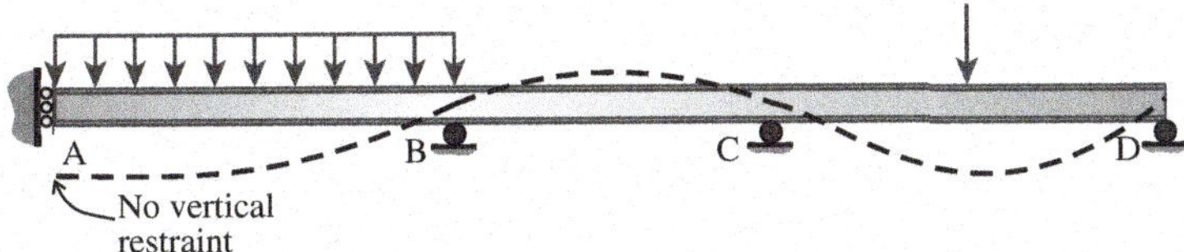

Now replace the vertical restraint at A and remove the vertical restraint at B to determine the direction of the vertical reaction at B. It is clear that an upward force must be applied to B to push the structure back into place.

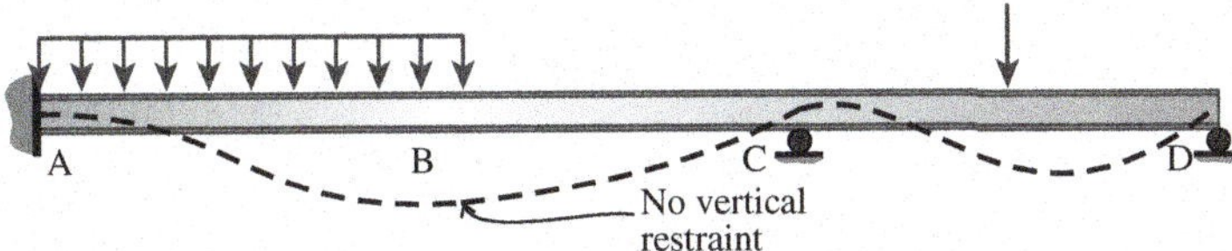

Do the same steps for joints C and D and the remaining reaction directions can be determined as shown.

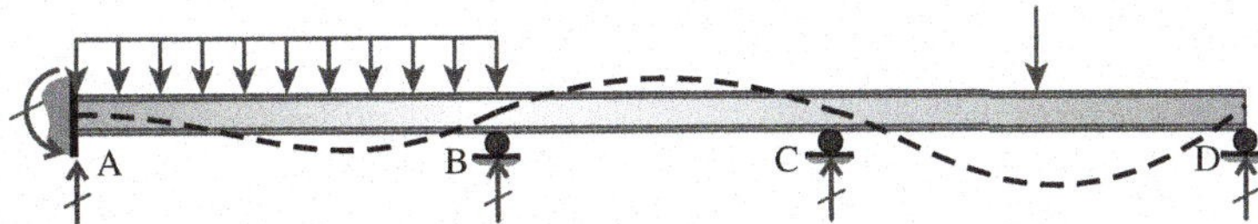

QUICK NOTE

Most of these reaction directions may be obvious from simple observation. However, it is not always obvious, such as for the structures in Figures 8.2(d) and (e). Can you follow the process outlined and confirm the reaction directions shown?

8.6 Elastic Beam Theory

Under changing loads, the neutral axis (N.A.) of a member changes in shape according to the positions and magnitudes of the loads. The *elastic curve*, denoted as y, of a member is the shape the neutral axis takes under temporary loads. A description of the elastic curve expresses the quantitative deflection of structure.

Expressions for the elastic curve are rooted in basic elastic beam theory, which describes the relationship between the internal moment within a member and the curvature of its deformed shape. To develop this relationship, the simple beam of Figure 8.3 is considered. Under the vertical loads shown, the beam deflects downward as indicated in the figure.

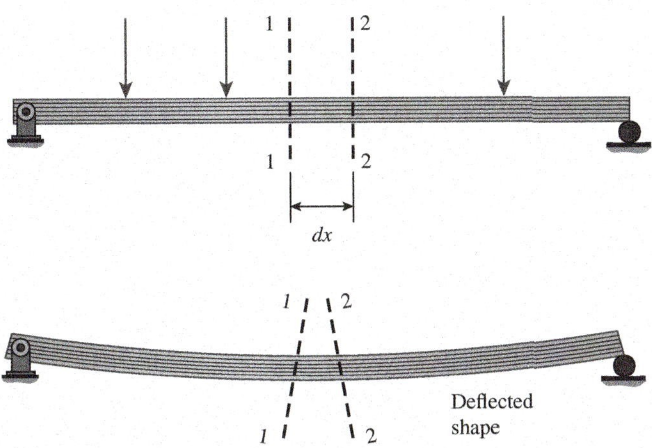

FIGURE 8.3 Deformed simply supported beams.

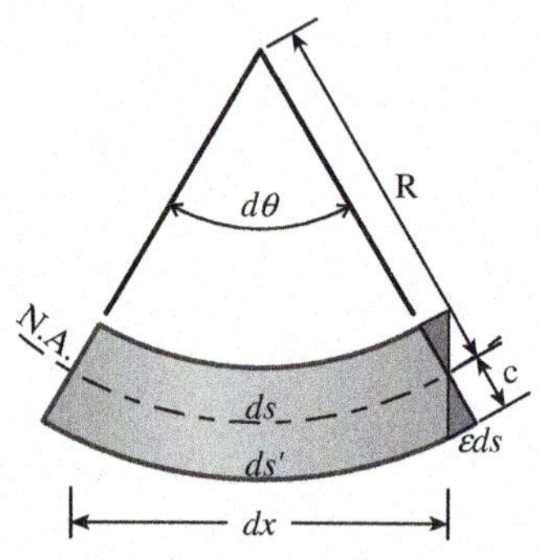

FIGURE 8.4 Infinitesimal deformed beam segment.

The segment of length dx, bounded on its end by line 1–1 and line 2–2, is shown in Figure 8.4. The size, degree of curvature, and distortion of the segment are tremendously exaggerated, so the slopes and deflections to be discussed can be easily seen. In Figure 8.4, the arc length of the neutral axis ds is calculated as

$$ds = Rd\theta$$

where R is the radius of the beam segment and $d\theta$ is the infinitesimal included angle. Through the process of deformation, the neutral axis does not undergo any strain (i.e., its length does not change.) However, as we travel up from the neutral axis, the beam segment experiences a shortening while the segment experiences an elongation as we travel down from the neutral axis. As shown in Figure 8.4, the bottom-most fiber of the beam segment undergoes the largest positive strain (ϵ). The new length for the bottom-most fiber is

$$ds' = (R + c)d\theta$$

where c is the distance from the neutral axis to the bottom-most fiber and R and $d\theta$ are as previously defined. This then makes the strain at the bottom-most fiber to be calculated as the change in length divided by the original length as shown.

$$\epsilon = \frac{(R + c)d\theta - Rd\theta}{Rd\theta} = \frac{c}{R}$$

Moving c to the other side of the equation yields:

$$\frac{1}{R} = \epsilon c$$

RECALL

Hooke's law for one-dimensional stress and the bending stress equations are given as

$$\epsilon = \frac{\sigma}{E} \qquad \sigma = \frac{Mc}{I}$$

where M is the internal bending moment, I is the moment of inertia of the cross section, and E is the modulus of elasticity of the material.

Using the well known Hooke's Law and bending stress equations as shown in the sidebar, we can rewrite the expression in terms of the internal moment M in the beam segment as

$$\frac{1}{R} = \frac{M}{EI}$$

The term $\frac{1}{R}$ is also known mathematically as the *curvature of the beam segment*. In summary, the curvature of the deformed segment is a function of the internal moment (M), a cross-sectional property (I), and the material behavior (E). Dipping into calculus, the expression for curvature can be rewritten in terms of the Cartesian coordinates x and y.

$$\frac{1}{R} = \frac{d^2y/dx^2}{(1 + (dy/dx)^2)^{(3/2)}} \approx \frac{d^2y}{dx^2}$$

DON'T GET CONFUSED

$$\frac{d^2y}{dx^2} \neq \left(\frac{dy}{dx}\right)^2$$

Therefore, the argument made for the denominator is not applicable to the numerator.

In this equation, y is the vertical deflection of the elastic curve, and x is the coordinate along the axis of the member. The justification for the approximate expression is that for small slopes—a condition present in most structures—the $(dy/dx)^2$ is very small compared to the 1 in the denominator. This yields the ever useful second-order differential equation giving a direct relation between the internal moment in a beam M and the deflection of the beam y.

$$\frac{d^2y}{dx^2} = \frac{M}{EI}$$

Understanding the relationship between bending moment and beam curvature can also help us to refine the qualitative deflected shapes discussed earlier in this chapter. One should recognize the following:

- A positive bending moment results in a positive curvature (i.e., concave up).
- A negative bending moment results in a negative curvature (i.e., concave down).
- A zero moment results in a zero curvature which is known as a *point of inflection* (PI).

If a moment diagram is available for the structure for which a deflected shape is to be sketched, the information in the moment diagram should direct how the curvatures are sketched. Figure 8.5

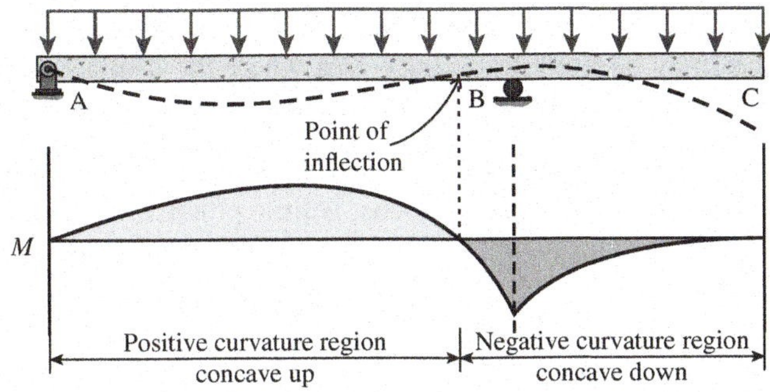

FIGURE 8.5 Relationship between deflected shape and the moment diagram.

revisits the beam shown in Example 8.2. If the moment diagram is known to be of the shape shown, not only can the regions of positive and negative curvature be found, but the point of inflection (PI) can be identified.

8.7 Deflection by Double Integration

The equations for the slope (θ) and the deflection (y) as a function of x can be developed by solving the second-order differential equation given in the previous section. The solution process for this equation is straightforward, since it can be solved by performing two successive integrations (i.e., *double integration*). The general steps for this process are outlined hereafter.

Procedure

1. Draw the deflected shape of the elastic curve (i.e., shape of the neutral axis).
   - Pay very close attention to the support conditions. Note any points of zero slope or zero displacement.
2. Establish the coordinate system(s).
   - Follow the guidance given in Section 6.3, watching for load discontinuities.
3. Develop an equation of moment for each segment (see Section 6.3).
4. Double integrate

$$y = \iint \frac{M}{EI} dx$$

   - Include constants of integration. This means that each beam segment will result in two integration constants.
   - Use known boundary conditions (BCs) to solve for the integration constants.
5. Substitute a value of x into the resulting equations to find the slope or deflection at a point of interest.

DID YOU KNOW?

Knowing that the PI represents a point of zero internal moment is a concept that will be exploited when approximate methods of analysis for statically indeterminate beams are discussed in Chapter 15.

Boundary Conditions

Double integration of the moment equation for each beam segment produces two integration constants per segment. This means that we need an equal number of boundary conditions to solve for these unknowns. Here are a few specifics related to the selection and use of boundary conditions.

- BCs can only be written in terms of displacement (y) or rotation (θ).
- Points of absolute displacement or rotation are preferred boundary conditions since they make the solution process a little easier. This refers to points on a beam segment where actual values of the slope or deflection are known.

COMMON MISTAKE

Students are very tempted to use BCs that are written in terms of moment or shear. The issue is that the successive integration presents equations of slope and displacement. Therefore, we must have knowledge of slope and displacement to solve of the constants within the equations.

- When there are not enough points where actual displacement/rotation are known, relative boundary conditions may be used. Look for points where the displacement/rotation in one segment is equal to the displacement/rotation in another segment.

The boundary conditions that are identified are dependent upon the coordinate system established for the development of the moment equations. A little practice selecting proper boundary conditions is warranted at this time.

EXAMPLE 8.6

A cantilever beam is shown with a point load being applied to its tip. Two possibilities are shown for selection of the coordinate system x. What boundary conditions should be used for each case (a) and (b)?

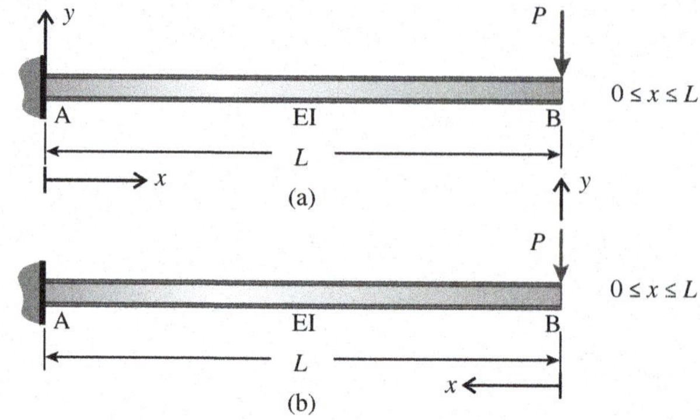

SOLUTION STRATEGY

There is only one beam segment. Look for any location on the beam where displacement and/or slope are already known.

SOLUTION

QUICK NOTE
Notice how the boundary conditions (i.e., y at A and θ at A) did not change. However, our description of these boundary conditions does depend on the coordinate system selected.

Point A is a fixed support. This means that the slope at point A is zero and the displacement at point A is also zero. This provides two boundary conditions that are sufficient:

Structure (a):

$$\underline{\underline{\theta(x=0)=0}} \qquad \underline{\underline{y(x=0)=0}}$$

These boundary conditions are read so that the rotation when x equals zero is zero and the displacement when x equals zero is zero.

Structure (b):

$$\underline{\underline{\theta(x=L)=0}} \qquad \underline{\underline{y(x=L)=0}}$$

These boundary conditions are read so that the rotation when x equals L is zero and the displacement when x equals L is zero.

Recognizing that the double integration will take place, it is often helpful to select a set of coordinates (see step 2) that keep the moment equations (see step 3) as short and simple as possible. Doing this however can complicate the description of the boundary conditions slightly. This next example identifies the BCs associated with different selections of coordinate systems.

EXAMPLE 8.7

The simply supported beam has a distributed load over half of its span. This means the beam is split into two segments for the development of deflection and slope equations because of the load discontinuity at B. Three possible scenarios for coordinate selection are shown. Identify the boundary conditions that could be used in each case.

SOLUTION STRATEGY

Since there are two beam segments (AB and BC), we need four boundary conditions. Scan the beam first for any points of known displacement or slope. If additional BCs are needed, look for a way to describe the slope and displacement where AB and BC meet.

SOLUTION

There are only two locations on the beam where the displacement is already known—namely, A and C. There are no points where we already know the slope, so that is not an option. We must now look at the point where the two segments meet (point B). We know that the displacement in segment AB at point B is the same as the displacement in BC at point B. We also know the same to be true for the rotations in each segment at point B.

Structure (a):

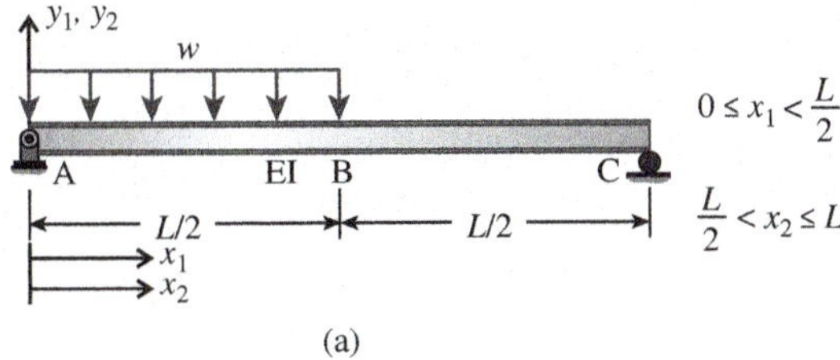

(a)

> **QUICK NOTE**
> Note that even though x_1 and x_2 start at the same location, they are valid for different domains. Don't try to use x_2 over the domain of $0 \leq x_2 < \frac{L}{2}$.

$$y_1(x_1 = 0) = 0 \qquad y_2(x_2 = L) = 0$$

$$y_1\left(x_1 = \tfrac{L}{2}\right) = y_2\left(x_2 = \tfrac{L}{2}\right) \qquad \theta_1\left(x_1 = \tfrac{L}{2}\right) = \theta_2\left(x_2 = \tfrac{L}{2}\right)$$

Structure (b):

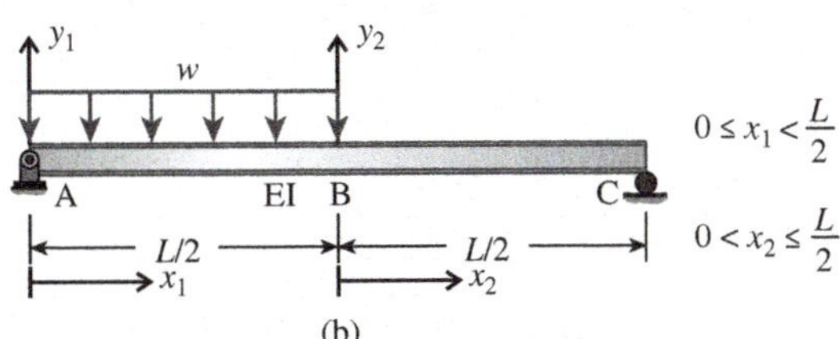

(b)

$$y_1(x_1 = 0) = 0 \qquad y_2\left(x_2 = \tfrac{L}{2}\right) = 0$$

$$y_1\left(x_1 = \tfrac{L}{2}\right) = y_2(x_2 = 0) \qquad \theta_1\left(x_1 = \tfrac{L}{2}\right) = \theta_2(x_2 = 0)$$

Structure (c):

> **QUICK NOTE**
> When a set of coordinates are opposite of one another as in case (c), the sign on the rotation reverses.
>
> 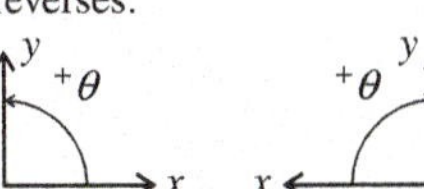
>
>
> This requires that the boundary condition describing the rotations at B must have opposite signs.

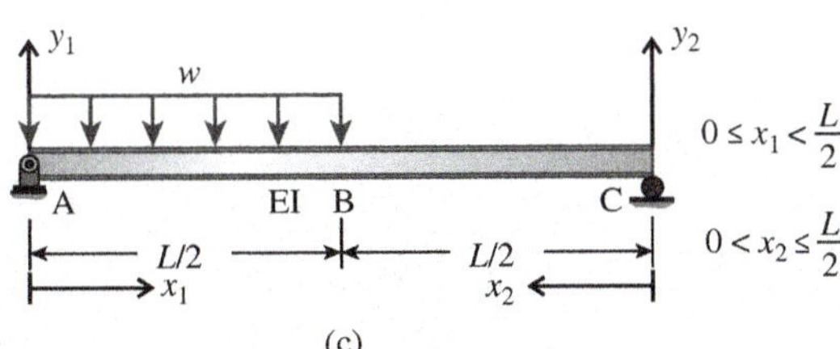

(c)

$$y_1(x_1 = 0) = 0 \qquad y_2(x_2 = 0) = 0$$

$$y_1\left(x_1 = \tfrac{L}{2}\right) = y_2\left(x_2 = \tfrac{L}{2}\right) \qquad \theta_1\left(x_1 = \tfrac{L}{2}\right) = -\theta_2\left(x_2 = \tfrac{L}{2}\right)$$

Double Integration

The procedure for conducting double integration has already been laid out. A couple of examples, considering the beams in Examples 8.6 and 8.7, are presented next. Pay particular attention to how the boundary conditions are evaluated to solve for the unknown constants. Students generally do quite well with performing the integration, but they tend to struggle with knowing what to do with the boundary conditions.

EXAMPLE 8.8

For the beam shown in Example 8.6 structure (a), the moment equation is given as

$$M(x) = P(x - L); \qquad 0 \leq x \leq L$$

Use the double integration method to find the equations for the slope and displacement of this beam. Identify what the displacement and the slope are at the tip of the beam (point B).

COMMON MISTAKE

Students are often tempted to perform a definite integration using the bounds for the domain of x as the integration limits. This will produce a scalar value and not an equation. You will always operate on an indefinite integral (i.e., no integration limits). This will require the integration constants previously discussed.

SOLUTION STRATEGY

The moment equation and boundary conditions have already been identified for this problem, so we need to pick up at step 4.

SOLUTION

Perform both integrations, making sure to add an integration constant each time.

$$\theta = \int \frac{M}{EI} dx = \int \frac{P(x-L)}{EI} dx = \frac{P}{EI}\left(\frac{1}{2}x^2 - Lx + C_1\right)$$

$$y = \int \theta dx = \int \frac{P}{EI}\left(\frac{x^2}{2} - Lx + C_1\right) dx$$

$$= \frac{P}{EI}\left(\frac{x^3}{6} - \frac{Lx^2}{2} + C_1 x + C_2\right)$$

Evaluate BC: $\theta(x = 0) = 0$

$$\theta(x=0) = \frac{P}{EI}\left(\frac{1}{2}(0)^2 - L(0) + C_1\right) = 0$$

$$\therefore \underline{\underline{C_1 = 0}}$$

We now use this known value for C_1 in future calculations.

Evaluate BC: $y(x = 0) = 0$

$$y(x=0) = \frac{P}{EI}\left(\frac{(0)^3}{6} - \frac{L(0)^2}{2} + (0)(0) + C_2\right) = 0$$

$$\therefore \underline{\underline{C_2 = 0}}$$

DID YOU KNOW?

By solving for these equations in terms of the variables P, L, E, and I, we can use the equations for many different scenarios without having to redo all of the work. This is the process for developing beam charts as found in Appendix D.

With the values of both constants found, the full expressions for the displacement, $y(x)$, and rotation, $\theta(x)$, are given as

$$\theta(x) = \frac{P}{EI}\left(\frac{1}{2}x^2 - Lx\right); \qquad 0 \leq x \leq L$$

$$y(x) = \frac{P}{EI}\left(\frac{x^3}{6} - \frac{Lx^2}{2}\right); \qquad 0 \leq x \leq L$$

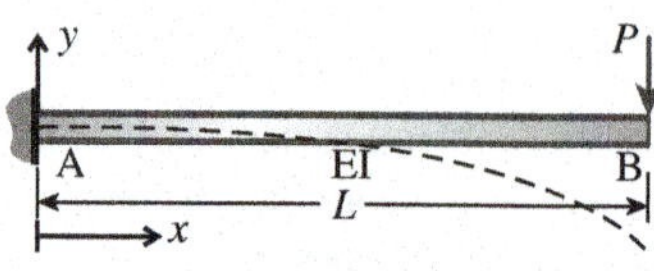

We need to now find the displacement and rotation at point B. We do this by evaluating the equations at $x = L$.

$$\theta(L) = \frac{P}{EI}\left(\frac{1}{2}L^2 - L(L)\right) = \underline{\underline{-\frac{PL^2}{2EI}}}$$

$$y(L) = \frac{P}{EI}\left(\frac{L^3}{6} - \frac{L(L)^2}{2}\right) = \underline{\underline{-\frac{PL^3}{3EI}}}$$

EXAMPLE 8.9

For the beam shown in Example 8.7 structure (c), the moment equations are given as

$$M(x_1) = -\frac{w}{8}\left(4x_1^2 - 3Lx_1\right); \qquad 0 \le x_1 < \frac{L}{2}$$
$$M(x_2) = \frac{wLx_2}{8}; \qquad 0 \le x_2 < \frac{L}{2}$$

Use the double integration method to find the equations for the displacement of this beam for both segments, y_1 and y_2.

SOLUTION STRATEGY

The moment equation and boundary conditions have already been identified for this problem, so we need to pick up at step 4.

SOLUTION

Perform both integrations making sure to add an integration constant each time. Do it for both beam segments, which will result in a total of four integration constants.

$$\theta_1 = \int \frac{M(x_1)}{EI}dx = \frac{1}{EI}\left[-\frac{w}{8}\left(\frac{4x_1^3}{3} - \frac{3Lx_1^2}{2}\right) + C_1\right]$$
$$\theta_2 = \int \frac{M(x_2)}{EI}dx = \frac{1}{EI}\left(\frac{wLx_2^2}{16} + C_2\right)$$

$$y_1 = \int \theta_1 dx = \frac{1}{EI}\left[-\frac{w}{8}\left(\frac{x_1^4}{3} - \frac{Lx_1^3}{2}\right) + C_1x_1 + C_3\right]$$
$$y_2 = \int \theta_2 dx = \frac{1}{EI}\left(\frac{wLx_2^3}{48} + C_2x_2 + C_4\right)$$

Evaluate BC: $y_1(x_1 = 0) = 0$

$$y_1(x_1 = 0) = \frac{1}{EI}\left[-\frac{w}{8}\left(\frac{(0)^4}{3} - \frac{L(0)^3}{2}\right) + C_1(0) + C_3\right] = 0 \qquad \therefore \underline{\underline{C_3 = 0}}$$

Evaluate BC: $y_2(x_2 = 0) = 0$

$$y_2(x_2 = 0) = \frac{1}{EI}\left(\frac{wL(0)^3}{48} + C_2(0) + C_4\right) = 0$$
$$\therefore \underline{\underline{C_4 = 0}}$$

Evaluate BC: $\theta_1\left(x_1 = \frac{L}{2}\right) = -\theta_2\left(x_2 = \frac{L}{2}\right)$

$$\frac{1}{EI}\left[-\frac{w}{8}\left(\frac{4}{3}\left(\frac{L}{2}\right)^3 - \frac{3}{2}L\left(\frac{L}{2}\right)^2\right) + C_1\right] = -\frac{1}{EI}\left[\frac{1}{16}wL\left(\frac{L}{2}\right)^2 + C_2\right]$$
$$\therefore C_2 = -C_1 - \frac{wL^3}{24}$$

Evaluate BC: $y_1\left(x_1 = \frac{L}{2}\right) = y_2\left(x_2 = \frac{L}{2}\right)$

$$\frac{1}{EI}\left[-\frac{w}{8}\left(\frac{1}{3}\left(\frac{L}{2}\right)^4 - \frac{1}{2}L\left(\frac{L}{2}\right)^3\right) + C_1\left(\frac{L}{2}\right)\right] = \frac{1}{EI}\left[\frac{1}{48}wL\left(\frac{L}{2}\right)^3 + \left(-C_1 - \frac{wL^3}{24}\right)\left(\frac{L}{2}\right)\right]$$

$$\therefore \underline{\underline{C_1 = -\frac{9wL^3}{384}}} \qquad \therefore \underline{\underline{C_2 = -\frac{7wL^3}{384}}}$$

COMMON MISTAKE

When performing the integrations, the students will assign integration constants of C_1 and C_2 to both beam segments. They then make the mistake of thinking that the integration constants for both segments are the same value. This is not the case. Rather, if you have two beam segments, you should end up with four integration constants $C_1 \ldots C_4$.

DON'T BE FOOLED

The integration constants don't always work out to be zero. It is dependent on beam configuration and the coordinate systems selected.

QUICK NOTE

Don't forget the negative sign in the boundary condition equation, since the coordinates are facing opposite directions.

With the values of all constants found, the full expressions for the displacement are given as

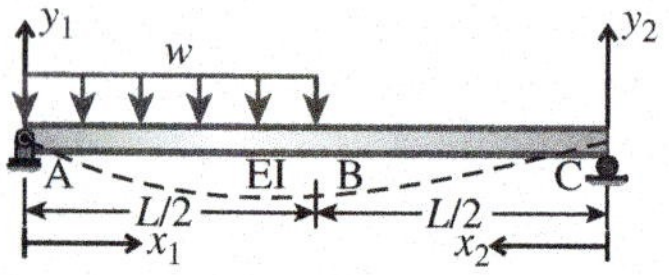

$$y_1(x_1) = \frac{w}{384EI}\left(-16x_1^4 + 24Lx_1^3 - 9L^3x_1\right); \qquad 0 \le x_1 < \frac{L}{2}$$

$$y_2(x_2) = \frac{w}{384EI}\left(8Lx_2^3 - 7L^3x_2\right); \qquad 0 \le x_2 < \frac{L}{2}$$

8.8 SAP2000 Computer Applications

As is clearly demonstrated within this chapter, the deflection of a beam is dependent upon the load being applied, the length of the spans, the geometry of the beam cross-section, and the type of material it is made of. It is no surprise then than SAP2000 requires that all of these characteristics be defined in every model that is developed.

It is often desirable to use SAP2000 to verify your hand calculations. There are a couple of common issues that must be done to a SAP2000 model to be able to make a consistent comparison. Below is a list of a few of the most common adjustments that need to be made.

- Turn off the self-weight of the structure.
- Turn off shear deformations.
- Neglect axial deformations in the structure. This keeps distance between the joints the same.
- Modeling a structure which doesn't have numerical values but rather is expressed in terms of P, w, L, E, and I need special consideration.

QUICK NOTE
Make sure you do a direct substitution for each variable. For example, don't put in a 1.0 for every point load that exists, rather plug 1.0 in for the variable P and input the resulting value. So if a point load on a structure is expressed as $P/2$, the magnitude put into SAP2000 should be 0.5.

The first two issues listed are covered in the SAP2000 tutorial shown in Appendix B. The latter two issues require a little more explanation. How to neglect axial deformations is further explained in Chapter 9.

To model structures like those presented in Examples 8.8 and 8.9, a simple procedure can be followed. Select a set of units in which to work (e.g., kips and inches, kips and feet, kilonewtons and meters, etc.). It doesn't matter what set you choose, as long as you are consistent throughout the model generation. For any variable present in the structure, input a value of 1.0. Set the shear area to be zero and set the cross-sectional area to be large—around 10,000. Any resulting computed displacements then will be comparable to the factor associated with the hand calculations. See the following exercise for a clarification of this point.

SAP2000 EXERCISE 8.1

QUICK NOTE
A fully functioning model of this example is found in the file *C8EX1.sdb*.

Model the beam shown in Example 8.6 structure (a) and verify the displacement results obtained for the tip of the cantilever found in Example 8.8. The easiest way to model this is using the model wizard and click on beam. We will select kips and inches for building this model. Set the following values.

Parameter	Value	Units
P	1.0	kips
L	1.0	inches
E	1.0	ksi
I	1.0	in^2
Shear area	0.0	in^2
A	10,000.0	in^2

Cross-section Input

Properties

Cross-section (axial) area	1.	Section modulus about 3 axis	1.
Torsional constant	1.	Section modulus about 2 axis	1.
Moment of Inertia about 3 axis	1.	Plastic modulus about 3 axis	1.
Moment of Inertia about 2 axis	1.	Plastic modulus about 2 axis	1.
Shear area in 2 direction	0.	Radius of Gyration about 3 axis	1.
Shear area in 3 direction	0.	Radius of Gyration about 2 axis	1.

Material Input

Isotropic Property Data

Modulus of Elasticity, E	1
Poisson's Ratio, U	0.3
Coefficient of Thermal Expansion, A	6.500E-06
Shear Modulus, G	0.3846

The deflected shape of the beam is as shown and the joint output is given in the table.

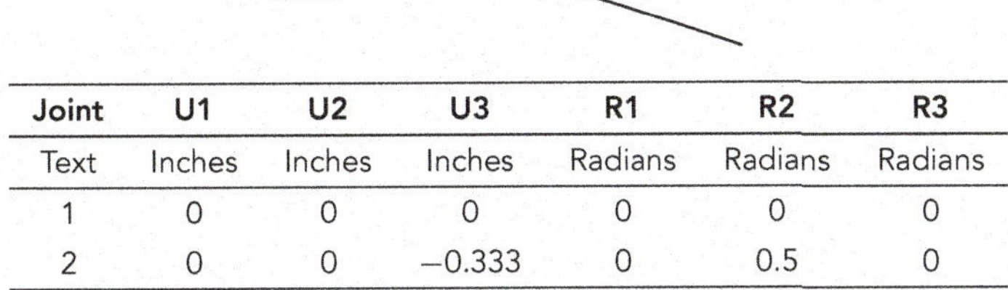

Joint	U1	U2	U3	R1	R2	R3
Text	Inches	Inches	Inches	Radians	Radians	Radians
1	0	0	0	0	0	0
2	0	0	−0.333	0	0.5	0

SPRING DEFINITION

Springs are defined under the *ASSIGN/JOINT* menu. For a planar model created in the XZ plane, a rotational spring should be defined about the 2-*axis*.

Spring Direction

Coordinate System: Local

Spring Stiffness

Translation 1	0.
Translation 2	0.
Translation 3	0.
Rotation about 1	0.
Rotation about 2	10.
Rotation about 3	0.

Options

- Add to Existing Springs
- Replace Existing Springs
- Delete Existing Springs

Advanced...

OK Cancel

SUMMARY

The expressions for tip displacement and rotation, as found in Example 8.8, are $\Delta = -\frac{PL^3}{3EI}$ and $\theta = \frac{PL^2}{2EI}$. If $L = P = E = I = 1.0$, then $\Delta = -0.333$ and $\theta = 0.5$. This is in direct agreement with the results from SAP2000. Thus, the results have been validated.

SAP2000 EXERCISE 8.2

Model the given frame and find the horizontal displacement at C. Do this for four different values of rotational spring stiffness: $k_R = 10$ (k·in)/rad, $k_R = 10{,}000$ (k·in)/rad, $k_R = 100{,}000$ (k·in)/rad, and $k_R = 1{,}000{,}000$ (k·in)/rad. Create a plot of the stiffness of the rotational spring (k_R) versus the horizontal displacement at C (δ_{Cx}). Make an observation on the trend. Plot the displaced shape for the four cases and make an observation on the impact the spring stiffness has on the appearance of the displaced shape.

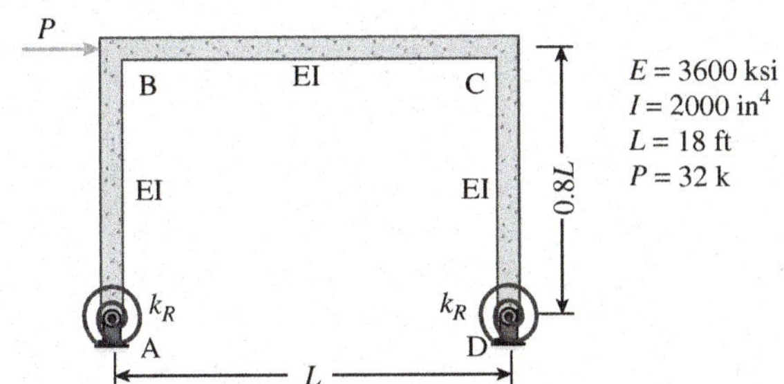

QUICK NOTE

A fully functioning model of this example is found in the file *C8EX2.sdb*.

SOLUTION

The following deflected shapes result.

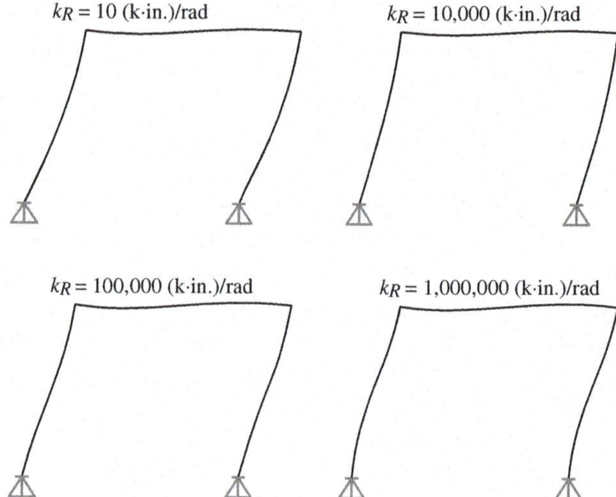

OBSERVATION

Notice how the rotation at the supports reduces as the rotational stiffness increases. The columns appear to be in single curvature when the stiffness is low but they are clearly in double curvature at the higher stiffness values.

The trend of the horizontal displacements at point C is demonstrated in the plot.

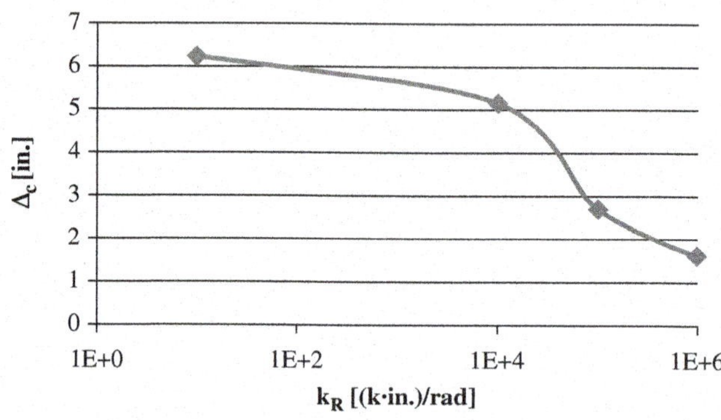

OBSERVATION

The horizontal displacement gradually decreases as the stiffness increases. Then there is a drastic decrease in displacement at the higher levels of stiffness. The flexible springs approximate a pin-supported condition whereas the stiff springs approximate a fixed condition. For your information, for a pin support Δ_{Cx} = 6.21 in. and for a fixed support Δ_{Cx} = 1.45 in.

8.9 Examples with Video Solutions

VE8.1 For the beam in Figure VE8.1, sketch the qualitative deflected shape and determine the direction of the resulting reactions.

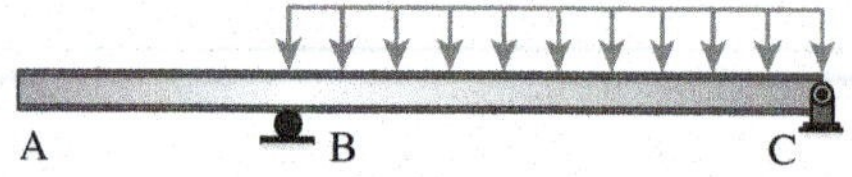

Figure VE8.1

VE8.2 For the beam in Figure VE8.2, sketch the qualitative deflected shape and determine the direction of the resulting reactions.

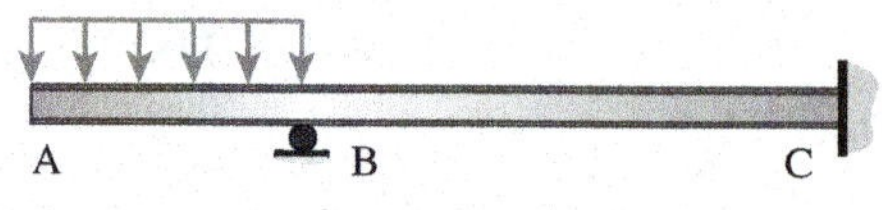

Figure VE8.2

VE8.3 For the frame in Figure VE8.3, sketch the qualitative deflected shape and determine the direction of the resulting reactions.

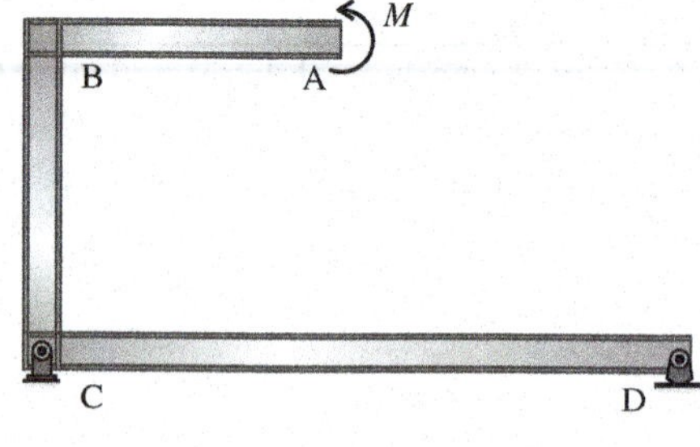

Figure VE8.3

VE8.4 For the beam in Figure VE8.4, identify the four boundary conditions that must be used for the coordinates given for each case.

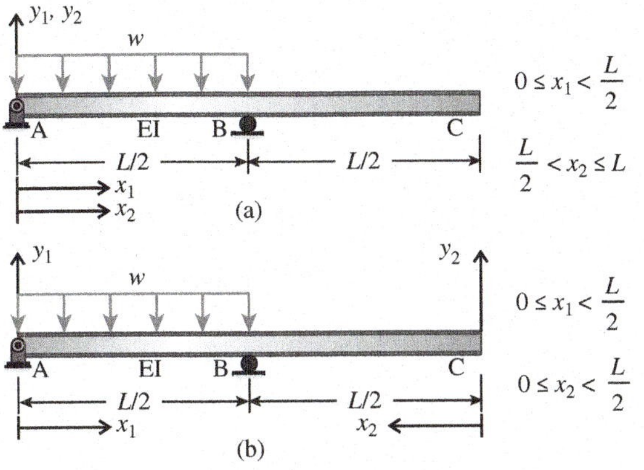

Figure VE8.4

VE8.5 For the beam and coordinate systems in Figure VE8.5, the moment equations are given for each segment. Develop the displacement equations ($y(x_1)$, $y(x_2)$) for the beam using the double integration method.

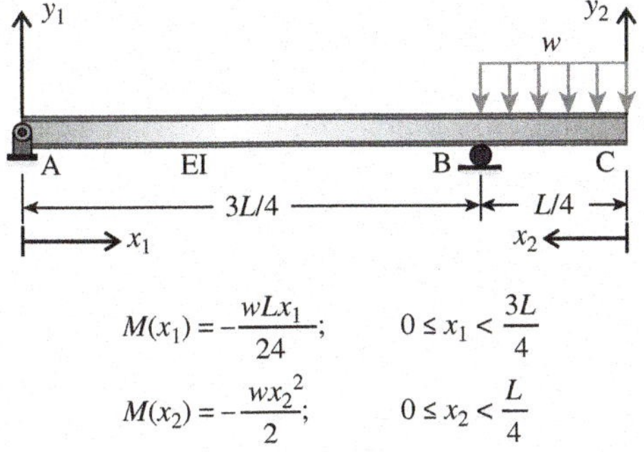

$$M(x_1) = -\frac{wLx_1}{24}; \qquad 0 \le x_1 < \frac{3L}{4}$$

$$M(x_2) = -\frac{wx_2^2}{2}; \qquad 0 \le x_2 < \frac{L}{4}$$

Figure VE8.5

8.10 Problems for Solution

Sections 8.4–8.5

For Problems 8.1 through 8.24, qualitatively sketch the deformed shape of the structures for the given loads. Also sketch the direction of all reactions.

P8.1 (Ans: $R_{By} \uparrow$, M_B ↺)

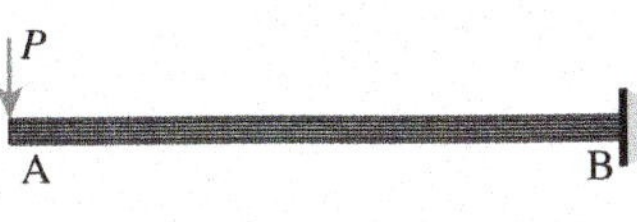

Problem 8.1

P8.2

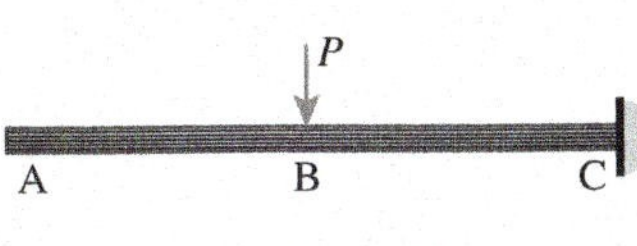

Problem 8.2

P8.3 (Ans: M_B ↻)

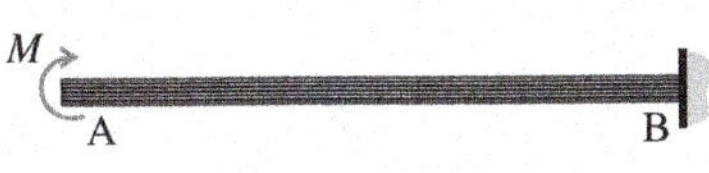

Problem 8.3

P8.4

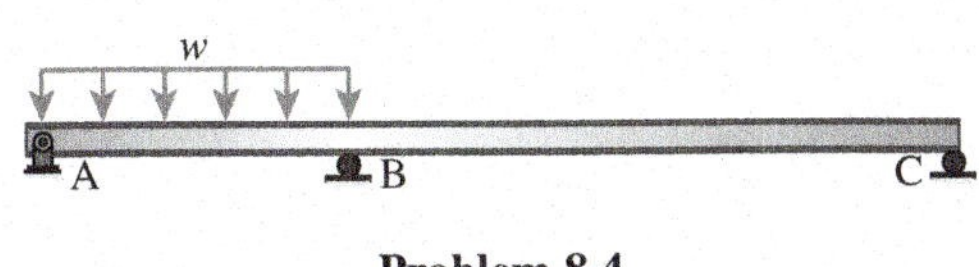

Problem 8.4

P8.5 (Ans: $R_{By} \uparrow$, $R_{Cy} \uparrow$)

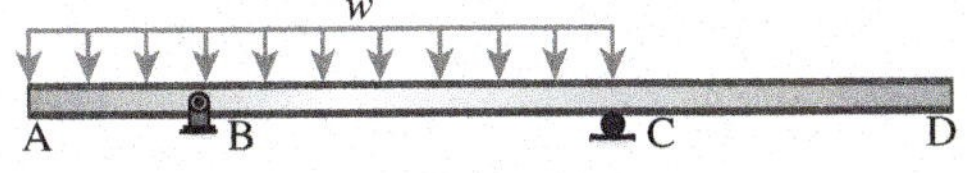

Problem 8.5

P8.6

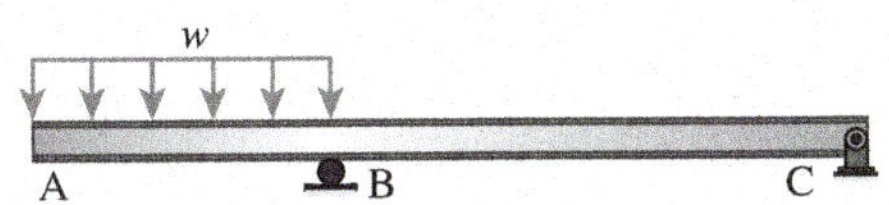

Problem 8.6

P8.7 (Ans: $R_{By} \downarrow$, $R_{Cy} \uparrow$)

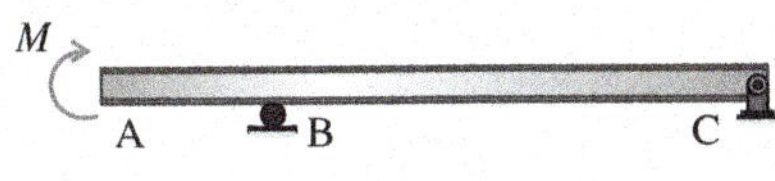

Problem 8.7

P8.8

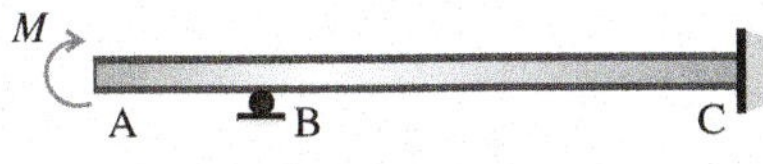

Problem 8.8

P8.9 (Ans: $R_{Ay} \uparrow$, $R_{By} \uparrow$, $R_{Cy} \uparrow$, $R_{Dy} \uparrow$)

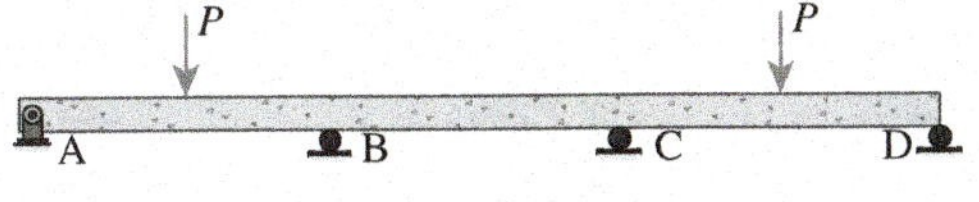

Problem 8.9

P8.10

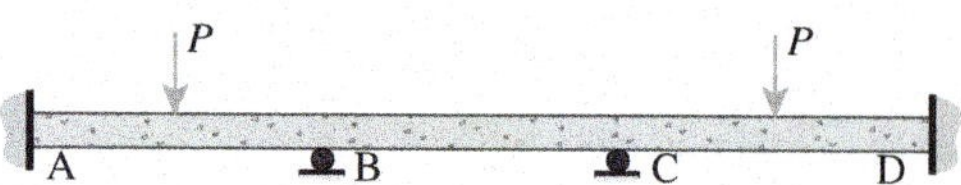

Problem 8.10

P8.11 (Ans: $R_{Ay} \downarrow$, $R_{By} \uparrow$, $R_{Cy} \uparrow$, $R_{Dy} \uparrow$, $R_{Ey} \uparrow$)

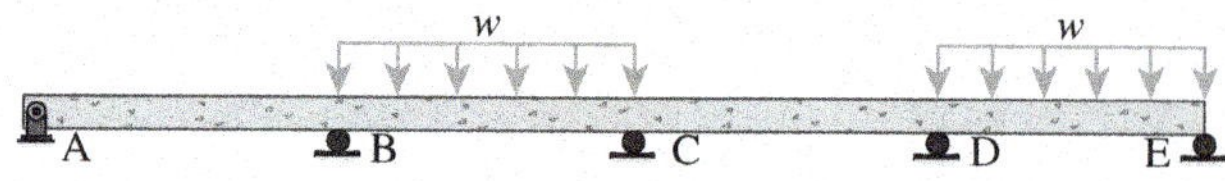

Problem 8.11

P8.12

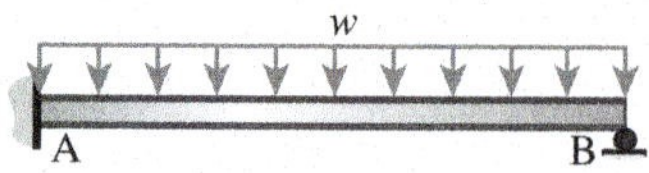

Problem 8.12

P8.13 (Ans: $R_{Ay} \uparrow$, $M_A \circlearrowleft$, $R_{By} \uparrow$)

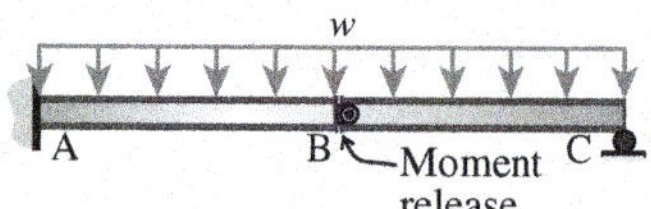

Problem 8.13

P8.14

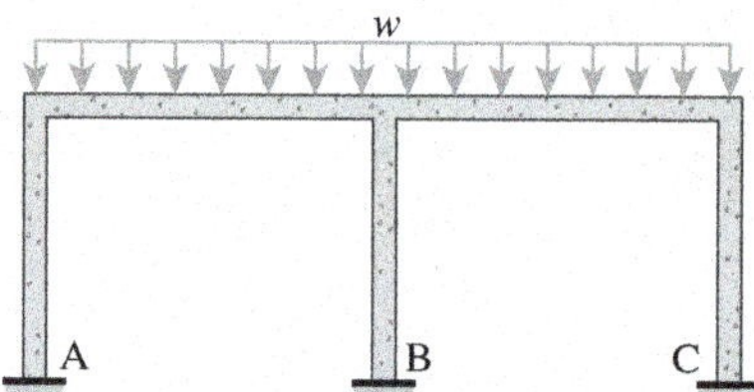

Problem 8.14

P8.15 (Ans: $R_{Ay} \uparrow$, $M_A \circlearrowleft$, $R_{Ax} \rightarrow$)

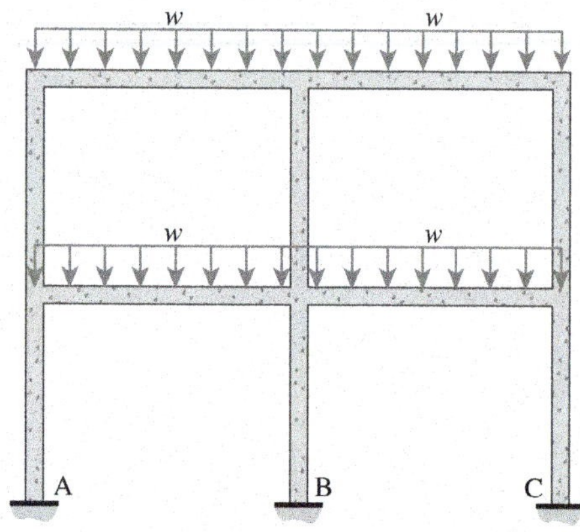

Problem 8.15

P8.16

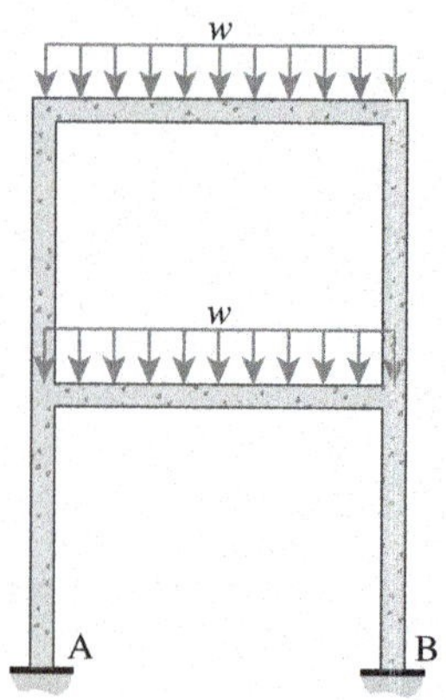

Problem 8.16

P8.17 (Ans: $R_{By} \uparrow$, $M_B \circlearrowleft$, $R_{Bx} \leftarrow$)

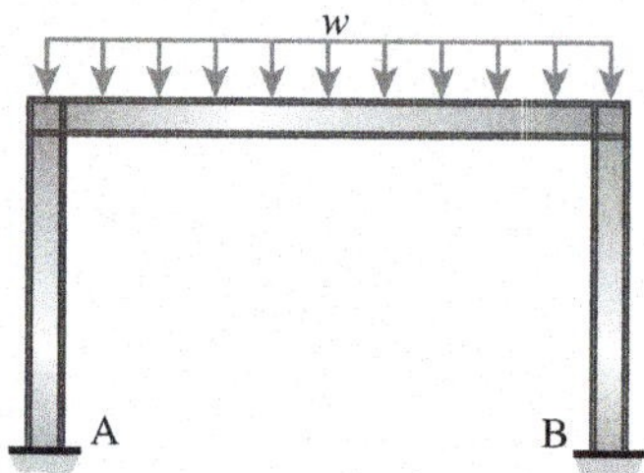

Problem 8.17

P8.18

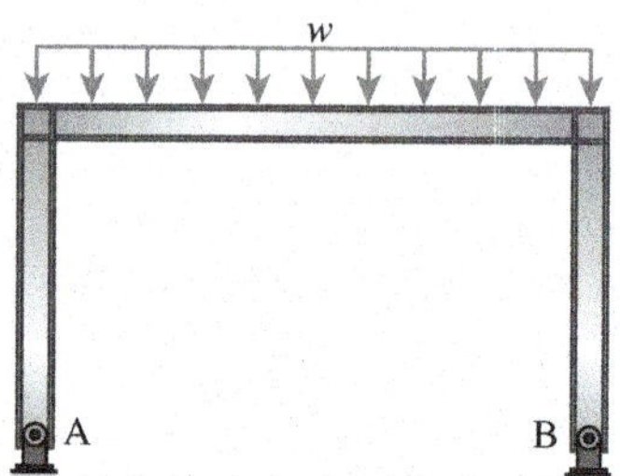

Problem 8.18

P8.19 (Ans: $R_{Ay} \downarrow$, $R_{Ax} \leftarrow$, $R_{By} \uparrow$, $R_{Bx} \leftarrow$)

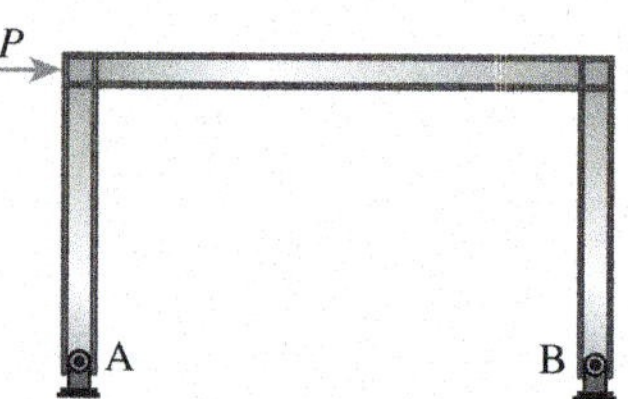

Problem 8.19

P8.20

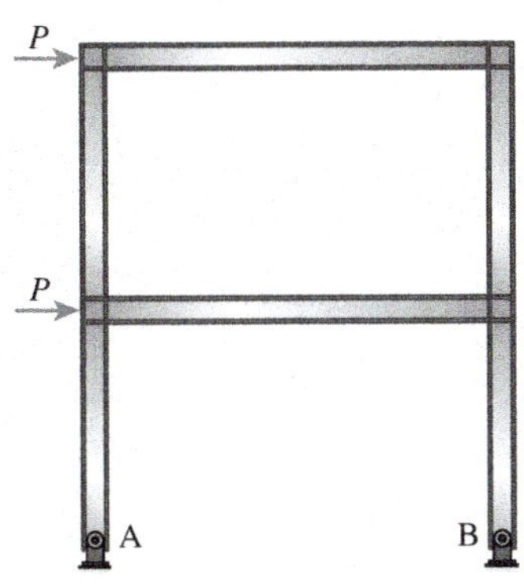

Problem 8.20

P8.21 (Ans: R_{Cy} ↑, R_{Dy} ↑)

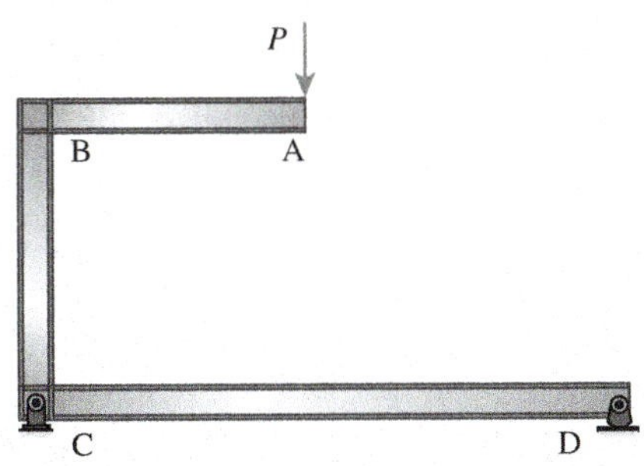

Problem 8.21

P8.22

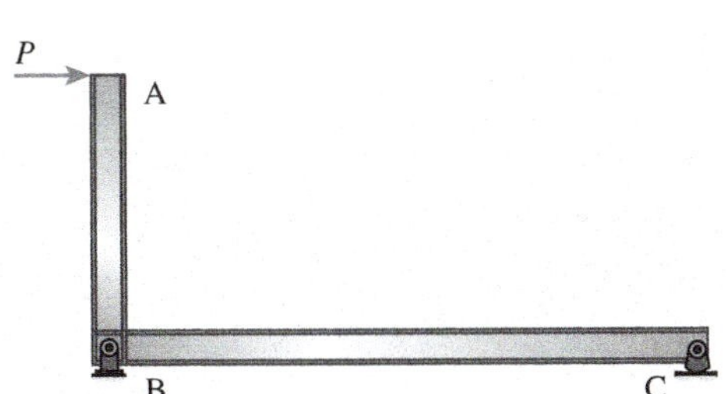

Problem 8.22

P8.23 (Ans: R_{By} ↑, R_{Cy} ↑)

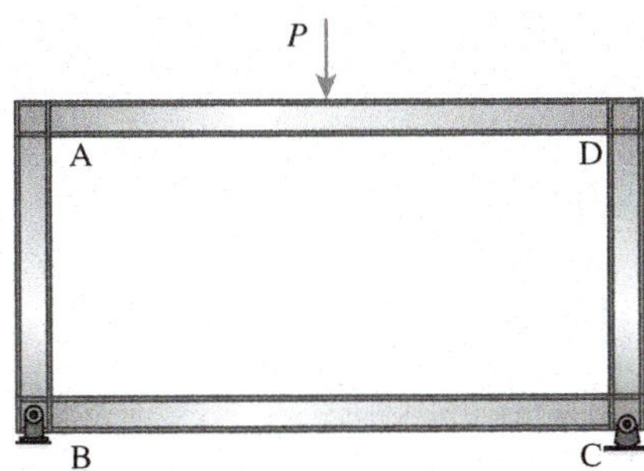

Problem 8.23

P8.24

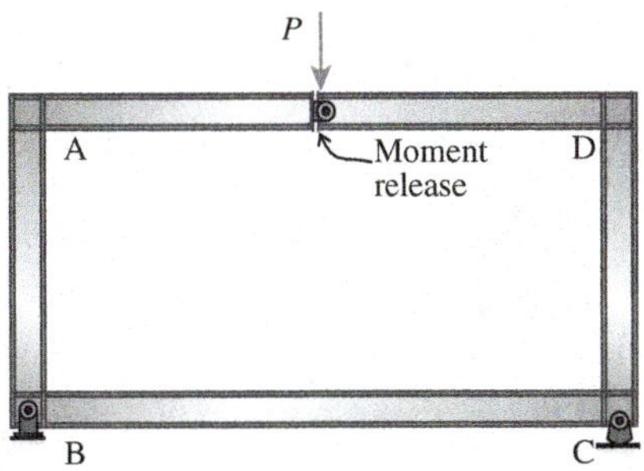

Problem 8.24

Section 7.6

For Problems 8.25 through 8.30 the qualitative moment diagrams are given for each beam. Use the moment diagram to help sketch the qualitative deformed shapes. Identify the direction of all reactions.

P8.25 (Ans: R_{Ay} ↓, M_A ↻, R_{Cy} ↑)

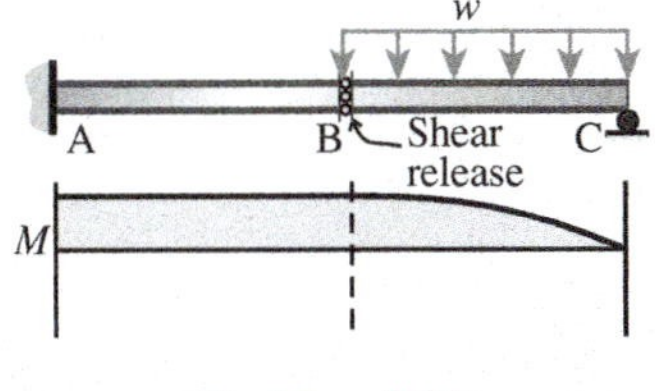

Problem 8.25

P8.26

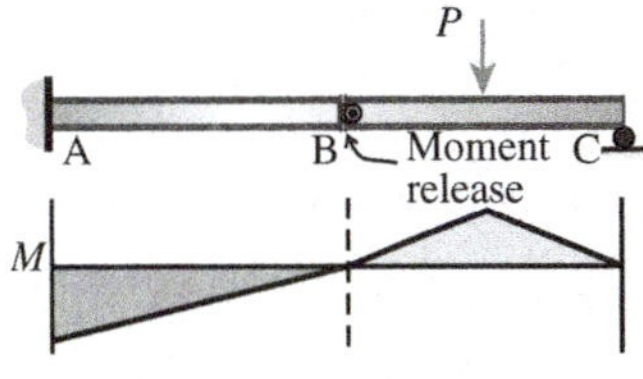

Problem 8.26

P8.27 (Ans: R_{Ay} ↑, M_A ↻)

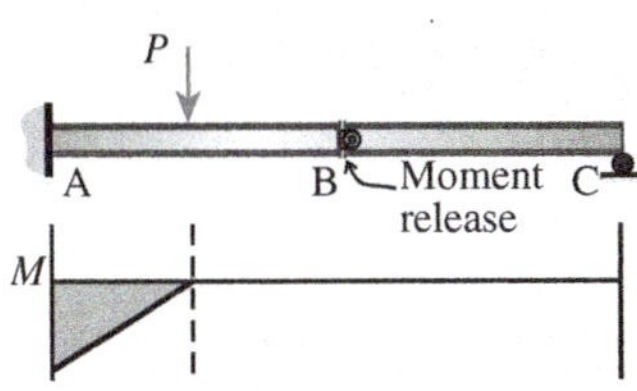

Problem 8.27

P8.28

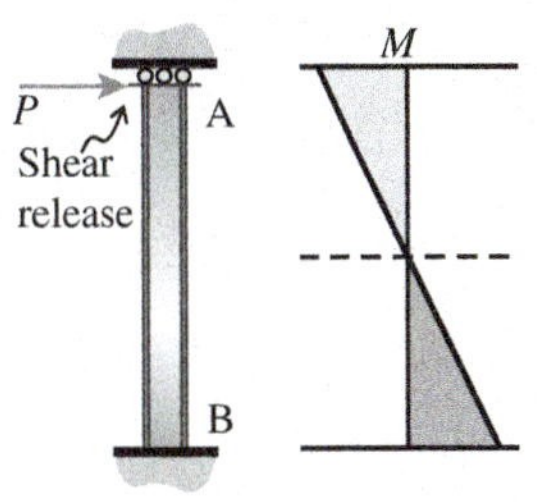

Problem 8.28

P8.29 (Ans: R_{Ay} ↑, M_A ↺, R_{By} ↑, M_B ↻)

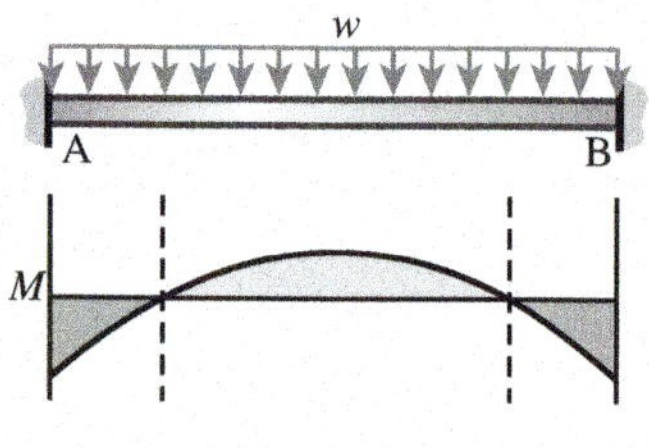

Problem 8.29

P8.30

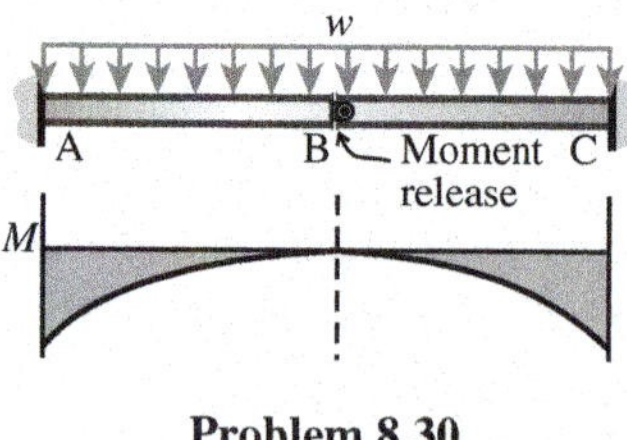

Problem 8.30

Section 8.7

For Problems 8.31 through 8.38 use the double integration method to solve for the deflection equations in terms of the coordinate system shown.

P8.31 $\left[\text{Ans: } y(x) = \frac{Mx^2}{EI}\right]$

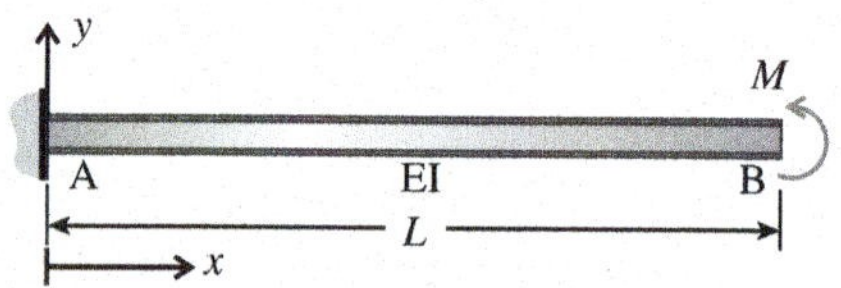

Problem 8.31

P8.32

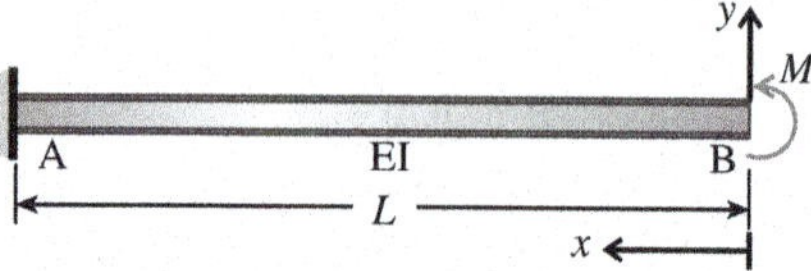

Problem 8.32

P8.33 $\left[\text{Ans: } y_1(x) = \frac{1}{EI}\left(\frac{Px^3}{3} - \frac{3PLx^2}{4}\right)\right]$

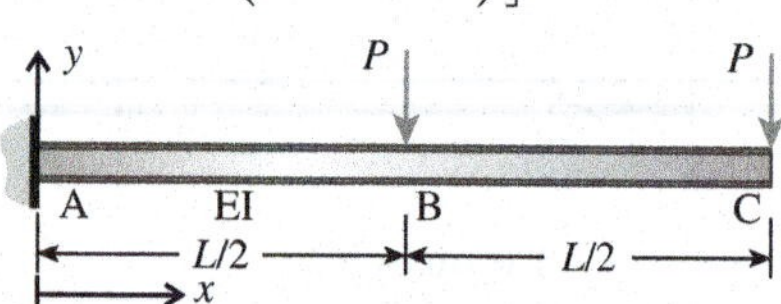

Problem 8.33

P8.34

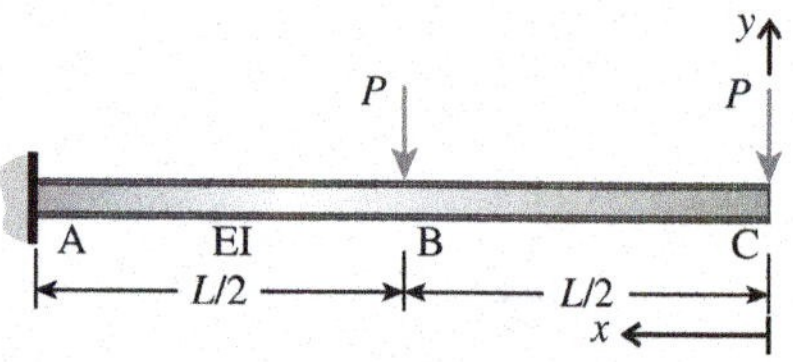

Problem 8.34

P8.35 $\left[\text{Ans: } y(x) = \frac{w}{2EI}\left(\frac{-x^4}{12} + \frac{L^3x}{3} - \frac{L^4}{4}\right)\right]$

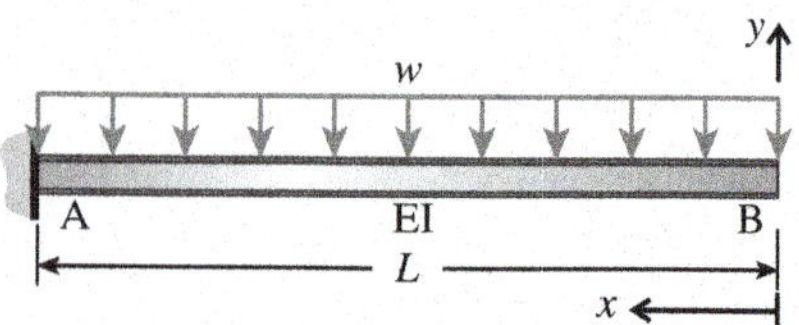

Problem 8.35

P8.36

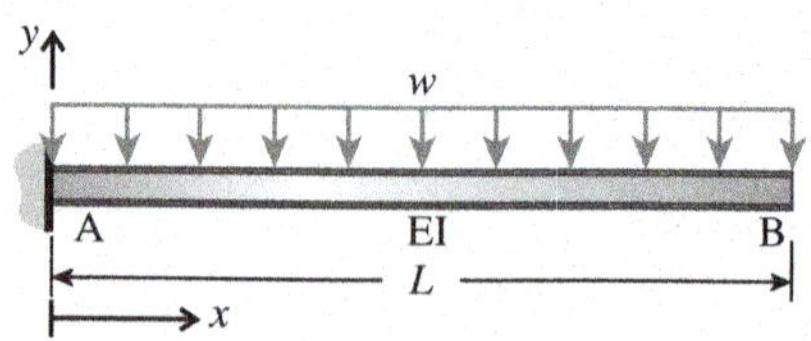

Problem 8.36

P8.37 $\left[\text{Ans: } y_1(x) = \frac{wx^2}{4EI}\left(\frac{-x^2}{12} + \frac{Lx}{3} - \frac{L^2}{2}\right)\right]$

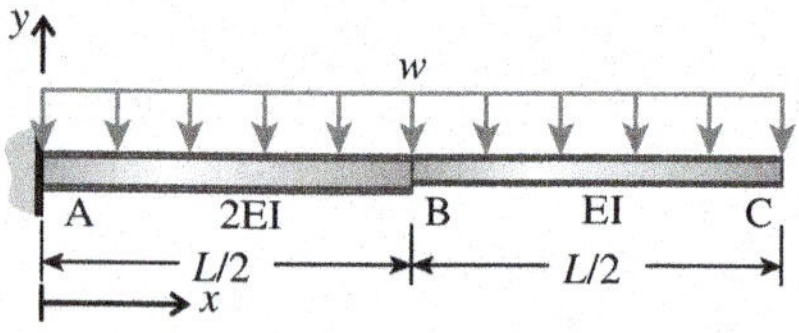

Problem 8.37

P8.38

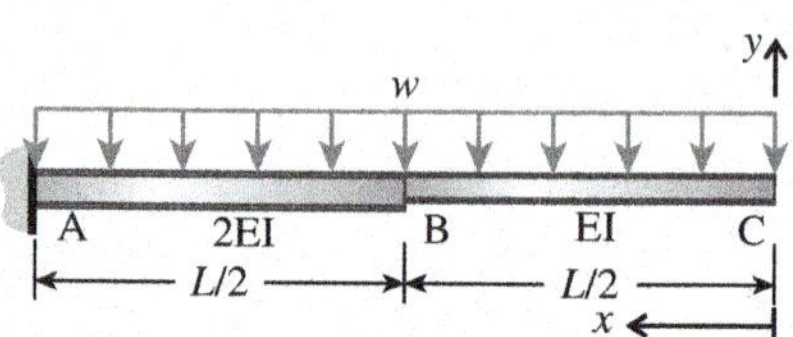

Problem 8.38

For Problems 8.39 through 8.52 use the double integration method to solve for the requested quantities. (Use whatever coordinate system you desire for the generation of the equations. You will then use your equations to solve for the quantities at the specific locations.)

P8.39 Determine for θ_A, θ_B, Δ_A, and Δ_B where $E = 29 \cdot 10^6$psi and $I = 1200$ in^4. (Ans: $\Delta_A = -2.648$ in., $\theta_B = 0.0124$ rad)

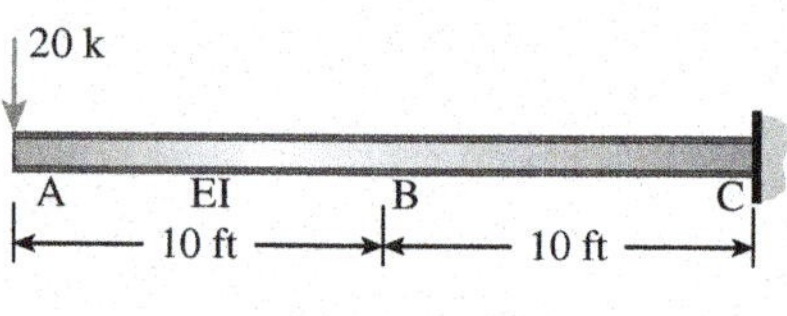

Problem 8.39

P8.40 Determine for θ_A, θ_B, Δ_A, and Δ_B where $E = 29 \cdot 10^6$ psi and $I = 2500$ in^4.

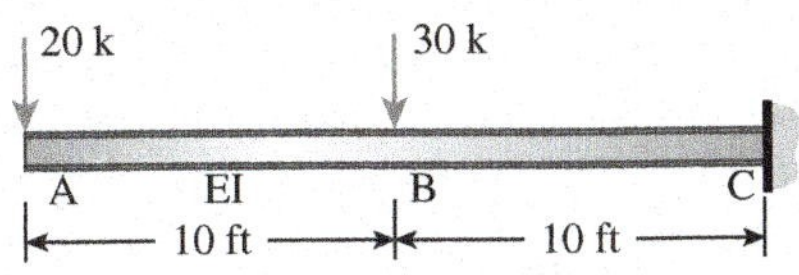

Problem 8.40

P8.41 Determine for θ_A, θ_B, Δ_A, and Δ_B where $E = 29 \cdot 10^6$ psi and $I = 1500$ in^4. (Ans: $\Delta_A = -0.579$ in., $\theta_B = 0.00414$ rad)

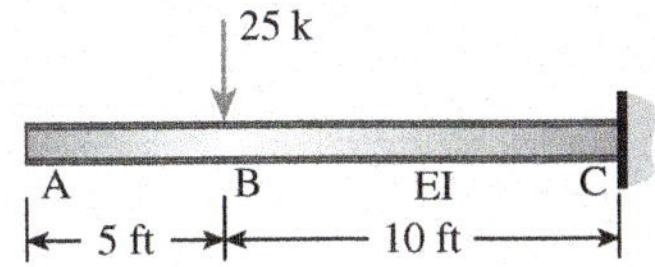

Problem 8.41

P8.42 Determine for θ_A and Δ_A where $E = 29 \cdot 10^6$ psi and $I = 1140$ in^4.

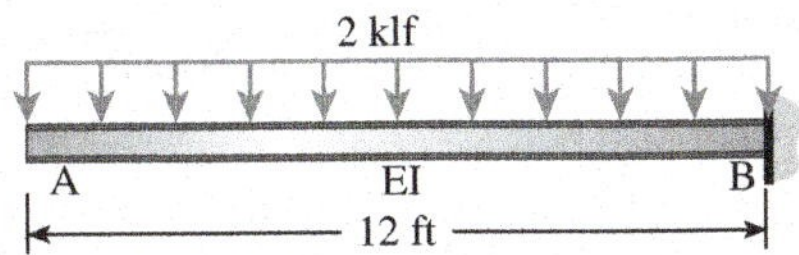

Problem 8.42

P8.43 Determine for Δ_B and Δ_C where $E = 29 \cdot 10^6$ psi and $I = 1000$ in^4. (Ans: $\Delta_B = -0.618$ in., $\Delta_C = -1.442$ in.)

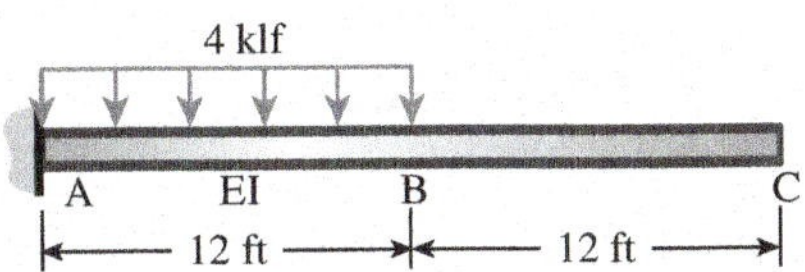

Problem 8.43

P8.44 Determine for Δ_B and Δ_C where $E = 29 \cdot 10^6$ psi and $I = 2100$ in^4.

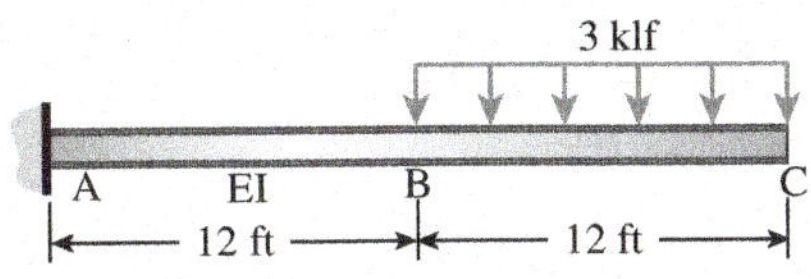

Problem 8.44

P8.45 Determine for Δ_B and Δ_C where $E = 29 \cdot 10^6$ psi and $I = 5200$ in^4. (Ans: $\Delta_B = -0.296$ in., $\Delta_C = -0.893$ in.)

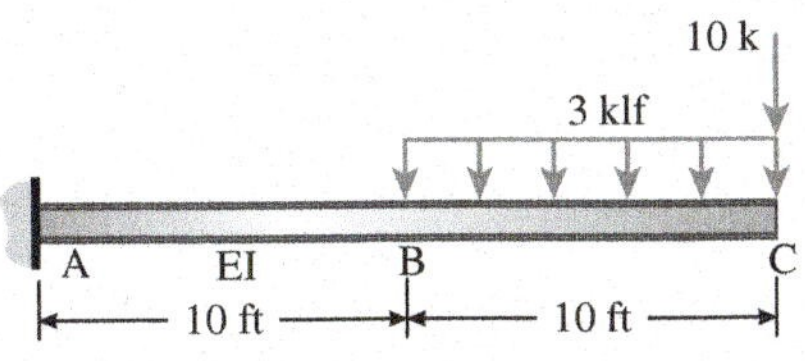

Problem 8.45

P8.46 Determine for Δ_B and θ_B where $E = 29 \cdot 10^6$ psi and $I = 2100$ in^4.

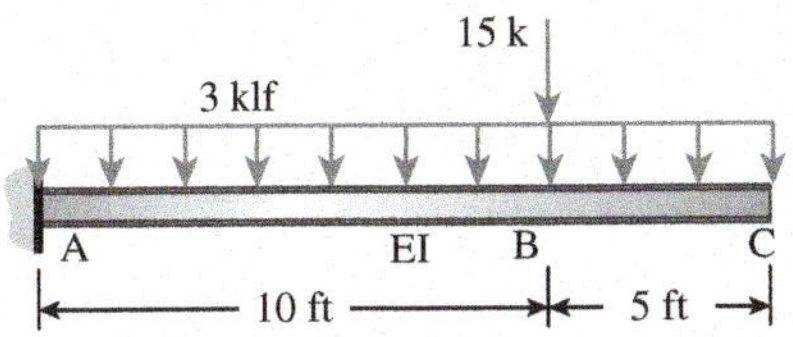

Problem 8.46

P8.47 Determine for Δ_B and θ_B where $E = 29 \cdot 10^6$ psi and $I = 600$ in^4. (Ans: $\Delta_B = -0.335$ in., $\theta_B = 0.00233$ rad)

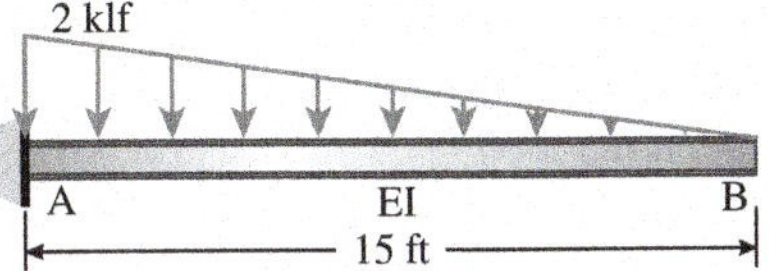

Problem 8.47

P8.48 Determine for Δ_B and θ_C where $E = 10 \cdot 10^6$ psi and $I = 3200$ in^4.

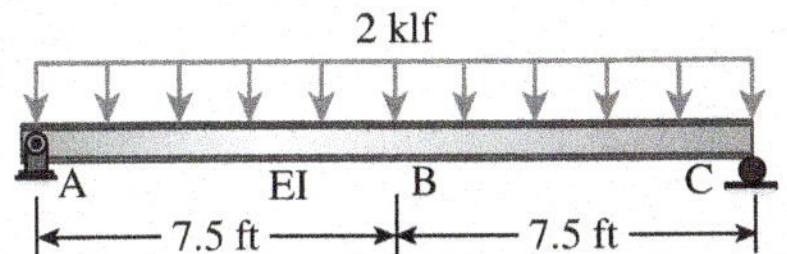

Problem 8.48

P8.49 Determine for Δ_B and θ_C where $E = 10 \cdot 10^6$ psi and $I = 1600$ in^4. (Ans: $\Delta_B = -0.0712$ in., $\theta_C = -0.00118$ rad)

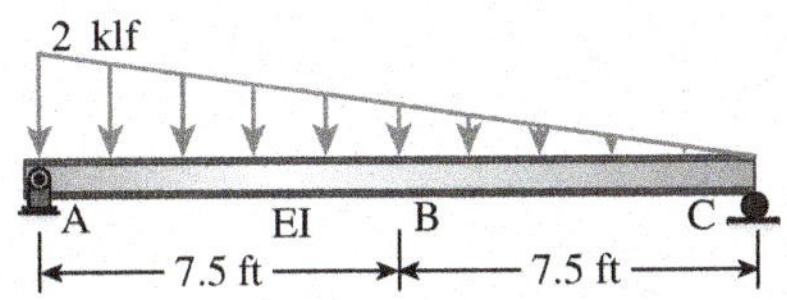

Problem 8.49

P8.50 Determine for Δ_B and θ_C where $E = 1.9 \cdot 10^6$ psi and $I = 1200$ in^4.

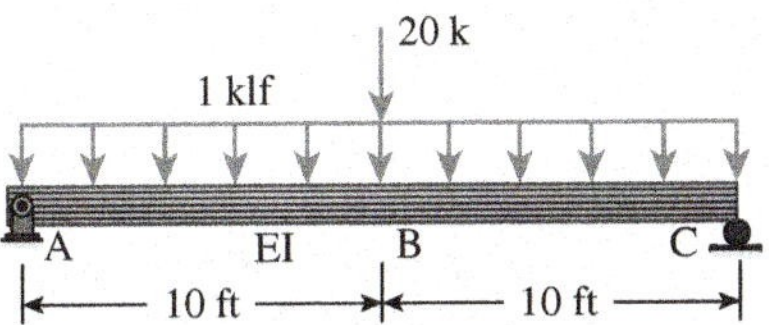

Problem 8.50

P8.51 Determine for Δ_C and θ_B where $E = 1.9 \cdot 10^6$ psi and $I = 950$ in^4. (Ans: $\Delta_C = 0.2992$ in., $\theta_B = -0.00623$ rad)

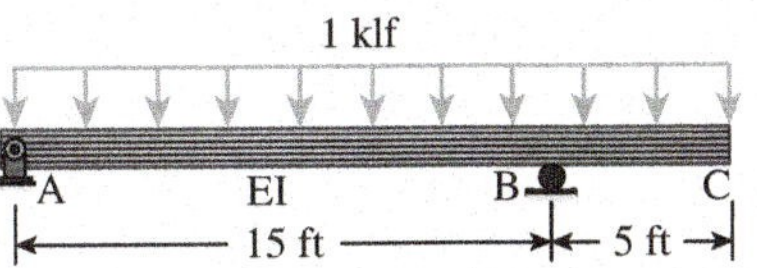

Problem 8.51

P8.52 Determine for Δ_C and θ_B where $E = 1.9 \cdot 10^6$ psi and $I = 1500$ in^4.

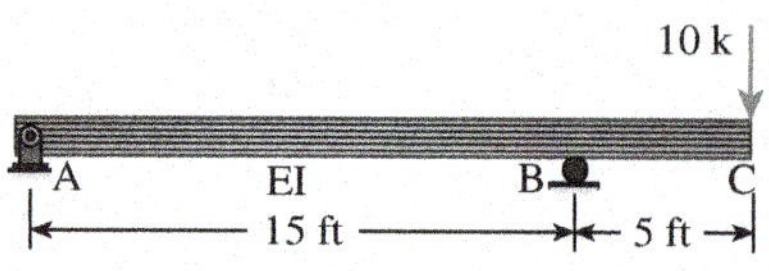

Problem 8.52

Section 8.8 SAP2000

P8.53 Model the frame shown in Problem 8.15 and plot its deformed shape. Assume the length of each member is 10 ft and that $w = 3$ klf. Set $E = 3600$ ksi, $I = 2000$ in^4, and $A = 10,000$ in^2 for each member. Now plot the deformed shape of the structure again if the supports at A, B, and C are all changed to pin connections. Describe the observed difference seen between the two cases.

P8.54 Model the given beam in SAP2000. Set $E = 1900$ ksi, $I = 900$ in^4, and $A = 10,000$ in^2 for each span. Plot the deformed shape. Now change the value of I for member ABC to 2900 in^4. Plot the deformed shape. What impact did the changing of the moment of inertia have on the overall deformed shape?

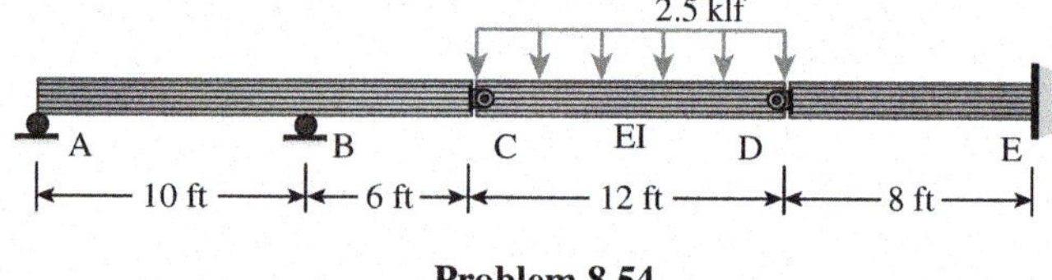

Problem 8.54

P8.55 Model the given beam in SAP2000. Set $E = 3600$ ksi, $I = 1500$ in^4, and $A = 10,000$ in^2. Plot the deformed shape and identify the magnitude and direction of all reactions. Do this for the following values of x (8 ft, 12 ft, 16 ft, and 20 ft). Indicate whether the net deflection at C is up or down and indicate the direction of the reaction at A for each case. What influence does the location of the roller have on these responses? (*Hint*: You can quickly change the location of B by selecting the node at B and using the *MOVE* command found under the *EDIT* pull-down menu.)

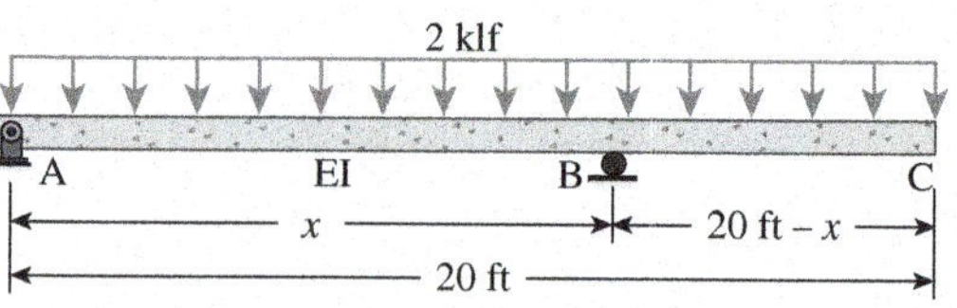

Problem 8.55

P8.56 Model the given beam in SAP2000. Set $E = 3600$ ksi, $I_{AB} = 1500$ in^4, and $A = 10,000$ in^2. Plot the deformed shape for each of the following cases: $I_{BC} = 750$ in^4, $I_{BC} = 1500$ in^4, and $I_{BC} = 3000$ in^4. What impact does changing the moment of inertia of the right span have on the overall deformed shape of the beam?

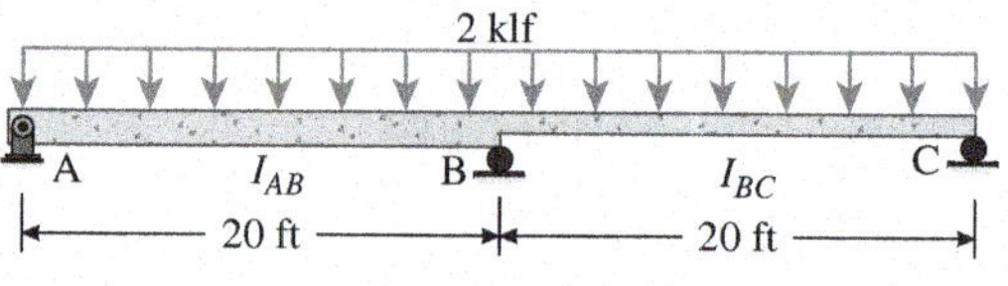

Problem 8.56

Deflection and Angle Changes Using Virtual Work

9

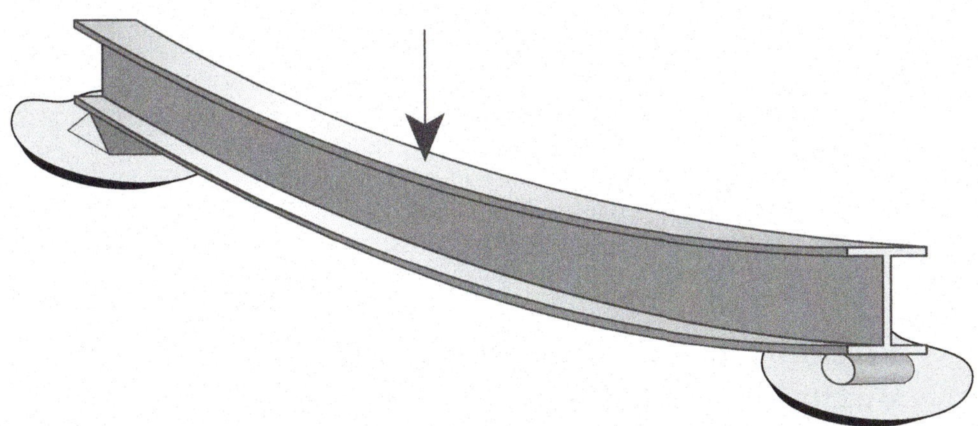

9.1 Introduction to Energy Methods

In this chapter, it is shown that the deflections and angle changes may be determined using the principle of conservation of energy. Only one of the common energy methods, *virtual work*, is presented here. This is to allow for a thorough coverage of this versatile method within this text. One should be aware that another powerful energy-based method that uses *Castigliano's second theorem* exists and also can be useful. However, for the sake of brevity, it is not discussed in this textbook.

Virtual work can be used for both simple and complicated structures. Some argue that geometric methods—mentioned in Chapter 8—are easier to use than virtual work and thus should be used for simple structures, saving virtual work for more complicated structures. This is a traditional approach to teaching structural analysis. However, the authors contend that by learning virtual work well and by using a geometric approach to solving the virtual work equation, it becomes a very powerful technique that can compete very well with the geometric methods when it comes to speed and simplicity. In addition, the energy methods are applicable to more types of structures.

DID YOU KNOW?
Virtual work can be used for calculating deflections in trusses in addition to beams and frames. This is a claim that cannot be made regarding the geometric approaches mentioned in Chapter 8.

9.2 Conservation of Energy Principle

Before the conservation of energy principle is introduced, a few comments are presented concerning the concept of work. For this discussion, *work* is defined as the product of a force and its displacement in the direction in which the force is acting. Should the force be constant during the displacement, as seen in Figure 9.1, the work will equal the force times the total displacement as

$$W = P\Delta$$

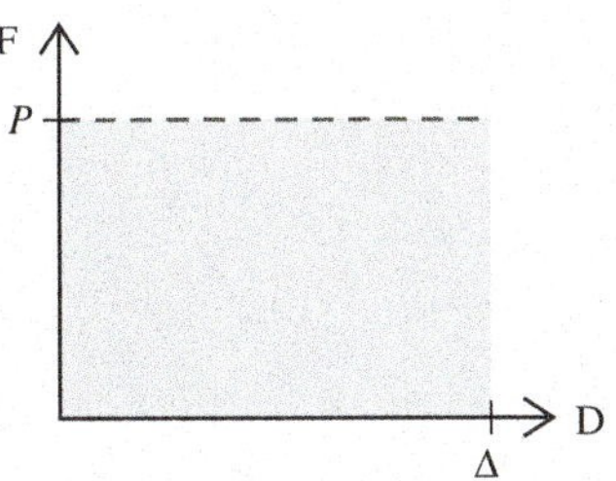

FIGURE 9.1
Force–deformation plot depicting work done for a constant force *P*.

On many practical occasions, the force changes in magnitude as does the deformation. For such a situation, it is necessary to compute the area under the force–deformation plot. This is done mathematically by summing up the small increments of work for which the force is assumed

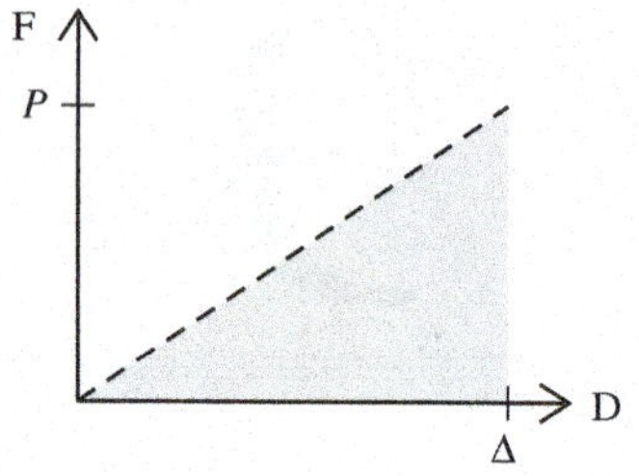

FIGURE 9.2
Force–deformation plot depicting work done for a linearly increasing force *P*.

to be constant. If a linear relationship exists between the force and displacement as shown in Figure 9.2, this integral produces

$$W = \int P d\Delta = \frac{1}{2} P\Delta$$

The *conservation of energy principle* is the basis of all energy methods. When a set of external loads is applied to a deformable structure, the points of load application move with the result that the members or elements making up the structure become deformed. According to the conservation of energy principle, the work done by the external loads, W_e, equals the work done by the internal forces acting in the elements of the structure, W_i. Stated mathematically, this principle takes the form:

$$W_e = W_i$$

As the deformation of a structure takes place, the internal work is stored within the structure as potential energy. If the elastic limit of the material is not exceeded, the elastic strain energy will be sufficient to return the structure to its original undeformed state when the loads are removed. Should a structure be subjected to more than one load, the total energy stored in the body will equal the sum of the energies stored in the structure by each of the loads.

It is to be clearly noted that the conservation of energy principle will be applicable only when static loads are applied to elastic systems. If the loads are not applied gradually, acceleration may occur, and some of the external work will be transferred into kinetic energy. If inelastic strains are present, some of the energy will be lost in the form of heat—among other things.

DID YOU KNOW?
Internal work is commonly referred to as *strain energy*. These terms are used interchangeably within this text.

9.3 Virtual Work or Complementary Virtual Work Method

With the virtual work principle, a system of forces in equilibrium is related to a compatible system of displacements in a structure. The virtual quantities discussed in this chapter do not exist in a real sense. When we speak of a virtual displacement, we are speaking of a fictitious displacement imposed on a structure. *The work performed by a set of real forces during a virtual displacement is called virtual work.*

NOMENCLATURE
The term *virtual* means to exist in essence but not in actual fact.

We see in the discussion to follow that for each virtual work theorem there is a corresponding theorem that is based on complementary virtual work. As a result, this method is sometimes referred to as *complementary virtual work.*

Virtual work is based on the principle of virtual velocities, which was introduced by Johann Bernoulli of Switzerland in 1717. The virtual work theorem can be stated as follows.

If a displacement is applied to a deformable body that is in equilibrium under a known load or loads, the external work performed by the existing load or loads due to this new displacement will equal the internal work performed by the stresses existing in the body that were caused by the original load or loads.

The virtual work or complementary work method is also referred to as the method of work or the dummy unit–load method. The name *dummy unit–load method* is given because a fictitious or dummy unit load is used in the solution.

Virtual work is based on the law of conservation of energy. To make use of this law in the derivation to follow, it is necessary to make the following assumptions.

1. The external and internal forces are in equilibrium.
2. The elastic limit of the material is not exceeded.
3. There is no movement of the supports.

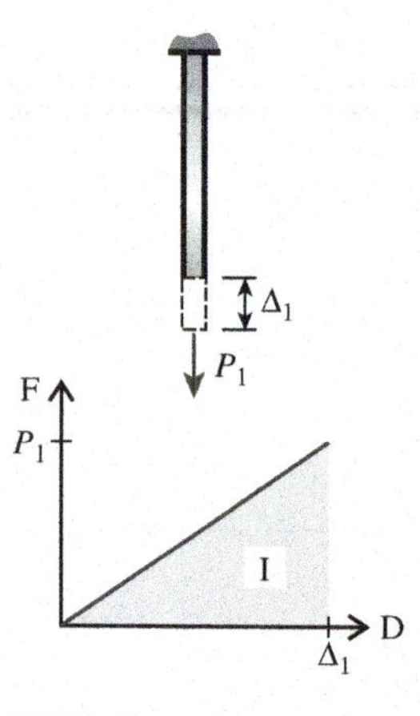

FIGURE 9.3 Deformation of axial bar.

Let a load P_1 be gradually applied to the bar shown in Figure 9.3 so that the load and the bar's deformation (or change in length) increase together from zero to their total values (P_1 and Δ_1). The external work done will equal the area (region I) under the force–displacement curve shown in Figure 9.3.

$$W_e = \frac{1}{2} P_1 \Delta_1$$

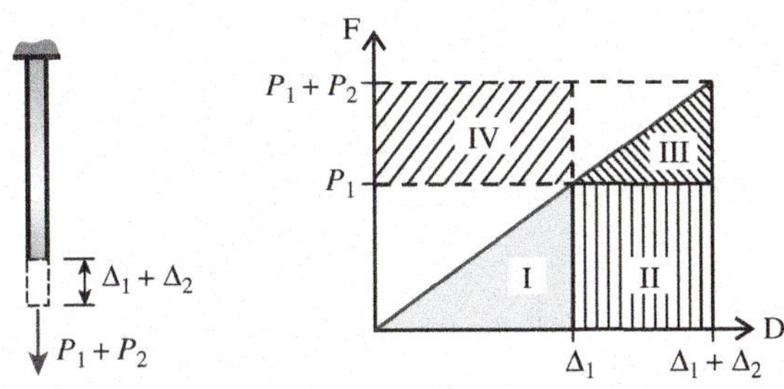

FIGURE 9.4
Force–displacement relationship for an elastic bar subjected to multiple loads.

Let another gradually applied force P_2 be added to the bar, causing an additional displacement Δ_2. This causes additional external work to occur (see Figure 9.4). The force P_1 will be entirely present for all of the Δ_2 displacement, and the work associated with that will be $P_1\Delta_2$. This work is represented in Figure 9.4 by region II. The gradually applied force P_2 will perform an additional amount of work equal to $1/2P_2\Delta_2$ denoted by region III. Thus, the total external work done by both forces is

$$W_e = \mathrm{I} + \mathrm{II} + \mathrm{III} = 1/2P_1\Delta_1 + P_1\Delta_2 + 1/2P_2\Delta_2$$

This expression for external work represents the total area beneath the sloped line in Figure 9.4. Another type of work is known as *complementary work*, W^*. It is represented by the total area above the curve. One may readily deduce then that each region (I, II, III) has an equal counterpart. For example, the area (i.e., work) of region II is equal to the complementary work shown in region IV. Complementary work does not really have a physical meaning as does the external work W_e, but it can provide a convenient way of looking at deflections in structures. Referencing the note in the side bar, the work represented in region IV can be called the *virtual external work* (δW_e^*), since it is the work done by a virtual load acting over the real displacement (Δ_1). The objective of the virtual work approach is to solve for the unknown real displacement (e.g., Δ_1) in a structure. Therefore, this description of the virtual work term (δW_e^*) is used hereafter.

QUICK NOTE

If the load P_1 is a real load and P_2 is a virtual load, the following interpretations can be made.

Region I The work done by the real load acting over the real displacement.

Region II The work done by the real load acting over a virtual displacement.

Region III The work done by the virtual load acting over a virtual displacement.

Region IV The work done by the virtual load acting over the real displacement.

9.4 Truss Deflections by Virtual Work

Since the conservation of energy principle says that the external work must be equal to the internal work (strain energy), the same thing could be said about virtual work. Such an interpretation would be

$$\delta W_e^* = \delta W_i^*$$

$$P_2\Delta_1 = \delta W_i^*$$

This relationship states that, if a virtual load is applied to a structure and the internal virtual strain energy is computed, the deflection in the structure can be directly determined. Thus, the focus of this section will be on deriving the strain energy term for a truss bar (axial member) and using it to calculate deflections in trusses.

Internal Virtual Strain Energy

The internal strain energy is easily derived in the same fashion that the external work was derived. Figure 9.5 shows the same sequence of loading as was described for Figure 9.4. The only difference is that the force–deformation plot is written in terms of the internal axial forces, N and n, instead of the external forces, P_1 and P_2.

Using Hooke's law in conjunction with well-known mechanics equations for stress ($\sigma = N/A$) and strain ($\epsilon = \Delta/L$) in an axial bar, the internal work equation can be written as

$$W_i = \mathrm{I} + \mathrm{II} + \mathrm{III}$$

$$= \frac{1}{2}(\sigma_1 A)(\epsilon_1 L) + (\sigma_1 A)(\epsilon_2 L) + \frac{1}{2}(\sigma_2 A)(\epsilon_2 L)$$

QUICK NOTE

N = internal axial force in the member due to the applied real load P_1

n = internal axial force in the member due to the applied virtual load P_2

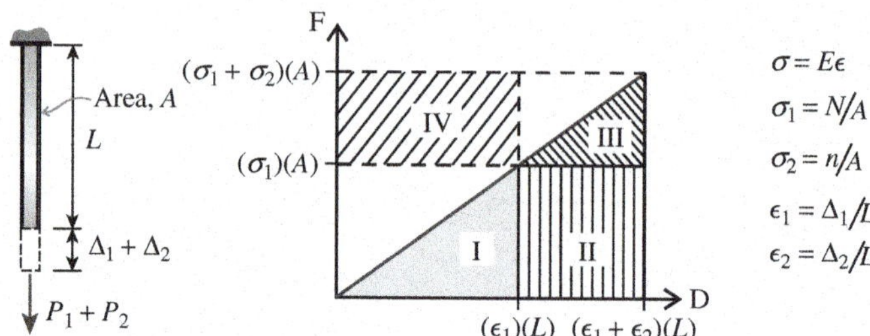

FIGURE 9.5 Depiction of internal strain energy in an axial bar.

This would mean that the complementary virtual strain energy represented by region IV is given as

$$\delta W_i^* = (\sigma_2 A)(\epsilon_1 L)$$

Back substituting expressions for σ, ϵ, and Hooke's law, the expression can further be modified into the following form.

$$\delta W_i^* = \left(\frac{n}{A}\right)(A)\left(\frac{\sigma_1}{E}\right)(L) = \left(\frac{n}{A}\right)(A)\left(\frac{N/A}{E}\right)(L)$$

$$\delta W_i^* = \frac{nNL}{AE}$$

QUICK NOTE

The expression for internal strain energy looks very familiar to the well-known mechanics expression that states that the deflection in an axial bar may be calculated by

$$\Delta = \frac{NL}{AE}$$

Students often desire to stick with this formulation instead of using the virtual work formulation. For a single truss member, this may be wise. However, virtual work really improves the ease of calculation when multiple members are considered, as is the case in a truss.

With an expression for the internal virtual strain energy in hand, the conservation of energy principle may be used to complete the virtual work method for trusses. The derivation of the internal strain energy was done for a single truss member. However, it is readily applied to an entire truss by considering the strain energy in all truss members. This is done by summing up all of the effects.

$$\delta W_e^* = \delta W_e^*$$

$$P_2\Delta_1 = \sum_{\text{all } i} \frac{n_i N_i L_i}{A_i E_i}$$

where the subscript i denotes the truss member and N, n, L, A, and E are the real bar force, virtual bar force, bar length, bar cross-sectional area and modulus of elasticity, respectively.

9.5 Application of Virtual Work to Trusses

The general form of the virtual work equation allows for the calculation of deflection at any location and in any direction within a truss. A basic procedure is outlined here.

DID YOU KNOW?

The virtual load P_2 can be any value because it is divided out in the last step. However, it is most common and convenient to assume the virtual load to be 1.0. This is why the method is sometimes referred to as the *unit dummy load method.* All examples and solutions in this text will assume the virtual load to be 1.0.

Procedure

1. Calculate the internal axial force in *EVERY* member of the truss due to the real loads applied on the truss (N_i).
2. Remove all of the real loads from the truss and place a unit virtual load (1.0) at the joint and in the direction of the desired deflection.
3. Calculate the axial force in *EVERY* member due to the applied virtual load (n_i).
4. Use the equation of virtual work to calculate the deflection.

$$\Delta = \sum_{\text{all } i} \frac{n_i N_i L_i}{A_i E_i} / (1.0)$$

Examples 9.1 and 9.2 illustrate the application of virtual work to trusses. To simplify the numerous multiplications, a table is used. The modulus of elasticity is carried through as a constant until the summation is made for all of the members, at which time its numerical value is used. Should there be members of different E values, it is necessary that their actual or their relative values be used for the individual multiplications. A positive value of $\sum(nNL/AE)$ indicates a deflection in the direction of the unit load.

EXAMPLE 9.1

Use the method of virtual work to determine the horizontal (Δ_{Ex}) and vertical (Δ_{Ey}) components of deflection at joint E in the truss shown. The cross-sectional areas for each bar are given and E = 29,000 ksi.

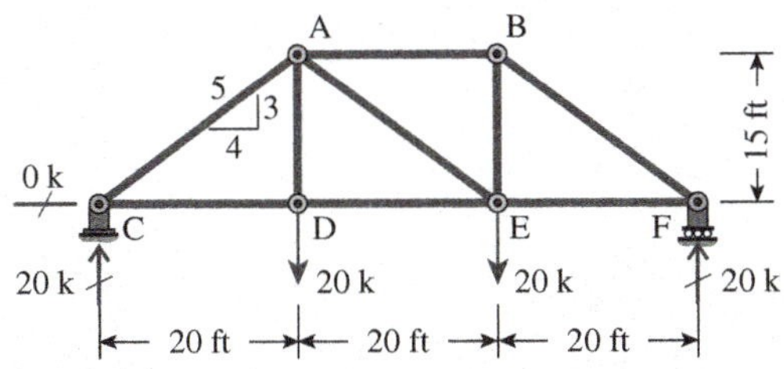

Cross-Sectional Areas:
All horizontal bars = 4.0 in^2
All vertical bars = 2.0 in^2
All diagonal bars = 3.0 in^2

SOLUTION STRATEGY

Analyze the truss for the given loads using the method of joints. This is done because every truss force must be known. Do this for the real (actual) and virtual loads following the procedure as outlined previously. Use a table to make computations easier—a spreadsheet is handy.

SOLUTION

The following figure shows the internal bar forces due to the external loads.

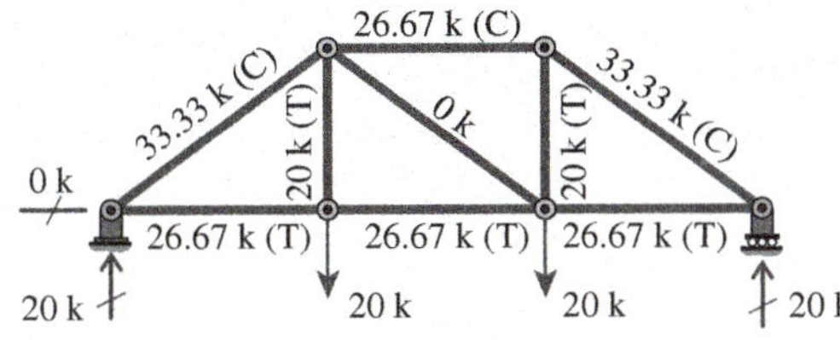

To find the vertical deflection at E (Δ_{Ey}), all of the real loads need to be removed, and a unit virtual load must be applied where the deflection value is wanted—in this case joint E. The following shows the internal axial loads due to the vertical virtual load.

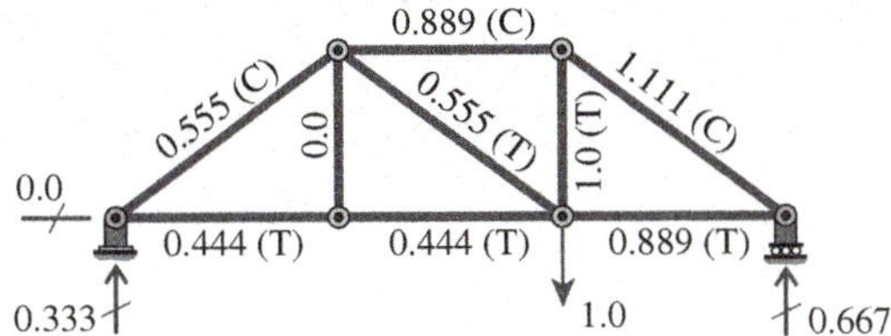

To find the horizontal deflection at E (Δ_{Ex}), all loads need to be removed and a unit virtual load must be applied in the horizontal direction at joint E. The following shows the internal axial loads due to the horizontal load.

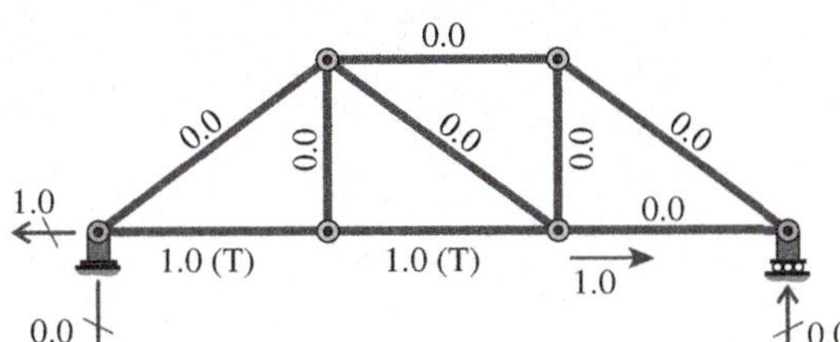

Now place all of the computed values in a table to facilitate the computation of the deflections. The value of E is left out of the table, since it is common to all terms.

DID YOU KNOW?
When placing the unit load at E, the analyst has the choice to point it upward or downward. The typical approach is to point it in the direction where the analyst thinks it will deflect. However, it is not always obvious in which direction it will deflect. Just assume a direction, and the sign on the answer will indicate whether the assumption was correct or not.

FRIENDLY ADVICE
Truss deflection analysis can be quite tedious due to the number of trusses that must be analyzed. It is really to your advantage to look for any zero force members using the guidance found in Chapter 7. For example, all but two of the members in this truss have zero forces and can be identified just by observation—thus saving a lot of calculation time.

Member	L (in.)	A (in^2)	N (k)	n_y	$n_y NL/A$ (k/in.)	n_x	$n_x NL/A$ (k/in.)
CA	300	3.0	−33.33	−0.555	1850	0	0
CD	240	4.0	26.67	0.444	710	1.000	1600
AD	180	2.0	20.00	0.000	0	0	0
AB	240	4.0	−26.67	−0.889	1423	0	0
AE	300	3.0	0	0.555	0	0	0
DE	240	4.0	26.67	0.444	710	1.000	1600
BE	180	2.0	20.00	0.667	1201	0	0
BF	300	3.0	−33.33	−1.111	3703	0	0
EF	240	4.0	26.67	0.889	1423	0	0
					$\sum = 11{,}020$		$\sum = 3200$

COMMON MISTAKE

Students often fail to use consistent units when assembling the table—particularly the length.

$$\Delta_{Ey} = \frac{11{,}020 \text{ k/in.}}{E} = \frac{11{,}020 \text{ k/in.}}{29{,}000 \text{ ksi}} = +0.38 \text{ in.} = \underline{\underline{0.38 \text{ in.} \quad \downarrow}}$$

$$\Delta_{Ex} = \frac{3200 \text{ k/in.}}{E} = \frac{3200 \text{ k/in.}}{29{,}000 \text{ ksi}} = +0.11 \text{ in.} = \underline{\underline{0.11 \text{ in.} \quad \rightarrow}}$$

SUMMARY

Since both displacements are computed to be positive, they are in the direction of the applied unit loads.

EXAMPLE 9.2

Use the method of virtual work to determine the vertical component of deflection of joint E in the truss shown. The cross-sectional areas of each member are listed, and the modulus of each member is taken as $E = 29{,}000$ ksi.

Members	Area (in^2)
AB, DE, CG	3.0
AF, FG	2.0
BF, DH, GH, HE	1.0
BC, CD	4.0
BG, GD	1.5

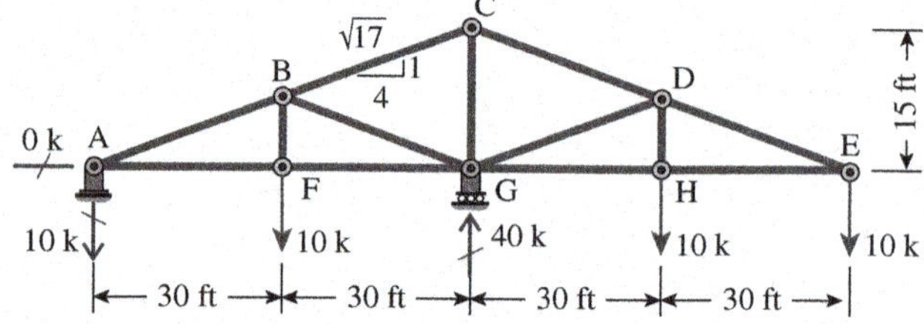

SOLUTION STRATEGY

Analyze the truss for the given loads using the method of joints. This is done because every truss force must be known. Do this for the real (actual) and virtual loads following the procedure as outlined previously. Use a table to make computations easier—a spreadsheet is handy for this.

SOLUTION

The following figure shows the internal bar forces due to the external loads.

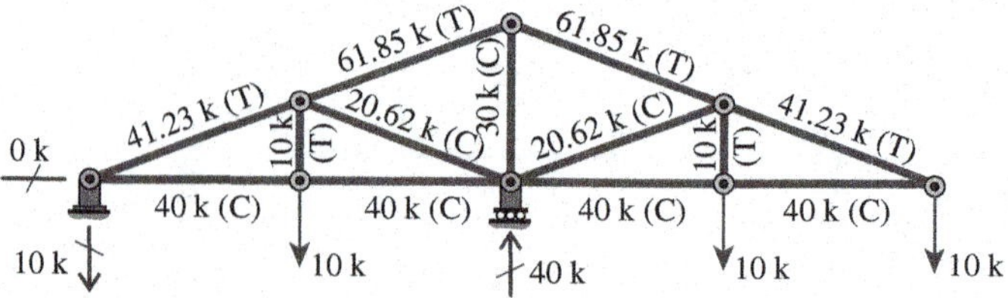

Remove the actual (real) loads from the truss and place a unit load in the vertical direction at joint E.

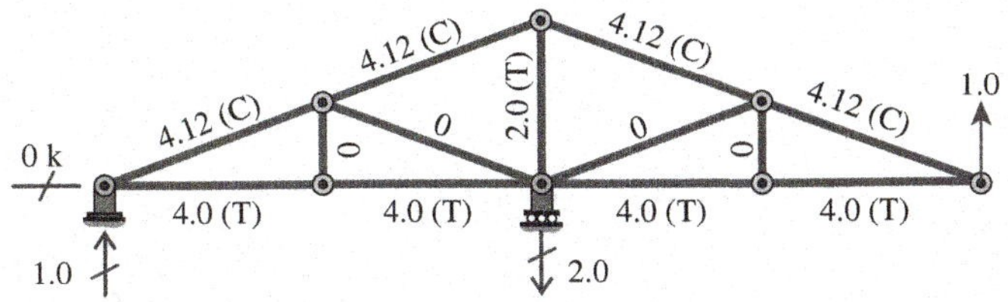

Member	L (in.)	A (in^2)	N (k)	n	nNL/A (k/in.)
AB	371.1	3.0	41.23	−4.12	−21013
BC	371.1	4.0	61.85	−4.12	−23641
CD	371.1	4.0	61.85	−4.12	−23641
DE	371.1	3.0	41.23	−4.12	−21013
AF	360.0	2.0	−40.00	4.00	−28800
FG	360.0	2.0	−40.00	4.00	−28800
GH	360.0	1.0	−40.00	4.00	−57600
HE	360.0	1.0	−40.00	4.00	−57600
BF	90.0	1.0	10.00	0.00	0
BG	371.1	1.5	−20.62	0.00	0
CG	180.0	3.0	−30.00	2.00	−3600
GD	371.1	1.5	−20.62	0.00	0
DH	90.0	1.0	10.00	0.00	0
					$\sum = -265{,}707$

Assemble all computed values into a table to facilitate the calculations.

$$\Delta_{Ey} = \frac{-265{,}707\text{ k/in.}}{E} = \frac{-265{,}707\text{ k/in.}}{29{,}000\text{ ksi}} = -9.16\text{ in.} = \underline{\underline{9.16\text{ in.}}} \downarrow$$

SUMMARY

Since the resulting deflection is negative, the assumed direction for the virtual load was incorrect.

DID YOU KNOW?

The virtual load applied to the structure does not have any units. However, units may be used because when solving the deflection equation one must divide again by the virtual load—including the units. Hence, it all works out in the end.

QUICK NOTE

In this example, notice that all of the numbers in the right-most column have the same sign. This does not have to be the case. It is possible to have some energy terms that add to the total deflection while other energy terms subtract from the total deflection. Just be methodical with the analysis of the truss and simply trust the signs on the numbers.

9.6 Deflections and Angle Changes of Beams and Frames

The principle of virtual work may be readily extended beyond trusses to other kinds of structures such as beams and frames. Since the expression and notion of external work does not depend on structure type, the only change to the virtual work equation previously introduced is the expression for internal virtual strain energy. Displacements within beams and frames are dominated by the deformations in the members due to bending (i.e., moment). Shear and axial deformations can, and often do, exist simultaneously with bending deformations. However, for the majority of beams and frames, the contribution is minimal when compared with bending deformations. Thus, this section introduces the strain energy term for bending deformations.

Consider the beam subjected to an arbitrary loading shown in Figure 9.6(a). Due to the applied load, the beam will deform. This deformation will then cause a deflection to occur at point C. The deformation of the beam is more easily seen if an infinitesimal element of the beam of length dx is cut out from the beam as shown in Figure 9.6(b). The internal moments that exist on the element are shown along with the resulting curvature. The relative rotation between the two moments is taken as $d\theta$. In Section 8.6, the relationship between M and $d\theta$ was developed and is shown again as

$$\frac{d^2y}{dx^2} = d\theta = \frac{M}{EI}$$

This expresses the real deformation within the beam member for the segment of length dx.

DID YOU KNOW?

Axial deformations in frames generally don't play an important part. They may be included, however, in the same fashion as for trusses.

For constant axial force,

$$\delta W_i^* = \frac{nNL}{AE}$$

For varying axial force,

$$\delta W_i^* = \int \frac{n(x)N(x)}{AE}dx$$

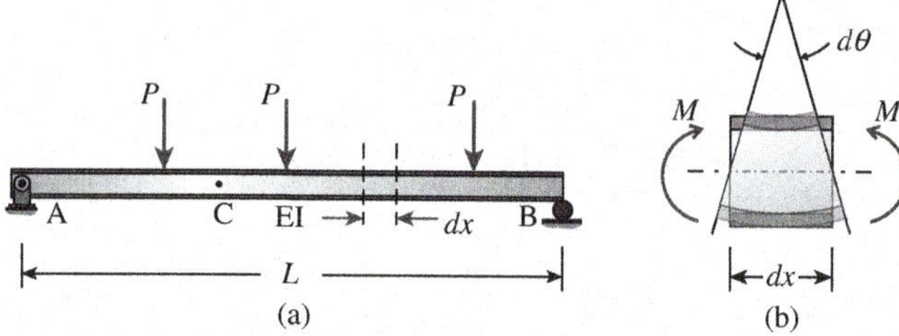

FIGURE 9.6 Beam subjected to an arbitrary real loading.

QUICK NOTE

Just as before, the virtual load is taken as a unit load and is added to the structure at the location and in the direction of the desired deflection.

QUICK NOTE

The well-known expression for calculating work is taking a force times a distance ($F\Delta$). A less known—but equally valid—expression is taken as a moment times a rotation ($M\theta$).

DID YOU KNOW?

To find the rotation at a location on a beam or frame, one simply needs to apply a unit moment at the point and in the direction of the desired rotation, so

$$1 \cdot \theta = \int_0^L \frac{mM}{EI} dx$$

where m in this case is the internal moment in the structure due to the virtual unit moment applied at the point of interest.

CAUTION

Moment equations must be developed in the same fashion as presented in Section 6.3. These equations are often piecewise, depending on the presence of load discontinuities. **Do NOT integrate over discontinuities in the moment equations.** Know the valid domain of each equation.

Now consider that the deflection at point C is desired. If a unit virtual load is applied at point C, as shown in Figure 9.7(a), the resulting internal moment at the segment dx is taken as m. The internal virtual strain energy $[d(\delta W_i^*)]$ is found for segment dx by multiplying the real deformation $M/(EI)$ by the virtual moment at that location m, giving the expression

$$d(\delta W_i^*) = (m)\left(\frac{M}{EI}\right)$$

where m is the internal moment due to the virtual load $P_2 = 1$ and M is the internal moment due to the actual load. The total internal strain energy is found by considering the contribution of strain energy along the entire length of the beam. This results in the integral as

$$\delta W_i^* = \int d(\delta W_i^*) = \int (m)\left(\frac{M}{EI}\right) dx$$

where m and M are both functions of x.

With the expression for internal strain energy due to bending, the virtual work equation can be completed. Thus,

$$1 \cdot \Delta = \int_0^L \frac{mM}{EI} dx$$

where $\int_0^L$ means to integrate over the entire structure. To do this one may need to perform piecewise integration.

Procedure

1. Identify appropriate coordinate system(s) for each beam/frame segment. Select these to keep the moment equations as simple as possible.
2. Calculate the internal moment equation due to the actual load, $M(x)$. This should be an equation of x.
3. Remove the actual load and place a unit load/moment at the location and in the direction of the desired displacement/rotation.
4. Calculate the internal moment equation, $m(x)$, due to the virtual load/moment that was applied.
5. Use the virtual work equation for finding deflection/rotation.

The next set of examples demonstrates the calculation of both deflections and rotations within beams and frames. Pay attention to the coordinate systems indicated. These are generally selected to keep the moment equations short and the subsequent integrations simple.

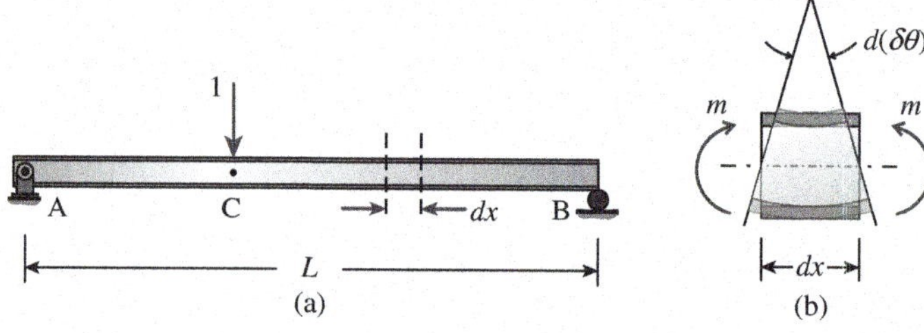

FIGURE 9.7 Beam subjected to a virtual load at C.

EXAMPLE 9.3

Use virtual work to determine the deflection (Δ_B) and rotation (θ_B) at point B for the beam shown here. $E = 29{,}000$ ksi and $I = 5000$ in^4.

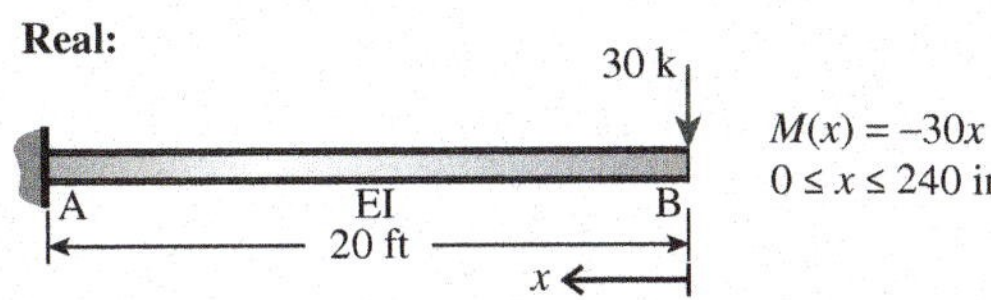

$M(x) = -30x$
$0 \le x \le 240$ in.

QUICK NOTE

If the origin of x is placed at A, the moment equation would be

$M(x) = -7200 + 30x$ (units: k·in.)

The origin selection does not have any influence on the final answer—just the intermediate steps.

SOLUTION STRATEGY

Select the right end of the beam as the origin of the moment equations. Use the procedure outlined in Chapter 6 for finding the moment equations. Follow the procedure outlined previously, making sure consistent units are used throughout.

SOLUTION

With the moment equations already given, remove the real load, apply a unit vertical load at point B, and solve for the moment equation m. Repeat the process for a unit moment being placed at B.

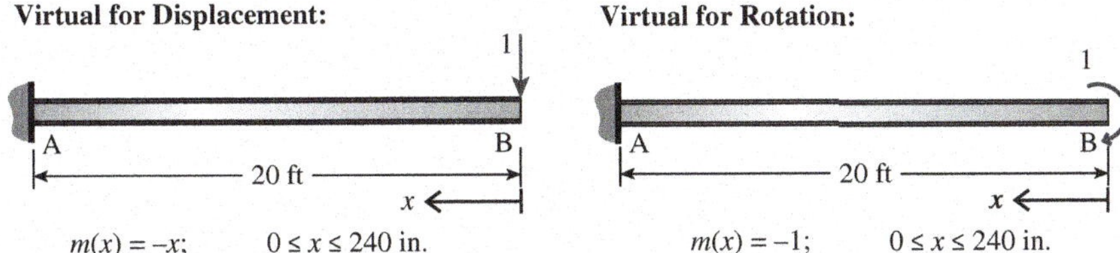

$m(x) = -x$; $0 \le x \le 240$ in. $m(x) = -1$; $0 \le x \le 240$ in.

Use the virtual work equations to solve for each requested quantity.

$$1 \cdot \Delta_B = \int_0^L \frac{mM}{EI}dx = \int_0^{240} \frac{(-x)(-30x)}{EI}dx = \left[\frac{10x^3}{EI}\right]_0^{240}$$

$$= \frac{(10\text{ k})(240\text{ in.})^3}{(29{,}000\text{ ksi})\left(5000\text{ in}^4\right)} = +0.953\text{ in.} = \underline{\underline{0.953\text{ in.}\ \downarrow}}$$

$$1 \cdot \theta_B = \int_0^L \frac{mM}{EI}dx = \int_0^{240} \frac{(-1)(-30x)}{EI}dx = \left[\frac{15x^2}{EI}\right]_0^{240}$$

$$= \frac{(15\text{ k})(240\text{ in.})^2}{(29{,}000\text{ ksi})\left(5000\text{ in}^4\right)} = +0.006\text{ rad} = \underline{\underline{0.006\text{ rad}\ \circlearrowright}}$$

QUICK NOTE

The same origin for x must be used for both the real and virtually loaded structures.

EXAMPLE 9.4

Use virtual work to find the deflection at point B in the beam shown. $E = 29{,}000$ ksi, $I = 1000$ in^4.

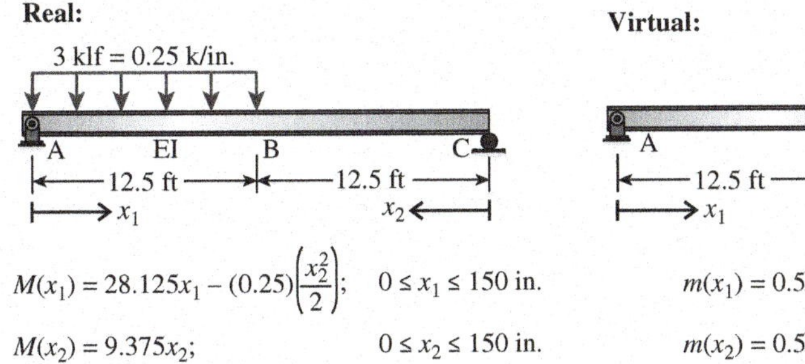

Virtual:

1
A B C
12.5 ft 12.5 ft
x_1 x_2

$M(x_1) = 28.125x_1 - (0.25)\left(\frac{x_1^2}{2}\right)$; $0 \le x_1 \le 150$ in.

$M(x_2) = 9.375x_2$; $0 \le x_2 \le 150$ in.

$m(x_1) = 0.5x_1$; $0 \le x_1 \le 150$ in.

$m(x_2) = 0.5x_2$; $0 \le x_2 \le 150$ in.

SOLUTION STRATEGY

Follow the procedure outlined previously. This beam has two segments: (AB) and (BC). Look for coordinates that keep the moment equations simple.

COMMON MISTAKE

Students who don't have all of their input (e.g., moment equations, integration limits, E, and I) put into a consistent set of units (e.g., kips and inches) at the beginning of the problem have a very high probability of performing an incorrect conversion at the end of the problem. Choose your units up front and use those units for everything.

SOLUTION

It is necessary to write one expression of M for segment AB and one for segment BC. Select A and C to be the origins, respectively. Remove the real load and replace it with a unit virtual load at B, then find the expressions for m.

Use the virtual work equation to solve for the unknown displacement at B.

$$1 \cdot \Delta_B = \int_0^{150} \frac{(0.5x_1)\left(28.125x_1 - 0.125x_1^2\right)}{(29{,}000\text{ ksi})(1000\text{ in}^4)} dx_1 + \int_0^{150} \frac{(0.5x_2)(9.375x_2)}{(29{,}000\text{ ksi})(1000\text{ in}^4)} dx_2$$

$$\therefore \Delta_B = +0.455\text{ in.} = \underline{\underline{0.455\text{ in.}}} \quad \downarrow$$

EXAMPLE 9.5

Use virtual work to determine the horizontal deflection at D (Δ_{Dx}) and the rotation at B (θ_B). Assume that each member has I = 1250 in^4 and E = 29,000 ksi.

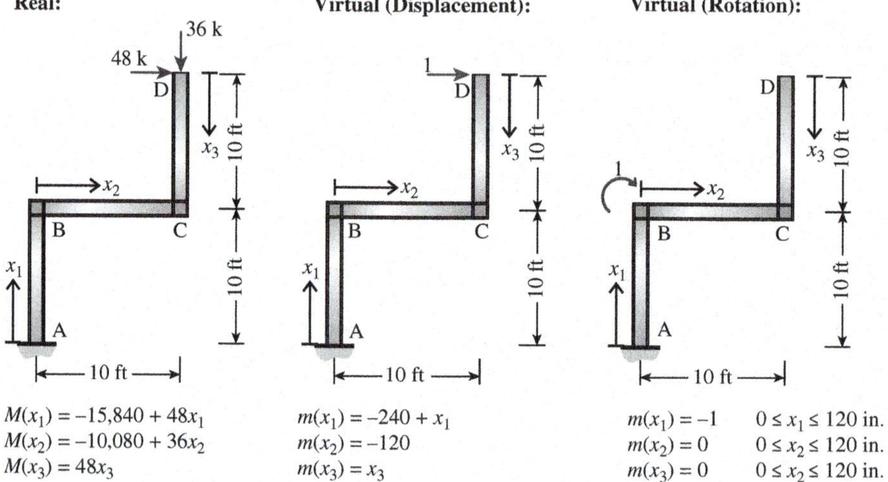

$M(x_1) = -15{,}840 + 48x_1$
$M(x_2) = -10{,}080 + 36x_2$
$M(x_3) = 48x_3$

$m(x_1) = -240 + x_1$
$m(x_2) = -120$
$m(x_3) = x_3$

$m(x_1) = -1$ $0 \le x_1 \le 120$ in.
$m(x_2) = 0$ $0 \le x_2 \le 120$ in.
$m(x_3) = 0$ $0 \le x_2 \le 120$ in.

SOLUTION STRATEGY

Review the analysis of frames discussed in Section 6.7 and the process for finding the moment equation in Section 6.3. Draw a separate FBD for each member and use the FBD to develop the equations.

QUICK NOTE

The total deflection at D is due to deformation in all three members. The intermediate evaluation of the virtual work equation as shown highlights the contribution each member makes. This reveals that, if one needs to reduce the overall deflection, increasing the I_{AB} value will have the biggest impact.

SOLUTION

Use the virtual work equations to solve for the requested displacement and rotation. All units are in kips and inches.

$$1 \cdot \Delta_{Dx} = \int_0^{120} \frac{(-240 + x_1)(-15{,}840 + 48x_1)}{EI} dx_1$$

$$+ \int_0^{120} \frac{(-120)(-10{,}080 + 36x_2)}{EI} dx_2 + \int_0^{120} \frac{(x_3)(48x_3)}{EI} dx_3$$

$$= 7.913\text{ in.} + 3.146\text{ in.} + 0.763\text{ in.} = \underline{\underline{11.822\text{ in.}}} \quad \rightarrow$$

$$1 \cdot \theta_B = \int_0^{120} \frac{(-1)(-15{,}840 + 48x_1)}{EI} dx_1$$

$$+ \int_0^{120} \frac{(0)(-10{,}080 + 36x_2)}{EI} dx_2 + \int_0^{120} \frac{(0)(48x_3)}{EI} dx_3$$

$$= 0.0\text{ rad} + 0.0\text{ rad} + 0.043\text{ rad} = \underline{\underline{0.043\text{ rad}}} \quad \circlearrowright$$

FRIENDLY ADVICE

Review the principles of statics and frame analysis. When the virtual moment is applied to joint B, many are surprised to be told that the moment in members BC and CD are zero. However, statics reveals this to be the case.

9.7 Application of Virtual Work Using Visual Integration

While virtual work is a very versatile method for calculating deflections, some see it as being an unfriendly method due to the need to develop moment equations and perform integrations. For this reason, many still encourage the learning of some of the geometric methods mentioned in Chapter 8 (such as moment-area and conjugate beam). The authors maintain that it is more useful to focus on the one method of virtual work and learn it well. However, the virtual work procedure can be sped up in many instances for moderate to simple beams and frames by using a different technique for integration.

Visual integration is a geometric technique for performing integrations. With a basic understanding of simple geometric shapes, one can avoid the need to develop moment equations and then perform a *conventional* integration. The concept of visual integration is not new. For example, consider that you were asked to calculate the integral of the force–displacement curve between the bounds of 0 and Δ shown in Figure 9.8. The resulting equation would look like

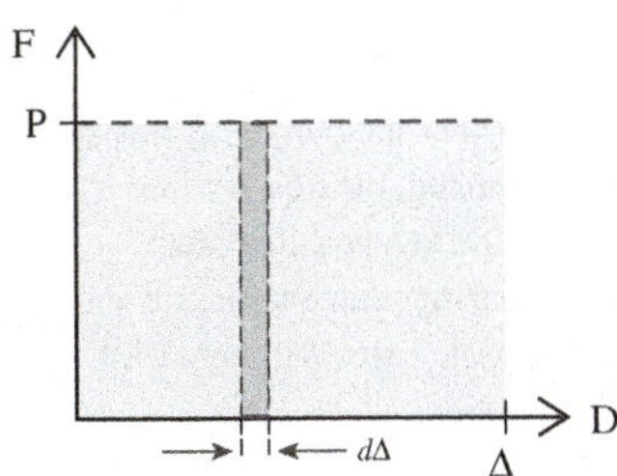

FIGURE 9.8
Force–displacement curve.

$$A = \int_0^{\Delta} P d\Delta = P\Delta$$

However, many do not have to go to such great effort to perform the integral. Rather, knowing that the bounded integral provides nothing more than the area under the curve between 0 and Δ, one may simply multiply the base (Δ) by the height (P) of the rectangle. This results in the same answer as previously—$P\Delta$—but done in a much quicker fashion. In other words, the integral was visually (i.e., geometrically) integrated. It is this same concept that is applied when looking at the virtual work equation.

As a reminder, the virtual work equation that considers the effects of the bending moment is

$$1 \cdot \Delta = \int_0^L \frac{mM}{EI} dx$$

The visual (geometric) integration of this is a little more involved than the previous illustration, but with a little practice, it is readily remembered and applied—saving much time over conventional integration in many cases. The virtual work equation using visual integration is

$$1 \cdot \Delta = \sum A_{\frac{M}{EI}} \left[m\left(\bar{x}\right) \right]$$

where $A_{\frac{M}{EI}}$ = the area(s) under the $\frac{M}{EI}$ diagram
$m(\bar{x})$ = the value of m located at the centroid of the $\frac{M}{EI}$ diagram
$\sum$ indicates to sum the terms for as many areas as exist

QUICK NOTE

The same equation works when looking for the rotation. The only difference is that m is the moment that results from a unit moment and not a unit force.

Procedure

1. Draw the $\frac{M}{EI}$ diagram for the structure subjected to the actual loading. This is done by generating moment diagrams as usual and then dividing by the quantity EI, keeping in mind that different values of EI may exist in different regions of the structure.
2. Remove the real load and apply a unit force or unit moment at the location and in the direction of the desired deflection or rotation, respectively.
3. Draw the m diagram for the virtual structure.
4. Note any discontinuities in either the $\frac{M}{EI}$ or m diagrams.
5. Calculate the area(s) under the sections of the $\frac{M}{EI}$ bounded by the discontinuities identified in the previous step. This is facilitated by referencing the basic geometric properties shown in Appendix D.
6. Locate the centroids ($\bar{x}$) of each of the areas found in the previous step. This is facilitated by referencing the basic geometric properties shown in Appendix D.

FURTHER GUIDANCE

- Remember that you cannot integrate over discontinuities using conventional integration. The rules don't change when doing visual integration. *Be conscientious* of the discontinuities.
- Calculating areas and centroid locations is made easier if the geometric shapes are kept simple. One way to keep the shapes simple is to generate the $\frac{M}{EI}$ diagrams using the *superposition method* introduced in Section 6.8.

7. Calculate the value of m at the location of each $\bar{x}$.
8. Use the summation equation previously introduced to find the displacement or rotation.

EXAMPLE 9.6

Rework the problem given in Example 9.3. This time use visual integration instead of conventional integration for the evaluation of the virtual work problem.

SOLUTION STRATEGY

Moment diagrams for cantilever beams should be generated easily. Refer back to Section 6.8 where moment diagrams for basic loadings on a cantilever beam are given. Follow the procedure outlined above.

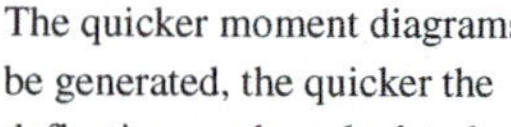

QUICK NOTE
The quicker moment diagrams can be generated, the quicker the deflection can be calculated. Memorizing moment diagrams for a few basic cases will help a lot. See Section 6.8.

SOLUTION

The moment diagrams for both the real load and the virtual loads shown in Example 9.3 are shown here with all of the appropriate areas and centroids labeled.

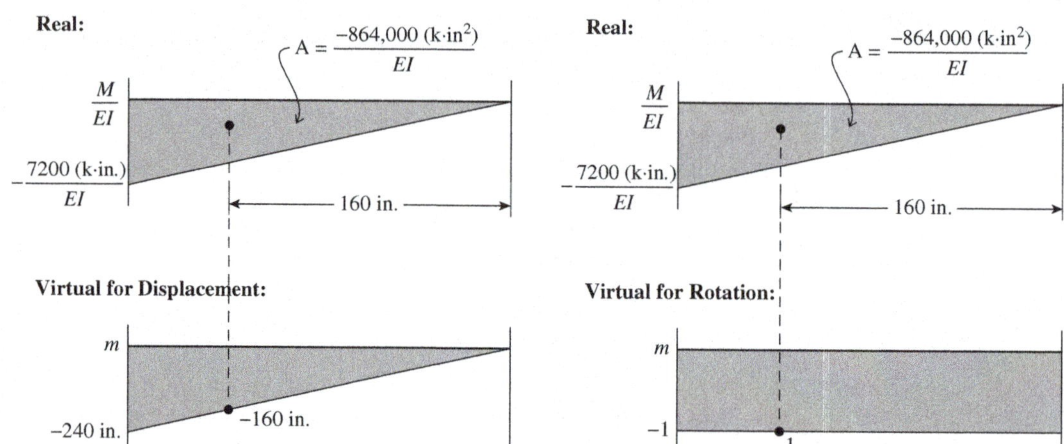

QUICK NOTE
The units of the little m diagrams look strange at first glance because they don't seem to agree with our understanding of moment. One must remember that the unit force and the unit moment applied in each case are dimensionless. Therefore, calculating the moment at A due to the unit force results in (1)(240 in.) = 240 in.

To calculate the deflection at B, there is only one area to consider because there are no load discontinuities between A and B. The area under the $\frac{M}{EI}$ diagram is

$$A_{\frac{M}{EI}} = -\frac{1}{2}\left(\frac{7200\ (\text{k·in.})}{EI}\right)(240\ \text{in.}) = -\frac{864{,}000\ (\text{k·in}^2)}{EI}$$

Thus, the deflection at B is calculated as

$$1 \cdot \Delta_B = \left(\frac{-864{,}000\ (\text{k·in}^2)}{(29{,}000\ \text{ksi})\ (5000\ \text{in}^4)}\right)(-160\ \text{in.}) = +0.953\ \text{in.} = \underline{\underline{0.953\ \text{in.}\ \downarrow}}$$

DID YOU KNOW?
The $\frac{M}{EI}$ diagram for the real load is used for both the deflection and rotation calculations. Usually only one sketch is made for this diagram. Two are presented here to make clear how the centroid of the $\frac{M}{EI}$ diagram maps to the m value. Future examples will not do this.

The rotation at B can be calculated in a similar fashion as

$$1 \cdot \theta_B = \left(\frac{-864{,}000\ (\text{k·in}^2)}{(29{,}000\ \text{ksi})\ (5000\ \text{in}^4)}\right)(-1) = +0.006\ \text{rad} = \underline{\underline{0.006\ \text{rad}\ \circlearrowright}}$$

SUMMARY

The results here are the same as in Example 9.3, yet no equations for the moments were needed.

EXAMPLE 9.7

Rework Example 9.4 using visual integration instead of conventional integration.

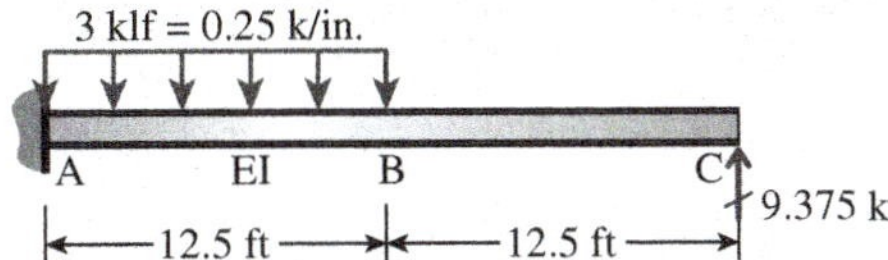

RECALL

In the superposition method, one must solve for all reactions and then choose one point on the beam/frame member to fix. Thus, $R_{Cy} = 9.375$ k for the beam of Example 9.4. Since A is selected as the point of fixity, R_{Cy} becomes an applied load at C.

SOLUTION STRATEGY

When sketching the $\frac{M}{EI}$ diagram, try to keep the moment diagrams as simple as possible so that areas and centroids can be identified easily. Use the principle of superposition presented in Section 6.8. Follow the rest of the procedure as outlined, and be careful to identify any discontinuities.

QUICK NOTE

Notice how the discontinuity at B in the little *m* diagram is projected to all other diagrams. This is to ensure that the areas are divided up according to this discontinuity.

SOLUTION

For the real load, use superposition by cantilevering at point A. This will result in two $\frac{M}{EI}$ diagrams—one for the distributed load and one for the force (reaction) at C.

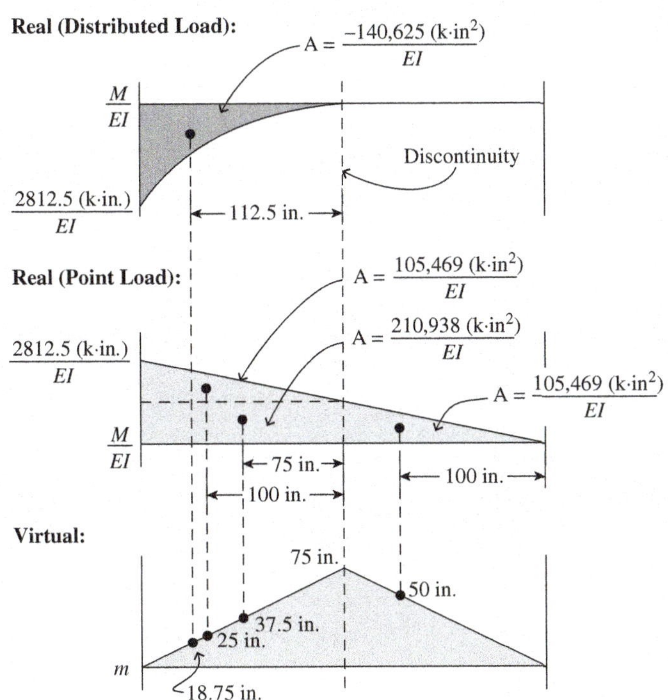

DID YOU KNOW?

One may look up formulas for finding areas and centroids of trapezoids. However, the authors recommend that if trapezoidal shapes are encountered they be broken up into rectangles and triangles. An alternative way of breaking up the one in this example is shown here.

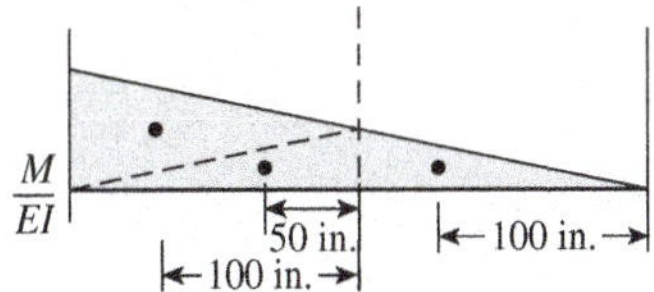

There are four separate areas that must be considered. This is where the summation comes into play.

$$1 \cdot \Delta_B = \left(\frac{-140{,}625\ (\text{k·in}^2)}{(29{,}000\ \text{ksi})(1000\ \text{in}^4)}\right)(18.75\ \text{in.})$$

$$+\left(\frac{210{,}938\ (\text{k·in}^2)}{(29{,}000\ \text{ksi})(1000\ \text{in}^4)}\right)(25\ \text{in.})$$

$$+\left(\frac{421{,}875\ (\text{k·in}^2)}{(29{,}000\ \text{ksi})(1000\ \text{in}^4)}\right)(37.5\ \text{in.})+\left(\frac{210{,}938\ (\text{k·in}^2)}{(29{,}000\ \text{ksi})(1000\ \text{in}^4)}\right)(50\ \text{in.})$$

$$= +0.455\ \text{in.} = \underline{\underline{0.455\ \text{in.}\ \downarrow}}$$

QUICK NOTE

When verbally communicating, the terms *big M* and *little m* are often used to denote the $\frac{M}{EI}$ and virtual diagram, respectively.

FRIENDLY ADVICE

Brushing up on the math skill of working with similar triangles can make the calculation of $m(\bar{x})$ very quick and easy. If you struggle working with similar triangles, you may find visual integration less appealing.

SUMMARY

The results obtained by using visual integration are identical to results found using conventional integration. At first glance, the visual integration approach may look like a cumbersome method, but with a little practice, you will find it is very easy and quick to use.

EXAMPLE 9.8

Using virtual work with the visual integration technique, find the rotation at D for the frame shown. $E = 29{,}000$ ksi, $I_1 = 2000\ \text{in}^4$, and $I_2 = 1500\ \text{in}^4$.

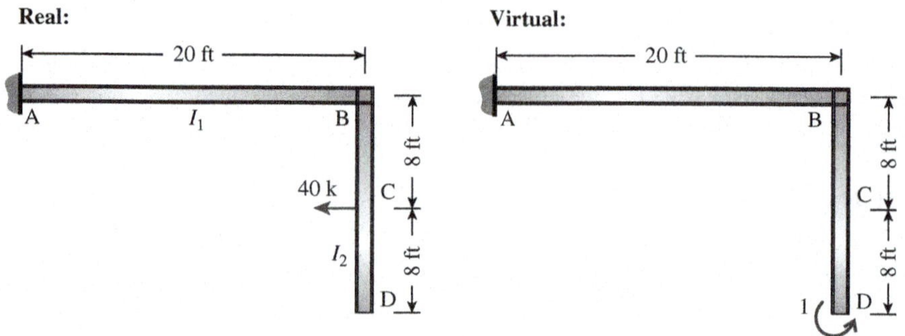

SOLUTION STRATEGY

The biggest challenge of this problem is to develop the moment diagrams. Break up the frame into members AB and BD. The moment diagrams are readily developed from this point. Follow the procedure as previously outlined.

SOLUTION

Since the rotation at D is desired, a unit moment is placed at point D on the virtual structure. We assume it rotates counterclockwise. The resulting moment diagrams, areas and centroidal values are shown below.

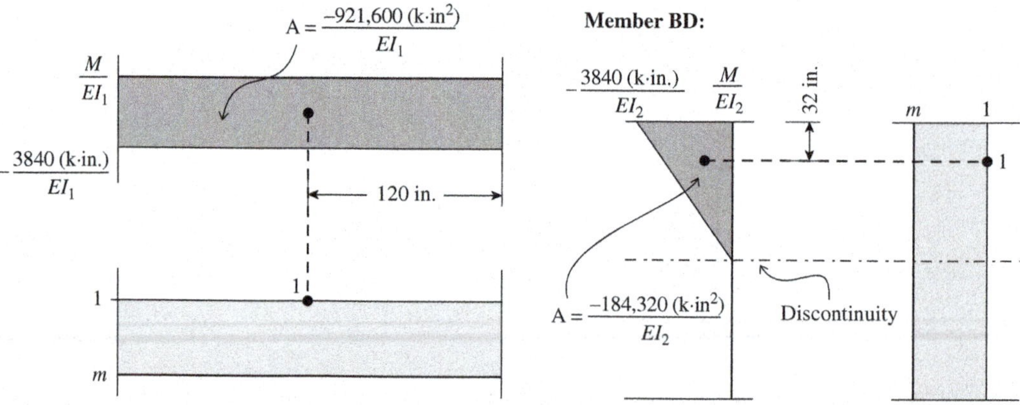

Sum up the areas multiplied by their respective m value.

QUICK NOTE

Note the different values of I used in the calculations.

$$1 \cdot \theta_D = \left(\frac{-921{,}600\ (\text{k·in}^2)}{(29{,}000\ \text{ksi})(2000\ \text{in}^4)}\right)(1) + \left(\frac{-184{,}320\ (\text{k·in}^2)}{(29{,}000\ \text{ksi})(1500\ \text{in}^4)}\right)(1)$$

$$\theta_D = -0.0201\ \text{rad} = \underline{\underline{0.0201\ \text{rad}\ \circlearrowright}}$$

SUMMARY

The negative number indicates that the assumed direction of rotation (counterclockwise) was incorrect.

9.8 Application of Virtual Work to Springs

Times may arise when the ideal assumptions dictated by roller, pin, and fixed supports may not be adequate to capture the structural behavior. Elastic springs are often a useful way to capture the fact that a support or connection is not perfectly rigid or perfectly unrestrained.

Figure 9.9 shows the symbols that are used in this text. The force–deformation relationship for a translational spring is well known, but the moment rotation relationship for a rotational spring is not. Both relationships are presented here.

$$N = (k_T)(\Delta) \qquad M = (k_R)(\theta)$$

Where N = the axial force in the translational spring
M = the internal moment in the rotational spring
Δ = extension/contraction of the translational spring
θ = relative rotation between both ends of the rotational spring
k_T = stiffness of translational spring (units: force/length)
k_R = stiffness of rotational spring (units: (force·length)/rad)

Translational spring:

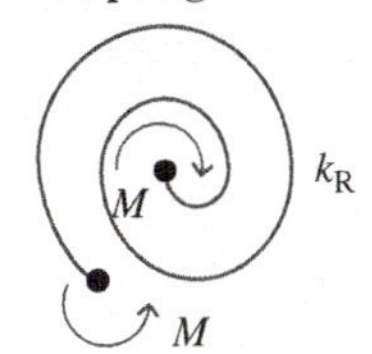

Rotational spring:

FIGURE 9.9 Illustrations of two types of springs.

To be able to include the contribution of springs to the overall deflection of a structure, we need to know what the internal strain energy terms are. Without any derivation these terms are given as

$$\delta W_i^* = \frac{n_{sp}N_{sp}}{k_T}$$

$$\delta W_i^* = \frac{m_{sp}M_{sp}}{k_R}$$

where N_{sp} = the axial force in the spring due to real load
n_{sp} = the axial force in the spring due to virtual load
M_{sp} = the internal moment in the spring due to real load
m_{sp} = the internal moment in the spring due to virtual load

QUICK NOTE

The quantities, N_{sp}, n_{sp}, M_{sp}, and m_{sp} are all scalar quantities. That means they are NOT equations and hence there is no integration term in the strain energy expression.

To calculate the deflection of a beam or frame that has one or more springs attached, one only needs to add the additional strain energy terms of the springs to the overall virtual work equation. The next two examples demonstrate how these spring terms are included in the expression.

EXAMPLE 9.9

Use virtual work to calculate the displacement at point C for the beam shown. $E = 4000$ ksi, $I = 1200$ in^4, and $k_T = 10$ k/in.

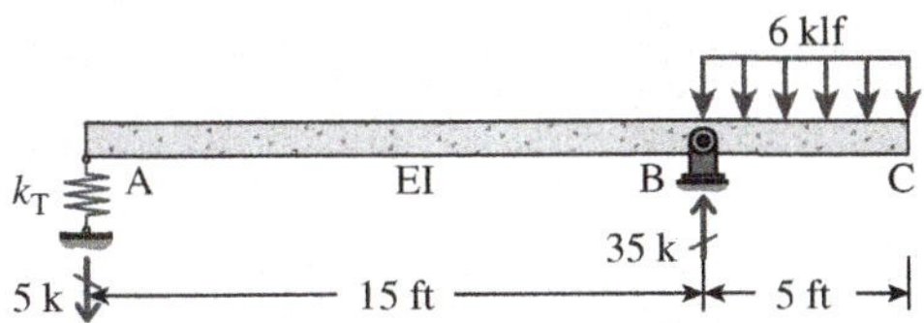

QUICK NOTE

The same rules of statics apply to springs as to they do to other structures. Don't let the squiggly lines scare you. Remember that translational springs can only take an axial load, and a rotational spring can only take a moment.

SOLUTION STRATEGY

Use visual integration for including the beam contribution to the virtual work equation. Solve the beam as you would before. Since there is only one spring, the virtual work equation will only have one additional term.

Virtual:

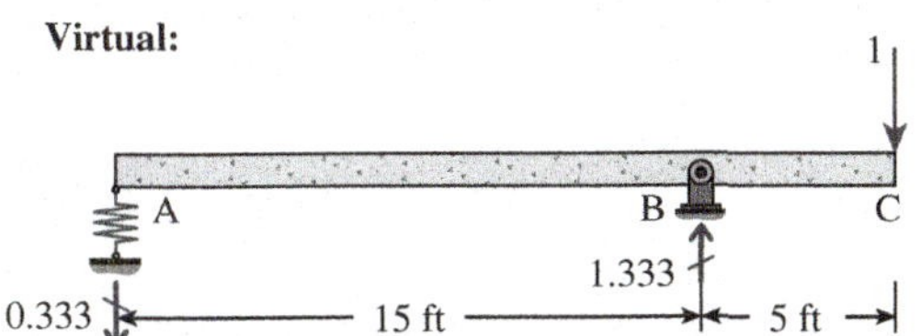

DID YOU KNOW?

If the question had asked for the rotation at C, then a unit moment would be applied at C. This would still produce an axial value in the spring, n_{sp}. Students often think that only applied forces can influence translational springs and only applied moments can influence rotational springs. This is not so.

Generate the $\frac{M}{EI}$ and the m diagrams. Compute the axial force in the translational spring for both cases—N_{sp} and n_{sp}.

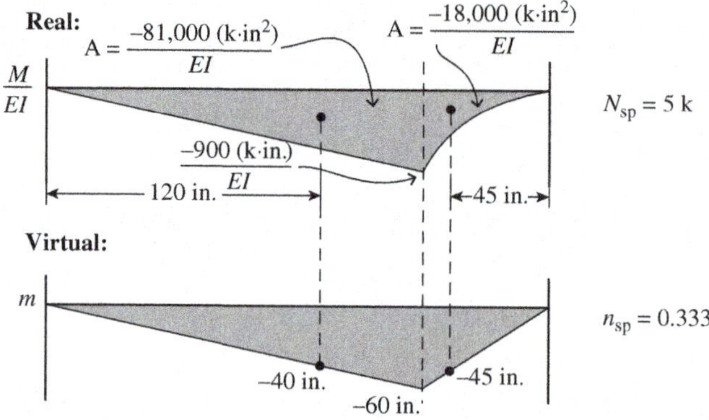

COMMON MISTAKES

- Neglecting the virtual work term for the spring.
- Not accounting for both segments of the beam—AB and BC.
- Substituting the moment equations for the beam into N_{sp} and n_{sp} terms instead of using the internal axial force values.

QUICK NOTE

Notice how the last term in the virtual work equation accounts for the strain energy in the spring. The terms in the numerator will always be scalars. The form of the spring term does not change—*regardless of whether conventional or visual integration is used.*

Use the virtual work equation to solve for the displacement at C.

$$1 \cdot \Delta_C = \left(\frac{-81{,}000\ (\text{k}\cdot\text{in}^2)}{(4000\ \text{ksi})(1200\ \text{in}^4)}\right)(-40\ \text{in.})$$

$$+\left(\frac{-18{,}000\ (\text{k}\cdot\text{in}^2)}{(4000\ \text{ksi})(1200\ \text{in}^4)}\right)(-45\ \text{in.}) + \frac{(5\ \text{k})(0.333)}{10\ \text{k/in.}}$$

$$= 0.675\ \text{in.} + 0.169\ \text{in.} + 0.167\ \text{in.} = +1.01\ \text{in.} = \underline{\underline{1.01\ \text{in.}}} \quad \downarrow$$

SUMMARY

The total deflection at C is a combination of deformation in members AB, BC, and the spring. The spring contribution represented approximately 16% of the total deflection while member AB contributed 67%. Therefore, if one wanted to reduce the total deflection, increasing E or I of member AB would be the most effective approach.

EXAMPLE 9.10

Use virtual work to solve for the vertical deflection at point A for the given beam. $E = 29{,}000$ ksi, $I = 500$ in^4, and $k_R = 40{,}000$ (k·in.)/rad.

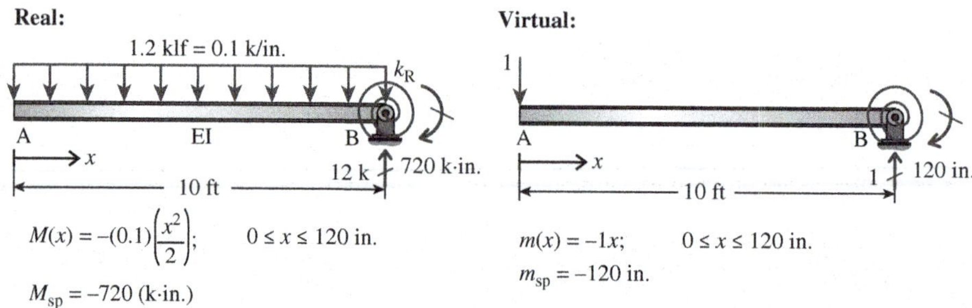

SOLUTION STRATEGY

Use conventional integration for including the beam contribution to the virtual work equation. Solve the beam as you would before. Since there is only one spring, the virtual work equation will only have one additional term.

SOLUTION

For the real load, find the moment equation for the beam, $M(x)$ and also the moment in the spring, M_{sp}. Remove the real load and add a unit virtual force at A. Derive the moment equation for the beam, $m(x)$, and also the internal moment in the spring, m_{sp}. Then use the virtual work equation to solve for the displacement.

$$1 \cdot \Delta_A = \int_0^{120} \frac{(-x)\,(-0.05x^2)}{(29{,}000\text{ ksi})(500\text{ in}^4)}dx + \frac{(-120\text{ in.})(-720\text{ (k·in.)})}{40{,}000\text{ (k·in.)/rad}}$$

$$= 0.18\text{ in.} + 2.16\text{ in.} = +2.34\text{ in.} = \underline{\underline{2.34\text{ in.}}} \quad \downarrow$$

SUMMARY

The displacement at A is due to deformation of the beam and also the deformation of the spring. However, 92% of the displacement is due to the rotational spring. This indicates that stiffening up the spring would have a far greater impact than stiffening up the beam.

COMMON MISTAKE

The virtual work term associated with the spring is often left out of the deflection equation. This is often done because of a lack of comfort the students have in dealing with springs. In this example, if the spring term had been left off, it would have produced a very large error in the estimation of deflection.

9.9 Consideration of Shear Deformations

As mentioned in Section 9.6, bending deformation is generally the largest contributor to overall deflection. However, for any member that is subjected to internal shear, there is a deformation associated with that shear. While the contribution is found to be insignificant for many cases, it can start to have an appreciable influence as the depth of the beam increases relative to the span length of the members.

One may include the effects of shear deformation by integrating the internal strain energy term for shear. The internal strain energy due to shear is given as

$$\delta W_i^* = \int_0^L \frac{vVK}{GA}dx$$

where
- V = the internal shear force due to the real load
- v = the internal shear force due to the virtual load
- G = shear modulus
- A = cross-sectional area of the member
- K = a shape factor which accounts for the shear stress variation on the cross section[1]

SHAPE FACTORS[1]

Shape of Cross Section	K
Solid rectangle	1.20
Solid circle	≈ 1.11
I-section	1.0[a]

[a] A should be used with just the area of the web.

EXAMPLE 9.11

For the cantilever beam, find the deflection at point A using the principle of virtual work. Include the effects of both bending and shear deformations. Assume the cross section to be rectangular. $E = 3600$ ksi, $G = 1500$ ksi, $I = 1150$ in^4, and $A = 40$ in^2.

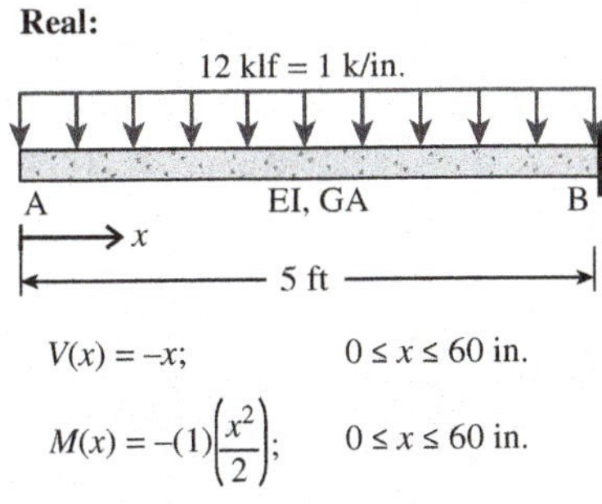

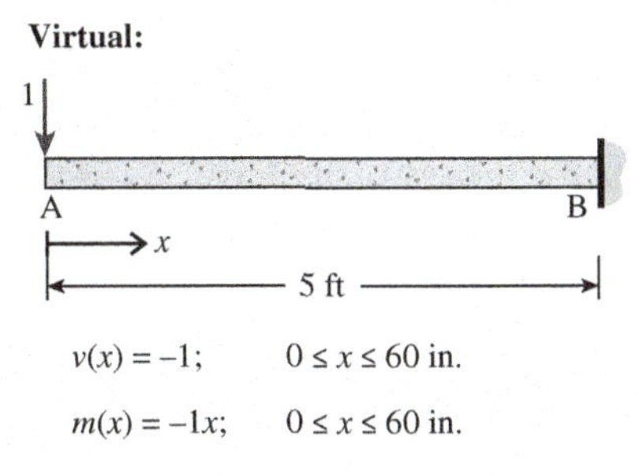

SOLUTION STRATEGY

Instead of just needing the moment equations, the shear equations also need to be developed for both the real and the virtual loads. Include both the bending moment strain energy and the shear strain energy terms in the virtual work equation.

[1] R. D. Cook and W. C. Young, *Advanced Mechanics of Materials*, Prentice Hall, New Jersey, 1999.

QUICK NOTE
The shear deformation contributed 8% to the overall deflection of this beam. As the beam gets shorter, the percentage of shear contribution will increase.

SOLUTION

The shape factor for a rectangle is 1.20. Write the virtual work equation using conventional integration.

$$1 \cdot \Delta_A = \int_0^{60} \frac{(-x)\left(-\frac{x^2}{2}\right)}{(3600\text{ ksi})(1150\text{ in}^4)}dx + \int_0^{60} \frac{(-1)(-x)(1.2)}{(1500\text{ ksi})(40\text{ in}^2)}dx$$

$$= 0.391\text{ in.} + 0.034\text{ in.} = +0.426\text{ in.} = \underline{\underline{0.426\text{ in.}}} \downarrow$$

9.10 SAP2000 Computer Applications

Computer analysis packages like SAP2000 require the user to provide the geometry of the structure, information on loading, cross sectional properties, and material properties—even for the most simple of problems. A natural consequence of this is that not only will it report member forces and reactions but it will also report deflections. It will draw a deflected shape, but it will only report numeric values of deflection and rotation at the nodes. There are two ways to access these numeric values. Once the deflected shape is shown on the screen, place the cursor over one of the nodes. A pop-up window will show the displacements and rotation for that node. The second way is to work through the *DISPLAY/SHOW TABLE* menu.

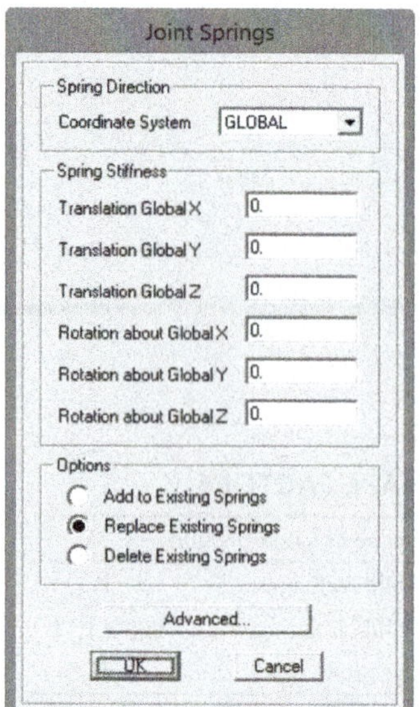

FIGURE 9.10 Spring assignment window in SAP2000.

Springs

The modeling of springs within SAP2000 is done by selecting the joint where the spring is to be added. Then go to the pull-down menu *ASSIGN/JOINT/SPRINGS*. There is an option to place an elastic spring in any of six directions. Three options are for translational springs, and three are for rotational springs (see Figure 9.10).

If a planar structure is modeled in the XZ plane, as is recommended, then translational springs could be assigned in the X or the Z axis. A rotational spring is assigned about the Y axis. The user must be conscientious of default units. Remember that a translational spring has units of (force/length) and a rotational spring has units of (force·length)/rad.

Shear and Axial Deformation

Shear deformations in SAP2000 are included whenever the *shear area* assigned to the cross section is nonzero. When defining the cross-sectional properties of a section, the user is required to set a cross-sectional area (A) and a shear area (SA). SA is calculated from the cross-sectional area and the shape factor (K).

$$SA = \frac{A}{K}$$

To neglect axial deformations within a frame structure, the cross-sectional area for each member may be set to be very large. As a rule of thumb, set the area to about 10,000 times the actual area. Don't go much larger than this or numerical instabilities could occur, causing invalid numerical results.

SAP2000 EXERCISE 9.1

The file *C9EX1.sdb* is a fully functioning model of the truss presented in Example 9.2. Plot the deflected shape and verify that the displacement at joint E = 9.16 in.

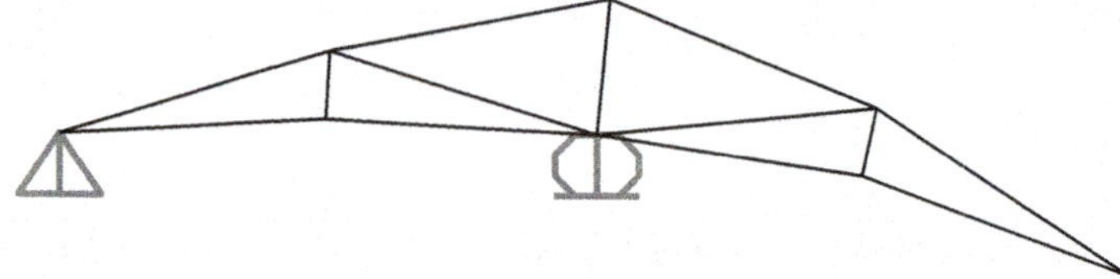

QUICK NOTE
The deflected shape can be plotted by clicking the following icon.

It may also be plotted by going to the *DISPLAY* menu.

Deflection at E in Units of Inches:

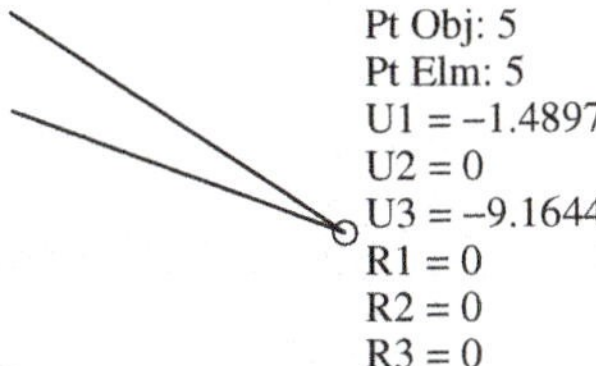

Now change the areas of bars BC and CD to 2.0 in^2 and find the displacement at E? (Δ_{Ey} = 10.79 in.).
Finally, change the areas of bars BG and GD to 4.0 in^2 and find the deflection at E. (Δ_{Ey} = 10.79 in.).

POINT TO PONDER

Changing the areas of bars BC and CD changed the deflection at point E, but changing the areas of bars BG and GD did not change the deflection. Does the virtual structure in Example 9.2 give any clue?

SAP2000 EXERCISE 9.2

The file *C9EX2.sdb* is a fully functioning model of the beam shown in Example 9.10. Plot the deflected shape and verify the deflection at A is 2.34 in.

Deflection at A in Units of Inches:

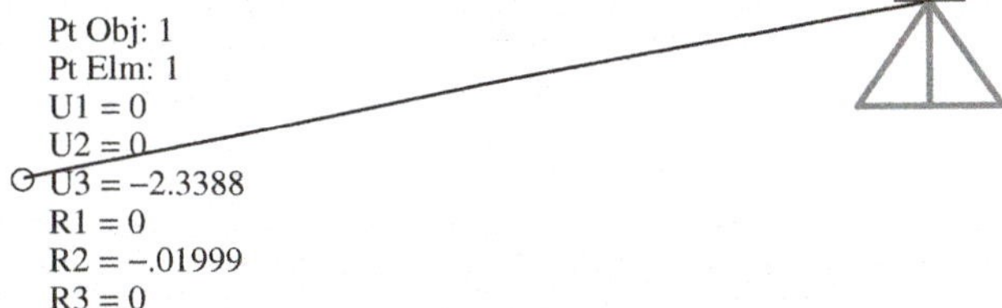

Now change the stiffness of the rotational spring at B to 400,000 (k·in.)/rad. What is the new deflection?

After Changing Rotational Stiffness Δ_A = 0.395 in.:

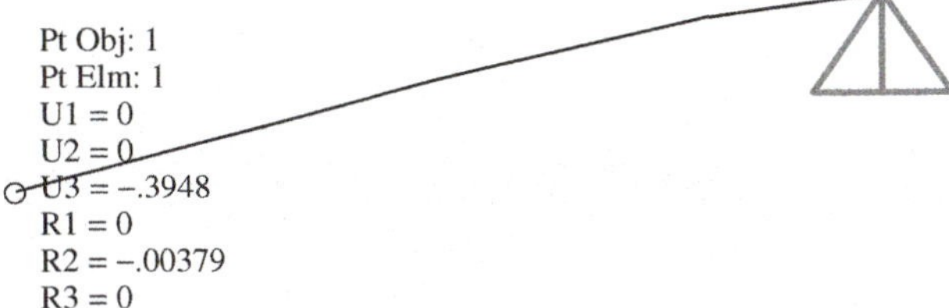

Now change the shear area of the beam to be 1 in^2. What is the new deflection?

After Changing the Shear Area Δ_A = 0.459 in.:

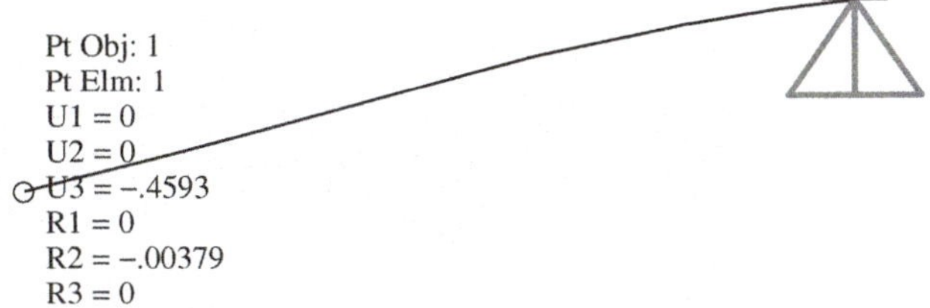

QUICK NOTE

The basic principle is that when stiffness of any structural member increases the deflection decreases, as seen with the rotational spring. Whenever the stiffness decreases—like when shear area is set small—the deflection increases.

9.11 Examples with Video Solutions

VE9.1 For the truss shown in Figure VE9.1, use virtual work to calculate the vertical deflection at joint C and the horizontal deflection at joint E. Assume all members have $E = 1900$ ksi and $A = 5.25$ in^2.

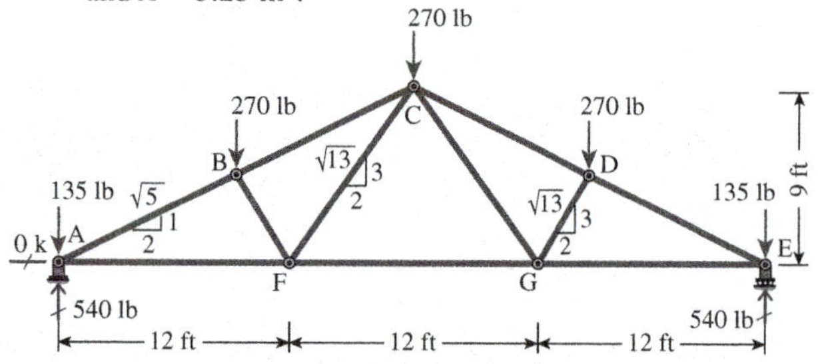

Figure VE9.1

VE9.2 For the beam shown in Figure VE9.2, use virtual work and conventional integration to solve for the vertical deflection at C and rotation at A. E and I are constant for the entire length. Solve for the answer symbolically.

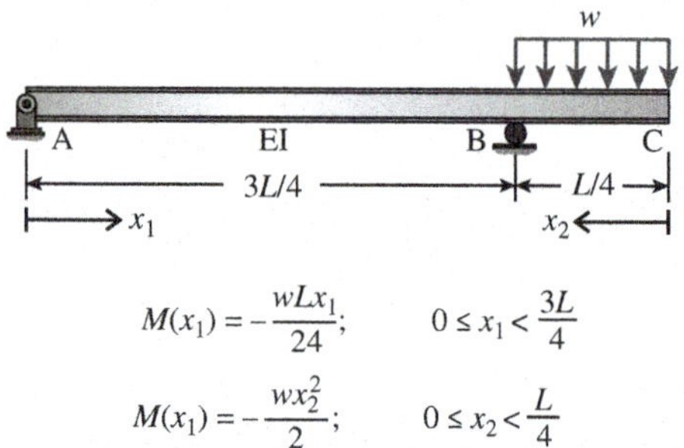

Figure VE9.2

VE9.3 Rework VE9.2, only use visual integration instead of conventional integration.

VE9.4 For the frame in Figure VE9.4, use virtual work and visual integration to calculate the horizontal deflection at D. $E = 29{,}000$ ksi and $I = 2000$ in^4.

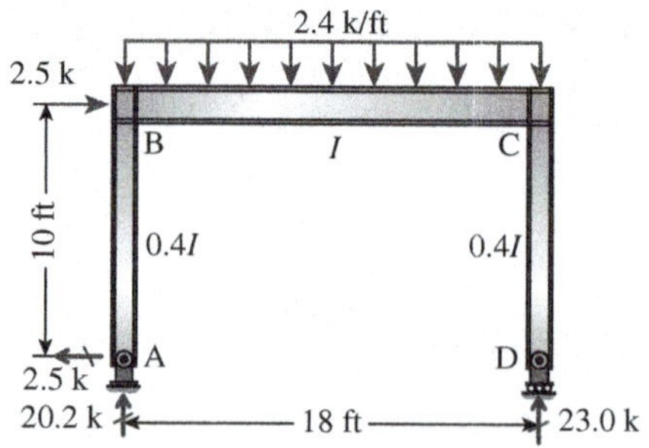

Figure VE9.4

VE9.5 The beam shown in Figure VE9.5 is supported on elastic springs with $k_T = 10$ k/in. Use virtual work with visual integration to solve for the vertical deflection at B. $E = 3600$ ksi and $I = 1500$ in^4.

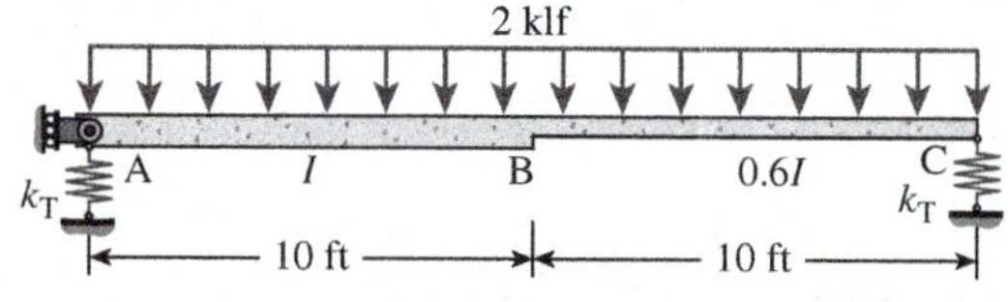

Figure VE9.5

9.12 Problems for Solution

Section 9.5

For Problems 9.1 through 9.6, use the virtual work method to determine the deflection of each of the joints indicated. $E = 29{,}000$ ksi for all members unless otherwise indicated. The cross section of each member is given as in^2 unless noted otherwise.

P9.1 Find Δ_{Cx} and Δ_{Cy}. Bar areas: AB = 7; BC = DE = EC = 2; AD = BE = 4; and DB = 6. (Ans: Δ_{Cy} = 2.148 in. ↓, Δ_{Cx} = 0.394 in. →)

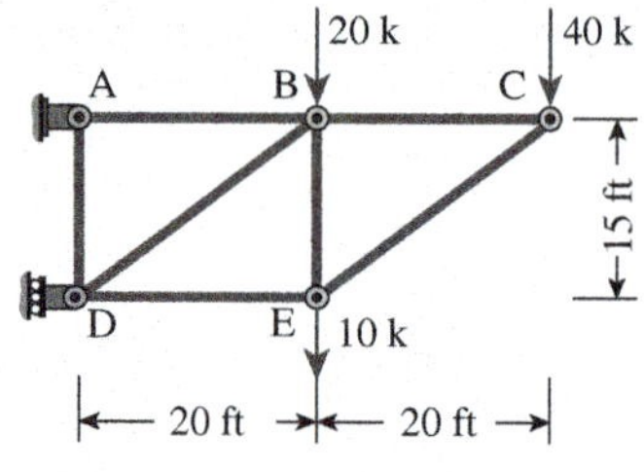

Problem 9.1

P9.2 Find Δ_{Ex} and Δ_{Dy}. Bar areas: AC = AB = 2 and all other members = 3.

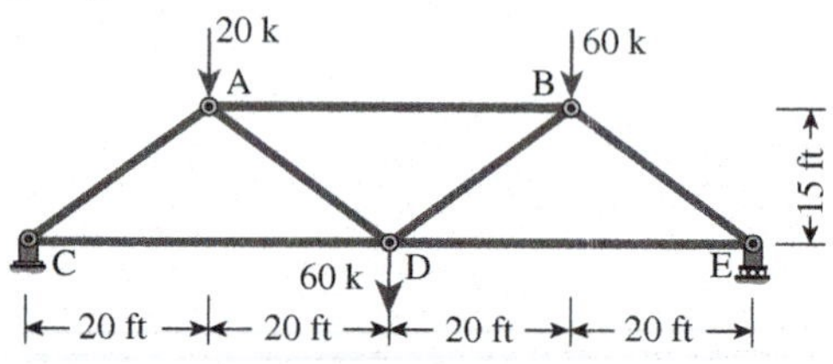

Problem 9.2

P9.3 Find Δ_{Bx} and Δ_{Cy}. (Ans: Δ_{Bx} = 13.6 mm →, Δ_{Cy} = 37.8 mm ↓)

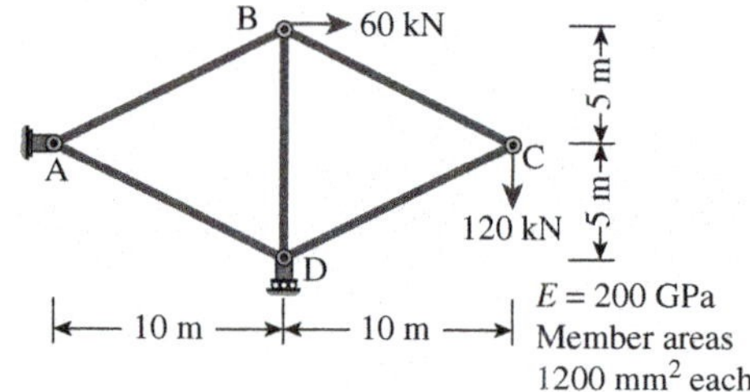

Problem 9.3

P9.4 Find Δ_{Ex} and Δ_{Dy}. Bar areas: top and bottom chords = 4; web members = 2.

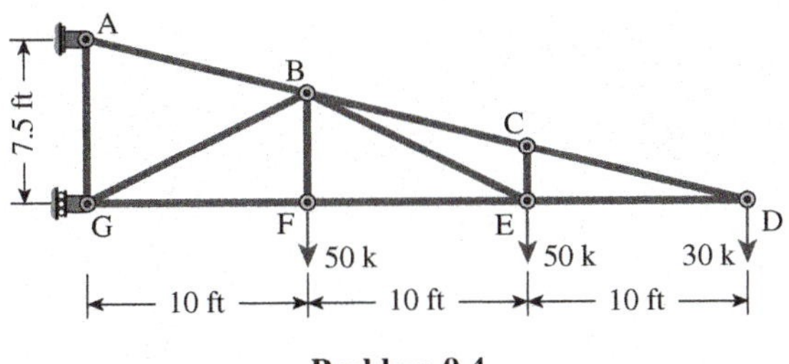

Problem 9.4

P9.5 Find Δ_{By} and Δ_{Cx}. Bar areas: top and bottom chords = 2.5; BD = 5. (Ans: Δ_{By} = 0.185 in. ↑, Δ_{Cx} = 0.370 in. ←)

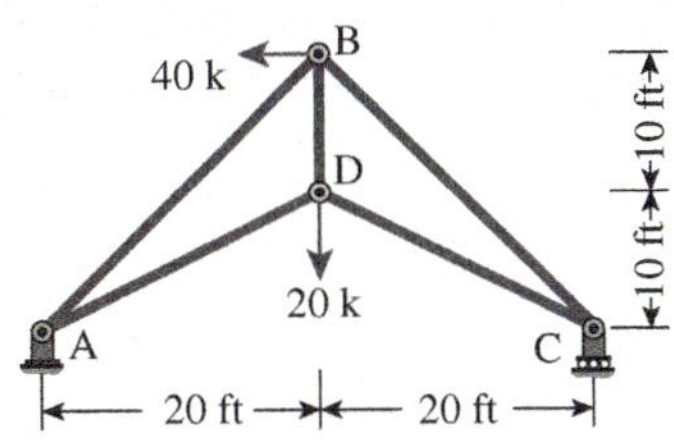

Problem 9.5

P9.6 Find Δ_{By} and Δ_{Cx}. Bar areas: top chords = 4; bottom chords = 2.5; BD = 3.

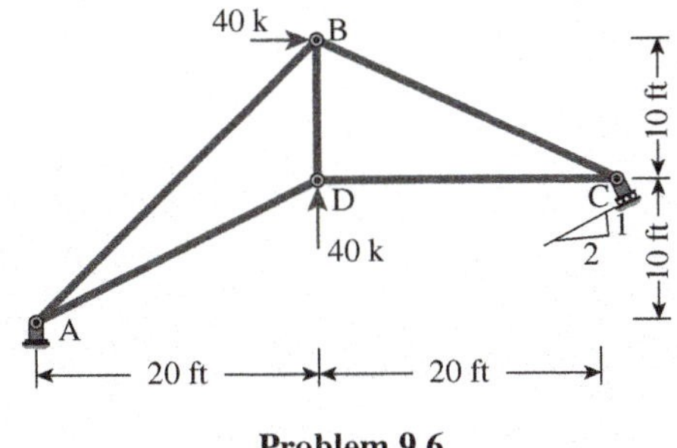

Problem 9.6

Section 9.6

For Problems 9.7 through 9.29, use the virtual work method using conventional integration and only bending deformations to find the indicated deflections/rotations in each structure. E = 29,000 ksi for all members unless otherwise indicated.

P9.7 Determine the vertical deflection at points A and B of the beam shown. I = 2000 in^4. (Ans: Δ_{Ay} = 5.49 in. ↓, Δ_{By} = 1.72 in. ↓)

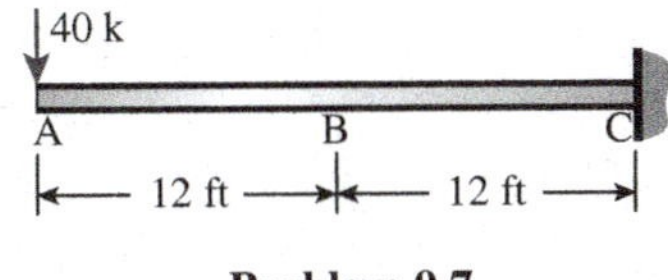

Problem 9.7

P9.8 Determine the vertical deflection at points B and C of the beam shown. I = 3500 in^4.

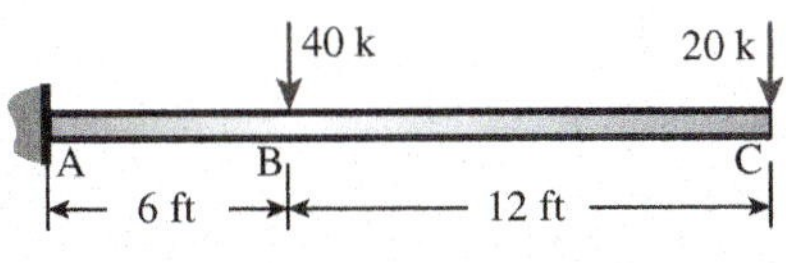

Problem 9.8

P9.9 Determine the vertical deflection and rotation at point B of the beam shown. I = 5000 in^4. (Ans: Δ_{By} = 0.602 in. ↓, θ_B = 0.0045 rad ↻)

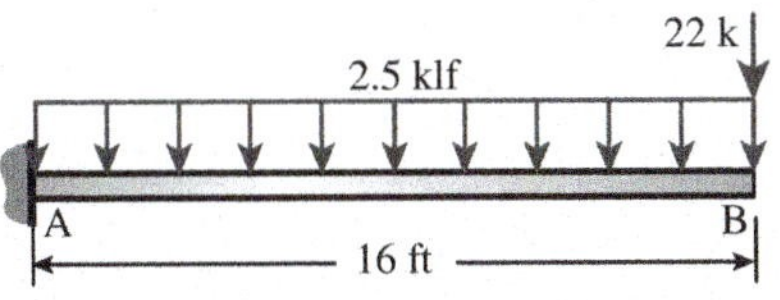

Problem 9.9

P9.10 Determine the vertical deflection and rotation at point A of the beam shown. I = 3500 in^4.

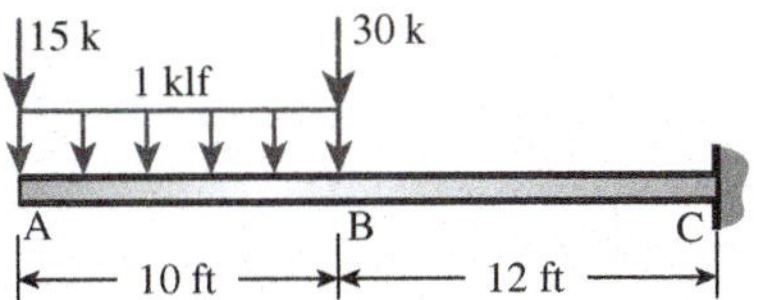

Problem 9.10

P9.11 Determine the vertical deflection and rotation at point A of the beam shown. I = 1140 in^4. (Ans: Δ_{Ay} = 0.271 in. ↓, θ_B = 0.0025 rad ↻)

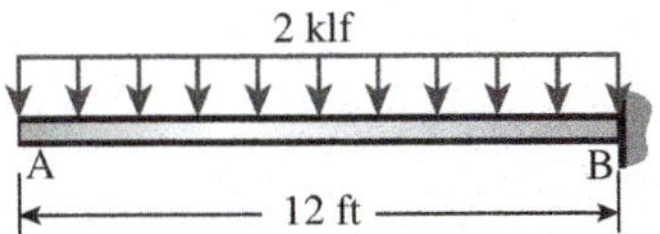

Problem 9.11

P9.12 Determine the vertical deflection and rotation at point C of the beam shown. I = 850 in^4.

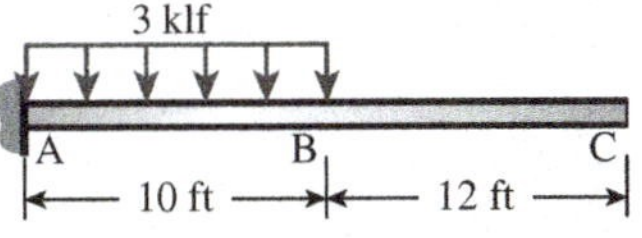

Problem 9.12

P9.13 Determine the vertical deflection at points C and D of the beam shown. I = $3{\cdot}10^8$ mm^4 and E = 200 GPa. (Ans: Δ_{Cy} = 74 mm ↓, Δ_{Dy} = 126 mm ↓)

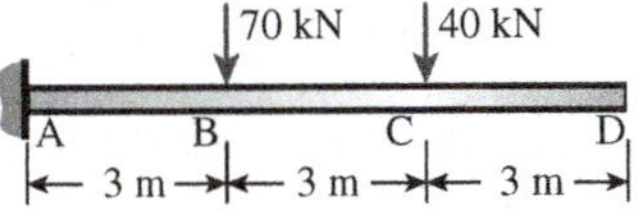

Problem 9.13

P9.14 Calculate the vertical deflection and rotation at point B of the beam shown. $I = 2.5 \cdot 10^8$ mm^4 and $E = 200$ GPa.

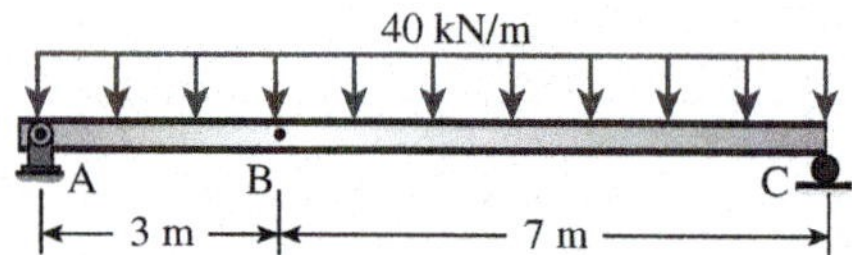

Problem 9.14

P9.15 Calculate the vertical deflection at point D and rotation at point B of the beam shown. $I = 1320$ in^4. (Ans: $\Delta_{Dy} = 2.457$ in. ↓, $\theta_B = 0.00731$ rad ↻)

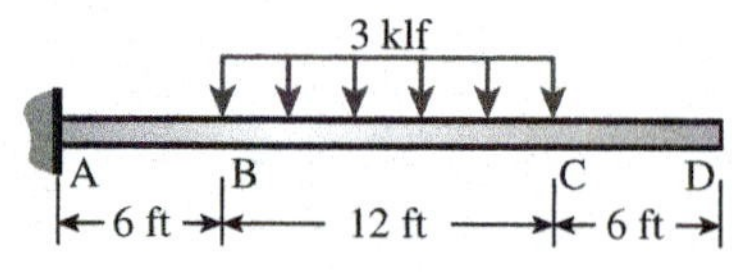

Problem 9.15

P9.16 Calculate the vertical deflection and rotation at point A of the beam shown.

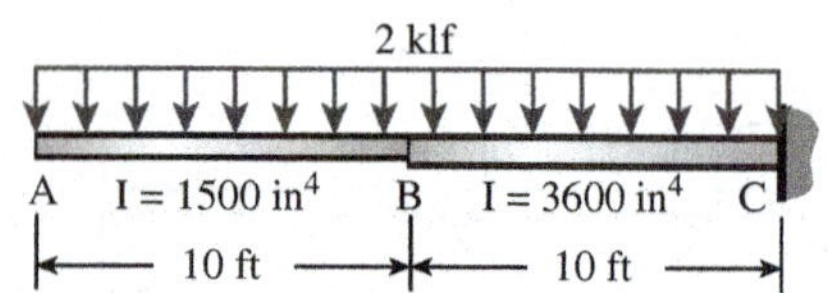

Problem 9.16

P9.17 Calculate the vertical deflection and rotation at point B of the beam shown. $I = 1500$ in^4. (Ans: $\Delta_{By} = 0.915$ in. ↓, $\theta_B = 0.00318$ rad ↻)

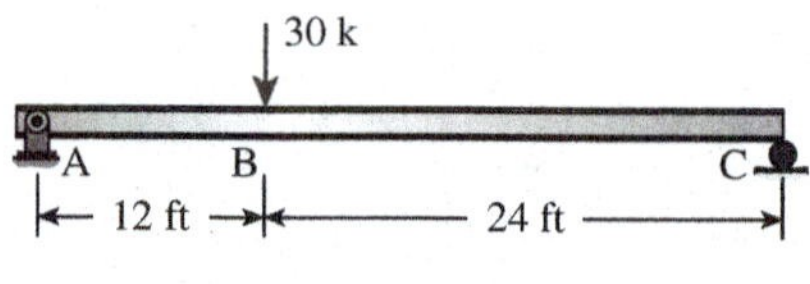

Problem 9.17

P9.18 Determine the vertical deflection at D and the rotation at C. $I = 2100$ in^4.

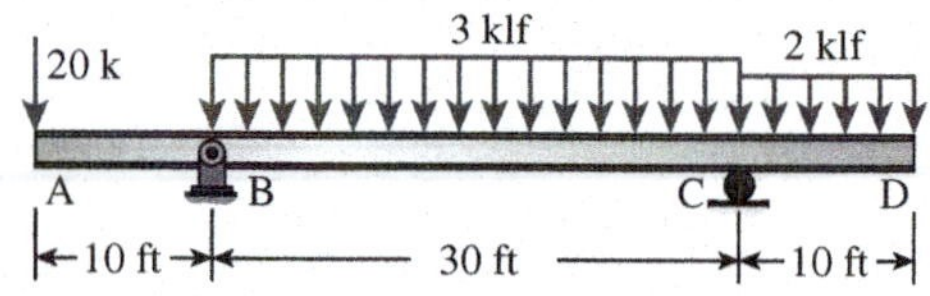

Problem 9.18

P9.19 Calculate the vertical deflection at point C. (Ans: $\Delta_{Cy} = 2.12$ in. ↓)

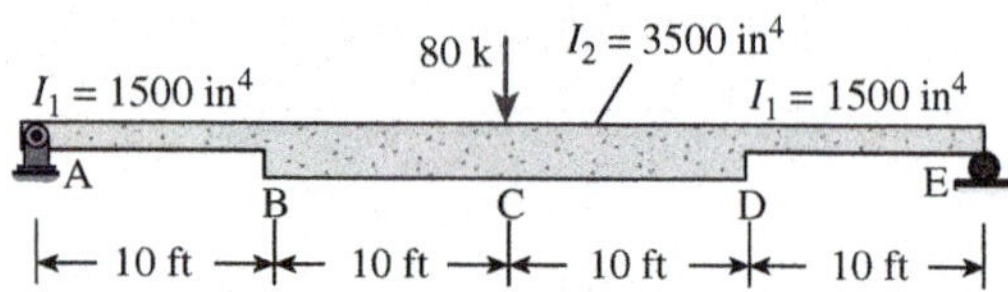

Problem 9.19

P9.20 Calculate the vertical deflection and rotation at point B.

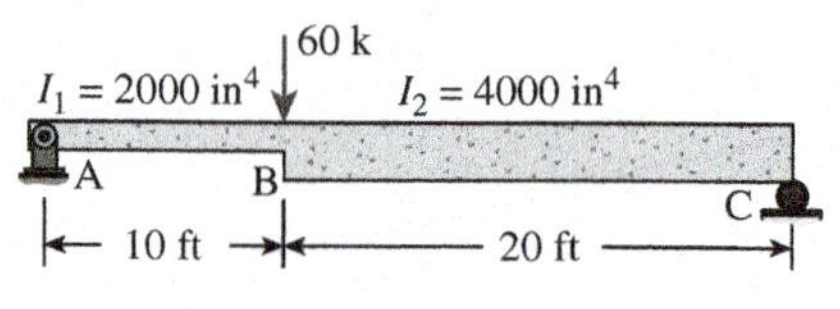

Problem 9.20

P9.21 Calculate the deflection at point D and the rotation at point B. $I = 1.46 \cdot 10^9$ mm^4 and $E = 200$ GPa. (Ans: $\Delta_{Dy} = 35.7$ mm ↓, $\theta_B = 0.00728$ rad ↻)

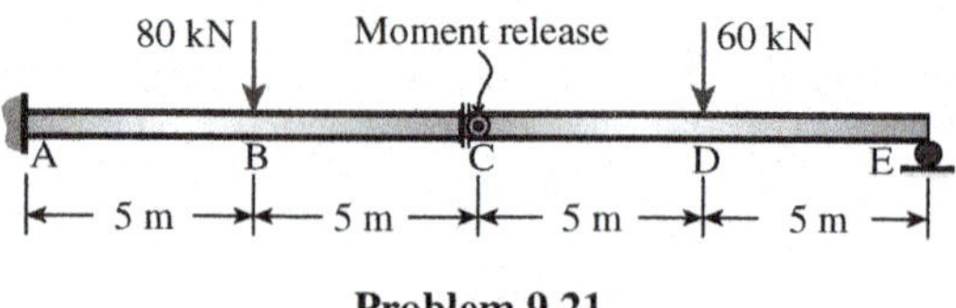

Problem 9.21

P9.22 Calculate the vertical deflection at point B. $I = 2250$ in^4.

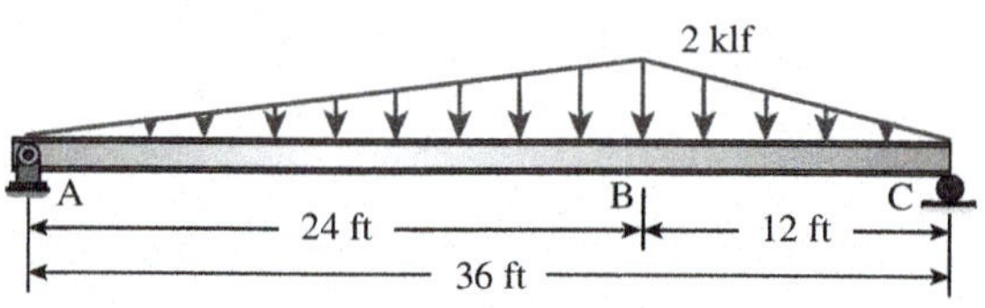

Problem 9.22

P9.23 Calculate the vertical deflection at point D. $I = 4250$ in^4. (Ans: $\Delta_{Dy} = 0.67$ in. ↑)

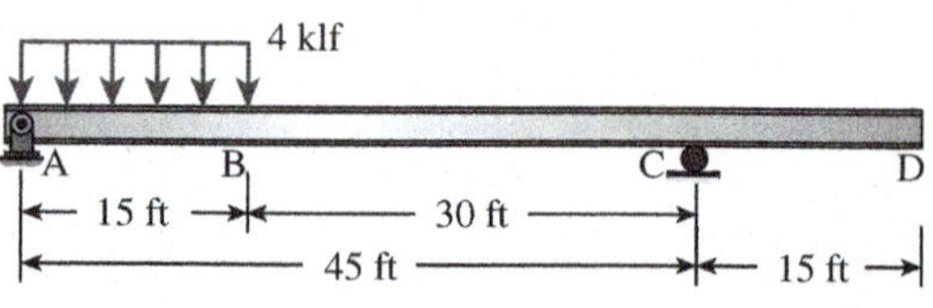

Problem 9.23

P9.24 Calculate the vertical deflection at point A and the horizontal deflection at point C. $I = 3000$ in^4.

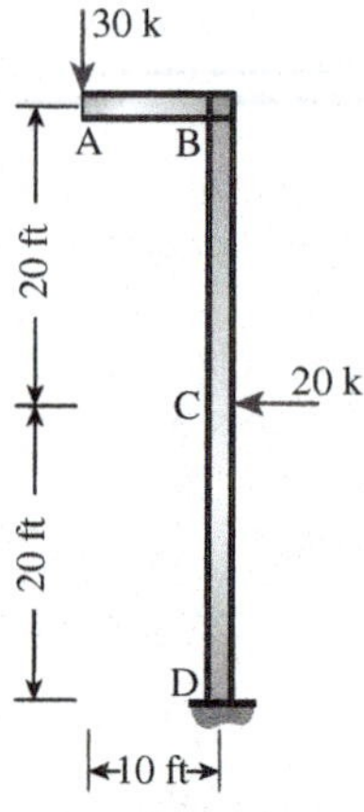

Problem 9.24

P9.25 Determine the horizontal deflection at point B and the vertical deflection at point D. (Ans: Δ_{Bx} = 1.208 in. →, Δ_{Dy} = 0.199 in. ↓)

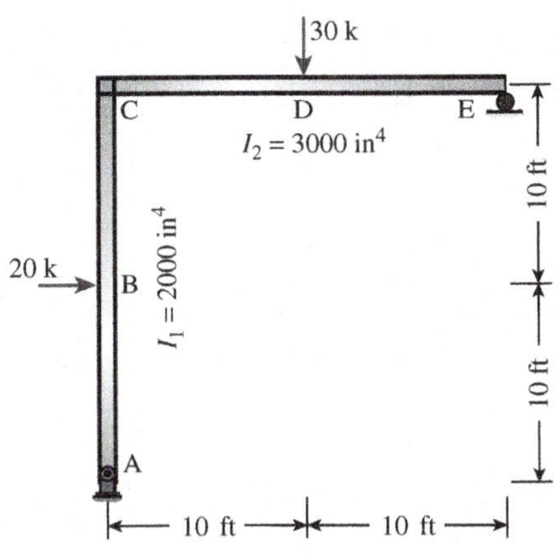

Problem 9.25

P9.26 Determine the vertical and horizontal deflection at point D. $I = 2.60{\cdot}10^9\ mm^4$ and E = 200 GPa.

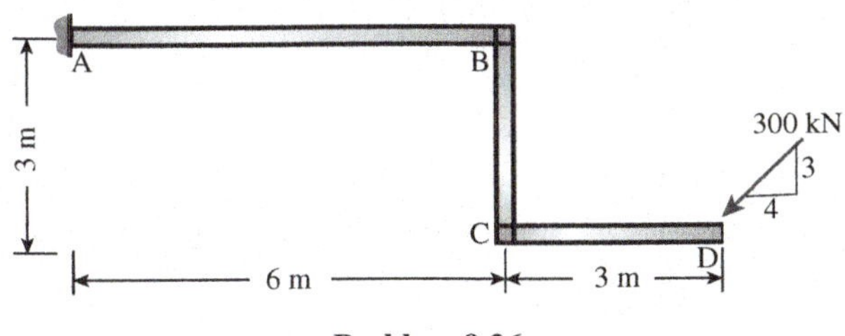

Problem 9.26

P9.27 Determine the horizontal deflection at point D and the rotation at point C. $I = 1200\ in^4$. (Ans: Δ_{Dx} = 7.72 in. ←, θ_C = 0.0358 rad ↺)

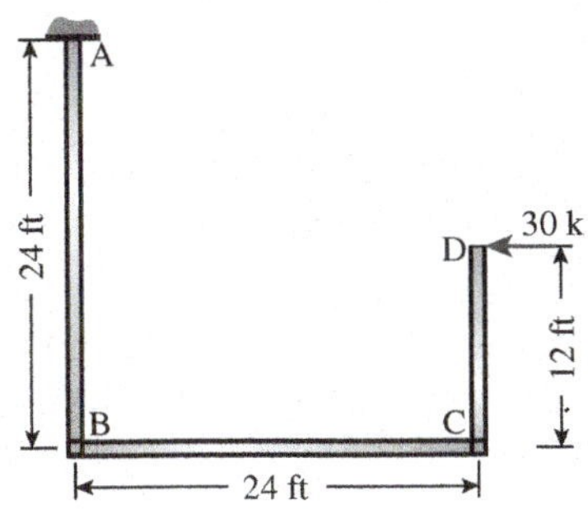

Problem 9.27

P9.28 Determine the horizontal deflection at point E and the rotation at point D. $I = 1850\ in^4$.

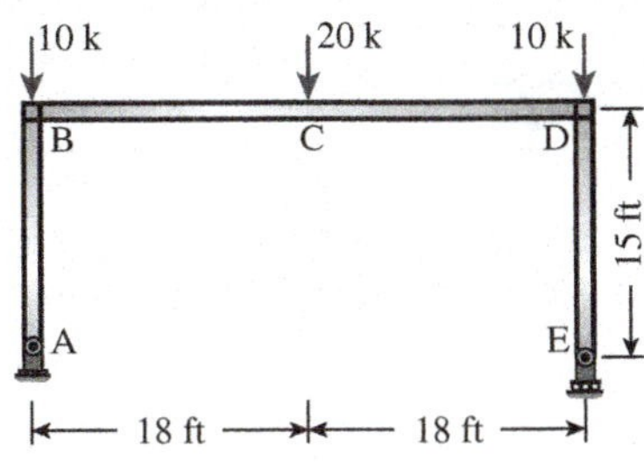

Problem 9.28

P9.29 Determine the horizontal and vertical deflection at point D. $I = 2750\ in^4$. (Ans: Δ_{Dy} = 1.156 in. ↓, Δ_{Dx} = 0.939 in. ←)

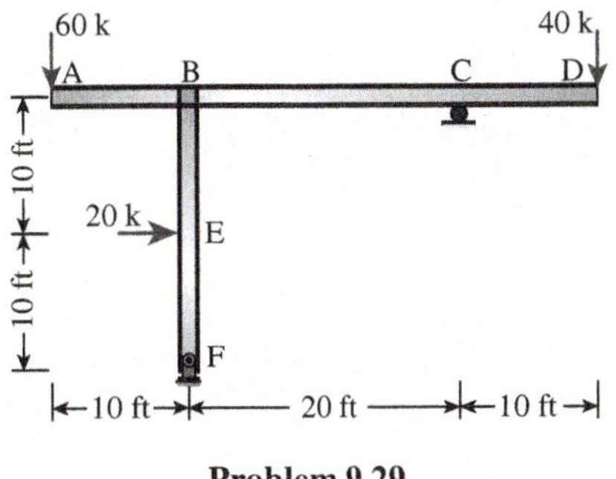

Problem 9.29

Section 9.7

For Problems 9.30 through 9.47, use the virtual work method using visual integration and only bending deformations to find the indicated deflections/rotations in each structure. E = 29,000 ksi for all members unless otherwise indicated.

P9.30 Determine the vertical deflections at points A and B. $I = 3250\ in^4$.

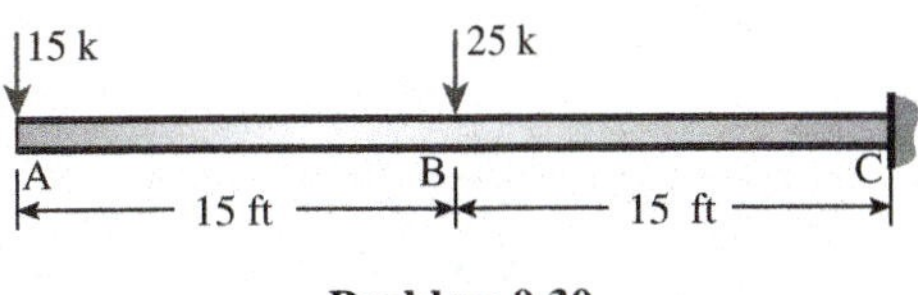

Problem 9.30

P9.31 Determine the vertical deflection and rotation at point C. $I = 1750\ in^4$. (Ans: Δ_{Cy} = 1.915 in. ↓, θ_C = 0.00638 rad ↻)

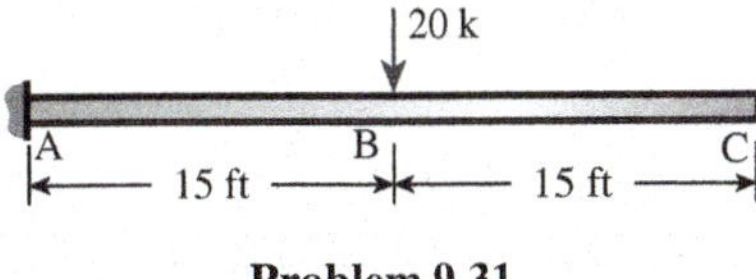

Problem 9.31

P9.32 Determine the vertical deflection and rotation at point B. $I = 5500\ in^4$.

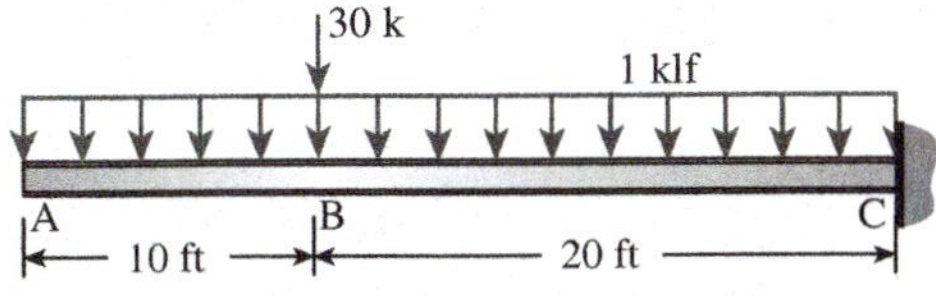

Problem 9.32

P9.33 Determine the vertical deflection at joints A and B. $I = 1250\ in^4$. (Ans: Δ_{Ay} = 2.443 in. ↓, Δ_{By} = 0.834 in. ↓)

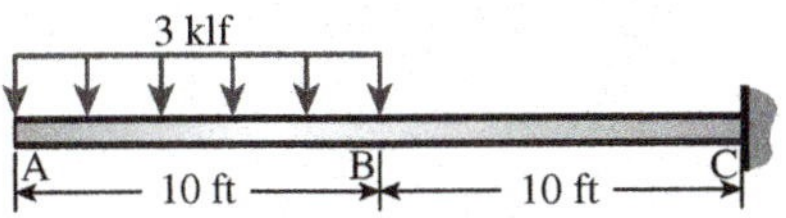

Problem 9.33

P9.34 Determine the vertical deflection at joints A and B. $I = 4.0{\cdot}10^8$ mm^4 and $E = 200$ GPa.

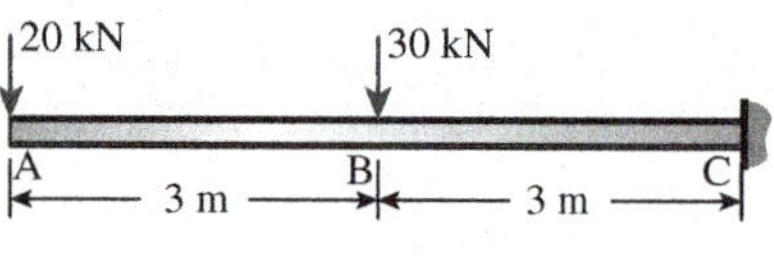

Problem 9.34

P9.35 Determine the vertical deflection and rotation at joint B. $I = 2.0{\cdot}10^8$ mm^4 and $E = 200$ GPa. (Ans: $\Delta_{By} = 23.4$ mm ↓, $\theta_B = 0.00313$ rad ↺)

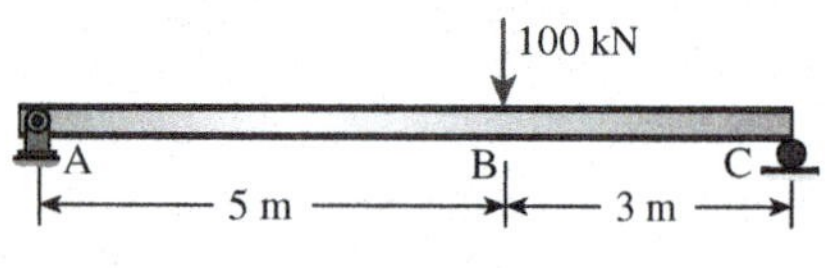

Problem 9.35

P9.36 Determine the vertical deflection at joints B and C. $I = 2250$ in^4.

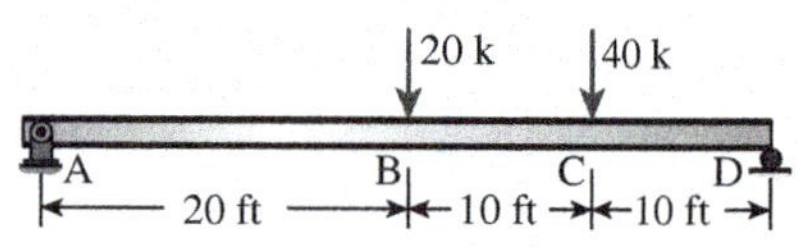

Problem 9.36

P9.37 Determine the vertical deflection and rotation at joint B. $I = 3100$ in^4. (Ans: $\Delta_{By} = 0.048$ in. ↓, $\theta_B = 0.0004$ rad ↺)

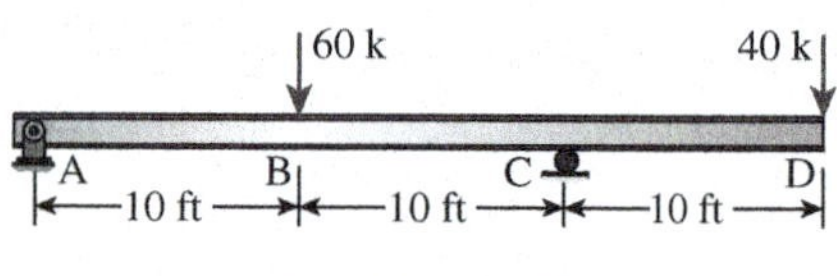

Problem 9.37

P9.38 Determine the horizontal deflection at joint B. $I = 1500$ in^4.

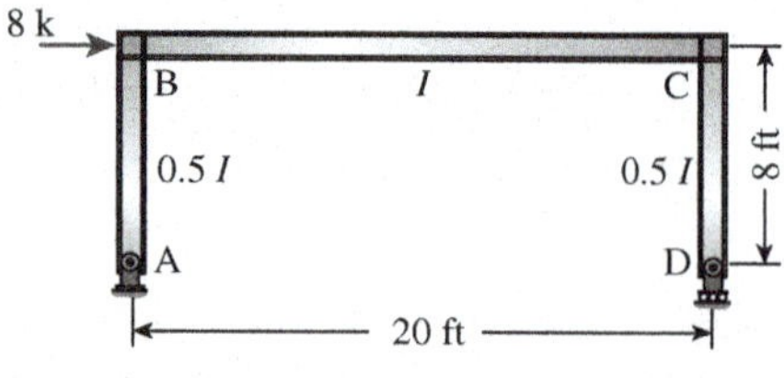

Problem 9.38

P9.39 Determine the vertical and horizontal deflection at joint C. $I = 2500$ in^4. (Ans: $\Delta_{Cy} = 0.134$ in. ↓, $\Delta_{Bx} = 0.0824$ in. →)

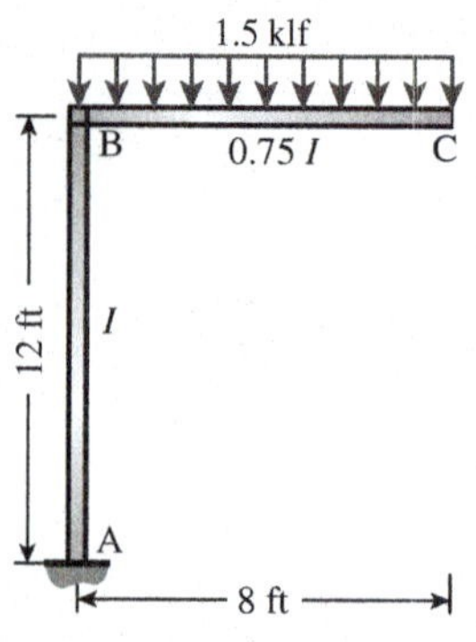

Problem 9.39

P9.40 Rework Problem 9.12.

P9.41 Rework Problem 9.21. (Ans: $\Delta_D = 35.7$ mm ↓, $\theta_B = 0.00728$ rad ↻)

P9.42 Rework Problem 9.24.

P9.43 Rework Problem 9.25. (Ans: $\Delta_{Bx} = 1.208$ in. →, $\Delta_{Dy} = 0.199$ in. ↓)

P9.44 Rework Problem 9.26.

P9.45 Rework Problem 9.27. (Ans: $\Delta_{Dx} = 7.72$ in. ←, $\theta_C = 0.0358$ rad ↺)

P9.46 Rework Problem 9.28.

P9.47 Rework Problem 9.29. (Ans: $\Delta_{Dy} = 1.156$ in. ↓, $\Delta_{Dx} = 0.939$ in. ←)

Section 9.8

For Problems 9.48 through 9.53, use the virtual work method considering bending and spring energy to find the indicated deflections/rotations in each structure. $E = 29{,}000$ ksi for all members unless otherwise indicated.

P9.48 Determine the vertical deflection at joints A and C. $I = 500$ in^4 and $k_R = 300{,}000$ (k·in.)/rad.

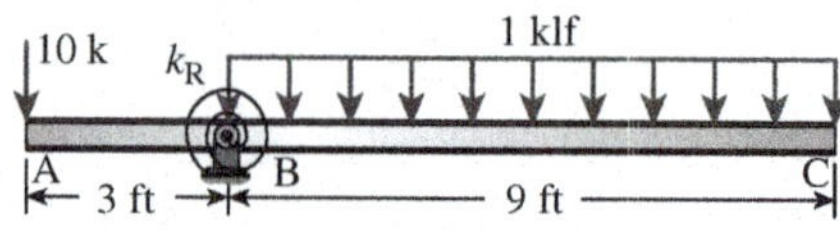

Problem 9.48

P9.49 Determine the vertical deflection at joint B and the rotation at joint C. I = 900 in^4 and k_R = 500,000 (k·in.)/rad. (Ans: Δ_{By} = 0.2289 in. ↓, θ_C = 0.00325 rad ↺)

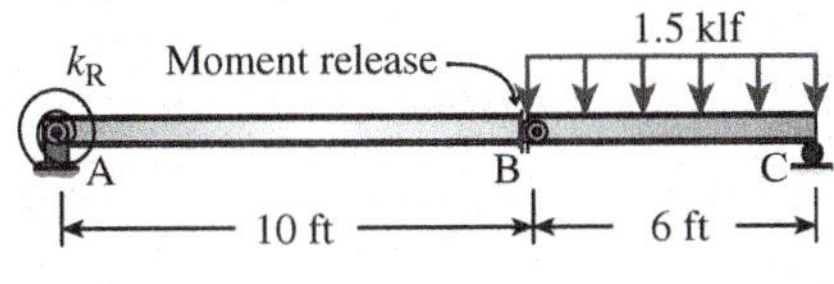

Problem 9.49

P9.50 Determine the vertical deflection just to the right of joint B and the rotation at joint C. I = 900 in^4 and k_R = 500,000 (k·in.)/rad.

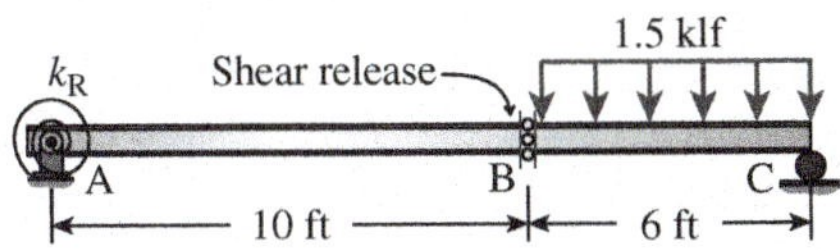

Problem 9.50

P9.51 Determine the vertical deflections at joints B and C. I = 1000 in^4 and k_T = 8 k/in. (Ans: Δ_{By} = 0.0894 in. ↓, Δ_{Cy} = 0.563 in. ↓)

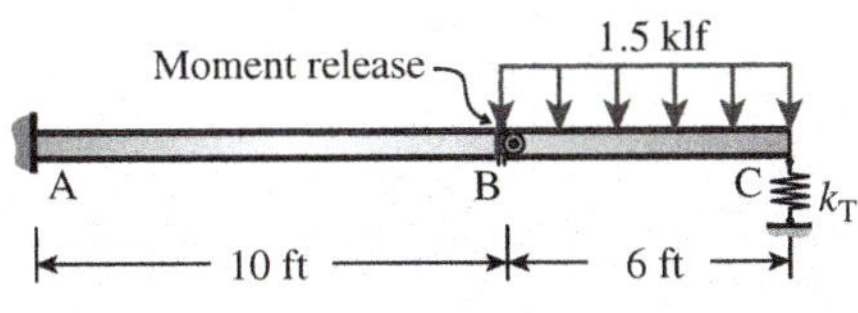

Problem 9.51

P9.52 Determine the vertical deflection at joint C and horizontal deflection at joint B. I = 1500 in^4 and k_R = 60,000 (k·in.)/rad.

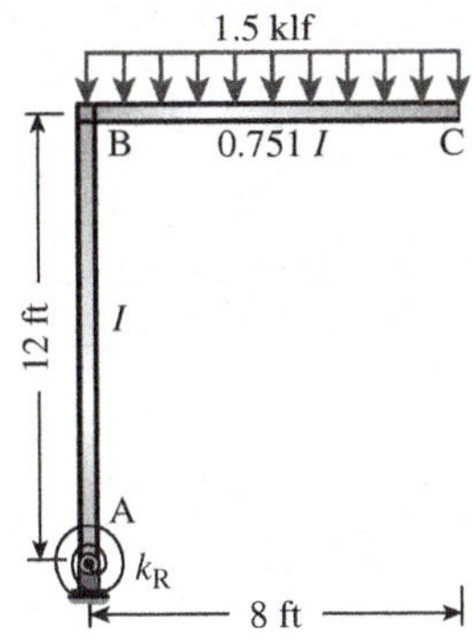

Problem 9.52

P9.53 Determine the vertical deflection and horizontal deflection at joint B. (*Hint*: treat the system of springs as a truss.) k_{T1} = 20 k/in. k_{T2} = 30 k/in. (Ans: Δ_{By} = 3.99 in. ↓, Δ_{Bx} = 1.47 in. ←)

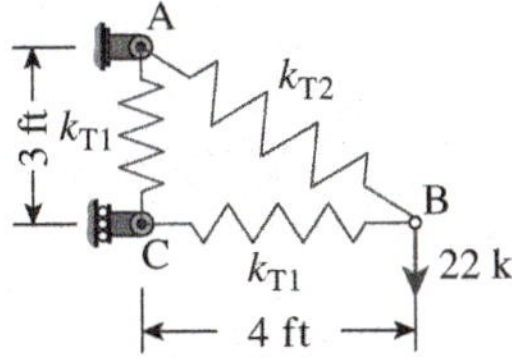

Problem 9.53

Section 9.9

For Problems 9.54 through 9.63, use the method of virtual work to compute the requested deflections. Include both bending and shear deformations. E = 29,000 ksi and G = 11,000 ksi for all members unless otherwise indicated.

P9.54 Rework Problem 9.7. K = 1.0 and A = 2.5 in^2.

P9.55 Rework Problem 9.9. K = 1.11 and A = 1.25 in^2. (Ans: Δ_{By} = 1.254 in. ↓, θ_B = 0.0045 rad ↻)

P9.56 Rework Problem 9.12. K = 1.5 and A = 4.0 in^2.

P9.57 Rework Problem 9.15. K = 1.0 and A = 2.0 in^2. (Ans: Δ_{Dy} = 2.693 in. ↓, θ_B = 0.00731 rad ↻)

P9.58 Rework Problem 9.17. K = 1.11 and A = 3.5 in^2.

P9.59 Rework Problem 9.23. K = 1.11 and A = 2.25 in^2. (Ans: Δ_{Dy} = −0.67 in. ↓)

P9.60 Rework Problem 9.30. K = 1.0 and A = 1.25 in^2.

P9.61 Rework Problem 9.33. K = 1.5 and A = 1.5 in^2. (Ans: Δ_{Ay} = 2.934 in. ↓, Δ_{By} = 1.161 in. ↓)

P9.62 Rework Problem 9.36. K = 1.0 and A = 2.0 in^2.

P9.63 Rework Problem 9.37. K = 1.5 and A = 3.0 in^2. (Ans: Δ_{By} = 0.212 in. ↓, θ_B = 0.00028 rad ↻)

Section 9.10 SAP2000

P9.64 Model the beam shown below in SAP2000 and find the vertical displacement at point C. E = 29,000 ksi, I = 1500 in^4. Do so for the following values of L: (20 ft, 25 ft, 30 ft, 35 ft, 40 ft, 42 ft, and 45 ft). Generate a plot of the location versus the deflection. Make an observation regarding the direction of the deflection of point C relative to the position of the roller at B. Using the generated plot, identify the approximate values of L that would cause there to be a net-zero displacement to occur at C. (*Hint*: The easiest way to move the location of the roller is to use the *Move* command under the *Edit* menu.)

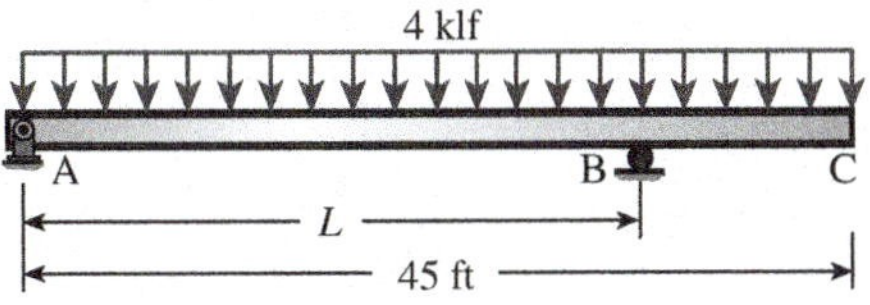

Problem 9.64

P9.65 Model the beam of Problem 9.22 in SAP2000 and plot the displaced shape. Move the roller from point C to point B and plot the displaced shape again. Report the rotation at A. Make an observation regarding the change in shape.

P9.66 Model the frame of Problem 9.28 in SAP2000 and find the vertical displacement at points B and C. Change the cross-sectional areas for all members and repeat the analysis. Do the analysis for the following cross-sectional areas and generate a plot of cross-sectional area versus deflection: (0.5 in^2, 1.0 in^2, 1.5 in^2, 2.0 in^2, 5.0 in^2, and 10.0 in^2). What impact does the cross-sectional area play in the displacements of both points?

P9.67 Model the beam of Problem 9.49 in SAP2000 and find the vertical displacement at point B. Change the rotational spring stiffness and repeat the analysis. Do the analysis for the following stiffness values (units of (k·in.)/rad): (500; 5000; 50,000; 500,000; and 5,000,000). Generate a plot of spring stiffness versus displacement. (Use a logarithmic scale for the plot of spring stiffness.) What impact does the spring stiffness play in the displacement? Is it a linear relationship?

P9.68 Model the beam of Problem 9.57 in SAP2000 and find the vertical displacement at point D. Be sure to include shear deformations by including the shear area which is A/K. Find the displacement for the following values of A: (0.05 in^2, 0.1 in^2, 0.5 in^2, 1.0 in^2, 2.0 in^2, and 5.0 in^2). Generate a plot of shear area versus displacement. What impact does the shear area play in the displacement? Is it a linear relationship?

PART 2

INDETERMINATE STRUCTURES

10 Introduction to Statically Indeterminate Structures

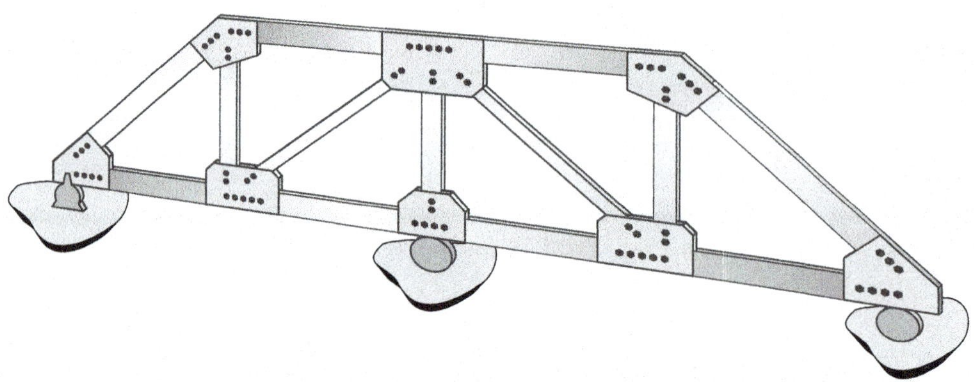

10.1 Introduction

When a structure has too many external reactions and/or internal forces than can be determined using only equations of static equilibrium, it is said to be *statically indeterminate*. A load placed on one part of a statically indeterminate or continuous structure will cause shearing forces, bending moments, and deflections in other parts of the structure. In other words, loads applied to a column can also affect the beams, slabs, and other columns and vice versa. This is often, but not necessarily, true with *statically determinate* structures.

Up to this point, the text has been so completely devoted to statically determinate structures that the reader may have been falsely led to believe that statically determinate beams and trusses are the rule in modern structures. In truth, it is difficult to find an ideal, simply supported beam. Probably the best place to look for one would be in a textbook on structures. This is because bolted or welded beam-to-column connections, as seen in Figure 10.1, do not produce ideal simple supports with zero moments as is often assumed.

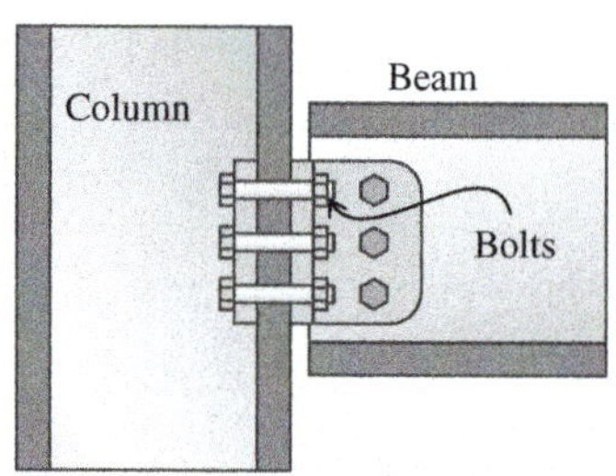

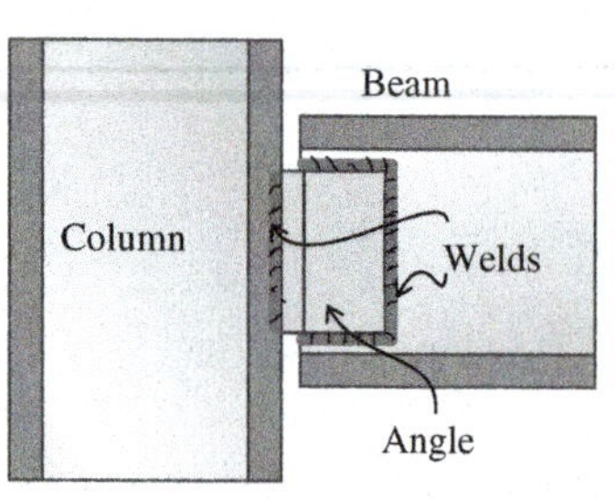

FIGURE 10.1 Typical steel frame connections.

The same holds for statically determinate trusses. Bolted or welded joints are not frictionless pins, as previously assumed. The other assumptions made about trusses in the earlier chapters of this book are not altogether true either. Thus, in a strict sense, all trusses are statically indeterminate because they have some bending and secondary forces.

Almost all cast-in-place, reinforced-concrete structures are statically indeterminate. The concrete for a large part of a concrete floor, including the support beams, and girders—and perhaps parts of the columns—may be placed at the same time. The reinforcing bars extend from member to member, as from one span of a beam into the next. When there are construction joints, the reinforcing bars are left protruding from the older concrete so they may be lapped or spliced to the bars in the newer concrete. In addition, the old concrete is cleaned and perhaps roughened, so the newer concrete will bond to it as well as possible. The result of all these facts is that reinforced-concrete structures are generally monolithic (i.e., one piece) or continuous and thus are statically indeterminate.

About the only way a statically determinate reinforced-concrete structure could be built would be with individual sections precast at a concrete plant and assembled on the job site. Even these structures could have some continuity in their joints. Until the early part of the 20th century, statically indeterminate structures were avoided as much as possible by most American engineers. However, three great developments completely changed the picture. These developments were monolithic reinforced-concrete structures, arc welding of steel structures, and modern methods of analysis that use the digital computer.

10.2 Continuous Structures

As the spans of simple structures become longer, the bending moments increase rapidly. If the weight of a structure per unit of length remained constant, regardless of the span, the dead-load moment would vary in proportion to the square of the span length ($M_{max} = wL^2/8$). This proportion, however, is not correct, because the weight of structures must increase with longer spans to be strong enough to resist the increased bending moments. Therefore, the dead-load moment increases at a greater rate than does the square of the span.

For economy, it pays in long spans to introduce types of structures that have smaller moments than the tremendous ones that occur in long-span, simply supported structures. One type of structure that considerably reduces bending moments is called cantilever-type construction, which was introduced in Section 5.10. Two other moment-reducing structures are discussed in the following paragraphs.

In some locations, a beam with fixed ends, rather than one with simple supports, may be possible. A comparison of the moments developed in a uniformly loaded simple beam with those in a uniformly loaded, fixed-ended beam is made in Figure 10.2.

The maximum bending moment in the fixed-ended beam is only two thirds of that in the simply supported beam. Usually, it is difficult to fix the ends, particularly in the case of bridges. For this reason, flanking spans are often used, as illustrated in Figure 10.3. These spans will partially restrain the interior supports, thus tending to reduce the moment in the center span. This figure compares the bending moments that occur in three uniformly loaded simple beams (spans of 60, 100, and 60 ft) with the moments of a uniformly loaded beam continuous over the same three spans.

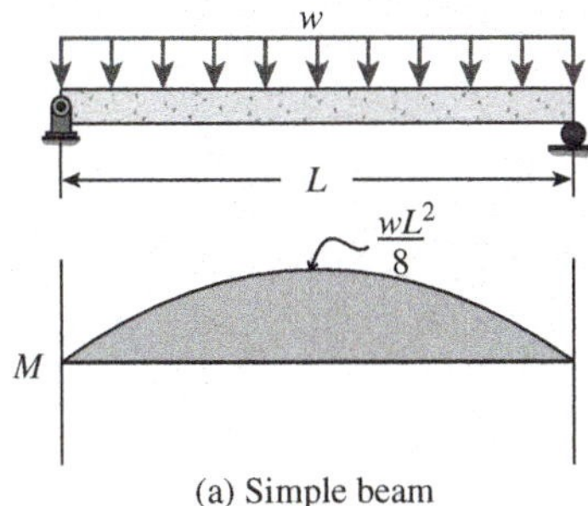

(a) Simple beam

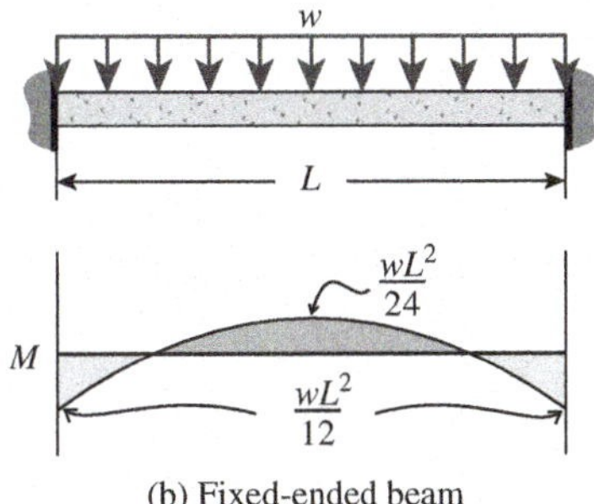

(b) Fixed-ended beam

FIGURE 10.2 A comparison of beam behaviors.

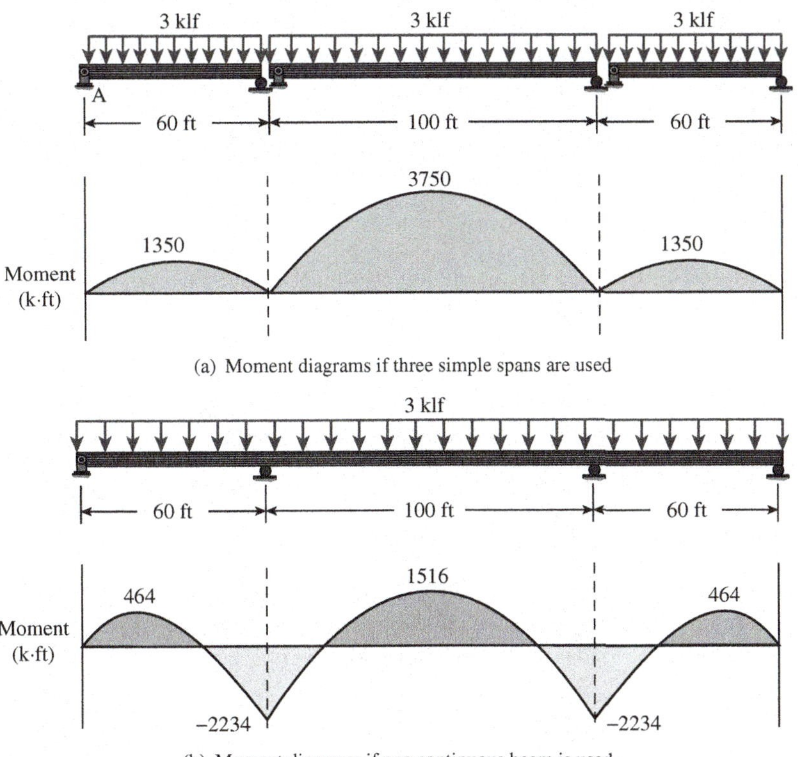

(a) Moment diagrams if three simple spans are used

(b) Moment diagrams if one continuous beam is used

FIGURE 10.3 Comparison of bending moments in three simple beams versus one continuous beam.

DID YOU KNOW?
The maximum deflection in the simple beam is

$$\Delta_{max} = (5wL^4)/(384EI)$$

The maximum deflection in the fixed-ended beam is:

$$\Delta_{max} = (wL^4)/(384EI)$$

This is a difference of five times.

DID YOU KNOW?
For a constant uniform load over the three spans, the smallest moment occurs when the flanking spans are from 0.3 to 0.4 times as long as the center span.

QUICK NOTE
As can be clearly shown, a statically indeterminate beam or frame normally will not collapse when the ultimate moment capacity is reached at just one location. Instead, there is a redistribution of the moments in the structure. This behavior is quite similar to the case in which three men are walking along with a log on their shoulders and one of the men gets tired and lowers his shoulder just a little. The result is a redistribution of load to the other men with corresponding changes in the reactions, shearing forces, and bending moments along the log. The building in Figure 10.4 did not have this advantage.

The maximum bending moment in the continuous beam is approximately 40% less than that for the simple beams. Unfortunately, there will not be a corresponding 40% reduction in total cost. The cost-reduction probably is only a small percentage of the total cost of the structure, because such items as foundations, connections, and floor systems are not reduced a great deal by the moment reduction. Varying the lengths of the flanking spans will change the magnitude of the largest moment occurring in the continuous member.

In the previous discussion, the maximum moments developed in beams were reduced appreciably by continuity. This reduction occurs when beams are rigidly connected to each other or where beams and columns are rigidly connected. There is a continuity of action in resisting a load applied to any part of a continuous structure because the load is resisted by the combined efforts of all of the members of the frame.

10.3 Advantages of Statically Indeterminate Structures

When comparing statically indeterminate structures with statically determinate structures, the first consideration for most engineers is likely the cost. However, making a general economic statement favoring one type of structure over another is impossible without reservation. Each structure presents a different and unique situation, and all factors must be considered—economic or otherwise. In general, statically indeterminate structures have the following advantages.

Savings in Materials

The smaller bending moments developed often enable the engineer to select smaller members for the structural components. The material savings could possibly be as high as 10 to 20% for highway bridges. Because of the large number of force reversals that occur in railroad bridges, the cost savings are closer to 10%.

A structural member of a given size can support more load if it is part of a continuous structure than if it is simply supported. The continuity permits the use of smaller members for the same loads and spans or increased spacing of supports for the same sized members. The possibility of fewer columns in buildings—or fewer piers in bridges—may permit a reduction in overall costs.

Continuous concrete or steel structures are cheaper without the joints, pins, and so on required to make them statically determinate, as was frequently the practice in past years. Monolithic reinforced-concrete structures are erected so that they are naturally continuous and statically indeterminate. Installing the hinges and other devices necessary to make them statically determinate is a difficult construction problem, and very expensive. Furthermore, if a building frame consisted of columns and simply supported beams, objectionable diagonal bracing between the joints would be necessary to provide a stable frame with sufficient rigidity.

Larger Safety Factors

Statically indeterminate structures often have higher safety factors than do statically determinate structures. When parts of statically indeterminate steel or reinforced-concrete structures are overstressed, they often have the ability to redistribute portions of those stresses to less-stressed areas. Statically determinate structures generally do not have this ability.[1] If the bending moments in a component of a statically determinate structure reach the ultimate moment capacity of that component, the structure will fail. This is not the case for statically indeterminate structures, since the load may be redistributed to other parts of the structure.

Greater Rigidity and Smaller Deflections

Statically indeterminate structures are more rigid and have smaller deflections than comparable statically determinate structures. Because of their continuity, they are stiffer and have greater stability against all types of loads (horizontal, vertical, moving, etc.).

[1] J. C. McCormac and S. F. Csernak. *Structural Steel Design* 5th ed. (Upper Saddle River, N.J.; Prentice Hall 2012), 243–254.

Courtesy of Daniel Metz

FIGURE 10.4 Collapse of a statically determinate steel frame building under construction.

More Attractive Structures

It is difficult to imagine statically determinate structures having the gracefulness and beauty of many statically indeterminate arches and rigid frames being erected today.

Adaptation to Cantilever Erection

The cantilever method of erecting bridges is of particular value where conditions underneath (probably marine traffic or deep water) hinder the erection of falsework. Continuous statically indeterminate bridges and cantilever-type bridges are conveniently erected by the cantilever method.

NOMENCLATURE
Falsework is another term for temporary shoring that must be put in place while a structure is under construction.

10.4 Disadvantages of Statically Indeterminate Structures

A comparison of statically determinate and statically indeterminate structures shows the latter have several disadvantages that make their use undesirable on many occasions. These disadvantages are discussed in the following paragraphs.

Support Settlement

Statically indeterminate structures are not desirable where foundation conditions are poor, because seemingly minor support settlements or rotations may cause major changes in the bending moments, shearing forces, reaction forces, and member forces. Where statically indeterminate bridges are used despite the presence of poor foundation conditions, it is occasionally necessary to physically measure the dead-load reactions. The supports of the bridge are jacked up or down until the calculated reaction forces are obtained. The support is then built to that elevation.

Development of Other Stresses

Support settlement is not the only condition that causes stress variations in statically indeterminate structures. Variation in the relative positions of members caused by temperature changes, poor fabrication, or internal deformation of members in the structure under load may cause significant force changes throughout the structure.

TEMPERATURE EFFECTS
Consider the determinate beam shown. If its temperature increases, the beam will want to elongate—which it is free to do.

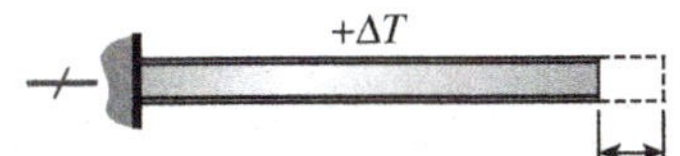

However, if the structure is fixed on each end (i.e., is indeterminate), it wants to elongate when it is heated up—but it is not allowed to do so. This produces both reactions and internal forces in the beam.

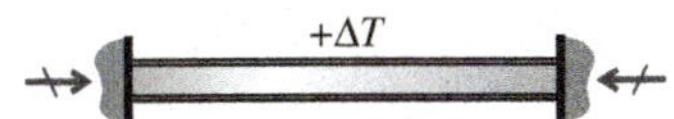

Difficulty of Analysis and Design

The forces in statically indeterminate structures depend not only on their dimensions but also on their elastic and cross-sectional properties (moduli of elasticity, moments of inertia, and areas). This situation presents a design difficulty: The forces cannot be determined until the member sizes are known, and the member sizes cannot be determined until their forces are known. The problem is handled by assuming member sizes and computing the forces, designing the members for these forces, and computing the forces for the new sizes, and so on, until the final design is obtained. Design by this method—*the method of successive approximations*—takes more time than the design of a comparable statically determinate structure, but the extra cost is only a small part of the total cost of the structure. Such a design is best done by interaction between the designer and the computer. Interactive computing is now used extensively in the aircraft and automobile industries.

Stress Reversals

Generally, more force reversals occur in statically indeterminate structures than in statically determinate structures. Additional material(s) may be required at certain sections to resist the different force conditions and to prevent fatigue failures.

10.5 Methods of Analyzing Statically Indeterminate Structures

DID YOU KNOW?

To obtain a unique and complete solution to any structural analysis problem, four conditions must be satisfied.

Equilibrium Forces and moments will sum to be zero in any direction.

Compatibility Also known as *consistent displacements* and means that the displacements within the structure must agree (i.e., the structure doesn't break).

Force-Deformation Relations There is a relationship between the force and the deformation on a structure. A full solution cannot be found without this information, including material behavior.

Boundary Conditions The calculated forces and displacements require a knowledge of the support conditions demonstrated in Chapter 8.

Statically indeterminate structures contain more unknown forces than there are equations of static equilibrium. As such, they cannot be analyzed using only the equations of static equilibrium; additional equations are needed. Forces beyond those needed to maintain a stable structure are *redundant forces*. The redundant forces can be reaction forces or forces in members that make up the structure. There are two general approaches used to find the magnitude of these redundant forces: *force methods* and *displacement methods*. The bases of these methods are discussed in this section.

Force Methods

With force methods, equations of condition involving displacement (i.e., displacement compatibility) at each of the redundant forces in the structure are introduced to provide the additional equations necessary for solution. Equations for displacement at and in the direction of the redundant forces are written in terms of the redundant forces; one equation is written for the displacement condition at each redundant force. The resulting equations solve for the redundant forces, which must be sufficiently large to satisfy the boundary conditions. As we will soon see, the boundary conditions do not necessarily have to be zero displacement. Force methods are also called *flexibility methods* or *compatibility methods*.

In 1864, James Clerk Maxwell published the first consistent force method for analyzing statically indeterminate structures. His method was based on a consideration of deflections, but the presentation (which included the reciprocal deflection theorem) was rather brief and attracted little attention. Ten years later, Otto Mohr independently extended the theory to almost its present stage of development. Analysis of redundant structures with the use of deflection computations is often referred to as the *Maxwell-Mohr method* or the *method of consistent distortions*.[2,3]

The force methods of structural analysis are somewhat useful for analyzing beams, frames, and trusses that are statically indeterminate to the first or second degree. They are also convenient for some single-story frames with unusual dimensions. For structures that are highly statically indeterminate, such as multistory buildings and large complex trusses, other methods are more appropriate and useful. These methods, which include moment distribution and the matrix

[2] J. I. Parcel and R. B. B. Moorman, *Analysis of Statically Indeterminate Structures* (New York: Wiley, 1955), 48.
[3] J. S. Kinney, *Indeterminate Structural Analysis* (Reading, Mass.: Addison-Wesley, 1957), 12–13.

methods, are more satisfactory and are introduced in later chapters. As such, the force methods have been almost completely superseded by the methods of analysis described in Chapters 19–21 of this text. Nevertheless, a study of the force methods provide an understanding of the behavior of statically indeterminate structures that might not otherwise be obtained.

Displacement or Stiffness Methods

In the displacement methods of analysis, the displacement of the joints (rotations and translations) necessary to describe fully the deformed shape of the structure are used in the equations instead of the redundant forces used in the force methods. When the simultaneous equations that result are solved, these displacements are determined and then substituted into the original equations to determine the various internal forces. The most commonly used displacement method is the matrix stiffness method discussed in Chapters 19 through 21.

10.6 Looking Ahead

In Chapters 11 and 12 the classical method of analyzing statically indeterminate structures using consistent distortions (flexibility method) is presented. This force-based method is primarily of historical interest and is almost never used in practice. However, it does help to establish an understanding of structural behavior, and it does form the basis for modern methods of analysis.

Although computer-based matrix methods (see Chapters 19 through 21) have become the mainstay for the structural engineer, many engineers still find it useful to have some hand-based techniques that can be used as a sanity check for computer output. Thus, the moment distribution method is presented in Chapters 13 and 14, and some quick yet approximate approaches are presented in Chapter 15.

QUICK NOTE

Truthfully speaking, computer analyses based on matrix methods are almost the "only game in town" today for achieving production.

11 Force Method for Statically Indeterminate Structures

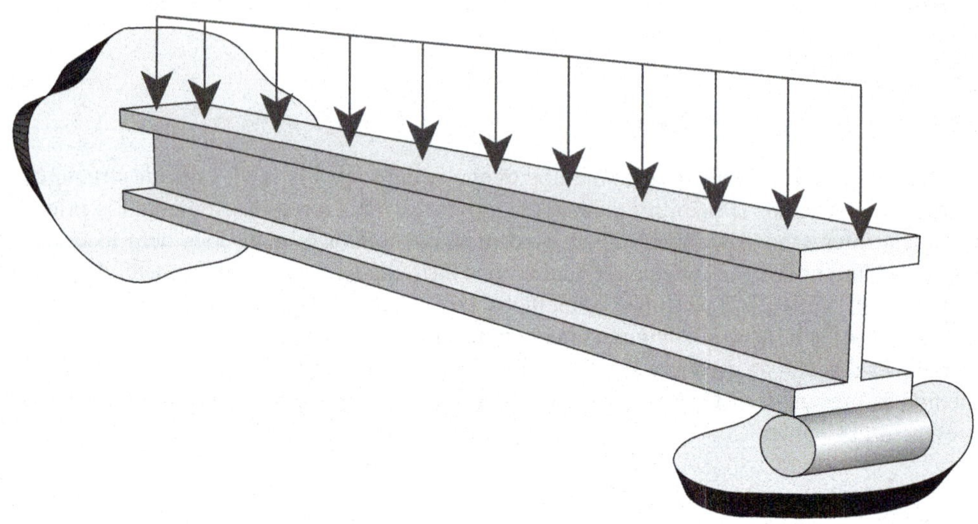

11.1 Beams and Frames with One Redundant

QUICK NOTE
The deflection at B in Figure 11.1(b) can be calculated using any number of methods, including virtual work or using a beam chart as discussed in Section 11.2.

The force method of analyzing statically indeterminate structures is often referred to as the *method of consistent distortions* or the *flexibility method.* For a first illustration of the consistent distortions, the two-span beam of Figure 11.1(a) is considered. The beam is assumed to consist of a material that follows Hooke's law. This statically indeterminate structure supports the loads P_1, and P_2 and is in turn supported by reaction components at points A, B, and C. Removal of support B would leave a statically determinate beam, proving the structure to be statically indeterminate to the first degree. With this support removed, it is a simple matter to find the deflection at B, which is Δ_B in Figure 11.1(b) and is caused by the external loads.

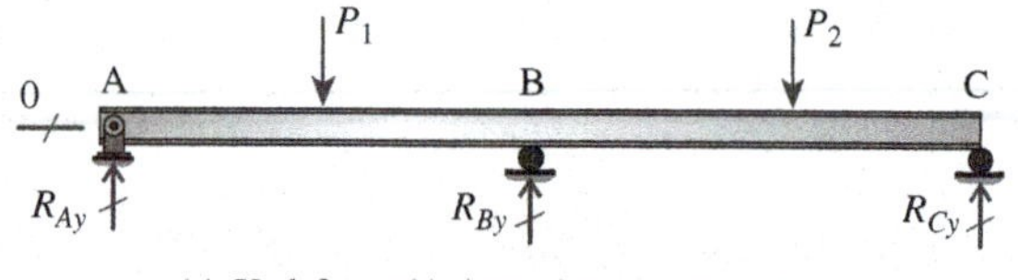

(a) Undeformed indeterminate beam—original

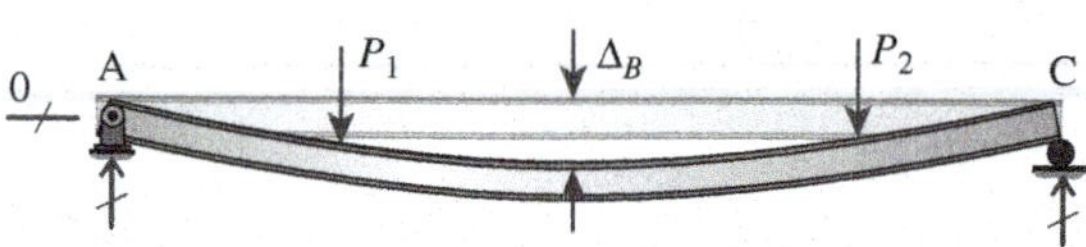

(b) Support B removed—determinate structure

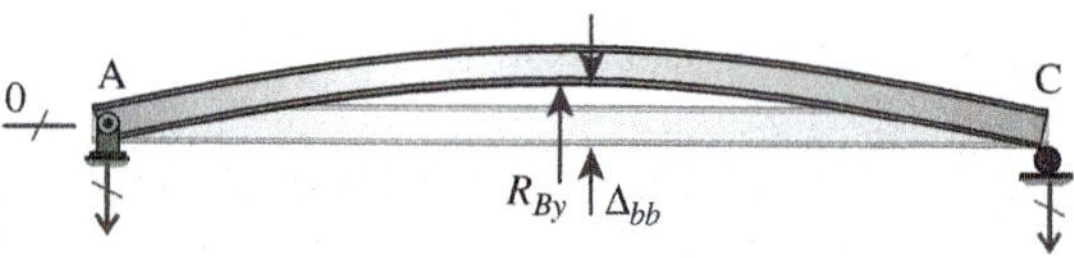

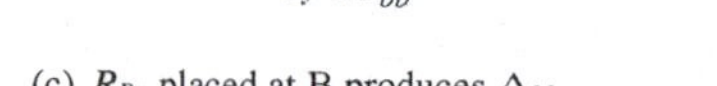

(c) R_{By} placed at B produces Δ_{bb}

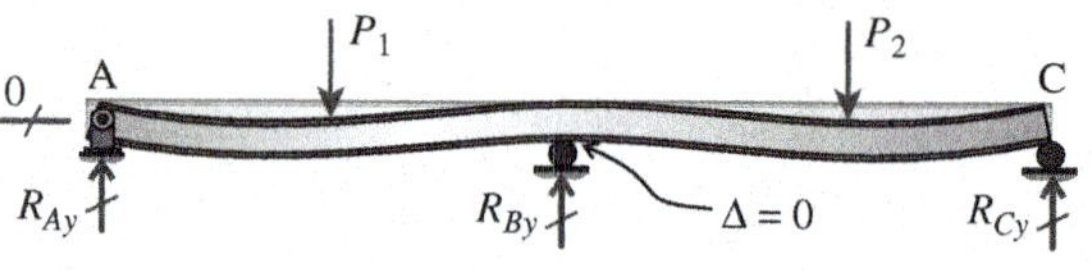

(d) Structures (b) and (c) superimposed to get displacement at B to be the same as in the original structure (a)

FIGURE 11.1
Demonstration of thought process for solving indeterminate structures using the force method.

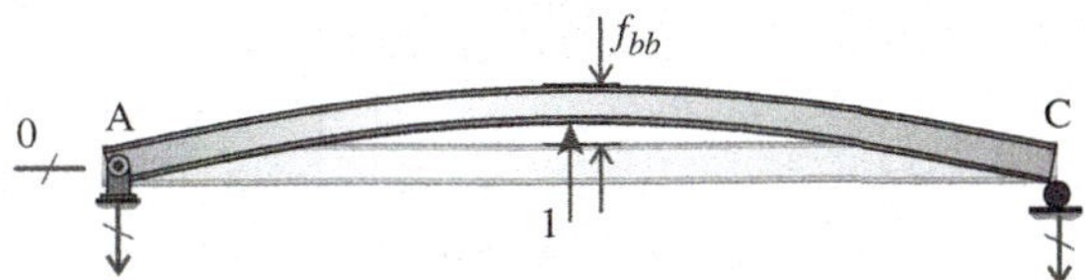

FIGURE 11.2 Unit loaded beam at B.

If the external loads are removed from the beam and the force R_{By} is placed at B, a deflection at B equal to Δ_{bb} will be developed, as indicated in Figure 11.1(c). Deflections due to external loads are denoted with capital letters herein. The deflection at point B on a beam due to external loads would be Δ_B. Deflections due to the *redundant load—R_{By}—*is denoted with two small letters. The first letter indicates the location of the deflection and the second letter indicates the location of the redundant load. For example, the deflection of a structure at E caused by a load at B would be written as Δ_{eb}.

Support B is unyielding, and its removal is merely a convenient assumption so that a determinate beam may be evaluated. An upward force is present at B, and it is sufficient to prevent any deflection. Another way of looking at it is that there is a force at B labeled R_{By} that is large enough to push point B back to its original non-deflected position. Said another way—the force R_{By} at B needs to produce a deflection Δ_{bb} at B that is equal in magnitude to Δ_B. This is stated mathematically using a *compatibility equation*:

$$\Delta_B + \Delta_{bb} = 0$$

The deflection Δ_B at B due to P_1 and P_2 in Figure 11.1(b) is easily found using previously learned techniques. However, the deflection at B due to the redundant force, R_{By} in Figure 11.1(c) is not so easily found. Not because the loading scenario is difficult to evaluate but because the value of R_{By} is unknown.

Consider the beam in Figure 11.2 that has a unit load at B causing a deflection at B equal to f_{bb}. If the deflection at B due to a unit load is f_{bb}, then the deflection at B due to a 10-kip load can be written as $10f_{bb}$, where the magnitude of the applied load is simply a scale factor for the flexibility coefficient f_{bb}. Therefore, the deflection at B due the redundant force R_{By} may be conveniently written as

$$\Delta_{bb} = f_{bb}R_{By}$$

Now we can substitute this known expression for Δ_{bb} into the compatibility equation which results in

$$\Delta_B + f_{bb}R_{By} = 0$$

The compatibility equation may be reworked to solve for the unknown redundant force R_{By}. So

$$R_{By} = -\frac{\Delta_B}{f_{bb}}$$

Thus, the unknown redundant force can be calculated by first solving for the two displacements Δ_B and f_{bb}. This whole equation was developed simply by ensuring that the displacement at B is equal to zero. This is a known quantity because of the roller support that is there in the original beam, as seen in Figure 11.1(d).

If we keep track of signs on the computed deflections, the sign on the final answer will indicate whether the assumed direction for R_{By} was correct. A plus sign indicates a correct assumption, and a minus sign means the assumption was incorrect.

Procedure

1. Identify the degree of statical indeterminacy (n).
2. Identify and remove the n unknown redundant forces and/or moments. The goal here is to reduce the structure to a determinate structure.
3. Construct the *PRIMARY* structure with the actual loads present but have the redundant forces that were identified in the previous step removed.

NOMENCLATURE

A **redundant force** is any resulting force—either internal force or support reaction—which is more than what would be required for the structure to be stable. For example, R_{By} in Figure 11.1 is taken as the redundant force because once it is removed the structure is still stable.

QUICK NOTE

The **compatibility equation** is a statement regarding known deformations on the structure. Such an equation provides the additional information beyond the equilibrium equations needed to solve indeterminate structures.

DID YOU KNOW?

The displacement on a structure due to a single unit load is known as the *flexibility coefficient* and is denoted in this text as f with two subscripts. For example, the flexibility coefficient f_{eb} would be read as the *deflection at point e due to a **unit load** at point b.*

VARIOUS STRUCTURES

For convenience, the following names are given to the various structures which are to be sketched.

PRIMARY is shown in Figure 11.1(b).

REDUNDANT is shown in Figure 11.1(c).

FLEXIBILITY is shown in Figure 11.2.

The primary and redundant structures are used to write the compatibility equation. The only deflections that must be calculated are found on the primary and flexibility structures.

4. Draw the deflected shape and identify the displacement/rotation at the points of each redundant force/moment. (e.g., Δ_B, θ_C, etc.).
5. Construct the *REDUNDANT* structure(s) with no other loads on it other than the redundant force/moment. A separate redundant structure for each redundant force/moment is needed. This should produce *n* redundant structures.
6. Sketch the deflected shape for each redundant structure and identify the displacement/rotation at the point of each redundant force/moment. (e.g., Δ_{bb}, θ_{ec}, etc.)
7. Write a compatibility equation for each location on the structure where a redundant force exists. This means to sum up all of the displacements at a given point from the primary and redundant structures. Equate this summation to a known displacement at that point. Remember that these displacements may be written in terms of their flexibility coefficients.
8. Determine the values of the deflections and flexibility coefficients. (e.g., Δ_B and f_{bb} using virtual work, beam charts, or any other method.)
9. Substitute computed displacements/rotations into the compatibility equation(s) and solve for the unknown redundant forces and moments.
10. Use the equations of equilibrium to solve for the remaining reactions and internal forces/moments.

FRIENDLY ADVICE

Experience has shown that those who do not sketch all of the indicated structures (primary, redundant, flexibility) during the learning process will persistently struggle with knowing what they should do. As experience is gained, one may abbreviate the outlined procedure and not make so many sketches.

The following examples illustrate the method of consistent distortions for both beams and frames. The key is to be able to set up the problem correctly by being able to assemble appropriate compatibility equations. The examples will not show the computation of the needed displacements but will rather indicate the method used to calculate them. The two primary methods the reader should consider is the use of *beam deflection charts* found in Appendix D or the use of *virtual work*. Note that once all reactions are known the shear and moment diagrams may be sketched.

EXAMPLE 11.1

Determine the reactions for the two-span beam shown using the method of consistent distortions (force method). $E = 29{,}000$ ksi and $I = 1000$ in^4.

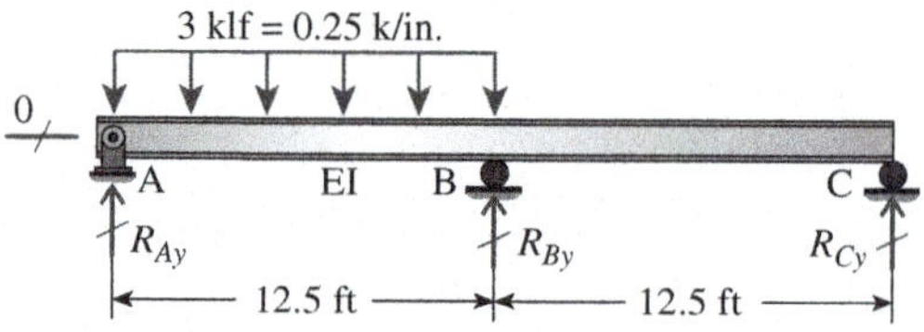

SOLUTION STRATEGY

Follow every step as outlined in the previous procedure, including making all of the necessary sketches. Do not begin calculating any deflections until the compatibility equation is constructed.

SELECTING A REDUNDANT

Only forces that cannot be found directly from equilibrium are eligible for selection. For example, the reaction in the *x*-direction can be found by $\sum F_X$ without any complications. Therefore it is not a valid option. See Example 11.2 for further guidance on redundant selection.

SOLUTION

Find the determinacy of the beam.

$$F = 4 \qquad n = 1 \qquad N = (1)(3) = 3$$
$$\text{SI} = 4 - 3 = \text{SI } 1^\circ$$

This requires that one redundant force is selected and removed and a deflected shape is sketched. We will select R_{By}.

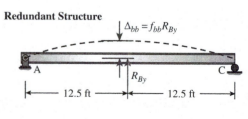

Primary Structure

3 klf = 0.25 k/in. Δ_B A C 12.5 ft 12.5 ft

Redundant Structure

$\Delta_{bb} = f_{bb}R_{By}$ A R_{By} C 12.5 ft 12.5 ft

The redundant structure is sketched with all of the original loads removed and the redundant force R_{By} applied at B. The compatibility equation now may be constructed by summing up the displacements at B and setting them equal to the actual displacement at B in the real structure, which is 0 inches in this case.

Compatibility Equation:

$$\Delta_B + \Delta_{bb} = 0 \Rightarrow \Delta_B + f_{bb}R_{By} = 0$$

$$\therefore R_{By} = -\frac{\Delta_B}{f_{bb}}$$

To solve for the unknown redundant force R_{By}, one needs only to solve for two displacements—Δ_B and f_{bb}. A sketch of the flexibility structure is made to clarify the meaning of f_{bb}, which is a deflection at B due to a unit load at B.

Flexibility Structure:

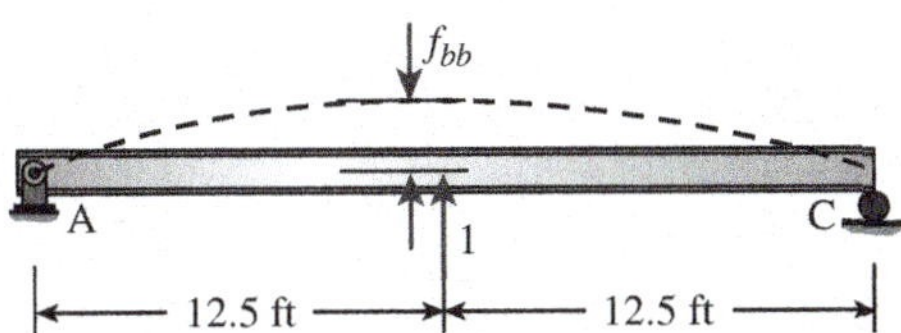

The process of calculating deflections was established in Chapters 8 and 9. For calculating Δ_B, the method of virtual work is used. This calculation was done in Example 9.4 and found to be 0.455 in. ↓. Since it is deflecting in the opposite direction of the assumed R_{By}, the deflection is considered to be negative.

$$\Delta_B = -0.455 \text{ in.}$$

The calculation of f_{bb} is easily handled by using the beam deflection charts provided in Appendix D. Beam 6 applies in this case and provides the following equation for the deflection at mid-span.

$$f_{bb} = \frac{PL^3}{48EI} = \frac{(1)(300 \text{ in.})^3}{48(29{,}000 \text{ ksi})(1000 \text{ in}^4)} = 0.0194 \text{ in./k}$$

A substitution of these displacements into the compatibility equation yields

$$R_{By} = -\frac{-0.455 \text{ in.}}{0.0194 \text{ in./k}} = +23.45 \text{ k} = \underline{\underline{23.45 \text{ k} \quad \uparrow}}$$

With R_{By} found, basic statics using $\sum M_A$ and $\sum F_y$ will find the remaining vertical reactions. Thus,

$$R_{Ay} = \underline{\underline{16.4 \text{ k} \quad \uparrow}} \qquad R_{Cy} = \underline{\underline{2.35 \text{ k} \quad \downarrow}}$$

DID YOU KNOW?

If Δ_B had not already been calculated by virtual work in Example 9.4, then it could have been calculated easily using beam 10 from the beam charts found in Appendix D.

QUICK NOTE

f_{bb} is positive because it displaces in the assumed direction of the redundant force R_{By}.

EXAMPLE 11.2

Determine the reactions for the propped cantilever beam shown using the method of consistent distortions. E and I are constant along the length.

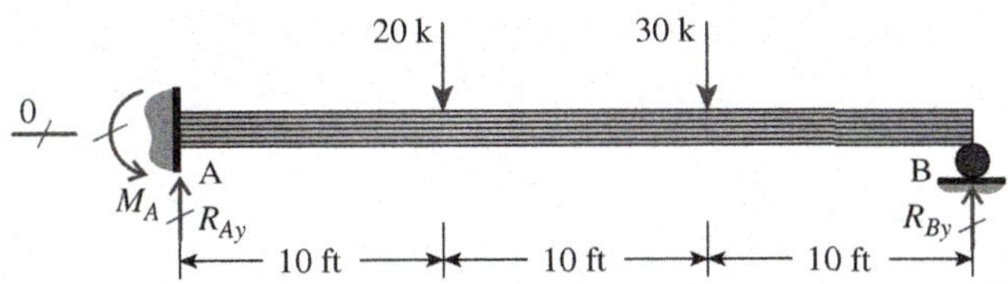

SOLUTION STRATEGY

Follow every step as outlined in the previous procedure, including making all of the necessary sketches. Do not begin calculating any deflections until the compatibility equation is constructed.

SELECTING A REDUNDANT

Try to select a redundant force that when removed leaves simple structures. For example, selecting R_{By} and removing it will result in a cantilever beam. Beam charts are available for cantilever beams.

SOLUTION

Find the determinacy of the beam.

$$F = 4 \qquad n = 1 \qquad N = (1)(3) = 3$$
$$\text{SI} = 4 - 3 = \text{SI } 1^\circ$$

This will require that one redundant force is selected and removed and a deflected shape is sketched. We will select R_{By}.

Primary Structure

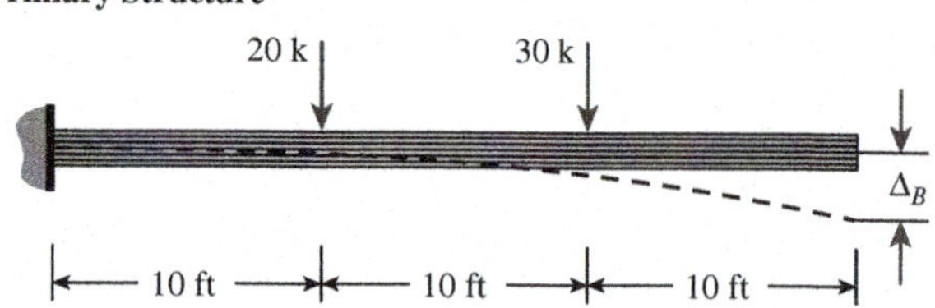

Redundant Structure

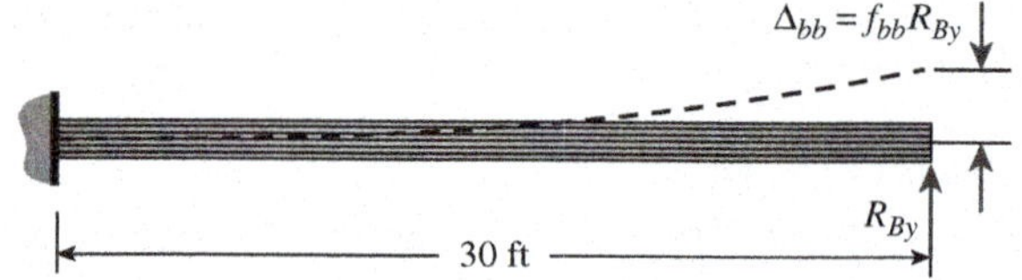

The redundant structure is sketched with all of the original loads removed and the redundant force R_{By} applied at B. Write the compatibility equation by summing displacements at B.

Compatibility Equation:

$$\Delta_B + \Delta_{bb} = 0 \Rightarrow \Delta_B + f_{bb}R_{By} = 0$$
$$\therefore R_{By} = -\frac{\Delta_B}{f_{bb}}$$

Solve for the two displacements Δ_{By} and f_{bb}.

Flexibility Structure:

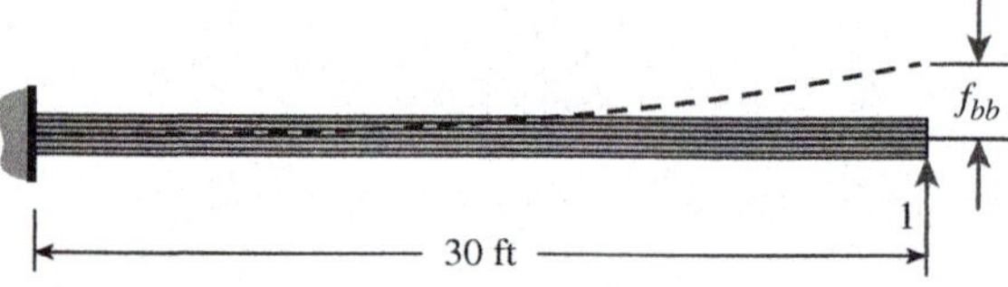

QUICK NOTE

Solving for Δ_B requires the superposition of the effects of each point load. Analyze the beam for 20 k, and then analyze it again for 30 k.

Since the primary and flexibility structures are cantilever beams, the beam deflection tables in Appendix D easily may be used for deflection calculations. Use beam 2.

$$\Delta_B = -\frac{(20\text{ k})(10\text{ ft})^2}{6EI}[(3)(30\text{ ft}) - 10\text{ ft}]$$
$$-\frac{(30\text{ k})(20\text{ ft})^2}{6EI}[(3)(30\text{ ft}) - 20\text{ ft}] = -\frac{1{,}000{,}000}{6EI}$$

The calculation of f_{bb} is just as easily handled by using the beam deflection charts. Beam 1 applies in this case and provides the following equation for the deflection at point B.

$$f_{bb} = \frac{PL^3}{3EI} = \frac{(1)(30)^3}{3EI} = \frac{9000}{EI}$$

QUICK NOTE

Make all calculations using consistent units from the beginning to the end of the problem. This way units conversion mistakes may be avoided. This example used kips and feet.

A substitution of these displacements into the compatibility equation yields

$$R_{By} = -\frac{-(1{,}000{,}000/6EI)}{(9000/EI)} = +18.52\text{ k} = \underline{\underline{18.52\text{ k} \quad \uparrow}}$$

With R_{By} found, basic statics using $\sum M_A$ and $\sum F_y$ will find the remaining vertical reactions. So

$$R_{Ay} = \underline{\underline{31.48\text{ k} \quad \uparrow}} \qquad M_A = \underline{\underline{244.4\text{ k}\cdot\text{ft} \quad \circlearrowleft}}$$

EXAMPLE 11.3

Rework Example 11.3 by using M_A at the fixed end as the redundant.

SOLUTION STRATEGY

Follow the same approach as has been used in Example 11.2.

SOLUTION

This beam is statically indeterminate to the first degree. Remove M_A and sketch both the primary and redundant structures.

Primary Structure

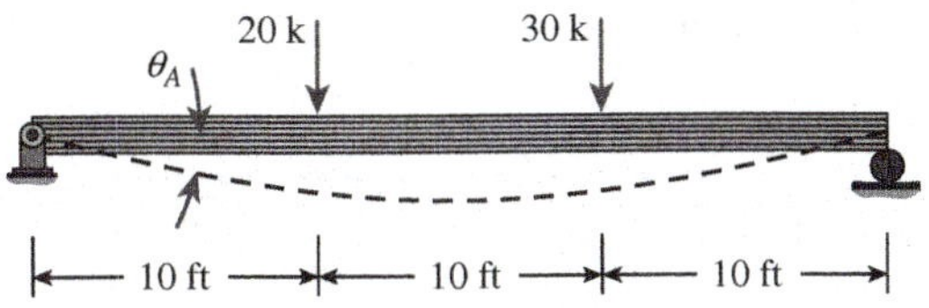

Redundant Structure

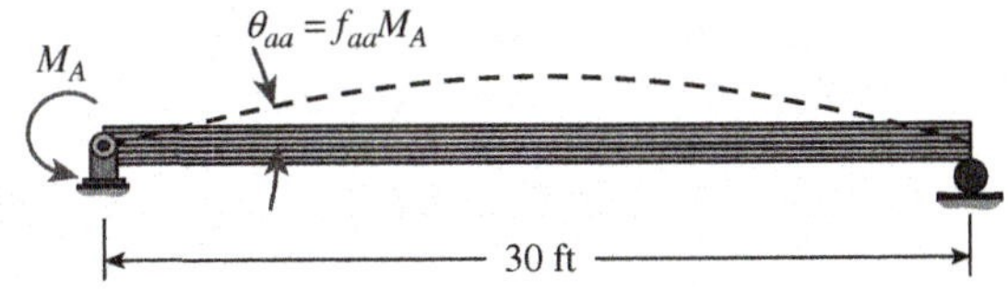

The redundant structure is sketched with all of the original loads removed and the redundant moment M_A applied at A. Write the compatibility equation by summing rotations at A.

Compatibility Equation:

$$\theta_A + \theta_{aa} = 0 \Rightarrow \theta_A + f_{aa}M_A = 0$$

$$\therefore M_A = -\frac{\theta_A}{f_{aa}}$$

Solve for the two rotations θ_A and f_{aa}.

Flexibility Structure:

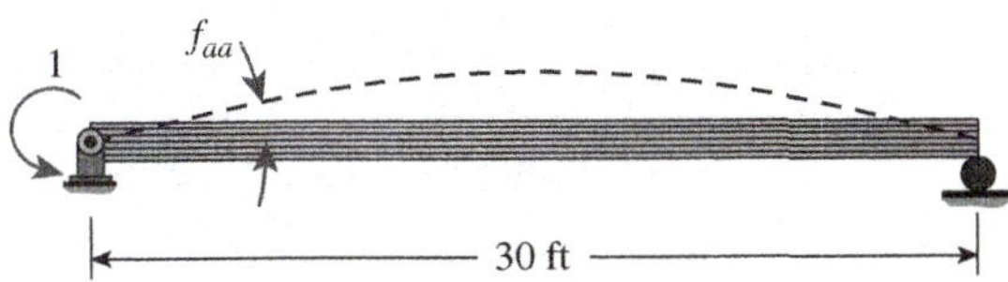

Since the primary and flexibility structures are simply supported beams, the beam deflection tables may be used for rotation calculations. Use beam 7.

$$\theta_A = -\frac{(20\text{ k})(20\text{ ft})\left[(30\text{ ft})^2 - (20\text{ ft})^2\right]}{6(30\text{ ft})EI} - \frac{(30\text{ k})(10\text{ ft})\left[(30\text{ ft})^2 - (10\text{ ft})^2\right]}{6(30\text{ ft})EI} = -\frac{22{,}000}{9EI}$$

QUICK NOTE

Solving for θ_A requires the superposition of the effects of each point load. Analyze the beam for 20 k, and then analyze it for 30 k.

The calculation of f_{aa} is just as easily handled by using the beam deflection charts. Beam 8 applies in this case and provides the following equation for the rotation at point A.

$$f_{aa} = \frac{ML}{3EI} = \frac{(1)(30\text{ ft})}{3EI} = \frac{10}{EI}$$

A substitution of these rotations into the compatibility equation yields

$$M_A = -\frac{-(22{,}000/9EI)}{(10/EI)} = +244.4\text{ k·ft} = \underline{\underline{244.4\text{ k·ft}\ \circlearrowleft}}$$

SUMMARY

Selection of M_A as a redundant produces the same results as found in Example 11.2. The positive sign on the answer indicates that the correct assumption was made for the reaction.

EXAMPLE 11.4

Rework Example 11.1 by using the internal moment M_B at B as the redundant.

SOLUTION STRATEGY

Consider the impact that removing the internal moment at B has on the behavior of the beam. This effectively places a moment release at B right above the roller support. All rotations and compatibility equations should be written in terms of relative rotations between spans AB and BC.

SOLUTION

This beam is statically indeterminate to the first degree. Remove M_B and sketch both the primary and redundant structures. Notice how the left and right spans of the beam now can rotate relative one to another.

Primary Structure

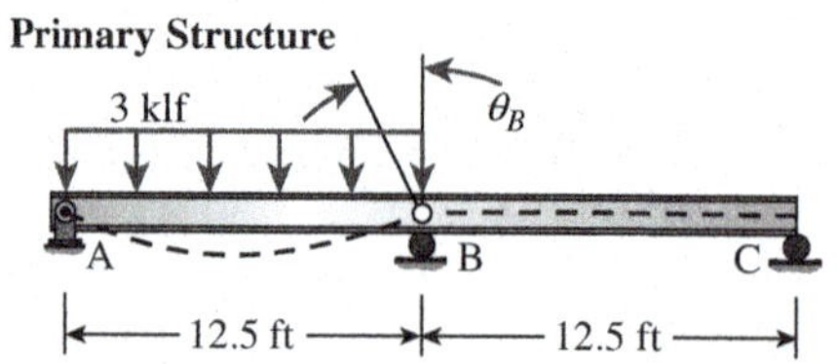

Redundant Structure

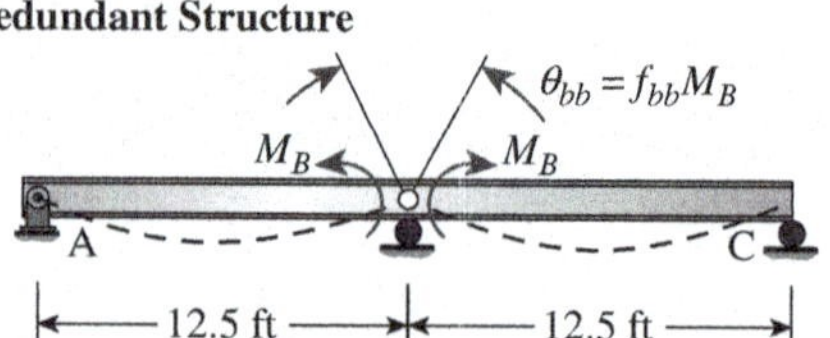

QUICK NOTE
The fact that the compatibility equation is set equal to zero does not say that the rotation at B in the original structure is zero. Rather, it is a statement that says that the relative rotation between spans AB and BC is zero.

The redundant structure is sketched with all of the original loads removed and the redundant moment M_B applied at B. Since M_B is an internal moment, it is represented as equal and opposite moments on either side of the moment release. Write the compatibility equation by summing the relative rotations at B.

Compatibility Equation:

$$\theta_B + \theta_{bb} = 0 \Rightarrow \theta_B + f_{bb}M_B = 0$$

$$\therefore M_B = -\frac{\theta_B}{f_{bb}}$$

Solve for the two rotations θ_B and f_{bb}.

QUICK NOTE
In the primary structure, span AB is loaded and will rotate at B, but span BC will not rotate. For the flexibility structure, both spans are loaded with a unit moment so the total rotation at B (f_{bb}) is a summation of the rotation of both beams.

Flexibility Structure:

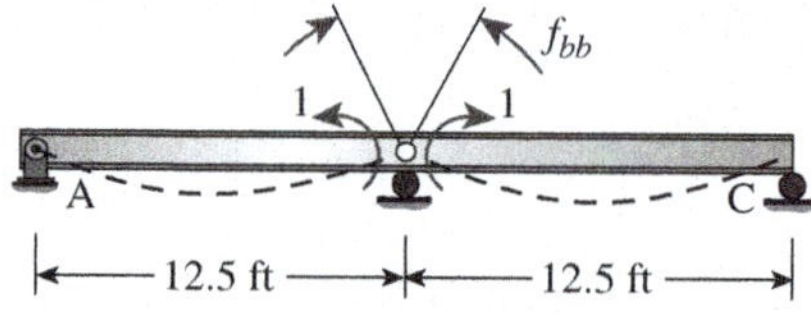

Each span of the primary and flexibility structures are effectively simple supported beams, so the beam deflection tables may be used for rotation calculations. Use beams 8 and 9.

$$\theta_B = \frac{wL^3}{24EI} = \frac{(3klf)(12.5\text{ ft})^3}{24EI} = \frac{244.1}{EI}$$

DID YOU KNOW?
The 2 in the f_{bb} equation accounts for the contribution of both beam segments to the overall rotation.

The calculation of f_{bb} is found from using a simple supported beam with a moment applied to the end (beam 8).

$$f_{bb} = \frac{ML}{3EI} - \frac{-ML}{3EI} = (2)\frac{(1)(12.5\text{ ft})}{3EI} = \frac{25}{3EI}$$

A substitution of these rotations into the compatibility equation yields

$$M_B = -\frac{(244.1/EI)}{(25/3EI)} = \underline{\underline{-29.3\text{ k·ft}}}$$

QUICK NOTE
The internal moment at B was assumed to be positive. Since the number turned out negative, we know that there is a negative moment at B.

Make a cut just to the right of B and draw a FBD. Then solve for R_{Cy}, as

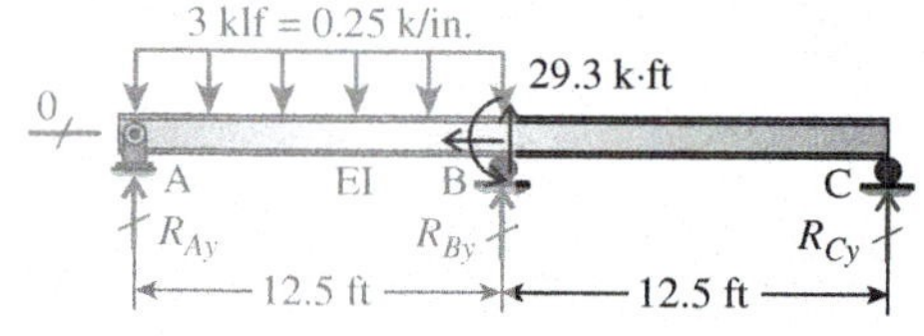

$$\circlearrowleft^{+} \sum M_B = (R_{Cy})(12.5\text{ ft}) + 29.3\text{ k·ft} = 0$$

$$\therefore R_{Cy} = -2.35\text{ k} = \underline{\underline{2.35\text{ k} \quad \downarrow}}$$

SUMMARY

This produces the same results as found in Example 11.1. It demonstrates that internal forces may be used as redundant forces, but it can be just a little more difficult for a beginner to visualize the displacements/rotations.

EXAMPLE 11.5

Determine the reactions for the frame shown using the method of consistent distortions. E and I are the same for both members.

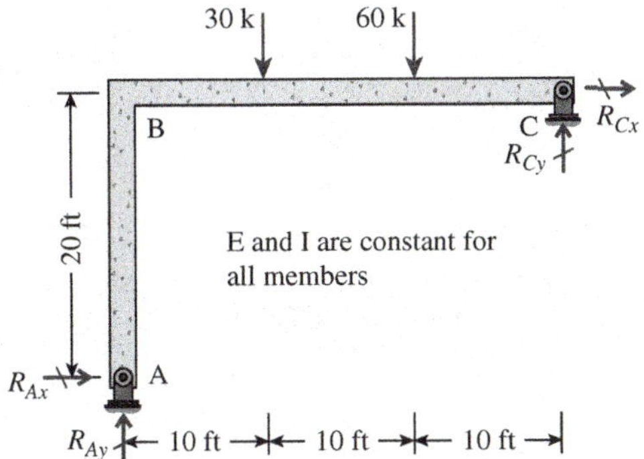

SELECTING A REDUNDANT

In this example, none of the reactions may be computed using just equilibrium. As such, any may be selected as the redundant force with no strong advantage of selecting one over the other.

SOLUTION STRATEGY

The approach for a frame is exactly identical to the approach used for the beam. The only difference comes when deflections are calculated.

SOLUTION

Find the determinacy of the frame.

$$F = 4 \qquad n = 1 \qquad N = (1)(3) = 3$$

$$\text{SI} = 4 - 3 = \text{SI } 1^\circ$$

This will require that one redundant force is selected and removed and a deflected shape is sketched. We will select R_{Ax}. The redundant structure is sketched with all of the original loads removed and the redundant force R_{Ax} applied at A. The compatibility equation may now be constructed by summing up the displacements at A and setting it equal to the actual displacement at A in the real structure which is 0 inches.

Primary Structure

30 k 60 k B C A 20 ft Δ_A 10 ft 10 ft 10 ft

Redundant Structure

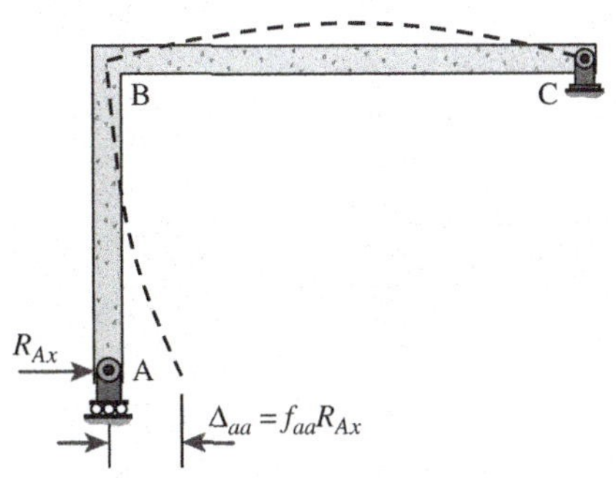

Compatibility Equation:

$$\Delta_A + \Delta_{aa} = 0 \;\Rightarrow\; \Delta_A + f_{aa}R_{Ax} = 0$$

$$\therefore R_{Ax} = -\frac{\Delta_A}{f_{aa}}$$

COMMON MISTAKE

The method of virtual work uses a unit-loaded structure to find deflection at a certain location. The flexibility structure is also a unit loaded structure. The roles of these structures are often confused. The unit load on the flexibility structure is a real load, whereas the unit load on the virtual structure is a virtual load used for computing m. These structures often look the same, but their roles are distinctly different.

This is why it is best to get the complete compatibility equation before any deflection calculations are performed.

To solve for the unknown displacements Δ_A and f_{aa}, the virtual work method introduced in Chapter 9 is used. The resulting displacements are computed as

$$\Delta_A = -\frac{86{,}600}{EI}$$

$$f_{aa} = \frac{6667}{EI}$$

Flexibility Structure:

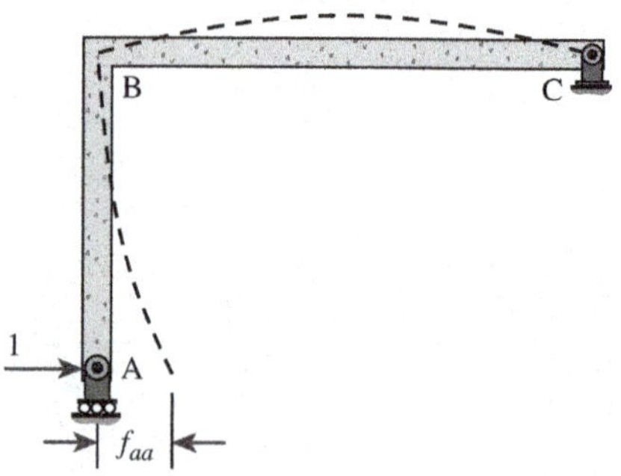

A substitution of these displacements into the compatibility equation yields

$$R_{Ax} = -\frac{-86{,}600/EI}{6667/EI} = +13\text{ k} = 13\text{ k} \rightarrow$$

With R_{Ax} found, basic statics using $\sum M_C$, $\sum F_y$, and $\sum F_x$ find the remaining three reactions:

$$R_{Cx} = 13\text{ k} \leftarrow \qquad R_{Ay} = 48.7\text{ k} \uparrow \qquad R_{Cy} = 41.3\text{ k} \uparrow$$

11.2 Maxwell's Law of Reciprocal Deflections

In dealing with this force method of analysis, many deflections must be computed to solve for unknown forces/moments. There is a surprising but useful relationship between the deflections at two points on a structure. This relationship was first published by James Clerk Maxwell in 1864. Maxwell's law may be stated as follows.

The deflection at one point (A) in a structure due to a load applied at another point (B) is exactly the same as the deflection at B if the same load is applied at A.

The rule is perfectly general and applies to any type of structure, whether it is a truss, beam, or frame that is made up of elastic materials following Hooke's law. The displacements may be caused by flexure, shear, or torsion.

Maxwell's law is not only applicable to the deflections in all of these types of structures but is also applicable to rotations. For instance, a unit couple at A will produce a rotation at B equal to the rotation caused at A if the same couple is applied at B.

Example 11.6 demonstrates that this law is correct for a simple cantilevered beam where the deflections at two points are determined using the beam deflection chart found in Appendix D.

QUICK NOTE

Maxwell's law is frequently applied when preparing influence lines for continuous structures, when analyzing statically indeterminate structures (as is done in this chapter), and with model-analysis problems.

EXAMPLE 11.6

Demonstrate Maxwell's law of reciprocal deflections by comparing the deflection at point B caused by a 10-kip load applied first at C (Δ_{BC}) with the deflection at C caused by a 10-kip load at B (Δ_{CB}). The two loading conditions and deflections are shown here. Assume E and I are constant along the length.

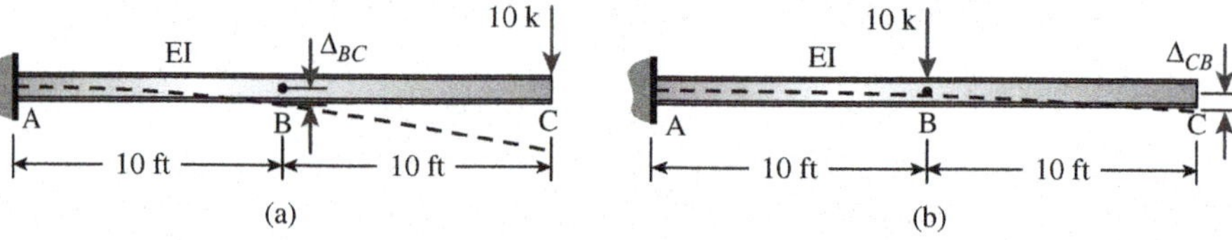

SOLUTION STRATEGY

Look in Appendix D and identify the beams that will be relevant. Beam (a) will use beam 1 from the deflection table, and beam (b) will use beam 2 from the deflection table.

SOLUTION

Compute Δ_{BC} first. Use the deflection equation given in the beam charts (beam 1) where $P = 10$ k, $L = 20$ ft, and $x = 10$ ft.

$$\Delta_{BC} = -\frac{Px^2}{6EI}(3L - x)$$

$$= -\frac{(10\text{ k})(10\text{ ft})^2}{6EI}[3(20\text{ ft}) - 10\text{ ft}] = -\frac{8333\text{ ft}^3\cdot\text{k}}{EI}$$

Now compute Δ_{CB}. Use the deflection equation given in the beam charts (beam 2) where $P = 10$ k, $L = 20$ ft, $a = 10$ ft, and $x = 20$ ft.

$$\Delta_{CB} = -\frac{Pa^2}{6EI}(3x - a)$$

$$= -\frac{(10\text{ k})(10\text{ ft})^2}{6EI}[3(20\text{ ft}) - 10\text{ ft}] = -\frac{8333\text{ ft}^3\cdot\text{k}}{EI}$$

SUMMARY

These two computed deflections are the same, which demonstrates Maxwell's law of reciprocal deflections: The deflection at B caused by a load at C is equal to the deflection at C caused by the same load at B.

11.3 Beams and Frames with Two or More Redundants

The force method of analyzing beams and frames with one redundant may be extended to beams and frames having two or more redundants. It will require that a separate compatibility equation is written for each redundant location. The continuous beam of Figure 11.3, which has two redundant reactions, is considered here.

To make the beam statically determinate, it is necessary to remove two supports. The reactions at B and C are selected as the redundants and are removed. This results in the primary structure shown in Figure 11.4 where the deflections Δ_B and Δ_C are due to the external loads. The external loads are theoretically removed from the beam, and a redundant force of R_{By} is placed at B. This is done again at point C, which results in a total of two redundant structures as shown in Figure 11.5.

Now a compatibility equation can be written for each location at B and C. This is done by summing up all of the displacements at B and C, respectively.

$$\Delta_B + f_{bb}R_{By} + f_{bc}R_{Cy} = 0$$
$$\Delta_C + f_{cb}R_{By} + f_{cc}R_{Cy} = 0$$

Figure 11.6 shows the flexibility structures that are a result of a unit load applied at each location independently.

There are now two compatibility equations that are written in terms of six displacements—four of which are flexibility coefficients. They also contain the two unknown redundant forces R_{By} and R_{Cy}. Once these equations are solved simultaneously, equilibrium may be used to solve for the rest of the reactions.

DID YOU KNOW?

The force method of analyzing indeterminate structures may be applied to structures with more than one redundant force. However, the reality is that it is not very feasible to do so for structures that have more than two redundants. Other methods such as moment distribution and the matrix method should be considered for structures with more redundants.

QUICK NOTE

Four flexibility coefficients are required to be calculated for this beam. Application of Maxwell's law, which was introduced in the previous section, can be of benefit here. It simply says that $f_{bc} = f_{cb}$. Thus, one need only calculate three flexibility coefficients instead of four.

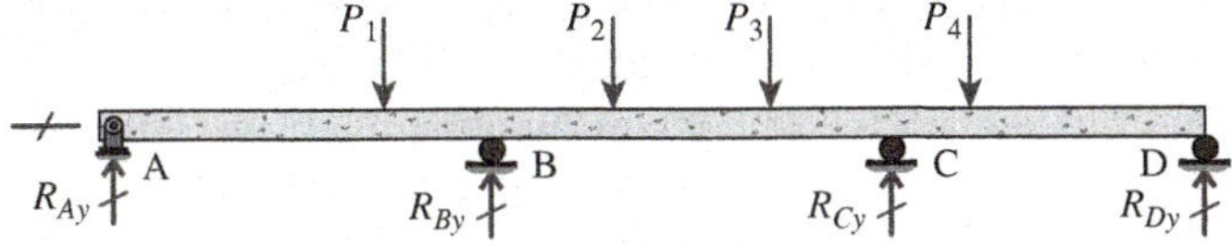

FIGURE 11.3 Multi-span beam with two redundant forces.

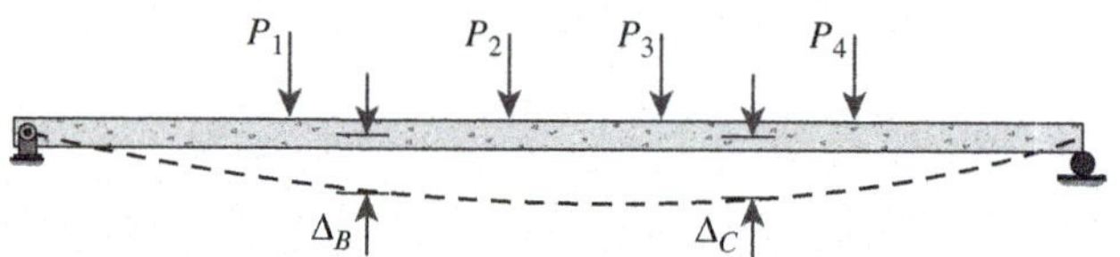

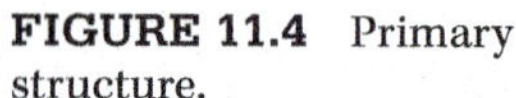

FIGURE 11.4 Primary structure.

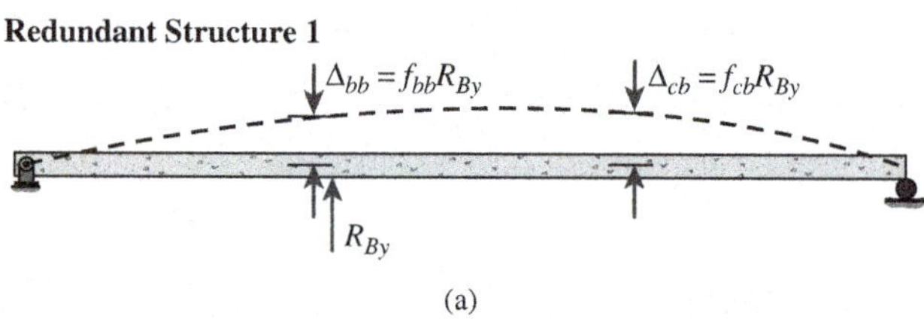

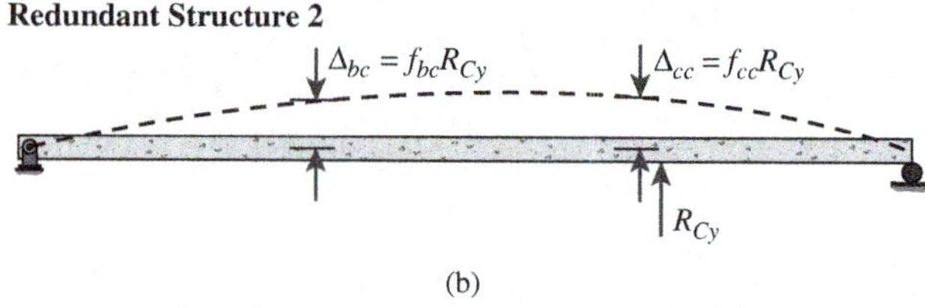

FIGURE 11.5 Redundant structures.

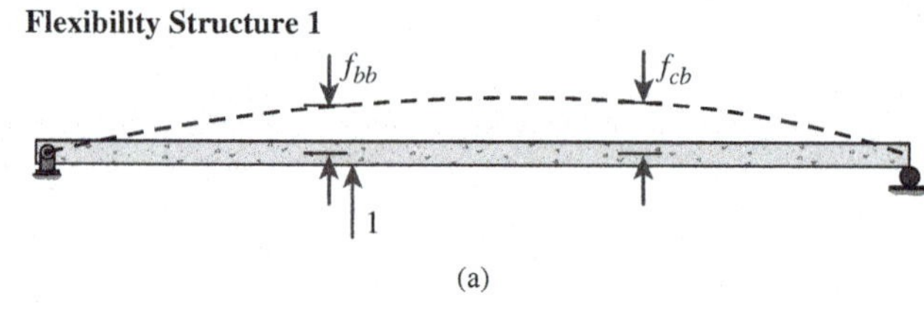

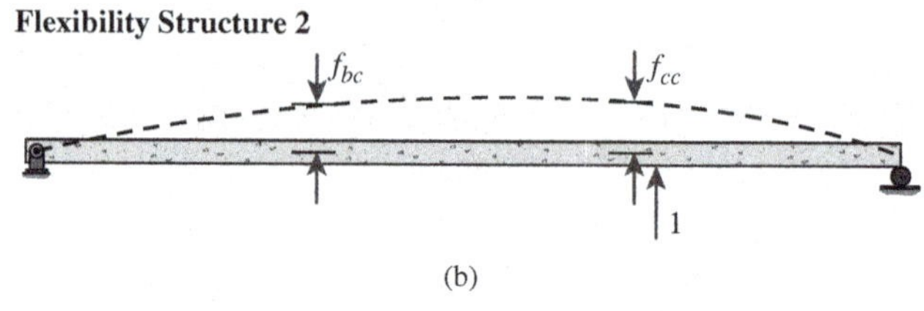

FIGURE 11.6 Flexibility structures.

The force method of computing redundant reactions may be extended indefinitely for structures with any number of redundants. The calculations become quite lengthy, however, if there are more than two redundants. For example, if three redundants existed, then three compatibility equations must be generated, which will result in 12 deflections needing to be calculated. This will be reduced to nine deflections if Maxwell's law is applied.

11.4 Support Settlement

Continuous beams with unyielding supports have been considered in the preceding sections. Should the supports settle or deflect from their theoretical positions, major changes may occur in the reactions, shears, moments, and stresses. Whatever the factors causing displacement (weak foundations, temperature changes, poor erection or fabrication, and so on), analysis may be made with the deflection expressions previously developed for continuous beams.

An expression for deflection at point B—the compatibility equation—in the two-span beam of Example 11.1 was previously written. The expression was developed on the assumption that support B was temporarily removed from the structure, allowing point B to deflect, after which the support was replaced. The reaction R_{By} at B was assumed to be of a sufficient magnitude to push B up to its original position of zero deflection. Should B actually settle 1.0 in., R_{By} will be smaller because it will only have to push B up a smaller amount. When settlement occurs, the only difference is to the right side of the compatibility equation.

QUICK NOTE

The right side of the compatibility equation should be set equal to the known settlement in the actual structure. It is best that the compatibility equation be written in terms of global axes. This means that if something moves downward, it is a negative number and if it moves upward, it is a positive number. Likewise, a movement to the right is positive and to the left is negative.

$$\Delta_B + f_{bb}R_{By} = -1 \text{ in.}$$

Consistent units must be used within the compatibility equation. The values of Δ_B and f_{bb} must be calculated in inches if the support movement is given in inches; they are calculated in feet if the support movement is given in feet, and so on. Example 11.7 illustrates the analysis of the two-span beam of Example 11.1 with the assumption of a $\frac{1}{4}$-in. settlement downward of the interior support (point B). The moment diagram is presented both for the case where no settlement occurs and for the case where settlement occurs.

EXAMPLE 11.7

Determine the reactions for the beam of Example 11.1 if support B settles downward by $\frac{1}{4}$ in. The beam is reproduced here for convenience. $E = 29{,}000$ ksi and $I = 1000$ in^4.

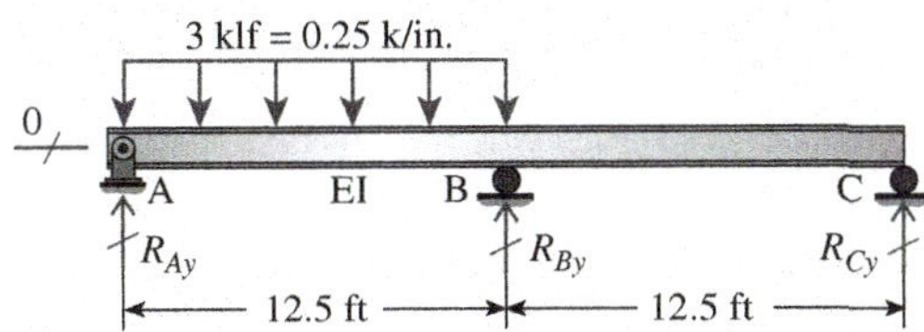

SOLUTION STRATEGY

Follow the exact same approach as demonstrated in Example 11.1. Draw the primary, redundant, and flexibility structures as have been shown. Sum up all the displacements at B to get the compatibility equation—only now set it equal to the settlement value instead of a zero displacement.

DID YOU KNOW?

If three people are walking along with a log on their shoulders (a statically indeterminate situation) and one of them lowers their shoulder slightly, they will not have to support as much of the total weight as before. They have, in effect, backed out from under the log and thrown more of its weight to the other people. The settlement of a support in a statically indeterminate continuous beam has the same effect.

SOLUTION

Using all of the primary, redundant, and flexibility sketches of Example 11.1, the compatibility equation is as shown.

QUICK NOTE

The negative sign assigned to the settlement value represents a downward motion.

Compatibility Equation:

$$\Delta_B + \Delta_{bb} = -0.25 \text{ in.} \Rightarrow \Delta_B + f_{bb}R_{By} = -0.25 \text{ in.}$$

$$\therefore R_{By} = -\frac{(\Delta_B + 0.25 \text{ in.})}{f_{bb}}$$

Previously, it was found that $\Delta_B = -0.455$ in. and $f_{bb} = 0.0194$ in./k. Using these, values R_{By} can be found as

$$R_{By} = -\frac{-0.455 \text{ in.} + 0.25 \text{ in.}}{0.0194 \text{ in./k}} = +10.57 \text{ k} = \underline{\underline{10.57 \text{ k} \uparrow}}$$

Setting up and evaluating equilibrium equations finds the remaining vertical reactions to be

$$R_{Ay} = \underline{\underline{22.84 \text{ k} \uparrow}} \qquad R_{Cy} = \underline{\underline{4.09 \text{ k} \uparrow}}$$

QUICK NOTE

Without a settlement at B, the reaction of C acts downward. With the settlement at B, the reaction not only changes magnitude but also changes direction.

The moment diagrams for both cases are shown for a comparison.

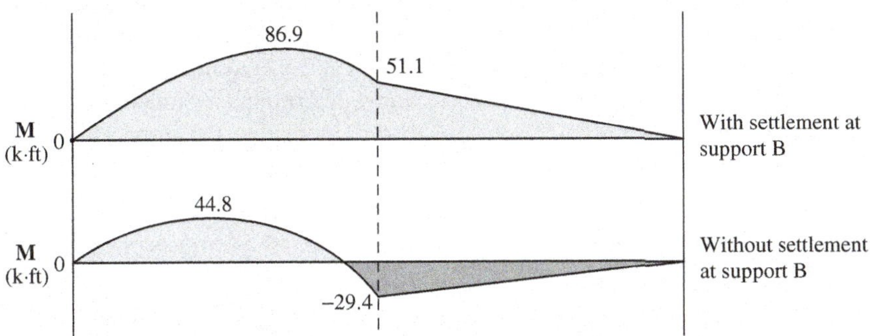

SUMMARY

The entire shape and magnitude of the moment diagram has changed between the two cases. This is all due to a settlement as small as 0.25 in. The moment diagram before settlement is repeated to illustrate the striking changes. The reader will now understand why structural engineers are reluctant to use statically indeterminate structures when foundation conditions are poor. Settlements may cause all sorts of changes and problems.

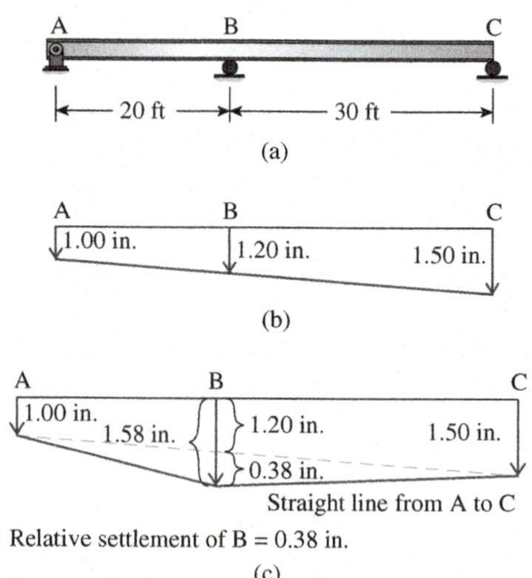

FIGURE 11.7

DID YOU KNOW?
The beam of Figure 11.7(a) has no applied loads to it. If a relative displacement occurs, stresses will develop within the beam. The Mandalay Bay casino in Las Vegas, Nevada, was built in 1998. Its center core was settling at different rate than the wings of the building. This caused visible distress in many locations in the structure.[1]

QUICK NOTE
The *EI* term takes into account cross-sectional and material properties associated with member bending. While not technically correct, this term is sometimes called a stiffness term for convenience. If *EI* is doubled, the stiffness doubles and any computed deflection is cut in half.

Multiple Support Settlement

When several or all of the supports are displaced, the analysis may be conducted on the basis of relative settlement values. For example, if all of the supports of the beam of Figure 11.7(a) were to settle 1.5 in., the stress conditions would theoretically be unchanged. If the supports settle different amounts but remain in a straight line, as illustrated in Figure 11.7(b), the situation theoretically is the same as before settlement.

Where inconsistent settlements occur and the supports no longer lie on a straight line, this causes the beam to distort because of the settlement that results in the development of internal stresses. The situation may be handled by drawing a line through the displaced positions of two of the supports, usually the end ones. The distances of the other supports from this line are determined and used in the calculations, as illustrated in Figure 11.7(c).

11.5 SAP2000 Computer Applications

In commercial computer analysis software packages like SAP2000, the analysis procedure for indeterminate structures is no different than for determinate structures. However, recall that one never needed to know cross-sectional or material properties (i.e., *EI*) to be able to calculate the reactions and internal forces for a determinate structure by hand. However, as shown in this chapter, one does need to know the relative cross-sectional and material properties to be able to analyze an indeterminate beam by hand. Since SAP2000 always requires the user to define these properties, it doesn't matter whether you are modeling a determinate or indeterminate structure.

The purpose of the following SAP exercise is to demonstrate the impact that the relative stiffness *EI* has on how loads are distributed in indeterminate structures.

SAP2000 EXERCISE 11.1

The beam of Example 11.1 has properties $E = 29{,}000$ ksi and $I = 1000$ in^4 and is modeled in the file *C11EX1.sdb*. In this exercise, the moment of inertia for each span can be adjusted separately. Leave $I_2 = 1000$ in^4 (see figure). Plot the moment diagram for the beam when $I_1 = I_2$. Then change I_1 so that it is a ratio of I_2, then find the maximum positive moment, the maximum negative moment, and the value of the reactions. Do this for the following ratios of I_1/I_2: 0.1, 0.5, 1.0, 2.0, and 10.0. Note the impact that the relative I values have on the how the forces are distributed in the beam.

[1]T. D. Richards and D. Kartofilis, "Micropile Underpinning of the Mandalay Bay Hotel & Casino." *54th Annual Geotechnical Engineering Conference*, St. Paul, Minnesota, February 2006.

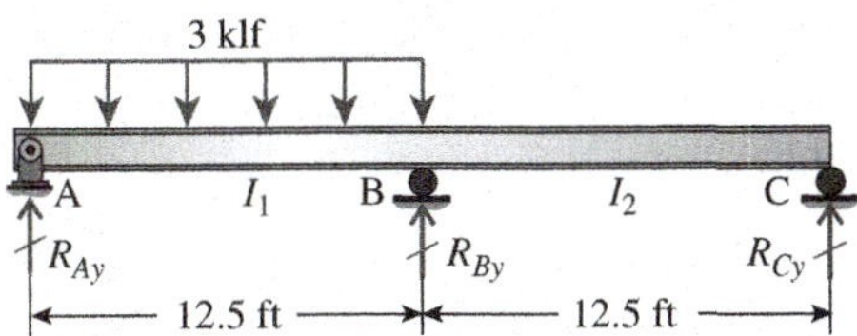

Here is the moment diagram when $I_1 = I_2$. The maximum positive moment is 44.9 k·ft, while the maximum negative moment is −29.3 k·ft.

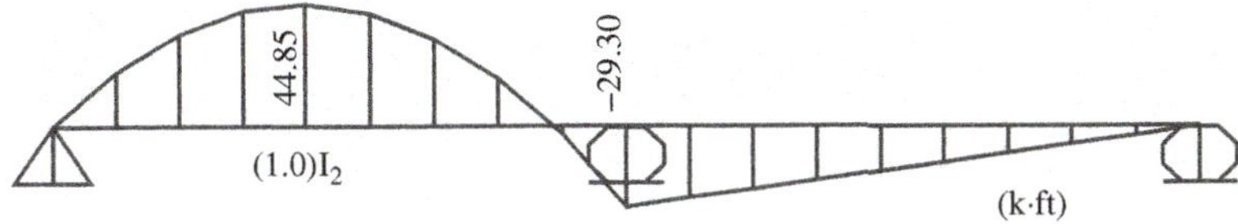

Now plot the maximum moment values for the various ratios of I_1/I_2.

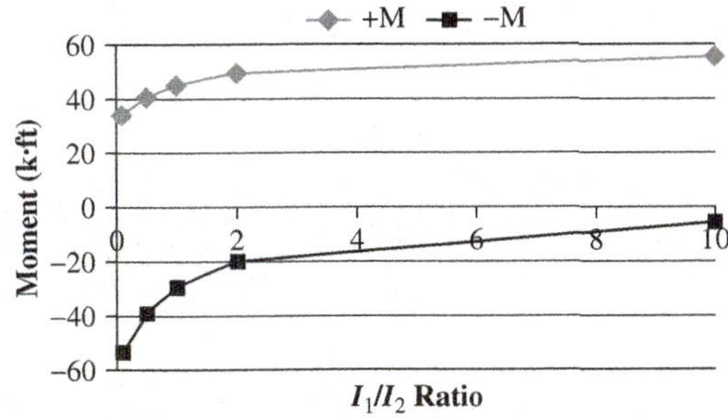

For this particular beam configuration and loading, the negative moment is most significantly impacted. As the moment of inertia of the left span increases relative to that of the right span, the left span takes more of the moment. Indeed, this is why the positive moment continues to increase as the ratio increases. The negative moment, which really is at the boundary of the left and right segments, approaches a zero asymptote, thus indicating that less load is traveling from the left span to the right.

Now plot the reaction values for the various ratios of I_1/I_2.

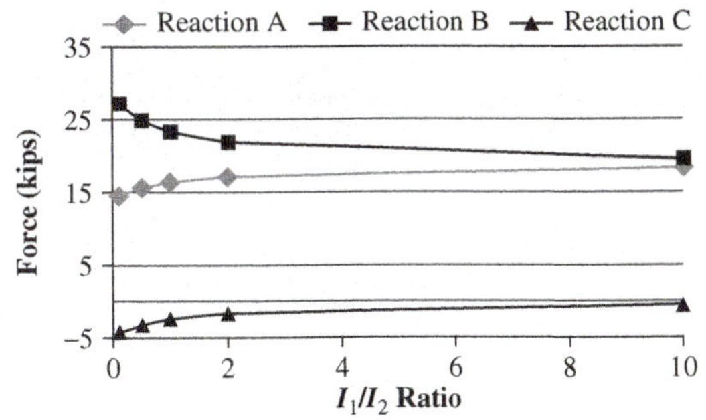

As the ratio increases, we can see a transfer of load from the right side of the beam (i.e., B and C) to the left side (i.e., A). Load is attracted to the more stiff portions of the structure. A larger I represents a larger stiffness.

QUICK NOTE

The moment of inertia of the left span may be changed by using the pull-down menu *DEFINE/SECTION PROPERTIES/FRAME SECTIONS*, selecting the section titled *Left Beam*, and clicking the modify button. Now select the set modifiers button. You may change the modifier for the moment about the 3-axis to 0.1, 0.5, 1.0, 2.0, or 10.0.

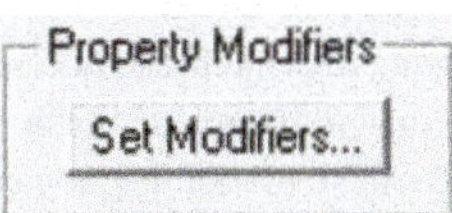

I_1/I_2	+M	−M
Ratio	(k·ft)	
0.1	34.33	−53.27
0.5	40.51	−39.06
1.0	44.85	−29.30
2.0	49.19	−19.53
10.0	55.5	−5.33

I_1/I_2	Reaction (k·ft)		
Ratio	A	B	C
0.1	14.49	27.27	−4.26
0.5	15.63	25.00	−3.13
1.0	16.41	23.44	−2.34
2.0	17.19	21.88	−1.56
10.0	18.32	19.60	−0.43

SAP2000 EXERCISE 11.2

The file *C11EX2.sdb* is a fully functioning model of the beam shown in Figure 11.7(a). Take $E = 29{,}000$ ksi and $I = 1250$ in^4.

Case 1: Apply a displacement at each support as noted in Figure 11.7(b) and confirm that no reactions or moment exists.

QUICK NOTE

Support displacements may be applied by selecting a node using the menu *ASSIGN/JOINT LOADS/DISPLACEMENTS*. In this case, the displacements will be applied in the Z-direction. Make sure your units are set to kips and inches for setting the displacement, and you switch back to kips and feet for plotting the moment diagrams.

QUICK NOTE

Applying the conditions of case 1, show that no internal moment develops. No plot of the moment diagram is provided here because it is zero. This is because there is no distortion of the beam.

Case 2: Impose support displacements as shown in Figure 11.7(c). Plot the resulting moment diagram. Confirm that this moment diagram is the same as results when the settlements at A and C are zero and the settlement at B is set to −0.38 in.

Case 3: Impose support displacements of −1.0 in., −1.5 in., and −1.58 in. at A, B, and C, respectively. Provide the moment diagram for this case.

Whether we apply the conditions shown in Figure 11.7(c) or only apply a displacement of −0.38 in. at B, the moment diagram is as shown.

Case 2 moment diagram

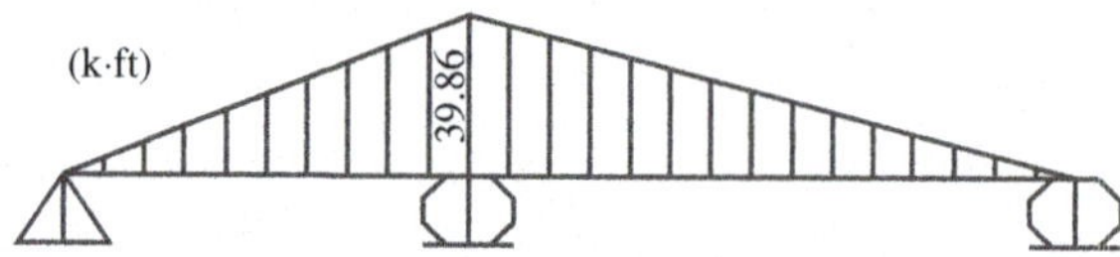

Case 3 moment diagram

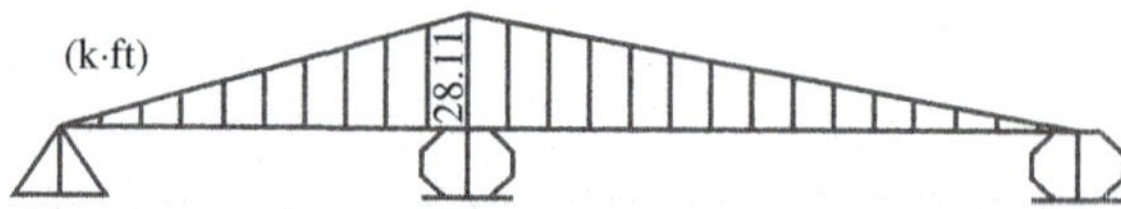

The switch-up of the support settlements causes the relative displacement to be decreased as shown. Instead of having a relative settlement at B of 0.38 in., there is only a relative settlement of 0.27 in. This is a 29% decrease in settlement, which results in a 29% decrease in the moment.

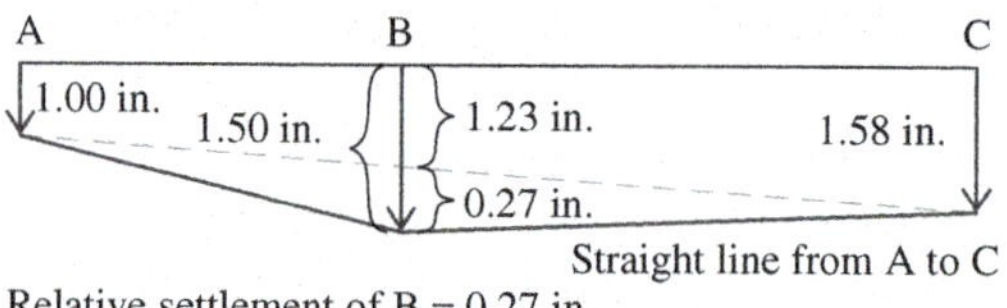

11.6 Examples with Video Solutions

VE11.1 For the beam shown in Figure VE11.1, use the force method of analysis to solve for all of the reactions including magnitude and direction. *EI* is constant along the entire length.

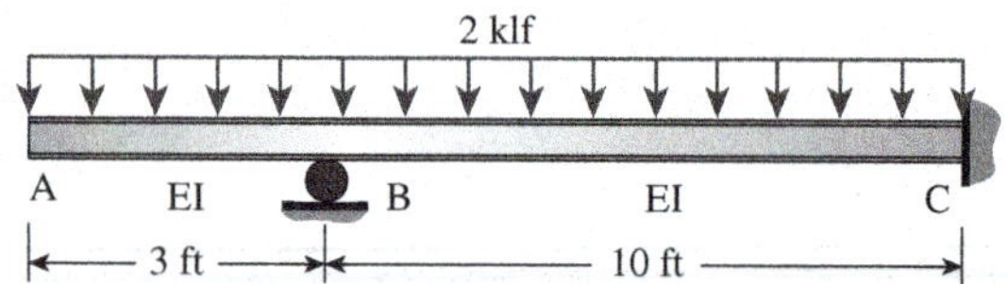

Figure VE11.1

VE11.2 For the frame shown in Figure VE11.2, use the force method of analysis to solve for all of the reactions. *E* is constant for all members but the moment of inertia is different for the columns and the beam.

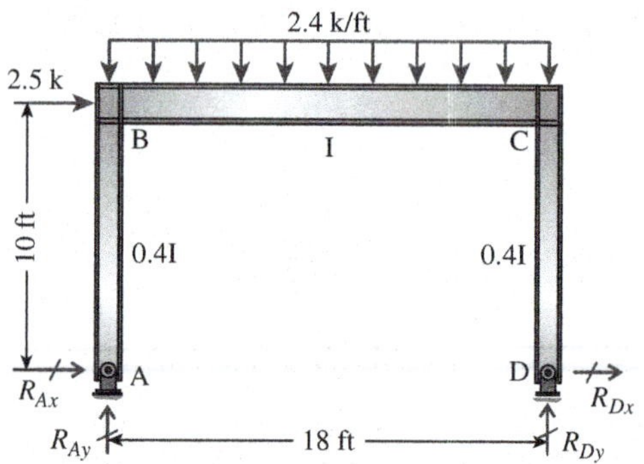

Figure VE11.2

VE11.3 For the beam in Figure VE11.3, use the force method of analysis to solve for all of the reactions. *E* and *I* are constant over the entire length.

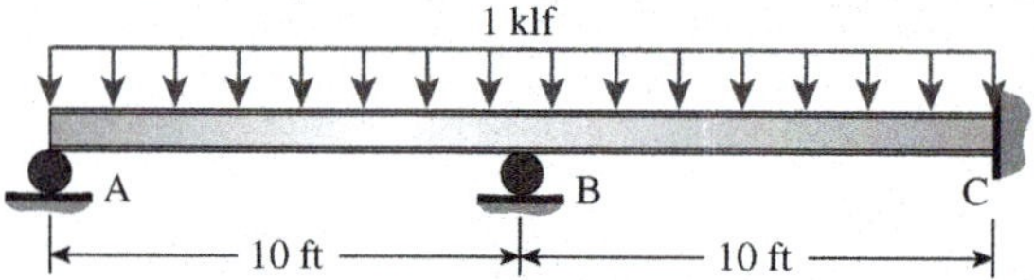

Figure VE11.3

11.7 Problems for Solution

Section 11.1

For Problems 11.1 through 11.16, compute the reactions for the beams and frames. E and I are constant unless noted otherwise. The force method (i.e., method of consistent distortions) is to be used.

P11.1 Select R_{By} as the redundant. (Ans: $R_{By} = 103.1$ k)

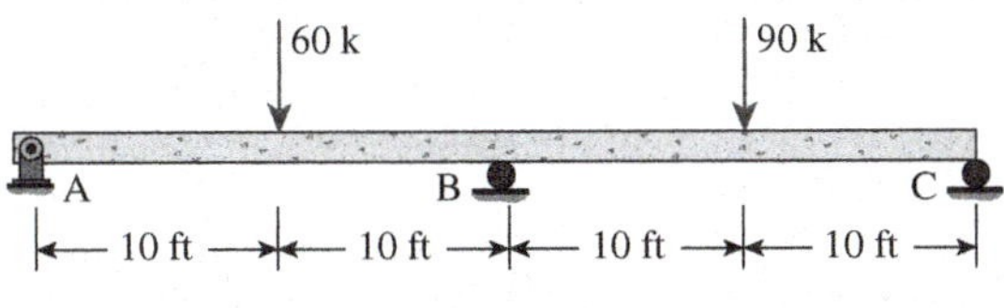

Problem 11.1

P11.2 Select R_{Ay} as the redundant.

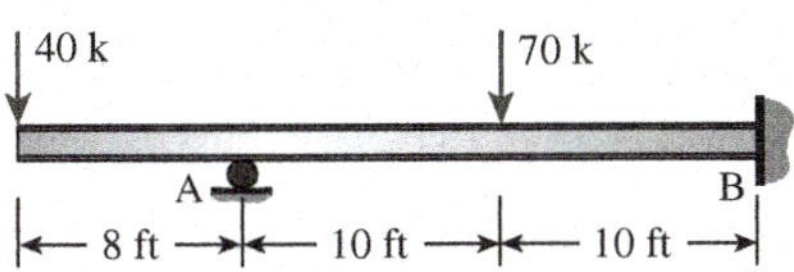

Problem 11.2

P11.3 Select R_{By} as the redundant. (Ans: $R_{By} = 123.7$ k)

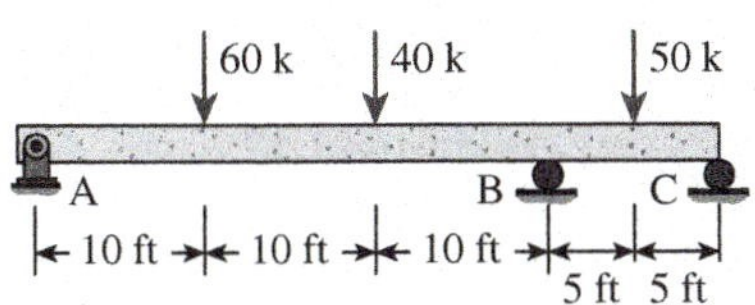

Problem 11.3

P11.4 Select R_{Ay} as the redundant.

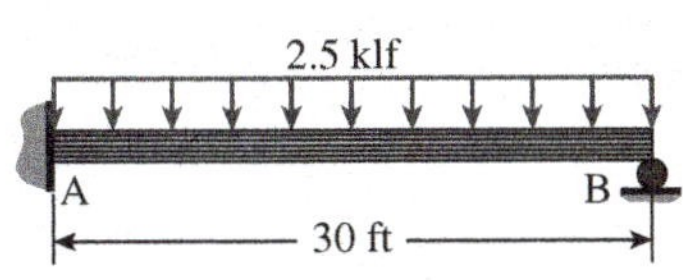

Problem 11.4

P11.5 Select R_{By} as the redundant. (Ans: $R_{By} = 82.5$ k)

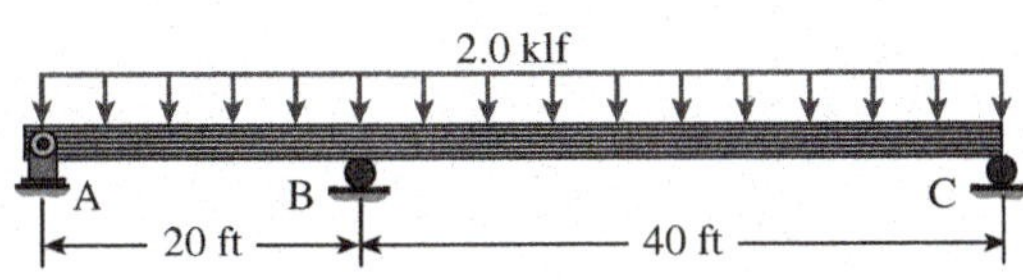

Problem 11.5

P11.6 Select R_{By} as the redundant.

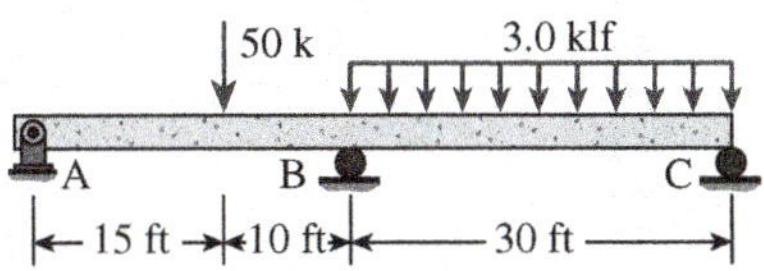

Problem 11.6

P11.7 Select R_{By} as the redundant. (Ans: $R_{By} = 92.8$ kN)

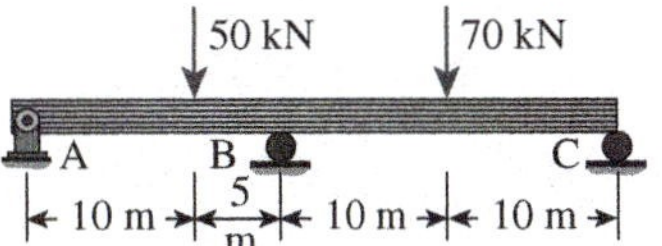

Problem 11.7

P11.8 Select R_{By} as the redundant.

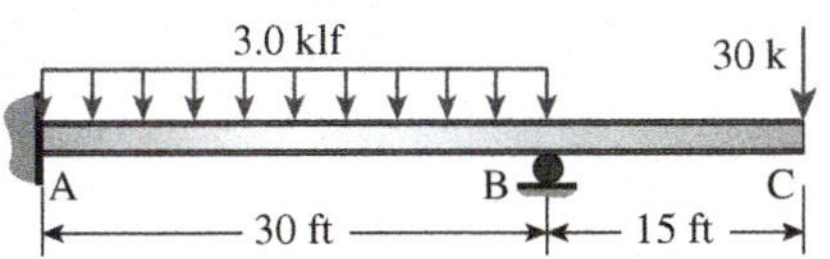

Problem 11.8

P11.9 Select R_{Cy} as the redundant. (Ans: $R_{Cy} = 6.3$ k)

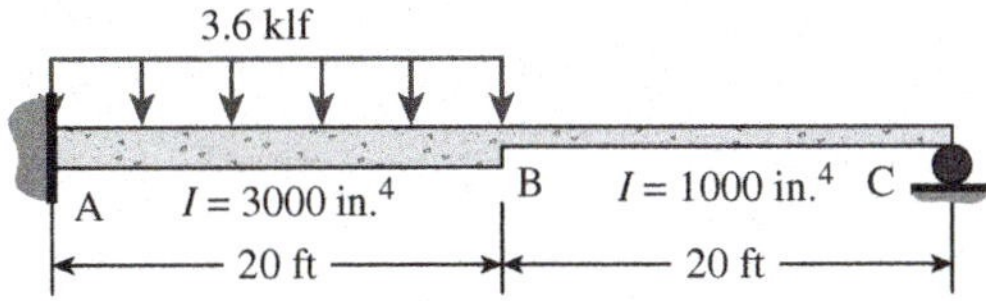

Problem 11.9

P11.10 Select R_{Cy} as the redundant.

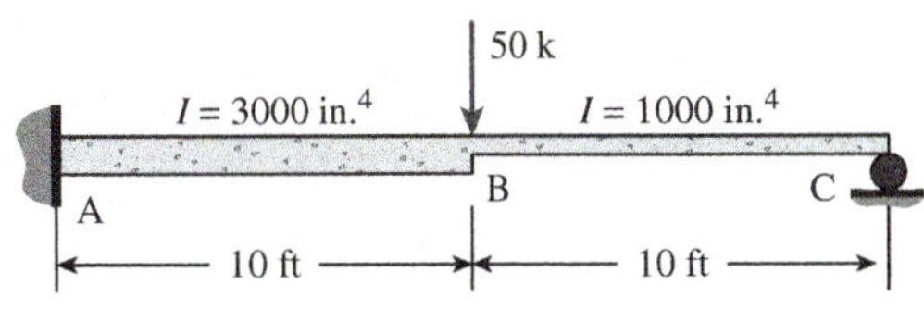

Problem 11.10

P11.11 Select R_{By} as the redundant. (Ans: $R_{By} = 43.9$ k)

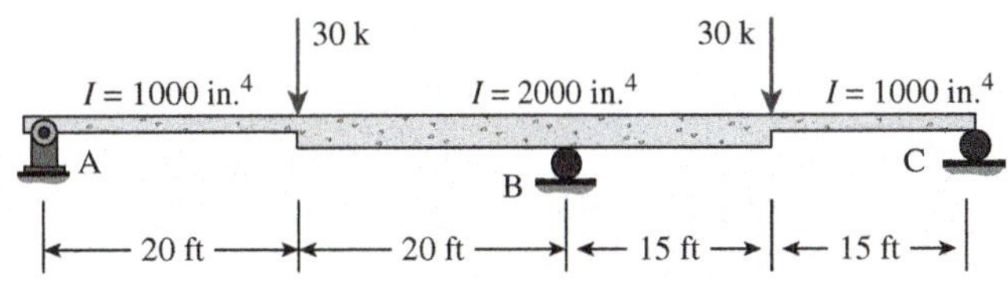

Problem 11.11

P11.12 Select R_{Cx} as the redundant.

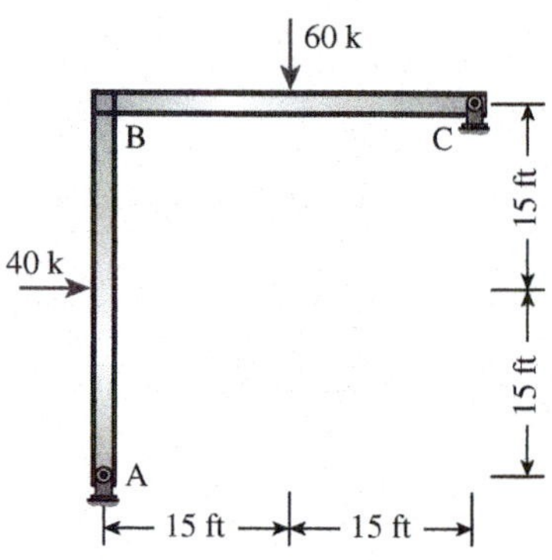

Problem 11.12

P11.13 Select R_{Cx} as the redundant. (Ans: $R_{Cx} = 22.37$ k)

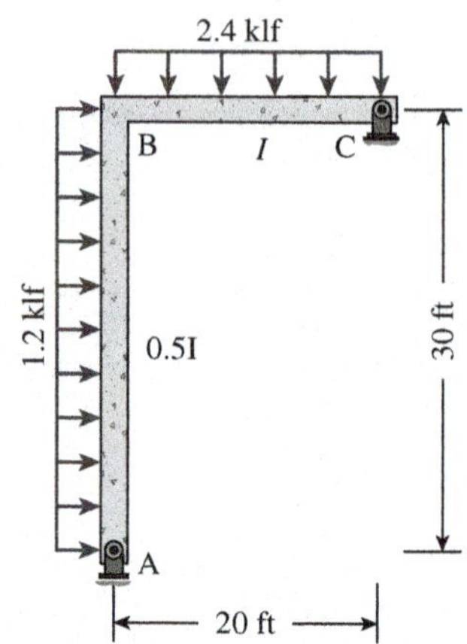

Problem 11.13

P11.14 Select R_{Dx} as the redundant.

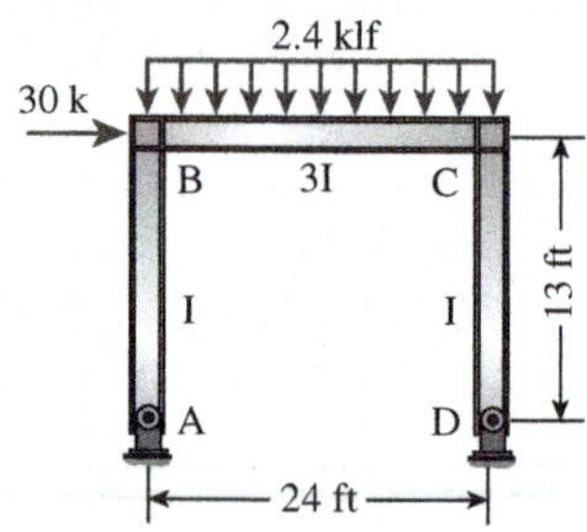

Problem 11.14

P11.15 Select R_{Cy} as the redundant. (Ans: $R_{Cy} = 22.5$ k)

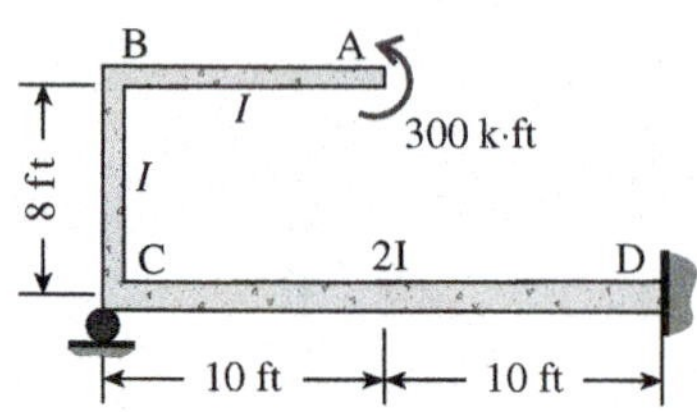

Problem 11.15

P11.16 Select R_{Cy} as the redundant.

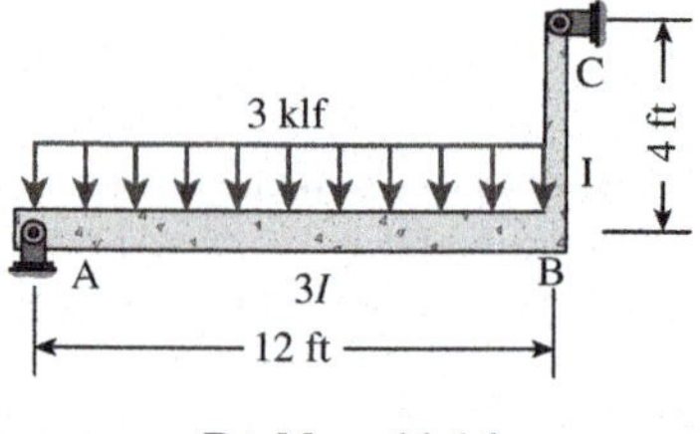

Problem 11.16

Section 11.3

For Problems 11.17 through 11.22, compute the reactions for the beams and frames. E and I are constant unless noted otherwise. The force method (i.e., method of consistent distortions) is to be used.

P11.17 Select R_{By} and R_{Cy} as the redundants. (Ans: $R_{By} = 42.59$ k, $R_{Cy} = 83.43$ k)

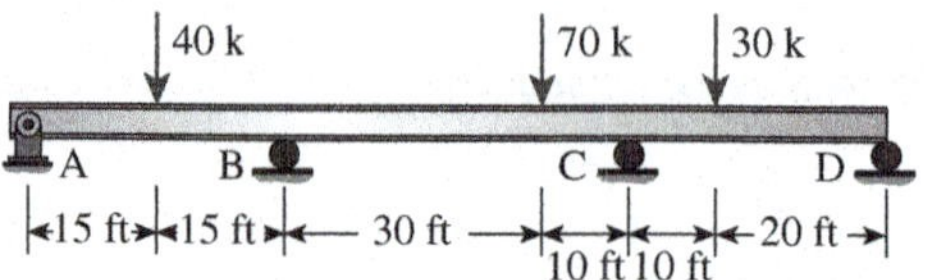

Problem 11.17

P11.18 Select R_{Ay} and R_{By} as the redundants.

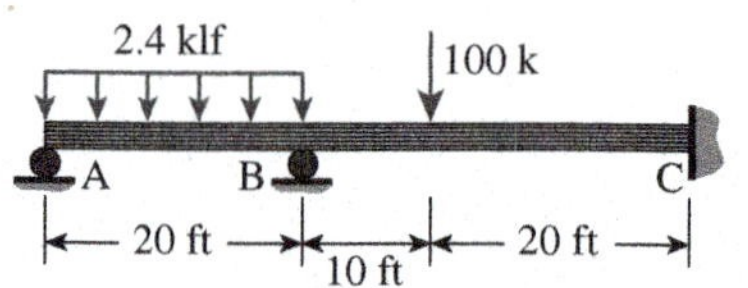

Problem 11.18

P11.19 Select R_{Ay} and R_{By} as the redundants. (Ans: $R_{Ay} = 66.8$ k, $R_{By} = 45.53$ k)

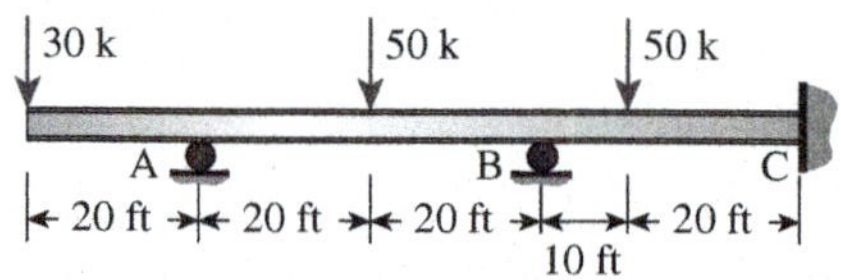

Problem 11.19

P11.20 Select R_{Dx} and R_{Dy} as the redundants.

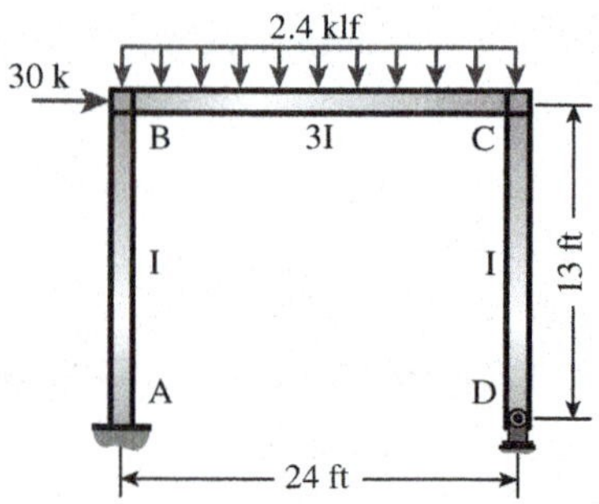

Problem 11.20

P11.21 Select R_{By} and R_{Cy} as the redundants. (Ans: $R_{By} = R_{Cy} = 24.75$ k)

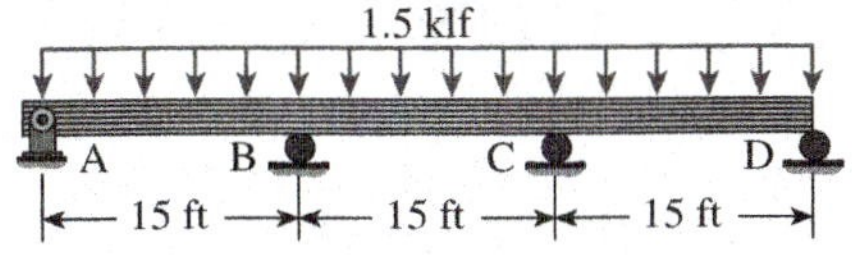

Problem 11.21

P11.22 Select R_{Ay} and R_{Dy} as the redundants.

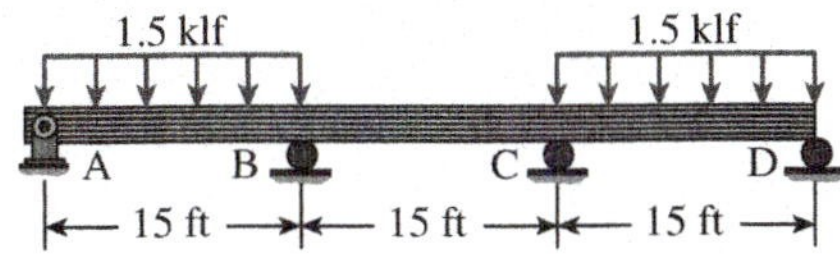

Problem 11.22

Section 11.4

For Problems 11.23 through 11.30, compute the reactions for the beams and frames. The force method (i.e., method of consistent distortions) is to be used.

P11.23 Analyze the beam of Problem 11.1, which is subjected to the indicated loads if support B settles 2.0 in. $E = 4000$ ksi and $I = 1200$ in^4. (Ans: $R_{By} = 98.96$ k)

P11.24 Analyze the beam of Problem 11.1 with the loads removed but if support B settles 1.5 in. $E = 4000$ ksi and $I = 1200$ in^4.

P11.25 Repeat Problem 11.24 but with support C settling 1.5 in. instead of support B. $E = 4000$ ksi and $I = 1200$ in^4. (Ans: $R_{By} = 1.56$ k)

P11.26 Repeat Problem 11.24, only all three supports settle the following amounts. A = 4.0 in., B = 2.0 in., and C = 3.5 in. $E = 4000$ ksi and $I = 1200$ in^4.

P11.27 Analyze the beam of Problem 11.4, which is subjected to the indicated loads if support B settles 4.0 in. $E = 1500$ ksi and $I = 1500$ in^4. (Ans: $R_{By} = 27.55$ k)

P11.28 Repeat Problem 11.27 where $E = 1500$ ksi and $I = 2500$ in^4.

P11.29 Analyze the frame of Problem 11.14, which is subjected to the indicated loads if support D moves 1.0 in. to the right. $E = 4500$ ksi and $I = 500$ in^4. (Ans: $R_{Dx} = 18.82$ k)

P11.30 Analyze the frame of Problem 11.16, which is subjected to the indicated loads if support C moves 0.5 in. down. $E = 29000$ ksi and $I = 500$ in^4.

Section 11.5 SAP2000

P11.31 Model the beam shown in Problem 11.11 and take E = 4000 ksi. For the middle 35 ft of the beam, set I equal to the values 200 in^4, 600 in^4, 1000 in^4, 1400 in^4, 1700 in^4, 2000 in^4, and 2300 in^4 and find the reaction at B. Generate a plot of $I_{\text{middle}}/I_{\text{outer}}$ versus reaction at B. What happens to the reaction at B as the moment of inertia increases for the middle portion relative to the outer sections? Now, set the I for the entire length to be 1800 in^4, set I for the entire beam to be 1000 in^4, and find the reaction at B for both cases. What conclusion can you draw about the influence of I?

P11.32 Model the beam given in Problem 11.5 in SAP2000. Set $E = 1900$ ksi and $I = 2500$ in^4. Find the reactions for the beam when the roller at B is set at different locations. Measuring from point A, place B at a distance of 5 ft, 15 ft, 25 ft, 35 ft, 45 ft, and 55 ft. Create a plot of the location of B versus the reaction at A and the reaction at C. What influence does the location of the roller have on these reactions? (*Hint*: You can quickly change the location of B by selecting the node at B and using the *MOVE* command found under the *EDIT* pull-down menu.)

P11.33 Model the primary structure given in Example 11.1 in SAP2000. Set $E = 29{,}000$ ksi and $I = 400$ in^4. What is the deflection at B? Next model the redundant structure given in the same example where $R_{By} = 23.45$ k. What is the deflection at B? Finally, add the reactions from the primary structure to the reactions from the redundant structure and compare with the results of the original structure in the example. What conclusions can be made about the summation of the primary and redundant structures?

P11.34 Use SAP2000 to model the two beams given where $E = 29{,}000$ ksi and $I = 1200$ in^4. In both cases, find the deflection at points A and B. What conclusions can you make regarding Maxwell's law?

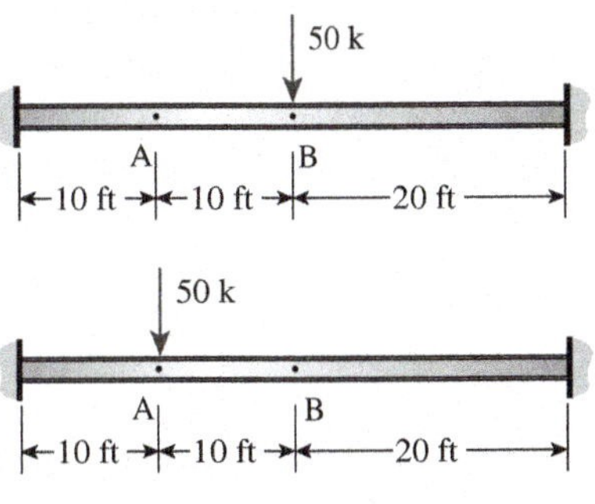

Problem 11.34

12 Force Method for Statically Indeterminate Structures Continued

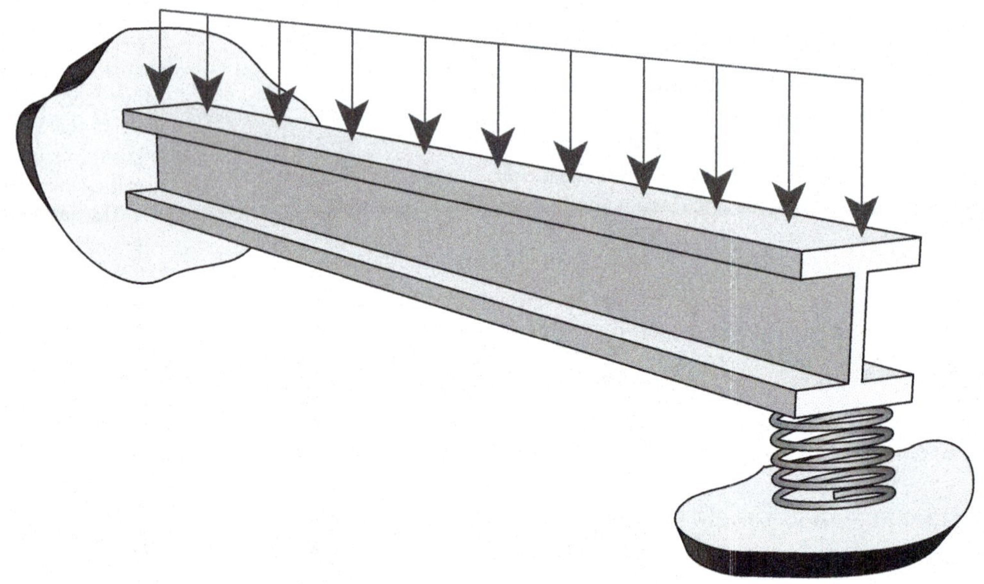

12.1 Analysis of Externally Redundant Trusses

Trusses may be statically indeterminate because of redundant reactions, redundant members, or a combination of redundant reactions and members. Externally redundant trusses will be considered initially, and they will be analyzed on the basis of deflection computations in a manner closely related to the procedure used in the preceding chapter for statically indeterminate beams and frames.

This process will require a primary structure, one or more redundant structures, and one or more flexibility structures. The forces in the truss members in the primary structure are not the correct final forces and are referred to herein as the N forces. Recall that the deflection in a truss is found using virtual work which gives

QUICK NOTE

The procedure for analyzing trusses having external redundants is exactly the same as was used for beams. The only difference that one experiences is how the deflection calculations are made.

$$1 \cdot \Delta = \sum_{\text{all } i} \frac{n_i N_i L_i}{A_i E_i}$$

n = axial bar forces due to a unit virtual load

N = axial bar forces due to the real load

L = length of the bar

A = cross-sectional area of the bar

E = modulus of elasticity of the bar

You will also need to find the deflection(s) in the flexibility structure(s). The flexibility structure has an applied unit load, and so does the virtual structure. For convenience, the forces in the flexibility structure and also in the virtual structure are represented as n. This means that the deflection at the location of the unit load can be found by

$$1 \cdot f = \sum_{\text{all } i} \frac{n_i^2 L_i}{A_i E_i}$$

Once the unknown redundant force is found, another full analysis of the truss may be done to find the actual bar forces. There is a more efficient way to handle this though without repeating many calculations. Recall that the original structure and all of its responses is the summation of the primary structure and the redundant structure(s). This includes all deflections and bar forces. Therefore, it is convenient to calculate the bar forces N_f by scaling up n by the value of the redundant R and adding it to N as shown.

QUICK NOTE

Spreadsheets offer convenience, speed, and less likelihood of calculation error. Analyzing indeterminate trusses is greatly aided by the use of these spreadsheets.

$$N_f = N + (n)(R)$$

Example 12.1 demonstrates the entire process as it applies to externally redundant trusses.

EXAMPLE 12.1

Compute the reactions and the member forces for the two-span continuous truss shown. The cross-sectional area of all the bars is 4 in^2—except for the internal bars AE, AF, BF, FC, and CG that have an area of 3 in^2. E is the same for all members.

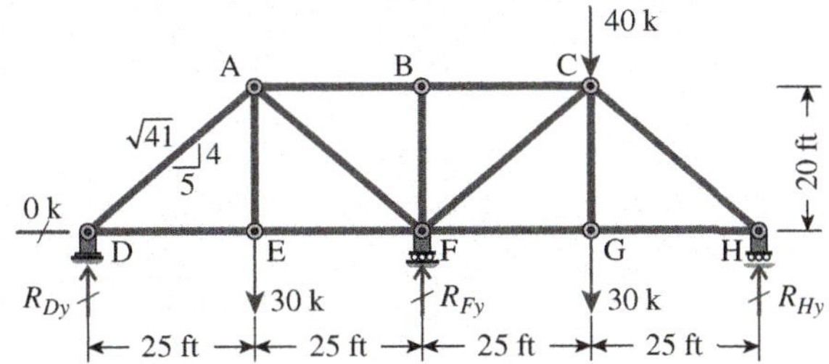

SOLUTION STRATEGY

Follow the procedure outlined in Chapter 11 for indeterminate structures. Once the compatibility equation is assembled, use a table to assemble all bar forces and calculate the requisite deflections.

REDUNDANT SELECTION

This truss is one-degree statically indeterminate, so it only has one redundant force. One cannot solve for the vertical reactions using just equilibrium. Therefore, the redundant may be selected as any of the vertical reactions.

SOLUTION

Calculate the determinacy of the truss. $m = 13$, $j = 8$, and $r = 4$.

$$(13 + 4) - 2(8) = 1 \Rightarrow \text{SI } 1^\circ$$

Only one redundant is present in this truss. R_{Fy} will be selected. Sketch the primary and redundant structures and then write the compatibility equation.

Primary Structure (N)

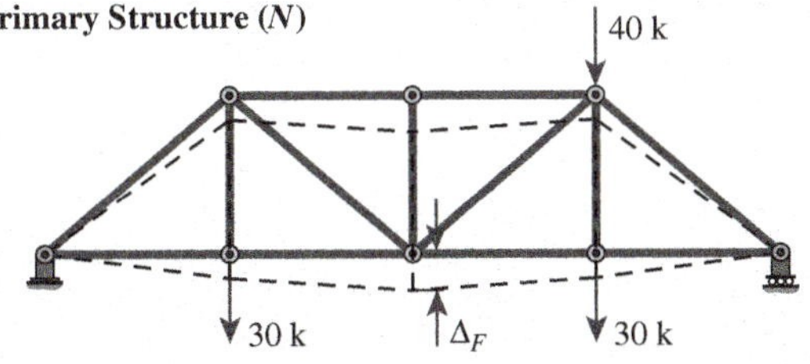

Redundant Structure

Compatibility Equation:

$$\Delta_F + \Delta_{ff} = \Delta_F + f_{ff} R_{Fy} = 0 \Rightarrow R_{Fy} = -\frac{\Delta_F}{f_{ff}}$$

The next step is to compute the deflections Δ_F and f_{ff} using virtual work. This requires a virtual unit load to be applied to joint F. The following structure serves both as the flexibility structure and the virtual structure and produces the axial bar forces n. Place all computed bar forces in the table.

Flexibility and Virtual Structures (n)

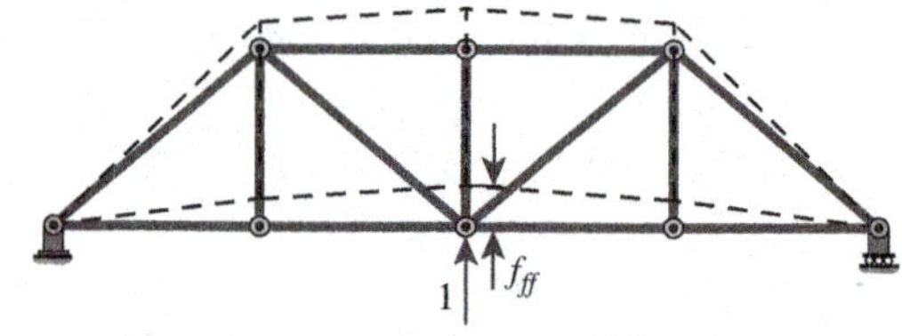

Member	L (in.)	A (in^2)	N (k)	n	$\Delta_F = \frac{nNL}{AE}$	$f_{ff} = \frac{n^2L}{AE}$	$N_f = N + R_{Fy}n$ (k)
DA	384	4.0	−64	0.8	−4915/E	61.4/E	−19.2
DE	300	4.0	50	−0.625	−2344/E	29.3/E	15.0
AE	240	3.0	30	0	0/E	0/E	30.0
AB	300	4.0	−62.5	−1.25	−5859/E	117.2/E	7.5
AF	384	3.0	16	−0.8	−1638/E	81.9/E	−28.8
EF	300	4.0	50	−0.625	−2344/E	29.3/E	15.0
BF	240	3.0	0	0	0/E	0/E	0
BC	300	4.0	−62.5	1.25	−5859/E	117.2/E	7.5
FC	384	3.0	−16	−0.8	1638/E	81.9/E	−60.8
FG	300	4.0	75	−0.625	−3516/E	29.3/E	40.0
CG	240	3.0	30	0	0/E	0/E	30.0
CH	384	4.0	−96	0.8	−7373/E	61.4/E	−51.2
GH	300	4.0	75	−0.625	−3516/E	29.3/E	40.0
				$\sum$	−35,726/E	638/E	

The calculated deflections are now available and may be inserted into the compatibility equation to solve for R_{Fy}. So

$$R_{Fy} = -\frac{-35{,}726/E}{638/E} = +56.0 \text{ k} = \underline{\underline{56.0 \text{ k} \quad \uparrow}}$$

The remaining reactions may be found using statics. Thus,

$$R_{Ay} = \underline{\underline{12.0 \text{ k} \quad \uparrow}} \qquad R_{Cy} = \underline{\underline{32.0 \text{ k} \quad \uparrow}}$$

All of the bar forces can be found using basic statics and truss analysis. However, they easily can be found by superimposing the primary and redundant structures. All results are shown in the table, but the calculation for member DA is shown here for illustration.

$$N_{DA} = -64 \text{ k} + (56 \text{ k})(0.8) = -19.2 \text{ k} = \underline{\underline{19.2 \text{ k} \quad (\text{C})}}$$

It should be evident that the procedure just demonstrated may be followed for trusses with more than one redundant. Remember that a separate redundant structure and compatibility equation must be sketched for each redundant removed. Furthermore, Maxwell's law of reciprocal displacements does apply to trusses, so this can help when finding flexibility coefficients. Refer to Sections 11.2 and 11.3 for further guidance.

12.2 Analysis of Internally Redundant Trusses

Internally redundant trusses may be analyzed in a manner closely related to the process used for externally redundant trusses. The force in one member is assumed to be the redundant and is theoretically cut. The remaining members must form a statically determinate and stable truss. The following procedure should be followed.

QUICK NOTE
The bars that are cut in the primary structure should have a zero axial force, $N = 0$.

Procedure

1. Identify the degree of statical indeterminacy (n). Recall SI $= m + r - 2j$.
2. Identify and CUT n unknown redundant bars. The goal here is to reduce the structure to a determinate structure.

3. Construct the *PRIMARY* structure with the actual loads present but having the bars cut.
4. Draw the deflected shape and identify the displacement at the point of each cut. Note that the displacement that should be identified is the relative displacement between the cut ends.
5. Construct the *REDUNDANT* structure(s) with no other loads on it other than the redundant axial force(s) applied to the cut bar(s). A separate redundant structure is needed for each redundant force. This should produce n redundant structures.
6. Sketch the deflected shape for each redundant structure and identify the displacement at the point of each redundant force.
7. Write a compatibility equation for each location on the structure where a redundant force exists. This will sum up all of the relative displacements at the cut bar from the primary and redundant structures. Equate this summation to a known relative displacement in the bar. This will usually be equal to zero, because the bars don't break in the real structure. Remember that these displacements may be written in terms of their flexibility coefficients.
8. Determine the values of the deflections and flexibility coefficients using virtual work. These equations are

$$1 \cdot \Delta = \sum_{\text{all } i} \frac{n_i N_i L_i}{A_i E_i} \qquad 1 \cdot f = \sum_{\text{all } i} \frac{n_i^2 L_i}{A_i E_i}$$

9. Substitute computed displacements into the compatibility equation(s) and solve for the unknown redundant forces.
10. Use the equations of equilibrium to solve for the remaining reactions and internal forces. The bar forces easily can be found using the procedure introduced in the previous section.

QUICK NOTE

When sketching the deflected shape for trusses, we should always assume the cut bars are in tension. This takes some thought. To get this right, you need to answer the question on where the load is being applied to the bar.

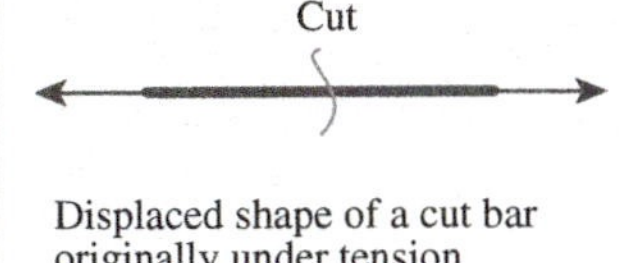

Displaced shape of a cut bar originally under tension

Displaced shape of a cut bar having an internal tensile force applied

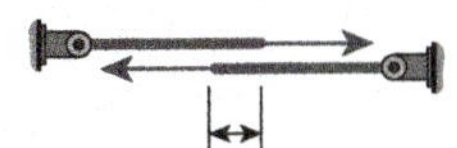

EXAMPLE 12.2

Determine the forces in the internally redundant truss shown. Members AD, AE, DB, and BE have areas of 1 in^2, and all of the rest have areas of 2 in^2. E = 29,000 ksi.

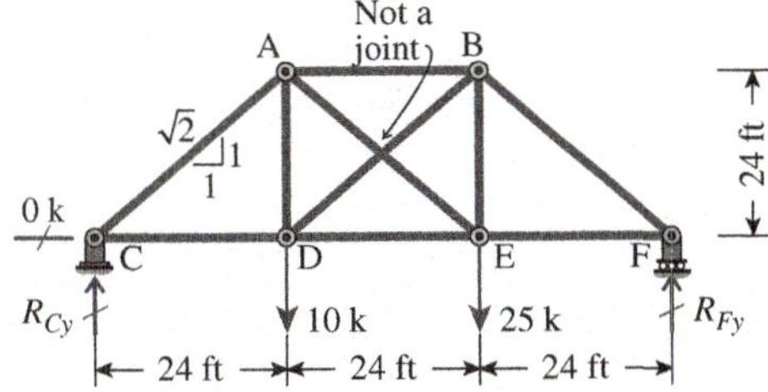

SOLUTION STRATEGY

Choose a redundant bar such that, when it is cut, a regular looking truss is left as the primary truss. Make the sketches by clearly assuming all cut bars were in tension, even if you know it is not.

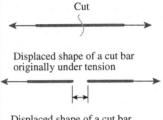

REDUNDANT SELECTION

Vertical reactions are easy to find using just equilibrium. Therefore, they cannot be selected as the redundant. Bars CA, CD, BF, and EF can also be found using just equilibrium. Any of the remaining bars are eligible for selection as a redundant.

SOLUTION

Calculate the determinacy of the truss. m = 10, j = 6, and r = 3.

$$(10 + 3) - 2(6) = 1 \Rightarrow \text{SI } 1^\circ$$

Only one redundant is present in this truss. N_{DB} will be selected. Sketch the primary and redundant structures and then write the compatibility equation.

Primary Structure (N)

A B C D E F Δ_{DB} 10 k 25 k

Redundant Structure

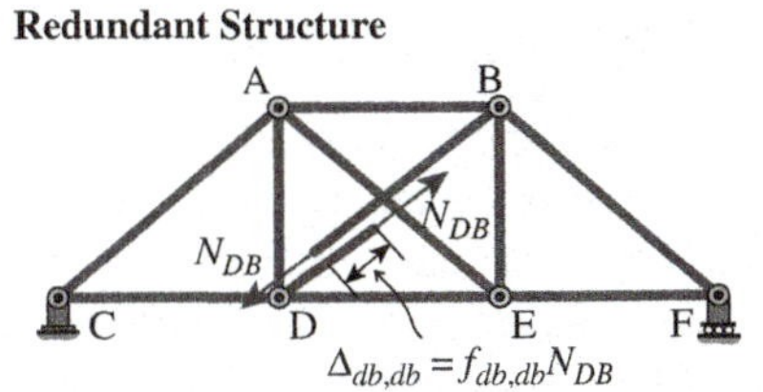

Compatibility Equation:

$$\Delta_{DB} + \Delta_{db,db} = \Delta_{DB} + f_{db,db} N_{DB} = 0 \Rightarrow N_{DB} = -\frac{\Delta_{DB}}{f_{db,db}}$$

COMMON MISTAKE

Students often struggle with the analysis of the flexibility structure. Remember that the rules of statics don't change. If a sum of moments equation is written about point C, the reaction at F is found to be zero. The other reactions are also zero. An internally applied unit load did not cause any external reactions.

The next step is to compute the deflections Δ_{DB} and $f_{db,db}$ using virtual work. This requires an internal virtual unit load to be applied to member DB. The following structure serves both as the flexibility structure and the virtual structure and produces the axial bar forces n. Place all computed bar forces in the table.

Flexibility and Virtual Structure (*n*)

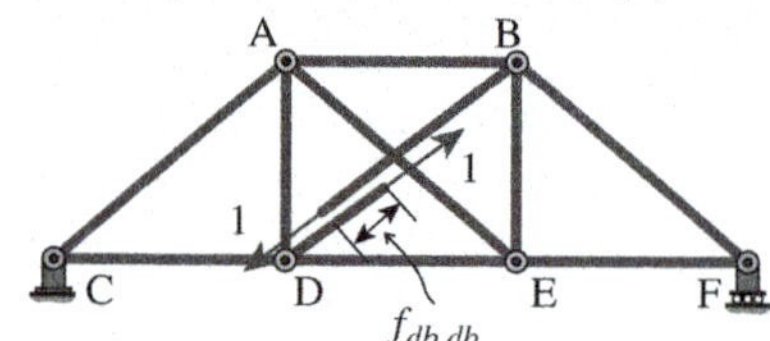

Member	L (in.)	A (in^2)	N (k)	n	$\Delta_{DB} = \frac{nNL}{AE}$	$f_{db,db} = \frac{n^2L}{AE}$	$N_f = N + N_{DB}n$ (k)
CA	407	2.0	−21.2	0	0	0.0	−21.2
CD	288	2.0	15.0	0	0	0.0	15.0
AD	288	1.0	10.0	−0.707	−2036	144	8.5
AB	288	2.0	−20.0	−0.707	2036	72.0	−21.5
AE	407	1.0	7.1	1	2890	407.0	9.3
DB	407	1.0	0	1	0	407.0	2.2
DE	288	2.0	15.0	−0.707	−1527	72	13.5
BE	288	1.0	20.0	−0.707	−4072	144.0	18.5
BF	407	2.0	−28.3	0	0	0.0	−28.3
EF	288	2.0	20.0	0	0	0.0	20.0
				Σ	−2710	1246	

The calculated deflections are now available and may be inserted into the compatibility equation to solve for N_{DB}. So

$$N_{DB} = -\frac{-2710/E}{1246/E} = +2.2 \text{ k} = \underline{\underline{2.2 \text{ k} \quad (T)}}$$

All of the remaining bar forces can be found using basic statics and truss analysis. However, they easily can be found by superimposing the primary and redundant structures. All results are shown in the table, but the calculation for member AB is shown here for illustration.

$$N_{AB} = -20 \text{ k} + (2.2 \text{ k})(-0.707) = -21.5 \text{ k} = \underline{\underline{21.5 \text{ k} \quad (C)}}$$

For trusses that have more than one redundant internally, multiple compatibility equations are required including multiple redundant structures. A simultaneous solution of these equations will provide the values of the redundant forces.

12.3 Analysis of Composite Structures

Previous sections and Chapter 11 dealt with indeterminate beams, frames, and trusses. The force method requires the calculation of deflections. The beams and frames consider the contribution of bending deformations while the trusses consider only axial deformations. However, some structures have both types of members present at the same time. These are known as composite structures and may include beam/frame members, truss members, and springs in any combination.

These types of structures are handled in the same way that previous structures were handled. It is the calculation of deflections where one must be careful to include the effects of all the members in the structure. The next two examples illustrate the process when beam members and truss members or springs are also present.

DID YOU KNOW?

A spring, whether rotational or translational, may be handled in the same way that truss bars were handled in the previous section. Recall the deflection equations from Chapter 9 as

$$1 \cdot \Delta = \frac{nN}{k_T} + \frac{mM}{k_R}$$

Review the analysis of springs in Section 9.8.

EXAMPLE 12.3

Using the force method, calculate the force in the steel rod of the composite structure shown.

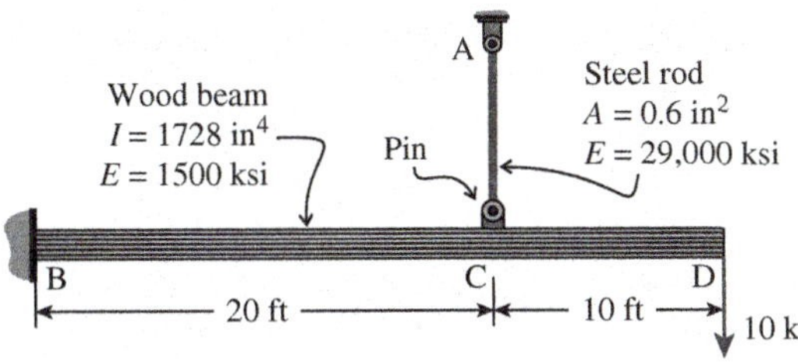

QUICK NOTE

Remember that to find the determinacy you must make a cut at any internal pin. The circled numbers indicate the number of unknown forces at that location after it has been cut apart at the pin.

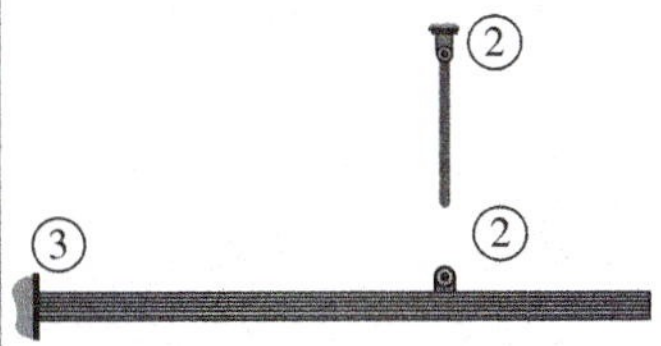

SOLUTION STRATEGY

Recognize that there is one truss bar and one beam present in this structure. Find the determinacy and select an appropriate number of redundants. When calculating deflections, make sure to include the contribution of both the truss bar and the beam.

SOLUTION

Calculate the determinacy of the structure (refer to the sidebar note) where $F = 7$, $n = 2$, and $N = (2)(3) = 6$.

$$7 - 6 = 1 \Rightarrow \text{SI } 1^\circ$$

Only one redundant is present in this composite structure. N_{AC} is selected as the redundant. Cut but *do not remove* AC, sketch the primary and redundant structures, and write the compatibility equation.

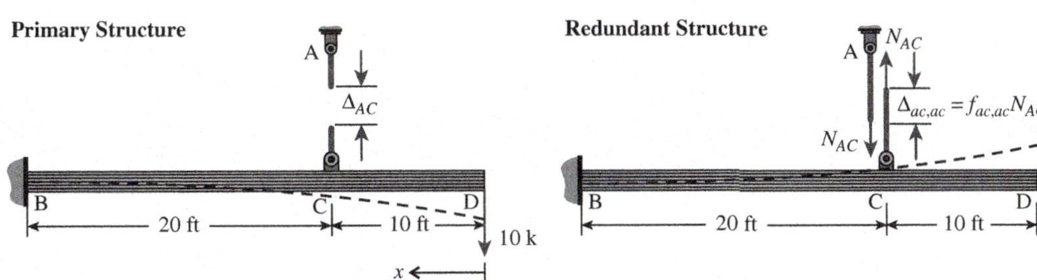

Compatibility Equation:

$$\Delta_{AC} + \Delta_{ac,ac} = \Delta_{AC} + f_{ac,ac}N_{AC} = 0 \Rightarrow N_{AC} = -\frac{\Delta_{AC}}{f_{ac,ac}}$$

Flexibility and Virtual Structure

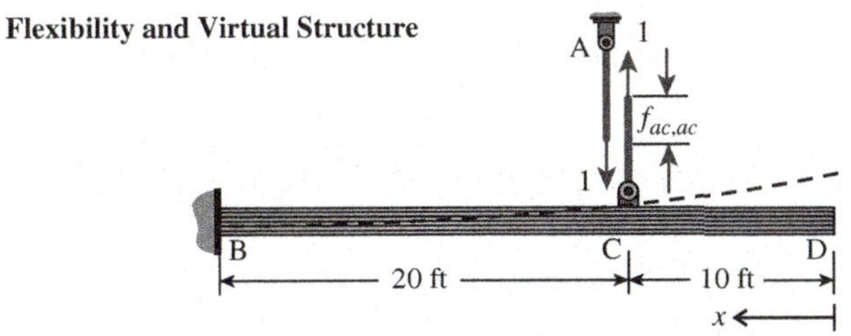

COMMON MISTAKE

Instead of cutting bar AC, students will often *remove* bar AC. This is not technically correct, but it won't impact the calculation of Δ_{AC} if this omission is made. However, take a look at the calculation for the flexibility coefficient, and it becomes clear that it does affect this value if bar AC had been removed.

The next step is to compute the deflections Δ_{AC} and $f_{ac,ac}$ using virtual work. This requires an internal virtual unit load to be applied to member AC. The previous structure serves both as the flexibility structure and the virtual structure and produces the m equation for BCD and the axial bar force n for AC. Referring to the coordinate system shown on the primary structure and the virtual structure, the following moment equations and axial bar forces are computed using units of kips and inches.

$$M(x) = -10x; \quad 0 \leq x \leq 360 \text{ in.}$$

$$m(x) = 0; \quad 0 \leq x < 120 \text{ in.}$$

$$m(x) = 1(x-120); \quad 120 \text{ in.} \leq x \leq 360 \text{ in.}$$

$$N = 0; \quad \text{comes from primary structure}$$

$$n = 1; \quad \text{comes from virtual structure}$$

The deflection calculation equations are

$$1 \cdot \Delta_{AC} = \int_0^{120} \frac{(0)(-10x)}{EI} dx + \int_{120}^{360} \frac{(x-120)(-10x)}{EI} dx + \frac{(1)(0)(120)}{AE} = 0 + \frac{8.07 \cdot 10^7 \text{ k}\cdot\text{in}^3}{EI} + \frac{0}{AE}$$

$$= \frac{8.07 \cdot 10^7 \text{ k}\cdot\text{in}^3}{(1500 \text{ ksi})(1728 \text{ in}^4)} = -31.1 \text{ in.}$$

$$1 \cdot f_{ac,ac} = \int_0^{120} \frac{(0)(0)}{EI} dx + \int_{120}^{360} \frac{(x-120)(x-120)}{EI} dx + \frac{(1)(1)(120)}{AE} = 0 + \frac{4.61 \cdot 10^6 \text{ in}^3}{EI} + \frac{120}{AE}$$

$$= \frac{4.61 \cdot 10^6 \text{ in}^3}{(1500 \text{ ksi})(1728 \text{ in}^4)} + \frac{120}{(0.6 \text{ in}^2)(29{,}000 \text{ ksi})}$$

$$= 1.78 \text{ in./k}$$

DID YOU KNOW?

Now that the force in bar AC has been found, basic statics can solve for the other reactions; then shear and moment diagrams may be sketched.

The calculated deflections are now available and may be inserted into the compatibility equation to solve for N_{AC}. So

$$N_{AC} = -\frac{-31.1 \text{ in.}}{1.785 \text{ in./k}} = +17.4 \text{ k} = \underline{\underline{17.4 \text{ k} \quad (\text{T})}}$$

Another form of composite structure is when springs—either translational or rotational—are attached. The process is exactly the same as shown in Example 12.3. Indeed, Example 12.3 is reworked in Example 12.4, only the truss bar AC has been replaced with a translational spring.

EXAMPLE 12.4

Using the force method, calculate the force in the spring of the composite structure shown.

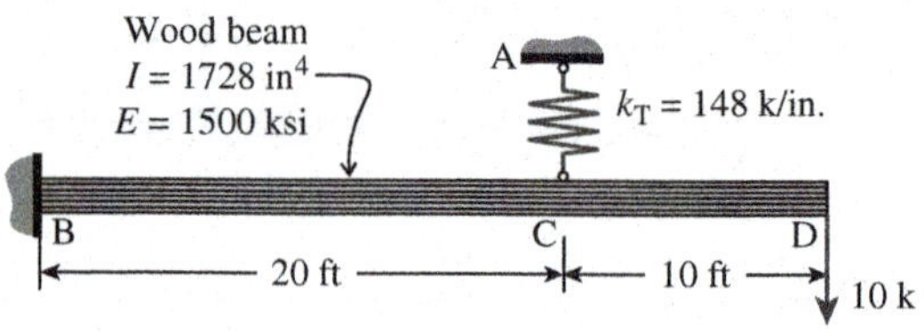

QUICK NOTE

Recognize that a translational spring only has one reaction.

Wood beam
$I = 1728$ in^4
$E = 1500$ ksi
A
$k_T = 148$ k/in.
B
C
D
20 ft
10 ft
10 k

SOLUTION STRATEGY

Recognize that there is one spring and one beam present in this structure. Find the determinacy and then select an appropriate number of redundants. When calculating deflections, make sure to include the contribution of both the spring and the beam.

SOLUTION

Calculate the determinacy of the structure (refer to the sidebar note) where $F = 4$, $n = 1$, and $N = (1)(3) = 3$.

$$4 - 3 = 1 \Rightarrow \text{SI } 1^\circ$$

Only one redundant is present in this composite structure. N_{AC} is selected as the redundant, because if spring AC is cut, the remaining basic cantilever beam is easier to deal with. Cut but *do not remove* AC, sketch the primary and redundant structures, and write the compatibility equation.

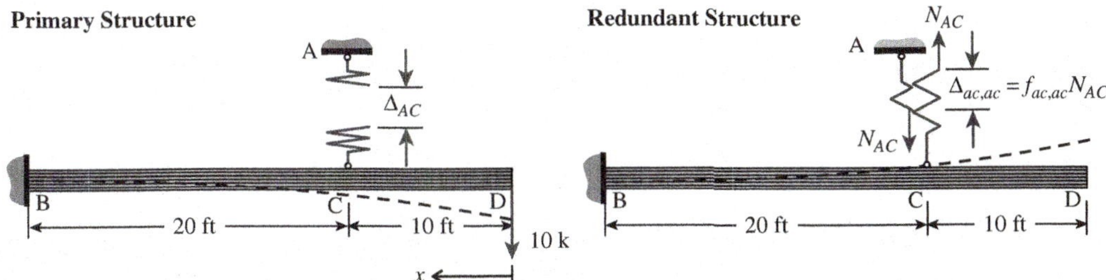

Compatibility Equation:

$$\Delta_{AC} + \Delta_{ac,ac} = \Delta_{AC} + f_{ac,ac}N_{AC} = 0 \Rightarrow N_{AC} = -\frac{\Delta_{AC}}{f_{ac,ac}}$$

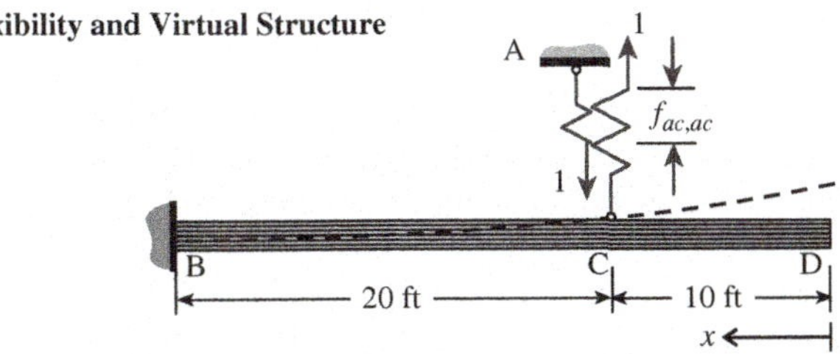

The next step is to compute deflections Δ_{AC} and $f_{ac,ac}$ using virtual work. This requires an internal virtual unit load to be applied to spring AC. The previous structure serves both as the flexibility structure and the virtual structure. The moment equations and axial forces, as given in Example 12.3, apply here. Therefore, the displacement equations may be written as

$$1 \cdot \Delta_{AC} = \int_0^{120} \frac{(0)(-10x)}{EI}dx + \int_{120}^{360} \frac{(x-120)(-10x)}{EI}dx + \frac{(1)(0)}{k_T} = 0 + \frac{8.07 \cdot 10^7 \text{ k·in}^3}{EI} + \frac{0}{k_T}$$

$$= \frac{8.07 \cdot 10^7 \text{ k·in}^3}{(1500 \text{ ksi})(1728 \text{ in}^4)} = -31.1 \text{ in.}$$

$$1 \cdot f_{ac,ac} = \int_0^{120} \frac{(0)(0)}{EI}dx + \int_{120}^{360} \frac{(x-120)(x-120)}{EI}dx + \frac{(1)(1)}{k_T} = 0 + \frac{4.61 \cdot 10^6 \text{ in}^3}{EI} + \frac{1}{k_T}$$

$$= \frac{4.61 \cdot 10^6 \text{ in}^3}{(1500 \text{ ksi})(1728 \text{ in}^4)} + \frac{1}{148 \text{ k/in.}} = 1.785 \text{ in./k}$$

COMMON MISTAKE

Students tend to view springs as supports and not as structural members. Don't remove the spring; just cut it if that is the redundant selected.

DID YOU KNOW?

A truss bar or any other structural member may be viewed as a spring. One only needs to find the associated stiffness value.

The calculated deflections are now available and may be inserted into the compatibility equation to solve for N_{AC}.

$$N_{AC} = -\frac{-31.1\text{ in.}}{1.785\text{ in./k}} = +17.4\text{ k} = \underline{\underline{17.4\text{ k}\quad(\text{T})}}$$

SUMMARY

The results of this example are the same as the results of Example 12.3. This is because the spring stiffness of this problem was set equal to the stiffness of the truss bar in the previous example where $k_T = (AE)/L$. Notice how the last term in the deflection equation becomes the same for both examples:

$$\frac{nNL}{AE} = \frac{nN}{k_T} \quad \text{if} \quad k_T = \frac{AE}{L}$$

DID YOU KNOW?
If the spring stiffness $k_T = 14.8$ k/in., the force in the spring would be 16.9 k and if the spring stiffness was 1.48 k/in., the force in the spring would be 12.7 k. The more flexible the spring becomes, the less load it will take.

12.4 Temperature Changes, Shrinkage, Fabrication Errors, and So On

Structures are subject to deformations due not only to external loads but also to temperature changes, support settlements, inaccuracies in fabrication dimensions, shrinkage in reinforced concrete members caused by drying and plastic flow, and so forth. Such deformations in statically indeterminate structures can cause the development of large additional forces in the members. As an illustration, it is assumed that the top-chord members of the truss in Figure 12.1 are exposed to the sun much more than are the other members. As a result, on a hot sunny day, they may have much higher temperatures than the other members, and the member forces may undergo some appreciable changes.

Problems such as these may be handled exactly as were the previous problems of this chapter. The only difference one experiences is how the deflections in the primary structure are calculated. Recall that for trusses the virtual work equation looks like

$$1 \cdot \Delta = \sum \frac{n_i N_i L_i}{A_i E_i}$$

When the deflection due to thermal expansion is experienced, the NL/AE term is replaced with $\alpha(\Delta T)L$ where

α = coefficient of thermal expansion for the material

ΔT = change in temperature. (+) is an increase in temperature and (−) is a decrease

L = length of the member

QUICK NOTE
The effect of support settlement on structural response has been demonstrated in Example 11.7 and SAP Exercise 11.2.

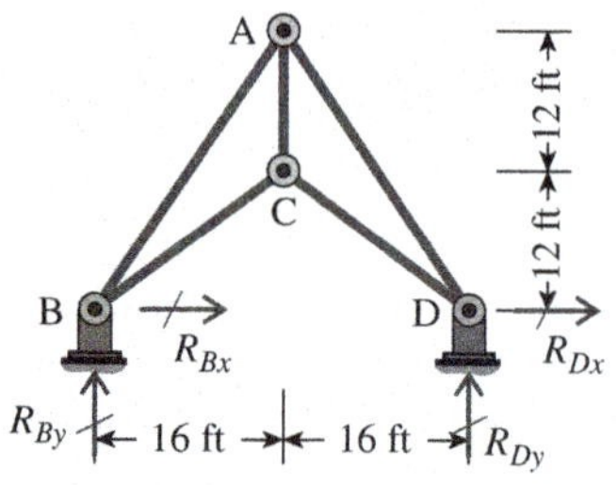

FIGURE 12.1 Truss used for discussion of effects of temperature changes.

VIRTUAL WORK EQUATION
The equation for calculating deflections in a truss due to thermal change in some of the members is

$$1 \cdot \Delta = \sum n_i \alpha_i (\Delta T_i) L_i$$

EXAMPLE 12.5

The top-chord members (BA and AD) of the statically indeterminate truss of Figure 12.1 are assumed to increase in temperature by 60°F. If E = 29,000 ksi and α = 0.0000065/°F, determine the forces induced in each of the truss members. The areas of all members are 10 in^2—except for AC, which is 5 in^2.

SOLUTION STRATEGY

This structure is SI 1°. Assume R_{Dx} is the redundant. Follow the procedure as previously used for trusses.

SOLUTION

Sketch the primary, redundant, and flexibility structures; then assemble the compatibility equation.

Primary Structure

Redundant Structure

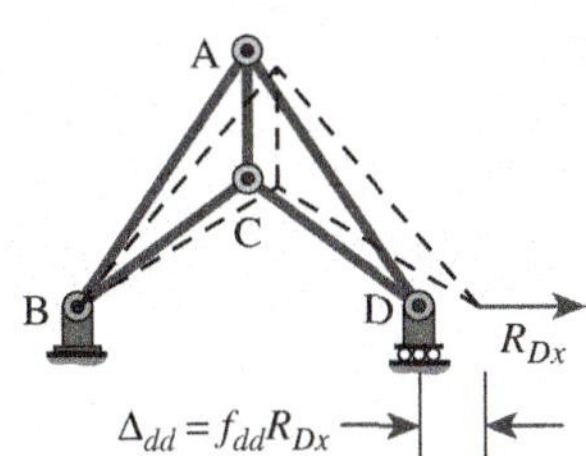

QUICK NOTE

Remember that the displacement in the primary structure is due to temperature changes in members BA and AD. That is why no loads appear on the sketch but there is a displacement. The direction of the unit load on the virtual structure is the direction assumed for Δ_D.

Compatibility Equation:

$$\Delta_D + \Delta_{dd} = \Delta_D + f_{dd}R_{Dx} = 0 \Rightarrow R_{Dx} = -\frac{\Delta_D}{f_{dd}}$$

Flexibility and Virtual Structure (gives *n*)

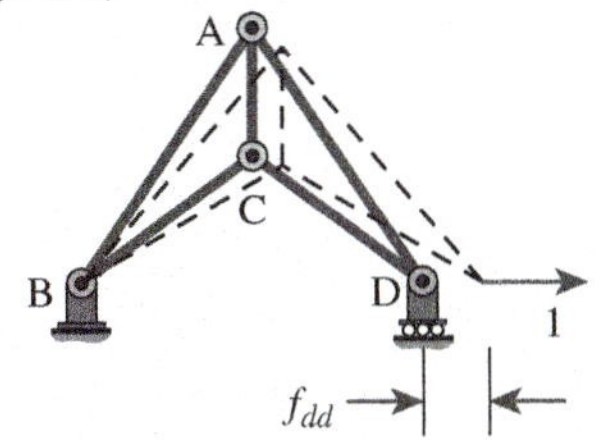

As in Example 12.1, a table is assembled for the calculation of both Δ_D and f_{dd}. However, there is no $(nNL)/(AE)$ term. Rather, it is replaced with a $\alpha(\Delta T)(L)$ term. A sample calculation for member BA is shown.

$$\alpha(\Delta T)(L)_{BA} = (0.0000065/^\circ\text{F})(60^\circ)(346\text{ in.}) = 0.135\text{ in.}$$

Member	L (in.)	A (in^2)	α (ΔT)L(in.)	n	(n)(α)(ΔT)(L)	n^2L/AE	$N_f = nR_{Dx}$ (k)
BA	346	10.0	0.135	−1.80	−0.243	0.0039	32.5
BC	240	10.0	0	2.50	0.000	0.0052	−45.0
AC	144	5.0	0	3.00	0.000	0.0089	−54.0
AD	346	10.0	0.135	−1.80	−0.243	0.0039	32.5
CD	240	10.0	0	2.50	0.000	0.0052	−45.0
				Σ	−0.487 in.	0.027 in./k	

Now substitute $\Delta_D = -0.487$ in. and $f_{dd} = 0.027$ in./k into the compatibility equation.

$$R_{Dx} = -\frac{-0.487\text{ in.}}{0.027\text{ in./k}} = 18\text{ k} = \underline{\underline{18\text{ k}}} \quad \rightarrow$$

The final forces in the truss members due to the temperature change can be found simply by multiplying n by R_{Dx}, since there are no axial forces in the primary structure. These are shown in the last column of the provided table.

12.5 SAP2000 Computer Applications

Temperature loads such as those addressed in Example 12.5 can be readily handled within most commercial analysis packages, including SAP2000. To assign such loads, you first need to select the members on which the loads are to be applied. The temperature loads are then found under the menu *ASSIGN/FRAME LOADS/TEMPERATURE*. The dialogue box shown in Figure 12.2 appears. The load is applied as a temperature change to the member.

The user may apply a temperature change uniformly across the entire depth of the member. They also have the option to apply a gradient temperature through the depth of the member. For example, the top surface of a bridge girder may heat up much quicker than the bottom surface because the top receives direct sunlight whereas the bottom does not. This will cause a non-uniform temperature to develop in the member. Such a temperature gradient will cause the girder to develop curvature, which can produce internal loads if it is an indeterminate girder.

QUICK NOTE

You need to be aware of the temperature units that are being used before you assign a temperature change in the dialog box. Enter a positive number if it is an increase in temperature and a negative number if it is a decrease in temperature.

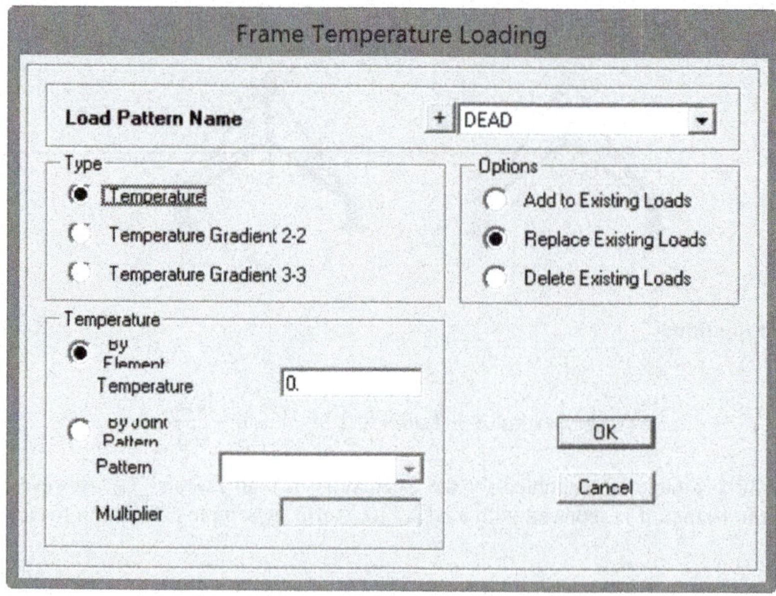

FIGURE 12.2 Dialogue box for applying temperature loads.

SAP2000 EXERCISE 12.1

The file *C12EX1.sdb* contains the truss presented in Example 12.5 except it doesn't have any loads applied. Apply a temperature change of 60°F to members BA and AD and confirm that the only reactions that develop are the horizontal reactions with a magnitude of 18 k.

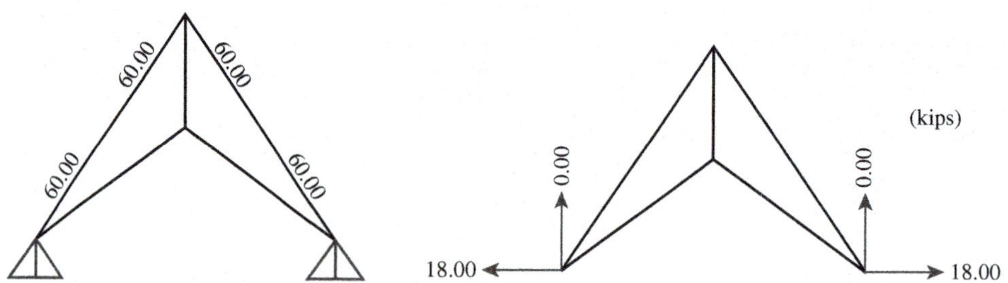

Remember that bar AC had a different cross-sectional area than all of the other bars. Change the area of AC to 10 in^2 like all of the other bars. What does this do to the reactions?

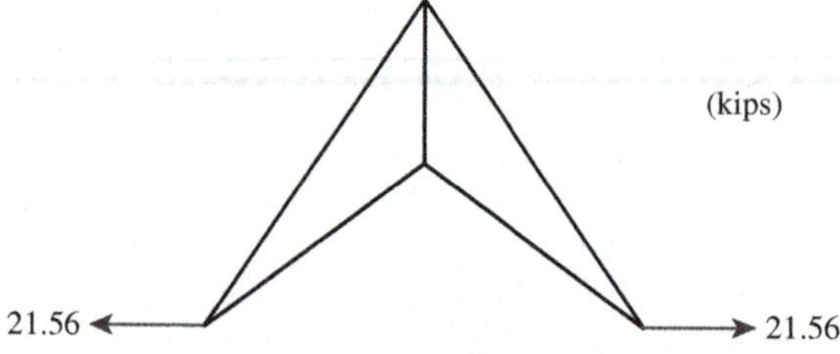

Refer back to Example 12.5 and note that the displacement in the primary structure was not dependent upon the cross-sectional areas of the members. However, the flexibility coefficient is dependent upon cross-sectional areas and will thus change when the area of AC changes. This then directly impacts the value of the reaction.

Now change the restraint at joint D to be a roller instead of a pin. Confirm that the horizontal displacement at D is 0.487 in. ←.

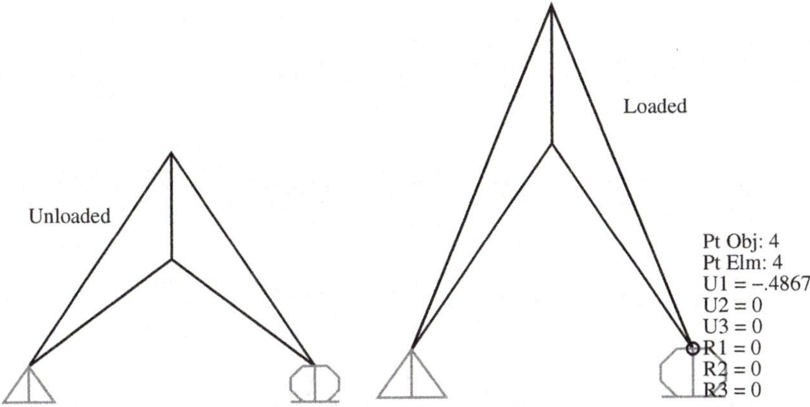

SAP2000 EXERCISE 12.2

The SAP2000 file *C12EX2.sdb* contains a fully functioning model of the beam/spring structure given in Example 12.4. Confirm that the reaction at the spring is indeed 17.4 k.

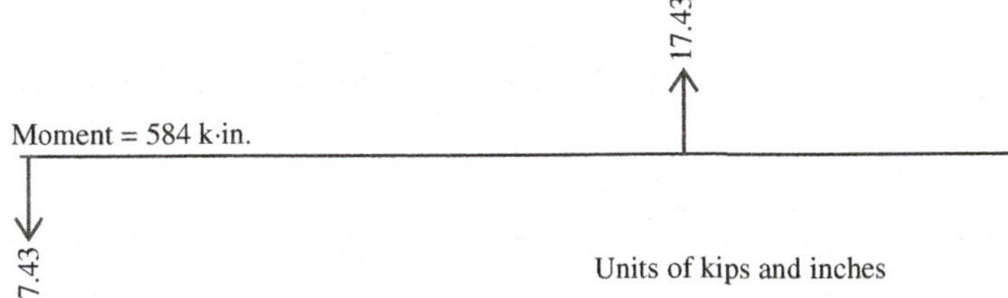

Change the value of the spring stiffness to the values 0 k/in., 0.1 k/in., 0.5 k/in., 1 k/in., 2 k/in., 5 k/in., and 10 k/in. Generate a plot of the stiffness of the spring versus the force in the spring and also a plot of the spring stiffness versus the moment at B.

SUMMARY

The general trend is that as the spring stiffness increases it attracts more of the load. In contrast, the moment at B rapidly decreases as the spring stiffness increases. This is because the load is shifting from B to the spring at A. Indeed, we see that the moment goes from negative into the positive range.

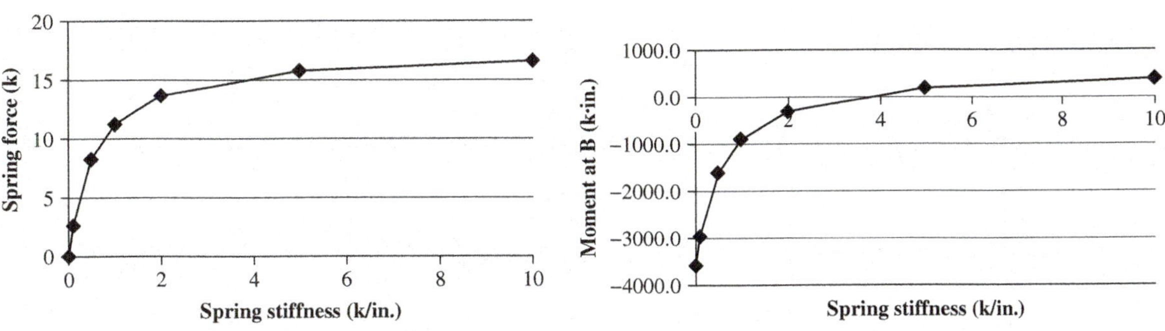

12.6 Examples with Video Solutions

VE12.1 Use the force method to solve for the bar forces in this indeterminate truss. Members AD and BD have an area of 2 in^2 and bar CD has an area of 1 in^2. E is the same for all members.

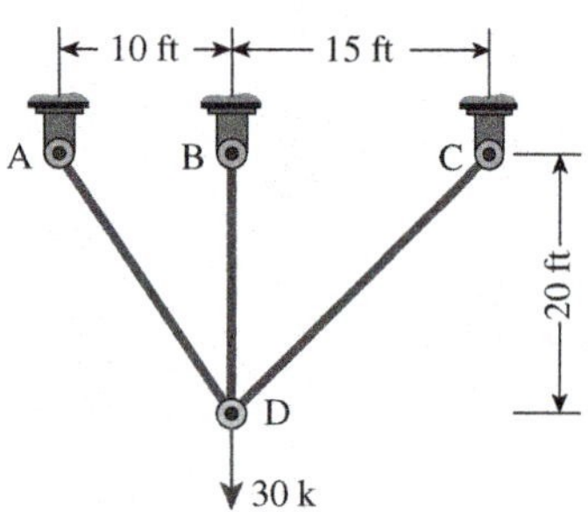

Figure VE12.1

VE12.2 For the composite structure shown in Figure VE12.2, use the force method of analysis to solve for the force in the steel rods.

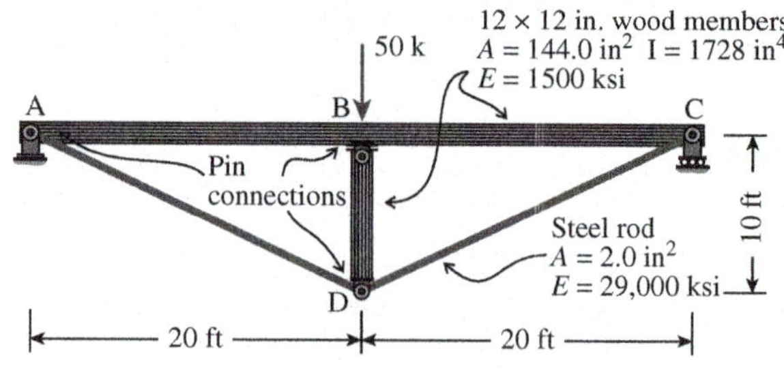

Figure VE12.2

VE12.3 For the beam/spring structure given in Figure VE12.3, use the force method of analysis to solve for the force in the spring. E = 29,000 ksi and I = 1000 in^4 for the beam.

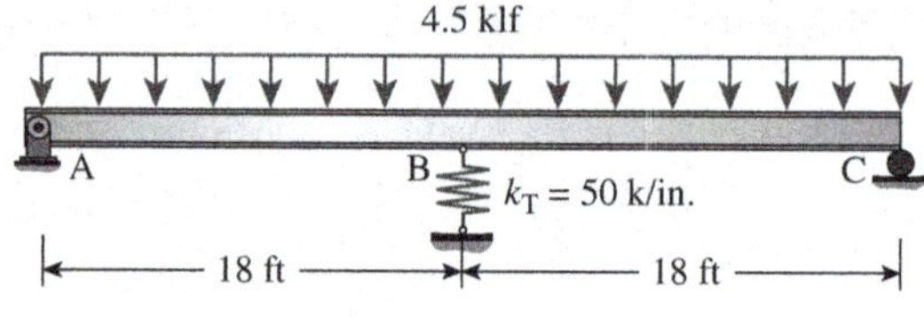

Figure VE12.3

12.7 Problems for Solution

Sections 12.1–12.2

For Problems 12.1 through 12.17 use the force method (method of consistent distortions) to find the value of the redundant force indicated for each truss. E is constant for all members.

P12.1 Find redundant N_{CD}. Bar areas given in in^2 are AD = 3, BD = 6, and CD = 4. (Ans: N_{CD} = 12.93 k)

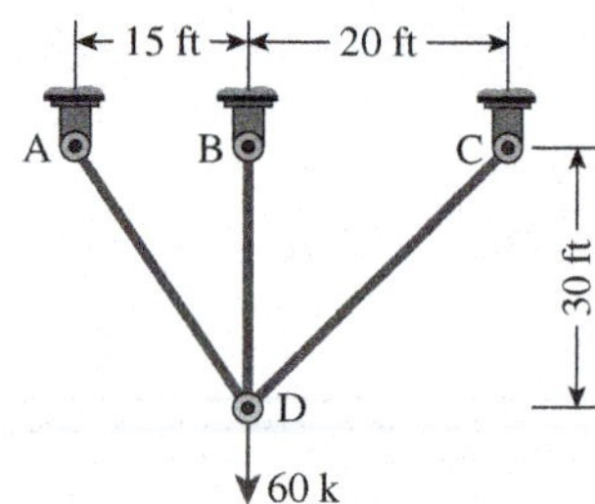

Problem 12.1

P12.2 Find redundant R_{Cy}. All bar areas are the same.

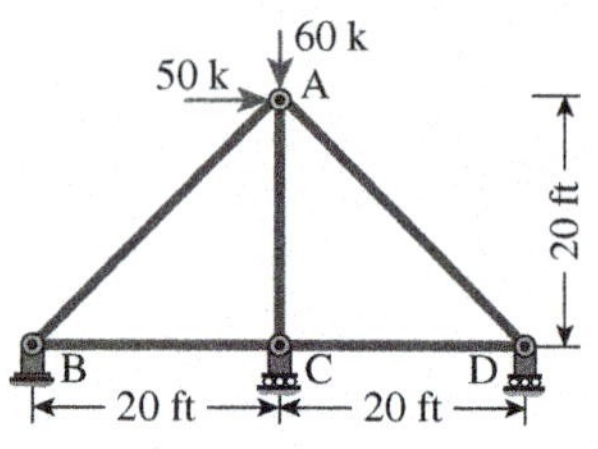

Problem 12.2

P12.3 Find redundant R_{Ey}. Areas of the upper and lower chords are 4.0 in^2 and the internal members are 2.0 in^2. (Ans: R_{Ey} = 3.79 k)

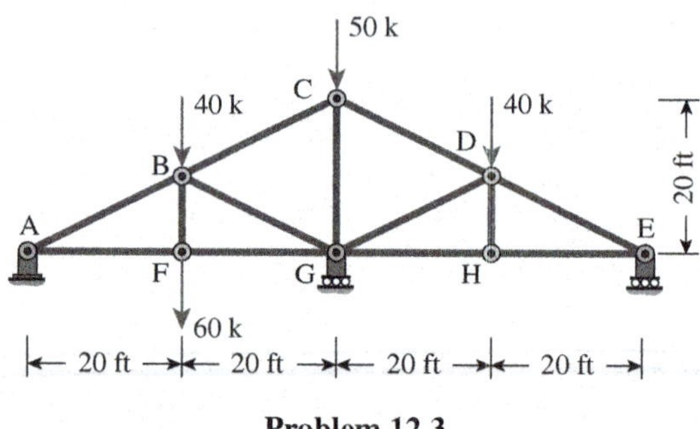

Problem 12.3

P12.4 Find redundant R_{Fx}. All areas are equal.

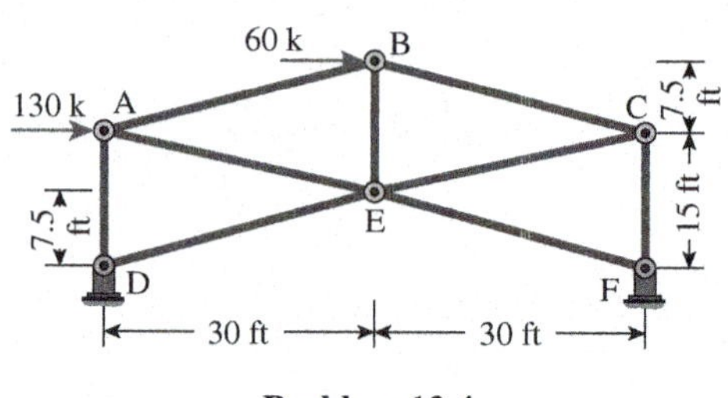

Problem 12.4

P12.5 Find redundant R_{Hy}. All areas are equal. (Ans: R_{Hy} = 149.17 k)

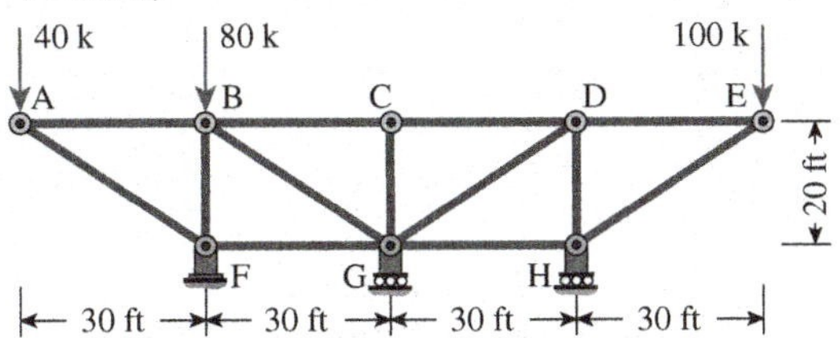

Problem 12.5

P12.6 Find redundant N_{AC}. All bar areas are 3000 mm^2—except AC, which is equal to 2000 mm^2.

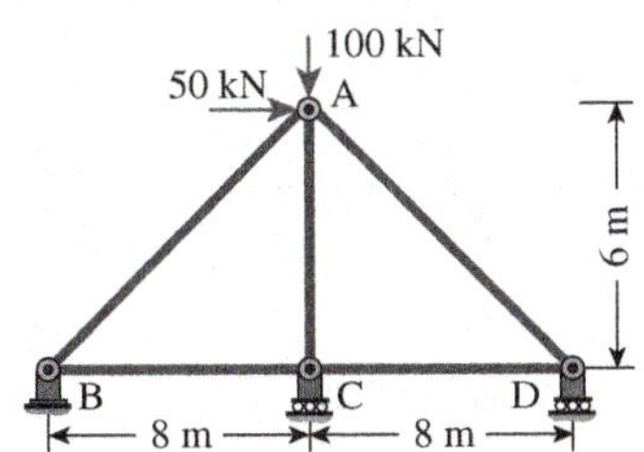

Problem 12.6

P12.7 Find redundant R_{Ey}. All bar areas equal. (Ans: R_{Ey} = 45.44 k)

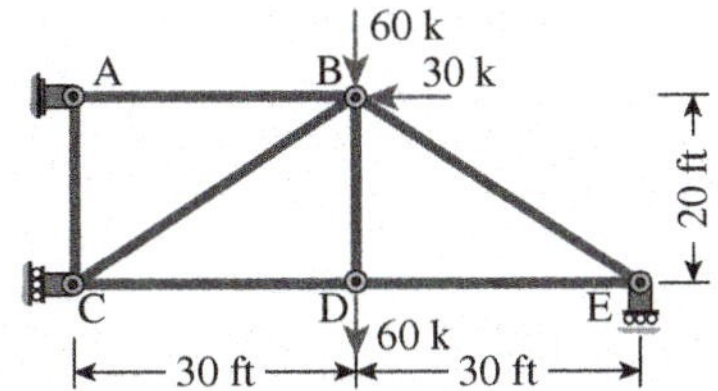

Problem 12.7

P12.8 Find redundant R_{Hx}. Bar areas are 4 in^2—except AF = BF = CF = CG = CH = 3 in^2.

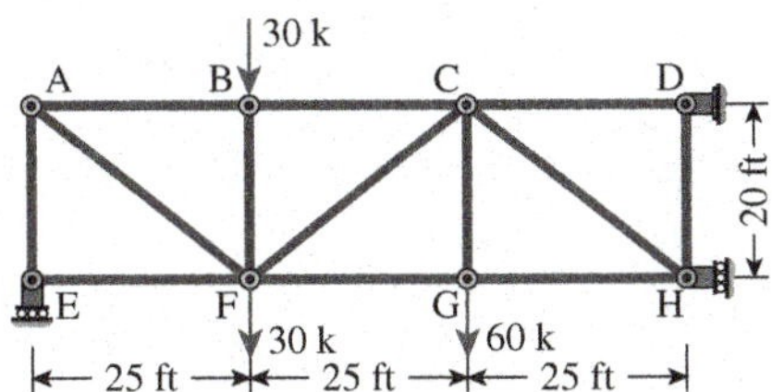

Problem 12.8

P12.9 Find redundant N_{BD}. Bar areas are 5 in^2—except diagonals are 2 in^2. (Ans: N_{BD} = −33.3 k)

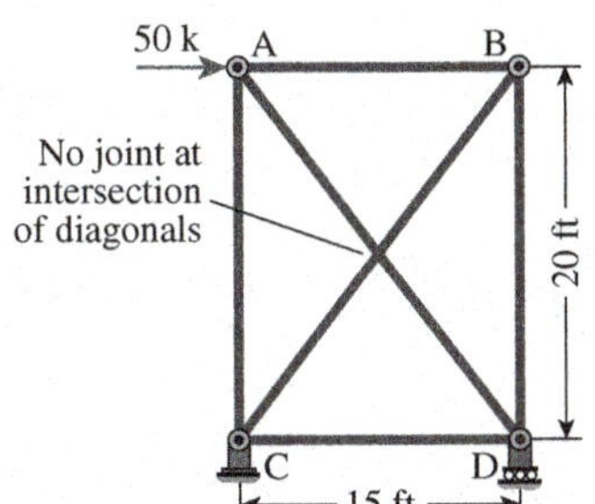

Problem 12.9

P12.10 Find redundant N_{AE}. All bar areas are equal.

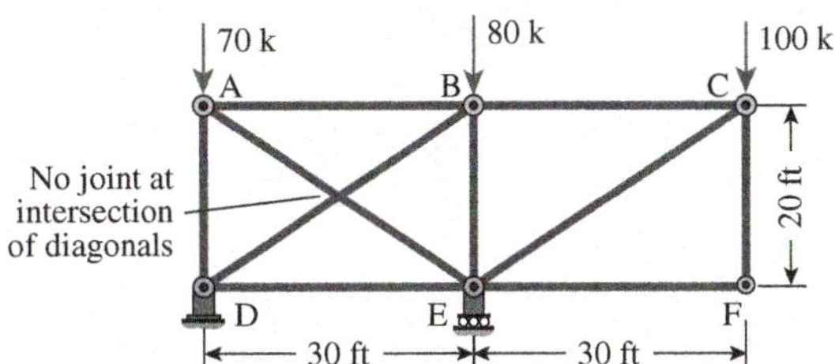

Problem 12.10

P12.11 Find redundant N_{AC}. All exterior bar areas are 40 mm^2. All interior bar areas are 50 mm^2. (Ans: N_{AC} = 126.52 kN)

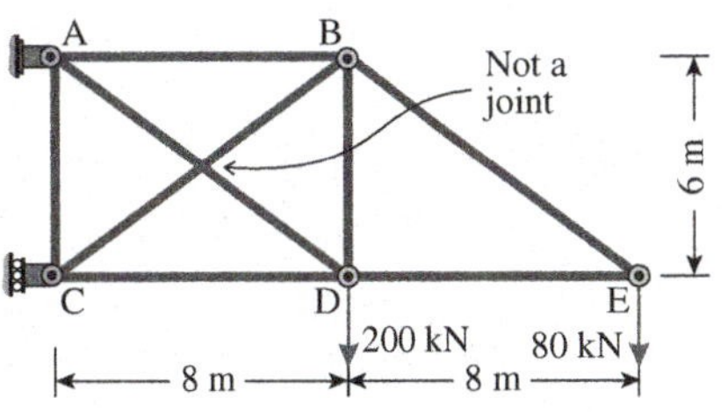

Problem 12.11

P12.12 Find redundant N_{DC}. Bar areas are AB = BC = 4 in^2, AD = BD = DC = 3 in^2, AE = CF = 6 in^2, and ED = DF = 5 in^2.

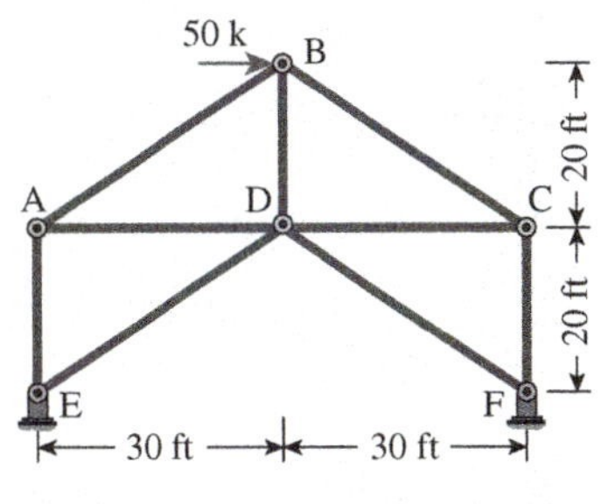

Problem 12.12

P12.13 Find redundant R_{Fy}. Bar areas are equal. (Ans: R_{Fy} = 67.86 kN)

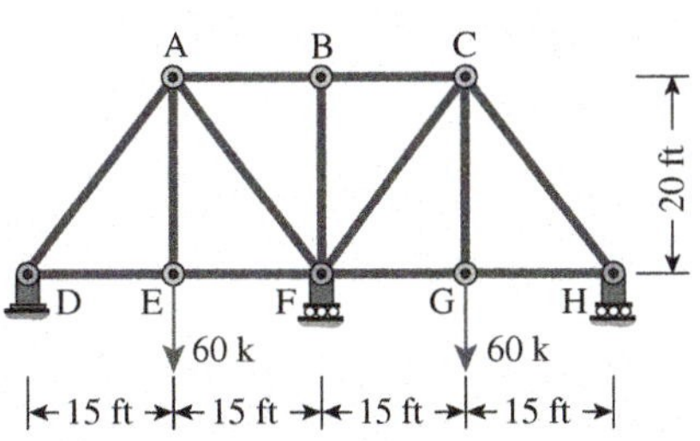

Problem 12.13

P12.14 Find redundant R_{Cx}. Bar areas are equal.

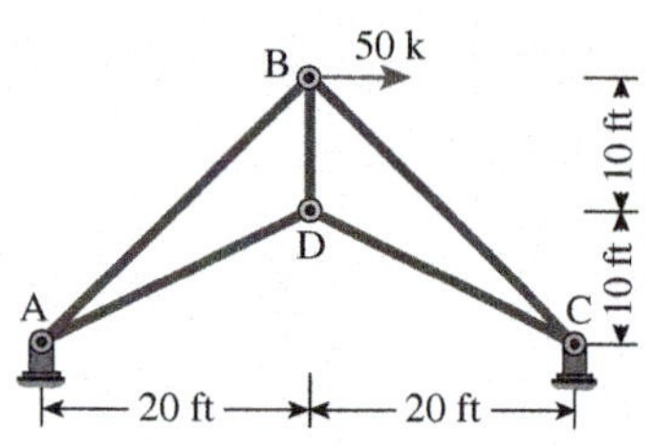

Problem 12.14

P12.15 Find redundant N_{EC}. Bar areas are equal. (Ans: $N_{EC} = -4.16$ k)

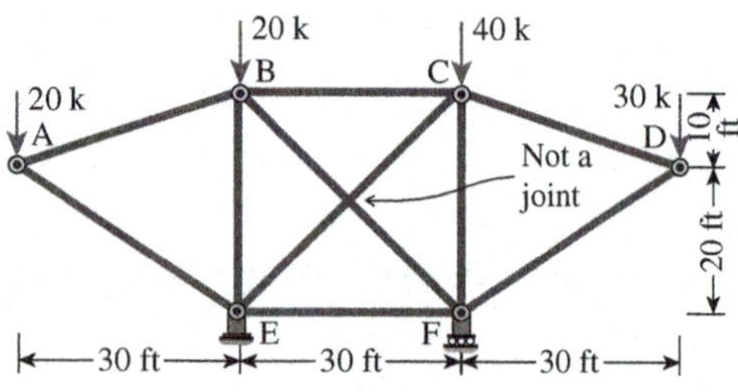

Problem 12.15

P12.16 Find redundant N_{AB}. All bar areas are 4 in^2—except the diagonals that are 5 in^2.

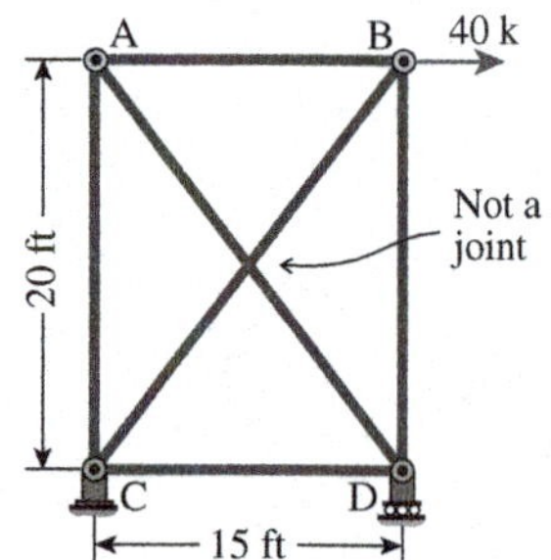

Problem 12.16

P12.17 Find redundant N_{DB}. All bar areas are equal. (Ans: $N_{DB} = 60.94$ kN)

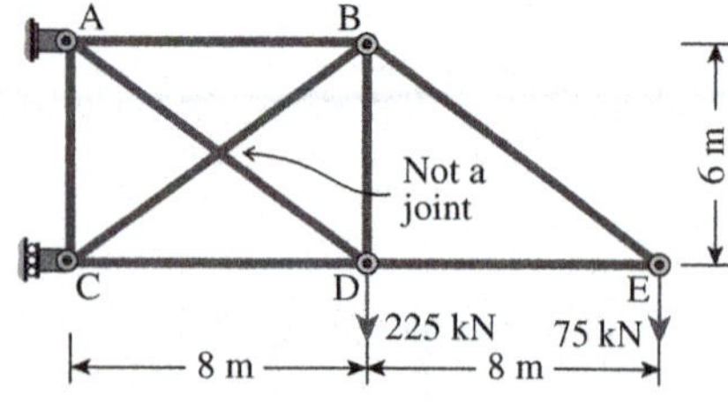

Problem 12.17

Section 12.3

For Problems 12.18 through 12.25, use the force method (method of consistent distortions) to find the value of the redundant force indicated for each structure shown. All members have $E = 29{,}000$ ksi unless noted otherwise.

P12.18 Find redundant R_{By}.

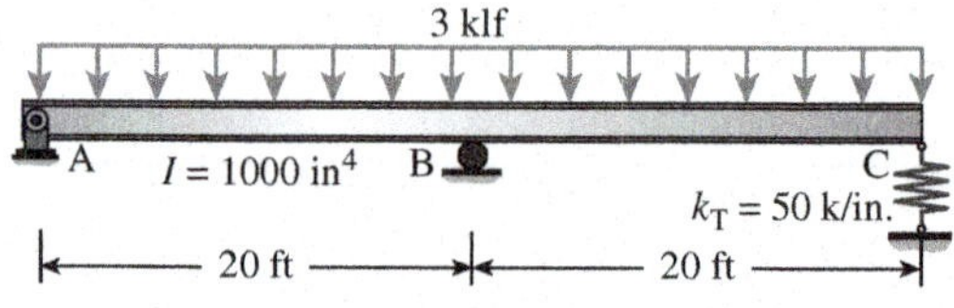

Problem 12.18

P12.19 Find redundant R_{By}. (Ans: $R_{By} = 64.88$ k)

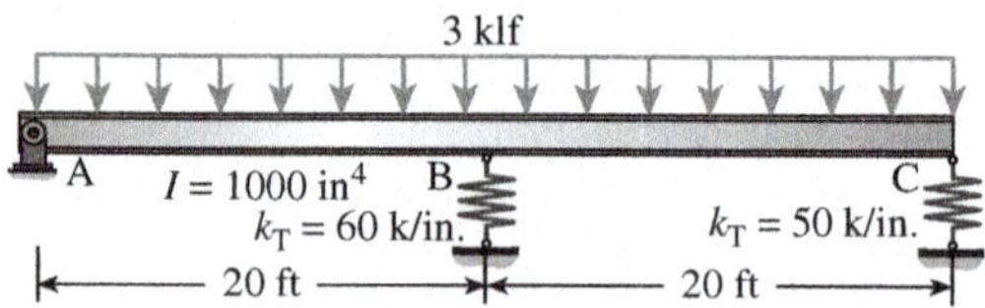

Problem 12.19

P12.20 Find redundant M_A.

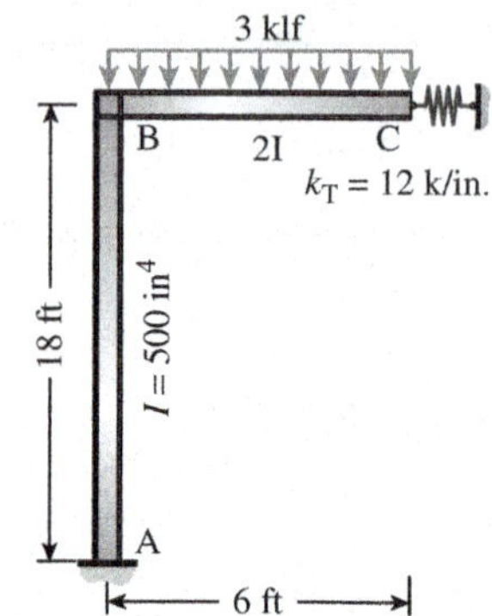

Problem 12.20

P12.21 Find redundant N_{BE}. (Ans: $N_{BE} = -24.5$ k)

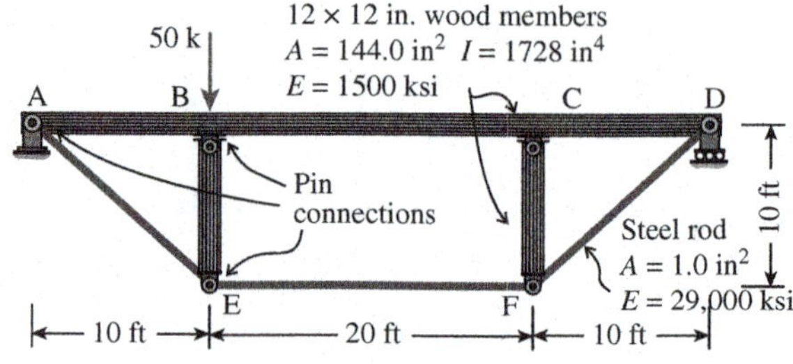

Problem 12.21

P12.22 Find redundant N_{FD}.

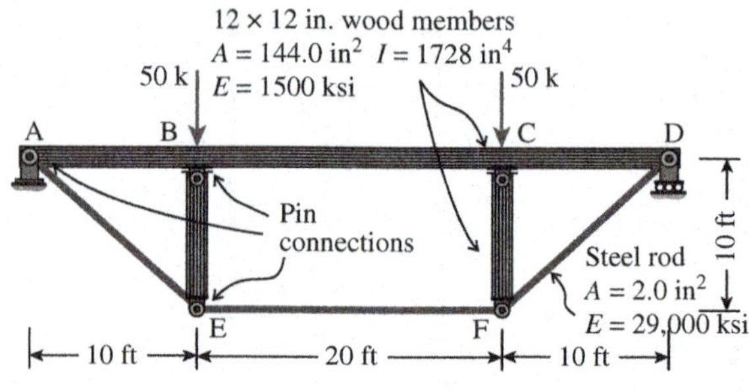

Problem 12.22

P12.23 Find redundant N_{AB}. (Ans: N_{AB} = 44.07 k)

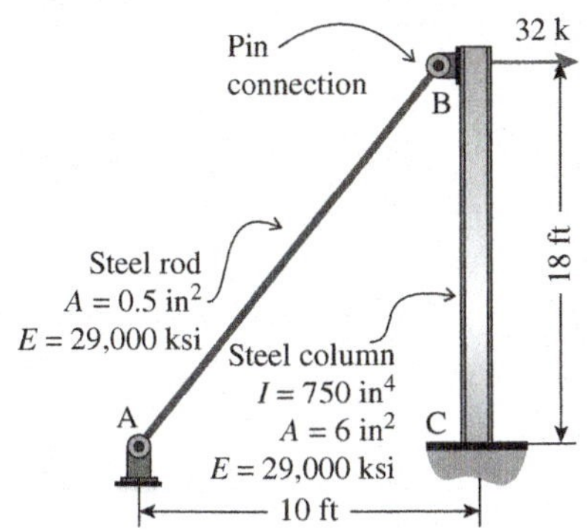

Problem 12.23

P12.24 Find redundant M_C.

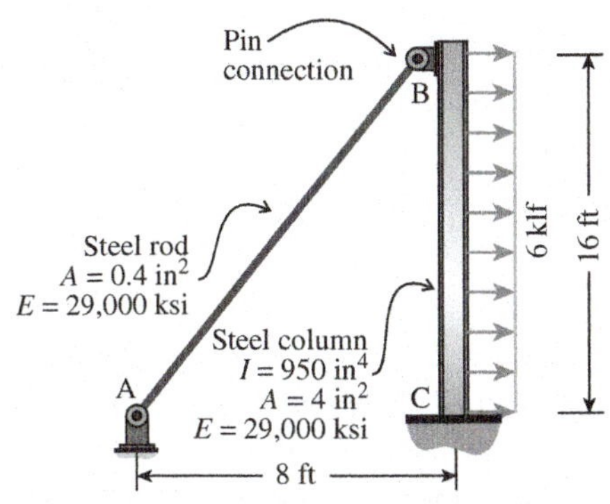

Problem 12.24

P12.25 Find redundant N_{CB}. (Ans: N_{CB} = 18.74 k)

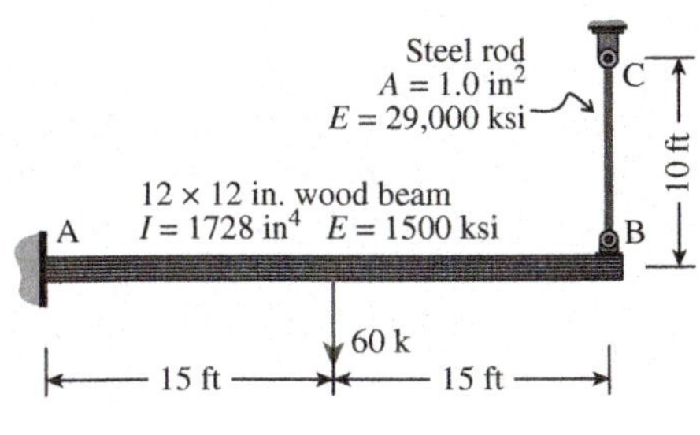

Problem 12.25

Section 12.4

For Problems 12.26 through 12.19, determine the indicated redundant force for the truss shown. Remove any loads on the truss and apply only the specified thermal change. Assume E = 29,000 ksi and the coefficient of thermal expansion $\alpha = 6.5 \times 10^{-6}/°F$. Apply these properties to all members of each truss.

P12.26 Find the redundant force R_{Fx}. Members AB and BC experience a temperature increase of 75°F. All exterior members each have a cross-sectional area of 8 in^2. The interior members AD, BD, and CD have a cross-sectional area of 6 in^2.

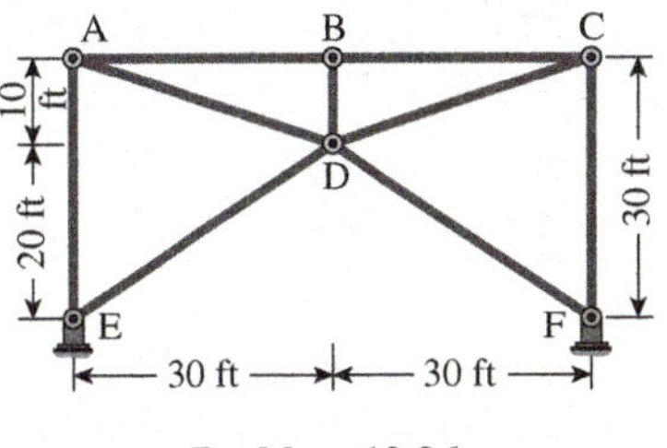

Problem 12.26

P12.27 Find the redundant force R_{Cx}. Consider the truss of Problem 12.7. Member BE experiences a temperature decrease of 100°F. All members have a cross-sectional area of 3 in^2. (Ans: R_{Cx} = 16.73 k)

P12.28 Find the redundant force R_{Fx}. Consider the truss of Problem 12.12. Members ED and DF experience a temperature decrease of 150°F.

P12.29 Find the redundant force N_{AB}, Consider the truss of Problem 12.16. Member BD experiences a temperature increase of 65°F. (Ans: N_{AB} = −6.15 k)

Section 12.5 SAP2000

P12.30 Model the truss shown in Problem 12.1 using SAP2000 and find the bar forces. Do this multiple times while changing the cross-sectional area of bar BD. Areas are given as (in^2) 0.2, 0.5, 0.8, 1.0, 1.5, 2.0, 4.0, 6.0, 10.0, and 20.0. Create a plot of the cross-sectional area versus the force in bar BD. Create another plot of the cross-sectional area of BD versus the force in bar DC. Make an observation of the trend observed in both plots.

P12.31 Model the beam shown in Problem 12.18 using SAP2000 and find the reactions. Do this multiple times while changing the stiffness of the spring. Stiffnesses are (k/in.) 2, 5, 10, 20, 40, 50, 70, 90, and 120. Create a plot of the spring stiffness versus the reaction at B and one for the spring stiffness versus the reaction at C. Make an observation of the trend observed in both plots.

P12.32 Use SAP2000 to model the truss in Problem 12.26. Model the temperature increase as indicated then apply the temperature change only to member AB. Repeat but only apply the temperature change to member BC. What change, if any, is observed in the bar forces for the three cases?

13 Moment Distribution for Beams

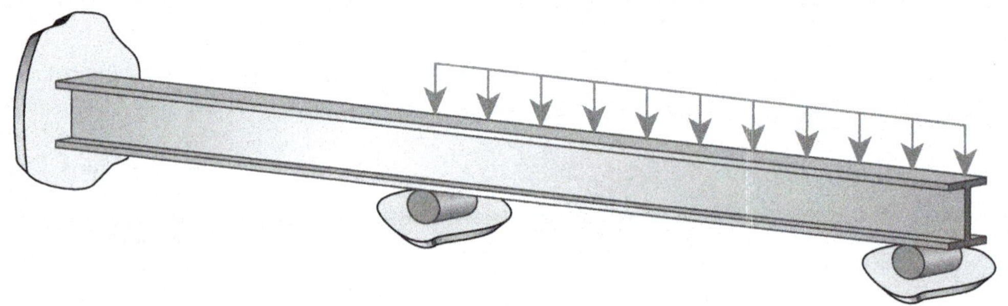

13.1 Introduction

QUICK NOTE

The *moment-distribution method* is also known as the *Hardy Cross method*.

The late Hardy Cross wrote papers about the *moment-distribution method* in 1929[1,2] and 1930[3] after having taught the subject to students at the University of Illinois since 1924. His papers began a new era in the analysis of statically indeterminate frames and gave added impetus to their use. The moment-distribution method of analyzing continuous beams and frames involves a little more labor than the approximate methods (covered in Chapter 15), but this method yields accuracy equivalent to that obtained from the infinitely more laborious classical "exact" methods.

The analysis of statically indeterminate structures in the preceding chapters may often require the solution of simultaneous equations. Simultaneous equations are not necessary in solutions by moment distribution—except in a few rare situations for complicated frames. The method that Cross developed involves successive cycles of computation with each cycle drawing closer to the "exact" answers. When these advantages are considered in light of the fact that the accuracy obtained by the lengthy "classical" methods is often questionable, the true worth of this quick and practical method is understood.

DID YOU KNOW?

The calculations may be stopped after two or three cycles, giving a very good approximate analysis, or they may be carried on to whatever degree of accuracy is desired.

From the 1930s until the 1960s, moment distribution was the dominant method used for the analysis of continuous beams and frames. Since the 1960s, however, there has been a continually increasing use of computers for analyzing all types of structures. Computers are extremely efficient for solving the simultaneous equations that are generated by other methods of analysis. Generally, the software is developed from the matrix-analysis procedures described in Chapters 19 through 21 of this book.

Even with the computer software available, moment distribution continues to be an important hand-calculation method for this analysis of continuous beams and frames. Structural engineers can use moment distribution to quickly make approximate analyses for preliminary designs and check computer results, which is very important. In addition, moment distribution may be solely used for the analysis of small structures.

[1]Hardy Cross, "Continuity as a Factor in Reinforced Concrete Design," *Proceedings of the American Concrete Institute*, Vol. 25 (1929), 669–708.

[2]Hardy Cross, "Simplified Rigid Frame Design," Report of Committee 301, *Proceedings of the American Concrete Institute*, Vol. 26 (1929), 170–183.

[3]Hardy Cross, "Analysis of Continuous Frames by Distributing Fixed-End Moments," *Proceedings of the American Society of Civil Engineers*, Vol. 56, No. 5 (May 1930): 919–928.

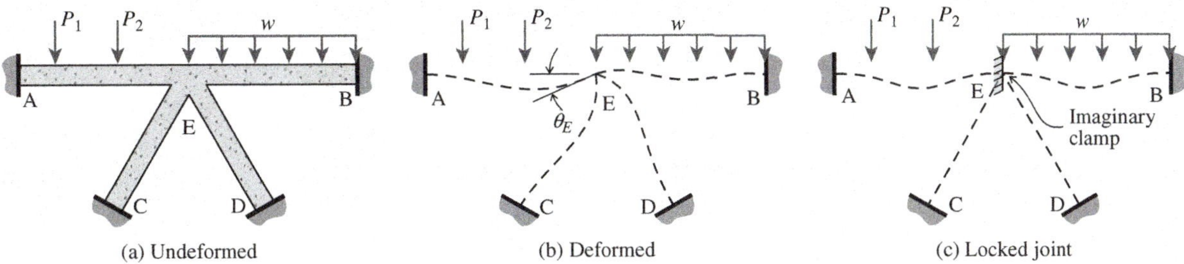

FIGURE 13.1 Behavior of an indeterminate frame.

In the discussion that follows, certain assumptions have been made. These assumptions are

1. The structures have members of constant cross section throughout their respective lengths. That is, the members are prismatic.
2. The joints at which two or more members frame together do not translate. This means they cannot displace.
3. The joints to which members are connected can rotate, but the ends of all members connected to a joint rotate the same amount as the joint. At a joint, there is no rotation of the ends of members relative to each other or to the joint.
4. Axial deformation of members is neglected.

Considering the frame of Figure 13.1(a), joints A to D are seen to be fixed. Joint E, however, is not fixed, and the loads on the structure will cause it to rotate slightly, as represented by the angle θ_E in Figure 13.1(b).

If an imaginary clamp is placed at E, fixing it so that it cannot be rotated, the structure will take the shape of Figure 13.1(c) when it is under load. For this situation, with all ends fixed, the fixed-end moments can be calculated with little difficulty using the table found in Appendix D.

If the clamp at E is removed, the joint will rotate slightly, rotating the ends of the members meeting there and causing a redistribution of the moments in the member ends. The changes in the moments or rotations at the E ends of members AE, BE, CE, and DE cause some effect at their other ends. When a moment is applied to one end of a member with the other end being fixed, there is some effect or *carryover* to the fixed end.

After the fixed-end moments are computed, the problem to be handled may be briefly stated as consisting of the calculation of (1) the moments caused in the E ends of the members by the rotation of joint E, (2) the magnitude of the moments carried over to the other ends of the members, and (3) the addition or subtraction of these latter moments to the original fixed-end moments. A complete procedure for this method is presented once some basic definitions and concepts have been established.

QUICK NOTE

The beauty of moment distribution lies in the simplicity of its theory and application. Readers will be able to grasp quickly the principles involved and will clearly understand what they are doing and why they are doing it.

DID YOU KNOW?

When one uses the **moment-distribution method** for solving indeterminate structures, they are directly solving for the internal moments that exist at a joint. Remember that the **force-based method** solves for redundant forces that could be reactions or internal forces/moments. In both methods, basic statics is used to solve for the remaining responses.

13.2 Sign Convention

The moment-distribution method requires that a different sign convention be used for internal moments than the traditional (beam sign convention) introduced in Chapter 6. The sign convention that will be used in this chapter states that if a moment on the end of a member acts in the clockwise direction it will be considered positive. This is illustrated in Figure 13.2. Contrast this with the traditional sign convention used for generating moment diagrams in Chapter 6, which is also shown in Figure 13.2.

QUICK NOTE

Students will often express a dislike of having to use a different sign convention and would rather not change from the traditional beam sign convention. However, it is necessary to do so. Indeed, as is discussed later in this chapter, it becomes necessary to be able to switch back and forth, so you are advised to become comfortable with it now.

Moment Distribution

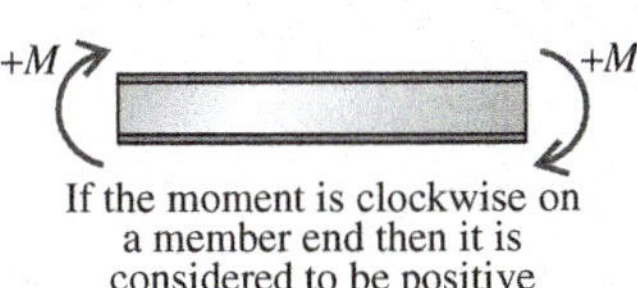

If the moment is clockwise on a member end then it is considered to be positive

Traditional

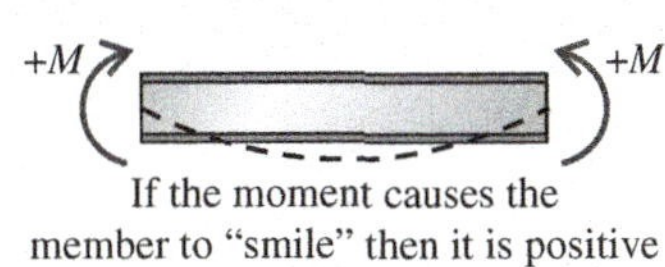

If the moment causes the member to "smile" then it is positive

FIGURE 13.2 Sign convention for two different uses.

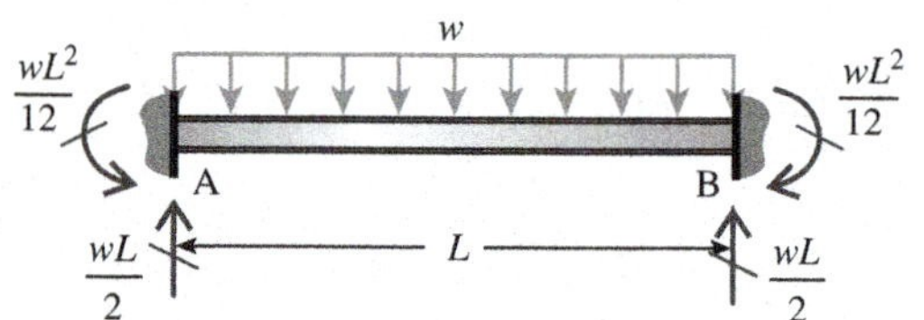

FIGURE 13.3 Uniformly distributed load on a fixed-fixed beam.

13.3 Basic Concepts and Definitions

Introducing a basic set of concepts and definitions will greatly aid in the understanding of the moment-distribution process. With only the basic introduction given in Section 13.1, it may be difficult at this time to fully understand the relevance of the concepts to be presented hereafter. However, the reader is encouraged to focus on the individual concepts at this stage. The relevance of each concept should become apparent once the moment-distribution method is formally laid out.

Fixed-End Moments (FEM)

QUICK NOTE
Beam elements that are fixed at both ends are statically indeterminate to the second degree. The reader may use the force method or other methods not introduced in this text to solve for these fixed-end moments. However, certain loading cases are so common that pre-solving for these few cases will save a lot of work in the long run.

When all of the joints of a structure are clamped to prevent any joint rotation, the external loads produce certain moments at the ends of the members to which they are applied. These moments are referred to as fixed-end moments (FEM). The table in Appendix D presents fixed-end moments for various common loading conditions.

Having the cases represented by the variables w, P, and L allows for many different loading scenarios to be analyzed with a single beam diagram. For example, a fixed-fixed beam with a length of 20 ft is subjected to a uniformly distributed load of 3 klf. Referring to Figure 13.3 or the chart in Appendix D, the fixed-end moments (FEMs) are found to be

$$\text{FEM} = \frac{wL^2}{12} = \frac{(3\ \text{klf})(20\ \text{ft})^2}{12} = 100\ \text{k·ft}$$

In this case, the FEMs are the same on both ends, but this will not always be the case. To distinguish between the moments at either end, a set of subscripts is often assigned to the FEM label. The first subscript denotes the joint at which the moment is located. The first and second subscripts indicate the member that experiences the moment. For example, the FEM on the left end of the beam in Figure 13.3 would be $\text{FEM}_{AB} = -100$ k·ft, and the moment on the right end would be $\text{FEM}_{BA} = 100$ k·ft. Both of these moments occur on member AB, but they occur at different joints. Also note the difference in the signs that follow the sign convention previously discussed.

Member Stiffness Factor (k)

DID YOU KNOW?
A stiffness factor may be described as the force or moment required to impose a unit displacement or rotation, respectively. For example, if $\theta = 1$ then $M = k_R$.

Every structure or structural element may be described by a stiffness factor. Such a factor describes the force–deformation behavior. This is most clearly seen when looking at springs—both translational and rotational. These relationships are shown in Figure 13.4. Notice that the relation between force P and displacement Δ is the stiffness factor k_T. One should also be comfortable with the notion that when a moment is applied to a rotational spring a rotation results. This relationship is described by the rotational stiffness factor k_R.

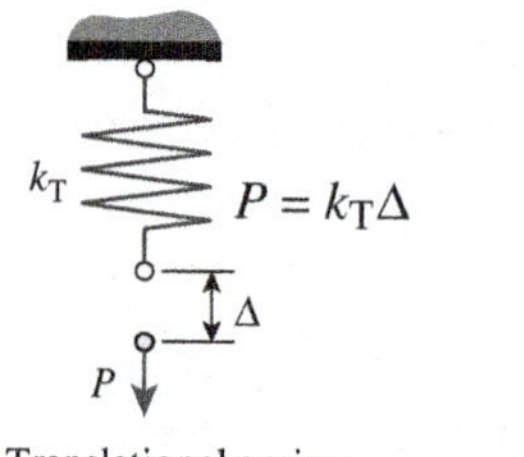

(a) Translational spring

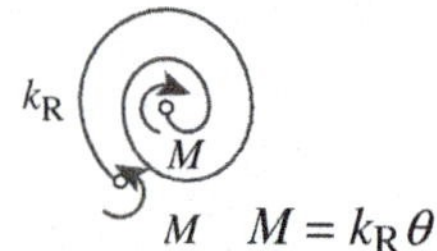

(b) Rotational spring

FIGURE 13.4 Typical spring types and their behavior.

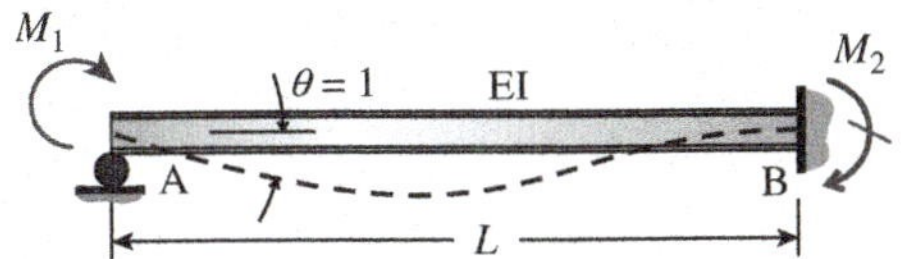

$$M_1 = k_{AB}\theta = k_{AB} = \frac{4EI}{L}$$

$$M_2 = \frac{M_1}{2} = \frac{2EI}{L}$$

FIGURE 13.5 Stiffness for fixed connection at B.

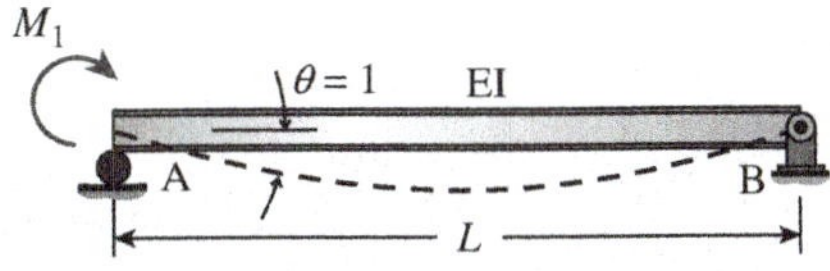

$$M_1 = k_{AB}\theta = k_{AB} = \frac{3EI}{L}$$

FIGURE 13.6 Stiffness for pinned connection at B.

The moment-distribution method is concerned with the rotational stiffness of beam members. Consider the propped cantilever beam shown in Figure 13.5. It has a moment applied at point A. The beam does rotate at point A, so logically we understand that the beam at point A has a rotational stiffness (k_{AB}). Without derivation, the stiffness for this beam is given as

$$k_{AB} = \frac{4EI}{L}$$

The stiffness value of the beam changes if the support at point B is changed from being fixed to being pinned. Figure 13.6 shows the results of such a change. Note that the moment required to rotate A one unit rotation is less than before. Simply put, stiffness will drop if the amount of support is reduced.

QUICK NOTE

The subscripts on the stiffness factor are interpreted the same way as for the FEMs. The first subscript is for the location of the stiffness and both subscripts indicate the contributing member.

DID YOU KNOW?

The stiffness value for the beam in Figure 13.5 may be found using the force method or other methods not discussed in this book (e.g., moment area). However, for the moment-distribution method, the given stiffness simply needs to be memorized and not derived again.

Distribution Factor (DF)

When multiple members come together at a joint where a moment is being applied, all of them will undergo some rotation. As such, they will share the moment between them. The *distribution factor* (DF) of a member indicates what portion of the applied moment the member will take. Consider the structure shown in Figure 13.7. It has three members coming together at joint A. If a moment M is applied at joint A, all three members will deform and all three members will take a portion of that moment. The next section develops the numerical evaluation of member distribution factors.

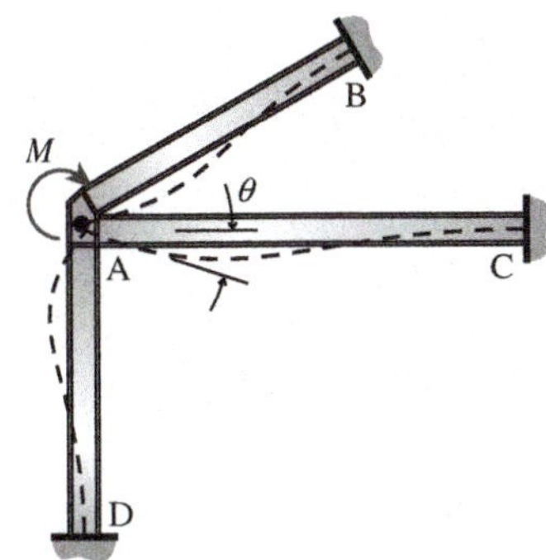

FIGURE 13.7 Deformed shape of frame due to a joint rotation.

Unbalanced Moments

Initially, the joints in a structure are considered to be clamped (see Figure 13.1). When a joint is released, it rotates if the sum of the fixed-end moments at the joint is not zero. The difference between zero and the actual sum of the end moments is known as the *unbalanced moment*.

Distributed Moments (DM)

After the clamp at a joint is released, the unbalanced moment causes the joint to rotate. The rotation twists the ends of the members at the joint and changes their moments. In other words, rotation of the joint is resisted by the members, and resisting moments are built up in the members as they are twisted. Rotation continues until equilibrium is reached—when the resisting moments equal the unbalanced moment—at which time the sum of the moments at the joint is equal to zero. The moments developed in the members resisting rotation are the *distributed moments* (DM).

Carryover Moments (CO)

The distributed moments in the ends of the members cause moments in the other ends, which are assumed to be fixed. These are the *carryover moments* (CO). Examine Figure 13.5. When the moment M_1 is applied at A, a moment reaction at B develops with a magnitude of exactly half of M_1. For the beam in Figure 13.6, no moment develops at point B. In the first case, some of the applied moment carries over to the other side, whereas there is no moment that carries over in the other case. Thus, the carryover factor is either 1/2 or 0, depending on the support conditions.

13.4 Distribution Factors

COMMON MISTAKE

Students will often come up with a set of distribution factors at a joint that doesn't sum up to 1.0. No matter how many or how few members meet at the joint, the sum of the distribution factors at that joint must not be more or less than 1.0.

In the previous section, the idea of distribution factors was introduced. For instance, the total moment applied at joint A in Figure 13.7 must be taken by the members that meet at that joint. What portion of the moment M is taken by member AB? This portion may be described by a distribution factor (DF_{AB}), which must be less than 1.0. The same can be said about members AC (DF_{AC}) and AD (DF_{AD}). As such, all of the distribution factors at a given joint must sum up to be 1.0, which says that all of the applied moment M is accounted for.

$$DF_{AB} + DF_{AC} + DF_{AD} = 1.0$$

Distribution factors are easily calculated using the fundamental principle that a moment at the joint will distribute according to the relative stiffness of the members framing into the joint. This principle is most clearly seen by looking at a set of translational springs shown in Figure 13.8. When the structure is subjected to the load P, it will deflect an amount Δ. Because of the given configuration, both springs will deflect the same amount: $\Delta = \Delta_1 = \Delta_2$ where the subscripts 1 and 2 refer to the left and right spring, respectively. This is a compatibility equation, as was seen in Chapter 11. Remembering that the displacement can be rewritten in terms of the spring stiffness and the spring force, the compatibility equation may be rewritten as

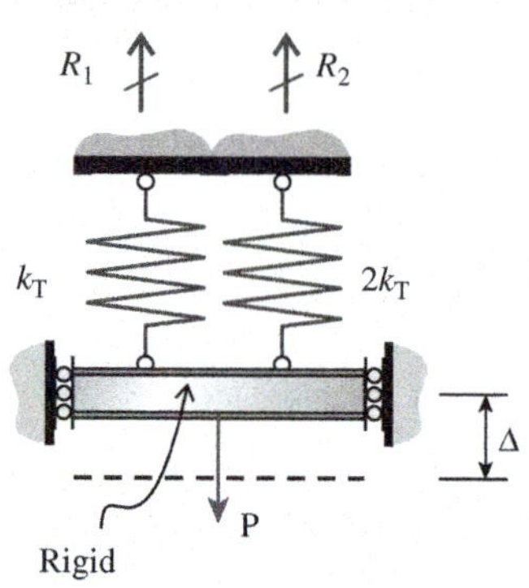

FIGURE 13.8 Parallel spring configuration.

$$\frac{R_1}{k_T} = \frac{R_2}{2k_T}$$

An equilibrium equation may be written for the same structure as

$$+\uparrow \sum F_y = R_1 + R_2 - P = 0$$

The simultaneous solution of these two equations reveals:

$$R_1 = \frac{P}{3} \qquad R_2 = \frac{2P}{3}$$

One may now observe that the force in the left spring is half of the force in the right spring. Also, the stiffness in the left spring is half of the stiffness in the right spring. Hence, the force distributes according to the relative stiffness of the springs. The distribution factors for this structure could be written as $DF_1 = 1/3$ and $DF_2 = 2/3$. Notice that $\sum DF = 1.0$.

The distribution factors are easily calculated by

$$DF_i = \frac{k_i}{\sum k}$$

QUICK NOTE

Another way of looking at the DF calculations is as member stiffness at the joint divided by the total joint stiffness.

Where k_i is the stiffness of member i and $\sum k$ is the sum of the stiffness of all members coming into the joint. For example, the DF for the left spring is calculated as

$$DF_1 = \frac{k_T}{k_T + 2k_T} = \frac{1}{3}$$

We now know that the left spring will take one third of the applied load P.

This concept of distribution according to relative stiffness is readily applied to members undergoing rotation, as shown in Example 13.1.

EXAMPLE 13.1

For the given beam, calculate the distribution factors for joint B.

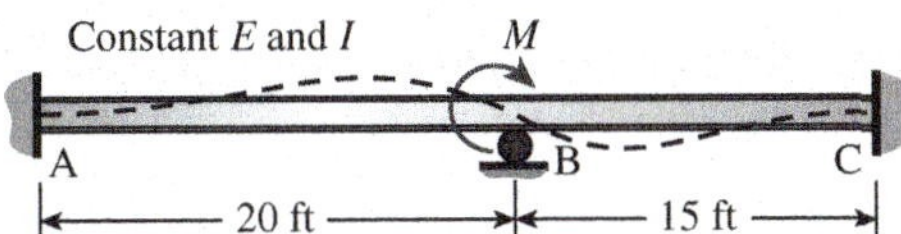

SOLUTION STRATEGY

Find the rotational stiffness value for each member meeting at joint B. Then use the distribution factor equation previously given.

> **QUICK NOTE**
> The appropriate stiffness factor ($\frac{4EI}{L}$ or $\frac{3EI}{L}$) is most easily identified by looking at the other end of the member in question. If the other end is fixed, the stiffness will be the 4, but if it is a pin or roller, then it will be the 3.

SOLUTION

The stiffness values for members BA and BC are

$$k_{BA} = \frac{4EI}{L} = \frac{4EI}{20 \text{ ft}}$$

$$k_{BC} = \frac{4EI}{L} = \frac{4EI}{15 \text{ ft}}$$

The distribution factors are found as

$$\text{DF}_{BA} = \frac{\frac{4EI}{20 \text{ ft}}}{\frac{4EI}{20 \text{ ft}} + \frac{4EI}{15 \text{ ft}}} = \underline{\underline{0.43}}$$

$$\text{DF}_{BC} = \frac{\frac{4EI}{15 \text{ ft}}}{\frac{4EI}{20 \text{ ft}} + \frac{4EI}{15 \text{ ft}}} = \underline{\underline{0.57}}$$

> **QUICK NOTE**
> Whenever E, I, 4, or 3 is constant for all members, it is often conveniently left off because they will cancel out in the end. For example, DF_{BA} may have been written as
>
> $$\text{DF}_{BA} = \frac{\frac{1}{20}}{\frac{1}{20} + \frac{1}{15}} = \underline{\underline{0.43}}$$

SUMMARY

The distribution factors for joint B sum up to be 1.0. They indicate that 43% of the applied moment M is taken by member BA and 57% of the applied moment is taken by member BC.

EXAMPLE 13.2

For the given beam, calculate the distribution factors for joint B.

SOLUTION STRATEGY

Find the rotational stiffness value for each member meeting at joint B. Then use the distribution factor equation previously given.

> **QUICK NOTE**
> Joint C is a roller and not fixed. This means that k_{BC} will use $\frac{3EI}{L}$.

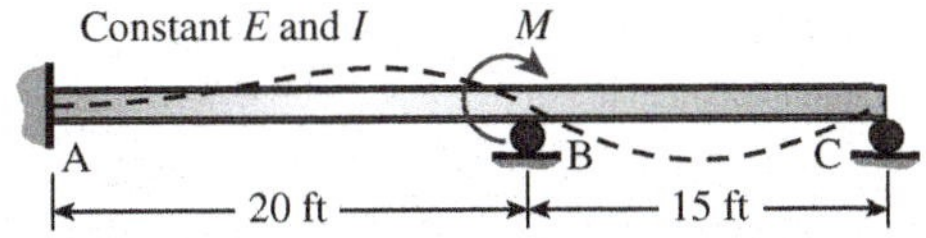

SOLUTION

The stiffness values for members BA and BC are

$$k_{BA} = \frac{4EI}{L} = \frac{4EI}{20 \text{ ft}} = \frac{EI}{5 \text{ ft}}$$

$$k_{BC} = \frac{3EI}{L} = \frac{3EI}{15 \text{ ft}} = \frac{EI}{5 \text{ ft}}$$

> **QUICK NOTE**
> The change of the support conditions changes the stiffness of member BC. Since it lost some support, its stiffness is reduced, which also affects its stiffness relative to member BA.

QUICK NOTE

To find DF_{BC}, we could also calculate it as

$$DF_{BC} = 1.0 - DF_{BA}$$

The distribution factors are found as

$$DF_{BA} = \frac{\frac{1}{5}}{\frac{1}{5} + \frac{1}{5}} = \underline{\underline{0.50}}$$

$$DF_{BC} = \frac{\frac{3EI}{15\text{ ft}}}{\frac{4EI}{20\text{ ft}} + \frac{3EI}{15\text{ ft}}} = \underline{\underline{0.50}}$$

SUMMARY

The distribution factors for joint B sum up to be 1.0. They indicate that 50% of the applied moment M is taken by member BA and 50% of the applied moment is taken by member BC.

13.5 Application of Moment Distribution

The few tools needed for applying moment distribution are now available and the procedure is given here.

Procedure

REMEMBER

This procedure solves for the internal moments at each joint in a structure due to a set of given loads.

1. Identify all joints. This includes the point of intersection of two or more members and support locations.
2. Lock all of the joints against rotation. This will cause the beam to be temporarily out of equilibrium but will reduce the beam to a set of fixed-fixed beams.
3. Calculate the member stiffness factors (k_i) based on the locked joints.
4. Calculate the member distribution factors (DF_i).
5. Determine the fixed-end moments (FEMs) for each member due to the applied load. Be sure to use the new sign convention to interpret these moments.
6. Select a joint that is out of equilibrium and *determine the amount of moment* (i.e., unbalanced moment) that needs to be added to that joint to *bring it into equilibrium*. This is done by summing up all moments at the joint and identifying how far the total is from zero.
7. Unlock the joint in question and apply the unbalanced moment. Distribute the unbalanced moment to the adjacent spans according to the distribution factors.
8. Carry moments to the opposite end of each affected member according to the CO moment. The factor will either be 1/2 or 0.
9. Relock the joint and repeat steps 6 through 8, focusing on another joint that is out of equilibrium until the desired accuracy is obtained.
10. Sum up each column in the table to get the final moments.

QUICK NOTE

The moment-distribution method is an iterative method and will converge to the "exact" solution with more iterations. The analyst may choose the level of accuracy desired, depending on their needs. The authors recommend that convergence is sufficient when the unbalanced moment is 1 to 2% of the moment at the joint.

The computations associated with the moment-distribution method are most easily handled in tabular form. Example 13.3 demonstrates this use of a table.

EXAMPLE 13.3

Use the moment-distribution method to solve for the internal moments at the joints of the given beam. E and I are constant over the entire length.

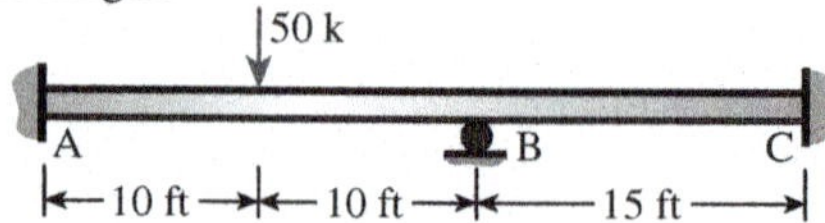

SOLUTION STRATEGY

The stiffness and distribution factors have already been calculated for this beam configuration in Example 13.1. Follow the outlined procedure.

SOLUTION

There are three joints—two of which are already fixed. Clamp joint B against rotation as shown.

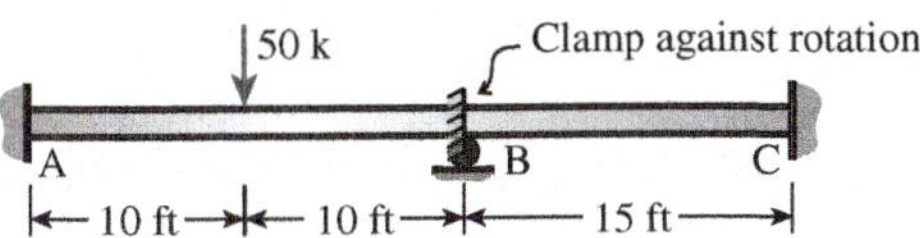

The next step is to calculate the fixed-end moments for members AB and BC. The FEMs for BC are zero because span BC has no loads placed on it. The FEMs for span AB are identified and computed using the charts in Appendix D.

$$\text{FEM}_{AB} = -\frac{PL}{8} = -\frac{(50\text{ k})(20\text{ ft})}{8} = -125\text{ k·ft}$$

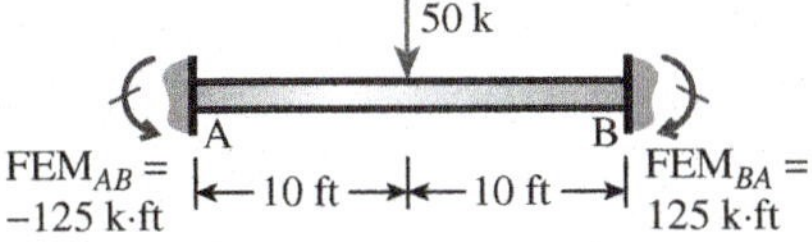

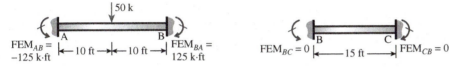

Place the distribution factors, as calculated in Example 13.1, and the fixed-end moments in a table and follow the procedure for balancing the moment.

A		B		C	Joints
AB		BA	BC	CB	
0.00		0.43	0.57	0.00	Distribution Factors
−125.0		125.0	0.0	0.0	Fixed-End Moment
		−53.8	−71.2		Distribution
−26.9	⟵		⟶	−35.6	Carryover
−151.9		71.2	−71.2	−35.6	Final Moments

Sum up the moments at joint B to get unbalanced moment = 125.0 + 0.0. This means a moment of −125.0 k·ft needs to be added to joint B to make the sum zero. This must be distributed to the adjacent members using the distribution factors.

$$(-125)(0.43) = -53.8\text{ k·ft}$$
$$(-125)(0.57) = -71.2\text{ k·ft}$$

Since both joints A and C are fixed, they have a carryover of 1/2. Once the carryover is complete, draw a horizontal line and sum each column in the table. No other joints are out of balance, so the analysis can stop here.

When the distribution has reached the accuracy desired, a double line is drawn under each column of figures. The final moment in the end of a member equals the sum of the moments at its position in the table. Unless a joint is fixed, the sum of the final end moments in the ends of the members meeting at the joint must total zero. A single joint may be balanced at a time or all joints may be balanced simultaneously. In this textbook, we adopt the simultaneous approach.

EXAMPLE 13.4

Using moment distribution, determine the member end moments for the three-span beam shown. E is constant over the entire length.

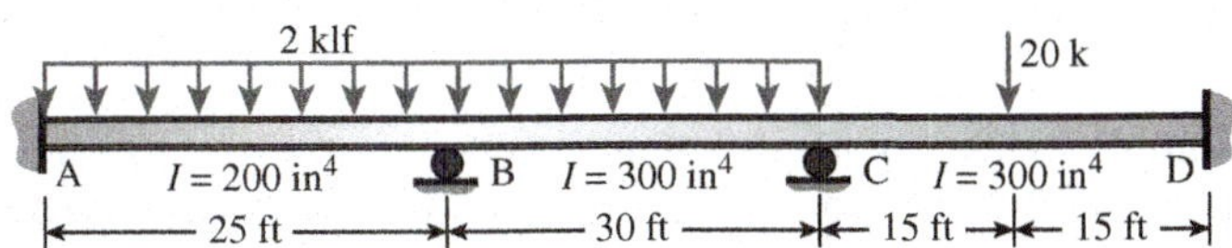

SOLUTION STRATEGY

Lock joints B and C against rotation. Solve for the FEMs and DFs using the same procedure as before. When filling out the table, balance all joints simultaneously.

SOLUTION

Since all beam segments are fixed-fixed, the $\frac{4EI}{L}$ stiffness factor will apply to all cases; thus, the 4 and E may be omitted from the DF equation.

$$DF_{BA} = \frac{\frac{200}{25}}{\frac{200}{25} + \frac{300}{30}} = 0.44 \qquad DF_{BC} = 1 - DF_{BA} = 0.56$$

$$DF_{CB} = \frac{\frac{300}{30}}{\frac{300}{30} + \frac{300}{30}} = 0.50 \qquad DF_{CD} = 1 - DF_{CB} = 0.50$$

QUICK NOTE

Joints B and C need to be clamped against rotation. This leaves three beam spans that are fixed-fixed. From the beam chart, the FEMs for a uniformly loaded fixed-fixed beam is

$$FEM = \frac{wL^2}{12}$$

For a point load placed mid-span, the FEM is

$$FEM = \frac{PL}{8}$$

The FEMs for each span are computed according to the equations given in the Quick Note sidebar. For example,

$$FEM_{BA} = \frac{(2\text{ klf})(25\text{ ft})^2}{12} = 104.2\text{ k·ft}$$

A		B		C			D	Joints
AB		BA	BC	CB	CD		DC	
0.00		0.44	0.56	0.50	0.50		0.00	DFs
−104.2		104.2	−150.0	150.0	−75.0		75.0	FEMs
		20.2	25.6	−37.5	−37.5			Dist 1
10.1	⟵		−18.8 ⟷	12.8		⟶	−18.8	CO 1
		8.3	10.5	−6.4	−6.4			Dist 2
4.1	⟵		−3.2 ⟷	5.3		⟶	−3.2	CO 2
		1.4	1.8	−2.6	−2.6			Dist 3
0.7	⟵		−1.3 ⟷	0.9		⟶	−1.3	CO 3
		0.6	0.7	−0.4	−0.4			Dist 4
0.3	⟵		−0.2 ⟷	0.4		⟶	−0.2	CO 4
		0.1	0.1	−0.2	−0.2			Dist 5
−89.0		134.7	−134.7	122.2	−122.2		51.5	Final

QUICK NOTE

Since all joints are balanced at each step, the new unbalanced moment at any joint is equal to the carryover from the previous step. Be sure to sum the entire column when the iterations are done.

The moments at joint B are summed as 104.2 − 150.0 = −45.8. The negative of this value is distributed to the adjacent spans using the distribution factors, so 0.44(45.8) = 20.2 and 0.56(45.8) = 25.6. Since all joints are fixed, the carryover from joint B to joints A and C is one-half of the total. So the carryover is $\frac{1}{2}(20.2) = 10.1$ and $\frac{1}{2}(25.6) = 12.8$.

Joint C is balanced simultaneously. The moments at joint C are summed as 150.0 − 75.0 = 75.0. The negative of this value is distributed to the adjacent spans using the distribution factors, so 0.50(−75.0) = −37.5 and 0.50(−75.0) = −37.5. Since all joints are fixed, the carryover from joint C to joints B and D is one-half of the total. So the carryover is $\frac{1}{2}(-37.5) = -18.8$ in both directions.

SUMMARY

The process described above for the *Dist 1* and *CO 1* is carried out repeatedly. Five distributions are carried out in this example. However, the error is less than 2% after only three distributions, which is a reasonable error.

Beginning with Example 13.5, a slightly different procedure is used for distributing the moments. Only one joint at a time is balanced, and the required carryovers are made from that joint. Generally speaking, it is desirable (but not necessary) to balance the joint that has the largest imbalance; make the carryovers; balance the joint with the next largest imbalance, and so on, because such a process will result in the quickest convergence. This procedure is quicker and provides fewer computations than the tabular method used for Examples 13.3 and 13.4 and more closely agrees with the procedure previously outlined.

EXAMPLE 13.5

Compute the end moments in the beam shown using the moment-distribution method.

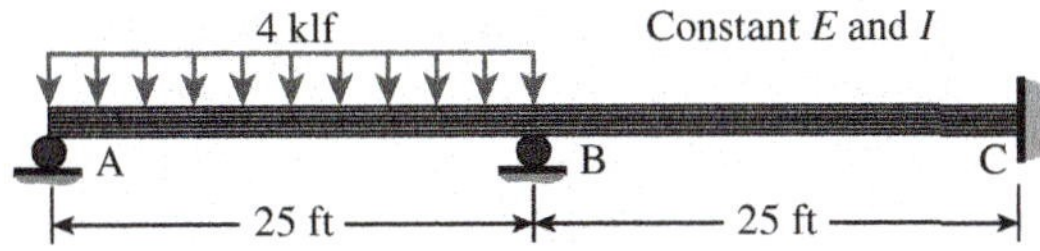

QUICK NOTE

Since all joints are locked against rotation, all FEMs, member stiffness factors, and DFs are based on the fixed-fixed condition.

$DF_{BA} = 0.5 \quad DF_{BC} = 0.5$

$FEM_{AB} = -208.3$ k·ft

$FEM_{BA} = 208.3$ k·ft

$FEM_{BC} = FEM_{CB} = 0$

SOLUTION STRATEGY

Lock joints A and B against rotation. Solve for the FEMs and DFs using the same procedure as before. In this example, the FEMs and DFs will simply be given, and the focus will be on balancing a single joint at a time.

SOLUTION

Place all the FEMs and DFs in a table as for previous examples. Start balancing the joint with the largest imbalance. Either joint A or B would meet this criterion.

DON'T FORGET

Joint A in the original beam is a roller support and thus has a zero moment. Hence, anytime there is a moment present in the table at joint A, that moment represents the unbalanced moment.

A		B			C	Joints
AB		BA	BC		CB	
1.00		0.50	0.50		0.00	DFs
−208.3		208.3	0.0		0.0	FEMs
208.3	⟶	104.2				Dist and CO for A
−78.1	⟵	−156.2	−156.2	⟶	−78.1	Dist and CO for B
78.1	⟶	39.1				Dist and CO for A
−9.8	⟵	−19.5	−19.5	⟶	−9.8	Dist and CO for B
9.8	⟶	4.9				Dist and CO for A
−1.2	⟵	−2.4	−2.4	⟶	−1.2	Dist and CO for B
1.2	⟶	0.6				Dist and CO for A
−0.2	⟵	−0.3	−0.3	⟶	−0.2	Dist and CO for B
0.2						Dist for A
0.0		178.5	−178.5		−89.2	Final Moments

To begin, the imbalanced load at A is −208.3. Therefore, a +208.3 needs to be added to that joint. Note the value of $DF_{AB} = 1.0$. This is because any moment added to joint A must be taken fully by member AB. All joints have been fixed, so the CO is one-half everywhere. Thus, 0.5(208.3) = 104.2 is carried over to joint B.

The first time joint B is balanced, the unbalanced moment is computed, including the FEMs and the CO from A, so 208.3 + 0.0 + 104.2 = 312.5. For the remaining iterations, it will only need to include the CO from A. The −312.5 now needs to be distributed at joint B according to the DFs and then carried over to joints A and C. This process continues until the CO from any joint is quite small relative to the FEMs.

Lastly, the numbers in each column should be summed to get the final moments at each joint.

COMMON MISTAKE

The first couple of times this moment-distribution method is used, students become very confused on what numbers need to be distributed and what numbers need to be carried over. It is important that the table is studied, but it is also important to read the descriptive text associated with the examples to ensure the calculations are understood.

In Chapter 15, several methods for approximately analyzing statically indeterminate structures is introduced. Moment distribution is one of the "exact" methods of analysis if it is carried out until the moments to be distributed and carried over become quite small. However, it also can be used as a superb approximate method for statically indeterminate structures if only a few cycles of distribution are made.

Consider the beam of Example 13.4. In this problem, each cycle of distribution is said to end when the unbalanced moments are balanced. Table 13.1 shows the ratios of the total moments up through each cycle at joints A and C to the final moments at those joints after all of the cycles are completed. These ratios should give an idea of how good partial moment distribution can be as an approximate method.

DID YOU KNOW?
For many cases where the structure and/or the loads are very unsymmetrical, there will be a few cycles of major adjustment of the moments throughout the structure. Until these adjustments are substantially made, moment distribution will not serve as a very good approximate method.

Table 13.1 Accuracy of moment distribution after various cycles of balancing for the beam of Example 13.4.

Values Given After Cycle No.	Moments at Support A (k·ft)	Ratio of Approximate Moment to "Exact" Moment at Support A	Moments at Support C (k·ft)	Ratio of Approximate Moment to "Exact" Moment at Support C
1	−104.2	1.171	−112.5	0.921
2	−94.1	1.057	−118.9	0.973
3	−90.0	1.011	−121.5	0.994
4	−89.3	1.003	−121.9	0.998
5	−89.0	1.000	−122.2	1.000

13.6 Modification of Stiffness and FEM for Simple Ends

DID YOU KNOW?
The term "simple end" refers to a support that doesn't resist moment (i.e., pin or roller). This means that the same stiffness and FEMs apply to a fixed-pinned case just as they do to a fixed-roller case.

In working Example 13.5, the roller support was considered to be clamped; the carryover was made to the end; and the joint was freed and balanced back to zero. Then one-half of that balancing moment was carried over to the adjacent support. This procedure, when repeated over and over, is absolutely correct, but it involves a little unnecessary work that may be eliminated by being aware of stiffness, carryover, and fixed-end moment information for beams other than those that are fixed-fixed.

Figure 13.5 shows that when joint B is fixed the stiffness is $(4EI)/L$ and the carryover moment is one-half. However, when the joint B is pinned (or a roller), as shown in Figure 13.6, the stiffness drops to $(3EI)/L$, and there is no carryover moment. Furthermore, the beam charts found in Appendix D also contain information for beams that are fixed-pinned. Example 13.6 reworks the beam given in Example 13.5, only this time joint A is not locked against rotation.

EXAMPLE 13.6

Determine the end moments for the beam in Example 13.5—only *do not lock the joint* at A.

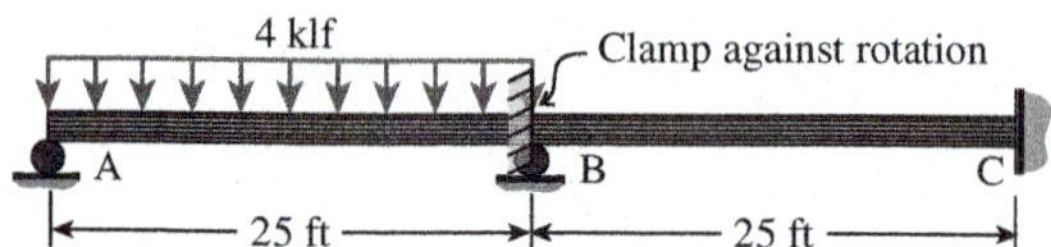

SOLUTION STRATEGY

With B being the only joint that is locked, the beams are generated as a roller-fixed and also a fixed-fixed. Use the appropriate stiffness factor and the correct FEM chart for each case.

QUICK NOTE
To determine which stiffness factor needs to be used, $4EI/L$ or $3EI/L$, look at the support at the end away from the joint of interest. For example, trying to find the stiffness factors for joint B, one would need to look at joint A and joint C. If the opposite joint is a roller or pin, use $3EI/L$, and if the other end is fixed, use $4EI/L$. Thus, $3EI/L$ applies to k_{BA}, and $4EI/L$ applies to k_{BC}.

SOLUTION

The FEMs are calculated using the beam chart for a fixed-roller beam, which says that FEM $= \frac{wL^2}{8} =$ 312.5 k·ft.

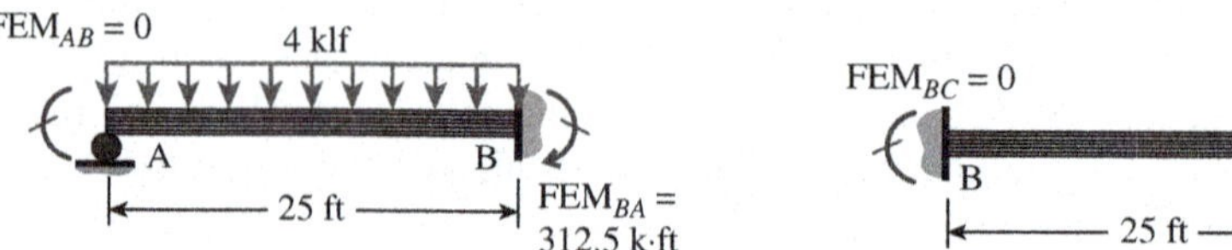

Calculate the stiffness factors for BA and BC. Since E and I are constant, they may be omitted from the stiffness calculation.

$$k_{BA} = \frac{3}{25} \qquad k_{BC} = \frac{4}{25}$$

The distribution factors are readily calculated for joint B. No distribution factor is needed for joints A or C.

$$DF_{BA} = \frac{\frac{3}{25}}{\frac{3}{25} + \frac{4}{25}} = 0.43 \qquad DF_{BC} = 1 - DF_{BA} = 0.57$$

A	B			C	Joints
AB	BA	BC		CB	
0.00	0.43	0.57		0.00	DFs
0.0	312.5	0.0		0.0	FEMs
	−134.4	−178.1	⟶	−89.1	Dist and CO for B
0.0	178.1	−178.1		−89.1	Final Moments

QUICK NOTE

There is no carryover from B to A, but there is carryover from B to C. One may easily identify the required carryover by examining the joint opposite the joint in question. For example, joint C is fixed, so it will receive carryover from B, but joint A is a roller, so it will not receive carryover from B.

SUMMARY

The same beam that took seven balancing steps in Example 13.5 now takes only one balancing step. This is because there was not an imbalance at A, and we weren't working with an artificial set of constraints at A that needed to be overcome.

13.7 Shearing Force and Bending Moment Diagrams

Drawing shear and moment diagrams is an excellent way to check the final moments computed by moment distribution and to obtain an overall picture of the stress condition in the structure. Before preparing the diagrams, it is necessary to revisit the sign convention introduced earlier in this chapter. Referring to Figure 13.2, the *moment-distribution* sign convention must be used to translate the moments found in the moment-distribution table onto free-body diagrams of each beam segment. The traditional sign convention then must be used to solve for the unknown internal shears and draw the moment diagrams.

While there is more than one way to generate shear and moment diagrams using the results of moment distribution, the following procedure is recommended to avoid confusion.

Procedure

1. Draw a separate free body diagram for each beam section.
   (a) Sketch a separate element for each span.
   (b) Place the actual loads on the element.
   (c) Use the moment-distribution sign convention to transfer the values and direction of the moment arrows to the element. Use the sign in the table to sketch the appropriate direction, but when writing the values of moment on the free body diagram, *do not use any negative signs.*
   (d) Sketch the unknown internal shears on the diagram, assuming a positive shear according to the sign convention given in Chapter 6.
   (e) Use equilibrium to solve for the values of shear.
2. Sketch the shear and moment diagrams for each beam span. Treat each span as an individual beam.

REMEMBER

The sign convention used for shear is

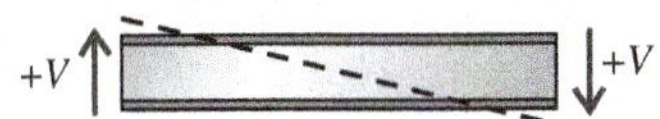

If the shear causes a clockwise rotation of the member it is positive

EXAMPLE 13.7

For the beam shown in Example 13.3, generate the shear and moment diagrams using the results from the moment-distribution table.

SOLUTION STRATEGY

Cut the beam just to the left of joint B and also just to the right of joint B. This will give a FBD for AB and one for BC. The FBD for the joint is not needed.

SOLUTION

Sketch the FBD of both segments AB and BC.

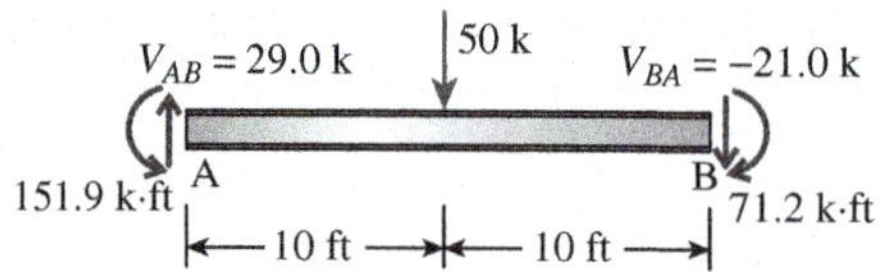

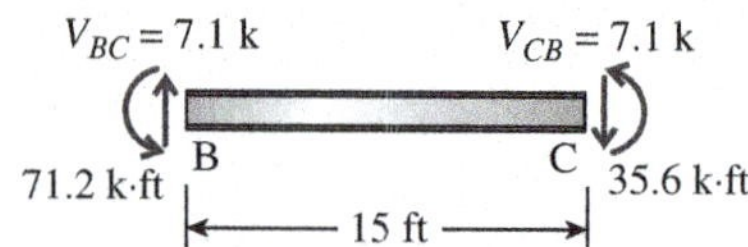

QUICK NOTE

For segment AB, one can sum moments about point A to solve for V_{BA} and then sum forces in the y direction to get V_{AB}. For segment BC, one can sum moments about point B to solve for V_{CB} and then sum force in the y direction to get V_{BC}.

The table indicates that at point A the moment is −151.9 k·ft. According to the moment-distribution sign convention, a negative moment is to be sketched counterclockwise. The table indicates that just to the left of point B the moment is +71.2 k·ft, which will be sketched as a clockwise moment. The same thought process is used to sketch the moments on element BC.

Use equilibrium to get the four values of shear.

SUMMARY

When drawing the shear and moment diagrams for the beam, it is convenient to draw them separately for each segment AB and BC—then join them together. For example, try sketching the shear and moment diagrams for segment AB. What does it look like? Do the same for segment BC. What does it look like? You may then combine them into one diagram, as shown here.

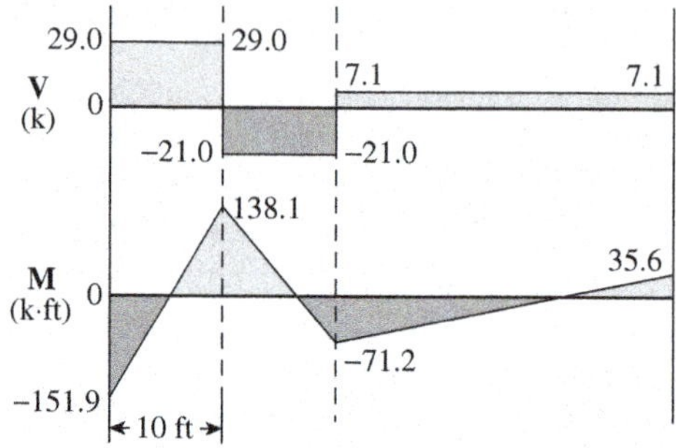

EXAMPLE 13.8

For the beam shown in Example 13.4, generate the shear and moment diagrams using the results from the moment-distribution table.

SOLUTION STRATEGY

Cut the beam just to the left and just to the right of joint B. Do the same for joint C. This will give three separate FBDs for beam segments: AB, BC, and CD. Treat each segment separately for analysis and diagram sketching.

SOLUTION

Sketch the FBD of each segment.

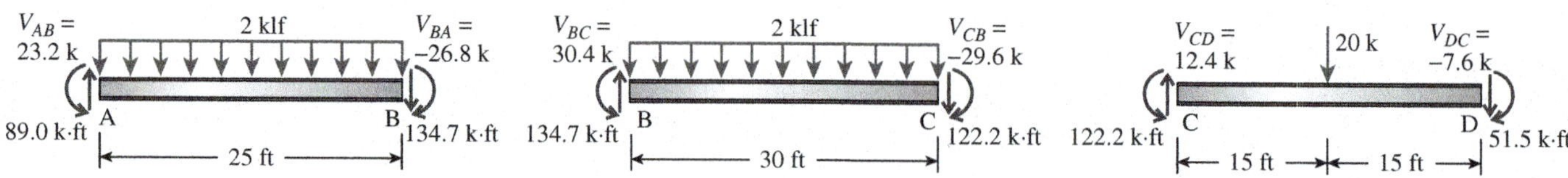

The table indicates that at point A the moment is −89.0 k·ft. According to the moment-distribution sign convention, a negative moment is to be sketched counterclockwise. The table indicates that just to the left of point B the moment is +134.7 k·ft, which will be sketched as a clockwise moment. The same thought process is used to sketch the moments on elements BC and CD.

Use equilibrium to get the four values of shear.

QUICK NOTE

For segment AB, one can sum moments about point A to solve for V_{BA} and sum forces in the y direction to get V_{AB}. The same process is followed for the other segments.

SUMMARY

The moment-distribution method is considered to be an "exact" method. However, the results are dependent not only upon the number of iterations but also upon the number of decimal places carried in the DFs and FEMs. If the values of the moment diagram are verified against the shear diagram, there is a slight discrepancy. This is due to rounding. However, the values of moment at the joints should agree between the table and the diagram.

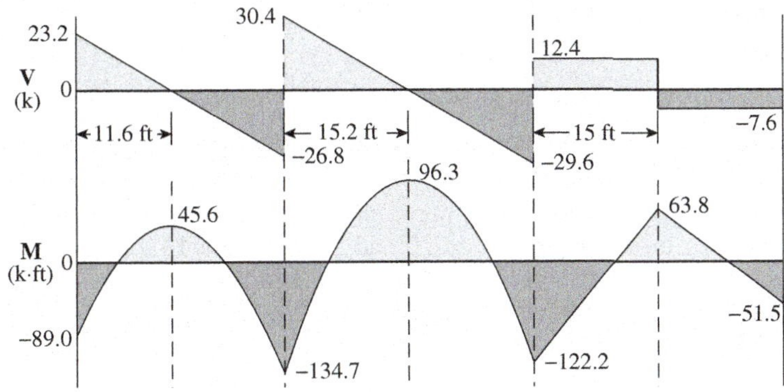

13.8 Spreadsheet Computer Applications

The moment-distribution method is an "exact" method of analysis for finding the internal moments at the joints of a beam. However, the "exactness" of the solution is dependent upon the number of significant digits used in the DFs, distributed moments, and carryover moments. It is also dependent on how many iterations are used in the analysis.

Both the significant digits and the iteration issues are very easily overcome through the use of spreadsheets. With very little effort, a few generic spreadsheets readily handle the analysis of many beam configurations. An Excel spreadsheet titled *Moment Distribution Worksheet.xls* is provided on this textbook website. This spreadsheet is set up to analyze generic two-span, three-span, and four-span beams.

Part of the spreadsheet for the beam of Example 13.4 is given in Figure 13.9. The formulas used in a few of the cells are shown in Table 13.2 to give the student some of the logic used to develop the spreadsheet.

DID YOU KNOW?

The approach presented in Examples 13.3 and 13.4 is most easily programmed into a spreadsheet. This is because every two rows in the table may be repeated as many times as desired.

	A	B	C	D	E	F	G	H	I	J	K	L	M
1				Span Length =	25		Span Length =	30			Span Length =	30	
2				E =	1		E =	1			E =	1	
3	Instructions:			I =	200		I =	300			I =	300	
4			Connection at A =		fixed						Connection at D =	fixed	
5	The only												
6	information			A		B			C			D	Joints
7	which should be			AB		BA	BC		CB	CD		DC	
8	provided by the user is:			0.000		0.444	0.556		0.500	0.500		0.000	DFs
9				-104.2		104.2	-150.0		150.0	-75.0		75.0	FEMs
10	* span lengths					20.4	25.4		-37.5	-37.5			Distribution
11	* E			10.2			-18.8		12.7			-18.8	Carryover
12	* I					8.3	10.4		-6.4	-6.4			Distribution
13	* end support conditions			4.2			-3.2		5.2			-3.2	Carryover
14	* FEMs					1.4	1.8		-2.6	-2.6			Distribution

FIGURE 13.9 Snapshot of moment distribution spreadsheet.

REMEMBER

The distribution factor is simply a ratio of the stiffness of a member divided by the the total stiffness of both members at the joint.

$$DF_{BA} = \frac{\frac{4*E3*E2}{E1}}{\frac{4*E3*E2}{E1} + \frac{4*H3*H2}{H1}}$$

However, the fours cancel out.

Table 13.2 Typical spreadsheet formulas for moment distribution.

Cell	Formula
F8	=E3*E2/E1/(E3*E2/E1+H3*H2/H1)
G8	=1-F8
F10	=-SUM(F9:G9)*F8
G10	=-SUM(F9:G9)*G8
D11	=0.5*F10
D8	=IF(E4="fixed",0,1)
D10	=IF(D8=0," ",-D9)
F11	=IF(ISNUMBER(D10),D10*0.5," ")

QUICK NOTE

With the spreadsheet available, it becomes very easy for the reader to investigate the influence that the different beam parameters E, I, and L have on the way moment is distributed within the beam.

With the generic spreadsheet developed, the analyst need only modify the FEMs for different loadings and also modify E, I, and L for the various spans. While the approach that is used in the spreadsheets is not the most efficient form of the moment-distribution method table, it is the easiest to program. For instance, once rows 10 and 11 in Figure 13.9 have been programmed, they may be copied and pasted as many times as needed to the rows below to get a converged answer. Once the distributed moment becomes acceptably small, each column is summed up to get the final moment in each joint.

13.9 Examples with Video Solutions

VE13.1 Use the moment-distribution method to find the internal moments at the joints of the beam in Figure VE13.1. Balance all joints simultaneously.

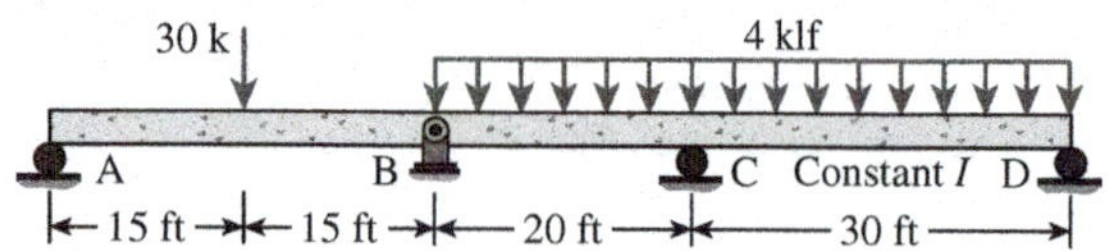

Figure VE13.1

VE13.2 For the beam shown in VE13.1, use the moment-distribution method to solve for the internal moments of the joints. Balance one joint at a time.

VE13.3 For the beam shown in VE13.1, use the moment-distribution method to solve for the internal moments of the joints. Use the modified stiffness and FEM approach, then balance one joint at a time.

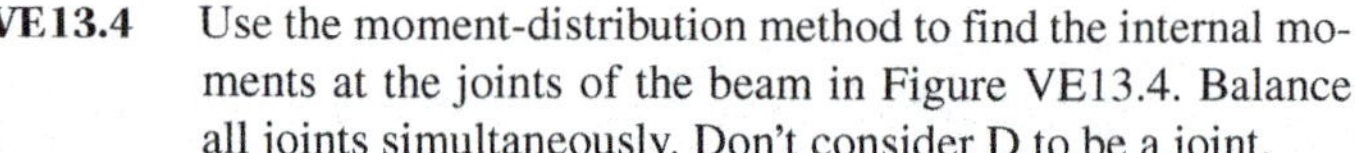

VE13.4 Use the moment-distribution method to find the internal moments at the joints of the beam in Figure VE13.4. Balance all joints simultaneously. Don't consider D to be a joint.

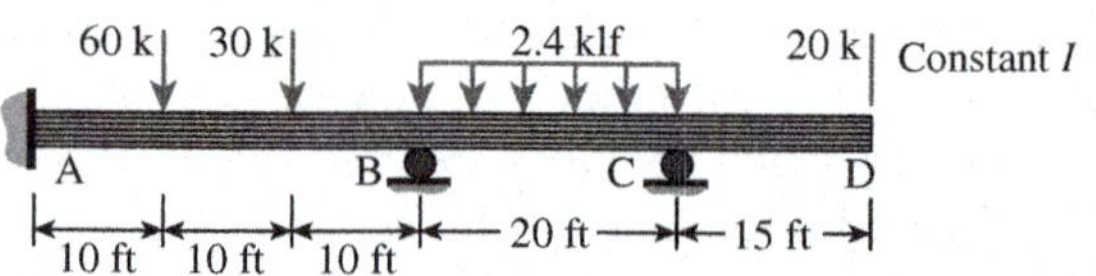

Figure VE13.4

VE13.5 Use the results of VE13.1 to generate the shear and moment diagrams for the beam shown.

VE13.6 Use the results of VE13.4 to generate the shear and moment diagrams for the beam shown.

13.10 Problems for Solution

Sections 13.1–13.5

For Problems 13.1 through 13.15, analyze the beams using the moment-distribution method and draw the shear and moment diagrams (see Section 13.7). EI is constant for all spans unless noted otherwise.

P13.1 (Ans: M_B = 185.4 k·ft)

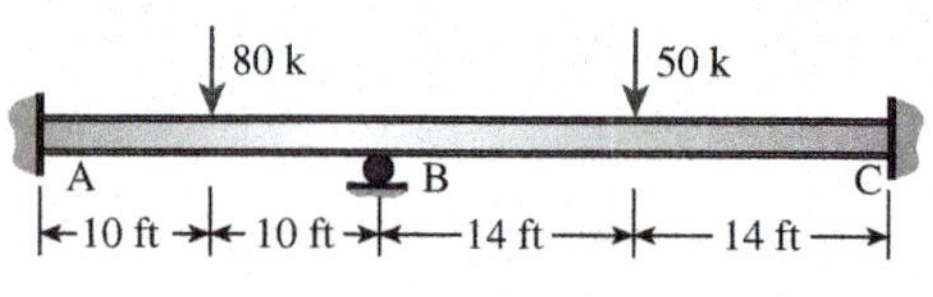

Problem 13.1

P13.2

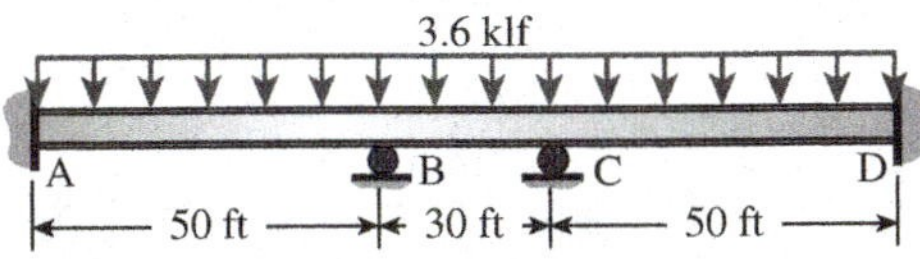

Problem 13.2

P13.3 (Ans: M_B = 332.3 k·ft)

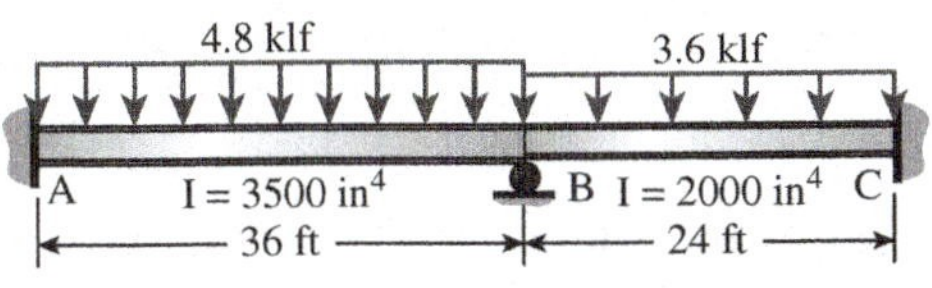

Problem 13.3

P13.4 Rework Problem 13.3 if a 60-k load is added at the centerline of span BC.

P13.5 (Ans: M_B = 362.6 k·ft, M_D = 336.3 k·ft)

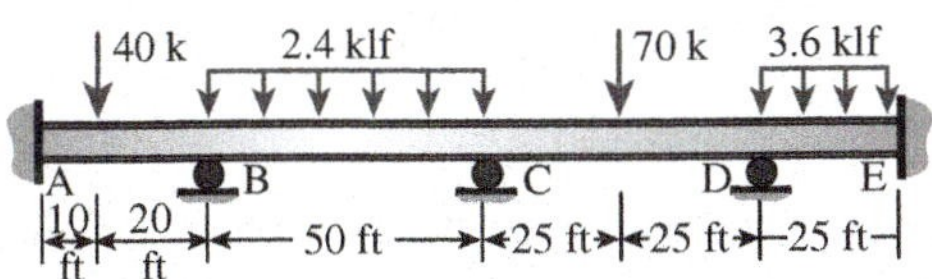

Problem 13.5

P13.6

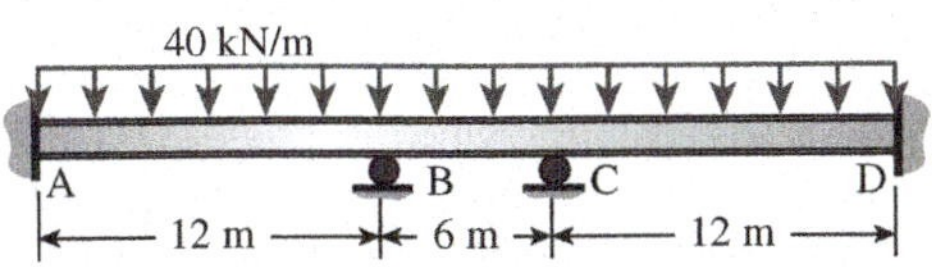

Problem 13.6

P13.7 (Ans: M_B = 130.9 kN·m)

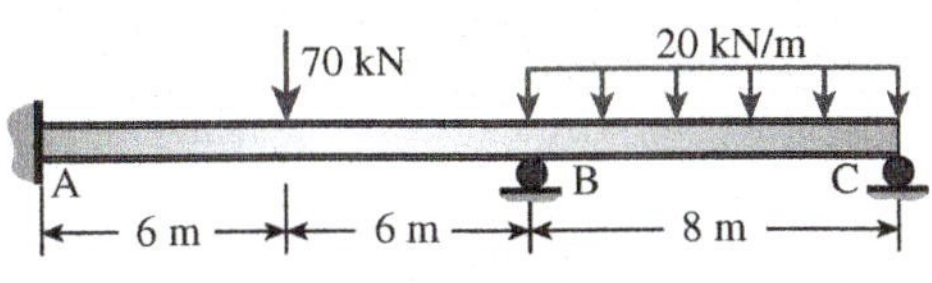

Problem 13.7

P13.8

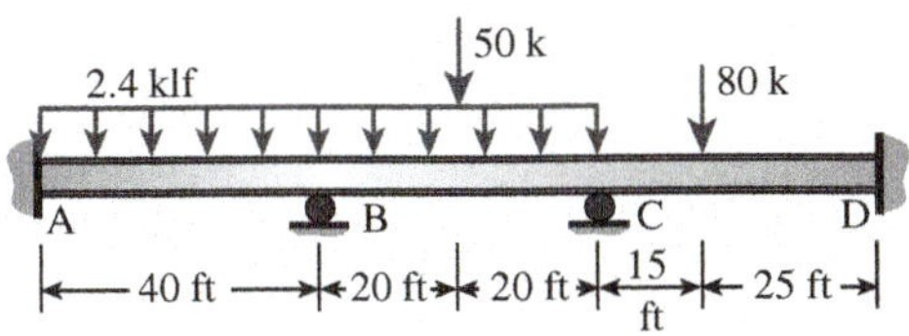

Problem 13.8

P13.9 (Ans: M_C = 186.9 k·ft, M_D = 882.2 k·ft)

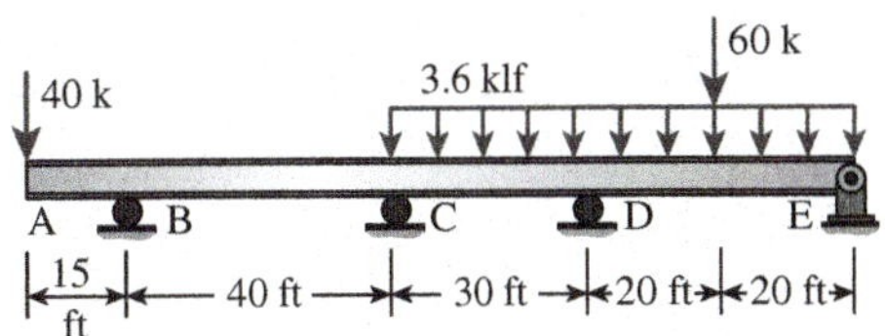

Problem 13.9

P13.10

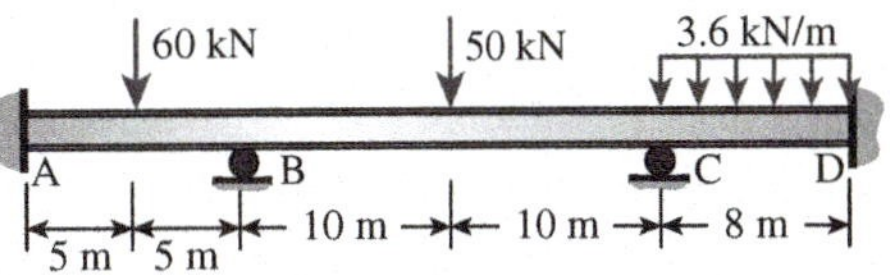

Problem 13.10

P13.11 (Ans: M_B = 148.0 k·ft, M_C = 202.0 k·ft)

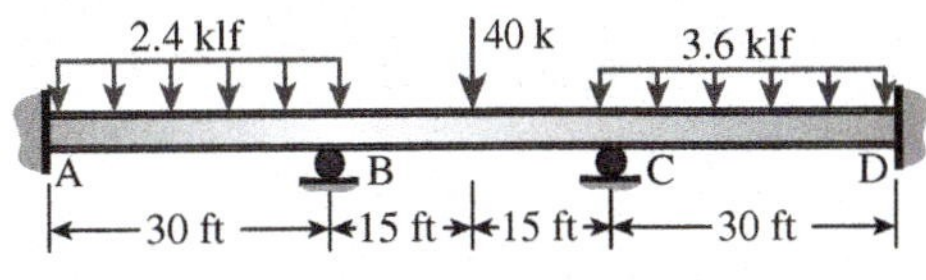

Problem 13.11

P13.12

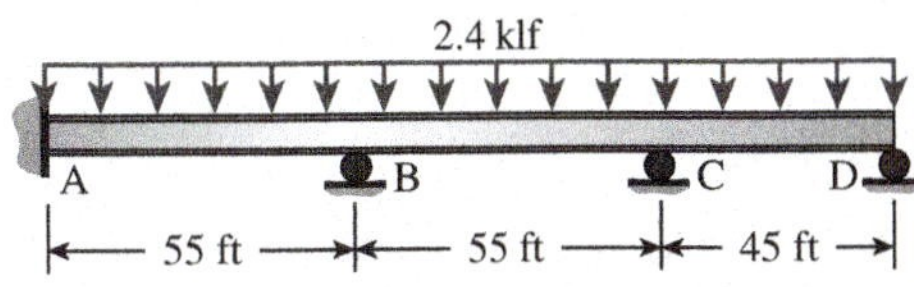

Problem 13.12

P13.13 (Ans: M_B = 220.7 k·ft, M_C = 275.8 k·ft)

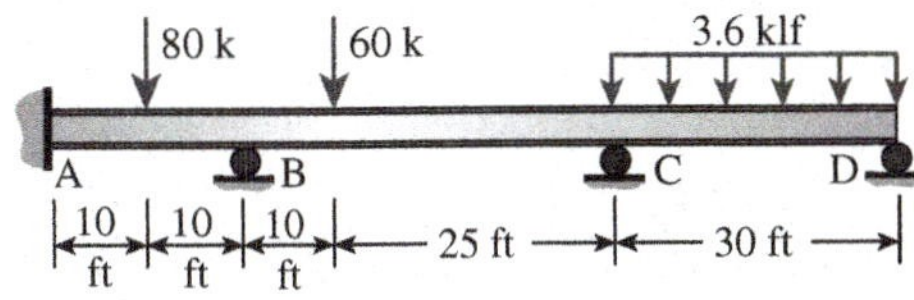

Problem 13.13

P13.14

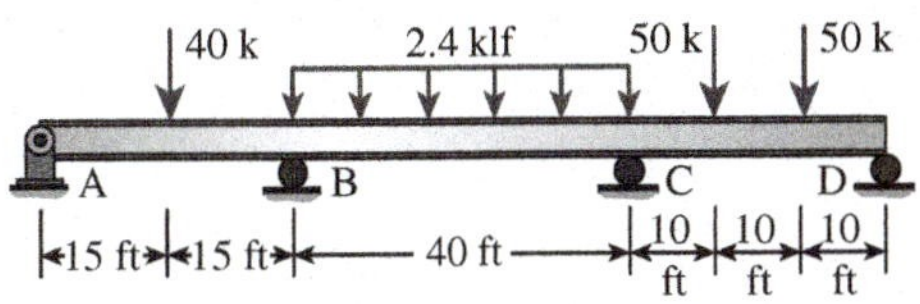

Problem 13.14

P13.15 Repeat Problem 13.2 if the end supports A and D are made simple supports. (Ans: M_B = 720.0 k·ft, M_C = 720.0 k·ft)

Sections 13.6–13.7

For Problems 13.16 through 13.21, analyze the beams using the moment-distribution method and draw the shear and moment diagrams. *EI* is constant for all spans unless noted otherwise. Use the modification of stiffness and FEM approach.

P13.16

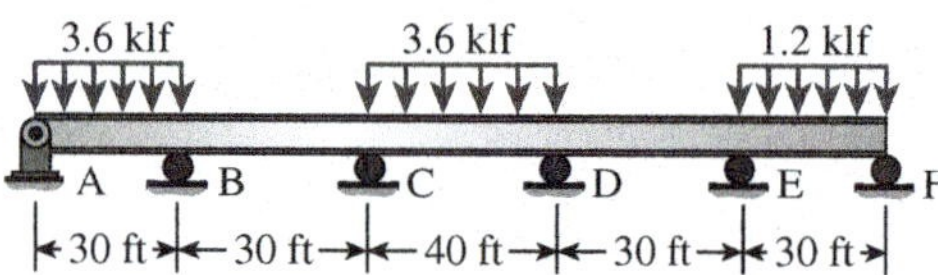

Problem 13.16

P13.17 (Ans: $M_B = 494.2$ k·ft, $M_C = 664.9$ k·ft)

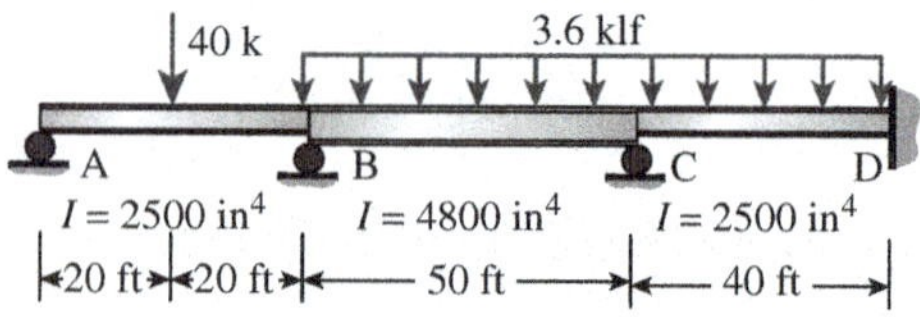

Problem 13.17

P13.18

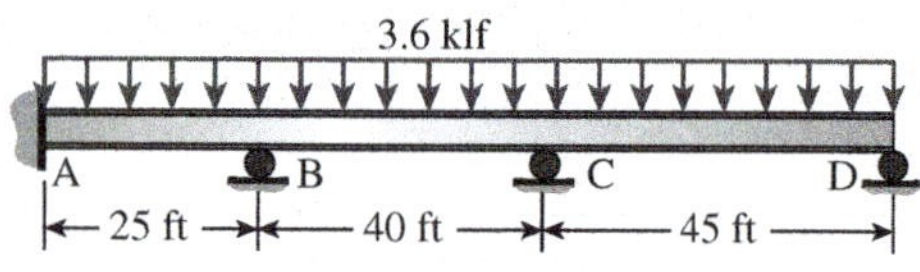

Problem 13.18

P13.19 (Ans: $M_C = 307.6$ k·ft)

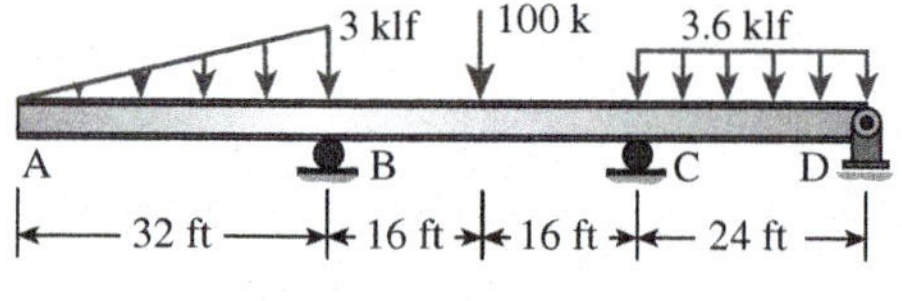

Problem 13.19

P13.20

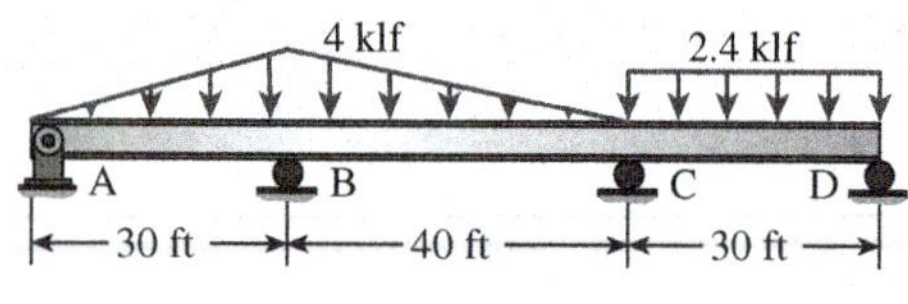

Problem 13.20

P13.21 (Ans: $M_B = 298.4$ k·ft, $M_C = 245.7$ k·ft)

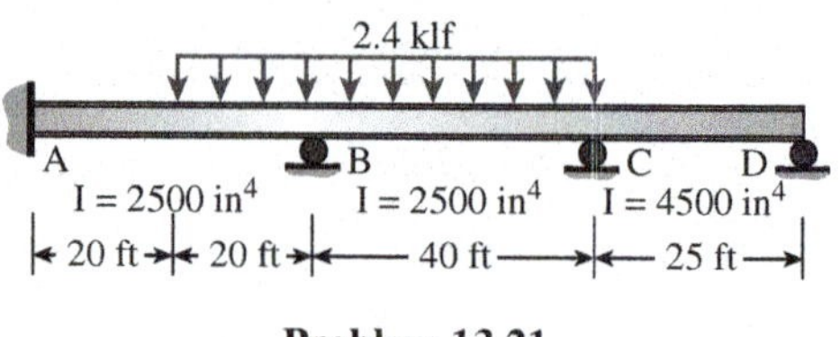

Problem 13.21

Section 13.8 SPREADSHEET

For the following problems, create a spreadsheet or use the one provided with this textbook - *Moment Distribution Worksheet.xls*.

P13.22 Analyze the beam presented in Problem 13.5 and find the moments at the joints. Now change the moment of inertia for span CD so that the ratios 0.1, 0.5, 1.0, 1.5, 2.0, 5.0, and 10.0 are compared to its original value. Generate a plot of the ratio versus the moments at the joints. How does the distribution of moments change when the moment of inertia of CD changes?

P13.23 Analyze the beam presented in Problem 13.6 and find the moments at the joints. Now change the moment of inertia for span AB so that the ratios 0.1, 0.5, 1.0, 1.5, 2.0, 5.0, and 10.0 are compared to its original value. Generate a plot of the ratio versus the moments at the joints. How does the distribution of moments change when the moment of inertia of AB changes?

P13.24 Analyze the beam presented in Problem 13.8 and find the moments at the joints. Now reanalyze the beam by removing the point loads. How does this change in loading affect the distribution?

P13.25 Analyze the beam presented in Problem 13.16 and find the moments at the joints. Now change the span length of member BC to the values (given in feet) 10, 20, 30, 40, 50, 60, 70, and 80 and reanalyze the beam. Create a plot of the span length of BC versus the internal moment at C and also a plot of the span length versus the internal moment at E. What influence does the span length of BC have on the internal moments at points C and E?

Moment Distribution for Frames

14

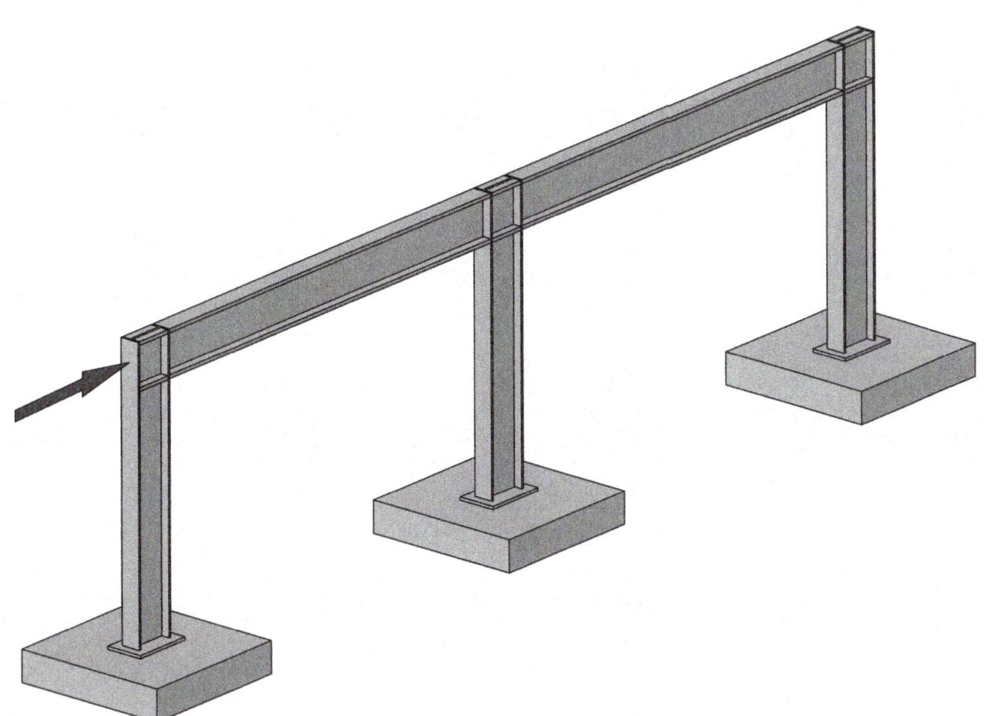

14.1 Frames with Sidesway Prevented

Moment distribution for frames where sidesway or lateral movement cannot occur is handled in the same manner as it is for beams. Analysis of frames without sidesway is illustrated by Examples 14.1 and 14.2. Where sidesway is possible, however, it must be taken into account because joint displacements cause rotations at the ends of members connected to the joints. This affects the moments in all of the members.

NOMENCLATURE

Sidesway refers to the condition where one or more of the joints in the structure may translate as shown in Figure 14.1(a) where joints B and C translate to the right. Contrast this to Figure 14.1(b) where the frame deflects but the joints do not translate to a new location.

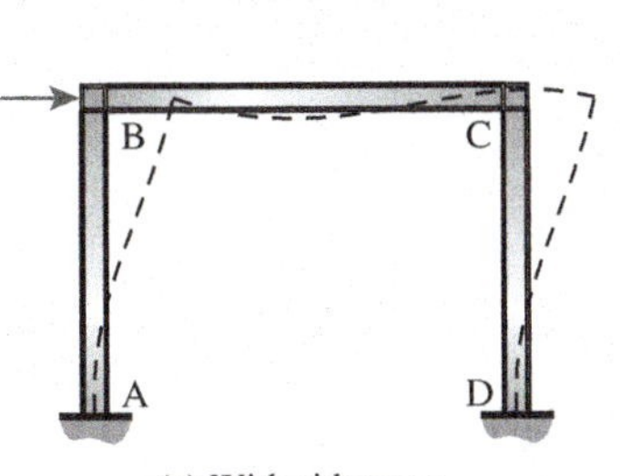

(a) With sidesway

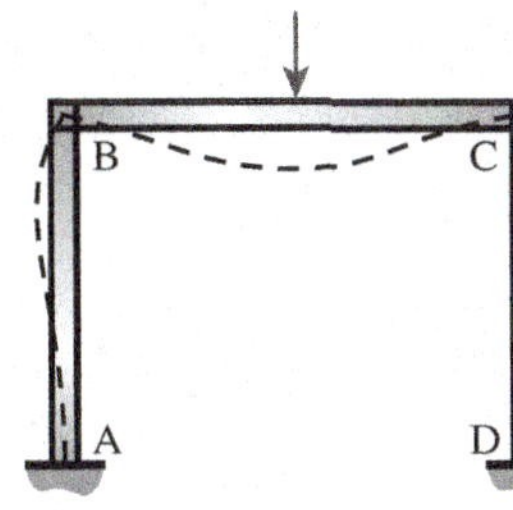

(b) Without sidesway

FIGURE 14.1
Demonstration of frames with and without sidesway.

EXAMPLE 14.1

Determine the end moments for the frame shown using the moment-distribution method. The relative value of I is shown for each member of the frame.

DID YOU KNOW?
One of the underlying assumptions in the moment-distribution method is that the length of a member doesn't change when the structure is loaded. Thus, since members AB and BC don't get shorter or longer, joint B cannot change location (i.e., cannot translate). This means the frame has no sway.

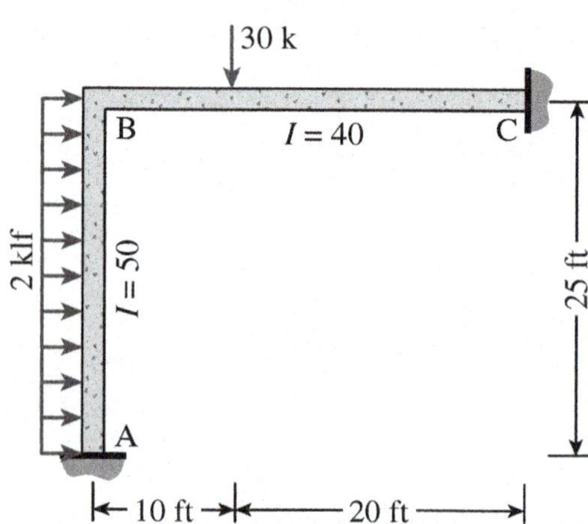

SOLUTION STRATEGY

Scan the structure for any joints that need to be clamped against rotation and clamp them. Follow the same process as used for beams. Just because member AB is oriented 90° relative to what was seen in the past, it is still handled in the same way.

SOLUTION

Clamp joint B against rotation. The resulting calculations are

$$k_{BA} = \frac{4E(50)}{25} = 8E \qquad k_{BC} = \frac{4E(40)}{30} = \frac{16E}{3}$$

$$DF_{BA} = \frac{8E}{8E + 16E/3} = 0.6 \qquad DF_{BC} = 1 - 0.6 = 0.4$$

Now assemble and complete the table.

FIXED-END MOMENTS

$$FEM_{BA} = -FEM_{AB} = \frac{(2\text{ klf})(25\text{ ft})^2}{12} = 104.2\text{ k·ft}$$

$$FEM_{BC} = -\frac{(30\text{ k})(10\text{ ft})(20\text{ ft})^2}{(30\text{ ft})^2} = -133.3\text{ k·ft}$$

$$FEM_{CB} = \frac{(30\text{ k})(10\text{ ft})^2(20\text{ ft})}{(30\text{ ft})^2} = 66.7\text{ k·ft}$$

A		B			C	Joints
AB		BA	BC		CB	
0.00		0.60	0.40		0.00	Distribution Factors
−104.2		104.2	−133.3		66.7	Fixed-End Moments
		17.5	11.6			Distribution
8.7	⟵			⟶	5.8	Carryover
−95.5		121.7	−121.7		72.5	Final Moments

EXAMPLE 14.2

Determine the end moments for the frame shown using the moment-distribution method. EI values are constant for all members in the structure.

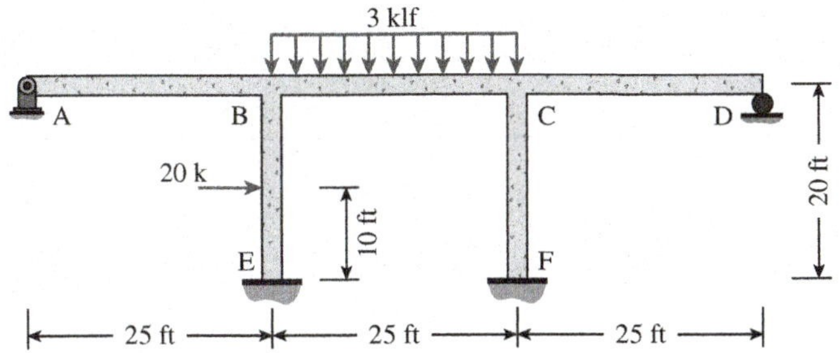

QUICK NOTE

Since the moment distribution does not rely on axial stiffness, the structure shown may be seen as a symmetric structure even though A is a pin and D is a roller. As such, the stiffness $k_{BA} = k_{CD}$ and so forth. This will make the distribution factors for joint B identical to the distribution factors for joint C.

SOLUTION STRATEGY

Use the modified stiffness factors for BA and CD. Only lock joints B and C against rotation. The pin at A ensures that no joints may translate. With three members coming into joint B and three into joint C, make sure that all of the distribution factors at a particular joint sum up to be one.

SOLUTION

The resulting stiffness and distribution factor calculations are

$$k_{BA} = \frac{3EI}{25} \qquad k_{BE} = \frac{4EI}{20} \qquad k_{BC} = \frac{4EI}{25}$$

$$\text{DF}_{BA} = \frac{3/25}{3/25 + 4/20 + 4/25} = 0.25$$

$$\text{DF}_{BE} = \frac{4/20}{3/25 + 4/20 + 4/25} = 0.42$$

$$\text{DF}_{BC} = 1 - 0.25 - 0.42 = 0.33$$

FIXED-END MOMENTS

$$\text{FEM}_{BA} = \text{FEM}_{AB} = \text{FEM}_{CD} = \text{FEM}_{DC} = \text{FEM}_{CF} = \text{FEM}_{FC} = 0$$

$$\text{FEM}_{BC} = -\frac{(3 \text{ klf})(25 \text{ ft})^2}{12} = -156.3 \text{ k·ft}$$

$$\text{FEM}_{BE} = \frac{(20 \text{ k})(20 \text{ ft})}{8} = 50 \text{ k·ft}$$

Note how all the distribution factors at joint B sum up to be 1.0. If they do not sum up to 1.0 then something was done wrong in the calculation. Now assemble and complete the table.

E	A	B			C			D	F	Joint
EB	AB	BA	BE	BC	CB	CF	CD	DC	FC	
0.00	**0.00**	**0.25**	**0.42**	**0.33**	**0.33**	**0.42**	**0.25**	**0.00**	**0.00**	DF
−50.0	0.0	0.0	50.0	−156.3	156.3	0.0	0.0	0.0	0.0	FEM
		26.6	44.6	35.1	−51.6	−65.6	−39.1			Dist 1
22.3				−25.8	17.5				−32.8	CO 1
		6.4	10.8	8.5	−5.8	−7.4	−4.4			Dist 2
5.4				−2.9	4.3				−3.7	CO 2
		0.7	1.2	1.0	−1.4	−1.8	−1.1			Dist 3
0.6				−0.7	0.5				−0.9	CO 3
		0.2	0.3	0.2	−0.2	−0.2	−0.1			Dist 4
−21.7	0.0	33.9	107.0	−140.9	119.6	−75.0	−44.6	0.0	−37.4	Final

SUMMARY

Moments are carried over from joint B to joints E and C. They also are carried over from joint C to joints B and F. The carryover arrows have not been placed in this table to avoid cluttering it.

A balanced joint, like joint B, will have moments that sum to zero. For example, $33.9 + 107.0 - 140.9 = 0$.

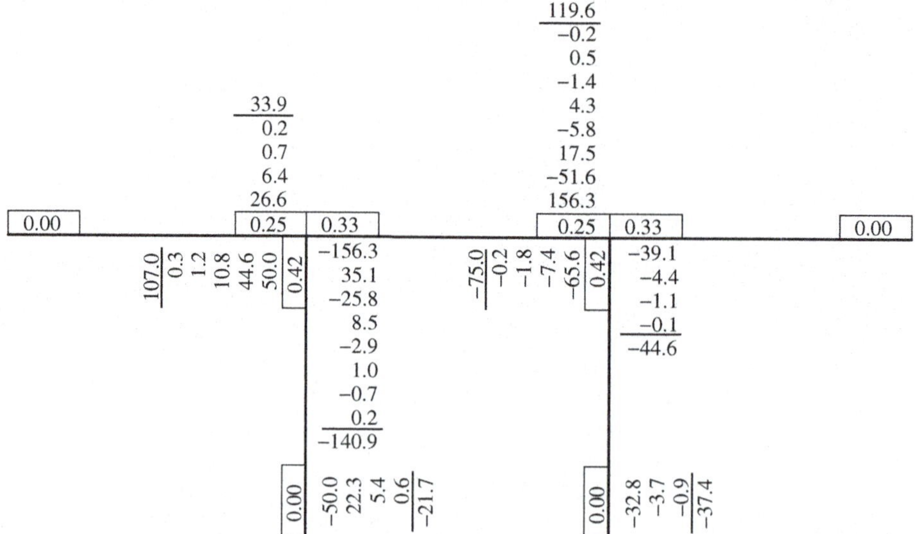

FIGURE 14.2 Alternate form of moment distribution tables for frames

In lieu of constructing a table for carrying out the moment distribution, some analysts find it convenient to place the numbers directly on the structure itself. The authors don't share this opinion. However, this alternate form is presented in Figure 14.2 without explanation in case the reader also finds it useful.

14.2 Sway Frames with Point Loads at Joints

As we begin our discussion of frames with sway, it is useful to introduce the analysis of frames subjected to lateral loads at a joint, as shown in Figure 14.3(a). With the 21-k load applied, joints B and C both translate to the right the same distance Δ. This joint translation is known as either sway or sidesway.

In Chapter 13, the moment-distribution method was introduced and required the analyst to first lock the joints against rotation and then calculate the FEMs. However, those FEMs were all based on joints that do not translate. Indeed, when joints do translate, the FEMs are modified from what is presented in the FEM chart by a factor relative to the *degree of translation* Δ. Unfortunately, the magnitude of Δ is not known ahead of time, so this requires the slightly different approach described here.

Procedure

- Lock the joints against rotation.
- Assume that $\Delta = 1$. Use appropriate units for the problem.
- Calculate FEMs based upon the imposed unit displacement. The FEMs for a fixed-fixed beam are shown in Figure 14.4.

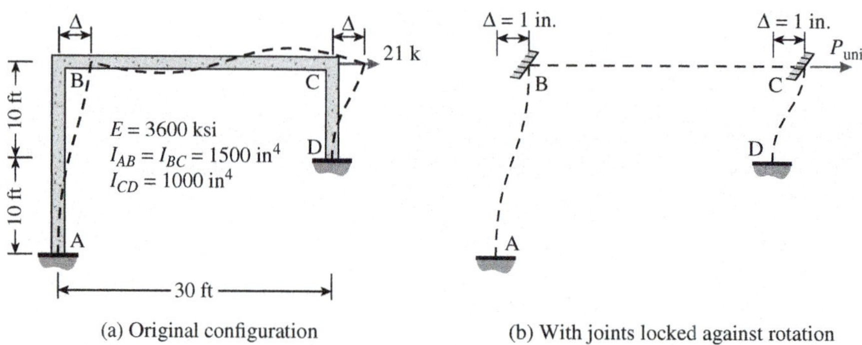

FIGURE 14.3 Frame with sidesway.

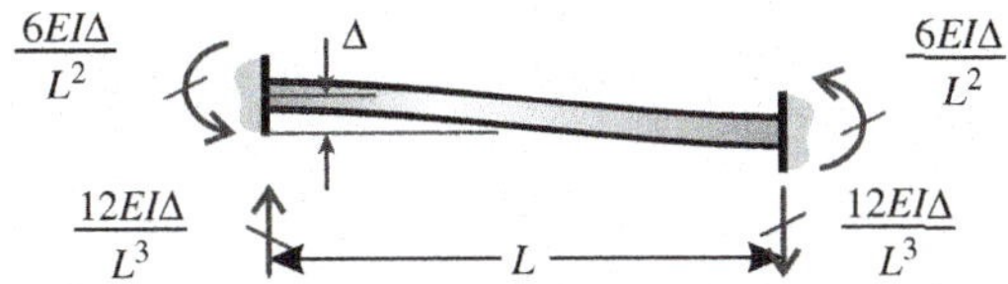

FIGURE 14.4 Fixed-end moments and fixed-end shears due to a displaced end.

- Calculate DFs and complete the moment-distribution table.
- Use the internal moments and FBDs to determine the magnitude of the *applied point load* (P_{unit}) needed to produce the unit displacement [see Figure 14.3(b)].
- Compute a scale factor S, that is the ratio of the actual load P to the load P_{unit}.
- Find the actual internal moments, reactions, displacement etc., due to the actual load by multiplying the results from the unit displaced structure by the scale factor S.

EXAMPLE 14.3

For the frame shown in Figure 14.3(a), use the moment-distribution method to calculate the internal moments at the joints, the horizontal reactions at the supports, and the horizontal displacement at joint C.

SOLUTION STRATEGY

Lock the necessary joints and calculate the member stiffness factors and associated DFs before calculating the FEMs. Note in Figure 14.3(b) that member BC doesn't deform, and as such, it does not develop any FEMs. Remember to use the sign-convention for interpreting the FEMs.

SOLUTION

Lock joints B and C against rotation only and calculate the stiffness factors and DFs.

$$k_{BA} = \frac{4E(1500)}{20} = 300E$$

$$k_{BC} = k_{CB} = \frac{4E(1500)}{30} = 200E$$

$$k_{CD} = \frac{4E(1000)}{10} = 400E$$

$$\text{DF}_{BA} = \frac{300E}{300E + 200E} = 0.6 \qquad \text{DF}_{BC} = 1 - 0.6 = 0.4$$

$$\text{DF}_{CB} = \frac{200E}{200E + 400E} = 0.333 \qquad \text{DF}_{BC} = 1 - 0.333 = 0.667$$

Now calculate the FEMs for members AB and CD where $\Delta = 1$ in. Refer to Figures 14.3(b) and 14.4.

$$\text{FEM}_{AB} = \text{FEM}_{BA} = -\frac{6(3600\text{ ksi})(1500\text{ in}^4)(1\text{ in.})}{(240\text{ in.})^2} = -562.5\text{ k·in.}$$

$$\text{FEM}_{CD} = \text{FEM}_{DC} = -\frac{6(3600\text{ ksi})(1000\text{ in}^4)(1\text{ in.})}{(120\text{ in.})^2} = -1500\text{ k·in.}$$

QUICK NOTE

Make sure that consistent units are used when calculating the FEMs. In this example, converting everything to kips and inches is most convenient.

A		B			C			D	Joints
AB		BA	BC		CB	CD		DC	
0.00		0.600	0.400		0.333	0.667		0.00	DFs
−562.5		−562.5	0.0		0.0	−1500.0		−1500.0	FEMs
		337.5	225.0		499.5	1000.5			Dist 1
168.8	←		249.8	↔	112.5		→	500.3	CO 1
		−149.9	−99.9		−37.5	−75.0			Dist 2
−74.9	←		−18.7	↔	−50.0		→	−37.5	CO 2
		11.2	7.5		16.6	33.3			Dist 3
5.6	←		8.3	↔	3.7		→	16.7	CO 3
		−5.0	−3.3		−1.2	−2.5			Dist 4
−2.5	←		−0.6	↔	−1.7		→	−1.2	CO 4
		0.4	0.2		0.6	1.1			Dist 5
−465.6		−368.2	368.2		542.6	−542.6		−1021.9	Final

RECALL

The moments in this table are the result of a load, P_{unit}, applied to joint C that produces a displacement of 1 in. at joint C. The next task is to solve for the value of P_{unit}.

The moments in the table are transferred to the FBDs, and the internal shears are found. These then can be transferred to the FBD of the frame, and P_{unit} is easily found.

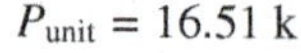

$$P_{unit} = 16.51 \text{ k}$$

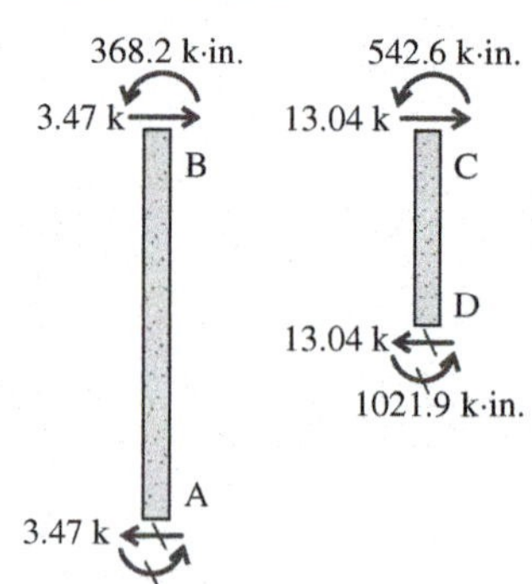

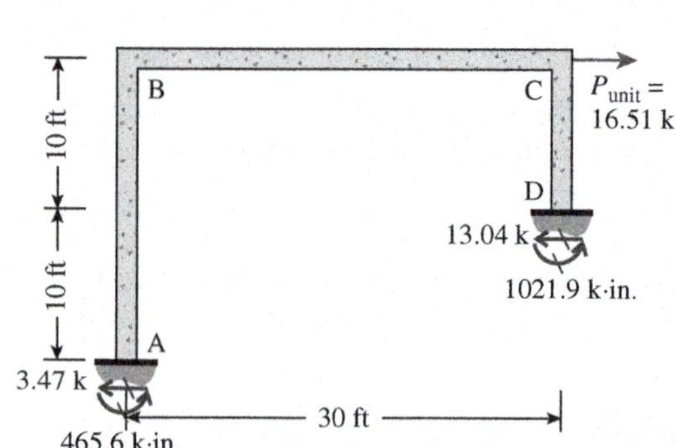

QUICK NOTE

The moments at B and C are internal moments, so they are reported as + or −. The moments at A and D are reactions and are reported with direction arrows.

The actual applied load to the frame is 21 k. The scale factor is calculated as

$$S = \frac{21 \text{ k}}{16.51 \text{ k}} = 1.27$$

The final step is to find reactions and displacements by taking the results from the P_{unit} structure and multiplying it by S.

$$\Delta_C = (1.27)(1.00 \text{ in.}) = \underline{\underline{1.27 \text{ in.} \quad \rightarrow}}$$

$$M_B = (1.27)(368.2 \text{ k·in.}) = \underline{\underline{+468.3 \text{ k·in.}}}$$

$$M_C = (1.27)(542.6 \text{ k·in.}) = \underline{\underline{-690.1 \text{ k·in.}}}$$

$$M_A = (1.27)(465.6 \text{ k·in.}) = \underline{\underline{592.1 \text{ k·in.} \quad \circlearrowleft}}$$

$$M_D = (1.27)(1021.9 \text{ k·in.}) = \underline{\underline{1299.7 \text{ k·in.} \quad \circlearrowleft}}$$

$$R_{Ax} = (1.27)(3.47 \text{ k}) = \underline{\underline{4.41 \text{ k} \quad \leftarrow}}$$

$$R_{Dx} = (1.27)(13.04 \text{ k}) = \underline{\underline{16.59 \text{ k} \quad \leftarrow}}$$

SUMMARY

Basic statics can be used to develop shear and moment diagrams. By developing the response for the unit displacement first, the response for any magnitude of point load is very easily found through the use of a scale factor. This idea of the unit displacement response is the root of the computer-based matrix methods discussed in Chapters 19 through 21.

Examples 14.4 and 14.5 are given to further familiarize the reader with the moment-distribution process. However, they will be presented in a more abbreviated fashion by *not* showing some of the calculations that the reader should be comfortable with at this stage—like stiffness factors and DFs.

EXAMPLE 14.4

Use the moment-distribution method to calculate the horizontal displacement at C, the horizontal reactions, and the internal moments in the frame shown.

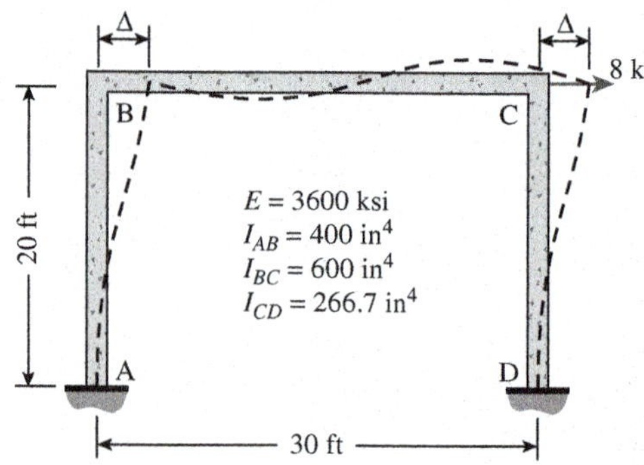

When joints B and C are locked against rotation, the following FEMs are calculated:

$$FEM_{AB} = FEM_{BA} = \frac{6(3600 \text{ ksi})(400 \text{ in}^4)}{(240 \text{ in.})^2} = -150 \text{ k·in.}$$

$$FEM_{CD} = FEM_{DC} = \frac{6(3600 \text{ ksi})(266.7 \text{ in}^4)}{(240 \text{ in.})^2} = -100 \text{ k·in.}$$

SOLUTION STRATEGY

Follow the exact same process given in Example 14.3.

SOLUTION

Set $\Delta = 1.0$ in. and solve for the point load at C that would be required to produce the 1.0-in. displacement.

A		B			C			D	Joints
AB		BA	BC		CB	CD		DC	
0.00		0.500	0.500		0.600	0.400		0.00	DFs
−150.0		−150.0	0.0		0.0	−100.0		−100.0	FEMs
		75.0	75.0		60.0	40.0			Dist 1
37.5	←		30.0	↔	37.5		→	20.0	CO 1
		−15.0	−15.0		−22.5	−15.0			Dist 2
−7.5	←		−11.3	↔	−7.5		→	−7.5	CO 2
		5.6	5.6		4.5	3.0			Dist 3
2.8	←		2.3	↔	2.8		→	1.5	CO 3
		−1.1	−1.1		−1.7	−1.1			Dist 4
−0.6	←		−0.8	↔	−0.6		→	−0.6	CO 4
		0.4	0.4		0.3	0.2			Dist 5
−117.8		−85.1	85.1		72.9	−72.9		−86.6	Final

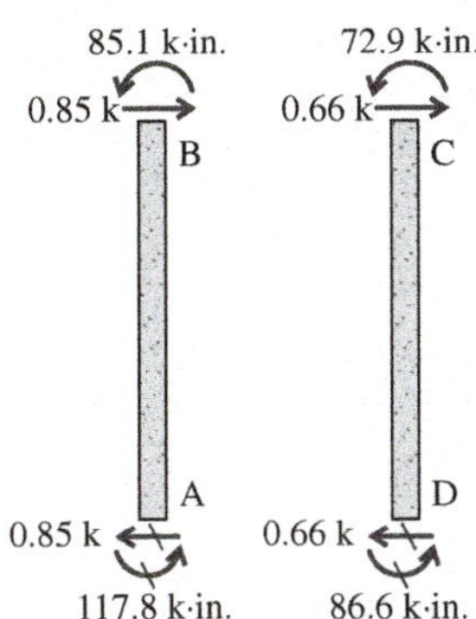

The applied force required to cause the 1.0-in. displacement is found by summing up the shears in both columns:

$$P_{\text{unit}} = 0.85\text{ k} + 0.66\text{ k} = 1.51\text{ k}$$

The scale factor required to shift the applied load up to 8.0 k is

$$S = \frac{8.0\text{ k}}{1.51\text{ k}} = 5.30$$

The final step is to find reactions and displacements by taking the results from the P_{unit} structure and multiplying it by S. Thus,

$$\Delta_C = (5.30)(1.00\text{ in.}) = \underline{\underline{5.30\text{ in.}\quad\rightarrow}}$$

$$M_B = (5.30)(85.1\text{ k·in.}) = \underline{\underline{+450.9\text{ k·in.}}}$$

$$M_C = (5.30)(72.9\text{ k·in.}) = \underline{\underline{-386.3\text{ k·in.}}}$$

$$M_A = (5.30)(117.8\text{ k·in.}) = \underline{\underline{624.0\text{ k·in.}\quad\circlearrowleft}}$$

$$M_D = (5.30)(86.6\text{ k·in.}) = \underline{\underline{458.7\text{ k·in.}\quad\circlearrowleft}}$$

$$R_{Ax} = (5.30)(0.85\text{ k}) = \underline{\underline{4.51\text{ k}\quad\leftarrow}}$$

$$R_{Dx} = (5.30)(0.66\text{ k}) = \underline{\underline{3.49\text{ k}\quad\leftarrow}}$$

QUICK NOTE

Notice how column AB, which has a larger moment of inertia, attracts more of the 8-k load than does column CD.

EXAMPLE 14.5

When joints B, C, and E are locked against rotation, the following FEMs are calculated:

$$FEM_{AB} = FEM_{BA} = \frac{6(3600\text{ ksi})(300\text{ in}^4)}{(360\text{ in.})^2} = -50.0\text{ k·in.}$$

$$FEM_{CD} = FEM_{DC} = \frac{6(3600\text{ ksi})(400\text{ in}^4)}{(360\text{ in.})^2} = -66.7\text{ k·in.}$$

$$FEM_{EF} = FEM_{FE} = \frac{6(3600\text{ ksi})(200\text{ in}^4)}{(240\text{ in.})^2} = -75\text{ k·in.}$$

Use the moment-distribution method to calculate the horizontal displacement at E and the horizontal and moment reactions for the frame shown.

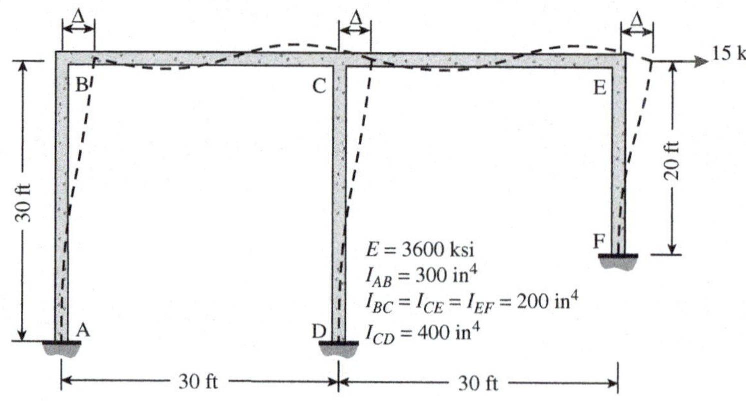

SOLUTION STRATEGY

Follow the same process given previously. Ensure that the table is constructed carefully to account for three members joining at joint C. Also be sure to account for carryover where appropriate.

SOLUTION

Set $\Delta = 1.0$ in. and solve for the point load at E that would be required to produce the 1.0-in. displacement.

A	B		C			E		D	F	Joint
AB	BA	BC	CB	CD	CE	EC	EF	DC	FE	
0.00	0.60	0.40	0.25	0.50	0.25	0.40	0.60	0.00	0.00	DF
−50.0	−50.0	0.0	0.0	−66.7	0.0	0.0	−75.0	−66.7	−75.0	FEM
	30.0	20.0	16.7	33.4	16.7	30.0	45.0			Dist 1
15.0		8.3	10.0		15.0	8.3		16.7	22.5	CO 1
	−5.0	−3.3	−6.3	−12.5	−6.3	−3.3	−5.0			Dist 2
−2.5		−3.1	−1.7		−1.7	−3.1		−6.3	−2.5	CO 2
	1.9	1.3	0.8	1.7	0.8	1.3	1.9			Dist 3
0.9		0.4	0.6		0.6	0.4		0.8	0.9	CO 3
	−0.3	−0.2	−0.3	−0.6	−0.3	−0.2	−0.3			Dist 4
−0.1		−0.2	−0.1		−0.1	−0.2		−0.3	−0.1	CO 4
	0.1	0.1		0.1		0.1	0.1			Dist 5
−36.7	−23.3	23.3	19.9	−44.7	24.9	33.3	−33.3	−55.8	−54.2	Final

The moments are transferred from the table to the FBD of the columns. Statics is used on each FBD to find the unknown column shears.

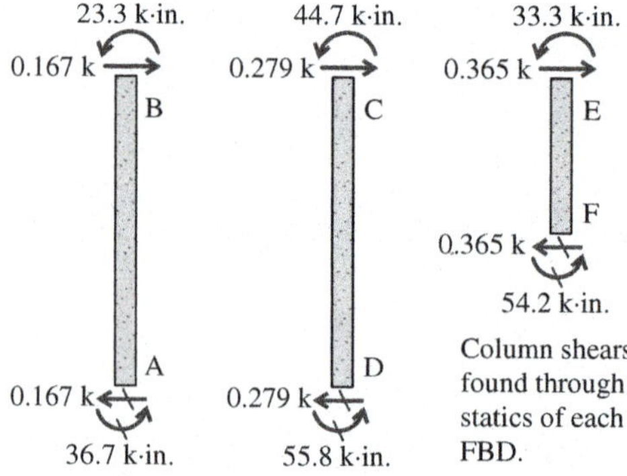

Column shears found through statics of each FBD.

The applied force required to cause the 1.0-in. displacement is found by summing up the shears in all columns:

$$P_{unit} = 0.167\text{ k} + 0.279\text{ k} + 0.365\text{ k} = 0.811\text{ k}$$

The scale factor required to shift the applied load up to 15.0 k is

$$S = \frac{15.0\text{ k}}{0.811\text{ k}} = 18.50$$

The final step is to find reactions and displacements by taking the results from the P_{unit} structure and multiplying it by S. Thus,

$$\Delta_E = (18.50)(1.00\text{ in.}) = \underline{\underline{18.50\text{ in.}\quad \rightarrow}}$$

$$M_A = (18.50)(36.7\text{ k·in.}) = \underline{\underline{679.0\text{ k·in.}\quad \circlearrowleft}}$$

$$M_D = (18.50)(55.8\text{ k·in.}) = \underline{\underline{1032.3\text{ k·in.}\quad \circlearrowleft}}$$

$$M_F = (18.50)(54.2\text{ k·in.}) = \underline{\underline{1002.7\text{ k·in.}\quad \circlearrowleft}}$$

$$R_{Ax} = (18.50)(0.167\text{ k}) = \underline{\underline{3.09\text{ k}\quad \leftarrow}}$$

$$R_{Dx} = (18.50)(0.279\text{ k}) = \underline{\underline{5.16\text{ k}\quad \leftarrow}}$$

$$R_{Fx} = (18.50)(0.365\text{ k}) = \underline{\underline{6.75\text{ k}\quad \leftarrow}}$$

14.3 General Frames with Sidesway

Examine the frame shown in Figure 14.5. At first glance, it does not appear to be subject to sidesway, and the reader is tempted to analyze the frame in the same way shown in Section 14.1. The reactions shown in Figure 14.5 are those computed if such an approach is taken without consideration of the sidesway. The clear indicators that this method does not work are the horizontal reactions. We readily note that equilibrium is not satisfied, and hence, this cannot be a legitimate solution.

Ask yourself, "What causes the imbalance of forces in this frame?" This question is most readily answered by re-examining Figure 14.4. The sway that is shown in the beam member produces moments at the fixed end. These moments are known as *sidesway moments* and must be considered in addition to the moments produced by the applied load. When both sources of moments are considered, equilibrium will be satisfied.

DID YOU KNOW?

A nonsymmetric load can cause sway to occur in a frame—even if the frame configuration is symmetric.

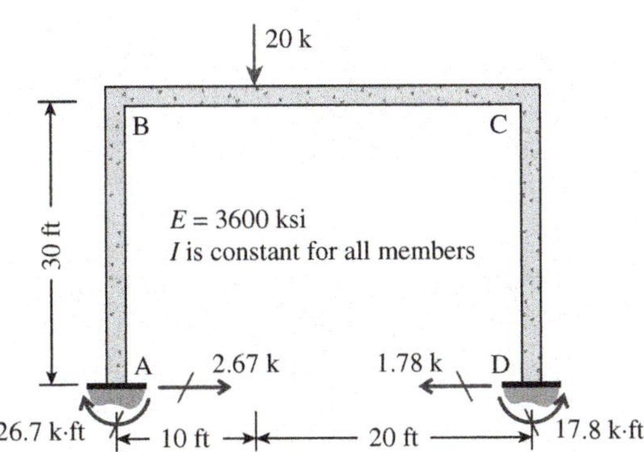

FIGURE 14.5 Computed reactions for a frame if sidesway is ignored when using moment distribution.

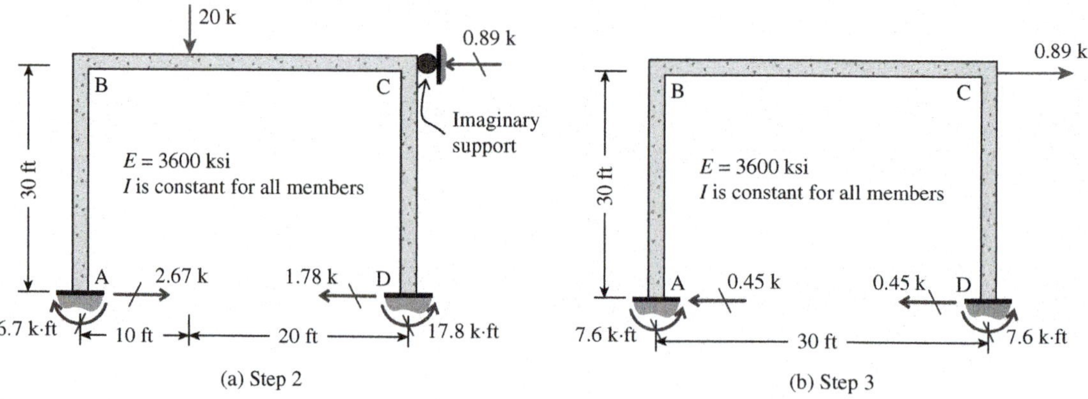

FIGURE 14.6 Load scenarios for the intermediate steps of moment distribution.

The process for analyzing frames subjected to sidesway is quite straightforward now that several tools for moment distribution are available.

Procedure

QUICK NOTE
The approach for handling frames with sway utilizes the principle of superposition. As such, all of the usual assumptions required to use superposition are necessary—including linear-elastic material and small displacements.

1. Lock the frame against sway by adding an imaginary support at one of the translating joints.
2. Use the moment-distribution method to calculate the moments and reactions—including at the imaginary support [see Figure 14.6(a)]. This uses the method discussed in Section 14.1.
3. Take the value of the reaction at the imaginary support, reverse the direction, and then apply it to a separate structure of the same configuration [see Figure 14.6(b)]. Solve for all moments and reactions. This uses the method introduced in Section 14.2.
4. Superimpose the structure of step 2 and the structure of step 3 to obtain the final results (see Figure 14.7).

QUICK NOTE
When the two structures of Figure 14.6 are superimposed, the point load at joint C disappears—as it should since it was a fictitious load anyway.

The demonstration given in Figures 14.5 through 14.7 shows how to accommodate sway in the frame through superposition. The horizontal and moment reactions are the only response quantities shown in this demonstration, but one should recognize that all response quantities in both frames of Figure 14.6 can and should be superimposed. This includes reactions, internal forces, and displacements.

The next several examples demonstrate the process just discussed. The results from Examples 14.3 and 14.5 will be useful in completing the following examples.

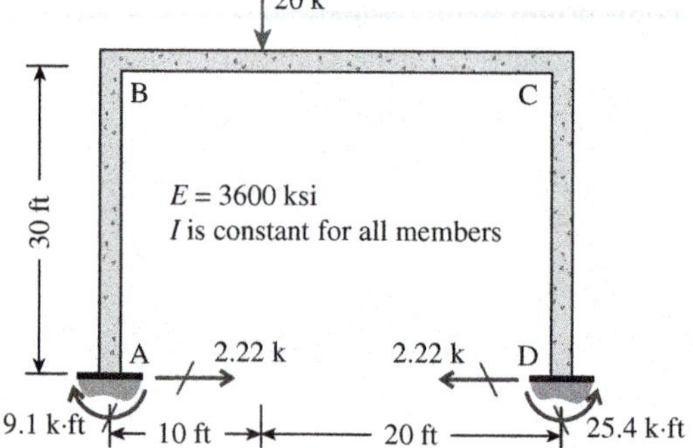

FIGURE 14.7 Correct reactions for the frame of Figure 14.5.

EXAMPLE 14.6

Using moment distribution, find all of the member end moments in the structure shown.

SOLUTION STRATEGY

Follow the process presented previously. The frame of Example 14.3 is the same configuration as this example. Use the results of that example to complete step 3.

SOLUTION

Joints B and C can translate side to side. Place a roller at joint C to prevent this translation, then use moment distribution to solve for the internal moments in the frame and also for the horizontal reaction at C. The following table results from the case where joint C is locked.

QUICK NOTE

At this stage, the calculation of FEMs and DFs is considered to be routine and the explicit calculations are no longer shown in the examples.

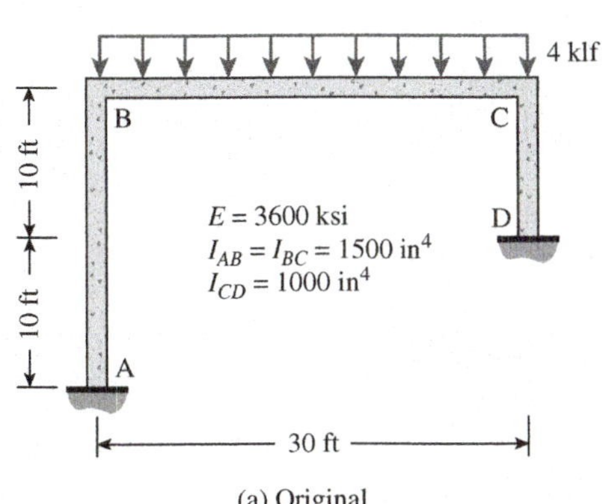

(a) Original

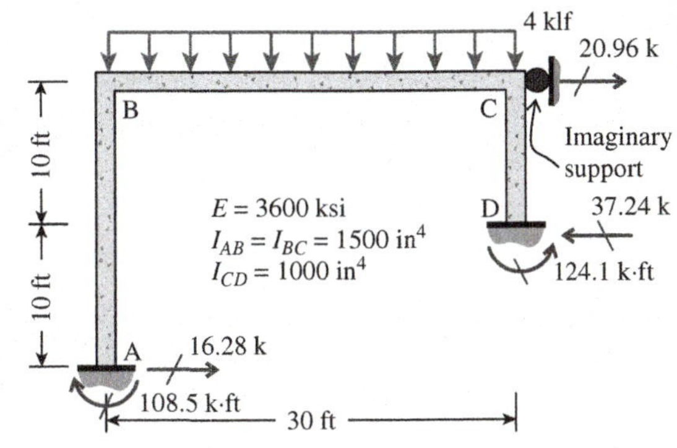

(b) With joint C locked against sway

A		B			C			D	Joints
AB		BA	BC		CB	CD		DC	
0.00		0.600	0.400		0.333	0.667		0.00	DFs
0.0		0.0	−300.0		300.0	0.0		0.0	FEMs
		180.0	120.0		−99.9	−200.1			Dist 1
90.0	←		−50.0	↔	60.0		→	−100.1	CO 1
		30.0	20.0		−20.0	−40.0			Dist 2
15.0	←		−10.0	↔	10.0		→	−20.0	CO 2
		6.0	4.0		−3.3	−6.7			Dist 3
3.0	←		−1.7	↔	2.0		→	−3.3	CO 3
		1.0	0.7		−0.7	−1.3			Dist 4
0.5	←		−0.3	↔	0.3		→	−0.7	CO 4
		0.2	0.1		−0.1	−0.2			Dist 5
108.5		217.2	−217.2		248.3	−248.3		−124.1	Final

SAMPLE CALCULATION

The column shears are found using basic statics:

$$V_{AB} = \frac{108.5\text{ k·ft} + 217.2\text{ k·ft}}{20\text{ ft}} = 16.28\text{ k}$$

$$V_{CD} = \frac{-124.1\text{ k·ft} - 248.3\text{ k·ft}}{10\text{ ft}} = -37.24\text{ k}$$

Step 3 requires that the 20.96-k reaction at C be reversed and applied at C. This would produce a structure that is shown in Figure 14.3(a), only the 21-k load is reversed the other direction. Refer to Example 14.3 to find the solution of this frame. The internal moments at each of the joints from this step are given as

$$M_B = (-1.27)(30.68\text{ k·ft}) = -38.97\text{ k·ft}$$

$$M_C = (-1.27)(45.22\text{ k·ft}) = +57.43\text{ k·ft}$$

$$M_A = (-1.27)(38.80\text{ k·ft}) = 49.28\text{ k·ft} \circlearrowright$$

$$M_D = (-1.27)(85.16\text{ k·ft}) = 108.15\text{ k·ft} \circlearrowright$$

CAUTION

It is very easy to make mistakes in the units here. The results from Example 14.3 are given in units of (k·in.), but they will be used with the results from step 2, which are in units of (k·ft). We present the internal moments in the consistent units of (k·ft).

The scale factor calculated in Example 14.3 is given as a negative in this example because the load is the opposite direction from what was assumed in the previous example.

The final internal moments at the joints are found by superimposing the results from step 2 and step 3. The signs on the moments at B and C are given using beam sign convention.

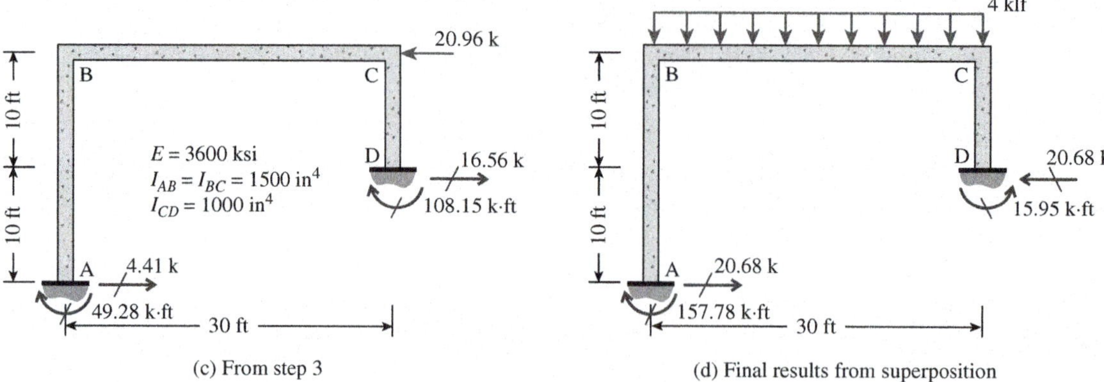

(c) From step 3

(d) Final results from superposition

$$M_B = -38.97 \text{ k·ft} - 217.2 \text{ k·ft} = \underline{\underline{-256.17 \text{ k·ft}}}$$

$$M_C = 57.43 \text{ k·ft} - 248.3 \text{ k·ft} = \underline{\underline{-190.87 \text{ k·ft}}}$$

$$M_A = 49.28 \text{ k·ft} + 108.5 \text{ k·ft} = \underline{\underline{157.78 \text{ k·ft} \circlearrowright}}$$

$$M_D = 108.15 \text{ k·ft} - 124.1 \text{ k·ft} = \underline{\underline{15.95 \text{ k·ft} \circlearrowleft}}$$

EXAMPLE 14.7

Using moment distribution, find the horizontal and moment reactions for the structure shown here.

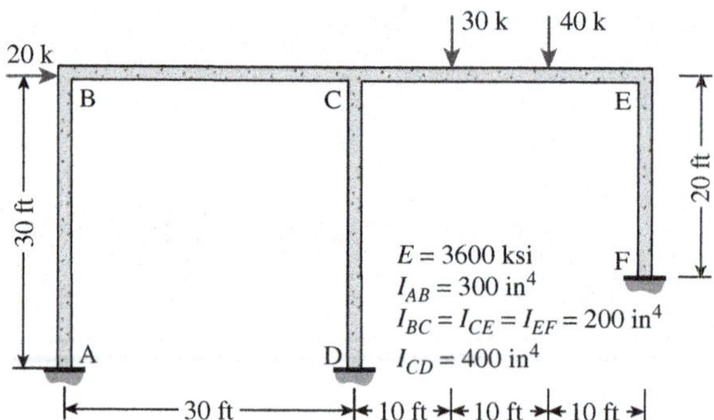

SOLUTION STRATEGY

Follow the process presented previously. The frame of Example 14.5 is the same configuration as this example. Use the results of that example to complete step 3.

SOLUTION

Joints B, C, and E can translate side to side. Place a roller at joint E to prevent this translation, then use moment distribution to solve for the internal moments in the frame and also for the horizontal reaction at E. The following table results from the case where joint E is locked.

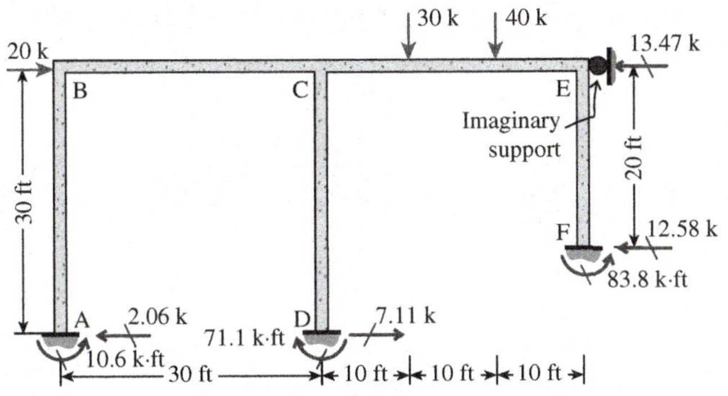

SAMPLE CALCULATION

The column shears are found using basic statics:

$$V_{AB} = \frac{-10.6\text{ k·ft} - 21.3\text{ k·ft}}{30\text{ ft}} = -1.06\text{ k}$$

$$V_{CD} = \frac{71.1\text{ k·ft} + 142.5\text{ k·ft}}{30\text{ ft}} = 7.11\text{ k}$$

$$V_{EF} = \frac{-83.8\text{ k·ft} - 167.7\text{ k·ft}}{20\text{ ft}} = -12.58\text{ k}$$

A	B		C			E		D	F	Joint
AB	BA	BC	CB	CD	CE	EC	EF	DC	FE	
0.00	0.60	0.40	0.25	0.50	0.25	0.40	0.60	0.00	0.00	DF
0.0	0.0	0.0	0.0	0.0	−222.0	244.0	0.0	0.0	0.0	FEM
	0.0	0.0	55.5	111.0	55.5	−97.6	−146.4			Dist 1
0.0		27.8	0.0		−48.8	27.8		55.5	−73.2	CO 1
	−16.7	−11.1	12.2	24.4	12.2	−11.1	−16.7			Dist 2
−8.3		6.1	−5.6		−5.6	6.1		12.2	−8.3	CO 2
	−3.7	−2.4	2.8	5.6	2.8	−2.4	−3.7			Dist 3
−1.8		1.4	−1.2		−1.2	1.4		2.8	−1.8	CO 3
	−0.8	−0.6	0.6	1.2	0.6	−0.6	−0.8			Dist 4
−0.4		0.3	−0.3		−0.3	0.3		0.6	−0.4	CO 4
	−0.2	−0.1	0.1	0.3	0.1	−0.1	−0.2			Dist 5
−10.6	−21.3	21.3	64.2	142.5	−206.6	167.7	−167.7	71.1	−83.8	Final

Step 3 requires that the 13.47-k reaction at E be reversed and applied at E. This produces a structure like the one shown in Example 14.5—only the magnitude of the load is different. For Example 14.5, $P_{\text{unit}} = 0.811$ k. The scale factor is then found to be

$$S = \frac{13.47\text{ k}}{0.811\text{ k}} = 16.61$$

To find the reactions and displacements for step 3, take the results from the P_{unit} structure of Example 14.5 and multiply them by S.

$$M_A = (16.61)(3.06\text{ k·ft}) = 50.8\text{ k·ft} \quad \circlearrowleft$$

$$M_D = (16.61)(4.65\text{ k·in.}) = 77.2\text{ k·ft} \quad \circlearrowleft$$

$$M_F = (16.61)(4.52\text{ k·in.}) = 75.0\text{ k·ft} \quad \circlearrowleft$$

$$R_{Ax} = (16.61)(0.167\text{ k}) = 2.77\text{ k} \quad \leftarrow$$

$$R_{Dx} = (16.61)(0.279\text{ k}) = 4.63\text{ k} \quad \leftarrow$$

$$R_{Fx} = (16.61)(0.365\text{ k}) = 6.06\text{ k} \quad \leftarrow$$

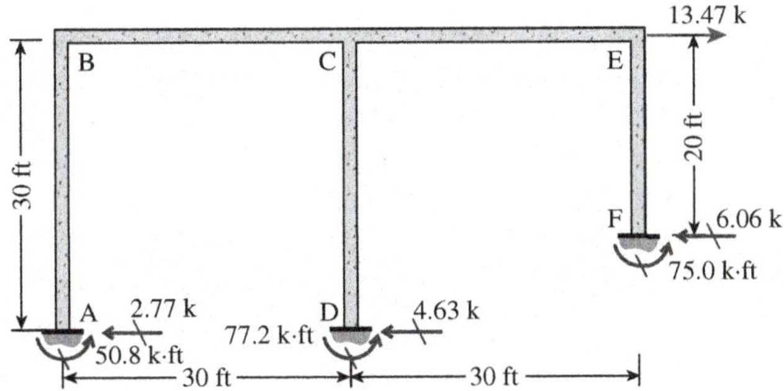

QUICK NOTE

Rounding throughout the process causes a slight error to occur in the analysis. A summation of the horizontal reactions should equal 20 k but in this case, it sums up to be 19.94 k. This constitutes a 0.3% error, which is acceptable.

The final reactions are found by superimposing the structure from step 2 with the structure from step 3.

$$M_A = 10.6\text{ k·ft} + 50.8\text{ k·ft} = \underline{\underline{61.4\text{ k·ft}}} \circlearrowleft$$

$$M_D = -71.1\text{ k·ft} + 77.2\text{ k·ft} = \underline{\underline{6.1\text{ k·ft}}} \circlearrowleft$$

$$M_F = 83.8\text{ k·ft} + 75.0\text{ k·ft} = \underline{\underline{158.8\text{ k·ft}}} \circlearrowleft$$

$$R_{Ax} = -1.06\text{ k} - 2.77\text{ k} = \underline{\underline{3.83\text{ k}}} \leftarrow$$

$$R_{Dx} = 7.11\text{ k} - 4.63\text{ k} = \underline{\underline{2.48\text{ k}}} \rightarrow$$

$$R_{Fx} = -12.58\text{ k} - 6.06\text{ k} = \underline{\underline{18.59\text{ k}}} \leftarrow$$

SUMMARY

QUICK NOTE

Notice how the shorter, stiffer column EF has a much larger shear and moment than do the other columns. This is because shorter columns are more stiff than longer columns for a given cross section. As such, the stiffer column will attract more of the applied load.

Remember that all structural responses should be superimposed. This example just showed the reactions, but all internal forces and displacements should be superimposed.

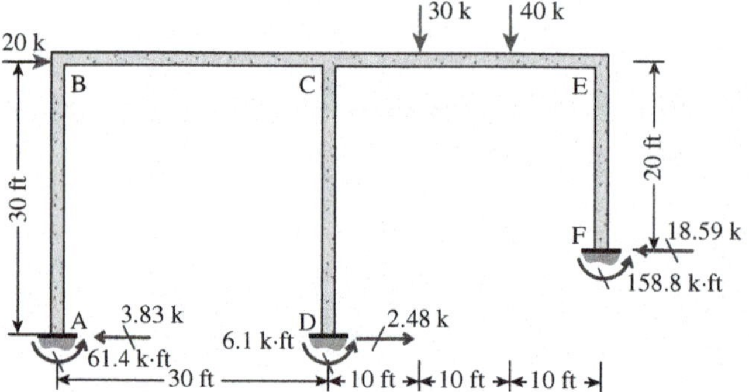

14.4 Frames with Sloping Legs

QUICK NOTE

Since $6E\Delta$ was a constant for the frames previously discussed, the resulting fixed-end moments were really proportional to their I/L^2 values.

The frames considered up to this point have been made up of vertical and horizontal members. It was presented earlier in this chapter that when sidesway occurs in such frames it causes fixed-end moments in the columns proportional to their $6EI\Delta/L^2$ values. However, lateral swaying did not produce fixed-end moments in the beams.

Not all frames are made of members that are only vertical and horizontal. In reality, the sloping legs of such a frame may be analyzed in the same way that frames with only vertical and horizontal members were analyzed. The only additional consideration the analyst must address is the calculation of the fixed-end moments.

DID YOU KNOW?

Frames that have sloping members but are not subject to sidesway require no special treatment for analysis. They can be analyzed using the approach discussed in Section 14.1.

As before, when joints are locked against rotation and subjected to sidesway, fixed-end moments develop at a value of $6EI\Delta/L^2$. The difference in frames with sloped members is that not every member will undergo the same degree of translation, Δ. The task lies in finding what the value of Δ is for each member.

Consider the frame shown in Figure 14.8(a). Because this frame is non-symmetric with a nonsymmetric load, joints B and C experience translation. Figure 14.8(b) shows the deformed shape of the structure if joints B and C are locked against rotation and then a translation is imposed. In this demonstration, joint C is displaced by an arbitrary four units, which means that joint B must displace horizontally by the same amount. To ensure that the length of member AB doesn't increase, joint B must also translate in the down direction as shown.

An underlying principle to sketching a translated joint is to remember that, if the length of a member is not to change, the joint can only move in a direction orthogonal to the axis of the member. A joint can only move in the direction of the member's axis if the entire member is translating as seen for member BC. First take note of joint C: It is only moving horizontally (orthogonal to CD) because any other movement would cause the length of CD to change. Now

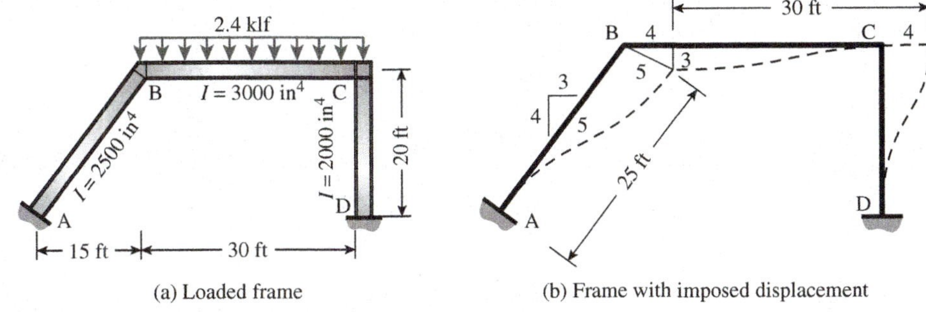

FIGURE 14.8 Sloped member frame.

examine joint B. Its new position is based on a movement orthogonal to member AB and also a movement orthogonal to member BC. In this case, the length of AB remains 25 ft, the length of BC remains 30 ft, and the length of CD remains at 20 ft.

The lateral movement of joint B by four units would then require the vertical movement of joint B by three units and the movement perpendicular to member AB of five units. These movements are based on geometry and the slope of member AB.

Thinking back to the analysis procedure used in Section 14.2, we will eventually need to impose a unit horizontal displacement at C. The resulting sway in each frame member is

$$\Delta_{BA} = \frac{5}{4}(1 \text{ in.}) = 1.25 \text{ in.}$$

$$\Delta_{BC} = \frac{3}{4}(1 \text{ in.}) = 0.75 \text{ in.}$$

$$\Delta_{CD} = \frac{4}{4}(1 \text{ in.}) = 1 \text{ in.}$$

If the frame members have a modulus of elasticity of E = 29,000 ksi, the fixed-end moments for each member are

$$\text{FEM}_{BA} = -\frac{6(29{,}000 \text{ ksi})(2500 \text{ in}^4)}{(300 \text{ in.})^2}(1.25 \text{ in.}) = -6042 \text{ k·in.} = -503.4 \text{ k·ft}$$

$$\text{FEM}_{BC} = \frac{6(29{,}000 \text{ ksi})(3000 \text{ in}^4)}{(360 \text{ in.})^2}(0.75 \text{ in.}) = 3021 \text{ k·in.} = 251.7 \text{ k·ft}$$

$$\text{FEM}_{CD} = -\frac{6(29{,}000 \text{ ksi})(2000 \text{ in}^4)}{(240 \text{ in.})^2}(1.0 \text{ in.}) = -6042 \text{ k·in.} = -503.4 \text{ k·ft}$$

With the methodology established for finding the fixed-end moments, one may then follow the moment-distribution procedure found in Section 14.3 to solve this frame.

DID YOU KNOW?

In the moment-distribution method, the underlying assumption is that the length of a member does not change under load. Therefore, all calculations and sketches are made assuming the distance between two points doesn't change. Don't let the exaggerated sketches of the deformed shapes convince you otherwise. These deformations are small in comparison with the overall dimensions of the structure. For example, the horizontal distance between point B and C remains at 30 ft in Figures 14.8(a) and 14.8(b).

QUICK NOTE

The signs shown in the moments are based on the moment-distribution sign convention. Refer to the beam chart in Appendix D to help identify whether the fixed-end moment is clockwise (positive) or counterclockwise (negative).

EXAMPLE 14.8

Determine the member end moments for the frame shown in Figure 14.8(a).

SOLUTION STRATEGY

Follow the exact same procedure used in the previous section of this text. Start by placing a temporary support at joint C. For step 3 of the process, use the fixed-end moments already calculated.

SOLUTION

Place an imaginary support at point C to prevent lateral sway. Carry out the moment distribution for this structure, then calculate all reactions.

A		B			C			D	Joints
AB		BA	BC		CB	CD		DC	
0.00		0.500	0.500		0.500	0.500		0.00	DFs
0.0		0.0	−180.0		180.0	0.0		0.0	FEMs
		90.0	90.0		−90.0	−90.0			Dist 1
45.0	←		−45.0	↔	45.0		→	−45.0	CO 1
		22.5	22.5		−22.5	−22.5			Dist 2
11.3	←		−11.3	↔	11.3		→	−11.3	CO 2
		5.6	5.6		−5.6	−5.6			Dist 3
2.8	←		−2.8	↔	2.8		→	−2.8	CO 3
		1.4	1.4		−1.4	−1.4			Dist 4
0.7	←		−0.7	↔	0.7		→	−0.7	CO 4
		0.4	0.4		−0.4	−0.4			Dist 5
59.8		119.9	−119.9		119.9	−119.9		−59.8	Final

SAMPLE CALCULATION—COLUMN AB

$$+\circlearrowleft \sum M_B = -59.8 \text{ k·ft} - 119.9 \text{ k·ft}$$
$$-R_{Ay}(15 \text{ ft}) + R_{Ax}(20 \text{ ft}) = 0$$

Two unknowns are present. This will require an analysis of member BC first to solve for B_y and C_{y1}.

The same procedure used for determining the horizontal reaction components at the bases of the vertical columns is used for the sloping columns. (That is, each column is considered to be a free body, and moments are taken at its top to determine the horizontal reaction at its base.) For vertical columns, the vertical reactions pass through the points where moments are taken and thus may be neglected. *This is not the case for sloping columns, and the vertical reactions will appear in the moment equations.*

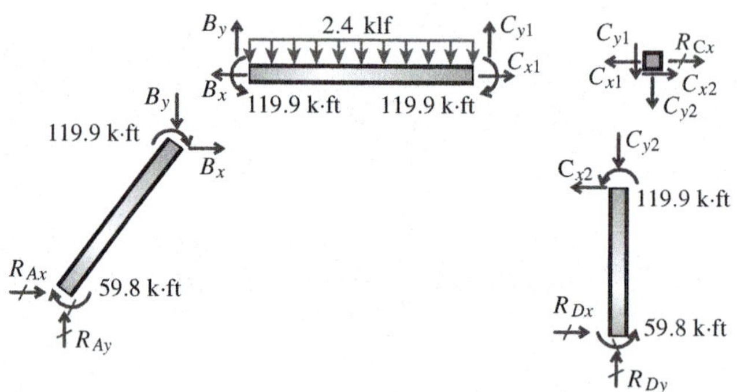

COMPUTED REACTIONS—STEP 2

$R_{Ay} = 36.0$ k ↑

$R_{Ax} = 36.0$ k →

$R_{Dy} = 36.0$ k ↑

$R_{Dx} = 9.0$ k ←

$R_{Cx} = 27.0$ k ←

Note that it is necessary to compute R_{Ay} before the equation can be solved for R_{Ax}. The authors find it convenient to remove the beam BC as a free body and compute the vertical component at each end by the columns. Once the value at B is determined, the sum of the vertical forces on column AB can be used to find R_{Ay}. The same procedure is followed after the assumed sidesway moments are distributed in step 3.

The next step is to analyze the condition shown in Figure 14.8(b). The following table shows this analysis. Once the internal moments are found, the reactions are calculated in the same process done previously.

A		B			C			D	Joints
AB		BA	BC		CB	CD		DC	
0.00		0.500	0.500		0.500	0.500		0.00	DFs
−503.4		−503.4	251.7		251.7	−503.4		−503.4	FEMs
		125.9	125.9		125.9	125.9			Dist 1
62.9	←		62.9	↔	62.9		→	62.9	CO 1
		−31.5	−31.5		−31.5	−31.5			Dist 2
−15.7	←		−15.7	↔	−15.7		→	−15.7	CO 2
		7.9	7.9		7.9	7.9			Dist 3
3.9	←		3.9	↔	3.9		→	3.9	CO 3
		−2.0	−2.0		−2.0	−2.0			Dist 4
−1.0	←		−1.0	↔	−1.0		→	−1.0	CO 4
		0.5	0.5		0.5	0.5			Dist 5
−453.3		−402.6	402.6		402.6	−402.6		−453.3	Final

The next step is to calculate the scale factor that should be applied to the results of the second table. This is the ratio of values of R_{Cx}:

$$S = \frac{27.0\text{ k}}{105.7\text{ k}} = 0.255$$

The final step is to superimpose the results of the first table with the results of the second table when it is multiplied by S. This will be done using beam-sign convention for the internal moments.

$$M_A = 59.8\text{ k·ft} + (0.255)(-453.3\text{ k·ft}) = \underline{\underline{-55.8\text{ k·ft}}}$$

$$M_B = -119.9\text{ k·ft} + (0.255)(402.6\text{ k·ft}) = \underline{\underline{-17.2\text{ k·ft}}}$$

$$M_C = -119.9\text{ k·ft} + (0.255)(-402.6\text{ k·ft}) = \underline{\underline{-222.3\text{ k·ft}}}$$

$$M_D = 59.8\text{ k·ft} + (0.255)(453.3\text{ k·ft}) = \underline{\underline{175.4\text{ k·ft}}}$$

COMPUTED REACTIONS—STEP 3

$R_{Ay} = \underline{\underline{26.8\text{ k}}} \downarrow$

$R_{Ax} = \underline{\underline{62.9\text{ k}}} \leftarrow$

$R_{Dy} = \underline{\underline{26.8\text{ k}}} \uparrow$

$R_{Dx} = \underline{\underline{42.8\text{ k}}} \leftarrow$

$R_{Cx} = \underline{\underline{105.7\text{ k}}} \rightarrow$

14.5 Multistory Frames

There are two possible ways in which the frame of Figure 14.9 may sway. The loads P_1 and P_2 obviously will cause both floors of the structure to sway to the right, but it is not known how much of the swaying is going to occur in the top floor (x condition) or how much will occur in the bottom floor (y condition). There are two sidesway conditions that need to be considered.

To analyze the frame by the usual sidesway procedure would involve:

1. An assumption of moments in the top floor for the x condition and the distribution of the moments throughout the frame.
2. An assumption of moments in the lower floor for the y condition and the distribution of those moments throughout the frame.

One equation could be written for the top floor by equating x times the horizontal forces caused by the x moments plus y times the horizontal forces caused by the y moments to the actual total shear on the floor, P_1. A similar equation could be written for the lower floor by equating the horizontal forces caused by the assumed moments to the shear on that floor, $P_1 + P_2$. Simultaneous solution of the two equations would yield the values of x and y. The final moments in the frame equal x times the x distributed moments plus y times the y distributed moments.

The sidesway method is not difficult to apply for a two-story frame, but for multistory frames it becomes unwieldy because each additional floor introduces another sidesway condition and another simultaneous equation. There are other options, like the classical *successive correction* method introduced by Professor C. T. Morris in 1932, which this text will not present. Rather, the reader is referred to Chapters 19 through 21 for an introduction to the stiffness method, which is capable of efficiently handling multistory frames of many heights and configurations.

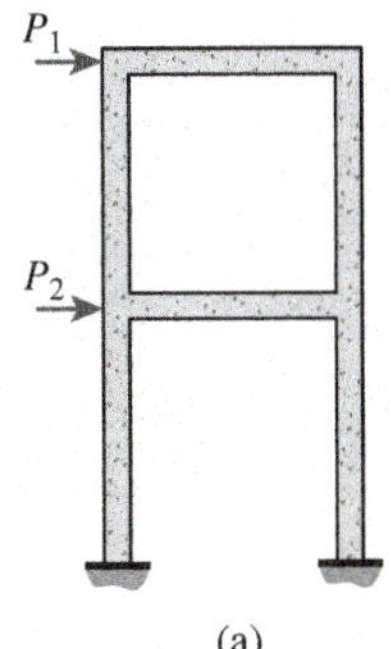

(a)

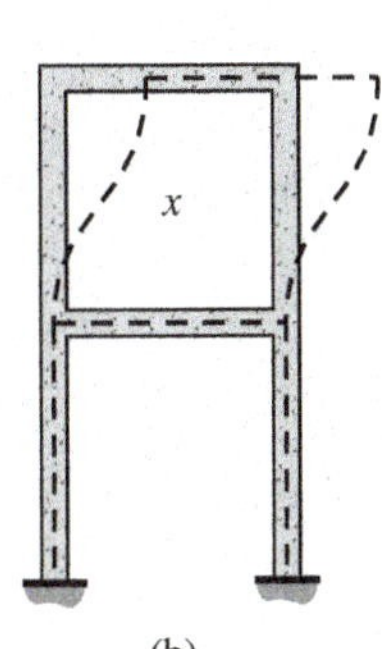

(b)

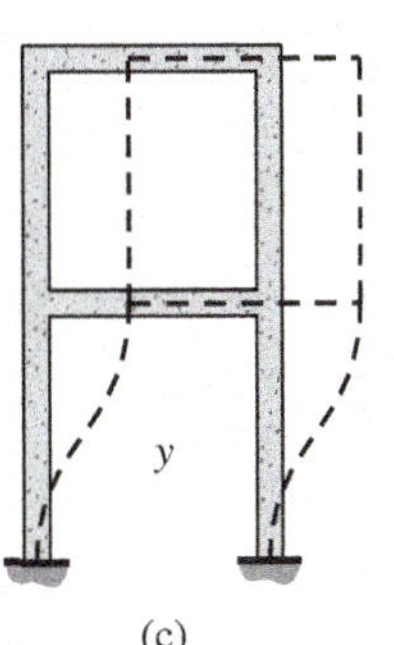

(c)

FIGURE 14.9 Story sway in a two story frame.

14.6 Examples with Video Solutions

VE14.1 Use the moment-distribution method to find the internal moments at the joints of the frame shown in Figure VE14.1.

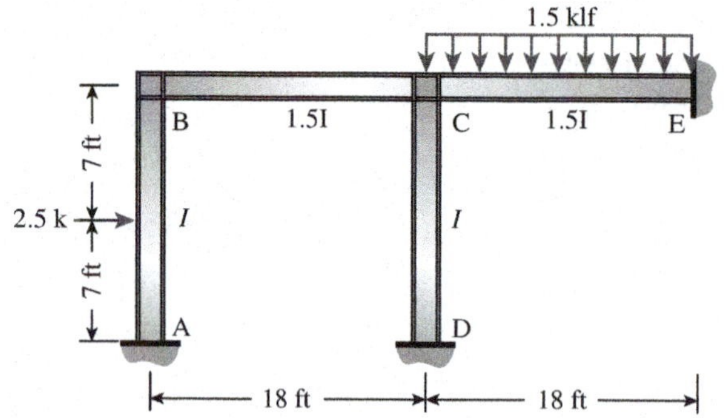

Figure VE14.1

VE14.2 Use the moment-distribution method to find the internal moments at the joints of the frame shown in Figure VE14.2. $I = 2400$ in^4.

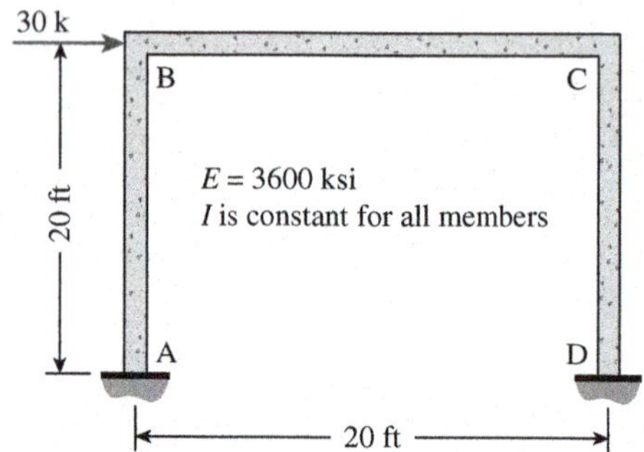

Figure VE14.2

VE14.3 Use the moment-distribution method to find the internal moments at the joints of the frame shown in Figure VE14.3. $I = 2400$ in^4.

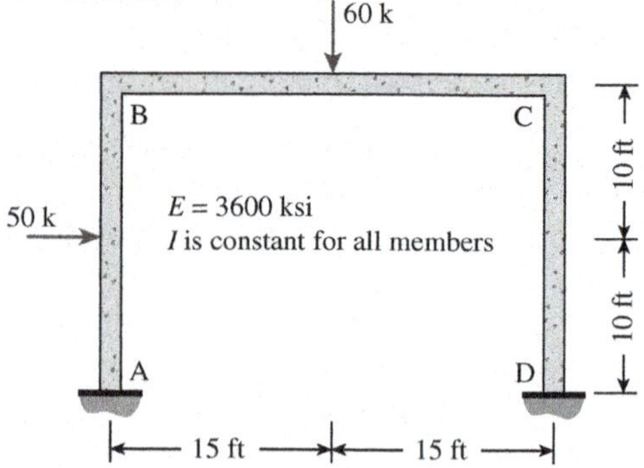

Figure VE14.3

14.7 Problems for Solution

Section 14.1

For Problems 14.1 through 14.6, use the moment distribution method to find the internal moments at all joints, then find the reactions. E and I are constant for all members unless otherwise noted.

P14.1 (Ans: M_{BA} = 71.25 k·ft, M_{BD} = 5.0 k·ft, M_{BC} = 76.25 k·ft)

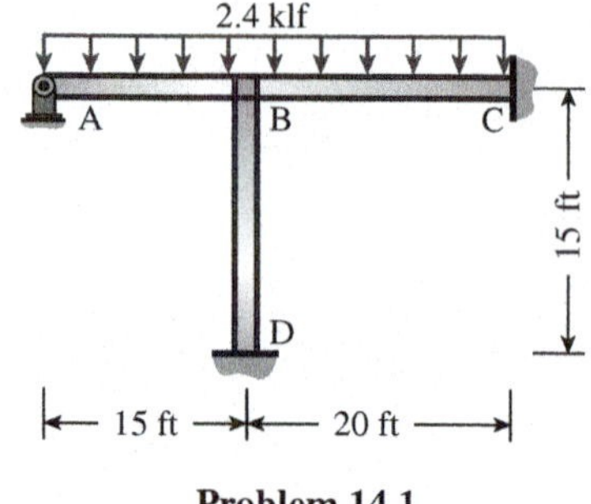

Problem 14.1

P14.2

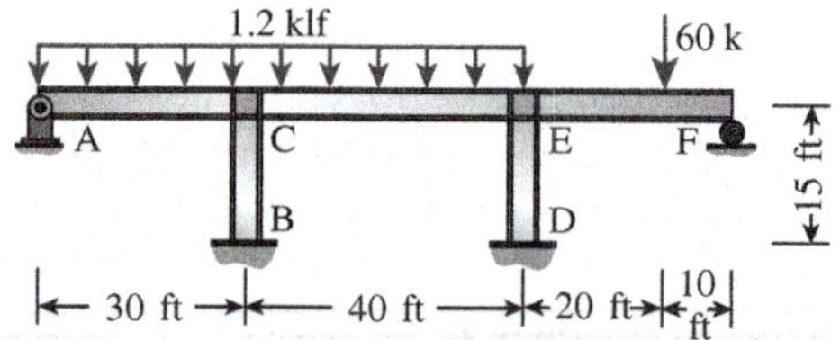

Problem 14.2

P14.3 (Ans: $M_{BA} = M_{CD}$ = 522.2 k·ft)

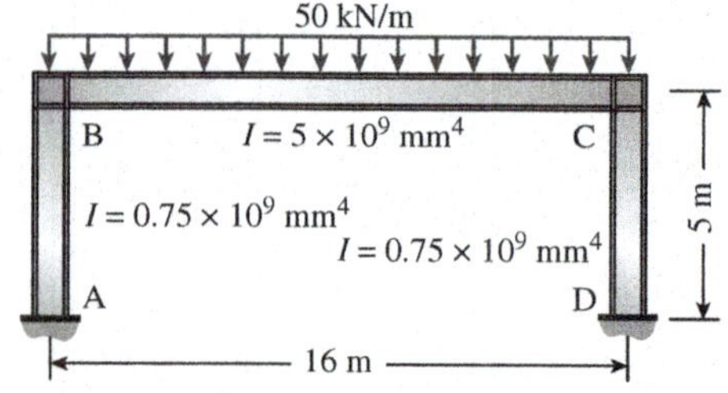

Problem 14.3

P14.4

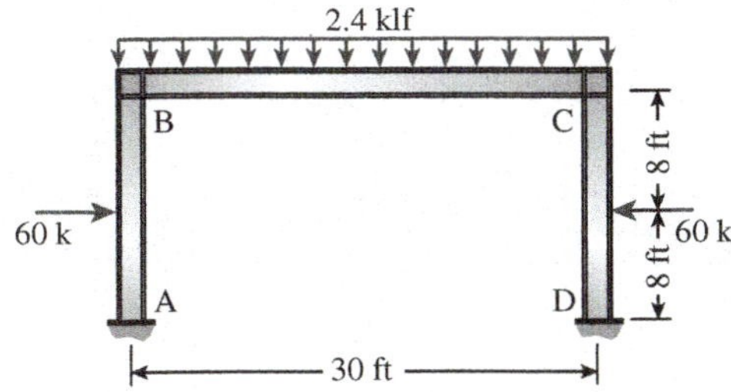

Problem 14.4

P14.5 (Ans: $M_{DB} = 211.3$ k·ft, $M_{DC} = 17.8$ k·ft, $M_{DF} = 193.3$ k·ft)

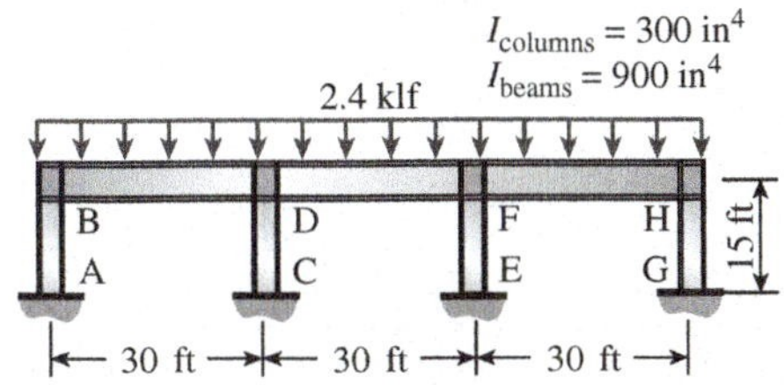

Problem 14.5

P14.6

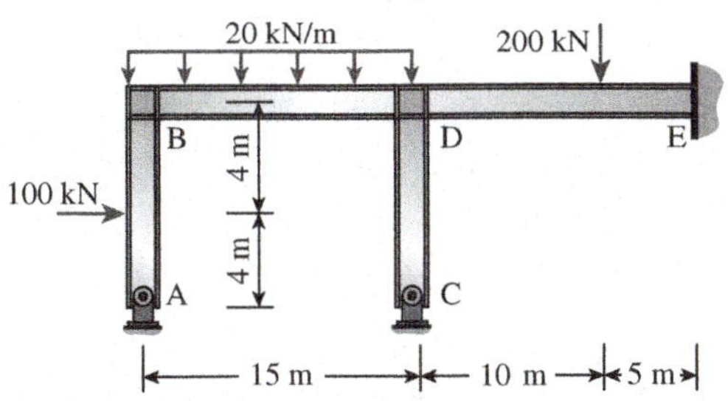

Problem 14.6

Section 14.2

For Problems 14.7 through 14.11, use the moment distribution method to find the internal moments at all joints. E = 29,000 ksi and I are constant for all members unless otherwise noted.

P14.7 Take $I = 1000$ in^4. (Ans: $M_B = M_C = 160$ k·ft)

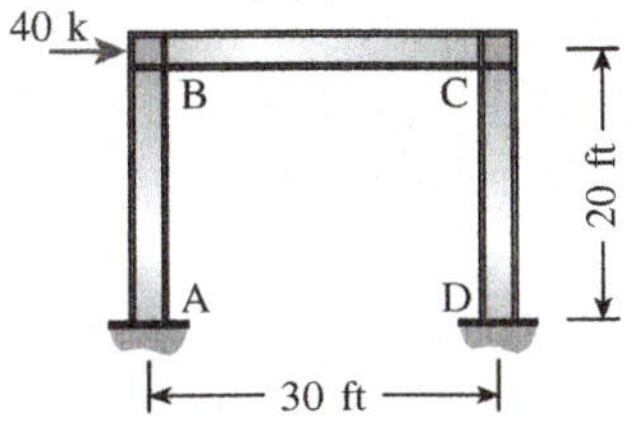

Problem 14.7

P14.8 Take $I = 1200$ in^4.

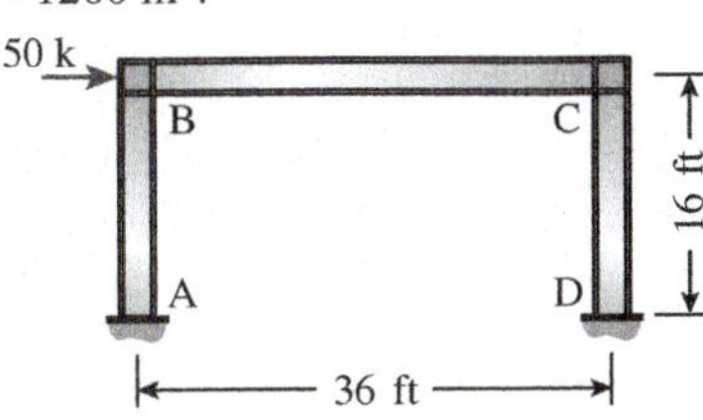

Problem 14.8

P14.9 Take $I = 1200$ in^4. (Ans: $M_B = 112.9$ k·ft, $M_C = 92.2$ k·ft)

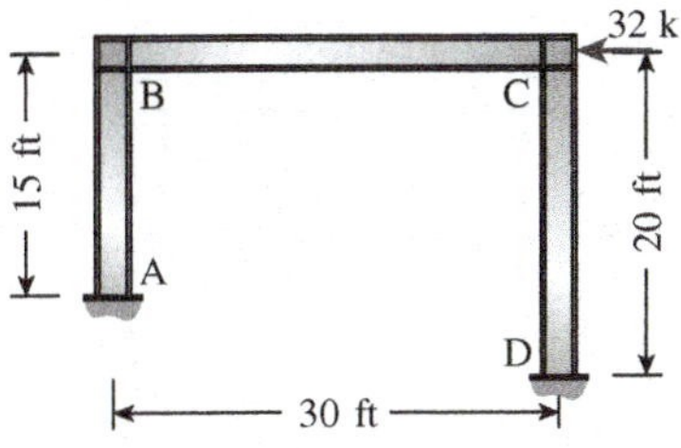

Problem 14.9

P14.10 Take $I = 800$ in^4.

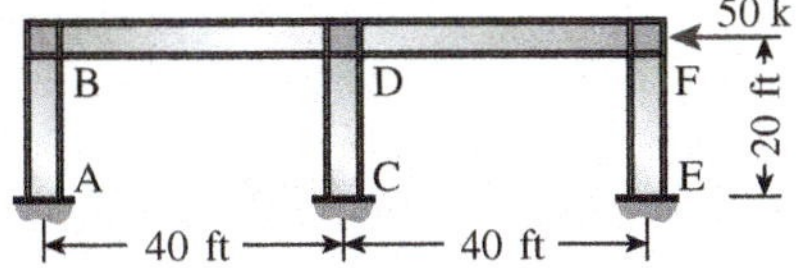

Problem 14.10

P14.11 Take $I = 1800$ in^4. (Ans: $M_{DB} = 33.1$ k·ft, $M_{DC} = 51.7$ k·ft, $M_{DF} = 18.7$ k·ft)

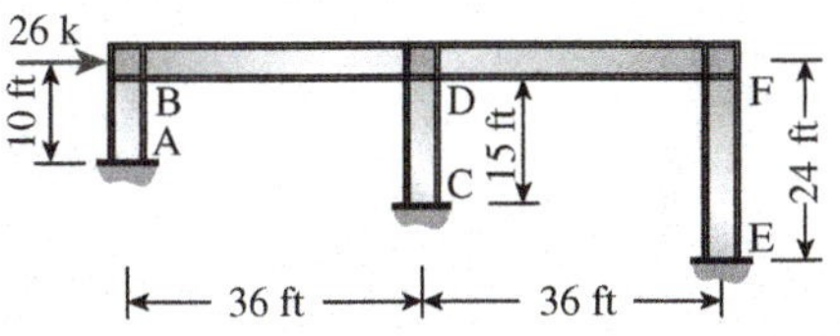

Problem 14.11

Section 14.3

For Problems 14.12 through 14.23, use the moment distribution method to find the horizontal and moment reactions. E = 29,000 ksi and $I = 1000$ in^4 are constant for all members unless otherwise noted.

P14.12

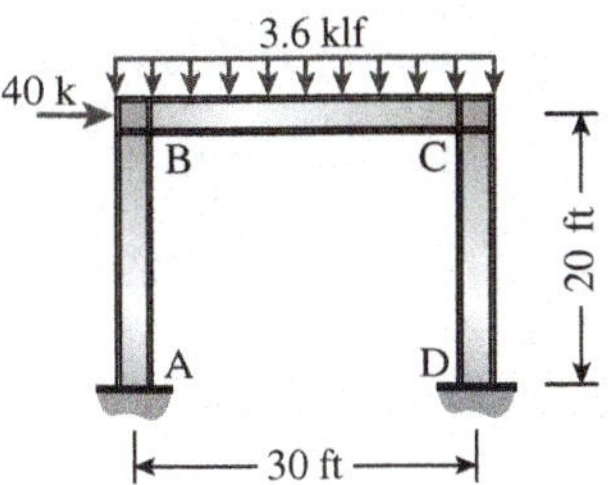

Problem 14.12

P14.13 (Ans: $R_{Ax} = 7.63$ k →, $M_A = 35.2$ k·ft ↺)

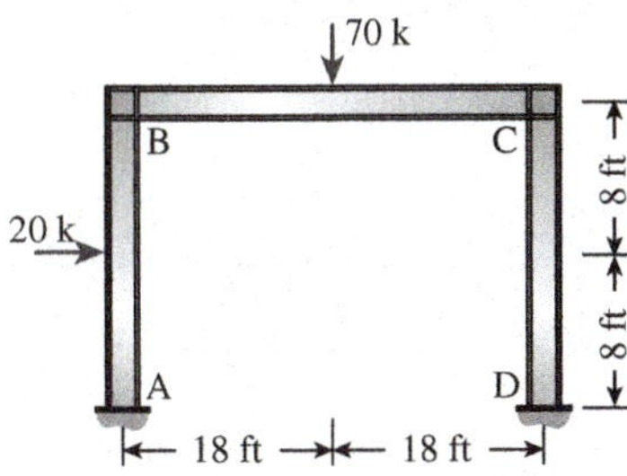

Problem 14.13

P14.14

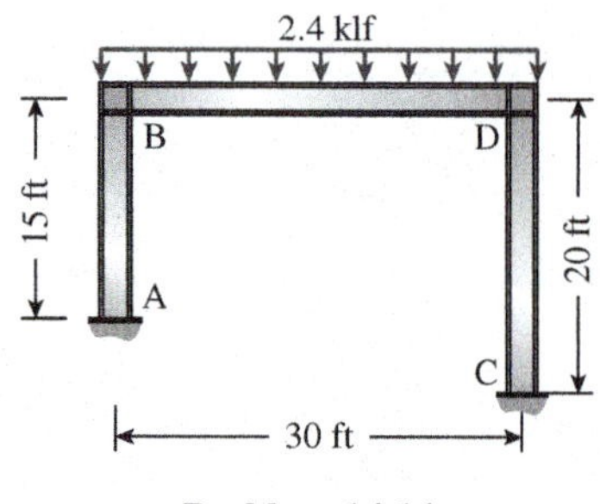

Problem 14.14

P14.15 Rework Problem 14.12 if the column bases are pinned. (Ans: R_{Ax} = 10.66 k ←, R_{Dx} = 29.34 k ←)

P14.16 Rework Problem 14.13 if column CD is pinned at the base.

P14.17 (Ans: R_{Ax} = 3.71 k →, M_A = 96.9 k·ft ↻)

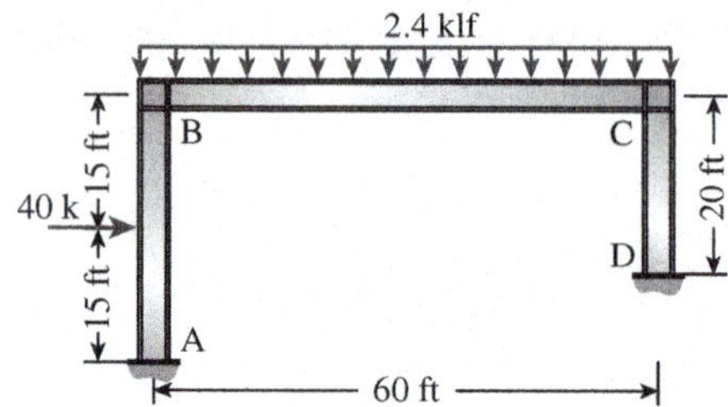

Problem 14.17

P14.18 Take E = 200 GPa and I = 1,800,000 mm^4.

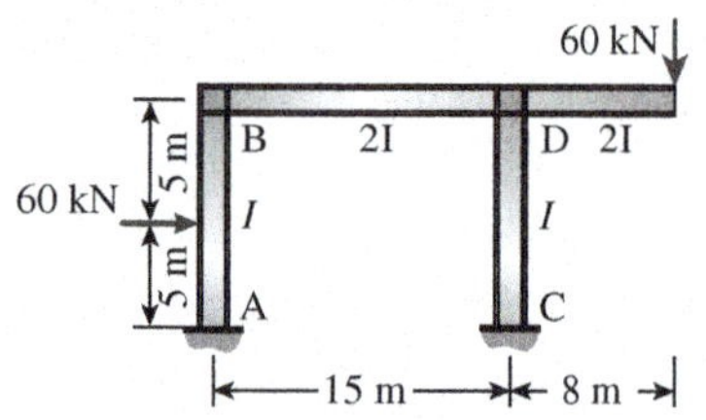

Problem 14.18

P14.19 (Ans: R_{Ax} = 38.14 k →, R_{Cx} = 20.26 k →, R_{Ex} = 8.4 k ←)

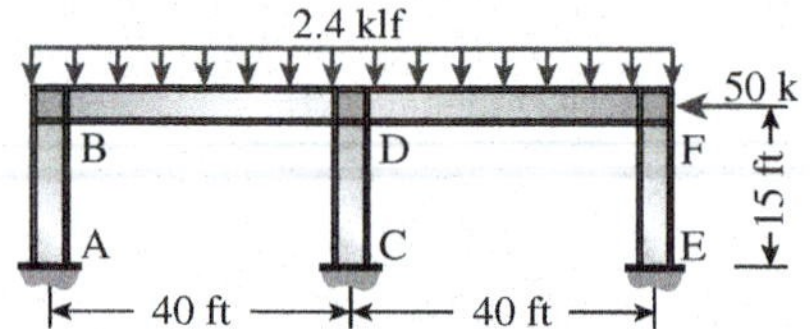

Problem 14.19

P14.20

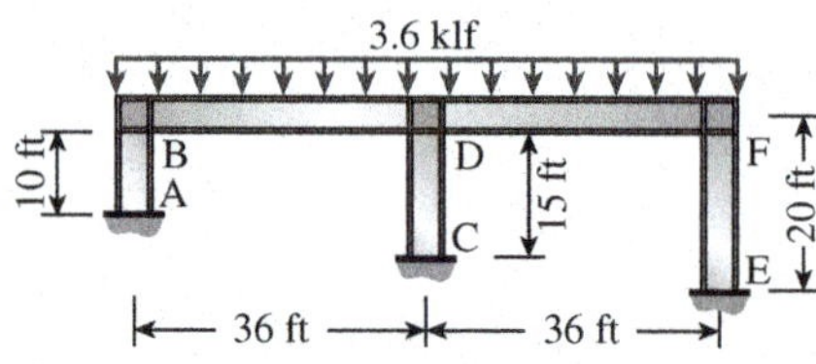

Problem 14.20

P14.21 (Ans: R_{Ax} = 25.06 k →, R_{Fx} = 14.94 k →)

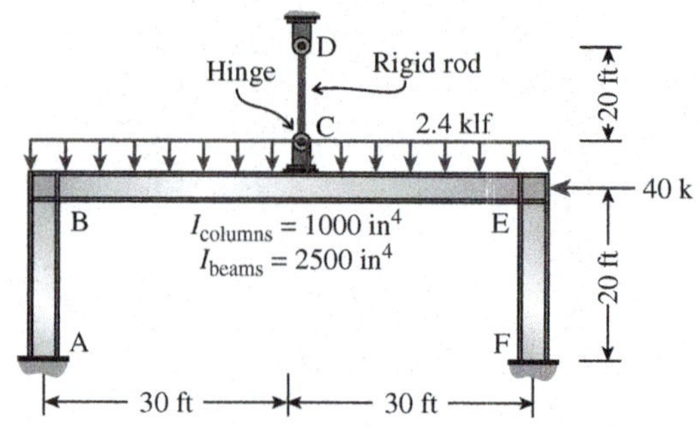

Problem 14.21

P14.22

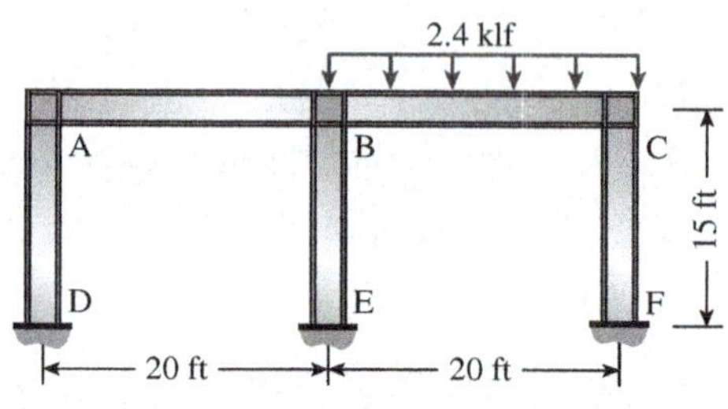

Problem 14.22

P14.23 Take E = 200 GPa and I = $2.0 \cdot 10^{-7}$ m^4. (Ans: R_{Ax} = 144.97 kN →, R_{Cx} = 32.33 kN ←, R_{Ex} = 192.63 kN ←)

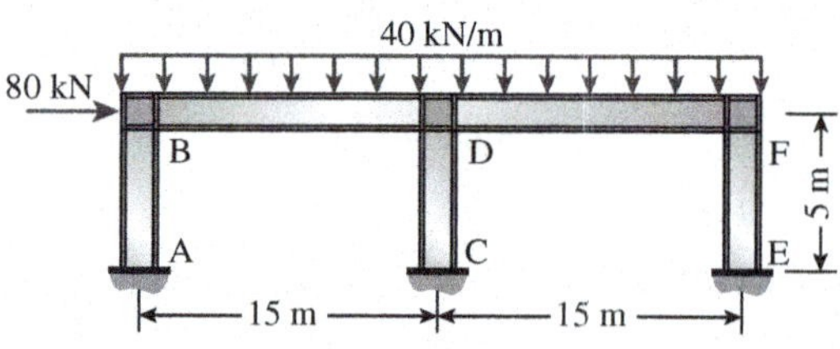

Problem 14.23

Section 14.4

For Problems 14.24 through 14.26, use the moment distribution method to find the reactions. E = 3600 ksi and is constant for all members.

P14.24

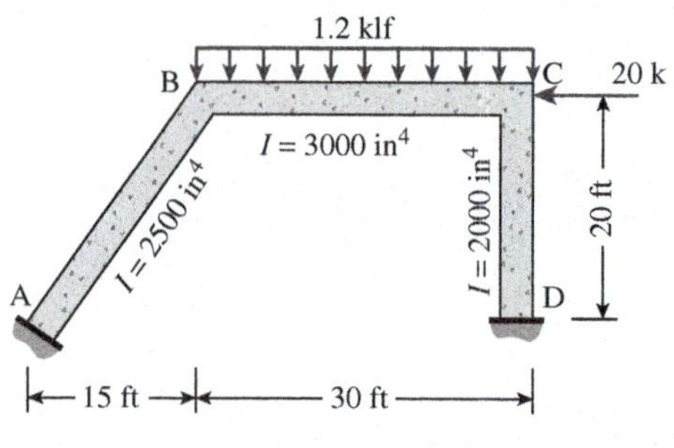

Problem 14.24

P14.25 (Ans: $R_{Dx} = 20$ k →, $M_A = 163$ k·ft ↻, $M_D = 228.6$ k·ft ↻)

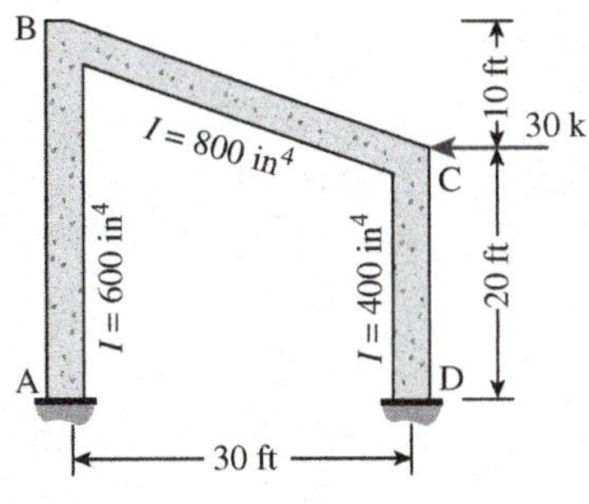

Problem 14.25

SAP2000 Problems

P14.26 Model the frame given in Problem 14.13 in SAP2000 but remove the 20-k load. Model all frame members as having $I = 1500$ in^4, $A = 10{,}000$ in^2, and $E = 29{,}000$ ksi. Remember to turn off self-weight and set the shear area to zero. Place the 70-k point load at joint B and find the horizontal displacement at C due to this load. Now shift the 70-k load 2 ft to the right and repeat the analysis. Keep shifting the load to the right in 2-ft increments and track the horizontal displacements for each location. Create a figure by plotting the distance from location B versus the horizontal displacement of C. Make comment on the relationship between the load location and frame sway (if any).

P14.27 Model the frame given in Problem 14.24 in SAP2000. Model all frame members as having I as given, $A = 10{,}000$ in^2, and $E = 3600$ ksi. Remember to turn off self-weight and set the shear area to zero. Change the moment of inertia of member AB to $I_{AB} = 500$ in^4, 1000 in^4, 1500 in^4, 2000 in^4, 2500 in^4, 3000 in^4, and 3500 in^4. Track the moment at A and the moment at C. Generate a plot of I_{AB}/I_{BC} versus the moment at A and also versus the moment at C. Make comment about the observed trends.

15 Approximate Analysis of Statically Indeterminate Structures

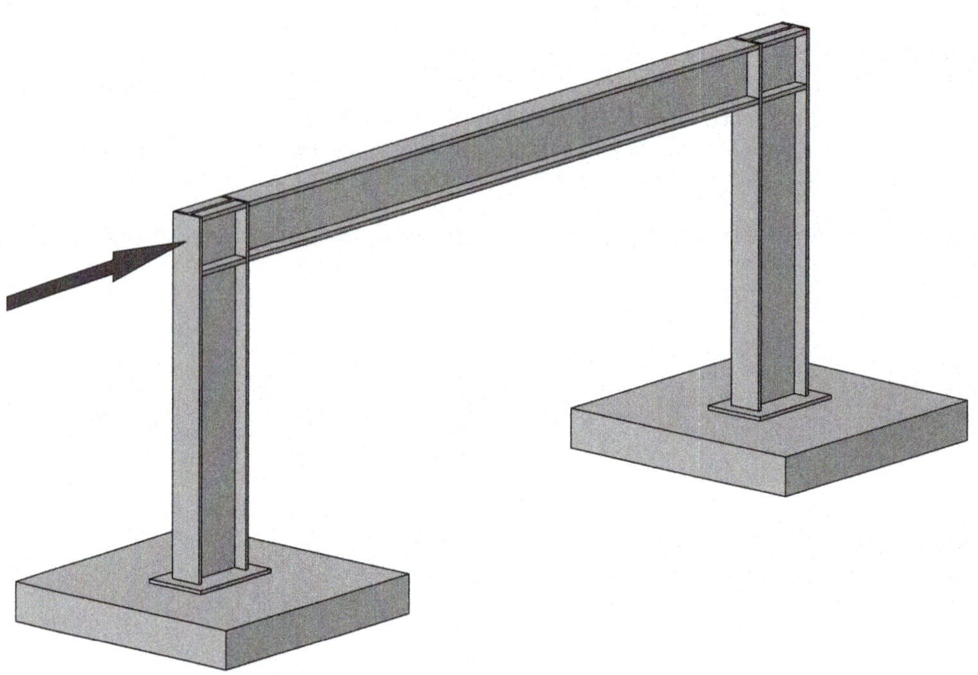

15.1 Introduction

The approximate methods presented in this chapter for analyzing statically indeterminate structures could very well be designated as classical methods. The same designation could be made for the moment-distribution method presented in Chapters 13 and 14. The methods discussed in this and the previous two chapters will occasionally be seen and used by an engineer in the course of everyday design as they are the methods of analysis sometimes used in current engineering practice.

Statically indeterminate structures may be analyzed "exactly" or "approximately." Several "exact" methods, which are based on elastic distortions, were discussed in the previous chapters. Approximate methods, which involve the use of simplifying assumptions, are presented in this chapter. These methods have many practical applications.

QUICK NOTE
With the proliferation of computers and high-end software, many of the reasons cited here for the benefit of knowing approximate methods are minimized. However, the authors maintain that the ability to perform "sanity" checks on computer results is necessary (see #4). Additionally, approximate methods aid in understanding structural behavior (see #6).

1. When costs are being estimated for alternative structural configurations and design concepts, approximate analyses are often very helpful. Approximate analyses and approximate designs of the various alternatives can be made quickly and used for initial cost estimates.

2. To analyze a statically indeterminate structure, an estimate of the member sizes must be made before the structure can be analyzed using an exact method. This is necessary because the analysis of a statically indeterminate structure is based on the elastic properties of the members. An approximate analysis of the structure will yield forces from which reasonably good initial estimates can be made of member sizes.

3. Today, computers are available with which exact analyses and designs of highly indeterminate structures can be made quickly and economically. To make use of computer programs, preliminary estimates of the size of the members should be made. If an approximate analysis of the structure has been done, very reasonable estimates of member sizes are possible. The result will be appreciable savings of both computer time and design hours.
4. Approximate analyses are quite useful for checking computer solutions, which is a very important matter.
5. An exact analysis may be too expensive for small noncritical systems, particularly when preliminary designs are being made. An acceptable and applicable approximate method is very appropriate for such a situation.
6. An additional advantage of approximate methods is that they provide the analyst with an understanding of the actual behavior of structures under various loading conditions. This important ability probably will not be developed directly from computer solutions.

To make an exact analysis of a complicated, statically indeterminate structure, a qualified analyst must model the structure; that is, the analyst must make certain assumptions about the behavior of the structure. For instance, the joints are assumed to be simple, rigid, or semi-rigid. Characteristics of material behavior and loading conditions must be assumed, and so on. The result of these assumptions is that all analyses are approximate. *We could say that we apply an "exact" analysis method to a structure that does not really exist.*

DID YOU KNOW?
All analysis methods are approximate in the sense that every structure is constructed within certain tolerances—no structure is perfect—and its behavior cannot be determined "exactly."

Many different methods are available for making approximate analyses. A few of the more common ones are presented here with consideration being given to trusses, continuous beams, and building frames. The approximate methods described in this chapter hopefully will provide you with a general knowledge about a wide range of statically indeterminate structures. Not all types of statically indeterminate structures are considered in this chapter. However, based on the ideas presented, you should be able to make reasonable assumptions when other types of statically indeterminate structures are encountered.

To be able to analyze a structure using the equations of static equilibrium, there must be no more unknowns than there are available equations of static equilibrium. If a truss or frame has 10 more unknowns than equations of equilibrium, it is statically indeterminate to the 10th degree. To analyze it by an approximate method, one assumption for each degree of indeterminacy—a total of 10 assumptions—must be made. Each assumption effectively provides another piece of information to use in the calculations.

15.2 Trusses with Two Diagonals in Each Panel

Diagonals Having Little Stiffness

The truss shown in Figure 15.1 has two diagonals in each panel. If one of the diagonal members were removed from each of the six panels, the truss would become statically determinate. Therefore, the original truss is statically indeterminate to the sixth degree and thus requires that an assumption must be made that provides six additional pieces of information.

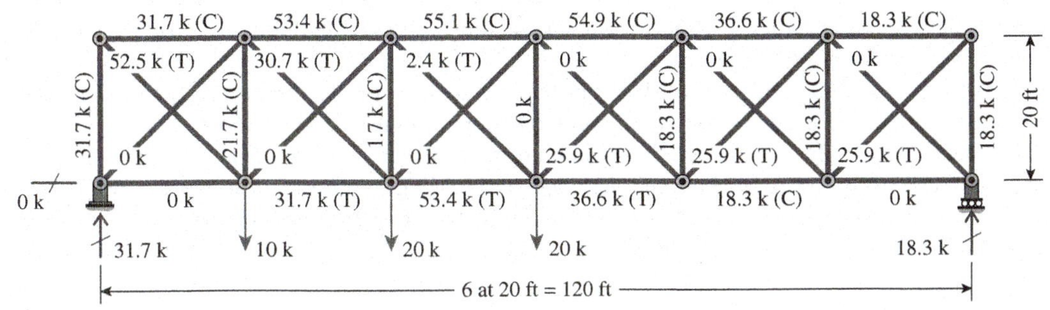

FIGURE 15.1 A truss analyzed assuming diagonals act only in tension.

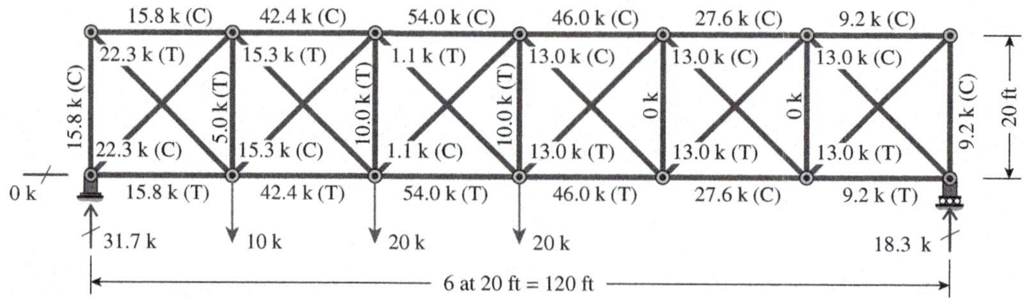

FIGURE 15.2 Approximate analysis of a truss assuming that each diagonal carries 50% of the panel shearing force.

ASSUMPTION
Only diagonal members in tension can take load. This means that members that would be in compression are assumed to have a zero load.

Frequently, the diagonals in a truss are relatively long and slender—often being made of a pair of small steel angles. They can carry reasonably large tensile forces but have negligible capacity in compression. For this situation, it is logical to assume that the shearing force in each panel is carried entirely by the diagonal that would be in tension for that sense of the shearing force (positive or negative). The other diagonal is assumed to have no force. Making this assumption in each panel effectively provides the forces (0 k) in six truss members. The forces in the remaining members can be evaluated with the equations of static equilibrium cast as the method of joints or the method of sections. The forces in Figure 15.1 were obtained on this basis.

Diagonals Having Considerable Stiffness

ASSUMPTION
One possible assumption is that both diagonal members in a panel take equal magnitude loads. One diagonal will be in tension while the other will be in compression.

In some trusses, the diagonals are constructed with sufficient stiffness to resist significant compressive loads. In panels with two substantial diagonals, the shearing force is carried by both diagonals. The division of shear causes one diagonal to be in tension and the other to be in compression. The usual approximation made is that each diagonal carries 50% of the shearing force in the panel; other divisions of the shearing force are also possible. Another typical division is that one-third of shearing force is carried by the diagonal acting in compression and two-thirds are carried by the diagonal in tension.

The forces calculated for the truss in Figure 15.2 are based on a 50% division of the shearing force in each panel.

15.3 Continuous Beams

Before beginning an exact analysis of a building frame, the sizes of the members in the frame must be estimated. Preliminary beam sizes can be obtained by considering their approximate moments. Frequently, a portion of the building can be removed and analyzed separately from the rest of the structure. For instance, one or more beam spans may be taken out as a free body and assumptions made as to the moments in those spans. To facilitate such an analysis, moment diagrams are shown in Figure 15.3 for several different uniformly loaded beams.

QUICK NOTE
Loads in an indeterminate structure are distributed according to the relative stiffness of the members that compose the structure. This is a problem because member sizes are designed based on loads, but the loads depend on the member sizes. Therefore, a reasonable estimate of loads without knowing member sizes is useful to have.

It is obvious from the figure that the assumed types of supports can have a tremendous effect on the magnitude of the calculated moments. For instance, the uniformly loaded simple beam in Figure 15.3(a) will have a maximum moment equal to $wL^2/8$. On the other hand, the uniformly loaded single-span fixed-ended beam of Figure 15.3(b) will have a maximum moment equal to $wL^2/12$. For a continuous uniformly loaded beam, the engineer may very well decide to estimate a maximum moment somewhere between the preceding values, at perhaps $wL^2/10$, and use that value for approximating the member size.

A very common method used for the approximate analysis of continuous reinforced-concrete structures involves the use of the American Concrete Institute's bending moment and shearing force coefficients.[1] These coefficients, which are reproduced in Table 15.1, provide estimated maximum shearing forces and bending moments for buildings of normal proportions. The values calculated in this manner usually will be somewhat larger than those obtained with an exact analysis. Consequently, appreciable economy can normally be obtained by taking the time or

[1] *Building Code Requirements for Reinforced Concrete*, ACI 318-11 (Detroit: American Concrete Institute) Section 8.3.3, 108–109.

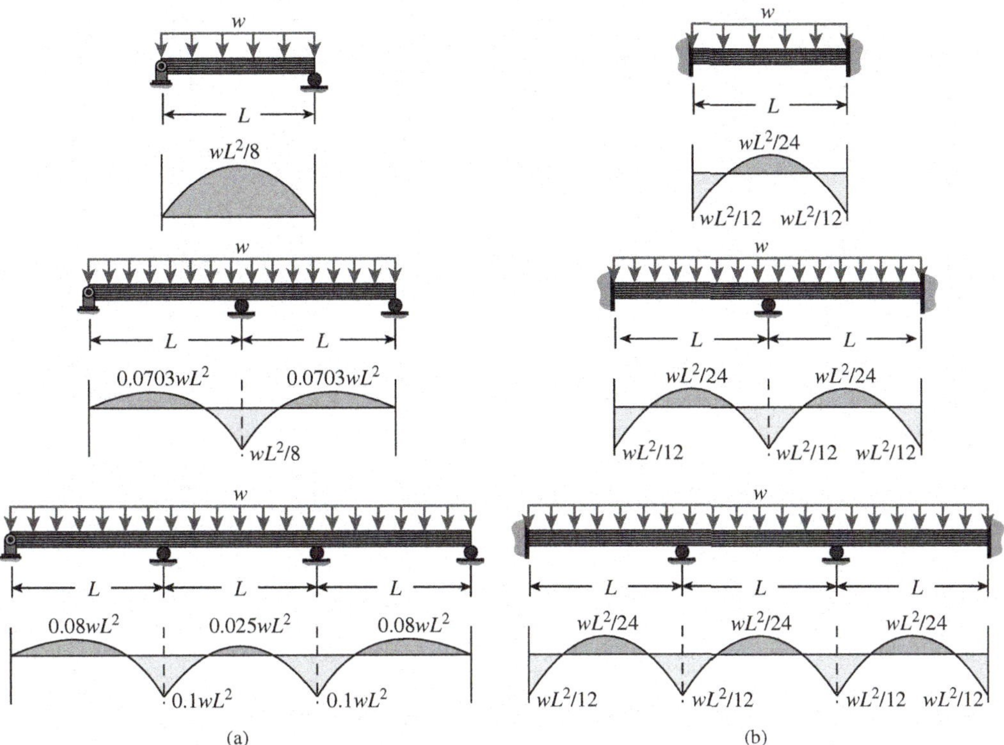

FIGURE 15.3 Moment diagrams for some typical continuous beams.

effort to make an "exact" analysis. In this regard, the engineer should realize that the ACI coefficients are best considered to apply to continuous frames having more than three or four continuous spans.

In applying the coefficients, w is the design load per unit of length, while L_n is the clear span for calculating positive bending moments and the average of the adjacent clear spans for calculating negative bending moments. These values were developed for members with approximately equal spans—the larger of two adjacent spans does not exceed the smaller by more than 20%—and

Table 15.1 ACI Moment Coefficients

Positive moment	
End spans	
If discontinuous end is restrained	$\frac{1}{11}wL_n^2$
If discontinuous end is integral with the support	$\frac{1}{14}wL_n^2$
Interior spans	$\frac{1}{16}wL_n^2$
Negative moment at the exterior face of the first interior support	
Two spans	$\frac{1}{9}wL_n^2$
More than two spans	$\frac{1}{10}wL_n^2$
Negative moment at other faces of interior supports	$\frac{1}{11}wL_n^2$
Negative moment at face of all supports for (a) slabs with spans not exceeding 10 ft and (b) beams and girders where ratio of sum of columns stiffness to beam stiffness exceeds 8 at each end of the span	$\frac{1}{11}wL_n^2$
Negative moment at interior faces of exterior supports for members built integrally with their supports	
Where the support is a spandrel beam or girder	$\frac{1}{24}wL_n^2$
Where the support is a column	$\frac{1}{16}wL_n^2$
Shear in end members at face of first interior support	$\frac{1.15}{2}wL_n$
Shear at face of all other supports	$\frac{1}{2}wL_n$

Source: American Concrete Institute ACI 318-11.

QUICK NOTE

In developing the ACI coefficients:

- The negative moment values were reduced to take into account the usual support widths and some moment redistribution before collapse.
- The positive moment values have been increased somewhat to account for the moment redistribution.
- The coefficients account for the fact that, in monolithic construction, the supports are not simple and moments are present at end supports where those supports are beams or columns.

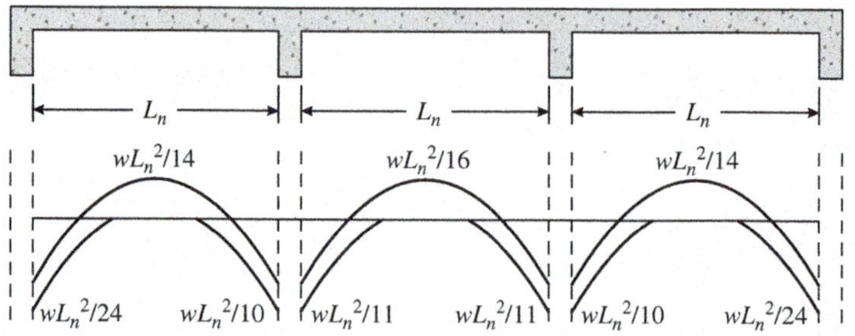

FIGURE 15.4 Moment envelopes for a continuous slab constructed integrally with exterior supports that are spandrel girders.

NOMENCLATURE

Spandrel Beam A beam on the exterior of a building that spans from column to column and marks the floor level between stories.

DID YOU KNOW?

It is not possible to produce maximum negative moments at both ends of a span simultaneously. It takes one placement of the live loads to produce maximum negative moment at one end of the span and another placement to produce maximum negative moment at the other end. The assumption of both maximums occurring at the same time is on the safe side.

DID YOU KNOW?

The motivation for setting the PIs to be at one-tenth points is because this is approximately halfway between the locations of zero moments for a simply supported beam and a fixed-fixed beam. In a regular frame, the end joints of the beams are not completely free to rotate as in a simply supported beam, but they are also not perfectly fixed against rotation as in a fixed-fixed beam (see Figure 15.6).

for cases where the ratio of the uniform service live load to the uniform service dead load is not greater than 3. In addition, the values are not applicable to prestressed-concrete members. Should these limitations not be met, a more precise method of analysis must be used.

For the design of a continuous beam or slab, the bending moment coefficients in effect provide two sets of moment diagrams for each span of the structure. One diagram is the result of placing the live loads so that they will cause maximum positive moment out in the span. The other is the result of placing the live loads to cause maximum negative moments at the supports.

The ACI coefficients give maximum values for a bending moment envelope for each span of a continuous frame. Typical envelopes are shown in Figure 15.4 for a continuous slab that is constructed integrally with its exterior supports, which are spandrel girders.

On some occasions, the analyst will take out a portion of a structure that includes not only the beams but also the columns for the floor above and the floor below, as shown in Figure 15.5. This procedure, usually called the *equivalent frame method*, is applicable only for gravity loads. The sizes of the members are estimated, and an analysis is made using one of the "exact" methods of analysis we have discussed.

15.4 Analysis of Building Frames for Vertical Loads

One approximate method for analyzing building frames for vertical loads involves the estimation of the location of the points of zero moment in the girders. These points, which occur where the moment changes from one sign to the other, are commonly called *points of inflection* (PIs) or *points of contraflexure*. One common practice is to assume the PIs are located at the one-tenth points from each girder end. In addition, the axial forces in the girders are assumed to be zero.[2]

These assumptions have the effect of creating a simple beam between the points of inflection, and the positive moments in the beam can be determined using simple equations of static equilibrium. Negative moments occur in the girders between their ends and the points of inflection. The negative moments are computed by treating the portion of the beam from the end to the point of inflection as a cantilever beam.

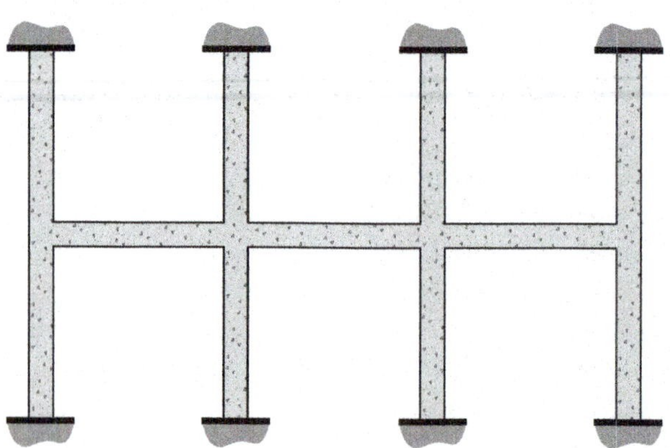

FIGURE 15.5 A portion of a building frame to be analyzed by the equivalent frame method.

[2]C. H. Norris, J. B. Wilbur, and S. Utku, *Elementary Structural Analysis*, 3rd ed. (New York: McGraw-Hill, 1976), 200–201.

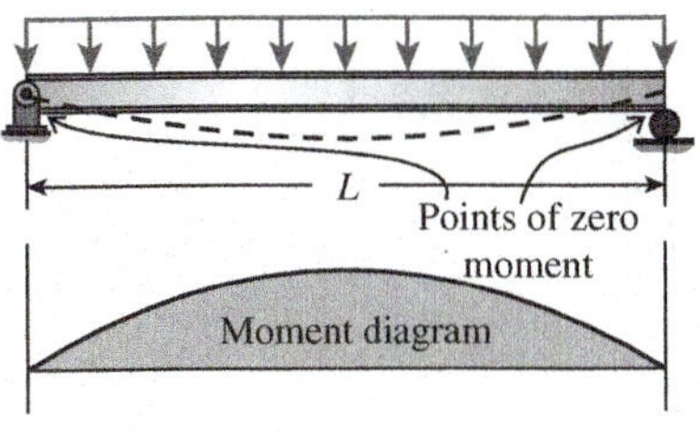

(a) Simply supported beam

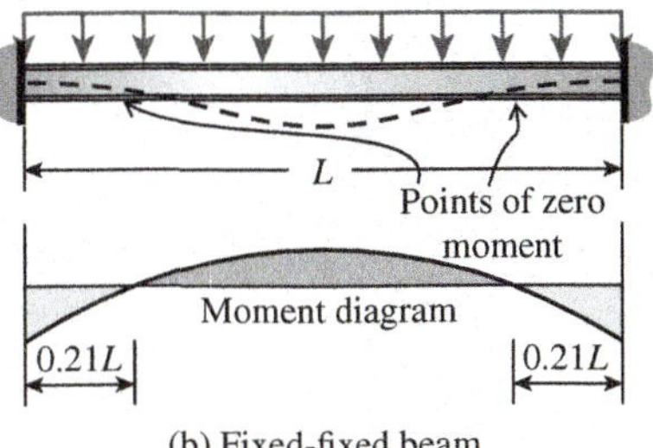

(b) Fixed-fixed beam

FIGURE 15.6 Points of zero moment (PIs) in ideal beams.

The shearing force at the end of the girders contributes to the axial forces in the columns. Similarly, the negative moments at the ends of the girders are transferred to the columns. At interior columns, the moments in the girders on each side oppose each other and may cancel. Exterior columns have moments caused by the girders framing into them on only one side; these moments need to be considered in design.

EXAMPLE 15.1

Determine the moment in all of the beams and the moment and reactions for column AEIM by using the approximate vertical analysis for the building frame shown.

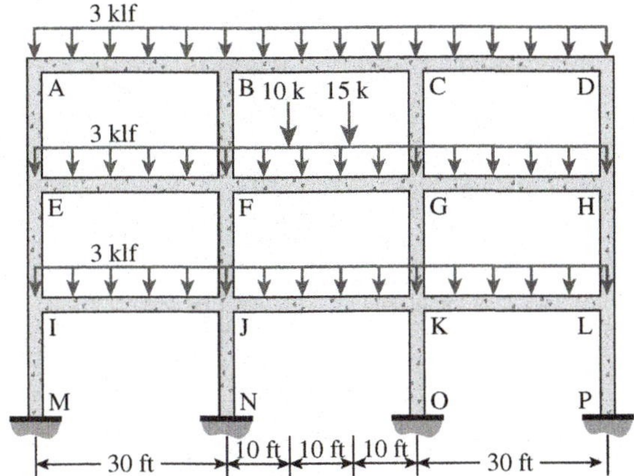

QUICK NOTE

Using the guidance found in Chapter 5, the statical indeterminacy of this frame is 27°. This means that to be able to completely solve for all forces in the frame, 27 additional pieces of information must be found. The approximate vertical analysis approach assumes 2 points of zero moment (i.e., internal hinges) per beam. This gives a total of 18 additional pieces of information. The axial force in each beam is assumed to be zero, giving another nine pieces of information. Hence, the number of assumption, equals the degree of indeterminacy.

SOLUTION STRATEGY

Assume that the PIs (i.e., points of zero moment) are located at one-tenth points and the beam ends are fixed supports. Furthermore, assume that no axial force develops in the beam. Draw appropriate FBDs of the beam. Recognize that all beams have the same span and the same loading except for beam FG.

SOLUTION

First solve for the moment in beam FG. A detailed drawing of the beam and the location (one-tenth span) of the assumed points of inflection (zero moment) are shown in the next figure.

Beam FG

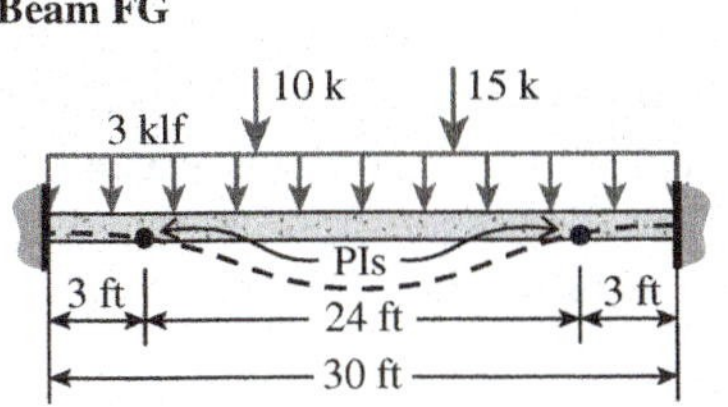

All Other Beams

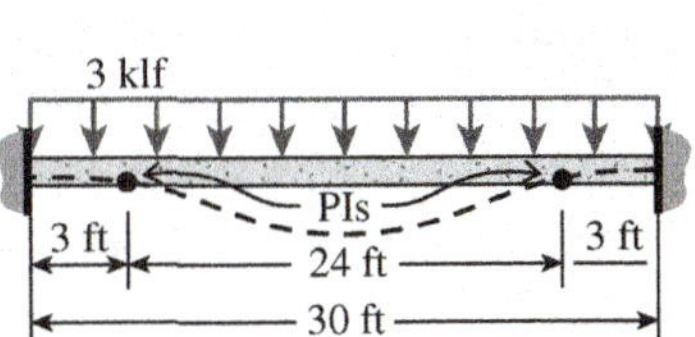

We can now split the beam at the PIs (hinges) into the simple span and the two cantilever spans. Solve for the reactions and moment diagram for the simple beam first. Transfer the reactions from the simple beam to the cantilever beams and then solve for their moments. The composite moment diagram shown is developed from connecting the two cantilevers with the simply supported beam results.

QUICK NOTE

Statics and the process for calculating the reactions and the moment diagrams are not shown here. Rather, the reader may refer to Chapters 5 and 6 if needed.

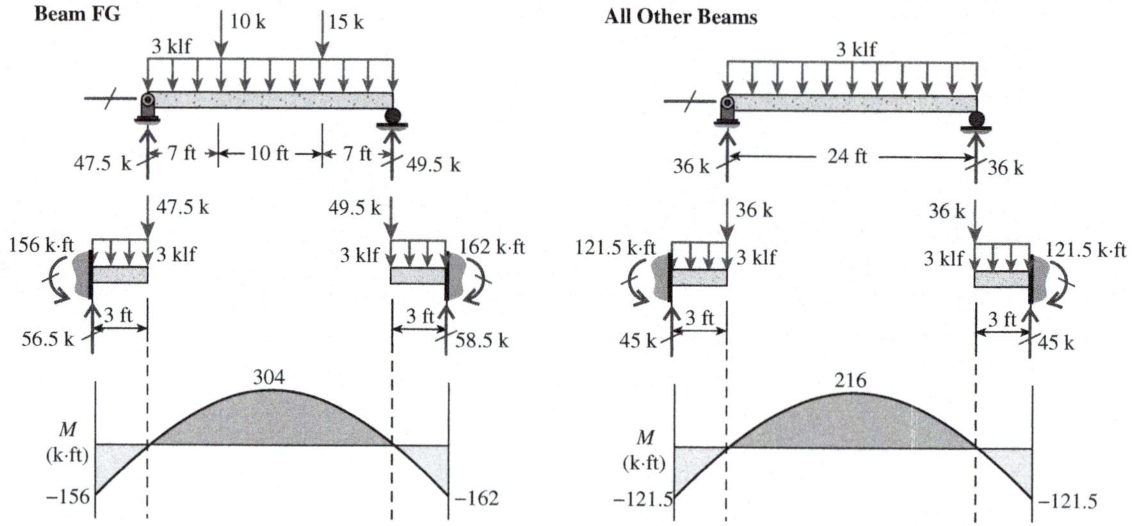

To find the moment at A, a FBD of the column should be sketched. It is useful to first analyze the beam, as has previously been shown. This will help the analyst to know what point loads should show up on the column FBD. Basic statics is used to solve for the reactions.

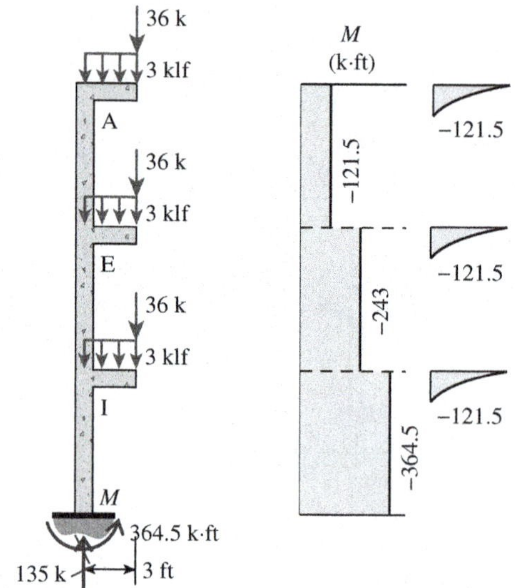

DID YOU KNOW?
The approximate analysis of vertical loads on frames does not produce any horizontal reactions. However, exact analysis will show the development of horizontal reactions.

DID YOU KNOW?
The actual location of the PIs in a beam is dependent upon the relative stiffness between the beam and the columns. The more stiff a column is, the closer the beam will approximate a fixed-fixed condition.

The one-tenth rule-of-thumb is a crude approximation of the actual locations of the PIs. The analyst may want to get a little better approximation of this location. Sketching the deflected shape often helps to make reasonable estimates for the locations of the PIs. For illustration, a continuous beam is shown in Figure 15.7(a), and a sketch of its estimated deflected shape for the loads shown is given in part (b). From the sketch, an approximate location of the PIs is estimated.

It might be useful for the reader to see where PIs occur for a few types of statically indeterminate beams. These may be helpful in estimating where PIs will occur in other structures. Moment diagrams are shown for several beams in Figure 15.8. The PIs obviously occur where the moment diagrams change from positive to negative moments or vice versa.

QUICK NOTE
The beam in Figure 15.7(a) has four PIs. However, we only need to estimate the location of two of them because the beam is only SI 2°.

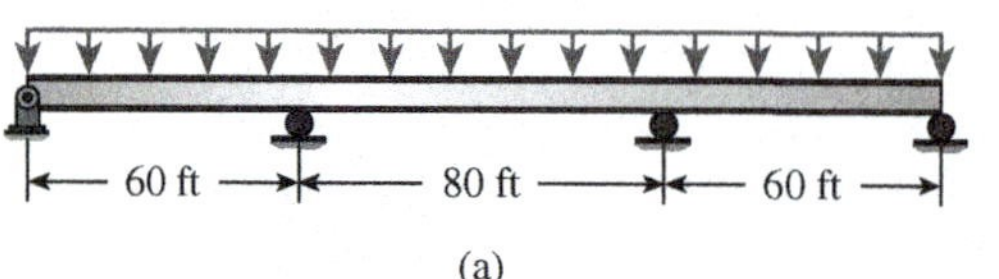

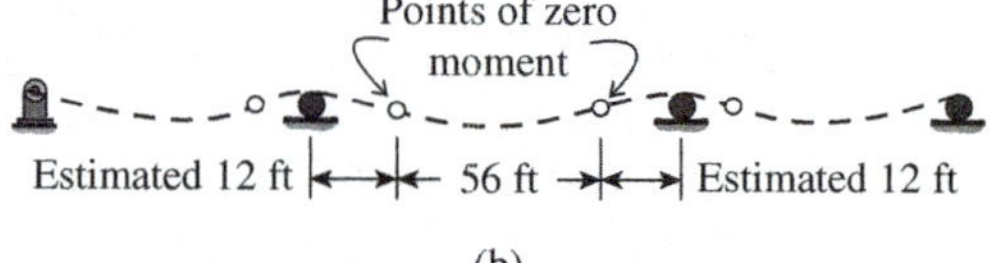

FIGURE 15.7 Estimated points of inflection in a continuous beam.

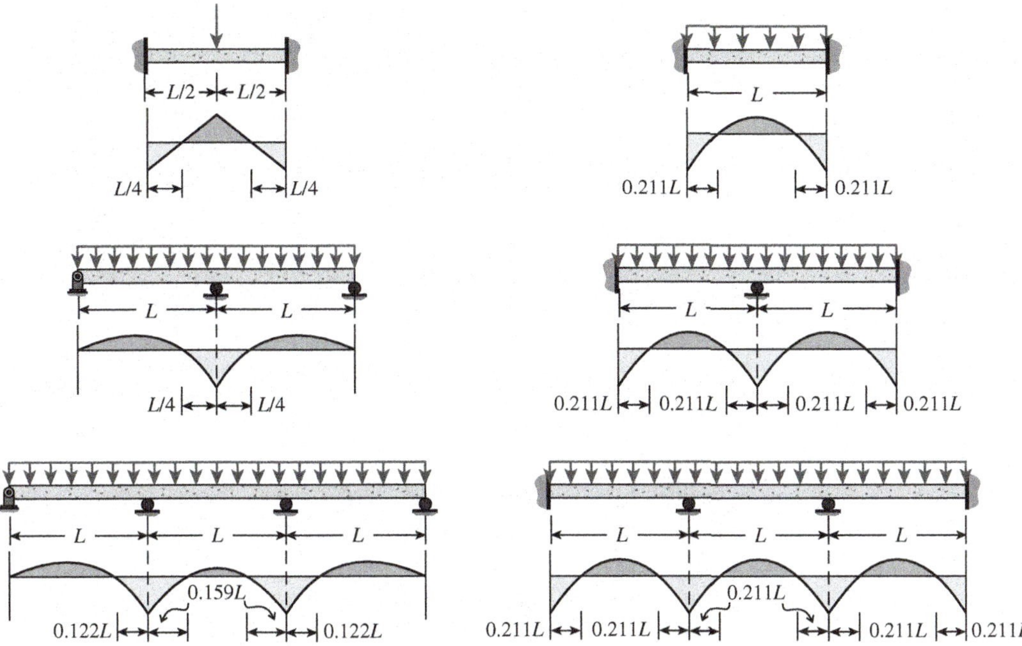

FIGURE 15.8 Location of points of inflection in several typical beams.

15.5 Analysis of Portal Frames

Portal frames of the type shown in Figures 15.9 and 15.10 may be fixed at their column bases, may be simply supported, or may be partially fixed. The columns of the frame of Figure 15.9(a) are assumed to have their bases fixed. Consequently, there will be three unknown forces of reaction at each support for a total of six unknowns. The structure is statically indeterminate to the third degree, and to analyze it by an approximate method, three assumptions must be made.

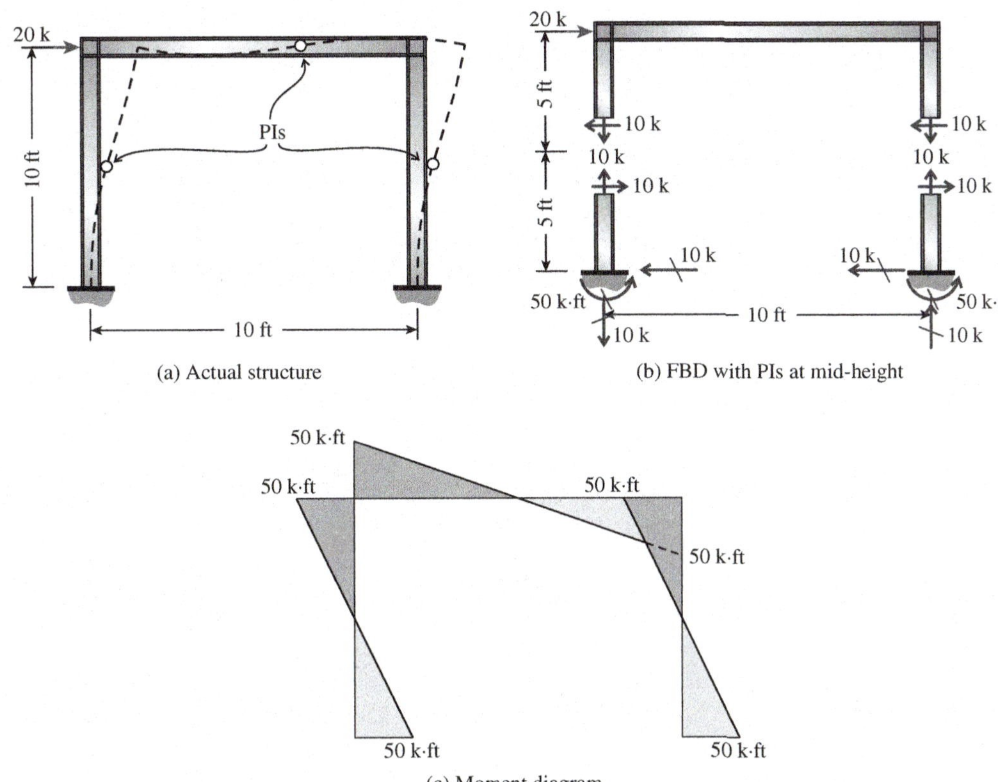

FIGURE 15.9 Approximate analysis of a portal frame with fixed column bases.

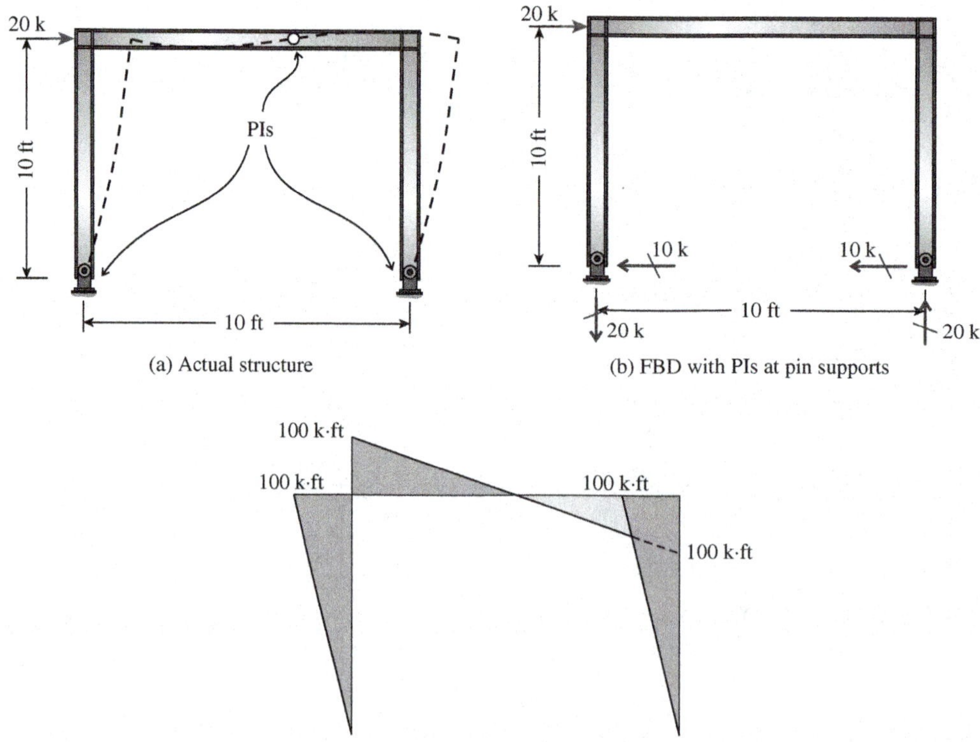

FIGURE 15.10 Approximate analysis of a portal frame with pinned column bases.

ASSUMPTIONS

Two assumptions are made for the analysis of portal frames.

1. Each column takes half of the shear load.
2. PIs (i.e., points of zero moment) are assumed to be at the mid-height of the columns if the bases are fixed. The points of zero moment are at the base of the columns if they are pinned.

QUICK NOTE

Notice that the moment is zero at the mid-span of the beam. Observing this and the deflected shape in Figure 15.9(a), it is seen that an implicit assumption is that a PI exists at the mid-span of the beam.

When a column is rigidly attached to its foundation, there can be no rotation at the base. Although the frame is subjected to wind loads causing the columns to bend laterally, a tangent to the column at the base will remain vertical. If the beam at the top of the columns is very stiff and rigidly fastened to the columns, the tangents to the columns at the junctions will remain vertical. A column rigidly fixed at the top and bottom will assume the shape of an S-curve when subjected to lateral loads, as shown in Figure 15.9(a).

At a point midway from the column base to the beam, the moment will be zero because it changes from a moment causing tension on one side of the column to a moment causing tension on the other side. If points of inflection are assumed in each column, two of the necessary three assumptions will have been made.

The third assumption usually made is that the horizontal shear is divided equally between the two columns at the plane of contraflexure. An "exact" analysis proves this is a very reasonable assumption as long as the columns are approximately the same size. If they are not similar in size, an assumption may be made that the shear is divided between them in a slightly different proportion with the stiffer column carrying the larger proportion of the shear. A good assumption for such cases is to distribute the shear to the columns in proportion to their EI/L^3 values.

To analyze the frame in Figure 15.9, the 20-kip lateral load is divided into 10-kip shearing forces at the column PIs as shown in Figure 15.9(b). The bending moments at the top and bottom of each column can be computed and are found to be equal to (10 k)(5 ft) = 50 k·ft. Moments then are taken about the left column PI to determine the right-hand vertical reaction as

$$(20\text{ k})(5\text{ ft}) - (10\text{ ft})R_{Ry} = 0$$

$$R_{Ry} = \underline{\underline{10\text{ k} \quad \uparrow}}$$

Finally, the moment diagrams are drawn with the results shown in Figure 15.9(c).

Should the column bases be assumed to be simply supported—that is, pinned or hinged—the points of contraflexure will occur at those hinges. Assuming the columns are the same size, the horizontal shear is split equally between the columns, and the other values are determined as shown in Figure 15.10.

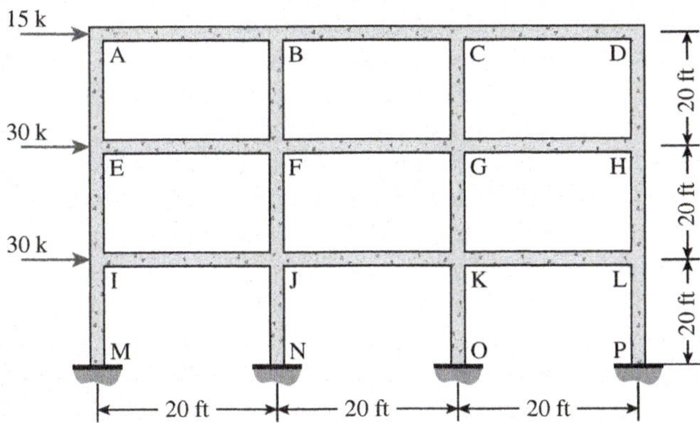

FIGURE 15.11 Three-story building frame.

The principles that have been discussed here can be applied to other framed structures regardless of the number of bays or the number of stories. Except for preliminary analysis, most structures are not analyzed using approximate methods. The ready availability of structural analysis software and the rapidity with which it can be used have now caused most analysis to be conducted with computers.

15.6 Analysis of Building Frames for Lateral Loads

Building frames are subjected to lateral loads as well as to vertical loads. The necessity for careful attention to these forces increases as buildings become taller. Not only must a building have sufficient lateral resistance to prevent failure, it also must have sufficient resistance to deflections to prevent injury to its various parts.

Lateral loads can be taken care of by means of X or other types of bracing, by shear walls, or by moment-resisting wind connections. Only the last of these three methods is considered in this chapter.

Large, rigid-frame buildings are highly indeterminate, and until the 1960s, their analysis by the so-called "exact" methods usually was impractical because of the multitude of equations involved. The total degree of indeterminacy of a rigid-frame building (internal and external) can be determined by considering it to consist of separate portals. One level of the rigid frame of Figure 15.11 is broken down into a set of portals in Figure 15.12. Each of the portals is statically indeterminate to the third degree, and the total degree of indeterminacy of a rigid-frame building equals three times the number of individual portals in the frame. For the frame in Figure 15.11, the degree of indeterminacy is (9 frames)(3°/frame) = 27°.

Another method of obtaining the degree of indeterminacy is to use the approach presented in Chapter 5 of this text, which is to assume that each of the girders is cut by an imaginary section. This would give $F = 39, n = 4, N = (4)(3) = 12$, and $\text{SI} = 39 - 12 = 27°$.

The building frame of Figure 15.11 is analyzed by two approximate methods in the pages to follow, and the results are compared with those obtained with the computer program SAP2000. The dimensions and loading of the frame are selected to illustrate the methods involved while keeping the computations as simple as possible. Since the structure is 27° statically indeterminate, there must be at least 27 assumptions made to permit an approximate solution.

The reader should be aware that with the availability of digital computers today it is feasible to make exact analyses in appreciably less time than that required to make approximate analyses without the use of computers. The more accurate values obtained permit the use of smaller members. The results of computer usage save money both in analysis time and in the use of smaller members. It is possible to analyze—in minutes—structures (such as tall buildings) that are statically indeterminate with hundreds or even thousands of redundants via the displacement

DID YOU KNOW?
Another item of importance is the provision of sufficient lateral rigidity to give the occupants a feeling of safety. There have been actual instances of occupants of the upper floors of tall buildings complaining of motion sickness on very windy days.

DID YOU KNOW?
Today such structures usually are analyzed "exactly" with modern computers, though approximate methods are used occasionally for preliminary analysis and sizing of members. Approximate methods also provide checks on computer solutions and give the analyst a better "feel" for a structure's behavior than they could obtain by examining the seemingly endless printout of a computer.

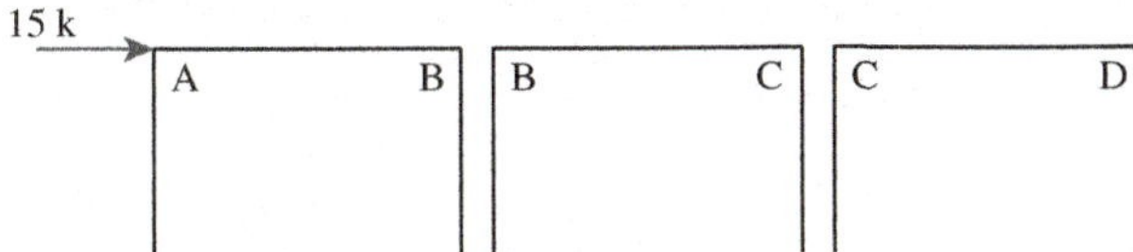

FIGURE 15.12 Top level of the frame of Figure 15.11.

method. The results for these highly indeterminate structures are far more accurate and more economical than those obtained with approximate analyses.

The two approximate methods considered here are the *portal method* and the *cantilever method.* These methods were used in so many successful building designs that they were almost the unofficial standard procedure for the design profession before the advent of modern computers. No consideration is given in either of these methods to the elastic properties of the members. These omissions can be very serious in asymmetrical frames and in very tall buildings.

If the height of a building is roughly five or more times its least lateral dimension, it is generally felt that a more precise method of analysis should be used than the portal or cantilever methods. There are several excellent approximate methods that make use of the elastic properties of the structures and that give values closely approaching the results of the exact methods. These include the *Factor method*,[3] the *Witmer method* of *K percentages*,[4] and the *Spurr method*.[5] Should an "exact" hand-calculation method be desired, the moment-distribution procedure of Chapters 13 and 14 is convenient.

DID YOU KNOW?
In both the portal and cantilever methods, the entire wind loads are assumed to be resisted by the building frames with no stiffening assistance from the floors, walls, and partitions. Changes in length of girders and columns are assumed to be negligible. However, they are not negligible in tall slender buildings whose height is five or more times the least horizontal dimension.

The Portal Method

The most common approximate method of analyzing building frames for lateral loads was at one time the portal method. Due to its simplicity, it probably was used more than any other approximate method for determining wind forces in building frames. This method, which was presented by Albert Smith in the *Journal of the Western Society of Engineers* in April 1915, was supposedly satisfactory for most buildings up to 25 stories in height.[6]

At least three assumptions must be made for each individual portal or for each girder. In the portal method, the frame is theoretically divided into independent portals (Figure 15.12) and the following three assumptions are made.

1. The columns bend in such a manner that there is a point of inflection at mid-height (see Figure 15.9).
2. The girders bend in such a manner that there is a point of inflection at their mid-span.
3. The horizontal shears on each level are arbitrarily distributed between the columns. One commonly used distribution (and the one illustrated here) is to assume the shear divides among the columns in the ratio of one part to exterior columns and two parts to interior columns. The reason for this ratio can be seen in Figure 15.12. Each of the interior columns is serving two portal frames, whereas the exterior columns are serving only one.

QUICK NOTE
Another common distribution of shear in the columns is to assume that the shear V taken by each column is in proportion to the floor area it supports. The shear distribution by the two procedures would be the same for a building with equal bays, but for one with unequal bays, the results would differ with the floor area method, probably giving more realistic results.

For this frame (Figure 15.13), there are 27 redundants; to obtain their values, one assumption as to the location of the point of inflection (i.e., points of zero moment) has been made for each

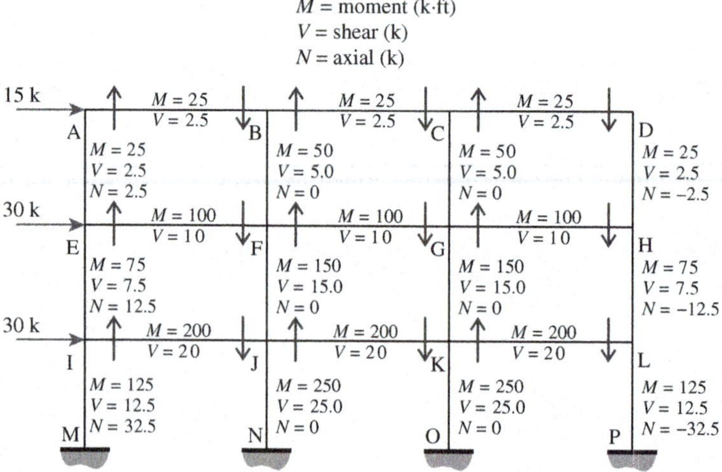

FIGURE 15.13 Building frame analyzed by the portal method for lateral loads.

[3] C.H. Norris, J.B. Wilbur, and S. Utk, *Elementary Structural Analysis*, 3rd ed. (New York: McGraw-Hill, 1976), 207–212.
[4] "Wind Bracing in Steel Buildings," *Transactions of the American Society of Civil Engineers* 105 (194): 1725–1727.
[5] *Ibid*, 1723–1725.
[6] *Ibid.*, 1723.

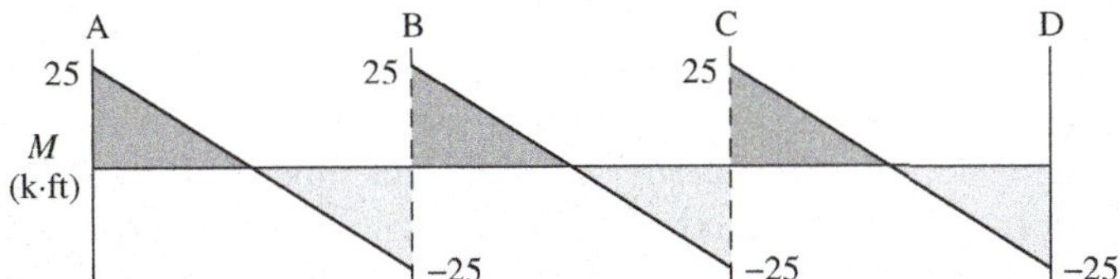

FIGURE 15.14 Moment diagram for beam ABCD by portal method.

of the 21 columns and girders. Three assumptions are made on each level as to the shear split in each individual portal, or the number of shear assumptions equals one less than the number of columns on each level. For the frame, nine shear assumptions are made, giving a total of 30 assumptions and only 27 redundants. More assumptions are made than necessary, but they are consistent with the solution (that is, if only 27 of the assumptions were used and the remaining values were obtained by statics, the results would be identical).

COMMON MISTAKE

Students will often assign the total shear in the middle level to be only 30 k and the total shear at the bottom level to be only 30 k. It is true that these are the lateral loads applied to each level, but the shears in the columns of each level must resist all of the lateral forces above it. Therefore, the total shear in the bottom level is
15 k + 30 k + 30 k = 75 k.

Frame Analysis

The frame of Figure 15.11 is analyzed on the basis of these assumptions. The arrows shown on the figure give the direction of the girder shears. The reader can visualize the stress condition of the frame if they assume the wind is tending to push it over from the left to right, stretching the left exterior columns and compressing the right exterior columns. The moment diagrams for a sample girder and column are given in Figures 15.14 and 15.15, respectively. Briefly, the calculations were made as follows.

Column Shears The shears in each column on the various levels were first obtained. The total shear on the top level is 15 kips. Because there are two exterior and two interior columns the expression may be written as

$$x + 2x + 2x + x = 15 \text{ k}$$
$$x = 2.5 \text{ k}$$
$$2x = 5.0 \text{ k}$$

The shear in column AE is 2.5 kips, in BF it is 5.0 kips, and so on. Similarly, the shears were determined for the columns on the first and second levels, where the total shears are 75 and 45 kips, respectively.

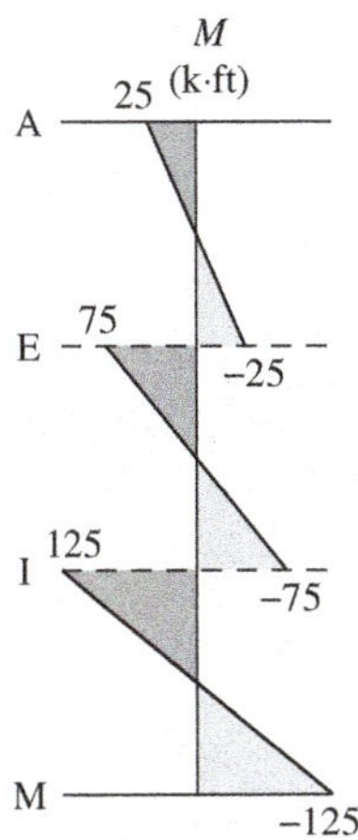

FIGURE 15.15 Moment diagram for column AEIM by portal method.

Column Moments The columns are assumed to have points of inflection at their mid-heights. Thus, cutting the column at mid-height and sketching a FBD will help in the analysis of the internal forces (see Figure 15.16). The moments in a column—top and bottom—equal the column shear times half the column height. For column AE, the calculation is

$$M_{AE} = (2.5 \text{ k})(10 \text{ ft}) = \underline{\underline{25 \text{ k·ft}}}$$

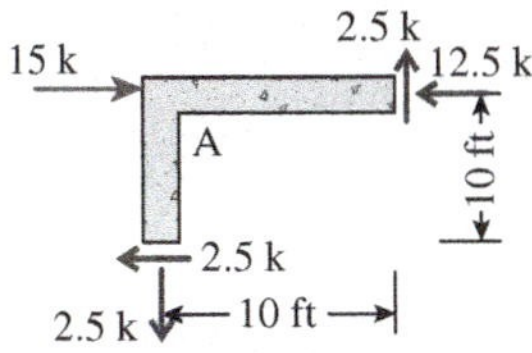

FIGURE 15.16 FBD of joint A. Only the shear in column AE was initially known. Basic statics was used to find the other three forces.

Girder Moments and Shears At any joint in the frame, the sum of the moments in the girders equals the sum of the moments in the columns. The column moments have been previously determined. By beginning at the upper left-hand corner of the frame (Figure 15.16) and working across from left to right, adding or subtracting the moments as the casefootno may be, the girder moments were found in this order: AB, BC, CD, EF, FG, and so on. It follows that, with points of inflection at girder centerlines, the girder shears equal the girder moments divided by half-girder lengths. For example, the moment at A is 25 k·ft, therefore the shear in girder AB is found by

$$V_{AB} = \frac{25 \text{ k·ft}}{10 \text{ ft}} = \underline{\underline{2.5 \text{ k}}}$$

Column Axial Forces The axial forces in the columns may be directly obtained from the girder shears. Starting at the upper left-hand corner, the column axial force in AE is numerically equal to the shear in girder AB. The axial force in column BF is equal to the difference between the two girder shears AB and BC, which equals zero in this case. (If the width of each of the portals is the same, the shears in the girder on one level will be equal, and the interior columns will have no axial force, since only lateral loads are considered.)

The Cantilever Method

FRIENDLY ADVICE

Basic statics will always govern the analysis of the frame. Recognizing that we assume zero moments at the mid-height and mid-span of all columns and girders, FBDs can be drawn for each joint. Moving from joint to joint (like in truss analysis) is a very methodical way to handle the analysis.

QUICK NOTE

The assumption for axial-force distribution is equivalent to making a number of axial-force assumptions equal to one less than the number of columns on each level. Again, the structure has 27 redundants, and 30 assumptions are made (21 column and girder PI assumptions and nine column axial-force assumptions), but the extra assumptions are consistent with the solution.

QUICK NOTE

The same approach shown in Figure 15.18 can be used to find the axial forces in the columns at the other levels in the structure. The only difference is that the FBD that is used consists of the entire structure and applied loads above the mid-height of the columns being considered.

Another simple method of analyzing building frames for lateral forces is the cantilever method presented by A. C. Wilson in the *Engineering Record*, September 5, 1908. This method is said to be a little more desirable than the portal method for high narrow buildings. It was used satisfactorily for buildings with heights not in excess of 25 to 35 stories.[7] It was not as popular as the portal method.

Mr. Wilson's method makes use of the assumptions that the portal method uses as to locations of points of inflection in columns and girders, but the third assumption differs somewhat. Rather than assume the shear on a particular level to divide between the columns in some ratio, the axial stress in each column is considered to be proportional to its distance from the center of gravity of all the columns on that level. If the columns on each level are assumed to have equal cross-sectional areas, then their forces will vary in proportion to their distances from the center of gravity. As the wind loads will tend to overturn the building, the columns on the leeward side will be compressed, whereas those on the windward side will be put in tension. The greater the distance a column is from the center of gravity of its group of columns, the greater will be its axial stress.

Frame Analysis

The frame previously analyzed by the portal method is now analyzed by the cantilever method in Figure 15.17. Briefly, the calculations are made as follows.

Column Axial Forces Considering first the top level, moments are taken about the *center of gravity* (cg) indicated in Figure 15.18. According to the third assumption, the axial force in BF will be only one-third of that in AE, since BF is only a third of the distance from the column center of gravity as is column AE. The axial forces in BF and AE will be tensile, whereas those in CG and DH will be compressive. The following expression is written with respect to the center of gravity shown in Figure 15.18 to determine the values of the column axial forces on the top level.

$$\overset{+}{\circlearrowleft} \sum M_{cg} = -(15\text{ k})(10\text{ ft}) + (3N)(30\text{ ft}) + (N)(10\text{ ft}) + (N)(10\text{ ft}) + (3N)(30\text{ ft}) = 0$$

$$\therefore \quad N = \underline{\underline{0.75\text{ k}}} \qquad 3N = \underline{\underline{2.25\text{ k}}}$$

The axial force in AE is 2.25 kips, and in BF is 0.75 kip, and so on. Similar moment calculations are made for each level to obtain the column axial forces.

Girder Shears The next step is to obtain the girder shears from the column axial forces. These shears are obtained by starting at the top left-hand corner and working across the top level,

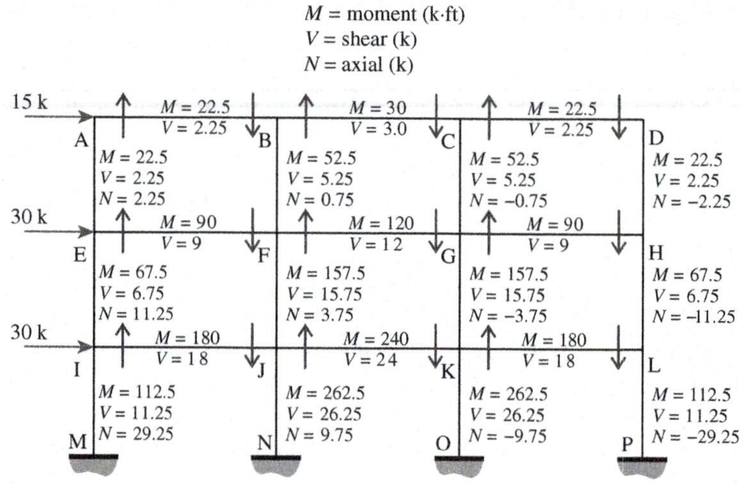

FIGURE 15.17 Building frame analyzed by cantilever method for lateral forces.

[7] *Ibid*, 1723.

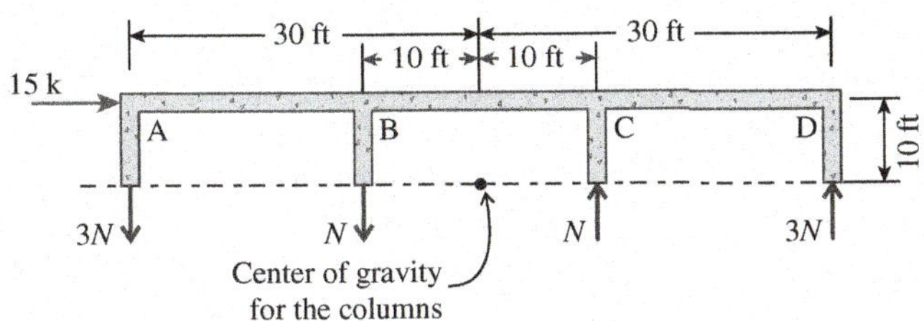

FIGURE 15.18 FBD showing only unknown axial forces when frame is cut at mid-height of the upper columns where internal moments are zero.

adding or subtracting the axial forces in the columns according to their signs. This procedure is similar to the method of joints used for finding truss forces.

Column and Girder Moments and the Column Shears The final steps can be quickly summarized. The girder moments—as before—are equal to the girder shears times the girder half-lengths. The column moments are obtained by starting at the top left-hand corner and working across each level in succession, adding or subtracting the previously obtained column and girder moments as indicated. The column shears are equal to the column moments divided by half the column heights.

FRIENDLY ADVICE

As you work from joint to joint solving for girder shears and internal moments, sketch a complete FBD of each joint with the known axial forces in the columns already identified. Use basic statics to solve for the unknown shears.

15.7 Exact and Approximate Analysis Results Comparison

Unlike the approximate methods, the exact analysis provided by a computer program like SAP2000 requires the input of material and cross-sectional properties. Indeed, it is the relative stiffness between the beams and the columns that plays into how well the approximate analyses estimate the results from the exact analysis. The comparison of the portal and cantilever methods with SAP2000 results assumes that the moment of inertia for the beams is twice the moment of inertia for the columns. These results are presented in Table 15.2.

Table 15.2 Member End Moments (k·ft)

	Portal		Cantilever		SAP 2000		Maximum
Member	End *i*	End *j*	End *i*	End *j*	End *i*	End *j*	Difference
AB	25.0	25.0	22.5	22.5	35.5	28.0	37%
BC	25.0	25.0	22.5	22.5	20.5	20.5	18%
CD	25.0	25.0	22.5	22.5	28.0	35.5	37%
EF	100.0	100.0	90.0	90.0	124.9	100.8	28%
FG	100.0	100.0	120.0	120.0	76.7	76.7	36%
GH	100.0	100.0	90.0	90.0	100.8	124.9	28%
IJ	200.0	200.0	180.0	180.0	188.2	242.3	26%
JK	200.0	200.0	240.0	240.0	134.2	134.2	44%
KL	200.0	200.0	180.0	180.0	242.3	188.2	26%
AE	25.0	25.0	22.5	22.5	35.5	23.7	37%
BF	50.0	50.0	52.5	52.5	48.5	42.3	19%
CG	50.0	50.0	52.5	52.5	48.5	42.3	19%
DH	25.0	25.0	22.5	22.5	35.5	23.7	37%
EI	75.0	75.0	67.5	67.5	101.2	84.8	33%
FJ	150.0	150.0	157.5	157.5	135.2	128.8	18%
GK	150.0	150.0	157.5	157.5	135.2	128.8	18%
HL	75.0	75.0	67.5	67.5	101.2	84.8	33%
IM	125.0	125.0	112.5	112.5	157.5	190.4	41%
JN	250.0	250.0	262.5	262.5	193.6	208.5	26%
KO	250.0	250.0	262.5	262.5	193.6	208.5	26%
LP	125.0	125.0	112.5	112.5	157.5	190.4	41%

QUICK NOTE

The maximum difference indicated in the table represents the maximum difference between any moment computed using SAP2000 and either the portal or cantilever results—whichever controls.

DID YOU KNOW?

The moments at both ends of a member are the same whenever the PIs are at midpoint of the member. When the moments on both ends of a member are not equal—see SAP2000 results—the PI is closer to the end having the smaller moment.

Note that, for several members, the approximate results vary considerably from the results obtained by the exact method. Experience with the exact method for handling indeterminate building frames will show that the points of inflection will not occur exactly at the midpoints. Using more realistic locations for the assumed inflection points will greatly improve the results.

15.8 Analysis of Vierendeel Trusses

In previous chapters, a truss has been defined as a structure assembled with a group of ties and struts connected at their joints with frictionless pins so the members are subjected only to axial tension or axial compression. A special type of truss (although it is not really a truss by the preceding definition) is the Vierendeel truss. A typical example of a Vierendeel truss is in Figure 15.19. You can see that these trusses are actually rigid frames or, as some people say, they are girders with big holes in them. The Vierendeel truss was developed in 1893 by the Belgian engineer and builder, Arthur Vierendeel. They have been used frequently in Europe, particularly in Belgium, for highway and railroad bridges but only occasionally in the United States.

A Vierendeel truss is usually constructed with reinforced concrete, but also may be fabricated with structural steel. The external loads are supported by means of the flexural resistance of the short heavy members. The continuous moment-resisting joints cause the trusses to be highly statically indeterminate.

These highly statically indeterminate structures can be analyzed approximately by the portal and cantilever methods described in the preceding sections. The results of the truss in Figure 15.19 are obtained by applying the portal method and are shown in Figure 15.20. For this symmetrical truss, the cantilever method will yield the same results. To follow the calculations, you might like to turn the structure on its end, because the shear being considered for the Vierendeel is vertical, whereas it was horizontal for the building frames previously considered. Because the truss is symmetric, the results are only shown for one half of the truss.

For many Vierendeel trusses—particularly those that are several stories in height—the lower horizontal members may be much larger and stiffer than the other horizontal members. To obtain better results with the portal or cantilever methods, the non-uniformity of sizes should be taken into account by assuming that more shearing force is carried by the stiffer members.

DID YOU KNOW?
Vierendeel trusses are rather inefficient, but because of their aesthetics, they are used in situations in which their large clear openings are desirable. In addition, they are particularly convenient in buildings where they can be constructed with depths equal to the story heights.

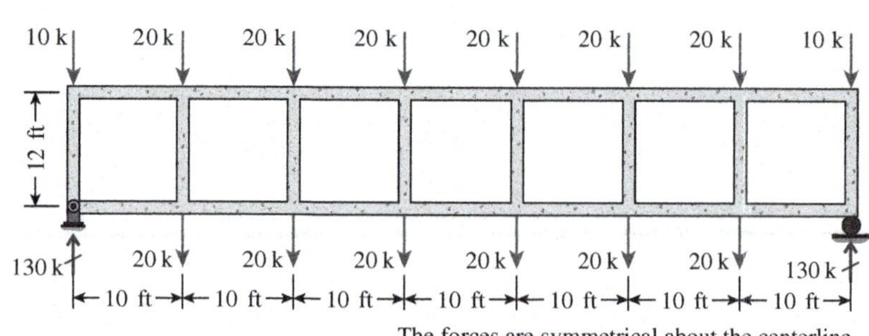

FIGURE 15.19 Vierendeel truss.

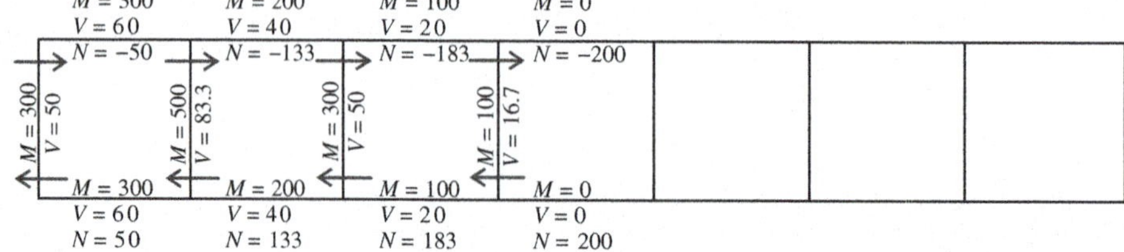

FIGURE 15.20 Results of a Vierendeel truss using the portal method.

15.9 Examples with Video Solution

VE15.1 Compute the forces in the diagonal and vertical members of the truss shown in Figure VE15.1, assuming that the diagonals that are in compression can only take one-fourth of the panel shear.

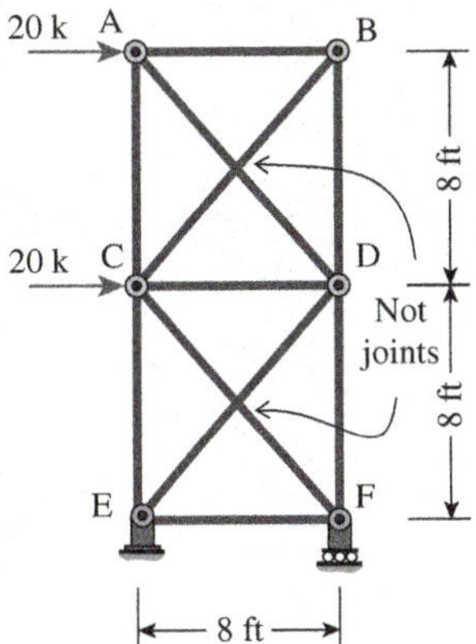

Figure VE15.1

VE15.2 Use the approximate vertical analysis to find the moment diagrams for the beams and columns of the frame shown in Figure VE15.2. Assume inflection points are located at one-tenth points.

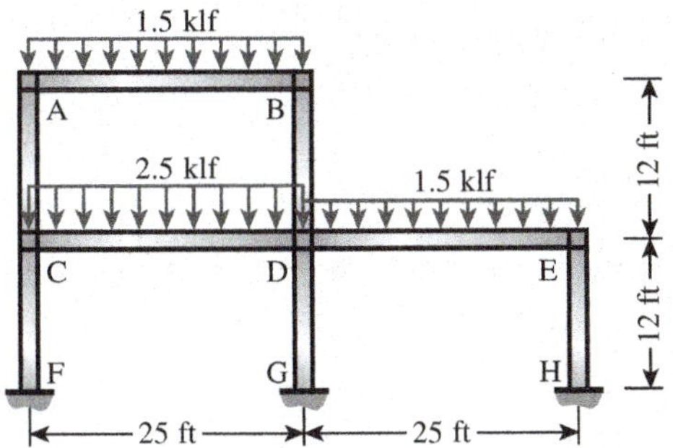

Figure VE15.2

VE15.3 Use the portal approximate method to generate the moment diagrams for the beams and columns of the frame shown in Figure VE15.3. Assume the interior column takes twice the shear that the exterior columns take.

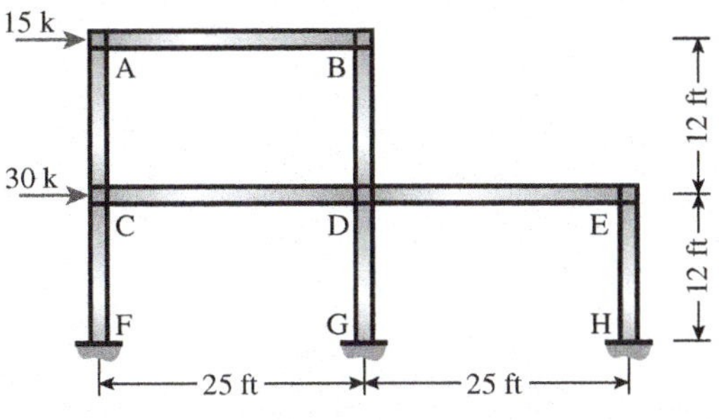

Figure VE15.3

VE15.4 Use the cantilever approximate method to generate the moment diagrams for the beams and columns of the frame shown in VE15.3.

15.10 Problems for Solution

Section 15.2

P15.1 Compute the forces in the diagonal members of the truss shown in the accompanying illustration if the diagonals are unable to carry compression. (Ans: N_{AI} = 113.14 k, N_{CK} = 28.28 k)

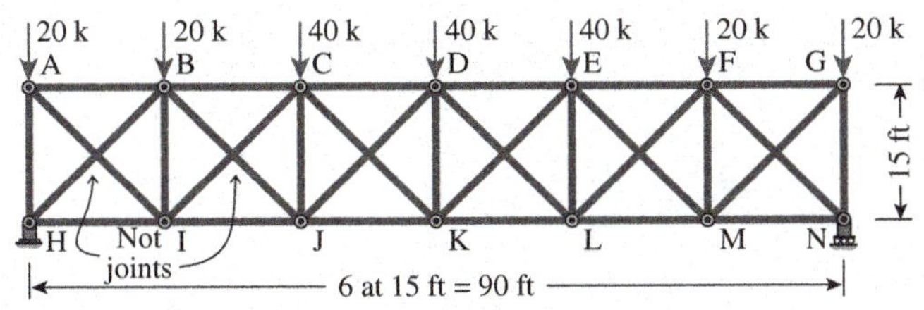

Problem 15.1

P15.2 Repeat Problem 15.1 if the diagonals that theoretically are in compression can resist half of the shear in each panel.

P15.3 Repeat Problem 15.1 if the diagonals that theoretically are in compression can resist one-third of the shear in each panel. (Ans: N_{AI} = 75.43 k, N_{CK} = 18.85 k)

P15.4 Compute the forces in the diagonal members of the cantilever truss shown if the diagonals are unable to carry any compression.

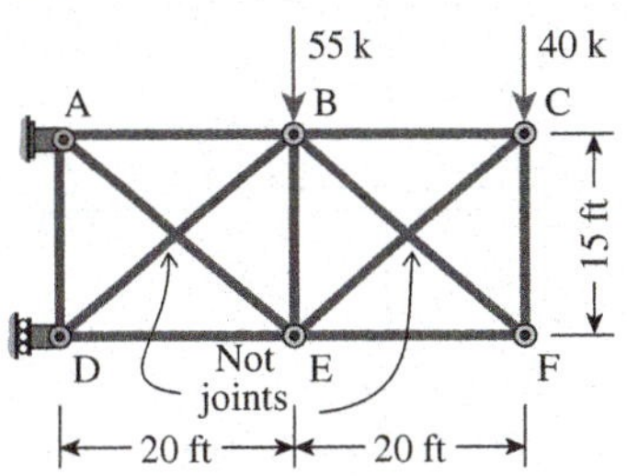

Problem 15.4

P15.5 Repeat Problem 15.4 if the diagonals that theoretically are in compression can resist only one-fourth of the shear in each panel. (Ans: $N_{BF} = 50.1$ k, $N_{BD} = -39.6$ k)

P15.6 Compute the forces in the members of the truss shown if the interior diagonals that would normally be in compression can resist no compression.

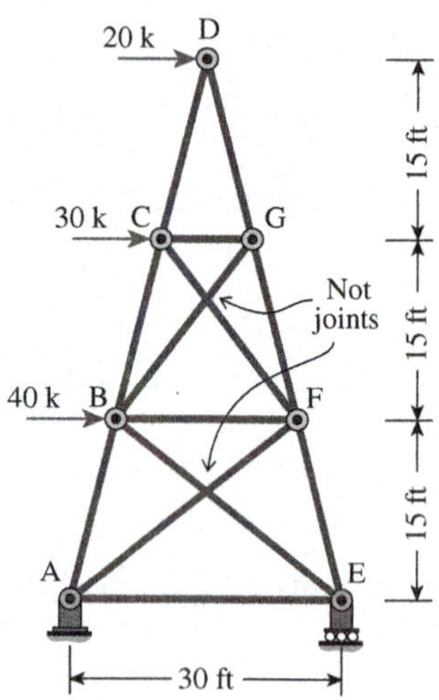

Problem 15.6

Section 15.3

P15.7 Prepare the shear and moment diagrams for the continuous beam shown using the ACI coefficients of Table 15.1. Assume the beam is constructed integrally with girders at all of its supports. Draw the moment diagrams as envelopes. (Ans: $M_{BA} = -23.2$ k·ft, $M_{DC} = -9.7$ k·ft)

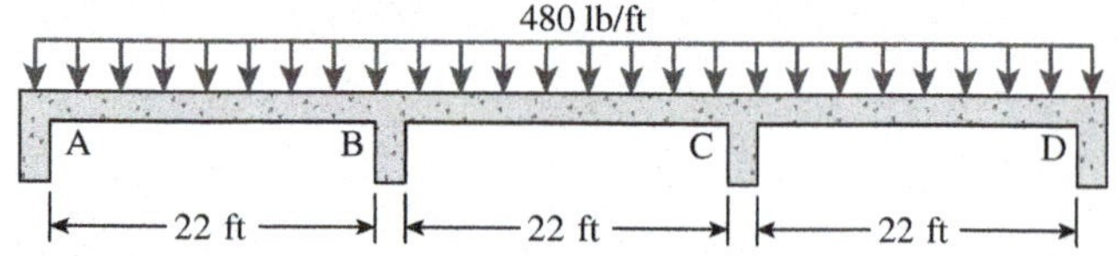

Problem 15.7

Section 15.4

P15.8 Use the vertical approximate analysis to generate the moment diagrams for all of the beams and also for column BDGJ. Assume the PIs are at one-tenth the length of the beams.

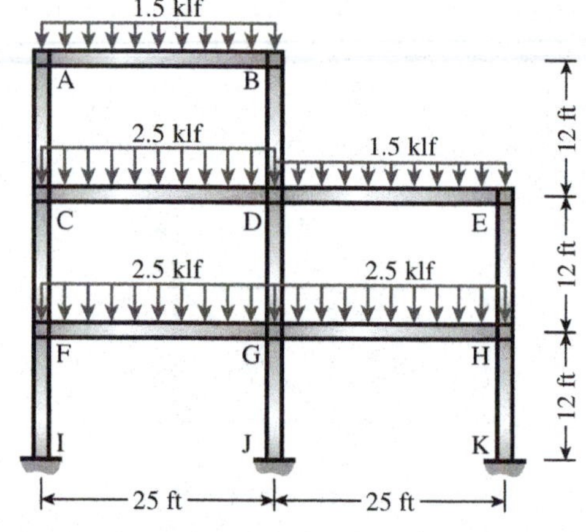

Problem 15.8

P15.9 Use the vertical approximate analysis to generate the moment diagrams for all of the beams and also for column BDGJ. Assume the PIs are at one-tenth the length of the beams. (Ans: $M_{BA} = -42.19$ k·ft, $M_{DG} = -70.31$ k·ft)

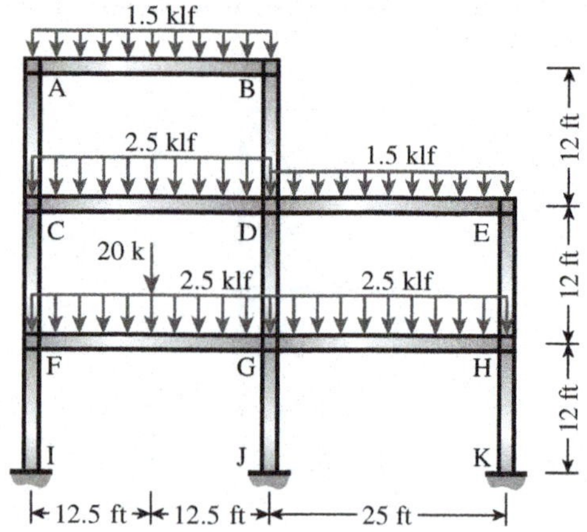

Problem 15.9

P15.10 Use the vertical approximate analysis to generate the moment diagrams for all of the beams and also for column BFJ. Assume the PIs for the outer beams are at one-tenth the length of the beams and that the PIs are at one-twentieth of the span length for the middle beams.

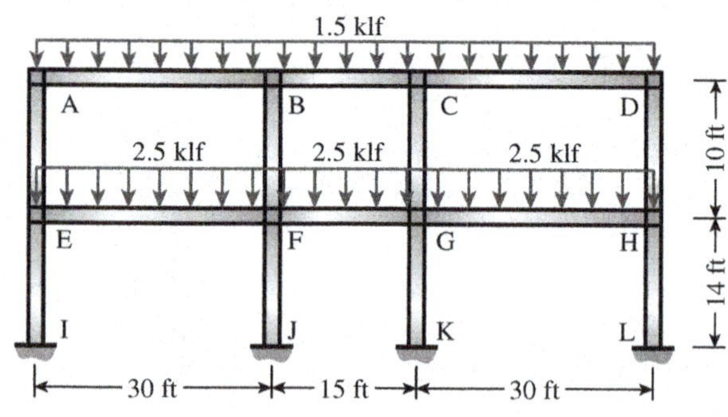

Problem 15.10

P15.11 Use the vertical approximate analysis to generate the moment diagrams for all of the beams and also for column BFJ. Assume the PIs for the outer beams are at one-tenth the length of the beams and that the PIs are at one-twentieth of the span length for the middle beams. (Ans: $M_{JF} = -170.63$ k·ft, $M_{BC} = -8.02$ k·ft)

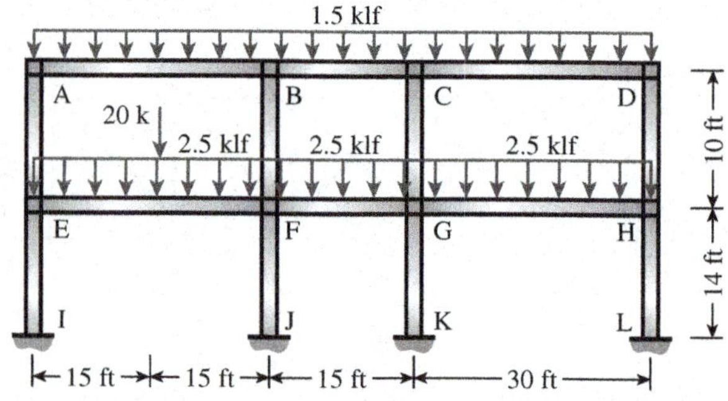

Problem 15.11

Section 15.5

P15.12 Analyze the portal shown if PIs are assumed to be located at column mid-heights and at mid-span of the beam. Sketch the moment diagrams.

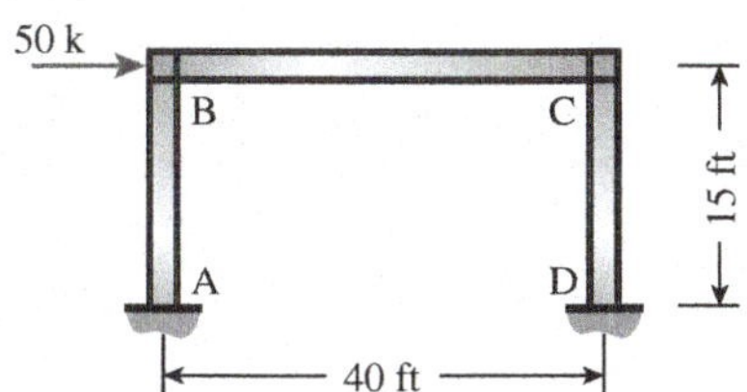

Problem 15.12

P15.13 Repeat Problem 15.12 if the column PIs are assumed to be located 5 ft above the column bases. (Ans: $M_{BA} = -250$ k·ft, $M_A = 125$ k·ft)

P15.14 Repeat Problem 15.12 if the column bases are assumed to be pinned and the PIs are located there.

Section 15.6

For Problems 15.15 through 15.18, compute the bending moments, shear forces, and axial forces for all of the members of the frames shown using the portal method. All joints are assumed to be rigid.

P15.15 (Ans: $M_{GF} = -210$ k·ft, $M_{FJ} = 280$ k·ft)

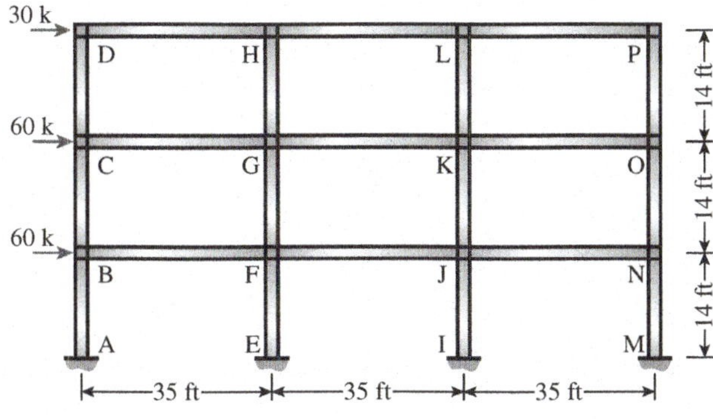

Problem 15.15

P15.16

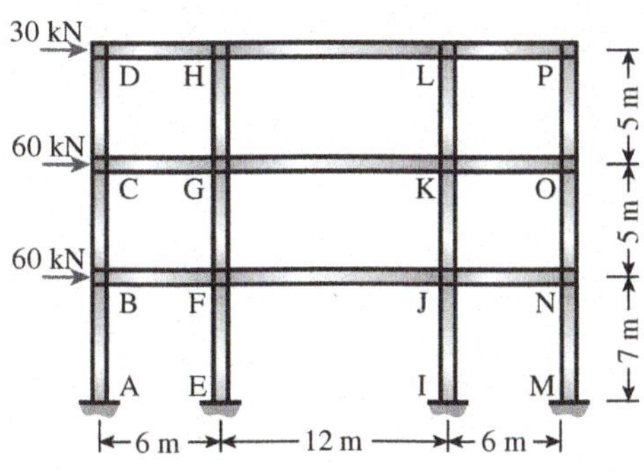

Problem 15.16

P15.17 (Ans: $M_{DG} = 300$ k·ft, $M_{DE} = -225$ k·ft)

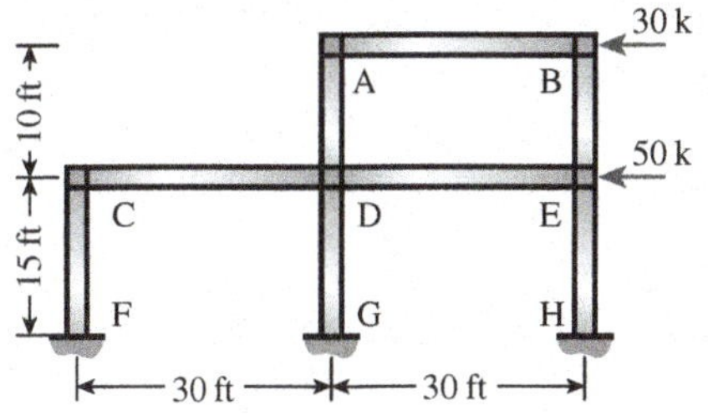

Problem 15.17

P15.18

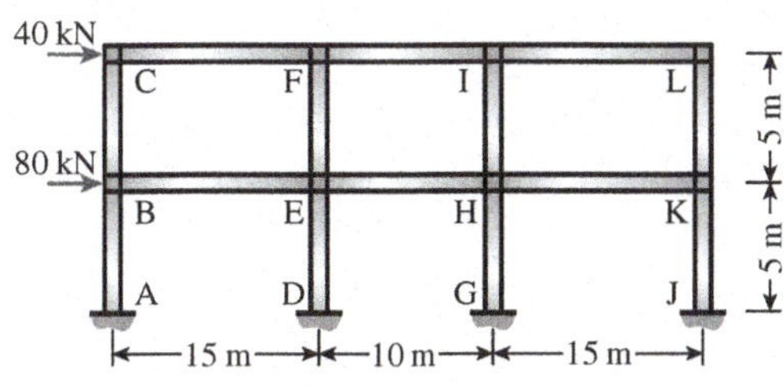

Problem 15.18

For Problems 15.19 through 15.22 compute the bending moments, shear forces, and axial forces for all of the members of the frames shown using the cantilever method. All joints are assumed to be rigid.

P15.19 Analyze the frame in Problem 15.15. (Ans: $M_{GF} = -220.5$ k·ft, $M_{FJ} = 336$ k·ft)

P15.20 Analyze the frame in Problem 15.16, assuming the bases are pinned.

P15.21 Analyze the frame in Problem 15.17. (Ans: $M_{DG} = 300$ k·ft, $M_{DE} = -150$ k·ft)

P15.22 Analyze the frame in Problem 15.18.

SAP2000 Problems

For Problems 15.23 through 15.30, define the cross-sectional area as 100,000 ft^2, set shear area equal to zero, and turn off self-weight.

P15.23 Use SAP2000 to model the frame in Problem 15.8. All members have the same modulus of elasticity and cross-sectional area. The beams have a moment of inertia twice that of the columns. Plot the moment diagram.

P15.24 Use SAP2000 to model the frame in Problem 15.8. All members have the same modulus of elasticity and cross-sectional area. The beams have a moment of inertia that is the same as that of the columns. Plot the moment diagram.

P15.25 Use SAP2000 to model the frame in Problem 15.11. All members have the same modulus of elasticity and cross-sectional area. The beams have a moment of inertia twice that of the columns. Plot the moment diagram.

P15.26 Use SAP2000 to model the frame in Problem 15.11. All members have the same modulus of elasticity and cross-sectional area. The beams have a moment of inertia that is the same as that of the columns. Plot the moment diagram.

P15.27 Use SAP2000 to model the frame in Problem 15.15. All members have the same modulus of elasticity and cross-sectional area. The beams have a moment of inertia twice that of the columns. Plot the moment diagram.

P15.28 Use SAP2000 to model the frame in Problem 15.15. All members have the same modulus of elasticity and cross-sectional area. The beams have a moment of inertia that is the same as that of the columns. Plot the moment diagram.

P15.29 Use SAP2000 to model the frame in Problem 15.17. All members have the same modulus of elasticity and cross-sectional area. The beams have a moment of inertia twice that of the columns. Plot the moment diagram.

P15.30 Use SAP2000 to model the frame in Problem 15.17. All members have the same modulus of elasticity and cross-sectional area. The beams have a moment of inertia that is the same as that of the columns. Plot the moment diagram.

Section 15.8

For Problems 15.31 and 15.32, compute the bending moments, shearing forces, and axial forces for all of the members of the rigid-jointed Vierendeel trusses using the portal method.

P15.31 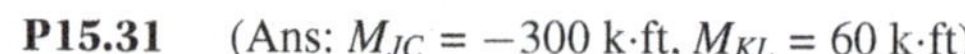
(Ans: $M_{JC} = -300$ k·ft, $M_{KL} = 60$ k·ft)

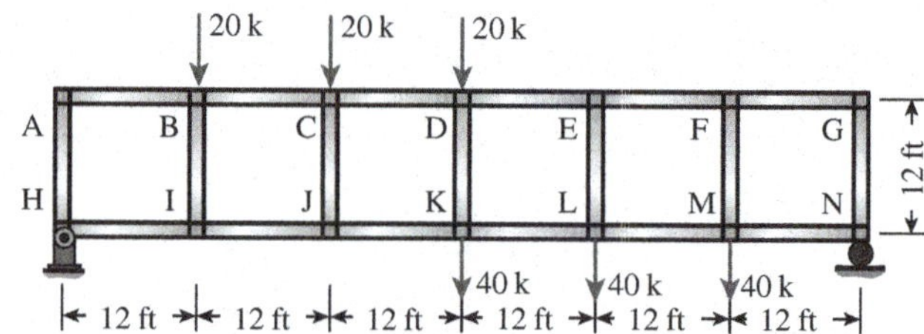

Problem 15.31

P15.32 Due to symmetry, only the left half needs to be analyzed.

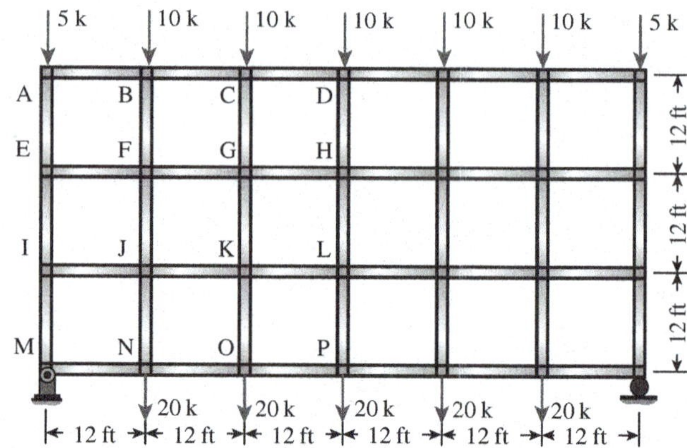

Problem 15.32

INFLUENCE LINES

PART 3

16 Influence Lines for Determinate Structures

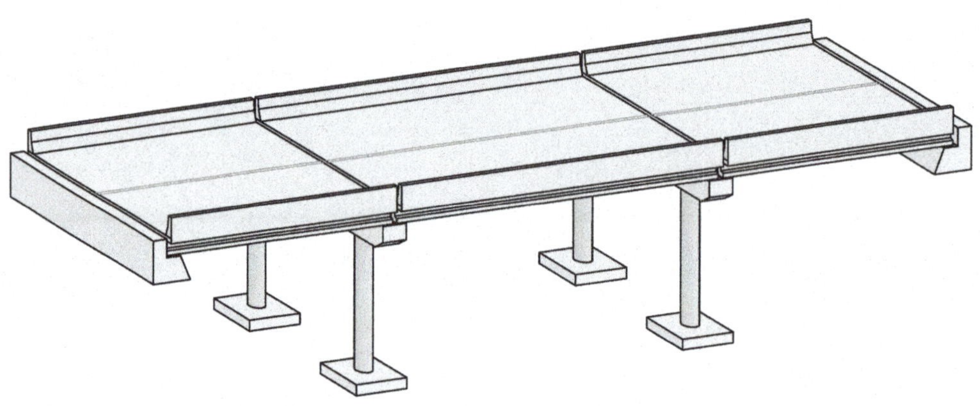

16.1 Introduction

DID YOU KNOW?
Each member of a structure must be designed for the most severe conditions that can possibly develop in that member. The designer must place the live loads at the positions where they will produce these conditions. **How does one know where to place the loads to produce this result?**

Structures supporting groups of loads fixed in one position have been discussed in the previous chapters. Regardless of whether beams, frames, or trusses were being considered and whether the responses sought were shears, reactions, or member forces, the loads were stationary. In practice, however, the engineer rarely deals with structures supporting only fixed loads. Nearly all structures are subject to loads moving back and forth across their spans. Perhaps bridges with their vehicular traffic are the most noticeable examples, but industrial buildings with traveling cranes, office buildings with furniture and human loads, and frames supporting conveyor belts fall in the same category.

The critical positions for placing live loads will not be the same for every member. For example, the maximum force in one member of a bridge truss may occur when a line of trucks extends from end to end of the bridge. The maximum force in some other member, however, may occur when the trucks extend only from that member to one end of the bridge. The maximum forces in certain beams and columns of a building occur when the live loads are concentrated in certain portions of the building. The maximum forces in other beams and columns occur when the live loads are placed elsewhere.

On some occasions, you can determine where to place the loads to give the most critical forces by inspection. On many other occasions, however, you will need to use certain criteria or diagrams to find the locations. The most useful of these devices is the *influence line* (IL).

16.2 The Influence Line Defined

The influence line, which was first used by Professor E. Winkler of Berlin in 1867, shows graphically how the movement of a unit load across a structure influences some response of the structure.[1] The responses that are typically represented include reactions, shears, moments, and deflections.

Influence lines are primarily used to determine where to place live loads to cause maximum forces. They also may be used to compute those forces. The procedure for drawing the diagrams is simply the plotting of values of the response under study as ordinates for various positions of the unit load along the span and then connecting those ordinates. You should mentally picture the load moving across the span and try to imagine what is happening to the function in question during the movement. This will be illustrated in the following section. The study of influence lines can immeasurably increase your knowledge of what happens to a structure under different loading conditions.

Studying the following sections should clearly fix in your mind what an influence line is. The actual mechanics of developing the diagrams are elementary once the definition is completely understood. No new fundamentals are introduced here; rather, a method of recording information in a convenient and useful form is given.

DEFINITION

An influence line may be defined as a diagram whose ordinates show the magnitude and character of some response of a structure as a unit load moves across it. Each ordinate of the diagram gives the value of the response when the load is at that ordinate.

16.3 Influence Lines for Simple Beam Reactions

Influence lines for the reactions of a simple beam are given in Figure 16.1. First consider the variation of the left-hand reaction, R_{Ay}, as a unit load moves from left to right across the beam. When the load is directly over the left support, $R_{Ay} = 1$. When the load is 2 ft to the right of the left support, $R_{Ay} = 18/20$ or 0.9. When the load is 4 ft to the right, $R_{Ay} = 16/20$ or 0.8, and so on.

Values of R_{Ay} are shown at 2-ft intervals as the unit load moves across the span. These values lie in a straight line because they change uniformly for equal intervals of the load. For every 2-ft interval, the ordinate changes 0.1. The values of R_{By} are plotted in a similar fashion in the same figure. For each position of the unit load, the sum of the ordinates of the two diagrams at any point equals (and for equilibrium certainly must equal) the unit load.

COMMON CONFUSION

At first glance, influence lines may look startling similar to the shear and moment diagrams that were previously introduced. They are not the same. An explanation of the difference will be given once the reader has been exposed to several influence lines.

16.4 Influence Lines for Simple Beam Shear Forces

Influence lines are plotted in Figure 16.2 for the shearing force at two sections in a simple beam. The beam sign convention for shear that was introduced in Chapter 6 and used for generating shear diagrams will continue to be used (refer to Figure 16.3). Another way of interpreting this:

> *A positive shearing force occurs when the sum of the transverse forces to the left of a section is up or when the sum of the forces to the right of the section is down.*

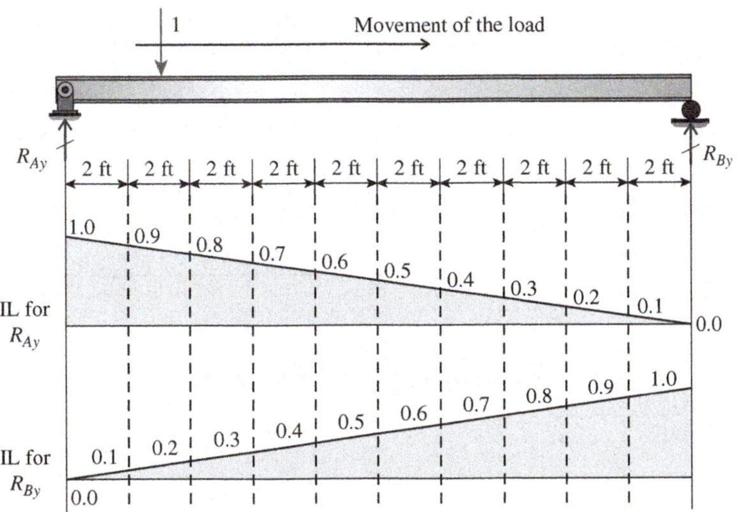

QUICK CHECK

Statics says that when the unit load is at mid-span then the value at both reactions will be 0.5. Notice the ordinate at the location of the mid-span on both influence lines (IL) to confirm this is true.

FIGURE 16.1 Influence lines for the reactions in a simple beam.

[1] J. S. Kinney, *Indeterminate Structural Analysis* (Reading, Mass.: Addison-Wesley, 1957), Chapter 1.

QUICK NOTE

Observe that the slope of the shearing force influence line to the left of the section in question is equal to the slope of the influence line to the right of the section. In Figure 16.2, for instance, for the influence line at section 1–1, the slope to the left is 0.2/4 = 0.05, while the slope to the right is 0.8/16 = 0.05. This information is very useful in drawing other shear influence lines.

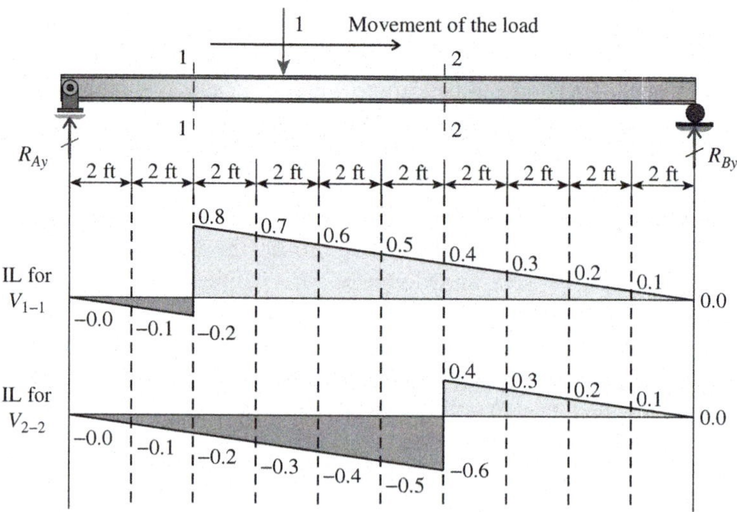

FIGURE 16.2 Two shearing force influence lines for a simple beam.

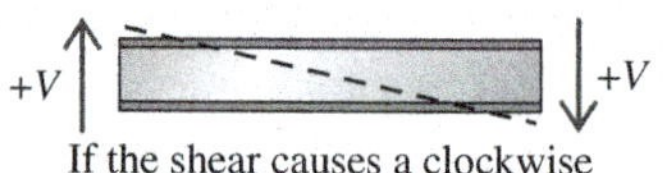

+V
+V
If the shear causes a clockwise rotation of the member it is positive

FIGURE 16.3 Shear force sign convention.

Placing the unit load over the left support causes no shearing force at either of the two sections. Moving the unit load 2 ft to the right of the left support results in $R_{Ay} = 0.9$. The sum of the forces to the left of section 1–1 is $0.9 - 1.0 = -0.1$ downward; thus the shearing force is 0.1. When the load is 4 ft to the right of the left support and an infinitesimal distance to the left of section 1–1, the shearing force to the left is $0.8 - 1.0 = -0.2$. If the load is moved a very slight distance to the right of section 1–1, the sum of the forces to the left of the section becomes 0.8 upward; thus the shearing force is +0.8. Continuing to move the load across the span toward the right support results in decreasing values of the shear at section 1–1. These values are plotted for 2-ft intervals of the unit load. The influence line for shear at section 2–2 is developed in the same manner.

16.5 Influence Lines for Simple Beam Moments

DID YOU NOTICE?

The unit load that is applied to the beam is dimensionless. As such, the shear influence line is dimensionless, and the moment influence line has the strange looking units of ft. If the student is more comfortable doing so, he may use the following units for shear/reaction and moments as k/k and (k·ft)/k, respectively. Either way, these units will make more sense once the method on how to use influence lines is introduced.

Influence lines for bending moments are plotted in Figure 16.4 at the same sections of the beam used in Figure 16.2 for the shearing force influence lines. To review, a positive moment causes tension in the bottom fibers of a beam. It occurs at a particular section when the sum of the external moments of all the forces to the left is clockwise or when the sum to the right is counterclockwise (see Figure 16.5). Moments are taken at each of the sections for 2-ft intervals of the unit load.

To illustrate how the calculations were made for the moment influence lines of this figure, the following illustrations are presented. If the unit load is 2 ft from the left end of the beam, R_{Ay} will be +0.9 and R_{By} is +0.1. The moment at section 1–1 may be determined by taking a

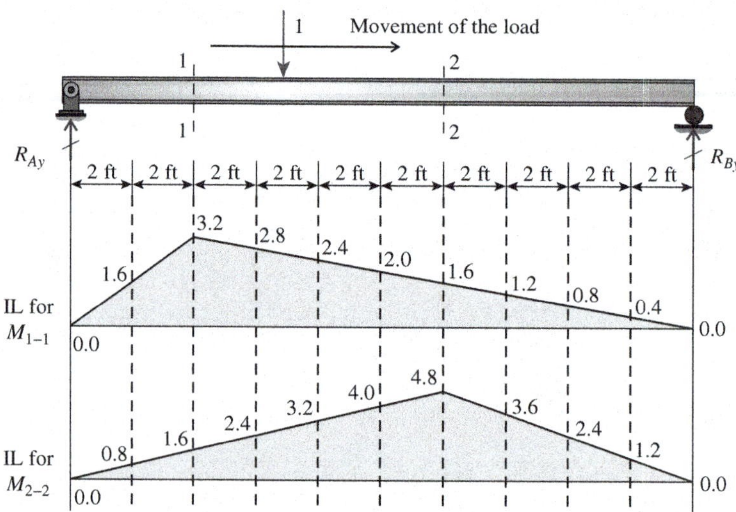

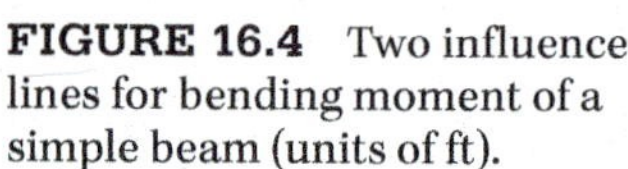

FIGURE 16.4 Two influence lines for bending moment of a simple beam (units of ft).

cut at section 1–1 and summing moments on either the left section or the right section. These calculations would be

$$M_{\text{left section}} = (0.9)(4\text{ ft}) - (1.0)(2\text{ ft}) = 1.6\text{ ft}$$
$$M_{\text{right section}} = (0.1)(16\text{ ft}) = 1.6\text{ ft}$$

If the load is at section 1–1, R_{Ay} will be +0.8 and $R_{By} = +0.2$. The moment at section 1–1 can be calculated using either the left or the right section as shown.

$$M_{\text{left section}} = (0.8)(4\text{ ft}) = 3.2\text{ ft}$$
$$M_{\text{right section}} = (0.2)(16\text{ ft}) = 3.2\text{ ft}$$

This process is continued by shifting the point load to the right in increments of 2 ft and then solving for the internal moment in section 1–1. The same is done for finding the moment in section 2–2.

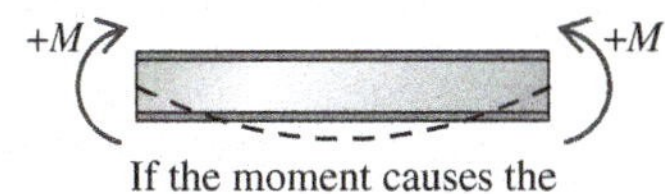

If the moment causes the member to "smile" then it is positive

FIGURE 16.5 Bending moment sign convention.

16.6 Influence Lines Using Quantitative Approach

Now that the reader has seen several influence lines, they may begin to appreciate the confusion that often arises regarding influence lines versus shear and moment diagrams.

Shear/Moment Diagrams Show the variation of shearing force or bending moment across an entire structure for loads fixed in one position.

Influence Lines Show the variation of a structural response (e.g., shear/moment) at one location in the structure caused by the movement of a unit load from one end of the structure to the other.

DID YOU KNOW?
Influence lines for statically determinate structures consist of a set of straight lines.

The quantitative approach for the generation of influence lines for determinate structures can require a lot of computations, as demonstrated in Figures 16.1,16.2, and 16.4. The unit load was placed at 11 different locations—moving from left to right—and the response quantities at the locations of interest (e.g., R_{Ay}, V_{1-1}, M_{2-2}, etc.) were then calculated and plotted. The process can be sped up by making a useful observation. Take a look at the IL for R_{Ay}. If the analyst knew the value at the left end of the beam and the right end of the beam, they would have sufficient information to construct the full IL. The intermediate nine points simply provide redundant information and are not necessary. Knowing ahead of time that the unit load only needed to be placed at the two locations would have saved tremendous time. The two locations for finding the IL for R_{Ay} would not have been sufficient to find the IL for V_{1-1}. A procedure for the generation of ILs is presented here with some guidance on selecting appropriate locations for the unit load.

Procedure

1. Place a dimensionless unit load at key locations along the beam where key locations are identified as
   - Beam supports
   - Beam ends
   - Moment or shear releases
   - Any *point of interest* (POI). Examples of POIs would be sections 1–1 and 2–2 shown in the previous demonstration. POIs are locations where the analyst wants to know a structural response quantity.
2. Calculate the *response quantity of interest* (e.g., *V*, *M*, *R*) and plot the value at the location where the unit load is placed. Basic statics is all that is required for this.
3. Connect the dots with straight lines if the structure is determinate. The dots will be connected with curved lines if the structure is indeterminate.

The following example demonstrates the quantitative approach. Notice how the analyses are only done at key locations.

QUICK NOTE
The POI for one IL may not necessarily be a POI for another IL. For example, the left support is the POI for the R_{Ay} IL, but section 1–1 is not. However, section 1–1 is the POI for the V_{1-1} IL.

BE AWARE
Notice the jump in the IL for shear, as seen in Figure 16.2, that occurs at the POI. To be able to capture this jump in shear ILs, you need to place the unit load just to the left of the POI and find the response quantity of interest. Then you need to place the unit load just to the right of the POI and again find the response quantity of interest.

EXAMPLE 16.1

For the beam with the double overhang shown, find the ILs for the response quantities using the quantitative approach: R_{By}, R_{Cy}, V_{1-1}, M_{1-1}, and V_{2-2}.

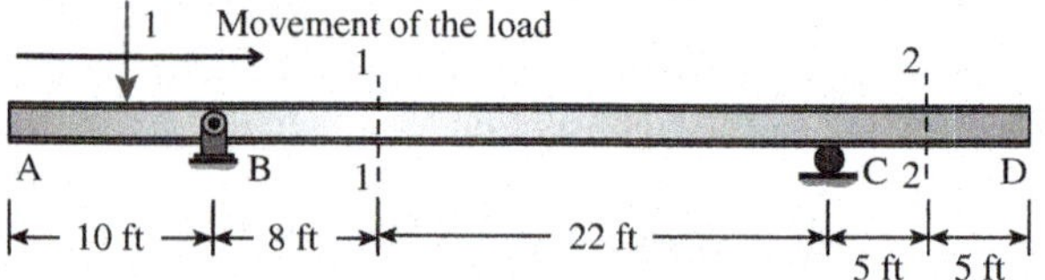

SOLUTION STRATEGY

Follow the procedure outlined previously. Note that the key locations for one IL may not necessarily be required for another IL. Before doing a single calculation, decide what statics process will be used to find the necessary quantities. Then follow this process no matter where the unit load is placed.

SOLUTION

Identify the key locations for each response quantity of interest. To assist in the learning process, the relevant key locations are explicitly listed here for each response quantity.

- R_{By} and R_{Cy}: Points A, B, C, and D
- V_{1-1} and M_{1-1}: Points A, B, C, and D and section 1–1
- V_{2-2}: Points A, B, C, and D and section 2–2

Outline the statics that will be used to solve for the response quantities of interest.

1. Sum moments about B to find R_{Cy}.
2. Sum forces in the y-direction to find R_{By}.
3. Cut the structure at section 1–1 and draw a FBD of the left section.
4. Sum forces in the y-direction to find V_{1-1}.
5. Sum moments about 1–1 to find M_{1-1}.
6. Cut the structure at section 2–2 and draw a FBD of the right section.
7. Sum the forces in the y-direction to find V_{2-2}.

All of the relevant FBDs are now shown in sequence of the calculations. Once the responses are found, plot them on the ILs.

QUICK NOTE

The outline of the statics used may appear to be a bit *over-the-top* and unnecessary. Generally, it doesn't need to be written down, but the analysts do want to get this straight in their minds before they start crunching numbers. Experience has shown that when the load starts moving locations, students often try to reinvent the statics halfway through the problem. Simply **identify the statics for the problem and stick with it** throughout the duration of the problem.

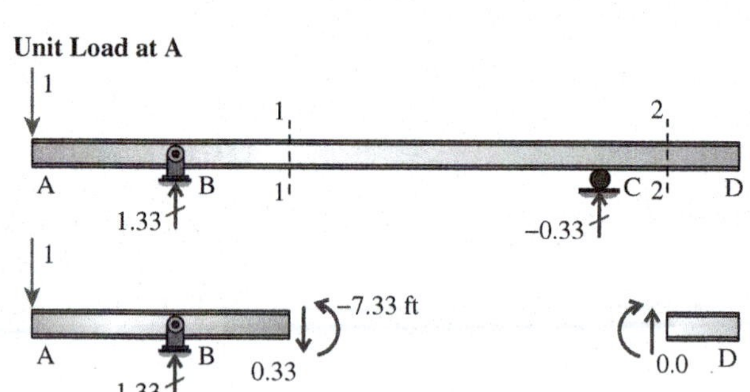

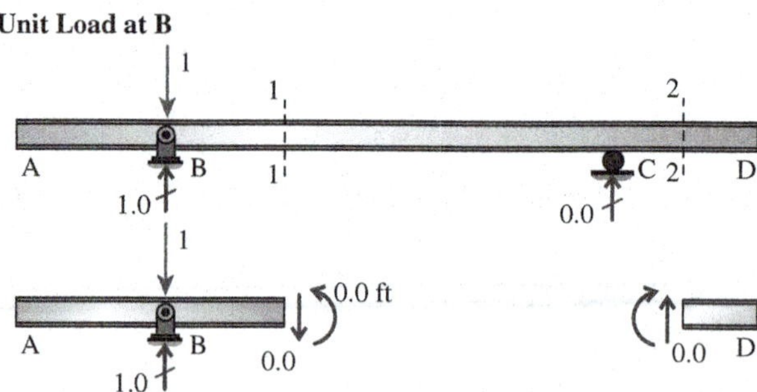

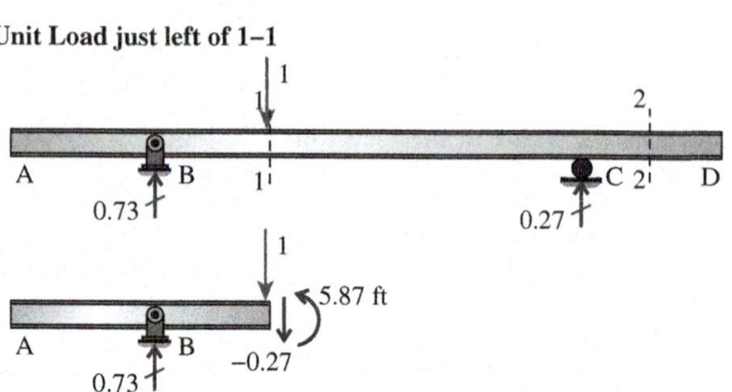

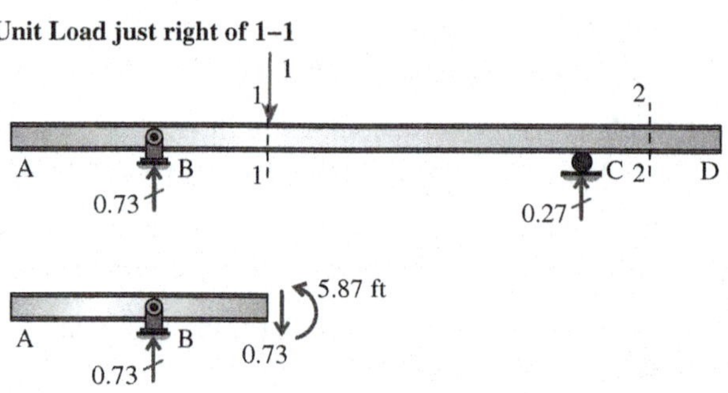

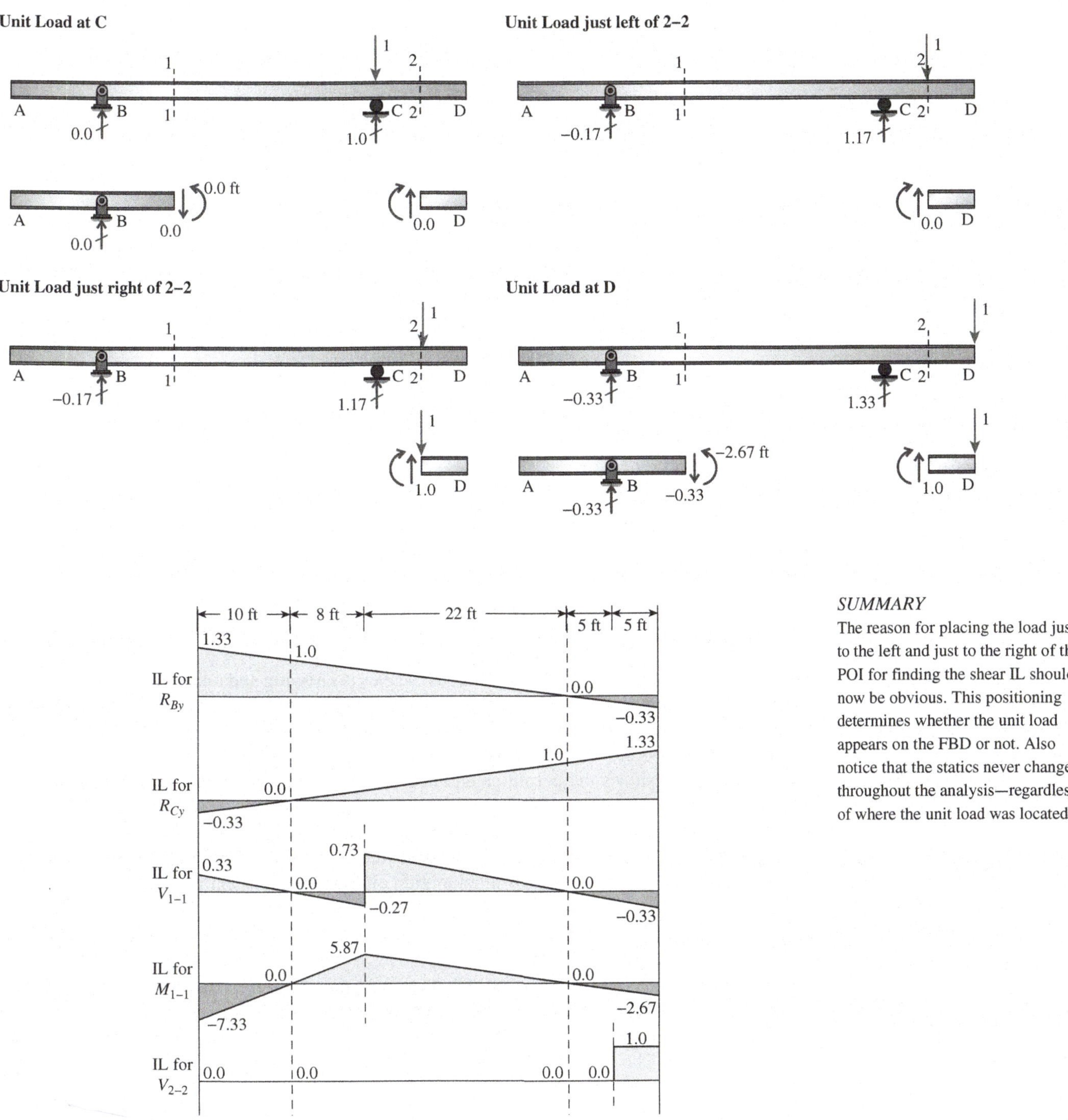

SUMMARY

The reason for placing the load just to the left and just to the right of the POI for finding the shear IL should now be obvious. This positioning determines whether the unit load appears on the FBD or not. Also notice that the statics never changed throughout the analysis—regardless of where the unit load was located.

16.7 Qualitative Influence Lines

Many structural analysis students initially have a great deal of trouble in preparing influence lines despite the simplicity of the mathematical calculations. The reason for this trouble is often rather perplexing to their professors. Perhaps the trouble occurs due to a lack of fully understanding the definition of influence lines. There is a very simple procedure that may help the students in their understanding and preparation of the diagrams. It involves the use of figures called qualitative influence lines. These diagrams enable the student to instantly draw the correct shape of influence lines without the need of any computations whatsoever.

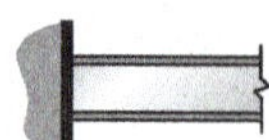

Fixed Support: Doesn't allow translation or rotation of the beam in any axis.

Pin Support: Doesn't allow translation of the beam in any axis but does permit rotation.

Roller Support: Doesn't allow translation in the vertical axis. Rotation is free to occur.

FIGURE 16.6 Permitted motion types for typical support types.

QUICK NOTE

Some uses of ILs only require you to know the shape of the IL and not the values. Qualitative ILs can be very useful for this purpose. However, occasions still arise where the numbers are required. The qualitative approach may be extended a little to provide this information. Even so, this approach is generally more efficient than the quantitative approach. **The presentation in this text will introduce the numerical extension to the qualitative approach.**

The influence lines drawn in the previous sections, for which numerical values were computed, are referred to as *quantitative influence lines*. It is possible, however, to make sketches of these diagrams with sufficient accuracy for many practical purposes without computing any numerical values. These latter diagrams are referred to as *qualitative influence lines*.

Qualitative influence lines are based on a principle introduced by the German Professor Heinrich Müller-Breslau.

> *The deflected shape of a structure represents, to some scale, the influence line for a structural response such as reaction, shear, or moment if the function in question is allowed to act through a unit displacement or rotation.*

In other words, *the structure draws its own influence line* when the proper displacement is applied.

Once learned, this method for generating influence lines for beams can be a very rapid process. The key to a successful IL generation is to learn how to sketch appropriate deflected shapes of the beams when a unit displacement or rotation is imposed upon it.

Sketching Deflected Shapes

The foundation to sketching accurate deflected shapes is knowing and understanding what movement the beam support conditions will and won't allow. Furthermore, one needs to understand what kind of relative displacement and/or rotation is permitted on either side of a moment release or a shear release. Figure 16.6 presents the permitted and non-permitted motion for typical support types. Figure 16.7 shows the permitted and non-permitted motion for internal moment releases and shear releases.

While the descriptions of allowable motions are correct, students have historically struggled in applying this guidance for sketching deflected shapes. As such, Figure 16.8 is presented here to demonstrate more readily the permitted motion of moment and shear releases.

Moment Release: Permits a relative rotation to occur on either side of the moment release (hinge). Does not permit relative displacement to occur.

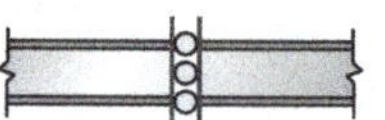

Shear Release: Permits a relative displacement to occur on either side of the shear release (slider). Does not permit relative rotation to occur.

FIGURE 16.7 Permitted motion types for typical internal releases.

ILs for Reactions

The IL for a selected reaction is generated by **imposing a positive unit displacement or rotation at the support**, which causes the reaction of interest. Sketch the deflected shape of the beam and honor all other support conditions. If the beam is determinate, all lines of the deflected shape should be straight. The resulting deflected shape **IS** the influence line for the reaction.

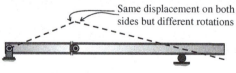

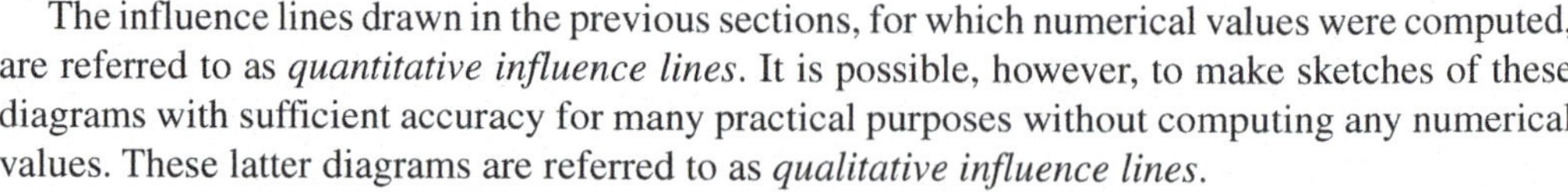

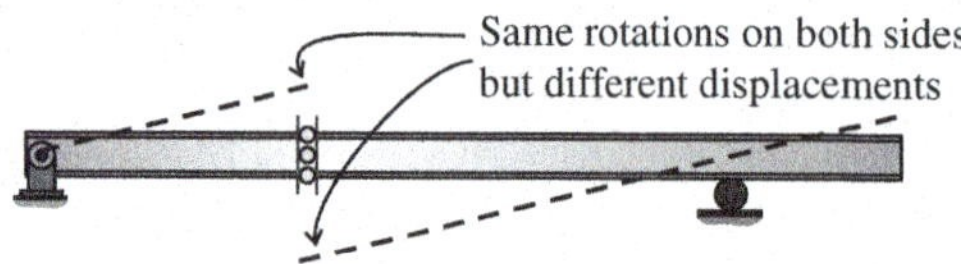

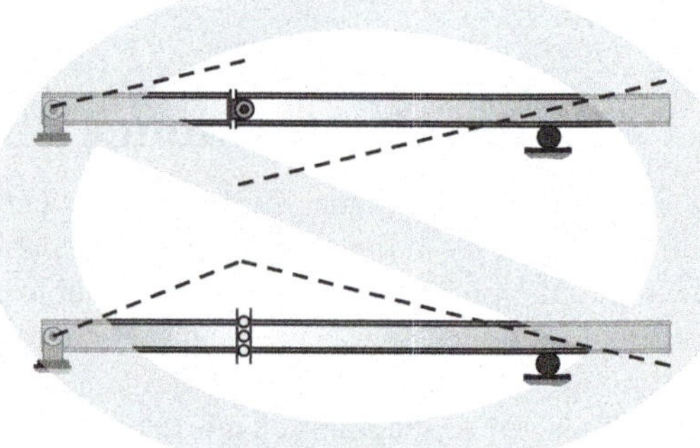

FIGURE 16.8 Correct and incorrect deflected shapes for beams with internal releases.

EXAMPLE 16.2

For the given beam, find the ILs for the reactions at A and B.

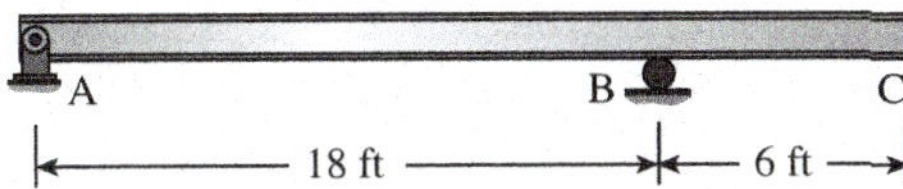

SOLUTION STRATEGY

Impose positive unit displacements at the appropriate support. Ensure that the other support condition is honored (see Figure 16.6). Once the deflected shape is sketched, use the principle of similar triangles to identify appropriate values.

SOLUTION

Impose a positive unit displacement at point A to find the IL for R_{Ay}. Because point B is a roller, it cannot translate up and down. The deflected shape (IL) is shown.
The value of the IL at point C is found using similar triangles. Thus,

$$\frac{-1}{18\text{ ft}} = \frac{C}{6\text{ ft}}$$

$$C = \underline{\underline{-0.33}}$$

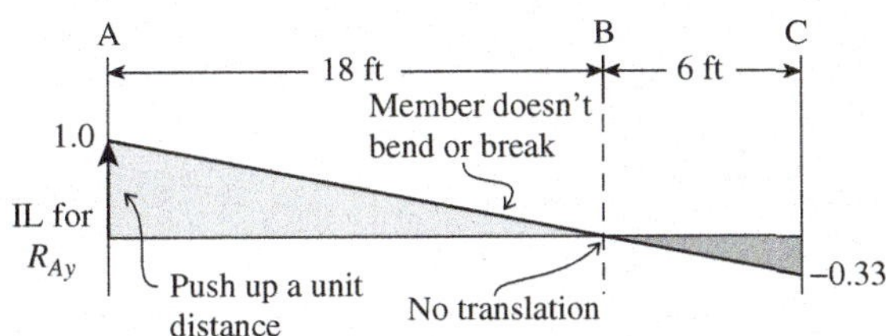

To find the IL for R_{By}, a positive unit displacement must be imposed at point B. Since point A is a pin support, it cannot translate, but it is free to rotate at that location.

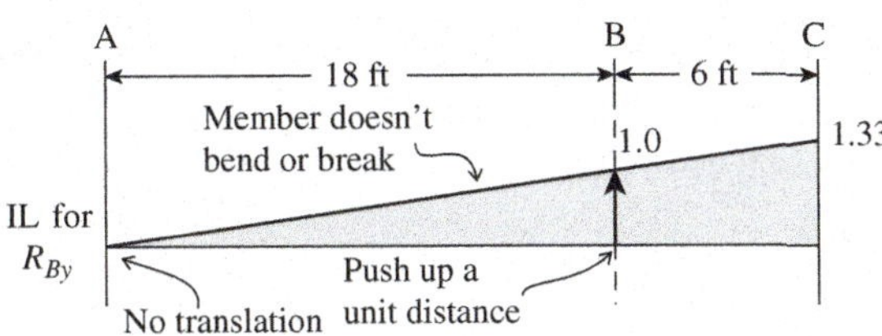

QUICK NOTE

Make sure that you are comfortable in using similar triangles. This concept is used extensively in finding values on ILs that are generated using Müller Breslau's approach.

SIMILAR TRIANGLES

The value at C is found by

$$\frac{1.0}{18\text{ ft}} = \frac{C}{18\text{ ft} + 6\text{ ft}}$$

$$C = \underline{\underline{1.33}}$$

ILs for Internal Shear

The IL for the internal shear is found by imposing a positive unit internal displacement at the location where the shear IL is desired. At first this may seem confusing, but the process is quite simple.

Procedure

1. Insert a shear release at the desired shear location.
2. Sketch a positive shear couple next to the shear release. Sketching the shear couple will assist in determining the direction of displacement (see sidebar).
3. Displace the beam on either side of the shear release such that a unit distance exists between the two beam ends.
4. Use similar triangles to calculate the other values on the deflected shape. **Remember that the rotation of the beam on each side of the shear release must be the same.**

POSITIVE SHEAR COUPLE

EXAMPLE 16.3

For the given beam, find the IL for the internal shear at D.

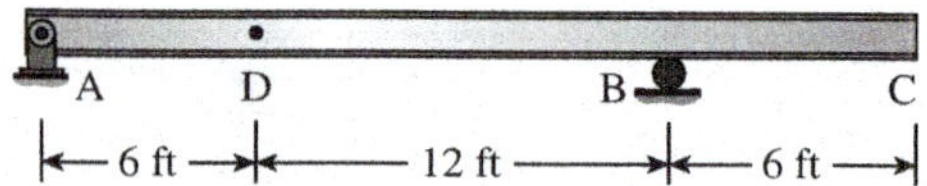

SOLUTION STRATEGY

Before imposing the unit displacement at D, make sure to place the shear release at D and sketch the positive shear couple. Also make sure that the support conditions are identified and understood before any other sketching is done.

SOLUTION

QUICK NOTE

This is a great time to identify the support conditions and what they mean. For example, points A and B cannot translate, but they can rotate because they are pin/roller.

Draw a set of axes for the beam; then sketch the release along with the shear couple as shown.

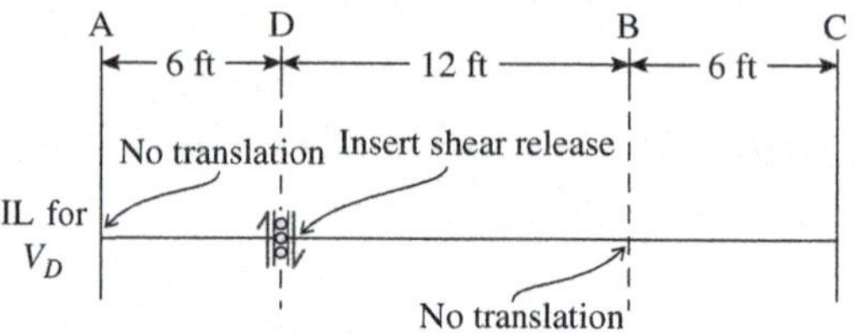

CALCULATIONS FOR IL INTENSITY

$$\frac{6 \text{ ft}}{6 \text{ ft} + 12 \text{ ft}} = 0.33$$

$$\frac{12 \text{ ft}}{6 \text{ ft} + 12 \text{ ft}} = 0.67$$

Push the end of the beams on either side of the shear release in the direction of the arrows. Be sure to maintain equal rotations on both sides.

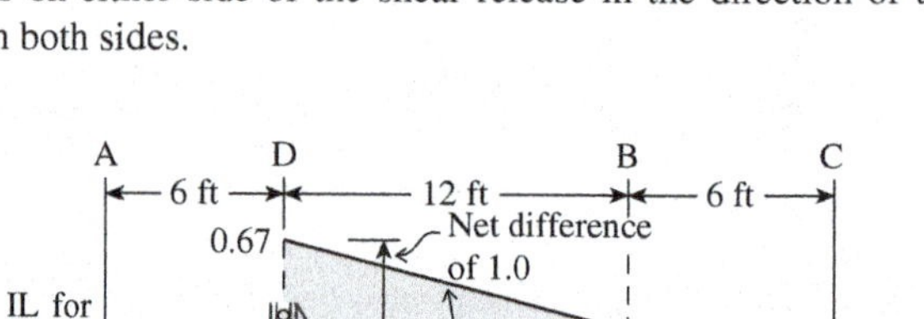

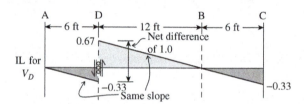

The relative magnitudes of the displacement on either side of the shear release are found by proportional distances on each beam section (6 ft and 12 ft). The displacement at C is found using similar triangles.

ILs for Internal Moment

POSITIVE MOMENT COUPLE

The IL for the internal moment is found by imposing a positive unit internal rotation at the location where the moment IL is desired. Similar as to the case for shear, an example may be needed for this statement to make full sense.

Procedure

1. Insert a moment release at the desired moment location.
2. Sketch a positive moment couple next to the moment release. Sketching the moment couple will assist in determining the direction of rotation and displacement (see sidebar).
3. Rotate the beam on either side of the moment release in the direction of the arrows such that a unit rotation exists between the two beam ends. The rotation of the members may cause the moment release to translate up or down. The member on each side of the moment release should rotate in the direction of the arrows—if it can rotate at all.
4. Use similar triangles and basic geometry to calculate the other values on the deflected shape.

EXAMPLE 16.4

For the beam given in Example 16.3, find the IL for the internal moment at D.

SOLUTION STRATEGY

Before imposing the unit rotation at D, make sure to place the moment release at D and sketch the positive moment couple. Also make sure that the support conditions are identified and understood before any other sketching is done.

SOLUTION

Draw a set of axes for the beam; then sketch the release along with the moment couple as shown.

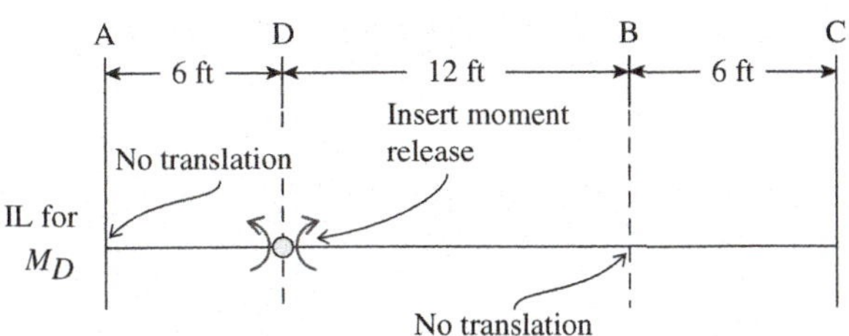

QUICK NOTE

As was the case in Example 16.3, points A and B cannot translate, but they can rotate.

Segment AD must rotate counterclockwise, and segment DBC must rotate clockwise. Out of necessity, this will require the moment release at D to translate upward.

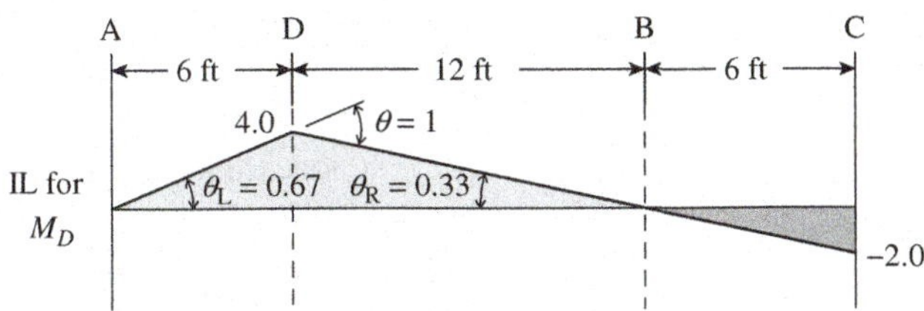

QUICK NOTE

The relative rotation between the adjacent segments AD and DBC must be 1.0. It is useful to recall the geometric property that says *the exterior angle is equal to the sum of the two interior opposite angles.* Thus, $\theta_L + \theta_R = 1.0$. The values of the interior angles are inversely proportional to the length of the segments.

The amplitudes of the displacements at D and at C are found using trigonometry and the small angle theory. Calculations are

$$D = (6\text{ ft})(\theta_L) = (6\text{ ft})(0.67) = \underline{\underline{4\text{ ft}}}$$

$$C = (6\text{ ft})(\theta_R) = (6\text{ ft})(-0.33) = \underline{\underline{-2\text{ ft}}}$$

EXAMPLE 16.5

For the given beam, draw the ILs for R_{Dy}, R_{By}, M_D, and V_{1-1}.

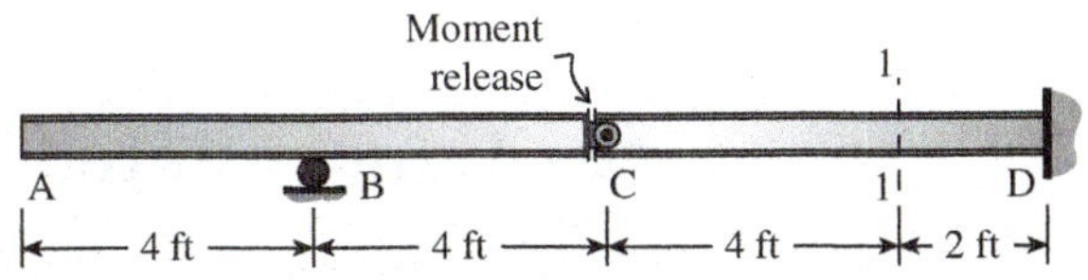

QUICK NOTE

Rather than simply presenting the resulting ILs using the qualitative approach, some effort is given here to provide an explanation of the thought process. A finalized IL does not need to contain all of the annotations shown in this example.

SOLUTION STRATEGY

Identify first the support conditions and make sure that a proper interpretation is rendered of these. Once this is done, the requisite displacement/rotation may be imposed.

SOLUTION

Generate the IL for R_{Dy}: The boundary conditions indicate that point B cannot translate and joint D cannot rotate. This means that member CD cannot rotate. With this fact identified, impose a positive unit displacement at D and then sketch member CD as being horizontal. Connect points B and C and extend to A. Use similar triangles to find the value at A.

COMMON MISTAKE

It is very easy to sketch a deflected shape that does not honor the support conditions. When an IL is completed, it is wise to ask the following questions:

- Have I rotated or translated the beam at a location where there is a support that shouldn't move?
- Have I allowed there to be different slopes on each side of a shear release?
- Have I allowed there to be different displacements on either side of a moment release?
- Have I broken or bent a member?

Generate the IL for R_{By}: Observe that joint D cannot rotate or translate. Therefore, member CD must remain horizontal and not undergo any translation. Now impose the unit displacement at joint B. Connect C with B; then extend to A. Use similar triangles to find the value at A.

Generate the IL for M_D: Observe that joints B and D cannot translate. Insert the moment release at joint D and observe that the member that is left of D must rotate counterclockwise. This must make a full unit rotation. Use the small angle theory to find the value at C:

$$C = (1)(-6\text{ ft}) = \underline{\underline{-6\text{ ft}}}$$

Use small angle theory to back calculate θ_L:

$$\theta_L = \frac{6\text{ ft}}{4\text{ ft}} = 1.5$$

Use similar triangles to find value at A.

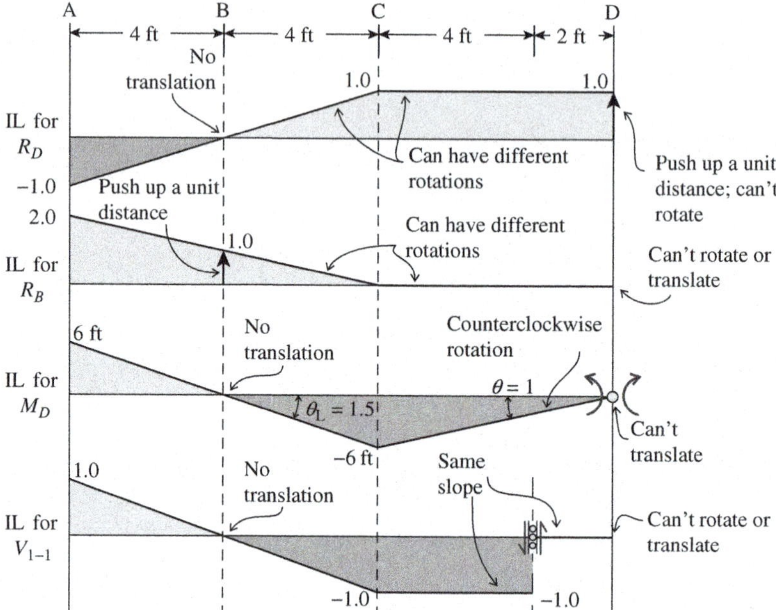

Generate IL for V_{1-1}: Note that point B cannot translate and joint D cannot translate or rotate. This means that the slope between 1−1 and D cannot move. Insert a shear release at 1−1 and impose a unit displacement. Since section 1−1 to D cannot move, all of the movement must come in section C to 1−1. Furthermore, this section cannot rotate because the slope on both sides of the shear release must be the same. Use similar triangles to find the value at A.

16.8 Uses of Influence Lines: Concentrated Loads

Influence lines are the plotted functions of structural responses for various positions of a unit load. Having an influence line for a particular response of a structure makes the value of the response immediately available for a concentrated load at any position on the structure. The beam of Figure 16.1 and the influence line for R_{Ay} are used to illustrate this statement. A concentrated 1-kip load placed 4 ft to the right of the left support would cause R_{Ay} to equal 0.8 kip. Should a concentrated load of 175 kips be placed in the same position, R_{Ay} would be 175 times as great, so $R_{Ay} = (0.8)(175\text{ k}) = 140\text{ k}$.

The value of a response due to a series of concentrated loads is quickly obtained by multiplying each concentrated load by the corresponding ordinate of the influence line for that response. Both a 150-kip load placed 6 ft from the left support and a 200-kip load placed 16 ft from the left support in Figure 16.1 would cause R_{Ay} to equal $(150\text{ k})(0.7) + (200\text{ k})(0.2) = 145\text{ k}$.

EXAMPLE 16.6

The influence lines for the left reaction, R_{Ay}, and the centerline moment, M_B, are shown for a simple beam. Determine the values of these functions for the several loads supported by the beam.

SOLUTION STRATEGY

Similar triangles may be readily used to solve for the ordinates of the ILs at the locations of the point loads.

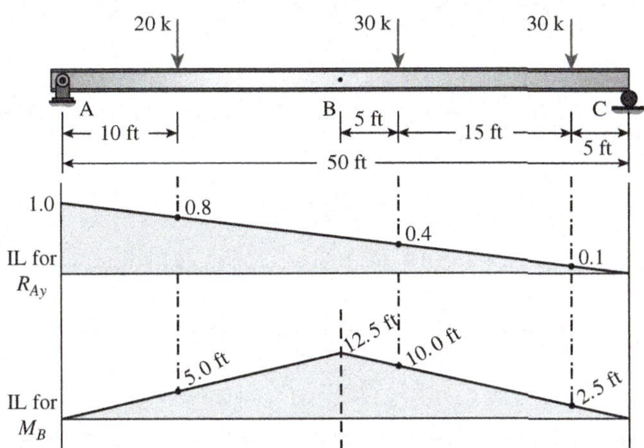

SOLUTION

Compute the magnitude of the left reaction, R_{Ay}. That value is equal to the summation of each load times the ordinate of the influence diagram at the location of the load. Thus,

$$R_{Ay} = (20\text{ k})(0.8) + (30\text{ k})(0.4) + (30\text{ k})(0.1) = \underline{\underline{31\text{ k}}}$$

Next, compute the magnitude of the moment at B. It is computed in the same manner as the left reaction:

$$M_B = (20\text{ k})(5.0\text{ ft}) + (30\text{ k})(10.0\text{ ft}) + (30\text{ k})(2.5\text{ ft}) = \underline{\underline{475\text{ k·ft}}}$$

QUICK NOTE

Notice the units that accompany the ILs. For the reaction, the IL is dimensionless, and for the moment, it has units of feet. However, once the ordinates are multiplied by the magnitude of the point loads, the units make sense.

16.9 Uses of Influence Lines: Uniform Loads

The value of a certain response of a structure may be obtained from the influence line when the structure is loaded with a uniform load by multiplying the area of that part of the influence line that corresponds with the loaded part of the member by the intensity of the uniform load. The following discussion proves this statement is correct.

A uniform load of intensity w lb/ft is equivalent to a continuous series of smaller loads of $(w)(0.1)$ lb on each 0.1 ft of the beam or $(w)(dx)$ lb on each length dx of the beam. Consider each length dx to be loaded with a concentrated load of magnitude $(w)(dx)$. The value of the function for one of these small loads is $[(w)(dx)](y)$ where y is the ordinate of the influence line at that point. The effect of all of these concentrated loads is equal to $\int(wy)dx$. This expression shows that the effect of a uniform load is equal to the intensity of the uniform load, w, times the area under the influence line, $\int ydx$, along the section of the structure covered by the uniform load.

DID YOU KNOW?

The area under an IL does not need to use an integration equation. Rather, well-known equations for calculating the area of rectangles, triangles, and trapezoids may be used quite readily.

EXAMPLE 16.7

Assume the beam in Example 16.6 is loaded with a uniform load of 3 klf. Determine the magnitude of R_{Ay} and the moment at B if the distributed load acts on the entire length of the beam. Then recompute these quantities if the load only acts on the left half of the beam.

SOLUTION STRATEGY

For each case, identify the region over which the distributed load acts. Only compute the areas under the ILs where the load is acting.

SOLUTION

Compute the values first for the load acting on the entire span of the beam.

$$R_{Ay} = (3\text{ klf})\left(\frac{(1.0)(50\text{ ft})}{2}\right) = (3\text{ klf})(25\text{ ft}) = \underline{\underline{75\text{ k}}}$$

$$M_B = (3\text{ klf})\left(\frac{(12.5\text{ ft})(50\text{ ft})}{2}\right) = (3\text{ klf})(625\text{ ft}^2) = \underline{\underline{1875\text{ k}\cdot\text{ft}}}$$

QUICK NOTE

Although the units associated with the moment IL look strange, the final units of the response work out when the area is multiplied by the intensity.

Now compute the same values for the uniform load when it acts only on the left half of the beam. The areas computed are for only the left half of the beam.

$$R_{Ay} = (3\text{ klf})\left(\frac{1.0+0.5}{2}(25\text{ ft})\right) = (3\text{ klf})(18.75\text{ ft}) = \underline{\underline{56.25\text{ k}}}$$

$$M_B = (3\text{ klf})\left(\frac{(12.5\text{ ft})(25\text{ ft})}{2}\right) = (3\text{ klf})(156.25\text{ ft}^2) = \underline{\underline{468.75\text{ k}\cdot\text{ft}}}$$

Should a structure support uniform and concentrated loads, the value of the function under study can be found by multiplying each concentrated load by its respective ordinate on the influence line and the uniform load by the area of the influence line corresponding to the section covered by the uniform load.

16.10 Determining Maximum Loading Effects Using Influence Lines

Beams must be designed to satisfactorily support the largest shearing forces and bending moments that can be caused by the loads to which they are subjected. As an example, consider the beam shown in Figure 16.9(a) and the influence line for bending moment at section 1–1

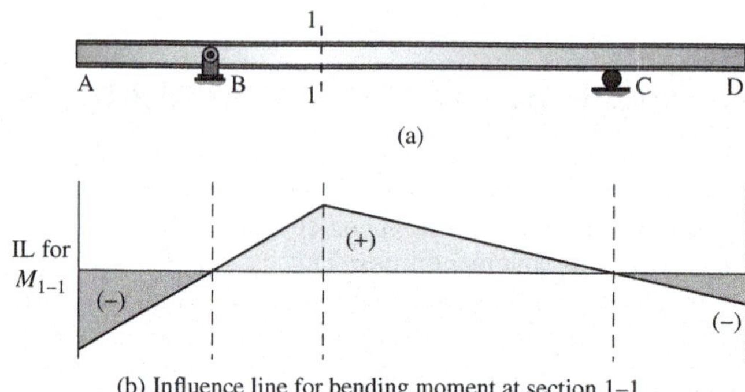

(b) Influence line for bending moment at section 1–1

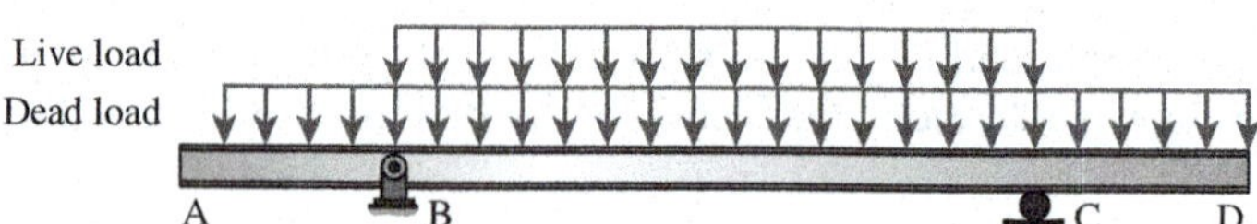

(c) Loads placed to cause maximum positive moment at section 1–1

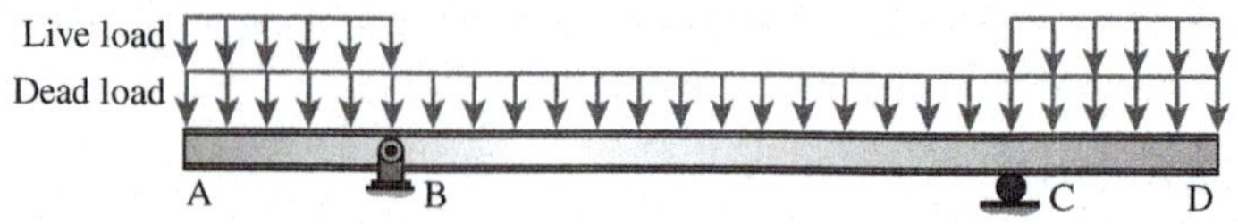

(d) Loads place to try to cause maximum negative moment at section 1–1

FIGURE 16.9 Placing loads to cause maximum forces.

shown in Figure 16.9(b). We now want to determine the maximum possible bending moments at section 1–1 for a uniform dead load and a uniform live load.

We first determine the loading to cause the maximum positive bending moment. The uniform dead load, which is the weight of the structure, is applied from one end of the beam to the other, as shown in Figure 16.9(c). The dead load is always acting on the entire structure. From the influence line, we see that the unit load causes a positive moment at section 1–1 only when it is located between the supports B and C. As such, the uniform live load is placed from B to C to determine the maximum positive bending moment, as shown in Figure 16.9(c). If there had been a concentrated live load acting with the uniform live load, it would have been placed at section 1–1, since the unit load causes the greatest positive moment when it is located there. The bending moment caused by these loads can be calculated using the ordinates on the influence line or by using the equations of static equilibrium.

QUICK NOTE

To create the maximum positive moment at 1–1, we should only place the live load where the IL is positive. Likewise, to create the maximum negative moment at 1–1, we should only place the live load where the IL is negative. This is a principle that may be applied to all ILs.

The maximum negative bending moment occurring at section 1–1 can be determined in a similar fashion. For this case, the loads would be placed as shown in Figure 16.9(d). When the unit load is placed on the cantilever portions of the beam, negative bending occurs at section 1–1. As such, the distributed live load is placed at those locations (A to B and C to D). If there had been a live concentrated load, it would have been placed at the left or right end of the beam—whichever had the largest negative ordinate on the influence line.

16.11 Maximum Loading Effects Using Beam Curvature

In the preceding section, an influence line was used to determine the critical positions for placing live loads to cause maximum bending moments. The same results can be obtained—perhaps obtained more easily in many situations—by considering the deflected shape or curvature of a member under load. If the live loads are placed so they cause the greatest curvature at a particular point, they will have bent the member to the greatest amount at that point. As such, the greatest bending moment will have been caused at that point.

For an illustration of this concept, let's determine the greatest positive bending moment at section 1–1 in the beam shown in Figure 16.10(a) due to the same loads considered in the last section. In Figure 16.10(b), the deflected shape of the beam is sketched, as it would be when a positive moment occurs at section 1–1. This deflected shape is shown by the dashed line. The dead load is placed all across the beam, while the live load is again placed from B to C; this location of the live load will magnify the deflected shape at section 1–1.

A similar situation is shown in Figure 16.10(c) to determine the maximum negative bending moment at section 1–1. The deflected shape of the beam, shown by the dashed line, is sketched consistent with a negative bending moment occurring at section 1–1. To magnify the negative or

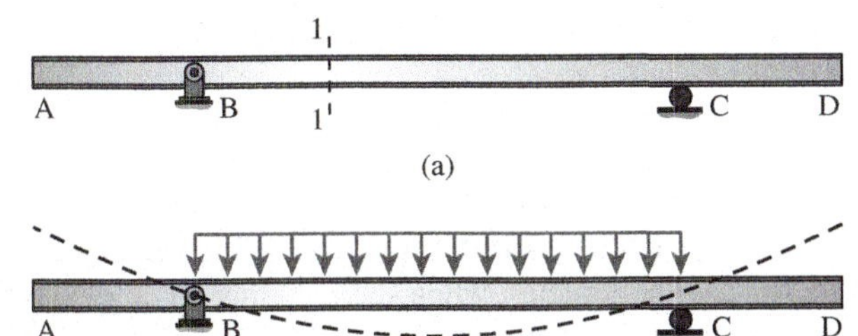

(b) Live loads placed to cause maximum positive moment at section 1–1

QUICK NOTE

The basic notion is that if a load is applied to counteract the deflected shape then the moment at section 1–1 will be reduced.

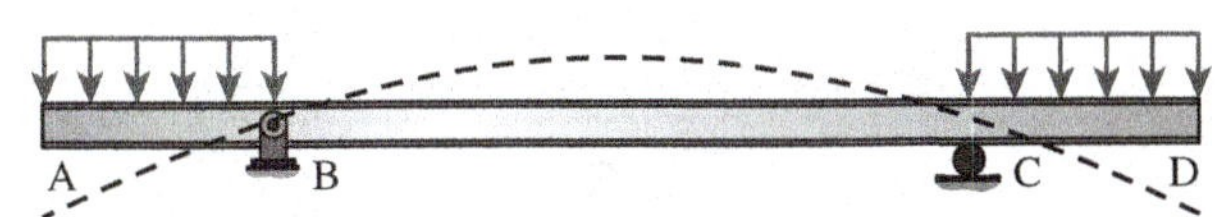

(c) Live loads placed to cause maximum negative moment at section 1–1

FIGURE 16.10 Obtaining maximum forces using curvature.

upward bending at section 1–1, the live load is placed on the cantilever portions of the beam—the parts outside the supports.

16.12 Maximum Values for Moving Loads

Highway and railway bridges are subjected to a series of point loads resulting from trucks and trains that move across them. (A description of such loads is presented in Sections 2.17 and 2.18.) To design a structure supporting moving loads such as these, the engineer must determine where to place the loads to cause maximum forces at various points in the structure.

QUICK NOTE

If one can place the loads at the positions causing maximum forces to occur, one need not worry about any other positions the loads might take on the structure. Such an identification can lead to real savings of time for the engineer.

If a structure is loaded with a uniform live load and not more than one or two moving concentrated loads, the critical positions for the loads will be obvious from the influence lines. If, however, the structure is to support a series of concentrated loads of varying magnitudes, such as groups of truck or train wheels, the problem is not as simple. The influence line provides an indication of the approximate positions for placing the loads, because it is reasonable to assume that the heaviest loads should be grouped near the largest ordinates of the diagram.

Space is not taken herein to consider all of the possible situations that might be faced in structural analysis. The authors feel, however, that the determination of the absolute maximum bending moment caused in a beam by a series of concentrated loads is so frequently encountered by the engineer that discussion of this topic is warranted.

DID YOU KNOW?

The bending moment diagram for a simple beam with concentrated loads will consist of straight lines regardless of the position of the loads. Therefore, the absolute maximum moment occurring during the movement of these loads across the span will occur at one of the loads. This is usually at the load nearest the center of gravity of the group of loads.

The absolute maximum bending moment in a simple beam usually is thought of as occurring at the beam centerline. A maximum bending moment does occur at the centerline if the beam is loaded with either a uniform load or a single concentrated load located at the centerline. A beam, however, may be required to support a series of varying moving concentrated loads—such as the wheels of a train—and the absolute maximum moment in all probability will occur at some position other than the centerline. To calculate the moment, it is necessary to find the point where it occurs and the position of the loads causing it. To assume that the largest moment is developed at the centerline of long-span beams is reasonable, but for short-span beams, this assumption may be considerably in error. It is therefore necessary to have a definite procedure for determining absolute maximum moment.

To examine the point of maximum moment, the beam shown in Figure 16.11 with the series of loads P_1, P_2, P_3, and so on is studied hereafter. The load P_3 is assumed to be the one nearest the center of gravity of the loads on the span. It is located a distance L_1 from P_R, which is the resultant of all the loads on the span. The left reaction, R_{Ay}, is located a distance x from P_R. In the following paragraphs, the maximum bending moment is assumed to occur at P_3, and a definite method is developed for placing this load to cause this maximum.

The bending moment at P_3 may be written by taking a cut at P_3 and summing moments about that cut as

$$M = R_{By}(L - x - L_1) - P_2(L_2 - L_1) - P_1(L_3 - L_1)$$

After substituting in the value of $R_{By} = P_R(x)/L$, we obtain

$$M = \left(\frac{P_R x}{L}\right)(L - x - L_1) - P_2(L_2 - L_1) - P_1(L_3 - L_1)$$

We want to find the value of x for which the moment at P_3 will be a maximum. The maximum bending moment at P_3, which occurs when the shearing force is zero, may be found by differentiating the bending moment equation with respect to x, setting the result equal to zero,

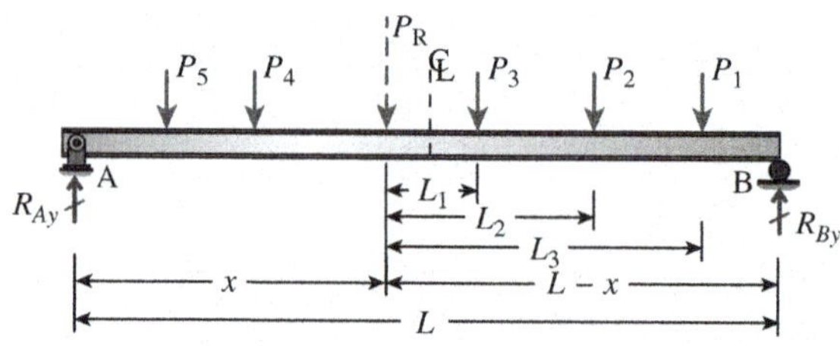

FIGURE 16.11 A simple beam with numerous applied loads.

and solving for x. Thus,

$$\frac{dM}{dx} = L - 2x - L_1 = 0$$

$$\therefore x = \frac{L}{2} - \frac{L_1}{2} = \underline{\underline{\frac{1}{2}(L - L_1)}}$$

From the preceding derivation, a general rule for absolute maximum moment is stated:

Maximum moment in a beam loaded with a moving series of concentrated loads usually will occur at the load nearest the center of gravity of the loads on the beam when the center of gravity of the loads on the beam is the same distance on one side of the centerline of the beam as the load nearest the center of gravity of the loads is on the other side.

Procedure

1. Calculate the resultant and center of gravity of the applied point loads.
2. Identify the point load which is nearest to the center of gravity.
3. Identify the location halfway between the center of gravity (P_R) and the nearest point load identified in step 2.
4. Position the series of point loads such that the location identified in step 3 lines up with the centerline of the beam, as shown in Figure 16.11.
5. Compute the moment under the point load closest to the beam centerline.

DID YOU KNOW?
Should the load nearest the center of gravity of the loads be relatively small, the absolute maximum moment may occur at some other load nearby. Occasionally, two or three loads have to be considered to find the greatest value.

The problem of finding the location is not a difficult one because of another moment criteria not described herein—*the average load to the left must be equal to the average load to the right*—must be satisfied. There will be little trouble determining which of the nearby loads will govern. Actually, it can be shown that the absolute maximum moment occurs under the load that would be placed at the centerline of the beam to cause maximum moment there, when that wheel is placed as far on one side of the beam centerline as the center of gravity of all the loads is on the other.[2]

EXAMPLE 16.8

Determine the absolute maximum moment that can occur in the 50-ft simple beam shown as the series of concentrated loads shown moves across the span.

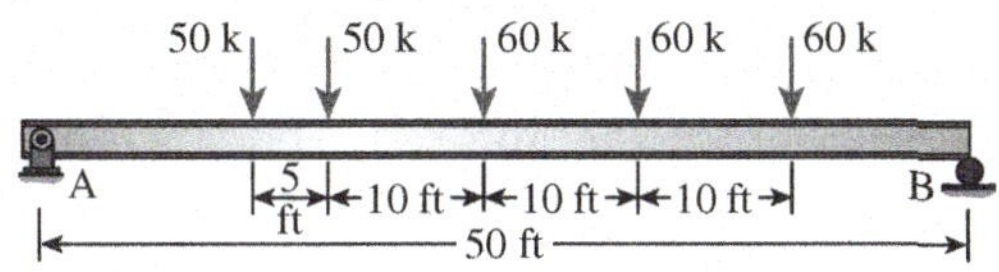

QUICK NOTE
Note that the beam figure only indicates the distance between the point loads but does not locate them on the beam. To find this location is the point of the problem.

SOLUTION STRATEGY

Focus on just the loads first. Once the center of gravity is found, the loads can be positioned as required.

SOLUTION

The center of gravity of the loads is determined measuring from the left-most load.

$$\frac{(50\text{ k})(0\text{ ft}) + (50\text{ k})(5\text{ ft}) + (60\text{ k})(15\text{ ft}) + (60\text{ k})(25\text{ ft}) + (60\text{ k})(35\text{ ft})}{(2)(50\text{ k}) + (3)(60\text{ k})}$$

$$= 16.96\text{ ft} \quad \text{(measured from the left load)}$$

DID YOU KNOW?
The location of the center of gravity is found using a simple weighted average.

[2]A. Jakkula, and H.K. Stephenson, *Fundamentals of Structural Analysis* (New York: Van Nostrand, 1953), 241–242.

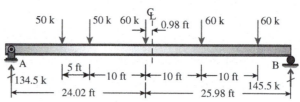

The distance from the first 60-k load and P_R is computed as

$$16.96 \text{ ft} - 15 \text{ ft} = 1.96 \text{ ft}$$

The location of the first 60-k load is then positioned on the beam such that it is located $(1.96 \text{ ft}/2) = 0.98$ ft left of the centerline.

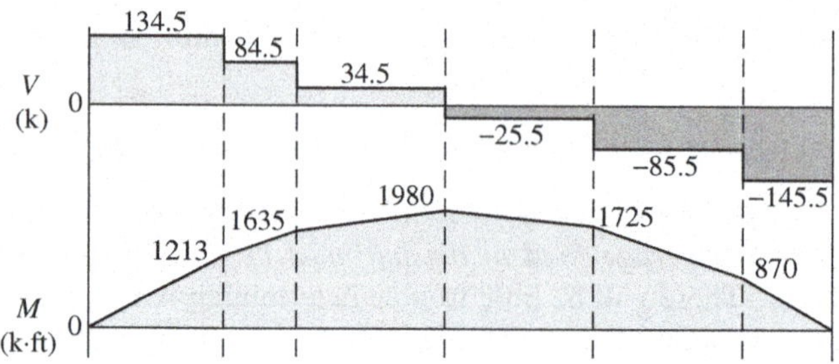

DID YOU KNOW?
Sketching the shear and moment diagrams for this beam quickly confirms that the maximum moment occurs under the left-most 60-k load.

QUICK NOTE
The absolutely largest moment that can occur for these loads is 1980 k·ft.

16.13 Influence Lines for Trusses

QUICK NOTE
Truss influence lines may be drawn and used for making force calculations, or they may be sketched roughly without computing the values of the ordinates and used only for placing the moving loads that cause maximum or minimum forces.

Beam-type bridges of structural steel, precast reinforced concrete, or prestressed concrete have almost completely taken over the short-span bridge market. Nevertheless, the drawing of influence lines for trusses is presented in this chapter for two reasons. First, the authors feel that the understanding of such diagrams gives the student a better understanding of the analysis of trusses and of the action of moving loads. Second, the information presented here serves as a background for the construction of influence lines for the longer span, statically indeterminate bridge trusses that are still economically competitive.

The procedure used for preparing influence lines for trusses is closely related to the quantitative approach used for beams. The exact manner of application of loads to a bridge truss from the bridge floor is described in the following subsections. A similar discussion could be made for the application of loads to roof trusses.

Arrangement of Bridge Floor Systems

The arrangement of the members of a bridge floor system should be carefully studied so that the manner of application of loads to the truss will be fully understood. Probably the most common type of floor system consists of a concrete slab supported by steel stringers running parallel to the trusses. The stringers run the length of each panel and are supported at their ends by floor beams that run transverse to the roadway and frame into the panel points or joints of the truss (see Figure 16.12).

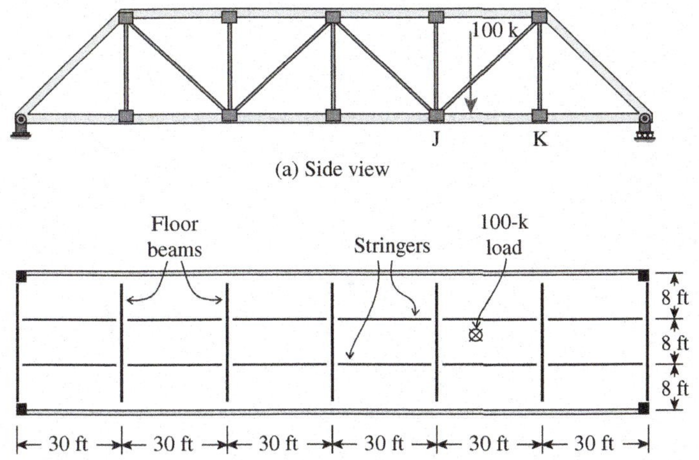

FIGURE 16.12 Typical truss bridge framing scheme.

The previous discussion apparently indicates that the stringers rest on the floor beams and the floor beams on the trusses. This method of explanation has been used to emphasize the manner in which loads are transferred from the pavement to the trusses, but the members usually are connected directly to each other on the same level. Stringers are conservatively assumed to be simply supported, but actually, there may be some continuity in their construction.

A 100-kip load is applied to the floor slab in the fifth panel of the truss of Figure 16.12. The load is transferred from the floor slab to the stringers, thence to the floor beams, and finally to joints J and K of the supporting trusses. The amount of load going to each stringer depends on the position of the load between the stringers: If halfway, each stringer would carry half. Similarly, the amount of load transferred from the stringers to the floor beams depends on the longitudinal position of the load.

Figure 16.13 shows a section of the bridge floor framing that includes the fifth panel. The calculations involved in figuring the transfer of a 100-kip load to the trusses are shown next. The final reactions shown for the floor beams represent the downward loads applied at the truss panel points and are shown in Figure 16.14. The computation of truss loads usually is a much simpler process than the one described here.

DID YOU KNOW?

The typical load path in a truss bridge system is as follows.

- Slab
- Stringers
- Floor beams
- Truss panels points (i.e., joints)

RECALL

When looking at a floor framing plan, the framing members are drawn so they don't touch each other when they are simply supported.

Load Transferred to Each Stringer:

$$R_A = \left(\frac{3}{8}\right)(100\text{ k}) = 37.5\text{ k}$$

$$R_B = \left(\frac{5}{8}\right)(100\text{ k}) = 62.5\text{ k}$$

Load Transferred from Stringer A to Floor Beams:

$$R_{4-4} = \left(\frac{20}{30}\right)(37.5\text{ k}) = 25\text{ k}$$

$$R_{5-5} = \left(\frac{10}{30}\right)(37.5\text{ k}) = 12.5\text{ k}$$

QUICK NOTE

Since the applied load is a point load, the procedure of calculating reactions by proportions is conveniently used (see Section 5.9).

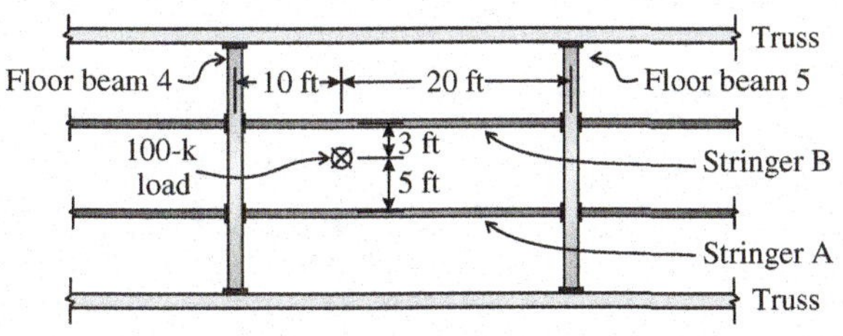

FIGURE 16.13 Bridge floor framing detail of panel 5.

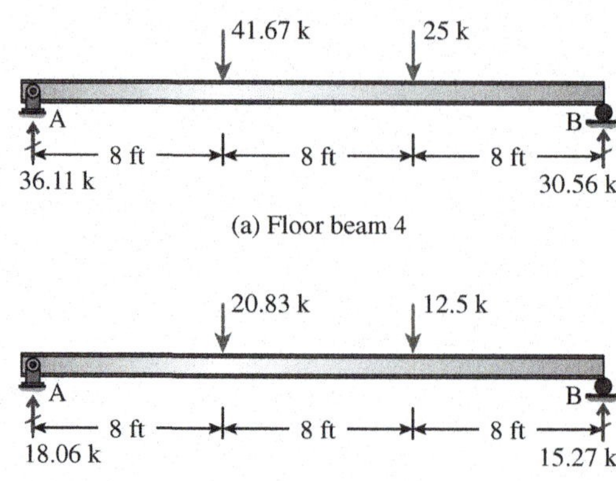

FIGURE 16.14 Floor beam loading diagrams.

Load Transferred from Stringer B to Floor Beams:

$$R_{4-4} = \left(\frac{20}{30}\right)(62.5\text{ k}) = 41.67\text{ k}$$

$$R_{5-5} = \left(\frac{10}{30}\right)(62.5\text{ k}) = 20.83\text{ k}$$

The floor beams are loaded as shown in Figure 16.14.

Procedure

The approach that must be taken for the development of influence lines for trusses is the quantitative approach. Recall that this approach requires that a unit load be placed at key locations along the structure and the response quantity of interest is found. The following presents a basic procedure which may be used for trusses.

1. Place a dimensionless unit load at key locations along the truss where key locations are identified as
   - Truss supports
   - Ends of the truss
   - Panel points on either side of the truss member of interest. One may place a unit load on each panel point of the truss, but this is simply more work and doesn't provide any more information to the analyst.
2. Calculate the response quantity of interest (e.g., N or R) and plot the value at the location where the unit load is placed. Basic statics is all that is required for this.
3. Connect the dots with straight lines.

QUICK NOTE
Influence lines for trusses are primarily found for reactions and/or axial member forces. (Remember that there is theoretically no shear or moment in a truss member.)

Influence Lines for Truss Reactions Influence lines for reactions of simply supported trusses are used to determine the maximum loads that may be applied to the supports. Although their preparation is elementary, they offer a good introductory problem in learning the construction of influence lines for truss members. Example 16.9 demonstrates the process and considers the truss in Figure 16.15.

QUICK NOTE
The reactions of the floor beams are what show up at the panel points (i.e., joints) of the truss. If the location of the 100-k load is changed, the magnitude of the floor beam reactions will change.

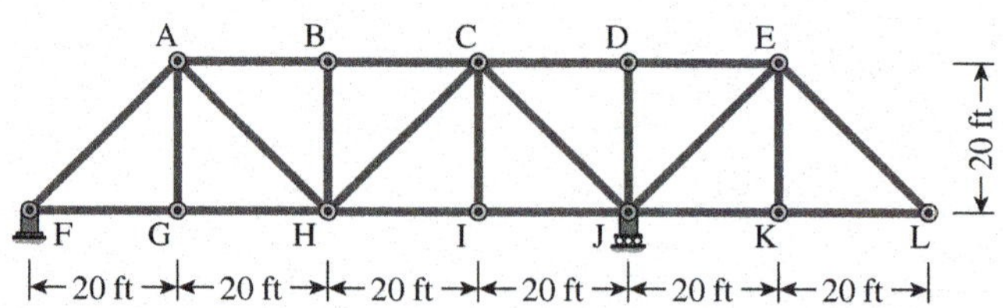

FIGURE 16.15 Parallel-chord truss.

EXAMPLE 16.9

Develop the influence lines for the vertical reactions at F and J of the truss shown in Figure 16.15.

SOLUTION STRATEGY

Identify the key locations relative to the influence lines that are requested. Remember that only a minimal number of locations for the unit load is required.

SOLUTION

The key locations are the truss ends and the supports. Joints F, J, and L are the only locations required. Go through the process as outlined in the procedure.

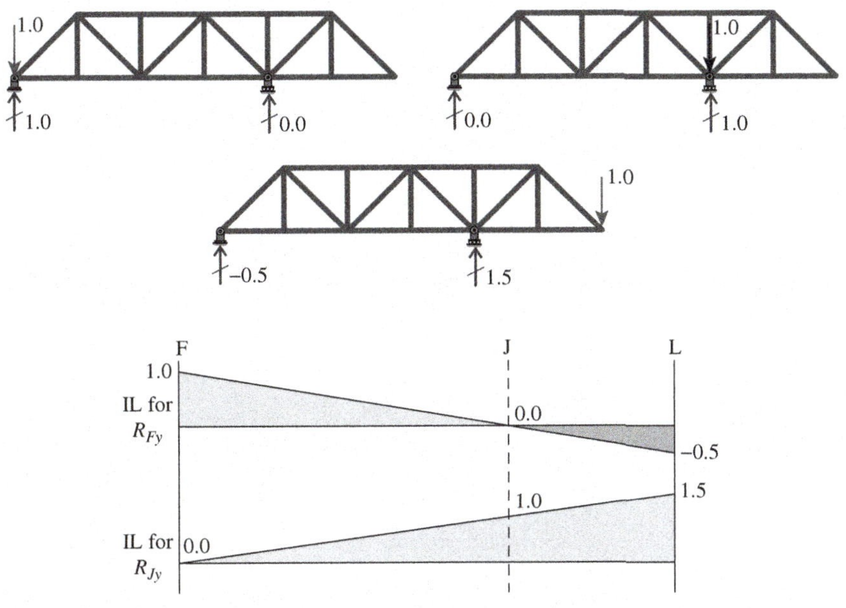

QUICK NOTE

Observation shows that the placement of the unit load at joint J was unnecessary. This kind of realization will more readily come as experience with truss influence lines is gained.

Influence Lines for Member Forces of Parallel-Chord Trusses Influence lines for forces in truss members may be constructed in the same manner as those for various beam responses. For trusses, the method of sections is often a very convenient form of analysis that is used to find the internal force in the member in question. This is done by passing an imaginary section through the truss, cutting the member in question, and taking moments at the intersection of the other members cut by the section. The resulting force in the member is equal to the moment divided by the lever arm. The force in the diagonal member of a panel is simply found using the sum of forces in the *y*-direction. This process is illustrated in Example 16.10.

EXAMPLE 16.10

Consider the truss given in Figure 16.15. Generate the influence lines for the following member forces: *GH*, *AH*, and *DE*.

SOLUTION STRATEGY

Use the method of sections to find the requisite internal forces. Once you figure out how to calculate the internal force for a member, do not change the analysis approach—even when the load changes location.

KEY POINTS

N_{GH} and N_{AH} require the load to be placed at F, G, H, J, and L.

N_{DE} requires the load to be placed at F, B, K, and L.

SOLUTION

The requested response quantities will require the unit load to be placed at six of the seven panel points. For demonstration purposes, the load will be placed at each panel point and the required internal forces will be listed. Pass a vertical cut through panels 2 and 5.

After finding the reactions for each load position, find N_{GH} by summing moments about point A. Find N_{AH} by summing the forces in the y-direction for the left section, and sum the moments about joint J for the right section to find N_{DE}.

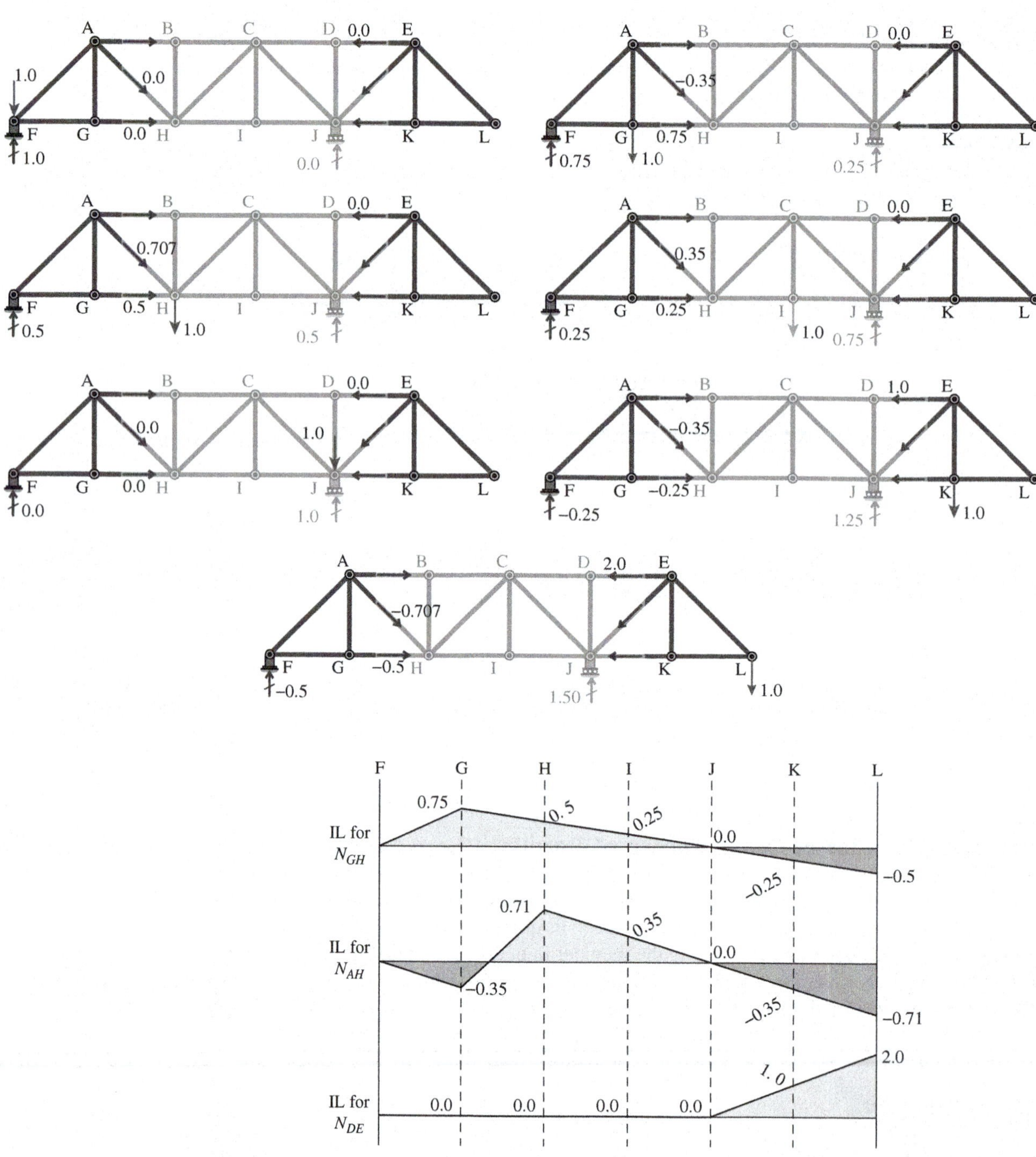

Influence Lines for K-Truss Figure 16.16 shows influence lines for several members of a K-truss. The approach necessary for preparing the diagrams for the chord members is equivalent to that used for the chords of any other parallel-chord truss. The ordinates for the force in member CD may be obtained by passing section 1–1 through the member in question and three other members (CJ, JQ, and QR). As the lines of action of these three forces intersect at Q, moments may be taken around Q to obtain the desired force in CD (N_{CD}).

The forces in the two diagonals of each panel may be obtained from the shear in the panel. By knowing that the horizontal components are equal and opposite, the relationship between

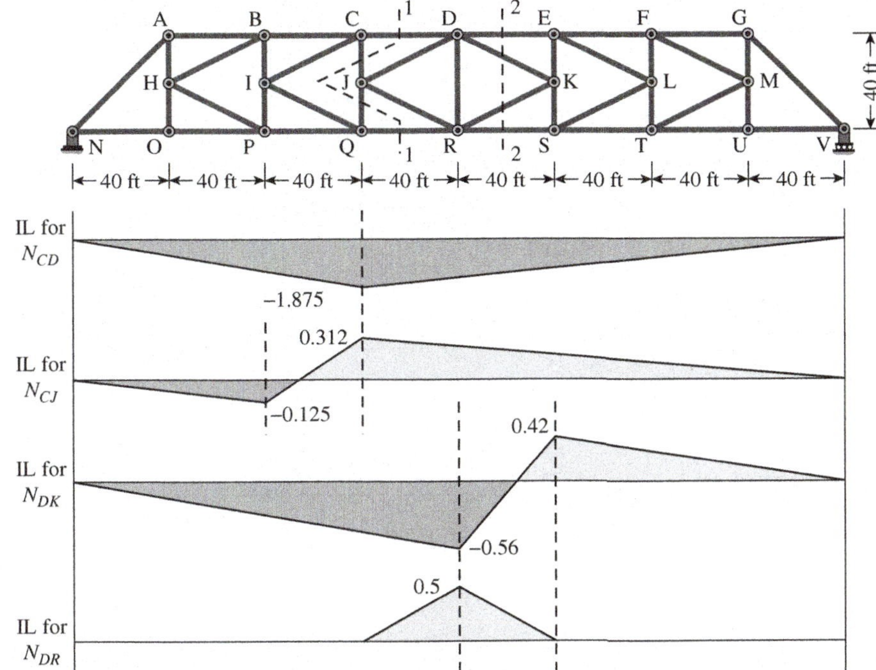

FIGURE 16.16 Influence lines for K-truss.

QUICK NOTE

It is always advisable to get the statics figured out before one attempts to construct the IL.

their vertical components can be found from their slopes. If the slopes are equal, the shear to be carried divides equally between the two. For example, one may cut at section 2–2 to gain access to members DK and RK. The IL for N_{DK} is given in Figure 16.16.

The influence lines for the verticals, such as CJ, may be determined from the influence lines for the adjoining diagonals (if available). On the other hand, the ordinates may be computed independently for various positions of the unit load. A possibility to find CJ would be to first find IQ and then make a cut through members CD, CJ, IQ, and PQ. Then a sum of the forces in the y-direction will find CJ. Another option would be to find IC and then do a joint analysis on joint C to find CJ.

The influence line for the mid-vertical DR can be developed by computing the vertical components of force in JR and RK (or in JD and DK) for each position of the unit load. The vertical components of force in each of these pairs of members will cancel each other unless the shear in panel 4 (CDRQ) is unequal to the shear in panel 5 (DESR), which is possible only when the unit load is at R.

Influence Lines for Member Forces of Nonparallel-Chord Trusses Influence line ordinates for the force in a chord member of a "curved-chord" truss may be determined by passing a vertical section through the panel and taking moments at the intersection of the diagonal and the other chord. The ordinates for the influence line for force in a diagonal may be obtained by passing a vertical section through the panel and taking moments at the intersection of the two chord members.

QUICK NOTE

Chapter 7 discusses the static analysis of trusses. "Curved-chord" trusses like the Parker truss were introduced and analyzed. Remember that with trusses such as these, the most convenient place to sum moments may not actually lie on the truss itself.

Determination of Maximum Forces in Trusses

Truss members are designed to resist the maximum forces that may be caused by any combination of the dead, live, and impact loads to which the truss may be subjected. The live load probably consists of a series of moving concentrated loads representing the wheel loads of the vehicles using the structure, but for convenience in analysis, an approximately equivalent uniform live load with only one or two concentrated loads often is used in their place. Live loads for which highway and railroad bridges are designed and common impact expressions are discussed in detail in Chapter 2.

The dead load, representing the weight of the structure and permanent attachments, extends for the entire length of the truss, but the uniform and concentrated live loads are placed at the

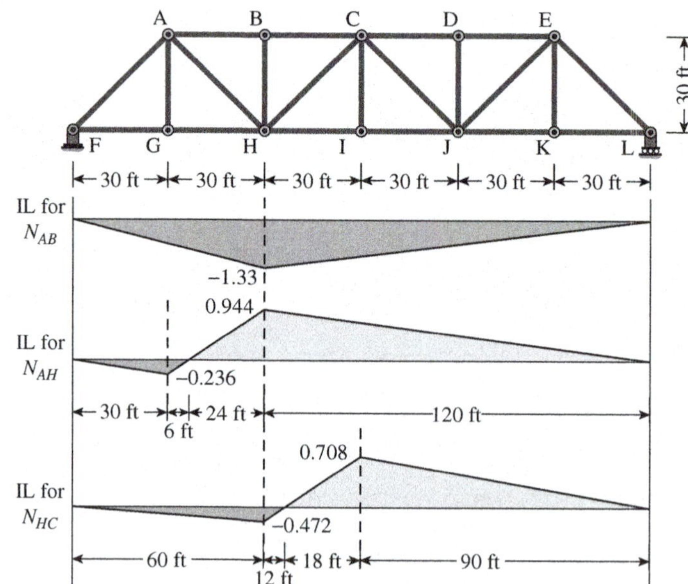

LOAD SCENARIO

1. Dead uniform load of 1.5 klf
2. Live uniform load of 2 klf
3. Moving concentrated load of 20 kips
4. Impact of 24.4%

FIGURE 16.17 Influence lines for a parallel-chord truss.

points on the influence line that cause maximum force of the character being studied. If tension is being studied, the live uniform load is placed along the section of the truss corresponding to the positive or tensile section of the influence line, and the live concentrated loads are placed at the maximum positive tensile ordinates on the diagram.

Members whose influence lines have both positive and negative ordinates may be in tension for one combination of loads and in compression for another (see member DK of Figure 16.16). A member subject to force reversal must be designed to resist both the maximum compressive and maximum tensile forces.

In the next few paragraphs, the maximum possible forces in several members of the truss of Figure 16.17 are determined due to the load scenario shown in the sidebar.

The influence lines are drawn, and the forces are computed by the exact method as described in the following paragraphs.

QUICK NOTE

Remember that one must multiply the area under the influence line by the magnitude of a distributed load to find the actual response to that load.

Member AB This member is in compression for every position of the unit load; therefore, the dead uniform load and the live uniform load are placed over the entire span. The moving concentrated load of 20 kips is placed at the maximum compression ordinate on the influence line. The impact factor is multiplied by the live load forces and added to the total.

$$DL = (1.5\text{ klf})(180\text{ ft})(-1.33)\left(\frac{1}{2}\right) = -180.0\text{ k}$$

$$LL = (2\text{ klf})(180\text{ ft})(-1.33)\left(\frac{1}{2}\right) = -240.0\text{ k}$$

$$+(20\text{ k})(-1.33) = -26.7\text{ k}$$

$$I = (0.244)(-240.0\text{ k} - 26.7\text{ k}) = -65.1\text{ k}$$

$$\text{total force} = \underline{\underline{-\ 511.8\text{ k}}}\quad (\text{C})$$

Member AH Examination of the influence line for AH shows that for some positions of the unit load the member is in compression, whereas for others it is in tension. The live loads should be placed over the positive portion of the diagram and the dead loads across the entire structure to obtain the largest possible tensile force. Similarly, the live loads should be placed over the negative portion of the diagram and the dead loads over the entire structure to obtain the largest possible compressive force.

Maximum Tension:

$$DL = (1.5\text{ klf})(144\text{ ft})(0.944)\left(\frac{1}{2}\right) = 102.0\text{ k}$$

$$+(1.5\text{ klf})(36\text{ ft})(-0.236)\left(\frac{1}{2}\right) = -6.4\text{ k}$$

$$LL = (2\text{ klf})(144\text{ ft})(0.944)\left(\frac{1}{2}\right) = 136.0\text{ k}$$

$$+(20\text{ k})(0.944) = 18.9\text{ k}$$

$$I = (0.244)(136.0\text{ k} + 18.9\text{ k}) = 37.8\text{ k}$$

$$\text{total force} = \underline{\underline{288.3\text{ k}}}\quad(\text{T})$$

Notice that when the maximum compression is calculated the answer is actually positive. This indicates that, for the given magnitude of the dead load and the given live loads, it is not possible for the member to be in compression. If the ratio of dead load to live load were to decrease, it may be possible that this member would experience compression.

Maximum Compression:

$$DL = (1.5\text{ klf})(144\text{ ft})(0.944)\left(\frac{1}{2}\right) = 102.0\text{ k}$$

$$+(1.5\text{ klf})(36\text{ ft})(-0.236)\left(\frac{1}{2}\right) = -6.4$$

$$LL = (2\text{ klf})(36\text{ ft})(-0.236)\left(\frac{1}{2}\right) = -8.5$$

$$+(20\text{ k})(-0.236) = -4.7\text{ k}$$

$$I = (0.244)(-8.5\text{ k} - 4.7\text{ k}) = -3.2\text{ k}$$

$$\text{total force} = \underline{\underline{79.2\text{ k}}}\quad(\text{T})$$

Member HC The calculations for AH proved that it could only have tensile forces—regardless of the positioning of the live loads given. The following calculations show force reversal may occur in member HC.

Maximum Tension:

$$DL = (1.5\text{ klf})(108\text{ ft})(0.708)\left(\frac{1}{2}\right) = 57.3\text{ k}$$

$$+(1.5\text{ klf})(72\text{ ft})(-0.472)\left(\frac{1}{2}\right) = -25.5\text{ k}$$

$$LL = (2\text{ klf})(108\text{ ft})(0.708)\left(\frac{1}{2}\right) = 76.4\text{ k}$$

$$+(20\text{ k})(0.708) = 14.2\text{ k}$$

$$I = (0.244)(76.4\text{ k} + 14.2\text{ k}) = 22.1\text{ k}$$

$$\text{total force} = \underline{\underline{144.5\text{ k}}}\quad(\text{T})$$

Maximum Compression:

$$DL = (1.5\text{ klf})(108\text{ ft})(0.708)\left(\frac{1}{2}\right) = 57.3\text{ k}$$

$$+(1.5\text{ klf})(72\text{ ft})(-0.472)\left(\frac{1}{2}\right) = -25.5\text{ k}$$

$$LL = (2.0\text{ klf})(72\text{ ft})(-0.472)\left(\frac{1}{2}\right) = -34.0\text{ k}$$

$$+(20\text{ k})(-0.472) = -9.4\text{ k}$$

$$I = (0.244)(-34.0\text{ k} - 9.4\text{ k}) = -10.6\text{ k}$$

$$\text{total force} = \underline{\underline{-22.2\text{ k}}}\quad \text{(C)}$$

16.14 Examples with Video Solution

VE16.1 Use the quantitative approach to generate the following ILs for the beam in Figure VE16.1. M_A, V_{1-1}, V_{2-2}, and R_{Cy}.

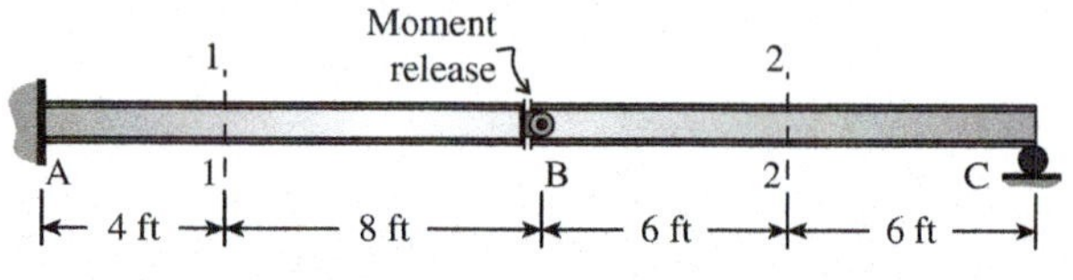

Figure VE16.1

VE16.2 Use the qualitative approach to generate the following ILs for the beam given in Figure VE16.1. M_A, V_{1-1}, V_{2-2}, and R_{Cy}.

VE16.3 Use the qualitative approach to generate the following ILs for the beam in Figure VE16.3. M_A, V_{1-1}, V_{2-2}, and R_{Cy}.

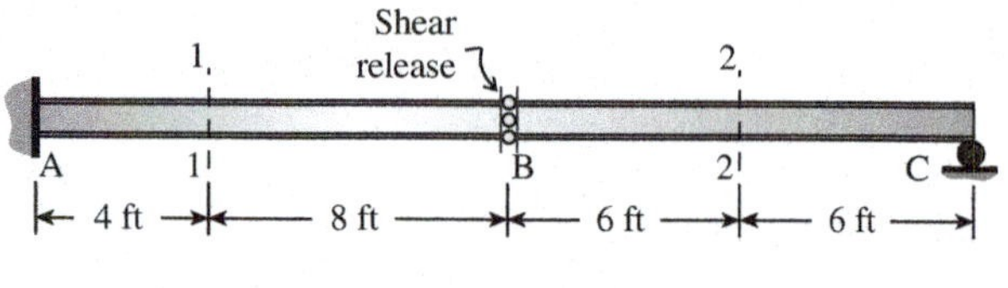

Figure VE16.3

VE16.4 Use the qualitative approach to generate the following ILs for the beam in Figure VE16.4. M_B, $V_{B-\text{left}}$, $V_{B-\text{right}}$, and R_{Dy}. Use the ILs to calculate M_B, $V_{B-\text{left}}$, $V_{B-\text{right}}$, and R_{Dy} due to the applied load.

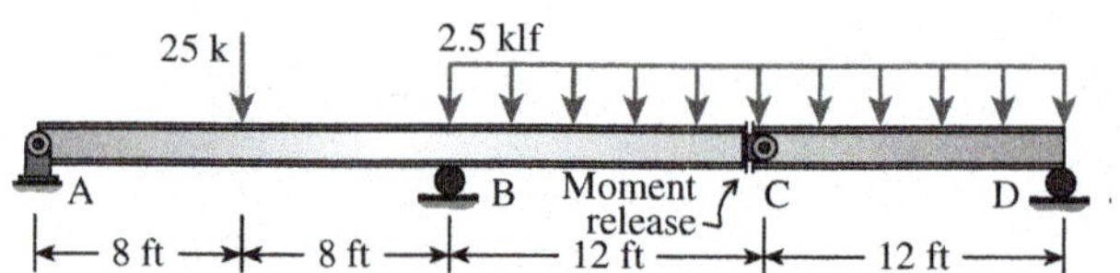

Figure VE16.4

VE16.5 For the truss shown in Figure VE16.5, draw the influence lines for the following bar forces: N_{BC}, N_{BG}, and N_{FG}.

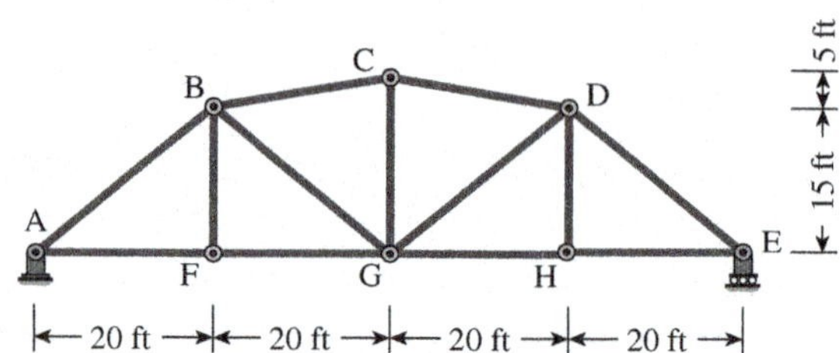

Figure VE16.5

16.15 Problems for Solution

Sections 16.1–16.6

For Problems 16.1 through 16.18, draw the quantitative influence lines for the situations listed.

P16.1 Both reactions and the shearing force and bending moment at section 1–1. (Ans: load @ 1–1: $V_{1-1} = -0.4, 0.6$, $M_{1-1} = 18$ m)

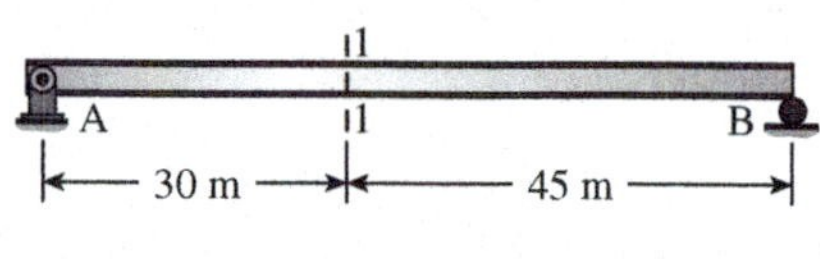

Problem 16.1

P16.2 Both reactions and the shearing force and bending moment at section 1–1.

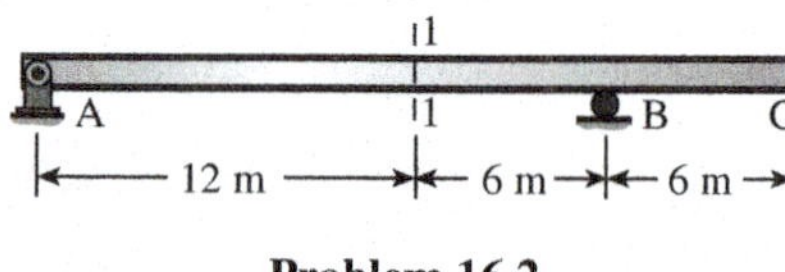

Problem 16.2

P16.3 Both reactions and the shearing force and bending moment at section 1–1. (Ans: load @ 1–1: $V_{1-1} = -0.333, 0.667$, $M_{1-1} = 10$ ft)

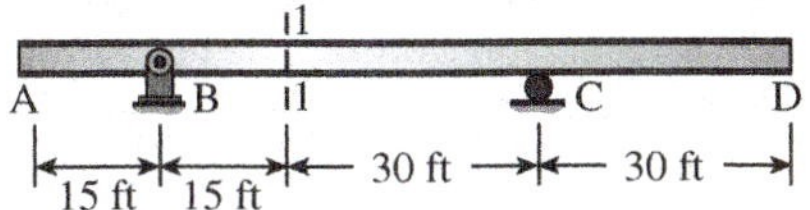

Problem 16.3

P16.4 Reaction at C, moment at C, and the shearing force and bending moment at section 1–1.

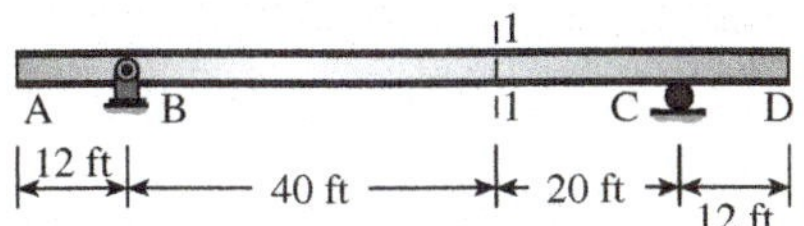

Problem 16.4

P16.5 Both reactions and the shearing force and bending moment at section 1–1. (Ans: load @ 1–1: $V_{1-1} = -0.6, 0.4$, $M_{1-1} = 12$ ft)

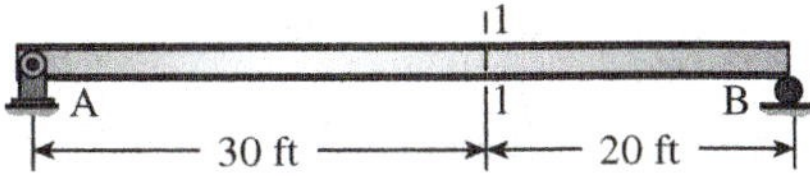

Problem 16.5

P16.6 Both reactions and the shearing force and bending moment at section 1–1.

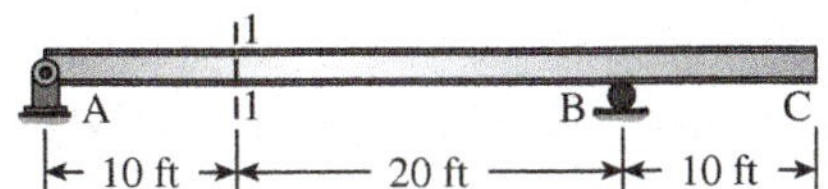

Problem 16.6

P16.7 Both reactions, as well as the shearing force and bending moment at section 1–1. (Ans: load @ 1–1: $V_{1-1} = -0.4, 0.6$, $M_{1-1} = 3$ ft)

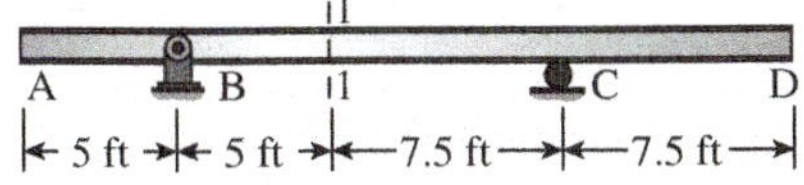

Problem 16.7

P16.8 Shearing force for sections 1–1, 2–2, and 3–3.

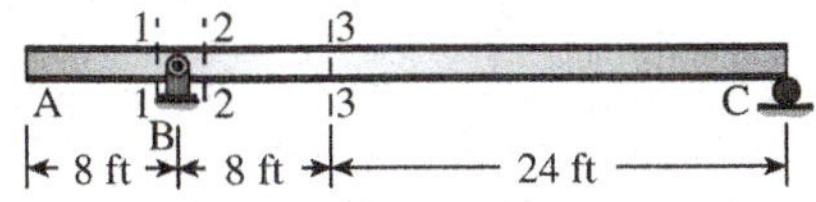

Problem 16.8

P16.9 Reaction at B, shear force at section 1–1, and bending moment at 2–2. (Ans: load @ A: $V_{1-1} = 0.375$, $M_{2-2} = -2.25$ ft)

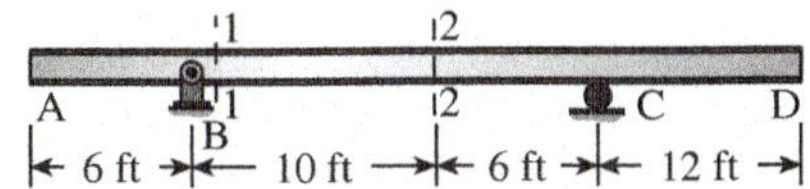

Problem 16.9

P16.10 Vertical reaction and moment at B; shearing force and bending moment at section 1–1.

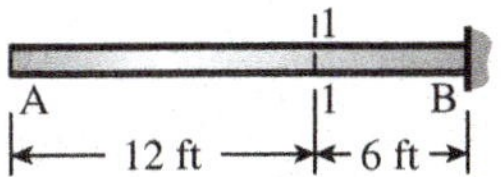

Problem 16.10

P16.11 Vertical reaction and moment at A; shearing force and bending moment at section 1–1. (Ans: load @ B: $V_{1-1} = 0.0, 1.0$, $M_A = 19$ ft)

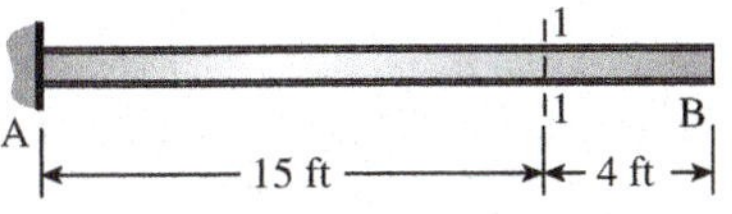

Problem 16.11

P16.12 Both reactions and the bending moment at B and C.

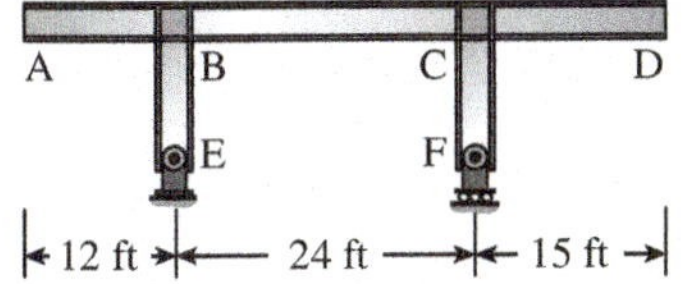

Problem 16.12

P16.13 Reaction at A, shear just left of C, and bending moment at C. (Ans: load @ B: $V_{C-\text{left}} = -1.0$, $M_C = -30$ ft)

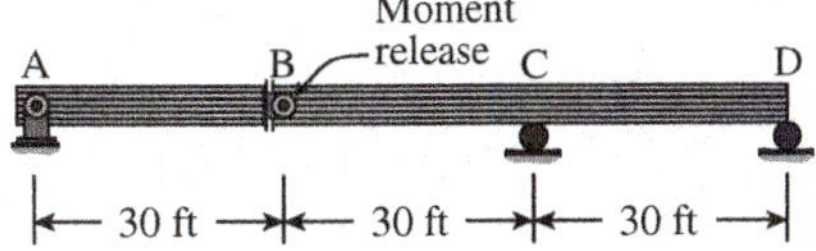

Problem 16.13

P16.14 Moment at D and B; shear force at C.

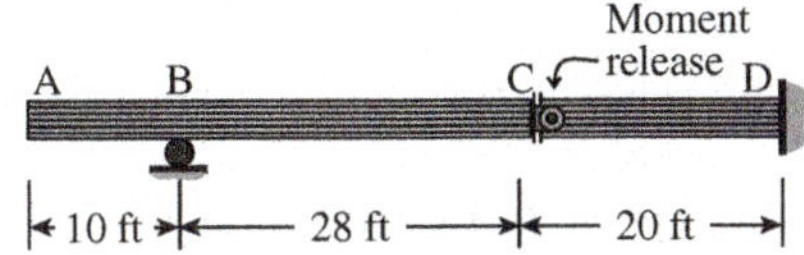

Problem 16.14

P16.15 Reaction at A, shear at section 1–1, and moment at C. (Ans: load @ B: $V_{1-1} = 0.0$, $M_C = -30$ ft)

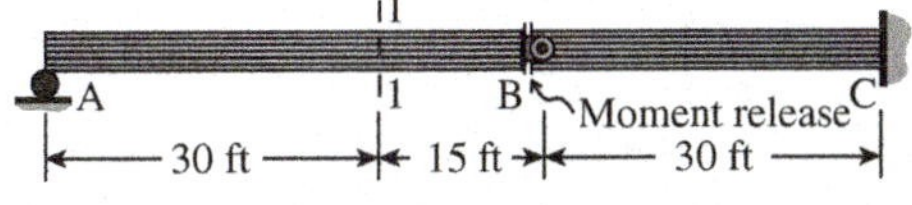

Problem 16.15

P16.16 Reaction at F, shear at D, and bending moment at B.

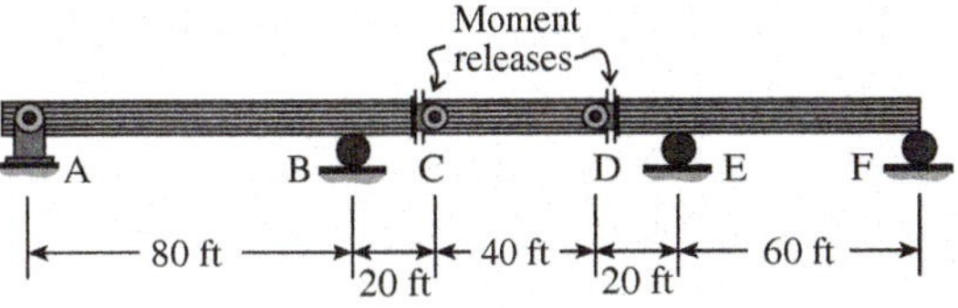

Problem 16.16

P16.17 Reaction at A, shear at section 1–1, and moment at C. (Ans: load @ B: $V_{1-1} = 1.0, 0.0$, $M_C = 45$ ft, -30 ft)

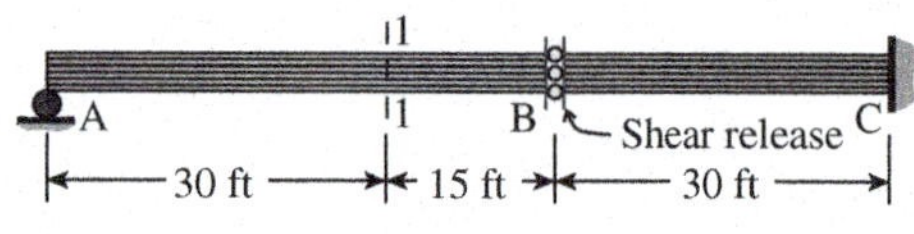

Problem 16.17

P16.18 Reaction at F, reaction at B, moment at B, and shear just to the left of D.

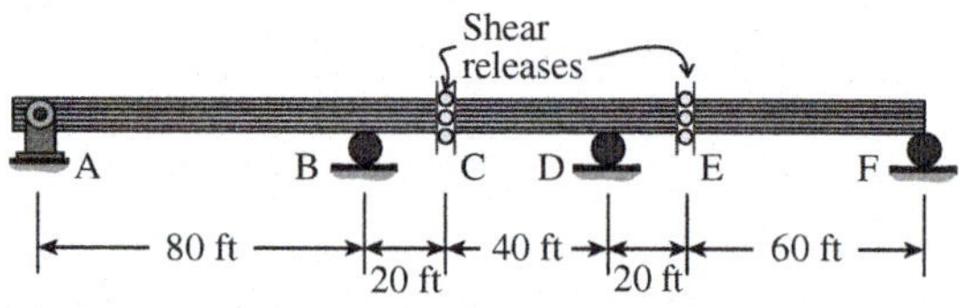

Problem 16.18

Section 16.7

For Problems 16.19 through 16.32, use the qualitative approach to work the following problems.

P16.19 Rework Problem 16.1. (Ans: load @ 1–1: $V_{1-1} = -0.4, 0.6$, $M_{1-1} = 18$ m)

P16.20 Rework Problem 16.2.

P16.21 Rework Problem 16.3. (Ans: load @ 1–1: $V_{1-1} = -0.333, 0.667$, $M_{1-1} = 10$ ft)

P16.22 Rework Problem 16.6.

P16.23 Rework Problem 16.7. (Ans: load @ 1–1: $V_{1-1} = -0.4, 0.6$, $M_{1-1} = 3$ ft)

P16.24 Rework Problem 16.8.

P16.25 Rework Problem 16.10. (Ans: load @ A: $V_{1-1} = -1.0$, $M_B = -18$ ft)

P16.26 Rework Problem 16.11.

P16.27 Rework Problem 16.13. (Ans: load @ B: $V_{C-\text{left}} = -1.0$, $M_C = -30$ ft)

P16.28 Rework Problem 16.14.

P16.29 Rework Problem 16.15. (Ans: load @ B: $V_{1-1} = 0.0$, $M_C = -30$ ft)

P16.30 Rework Problem 16.16.

P16.31 Rework Problem 16.17. (Ans: load @ B: $V_{1-1} = 1.0, 0.0$, $M_C = 45$ ft, -30 ft)

P16.32 Rework Problem 16.18.

Sections 16.8–16.10

For Problems 16.33 through 16.39, use influence lines to determine the requested quantities for a uniform dead load of 1.5 klf, a moving uniform live load of 2.5 klf, and a moving concentrated live load of 25 k.

P16.33 Find the maximum reaction at A, maximum positive shear, and moment at section 1–1. (Ans: $R_{Ay} = 125$ k, $V_{1-1} = 28.1$ k)

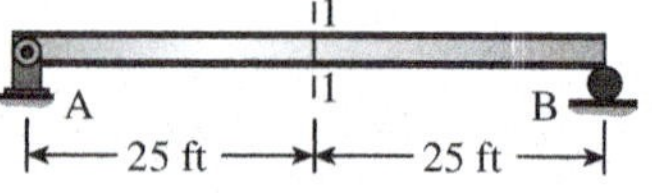

Problem 16.33

P16.34 Maximum reaction at B, maximum negative shear, and moment at section 1–1.

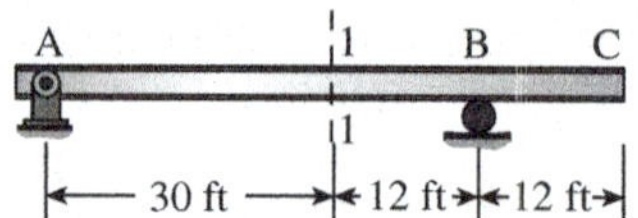

Problem 16.34

P16.35 For the beam in Problem 16.34, find the maximum negative value of shear and maximum positive moment at section 1–1, and the maximum reaction at A. (Ans: $R_{Ay} = 106.5$ k, $V_{1-1(\text{neg})} = -65$ k)

P16.36 Find the maximum negative shear and negative moment at section 1–1.

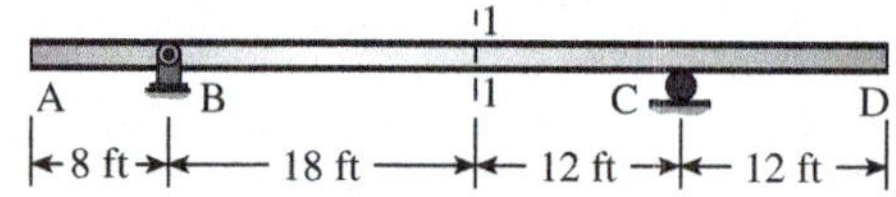

Problem 16.36

P16.37 Find the maximum upward reactions at A and D and the maximum positive moment at section 1–1. (Ans: $R_{Ay} = 75$ k, $M_{1-1} = 487.5$ k·ft)

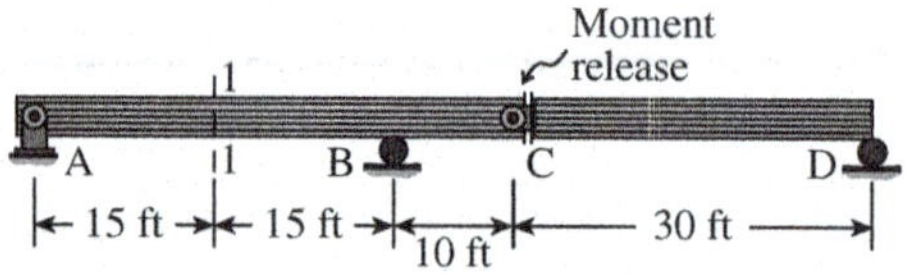

Problem 16.37

P16.38 For the beam in Problem 16.6, find the minimum reaction at A and the maximum positive moment and negative shear at section 1–1.

P16.39 For the beam in Problem 16.17, find the maximum positive moment at C and the maximum positive shear at section 1–1. (Ans: $M_C = 4500$ k·ft, $V_{1-1(\text{pos})} = 85$ k)

Section 16.14

P16.40 For the beam shown, determine the absolute maximum shear and moment due to the moving load as shown. It is anticipated that the maximum shear is found by the reader employing a trial-and-error process, while the maximum moment is found by the method outlined in Section 16.14.

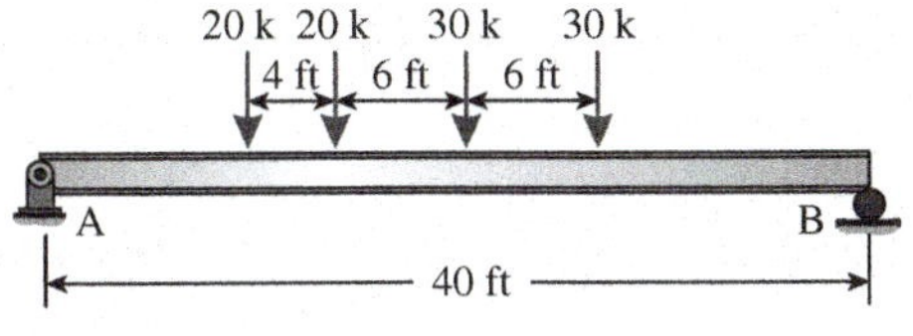

Problem 16.40

P16.41 A simple beam of 25-m span supports a pair of 80-kN moving concentrated loads which are 4-m apart. Compute the maximum possible moment at the centerline of the beam and also the absolute maximum moment in the beam. (Ans: M_{max} = 846.4 kN·m)

P16.42 The given loading scenarios describe the AASHTO design truck of Section 2.17. Compute the absolute maximum moment in a 100-ft simple beam for each load case.

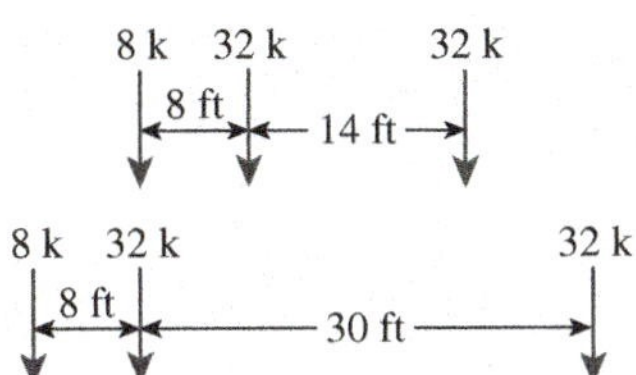

Problem 16.42

P16.43 The given loading scenario represents part of a train load as described in Section 2.18. Compute the absolute maximum moment in a 150-ft simple beam. (Ans: M_{max} = 8933 k·ft)

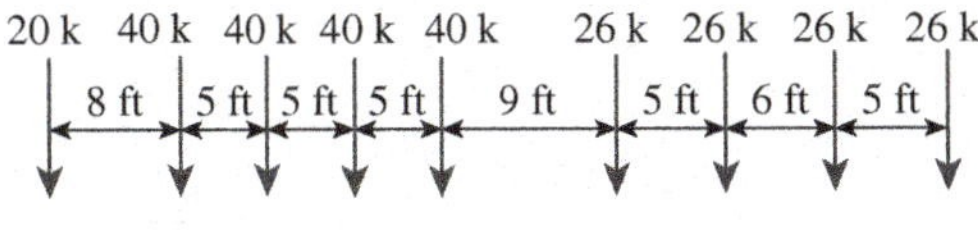

Problem 16.43

Section 16.15

For Problems 16.44 through 16.51, draw the axial force influence lines for the members indicated.

P16.44 AB, AF, and EF.

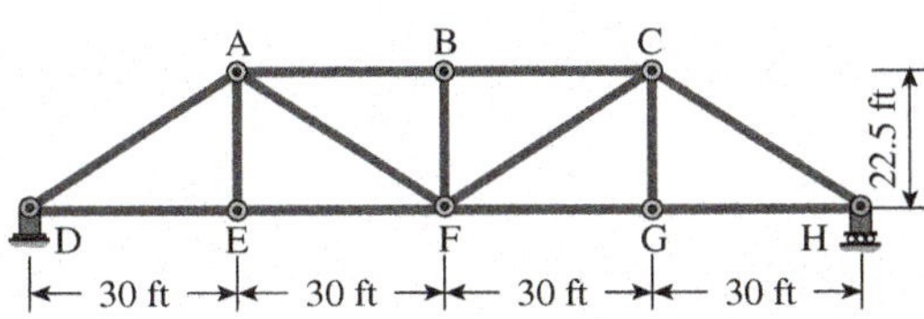

Problem 16.44

P16.45 BC, CH, HI, and CI. (Ans: load @ H: N_{BC} = −1.78, N_{CH} = 0.56)

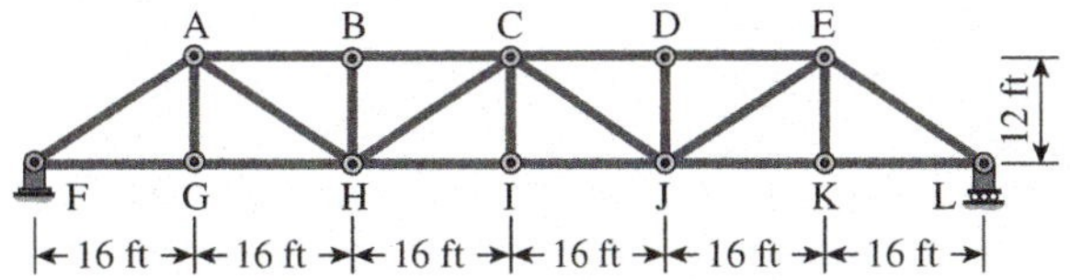

Problem 16.45

P16.46 BC, CF, and FG.

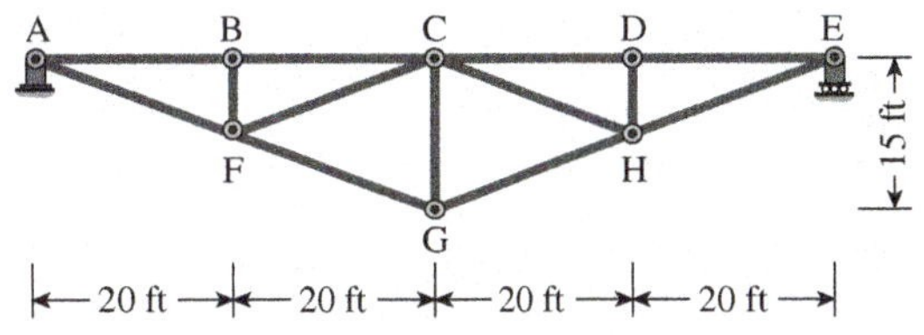

Problem 16.46

P16.47 AB, AF, and BF. (Ans: load @ G: N_{AB} = −0.97, N_{AF} = 0.83).

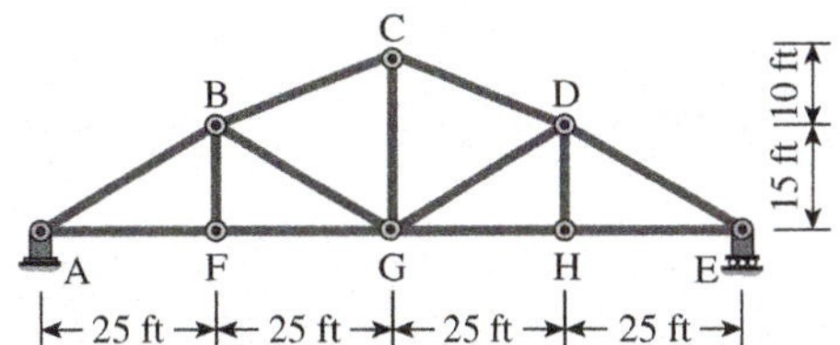

Problem 16.47

P16.48 AB, BE, EF, and GH.

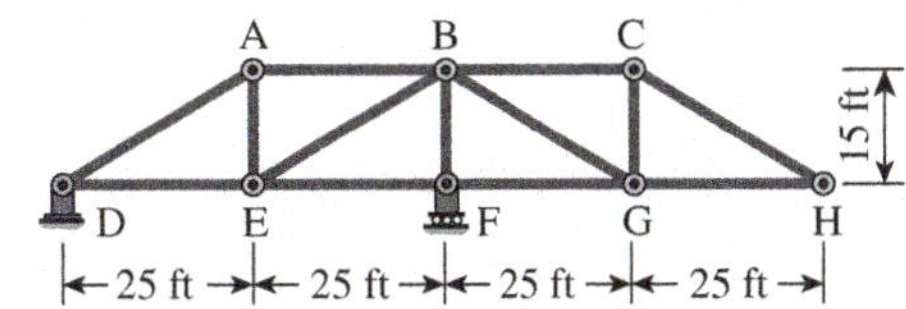

Problem 16.48

P16.49 BC, BH, and GH. (Ans: load @ F: N_{BC} = 1.69, N_{GH} = −1.33)

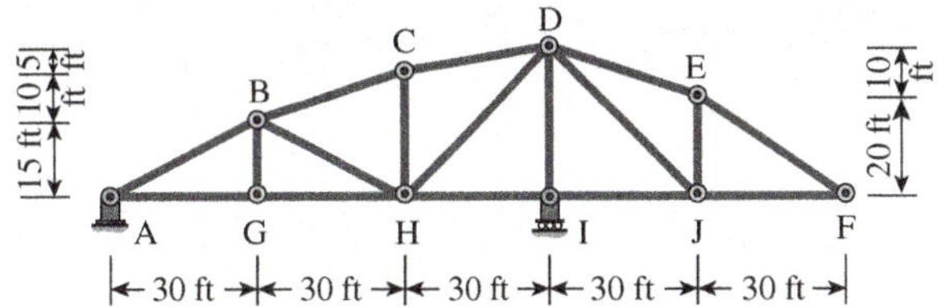

Problem 16.49

P16.50 BC, CG, and GH.

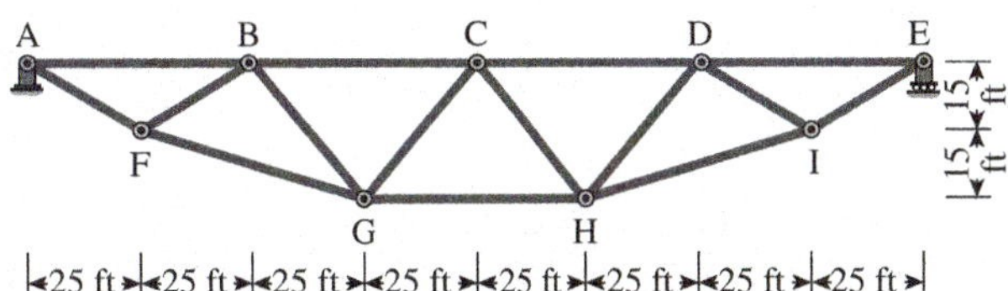

Problem 16.50

P16.51 BC, GC, GM, and LM. (Ans: load @ M: $N_{BC} = -1.0$, $N_{GC} = 0.56$)

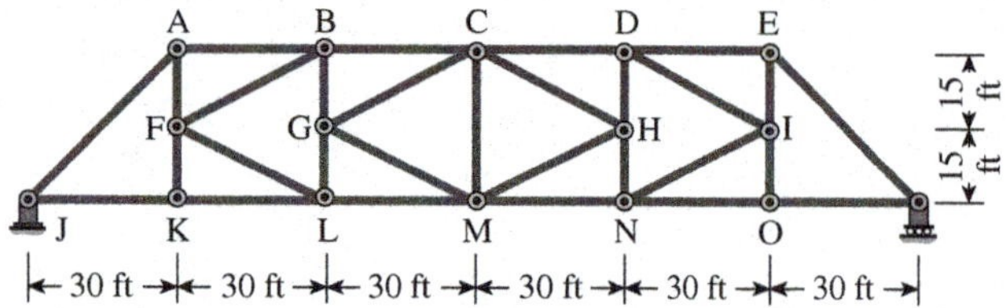

Problem 16.51

P16.52 Determine the maximum and minimum axial forces in member CH of the truss in Problem 16.45 given the following loading scenario: uniform dead load of 1 klf, a moving uniform live load of 2 klf, a moving concentrated load of 20 k, and an impact factor of 30%.

P16.53 Determine the maximum and minimum axial forces in member FG of the truss in Problem 16.46 given the following loading scenario: uniform dead load of 1.5 klf, a moving uniform live load of 1.5 klf, a moving concentrated load of 10 k, and an impact factor of 15%. (Ans: $N_{FG-\min} = 85.2$ k, $N_{FG-\max} = 199.5$ k)

P16.54 Determine the maximum and minimum axial forces in member BC of the truss in Problem 16.49 given the following loading scenario: uniform dead load of 1 klf, a moving uniform live load of 2 klf, a moving concentrated load of 10 k, and an impact factor of 10%.

Influence Lines for Statically Indeterminate Structures

17

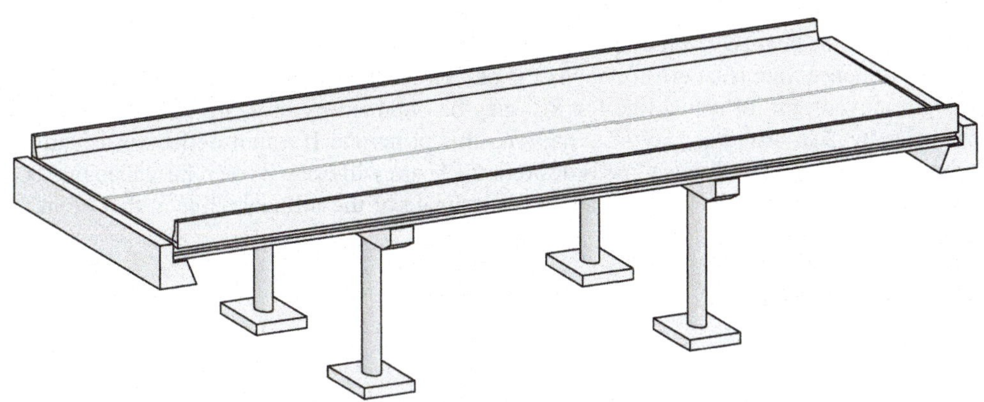

17.1 Influence Lines for Statically Indeterminate Beams

The uses of influence lines for statically indeterminate structures are the same as those for statically determinate structures. They enable the designer to locate the critical positions for placing live loads and to compute forces for various positions of the loads. For determinate structures, it is possible to compute the ordinates for a few controlling points by statics and to connect those values with a set of straight lines. Unfortunately, influence lines for continuous structures require the computation of ordinates at a large number of points because the diagrams are either curved or made up of a series of chords. The chord-shaped diagram occurs where loads can only be transferred to the structure at intervals—as is the case with trusses.

The problem of preparing the diagrams is not as difficult as the preceding paragraph seems to indicate. This is because a large percentage of the work may be eliminated by applying Maxwell's law of reciprocal deflections, as previously introduced in Chapter 11. The preparation of an influence line for the interior reaction of the two-span beam of Figure 17.1 is considered in the following paragraphs.

The procedure for calculating R_{By} in the past (see Chapter 11) has been to remove it from the beam and then compute Δ_B and Δ_{bb} and substitute their values into the compatibility equation:

$$R_{By} = -\frac{\Delta_B}{\Delta_{bb}}$$

The same procedure may be used in drawing an influence line for R_{By}. A unit load is placed at some point x, causing Δ_B to equal Δ_{bx} from which the expression is

$$R_{By} = -\frac{\Delta_B}{\Delta_{bb}} = -\frac{\Delta_{bx}}{\Delta_{bb}}$$

At first glance, it appears that the unit load will have to be placed at numerous points on the beam and the value of Δ_{bx} laboriously computed for each. A study of the deflections caused by

DID YOU KNOW?

Influence lines for statically indeterminate structures are not as simple to draw as they are for statically determinate structures. However, with a little background, the work can be minimized.

DID YOU KNOW?

The procedure described for generating quantitative influence lines for beams also can be applied to trusses. However, the deflection calculations needed for the trusses can be quite time consuming. As such, an explicit treatment of the analysis is not presented in this text. Rather, some of the exercises at the end of this chapter will use SAP2000 to help develop ILs for indeterminate trusses.

RECALL

The first subscript indicates the location of the computed deflection, and the second subscript indicates the location of the load.

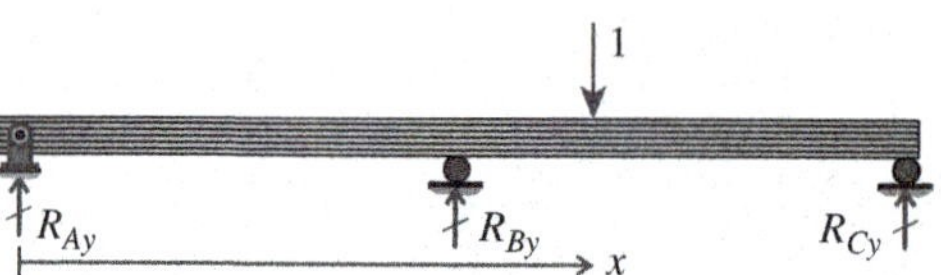

FIGURE 17.1 Two-span indeterminate beam.

a unit load at point x, however, proves these computations to be unnecessary. By Maxwell's law, the deflection at B due to a unit load at x (Δ_{bx}) is identical with the deflection at x due to a unit load at B (Δ_{xb}). Thus, the expression for R_{By} becomes

$$R_{By} = -\frac{\Delta_{xb}}{\Delta_{bb}}$$

QUICK NOTE
Essentially, the process for finding the IL for a reaction is to remove the reaction and replace it with a unit load. Calculate the deflection anywhere along the length of the beam, and then normalize these values by the displacement occurring under the unit load. This process produces a unit displacement at the location of the unit load.

It is now evident that the unit load need only be placed at B, and the deflections at various points across the beam can be computed. Dividing each of these values by Δ_{bb} gives the ordinates for the influence line. If a deflection curve is plotted for the beam for a unit load at B (support B being removed), an influence line for R_{By} may be obtained by dividing each of the deflection ordinates by Δ_{bb}. Another way of expressing this principle: If a unit deflection is caused at a support for which the influence line is desired, the beam will draw its own influence line because the deflection at any point in the beam is the ordinate of the influence line at that point for the reaction in question.

Procedure—SI 1°

1. Identify the redundant reaction force.
2. Remove the redundant force and replace it with a unit load.
3. Compute the deflection at regular intervals along the length of the beam due to the unit load that remains at the location where the redundant force had been.
4. Divide all of the computed deflections by the deflection located under the unit load. This should produce a deflection at the unit load of 1.0. This is the influence line for the redundant reaction.
5. If the influence line is desired for any other response quantity (R, V, and M), refer back to the original beam (i.e., before the redundant force was removed). Moving from left to right, place the unit load at regular intervals. Use the IL for the redundant reaction to solve for the redundant value for the unit load. Then use basic statics to solve for the desired response quantity.

QUICK NOTE
The given procedure is to generate a quantitative influence line for a beam that is statically indeterminate to the first degree. Once this is understood, the reader will more readily pick up the process for dealing with multiple degrees of indeterminacy.

The influence line for the reaction at the interior support of a two-span beam is presented in Example 17.1. Influence lines also are shown for the end reactions—the values for ordinates having been obtained by statics from those computed for the interior reaction.

EXAMPLE 17.1

Draw influence lines for the reactions at each support of the beam shown. Also find the IL for the shear just to the left of B (V_{B-L}) and the moment at B (M_B). Report the influence line ordinates at the locations shown.

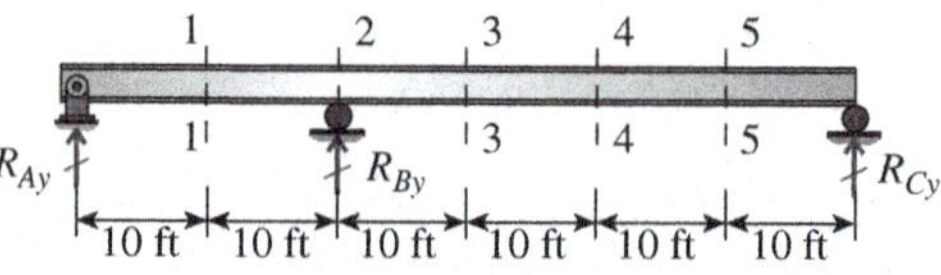

DID YOU KNOW?
There is no hard and fast rule governing the selection of the intervals used for calculating ordinates. The selection is made to ensure that a proper shape is sketched but minimize the number of ordinates that are calculated. A proper balance must be struck.

SOLUTION STRATEGY

Select R_{By} as the redundant force. Once the unit load is placed at B, the real work is in the computation of deflections. It is wise to look for any resources available. In this case, we use the beam charts found in Appendix D.

SOLUTION

Remove R_{By}, place a unit load at B, and compute the deflections occurring at 10-ft intervals.

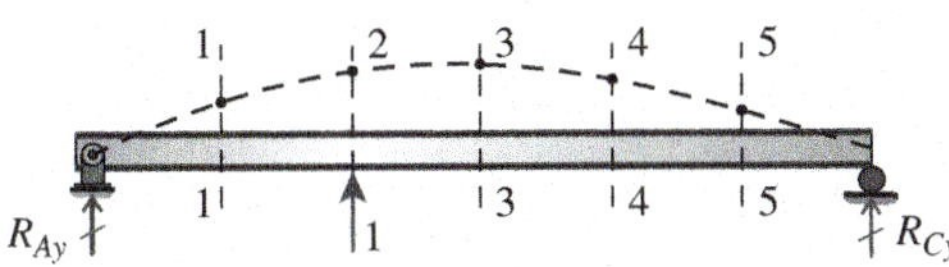

DEFLECTION EQUATION

The deflection equation for a simple supported beam subjected to a point load is given as

$$\Delta = \frac{Pbx}{6LEI}(L^2 - b^2 - x^2)$$
$$0 \le x \le a$$

$$\Delta_1 = \frac{(1)(40\text{ ft})(10\text{ ft})}{6(60\text{ ft})EI}\left[(60\text{ ft})^2 - (40\text{ ft})^2 - (10\text{ ft})^2\right] = \frac{2111}{EI}$$

$$\Delta_2 = \frac{(1)(40\text{ ft})(20\text{ ft})}{6(60\text{ ft})EI}\left[(60\text{ ft})^2 - (40\text{ ft})^2 - (20\text{ ft})^2\right] = \frac{3556}{EI}$$

For the calculation of Δ_1 and Δ_2, x is measured from the left end of the beam, but it is measured from the right end of the beam for all other deflections.

$$\Delta_3 = \frac{(1)(20\text{ ft})(30\text{ ft})}{6(60\text{ ft})EI}\left[(60\text{ ft})^2 - (20\text{ ft})^2 - (30\text{ ft})^2\right] = \frac{3833}{EI}$$

$$\Delta_4 = \frac{(1)(20\text{ ft})(20\text{ ft})}{6(60\text{ ft})EI}\left[(60\text{ ft})^2 - (20\text{ ft})^2 - (20\text{ ft})^2\right] = \frac{3111}{EI}$$

$$\Delta_5 = \frac{(1)(20\text{ ft})(10\text{ ft})}{6(60\text{ ft})EI}\left[(60\text{ ft})^2 - (20\text{ ft})^2 - (10\text{ ft})^2\right] = \frac{1722}{EI}$$

All of the computed deflections need to be normalized by the Δ_2. This will ensure that the deflection at 2–2 is 1.0.

QUICK NOTE

The normalized displacements are the ordinates of the influence line.

$$\frac{\Delta_1}{\Delta_2} = 0.594 \qquad \frac{\Delta_2}{\Delta_2} = 1.000$$

$$\frac{\Delta_3}{\Delta_2} = 1.078 \qquad \frac{\Delta_4}{\Delta_2} = 0.875$$

$$\frac{\Delta_5}{\Delta_2} = 0.484$$

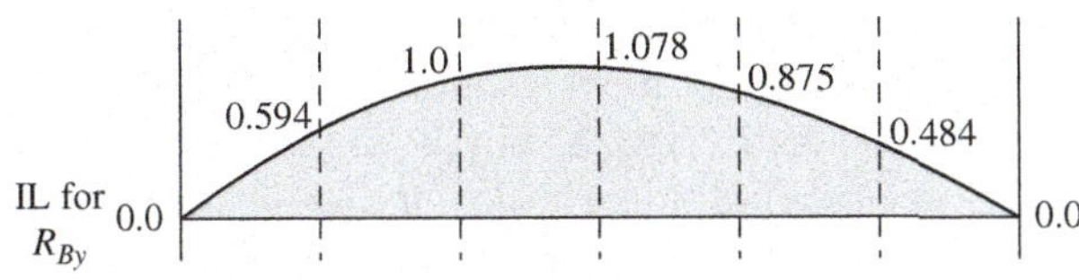

With the IL for R_{By} in hand, the remaining ILs may be found using the same approach as for determinate beams. That is, we will place a unit load at the locations of interest and solve for the quantities requested. For the sake of brevity, only the first location analysis is shown in detail.

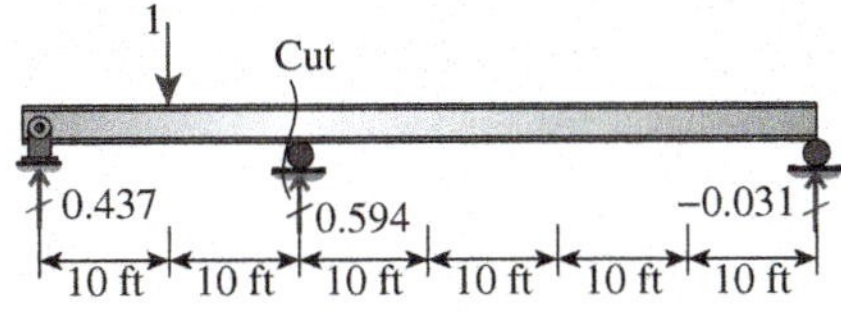

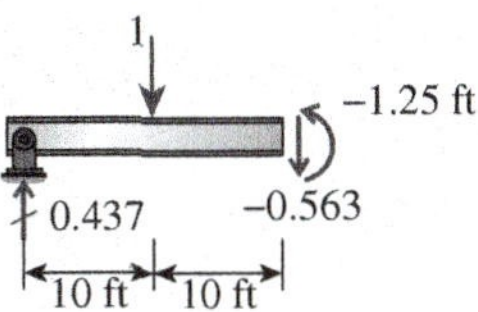

When the unit load is placed at 1–1, the IL says that $R_{By} = 0.594$. Summing moments about point A will then provide $R_{Cy} = -0.031$, and summing forces in the y-direction will give $R_{Ay} = 0.437$. Finally, a cut is made just to the left of point B, a FBD of the left section is made, and the shear and moment are computed to be $V_{B-L} = -0.563$ and $M_B = -1.25$ ft.

RECALL
When finding the IL for shear, you will need to place the unit load just to the left of the cut then just to the right of the cut.

Follow the same process for the other locations of the unit load (i.e., 2–2, 3–3, 4–4, 5–5, and C). This will produce the following ILs.

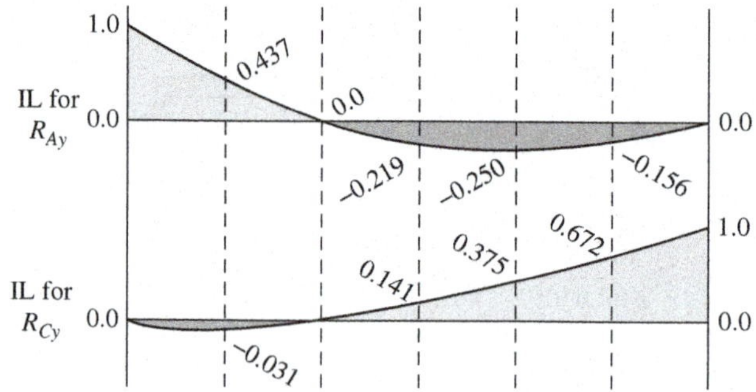

QUICK NOTE
Recall that, for determinate beams, there should be a unit jump in the shear IL at the location where the shear is being found. As seen here, there is no difference for indeterminate beams.

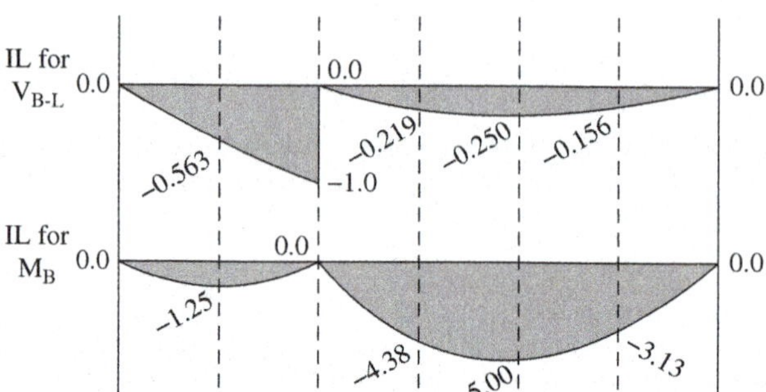

DID YOU KNOW?
The simultaneous equations are solved quickly, even though a large number of ordinates are being computed, because the only variables in the equations are Δ_{xb} and Δ_{xc}.

The next problem is to draw the influence lines for beams continuous over three spans that have two redundants. For this discussion, the beam of Figure 17.2 is considered, and the reactions R_{By} and R_{Cy} are assumed to be the redundants.

It will be necessary to remove the redundants and compute the deflections at various sections in the beam for a unit load at B and for a unit load at C.

By Maxwell's law, a unit load at any point x causes a deflection at B (Δ_{bx}) equal to the deflection at x due to a unit load at B (Δ_{xb}). Similarly, $\Delta_{cx} = \Delta_{xc}$. After computing Δ_{xb} and Δ_{xc} at the several sections, their values at each section may be substituted into the following simultaneous equations, whose solution will yield the values of R_{By} and R_{Cy}.

$$\Delta_{xb} + R_{By}\Delta_{bb} + R_{Cy}\Delta_{bc} = 0$$

$$\Delta_{xc} + R_{By}\Delta_{cb} + R_{Cy}\Delta_{cc} = 0$$

After the influence lines are prepared for the redundant reactions of a beam, the ordinates for any other response (moment, shear, and so on) can be determined by statics, as done in Example 17.1. Example 17.2 illustrates the calculations necessary for preparing influence lines for several responses of a three-span continuous beam.

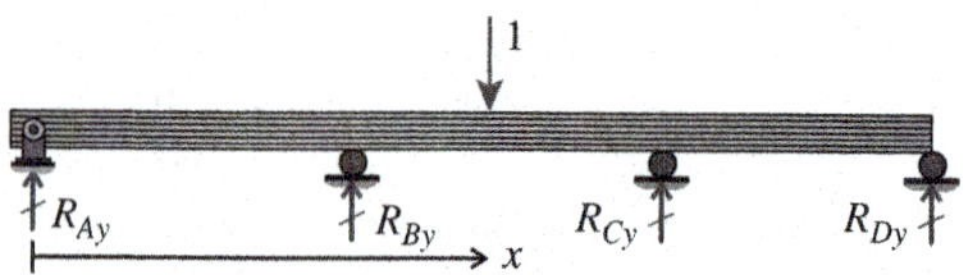

FIGURE 17.2 Three-span indeterminate beam.

EXAMPLE 17.2

Draw influence lines for R_{By}, R_{Cy}, and R_{Dy}; moment at section 7, M_7; and shear at section 6, V_6, in the beam shown.

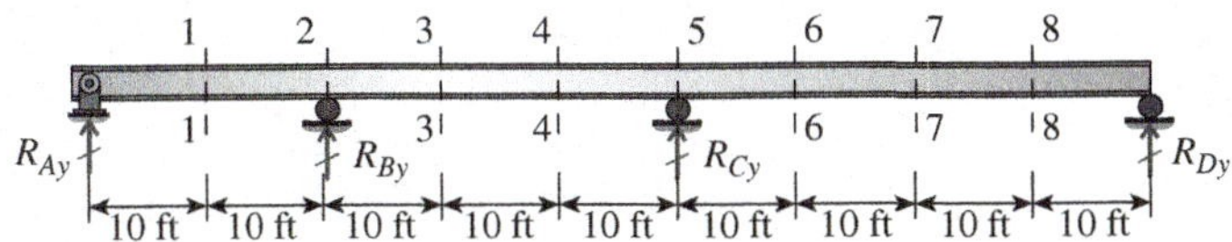

SOLUTION STRATEGY

Choose redundant forces such that you will be left with a beam for which deflections are easy to calculate. Selecting R_{By} and R_{Cy} as the redundants leaves a simply supported beam. The equations and procedures shown in Example 17.1 then may be followed.

QUICK NOTE

This process requires many repetitive calculations. A spreadsheet is a great way to perform these calculations without all of the tedium. Indeed, all of the calculated values given in the table in this example were done using a spreadsheet.

SOLUTION

Remove R_{By} and R_{Cy}. Draw two beams placing a unit load at B on one of them and a unit load at C on the other one. Use the deflection equation given in Example 17.1 to find the deflection at each location. These deflections should be in terms of EI. However, since it is constant, we don't need to include it.

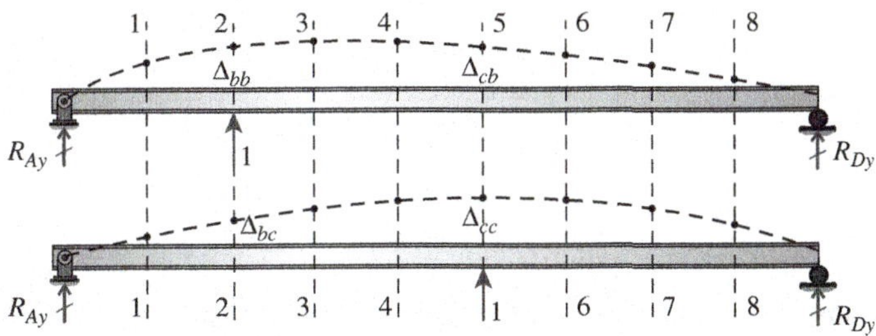

The following table includes the values which come from the deflection equation. These show up under Δ_{xb} and Δ_{xc}. The Δ_{xb} values are deflections in the first beam, and Δ_{xc} values are deflections in the second beam.

Location	Section	Δ_{xb}	Δ_{xc}	R_{By}	R_{Cy}	R_{Dy}	M_7	V_6
	1	4019	4741	0.645	−0.073	0.009	0.17	−0.009
B	2	7259 = Δ_{bb}	9037 = Δ_{bc}	1.000	0	0	0	0
	3	9111	12444	0.870	0.309	−0.032	−0.64	0.032
	4	9630	14519	0.443	0.710	−0.048	−0.97	0.048
C	5	9037 = Δ_{cb}	14815 = Δ_{cc}	0	1.000	0	0	0
	6	7556	13056	−0.234	1.024	0.150	3.00	−0.150
								0.850
	7	5407	9630	−0.267	0.813	0.385	7.71	0.615
	8	2815	5093	−0.167	0.446	0.678	3.57	0.322

The values of R_{By} and R_{Cy} are determined from solving simultaneous equations. A sample calculation for the first row in the table is given here.

$$\Delta_{1b} + R_{By}\Delta_{bb} + R_{Cy}\Delta_{bc} = 0$$
$$\Delta_{1c} + R_{By}\Delta_{cb} + R_{Cy}\Delta_{cc} = 0$$

Substitute known values into this equation and solve using any method you choose:

$$4019 + R_{By}(7259) + R_{Cy}(9037) = 0$$

$$4741 + R_{By}(9037) + R_{Cy}(14815) = 0$$

$$\therefore R_{By} = \underline{\underline{0.645}} \qquad R_{Cy} = \underline{\underline{-0.073}}$$

DID YOU KNOW?

A convenient way to solve simultaneous equations, which is conducive for programming into spread sheets, is to use the matrix formulation.

$$\begin{Bmatrix} R_{By} \\ R_{Cy} \end{Bmatrix} = \begin{bmatrix} \Delta_{bb} & \Delta_{bc} \\ \Delta_{cb} & \Delta_{cc} \end{bmatrix}^{-1} \begin{Bmatrix} \Delta_{xb} \\ \Delta_{xc} \end{Bmatrix}$$

With the reactions at B and C found for all locations of the unit load, we can now use basic statics to solve for R_{Dy}, M_7, and V_6. Once the table is complete, a plot of each influence line may be constructed.

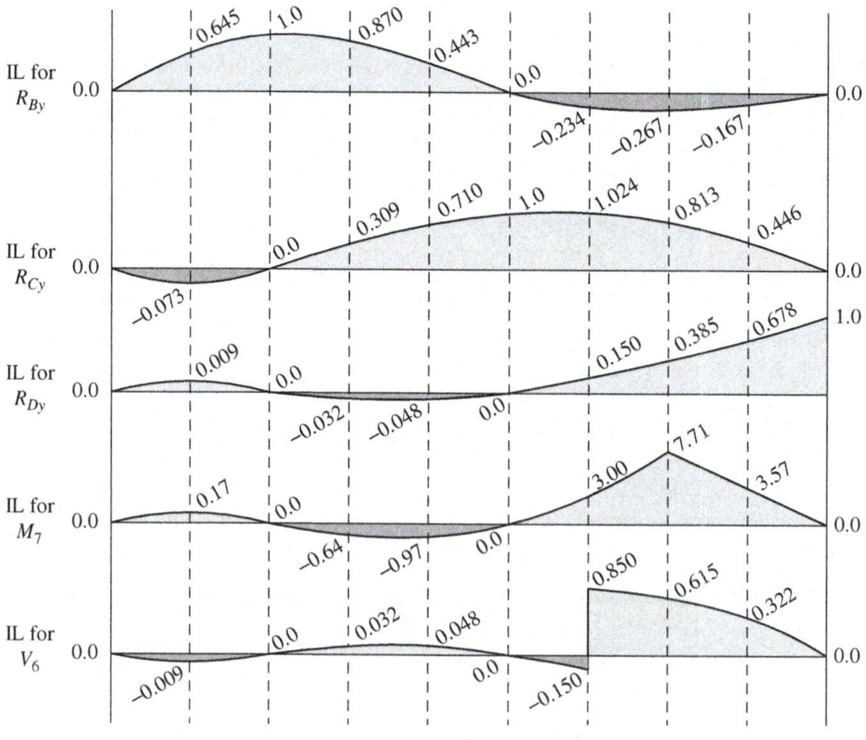

Influence lines for the reactions, shears, moments, and so on for frames can be prepared exactly as they are for the statically indeterminate beams just considered. The same notion may be applied to trusses as well. Space is not taken to present such calculations. The next section of this chapter is concerned with the preparation of qualitative influence lines. The purpose of this is to show the reader how to place live loads in beams and frames so as to cause maximum values.

17.2 Qualitative Influence Lines for Indeterminate Beams and Frames

Maxwell's presentation of his theorem in 1864 was so brief that its value was not fully appreciated until 1886 when Heinrich Müller-Breslau clearly showed its true worth, as demonstrated in the previous chapter.[1] Müller-Breslau's principle may be stated in detail as follows.

> *The deflected shape of a structure represents, to some scale, the influence line for a response such as a stress, shear, moment, or reaction component if the function is allowed to act through a unit displacement.*

This principle, which is applicable to statically determinate and indeterminate beams, frames, and trusses, is more fully explained hereafter.

Müller-Breslau's principle is based upon Castigliano's theorem of least work. This theorem can be expressed as follows.

> *When a virtual displacement is induced into a structure, the total virtual work done by all the active forces must equal zero.*

RECALL

Our introduction to virtual work was presented in Chapter 9 dealing with deflection calculations. In that chapter, virtual loads were applied that acted over a real distance, and work was calculated as force times a distance. The virtual work referred to in this chapter deals with a real load acting over a virtual displacement.

Proof of Müller-Breslau's principle can be made by considering the beam of Figure 17.3(b) when it is subjected to a moving unit load. To determine the magnitude of the reaction R_{By}, we can remove the support at B and allow R_{By} to move through a small virtual distance Δ_B, as shown in part (b) of the figure. Notice the beam's position is now represented by the line AB′ in the

[1] J.S. Kinney, *Indeterminate Structural Analysis* (Reading, Mass.: Addison-Wesley, 1957), 14.

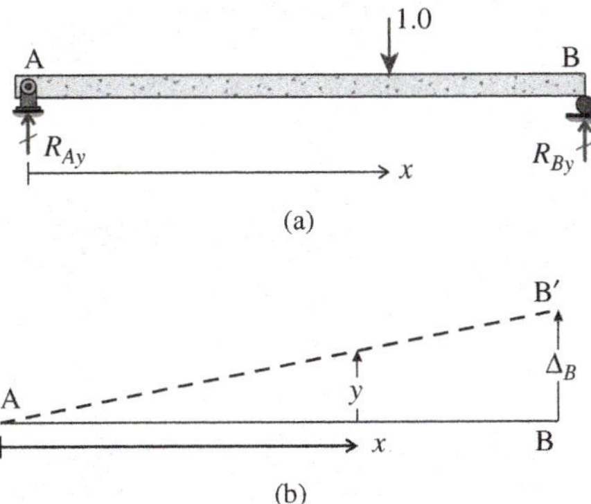

FIGURE 17.3
Demonstration of Müller-Breslau's principle.

figure, and the unit load has been moved through a positive virtual distance of y. We know that the virtual work must all be equal to zero, and as such, we can write

$$R_{By}(\Delta_B) + (-1.0)(y)$$
$$R_{By} = \frac{y}{\Delta_B}$$

QUICK NOTE

The negative in front of the 1.0 is present because the unit load is acting in the opposite direction of the virtual displacement y. This produces negative work.

If Δ_B is set equal to 1.0—as is done when generating a qualitative IL for R_{By}—then R_{By} is equal to y.

$$R_{By} = \frac{y}{1.0} = y$$

You can now see that y is the ordinate of the deflected beam at the location of the unit load. It is also the value of the reaction R_{By} due to the unit load located at x. Similar proofs can be developed for the influence lines for other functions of a structure, such as shear and moments.

Müller-Breslau's principle is of such importance that space is taken here to emphasize its value. The shape of the usual influence line needed for continuous structures is so simple to obtain from his principle that (in many situations) it is unnecessary to perform the rather tedious labor needed to compute the numerical values of the ordinates. It is possible to sketch the diagram roughly with sufficient accuracy to locate the critical positions for the live load for various responses of the structure. This possibility is of particular importance for building frames, as will be illustrated in subsequent paragraphs.

DID YOU KNOW?

The deflected beam position AB′ represents the influence line for R_{By} as evidenced by

$$R_{By} = y$$

Procedure

The process for developing qualitative influence lines for indeterminate beams and frames is really the same as it was for determinate beams, as introduced in Chapter 16. However, it can be difficult to calculate the ordinates of the resulting influence lines. Furthermore, just having the shape of the influence line is often good enough to be able to locate the loads at the critical locations. As such, the influence lines for indeterminate beams and frames will truly be qualitative in that they will have no numbers associated with them. The following provides a few key concepts which must be remembered when generating qualitative influence lines using the Müller-Breslau principle.

- For a reaction IL, impose a positive displacement—technically a unit—at the location of the reaction. Sketch the deflected shape of the beam/frame to reasonable scale.
- An IL for a positive moment requires that a moment release is inserted at the point of interest; then a positive moment couple is applied to both sides of the release. Sketch the deflected shape so that it is consistent with the positive moment couple.
- An IL for a negative moment requires that a moment release is inserted at the point of interest; then a negative moment couple is applied to both sides of the release. Sketch the deflected shape so that it is consistent with the negative moment couple.
- An IL for shear requires that a shear release is inserted at the point of interest; then a positive shear couple is applied to both sides of the release. Sketch the deflected shape so that it is consistent with the positive shear couple.

QUICK NOTE

A quick review of how to sketch deflected shapes—as presented in Chapter 8—is recommended at this time. Be sure to focus on the support conditions and joints in the frames.

EXAMPLE 17.3

For the continuous beam shown, sketch the qualitative influence lines for the following structural responses: R_{Ay}, R_{Cy}, positive M_{1-1}, and V_{2-2}.

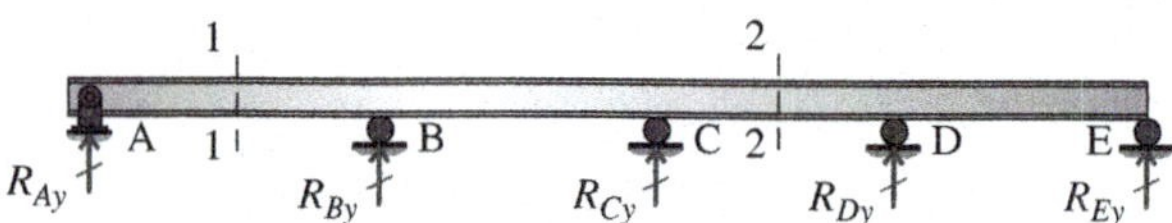

SOLUTION STRATEGY

Understand and follow the guidance given for each of the requested responses. When imposing the required displacement/rotation, ensure that all other support conditions are enforced.

SOLUTION

DESCRIPTION

The following is the action taken for each IL generation.

R_{Ay} Impose a positive displacement at A. Points B, C, D, and E don't translate but they do rotate.

R_{Cy} Impose a positive displacement at C. Points A, B, D, and E don't translate but they do rotate.

M_{1-1} Insert a moment release at section 1–1 and sketch a positive moment couple. Members to the left and right of the moment release rotate according to the moment couple applied.

V_{2-2} Insert a shear release at section 2–2 and sketch a positive shear couple. Members to the left and right of the shear release displace in the direction of the shear arrow.

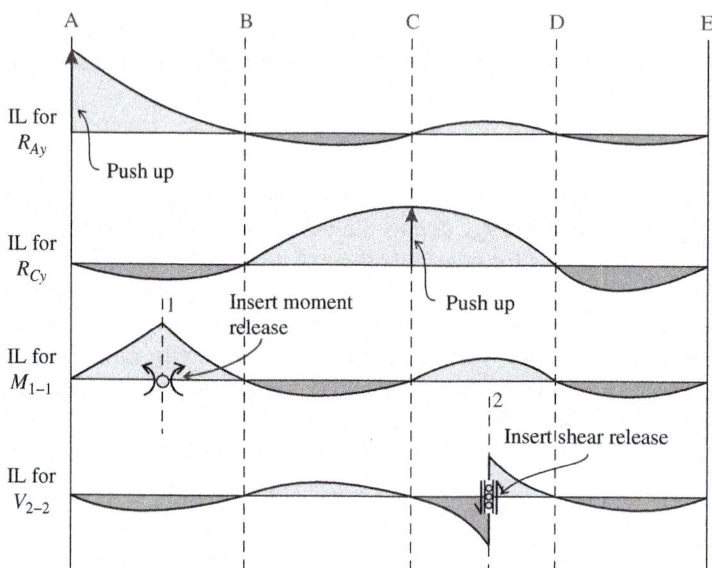

Sketching ILs for shear and moment at supports can be a bit challenging for many students. This challenge is not because of a lack of understanding of the procedure. Rather, the deflected shapes seem to be a little less intuitive. This difficulty is overcome—at least partially—by remembering the following points.

- A member will rotate in the direction of the applied moments. This is why we sketch the moment couple after the moment release is inserted. Simply rotate the members in the direction of the arrows.
- The slope on either side of a shear release must be the same. This will require that the displacement on each side of the release must be going in the opposite directions.
- If a moment release is inserted directly over a support, the release itself cannot move up or down. However, the beam segments will still rotate according to the direction arrows.
- If a shear release is placed immediately adjacent to a support, the beam cannot move up or down on the side of the release where the support is located. This will require that all of the differential movement between both sides of the shear release must come from one side.

The following example demonstrates these guidelines.

EXAMPLE 17.4

For the continuous beam shown in Example 17.3, sketch the qualitative influence lines for the following structural responses: negative moment over support C ($-M_C$) and shear just to the left of support D (V_{D-L}).

SOLUTION STRATEGY

Understand and follow the guidance given for each of the requested responses. When imposing the required displacement/rotation, ensure that all other support conditions are enforced.

SOLUTION

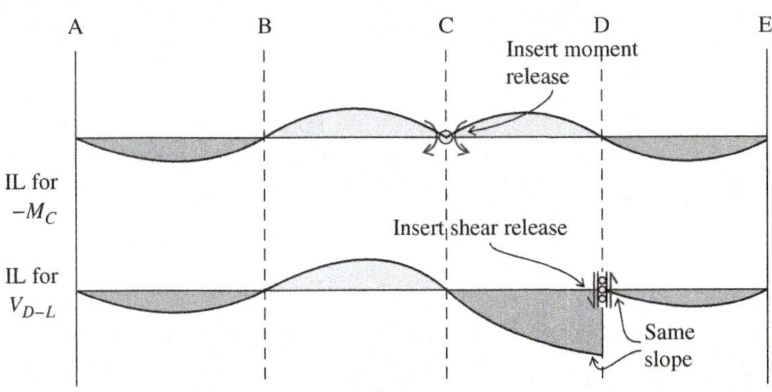

DESCRIPTION

The following are the actions taken for each IL generation.

M_C Insert a moment release just above support C. Since the problem is asking for negative moment, a negative moment couple is sketched. Since the release is above the support, it can't translate up or down. However, the members to the left and right of the moment release rotate according to the moment couple applied. All other support conditions are enforced.

V_{D-L} Insert a shear release just to the left of support D. Sketch a positive shear couple. The right side of the shear release is over the support, so it can't move up or down. The left side of the shear release displaces in the direction of the shear arrow. Remember that the slope on each side of the shear release must be the same.

Influence lines for buildings can be constructed using the same principles as outlined for continuous beams. Within frames the joints are free to rotate. However, each joint is assumed to be rigidly connected. The angles between the members in a joint do not change—even though the overall joint may rotate.

A warning should be given regarding qualitative influence lines. They should be drawn for responses near the center of spans or at the supports, but for sections near one-fourth points, they may not be sketched without a good deal of study. Near the one-fourth point of a span is a point called the fixed point at which the influence line changes in type. The subject of fixed points is discussed at some length in the book *Continuous Frames of Reinforced Concrete* by H. Cross and N. D. Morgan.[2]

QUICK NOTE

Refer to Chapter 8 for guidance on sketching deflected shapes for frames.

EXAMPLE 17.5

For the given three-story frame. Sketch the ILs for the positive moment in the midspan of beam IJ (M_{IJ}).

SOLUTION STRATEGY

Always start by sketching in the appropriate release and then sketch the span that is in question. Once this is done, the remaining spans and columns may be appropriately sketched.

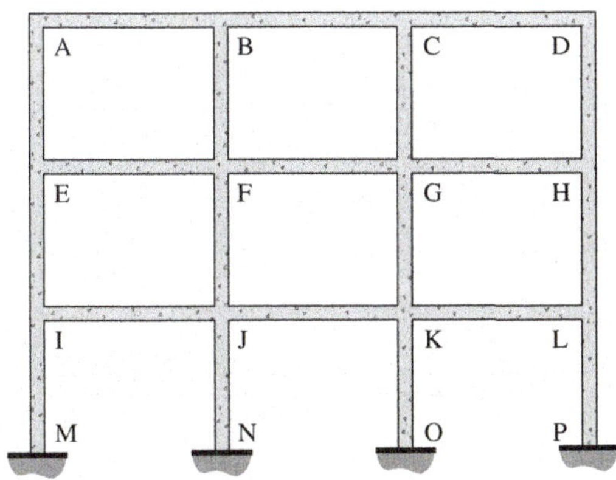

QUICK NOTE

Remember to enforce all of the support conditions. Joints M, N, O, and P don't allow any translation or rotation at their respective locations.

[2]H. Cross and N. D. Morgan, *Continuous Frames of Reinforced Concrete* (New York: Wiley, 1932).

SOLUTION

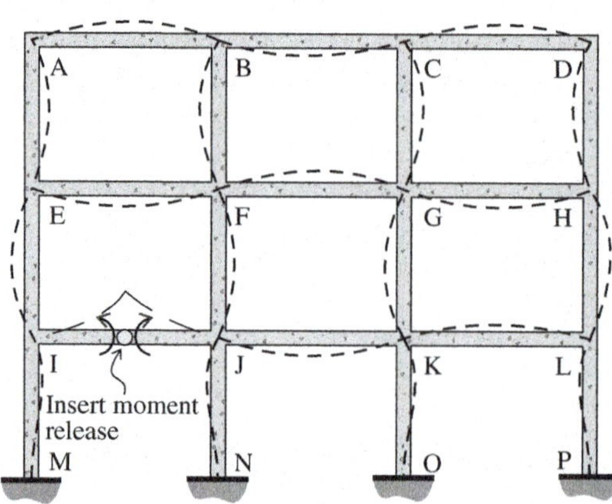

DID YOU KNOW?

Sketching ILs for structural responses near joints—such as the moment just to the left of joint G—can be a real challenge. It is not obvious what the deflected shape should look like. This is because some of the members are more of an S-shape, but this is difficult to identify qualitatively.

17.3 Influence Lines for Determining Loading Scenarios for Continuous Systems

RECALL

Chapter 16 presented how to use ILs to compute structural responses due to uniformly distributed loads. The approach is to compute the area under the IL and multiply it by the intensity of the distributed load.

The main benefit and purpose for sketching qualitative ILs for indeterminate beams and frames is the information they provide regarding the placement of lives loads for maximum effects. Maximum load effects are realized when the uniformly distributed live load is placed only over spans that have the same sign assigned to the IL. This means either place the live load only on the spans whose ILs are positive or only on the spans that are negative. Placing the live load concurrently on spans that are both positive and negative results in responses that are not maximum.

EXAMPLE 17.6

Consider the ILs presented in Examples 17.3 and 17.4. Determine where the live load should be placed to cause the maximum values of the following responses: R_{Ay}, R_{Cy}, $-M_C$, and $-V_{D-L}$.

SOLUTION STRATEGY

Only place the live load on the spans that add to the desired effect.

SOLUTION

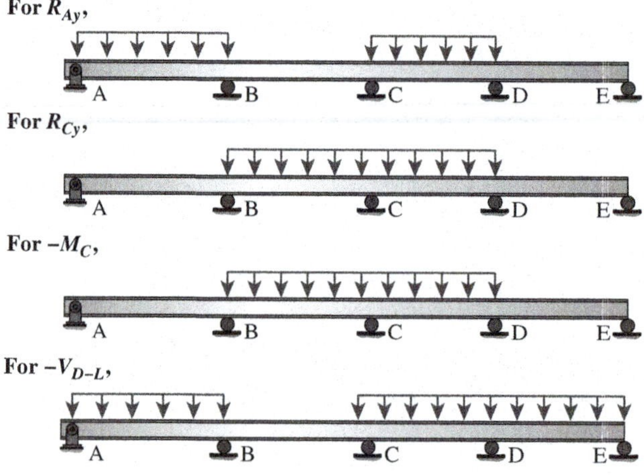

QUICK NOTE

- The IL for M_C was developed for the negative moment at C. Thus, the live load is placed wherever the IL is positive. This will give the maximum negative moment at C.
- The IL for V_{D-L} was developed for the positive shear just to the left of D. Therefore, if we want the maximum negative shear at D, we should place the live load only where the IL is negative.

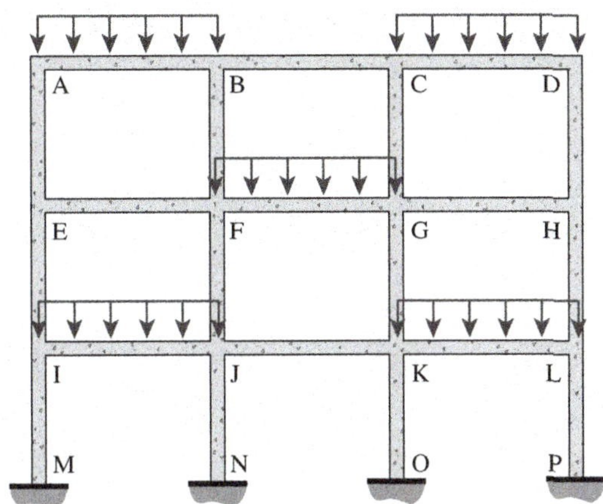

FIGURE 17.4 Live load position for maximum positive moment in beam IJ.

The placement of live loads to cause maximum effects is sometimes referred to as *skip loading* or *checkerboard loading*. This is because of the frequent placement of loads on alternating spans, as is clearly seen in Figure 17.4. In this figure, the live load placement that will cause the maximum positive moment in beam IJ of Example 17.5 is shown.

17.4 Examples with Video Solution

VE17.1 Use the quantitative approach to generate the ILs for the beam of Figure VE17.1. R_{Ay}, V_{1-1}, and M_{1-1}. Use 6-ft intervals.

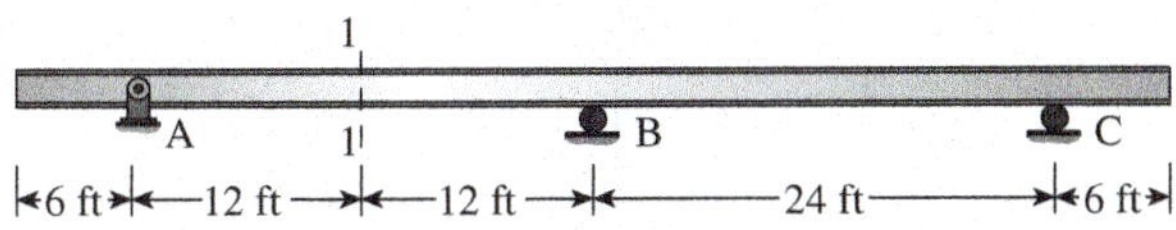

Figure VE17.1

VE17.2 Use the qualitative approach to generate the ILs for the beam given in Figure VE17.1. R_{By}, V_{C-L} (shear to the left of C), and $+M_B$. Use 6-ft intervals.

17.5 Problems for Solution

Section 17.1

For Problems 17.1 through 17.6, draw the quantitative influence lines for the situations listed.

P17.1 Reactions for all supports of the beam shown. Place the unit load at 10-ft intervals. (Ans: load @ 10 ft: $R_{By} = 0.688$, $R_{Cy} = -0.094$)

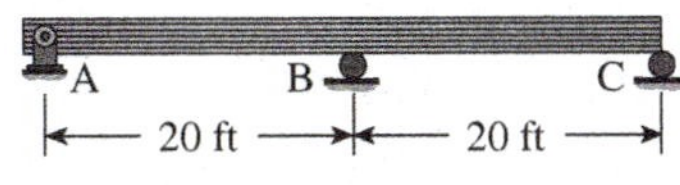

Problem 17.1

P17.2 The vertical reaction and moment at A. Place the unit load at 5-ft intervals.

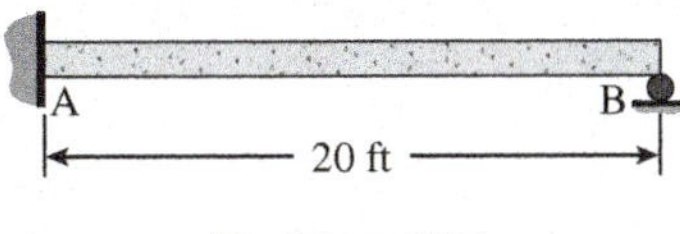

Problem 17.2

P17.3 Shear immediately to the left of support B and the moment at B. Place the unit load at 10-ft intervals. (Ans: load @ 20 ft: $R_{By} = 1.0$, $R_{B-\text{left}} = -1.0, 0.0$)

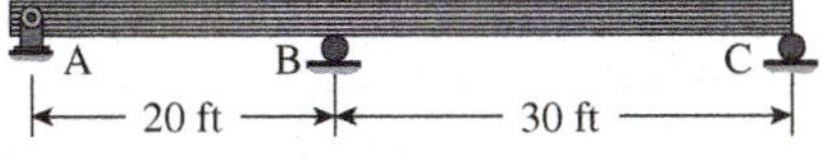

Problem 17.3

P17.4 Shear and moment at section 1–1. Place the unit load at 10-ft intervals.

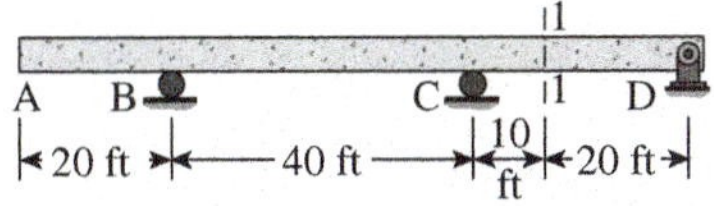

Problem 17.4

P17.5 Vertical reactions and moments at A and B. Place the unit load at 10-ft intervals. (Ans: load @ 40 ft: $R_{By} = 1.50$, $M_A = 5$ ft)

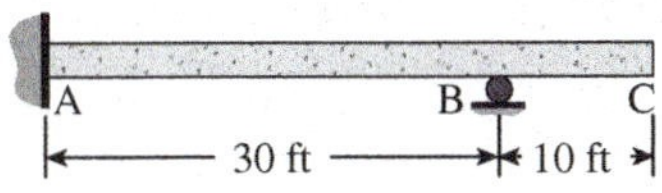

Problem 17.5

P17.6 Vertical reactions at A and B and the moment at section 1–1. Place the unit load at 10-ft intervals.

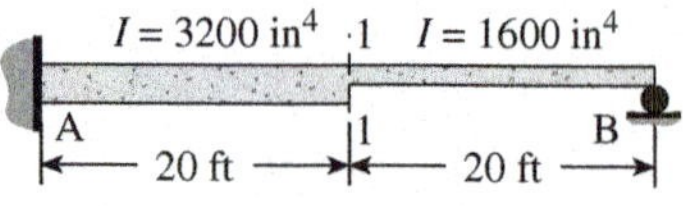

Problem 17.6

Section 17.2

For Problems 17.7 through 17.14, use Müller-Breslau's principle to draw the qualitative influence lines for the situations listed.

P17.7 Reactions at A and C, positive moment at section 1–1, and the positive shear at section 2–2.

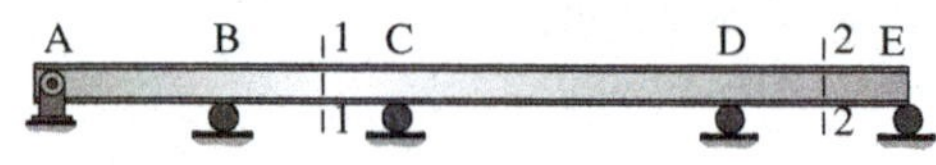

Problem 17.7

P17.8 Reaction at A, positive moment and positive shear at section 1–1, and positive shear just to the right of support D.

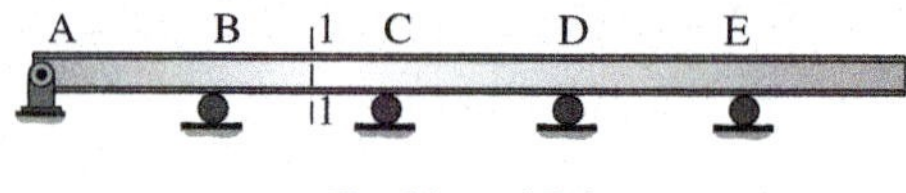

Problem 17.8

P17.9 Vertical reactions at A and C, negative moment at A, and positive shear just to the right of B.

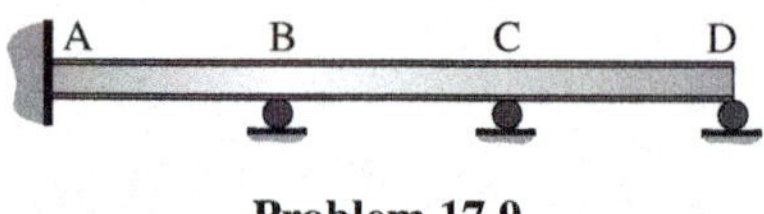

Problem 17.9

P17.10 Vertical and moment reaction at C, negative moment, and negative shear at section 1–1.

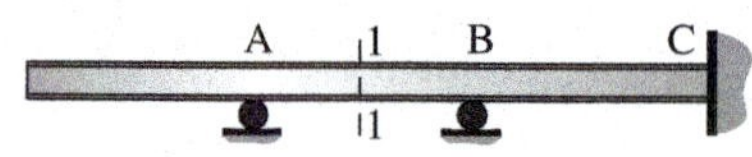

Problem 17.10

P17.11 Reaction at A, positive moment at section 1–1, positive shear at section 2–2, and negative shear at section 3–3.

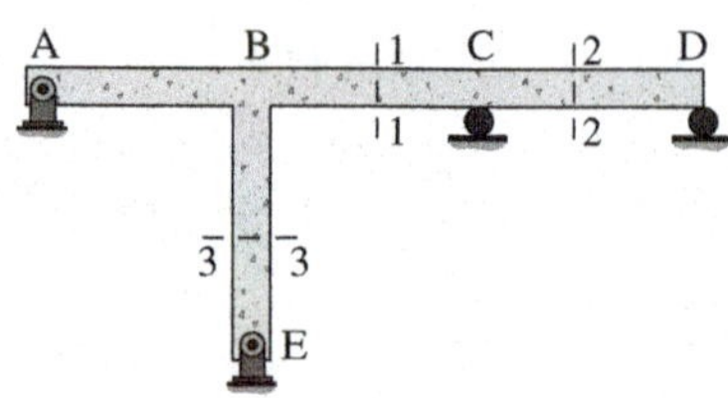

Problem 17.11

P17.12 Positive moment at section 1–1; negative shear just to the right of point 2.

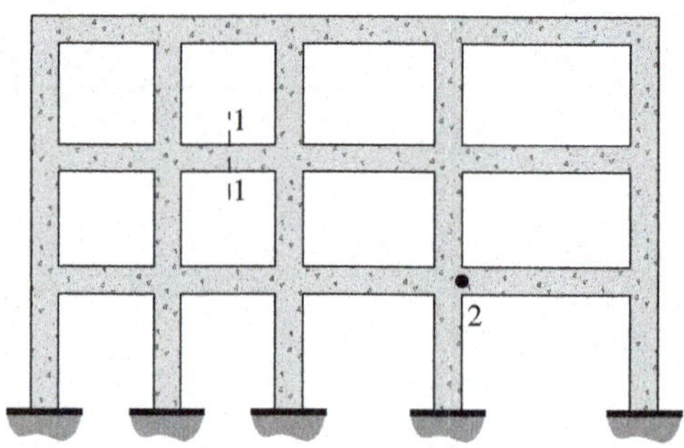

Problem 17.12

P17.13 Negative moment at section 1–1; positive shear just to the right of point 2.

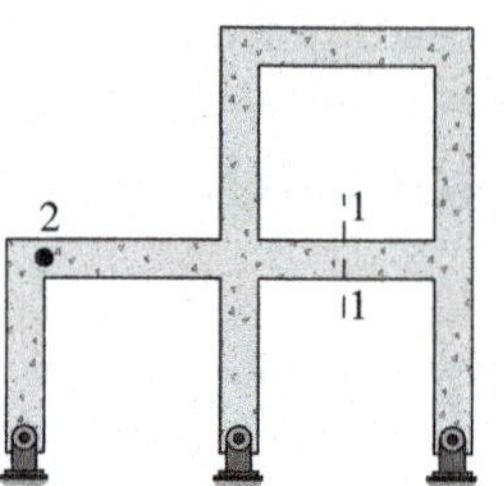

Problem 17.13

P17.14 Positive moment at section 1–1 and positive shear just to the right of point 2.

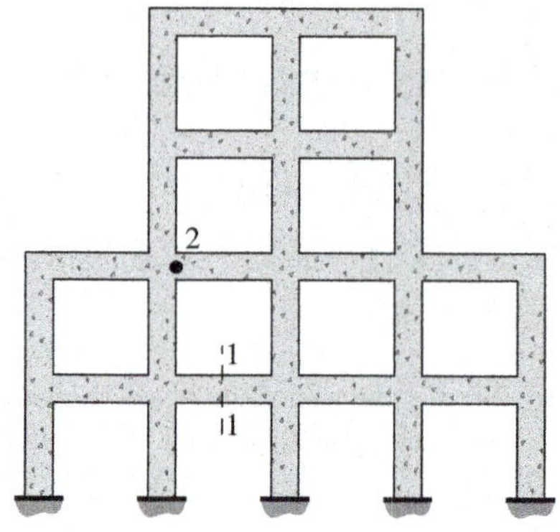

Problem 17.14

Section 17.3

For Problems 17.15 through 17.18, use qualitative influence lines to determine the loading locations for uniformly distributed live load to cause the maximum response for the situations listed.

P17.15 For the beam given in Problem 17.7, find the positive shear at section 1–1, positive moment at C, and the reaction at E.

P17.16 For the beam given in Problem 17.10, find the positive internal moment at C, the positive moment at section 1–1, and the shear just to the right of B.

P17.17 For the frame given in Problem 17.12, find the positive shear at section 1–1 and the positive moment just to the right of point 2.

P17.18 For the frame given in Problem 17.14, find the negative moment at section 1–1 and the positive shear just to the right of point 2.

SAP2000

P17.19 Use SAP2000 to model the truss shown where E = 29,000 ksi, The cross-sectional areas of the top and bottom chords are 2.0 in^2 each, while all other members are 1.0 in^2 each. Generate the IL for the force in member CI by moving a unit load across the lower joints in the truss. Now change the lower chord area to 1.0 in^2 and generate the IL again. Does this change the IL as compared to the original scenario?

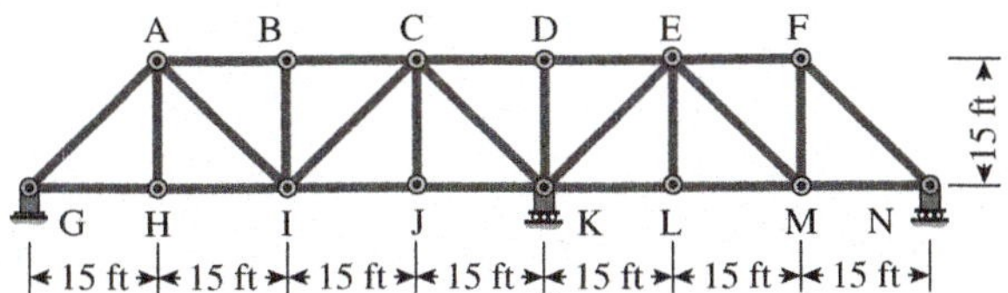

Problem 17.19

P17.20 Use SAP2000 to model the truss shown where E = 29,000 ksi. The cross-sectional area for each member is 1.0 in^2. Generate the IL for the force in member FG by moving a unit load across the lower joints in the truss. Now change the cross-sectional area of member BC to 2.0 in^2 and generate the IL again. Do it one more time where BC is 4.0 in^2. Does this change the IL as compared to the original scenario?

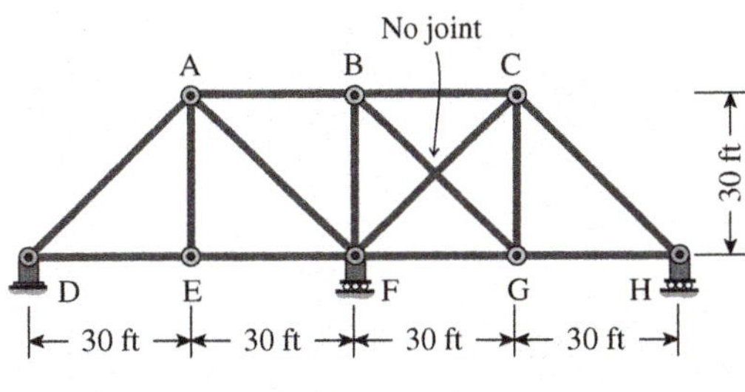

Problem 17.20

MATRIX METHODS FOR STRUCTURAL ANALYSIS

PART 4

18 Introduction to Matrix Methods

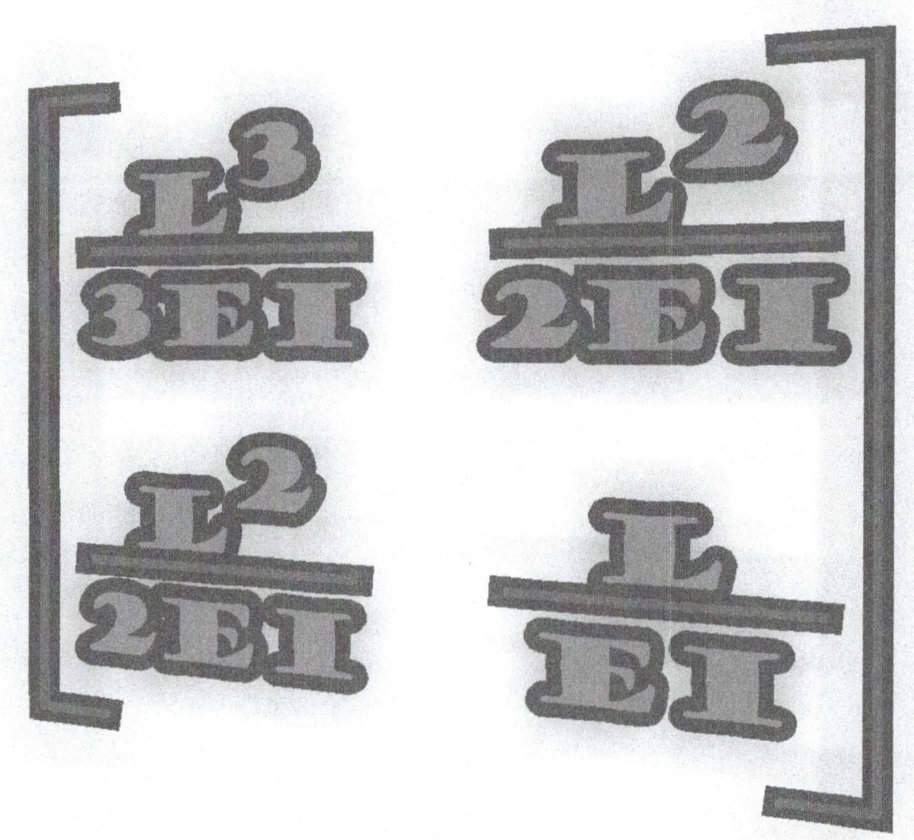

18.1 Structural Analysis Using the Computer

DID YOU KNOW?
A computer analysis of structures is only as good as the input information provided by the analyst. The student ought to see computer analysis as a way to minimize the need for tedious calculations, but it is not a replacement for the analyst's sound understanding of structural behavior and analysis.

During the past four decades, there have been tremendous changes in the structural analysis methods used in engineering practice. These changes have primarily occurred because of the great developments made with high-speed digital computers and the increasing use of very complex structures. Matrix methods of analysis provide a convenient mathematical language for describing complex structures, and the necessary matrix manipulations can easily be handled with computers.

The decreasing cost of personal computers has escalated this trend and has made impressive computing power available in almost every design office. It is therefore important for all students of structural engineering to understand the fundamentals of structural analysis as performed on the computer and to appreciate both the strengths and the weaknesses of this type of analysis.

Structural analysis, as performed on the computer, involves no new concepts of structural engineering. However, the organization of the work must be both versatile and precise. The computer is capable of extraordinary feats of arithmetic, but it can only do those tasks that can be described with simple, precise, and unambiguous instructions. In the following chapters, the authors have attempted to explain how structural analysis problems are organized for use by the computer and how instructions can be written that permit the computer to solve a large variety of problems at the bidding of the analyst.

18.2 Matrix Methods

Structural engineers have been attempting to handle analysis problems for a good many years by applying the methods used by mathematicians in linear algebra. Although many structures could be analyzed with the resulting equations, the work was extremely tedious—at least until large-scale computers became available. In fact, the usual matrix equations are not manageable with hand-held calculators unless the most elementary structures are involved.

Today, matrix analysis (using computers) has almost completely replaced the classical methods of analysis in engineering offices. As a result, engineering educators and writers of structural analysis textbooks are faced with a difficult decision. Should they require a thorough study of the classical methods followed by a study of modern matrix methods; should they require the students to study both at the same time in an integrated approach; or should they just present a study of the modern methods? The reader can see from the preceding chapters that the authors feel that an initial study of some of the classical methods followed by a study of the matrix methods will result in an engineer who has a better understanding of structural behavior.

QUICK NOTE

An engineer who has a sound understanding of structural behavior inherently possesses a strong foundation for the ever needed *engineering judgment*.

Any method of analysis involving linear algebraic equations can be put into matrix notation and matrix operations used to obtain their solution. The possibility of the application of matrix methods by the structural engineer is very important, because all linearly elastic, statically determinate, and indeterminate structures are governed by systems of linear equations.

The simple numerical examples presented in the next few chapters could be solved more quickly by classical methods using a pocket calculator rather than with a matrix approach. However, as structures become more complex and as more loading patterns are considered, matrix methods using computers become increasingly useful.

MATRIX ALGEBRA

Engineering students have some background in matrix algebra; however, the authors' students seem to need a little review of the subject before they study the material presented in this and the next three chapters. If you should fall into this class of students, you might very well like to look over the matrix algebra review sections presented in Appendix C.

Courtesy of Corinne Lawler

FIGURE 18.1 Science museum as part of the City of Arts and Sciences in Valencia, Spain.

18.3 Force and Displacement Methods of Analysis

The methods presented in earlier chapters for analyzing statically indeterminate structures can be placed in two general classes. These are the force methods and the displacement methods. Both of these methods have been developed to a stage where they can be applied to almost any structure—trusses, beams, frames, plates, shells, and so on. The displacement procedures, however, are much more commonly used today, since they can be more easily programmed for solution by computers (as described in Chapters 19 through 21).

Force Method of Analysis

DID YOU KNOW?
The *method of consistent distortions* introduced in Chapters 11 and 12 is a force method.

With the *force method*, also called the *flexibility* or *compatibility method*, redundants are selected and removed from the structure so that a stable and statically determinate structure remains. An equation of deformation compatibility is written at each location where a redundant has been removed. These equations are written in terms of the redundants, and the resulting equations are solved for the numerical values of the redundants. After the redundants are determined, statics can be used to compute all other desired internal forces, moments, and so on.

Displacement Method of Analysis

DID YOU KNOW?
Both the flexibility and the displacement methods can be analyzed using a matrix formulation. However, of the two methods, the displacement method is most easily generalized for many types and configurations of structures.

With the *displacement method* of analysis, also called the *stiffness* or *equilibrium method*, the displacements of the joints necessary to fully describe the deformed shape of the structure are used in a set of simultaneous equations. When the equations are solved for these displacements, they are substituted into the force-deformation relations of each member to determine the various internal forces.

Note that the number of unknowns in the displacement method is generally much greater than the number of unknowns in the force method. Despite this fact, the displacement method is of the greatest importance because, as will be shown, it is the matrix implementation of this method that can be computerized most easily for general usage. As a result, only the next section is devoted to the force method, while the remainder of this text is devoted to the displacement method.

18.4 Introduction to the Force or Flexibility Method

This method is actually the method of consistent distortions (initially introduced in Chapter 11) cast in matrix form. The steps involved in analyzing a statically indeterminate structure by this method are outlined here.

QUICK NOTE
The displacements due to the unit load are called *flexibility coefficients*. The actual displacement at joint 1 due to redundant R_1 is R_1 times the deflection caused by a unit load acting there, i.e., $(R_1)(\delta_{1,1})$; the displacement at joint 2 due to R_1 is $(R_1)(\delta_{2,1})$, and so on.

1. A sufficient number of redundants is chosen and removed from the structure to make it statically determinate. The remaining structure, which is often called the *primary structure* or the *released structure*, must be stable.
2. The primary structure is analyzed to determine the deformations located at and in the direction of the redundants that were removed.
3. A flexibility structure is created for each redundant force that was removed. This involves removal of all applied loads and then placing a unit load at the point and in the direction of one of the redundants. The deformation at that redundant location and also at each of the other redundant locations is determined. For instance, the deflection due to a unit load at point 1 is computed and labeled $\delta_{1,1}$, the deflection at point 2 due to a unit load at point 1 is labeled $\delta_{2,1}$, and so on. This same procedure is followed with a unit load applied at each of the other redundant locations.
4. Finally, simultaneous equations of deformation compatibility are written at each of the redundant locations. The unknowns in these equations are the redundant forces. The equations are expressed in matrix form and solved for the redundants.

To illustrate this procedure, the four-span beam of Figure 18.2 is considered. This beam is statically indeterminate to the third degree. The three support reactions R_1, R_2, and R_3 have been selected as the redundants and considered to be removed from the structure, as shown in part (b)

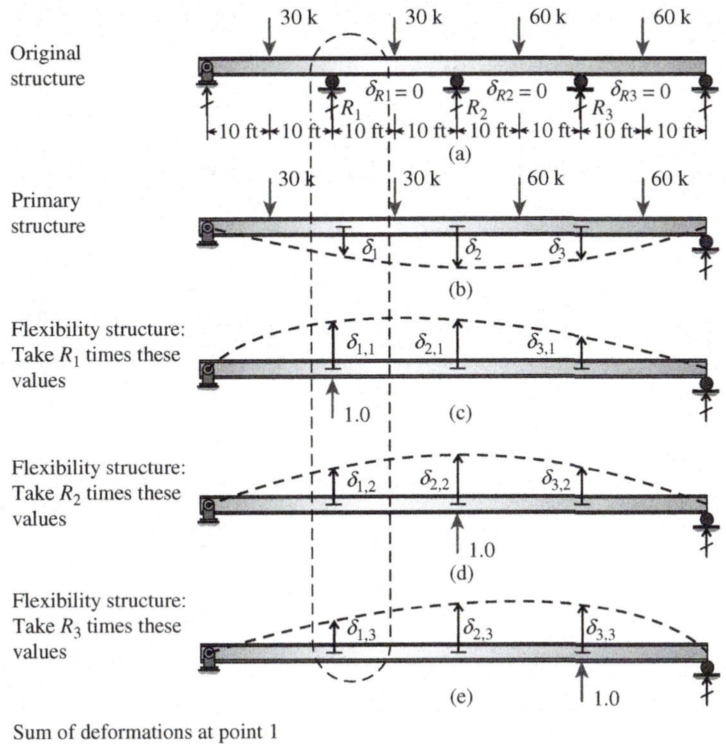

Sum of deformations at point 1
$= \delta_1 + R_1\delta_{1,1} + R_2\delta_{1,2} + R_3\delta_{1,3} = 0$

FIGURE 18.2
Demonstration of flexibility approach.

RECALL

The first subscript of the flexibility coefficient is the location of the computed displacement. The second subscript indicates the location where the unit load is applied.

QUICK NOTE

The results of the flexibility structures (e.g., $\delta_{1,1}$, $\delta_{2,1}$, $\delta_{3,1}$) must be scaled up by R_1, R_2, and R_3, respectively, before they can be added to the primary structure.

of the figure. The external loads will cause the beam to deflect downward by the values δ_1, δ_2, and δ_3, as shown.

In Figure 18.2(c), a unit load is applied at point 1 acting upward. It causes upward deflections at points 1, 2, and 3, respectively, equal to $\delta_{1,1}$, $\delta_{2,1}$, and $\delta_{3,1}$. Similarly, in Figure 18.2 (d) and (e), upward unit loads are applied at points 2 and 3, respectively, and the deformations at the three redundant points are determined.

QUICK NOTE

In the previous chapters of this text, joints in structures were labeled using alphabet characters. As we begin looking at matrix methods—with an intent to computerize—it is more convenient to label the joints with numbers.

Equations are then written for the total deformation at each of the points. It can be seen that these equations will be expressed in terms of all the redundants. This means that each redundant affects the displacement associated with each of the other redundant locations.

Now we can write an expression for the deformation at each of the joints. Here δ_{R1} is the total deflection at point 1, δ_{R2} is the total deflection at joint 2, etc. Since this beam is assumed to have unyielding supports, these values are each equal to zero.

$$\delta_1 + \delta_{1,1}R_1 + \delta_{1,2}R_2 + \delta_{1,3}R_3 = \delta_{R1}$$

$$\delta_2 + \delta_{2,1}R_1 + \delta_{2,2}R_2 + \delta_{2,3}R_3 = \delta_{R2}$$

$$\delta_3 + \delta_{3,1}R_1 + \delta_{3,2}R_2 + \delta_{3,3}R_3 = \delta_{R3}$$

These equations can be put into matrix form as

$$\begin{Bmatrix} \delta_1 \\ \delta_2 \\ \delta_3 \end{Bmatrix} + \begin{bmatrix} \delta_{1,1} & \delta_{1,2} & \delta_{1,3} \\ \delta_{2,1} & \delta_{2,2} & \delta_{2,3} \\ \delta_{3,1} & \delta_{3,2} & \delta_{3,3} \end{bmatrix} \begin{Bmatrix} R_1 \\ R_2 \\ R_3 \end{Bmatrix} = \begin{Bmatrix} \delta_{R1} \\ \delta_{R2} \\ \delta_{R3} \end{Bmatrix}$$

This equation actually shows that the displacements due to the loads plus the flexibility matrix times the redundants equals the final deformation at the supports. It can be written in compact symbolic form as

$$\{\delta_L\} + [F]\{R\} = \{\delta_R\}$$

DEFINED TERMS

$\{\delta_L\}$ is a vector of displacements due to the imposed loads.

$[F]$ F is a matrix of flexibility coefficients.

$\{R\}$ is a vector of unknown redundant forces.

$\{\delta_R\}$ is a vector of final deformations at the supports, (all equal to zero in this case).

This equation may be written in a somewhat altered form and solved for the redundants:

$$\{R\} = [F]^{-1} (\{\delta_R\} - \{\delta_L\})$$

The symbol $[F]^{-1}$ represents the inverse of matrix $[F]$. This matrix may be found in a formal way through a mathematical procedure described in Appendix C. However, in most practical problems, the inverse is not found explicitly, but the set of algebraic equations is solved simultaneously by some other procedure, e.g., the Gauss method. The inverse notation will continue to be used in this text to represent the solution symbolically to the set of algebraic equations.

In Example 18.1, the redundant reactions are determined for the beam of Figure 18.2. In this problem, it is necessary to compute three support deformations and nine flexibility coefficients. These rather lengthy calculations, which can be handled by methods previously presented in this text, are not shown here. Application of Maxwell's law of reciprocal deflections shortens the work somewhat in that $\delta_{1,2} = \delta_{2,1}$, $\delta_{1,3} = \delta_{3,1}$, and $\delta_{2,3} = \delta_{3,2}$. In addition, because of symmetry for this particular beam, $\delta_{1,1} = \delta_{3,3}$.

EXAMPLE 18.1

Determine the values of reactions R_1, R_2, and R_3 of the continuous beam of Figure 18.2 using a matrix approach and the force or flexibility method.

SOLUTION STRATEGY

The calculation of deflections can be tedious, but the selection of supports 1 through 3 as the redundants permits the use of beam deflection charts.

SOLUTION

The following deflections are obtained using the beam deflection chart for a simple supported beam.

QUICK NOTE

Virtual work also may be used to calculate the requisite deflections.

Primary Structure:

$$\delta_1 = -\frac{850{,}000}{EI}\ \text{k·ft}^3 \qquad \delta_2 = -\frac{1{,}230{,}000}{EI}\ \text{k·ft}^3$$

$$\delta_3 = -\frac{905{,}000}{EI}\ \text{k·ft}^3$$

Flexibility Structures:

$$\delta_{1,1} = \frac{6000}{EI}\text{ft}^3 \qquad \delta_{1,2} = \delta_{2,1} = \frac{7333}{EI}\text{ft}^3$$

$$\delta_{1,3} = \delta_{3,1} = \frac{4667}{EI}\text{ft}^3 \qquad \delta_{2,2} = \frac{10{,}667}{EI}\text{ft}^3$$

$$\delta_{2,3} = \delta_{3,2} = \frac{7333}{EI}\text{ft}^3 \qquad \delta_{3,3} = \frac{6000}{EI}\text{ft}^3$$

QUICK NOTE

For this structure and most structures that are analyzed, the actual displacements at the redundant supports are zero.

$$\begin{Bmatrix} \delta_{R1} \\ \delta_{R2} \\ \delta_{R3} \end{Bmatrix} = \begin{Bmatrix} 0 \\ 0 \\ 0 \end{Bmatrix} \text{ft}$$

Sum up the displacements at the location of each redundant and set equal to the actual displacement at each support.

$$\frac{1}{EI}[-850{,}000 + 6000R_1 + 7333R_2 + 4667R_3] = 0$$

$$\frac{1}{EI}[-1{,}230{,}000 + 7333R_1 + 10{,}667R_2 + 7333R_3] = 0$$

$$\frac{1}{EI}[-905{,}000 + 4667R_1 + 7333R_2 + 6000R_3] = 0$$

Rewrite the equations into matrix form:

$$\frac{1}{EI}\begin{bmatrix} 6000 & 7333 & 4667 \\ 7333 & 10{,}667 & 7333 \\ 4667 & 7333 & 6000 \end{bmatrix} \begin{Bmatrix} R_1 \\ R_2 \\ R_3 \end{Bmatrix} = \frac{1}{EI}\begin{Bmatrix} 850{,}000 \\ 1{,}250{,}000 \\ 905{,}000 \end{Bmatrix}$$

Recognizing that the $1/(EI)$ term cancels from both sides, the unknown forces may be found by evaluating the following equation.

$$\begin{Bmatrix} R_1 \\ R_2 \\ R_3 \end{Bmatrix} = \begin{bmatrix} 6000 & 7333 & 4667 \\ 7333 & 10{,}667 & 7333 \\ 4667 & 7333 & 6000 \end{bmatrix}^{-1} \begin{Bmatrix} 850{,}000 \\ 1{,}250{,}000 \\ 905{,}000 \end{Bmatrix}$$

$$\therefore R_1 = 34.0 \text{ k} \uparrow$$

$$R_2 = 40.2 \text{ k} \uparrow$$

$$R_3 = 75.3 \text{ k} \uparrow$$

QUICK NOTE

The procedure to evaluate this matrix-based equation is shown in Appendix C. Many of the calculators students have today would have no problem performing this calculation. However, Section 19.10 also shows how to do this calculation using spreadsheets.

In Example 18.2, only axial deformations are considered in order to simplify the numerical calculations required while still illustrating the principles involved in the method. Similar problems are worked by the stiffness method (i.e., displacement based method) in later chapters.

EXAMPLE 18.2

The axial member shown here is supported at each end and in the middle and is loaded with axial forces at intermediate locations. Determine the reactions at the supports.

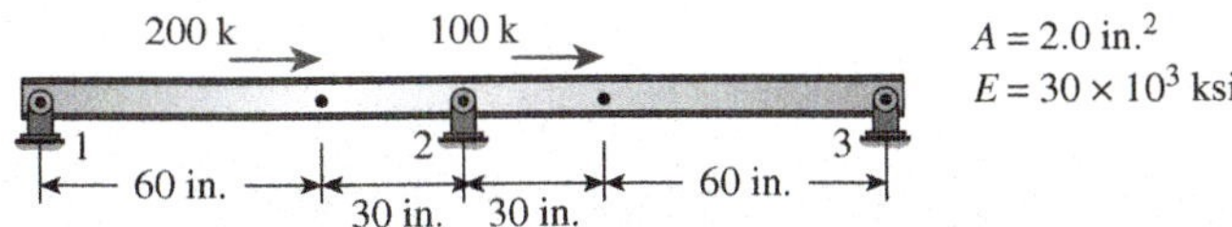

$A = 2.0$ in.^{2}
$E = 30 \times 10^3$ ksi

DID YOU KNOW?

The axial deformation in an axial bar is conveniently calculated as derived from basic mechanics:

$$\delta = \frac{NL}{AE}$$

where N is the internal axial force, L is the segment length, A is the cross-sectional area, and E is the modulus of elasticity.

SOLUTION STRATEGY

Find the number of redundants required to be removed. It is helpful to remember that axially loaded members only have one relevant equation of equilibrium. Follow the same process as in the previous example.

SOLUTION

The number of reactions in the horizontal direction is three. There is only one equation of equilibrium $\sum F_x$. This means that the axial member is SI 2°. Select R_2 and R_3 as the redundants.

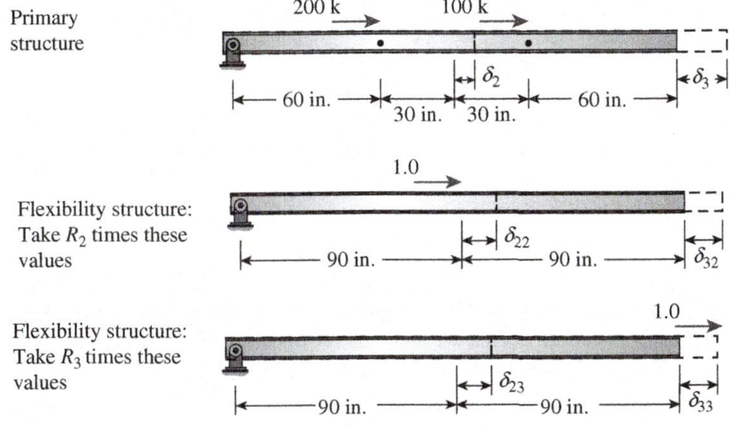

AXIAL FORCES

There are three relevant segments in the primary structure. They have the following internal axial forces working from left to right: $N_{left} = 300$ k, $N_{center} = 100$ k, and $N_{right} = 0$ k.

For the first flexibility structure, the internal axial forces for the two segments are $N_{12} = 1.0$ and $N_{23} = 0.0$.

For the second flexibility structure, the internal axial forces for the two segments are $N_{12} = 1.0$ and $N_{23} = 1.0$.

Deflection Calculations:

$$\delta_2 = \frac{1}{(2.0 \text{ in}^2)(30{,}000 \text{ ksi})}[(300 \text{ k})(60 \text{ in.}) + (100 \text{ k})(30 \text{ in.})]$$

$$= 0.35 \text{ in.}$$

$$\delta_3 = \frac{1}{(2.0\ \text{in}^2)(30{,}000\ \text{ksi})}[(300\ \text{k})(60\ \text{in.}) + (100\ \text{k})(60\ \text{in.}) + (0\ \text{k})(60\ \text{in.})]$$

$$= 0.40\ \text{in.}$$

$$\delta_{22} = \frac{1}{(2.0\ \text{in}^2)(30{,}000\ \text{ksi})}[(1.0)(90\ \text{in.})]$$

$$= 0.0015\ \text{in./k}$$

$$\delta_{32} = \delta_{23} = \frac{1}{(2.0\ \text{in}^2)(30{,}000\ \text{ksi})}[(1.0)(90\ \text{in.}) + (0.0)(90\ \text{in.})]$$

$$= 0.0015\ \text{in./k}$$

$$\delta_{33} = \frac{(1.0)(180\ \text{in.})}{(2\ \text{in}^2)(30{,}000\ \text{ksi})} = 0.003\ \text{in./k}$$

DID YOU KNOW?
With a little experience, it is possible to assemble the matrix form of the equations without first writing the individual compatibility equations.

Assemble the equations by summing up the displacements at each redundant location and set equal to zero. Thus,

$$0.35\ \text{in.} + (0.0015\ \text{k/in.})R_2 + (0.0015\ \text{in./k})R_3 = 0$$

$$0.40\ \text{in.} + (0.0015\ \text{k/in.})R_2 + (0.0030\ \text{in./k})R_3 = 0$$

Matrix Form:

$$\begin{bmatrix} 0.0015\ (\text{in./k}) & 0.0015\ (\text{in./k}) \\ 0.0015\ (\text{in./k}) & 0.0030\ (\text{in./k}) \end{bmatrix} \begin{Bmatrix} R_2 \\ R_3 \end{Bmatrix} = \begin{Bmatrix} -0.35 \\ -0.40 \end{Bmatrix}\ \text{in.}$$

Rework the matrix equation and solve for the unknown axial forces.

QUICK NOTE
Take note of the units associated with the flexibility coefficients. They are a length per unit force.

$$\begin{Bmatrix} R_2 \\ R_3 \end{Bmatrix} = \begin{bmatrix} 0.0015\ (\text{in./k}) & 0.0015\ (\text{in./k}) \\ 0.0015\ (\text{in./k}) & 0.0030\ (\text{in./k}) \end{bmatrix}^{-1} \begin{Bmatrix} -0.35 \\ -0.40 \end{Bmatrix}\ \text{in.}$$

$$\therefore R_2 = -200\ \text{k} = \underline{\underline{200\ \text{k}\quad \leftarrow}}$$

$$R_3 = -33.3\ \text{k} = \underline{\underline{33.3\ \text{k}\quad \leftarrow}}$$

Statics then shows that $R_1 = \underline{\underline{66.7\ \text{k}\quad \leftarrow}}$.

QUICK NOTE
The flexibility approach shown sometimes requires the analyst to calculate many different deflection numbers. This can be quite tedious and is difficult to generalize for many different structural configurations. The stiffness method discussed in the next chapter handles this much easier.

An inherent difficulty with the use of the force method is that the analyst must know at the outset how many redundants there are in the structure to be analyzed. Furthermore, they must choose which members or reactions are to serve as redundants. The choice of redundants, which is not unique, may affect the complexity involved in solving the problem at hand. However, with the stiffness method introduced in Chapter 19, it is much easier to prepare a general program applicable to all types of statically determinate or indeterminate structures.

A full discussion of the force method is not included in this text because current structural analysis trends are decidedly in the direction of the displacement or stiffness method. It would be possible to expand the force method described herein to include other items, such as the development of the flexibility coefficients using matrix methods, a consideration of different types of structures, and so on.

18.5 Examples with Video Solution

VE18.1 Use the flexibility approach to find the reactions for the beam shown in Figure VE18.1. Choose the moments at 1 and 3 and the reaction at 2 as the redundants. Beam charts may be readily used for calculating the necessary deflections.

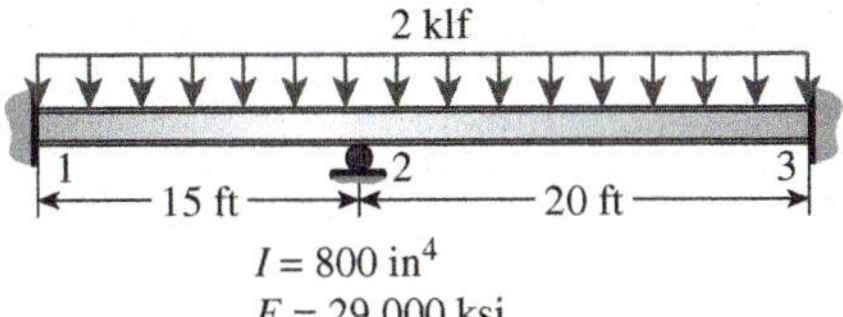

Figure VE18.1

VE18.2 Rework the problem given in Figure VE18.1, only now assume that there is a 0.5-in. settlement (downward) at support 2.

VE18.3 Use the flexibility approach to solve for the reactions of the indeterminate axial bar shown in Figure VE18.3.

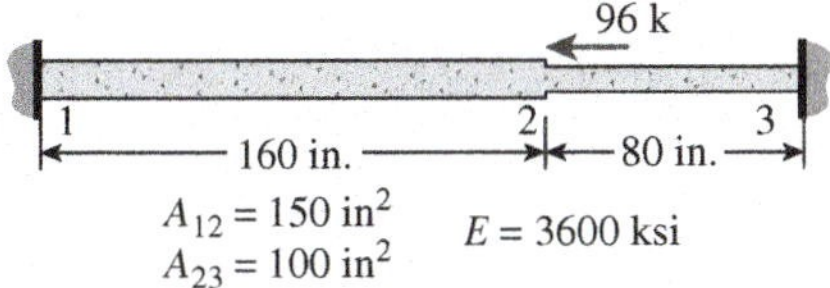

Figure VE18.3

18.6 Problems for Solution

Section 18.4

P18.1 Find the flexibility coefficients associated with joint loads applied at joint 2 (i.e., find the coefficients of matrix $[F]$) in

$$\begin{Bmatrix} \Delta_2 \\ \theta_2 \end{Bmatrix} = [F] \begin{Bmatrix} Q_2 \\ M_2 \end{Bmatrix}$$

(Ans: $\delta_{\Delta\Delta} = \frac{L^3}{3EI}$, $\delta_{\theta\theta} = \frac{L}{EI}$)

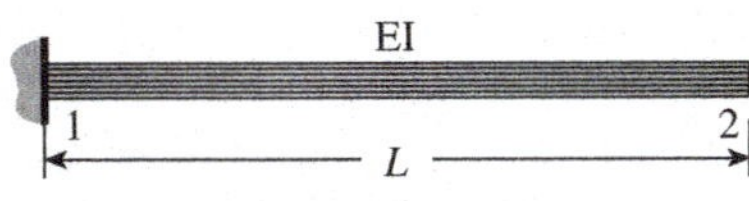

Problem 18.1

P18.2 Use the flexibility method to determine the reactions at joint 2 of the fixed-end beam that supports a uniformly distributed load, w. You may want to use beam deflection charts and also the flexibility coefficients found in Problem 18.1.

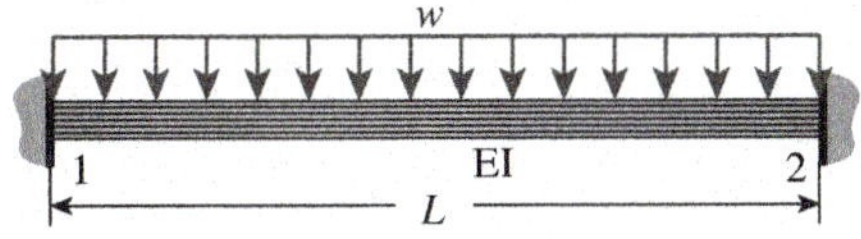

Problem 18.2

P18.3 Determine the flexibility coefficients associated with the continuous beam shown. Use the moments at joints 2 and 3 as redundants; the released structure will then appear as two simply supported beams, as shown in the figure. The flexibility coefficients will be part of matrix $[F]$ in

$$\begin{Bmatrix} \theta_{2,1} - \theta_{2,3} \\ \theta_3 \end{Bmatrix} = [F] \begin{Bmatrix} M_2 \\ M_3 \end{Bmatrix}$$

(Ans: $\delta_{22} = \frac{2L}{3EI}$, $\delta_{33} = \frac{L}{3EI}$)

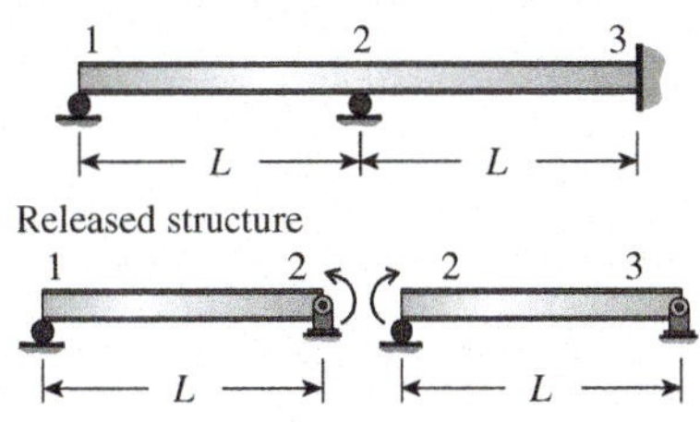

Problem 18.3

P18.4 Use the flexibility method to find the internal moments at joints 2 and 3 for the continuous beam shown. The beam is loaded with a uniformly distributed load over its entire length. Consider using the flexibility coefficients determined in Problem 18.3.

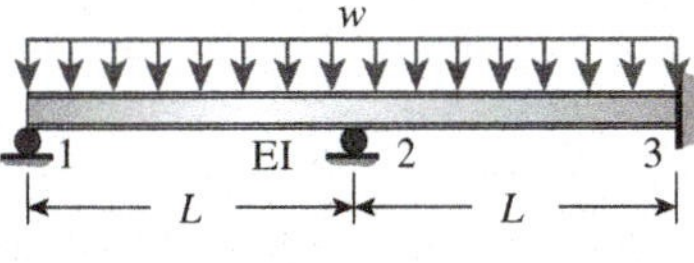

Problem 18.4

P18.5 Use the flexibility coefficients found in Problem 18.3 to determine the moments at joints 2 and 3 for the continuous beam loaded with a concentrated load as shown. (Ans: $M_2 = \frac{PL}{7}$, $M_3 = \frac{-PL}{14}$)

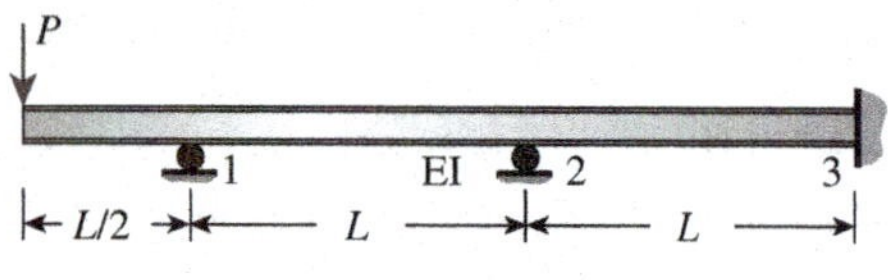

Problem 18.5

P18.6 Determine the flexibility coefficients associated with the continuous beam shown in Problem 18.3. Use the reactions at joints 1 and 2 as redundants; the released structure is also shown. The flexibility coefficients will be part of matrix $[F]$ in

$$\begin{Bmatrix} \Delta_1 \\ \Delta_2 \end{Bmatrix} = [F] \begin{Bmatrix} Q_1 \\ Q_2 \end{Bmatrix}$$

Released structure

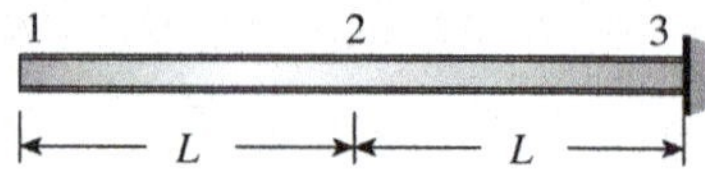

Problem 18.6

P18.7 Determine the reactions at joints 1 and 2 for the continuous beam shown in Problem 18.6, assuming that a uniformly distributed load acts over the entire length of the beam, as shown in Problem 18.4. (Ans: $R_1 = \frac{11wl}{28}$, $R_2 = \frac{8wL}{7}$)

P18.8 Determine the reactions at joints 1 and 2 for the continuous beam shown in Problem 18.6, assuming that a uniformly distributed load acts over the span between joints 2 and 3 only.

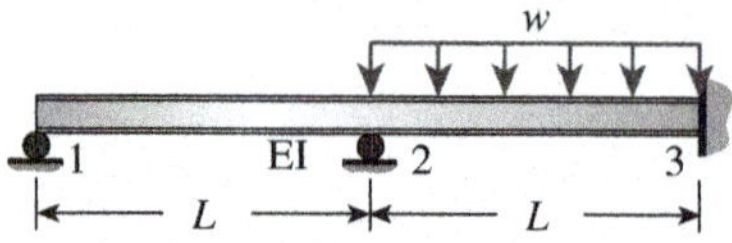

Problem 18.8

P18.9 Determine the end moments and moment at joint 2 for the continuous beam shown. The loading carried by the beam consists of a uniformly distributed load over the left span. Use the moments at the joints as redundants; the released structure is also shown. (Ans: $M_1 = \frac{5wL^2}{48}$, $M_2 = \frac{-wL^2}{24}$)

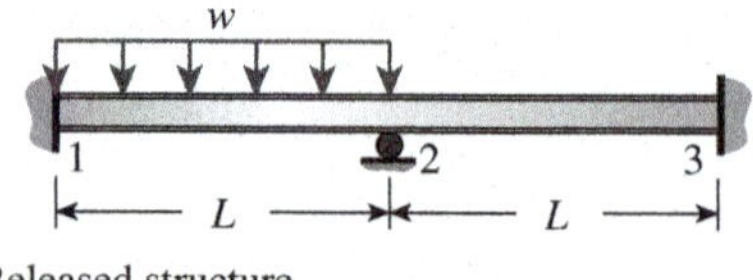

Released structure

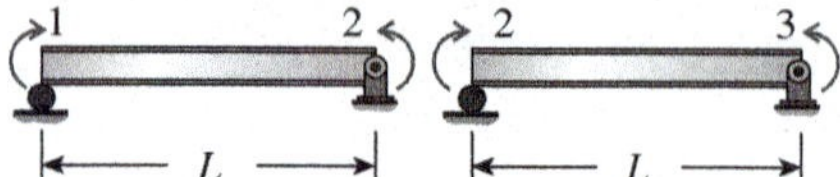

Problem 18.9

P18.10 An axial force member consisting of three segments having different cross-sectional areas is supported at each end and loaded with an intermediate 100-k load. Use the flexibility method to find the support reactions.

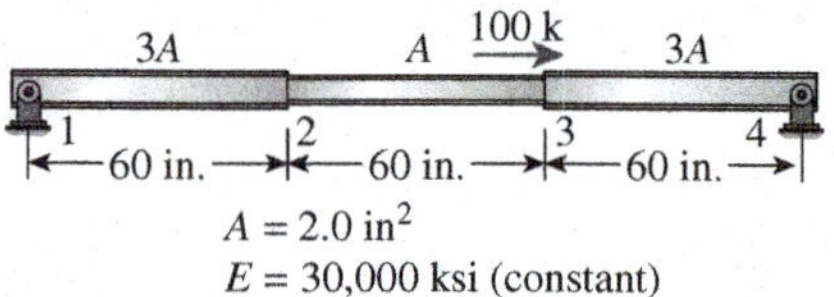

Problem 18.10

SAP2000

For the following problems, assume unity for all given variables, such as L, E, I, w, and P.

P18.11 Consider the beam shown in Problem 18.1. Model this beam in SAP2000 and use it to calculate the flexibility coefficients for joint 2. This includes both Δ_2 and θ_2. Assemble the flexibility matrix $[F]$.

P18.12 Consider the beam shown in Problem 18.6. Model this beam in SAP2000 and use it to calculate the flexibility coefficients for joints 1 and 2. This includes both Δ_1 and Δ_2. Assemble the flexibility matrix $[F]$. Do the same thing again, only this time let the left span be equal to $L/2$. When you compare both flexibility matrices, what influence did the span length have on the results?

P18.13 Consider the axial member shown in Problem 18.10. Model this member in SAP2000 and find the axial force in each of the three segments. Rerun the analysis when segment 1–2 has the following cross-sectional areas: $0.5A$, A, $2A$, $3A$, $5A$, and $10A$. Generate a plot which shows the axial force in each bar versus the cross-sectional area. What influence does the cross-sectional area of segment 1–2 have on the internal forces in the individual segments?

P18.14 Consider the axial member shown in Problem 18.10. Model this member in SAP2000 and find the axial force in each of the three segments. Run the analysis again when joint 2 also has a pin support. What impact does the additional support have on the axial forces in the members? What rationale can you give for the internal force computed for segment 1–2?

Direct Stiffness Method for Trusses

19

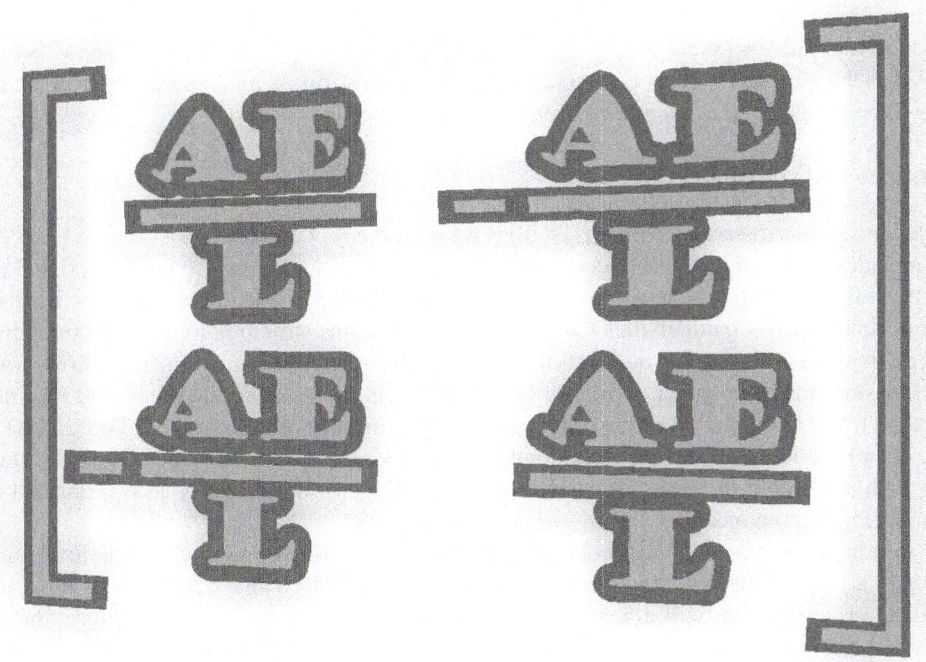

19.1 Introduction

In the previous chapter, the flexibility approach was introduced in its matrix form. This required that the analyst first wrote a set of compatibility equations that were then used to solve for unknown forces. In contrast, when a structure is being analyzed with the displacement or stiffness method, the joint displacements (translations and rotations) are treated as unknowns. Equilibrium equations are written for each joint of the structure in terms of (1) the applied loads, (2) the properties of the members framing into the joint, and (3) the unknown joint displacements. The result is a set of linear, algebraic equations that can be solved simultaneously for the joint displacements. These displacements are then used to determine the individual internal member forces and support reactions.

The stiffness method can be used with equal ease for the analysis of statically determinate or statically indeterminate structures. The analyst does not have to make a choice of redundants and does not need to specify or even know whether the structure is statically determinate or indeterminate. Furthermore, if the structure is unstable, no solution can be determined, and the analyst is thereby warned of the instability.

In this chapter, the authors develop the fundamentals of the stiffness method for truss bars and truss assemblies. By the end of the chapter, the reader should be able to fully analyze a truss using the stiffness approach, including all joint displacements, reactions, and bar forces. Chapter 20 will extend the concepts learned in this chapter to the analysis of beams and frames.

SUMMARY

The **flexibility approach** uses a set of compatibility equations to solve for unknown redundants and then uses statics to find the remaining forces. The **stiffness approach** uses a set of equilibrium equations to solve for unknown displacements. It then uses these displacements to solve for the remaining forces.

Elements are linear

Need only X and Y coordinates

P

Movement of a Truss Joint

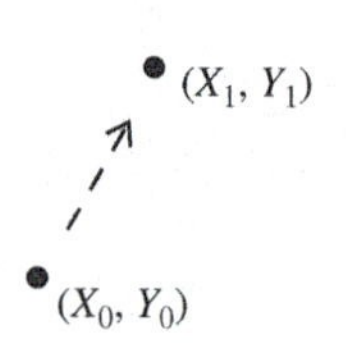

FIGURE 19.1 Illustration of DOFs for a planar truss joint.

19.2 Definitions and Concepts

The direct stiffness approach, discussed in this and the following chapters, is really a very methodical approach to the analysis of structures. As such, it is wise to establish a basis from which all ensuing concepts may be presented. To assist in this effort, a number of definitions and concepts are initially presented to the reader for their consideration. At this stage, it may not be obvious what these concepts have to do with the direct stiffness method, but later in the chapter it will be necessary for this information to be understood.

DID YOU KNOW?
A DOF may be seen as a permissible joint movement—whether translation or rotation.

Degree of Freedom (DOF)

The concept of a *degree of freedom* (DOF) is found at the root of the stiffness method. It is defined as an independent displacement coordinate located on a structure used to help describe the structure's deflected shape. A structure must have enough DOFs to adequately describe its displaced shape. In the matrix method, the DOFs are just defined at the structural joints. Structural displacements at locations other than the joints must be found by interpolation, as described in Section 20.5.

TYPICAL DOFs
The following is a list of considered DOFs per joint in various types of structures.

Planar Truss Two DOFs (2-translational).

3D Truss Three DOFs (3-translational).

Beam Two DOFs (1-translational, 1-rotational).

Planar Frame Three DOFs (2-translational, 1-rotational).

3D Frame Six DOFs (3-translational, 3-rotational).

Consider the truss shown in Figure 19.1. When it is subjected to the point load P, it deforms as shown. All of the members are only loaded axially, so they remain straight. Thus, if one knows the X and Y coordinates of all of the joints, they could describe the displaced shape of the truss. As such, each joint in a truss is said to have two DOFs (or in other words, is described by two translational components).

A little more information is required to describe the motion of a joint in a frame element. If a joint in a truss rotated, it would not change what the displaced shape looks like. However, in a frame, a rotating joint will affect the deflected shape (see Figure 19.2). Therefore, the joint in a planar frame requires three DOFs (2-translational and 1-rotational) to describe the displaced shape. The margin note indicates typical DOFs for each joint/node type.

DID YOU KNOW?
In the matrix analysis of structures, the term **node** is used interchangeably with the word **joint**. For line elements such as truss bars, beams, and frames, there are two joints/nodes per member—one at each end.

When the stiffness method is used for analysis, it is convenient to label the DOFs associated with each structural element (e.g., beam, truss, frame member) and also with the entire structure (e.g., assemblage of multiple members). As shown in the side margin, an elongated arrow head will be used to symbolize the location and assumed positive direction of a DOF. *Do not confuse these with actual forces.*

The DOF labeling scheme used in this textbook is based on whether the DOFs are being labeled in the local or global coordinate systems (see the next subsection). In the local coordinate system, alphabetic characters (e.g., a, b, c, etc.) will be used. In the global coordinate systems, the DOFs will be labeled using numbers (e.g., 1, 2, 3, etc.). Depending on the orientation of the member, the local and global DOFs may indeed be the same. In these instances, the global notation will be adopted.

DOF NOTATION

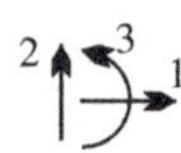

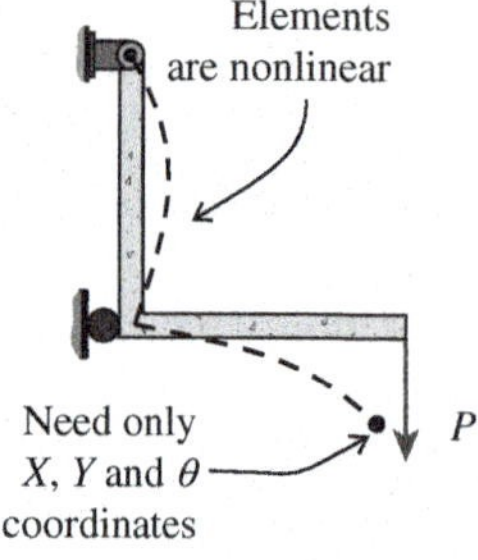

Movement of a Frame Joint

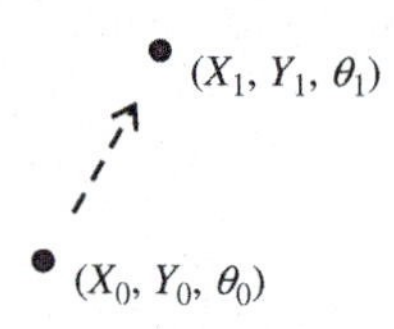

FIGURE 19.2 Illustration of DOFs for a planar frame joint.

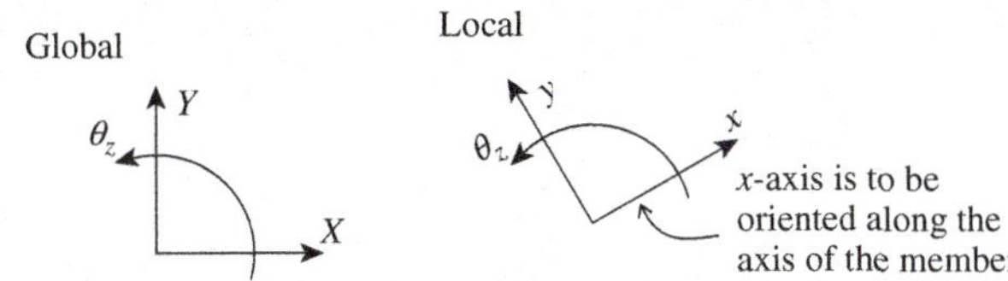

FIGURE 19.3 Two basic coordinate systems defined.

Coordinate System

An understanding of two types of coordinate systems is required for the matrix formulation of a structural problem. One coordinate system is needed to describe the entire structure—*global coordinates*—and a separate coordinate system is required for each individual member in the structure—*local coordinates*. As demonstrated in Figure 19.3, the right-handed Cartesian coordinate system is used for both. Uppercase letters X and Y will be used to represent the global coordinates, and the lowercase letters x and y will be used to describe the local coordinates.

The reason for having multiple coordinate systems is that some information makes more sense relative to the whole structure (i.e., global coordinates), such as displacements and reactions. Other information, such as member internal forces, makes much more sense when reported in a local coordinate system such as shear and axial force.

QUICK NOTE

This textbook will only consider the analysis of planar structures and will thus only necessitate a 2-D coordinate system. One should know, however, that this method readily can be extended to deal with 3-D structures with a little more effort and understanding.

Notation Throughout the next few chapters, there will be a consistent need to transform forces, displacements, and stiffness factors from the local to the global coordinate system and back again. As such, a certain notation for these terms is adopted to assist in the clarity of the analysis and results.

Term	Local	Global
Forces	Q	F
Displacements	u	Δ
Stiffness	k	K

DID YOU KNOW?

Every member in the structure will have its own local coordinate system, as shown here.

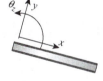

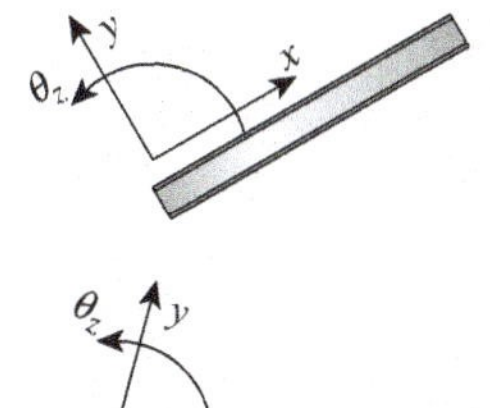

Stiffness Factors

The stiffness factor (or factors) are a way of describing the relationship between the internal forces in the member and the deformation of that member. Consider a member that is subjected to an axial internal load N. The internal normal stress (σ) is calculated by dividing the normal force by the cross-sectional area of the member, A. So

$$\sigma = \frac{N}{A}$$

The strain in the member (ϵ) is calculated by using the change in length of the member (Δ) and the original length of the member, L, as

$$\epsilon = \frac{\Delta}{L}$$

If the material remains linear, Hooke's law can be used to relate the stress and the strain in the axial member. Thus,

$$\sigma = E\epsilon$$

where E is the modulus of elasticity. Substitute in the known expressions for stress and strain and then solve for N:

$$\frac{N}{A} = E\frac{\Delta}{L}$$

$$N = \frac{AE\Delta}{L} = \left(\frac{AE}{L}\right)\Delta$$

QUICK NOTE
One should take note that the stiffness factor k is a function of a cross-sectional property (A), a material property (E), and member geometry (L). This will be a characteristic of all of the stiffness factors presented.

Notice that N is related to Δ through the term $(AE)/L$. This term is known as the stiffness of an axial member and is often represented using the letter k.

$$k = \frac{AE}{L}$$

The stiffness factor is defined: *The force required to generate a unit displacement.* Note that when $\Delta = 1$, then $N = k$. The stiffness coefficient is actually the reciprocal of the flexibility coefficient, which is the displacement resulting from a unit force $\delta = 1/k$.

Structural Analysis Requirements

The complete solution to any structural analysis problem must satisfy three basic requirements. The formulation of the stiffness method requires that all three of these conditions are met explicitly.

QUICK NOTE
As the matrix formulation is presented, look for each of these three components to be visible in the analysis process.

Equilibrium All forces and moments must sum to zero, otherwise acceleration would occur. The equations of equilibrium are found in this format:

$$\{F\} - [K]\{\Delta\} = \{0\}$$

This equation is slightly reworked and is most often seen in the form:

$$\{F\} = [K]\{\Delta\}$$

Compatibility Deformations in the structure must occur such that the structure remains connected in a manner consistent with the supports and connections.

Force-Displacement Relations The relationship between force and displacement in the structure must be described and honored. This is achieved in the form of the stiffness factor that was introduced before and is based on material properties, cross-sectional properties, and member geometry.

19.3 Kinematic Determinacy

KINEMATIC INDETERMINACY
The total number of additional DOFs that must be restrained to prevent any displacement of the structure.

When using the classical approaches to solving indeterminate structures, the concept of *statical indeterminacy* was used extensively. The degree of statical indeterminacy expressed the number of unknown redundant forces in a structure. The stiffness method, however, does not require knowledge of this type of determinacy.

The concept of kinematic determinacy is of more relevance in the stiffness approach. *Kinematic indeterminacy* is simply the total number of displacements/rotations that must be known to fully describe the displaced shape of the structure. In other words, this is the number of unknowns that the analyst must find to fully analyze the structure.

Procedure

QUICK NOTE
Sometimes in modeling frames, the analyst may reasonably assume that no axial deformation occurs in the members. This means that the distance between any two joints will not change. The benefit of this is a reduction in the number of DOFs, as illustrated in Example 19.2

A simple process may be used to identify the degree of kinematic indeterminacy and is outlined as follows. Note that the resulting number has nothing to do with the applied loads but rather is associated with possible displacements.

1. Label all the unrestrained DOFs on the structure.
   - For each joint in a planar truss, there should be two orthogonal DOFs.
   - For each joint in a planar frame, there should be two orthogonal and one rotational DOF.
2. Count the number of DOFs that have been labeled. This number is the degree of kinematic indeterminacy (KI°).

EXAMPLE 19.1

For the given truss, identify the degree of kinematic indeterminacy.

SOLUTION STRATEGY

Recognize that every joint in the truss has two DOFs. Label and count only the DOFs that don't correspond to a support.

SOLUTION

In this case, there are eight joints, which means the truss has 16 DOFs. However, three of the DOFs are currently restrained. This leaves 13 DOFs unrestrained.

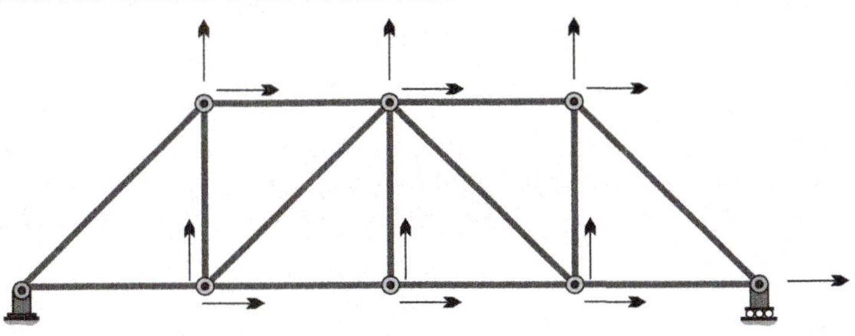

∴ Kinematic indeterminacy is KI 13°

QUICK NOTE

Only the unrestrained degrees of freedom are labeled.

EXAMPLE 19.2

For the given frame, identify the degree of kinematic indeterminacy.

SOLUTION STRATEGY

Recognize that every joint in the frame has three DOFs. Label and count only the DOFs that don't correspond to a support.

SOLUTION

In this case, there are 16 joints, which means the frame has a total of 48 DOFs. Two of the joints have all three DOFs restrained (total of 6) and two joints where only the translational DOFs are restrained (total of 4). This leaves 38 DOFs unrestrained.

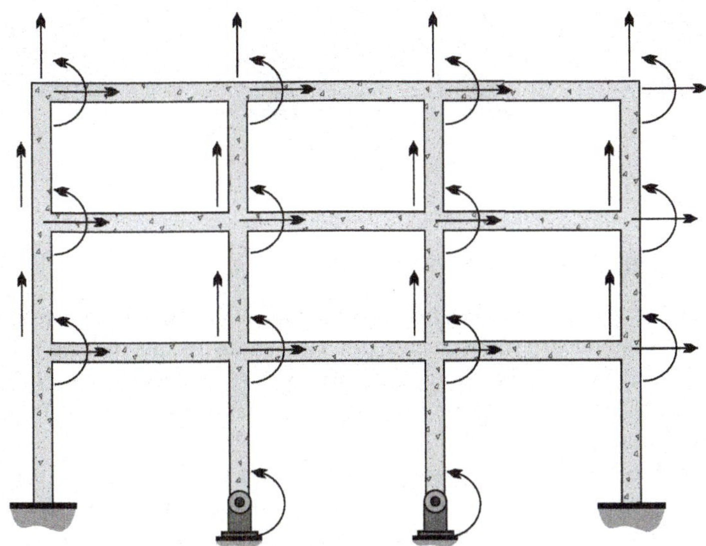

∴ Kinematic indeterminacy is KI 38°

QUICK NOTE

38 displacement/rotation components would need to be restrained in this frame to make it kinematically determinate.

Occasionally, the description of a displaced shape of a structure may be simplified by making certain assumptions. One common assumption made in the analysis of frames is that there is no axial deformation that occurs in the frame members. The implication of this assumption is that the relative distance between the joints at the ends of a member doesn't change. With such an assumption, many of the translational DOFs in a frame become redundant and uninformative. Therefore, any DOFs that are inline with a given member or members may be eliminated and represented with only one DOF.

EXAMPLE 19.3

For the frame of Example 19.2, identify the degree of kinematic indeterminacy if the assumption is made that no axial deformation can occur in any of the members.

DID YOU KNOW?
No vertical unrestrained DOFs exist in this frame. Since the length of the members can't change and the supports don't allow for vertical translation, none of the joints can vertically translate.

SOLUTION STRATEGY

Recognize that every joint in the frame has three DOFs. However, if multiple translational DOFs lie in a line, replace them all with only a single translational DOF. Pay attention to translational DOFs that are in line with a support. Label and count only the DOFs that don't correspond to a support.

SOLUTION

Refer to all of the DOFs labeled in Example 19.2. If no axial deformation occurs in any of the frame members, all the joints at a given floor level must move the same amount. This means that the four horizontal DOFs at each level may be reduced to one.

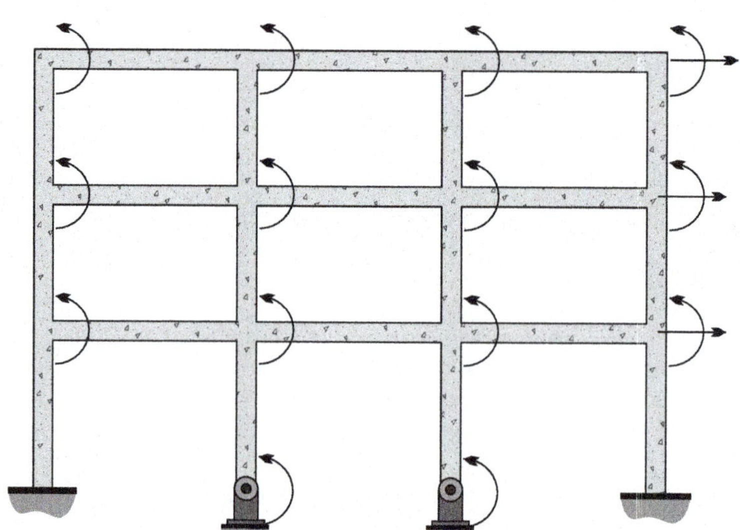

KINEMATIC INDETERMINACY
Count the number of labeled DOFs.

KI 17°

The simple assumption of no axial deformations reduced the degree of kinematic indeterminacy from 38 in Example 19.2 to 17.

19.4 Stiffness Method

The stiffness method is really a set of equilibrium equations that are written in terms of an applied set of forces (Q) and unknown displacements (u). The details of this method are introduced by illustration. Consider that a truss bar is subjected to a force Q_1 at one end and Q_2 at the other end, as shown in Figure 19.4(a). We desire to find the horizontal displacements at each end (u_1, u_2) due to these applied loads.

The approach taken to find the unknown displacements (u_1 and u_2) is to use superposition. Both DOFs first will be locked against translation. Then a displacement at DOF 1 will be imposed in the positive direction (u_1). The reactions at both DOFs are calculated. In this case, the force in the truss bar at DOF 1 is taken as

$$Q_{11} = \frac{AE}{L} u_1$$

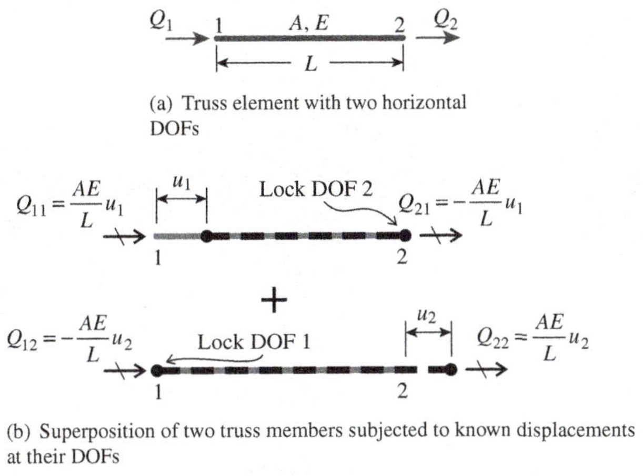

(a) Truss element with two horizontal DOFs

(b) Superposition of two truss members subjected to known displacements at their DOFs

$$Q_{11}=\frac{AE}{L}u_1 \quad \text{Lock DOF 2} \quad Q_{21}=-\frac{AE}{L}u_1$$

$$Q_{12}=-\frac{AE}{L}u_2 \quad \text{Lock DOF 1} \quad Q_{22}=\frac{AE}{L}u_2$$

FIGURE 19.4 Illustration of the stiffness method equations for a truss member.

QUICK NOTE

For the sake of this discussion, assume that only the horizontal DOFs of the truss bar are considered at joints 1 and 2. For clarity, we label each DOF as 1 and 2, respectively.

which is a well-known solution of basic mechanics. This results in a reaction at DOF 2 equal to

$$Q_{21} = -\frac{AE}{L}u_1$$

This entire process is repeated again; only now the displacement is imposed at DOF 2 and the ensuing reactions are recorded as shown in Figure 19.4(b). The original member of Figure 19.4(a) may be easily found by superimposing the two cases in 19.4(b). This results in

$$Q_1 = Q_{11} + Q_{12} = \tfrac{AE}{L}u_1 - \tfrac{AE}{L}u_2$$

$$Q_2 = Q_{21} + Q_{22} = -\tfrac{AE}{L}u_1 + \tfrac{AE}{L}u_2$$

DID YOU KNOW?

The subscripts associated with the Q terms may be interpreted as Q_{ij} where

i = location of the resulting force

j = location of the imposed displacement

The force at i is due to an imposed displacement at j.

A system of simultaneous linear equations may be solved readily using the matrix formulation. This requires the analyst to rewrite the equations in terms of the equation coefficients. Thus,

$$\begin{Bmatrix} Q_1 \\ Q_2 \end{Bmatrix} = \begin{bmatrix} \frac{AE}{L} & -\frac{AE}{L} \\ -\frac{AE}{L} & \frac{AE}{L} \end{bmatrix} \begin{Bmatrix} u_1 \\ u_2 \end{Bmatrix}$$

QUICK NOTE

Both forces, Q_{11} and Q_{21}, are a function of the imposed displacement, u_1.

The matrix containing the A, E, and L terms is known as the stiffness matrix $[k]$. The unknown displacements may be found following basic matrix math, which includes taking the inverse of a matrix.

$$\begin{Bmatrix} u_1 \\ u_2 \end{Bmatrix} = \begin{bmatrix} \frac{AE}{L} & -\frac{AE}{L} \\ -\frac{AE}{L} & \frac{AE}{L} \end{bmatrix}^{-1} \begin{Bmatrix} Q_1 \\ Q_2 \end{Bmatrix}$$

FRIENDLY ADVICE

Basic matrix math and direction on how to perform this math is introduced in Section 19.10. Take the time now to review these basic math concepts.

If one gives some numbers to A, E, L, and Q and attempts to solve for the unknown displacements, they will find that the stiffness matrix is singular and the inverse cannot be found. Each structure must have adequate supports, as discussed in Chapter 5 in the presentation on stability.

Consider the truss element again; only now a support is provided at DOF 2 (see Figure 19.5). The matrix formulation looks the same, but u_2 is a known value of 0.0. This means the two unknowns are u_1 and the reaction R_2 (formerly called Q_2). The solution first requires the matrix to be partitioned, so the the unrestrained DOFs may be independently analyzed. Thus,

$$\left\{\begin{array}{c} Q_1 \\ \hline R_2 \end{array}\right\} = \left[\begin{array}{c|c} \frac{AE}{L} & -\frac{AE}{L} \\ \hline -\frac{AE}{L} & \frac{AE}{L} \end{array}\right] \left\{\begin{array}{c} u_1 \\ \hline 0 \end{array}\right\}$$

SINGULAR MATRIX

A singular matrix is one where the determinant is zero.

$$\det\left|\frac{AE}{L}\begin{bmatrix} 1 & -1 \\ -1 & 1 \end{bmatrix}\right| = 0$$

If the determinant is zero, the inverse of a matrix cannot be calculated. A singular stiffness matrix is an indication that the structure being analyzed is unstable. This means that additional restraints must be added before analysis may ensue.

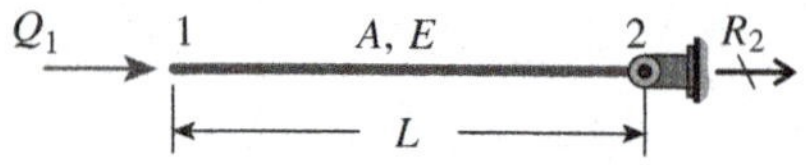

Q_1 1 A, E 2 R_2 L

FIGURE 19.5 Truss member with proper restraint.

The expression for u_1 is found by multiplying out the first equation, resulting in

$$Q_1 = \left(\frac{AE}{L}\right)u_1 - \left(\frac{AE}{L}\right)(0) \Rightarrow u_1 = Q_1\left(\frac{L}{AE}\right)$$

QUICK NOTE

Take note that this process first solves for the unknown displacement(s) and then uses those displacement(s) to solve for the unknown reactions.

With the expression for u_1 found, the second equation in the matrix may be multiplied out to solve for the reaction R_2. So

$$R_2 = -\left(\frac{AE}{L}\right)\left[Q_1\left(\frac{L}{AE}\right)\right] + \left(\frac{AE}{L}\right)(0) \Rightarrow R_2 = -Q_1$$

Matrix Procedure for Solution

NOTATION

u corresponds with **unrestrained DOFs** (i.e., DOFs that are free to move).

r corresponds with **restrained DOFs** (i.e., DOFs located at supports).

In the previous discussion, the stiffness matrix was partitioned and then the equations having unknown displacements were solved. Take note that, during the whole process of getting unknown displacements and unknown reactions, the only terms in the stiffness matrix that were needed were the upper-left and lower-left quadrants. This observation can be generalized with the notation:

$$\left\{\begin{array}{c} F_u \\ \hline R_r \end{array}\right\} = \left[\begin{array}{c|c} K_{uu} & K_{ur} \\ \hline K_{ru} & K_{rr} \end{array}\right]\left\{\begin{array}{c} \Delta_u \\ \hline \Delta_r \end{array}\right\}$$

The presentation here is represented in its most general form where the stiffness matrix, displacement vector, and force vector are given using the notation for global coordinates. Sometimes, the local coordinates of a member will correspond with the global coordinates of the overall structure. This was the case for the previous illustration, which is why the terms Q and F were used interchangeably—as were u and Δ.

DID YOU KNOW?

When going through the process of developing a stiffness matrix, it can often be difficult to tell if a reaction develops at a given DOF. An easy way to tell is that, if a DOF is attached to a *deformed* member, the DOF should have a resulting reaction. Otherwise, the reaction is zero.

QUICK NOTE

The stiffness term k_{12} is read as the reaction that develops at DOF 1 due to a unit displacement at DOF 2. This stiffness term is placed in row 1, column 2 of the stiffness matrix.

Stiffness Matrix Development

1. Label and number all DOFs on the structure starting from 1 and working up. Make sure to number all of the unrestrained (i.e., unsupported) DOFs first and then label the restrained DOFs next.
2. Lock all DOFs in the structure against movement. This essentially means you are making the structure kinematically determinate.
3. Start with DOF 1 and impose a *positive unit* displacement at this DOF while not allowing any of the others to move. It is often convenient to sketch the deflected shape of this displacement. Calculate the reactions that develop at each DOF (these reactions are the stiffness factors).
4. Place the stiffness factors into the stiffness matrix according to their DOF numbers K_{ij}. Here i is the DOF where the resulting reaction occurs and also indicates the row index in the stiffness matrix. j is the DOF where the unit displacement was imposed and represents the column index in the matrix.
5. Repeat steps 3 through 5 for all DOFs in the structure.

EXAMPLE 19.4

An assembly of three truss bars is shown here. Each bar is identified with a boxed number. The E value for all bars is the same at 29,000 ksi. The cross-sectional areas for each bar are $A_1 = A_3 = 1.5\ \text{in}^2$ and $A_2 = 3.0\ \text{in}^2$. For the DOF numbering scheme given, assemble the stiffness matrix for this assembly.

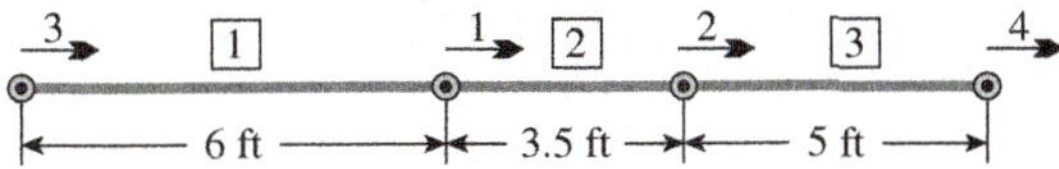

SOLUTION STRATEGY

Follow the procedure outlined previously and start with DOF 1. It is wise to sketch the deformed shapes in each case. Knowing that $(AE)/L$ terms are going to be needed, it may be convenient to calculate these values before starting the stiffness matrix procedure.

SOLUTION

Lock all DOFs against movement and then impose a positive unit displacement at DOF 1. Notice that this causes bar 1 to get longer by a unit and bar 2 to get shorter by a unit while bar 3 doesn't change length. Using the $(AE)/L$ values in the side margin, the reactions at each DOF are readily computed.

CALCULATIONS

$$\left(\frac{AE}{L}\right)_1 = \frac{(1.5\ \text{in}^2)(29{,}000\ \text{k/in}^2)}{72\ \text{in.}} = \underline{\underline{604\ \text{k/in.}}}$$

$$\left(\frac{AE}{L}\right)_2 = \frac{(3.0\ \text{in}^2)(29{,}000\ \text{k/in}^2)}{42\ \text{in.}} = \underline{\underline{2071\ \text{k/in.}}}$$

$$\left(\frac{AE}{L}\right)_3 = \frac{(1.5\ \text{in}^2)(29{,}000\ \text{k/in}^2)}{60\ \text{in.}} = \underline{\underline{725\ \text{k/in.}}}$$

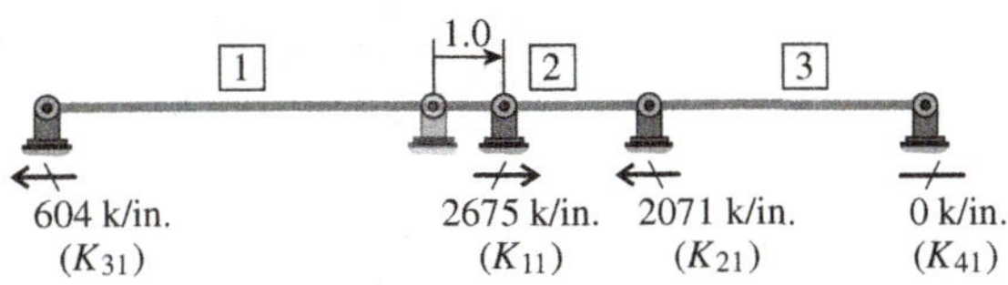

These reactions (i.e., stiffness values) are dropped into the stiffness matrix as

$$K = \begin{array}{c} \\ 1 \\ 2 \\ 3 \\ 4 \end{array} \begin{array}{c} \begin{array}{cccc} 1 & 2 & 3 & 4 \end{array} \\ \begin{bmatrix} 2675 & - & - & - \\ -2071 & - & - & - \\ -604 & - & - & - \\ 0 & - & - & - \end{bmatrix} \end{array} \text{k/in.}$$

Doing the same thing for the other three DOFs fills out the remaining slots in the stiffness matrix.

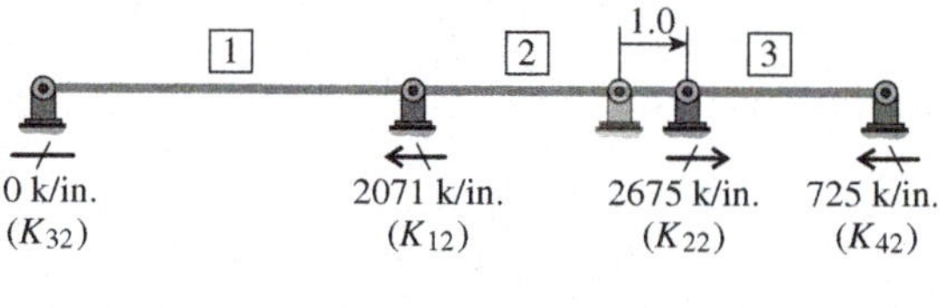

QUICK NOTE

Only one DOF is to be displaced at at time, and only the DOFs that are connected to deformed bars develop a reaction forces. For example, when DOF 3 is given a unit displacement, only bar 1 deforms. The only DOFs that are connected to bar 1 are DOF 3 and DOF 1. Hence, the reactions at the other DOFs are necessarily zero.

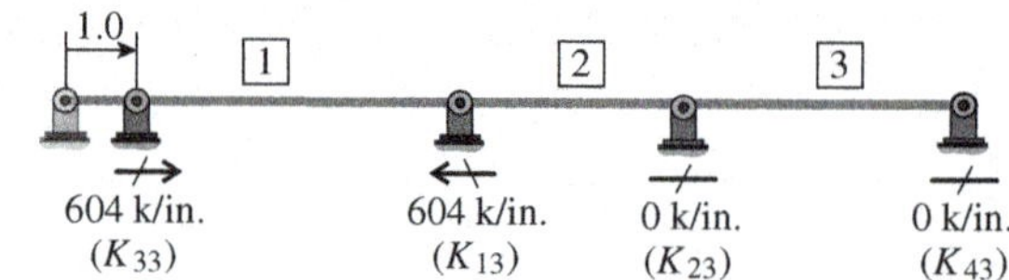

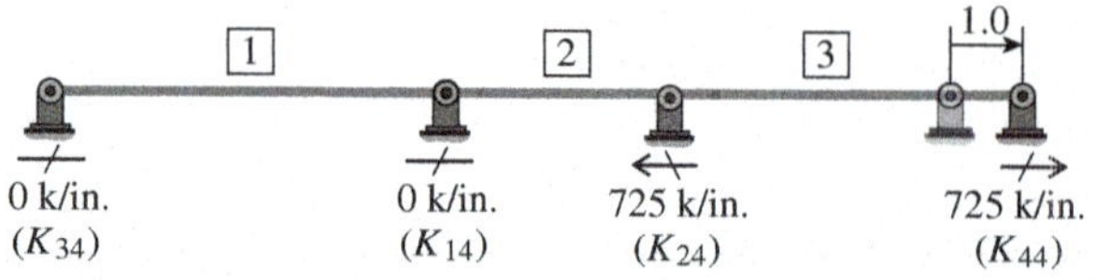

DID YOU NOTICE?

$K_{12} = K_{21}$, $K_{13} = K_{31}$, $K_{14} = K_{41}$, etc. The stiffness factors with the same subscripts have the same value. This produces a *symmetric stiffness matrix*. For linear elastic structures, the stiffness matrix will *ALWAYS* be symmetric. This knowledge can expedite the assembly of the matrix.

$$K = \begin{array}{c} \\ 1 \\ 2 \\ 3 \\ 4 \end{array} \begin{array}{c} \begin{array}{cccc} 1 & 2 & 3 & 4 \end{array} \\ \begin{bmatrix} 2675 & -2071 & -604 & 0 \\ -2071 & 2796 & 0 & -725 \\ -604 & 0 & 604 & 0 \\ 0 & -725 & 0 & 725 \end{bmatrix} \end{array} \text{ k/in.}$$

DID YOU KNOW?

The $[K_{uu}]$ matrix is a square matrix whose size is equal in size to the degree of kinematic indeterminacy that was previously discussed. For example, if a structure is KI 6° then $[K_{uu}]$ will be a 6 × 6 square matrix.

Analysis of a Structure Once the stiffness matrix is found, the following procedure may be used for finding displacements and reaction forces.

1. Assemble a force column vector that has the known applied forces at each DOF and the unknown reaction forces. Place them in the same DOF order as used for the stiffness matrix. If there is no applied force at an unrestrained DOF, insert a 0.
2. Assemble a displacement column vector that has the unknown displacements of the unrestrained DOFs and also the known displacements at the supports (for now these should be zero).
3. Partition the stiffness matrix, force vector, and displacement vector such that the unrestrained DOFs are grouped and the restrained DOFs are grouped where $\{\Delta_r\} = \{0\}$. Thus,

$$\left\{ \begin{array}{c} F_u \\ \hline R_r \end{array} \right\} = \left[\begin{array}{c|c} K_{uu} & K_{ur} \\ \hline K_{ru} & K_{rr} \end{array} \right] \left\{ \begin{array}{c} \Delta_u \\ \hline \Delta_r \end{array} \right\}$$

4. Solve for the unknown displacements by

$$\{\Delta_u\} = [K_{uu}]^{-1}\{F_u\}$$

5. Once $\{\Delta_u\}$ is found, the reaction forces may be found by

$$\{R_r\} = [K_{ru}]\{\Delta_u\}$$

QUICK NOTE

The matrix math shown here can be done by hand for very small matrices. However, readers will likely want to use calculators or some other math software to do these calculations. At the end of this chapter, we will introduce how to do these calculations using a spreadsheet.

EXAMPLE 19.5

Use the matrix formulation of the stiffness approach to find the joint displacements and the reactions for the following assembly of truss bars. This is the same assembly presented in Example 19.4, so the same DOF numbering and stiffness matrix can be used.

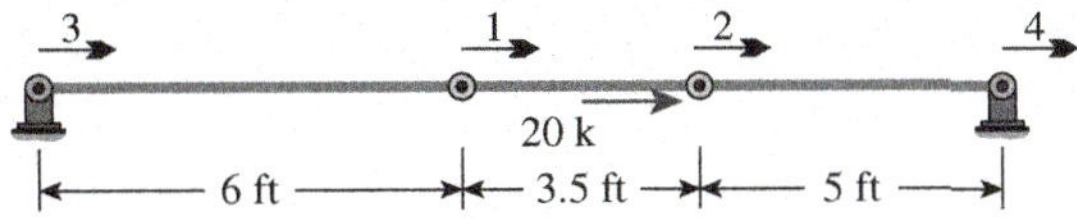

SOLUTION STRATEGY

Confirm that the stiffness matrix was assembled by considering the unrestrained DOFs first and the restrained DOFs second, then follow the process as outlined.

SOLUTION

The force vector is a 4 × 1 column vector because there are four DOFs in this structure. The force applied at DOF 1 $(F_1) = 0.0$ and at DOF 2 $(F_2) = 20$ k. The reactions R_3 and R_4 are currently unknown, so the total partitioned force vector looks like

$$\begin{Bmatrix} F_u \\ \hline R_r \end{Bmatrix} = \begin{Bmatrix} 0.0 \\ 20.0 \\ \hline R_3 \\ R_4 \end{Bmatrix} \text{k}$$

DISPLACEMENT VECTOR

The displacements at DOFs 1 and 2 (Δ_1 and Δ_2) are unknown, but the displacements at DOFs 3 and 4 are zero. This gives the following displacement vector.

$$\begin{Bmatrix} \Delta_u \\ \hline \Delta_r \end{Bmatrix} = \begin{Bmatrix} \Delta_1 \\ \Delta_2 \\ \hline 0.0 \\ 0.0 \end{Bmatrix} \text{in.}$$

The partitioned stiffness matrix taken from Example 19.4 is

$$K = \left[\begin{array}{c|c} K_{uu} & K_{ur} \\ \hline K_{ru} & K_{rr} \end{array}\right] = \begin{array}{c} \\ 1 \\ 2 \\ 3 \\ 4 \end{array} \begin{array}{c} \begin{array}{cccc} 1 & 2 & 3 & 4 \end{array} \\ \left[\begin{array}{cc|cc} 2675 & -2071 & -604 & -0 \\ -2071 & 2796 & 0 & -725 \\ \hline -604 & 0 & 604 & 0 \\ 0 & -725 & 0 & 725 \end{array}\right] \end{array} \text{k/in.}$$

The unknown displacements may be found by the following matrix operation.

$$\{\Delta_u\} = [K_{uu}]^{-1}\{F_u\} \Rightarrow \begin{Bmatrix} \Delta_1 \\ \Delta_2 \end{Bmatrix} = \begin{bmatrix} 2675 & -2071 \\ -2071 & 2796 \end{bmatrix}^{-1} \begin{Bmatrix} 0.0 \\ 20.0 \end{Bmatrix}$$

$$\begin{Bmatrix} \Delta_1 \\ \Delta_2 \end{Bmatrix} = \begin{bmatrix} 0.00088 & 0.00065 \\ 0.00065 & 0.00084 \end{bmatrix} \begin{Bmatrix} 0.0 \\ 20.0 \end{Bmatrix} = \begin{Bmatrix} 0.0130 \\ 0.0168 \end{Bmatrix} \text{in.}$$

The reactions of the assembly may now be found.

$$\{R_r\} = [K_{ru}]\{\Delta_u\}$$

$$\begin{Bmatrix} R_3 \\ R_4 \end{Bmatrix} = \begin{bmatrix} -604 & 0 \\ 0 & -725 \end{bmatrix} \begin{Bmatrix} 0.0130 \\ 0.0168 \end{Bmatrix} = \begin{Bmatrix} -7.84 \\ -12.16 \end{Bmatrix} \text{k}$$

QUICK NOTE

The displacement and reaction results are to be interpreted in terms of the global coordinate system (i.e., right is positive and left is negative).

The overall process of creating the stiffness matrix for a structure was introduced in the previous section. This involved locking DOFs and imposing unit displacements. While this approach is correct, it is not realistic to apply such an approach to structures with a large number of DOFs. A more direct and not so cumbersome approach is needed. This involves the creation of a stiffness

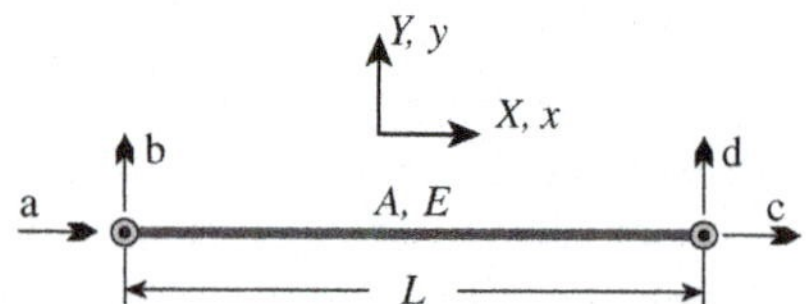

FIGURE 19.6 Axial force element.

matrix for each element (member) in the structure and then combining them in a meaningful way to represent the entire structure. The next three sections of this chapter are devoted to the development of this approach.

19.5 Stiffness Matrix for Axial Force Members

QUICK NOTE

The terms **member** and **element** will be used interchangeably just like the terms **joint** and **node** are used interchangeably.

The process for creating the stiffness matrix for an entire truss structure is more easily achieved if the stiffness matrix for individual axial members is first developed. In the previous section, the stiffness matrix for a single member was already introduced. However, it considered DOFs only along the member axis (i.e., two DOFs). For more general use, the element stiffness matrix needs to capture two orthogonal degrees of translation (i.e., two DOFs at each end).

RECALL

DOFs for an element in local coordinates will be labeled with letters, whereas numbers will be used for DOFs listed in the global coordinate system. For Figure 19.6, we could use either alpha or numeric characters. We will choose alpha characters for a reason that becomes obvious in the next section.

Consider the axial member shown in Figure 19.6. It has two orthogonal DOFs at each end (i.e., node), and the member local axis agrees with the global axis. The familiar procedure of restraining all the DOFs and then imposing unit displacements is used. Recall that the resulting reactions are the stiffness factors.

Figure 19.7 demonstrates the sequence of imposing the unit displacements in the members. The resulting stiffness matrix is shown hereafter. Most of the reactions that are shown make perfect sense, such as those in Figures 19.7(a) and 19.7(c). However, the reactions resulting from the displacements shown in Figures 19.7(b) and 19.7(d) require additional explanation. For example, when a unit displacement is imposed to the b DOF, the displacement and the resulting rotation of the member is assumed to be small enough that the net length of the member does not change. If the member lengths do not change, no axial force develops, and thus no reactions (i.e., stiffness factors) develop. Thus,

$$k = \begin{array}{c} \\ a \\ b \\ c \\ d \end{array} \begin{array}{c} \begin{array}{cccc} a & b & c & d \end{array} \\ \begin{bmatrix} \frac{AE}{L} & 0 & -\frac{AE}{L} & 0 \\ 0 & 0 & 0 & 0 \\ -\frac{AE}{L} & 0 & \frac{AE}{L} & 0 \\ 0 & 0 & 0 & 0 \end{bmatrix} \end{array}$$

DID YOU KNOW?

Considering the effects of large displacements in the derivation of the stiffness matrix is to include what are called *second-order effects*—also known as geometric nonlinearities. A presentation of such is beyond the scope of this introductory text but refer to *Matrix Structural Analysis* for a full presentation.[1]

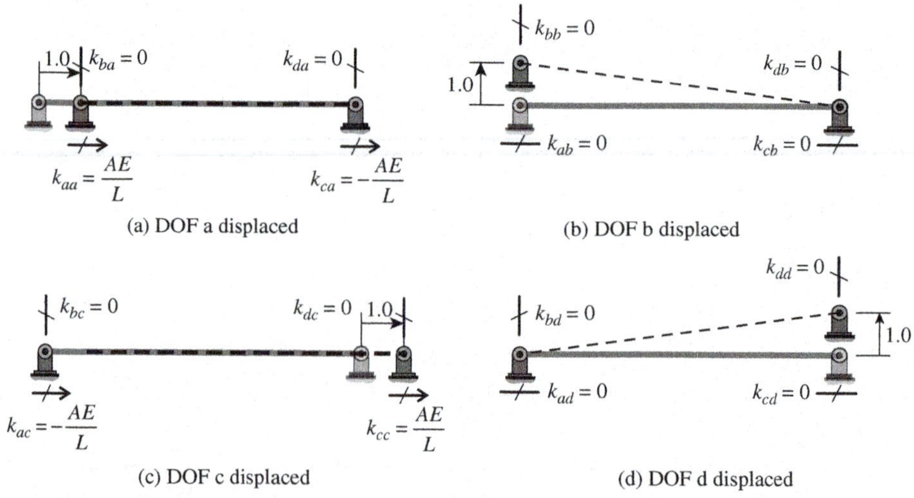

1.0 $k_{ba} = 0$ $k_{da} = 0$ $k_{aa} = \frac{AE}{L}$ $k_{ca} = -\frac{AE}{L}$

(a) DOF a displaced

$k_{bb} = 0$ 1.0 $k_{db} = 0$ $k_{ab} = 0$ $k_{cb} = 0$

(b) DOF b displaced

$k_{bc} = 0$ $k_{dc} = 0$ 1.0 $k_{ac} = -\frac{AE}{L}$ $k_{cc} = \frac{AE}{L}$

(c) DOF c displaced

$k_{dd} = 0$ $k_{bd} = 0$ 1.0 $k_{ad} = 0$ $k_{cd} = 0$

(d) DOF d displaced

FIGURE 19.7 Development of axial member stiffness matrix.

[1]W. McGuire, R.H. Gallagher, and R.D. Ziemian, *Matrix Structural Analysis, 2ed*, (New York: John Wiley & Sons, Inc.), 2000, 242–268.

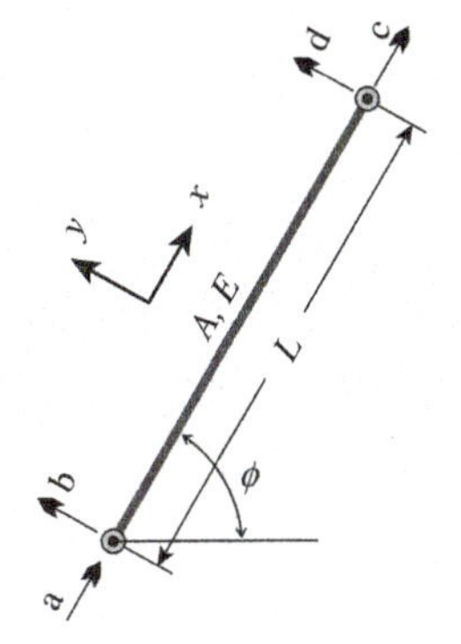

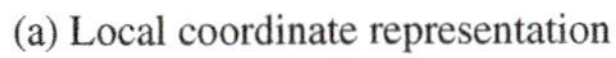
(a) Local coordinate representation

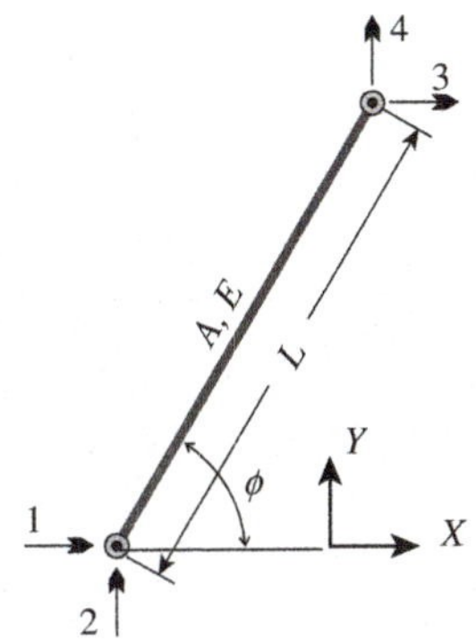

(b) Global coordinate representation

FIGURE 19.8 DOFs and coordinate system for inclined axial members.

19.6 Stiffness Matrix for Inclined Axial Force Members

Look at the basic truss in Figure 19.1. Note that not all of the axial members are oriented horizontally like the member in Figure 19.6. As such, a stiffness matrix must be developed that can accommodate an axial member having a different orientation other than horizontal.

First, we must re-examine the difference between the local and global coordinate systems. As demonstrated in Figure 19.8, the local axis is oriented the same way as the member, while the global system maintains a horizontal and vertical orientation. The DOFs at each end either can be described in local [Figure 19.8(a)] or global coordinates [Figure 19.8(b)]. The previous section already presented the stiffness matrix for an axial member based on its local coordinates. This section will present the stiffness matrix now in the global coordinate system.

QUICK NOTE

As will be shown in the next section, all element stiffness matrices must eventually be presented in global coordinates so that they all share the same coordinate system.

The derivation of the stiffness matrix for a rotated axial member is a little more challenging than had previously been experienced with the non-rotated member. This is because there are some additional geometry considerations that will also require some use of trigonometry.

Begin by restraining all DOFs and imposing a positive unit displacement at DOF 1 [see Figure 19.9(a)]. Assume that the displacement is small enough that any net change in the orientation angle ϕ can be neglected. Then the net change in the axial length of the member (e_1) is found as

$$e_1 = -(1.0)\cos\phi$$

where the negative sign indicates a shortening of the member and a resulting compression force. The axial force in the member is now found as

$$Q = \frac{AE}{L}e_1 = -\frac{AE}{L}\cos\phi$$

This then needs to be resolved into horizontal and vertical components that correspond to DOFs 1 and 2, respectively (i.e., K_{11} and K_{21}). This is shown in Figure 19.10. Using basic equilibrium equations, one can then solve for the remaining reactions (i.e., K_{31} and K_{41}).

TRIGONOMETRY REMINDER

The stiffness terms rely on being able to correctly identify the change in length of the member, for example e_1. The trigonometry is shown here in a scaled-up manner.

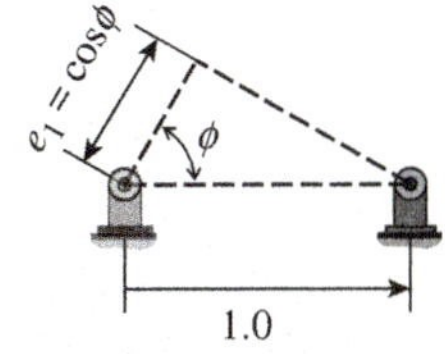

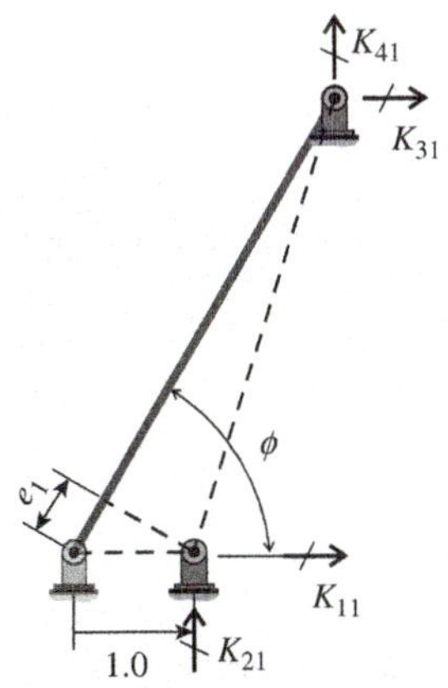

(a) Unit displacement at DOF 1

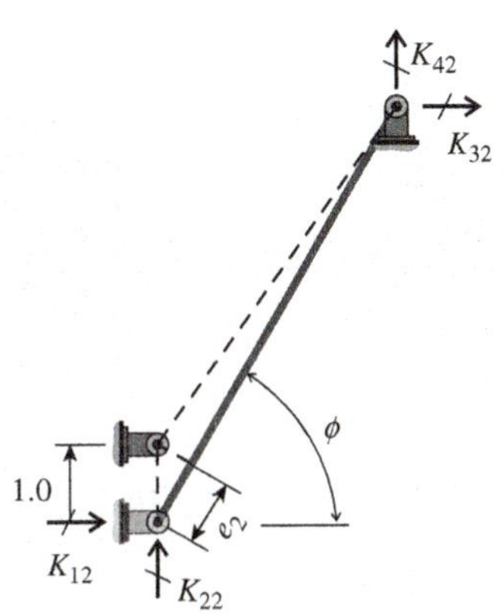

(b) Unit displacement at DOF 2

FIGURE 19.9 Displaced inclined axial members.

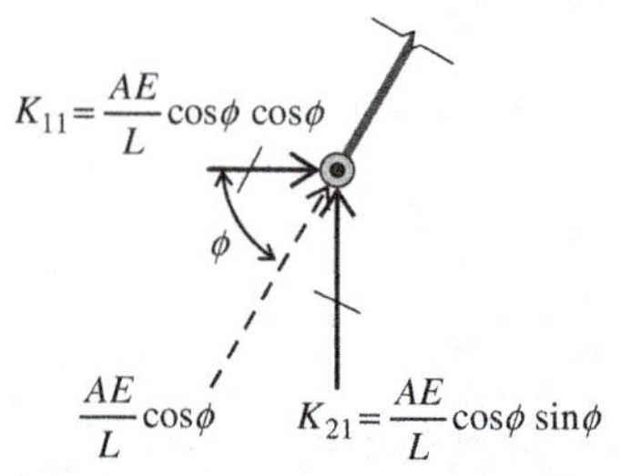

FIGURE 19.10 Resolving axial force into components.

$$K_{11} = \frac{AE}{L}\cos^2\phi \qquad K_{21} = \frac{AE}{L}\cos\phi\sin\phi$$

$$K_{31} = -\frac{AE}{L}\cos^2\phi \qquad K_{41} = -\frac{AE}{L}\cos\phi\sin\phi$$

We can now drop these stiffness values into the first column of the stiffness matrix. Take note that the $(AE)/L$ term is constant to all terms, so it is conveniently pulled out in front of the matrix.

$$K = \frac{AE}{L}\;\begin{array}{c}\\ 1\\ 2\\ 3\\ 4\end{array}\begin{array}{c}\begin{array}{cccc}1 & 2 & 3 & 4\end{array}\\ \left[\begin{array}{cccc}\cos^2\phi & - & - & -\\ \cos\phi\sin\phi & - & - & -\\ -\cos^2\phi & - & - & -\\ -\cos\phi\sin\phi & - & - & -\end{array}\right]\end{array}$$

QUICK NOTE

An abbreviated representation of the member stiffness matrix replaces $\cos\phi$ with c and $\sin\phi$ with s as shown here.

$$K = \frac{AE}{L}\begin{bmatrix} c^2 & cs & -c^2 & -cs\\ cs & s^2 & -cs & -s^2\\ -c^2 & -cs & c^2 & cs\\ -cs & -s^2 & cs & s^2\end{bmatrix}$$

The same approach may be used for the other three DOFs to get the remaining terms in the matrix. Figure 19.9(b) shows the unit displacement being imposed at DOF 2. The total element stiffness matrix is given as

$$K = \frac{AE}{L}\;\begin{array}{c}\\ 1\\ 2\\ 3\\ 4\end{array}\begin{array}{c}\begin{array}{cccc}1 & 2 & 3 & 4\end{array}\\ \left[\begin{array}{cccc}\cos^2\phi & \cos\phi\sin\phi & -\cos^2\phi & -\cos\phi\sin\phi\\ \cos\phi\sin\phi & \sin^2\phi & -\cos\phi\sin\phi & -\sin^2\phi\\ -\cos^2\phi & -\cos\phi\sin\phi & \cos^2\phi & \cos\phi\sin\phi\\ -\cos\phi\sin\phi & -\sin^2\phi & \cos\phi\sin\phi & \sin^2\phi\end{array}\right]\end{array}$$

Identifying the Angle ϕ

For some members, the angle of ϕ is readily apparent, yet for other members, it becomes a little less obvious and indeed is dependent upon the decision of the analyst. However, it is imperative that a consistent interpretation is made. For this reason, a formalized way of defining ϕ is given here.

1. Choose and identify which joint is the starting joint (called the i joint) and which is the ending joint (called the j joint). It doesn't matter which joint is assigned to be which.
2. An indicator arrow head—pointed away from the starting joint—is sketched on the member indicating the local x-axis of the member.
3. Being positioned at joint i, the angle ϕ is measured from the global X-axis and rotated counterclockwise to the member. This means that $0° \leq \phi \leq 360°$.

Figure 19.11 demonstrates this measurement for several different scenarios.

Direct Method For Computing $\cos\phi$ and $\sin\phi$

One often finds it convenient to avoid the direct calculation of ϕ and jump directly to calculating the values of $\cos\phi$ and $\sin\phi$. This is particularly useful when programming the matrix method into spreadsheets or other math-based software. It can be done based solely upon the coordinates

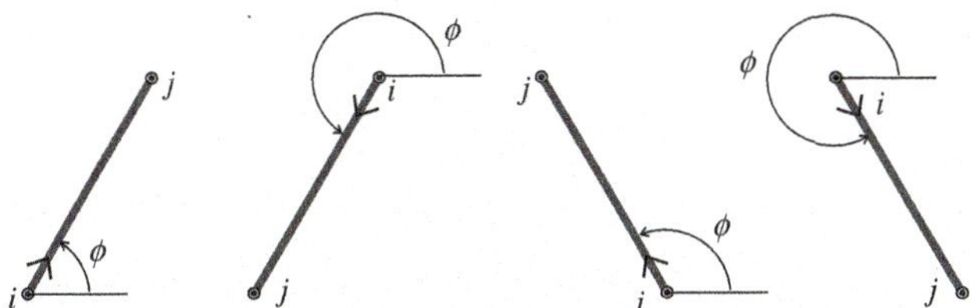

FIGURE 19.11 How to measure angle of ϕ.

of the joints located at the end of the axial member. The first calculation is to find the length, L, of the member using the Pythagorean theorem. Refer to Figure 19.12 for understanding of the notation.

$$L = \sqrt{(X_j - X_i)^2 + (Y_j - Y_i)^2}$$

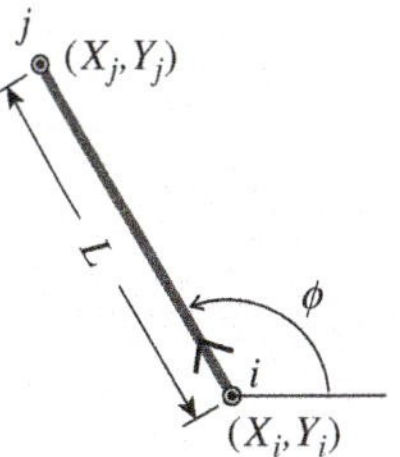

FIGURE 19.12 Coordinates of a member.

The $\cos\phi$ and $\sin\phi$ now can be calculated using the well-known trigonometric identities that are expressed in the terms of the end coordinates of the members.

$$\cos\phi = \frac{X_j - X_i}{L} \qquad \sin\phi = \frac{Y_j - Y_i}{L}$$

This type of calculation will give the correct signs to these terms in all cases and is easily programmed into a spreadsheet or other math software.

Transformation Matrix for Coordinate Conversion

Throughout the analysis of a structure, the analyst will have to convert member properties (i.e., stiffness values) from local to global coordinates. They will also find themselves needing to transform both the forces (Q, F) and the displacements (u, Δ) back and forth between coordinate systems. Closed-form matrices may be developed for this transformation—much like was done for the stiffness matrix of the inclined member presented previously. However, it is often convenient to handle the transformations using a transformation matrix.

The form of the transformation matrix is most easily seen with the following presentation. Assume that a set of forces F_1 and F_2 are given in a global coordinate system, XY. Now assume that these forces need to be transformed (Q_a and Q_b) into a local coordinate system, xy, which is oriented with an angle of ϕ from the horizontal (see Figure 19.13).

The horizontal and vertical forces, F_1 and F_2, are resolved into components that are oriented with the xy axes, as shown in Figure 19.13(b). To find the total forces in the xy coordinates (Q_a and Q_b), the resolved components shown in Figure 19.13(b) are summed up. This generates the two expressions as

$$Q_a = F_1 \cos\phi + F_2 \sin\phi$$

$$Q_b = -F_1 \sin\phi + F_2 \cos\phi$$

These two equations can be rewritten in matrix form as

$$\begin{Bmatrix} Q_a \\ Q_b \end{Bmatrix} = \begin{bmatrix} \cos\phi & \sin\phi \\ -\sin\phi & \cos\phi \end{bmatrix} \begin{Bmatrix} F_1 \\ F_2 \end{Bmatrix}$$

QUICK NOTE

The coordinate transformation matrix for two force components is represented as

$$[T] = \begin{bmatrix} \cos\phi & \sin\phi \\ -\sin\phi & \cos\phi \end{bmatrix}$$

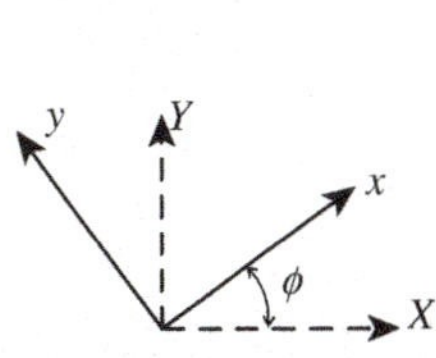

(a) Coordinate systems

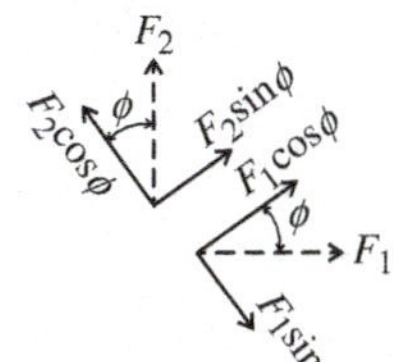

(b) Resolved global forces

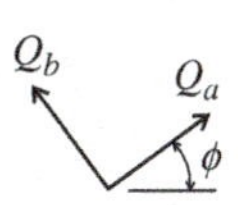

(c) Local forces

FIGURE 19.13 Transformation of forces from one coordinate system to another.

Referring to the *Quick Note* on the previous page, this simplified equation may be written as

$$\begin{Bmatrix} Q_a \\ Q_b \end{Bmatrix} = [T] \begin{Bmatrix} F_1 \\ F_2 \end{Bmatrix}$$

DID YOU KNOW?

The determinant of the transformation matrix is 1.0, which means that the inverse of the transformation matrix is nothing more than the transpose of the matrix.

$$[T]^{-1} = [T]^T$$

Since the transformation matrix is easier to calculate, we use $[T]^T$ anywhere the inverse should be used.

The transformation just demonstrated is for only two force components. A truss bar has two force components at each end, giving a total of four force components. Without a formal derivation, the transformation matrix is presented here for a truss bar—recognizing that the basic process for derivation has already been given.

$$[T] = \begin{bmatrix} \cos\phi & \sin\phi & 0 & 0 \\ -\sin\phi & \cos\phi & 0 & 0 \\ 0 & 0 & \cos\phi & \sin\phi \\ 0 & 0 & -\sin\phi & \cos\phi \end{bmatrix}$$

All of the relevant transformation equations may be summarized as follows.

Global to Local Transformation	**Local to Global Transformation**
$\{Q\} = [T]\{F\}$	$\{F\} = [T]^T\{Q\}$
$\{u\} = [T]\{\Delta\}$	$\{\Delta\} = [T]^T\{u\}$
$[k] = [T][K][T]^T$	$[K] = [T]^T[k][T]$

QUICK NOTE

Some individuals prefer to use the transformation matrix formulation, while others prefer to have the predefined matrices and vectors. As such, both forms have been given here.

The expanded global stiffness matrix for a truss element already has been given in this text. For convenience, the global force vector and the global displacement vectors are provided here in their expanded form. Refer to Figure 19.8 to understand the subscripts.

$$\{F\} = \begin{Bmatrix} Q_a\cos\phi - Q_b\sin\phi \\ Q_b\cos\phi + Q_a\sin\phi \\ Q_c\cos\phi - Q_d\sin\phi \\ Q_d\cos\phi + Q_c\sin\phi \end{Bmatrix} \qquad \{\Delta\} = \begin{Bmatrix} u_a\cos\phi - u_b\sin\phi \\ u_b\cos\phi + u_a\sin\phi \\ u_c\cos\phi - u_d\sin\phi \\ u_d\cos\phi + u_c\sin\phi \end{Bmatrix}$$

EXAMPLE 19.6

The three-member truss shown has all of the global DOFs identified, the selected origin of each member identified, and the members numbered. All bars have a cross-sectional area of 2.25 in^2 and E = 10,000 ksi. Find the global stiffness matrix for bar 2.

SOLUTION STRATEGY

Isolate bar 2, noting that the origin of the member coincides with the global DOFs 3 and 4, while the end member coincides with the global DOFs 1 and 2. Develop the transformation matrix and the local element stiffness matrix using the end coordinates of the member.

SOLUTION

Identify the coordinates of the starting joint (i) and the ending joint (j) of member 2. Use these to calculate L_2, $\cos\phi_2$, and $\sin\phi_2$ directly.

JOINT COORDINATES OF MEMBER 2

$$(X_i, Y_i) = (0, 0) \text{ in.}$$

$$(X_j, Y_j) = (108, 144) \text{ in.}$$

The lower left-hand joint in the truss was arbitrarily selected as the origin for the entire structure.

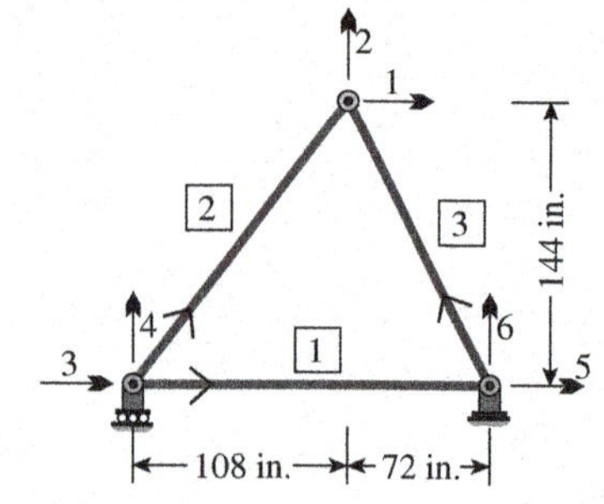

$$L_2 = \sqrt{(108\text{ in.} - 0\text{ in.})^2 + (144\text{ in.} - 0\text{ in.})^2} = 180\text{ in.}$$

$$\cos\phi_2 = \frac{108\text{ in.} - 0\text{ in.}}{180\text{ in.}} = 0.6$$

$$\sin\phi_2 = \frac{144\text{ in.} - 0\text{ in.}}{180\text{ in.}} = 0.8$$

The transformation matrix for member 2 is then assembled as

$$T_2 = \begin{array}{c} \\ a \\ b \\ c \\ d \end{array} \begin{array}{c} \begin{array}{cccc} 3 & 4 & 1 & 2 \end{array} \\ \begin{bmatrix} 0.6 & 0.8 & 0 & 0 \\ -0.8 & 0.6 & 0 & 0 \\ 0 & 0 & 0.6 & 0.8 \\ 0 & 0 & -0.8 & 0.6 \end{bmatrix} \end{array}$$

QUICK NOTE

The DOFs labeled on the rows of the transformation matrix represent the local coordinates and always will be *a* through *d* in order. The DOFs labeled on the columns of the transformation matrix represent the global coordinates associated with the *i* and *j* ends, respectively, of the member in question. At each end, the horizontal DOF is listed first, and the vertical DOF is listed second.

The stiffness term (AE/L) for member 2 is found as

$$\left(\frac{AE}{L}\right)_2 = \frac{(2.25\text{ in}^2)(10{,}000\text{ ksi})}{180\text{ in.}} = 125\text{k/in.}$$

The local element stiffness matrix then becomes

$$k^{(2)} = 125 \begin{array}{c} \\ a \\ b \\ c \\ d \end{array} \begin{array}{c} \begin{array}{cccc} a & b & c & d \end{array} \\ \begin{bmatrix} 1 & 0 & -1 & 0 \\ 0 & 0 & 0 & 0 \\ -1 & 0 & 1 & 0 \\ 0 & 0 & 0 & 0 \end{bmatrix} \end{array} \text{k/in.}$$

NOTATION

The stiffness matrices for individual elements are denoted with a superscript in parenthesis. For example, $k^{(2)}$ is the local element stiffness matrix for member 2, while $K^{(2)}$ is the global stiffness matrix for element 2. The reason for making the notation a superscript rather than a subscript—as seems natural—becomes obvious in the next section.

The element stiffness matrix for member 2—given in global coordinates—finally can be calculated using

$$K^{(2)} = [T]_2^T [k]^{(2)} [T]_2 = \begin{array}{c} \\ 3 \\ 4 \\ 1 \\ 2 \end{array} \begin{array}{c} \begin{array}{cccc} 3 & 4 & 1 & 2 \end{array} \\ \begin{bmatrix} 45.0 & 60.0 & -45.0 & -60.0 \\ 60.0 & 80.0 & -60.0 & -80.0 \\ -45.0 & -60.0 & 45.0 & 60.0 \\ -60.0 & -80.0 & 60.0 & 80.0 \end{bmatrix} \end{array} \text{k/in.}$$

RECALL

Refer to Appendix C for how to perform the specific matrix multiplication and matrix transpose. Spreadsheets (see Section 19.10) and some calculators can make this calculation quickly.

19.7 Assemblage of Structure-Level Stiffness Matrix for Planar Trusses

An assembly of truss bars was shown in Example 19.4. The structure stiffness matrix for this assemblage was found by locking all DOFs (four in this case), imposing a unit displacement at each DOF independently, and then solving for all ensuing reactions. One column in the stiffness matrix was generated for each independent DOF that was displaced. While this procedure worked well enough for small and simple structures, it becomes very cumbersome when the overall number of DOFs in the structure gets larger. For this reason, a more methodical and *easy-to-automate* approach is needed to assemble the structure-level stiffness matrix. This approach is illustrated for the truss bar assembly given in Example 19.4.

The development of the element stiffness matrix for each of the members 1 through 3 is left as an exercise for the reader. Rather, the three stiffness matrices are presented here with the DOFs labeled according to those shown in the example.

$$K^{(1)} = \begin{matrix} & 3 & 1 \\ 3 & 604 & -604 \\ 1 & -604 & 604 \end{matrix} \text{ k/in.} \quad K^{(2)} = \begin{matrix} & 1 & 2 \\ 1 & 2071 & -2071 \\ 2 & -2017 & 2071 \end{matrix} \text{ k/in.} \quad K^{(3)} = \begin{matrix} & 2 & 4 \\ 2 & 725 & -725 \\ 4 & -725 & 725 \end{matrix} \text{ k/in.}$$

Since the structure has four DOFs, the structure-level stiffness matrix is a 4 × 4 matrix. Each term in this matrix is found by

$$K_{i,j} = \sum_{n=1}^{\text{number of elements}} K_{i,j}^{(n)}$$

RECALL

The first subscript represents the row in the matrix and the second subscript represents the column in the matrix.

where n is the element number and i and j are integers between 1 and the total number of DOFs in the structure. Therefore, to find the stiffness term for $K_{1,1}$, the following calculation is performed. Take note that since, $K_{1,1}^{(3)}$ doesn't appear in any of the element matrices, its value is taken as zero.

$$K_{1,1} = K_{1,1}^{(1)} + K_{1,1}^{(2)} + K_{1,1}^{(3)} = 604 + 2071 + 0 = 2675 \text{ k/in.}$$

All combinations of i and j should be considered where $i = 1, ..., 4$ and $j = 1, ..., 4$. The first column of the stiffness matrix is filled out with these values:

$$K_{2,1} = K_{2,1}^{(1)} + K_{2,1}^{(2)} + K_{2,1}^{(3)} = 0 - 2071 + 0 = -2071 \text{ k/in.}$$
$$K_{3,1} = K_{3,1}^{(1)} + K_{3,1}^{(2)} + K_{3,1}^{(3)} = -604 + 0 + 0 = -604 \text{ k/in.}$$
$$K_{4,1} = K_{4,1}^{(1)} + K_{4,1}^{(2)} + K_{4,1}^{(3)} = 0 + 0 + 0 = 0 \text{ k/in.}$$

The total matrix then works out to be the same that was developed in Example19.4, as

$$K = \begin{matrix} & 1 & 2 & 3 & 4 \\ 1 & 2675 & -2071 & -604 & 0 \\ 2 & -2071 & 2796 & 0 & -725 \\ 3 & -604 & 0 & 604 & 0 \\ 4 & 0 & -725 & 0 & 725 \end{matrix} \text{ k/in.}$$

Procedure

QUICK NOTE

When numbering the DOFs in a structure, be sure to number the unrestrained DOFs first followed by the restrained DOFs.

1. Label all DOFs on the structure (two orthogonal DOFs per joint).
2. Label the members with a unique number. Any order will work.
3. Identify the x-axis of each member with the little arrow head. This will essentially define the i end and the j end of the member. Once again, this is an arbitrary selection and is the choice of the analyst.
4. Calculate the member length along with $\cos\phi$ and $\sin\phi$.
5. Assemble a separate local stiffness matrix for each element in the structure, $k^{(n)}$.
6. Generate a separate transformation matrix $[T]_n$ for each element and use them to calculate the global stiffness matrix for each individual element, $K^{(n)}$.
7. Assemble the global stiffness matrix K for the entire structure by summing all stiffness coefficients that have the same subscripts. Consider the contribution of all elements in the structure.

$$K_{i,j} = \sum_{n=1}^{\text{number of elements}} K_{i,j}^{(n)}$$

EXAMPLE 19.7

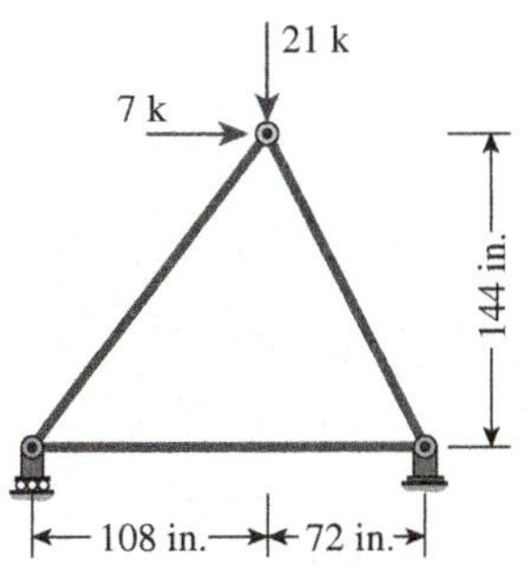

For the truss given in Example 19.6, assemble the structure stiffness matrix $[K]$. Then analyze the truss, finding the joint displacements and the reaction forces for the loading shown.

SOLUTION STRATEGY

Follow the procedure outlined for developing the global structure stiffness matrix, recognizing that some steps have already been completed. Review Example 19.5 for a review of the approach for calculating joint displacements and reactions.

SOLUTION

The global element stiffness matrix for member 2, $K^{(2)}$, was developed in Example 19.6, while the global stiffness matrices for the other two elements are developed later in Video Example 19.2. All three matrices are presented here for convenience.

$$K^{(1)} = \begin{array}{c} \\ 3 \\ 4 \\ 5 \\ 6 \end{array}\begin{array}{c} \begin{array}{cccc} 3 & 4 & 5 & 6 \end{array} \\ \begin{bmatrix} 125.0 & 0.0 & -125.0 & 0.0 \\ 0.0 & 0.0 & 0.0 & 0.0 \\ -125.0 & 0.0 & 125.0 & 0.0 \\ 0.0 & 0.0 & 0.0 & 0.0 \end{bmatrix} \end{array} \text{ k/in.}$$

QUICK NOTE

Steps 1 through 6 of the structure stiffness matrix generation procedure have already been completed in previous examples.

$$K^{(2)} = \begin{array}{c} \\ 3 \\ 4 \\ 1 \\ 2 \end{array}\begin{array}{c} \begin{array}{cccc} 3 & 4 & 1 & 2 \end{array} \\ \begin{bmatrix} 45.0 & 60.0 & -45.0 & -60.0 \\ 60.0 & 80.0 & -60.0 & -80.0 \\ -45.0 & -60.0 & 45.0 & 60.0 \\ -60.0 & -80.0 & 60.0 & 80.0 \end{bmatrix} \end{array} \text{ k/in.} \qquad K^{(3)} = \begin{array}{c} \\ 5 \\ 6 \\ 1 \\ 2 \end{array}\begin{array}{c} \begin{array}{cccc} 5 & 6 & 1 & 2 \end{array} \\ \begin{bmatrix} 28.0 & -55.9 & -28.0 & 55.9 \\ -55.9 & 111.8 & 55.9 & -111.8 \\ -28.0 & 55.9 & 28.0 & -55.9 \\ 55.9 & -111.8 & -55.9 & 111.8 \end{bmatrix} \end{array} \text{ k/in.}$$

Follow the process previously outlined, which says that all element stiffness values that share the same subscripts should be added together. Here is an example of a few of these.

$$K_{1,1} = K_{1,1}^{(1)} + K_{1,1}^{(2)} + K_{1,1}^{(3)} = 0.0 + 45.0 + 28.0 = 73.0 \text{ k/in.}$$

$$K_{2,1} = K_{2,1}^{(1)} + K_{2,1}^{(2)} + K_{2,1}^{(3)} = 0.0 + 60.0 - 55.9 = 4.1 \text{ k/in.}$$

$$K_{1,2} = K_{1,2}^{(1)} + K_{1,2}^{(2)} + K_{1,2}^{(3)} = 0.0 + 60.0 - 55.9 = 4.1 \text{ k/in.}$$

DID YOU KNOW?

All stiffness matrices of structures, which are linear-elastic in behavior, are symmetric. So

$$K_{i,j} = K_{j,i}$$

This useful recognition can expedite the assembly of the structure stiffness matrix.

This should be done for each term in the K matrix. Notice the side note about symmetry.
The entire structure matrix is given here already partitioned according to the unrestrained and the restrained DOFs.

$$K = \left[\begin{array}{c|c} K_{uu} & K_{ur} \\ \hline K_{ru} & K_{rr} \end{array}\right]$$

$$K = \begin{array}{c} \\ 1 \\ 2 \\ 3 \\ 4 \\ 5 \\ 6 \end{array}\begin{array}{c} \begin{array}{cccccc} 1 & 2 & 3 & 4 & 5 & 6 \end{array} \\ \left[\begin{array}{ccc|ccc} 73.0 & 4.1 & -45.0 & -60.0 & -28.0 & 55.9 \\ 4.1 & 191.8 & -60.0 & -80.0 & 55.9 & -111.8 \\ -45.0 & -60.0 & 170.0 & 60.0 & -125.0 & 0.0 \\ \hline -60.0 & -80.0 & 60.0 & 80.0 & 0.0 & 0.0 \\ -28.0 & 55.9 & -125.0 & 0.0 & 153.0 & -55.9 \\ 55.9 & -111.8 & 0.0 & 0.0 & -55.9 & 111.8 \end{array}\right] \end{array} \text{ k/in.}$$

RECALL

The matrix operation used to find DOF displacements is

$$\{\Delta_u\} = [K_{uu}]^{-1}\{F_u\}$$

The next step is to assemble the force vector for the unrestrained DOFs. Referring to the figure in Example 19.6 and also the figure in this example, we see that there is a positive 7-k load in DOF 1 and a negative 21-k load in DOF 2, while no load is applied in DOF 3.

$$\{F_u\} = \begin{Bmatrix} F_1 \\ F_2 \\ F_3 \end{Bmatrix} = \begin{Bmatrix} 7 \\ -21 \\ 0 \end{Bmatrix} \text{k}$$

DID YOU KNOW?

$$[K_{uu}]^{-1} = \begin{array}{c} \\ 1 \\ 2 \\ 3 \end{array}\begin{array}{c} \begin{array}{ccc} 1 & 2 & 3 \end{array} \\ \begin{bmatrix} 73.0 & 4.1 & -45.0 \\ 4.1 & 191.8 & -60.0 \\ -45.0 & -60.0 & 170.0 \end{bmatrix}^{-1} \end{array}$$

$$= \begin{array}{c} \\ 1 \\ 2 \\ 3 \end{array}\begin{array}{c} \begin{array}{ccc} 1 & 2 & 3 \end{array} \\ \begin{bmatrix} 0.0166 & 0.0011 & 0.0048 \\ 0.0011 & 0.0059 & 0.0024 \\ 0.0048 & 0.0024 & 0.0080 \end{bmatrix} \end{array}$$

The displacements at the unrestrained DOFs are found next.

$$\{\Delta_u\} = \begin{Bmatrix} \Delta_1 \\ \Delta_2 \\ \Delta_3 \end{Bmatrix} = \begin{bmatrix} 73.0 & 4.1 & -45.0 \\ 4.1 & 191.8 & -60.0 \\ -45.0 & -60.0 & 170.0 \end{bmatrix}^{-1} \begin{Bmatrix} 7 \\ -21 \\ 0 \end{Bmatrix} = \underline{\underline{\begin{Bmatrix} 0.0921 \\ -0.1167 \\ -0.0168 \end{Bmatrix} \text{in.}}}$$

The calculated displacements are used to now find the values at the reactions using the following relationship, which was presented earlier in this chapter.

$$\{R_r\} = [K_{ru}]\{\Delta_u\}$$

$$\{R_r\} = \begin{Bmatrix} R_4 \\ R_5 \\ R_6 \end{Bmatrix} = \begin{bmatrix} -60.0 & -80.0 & 60.0 \\ -28.0 & 55.9 & -125.0 \\ 55.9 & -111.8 & 0.0 \end{bmatrix} \begin{Bmatrix} 0.0921 \\ -0.1167 \\ -0.0168 \end{Bmatrix} = \underline{\underline{\begin{Bmatrix} 2.8 \\ -7.0 \\ 18.2 \end{Bmatrix} \text{k}}}$$

QUICK NOTE

The global coordinate system is used to interpret the signs on the displacement and reaction vectors.

19.8 Solving for Member End Forces

After finding both the joint displacements and the joint reactions, the member forces are the last major piece of required information to complete a truss analysis. The matrix formulation allows for the direct calculation of the forces that appear at the ends of each member. For truss bars, the axial forces in a bar are constant throughout the length of the bar. Thus, if the analyst knows the forces at the end of the bar, they know the force at any location along the length.

Finding member end forces is a two step process.

1. Convert the known global displacements at both ends of the nth truss bar, $\{\Delta\}^{(n)}$, into the local coordinate system for that bar, $\{u\}^{(n)}$. This is done through the use of the transformation matrix as

$$\{u\}^{(n)} = [T]_n \{\Delta\}^{(n)}$$

The $\Delta^{(n)}$ vector is a 4 × 1 (i.e., 4 rows and 1 column) vector that is populated with the global displacements associated with the bar of interest. The displacements should be positioned in the vector in this order.

(a) Horizontal at the *i* end

(b) Vertical at the *i* end

(c) Horizontal at the *j* end

(d) Vertical at the *j* end

QUICK NOTE

One may readily find the forces at the ends of the members in terms of the global coordinates shown here.

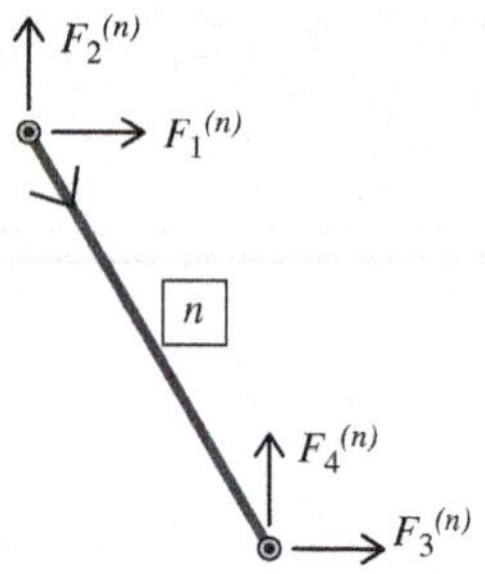
$F_2^{(n)}$
$F_1^{(n)}$
n
$F_4^{(n)}$
$F_3^{(n)}$

To have real meaning for interpretation, these forces should be converted to axes that are parallel and perpendicular to the member (i.e., axial and shear). This requires the member end forces to be reported in the local coordinate system.

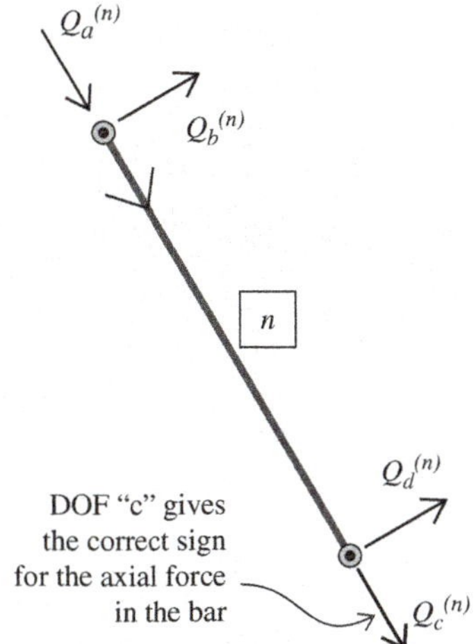

FIGURE 19.14
Interpretation of member end forces for trusses.

2. Compute the local member end forces $\{Q\}^{(n)}$ by using the local element stiffness matrix, $[k]^{(n)}$, and the local displacement vector, $\{u\}^{(n)}$.

$$\{Q\}^{(n)} = [k]^{(n)} \{u\}^{(n)}$$

Figure 19.14 shows the member end forces in the local coordinate system. $Q_a^{(n)}$ and $Q_c^{(n)}$ are the axial forces and should be equal in magnitude but of opposite sign. The forces $Q_b^{(n)}$ and $Q_d^{(n)}$ are shear forces and should work out to be zero for a truss bar.

DID YOU KNOW?

The DOF "c" in the local coordinate system gives the correct magnitude and the correct sign of the axial force in the member. This is because the assumed positive direction of the DOF corresponds to a tensile force in the bar.

EXAMPLE 19.8

Find the member axial forces for the truss presented and analyzed in Examples 19.6 and 19.7. Indicate whether they are in tension (T) or compression (C).

SOLUTION STRATEGY

All results are dependent upon the local element stiffness matrix for each member, the transformation matrix for each member, and also upon the global displacement vector for each member. Gather all of this information and then perform the required matrix operations.

SOLUTION

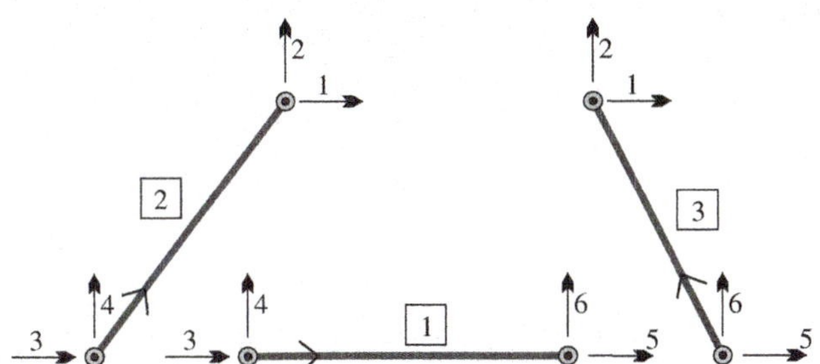

COMPLETE DISPLACEMENT VECTOR

$$\begin{Bmatrix} \Delta_1 \\ \Delta_2 \\ \Delta_3 \\ \Delta_4 \\ \Delta_5 \\ \Delta_6 \end{Bmatrix} = \begin{Bmatrix} 0.0921 \\ -0.1167 \\ -0.0168 \\ 0.0 \\ 0.0 \\ 0.0 \end{Bmatrix} \text{ in.}$$

$$\{\Delta\}^{(1)} = \begin{Bmatrix} \Delta_3 \\ \Delta_4 \\ \Delta_5 \\ \Delta_6 \end{Bmatrix} = \begin{Bmatrix} -0.0168 \\ 0.0 \\ 0.0 \\ 0.0 \end{Bmatrix} \qquad \{\Delta\}^{(2)} = \begin{Bmatrix} \Delta_3 \\ \Delta_4 \\ \Delta_1 \\ \Delta_2 \end{Bmatrix} = \begin{Bmatrix} -0.0168 \\ 0.0 \\ 0.0921 \\ -0.1167 \end{Bmatrix} \qquad \{\Delta\}^{(3)} = \begin{Bmatrix} \Delta_5 \\ \Delta_6 \\ \Delta_1 \\ \Delta_2 \end{Bmatrix} = \begin{Bmatrix} 0.0 \\ 0.0 \\ 0.0921 \\ -0.1167 \end{Bmatrix}$$

The transformation and local stiffness matrices are shown here for each member so that it is obvious to the reader what numbers are used in the calculation.

$$[k]^{(1)} = \begin{array}{c} \\ a \\ b \\ c \\ d \end{array} \begin{array}{c} \begin{array}{cccc} a & b & c & d \end{array} \\ \begin{bmatrix} 125.0 & 0.0 & -125.0 & 0.0 \\ 0.0 & 0.0 & 0.0 & 0.0 \\ -125.0 & 0.0 & 125.0 & 0.0 \\ 0.0 & 0.0 & 0.0 & 0.0 \end{bmatrix} \end{array} \text{k/in.} \qquad [k]^{(2)} = \begin{array}{c} \\ a \\ b \\ c \\ d \end{array} \begin{array}{c} \begin{array}{cccc} a & b & c & d \end{array} \\ \begin{bmatrix} 125.0 & 0.0 & -125.0 & 0.0 \\ 0.0 & 0.0 & 0.0 & 0.0 \\ -125.0 & 0.0 & 125.0 & 0.0 \\ 0.0 & 0.0 & 0.0 & 0.0 \end{bmatrix} \end{array} \text{k/in.}$$

$$[k]^{(3)} = \begin{array}{c} \\ a \\ b \\ c \\ d \end{array} \begin{array}{c} \begin{array}{cccc} a & b & c & d \end{array} \\ \begin{bmatrix} 139.8 & 0.0 & -139.8 & 0.0 \\ 0.0 & 0.0 & 0.0 & 0.0 \\ -139.8 & 0.0 & 139.8 & 0.0 \\ 0.0 & 0.0 & 0.0 & 0.0 \end{bmatrix} \end{array} \text{k/in.}$$

$$[T]_1 = \begin{array}{c} \\ a \\ b \\ c \\ d \end{array} \begin{array}{c} \begin{array}{cccc} 3 & 4 & 5 & 6 \end{array} \\ \begin{bmatrix} 1.00 & 0.00 & 0.00 & 0.00 \\ 0.00 & 1.00 & 0.00 & 0.00 \\ 0.00 & 0.00 & 1.00 & 0.00 \\ 0.00 & 0.00 & 0.00 & 1.00 \end{bmatrix} \end{array} \qquad [T]_2 = \begin{array}{c} \\ a \\ b \\ c \\ d \end{array} \begin{array}{c} \begin{array}{cccc} 3 & 4 & 1 & 2 \end{array} \\ \begin{bmatrix} 0.60 & 0.80 & 0.00 & 0.00 \\ -0.80 & 0.60 & 0.00 & 0.00 \\ 0.00 & 0.00 & 0.60 & 0.80 \\ 0.00 & 0.00 & -0.80 & 0.60 \end{bmatrix} \end{array}$$

$$[T]_3 = \begin{array}{c} \\ a \\ b \\ c \\ d \end{array} \begin{array}{c} \begin{array}{cccc} 5 & 6 & 1 & 2 \end{array} \\ \begin{bmatrix} -0.45 & 0.89 & 0.00 & 0.00 \\ -0.89 & -0.45 & 0.00 & 0.00 \\ 0.00 & 0.00 & -0.45 & 0.89 \\ 0.00 & 0.00 & -0.89 & -0.45 \end{bmatrix} \end{array}$$

The member end displacements in local coordinates now can be found for each member.

$$\begin{Bmatrix} u_a \\ u_b \\ u_c \\ u_d \end{Bmatrix}^{(1)} = [T]_1\{\Delta\}^{(1)} = \begin{Bmatrix} -0.0168 \\ 0.0 \\ 0.0 \\ 0.0 \end{Bmatrix} \text{in.} \qquad \begin{Bmatrix} u_a \\ u_b \\ u_c \\ u_d \end{Bmatrix}^{(2)} = [T]_2\{\Delta\}^{(2)} = \begin{Bmatrix} -0.0101 \\ 0.0134 \\ -0.0381 \\ -0.1437 \end{Bmatrix} \text{in.}$$

$$\begin{Bmatrix} u_a \\ u_b \\ u_c \\ u_d \end{Bmatrix}^{(3)} = [T]_3\{\Delta\}^{(3)} = \begin{Bmatrix} 0.000 \\ 0.000 \\ -0.146 \\ -0.030 \end{Bmatrix} \text{in.}$$

The final step is to use the local element stiffness matrix and the local displacement vector to find the member end forces.

$$\begin{Bmatrix} Q_a \\ Q_b \\ Q_c \\ Q_d \end{Bmatrix}^{(1)} = [k]^{(1)}\{u\}^{(1)} = \begin{Bmatrix} -2.100 \\ 0.000 \\ 2.100 \\ 0.000 \end{Bmatrix} \text{k} \qquad \begin{Bmatrix} Q_a \\ Q_b \\ Q_c \\ Q_d \end{Bmatrix}^{(2)} = [k]^{(2)}\{u\}^{(2)} = \begin{Bmatrix} 3.500 \\ 0.000 \\ -3.500 \\ 0.000 \end{Bmatrix} \text{k}$$

$$\begin{Bmatrix} Q_a \\ Q_b \\ Q_c \\ Q_d \end{Bmatrix}^{(3)} = [k]^{(3)}\{u\}^{(3)} = \begin{Bmatrix} 20.348 \\ 0.000 \\ -20.348 \\ 0.000 \end{Bmatrix} \text{k}$$

There are no forces in the b and d DOFs of all members. This is not a surprise, since truss members should have no shear. The sense of the axial force (i.e., tension or compression) is determined from the sign of Q_c. Bar 1 is in tension, while the other two bars are in compression.

EXAMPLE 19.9

Find the member end forces of all members of the truss assembly shown in Examples 19.4 and 19.5. Assume that the local element stiffness matrices are

$$K^{(1)} = \begin{matrix} & 3 & 1 \\ 3 & 604 & -604 \\ 1 & -604 & 604 \end{matrix} \text{ k/in.} \qquad K^{(2)} = \begin{matrix} & 1 & 2 \\ 1 & 2071 & -2071 \\ 2 & -2071 & 2071 \end{matrix} \text{ k/in.}$$

$$K^{(3)} = \begin{matrix} & 2 & 4 \\ 2 & 725 & -725 \\ 4 & -725 & 725 \end{matrix} \text{ k/in.}$$

DID YOU NOTICE?

The local and global coordinate systems agree for all members. This eliminates the need of a transformation matrix.

SOLUTION STRATEGY

Look to Example 19.5 to find the joint displacements. Generate a displacement vector for each of the members. There is no need to transform from global to local coordinates, so only one matrix operation is needed. Hence, $\{Q\} = [k]\{u\}$.

SOLUTION

The displacement vectors, given in local coordinates are

$$\begin{Bmatrix} \Delta_3 \\ \Delta_1 \end{Bmatrix} = \begin{Bmatrix} u_a \\ u_b \end{Bmatrix}^{(1)} = \begin{Bmatrix} 0.0000 \\ 0.0130 \end{Bmatrix} \text{in.} \qquad \begin{Bmatrix} \Delta_1 \\ \Delta_2 \end{Bmatrix} = \begin{Bmatrix} u_a \\ u_b \end{Bmatrix}^{(2)} = \begin{Bmatrix} 0.0130 \\ 0.0168 \end{Bmatrix} \text{in.}$$

$$\begin{Bmatrix} \Delta_2 \\ \Delta_4 \end{Bmatrix} = \begin{Bmatrix} u_a \\ u_b \end{Bmatrix}^{(3)} = \begin{Bmatrix} 0.0168 \\ 0.0000 \end{Bmatrix} \text{in.}$$

Using the element stiffness matrices and the displacement vectors, the local element forces are found.

$$\begin{Bmatrix} Q_a \\ Q_b \end{Bmatrix}^{(1)} = [K]^{(1)}\begin{Bmatrix} \Delta_3 \\ \Delta_1 \end{Bmatrix} = \begin{Bmatrix} -7.85 \\ 7.85 \end{Bmatrix} \text{k} \qquad \begin{Bmatrix} Q_a \\ Q_b \end{Bmatrix}^{(2)} = [K]^{(2)}\begin{Bmatrix} \Delta_1 \\ \Delta_2 \end{Bmatrix} = \begin{Bmatrix} -7.85 \\ 7.85 \end{Bmatrix} \text{k}$$

$$\begin{Bmatrix} Q_a \\ Q_b \end{Bmatrix}^{(3)} = [K]^{(3)}\begin{Bmatrix} \Delta_2 \\ \Delta_4 \end{Bmatrix} = \begin{Bmatrix} 12.15 \\ -12.15 \end{Bmatrix} \text{k}$$

FINDINGS

Bars 1 and 2 are in tension and bar 3 is in compression. A quick equilibrium check will show the axial bar forces are reasonable.

19.9 Characteristics of Stiffness Matrices

In the previous sections, stiffness matrices have been developed for truss bars. These stiffness matrices are all written in a special order, which should be carefully studied. The general stiffness equation is written symbolically as

$$\{F\} = [K]\{\Delta\}$$

The order of listing the joint forces in the matrix $[K]$ should be the same as the order of listing of the corresponding displacements in the vector $\{\Delta\}$. Thus, if the first listed joint force in $\{F\}$ is F_1, the first listed displacement in the vector $\{\Delta\}$ should be Δ_1, and so on. If this ordering of the equations is preserved throughout the analysis, all stiffness matrices will have the following characteristics.

1. They will be symmetric (i.e., $K_{i,j} = K_{j,i}$). A proof of this characteristic can be obtained from Maxwell's law of reciprocal deformations.
2. The full stiffness matrix for either a single element or for a general structure is singular, which means its determinant is zero and the inverse of that matrix does not exist. However, if sufficient boundary conditions are specified so that the structure is stable (and the stiffness matrix is partitioned to reflect these conditions), the resulting stiffness matrix is not singular.
3. The stiffness coefficients on the main diagonal are always positive. The reason for this characteristic stems from the fact that the stiffness coefficients on the main diagonal each represent the force at a joint required to produce a corresponding displacement at that joint. A negative force required to produce a positive displacement is contrary to observed physical behavior of structures.

19.10 Spreadsheet Computer Applications

The stiffness approach, as presented in this chapter, essentially breaks each structure down to a bunch of individual and simple building blocks (i.e., truss bars). The stiffness matrix for a single truss bar is well known and is easy to construct—even for an inclined member. Thus, very large and complex trusses can be constructed using these simple truss elements.

The challenge to this method is in the number of calculations and matrices that must be handled. However, personal computers can handle these calculations quickly and effectively. Using spreadsheets for these calculations can help in the learning process while assisting with the mathematics. It is also a tool to which most readers will have access.

QUICK NOTE
Many repetitive calculations are required as part of the stiffness approach. Many math-based software packages can readily handle these calculations. Spreadsheets, while not the most versatile tool, are used in this text to assist in these calculations. The reason for this is that the calculations are quite transparent—including the matrix math.

Basic Microsoft Excel Commands

Table 19.1 presents the four basic Excel commands used to perform matrix math.

QUICK NOTE
The =mdeterm command may be used to check whether a stiffness matrix is singular or not. A singular matrix has a determinant of zero.

Table 19.1 Matrix commands in Excel.

Command		Description
=mmult	⇒	matrix multiplication
=minverse	⇒	matrix inverse
=mdeterm	⇒	determinant of matrix
=transpose	⇒	transpose of matrix

A couple of key points need to be understood to be able to correctly use these commands.

1. Highlight the destination cells.
   - Appropriate vector/matrix dimensions of the answer should be known. Review Appendix C for a review of matrix math.
   - If you don't highlight enough cells, you will be missing part of the answer and will not know it.

- If you highlight too many cells, you will see all of the answers, but you will have a matrix that is the wrong size.

2. Type in the desired formula.
   - You may nest multiple commands in a single formula.
3. Press this group of keys to create the matrix range: **ctrl-shift-enter**.
   - If this has been correctly performed, the formula in any of the cells of the destination matrix should have curly braces (e.g., {= *transpose*(:)}).

Spreadsheet for Solution

The analysis of the truss presented in Examples 19.6 through 19.8 is presented in this section using Excel. Some screen captures of the basic setup are given here, but a fully functioning spreadsheet titled *Example 20-6.xlsx* is provided on the textbook website.

A local element matrix for each truss bar is set up where the length, $\cos\phi$, and $\sin\phi$ are calculated from the member end coordinates (X_i, Y_i) and (X_j, Y_j). This is shown in Figures 19.15 and 19.16. The calculation of the global stiffness matrix for element 1 requires the use of nested formulas. For example, highlight cells C26:F29. Then type the following in the formula bar: *=MMULT(TRANSPOSE(C19:F22),MMULT(C13:F16,C19:F22))*. Once this is typed, press **ctrl-shift-enter**, and Excel will place the formula in curly braces {} and fill out the entire matrix.

With the individual global-element stiffness matrices available, it is now time to generate the structure level stiffness matrix—making sure to partition it according to unrestrained DOFs and restrained DOFs. Concurrent with this is the assembly of the applied force vector and the unknown displacement vector. This is shown in Figures 19.17 and 19.18. For convenience, the assembly of the structure stiffness matrix is shown by giving the formulas used in each cell.

DID YOU KNOW?
Within the spreadsheet, great care should be used to set up the calculation for the first element. Once this is completely defined the remainder of the elements are very quickly generated using the *copy and paste* feature.

VBA Script for [*K*] Assembly The assembly of the global structure stiffness matrix can be a bit tedious and prone to error when the number of elements and the size of the matrix increases. To assist in this, the reader may use a simple *Visual Basic for Applications (VBA)* script, as shown hereafter. The file titled *Example 20-6.xlsm* demonstrates how this is used. Once the formula is defined for a single cell in the $[K]$ matrix, it may be readily copied to the other cells in the matrix without any effort.

	A	B	C	D	E	F	G	H	I	J	K	L	M	N
5				Element 1							Element 2			
6	Node Coordinates							Node Coordinates						
7	Xi =	0	in	Yi =	0	in		Xi =	0	in	Yi =	0	in	
8	Xj =	180	in	Yj =	0	in		Xj =	108	in	Yj =	144	in	
9	L_1 =	180	in	E =	10000	ksi		L_2 =	180	in	E =	10000	ksi	
10	cos(φ)=	1		A =	2.25	in²		cos(φ)=	0.6		A =	2.25	in²	
11	sin(φ)=	0						sin(φ)=	0.8					
12			a	b	c	d				a	b	c	d	
13		a	125.0	0.0	-125.0	0.0			a	125.0	0.0	-125.0	0.0	
14	k_1 =	b	0.0	0.0	0.0	0.0	k/in	k_2 =	b	0.0	0.0	0.0	0.0	k/in
15		c	-125.0	0.0	125.0	0.0			c	-125.0	0.0	125.0	0.0	
16		d	0.0	0.0	0.0	0.0			d	0.0	0.0	0.0	0.0	
17														
18			3	4	5	6				3	4	1	2	
19		a	1.00	0.00	0.00	0.00			a	0.60	0.80	0.00	0.00	
20	T_1 =	b	0.00	1.00	0.00	0.00		T_2 =	b	-0.80	0.60	0.00	0.00	
21		c	0.00	0.00	1.00	0.00			c	0.00	0.00	0.60	0.80	
22		d	0.00	0.00	0.00	1.00			d	0.00	0.00	-0.80	0.60	
23														
24														
25			3	4	5	6				3	4	1	2	
26		3	125.0	0.0	-125.0	0.0			3	45.0	60.0	-45.0	-60.0	
27	$K^{(1)}$ =	4	0.0	0.0	0.0	0.0	k/in	$K^{(2)}$ =	4	60.0	80.0	-60.0	-80.0	k/in
28		5	-125.0	0.0	125.0	0.0			1	-45.0	-60.0	45.0	60.0	
29		6	0.0	0.0	0.0	0.0			2	-60.0	-80.0	60.0	80.0	

FIGURE 19.15 Stiffness matrices for elements one and two.

Element 3

Node Coordinates

Xi =	180	in	Yi =	0	in
Xj =	108	in	Yj =	144	in
L_3 =	160.997	in	E =	10000	ksi
cos(ϕ)=	-0.44721		A =	2.25	in^2
sin(ϕ)=	0.89443				

$$\mathbf{k}_3 = \begin{array}{c} \\ a \\ b \\ c \\ d \end{array}\!\!\begin{array}{c} \begin{array}{cccc} a & b & c & d \end{array} \\ \begin{pmatrix} 139.8 & 0.0 & -139.8 & 0.0 \\ 0.0 & 0.0 & 0.0 & 0.0 \\ -139.8 & 0.0 & 139.8 & 0.0 \\ 0.0 & 0.0 & 0.0 & 0.0 \end{pmatrix} \end{array} \text{ k/in}$$

$$\mathbf{T}_3 = \begin{array}{c} \\ a \\ b \\ c \\ d \end{array}\!\!\begin{array}{c} \begin{array}{cccc} 5 & 6 & 1 & 2 \end{array} \\ \begin{pmatrix} -0.45 & 0.89 & 0.00 & 0.00 \\ -0.89 & -0.45 & 0.00 & 0.00 \\ 0.00 & 0.00 & -0.45 & 0.89 \\ 0.00 & 0.00 & -0.89 & -0.45 \end{pmatrix} \end{array}$$

$$\mathbf{K}^{(3)} = \begin{array}{c} \\ 5 \\ 6 \\ 1 \\ 2 \end{array}\!\!\begin{array}{c} \begin{array}{cccc} 5 & 6 & 1 & 2 \end{array} \\ \begin{pmatrix} 28.0 & -55.9 & -28.0 & 55.9 \\ -55.9 & 111.8 & 55.9 & -111.8 \\ -28.0 & 55.9 & 28.0 & -55.9 \\ 55.9 & -111.8 & -55.9 & 111.8 \end{pmatrix} \end{array} \text{ k/in}$$

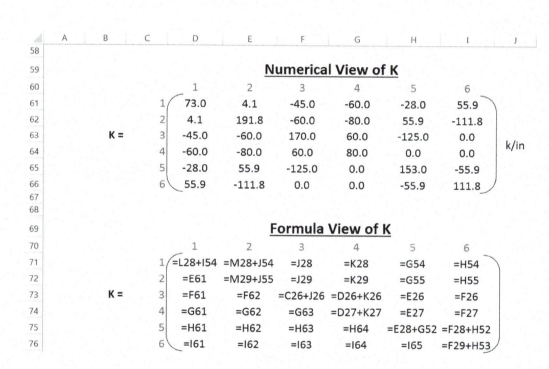

FIGURE 19.16 Element three stiffness matrix.

Numerical View of K

$$\mathbf{K} = \begin{array}{c} \\ 1 \\ 2 \\ 3 \\ 4 \\ 5 \\ 6 \end{array}\!\!\begin{array}{c} \begin{array}{cccccc} 1 & 2 & 3 & 4 & 5 & 6 \end{array} \\ \begin{pmatrix} 73.0 & 4.1 & -45.0 & -60.0 & -28.0 & 55.9 \\ 4.1 & 191.8 & -60.0 & -80.0 & 55.9 & -111.8 \\ -45.0 & -60.0 & 170.0 & 60.0 & -125.0 & 0.0 \\ -60.0 & -80.0 & 60.0 & 80.0 & 0.0 & 0.0 \\ -28.0 & 55.9 & -125.0 & 0.0 & 153.0 & -55.9 \\ 55.9 & -111.8 & 0.0 & 0.0 & -55.9 & 111.8 \end{pmatrix} \end{array} \text{ k/in}$$

Formula View of K

K =

	1	2	3	4	5	6
1	=L28+I54	=M28+J54	=J28	=K28	=G54	=H54
2	=E61	=M29+J55	=J29	=K29	=G55	=H55
3	=F61	=F62	=C26+J26	=D26+K26	=E26	=F26
4	=G61	=G62	=G63	=D27+K27	=E27	=F27
5	=H61	=H62	=H63	=H64	=E28+G52	=F28+H52
6	=I61	=I62	=I63	=I64	=I65	=F29+H53

FIGURE 19.17 Global structural stiffness matrix.

$$\begin{Bmatrix} F_u \\ \hline F_r \end{Bmatrix} = \begin{Bmatrix} F_1 \\ F_2 \\ F_3 \\ \hline R_4 \\ R_5 \\ R_6 \end{Bmatrix} = \begin{Bmatrix} 7 \\ -21 \\ 0 \\ \hline ? \\ ? \\ ? \end{Bmatrix} \text{ kips} \qquad \mathbf{K} = \begin{array}{c} \\ 1 \\ 2 \\ 3 \\ 4 \\ 5 \\ 6 \end{array}\!\!\begin{array}{c} \begin{array}{ccc|ccc} 1 & 2 & 3 & 4 & 5 & 6 \end{array} \\ \left(\begin{array}{ccc|ccc} 73.0 & 4.1 & -45.0 & -60.0 & -28.0 & 55.9 \\ 4.1 & 191.8 & -60.0 & -80.0 & 55.9 & -111.8 \\ -45.0 & -60.0 & 170.0 & 60.0 & -125.0 & 0.0 \\ \hline -60.0 & -80.0 & 60.0 & 80.0 & 0.0 & 0.0 \\ -28.0 & 55.9 & -125.0 & 0.0 & 153.0 & -55.9 \\ 55.9 & -111.8 & 0.0 & 0.0 & -55.9 & 111.8 \end{array}\right) \end{array} \text{ k/in}$$

$$\begin{Bmatrix} \Delta_u \\ \hline \Delta_r \end{Bmatrix} = \begin{Bmatrix} \Delta_1 \\ \Delta_2 \\ \Delta_3 \\ \hline \Delta_4 \\ \Delta_5 \\ \Delta_6 \end{Bmatrix} = \begin{Bmatrix} ? \\ ? \\ ? \\ \hline 0 \\ 0 \\ 0 \end{Bmatrix} \text{ inches} \qquad \mathbf{K} = \begin{array}{c} 1,2,3 \\ 4,5,6 \end{array}\!\!\begin{array}{c} \begin{array}{c|c} 1\;\;2\;\;3 & 4\;\;5\;\;6 \end{array} \\ \left(\begin{array}{c|c} K_{uu} & K_{ur} \\ \hline K_{ru} & K_{rr} \end{array}\right) \end{array}$$

FIGURE 19.18 Partitioned matrices.

```
Function global_K(k, dof)
Dim n, nn As Integer
Dim m, mm, i, j As Integer
stiff = k
d = dof
n = UBound(d, 2)
nn = UBound(stiff)
ReDim globalStiff(1 To n, 1 To n) As Double
For m = 1 To n
  For mm = 1 To n
  For i = 1 To nn - 1
    For j = 1 To nn - 1
      If (stiff(1, i + 1) = d(1, m) And stiff(1, j +
1) = d(1, mm)) Then
        globalStiff(m, mm) = stiff(i + 1, j + 1)
      End If
    Next j
  Next i
  Next mm
Next m
global_K = globalStiff
End Function
```

Here are instructions on how to use the VBA script.

1. Highlight the destination cells for the entire structure matrix.
2. Type: =global_K(k, dof).

   k = element stiffness matrix which includes the DOF labels. Thus, if your element stiffness matrix is a 4 × 4, you would highlight a 5 × 5 block of cells, including the row and column labels.

   dof = this should be a row vector that contains the DOF numbering for the entire structure. The length of this vector defines the final dimensions of the global stiffness matrix. Thus, if you highlight a row vector that has the labels for six DOFs, you will end up with a 6 × 6 global stiffness matrix.
3. To account for multiple elements, you should add a plus sign and then repeat step 2. This looks like: =global_K(k1, dof) + global_K(k2, dof) + ... (the DOF vector will be the same in all cases).
4. Press ctrl-shift-enter simultaneously.

Further Excel Operations Once the structure stiffness matrix and force vector are constructed and partitioned, the matrix math may be carried out to find first the joint displacements and then the support reactions. Figure 19.19 contains these calculations. The nested formula for

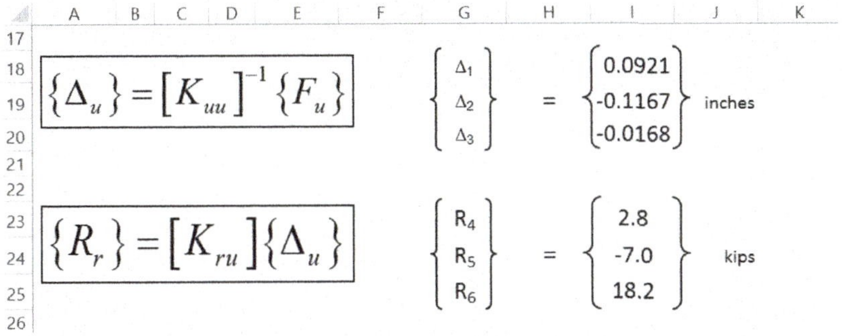

$$\{\Delta_u\} = [K_{uu}]^{-1}\{F_u\} \qquad \begin{Bmatrix} \Delta_1 \\ \Delta_2 \\ \Delta_3 \end{Bmatrix} = \begin{Bmatrix} 0.0921 \\ -0.1167 \\ -0.0168 \end{Bmatrix} \text{ inches}$$

$$\{R_r\} = [K_{ru}]\{\Delta_u\} \qquad \begin{Bmatrix} R_4 \\ R_5 \\ R_6 \end{Bmatrix} = \begin{Bmatrix} 2.8 \\ -7.0 \\ 18.2 \end{Bmatrix} \text{ kips}$$

FIGURE 19.19 Computed displacement and reaction vectors.

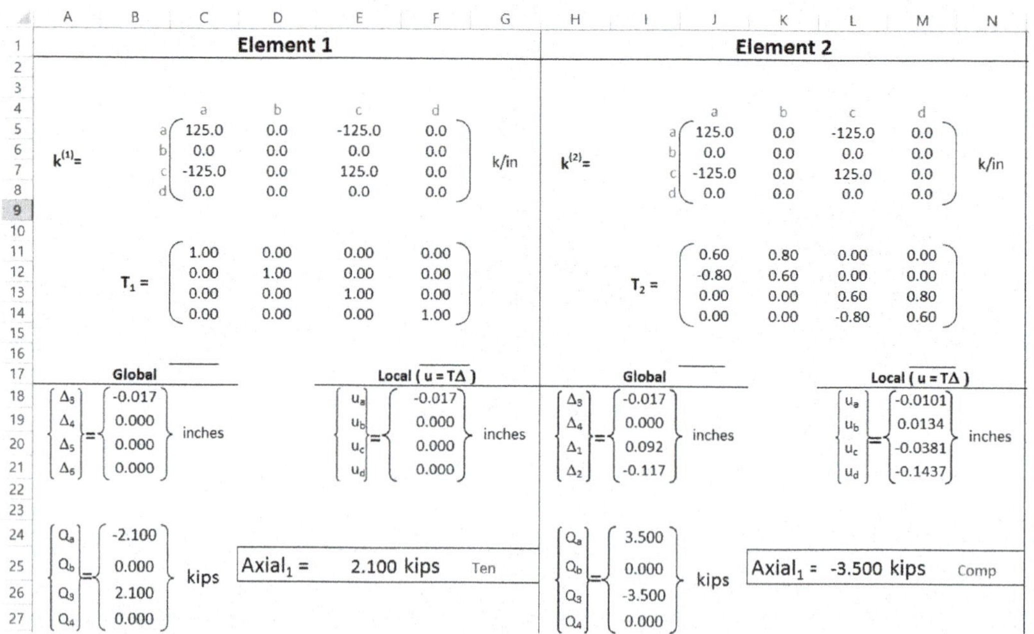

FIGURE 19.20 Member end forces for elements one and two.

calculating Δ is =*MMULT(MINVERSE(J2:L4),E2:E4)*. The formula for calculating the reactions is =*MMULT(J5:L7,I18:I20)*.

Lastly, the member end forces are found by first transforming the Δ vectors to local coordinates and then using the local stiffness matrix to get the member end forces (see Figures 19.20 and 19.21). Using member 1 as an example, the formula used for the transformation to local coordinates is =*MMULT (C11:F14,B18:B21)*. The formula used for getting the member end forces is =*MMULT(C5:F8,F18:F21)*.

Element 3

$k^{(3)}$ =

	a	b	c	d
a	139.8	0.0	-139.8	0.0
b	0.0	0.0	0.0	0.0
c	-139.8	0.0	139.8	0.0
d	0.0	0.0	0.0	0.0

k/in

T_3 =

-0.45	0.89	0.00	0.00
-0.89	-0.45	0.00	0.00
0.00	0.00	-0.45	0.89
0.00	0.00	-0.89	-0.45

Global: Δ_5, Δ_6, Δ_1, Δ_2 = 0.000, 0.000, 0.092, -0.117 inches

Local (u = TΔ): u_a, u_b, u_c, u_d = 0.000, 0.000, -0.146, -0.030 inches

Q_a, Q_b, Q_3, Q_4 = 20.348, 0.000, -20.348, 0.000 kips

Axial$_1$ = -20.348 kips Comp

FIGURE 19.21 Member end forces for element three.

19.11 Examples with Video Solution

VE19.1 For the truss given in Figure VE19.1, identify the degree of statical indeterminacy and also find the degree of kinematic indeterminacy.

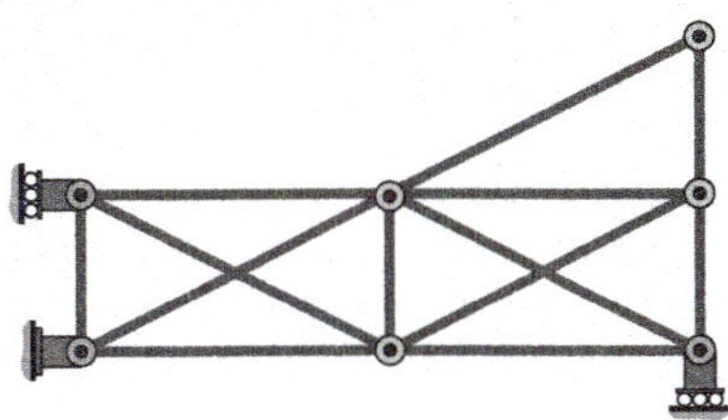

Figure VE19.1

VE19.2 For the truss given in Example 19.6, find the angles ϕ which should be used for both members 1 and 3, then develop their respective global element stiffness matrices.

VE19.3 An assembly of truss bars is given where $\left(\frac{AE}{L}\right)_1 = \left(\frac{AE}{L}\right)_4 =$ 32 k/in. and $\left(\frac{AE}{L}\right)_2 = \left(\frac{AE}{L}\right)_3 = 19$ k/in. By using the approach of applying unit displacements, derive the structure stiffness matrix using the DOFs as labeled in Figure VE19.3.

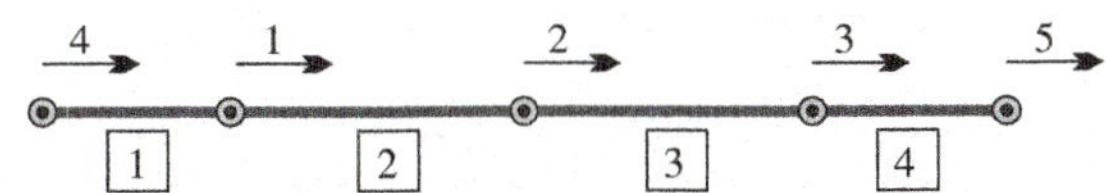

Figure VE19.3

VE19.4 Consider the truss assembly and stiffness matrix derived in VE19.3. For the support conditions and loading shown in Figure VE19.4, find all joint displacements, reaction forces, and member forces.

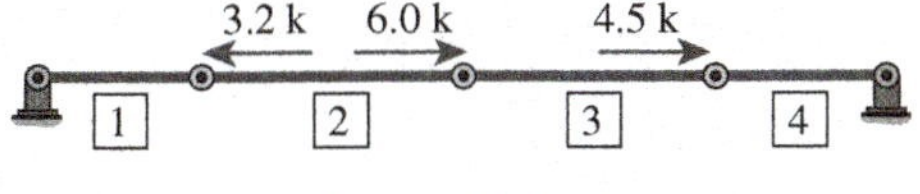

Figure VE19.4

VE19.5 Use the stiffness approach to solve for joint displacements, reactions, and member forces in the truss shown in Figure VE19.5. Use Excel to perform the computations. $E = 10{,}000$ ksi for all members. $A_1 = A_2 = 0.3$ in^2 and $A_3 = 2.0$ in^2.

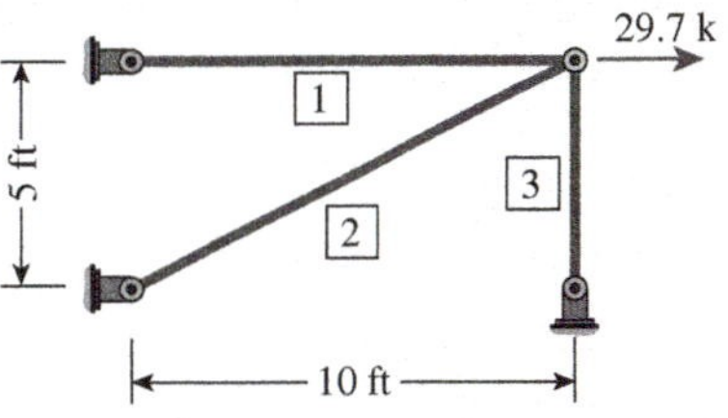

Figure VE19.5

19.12 Problems for Solution

Section 19.3

For Problems 19.1 through 19.6, determine the degree of statical and kinematic indeterminacy for the structure given.

P19.1 (Ans: SI 1°, KI 8°)

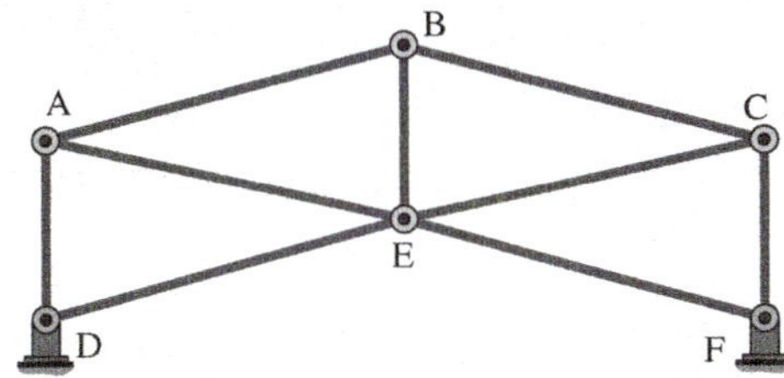

Problem 19.1

P19.2

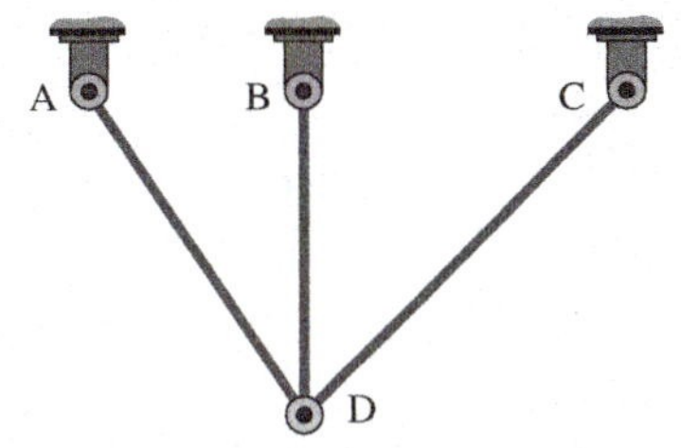

Problem 19.2

P19.3 (Ans: SI 6°, KI 25°)

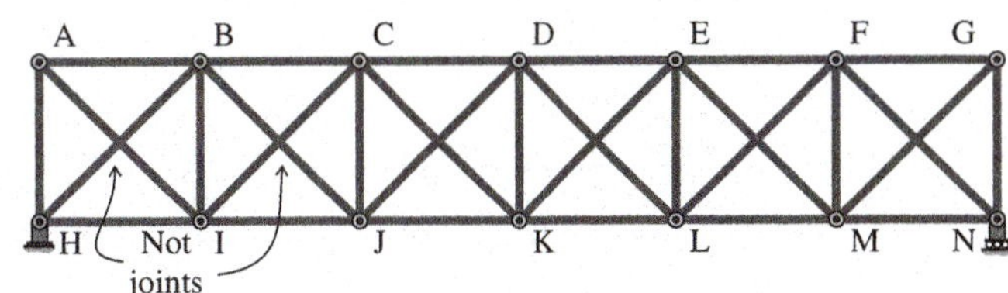

Problem 19.3

P19.4 For the kinematic indeterminacy, consider both when axial deformation is considered and when it is not considered.

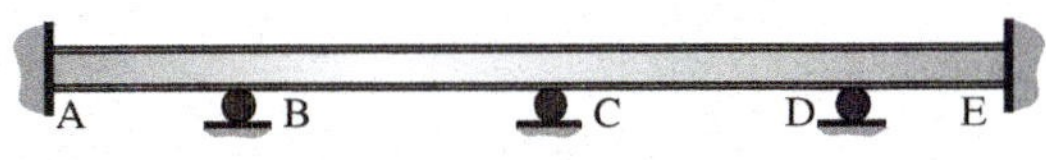

Problem 19.4

P19.5 For the kinematic indeterminacy, consider both when axial deformation is allowed and when it is not allowed. (Ans: SI 9°, KI 12°, KI 5°)

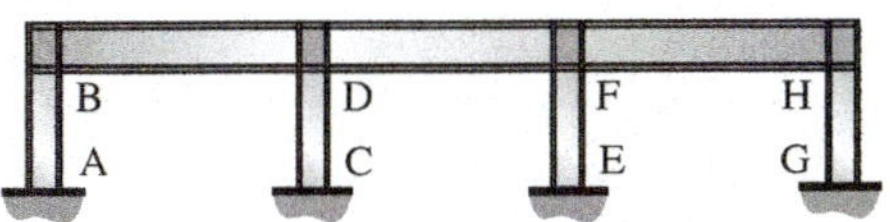

Problem 19.5

P19.6 For the kinematic indeterminacy, consider both when axial deformation is allowed and when it is not allowed.

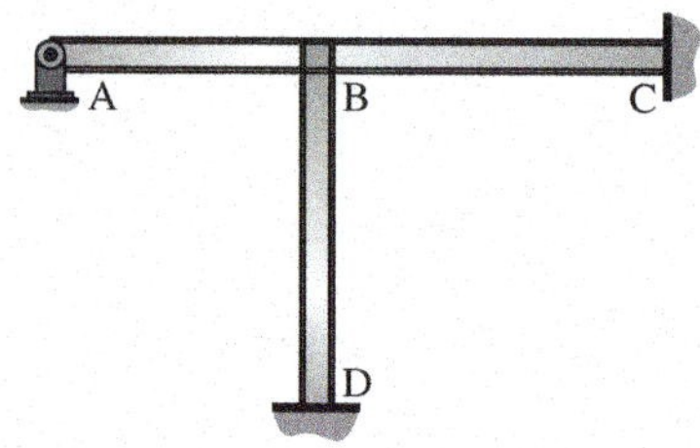

Problem 19.6

Section 19.4

For Problems 19.7 through 19.10, use the unit displacement approach to develop the structural stiffness matrix for the structure shown. Use the matrix formulation to solve for joint displacements and reaction forces. Use basic statics to find the internal member forces. All members are axial members, so only horizontal DOFs are to be considered.

P19.7

$$\left(\frac{AE}{L}\right)_1 = \left(\frac{AE}{L}\right)_2 = \left(\frac{AE}{L}\right)_3 = 150\ \text{k/in.}$$

(Ans: $N_1 = 20\ \text{k}$, $N_2 = 0.0\ \text{k}$, $N_3 = -20.0\ \text{k}$)

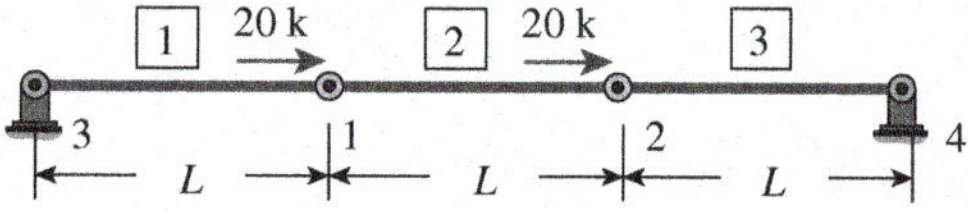

Problem 19.7

P19.8

$$\left(\frac{AE}{L}\right)_1 = \left(\frac{AE}{L}\right)_2 = \left(\frac{AE}{L}\right)_3 = 100\ \text{k/in.}$$

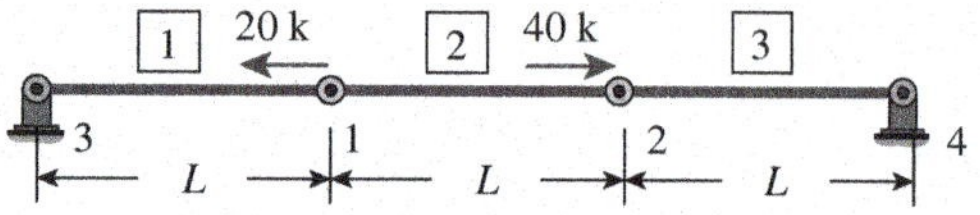

Problem 19.8

P19.9 Repeat Problem 19.8, only use the following stiffness values.

$$\left(\frac{AE}{L}\right)_1 = \left(\frac{AE}{L}\right)_3 = 300\ \text{k/in. and } \left(\frac{AE}{L}\right)_2 = 100\ \text{k/in.}$$

(Ans: $N_1 = -8\ \text{k}$, $N_2 = 12\ \text{k}$, $N_3 = -28\ \text{k}$)

P19.10

$$\left(\frac{AE}{L}\right)_1 = \left(\frac{AE}{L}\right)_2 = \left(\frac{AE}{L}\right)_3 = 100\ \text{k/in.}$$

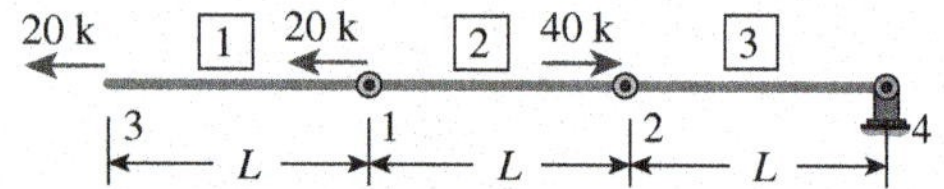

Problem 19.10

P19.11 For the structure shown, assume that the external force F_1, acting at node 1, produces a displacement $\Delta_1 = 0.02$ in. Determine F_1, the resulting displacement at node 2 (Δ_2), and reactions at nodes 3 and 4 (R_3 and R_4).

$$\left(\frac{AE}{L}\right)_1 = \left(\frac{AE}{L}\right)_2 = \left(\frac{AE}{L}\right)_3 = 100\ \text{k/in.}$$

Hint: Once you assemble the stiffness equation in matrix form, you may want to multiply it out and then solve first for Δ_2 then revert back to the matrix formulation for the remaining calculations. (Ans: $F_1 = 3\ \text{k}$, $\Delta_2 = 0.01$ in.)

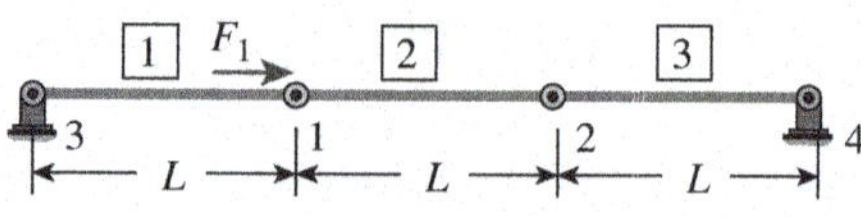

Problem 19.11

P19.12 For the structure shown, assume that the external force F_1 acting at node 1 and the known force F_2 acting at node 2 produce a displacement $\Delta_1 = 0.05$ in. Determine F_1, the resulting displacement at node 2 (Δ_2), and reactions at nodes 3 and 4 (R_3 and R_4).

$$\left(\frac{AE}{L}\right)_1 = \left(\frac{AE}{L}\right)_2 = \left(\frac{AE}{L}\right)_3 = 100\ \text{k/in.}$$

Hint: Once you assemble the stiffness equation in matrix form you may want to multiply it out and then solve first for Δ_2 then revert back to the matrix formulation for the remaining calculations.

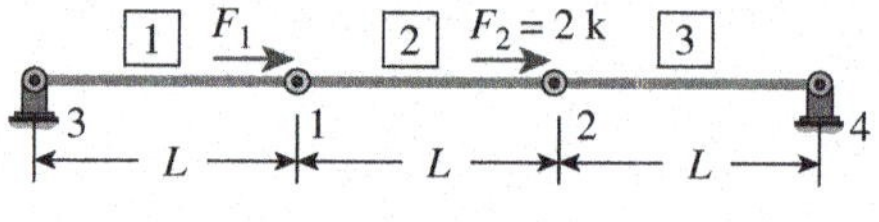

Problem 19.12

Section 19.6

For Problems 19.13 through 19.18, determine the global member stiffness matrix, $[K]$ for the truss element shown using the origin which is indicated by the arrow head. The coordinates for each node are given in the figure. Assume $E = 29{,}000$ ksi for all members.

P19.13 $A = 3\ \text{in}^2$ (Ans: $K_{11} = 92.8$ k/in., $K_{23} = -232.1$ k/in.)

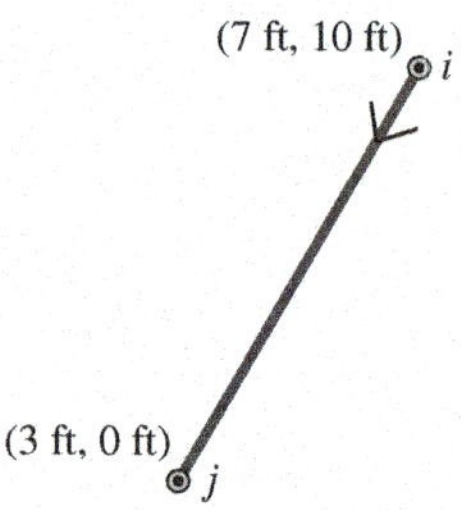

Problem 19.13

P19.14 $A = 3\text{ in}^2$

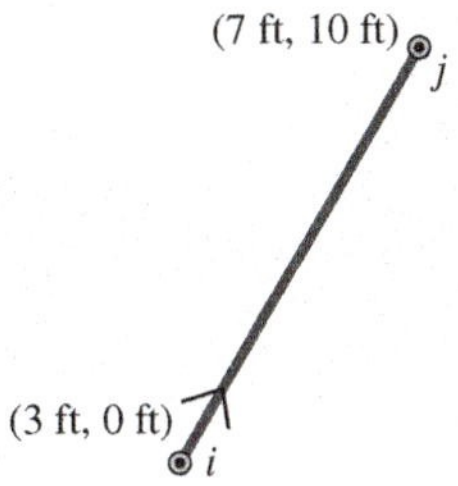

Problem 19.14

P19.15 $A = 2\text{ in}^2$ (Ans: $K_{11} = 92.8$ k/in., $K_{23} = 61.9$ k/in.)

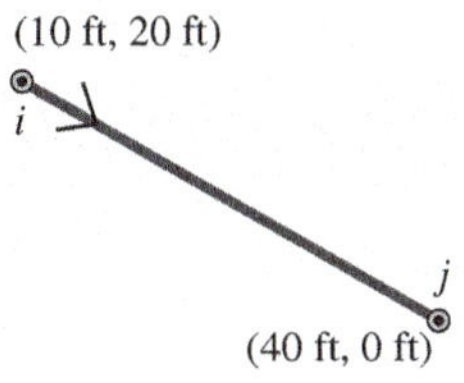

Problem 19.15

P19.16 $A = 2\text{ in}^2$

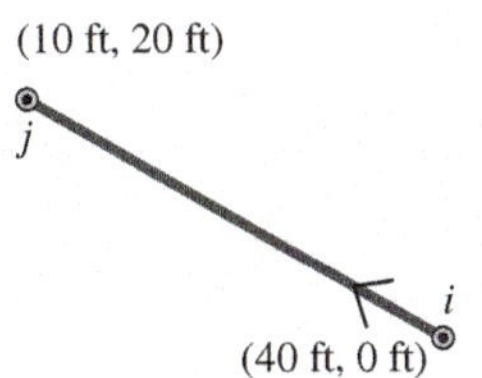

Problem 19.16

P19.17 $A = 1.5\text{ in}^2$ (Ans: $K_{11} = 0.0$ k/in., $K_{24} = -226.6$ k/in.)

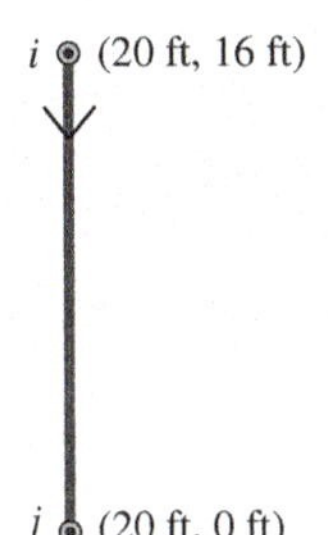

Problem 19.17

P19.18 $A = 1.5\text{ in}^2$

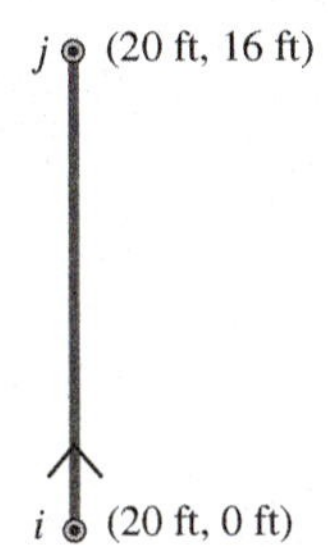

Problem 19.18

Sections 19.5 through 19.9

For Problems 19.19 through 19.24, use the matrix formulation of the stiffness method to compute all nodal displacements, reactions, and member end forces in the trusses shown. Assume $E = 29{,}000$ ksi for all members.

P19.19 All bar areas are $A = 1.5\text{ in}^2$. (Ans: $N_3 = 60.6$ k, $N_6 = 21.1$ k)

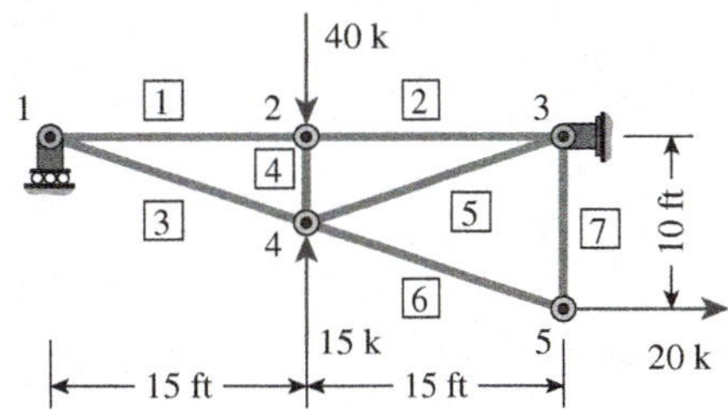

Problem 19.19

P19.20 All bar areas are $A = 0.75\text{ in}^2$.

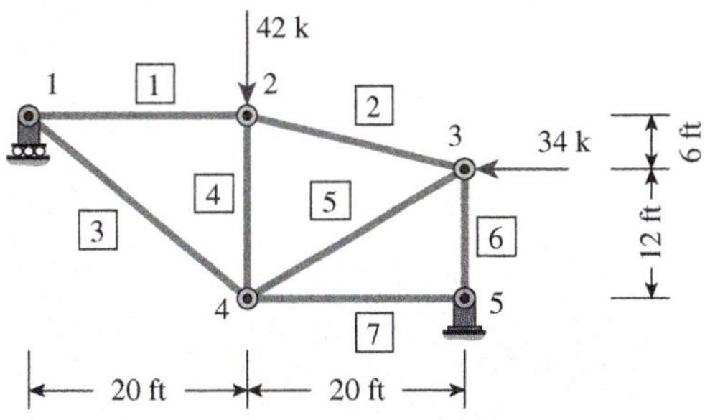

Problem 19.20

P19.21 All bar areas are $A = 0.5\text{ in}^2$. (Ans: $N_3 = -40.0$ k, $N_4 = 19.4$ k)

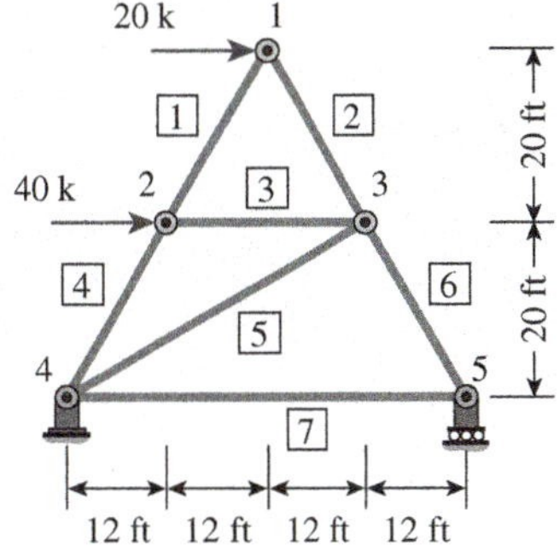

Problem 19.21

P19.22 All bar areas except bars 4 and 5 are $A = 1.5\text{ in}^2$. For bars 4 and 5, $A = 0.5\text{ in}^2$.

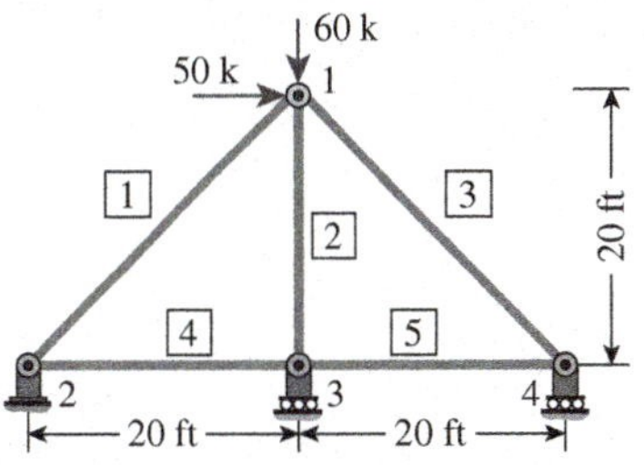

Problem 19.22

P19.23 All bar areas except bars 4 and 5 are $A = 1.5$ in^2. For bars 4 and 5, $A = 0.5$ in^2. (Ans: $N_3 = -35.4$ k, $N_2 = 0.0$ k)

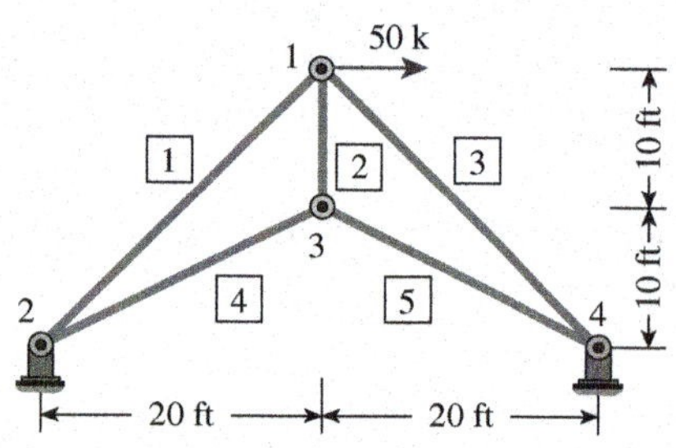

Problem 19.23

P19.24 All bar areas are $A = 1.0$ in^2.

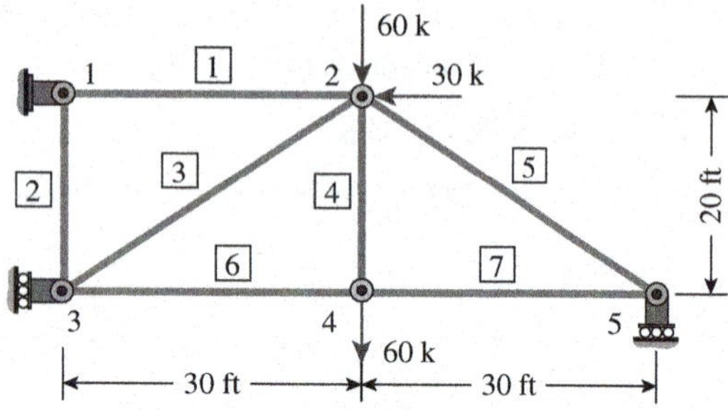

Problem 19.25

Section 19.10

P19.25 Model the truss in Problem 19.21 in a spreadsheet using the matrix approach. Analyze the truss multiple times where the cross-sectional area for bar 6 is set to the following values. $A_6 = 0.1$ in^2, 0.5 in^2, 0.75 in^2, 1.0 in^2, 2 in^2, and 5 in^2. Generate a plot of the area of bar 6 versus the horizontal displacements at joints 1 and 5. Make an observation of the trends.

P19.26 Model the truss in Problem 19.24 in a spreadsheet using the matrix approach. Analyze the truss multiple times where the vertical coordinate of joint 4, as measured from the bottom of the truss, has the following values: −10 ft, −5 ft, 0 ft, 5 ft, and 10 ft. Generate a plot of the vertical coordinate of joint 4 versus the vertical reaction at joint 5 and horizontal reaction of joint 3. Make an observation of the trends.

Direct Stiffness Method for Beams and Frames

20

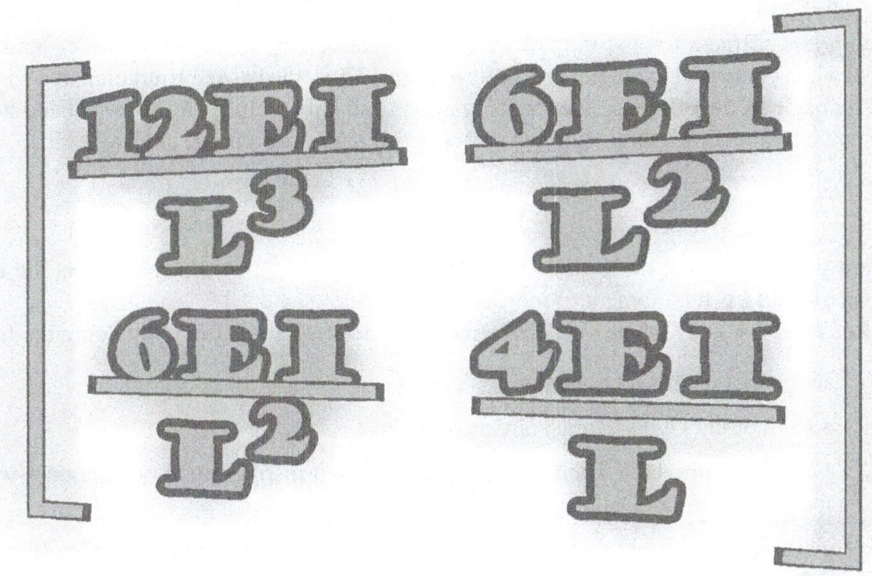

20.1 Introduction

The previous chapter introduced the matrix formulation of the stiffness approach. All of the basic principles and processes were demonstrated using truss bars and truss assemblies. Recall that truss members are, by their very definition, only subjected to axial forces that only result in axial deformations. This chapter looks to extend the application of this method to structures whose members are also subjected to shear and moment—namely beams and frames.

This chapter serves as a follow-up to the previous chapter and will not re-introduce basic definitions, concepts, and processes. We would recommend that the reader review and understand Chapter 19 prior to undertaking a study of the material presented herein.

QUICK NOTE

The presentation of the matrix formulation of the stiffness approach is restricted to the discussion of planar (two-dimensional) structures. Additional components of shear, moment, and torsion would be present for three-dimensional structures.

DID YOU KNOW?

A structure is nothing more than an assemblage of simple and basic structural elements. The type of structure you can assemble is dependent upon what elements you have available to you in your virtual *toolbox*. The first element you have placed in this toolbox is the *axial* element. This section will now add another element to this toolbox—the *beam element*—which increases the types of structures that may be analyzed.

20.2 Stiffness Matrix for Flexural (Beam) Elements

The stiffness approach requires you to already know something about the stiffness of the elements you use prior to implementing the method. Other methods, such as the flexibility approach, do not have such a restriction but are also not as easily implemented on large structures as is the stiffness approach. Indeed, the stiffness approach requires us to find element stiffness values using some other method before a structural stiffness matrix can be assembled. One may quickly see that this whole process could become extremely cumbersome and time consuming if every new structure required us to first use the flexibility approach to find the stiffness terms and then the stiffness approach to solve for all structural actions. For this reason, it is useful to view a structure (whether simple or complex) as nothing more than an assemblage of a few basic structural elements whose properties are already known.

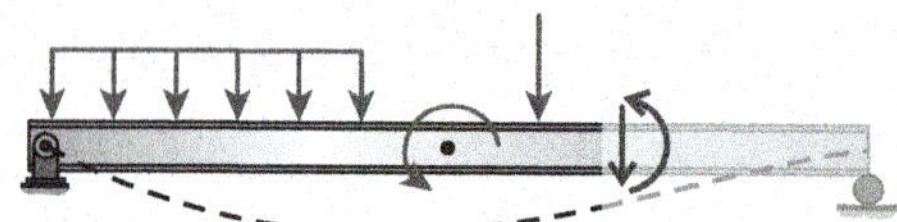

FIGURE 20.1 Typical beam action.

QUICK NOTE
A beam element is defined by four degrees-of-freedom—a single translation and rotation at each end. This will result in a 4 × 4 stiffness matrix.

In this section, we introduce the stiffness terms for a *beam element*. Referring to Figure 20.1, a beam element is an element where the load is applied in a direction perpendicular to the member or in the form of a moment. These kinds of applied forces only produce internal shears and internal moments. Further, the resulting deflected shape at any location along the beam can be fully described by a vertical displacement and a rotation. For this reason, the DOFs at each node on the beam element are a vertical translation and a rotation, as shown in Figure 20.2.

RECALL
In the discussion in Chapter 9 on computing deflections using virtual work, Example 9.11 demonstrated that both shear and flexural deformations contribute to the overall deflection of a beam. In this example, shear deformations constituted only 8% of the overall deflection.

In an actual beam like the one shown in Figure 20.1, the total deflected shape is a combination of deflection due to internal moment (flexural) and the deflection due to internal shear. Knowing the relationship between internal loads and the resulting deflection is key to developing the appropriate stiffness terms for an element. For most beams, shear deformations tend to be quite small when compared to the flexural deformations. Thus, the basic formulation of a beam stiffness matrix neglects the effects of shear deformations and only includes those associated with flexure, as shown by the differential equation that was originally introduced in Chapter 8:

$$\frac{d^2y}{dx^2} = \frac{M}{EI}$$

where y is the vertical deflection, M is the internal moment in the beam, E is the modulus of elasticity, and I is the moment of inertia.

QUICK NOTE
The consideration of shear deformations is addressed in the following chapter.

Referring back to Chapter 19, the element stiffness matrix is found by doing the following.

1. Label all DOFs for the member.
2. Lock all of the DOFs against any movement.
3. Select one of the DOFs and impose a positive unit displacement or rotation—whichever is appropriate for that DOF.
4. Calculate the reactions due to the imposed motion. Another method, like flexibility, is usually used to calculate these.
5. The reactions are known as the stiffness factors and should be placed in the relevant column of the matrix.
6. Repeat steps 3 through 5 for all DOFs on the element.

QUICK NOTE
The labels a, b, c, and d are used to represent the DOFs of a single element. The numbers 1, 2, 3, etc., are used to describe the DOFs at the structural level.

The stiffness matrix for the generic beam element shown in Figure 20.2 is developed through this procedure. Figure 20.3 demonstrates this sequence. The values of the reactions are given here without showing the actual calculations. Rather, this is left as an exercise for the reader to do using principles presented in earlier chapters of this textbook.

DID YOU NOTICE?
Several of these stiffness factors had previously been introduced in Chapter 13 dealing with moment distribution (refer to Figure 13.5).

$$k = \begin{array}{c} \\ a \\ b \\ c \\ d \end{array} \begin{array}{c} \begin{array}{cccc} \quad a \quad & \quad b \quad & \quad c \quad & \quad d \quad \end{array} \\ \begin{bmatrix} \frac{12EI}{L^3} & \frac{6EI}{L^2} & -\frac{12EI}{L^3} & \frac{6EI}{L^2} \\ \frac{6EI}{L^2} & \frac{4EI}{L} & -\frac{6EI}{L^2} & \frac{2EI}{L} \\ -\frac{12EI}{L^3} & -\frac{6EI}{L^2} & \frac{12EI}{L^3} & -\frac{6EI}{L^2} \\ \frac{6EI}{L^2} & \frac{2EI}{L} & -\frac{6EI}{L^2} & \frac{4EI}{L} \end{bmatrix} \end{array}$$

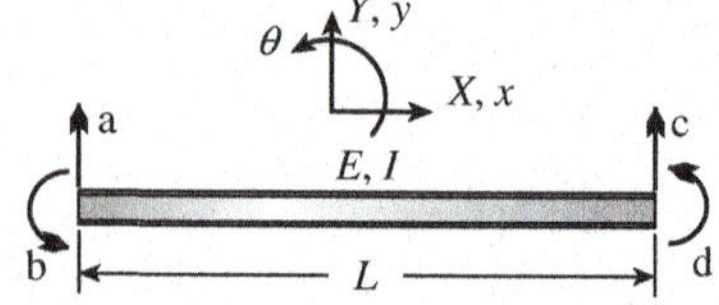

FIGURE 20.2 DOFs for a typical beam element.

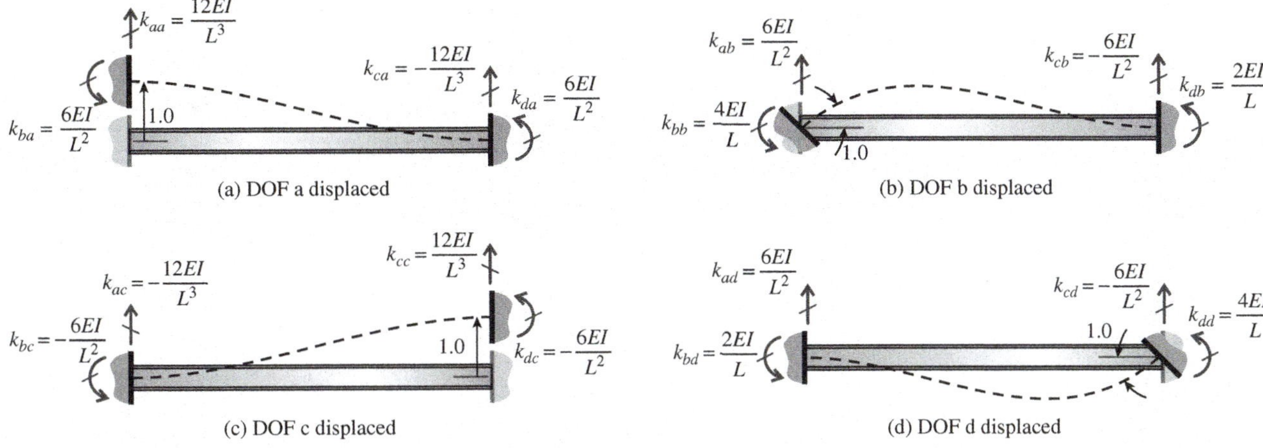

FIGURE 20.3 Development of beam member stiffness matrix.

At this point, it is probably worthwhile to mention the units associated with these stiffness terms. When dealing with axial members, all DOFs were translational, which resulted in all stiffness terms in the matrix having units of force per length (F/L). Beam elements have a mixture of translational and rotational DOFs. This causes various sets of units to exist in the same matrix. They are most readily identified by the terms $\frac{EI}{L^3}$, $\frac{EI}{L^2}$, and $\frac{EI}{L}$. Recognizing the using of E and I are $\frac{F}{L^2}$ and L^4, respectively, we may see that the basic units of each type of stiffness term are $\frac{F}{L}$, $\frac{F}{\text{rad}}$, and $\frac{F \cdot L}{\text{rad}}$. The rad unit shouldn't be too surprising, considering rotations were imposed as part of the stiffness development process.

QUICK NOTE

Because of the mixed units that show up in the matrix, the authors recommend that consistent units (e.g., kips and inches or kN and m) be defined at the beginning of any problem. Then one will easily know the proper units of the results without the difficulty in tracking units and handling conversions throughout the process. For example, don't input the element length in units of ft and then define I in units of in^4. Define both in the same set of units.

20.3 Matrix Stiffness Method Applied to Beams

The beginnings of the stiffness method applied to beams is the same as it was for trusses. A structural-level stiffness matrix must be generated, a force vector needs to be assembled, and the restrained DOFs need to be identified. The matrix-based computations outlined in Section 19.4 then will be carried out.

Direct Assemblage of Structural Stiffness Matrix

The structure-level stiffness matrix may be generated by taking the entire beam and locking all of its DOFs. A unit displacement or a unit rotation is applied to each DOF independently, and the resulting reactions are computed. Consider the beam assembly shown in Figure 20.4.

Figure 20.5 shows the deformed shape and resulting reactions for the cases where DOFs 1 through 3 are given unit displacements/rotations. Displacements/rotations for DOFs 4 through 6 are done in the same fashion. This then produces a 6×6 stiffness matrix $[K]$ for the beam structure.

QUICK NOTE

The numbering sequence of the DOFs seems a bit arbitrary at this stage. Examples shown later in this chapter will make this numbering sequence more obvious.

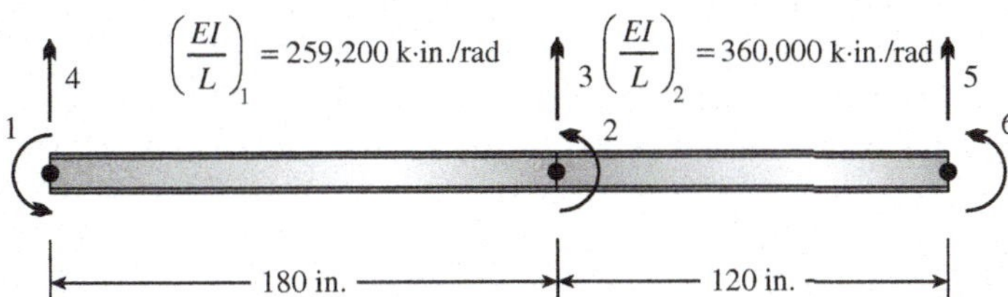

FIGURE 20.4 Two-member beam assembly.

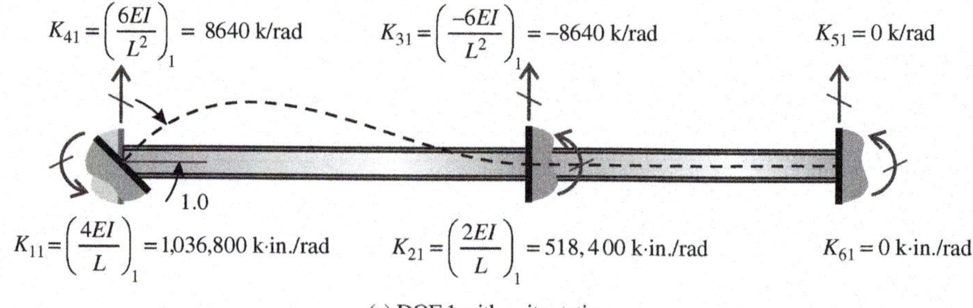

(a) DOF 1 with unit rotation.

$$K_{42} = \left(\frac{6EI}{L^2}\right)_1 = 8640 \text{ k/rad} \qquad K_{32} = -\left(\frac{6EI}{L^2}\right)_1 + \left(\frac{6EI}{L^2}\right)_2 = 9360 \text{ k/rad} \qquad K_{52} = -\left(\frac{6EI}{L^2}\right)_2 = -18{,}000 \text{ k/rad}$$

1.0

$$K_{12} = \left(\frac{2EI}{L}\right)_1 = 518{,}400 \text{ k·in./rad} \qquad K_{22} = \left(\frac{4EI}{L}\right)_1 + \left(\frac{4EI}{L}\right)_2 = 2{,}476{,}800 \text{ k·in./rad} \qquad K_{62} = \left(\frac{2EI}{L}\right)_2 = 720{,}000 \text{ k·in./rad}$$

(b) DOF 2 with unit rotation.

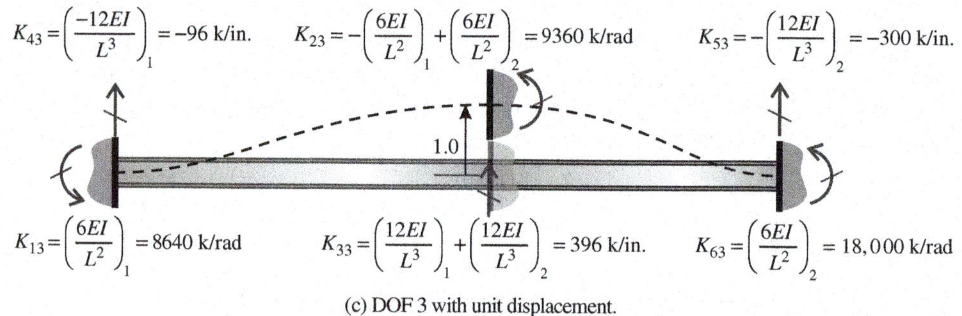

$K_{43} = \left(\frac{-12EI}{L^3}\right)_1 = -96$ k/in.
$K_{23} = -\left(\frac{6EI}{L^2}\right)_1 + \left(\frac{6EI}{L^2}\right)_2 = 9360$ k/rad
$K_{53} = -\left(\frac{12EI}{L^3}\right)_2 = -300$ k/in.
1.0
$K_{13} = \left(\frac{6EI}{L^2}\right)_1 = 8640$ k/rad
$K_{33} = \left(\frac{12EI}{L^3}\right)_1 + \left(\frac{12EI}{L^3}\right)_2 = 396$ k/in.
$K_{63} = \left(\frac{6EI}{L^2}\right)_2 = 18{,}000$ k/rad

(c) DOF 3 with unit displacement.

FIGURE 20.5 Unit displacements and resulting stiffness values.

DID YOU KNOW?
One may easily determine which members contribute to the overall stiffness at a DOF by observing the deformed shapes of the members. A particular DOF will only have contribution from a member if the member to which it is attached is deformed. Look at Figure 20.5(a). Only member 1 is deformed, so only DOFs directly attached to member 1 will have nonzero values.

RECALL
The terms in this matrix were developed using the units of kips and inches. But because the matrix contains both translational and rotational DOFs, not all terms have the same set of units. For example, K_{11} has units of (k · in.)/rad, K_{33} has units of k/in. and K_{31} has units of k/rad.

$$K = \begin{array}{c} \\ 1 \\ 2 \\ 3 \\ 4 \\ 5 \\ 6 \end{array} \begin{array}{c} \begin{array}{cccccc} 1 & 2 & 3 & 4 & 5 & 6 \end{array} \\ \begin{bmatrix} 1{,}036{,}800 & 518{,}400 & -8640 & 8640 & 0 & 0 \\ 518{,}400 & 2{,}476{,}800 & 9360 & 8640 & 18{,}000 & 720{,}000 \\ -8640 & 9360 & 396 & -96 & -300 & -18{,}000 \\ 8640 & 8640 & -96 & 96 & 0 & 0 \\ 0 & 18{,}000 & -300 & 0 & 300 & -18{,}000 \\ 0 & 720{,}000 & -18{,}000 & 0 & -18{,}000 & 1{,}440{,}000 \end{bmatrix} \end{array}$$

Be aware that the contribution of each of the members may counteract one another. For example, in Figure 20.5(c), a unit displacement is imposed at DOF 3. This causes a clockwise (i.e., negative) moment reaction to develop at DOF 2 because of member 1 but a counterclockwise (i.e., positive) moment reaction at DOF 2 because of member 2. Thus, the net reaction is

$$K_{23} = -\left(\frac{6EI}{L^2}\right)_1 + \left(\frac{6EI}{L^2}\right)_2 = -8640 + 18{,}000 = 9360 \text{ k/rad}$$

This process provides a direct and clear view of the contribution of each beam element to each DOF. However, this process can be very tedious for structures of any significant size. For this reason, it is generally preferable to use the approach that is discussed next.

Element-Based Assemblage of the Structural Stiffness Matrix

The structural stiffness matrix for a beam may be assembled by combining the stiffnesses of the individual elements in a very methodical manner. This procedure was previously introduced for trusses and is slightly modified and presented here with a focus on beam assemblies.

1. Label all DOFs on the structure (one vertical translation and one rotation per joint).
2. Label the members with a unique number. Any order will work.
3. Assemble a separate local stiffness matrix for each element in the structure, so $k^{(n)} = K^{(n)}$.
4. Assemble the global stiffness matrix K for the entire structure by summing all stiffness coefficients that have the same subscripts. Consider the contribution of all elements in the structure.

$$K_{i,j} = \sum_{n=1}^{\text{number of elements}} K_{i,j}^{(n)}$$

The individual element stiffness matrices for the beam assembly shown in Figure 20.4 are shown here.

$$K^{(1)} = \begin{array}{c} \\ 4 \\ 1 \\ 3 \\ 2 \end{array} \begin{array}{c} \begin{array}{cccc} 4 & 1 & 3 & 2 \end{array} \\ \begin{bmatrix} 96 & 8640 & -96 & 8640 \\ 8640 & 1{,}036{,}800 & -8640 & 518{,}400 \\ -96 & -8640 & 96 & -8640 \\ 8640 & 518{,}400 & -8640 & 1{,}036{,}800 \end{bmatrix} \end{array}$$

$$K^{(2)} = \begin{array}{c} \\ 3 \\ 2 \\ 5 \\ 6 \end{array} \begin{array}{c} \begin{array}{cccc} 3 & 2 & 5 & 6 \end{array} \\ \begin{bmatrix} 300 & 18{,}000 & -300 & -18{,}000 \\ 18{,}000 & 1{,}440{,}000 & 18{,}000 & 720{,}000 \\ -300 & 18{,}000 & 300 & -18{,}000 \\ -18{,}000 & 720{,}000 & -18{,}000 & 1{,}440{,}000 \end{bmatrix} \end{array}$$

The process of assembling the structural stiffness matrix from the element stiffness matrices is quite straightforward. The calculations for a few of the structural stiffness terms are shown here.

$$K_{33} = K_{33}^{(1)} + K_{33}^{(2)} = 96 + 300 = 396 \text{ k/in.}$$

$$K_{32} = K_{32}^{(1)} + K_{32}^{(2)} = -8640 + 18{,}000 = 9360 \text{ k/rad}$$

$$K_{66} = K_{66}^{(2)} = 1{,}440{,}000 \text{ k}\cdot\text{in./rad}$$

Similar calculations are done for each pair of subscript indices (i.e., i, j) in the structural stiffness matrix. This would be a total of $6 \times 6 = 36$ calculations. However, recalling that the stiffness matrix is symmetric can reduce this process to 21 calculations.

Analyzing Beams Using the Matrix Method

The matrix setup and operations were described at some length in the previous chapter. For convenience and reinforcement, the procedure for the matrix analysis of a structure is presented here.

1. Assemble a force column vector that has the known applied forces/moments at each DOF and the unknown reaction forces/moments. Place them in the same DOF order as was used for the stiffness matrix. If there is no applied force/moment at an unrestrained DOF, insert a 0.

DID YOU KNOW?

Since beams are horizontally oriented structures, the local axis of each individual element is in the same orientation as the global coordinate system. Thus, the following equality is appropriate to make.

$$k^{(n)} = K^{(n)}$$

QUICK NOTE

When numbering the DOFs in a structure, be sure to number the unrestrained DOFs first followed by the restrained DOFs.

SAMPLE CALCULATION

One of the stiffness terms for element 1:

$$K_{44}^{(1)} = \frac{12EI}{L^3}$$

$$= \frac{(12)\,[259{,}200 \text{ k}\cdot\text{in./rad}]}{(180 \text{ in.})^2} = 96 \text{ k/in.}$$

2. Assemble a displacement column vector that has both the unknown displacements/rotations of the unrestrained DOFs and the known displacements/rotations at the supports (for now these should be zero).

3. Partition the stiffness matrix, force vector, and displacement vector such that the unrestrained DOFs are grouped and the restrained DOFs are grouped where $\{\Delta_r\} = \{0\}$. Thus,

$$\left\{\begin{array}{c} F_u \\ \hline R_r \end{array}\right\} = \left[\begin{array}{c|c} K_{uu} & K_{ur} \\ \hline K_{ru} & K_{rr} \end{array}\right] \left\{\begin{array}{c} \Delta_u \\ \hline \Delta_r \end{array}\right\}$$

QUICK NOTE
Refer to Appendix C for a review of matrix math.

4. Solve for the unknown displacements/rotations by

$$\{\Delta_u\} = [K_{uu}]^{-1}\{F_u\}$$

5. Once $\{\Delta_u\}$ is found, the reaction forces/moments may be found by

$$\{R_r\} = [K_{ru}]\{\Delta_u\}$$

EXAMPLE 20.1

Use the matrix formulation of the stiffness approach to find the joint displacements/rotations and the reactions for the following beam. This is the same beam assembly as presented in Figure 20.4, so the previously developed stiffness matrix may be used.

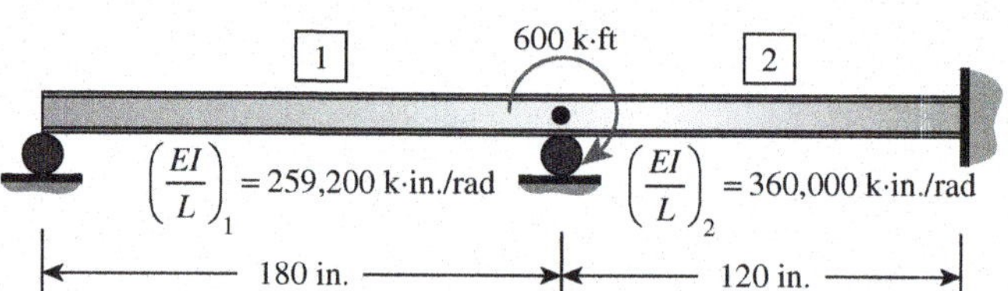

QUICK NOTE
Remember to be consistent in the units you use. The stiffness matrix was developed using kips and inches, so these same units will be used for the force vector.

SOLUTION STRATEGY

Confirm that the stiffness matrix was assembled by considering the unrestrained DOFs first and the restrained DOFs second, then follow the process as outlined.

SOLUTION

The force vector is a 6×1 column vector because there are six DOFs in this beam. However only two of these DOFs are unrestrained—1 and 2; both of which are rotational. By observation, $M_1 = 0$ k·in. and $M_2 = -600$ k·ft $= -7200$ k·in. The negative on the moment results from the sign convention established in Chapter 19, which says that clockwise rotations are negative. The reactions R_3, R_4, R_5, and M_6 are currently unknown, so the total partitioned force vector looks like

$$\left\{\begin{array}{c} F_u \\ \hline R_r \end{array}\right\} = \left\{\begin{array}{c} M_1 \\ M_2 \\ \hline R_3 \\ R_4 \\ R_5 \\ M_6 \end{array}\right\} = \left\{\begin{array}{c} 0.0 \\ -7200 \\ \hline R_3 \\ R_4 \\ R_5 \\ M_6 \end{array}\right\}$$

DISPLACEMENT VECTOR
The rotations at DOFs 1 and 2 (θ_1 and θ_2) are unknown, but the displacements/rotations at the other DOFs are zero. This gives the following displacement vector.

$$\left\{\begin{array}{c} \Delta_u \\ \hline \Delta_r \end{array}\right\} = \left\{\begin{array}{c} \theta_1 \\ \theta_2 \\ \hline \Delta_3 \\ \Delta_4 \\ \Delta_5 \\ \theta_6 \end{array}\right\} = \left\{\begin{array}{c} \theta_1 \\ \theta_2 \\ \hline 0.0 \\ 0.0 \\ 0.0 \\ 0.0 \end{array}\right\}$$

The structural stiffness matrix is partitioned between rows 2 and 3 and also columns 2 and 3. This is because there are only two unrestrained DOFs.

$$K = \left[\begin{array}{c|c} K_{uu} & K_{ur} \\ \hline K_{ru} & K_{rr} \end{array}\right] = K = \begin{array}{c} \\ 1 \\ 2 \\ 3 \\ 4 \\ 5 \\ 6 \end{array} \begin{array}{c} \begin{array}{cccccc} 1 & 2 & 3 & 4 & 5 & 6 \end{array} \\ \left[\begin{array}{cc|cccc} 1{,}036{,}800 & 518{,}400 & -8640 & 8640 & 0 & 0 \\ 518{,}400 & 2{,}476{,}800 & 9360 & 8640 & 18{,}000 & 720{,}000 \\ \hline -8640 & 9360 & 396 & -96 & -300 & -18{,}000 \\ 8640 & 8640 & -96 & 96 & 0 & 0 \\ 0 & 18{,}000 & -300 & 0 & 300 & -18{,}000 \\ 0 & 720{,}000 & -18{,}000 & 0 & -18{,}000 & 1{,}440{,}000 \end{array}\right] \end{array}$$

The unknown rotations may be found by the following matrix operation.

$$\{\Delta_u\} = [K_{uu}]^{-1}\{F_u\} \Rightarrow \begin{Bmatrix} \theta_1 \\ \theta_2 \end{Bmatrix} = \begin{bmatrix} 1{,}036{,}800 & 518{,}400 \\ 518{,}400 & 2{,}476{,}800 \end{bmatrix}^{-1} \begin{Bmatrix} 0.0 \\ -7200.0 \end{Bmatrix}$$

$$\begin{Bmatrix} \theta_1 \\ \theta_2 \end{Bmatrix} = \begin{bmatrix} 1.08 \cdot 10^{-6} & -2.25 \cdot 10^{-7} \\ -2.25 \cdot 10^{-7} & 4.51 \cdot 10^{-7} \end{bmatrix} \begin{Bmatrix} 0.0 \\ -7200.0 \end{Bmatrix} = \begin{Bmatrix} 0.0016 \\ -0.0032 \end{Bmatrix} \begin{matrix} \text{rad} \\ \text{rad} \end{matrix}$$

The reactions of the assembly may now be found.

$$\{R_r\} = [K_{ru}]\{\Delta_u\}$$

$$\begin{Bmatrix} R_3 \\ R_4 \\ R_5 \\ M_6 \end{Bmatrix} = \begin{bmatrix} -8640 & 9360 \\ 8640 & 8640 \\ 0 & -18{,}000 \\ 0 & 720{,}000 \end{bmatrix} \begin{Bmatrix} 0.0016 \\ -0.0032 \end{Bmatrix} = \begin{Bmatrix} -44.4 \\ -14.0 \\ 58.4 \\ -2338 \end{Bmatrix} \begin{matrix} \text{k} \\ \text{k} \\ \text{k} \\ \text{k}\cdot\text{in.} \end{matrix}$$

QUICK NOTE
The displacement and reaction results are to be interpreted in terms of the global coordinate system (i.e., up is positive and counterclockwise is positive).

EXAMPLE 20.2

Use the matrix formulation of the stiffness approach to find the joint displacements/rotations and the reactions for the following beam. This is the same beam assembly presented in Figure 20.4, so the previously developed stiffness matrix may be used.

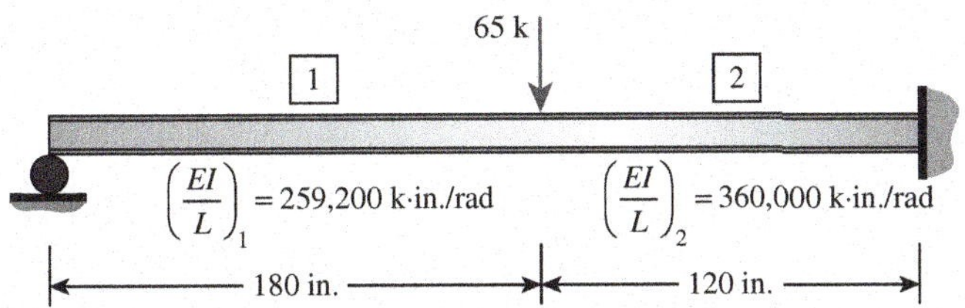

SOLUTION STRATEGY

Confirm that the stiffness matrix was assembled by considering the unrestrained DOFs first and the restrained DOFs second, then follow the process as outlined.

QUICK NOTE
This is the same beam as in Example 20.1, but now DOF 3 is also unrestrained. The sequence of DOF numbering in Figure 20.4 should now make sense.

SOLUTION

The force vector is a 6 × 1 column vector because there are six DOFs in this beam. The first three of these DOFs are now unrestrained. By observation $M_1 = M_2 = 0$ k·in. and $F_3 = -65$ k. The reactions R_4, R_5, and M_6 are currently unknown so the total partitioned force vector looks like

DISPLACEMENT VECTOR

The following is the appropriate displacement vector.

$$\begin{Bmatrix} \Delta_u \\ \hline \Delta_r \end{Bmatrix} = \begin{Bmatrix} \theta_1 \\ \theta_2 \\ \Delta_3 \\ \hline \Delta_4 \\ \Delta_5 \\ \theta_6 \end{Bmatrix} = \begin{Bmatrix} \theta_1 \\ \theta_2 \\ \Delta_3 \\ \hline 0.0 \\ 0.0 \\ 0.0 \end{Bmatrix}$$

$$\begin{Bmatrix} F_u \\ \hline R_r \end{Bmatrix} = \begin{Bmatrix} M_1 \\ M_2 \\ R_3 \\ \hline R_4 \\ R_5 \\ M_6 \end{Bmatrix} = \begin{Bmatrix} 0.0 \\ 0.0 \\ -65.0 \\ \hline R_4 \\ R_5 \\ M_6 \end{Bmatrix}$$

The structural stiffness matrix is partitioned between rows 3 and 4 and also columns 3 and 4. This is because there are now three unrestrained DOFs.

$$K = \left[\begin{array}{c|c} K_{uu} & K_{ur} \\ \hline K_{ru} & K_{rr} \end{array}\right] =$$

$$K = \begin{array}{c} \\ 1 \\ 2 \\ 3 \\ 4 \\ 5 \\ 6 \end{array} \begin{array}{c} \begin{array}{cccccc} 1 & 2 & 3 & 4 & 5 & 6 \end{array} \\ \left[\begin{array}{ccc|ccc} 1{,}036{,}800 & 518{,}400 & -8640 & 8640 & 0 & 0 \\ 518{,}400 & 2{,}476{,}800 & 9360 & 8640 & 18{,}000 & 720{,}000 \\ -8640 & 9360 & 396 & -96 & -300 & -18{,}000 \\ \hline 8640 & 8640 & -96 & 96 & 0 & 0 \\ 0 & 18{,}000 & -300 & 0 & 300 & -18{,}000 \\ 0 & 720{,}000 & -18{,}000 & 0 & -18{,}000 & 1{,}440{,}000 \end{array}\right] \end{array}$$

The unknown rotations and displacement may be found by the following matrix operation.

$$\{\Delta_u\} = [K_{uu}]^{-1}\{F_u\} \Rightarrow \begin{Bmatrix} \theta_1 \\ \theta_2 \\ \Delta_3 \end{Bmatrix} = \begin{bmatrix} 1{,}036{,}800 & 518{,}400 & -8640 \\ 518{,}400 & 2{,}476{,}800 & 9360 \\ -8640 & 9360 & 396 \end{bmatrix}^{-1} \begin{Bmatrix} 0.0 \\ 0.0 \\ -65.0 \end{Bmatrix}$$

$$\begin{Bmatrix} \theta_1 \\ \theta_2 \\ \Delta_3 \end{Bmatrix} = \begin{bmatrix} 1.62 \cdot 10^{-6} & -5.19 \cdot 10^{-7} & 4.77 \cdot 10^{-5} \\ -5.19 \cdot 10^{-7} & 6.10 \cdot 10^{-7} & -2.57 \cdot 10^{-5} \\ 4.77 \cdot 10^{-5} & -2.57 \cdot 10^{-5} & 4.17 \cdot 10^{-3} \end{bmatrix} \begin{Bmatrix} 0.0 \\ 0.0 \\ -65.0 \end{Bmatrix} = \underline{\underline{\begin{Bmatrix} -0.0031 \\ 0.0017 \\ -0.2713 \end{Bmatrix} \begin{array}{l} \text{rad} \\ \text{rad} \\ \text{in.} \end{array}}}$$

The reactions of the assembly may now be found.

$$\{R_r\} = [K_{ru}]\{\Delta_u\}$$

$$\begin{Bmatrix} R_4 \\ R_5 \\ M_6 \end{Bmatrix} = \begin{bmatrix} 8640 & 8640 & -96 \\ 0 & -18{,}000 & -300 \\ 0 & 720{,}000 & 18{,}000 \end{bmatrix} \begin{Bmatrix} -0.0031 \\ 0.0017 \\ -0.2713 \end{Bmatrix} = \underline{\underline{\begin{Bmatrix} 13.7 \\ 51.3 \\ -3678 \end{Bmatrix} \begin{array}{l} \text{k} \\ \text{k} \\ \text{k·in.} \end{array}}}$$

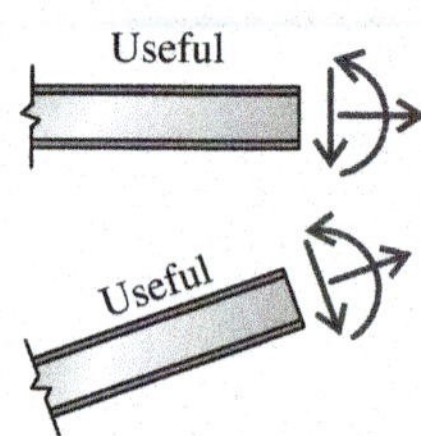

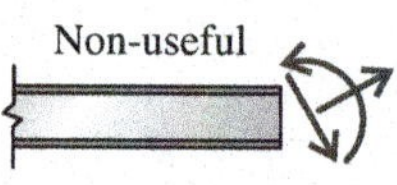

FIGURE 20.6 Various representations of member end forces.

20.4 Solving for Member End Forces

Member end forces are computed as previously discussed in Chapter 19 Section 8. Specifically, the member end forces need to be reported in member local coordinates so that they retain useful meaning for the analyst. For example, Figure 20.6 shows three scenarios for reporting member end forces. The first two scenarios are useful in that they report the values in terms of a normal force N, a shear force V, and a moment M. The third representation is not useful in that the two orthogonal forces are not parallel and perpendicular to member. It is interesting to note that

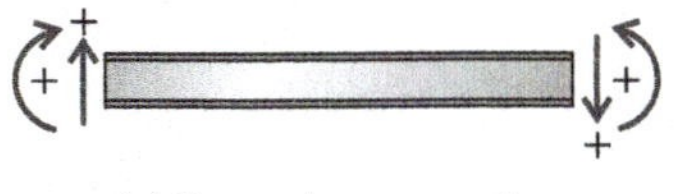
(a) Beam sign convention

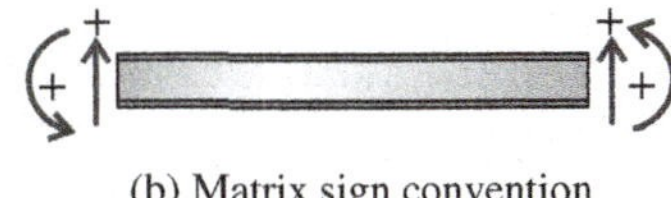
(b) Matrix sign convention

FIGURE 20.7 Sign conventions for internal forces.

the moment has the same physical meaning in all three scenarios—a characteristic that will be beneficial to note when dealing with frames later in this chapter.

The matrix operations for computing member end forces in the local coordinates $\{Q\}$ requires knowledge of the element stiffness matrix in the local coordinates $[k]$ and the member end displacements given in local coordinates $\{u\}$ as

$$\{Q\}^{(n)} = [k]^{(n)}\,\{u\}^{(n)}$$

Since beam members primarily have a horizontal position, this means that the local coordinates for the member match up with the global coordinates for the entire beam. The consequence is that the displacement vector and the stiffness matrix are the same in both coordinates and doesn't require any type of transformation as shown. So

$$[k]^{(n)} = [K]^{(n)}$$
$$\{u\}^{(n)} = \{\Delta\}^{(n)}$$

Thus, the member end forces for a single beam element can be carried out with the following matrix operation.

$$\{Q\}^{(n)} = [K]^{(n)}\,\{\Delta\}^{(n)}$$

Understanding Sign Convention

In addition to the magnitude of the member end forces, we also want to have a good interpretation of the signs associated with these forces. As thoroughly discussed in Chapter 6, the generation of shear and moment diagrams requires the end forces to be reported in *beam sign convention*. However, the matrix approach reports these values using a right-handed sign convention. Therefore, we need to be able to translate between the two sign conventions. Figure 20.7 shows the differences between these two conventions for internal moment and internal shear.

Note that the shear at DOF a and the moment at DOF d match between the two sign conventions. However, the shear at DOF c and the moment at DOF b have opposite directions between the two sign conventions. Therefore, the following mapping may be utilized where Q represents the matrix notation and V and M represent the beam sign convention.

RECALL

The generic DOF labeling adopted in this book for a beam element is shown here.

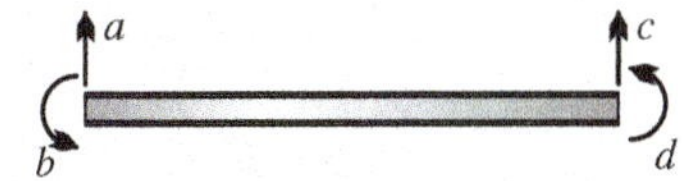

$$\begin{Bmatrix} Q_a \\ Q_b \\ Q_c \\ Q_d \end{Bmatrix} = \begin{Bmatrix} V_a \\ -M_b \\ -V_c \\ M_d \end{Bmatrix}$$

EXAMPLE 20.3

Use the results found in Example 20.1 to solve for the member end forces for each of the beam elements. Report the findings using beam sign convention and then sketch the shear and moment diagrams for each member.

SOLUTION STRATEGY

Collect the element stiffness matrices and the nodal displacements/rotations for the entire structure from Example 20.1. Use the matrix operation for computing member end forces and then transfer these forces over to a FBD using beam sign convention. Review Chapter 6 on how to generate shear and moment diagrams.

SOLUTION

The element stiffness matrices and the element displacement vectors are assembled from the information in Example 20.1.

$$K^{(1)} = \begin{array}{c} 4 \\ 1 \\ 3 \\ 2 \end{array} \begin{bmatrix} 96 & 8640 & -96 & 8640 \\ 8640 & 1{,}036{,}800 & -8640 & 518{,}400 \\ -96 & -8640 & 96 & -8640 \\ 8640 & 518{,}400 & -8640 & 1{,}036{,}800 \end{bmatrix} \quad (\text{columns } 4,1,3,2)$$

$$K^{(2)} = \begin{array}{c} 3 \\ 2 \\ 5 \\ 6 \end{array} \begin{bmatrix} 300 & 18{,}000 & -300 & -18{,}000 \\ 18{,}000 & 1{,}440{,}000 & 18{,}000 & 720{,}000 \\ -300 & 18{,}000 & 300 & -18{,}000 \\ -18{,}000 & 720{,}000 & -18{,}000 & 1{,}440{,}000 \end{bmatrix} \quad (\text{columns } 3,2,5,6)$$

$$\{\Delta\}^{(1)} = \begin{Bmatrix} \Delta_4 \\ \theta_1 \\ \Delta_3 \\ \theta_2 \end{Bmatrix} = \begin{Bmatrix} 0.0 \\ 0.0016 \\ 0.0 \\ -0.0032 \end{Bmatrix} \begin{matrix} \text{in.} \\ \text{rad} \\ \text{in.} \\ \text{rad} \end{matrix} \qquad \{\Delta\}^{(2)} = \begin{Bmatrix} \Delta_3 \\ \theta_2 \\ \Delta_5 \\ \theta_6 \end{Bmatrix} = \begin{Bmatrix} 0.0 \\ -0.0032 \\ 0.0 \\ 0.0 \end{Bmatrix} \begin{matrix} \text{in.} \\ \text{rad} \\ \text{in.} \\ \text{rad} \end{matrix}$$

The individual member end forces are now computed.

$$\{Q\}^{(1)} = [K]^{(1)}\{\Delta\}^{(1)} = \begin{Bmatrix} Q_4 \\ Q_1 \\ Q_3 \\ Q_2 \end{Bmatrix} = \begin{Bmatrix} -14.03 \\ 0.0 \\ 14.03 \\ -2524.7 \end{Bmatrix} \begin{matrix} \text{k} \\ \text{k·in.} \\ \text{k} \\ \text{k·in.} \end{matrix} \Rightarrow \begin{Bmatrix} V_4 \\ M_1 \\ V_3 \\ M_2 \end{Bmatrix} = \begin{Bmatrix} -14.03 \\ 0.0 \\ -14.03 \\ -2524.7 \end{Bmatrix} \begin{matrix} \text{k} \\ \text{k·in.} \\ \text{k} \\ \text{k·in.} \end{matrix}$$

$$\{Q\}^{(2)} = [K]^{(2)}\{\Delta\}^{(2)} = \begin{Bmatrix} Q_3 \\ Q_2 \\ Q_5 \\ Q_6 \end{Bmatrix} = \begin{Bmatrix} -58.44 \\ -4675.3 \\ 58.44 \\ -2337.7 \end{Bmatrix} \begin{matrix} \text{k} \\ \text{k·in.} \\ \text{k} \\ \text{k·in.} \end{matrix} \Rightarrow \begin{Bmatrix} V_3 \\ M_2 \\ V_5 \\ M_6 \end{Bmatrix} = \begin{Bmatrix} -58.44 \\ 4675.3 \\ -58.44 \\ -2337.7 \end{Bmatrix} \begin{matrix} \text{k} \\ \text{k·in.} \\ \text{k} \\ \text{k·in.} \end{matrix}$$

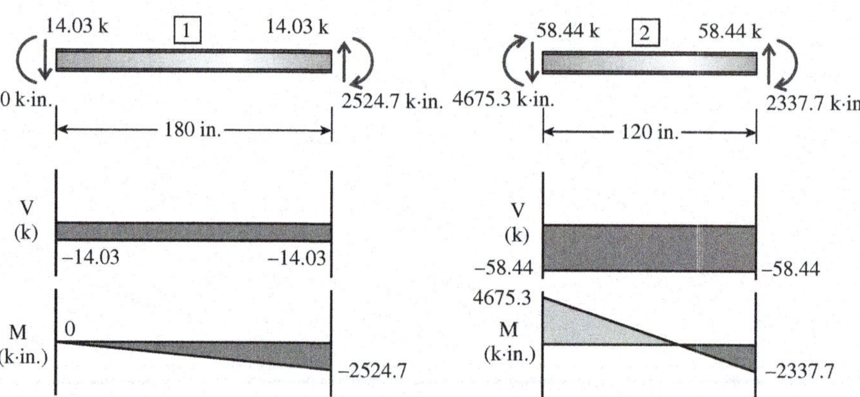

EXAMPLE 20.4

Use the results found in Example 20.2 to solve for the member end forces for each of the beam elements. Report the findings using beam sign convention and then sketch the shear and moment diagrams for each member.

SOLUTION STRATEGY

Collect the element stiffness matrices and the nodal displacements/rotations for the entire structure from Example 20.2. Use the matrix operation for computing member end forces and then transfer these forces over to a FBD using beam sign convention.

SOLUTION

The element stiffness matrices are shown in Example 20.3 and the element displacement vectors are assembled from the information in Example 20.2.

$$\{\Delta\}^{(1)} = \begin{Bmatrix} \Delta_4 \\ \theta_1 \\ \Delta_3 \\ \theta_2 \end{Bmatrix} = \begin{Bmatrix} 0.0 \\ -0.0031 \\ -0.2173 \\ 0.0017 \end{Bmatrix} \begin{matrix} \text{in.} \\ \text{rad} \\ \text{in.} \\ \text{rad} \end{matrix} \qquad \{\Delta\}^{(2)} = \begin{Bmatrix} \Delta_3 \\ \theta_2 \\ \Delta_5 \\ \theta_6 \end{Bmatrix} = \begin{Bmatrix} -0.2173 \\ 0.0017 \\ 0.0 \\ 0.0 \end{Bmatrix} \begin{matrix} \text{in.} \\ \text{rad} \\ \text{in.} \\ \text{rad} \end{matrix}$$

The individual member end forces are now computed.

$$\{Q\}^{(1)} = [K]^{(1)}\{\Delta\}^{(1)} = \begin{Bmatrix} Q_4 \\ Q_1 \\ Q_3 \\ Q_2 \end{Bmatrix} = \begin{Bmatrix} 13.74 \\ 0.0 \\ -13.74 \\ 2473.2 \end{Bmatrix} \begin{matrix} \text{k} \\ \text{k·in.} \\ \text{k} \\ \text{k·in.} \end{matrix} \Rightarrow \begin{Bmatrix} V_4 \\ M_1 \\ V_3 \\ M_2 \end{Bmatrix} = \begin{Bmatrix} 13.74 \\ 0.0 \\ 13.74 \\ 2473.2 \end{Bmatrix} \begin{matrix} \text{k} \\ \text{k·in.} \\ \text{k} \\ \text{k·in.} \end{matrix}$$

$$\{Q\}^{(2)} = [K]^{(2)}\{\Delta\}^{(2)} = \begin{Bmatrix} Q_3 \\ Q_2 \\ Q_5 \\ Q_6 \end{Bmatrix} = \begin{Bmatrix} -51.26 \\ -2473.2 \\ 51.26 \\ -3678.0 \end{Bmatrix} \begin{matrix} \text{k} \\ \text{k·in.} \\ \text{k} \\ \text{k·in.} \end{matrix} \Rightarrow \begin{Bmatrix} V_3 \\ M_2 \\ V_5 \\ M_6 \end{Bmatrix} = \begin{Bmatrix} -51.26 \\ 2473.2 \\ 51.26 \\ -3678.0 \end{Bmatrix} \begin{matrix} \text{k} \\ \text{k·in.} \\ \text{k} \\ \text{k·in.} \end{matrix}$$

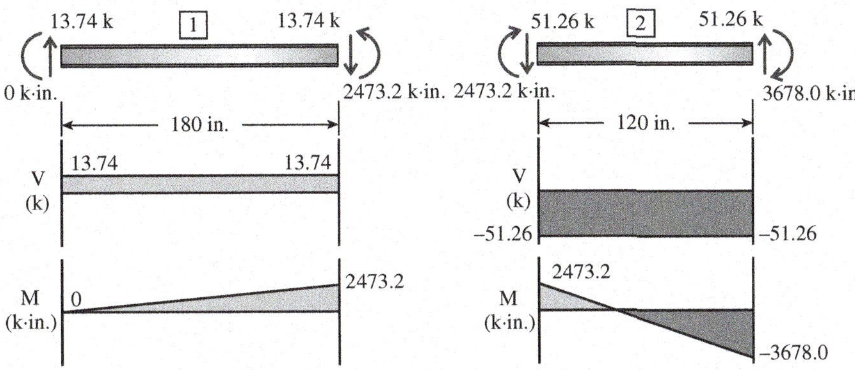

20.5 Plotting Deflections Using Beam Shape Functions

It doesn't take long utilizing the matrix method for one to recognize that it only provides response information to the analyst at the nodes (i.e., DOFs). To gain information about member responses at locations between the node, one needs to use some additional rules. In Examples 20.3 and 20.4, the member end forces were determined for each beam element. These member end forces were used in conjunction with the methods developed in Chapter 6 to find internal forces at any location along the length (i.e., generate shear and moment diagrams).

In a like fashion, nodal displacements may be used to identify member displacements at any location between the nodes. The tools used to do this are known as *shape functions*. Shape functions essentially describe the displaced shape of the element in terms of the member end displacements and rotations. This concept is demonstrated in Figure 20.8 where $u_y(x)$ is the displacement in a direction perpendicular to the axis of the member as a function of x.

The solution to this is found using the principle of superposition, which says that if we calculate the displacement at x due to each of the nodal displacements/rotations individually, then we can find the total response $u_y(x)$ by adding all of these effects.

$$u_y(x) = u_{y-a}(x) + u_{y-b}(x) + u_{y-c}(x) + u_{y-d}(x)$$

DID YOU KNOW?

The term *shape function* is sometimes referred to as an *interpolation function*. This is because it is a function used to interpolate displacements between the nodes.

NOTATION

$u_{y-a}(x)$ simply represents the vertical displacement at x due to a displacement at DOF a. Likewise $u_{y-b}(x)$ represents the vertical displacement at x due to a rotation at DOF b, and so forth.

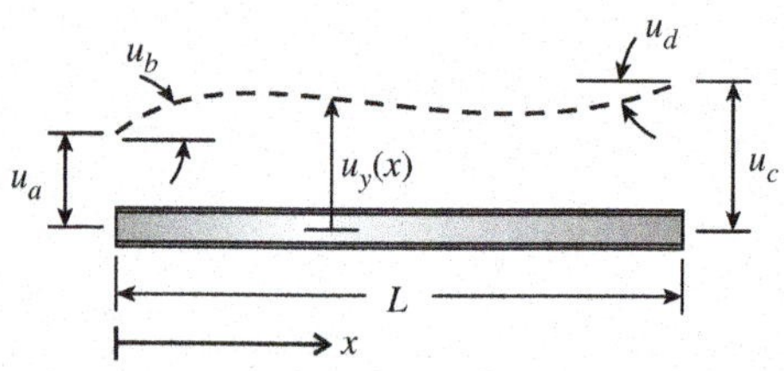

FIGURE 20.8 Illustration of beam element displacement as a function of nodal displacements.

Figure 20.9 represents the individual cases that get superimposed to obtain the condition in Figure 20.8. The task now is to find expressions for $u_{y-a}(x)$, $u_{y-b}(x)$, and so on in terms of u_a, u_b respectively. A full derivation will not be shown here, but the method of double integration may be used to show that the following equations describe these independent displacements.

$$u_{y-a}(x) = u_a\left[1-3\left(\tfrac{x}{L}\right)^2+2\left(\tfrac{x}{L}\right)^3\right] = u_aN_a(x)$$

$$u_{y-b}(x) = u_b\left[x\left(1-\tfrac{x}{L}\right)^2\right] = u_bN_b(x)$$

$$u_{y-c}(x) = u_c\left[3\left(\tfrac{x}{L}\right)^2-2\left(\tfrac{x}{L}\right)^3\right] = u_cN_c(x)$$

$$u_{y-d}(x) = u_d\left[\tfrac{x^2}{L}\left(-1+\tfrac{x}{L}\right)\right] = u_dN_d(x)$$

DID YOU KNOW?

A shape function N_i describes the displaced shape of an element when a unit displacement/rotation is imposed at DOF i (see Figure 20.10).

In looking over these equations, one may note that the terms in the square brackets have been assigned the symbol N. These terms are known as the shape functions and are presented concisely as

$$N_a(x) = 1-3\left(\tfrac{x}{L}\right)^2+2\left(\tfrac{x}{L}\right)^3 \qquad N_b(x) = x\left(1-\tfrac{x}{L}\right)^2$$

$$N_c(x) = 3\left(\tfrac{x}{L}\right)^2-2\left(\tfrac{x}{L}\right)^3 \qquad N_d(x) = \tfrac{x^2}{L}\left(-1+\tfrac{x}{L}\right)$$

This allows then for the displacement at x to be written in terms of these shape functions. Thus,

$$u_y(x) = u_aN_a(x)+u_bN_b(x)+u_cN_c(x)+u_dN_d(x)$$

QUICK NOTE

To find a value of u_y at a given value of x, first evaluate each of the shape functions at the x value, assemble these values into a row vector as shown, and then carry out the matrix multiplication.

In matrix form, this operation is

$$u_y(x) = \begin{bmatrix} N_a(x) & N_b(x) & N_c(x) & N_d(x) \end{bmatrix} \begin{Bmatrix} u_a \\ u_b \\ u_c \\ u_d \end{Bmatrix}$$

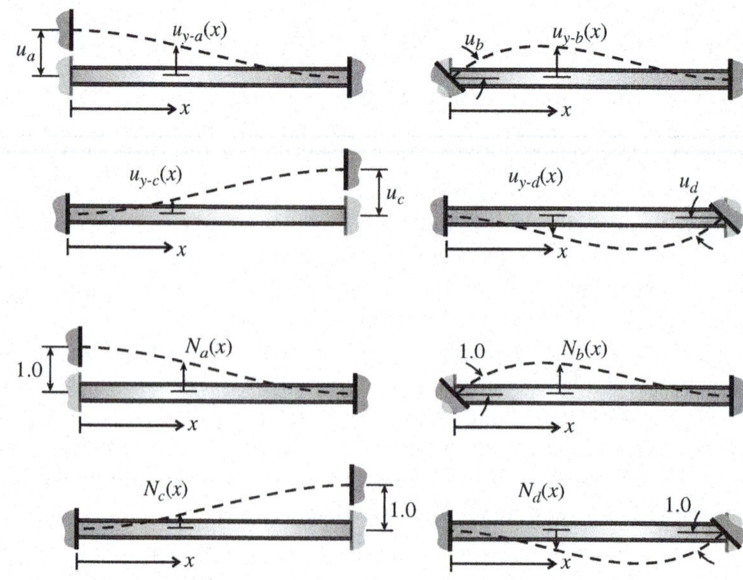

FIGURE 20.9 Independent displacement cases.

FIGURE 20.10 Shape functions illustrated.

EXAMPLE 20.5

With the use of shape functions, plot the deflected shape for the beam shown in Example 20.1.

SOLUTION STRATEGY

Pull the nodal displacement vector from Example 20.3. Decide how frequently to plot values along the length of the member. Set up a table showing the calculations of the shape functions for each member at differing values of x. Carry out the matrix operation $\lfloor N \rfloor \{u\}^{(n)}$ for each member.

QUICK NOTE

For this example, each member was divided into five segments. Generally, 5 to 10 segments will provide sufficient resolution in the plot.

SOLUTION

Collect the nodal displacements for each member from Example 20.1

$$\begin{Bmatrix} u_a \\ u_b \\ u_c \\ u_d \end{Bmatrix}^{(1)} = \begin{Bmatrix} \Delta_4 \\ \theta_1 \\ \Delta_3 \\ \theta_2 \end{Bmatrix} = \begin{Bmatrix} 0.0 \\ 0.0016 \\ 0.0 \\ -0.0032 \end{Bmatrix} \begin{matrix} \text{in.} \\ \text{rad} \\ \text{in.} \\ \text{rad} \end{matrix} \qquad \begin{Bmatrix} u_a \\ u_b \\ u_c \\ u_d \end{Bmatrix}^{(2)} = \begin{Bmatrix} \Delta_3 \\ \theta_2 \\ \Delta_5 \\ \theta_6 \end{Bmatrix} = \begin{Bmatrix} 0.0 \\ -0.0032 \\ 0.0 \\ 0.0 \end{Bmatrix} \begin{matrix} \text{in.} \\ \text{rad} \\ \text{in.} \\ \text{rad} \end{matrix}$$

For each element, set up a table that calculates the value of the shape functions for each value of x where $L_1 = 180$ in. and $L_2 = 120$ in.

Element 1

x (in.)	N_a	N_b	N_c	N_d	u_y (in.)
0	1	0	0	0	0
36	0.896	23.04	0.104	−5.76	0.0561
72	0.648	25.92	0.352	−17.28	0.0982
108	0.352	17.28	0.648	−25.92	0.1122
144	0.104	5.76	0.896	−23.04	0.0842
180	0	0	1	0	0

Element 2

x (in.)	N_a	N_b	N_c	N_d	u_y (in.)
0	1	0	0	0	0
24	0.896	15.36	0.104	−3.84	−0.0499
48	0.648	17.28	0.352	−11.52	−0.0561
72	0.352	11.52	0.648	−17.28	−0.0374
96	0.104	3.84	0.896	−15.36	−0.0125
120	0	0	1	0	0

By way of demonstration, the calculations for $x = 36$ in. for element 1 are shown here.

$$\begin{aligned}
N_a(36\text{ in.}) &= 1 - 3\left(\tfrac{36}{180}\right)^2 + 2\left(\tfrac{36}{180}\right)^3 &&= 0.896 \\
N_b(36\text{ in.}) &= 36\left(1 - \tfrac{36}{180}\right)^2 &&= 23.04 \\
N_c(36\text{ in.}) &= 3\left(\tfrac{36}{180}\right)^2 - 2\left(\tfrac{36}{180}\right)^3 &&= 0.104 \\
N_d(36\text{ in.}) &= \tfrac{(36)^2}{180}\left(-1 + \tfrac{36}{180}\right) &&= -5.76
\end{aligned}$$

$$u_y^{(1)}(36\text{ in.}) = \lfloor\ 0.896 \quad 23.04 \quad 0.104 \quad -5.76\ \rfloor \begin{Bmatrix} 0.0 \\ 0.0016 \\ 0.0 \\ -0.0032 \end{Bmatrix} = \underline{\underline{0.0561\text{ in.}}}$$

The plot of the deflected shape can now be generated.

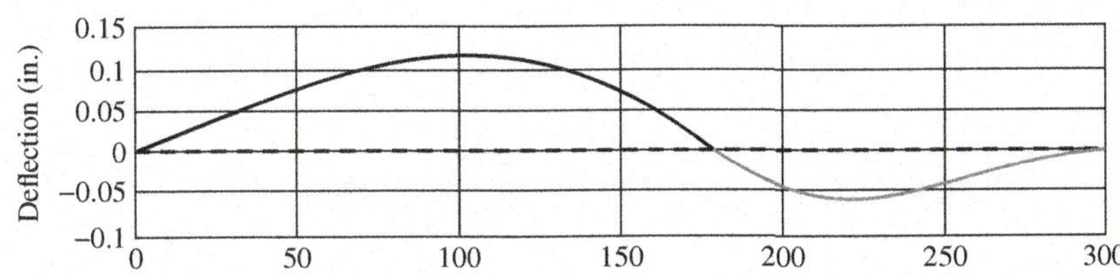

DID YOU NOTICE?

The x values used for the evaluation of each element are relative to the individual element. Thus, if you were to simply plot x versus u_y for both elements, they would plot on top of each other. To create a plot for the entire beam, the plotted values of x need to be offset based on member origin. For example, member 2 starts at 180 in. from the left of the beam. Thus, all x values need 180 in. added to them before you can plot member 2.

DID YOU KNOW?
The values of the shape functions are a function of just the element length and is not dependent on the type of load being applied. Rather, the nodal displacements/rotations are influenced by the loading, and this is what alters the deflected shape.

By way of interest, the deflected shape for the beam given in Example 20.2 is shown here. Its calculations are done in the exact same fashion as for the previous beam.

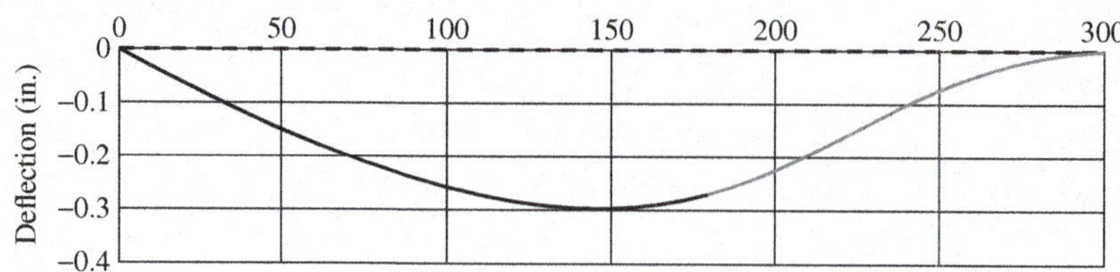

20.6 Loading Between Nodes (Statical Equivalency)

By now, the reader may have come to appreciate the beauty, finesse, and thoroughness of the matrix formulation for structural analysis. However, there is a notable shortcoming with the formulation the way it currently stands. The matrix-based equilibrium equations that have been developed and used are written for the DOFs in the structure. This means that when the force vector $\{F_u\}$ is assembled it is based on the external forces that exist at each of the unrestrained DOFs. So this begs the question: *What happens when the external forces are not applied at the DOFs in the actual structure?* Take a look at Figure 20.11. Note that it is made of three beam elements and has four unrestrained DOFs (all of which are rotational). None of the actual applied loads occur at any of the DOFs.

There are two general strategies for dealing with this issue.

QUICK NOTE
No one method must be used all of the time. There may be occasions where it is more straightforward to add another node and complete the analysis. In commercial structural analysis software, the user has the option to add extra nodes at the location of point loads or applied moments. If they do not do this, the program defaults to method 2. There is very little impact, if any, upon the end user.

Method 1 Place additional nodes (i.e., DOFs) at the locations of the applied loads.

- Very easy to implement.
- This will only address the issue with applied point loads and moments.
- Can quickly increase the size of the structure level stiffness matrix $[K]$.

Method 2 Replace member loads with *statically equivalent* nodal loads.

- Valid for all types of loading scenarios.
- Does not increase the size of the structure level stiffness matrix $[K]$.
- All final response quantities must be found by superposition.

Statically Equivalent Loads

DID YOU KNOW?
The given process will only ensure that the nodal displacements/rotations between the original structure and the statically equivalent structure agree. All other response quantities (displacements between nodes, reactions, and internal shear and moment) will generally not be equivalent in both structures.

By definition, *statically equivalent loads* are a set of nodal loads that produce the same displacements/rotations at all DOFs as would have been caused by the actual member loads. While this may initially sound complex, the process for finding the magnitude of such loads is fairly straightforward.

1. Lock all DOFs in the structure against rotation/translation.
2. Using beam charts (found in Appendix D), find the reactions—both forces and moments—that occur at the locked DOFs due to the member loads. These are known as fixed-end moments (FEMs) and fixed-end shears (FEVs) and are assembled into a force vector noted as

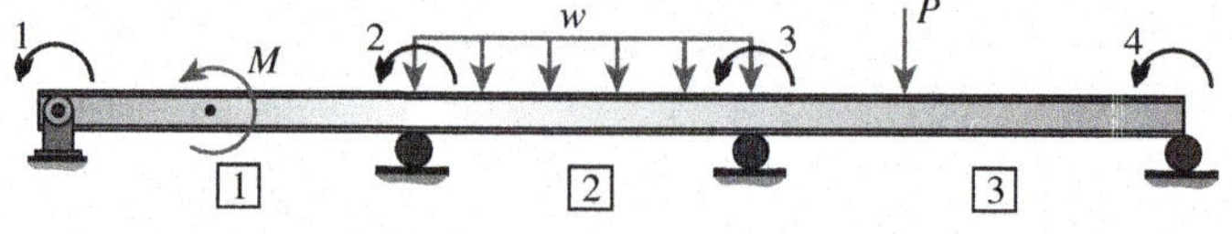

FIGURE 20.11 Illustration of loads not located at current DOFs.

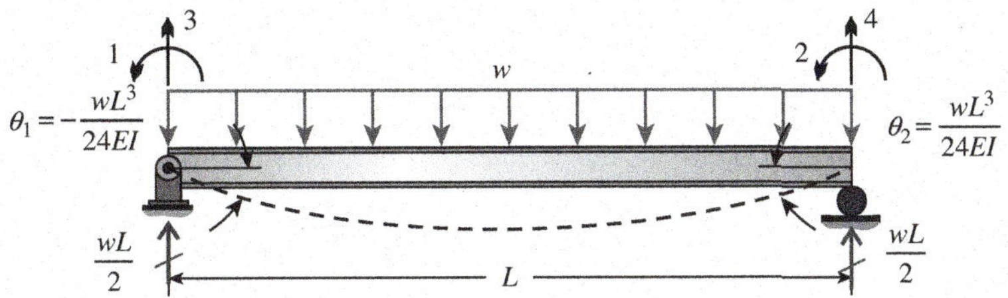

FIGURE 20.12 Solution for a uniformly loaded simply supported beam.

$\{F\}_{\text{FEF}}$ where the FEF is a generic term called fixed-end forces that include both FEMs and FEVs.

3. Assemble the force vector $\{F\}$ the way you normally would for any loads that were originally located at DOFs. The equilibrium equation takes the form:

$$\left\{\begin{array}{c} F_u \\ \hline F_r \end{array}\right\} = \left[\begin{array}{c|c} K_{uu} & K_{ur} \\ \hline K_{ru} & K_{rr} \end{array}\right]\left\{\begin{array}{c} \Delta_u \\ \hline \Delta_r \end{array}\right\} + \left\{\begin{array}{c} F_{\text{FEF}-u} \\ \hline F_{\text{FEF}-r} \end{array}\right\}$$

4. Move the vector $\{F_{\text{FEF}-u}\}$ to the left side of the equation and carry out all matrix operations as you normally would.

$$\{\Delta_u\} = [K_{uu}]^{-1}\{F_u - F_{\text{FEF}-u}\}$$

$$\{F_r\} = [K_{ru}]\{\Delta_u\} + \{F_{\text{FEF}-r}\}$$

QUICK NOTE

The net effect of moving the vector $\{F_{\text{FEF}-u}\}$ to the left side of the equation is that the FEMs/FEVs that appear on the fully fixed structure are reversed when they appear on the statically equivalent structure (see Figure 20.13).

This process will be illustrated using a simply supported beam that has a uniformly distributed load. Figure 20.12 shows the known solution for this beam and also has the selected DOFs labeled. The fixed-end forces and the statically equivalent beam are shown in Figure 20.13. The FEFs are taken from the beam table found in Appendix D.

Assemble the force vectors and the beam stiffness matrix.

$$\{F\} = \left\{\begin{array}{c} 0 \\ 0 \\ \hline R_3 \\ R_4 \end{array}\right\} \quad \{F_{\text{FEF}}\} = \left\{\begin{array}{c} \frac{wL^2}{12} \\ -\frac{wL^2}{12} \\ \hline \frac{wL}{2} \\ \frac{wL}{2} \end{array}\right\} \quad [K] = \left[\begin{array}{cc|cc} \frac{4EI}{L} & \frac{2EI}{L} & \frac{6EI}{L^2} & -\frac{6EI}{L^2} \\ \frac{2EI}{L} & \frac{4EI}{L} & \frac{6EI}{L^2} & -\frac{6EI}{L^2} \\ \hline \frac{6EI}{L^2} & \frac{6EI}{L^2} & \frac{12EI}{L^3} & -\frac{12EI}{L^3} \\ -\frac{6EI}{L^2} & -\frac{6EI}{L^2} & -\frac{12EI}{L^3} & \frac{12EI}{L^3} \end{array}\right]$$

As shown in step 4, the rotations for the unrestrained DOFs may be solved using

$$\{\Delta_u\} = \left\{\begin{array}{c} \theta_1 \\ \theta_2 \end{array}\right\} = \left[\begin{array}{cc} \frac{4EI}{L} & \frac{2EI}{L} \\ \frac{2EI}{L} & \frac{4EI}{L} \end{array}\right]^{-1}\left\{\begin{array}{c} 0 - \frac{wL^2}{12} \\ 0 + \frac{wL^2}{12} \end{array}\right\} = \underline{\underline{\left\{\begin{array}{c} -\frac{wL^3}{24EI} \\ \frac{wL^3}{24EI} \end{array}\right\}}}$$

QUICK NOTE

Recognize that in the original structure there are no forces applied at the DOFs. This means:

$$\{F_u\} = \left\{\begin{array}{c} 0 \\ 0 \end{array}\right\}$$

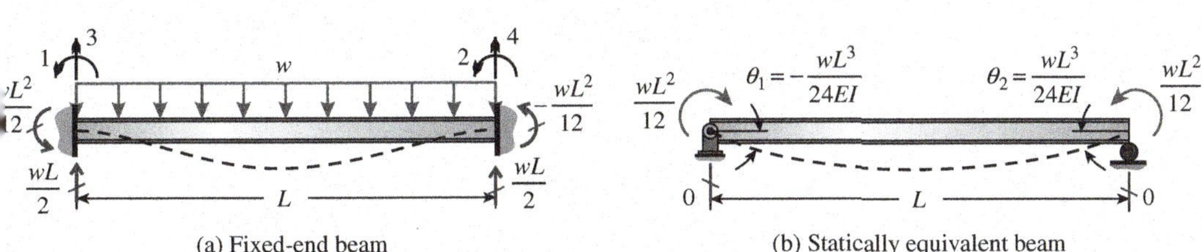

FIGURE 20.13 Statically equivalent forces for a uniformly loaded beam.

One should note at this point that the rotations in the statically equivalent structure did indeed come out to be equal to the known rotations of the original structure. The final step is to solve for the reactions. Note that the reactions for the statically equivalent structure worked out to be zero, as shown by the matrix operations and illustrated in Figure 20.13(b). However, when these are added back to the reactions from the fixed-fixed beam, the final reactions agree with the original beam in Figure 20.12.

$$\{F_r\} = \begin{Bmatrix} R_3 \\ R_4 \end{Bmatrix} = \begin{bmatrix} \frac{6EI}{L^2} & \frac{6EI}{L^2} \\ -\frac{6EI}{L^2} & -\frac{6EI}{L^2} \end{bmatrix} \begin{Bmatrix} -\frac{wL^3}{24EI} \\ \frac{wL^3}{24EI} \end{Bmatrix} + \begin{Bmatrix} \frac{wL}{2} \\ \frac{wL}{2} \end{Bmatrix}$$

$$= \begin{Bmatrix} 0 \\ 0 \end{Bmatrix} + \begin{Bmatrix} \frac{wL}{2} \\ \frac{wL}{2} \end{Bmatrix} = \begin{Bmatrix} \frac{wL}{2} \\ \frac{wL}{2} \end{Bmatrix}$$

EXAMPLE 20.6

Use the matrix method to solve for the nodal rotations/displacements and also the reactions for the given beam. Take E = 29,000 ksi and I = 500 in^4 for all beam spans.

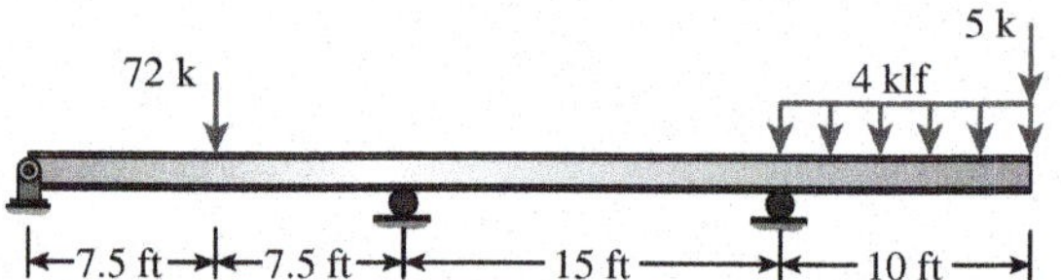

SOLUTION STRATEGY

Set up the stiffness matrices in the same fashion as done previously. Identify which loads are nodal loads and which are member loads. This will help with the assembly of both $\{F_u\}$ and $\{F_{FEF}\}$.

SOLUTION

The following DOF labeling is used throughout this example.

QUICK NOTE

Using standard procedures as previously introduced, the $[K_{uu}]$ and $[K_{ru}]$ matrices are found and presented here. These matrices have been developed using the units of kips and in.

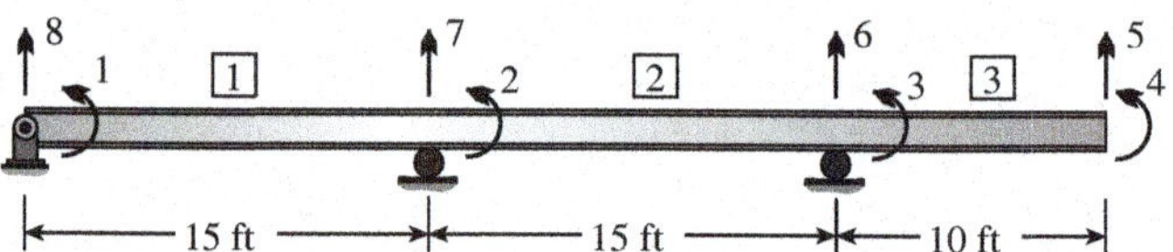

$$[K_{uu}] = \begin{array}{c} \\ 1 \\ 2 \\ 3 \\ 4 \\ 5 \end{array} \begin{array}{c} \begin{array}{ccccc} 1 & 2 & 3 & 4 & 5 \end{array} \\ \begin{bmatrix} 322{,}222 & 161{,}111 & 0 & 0 & 0 \\ 161{,}111 & 644{,}444 & 161{,}111 & 0 & 0 \\ 0 & 161{,}111 & 805{,}556 & 241{,}667 & -6042 \\ 0 & 0 & 241{,}667 & 483{,}333 & -6042 \\ 0 & 0 & -6042 & -6042 & 101 \end{bmatrix} \end{array}$$

$$[K_{ru}] = \begin{array}{c} \\ 6 \\ 7 \\ 8 \end{array} \begin{array}{c} \begin{array}{ccccc} 1 & 2 & 3 & 4 & 5 \end{array} \\ \begin{bmatrix} 0 & -2685 & 3356 & 6042 & -101 \\ -2685 & 0 & 2685 & 0 & 0 \\ 2685 & 2685 & 0 & 0 & 0 \end{bmatrix} \end{array}$$

The fixed-end forces and moments are found by using the beam charts found in Appendix D. Be careful of the units and make sure that they are consistent with how you have defined the stiffness matrices—in this case kips and in.

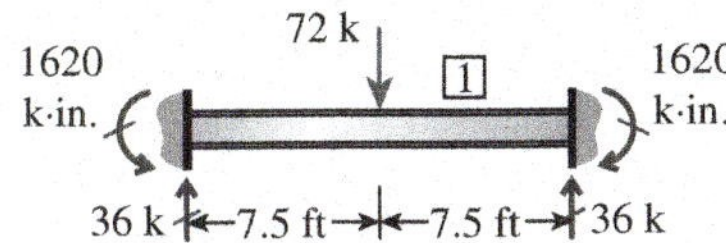

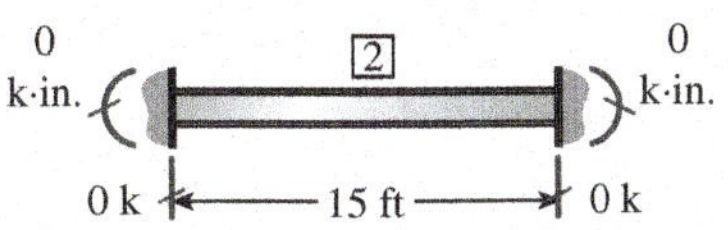

Using global sign-convention, we need to interpret all of the FEVs and FEMs and place them in the FEF vector as shown.

$$\begin{Bmatrix} F_{\text{FEF}-u} \\ \hline F_{\text{FEF}-r} \end{Bmatrix} = \begin{Bmatrix} 1620.0 \\ -1620.0 \\ 400.0 \\ -400.0 \\ 20 \\ \hline 20 \\ 36 \\ 36 \end{Bmatrix} \begin{matrix} \text{k·in.} \\ \text{k·in.} \\ \text{k·in.} \\ \text{k·in.} \\ \text{k} \\ \text{k} \\ \text{k} \\ \text{k} \end{matrix}$$

The force vector for the unrestrained DOFs is easily assembled by $F_u - F_{\text{FEF}-u}$, which produces the following vector and is effectively analyzing the structure shown.

$$\{F_u\} - \{F_{\text{FEF}-u}\} = \begin{Bmatrix} 0.0 \\ 0.0 \\ 0.0 \\ 0.0 \\ -5.0 \end{Bmatrix} - \begin{Bmatrix} 1620.0 \\ -1620.0 \\ 400.0 \\ -400.0 \\ 20.0 \end{Bmatrix} = \begin{Bmatrix} -1620.0 \\ 1620.0 \\ -400.0 \\ 400.0 \\ -25.0 \end{Bmatrix} \begin{matrix} \text{k·in.} \\ \text{k·in.} \\ \text{k·in.} \\ \text{k·in.} \\ \text{k} \end{matrix}$$

Following the matrix formulation previously introduced, the nodal displacements may be computed as

$$\begin{Bmatrix} \theta_1 \\ \theta_2 \\ \theta_3 \\ \theta_4 \\ \Delta_5 \end{Bmatrix} = \{\Delta_u\} = [K_{uu}]^{-1}\{F_u - F_{\text{FEF}-u}\} = [K_{uu}]^{-1} \begin{Bmatrix} -1620.0 \\ 1620.0 \\ -400.0 \\ 400.0 \\ -25.0 \end{Bmatrix} = \begin{Bmatrix} -0.0091 \\ 0.0081 \\ -0.0134 \\ -0.0225 \\ -2.400 \end{Bmatrix} \begin{matrix} \text{rad} \\ \text{rad} \\ \text{rad} \\ \text{rad} \\ \text{in.} \end{matrix}$$

FORCE VECTOR

The basic force vector $\{F\}$ is easily assembled by identifying any applied forces or reactions that are applied directly at a DOF.

$$\begin{Bmatrix} F_u \\ \hline F_r \end{Bmatrix} = \begin{Bmatrix} 0.0 \\ 0.0 \\ 0.0 \\ 0.0 \\ -5 \\ \hline R_6 \\ R_7 \\ R_8 \end{Bmatrix} \begin{matrix} \text{k·in.} \\ \text{k·in.} \\ \text{k·in.} \\ \text{k·in.} \\ \text{k} \\ \text{k} \\ \text{k} \\ \text{k} \end{matrix}$$

DID YOU KNOW?

To find the FEF at a particular DOF, you must add all of the effects from every member that shares that DOF. For example, DOF 2 is common to both members 1 and 2. Therefore, to find the FEM at DOF 2, we need to sum the FEMs from both members. In this case, member 2 doesn't have a load, so this concept is not overly obvious. The equation would really look like this:

$$\begin{aligned} \text{FEM}_2 &= -1620 \text{ k·in.} + 0.0 \text{ k·in.} \\ &= -1620 \text{ k·in.} \end{aligned}$$

Finally, the reactions are calculated using $[K_{ru}]$ and $\{\Delta_u\}$, but don't forget to superimpose $\{F_{\text{FEF}-r}\}$.

$$\begin{Bmatrix} R_6 \\ R_7 \\ R_8 \end{Bmatrix} = \{F_r\} = [K_{ru}]\{\Delta_u\} + \{F_{\text{FEF}-r}\} = [K_{ru}]\begin{Bmatrix} -0.0091 \\ 0.0081 \\ -0.0134 \\ -0.0225 \\ -2.400 \end{Bmatrix} + \begin{Bmatrix} 20\text{ k} \\ 36\text{ k} \\ 36\text{ k} \end{Bmatrix}$$

$$= \begin{Bmatrix} 39.1\text{ k} \\ -11.5\text{ k} \\ -2.6\text{ k} \end{Bmatrix} + \begin{Bmatrix} 20\text{ k} \\ 36\text{ k} \\ 36\text{ k} \end{Bmatrix} = \begin{Bmatrix} 59.1 \\ 24.5 \\ 33.4 \end{Bmatrix} \begin{matrix} \text{k} \\ \text{k} \\ \text{k} \end{matrix}$$

20.7 Superposition to Obtain Shear, Moment, and Deflection Diagrams

RECALL

Statically equivalent loads are a set of nodal loads that produce the same displacements/rotations at all DOFs as caused by the actual member loads.

In the previous section, an approach was introduced on how to deal with loads that are located between nodes on a beam element. The concept that was used required the computation of a set of statically equivalent loads—seen previously as $\{F_{\text{FEF}}\}$. These loads actually only produce correct displacements at the DOFs. All other responses, including reactions, member end forces, displacements between nodes, etc., must be found by superimposing the results of the fixed-end members with the results from the matrix analysis. This requirement is explicitly seen in

$$\{F_r\} = [K_{ru}]\{\Delta_u\} + \{F_{\text{FEF}-r}\}$$

Just as is the case with the reactions, member end forces must be computed by superimposing the results of the matrix analysis with the fixed-end forces.

EXAMPLE 20.7

RECALL

Member end forces are computed by the matrix operation:

$$\begin{Bmatrix} Q_a \\ Q_b \\ Q_c \\ Q_d \end{Bmatrix}^{(n)} = [k]^{(n)} \begin{Bmatrix} u_a \\ u_b \\ u_c \\ u_d \end{Bmatrix}^{(n)}$$

where (n) refers to the element number and the DOFs are in the order shown in Figure 20.2.

Consider the beam analyzed in Example 20.6. If the approach described in Section 20.4 is followed, the following member end forces are computed directly from the matrix analysis considering the equivalent static loads.

$$\{Q\}^{(1)} = \begin{Bmatrix} -2.58 \\ -1620.0 \\ 2.58 \\ 1155.0 \end{Bmatrix} \begin{matrix} \text{k} \\ \text{k·in.} \\ \text{k} \\ \text{k·in.} \end{matrix} \qquad \{Q\}^{(2)} = \begin{Bmatrix} -14.08 \\ 465.0 \\ 14.08 \\ -3000.0 \end{Bmatrix} \begin{matrix} \text{k} \\ \text{k·in.} \\ \text{k} \\ \text{k·in.} \end{matrix}$$

$$\{Q\}^{(3)} = \begin{Bmatrix} 25.00 \\ 2600.0 \\ -25.00 \\ 400.0 \end{Bmatrix} \begin{matrix} \text{k} \\ \text{k·in.} \\ \text{k} \\ \text{k·in.} \end{matrix}$$

Find the actual member end forces for these beam members.

SOLUTION STRATEGY

Superposition of the member end forces from the matrix analysis and from the fixed-end members must be made. Since we already have the force vectors from the matrix analysis, the next step is to assemble the force vectors for the fixed end members.

SOLUTION

The following force vectors can be assembled by examining the fixed-ended members shown in Example 20.6.

$$\{FEF\}^{(1)} = \begin{Bmatrix} 36.00 \\ 1620.0 \\ 36.0 \\ -1620.0 \end{Bmatrix} \begin{matrix} \text{k} \\ \text{k·in.} \\ \text{k} \\ \text{k·in.} \end{matrix} \quad \{FEF\}^{(2)} = \begin{Bmatrix} 0.00 \\ 0.0 \\ 0.00 \\ 0.0 \end{Bmatrix} \begin{matrix} \text{k} \\ \text{k·in.} \\ \text{k} \\ \text{k·in.} \end{matrix} \quad \{FEF\}^{(3)} = \begin{Bmatrix} 20.00 \\ 400.0 \\ 20.00 \\ -400.0 \end{Bmatrix} \begin{matrix} \text{k} \\ \text{k·in.} \\ \text{k} \\ \text{k·in.} \end{matrix}$$

With these force vectors now available, the superposition may proceed. Thus,

$$\begin{Bmatrix} Q_a \\ M_b \\ Q_c \\ M_d \end{Bmatrix}^{(1)} = \{Q\}^{(1)} + \{FEF\}^{(1)} = \begin{Bmatrix} 33.42 \\ 0.0 \\ 38.58 \\ -465.0 \end{Bmatrix} \begin{matrix} \text{k} \\ \text{k·in.} \\ \text{k} \\ \text{k·in.} \end{matrix}$$

$$\begin{Bmatrix} Q_a \\ M_b \\ Q_c \\ M_d \end{Bmatrix}^{(2)} = \{Q\}^{(2)} + \{FEF\}^{(2)} = \begin{Bmatrix} -14.08 \\ 465.0 \\ 14.08 \\ -3000.0 \end{Bmatrix} \begin{matrix} \text{k} \\ \text{k·in.} \\ \text{k} \\ \text{k·in.} \end{matrix}$$

$$\begin{Bmatrix} Q_a \\ M_b \\ Q_c \\ M_d \end{Bmatrix}^{(3)} = \{Q\}^{(3)} + \{FEF\}^{(3)} = \begin{Bmatrix} 45.00 \\ 3000.0 \\ -5.00 \\ 0.0 \end{Bmatrix} \begin{matrix} \text{k} \\ \text{k·in.} \\ \text{k} \\ \text{k·in.} \end{matrix}$$

RECALL

The member end forces computed in this way are reported in the matrix-based sign convention. You will need to convert to the beam sign convention before sketching any shear and moment diagrams (see Section 20.4). This mapping from matrix to beam sign convention is repeated here for convenience.

$$\begin{Bmatrix} Q_a \\ Q_b \\ Q_c \\ Q_d \end{Bmatrix}^{(n)} = \begin{Bmatrix} V_a \\ -M_b \\ -V_c \\ M_d \end{Bmatrix}^{(n)}$$

Diagrams for Shear, Moment, and Displacement

Between Section 20.6 and the current section, we have shown how the analyst may compute the correct values for nodal displacements, reactions, and member end forces when loads are applied between the nodes. While useful, these response quantities don't provide all of the information an analyst may desire to know. Specifically, one may also desire to know the internal shear, internal moment, and displacement at locations other than at the nodes. This requires the generation of shear, moment, and deflection diagrams, as illustrated in Sections 20.4 and 20.5. As with the other responses, to get correct diagrams, the results from the matrix analysis need to be superimposed with the results of the fixed-end members.

To assist in the plotting of shear, moment, and deflection diagrams, equations for these quantities for several fixed-ended members are provided in Figures 20.14 and 20.15. This will alleviate some workload and potential error if you derive these from principles learned earlier in this textbook.

QUICK NOTE

Generating shear and moment diagrams from member end forces is a straightforward process following basic procedures discussed in Chapter 6. Examples of typical diagrams are found in Examples 20.3 and 20.4. Solving for member displacements based on member end displacements is demonstrated in Example 20.5.

$$FEV_{AB} = \frac{wL}{2} \qquad FEV_{BA} = \frac{wL}{2} \qquad FEM_{AB} = \frac{wL^2}{12} \qquad FEM_{BA} = \frac{wL^2}{12}$$

$$V(x) = FEV_{AB} - wx; \quad 0 \le x \le L$$

$$M(x) = \frac{wx}{2}(L-x) - FEM_{AB}; \quad 0 \le x \le L$$

$$u_y(x) = \frac{x^2}{2EI}\left(-\frac{wx^2}{12} + \frac{FEV_{AB}\,x}{3} - FEM_{AB}\right); \quad 0 \le x \le L$$

FIGURE 20.14 Response of fixed-end beam with uniformly distributed load.

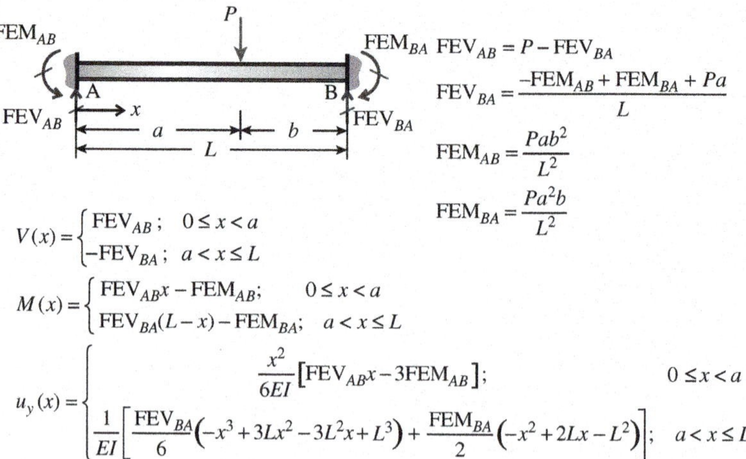

FIGURE 20.15 Response of fixed-end beam with point load.

QUICK NOTE

The beam responses of shear, moment, and displacement are shown in Figures 20.16, 20.17, and 20.18, respectively. Each of these figures shows three plots. The first plot shows the response of the beam to the statically equivalent loads at the joints (i.e., matrix analysis). The second shows the beam response when all joints are fixed against motion (i.e., responses described in Figures 20.13 and 20.14). The third is a superposition of the first two and represents the actual response in the original beam. *Note that all three plots use the same scale so that the reader may gain a sense of the relative contributions of the first and second parts.*

The equations given in Figures 20.14 and 20.15 can be used to plot the shear, moment, and deflection diagrams for the fixed-ended beams. Thus, the result can be superimposed with those results from the matrix analysis. Figures 20.16 through 20.18 show these diagrams for the beam analyzed in Examples 20.6 and 20.7. Note that each diagram shows the results from the matrix analysis, the fixed-end condition, and then the superimposed (i.e., final and correct) condition.

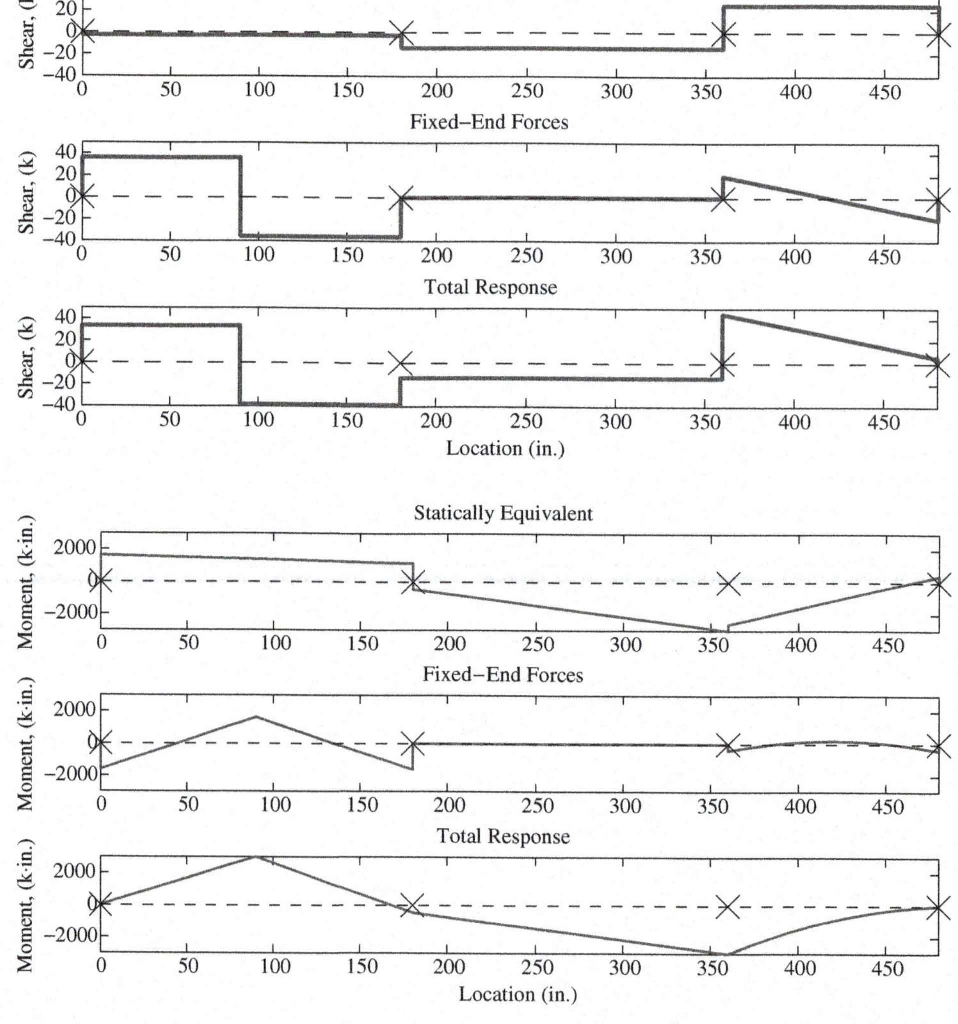

FIGURE 20.16 Superimposed responses to obtain the shear diagram.

FIGURE 20.17 Superimposed responses to obtain the moment diagram.

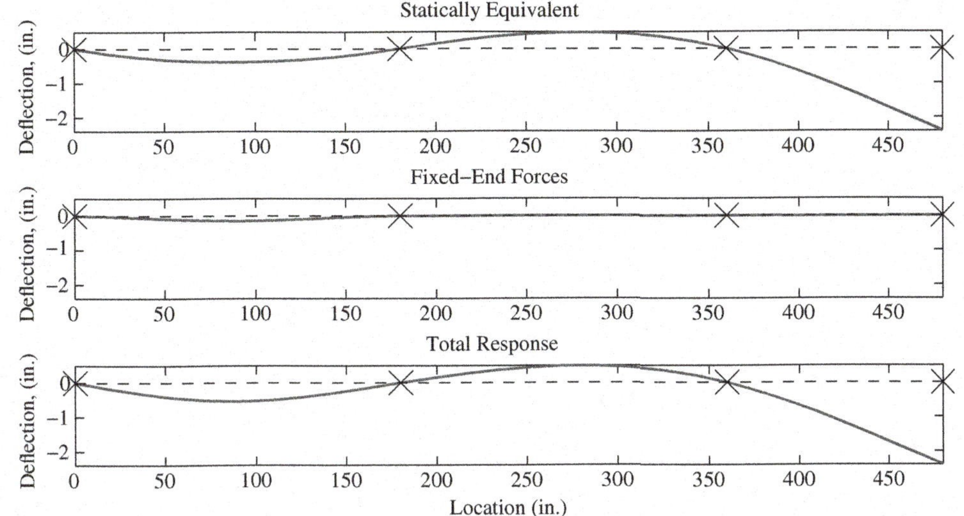

FIGURE 20.18 Superimposed responses to obtain the deflection diagram.

20.8 Stiffness Matrix for Combined Axial and Flexural (Frame) Elements

In Chapter 19, the assembly of a stiffness matrix for a truss element was generated. Section 20.2 of this chapter presented the stiffness matrix for a beam element. A more generic element, the *frame element*, easily can be developed by combining the behaviors of a truss and a beam element. It is helpful to note that a truss bar can only sustain axial load, while a beam element can only sustain transverse loads and applied moments. It stands to reason that, if a frame element is a combination of the truss and beam, the frame element can sustain internal axial force, shear force, and bending moment.

The stiffness matrix for a frame element given in local coordinates is essentially a superposition of the truss stiffness and the beam stiffness matrices previously given. This matrix, which is a 6 × 6, is given here.

RECALL

A *truss element* only has internal axial load, while a *beam element* only has internal shear and internal moment.

QUICK NOTE

Each joint in a frame has three DOFs—two translational and one rotational.

$$
k = \begin{array}{c} \\ a \\ b \\ c \\ d \\ e \\ f \end{array}
\begin{array}{c} \begin{array}{cccccc} a & b & c & d & e & f \end{array} \\
\begin{bmatrix}
\frac{AE}{L} & 0 & 0 & -\frac{AE}{L} & 0 & 0 \\
0 & \frac{12EI}{L^3} & \frac{6EI}{L^2} & 0 & -\frac{12EI}{L^3} & \frac{6EI}{L^2} \\
0 & \frac{6EI}{L^2} & \frac{4EI}{L} & 0 & -\frac{6EI}{L^2} & \frac{2EI}{L} \\
-\frac{AE}{L} & 0 & 0 & \frac{AE}{L} & 0 & 0 \\
0 & -\frac{12EI}{L^3} & -\frac{6EI}{L^2} & 0 & \frac{12EI}{L^3} & -\frac{6EI}{L^2} \\
0 & \frac{6EI}{L^2} & \frac{2EI}{L} & 0 & -\frac{6EI}{L^2} & \frac{4EI}{L}
\end{bmatrix} \end{array}
$$

DID YOU KNOW?

A *frame element* is sometimes referred to as a *beam-column* element. This reference is obvious if one recognizes that an ideal column is an axial member just like a truss.

DID YOU KNOW?

A careful observation of the frame-element stiffness matrix will show that no cross-terms exist between the axial and/or shear. This is in direct consequence of the assumption of small displacements.

Force and Displacement Vectors

The local force vector for a frame member may be presented as follows. Refer back to Figure 20.7 for a presentation and discussion on the conversion between matrix sign convention represented

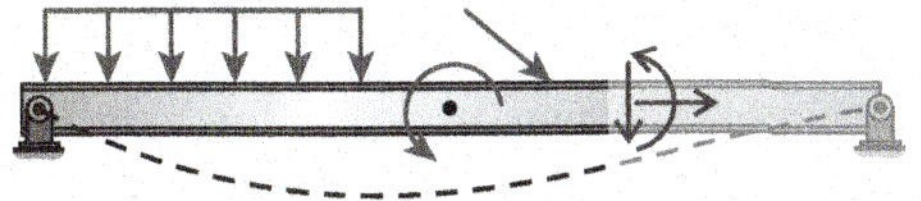

FIGURE 20.19 Typical frame element action.

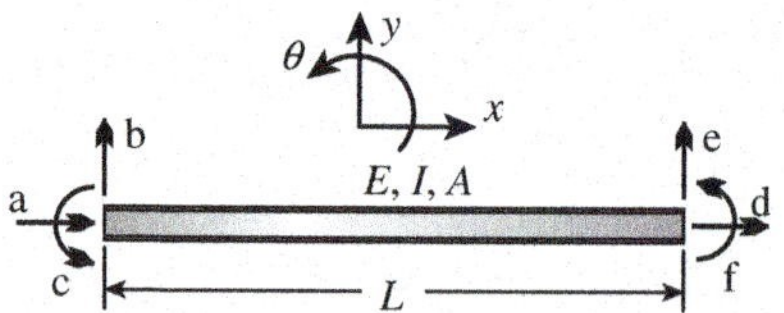

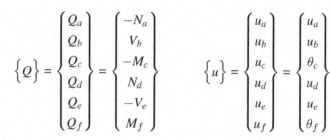

FIGURE 20.20 DOFs for a typical frame section given in local coordinates.

with a Q and beam sign convention represented with V, M, and N.

QUICK NOTE
For the member end forces, one needs to convert the results from the matrix operations into beam sign convention prior to creating shear and moment diagrams. The negative signs shown in the force vector reflect the requisite conversion.

$$\{Q\} = \begin{Bmatrix} Q_a \\ Q_b \\ Q_c \\ Q_d \\ Q_e \\ Q_f \end{Bmatrix} = \begin{Bmatrix} -N_a \\ V_b \\ -M_c \\ N_d \\ -V_e \\ M_f \end{Bmatrix} \qquad \{u\} = \begin{Bmatrix} u_a \\ u_b \\ u_c \\ u_d \\ u_e \\ u_f \end{Bmatrix} = \begin{Bmatrix} u_a \\ u_b \\ \theta_c \\ u_d \\ u_e \\ \theta_f \end{Bmatrix}$$

20.9 Transformation Matrix for Inclined Frame Element

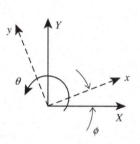

Not all frame members in a structure will be oriented in the horizontal direction. Just as was the case with truss elements in Section 20.6, frame elements must be able to be rotated to a new orientation. A complete derivation could be shown on how to obtain the transformation matrix for a frame element. However, little benefit is achieved by this, since this was already shown for trusses in Section 20.6. The only difference between the truss and frame element is that the frame element also has a rotational DOF at each node.

The rotational DOF does not require a transformation to change from the global to the local coordinate system and vice versa. An examination of Figure 20.21 illustrates that even when the coordinate system is rotated by the angle ϕ, this does not change the frame of reference for the angle θ. This says that the rotation θ is the same whether it is referencing the $x - y$ or the $X - Y$ coordinate system. This phenomenon shows up in the transformation matrix as a 1 being placed at $T_{c,c}$ and $T_{f,f}$. The actual transformation matrix for frames is given here.

FIGURE 20.21 Axes transformation.

RECALL
The Greek letter ϕ represents the orientation of a member relative to the global coordinate system. The Greek letter θ represents the rotation of a joint due to deformation.

$$[T] = \begin{array}{c} \\ 1 \\ 2 \\ 3 \\ 4 \\ 5 \\ 6 \end{array} \begin{array}{c} \begin{array}{cccccc} a & b & c & d & e & f \end{array} \\ \begin{bmatrix} \cos\phi & \sin\phi & 0 & 0 & 0 & 0 \\ -\sin\phi & \cos\phi & 0 & 0 & 0 & 0 \\ 0 & 0 & 1 & 0 & 0 & 0 \\ 0 & 0 & 0 & \cos\phi & \sin\phi & 0 \\ 0 & 0 & 0 & -\sin\phi & \cos\phi & 0 \\ 0 & 0 & 0 & 0 & 0 & 1 \end{bmatrix} \end{array}$$

$$\{\Delta\}^{(n)} = \begin{Bmatrix} \cos\phi \cdot u_a - \sin\phi \cdot ub \\ \sin\phi \cdot u_a + \cos\phi \cdot u_b \\ u_c \\ \cos\phi \cdot u_d - \sin\phi \cdot u_e \\ \sin\phi \cdot u_d + \cos\phi \cdot u_e \\ u_f \end{Bmatrix}$$

$$\{F\}^{(n)} = \begin{Bmatrix} \cos\phi \cdot Q_a - \sin\phi \cdot Q_b \\ \sin\phi \cdot Q_a + \cos\phi \cdot Q_b \\ Q_c \\ \cos\phi \cdot Q_d - \sin\phi \cdot Q_e \\ \sin\phi \cdot Q_d + \cos\phi \cdot Q_e \\ Q_f \end{Bmatrix}$$

With the transformation matrix in hand, the same matrix operations, as previously shown for trusses, will also apply to frames. These matrix operations are repeated here for convenience.

Global to Local Transformation	**Local to Global Transformation**
$\{Q\} = [T]\{F\}$	$\{F\} = [T]^T\{Q\}$
$\{u\} = [T]\{\Delta\}$	$\{\Delta\} = [T]^T\{u\}$
$[k] = [T][K][T]^T$	$[K] = [T]^T[k][T]$

Carrying out the matrix operations shown, a closed form of the element stiffness matrix, element end displacements, and element end forces in the global coordinates is given hereafter.

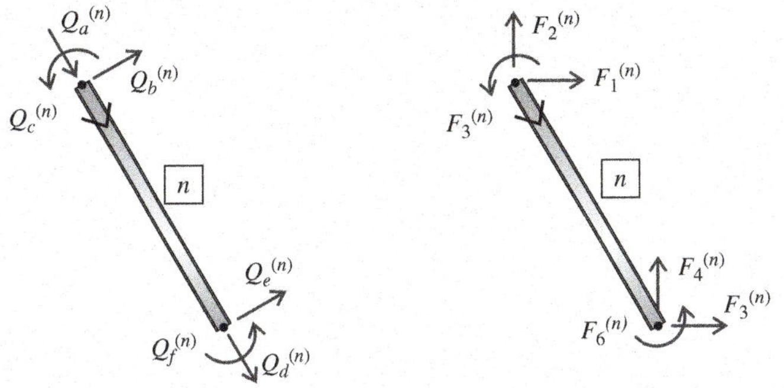

(a) In local coordinates (b) In global coordinates

FIGURE 20.22 Member end forces for a frame element.

RECALL

$$c^2 = \cos^2\phi$$
$$s^2 = \sin^2\phi$$
$$sc = \sin\phi\cos\phi$$

Figure 20.22 represents the effects of these transformations.

$$
\left[K\right]^{(n)} = \begin{array}{c@{}c}
 & \begin{array}{cccccc} 1 & 2 & 3 & 4 & 5 & 6 \end{array} \\
\begin{array}{c} 1 \\ 2 \\ 3 \\ 4 \\ 5 \\ 6 \end{array} &
\left[\begin{array}{cccccc}
c^2\frac{AE}{L} + s^2\frac{12EI}{L^3} & sc\frac{AE}{L} - sc\frac{12EI}{L^3} & -s\frac{6EI}{L^2} & -c^2\frac{AE}{L} - s^2\frac{12EI}{L^3} & -sc\frac{AE}{L} + sc\frac{12EI}{L^3} & -s\frac{6EI}{L^2} \\
sc\frac{AE}{L} - sc\frac{12EI}{L^3} & s^2\frac{AE}{L} + c^2\frac{12EI}{L^3} & c\frac{6EI}{L^2} & -sc\frac{AE}{L} + sc\frac{12EI}{L^3} & -s^2\frac{AE}{L} - c^2\frac{12EI}{L^3} & c\frac{6EI}{L^2} \\
-s\frac{6EI}{L^2} & c\frac{6EI}{L^2} & \frac{4EI}{L} & s\frac{6EI}{L^2} & -c\frac{6EI}{L^2} & \frac{2EI}{L} \\
-c^2\frac{AE}{L} - s^2\frac{12EI}{L^3} & -sc\frac{AE}{L} + sc\frac{12EI}{L^3} & s\frac{6EI}{L^2} & c^2\frac{AE}{L} + s^2\frac{12EI}{L^3} & sc\frac{AE}{L} - sc\frac{12EI}{L^3} & s\frac{6EI}{L^2} \\
-sc\frac{AE}{L} + sc\frac{12EI}{L^3} & -s^2\frac{AE}{L} - c^2\frac{12EI}{L^3} & -c\frac{6EI}{L^2} & sc\frac{AE}{L} - sc\frac{12EI}{L^3} & s^2\frac{AE}{L} + c^2\frac{12EI}{L^3} & -c\frac{6EI}{L^2} \\
-s\frac{6EI}{L^2} & c\frac{6EI}{L^2} & \frac{2EI}{L} & s\frac{6EI}{L^2} & -c\frac{6EI}{L^2} & \frac{4EI}{L}
\end{array}\right]
\end{array}
$$

20.10 Matrix Stiffness Method Applied to Frames

The stiffness method is applied to frames the same as it was applied to the trusses and beams in that it uses all of the same matrix operations. The next couple of examples demonstrate this application.

EXAMPLE 20.8

For the given frame, use the matrix formulation to solve for the nodal displacements, the reactions, and the member-end forces. The DOFs are labeled on the structure.

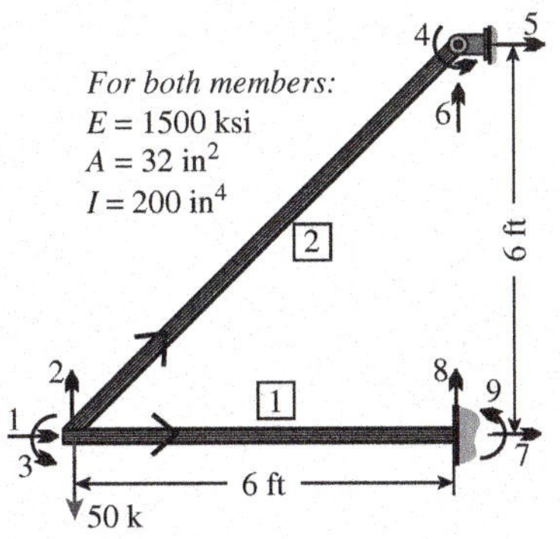

MEMBER INFORMATION

The following geometric member properties are derived.

Member 1:

$$L = 72 \text{ in.}$$
$$\phi = 0^\circ$$

Member 2:

$$L = 101.8 \text{ in.}$$
$$\phi = 45^\circ$$

SOLUTION STRATEGY

Use the labeled DOFs, the indicated member origins, and member properties to develop the local-element stiffness matrices, the transformation matrices, and the global-element stiffness matrices. The same procedure that was used for trusses and beams may be used here. *Get all units consistent* at the beginning of the analysis. Units of kips and inches will be used throughout this example.

$$k^{(1)} = \begin{array}{c} \\ a \\ b \\ c \\ d \\ e \\ f \end{array} \begin{array}{c} \begin{array}{cccccc} a & b & c & d & e & f \end{array} \\ \begin{bmatrix} 666.7 & 0.0 & 0.0 & -666.7 & 0.0 & 0.0 \\ 0.0 & 9.6 & 347.2 & 0.0 & -9.6 & 347.2 \\ 0.0 & 347.2 & 16666.7 & 0.0 & -347.2 & 8333.3 \\ -666.7 & 0.0 & 0.0 & 666.7 & 0.0 & 0.0 \\ 0.0 & -9.6 & -347.2 & 0.0 & 9.6 & -347.2 \\ 0.0 & 347.2 & 8333.3 & 0.0 & -347.2 & 16666.7 \end{bmatrix} \end{array}$$

QUICK NOTE

The element stiffness matrices are computed for their respective local coordinate systems in units of kips and inches.

$$k^{(2)} = \begin{array}{c} \\ a \\ b \\ c \\ d \\ e \\ f \end{array} \begin{array}{c} \begin{array}{cccccc} a & b & c & d & e & f \end{array} \\ \begin{bmatrix} 471.4 & 0.0 & 0.0 & -471.4 & 0.0 & 0.0 \\ 0.0 & 3.4 & 173.6 & 0.0 & -3.4 & 173.6 \\ 0.0 & 173.6 & 11785.1 & 0.0 & -173.6 & 5892.6 \\ -471.4 & 0.0 & 0.0 & 471.4 & 0.0 & 0.0 \\ 0.0 & -3.4 & -173.6 & 0.0 & 3.4 & -173.6 \\ 0.0 & 173.6 & 5892.6 & 0.0 & -173.6 & 11785.1 \end{bmatrix} \end{array}$$

$$T_1 = \begin{array}{c} \\ a \\ b \\ c \\ d \\ e \\ f \end{array} \begin{array}{c} \begin{array}{cccccc} 1 & 2 & 3 & 7 & 8 & 9 \end{array} \\ \begin{bmatrix} 1.00 & 0.00 & 0.00 & 0.00 & 0.00 & 0.00 \\ 0.00 & 1.00 & 0.00 & 0.00 & 0.00 & 0.00 \\ 0.00 & 0.00 & 1.00 & 0.00 & 0.00 & 0.00 \\ 0.00 & 0.00 & 0.00 & 1.00 & 0.00 & 0.00 \\ 0.00 & 0.00 & 0.00 & 0.00 & 1.00 & 0.00 \\ 0.00 & 0.00 & 0.00 & 0.00 & 0.00 & 1.00 \end{bmatrix} \end{array} \quad T_2 = \begin{array}{c} \\ a \\ b \\ c \\ d \\ e \\ f \end{array} \begin{array}{c} \begin{array}{cccccc} 1 & 2 & 3 & 5 & 6 & 4 \end{array} \\ \begin{bmatrix} 0.71 & 0.71 & 0.00 & 0.00 & 0.00 & 0.00 \\ -0.71 & 0.71 & 0.00 & 0.00 & 0.00 & 0.00 \\ 0.00 & 0.00 & 1.00 & 0.00 & 0.00 & 0.00 \\ 0.00 & 0.00 & 0.00 & 0.71 & 0.71 & 0.00 \\ 0.00 & 0.00 & 0.00 & -0.71 & 0.71 & 0.00 \\ 0.00 & 0.00 & 0.00 & 0.00 & 0.00 & 1.00 \end{bmatrix} \end{array}$$

DID YOU KNOW?

The ordering of the DOFs in the global element stiffness matrix should be made as follows. This is due to the way the transformation matrix $[T]$ was derived in Section 20.9.

- Starting node—horizontal
- Starting node—vertical
- Starting node—rotation
- Ending node—horizontal
- Ending node—vertical
- Ending node—rotation

For element 1, this would produce an order of 1, 2, 3, 5, 6, 4.

The global element stiffness matrices are found by using the matrix operation:

$$[K]^{(n)} = [T]_n^T [k]^{(n)} [T]_n$$

where n represents the nth element. Thus,

$$K^{(1)} = \begin{array}{c} \\ 1 \\ 2 \\ 3 \\ 7 \\ 8 \\ 9 \end{array} \begin{array}{c} \begin{array}{cccccc} 1 & 2 & 3 & 7 & 8 & 9 \end{array} \\ \begin{bmatrix} 667 & 0 & 0 & -667 & 0 & 0 \\ 0 & 10 & 347 & 0 & -10 & 347 \\ 0 & 347 & 16667 & 0 & -347 & 8333 \\ -667 & 0 & 0 & 667 & 0 & 0 \\ 0 & -10 & -347 & 0 & 10 & -347 \\ 0 & 347 & 8333 & 0 & -347 & 16667 \end{bmatrix} \end{array}$$

$$K^{(2)} = \begin{array}{c} \\ 1 \\ 2 \\ 3 \\ 5 \\ 6 \\ 4 \end{array} \begin{array}{c} \begin{array}{cccccc} 1 & 2 & 3 & 5 & 6 & 4 \end{array} \\ \begin{bmatrix} 237 & 234 & -123 & -237 & -234 & -123 \\ 234 & 237 & 123 & -234 & -237 & 123 \\ -123 & 123 & 11785 & 123 & -123 & 5893 \\ -237 & -234 & 123 & 237 & 234 & 123 \\ -234 & -237 & -123 & 234 & 237 & -123 \\ -123 & 123 & 5893 & 123 & -123 & 11785 \end{bmatrix} \end{array}$$

We could now assemble the entire structure level stiffness matrix at this point, but we know that we really only need $\left[K_{uu}\right]$ and $\left[K_{ru}\right]$. Thus, we assemble these matrices directly using the same approach as used in all previous examples.

$$K_{i,j} = \sum_{n=1}^{\text{number of elements}} K_{i,j}^{(n)}$$

Where there are two elements, the unrestrained DOFs are 1 through 4 and the restrained DOFs are 5 through 9.

$$K_{uu} = \begin{array}{c} 1 \\ 2 \\ 3 \\ 4 \end{array}\overset{\begin{array}{cccc} 1 & 2 & 3 & 4 \end{array}}{\begin{bmatrix} 904.1 & 234.0 & -122.8 & -122.8 \\ 234.0 & 247.1 & 470.0 & 122.8 \\ -122.8 & 470.0 & 28451.8 & 5892.6 \\ -122.8 & 122.8 & 5892.6 & 11785.1 \end{bmatrix}} \qquad K_{ru} = \begin{array}{c} 5 \\ 6 \\ 7 \\ 8 \\ 9 \end{array}\overset{\begin{array}{cccc} 1 & 2 & 3 & 4 \end{array}}{\begin{bmatrix} -237.4 & -234.0 & 122.8 & 122.8 \\ -234.0 & -237.4 & -122.8 & -122.8 \\ -666.7 & 0.0 & 0.0 & 0.0 \\ 0.0 & -9.6 & -347.2 & 0.0 \\ 0.0 & 347.2 & 8333.3 & 0.0 \end{bmatrix}}$$

The force vector for the unrestrained DOFs is assembled by observing the applied loads that exist on the structure. Then the displacement vector may be computed.

$$\left\{F_u\right\} = \begin{array}{c} 1 \\ 2 \\ 3 \\ 4 \end{array}\begin{Bmatrix} 0.0 \\ -50.0 \\ 0.0 \\ 0.0 \end{Bmatrix}\begin{array}{l} \text{k} \\ \text{k} \\ \text{k·in.} \\ \text{k·in.} \end{array} \qquad \left\{\Delta_u\right\} = \left[K_{uu}\right]^{-1}\left\{F_u\right\} = \begin{Bmatrix} \Delta_1 \\ \Delta_2 \\ \theta_3 \\ \theta_4 \end{Bmatrix} = \begin{Bmatrix} 0.0738 \\ -0.2819 \\ 4.69E-03 \\ 1.36E-03 \end{Bmatrix}\begin{array}{l} \text{in.} \\ \text{in.} \\ \text{rad} \\ \text{rad} \end{array}$$

The reactions are now found in the normal fashion.

$$\left\{F_r\right\} = \begin{Bmatrix} R_5 \\ R_6 \\ R_7 \\ R_8 \\ M_9 \end{Bmatrix} = \left[K_{ru}\right]\left\{\Delta_u\right\} = \begin{Bmatrix} 49.18 \\ 48.91 \\ -49.18 \\ 1.09 \\ -58.76 \end{Bmatrix}\begin{array}{l} \text{k} \\ \text{k} \\ \text{k} \\ \text{k} \\ \text{k·in.} \end{array}$$

The last issue to tackle is the computation of the member end forces. This requires the assemblage of a global displacement vector for each member. This is done by observing the DOFs associated with each member and either pulling the values from the vector Δ_u or from the known displacements of the restrained DOFs (i.e., 0.0).

$$\Delta^{(1)} = \begin{Bmatrix} \Delta_1 \\ \Delta_2 \\ \theta_3 \\ \Delta_7 \\ \Delta_8 \\ \theta_9 \end{Bmatrix} = \begin{Bmatrix} 0.0738 \\ -0.2819 \\ 4.69E-03 \\ 0.0 \\ 0.0 \\ 0.0 \end{Bmatrix}\begin{array}{l} \text{in.} \\ \text{in.} \\ \text{rad} \\ \text{in.} \\ \text{in.} \\ \text{rad} \end{array} \qquad \Delta^{(2)} = \begin{Bmatrix} \Delta_1 \\ \Delta_2 \\ \theta_3 \\ \Delta_5 \\ \Delta_6 \\ \theta_4 \end{Bmatrix} = \begin{Bmatrix} 0.0738 \\ -0.2819 \\ 4.69E-03 \\ 0.0 \\ 0.0 \\ 1.36E-03 \end{Bmatrix}\begin{array}{l} \text{in.} \\ \text{in.} \\ \text{rad} \\ \text{in.} \\ \text{in.} \\ \text{rad} \end{array}$$

These displacement vectors are transformed to the local coordinates. Then these are used in conjunction with the element stiffness matrices to find the member end forces.

$$\begin{Bmatrix} u_a \\ u_b \\ \theta_c \\ u_d \\ u_e \\ \theta_f \end{Bmatrix}^{(1)} = [T]_1\{\Delta\}^{(1)} = \begin{Bmatrix} 0.0738 \\ -0.2819 \\ 4.69E-03 \\ 0.0 \\ 0.0 \\ 0.0 \end{Bmatrix} \begin{matrix} \text{in.} \\ \text{in.} \\ \text{rad} \\ \text{in.} \\ \text{in.} \\ \text{rad} \end{matrix} \qquad \begin{Bmatrix} u_a \\ u_b \\ \theta_c \\ u_d \\ u_e \\ \theta_f \end{Bmatrix}^{(2)} = [T]_2\{\Delta\}^{(2)} = \begin{Bmatrix} -0.1471 \\ -0.2515 \\ 4.69E-03 \\ 0.0 \\ 0.0 \\ 1.36E-03 \end{Bmatrix} \begin{matrix} \text{in.} \\ \text{in.} \\ \text{rad} \\ \text{in.} \\ \text{in.} \\ \text{rad} \end{matrix}$$

$$\begin{Bmatrix} Q_a \\ Q_b \\ M_c \\ Q_d \\ Q_e \\ M_f \end{Bmatrix}^{(1)} = \begin{Bmatrix} 49.18 \\ -1.09 \\ -19.65 \\ -49.18 \\ 1.09 \\ -58.76 \end{Bmatrix} \begin{matrix} \text{k} \\ \text{k} \\ \text{k·in.} \\ \text{k} \\ \text{k} \\ \text{k·in.} \end{matrix} \qquad \begin{Bmatrix} Q_a \\ Q_b \\ M_c \\ Q_d \\ Q_e \\ M_f \end{Bmatrix} = \begin{Bmatrix} -69.36 \\ 0.19 \\ 19.65 \\ 69.36 \\ -0.19 \\ 0.00 \end{Bmatrix} \begin{matrix} \text{k} \\ \text{k} \\ \text{k·in.} \\ \text{k} \\ \text{k} \\ \text{k·in.} \end{matrix}$$

A graphical interpretation of the reactions and the member internal forces for this frame is given here. Recall that the directions are interpreted using the right-hand coordinate system.

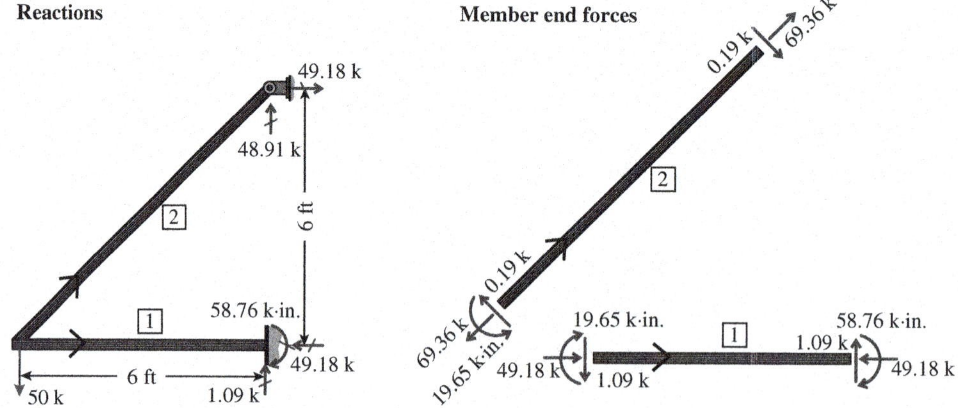

EXAMPLE 20.9

For the frame structure shown, compute the nodal displacements, reactions, and member end forces using the matrix method.

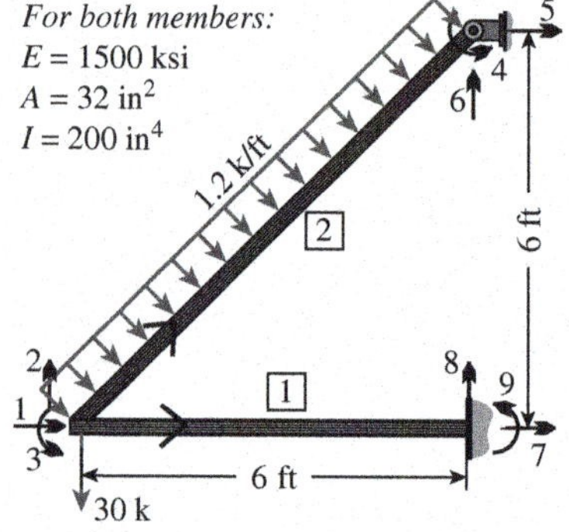

SOLUTION STRATEGY

This frame has the same configuration and the same properties as the one presented in Example 20.8. Use all of the previously developed matrices to solve this problem. The focus here is to come up with the statically equivalent loads F_{FEF} due to the distributed load. Follow the same procedure as was followed for beams in Section 20.6. The only difference is that the FEFs will need to be transformed from the local coordinate system to the global coordinate system before they can be used.

QUICK NOTE

Since we already have the matrices used for analysis, we can skip directly to finding the statically equivalent loads due to the load on member 2. The DOFs at each end are locked, and from the beam chart, the following values are obtained in the *local coordinate system*.

FEM:

$$\frac{wL^2}{12} = \frac{(0.1\text{ k/in.})(101.8\text{ in.})^2}{12} = 86.4\text{ k·in.}$$

FEV:

$$\frac{wL}{2} = \frac{(0.1\text{ k/in.})(101.8\text{ in.})}{2} = 5.09\text{ k}$$

$$Q^{(2)}_{FEF} = \begin{Bmatrix} 0.0 \\ 5.09 \\ 86.4 \\ 0.0 \\ 5.09 \\ -86.4 \end{Bmatrix} \begin{matrix} \text{k} \\ \text{k} \\ \text{k·in.} \\ \text{k} \\ \text{k} \\ \text{k·in.} \end{matrix}$$

1.2 k/ft, 2, 101.8 in., 5.09 k, 0 k, 86.4 k·in., 5.09 k, 0 k, 86.4 k·in.

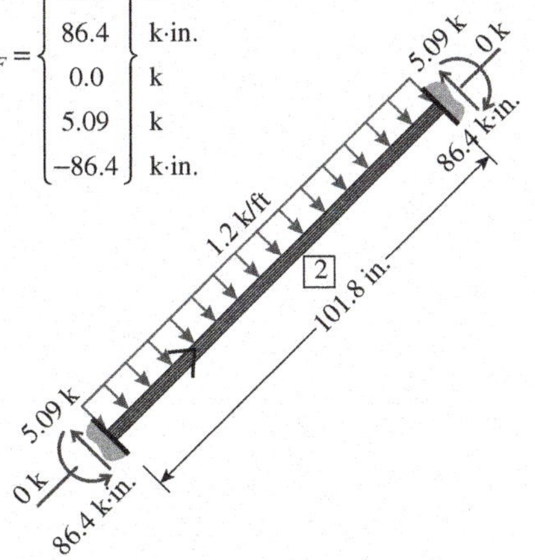

This force vector has been found in the local coordinate system, but it must be transformed to the global coordinate system prior to its use in the matrix formulation.

$$\{F_{\text{FEF}}\}^{(2)} = [T]_2^T Q^{(2)}_{\text{FEF}} = \begin{matrix} 1 \\ 2 \\ 3 \\ 5 \\ 6 \\ 4 \end{matrix} \begin{Bmatrix} -3.60 \\ 3.60 \\ 86.40 \\ -3.60 \\ 3.60 \\ -86.40 \end{Bmatrix} \begin{matrix} \text{k} \\ \text{k} \\ \text{k·in.} \\ \text{k} \\ \text{k} \\ \text{k·in.} \end{matrix}$$

The next step is to assemble the entire force vector for the unrestrained DOFs.

RECALL

The first force vector, F_u, represents the loads that are directly applied to DOFs, while the second force vector, $F_{\text{FEF}-u}$, represents the statically equivalent forces at the unrestrained DOFs.

$$\{F_u\} - \{F_{\text{FEF}-u}\} = \begin{matrix} 1 \\ 2 \\ 3 \\ 4 \end{matrix} \begin{Bmatrix} 0.0 \\ -30.0 \\ 0.0 \\ 0.0 \end{Bmatrix} - \begin{Bmatrix} -3.60 \\ 3.60 \\ 86.40 \\ -86.40 \end{Bmatrix} = \begin{Bmatrix} 3.60 \\ -33.60 \\ -86.40 \\ 86.40 \end{Bmatrix} \begin{matrix} \text{k} \\ \text{k} \\ \text{k·in.} \\ \text{k·in.} \end{matrix}$$

With the force vector in hand, the next step is to compute the nodal displacements and then use those displacements to compute the reactions.

RECALL

The K_{uu} and the K_{ru} matrices were found in Example 20.8

$$\{\Delta_u\} = \begin{Bmatrix} \Delta_1 \\ \Delta_2 \\ \theta_3 \\ \theta_4 \end{Bmatrix} = [K_{uu}]^{-1} \begin{Bmatrix} 3.60 \\ -33.60 \\ -86.40 \\ 86.40 \end{Bmatrix} = \begin{Bmatrix} 0.0541 \\ -0.1890 \\ -1.92E-03 \\ 1.08E-02 \end{Bmatrix} \begin{matrix} \text{in.} \\ \text{in.} \\ \text{rad} \\ \text{rad} \end{matrix}$$

DON'T FORGET

Remember that to find the true reactions of the structure we must superimpose these back with the FEFs located at the restrained DOFs, which in this case are DOFs 5 and 6.

$$\{F_r\} = \begin{Bmatrix} R_5 \\ R_6 \\ R_7 \\ R_8 \\ M_9 \end{Bmatrix} = [K_{ru}] \begin{Bmatrix} 0.0541 \\ -0.1890 \\ -1.92E-03 \\ 1.08E-02 \end{Bmatrix} + \begin{Bmatrix} -3.60 \\ 3.60 \\ 0.00 \\ 0.00 \\ 0.00 \end{Bmatrix} = \begin{Bmatrix} 28.87 \\ 34.71 \\ -36.06 \\ 2.49 \\ -81.64 \end{Bmatrix} \begin{matrix} \text{k} \\ \text{k} \\ \text{k} \\ \text{k} \\ \text{k·in.} \end{matrix}$$

The member end forces are found by taking the global displacements Δ, converting to local displacements u, and using the element local stiffness matrix as shown explicitly in Example 20.8. The unique step in this problem is that the member end forces found from the matrix operations then need to be superimposed with with the FEFs for each member given in local coordinates.

$$\left\{Q\right\}^{(1)} = [k]^{(1)}\{u\}^{(1)} + \left\{Q\right\}^{(1)}_{\text{FEF}} = \begin{Bmatrix} 36.06 \\ -2.49 \\ -97.68 \\ -36.06 \\ 2.49 \\ -81.64 \end{Bmatrix} + \begin{Bmatrix} 0.0 \\ 0.0 \\ 0.0 \\ 0.0 \\ 0.0 \\ 0.0 \end{Bmatrix} = \begin{Bmatrix} 36.06 \\ -2.49 \\ -97.68 \\ -36.06 \\ 2.49 \\ -81.64 \end{Bmatrix} \begin{matrix} \text{k} \\ \text{k} \\ \text{k}\cdot\text{in.} \\ \text{k} \\ \text{k} \\ \text{k}\cdot\text{in.} \end{matrix}$$

$$\left\{Q\right\}^{(2)} = [k]^{(2)}\{u\}^{(2)} + \left\{Q\right\}^{(2)}_{\text{FEF}} = \begin{Bmatrix} -44.95 \\ 0.96 \\ 11.28 \\ 44.95 \\ -0.96 \\ 86.40 \end{Bmatrix} + \begin{Bmatrix} 0.0 \\ 5.09 \\ 86.4 \\ 0.0 \\ 5.09 \\ -86.4 \end{Bmatrix} = \begin{Bmatrix} -44.95 \\ 6.05 \\ 97.68 \\ 44.95 \\ 4.13 \\ 0.00 \end{Bmatrix} \begin{matrix} \text{k} \\ \text{k} \\ \text{k}\cdot\text{in.} \\ \text{k} \\ \text{k} \\ \text{k}\cdot\text{in.} \end{matrix}$$

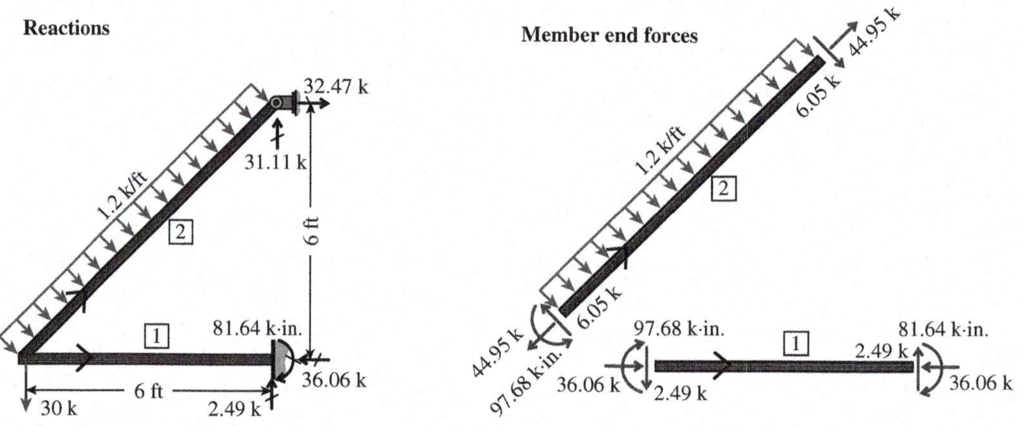

20.11 Spreadsheet Computer Applications

The procedure for implementing the matrix analysis procedure within a spreadsheet is fully explained in Section 19.10. The procedure was demonstrated on a truss. For completeness sake, the structure given in Example 20.9 is shown in Figures 20.23 through 20.27.

Element 1

x_{start} =	0	in	x_{end} =	72	in
y_{start} =	0	in	y_{end} =	0	in
L =	72	in	A =	32.0	in^2
E =	1500	ksi	I =	200	in^4

$k^{(1)}$ =

	a	b	c	d	e	f
a	666.7	0.0	0.0	-666.7	0.0	0.0
b	0.0	9.6	347.2	0.0	-9.6	347.2
c	0.0	347.2	16666.7	0.0	-347.2	8333.3
d	-666.7	0.0	0.0	666.7	0.0	0.0
e	0.0	-9.6	-347.2	0.0	9.6	-347.2
f	0.0	347.2	8333.3	0.0	-347.2	16666.7

T_1 =

	1	2	3	7	8	9
a	1.00	0.00	0.00	0.00	0.00	0.00
b	0.00	1.00	0.00	0.00	0.00	0.00
c	0.00	0.00	1.00	0.00	0.00	0.00
d	0.00	0.00	0.00	1.00	0.00	0.00
e	0.00	0.00	0.00	0.00	1.00	0.00
f	0.00	0.00	0.00	0.00	0.00	1.00

$K^{(1)}$ =

	1	2	3	7	8	9
1	667	0	0	-667	0	0
2	0	10	347	0	-10	347
3	0	347	16667	0	-347	8333
7	-667	0	0	667	0	0
8	0	-10	-347	0	10	-347
9	0	347	8333	0	-347	16667

Element 2

x_{start} =	0	in	x_{end} =	72	in
y_{start} =	0	in	y_{end} =	72	in
L =	101.8	in	A =	32.0	in^2
E =	1500	ksi	I =	200	in^4

$k^{(2)}$ =

	a	b	c	d	e	f
a	471.4	0.0	0.0	-471.4	0.0	0.0
b	0.0	3.4	173.6	0.0	-3.4	173.6
c	0.0	173.6	11785.1	0.0	-173.6	5892.6
d	-471.4	0.0	0.0	471.4	0.0	0.0
e	0.0	-3.4	-173.6	0.0	3.4	-173.6
f	0.0	173.6	5892.6	0.0	-173.6	11785.1

T_2 =

	1	2	3	5	6	4
a	0.71	0.71	0.00	0.00	0.00	0.00
b	-0.71	0.71	0.00	0.00	0.00	0.00
c	0.00	0.00	1.00	0.00	0.00	0.00
d	0.00	0.00	0.00	0.71	0.71	0.00
e	0.00	0.00	0.00	-0.71	0.71	0.00
f	0.00	0.00	0.00	0.00	0.00	1.00

$K^{(2)}$ =

	1	2	3	5	6	4
1	237	234	-123	-237	-234	-123
2	234	237	123	-234	-237	123
3	-123	123	11785	123	-123	5893
5	-237	-234	123	237	234	123
6	-234	-237	-123	234	237	-123
4	-123	123	5893	123	-123	11785

FIGURE 20.23 Element stiffness matrices.

K =

	1	2	3	4	5	6	7	8	9
1	904.1	234.0	-122.8	-122.8	-237.4	-234.0	-666.7	0.0	0.0
2	234.0	247.1	470.0	122.8	-234.0	-237.4	0.0	-9.6	347.2
3	-122.8	470.0	28451.8	5892.6	122.8	-122.8	0.0	-347.2	8333.3
4	-122.8	122.8	5892.6	11785.1	122.8	-122.8	0.0	0.0	0.0
5	-237.4	-234.0	122.8	122.8	237.4	234.0	0.0	0.0	0.0
6	-234.0	-237.4	-122.8	-122.8	234.0	237.4	0.0	0.0	0.0
7	-666.7	0.0	0.0	0.0	0.0	0.0	666.7	0.0	0.0
8	0.0	-9.6	-347.2	0.0	0.0	0.0	0.0	9.6	-347.2
9	0.0	347.2	8333.3	0.0	0.0	0.0	0.0	-347.2	16666.7

FIGURE 20.24 Global stiffness matrix.

$$\{F_{FEF}\} = [T]^T \{Q_{FEF}\}$$

$$\{Q_{FEF}\}^{(2)} = \begin{Bmatrix} 0.0 \text{ k} \\ 5.09 \text{ k} \\ 86.4 \text{ k-in} \\ 0.0 \text{ k} \\ 5.09 \text{ k} \\ -86.4 \text{ k-in} \end{Bmatrix} \qquad \{F_{FEF}\}^{(2)} = \begin{Bmatrix} -3.60 \text{ k} \\ 3.60 \text{ k} \\ 86.40 \text{ k-in} \\ -3.60 \text{ k} \\ 3.60 \text{ k} \\ -86.40 \text{ k-in} \end{Bmatrix}$$

FIGURE 20.25 Fixed-end forces resulting from the distributed load shown in local and global coordinates.

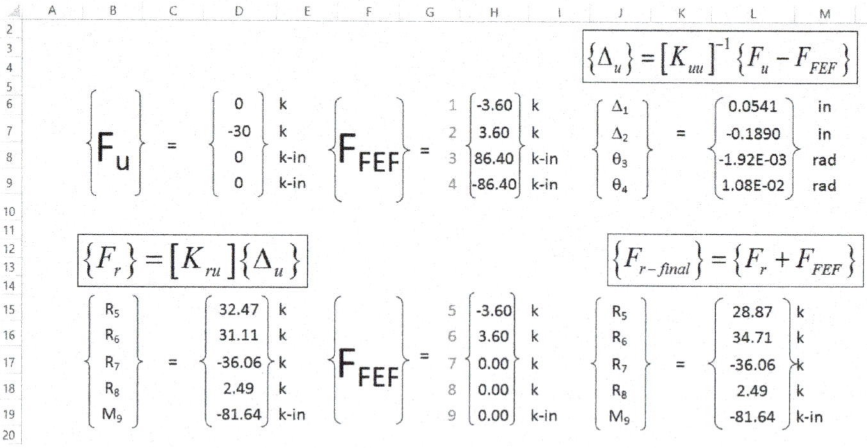

$$\{\Delta_u\} = [K_{uu}]^{-1}\{F_u - F_{FEF}\}$$

$$\{F_u\} = \begin{Bmatrix} 0 \\ -30 \\ 0 \\ 0 \end{Bmatrix} \begin{matrix} \text{k} \\ \text{k} \\ \text{k-in} \\ \text{k-in} \end{matrix} \qquad \{F_{FEF}\} = \begin{Bmatrix} -3.60 \\ 3.60 \\ 86.40 \\ -86.40 \end{Bmatrix} \begin{matrix} \text{k} \\ \text{k} \\ \text{k-in} \\ \text{k-in} \end{matrix} \qquad \begin{Bmatrix} \Delta_1 \\ \Delta_2 \\ \theta_3 \\ \theta_4 \end{Bmatrix} = \begin{Bmatrix} 0.0541 \\ -0.1890 \\ -1.92\text{E-03} \\ 1.08\text{E-02} \end{Bmatrix} \begin{matrix} \text{in} \\ \text{in} \\ \text{rad} \\ \text{rad} \end{matrix}$$

$$\{F_r\} = [K_{ru}]\{\Delta_u\} \qquad \{F_{r-final}\} = \{F_r + F_{FEF}\}$$

$$\begin{Bmatrix} R_5 \\ R_6 \\ R_7 \\ R_8 \\ M_9 \end{Bmatrix} = \begin{Bmatrix} 32.47 \\ 31.11 \\ -36.06 \\ 2.49 \\ -81.64 \end{Bmatrix} \begin{matrix} \text{k} \\ \text{k} \\ \text{k} \\ \text{k} \\ \text{k-in} \end{matrix} \qquad \{F_{FEF}\} = \begin{Bmatrix} -3.60 \\ 3.60 \\ 0.00 \\ 0.00 \\ 0.00 \end{Bmatrix} \begin{matrix} \text{k} \\ \text{k} \\ \text{k} \\ \text{k} \\ \text{k-in} \end{matrix} \qquad \begin{Bmatrix} R_5 \\ R_6 \\ R_7 \\ R_8 \\ M_9 \end{Bmatrix} = \begin{Bmatrix} 28.87 \\ 34.71 \\ -36.06 \\ 2.49 \\ -81.64 \end{Bmatrix} \begin{matrix} \text{k} \\ \text{k} \\ \text{k} \\ \text{k} \\ \text{k-in} \end{matrix}$$

FIGURE 20.26 Basic matrix operations required to solve for joint displacements and reactions.

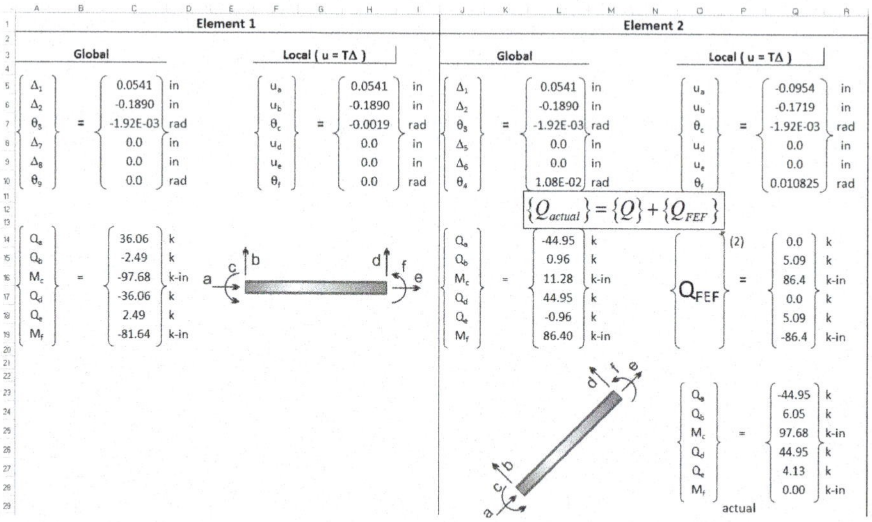

Element 1

Global:
$$\begin{Bmatrix} \Delta_1 \\ \Delta_2 \\ \theta_3 \\ \Delta_7 \\ \Delta_8 \\ \theta_9 \end{Bmatrix} = \begin{Bmatrix} 0.0541 \\ -0.1890 \\ -1.92\text{E-03} \\ 0.0 \\ 0.0 \\ 0.0 \end{Bmatrix} \begin{matrix} \text{in} \\ \text{in} \\ \text{rad} \\ \text{in} \\ \text{in} \\ \text{rad} \end{matrix}$$

Local (u = TΔ):
$$\begin{Bmatrix} u_a \\ u_b \\ \theta_c \\ u_d \\ u_e \\ \theta_f \end{Bmatrix} = \begin{Bmatrix} 0.0541 \\ -0.1890 \\ -0.0019 \\ 0.0 \\ 0.0 \\ 0.0 \end{Bmatrix} \begin{matrix} \text{in} \\ \text{in} \\ \text{rad} \\ \text{in} \\ \text{in} \\ \text{rad} \end{matrix}$$

$$\begin{Bmatrix} Q_a \\ Q_b \\ M_c \\ Q_d \\ Q_e \\ M_f \end{Bmatrix} = \begin{Bmatrix} 36.06 \\ -2.49 \\ -97.68 \\ -36.06 \\ 2.49 \\ -81.64 \end{Bmatrix} \begin{matrix} \text{k} \\ \text{k} \\ \text{k-in} \\ \text{k} \\ \text{k} \\ \text{k-in} \end{matrix}$$

Element 2

Global:
$$\begin{Bmatrix} \Delta_1 \\ \Delta_2 \\ \theta_3 \\ \Delta_5 \\ \Delta_6 \\ \theta_4 \end{Bmatrix} = \begin{Bmatrix} 0.0541 \\ -0.1890 \\ -1.92\text{E-03} \\ 0.0 \\ 0.0 \\ 1.08\text{E-02} \end{Bmatrix} \begin{matrix} \text{in} \\ \text{in} \\ \text{rad} \\ \text{in} \\ \text{in} \\ \text{rad} \end{matrix}$$

Local (u = TΔ):
$$\begin{Bmatrix} u_a \\ u_b \\ \theta_c \\ u_d \\ u_e \\ \theta_f \end{Bmatrix} = \begin{Bmatrix} -0.0954 \\ -0.1719 \\ -1.92\text{E-03} \\ 0.0 \\ 0.0 \\ 0.010825 \end{Bmatrix} \begin{matrix} \text{in} \\ \text{in} \\ \text{rad} \\ \text{in} \\ \text{in} \\ \text{rad} \end{matrix}$$

$$\{Q_{actual}\} = \{Q\} + \{Q_{FEF}\}$$

$$\begin{Bmatrix} Q_a \\ Q_b \\ M_c \\ Q_d \\ Q_e \\ M_f \end{Bmatrix} = \begin{Bmatrix} -44.95 \\ 0.96 \\ 11.28 \\ 44.95 \\ -0.96 \\ 86.40 \end{Bmatrix} \begin{matrix} \text{k} \\ \text{k} \\ \text{k-in} \\ \text{k} \\ \text{k} \\ \text{k-in} \end{matrix} \qquad \{Q_{FEF}\}^{(2)} = \begin{Bmatrix} 0.0 \\ 5.09 \\ 86.4 \\ 0.0 \\ 5.09 \\ -86.4 \end{Bmatrix} \begin{matrix} \text{k} \\ \text{k} \\ \text{k-in} \\ \text{k} \\ \text{k} \\ \text{k-in} \end{matrix}$$

$$\begin{Bmatrix} Q_a \\ Q_b \\ M_c \\ Q_d \\ Q_e \\ M_f \end{Bmatrix}_{actual} = \begin{Bmatrix} -44.95 \\ 6.05 \\ 97.68 \\ 44.95 \\ 4.13 \\ 0.00 \end{Bmatrix} \begin{matrix} \text{k} \\ \text{k} \\ \text{k-in} \\ \text{k} \\ \text{k} \\ \text{k-in} \end{matrix}$$

FIGURE 20.27 Computation of member end forces, including the adjustment for the FEFs.

20.12 SAP2000 Computer Applications

The handling of loads located between joints, along with the requisite superposition, is automatically handled behind the scenes in SAP2000. The user is only required to define the structure as described and then look at the final results. However, one needs to understand how information is presented within the software.

Both the graphical and tabular forms of the member responses rely on the information being reported at certain locations along the length of the member. These locations are referred to as *stations*. The program uses the following defaults when setting the reporting stations.

- Horizontal members will have stations spaced at no more than 2-ft apart.
- Non-horizontal members will have only three stations—one at each end and at mid-span.
- A station is automatically placed wherever there is a load discontinuity (i.e., point load or change in load intensity).

These defaults are often sufficient and don't require any special consideration. However, when non-horizontal members have distributed loads, the default is insufficient to represent the actual internal forces. For example, the structure discussed in Example 20.9 is modeled in SAP2000 and the moment diagram is generated. Figure 20.28 shows the default station assignments, nine stations per member, and 21 stations per member.

One will readily note that an increase in the number of stations for the horizontal member did not improve our understanding of what is happening in the horizontal member. However, this change did significantly affect our understanding of the internal moment in the inclined member.

RECALL

The matrix method directly solves for the structural response at the ends of members. Shape functions (see Section 20.5) along with shear and moment equations (see Section 20.7) are used to interpolate results between the member ends.

DID YOU KNOW?

A station location is noted as a hatch mark on the diagram which is drawn perpendicular to the member. Thus, the more hatch marks you see, then the more stations are being used to report results.

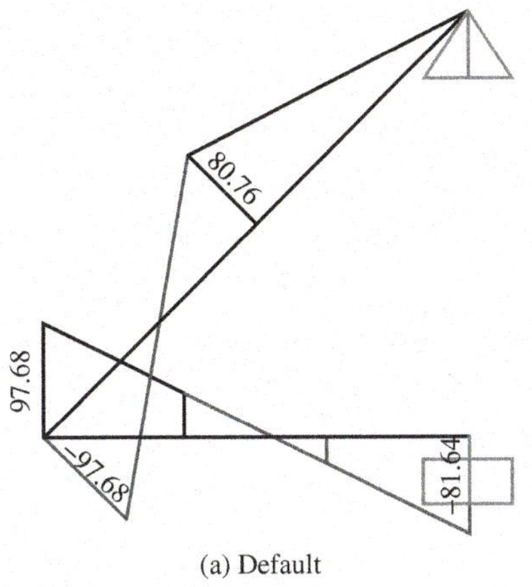

(a) Default

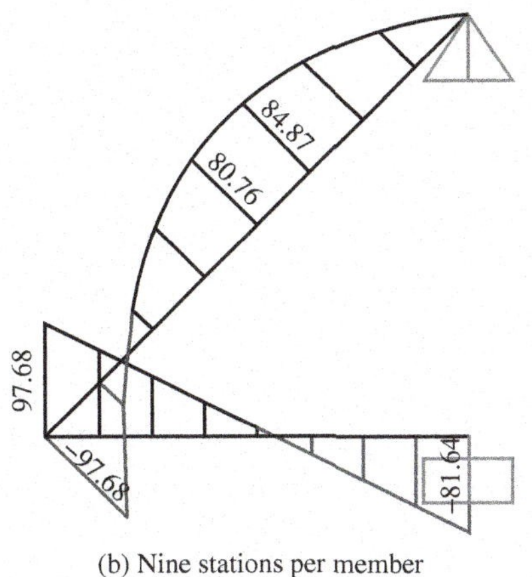

(b) Nine stations per member

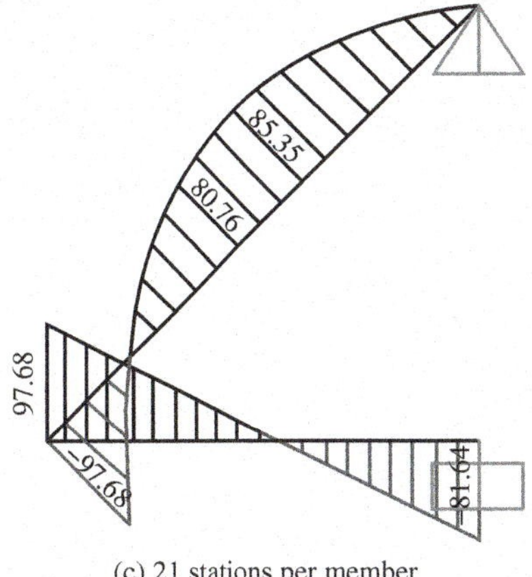

(c) 21 stations per member

FIGURE 20.28 Moment diagram representation using different numbers of stations (units of k·in.).

It is important to recognize that the number of stations does not affect the actual numerical results. Rather, it only affects our understanding of the results. For example, note that the moment at mid-span of the inclined member is 80.76 k·in. This value at mid-span is the same in all diagrams. The additional stations simply report more information to us. The lesson learned is that, if a member is subjected to a distributed load, it is wise to have around nine or more stations.

Tabular Output

The number of sections defined for a member will also influence the tabular output of the results. Figures 20.29 and 20.30 show the member forces as they are reported for the inclined members.

QUICK NOTE

Setting a member to have many stations unnecessarily will only generate more data that the analyst will need to sift through. For example, a truss bar will have the same axial force along its entire length. Adding more stations is simply uninformative.

Setting Stations

The number of stations may be adjusted for any member by placing the cursor over that member and right-clicking the mouse. This will generate a pop-up window like that shown in Figure 20.31. Then double-click on the cell that has been circled in Figure 20.31. This will produce another pop-up window, as shown in Figure 20.32. The user may then set the number of stations to be used for that member or set a minimum spacing to be used between stations.

TABLE: Element Forces - Frames

Frame	Station	OutputCase	CaseType	P	V2	M3
Text	in	Text	Text	Kip	Kip	Kip-in
2	0	DEAD	LinStatic	44.955	-6.05	-97.676
2	50.912	DEAD	LinStatic	44.955	-0.959	80.762
2	101.823	DEAD	LinStatic	44.955	4.132	-1.492E-13

FIGURE 20.29 Element forces for inclined member when three stations are used.

TABLE: Element Forces - Frames

Frame	Station	OutputCase	CaseType	P	V2	M3
Text	in	Text	Text	Kip	Kip	Kip-in
2	0	DEAD	LinStatic	44.955	-6.05	-97.676
2	12.728	DEAD	LinStatic	44.955	-4.778	-28.767
2	25.456	DEAD	LinStatic	44.955	-3.505	23.943
2	38.184	DEAD	LinStatic	44.955	-2.232	60.452
2	50.912	DEAD	LinStatic	44.955	-0.959	80.762
2	63.64	DEAD	LinStatic	44.955	0.314	84.871
2	76.368	DEAD	LinStatic	44.955	1.586	72.781
2	89.095	DEAD	LinStatic	44.955	2.859	44.49
2	101.823	DEAD	LinStatic	44.955	4.132	-1.292E-13

FIGURE 20.30 Element forces for inclined member when nine stations are used.

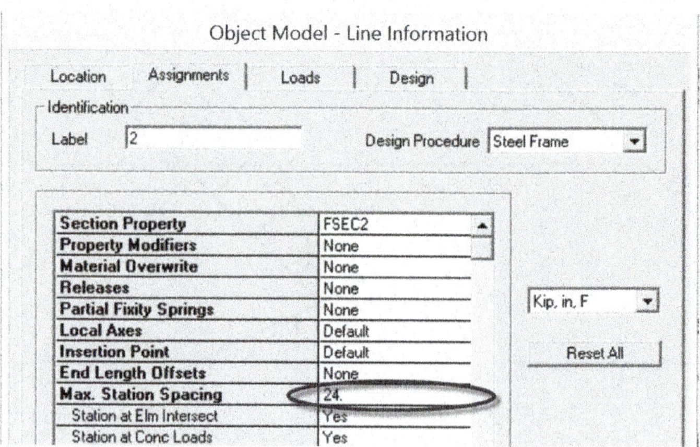

FIGURE 20.31 SAP2000 pop-up window for modifying element information.

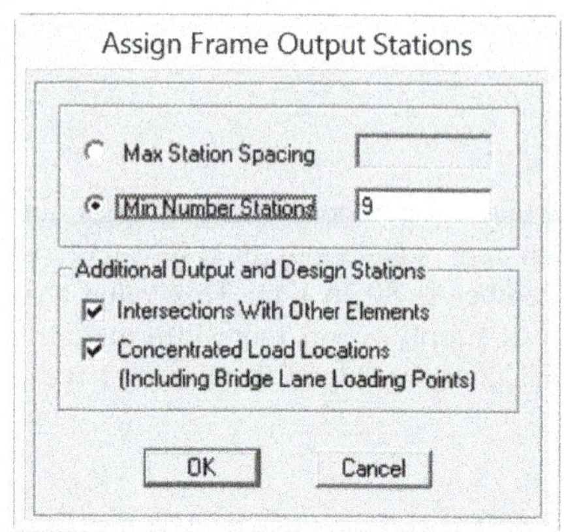

FIGURE 20.32 Setting station spacing.

20.13 Examples with Video Solution

VE20.1 Generate the structure stiffness matrix for the beam shown in Figure VE20.1. $E = 3500$ ksi and $I = 515$ in^4.

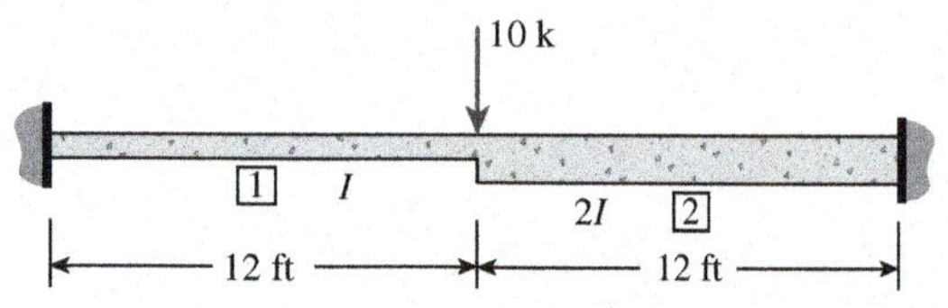

Figure VE20.1

VE20.2 For the beam shown in Figure VE 20.1, use the matrix approach to solve for the joint displacements, reactions, and member end forces.

VE20.3 For the beam analyzed in Figure VE 20.2, use a spreadsheet to plot the resulting shear and moment diagrams.

VE20.4 For the beam analyzed in Figure VE 20.2, use shape functions to plot its deflected shape.

VE20.5 Using the stiffness matrix developed in Figure VE 20.1, solve for the joint displacements, reactions, and member end forces for the loading condition shown in Figure VE20.5.

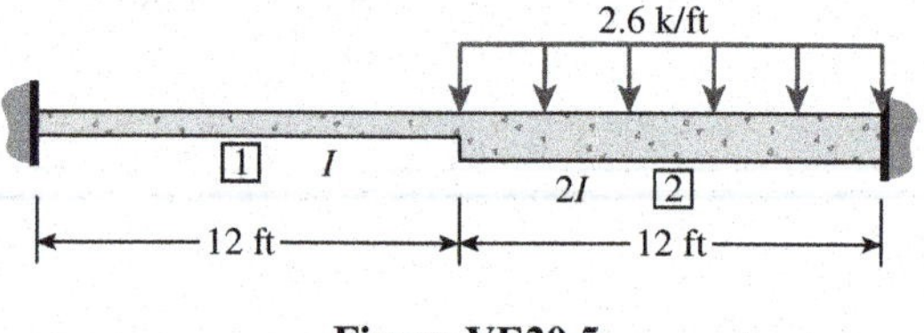

Figure VE20.5

20.14 Problems for Solution

Sections 20.3–20.4

For Problems 20.1 through 20.5, develop the structure stiffness matrix for the DOFs shown. Provide answers in units of k and in.

P20.1 $E = 1900$ ksi and $I = 1250$ in^4. (Ans: $K_{11} = 56{,}548$(k·in.)/rad, $K_{33} = 6.01$ k/in.)

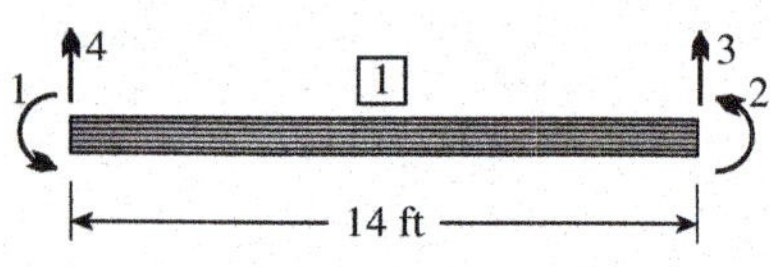

Problem 20.1

P20.2 $E = 5400$ ksi and $I = 5000$ in^4.

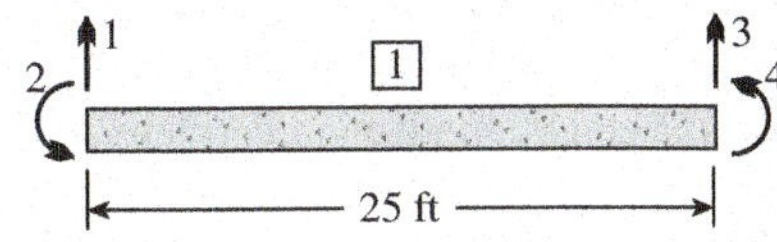

Problem 20.2

P20.3 $E = 29{,}000$ ksi and $I = 350$ in^4. (Ans: $K_{11} = 36.59$ k/in., $K_{22} = 395{,}589.7$(k·in.)/rad)

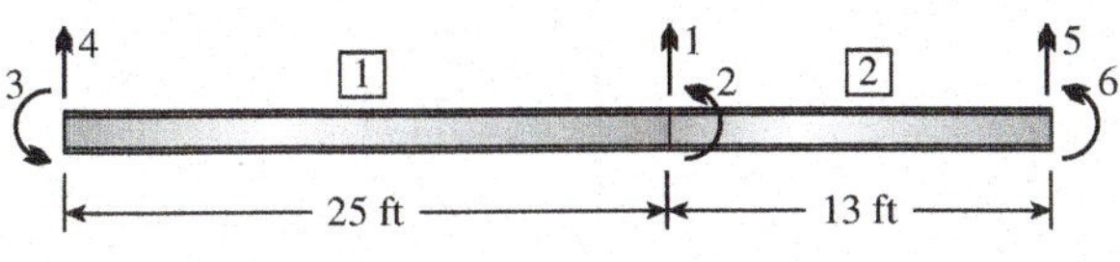

Problem 20.3

P20.4 Rework Problem 20.3 with the following moments of inertia. $I_1 = 1050$ in^4 and $I_2 = 350$ in^4.

P20.5 *EI* constant for entire beam. $E = 29{,}000$ ksi and $I = 225$ in^4. (Ans: $K_{11} = 255.09$ k/in., $K_{44} = 543{,}750$(k·in.)/rad)

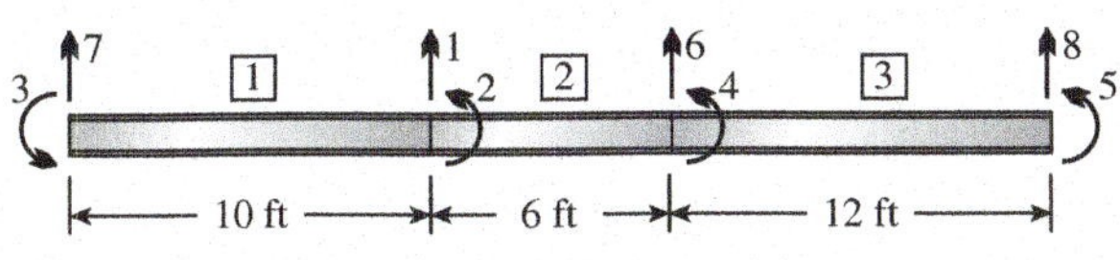

Problem 20.5

For Problems 20.6 through 20.16, use the matrix approach to solve for the joint displacements, reactions, and member end forces. Perform calculations in units of k and in.

P20.6 $E = 1900$ ksi and $I = 1250$ in^4.

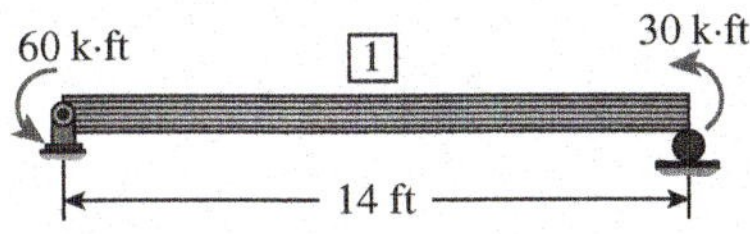

Problem 20.6

P20.7 $E = 1900$ ksi and $I = 1250$ in^4. (Ans: $V_{right} = 6.43$ k, $M_{right} = 360$ k·in.)

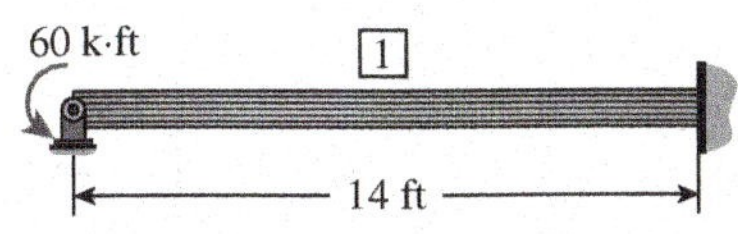

Problem 20.7

P20.8 $E = 5400$ ksi and $I = 5000$ in^4.

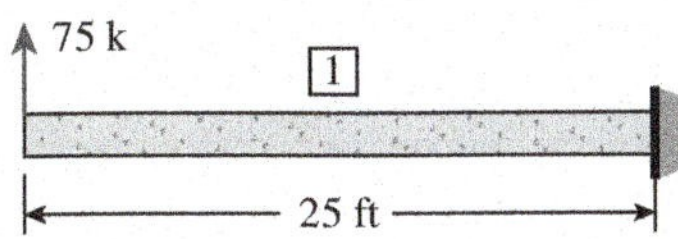

Problem 20.8

P20.9 $E = 5400$ ksi and $I = 5000$ in^4. (Ans: $V_{right} = -82.0$ k, $M_{right} = -12{,}300$ k·in.)

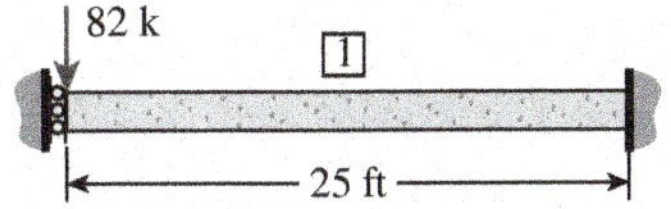

Problem 20.9

P20.10 $E = 29{,}000$ ksi and $I = 350$ in^4.

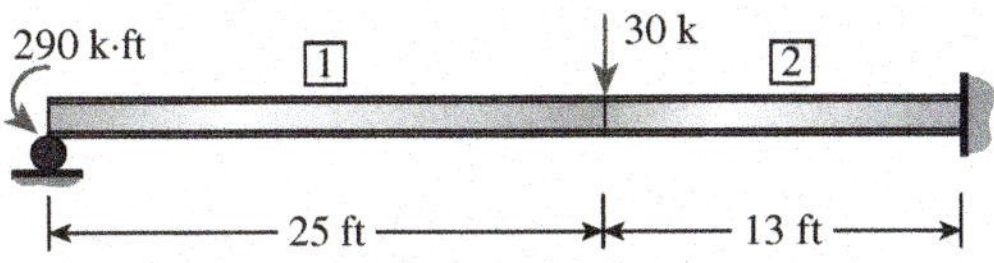

Problem 20.10

P20.11 $E = 29{,}000$ ksi and $I = 350$ in^4. (Ans: $V_{2-right} = -7.47$ k, $M_{1-right} = 2256.8$ k·in.)

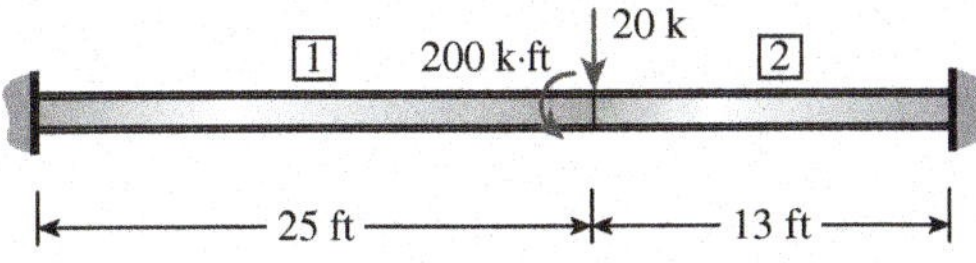

Problem 20.11

P20.12 Rework Problem 20.10 with the following moments of inertia. $I_1 = 1050$ in^4 and $I_2 = 350$ in^4.

P20.13 Rework Problem 20.11 with the following moments of inertia. $I_1 = 1050$ in^4 and $I_2 = 350$ in^4. (Ans: $V_{2-right} = -5.51$ k, $M_{1-right} = 2573.4$ k·in.)

P20.14 *EI* constant for entire beam. $E = 29{,}000$ ksi and $I = 225$ in^4.

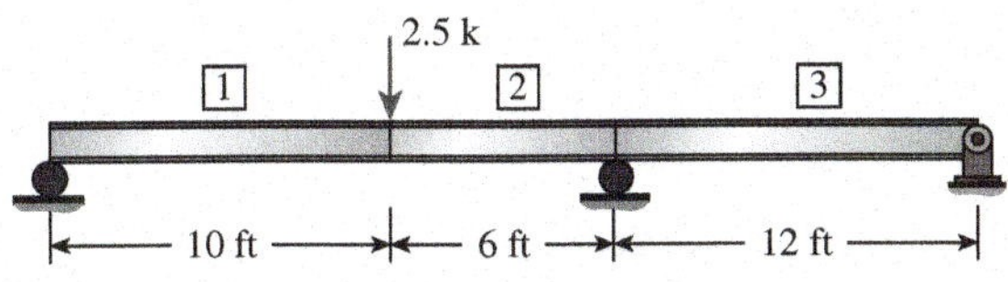

Problem 20.14

P20.15 *EI* constant for entire beam. $E = 29{,}000$ ksi and $I = 225$ in^4. (Ans: $V_{1-right} = 1.29$ k, $M_{3-left} = -185.1$ k·in.)

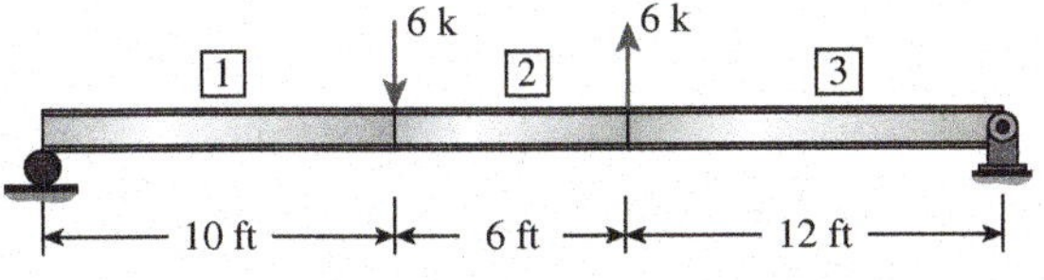

Problem 20.15

P20.16 *EI* constant for entire beam. $E = 29{,}000$ ksi and $I = 225$ in^4.

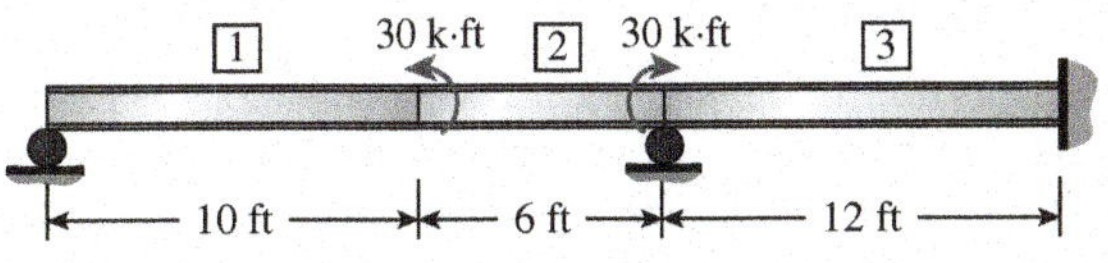

Problem 20.16

For Problems 20.17 through 20.20, use the matrix approach to solve for the member end forces. In each case the displacements at each DOF are given. Perform calculations in units of k and in.

P20.17 $E = 1900$ ksi, $I = 1250$ in^4, and

$$\{u\}^{(1)} = \begin{matrix} a \\ b \\ c \\ d \end{matrix} \begin{Bmatrix} 0.0 \\ 0.001 \\ 1.1 \\ -0.0015 \end{Bmatrix} \begin{matrix} \text{in.} \\ \text{rad} \\ \text{in.} \\ \text{rad} \end{matrix}$$

(Ans: $V_a = -6.86$ k, $M_B = 541.2$ k·in.)

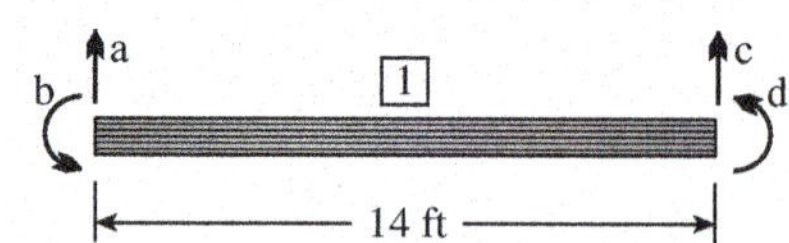

Problem 20.17

P20.18 Use the element shown in Problem 20.17. $E = 1900$ ksi, $I = 1250$ in^4, and

$$\{u\}^{(1)} = \begin{matrix} a \\ b \\ c \\ d \end{matrix} \begin{Bmatrix} -0.255 \\ -0.0005 \\ 0.652 \\ -0.0025 \end{Bmatrix} \begin{matrix} \text{in.} \\ \text{rad} \\ \text{in.} \\ \text{rad} \end{matrix}$$

P20.19 $E = 5400$ ksi, $I = 5000$ in^4, and

$$\{u\}^{(1)} = \begin{matrix} a \\ b \\ c \\ d \end{matrix} \begin{Bmatrix} -0.55 \\ 0.0021 \\ -1.5 \\ -0.0005 \end{Bmatrix} \begin{matrix} \text{in.} \\ \text{rad} \\ \text{in.} \\ \text{rad} \end{matrix}$$

(Ans: $V_a = 14.28$ k, $M_B = -2376.0$ k·in.)

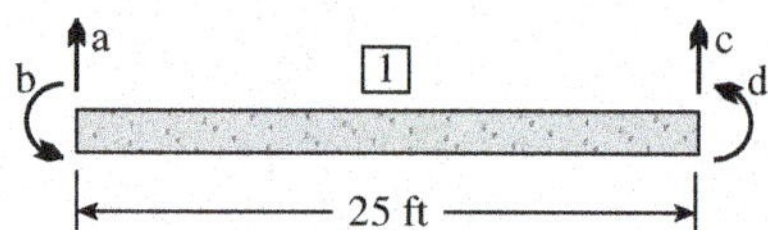

Problem 20.19

P20.20 Use the element shown in Problem 20.19. $E = 5400$ ksi and $I = 5000$ in^4.

$$\{u\}^{(1)} = \begin{matrix} a \\ b \\ c \\ d \end{matrix} \begin{Bmatrix} -0.355 \\ 0.0015 \\ 0.762 \\ -0.0035 \end{Bmatrix} \begin{matrix} \text{in.} \\ \text{rad} \\ \text{in.} \\ \text{rad} \end{matrix}$$

Section 20.5

For Problems 20.21 through 20.29, use shape functions to plot the displaced shape of each beam shown. In each case, the displacements at each DOF are given. Each beam element should be divided into 10 equal segments and the plots should be computer generated. Perform calculations in units of in.

P20.21 Consider the condition given in Problem 20.17. (Ans: displacement @ mid-span 0.603 in.)

P20.22 Consider the condition given in Problem 20.18.

P20.23 Consider the condition given in Problem 20.19. (Ans: displacement @ mid-span −0.928 in.)

P20.24 Consider the condition given in Problem 20.20.

P20.25

$$\{\Delta\} = \begin{matrix} 4 \\ 3 \\ 1 \\ 2 \\ 5 \\ 6 \end{matrix} \begin{Bmatrix} 0.0 \\ 0.02722 \\ -0.108 \\ -0.00416 \\ 0.0 \\ 0.0 \end{Bmatrix} \begin{matrix} \text{in.} \\ \text{rad} \\ \text{in.} \\ \text{rad} \\ \text{in.} \\ \text{rad} \end{matrix}$$

(Ans: displacement @ mid-span of member 1 = 1.123 in.)

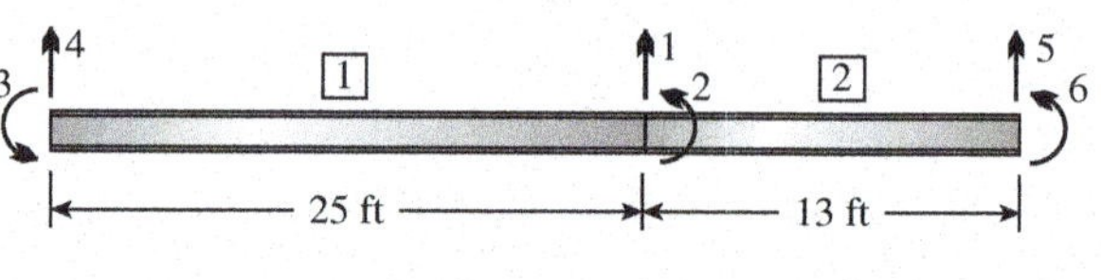

Problem 20.25

P20.26 For the beam in Problem 20.25,

$$\{\Delta\} = \begin{matrix} 4 \\ 3 \\ 1 \\ 2 \\ 5 \\ 6 \end{matrix} \begin{Bmatrix} 0.0 \\ 0.0 \\ -1.1019 \\ 0.01115 \\ 0.0 \\ 0.0 \end{Bmatrix} \begin{matrix} \text{in.} \\ \text{rad} \\ \text{in.} \\ \text{rad} \\ \text{in.} \\ \text{rad} \end{matrix}$$

P20.27

$$\{\Delta\} = \begin{matrix} 7 \\ 3 \\ 1 \\ 2 \\ 6 \\ 4 \\ 8 \\ 5 \end{matrix} \begin{Bmatrix} 0.0 \\ -0.00047 \\ 0.0 \\ 0.00095 \\ 0.0844 \\ 0.00078 \\ 0.0 \\ -0.00127 \end{Bmatrix} \begin{matrix} \text{in.} \\ \text{rad} \\ \text{in.} \\ \text{rad} \\ \text{in.} \\ \text{rad} \\ \text{in.} \\ \text{rad} \end{matrix}$$

(Ans: displacement @ mid-span of member 2 = 0.044 in.)

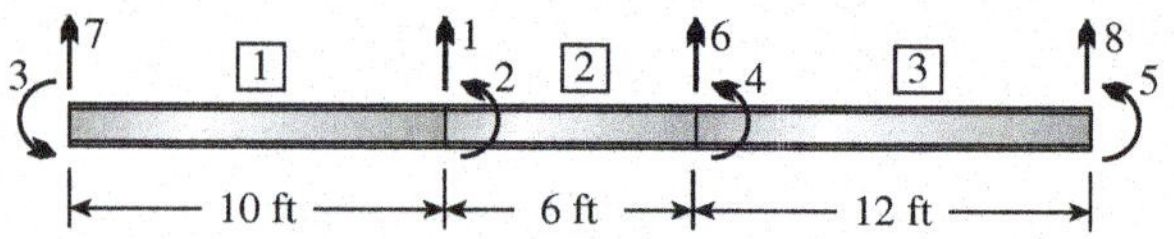

Problem 20.27

P20.28 For the beam in Problem 20.27,

$$\{\Delta\} = \begin{matrix} 7 \\ 3 \\ 1 \\ 2 \\ 6 \\ 4 \\ 8 \\ 5 \end{matrix} \begin{Bmatrix} 0.0 \\ -0.00029 \\ 0.0139 \\ 0.00092 \\ 0.0 \\ -0.00116 \\ 0.0 \\ 0.0 \end{Bmatrix} \begin{matrix} \text{in.} \\ \text{rad} \\ \text{in.} \\ \text{rad} \\ \text{in.} \\ \text{rad} \\ \text{in.} \\ \text{rad} \end{matrix}$$

P20.29 For the beam in Problem 20.27,

$$\{\Delta\} = \begin{matrix} 7 \\ 3 \\ 1 \\ 2 \\ 6 \\ 4 \\ 8 \\ 5 \end{matrix} \begin{Bmatrix} 0.0 \\ -0.0005 \\ -0.0309 \\ 0.00023 \\ 0.0 \\ 0.00038 \\ 0.0 \\ -0.00019 \end{Bmatrix} \begin{matrix} \text{in.} \\ \text{rad} \\ \text{in.} \\ \text{rad} \\ \text{in.} \\ \text{rad} \\ \text{in.} \\ \text{rad} \end{matrix}$$

(Ans: displacement @ mid-span of member 2 = −0.017 in.)

Sections 20.6 and 20.7

For Problems 20.30 through 20.40, use the matrix method to solve for the joint displacements, reactions, and member end forces. Use these results to plot the shear and moment diagrams by hand. Check to ensure that your diagrams do agree with the member end forces you calculated. Perform calculations in units of k and in.

P20.30 $E = 1900$ ksi and 1250 in^4.

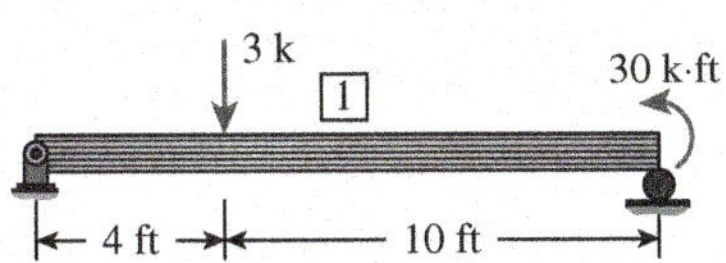

Problem 20.30

P20.31 $E = 1900$ ksi and 1250 in^4. (Ans: $R_{left} = 24.57$ k, $\theta_{right} = 0.0137$ rad)

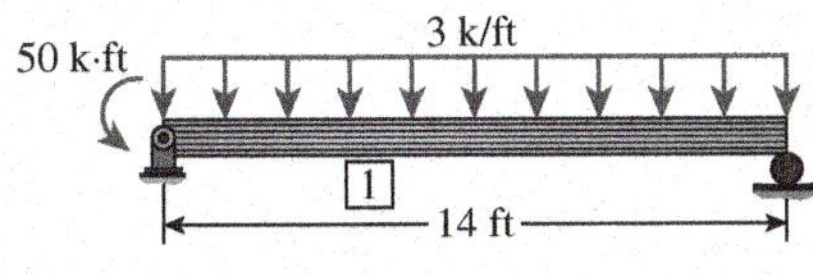

Problem 20.31

P20.32 $E = 1900$ ksi and 1250 in^4.

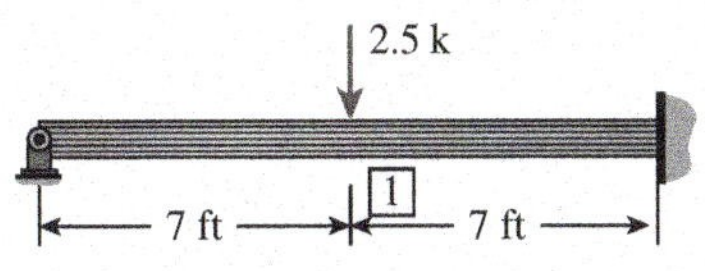

Problem 20.32

P20.33 $E = 5400$ ksi and 5000 in^4. (Ans: $\Delta_{left} = 5.30$ in., $M_{right} = 901.0$ k·in.)

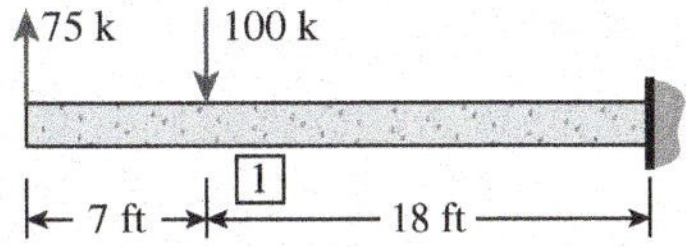

Problem 20.33

P20.34 $E = 5400$ ksi and 5000 in^4.

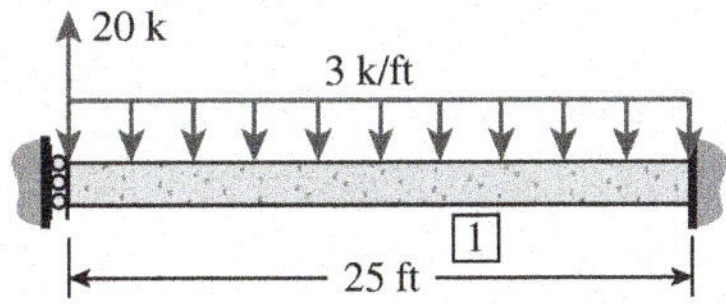

Problem 20.34

P20.35 $E = 29{,}000$ ksi and $I = 350$ in^4. (Ans: $V_{left\text{-}end} = 33.17$ k, $M_{right\text{-}end} = -6884$ k·in.)

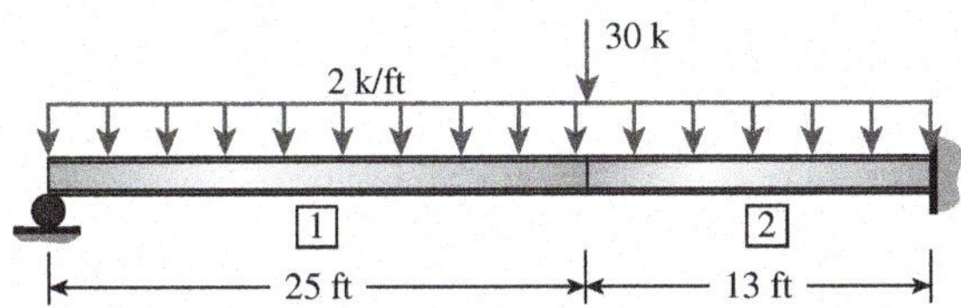

Problem 20.35

P20.36 $E = 29{,}000$ ksi and $I = 350$ in^4.

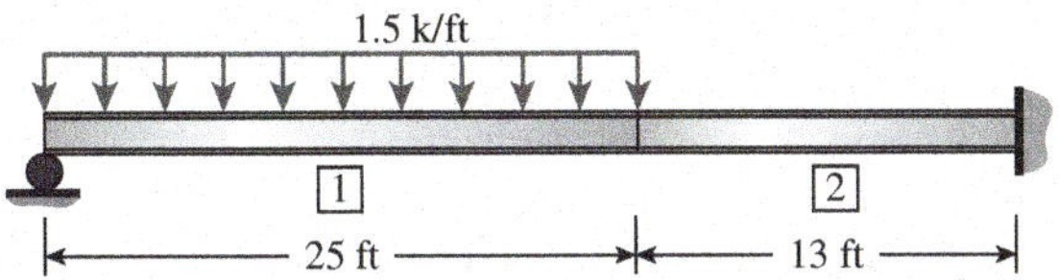

Problem 20.36

P20.37 $E = 29{,}000$ ksi and $I = 350$ in^4. (Ans: $V_{left\text{-}end} = 4.87$ k, $M_{right\text{-}end} = -3781$ k·in.)

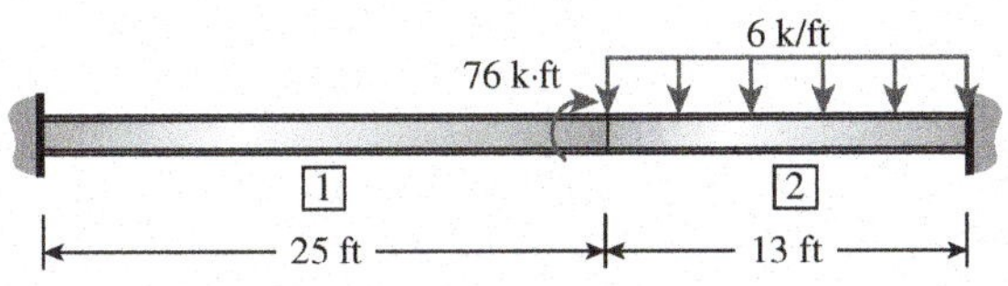

Problem 20.37

P20.38 $E = 29{,}000$ ksi and $I = 225$ in^4.

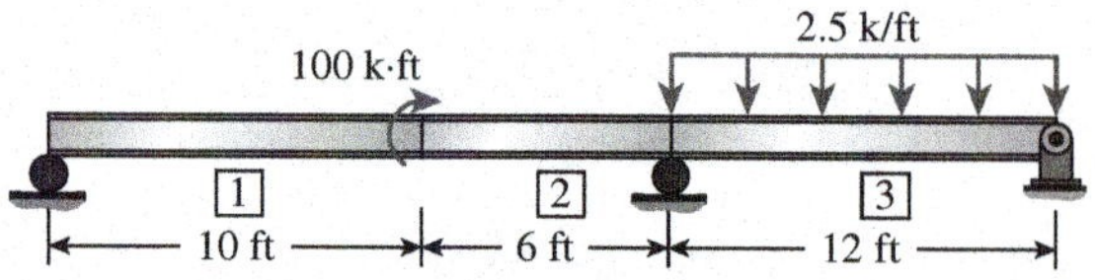

Problem 20.38

P20.39 $E = 29{,}000$ ksi and $I = 225$ in^4. (Ans: $V_{\text{left}-1} = 5.24$ k, $M_{\text{right}-1} = 509.2$ k·in.)

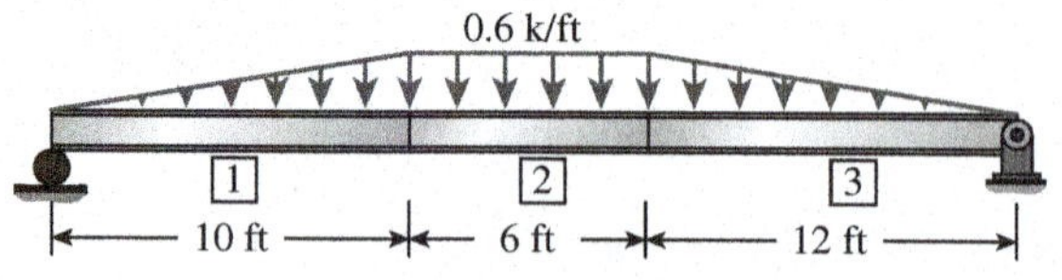

Problem 20.39

P20.40 $E = 29{,}000$ ksi and $I = 225$ in^4.

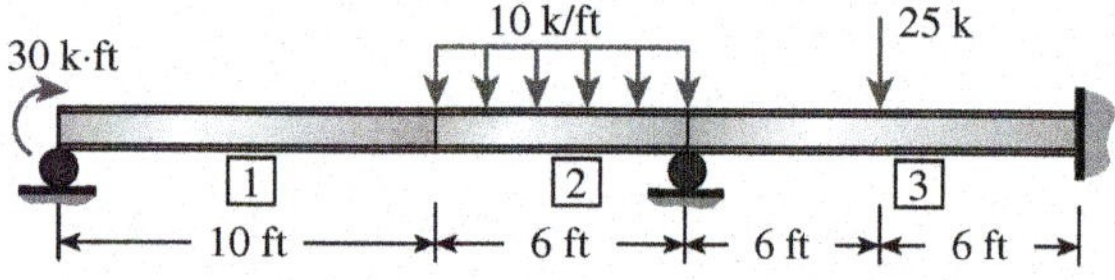

Problem 20.40

Sections 20.8 and 20.10

P20.41 Determine the global element stiffness matrix for the member shown. Use the member origin as indicated on the sketch. Leave answers in terms of A, I, E, and L. (Ans: $K_{11} = \frac{12EI}{L^3}$, $K_{22} = \frac{AE}{L}$)

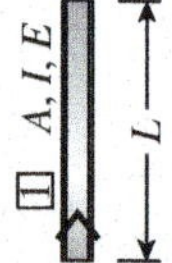

Problem 20.41

P20.42 Determine the global element stiffness matrix for the member shown. Use the member origin as indicated on the sketch. Leave answers in terms of A, I, E, and L.

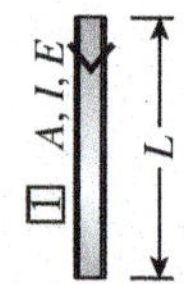

Problem 20.42

P20.43 For the inclined cantilever, use the matrix approach to solve for the joint displacements, reactions, and member end forces. Use the DOF labeling as given in the figure. (Ans: $\Delta_1 = 0.6708$ in., $M_6 = 254.6$ k·in.)

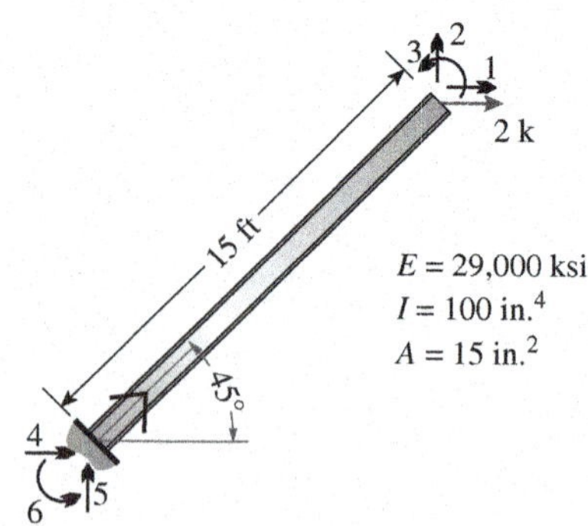

Problem 20.43

For Problems 20.44 through 20.53, use the matrix method to solve for the joint displacements, reactions, and member end forces. The origins of each member are indicated on the figure. Perform calculations in units of k and in.

P20.44

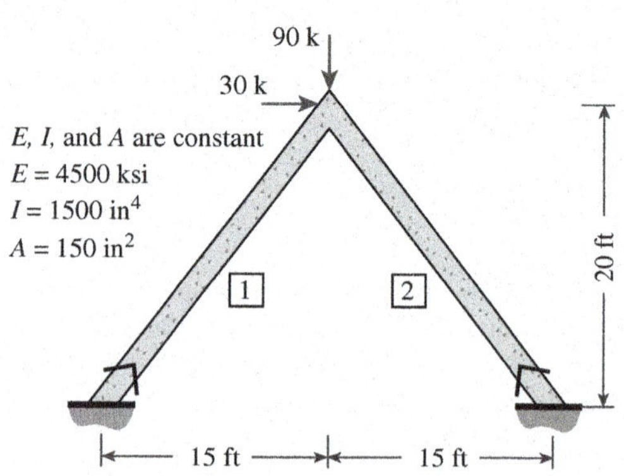

Problem 20.44

P20.45 (Ans: for member 2 $V_{\text{left}} = -42.21$ k, $M_{\text{left}} = -2348.4$ k·in.)

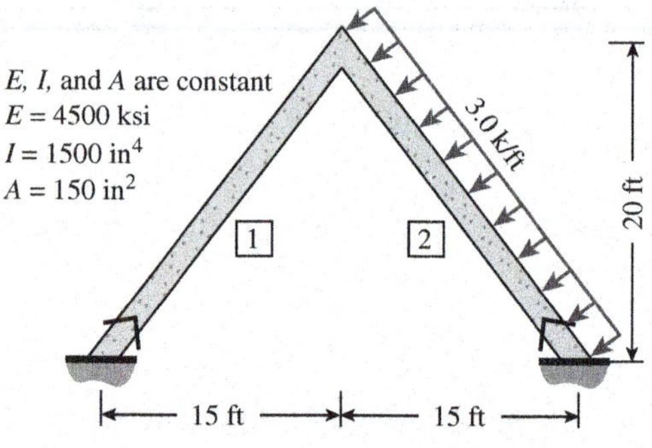

Problem 20.45

P20.46

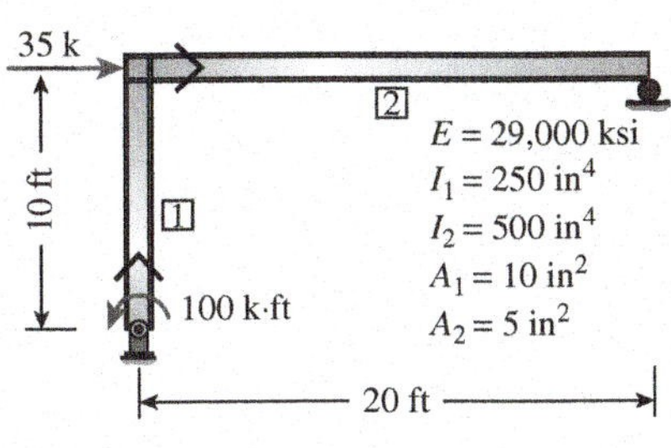

Problem 20.46

P20.47 (Ans: for member 2 $V_{\text{left}} = 6.50$ k, $M_{\text{left}} = -4200$ k·in.)

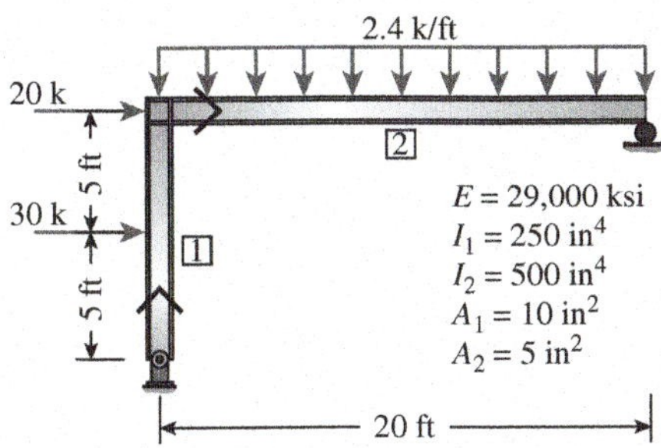

Problem 20.47

P20.48

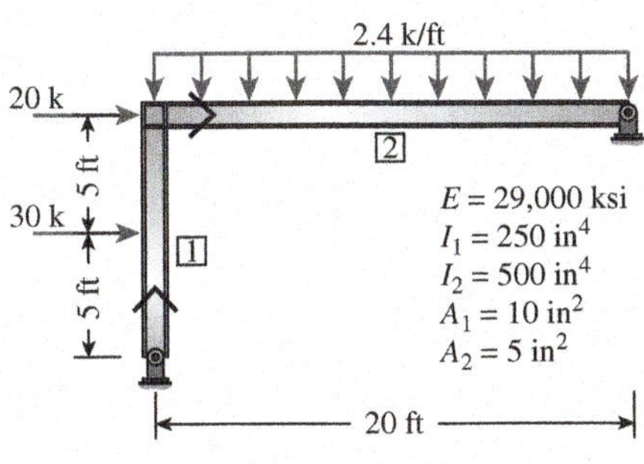

Problem 20.48

P20.49 (Ans: for member 1 $V_{\text{left}} = 10.92$ k, $M_{\text{left}} = 402$ k·in.)

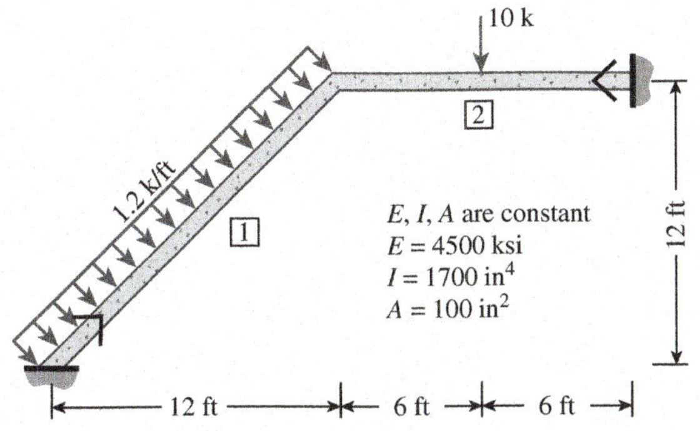

Problem 20.49

P20.50

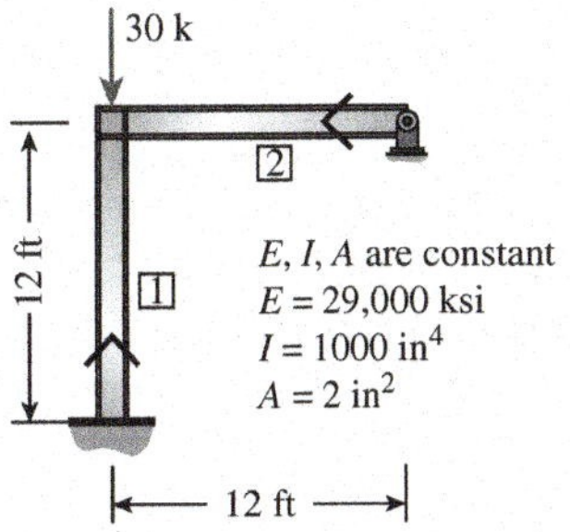

Problem 20.50

P20.51 (Ans: for member 1 $V_{\text{left}} = 0.72$ k, $M_{\text{left}} = 107.5$ k·in.)

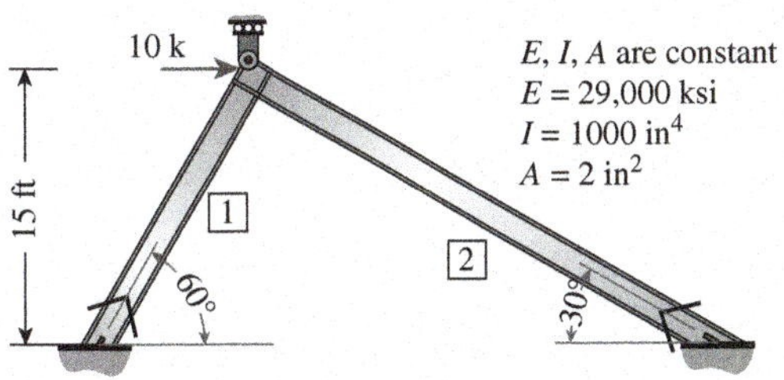

Problem 20.51

P20.52

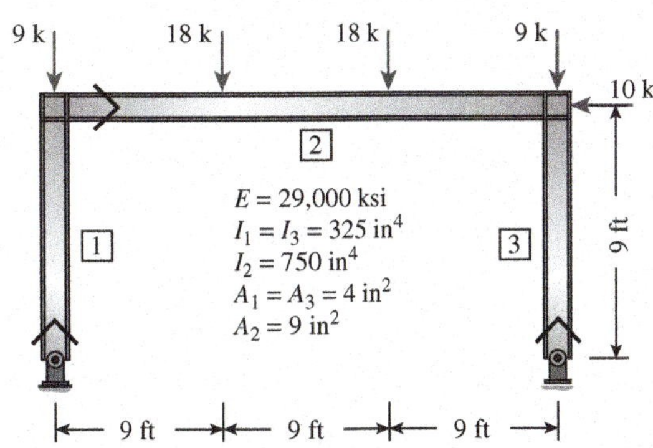

Problem 20.52

P20.53 (Ans: for member 1 $V_{\text{left}} = 19.67$ k, $M_{\text{right}} = -792.9$ k·in.)

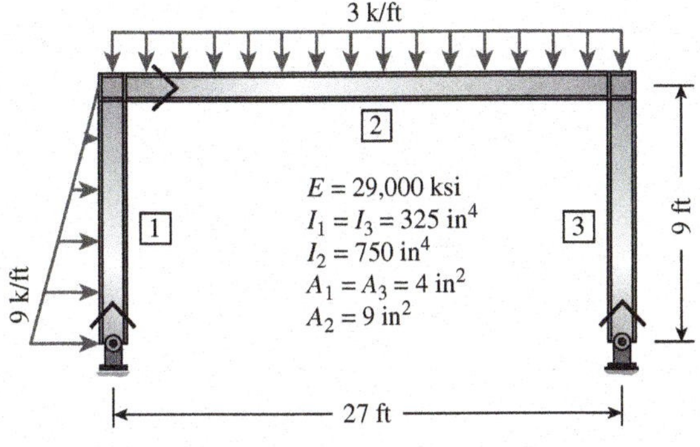

Problem 20.53

SAP2000

Please note that when you construct a SAP model in the $X-Z$ plane that a counterclockwise rotation has a negative sign as opposed to a positive sign when it is built in the $X-Y$ plane. Take this into account as you compare results in the following problems.

P20.54 Using SAP2000, model the frame described in Example 20.9. Report the displacements at the joints and plot the moment diagram. Next provide a fixed support at each joint and rerun the analysis. Report the moment diagram and the support reactions and compare them with the force vector $\{F_u - F_{\text{FEF}}\}$ as given in Example 20.9. Lastly, revert back to the original support conditions and remove all loads. Then apply only the statically equivalent loads as computed in Example 20.9, find the joint displacements, and plot the moment diagram. Show that the results of the latter two cases will superimpose to equal the original case.

P20.55 Compute the element stiffness matrix for the beam shown—use the DOFs as labeled. Now model the beam within SAP2000. Make sure to turn off shear deformations and provide a fixed support at each end and impose a 1-in. displacement at DOF 1. Find the reactions at the supports. Then remove the unit displacement at DOF 1, impose a unit rotation at DOF 2, and find the reactions. Repeat for DOFs 3 and 4. Make observations regarding the original stiffness matrix and the reactions you found using SAP2000.

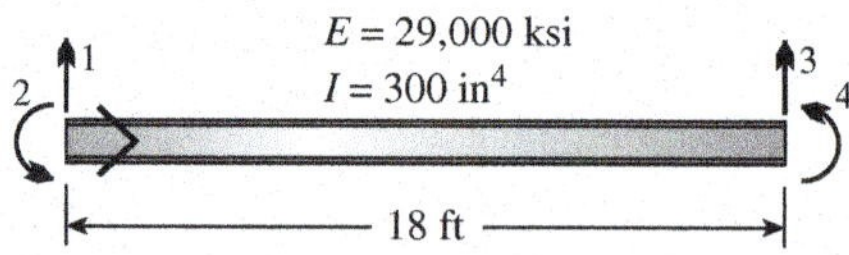

Problem 20.55

P20.56 Compute the element stiffness matrix for the frame shown—use the DOFs as labeled. Now model the frame within SAP2000. Make sure to turn off shear deformations and provide a fixed support at each end and impose a 1-in. displacement at DOF 1. Find the reactions at the supports. Then remove the unit displacement at DOF 1, impose a unit displacement at DOF 2, and then find the reactions. Finally, remove the other displacements and impose a unit rotation at DOF 3 then find the reactions. Make observations regarding the original stiffness matrix and the reactions you found using SAP2000.

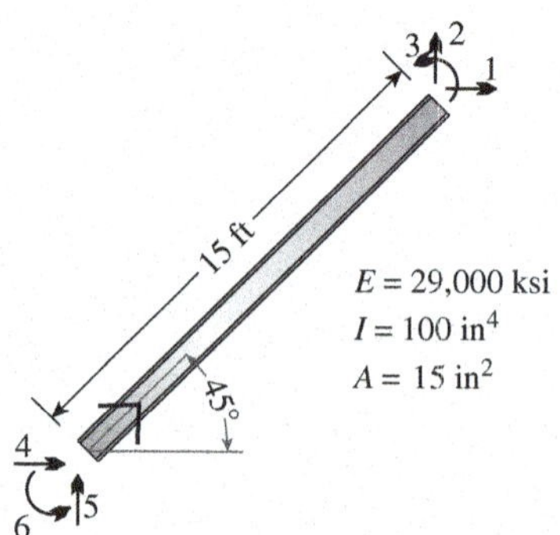

Problem 20.56

P20.57 Using SAP2000, model the frame described in Problem 20.50. Report all nodal displacements. Now change the cross-sectional areas for both members to 0 in^2 and report the nodal displacements again. Provide an explanation of the differences, if any.

Additional Topics for the Direct Stiffness Method

21

21.1 Introduction

In the previous two chapters, the direct stiffness method was demonstrated for trusses, beams, and frames. With this knowledge, many practical and useful analyses may be carried out for realistic structures. Indeed, the reader may note that their ability to deal with complex and heavily indeterminate structures has been enhanced through these chapters.

The purpose of the current chapter is to further demonstrate how the direct stiffness method set with a matrix formulation may be readily implemented to deal with useful—but less common—structural analysis problems. The reader is expected to be well versed in the methods discussed in Chapters 19 and 20.

21.2 Stiffness Formulation for Structures with Enforced Displacements

Although, in general, structures deflect in response to applied loads, there are occasions when a structure is forced to displace a specified amount. A continuous beam whose supports are not aligned properly is one example. The misalignment may be corrected by jacking the structure into position. The analyst may want to know what forces will be required to force the beam through the necessary displacement in order to make it fit.

An example of misaligned supports is shown in Figure 21.1, where the left-most support of the continuous beam is assumed to be misaligned by an amount Δ_0. The question that arises is

QUICK NOTE

Having structural supports misaligned is not an uncommon problem. Forcing the structure into place using sledge hammers or jacks is a common approach used by builders. It is important to note that this will induce internal forces within the structure—even before external loads have been applied.

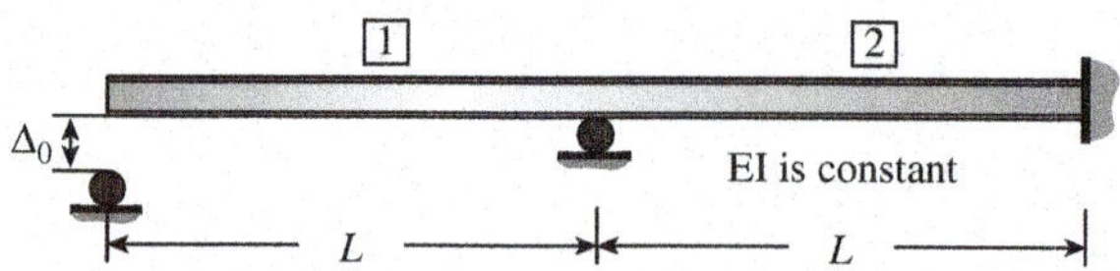

FIGURE 21.1 Continuous beam with misaligned support.

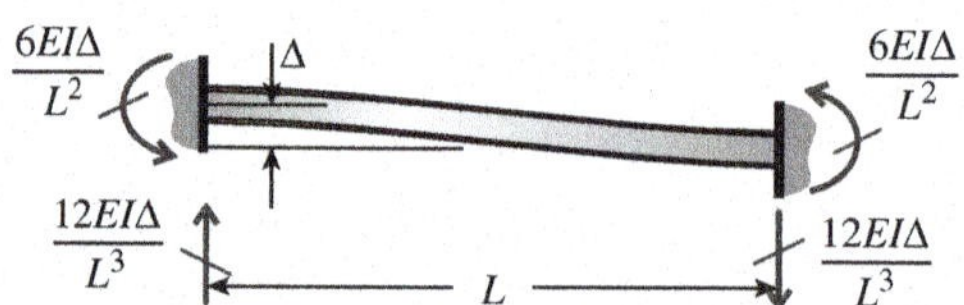

FIGURE 21.2 Fixed-end forces for support movement.

how one is to incorporate this effect into the matrix formulation. The key to this solution is found in Section 20.6, where the concept of *statical equivalency* was introduced. One simply needs to find the fixed-end forces $\{F_{\text{FEF}}\}$ that develop due to the specified support movement. Figure 21.2 shows a generic beam chart used for determining these forces.

The basic steps for this analysis are as follows.

1. Lock all DOFs.
2. Determine the forces required at the DOFs that are needed for compatibility—$\{F_{\text{FEF}}\}$. These FEFs should be due to the support movement.
3. Apply FEFs in the opposite direction at the appropriate DOFs and then analyze using the matrix procedure. Thus,

$$\{F_u - F_{\text{FEF}}\}$$

4. Resultant forces, reactions, and displacements from steps 2 and 3 are to be superimposed.

DID YOU KNOW?
The fixed-end forces due to support settlement/rotation are easily determined from the stiffness factors that were previously presented in Figure 20.3 of Section 20.2.

EXAMPLE 21.1

For the beam given in Figure 21.1, consider the left support has been set too low by 0.25 in. If $L = 240$ in. and $EI = 7.25 \times 10^6$ k·in^2, solve for the joint displacements, reactions, and member end forces that result from this misalignment.

SOLUTION STRATEGY

All stiffness matrices and force vectors must be assembled in the same fashion as done in the previous two chapters. Fixed-end forces due to the misalignment should be calculated using the values from Figure 21.2.

DID YOU KNOW?
The DOF, which is associated with the known misalignment, should be treated as a restrained DOF.

SOLUTION

The following DOF numbering is used and results in the element and structural stiffness matrices as shown.

QUICK NOTE
Remember that all structural responses (i.e., displacements, reactions, and member end forces) should be superimposed to obtain final results.

$$K^{(1)} = \begin{array}{c} \\ 3 \\ 1 \\ 4 \\ 2 \end{array} \begin{array}{c} \begin{array}{cccc} 3 & 1 & 4 & 2 \end{array} \\ \begin{bmatrix} 6.29 & 755 & -6.29 & 755 \\ 755 & 120{,}833 & -755 & 60{,}417 \\ -6.29 & -755 & 6.29 & -755 \\ 755 & 60{,}417 & -755 & 120{,}833 \end{bmatrix} \end{array} \qquad K^{(2)} = \begin{array}{c} \\ 4 \\ 2 \\ 5 \\ 6 \end{array} \begin{array}{c} \begin{array}{cccc} 4 & 2 & 5 & 6 \end{array} \\ \begin{bmatrix} 6.29 & 755 & -6.29 & 755 \\ 755 & 120{,}833 & -755 & 60{,}417 \\ -6.29 & -755 & 6.29 & -755 \\ 755 & 60{,}417 & -755 & 120{,}833 \end{bmatrix} \end{array}$$

This results in the following structure level stiffness matrices.

$$K_{uu} = \begin{matrix} & 1 & 2 \\ 1 & 120{,}833 & 60{,}417 \\ 2 & 60{,}417 & 241{,}667 \end{matrix} \qquad K_{ru} = \begin{matrix} & 1 & 2 \\ 3 & 755 & 755 \\ 4 & -755 & 0 \\ 5 & 0 & -755 \\ 6 & 0 & 60{,}417 \end{matrix}$$

QUICK NOTE

Units for translational and rotational DOFs are k/in. and (k·in.)/rad, respectively.

Because there are no externally applied loads, the applied force vector $\{F_u\}$ is populated with zeros.

The next step is to find the equivalent static forces due to a downward deflection of 0.25 in. at DOF 3. Since we need to lock all DOFs before applying the known displacement, only the DOFs associated with member 1 will develop FEFs. Use Figure 21.2 to get the FEF values.

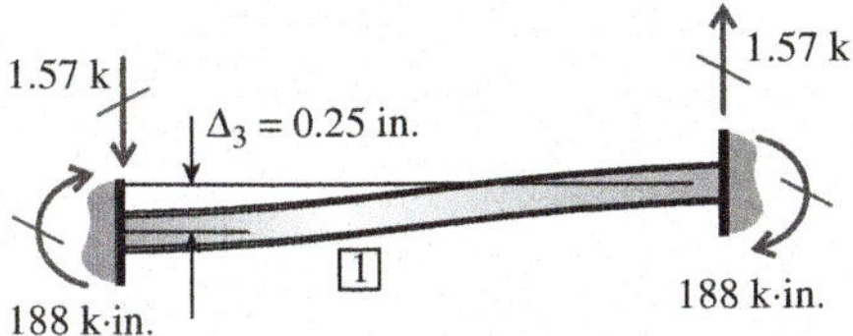

CALCULATIONS

$$FEV_3 = \frac{12(7,250,000)(-0.25)}{240^3} = -1.57 \text{ k}$$

$$FEM_1 = \frac{6(7,250,000)(-0.25)}{240^2} = -188.8 \text{ k·in.}$$

Therefore the FEFs for the unrestrained DOFs are

$$\{F_{FEF-u}\} = \begin{Bmatrix} -188.8 \\ -188.8 \end{Bmatrix} \begin{matrix} \text{k·in.} \\ \text{k·in.} \end{matrix}$$

The FEFs for the restrained DOFs for the entire structure are

$$\{F_{FEF-r}\} = \begin{Bmatrix} -1.57 \\ 1.57 \\ 0.0 \\ 0.0 \end{Bmatrix} \begin{matrix} \text{k} \\ \text{k} \\ \text{k} \\ \text{k·in.} \end{matrix}$$

The calculation of joint rotations may now be carried out.

$$\{\Delta_u\} = \begin{Bmatrix} \theta_1 \\ \theta_2 \end{Bmatrix} = [K_{uu}]^{-1} \left(\begin{Bmatrix} 0.0 \\ 0.0 \end{Bmatrix} - \begin{Bmatrix} -188.8 \\ -188.8 \end{Bmatrix} \right) = \begin{Bmatrix} 0.00134 \\ 0.00045 \end{Bmatrix} \begin{matrix} \text{rad} \\ \text{rad} \end{matrix}$$

The structure analyzed for the statically equivalent loads is shown in the following sketch.

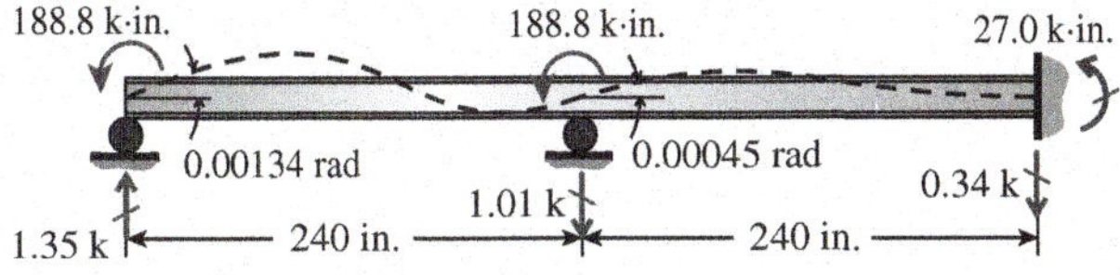

Reactions for the beam—including superposition—are

$$\begin{Bmatrix} R_3 \\ R_4 \\ R_5 \\ M_6 \end{Bmatrix} = [K_{ru}] \begin{Bmatrix} 0.00134 \\ 0.00045 \end{Bmatrix} + \{F_{FEF}\} = \begin{Bmatrix} 1.35 \\ -1.01 \\ -0.34 \\ 27.0 \end{Bmatrix} + \begin{Bmatrix} -1.57 \\ 1.57 \\ 0.0 \\ 0.0 \end{Bmatrix} = \begin{Bmatrix} -0.22 \\ 0.56 \\ -0.34 \\ 27.0 \end{Bmatrix} \begin{matrix} \text{k} \\ \text{k} \\ \text{k} \\ \text{k·in.} \end{matrix}$$

The structure representing the FEFs is shown in the following sketch.

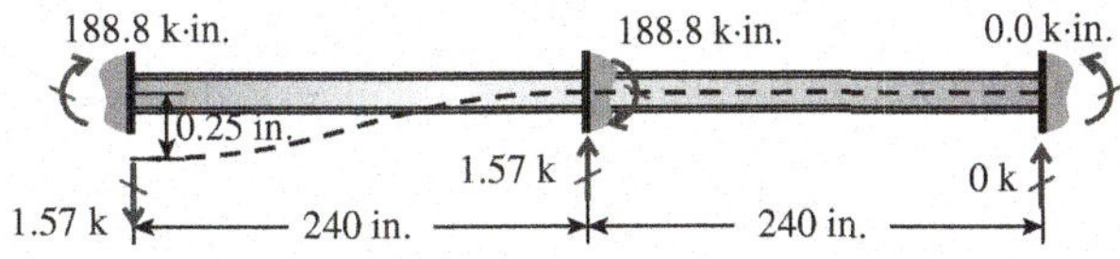

The results for the actual structure are given in the following sketch.

27.0 k·in.
0.25 in.
0.00134 rad
0.00045 rad
0.56 k
0.34 k
240 in.
240 in.
0.22 k

QUICK NOTE

Be aware that in the original structure there is a 0.25-in. displacement at the left roller. Superposition will account for this.

DID YOU KNOW?
The actual displacement at DOF 3 is −0.25 in. If you use this value in your displacement vector $u^{(n)}$, you will directly solve for the member end forces without the need for superposition. This only works for situations where there are no member loads applied that cause FEFs.

Member End Forces

The member end forces can be found by using the element stiffness matrix with the element DOF displacements of $\{Q\}^{(n)} = [K]^{(n)}\{u\}^{(n)} + \{Q_{\text{FEF}}\}^{(n)}$. Be sure to superimpose back with the FEFs.

$$\{Q\}^{(1)} = [K]^{(1)} \begin{Bmatrix} 0.0 \\ 0.00134 \\ 0.0 \\ 0.00045 \end{Bmatrix} + \begin{Bmatrix} -1.57 \\ -188.8 \\ 1.57 \\ -188.8 \end{Bmatrix} = \begin{Bmatrix} -0.22 \\ 0.0 \\ 0.22 \\ -53.9 \end{Bmatrix} \begin{matrix} \text{k} \\ \text{k·in.} \\ \text{k} \\ \text{k·in.} \end{matrix}$$

$$\{Q\}^{(2)} = [K]^{(2)} \begin{Bmatrix} 0.0 \\ 0.00045 \\ 0.0 \\ 0.0 \end{Bmatrix} + \begin{Bmatrix} 0.0 \\ 0.0 \\ 0.0 \\ 0.0 \end{Bmatrix} = \begin{Bmatrix} 0.34 \\ 53.9 \\ -0.34 \\ 27.0 \end{Bmatrix} \begin{matrix} \text{k} \\ \text{k·in.} \\ \text{k} \\ \text{k·in.} \end{matrix}$$

21.3 Stiffness Formulation for Structures Subjected to Temperature Changes

QUICK NOTE
When the sun rises each day, the temperature at the top of a bridge girder rises faster than it does at the bottom of the girder. This can cause the girder to crown. Anything that tries to prevent this crowning will require forces to develop.

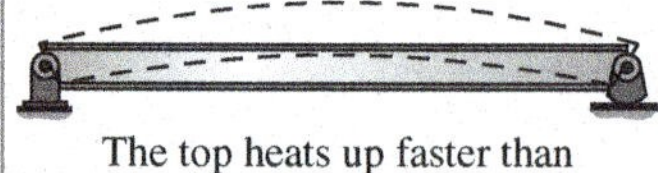

The top heats up faster than the bottom

Frequently, one or more members of a structure are subject to changes in temperature for various reasons. This will result in changes of length and/or curvature of those members. For a statically indeterminate structure, such changes will generally result in the development of internal forces. For example, if a truss bar heats up, it then wants to get longer. If the supports or other structural members try to prevent this elongation, forces are developed in the supports and members. If a structure is determinate, it is free to expand or contract without any forces trying to restrain it (see Figure 21.3). The analyst may need to know the forces and displacements caused by temperature changes.

FEFs for Truss Bars

It is well known that the extension that occurs in a bar due to a thermal change can be expressed as

$$\Delta = \alpha(\Delta T)L$$

RECALL
A brief introduction to thermal effects in truss bars was given in Section 12.4.

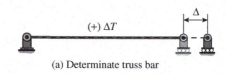

where α is the coefficient of thermal expansion, ΔT is the change in temperature, and L is the original length of the member.

The axial displacement subjected to an axial load N is also well established as

$$\Delta = \frac{NL}{AE}$$

where N is the internal axial force in the bar, L is the length, A is the cross-sectional area, and E is the modulus of elasticity.

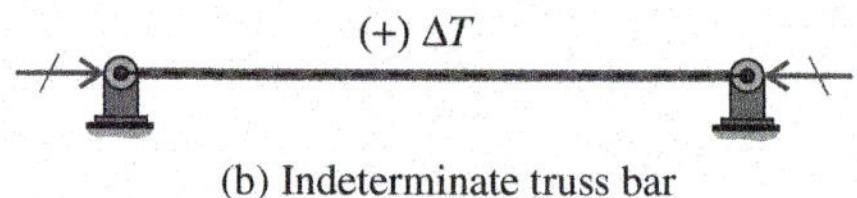

FIGURE 21.3 Comparison of a determinate and indeterminate truss bar subjected to a temperature increase.

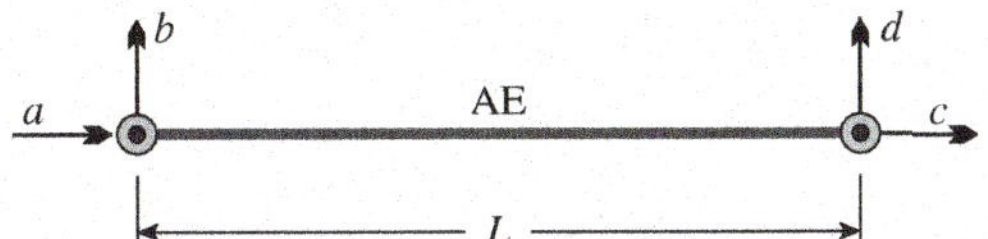

FIGURE 21.4 DOF labeling for a truss bar in local coordinates.

To be able to implement thermal changes into the matrix formulation, the FEFs due to a temperature change are required to be found. Recall that to find FEFs, all DOFs must first be locked—just like the second bar shown in Figure 21.3. The developed forces in the bar can be found by the following compatibility equation.

$$\alpha(\Delta T)L + \frac{NL}{AE} = 0$$

Solving for N, we get

$$N = -AE\alpha(\Delta T)$$

Thus, the FEF vector for a truss bar, as shown in Figure 21.4, is readily recognized as

$$\begin{Bmatrix} Q_{a-\text{FEF}} \\ Q_{b-\text{FEF}} \\ Q_{c-\text{FEF}} \\ Q_{d-\text{FEF}} \end{Bmatrix} = AE\alpha(\Delta T)\begin{Bmatrix} 1 \\ 0 \\ -1 \\ 0 \end{Bmatrix}$$

With the FEFs found, the analysis of the structure proceeds in the same fashion as it did for Example 21.1.

DID YOU KNOW?

One must convert the FEFs into the global coordinates to be able to assemble $\{F_{\text{FEF}}\}$. This easily can be accomplished with the transformation matrix $[T]$.

$$\{F_{\text{FEF}}\}^{(n)} = [T_n]^T \{Q_{\text{FEF}}\}^{(n)}$$

EXAMPLE 21.2

For the truss shown, bar number 2 experiences a temperature increase of 80°F. For all bars, $E = 10{,}000$ ksi, $A = 2.25\ \text{in}^2$, and $\alpha = 13.1 \times 10^{-6}/°\text{F}$. Use the matrix approach to find all joint displacements, reactions, and member forces. Use the DOF numbering given.

SOLUTION STRATEGY

Find all member and structural stiffness matrices as outlined in Chapter 19. You may specifically look at Example 19.7. The FEFs for member 2 must first be found in local coordinates and then converted to global coordinates to assemble the force vector. Remember to superimpose the FEFs with the statically equivalent structure to obtain final results.

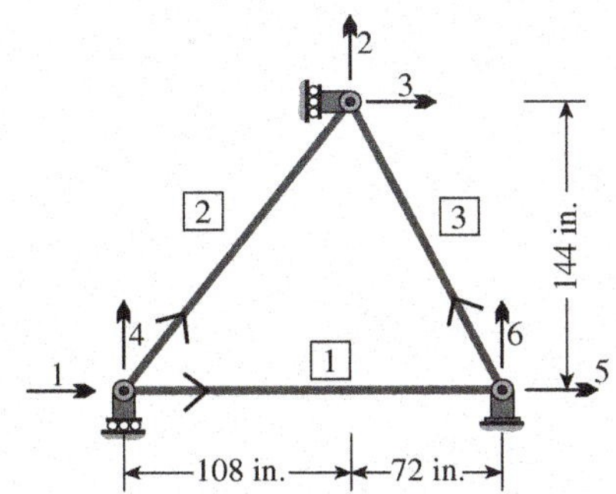

CALCULATION

The axial stiffness for each bar is computed as

$$\left(\frac{AE}{L}\right)_{1,2} = \frac{(2.25\ \text{in}^2)(10{,}000\ \text{ksi})}{180\ \text{in.}} = 125.0\ \text{k/in.}$$

$$\left(\frac{AE}{L}\right)_{3} = \frac{(2.25\ \text{in}^2)(10{,}000\ \text{ksi})}{161\ \text{in.}} = 139.8\ \text{k/in.}$$

SOLUTION

The element stiffness matrices in local coordinates are easily assembled using the stiffnesses calculated in the side bar. Thus, the full matrices are not shown here. Rather, the transformation matrices along with the global-element stiffness matrices are shown explicitly.

$$[T]_1 = \begin{array}{c} \\ a \\ b \\ c \\ d \end{array}\begin{array}{c} \begin{array}{cccc} 1 & 4 & 5 & 6 \end{array} \\ \begin{bmatrix} 1.00 & 0.00 & 0.00 & 0.00 \\ 0.00 & 1.00 & 0.00 & 0.00 \\ 0.00 & 0.00 & 1.00 & 0.00 \\ 0.00 & 0.00 & 0.00 & 1.00 \end{bmatrix} \end{array} \qquad K^{(1)} = \begin{array}{c} \\ 1 \\ 4 \\ 5 \\ 6 \end{array}\begin{array}{c} \begin{array}{cccc} 1 & 4 & 5 & 6 \end{array} \\ \begin{bmatrix} 125.0 & 0.0 & -125.0 & 0.0 \\ 0.0 & 0.0 & 0.0 & 0.0 \\ -125.0 & 0.0 & 125.0 & 0.0 \\ 0.0 & 0.0 & 0.0 & 0.0 \end{bmatrix} \end{array} \text{k/in.}$$

$$[T]_2 = \begin{array}{c} \\ a \\ b \\ c \\ d \end{array}\begin{array}{c} \begin{array}{cccc} 1 & 4 & 3 & 2 \end{array} \\ \begin{bmatrix} 0.60 & 0.80 & 0.00 & 0.00 \\ -0.80 & 0.60 & 0.00 & 0.00 \\ 0.00 & 0.00 & 0.60 & 0.80 \\ 0.00 & 0.00 & -0.80 & 0.60 \end{bmatrix} \end{array} \qquad K^{(2)} = \begin{array}{c} \\ 1 \\ 4 \\ 3 \\ 2 \end{array}\begin{array}{c} \begin{array}{cccc} 1 & 4 & 3 & 2 \end{array} \\ \begin{bmatrix} 45.0 & 60.0 & -45.0 & -60.0 \\ 60.0 & 80.0 & -60.0 & -80.0 \\ -45.0 & -60.0 & 45.0 & 60.0 \\ -60.0 & -80.0 & 60.0 & 80.0 \end{bmatrix} \end{array} \text{k/in.}$$

$$[T]_3 = \begin{array}{c} \\ a \\ b \\ c \\ d \end{array}\begin{array}{c} \begin{array}{cccc} 5 & 6 & 3 & 2 \end{array} \\ \begin{bmatrix} -0.45 & 0.89 & 0.00 & 0.00 \\ -0.89 & -0.45 & 0.00 & 0.00 \\ 0.00 & 0.00 & -0.45 & 0.89 \\ 0.00 & 0.00 & -0.89 & -0.45 \end{bmatrix} \end{array} \qquad K^{(3)} = \begin{array}{c} \\ 5 \\ 6 \\ 3 \\ 2 \end{array}\begin{array}{c} \begin{array}{cccc} 5 & 6 & 3 & 2 \end{array} \\ \begin{bmatrix} 28.0 & -55.9 & -28.0 & 55.9 \\ -55.9 & 111.8 & 55.9 & -111.8 \\ -28.0 & 55.9 & 28.0 & -55.9 \\ 55.9 & -111.8 & -55.9 & 111.8 \end{bmatrix} \end{array} \text{k/in.}$$

The following structure stiffness matrices now can be assembled.

$$K_{uu} = \begin{array}{c} \\ 1 \\ 2 \end{array}\begin{array}{c} \begin{array}{cc} 1 & 2 \end{array} \\ \begin{bmatrix} 170.0 & -60.0 \\ -60.0 & 191.8 \end{bmatrix} \end{array} \text{k/in.} \qquad K_{ru} = \begin{array}{c} \\ 3 \\ 4 \\ 5 \\ 6 \end{array}\begin{array}{c} \begin{array}{cc} 1 & 2 \end{array} \\ \begin{bmatrix} -45.0 & 4.1 \\ 60.0 & -80.0 \\ -125.0 & 55.9 \\ 0.0 & -111.8 \end{bmatrix} \end{array} \text{k/in.}$$

THERMAL FORCES

$AE\alpha(\Delta T) = (2.25 \text{ in}^2)(10{,}000 \text{ ksi})\cdot(13.1 \times 10^{-6})(80°F) = 23.6 \text{ k}$

The challenge in this problem is to assemble the force vector. Bar 2 is the only bar that experiences a thermal change, so we first find the FEFs for bar 2 in the local coordinates.

$$\{Q_{\text{FEF}}\}^{(2)} = \begin{array}{c} a \\ b \\ c \\ d \end{array}\begin{Bmatrix} 23.6 \\ 0.0 \\ -23.6 \\ 0.0 \end{Bmatrix}\begin{array}{c} \text{k} \\ \text{k} \\ \text{k} \\ \text{k} \end{array}$$

Now convert into global coordinates using $[T]_2$ that gives

$$\{F_{\text{FEF}}\}^{(2)} = [T_2]^T\{Q_{\text{FEF}}\}^{(2)} = \begin{array}{c} 1 \\ 4 \\ 3 \\ 2 \end{array}\begin{Bmatrix} 14.1 \\ 18{,}9 \\ -14.1 \\ -18.9 \end{Bmatrix}\begin{array}{c} \text{k} \\ \text{k} \\ \text{k} \\ \text{k} \end{array}$$

QUICK NOTE

Because there are no external loads applied to the truss, $\{F_u\}$ is

$$\{F_u\} = \begin{Bmatrix} 0.0 \\ 0.0 \end{Bmatrix}$$

If there were also joint loads, this is where they would be added.

The joint deflection now may be computed as

$$\Delta_u = \begin{Bmatrix} \Delta_1 \\ \Delta_2 \end{Bmatrix} = [K_{uu}]^{-1}(\{F_u\} - \{F_{\text{FEF}-u}\})$$

$$\Delta_u = \begin{Bmatrix} \Delta_1 \\ \Delta_2 \end{Bmatrix} = [K_{uu}]^{-1}\left(\begin{Bmatrix} 0.0 \\ 0.0 \end{Bmatrix} - \begin{Bmatrix} 14.1 \\ -18.9 \end{Bmatrix}\right) = \begin{Bmatrix} -0.0545 \\ 0.0813 \end{Bmatrix}\begin{array}{c} \text{in.} \\ \text{in.} \end{array}$$

Reactions come next:

$$F_r = \begin{Bmatrix} R_3 \\ R_4 \\ R_5 \\ R_6 \end{Bmatrix} = [K_{ru}]\{\Delta_u\} + \{F_{\text{FEF}-r}\}$$

$$F_r = \begin{Bmatrix} R_3 \\ R_4 \\ R_5 \\ R_6 \end{Bmatrix} = [K_{ru}]\begin{Bmatrix} -0.0545 \\ 0.0813 \end{Bmatrix} + \begin{Bmatrix} -14.1 \\ 18.9 \\ 0.0 \\ 0.0 \end{Bmatrix} = \begin{Bmatrix} -11.4 \\ 9.1 \\ 11.4 \\ -9.1 \end{Bmatrix} \begin{matrix} \text{k} \\ \text{k} \\ \text{k} \\ \text{k} \end{matrix}$$

The last step is to compute the bar forces, being sure to note that, since bar 2 had the temperature change, the final results are found through superposition.

RECALL

$$\{u\} = [T]\{\Delta\}$$

and

$$\{Q\} = [k]\{u\}$$

$$\{u\}^{(1)} = [T]_1 \begin{Bmatrix} -0.0545 \\ 0.0 \\ 0.0 \\ 0.0 \end{Bmatrix} = \begin{Bmatrix} -0.0545 \\ 0.0 \\ 0.0 \\ 0.0 \end{Bmatrix} \begin{matrix} \text{in.} \\ \text{in.} \\ \text{in.} \\ \text{in.} \end{matrix} \qquad \{u\}^{(2)} = [T]_2 \begin{Bmatrix} -0.0545 \\ 0.0 \\ 0.0 \\ 0.0813 \end{Bmatrix} = \begin{Bmatrix} -0.0327 \\ 0.0436 \\ 0.0650 \\ 0.0488 \end{Bmatrix} \begin{matrix} \text{in.} \\ \text{in.} \\ \text{in.} \\ \text{in.} \end{matrix}$$

$$\{u\}^{(3)} = [T]_3 \begin{Bmatrix} 0.0 \\ 0.0 \\ 0.0 \\ 0.0813 \end{Bmatrix} = \begin{Bmatrix} 0.0 \\ 0.0 \\ 0.0727 \\ -0.0364 \end{Bmatrix} \begin{matrix} \text{in.} \\ \text{in.} \\ \text{in.} \\ \text{in.} \end{matrix}$$

$$\{Q\}^{(1)} = [k]^{(1)} \begin{Bmatrix} -0.0545 \\ 0.0 \\ 0.0 \\ 0.0 \end{Bmatrix} = \begin{Bmatrix} -6.8 \\ 0.0 \\ 6.8 \\ 0.0 \end{Bmatrix} \begin{matrix} \text{k} \\ \text{k} \\ \text{k} \\ \text{k} \end{matrix} \qquad \{Q\}^{(3)} = [k]^{(3)} \begin{Bmatrix} 0.0 \\ 0.0 \\ 0.0727 \\ -0.0364 \end{Bmatrix} = \begin{Bmatrix} -10.2 \\ 0.0 \\ 10.2 \\ 0.0 \end{Bmatrix} \begin{matrix} \text{k} \\ \text{k} \\ \text{k} \\ \text{k} \end{matrix}$$

$$\{Q\}^{(2)} = [k]^{(2)} \begin{Bmatrix} -0.0327 \\ 0.0436 \\ 0.0650 \\ 0.0488 \end{Bmatrix} + \begin{Bmatrix} 23.6 \\ 0.0 \\ -23.6 \\ 0.0 \end{Bmatrix} = \begin{Bmatrix} 11.4 \\ 0.0 \\ -11.4 \\ 0.0 \end{Bmatrix} \begin{matrix} \text{k} \\ \text{k} \\ \text{k} \\ \text{k} \end{matrix}$$

FEFs for Beams

A beam can actually experience a temperature gradient across its depth (d). This means that the temperature at the top of the beam (T_t) and at the bottom (T_b) are different. This causes different strain rates to occur across the depth of the beam, which in turn induces curvature as shown in Figure 21.5.

Recognizing that the temperature gradient wants to cause curvature, we can quickly understand that moments will develop at the beam ends when the beam is fixed-fixed (i.e., restrained from motion). A full derivation of these fixed-end moments is not shown here but may be found

DID YOU KNOW?

The temperature in a beam is assumed to vary linearly across the depth of the beam. A well know mechanics equation states that the strain due to temperature can be computed as

$$\epsilon = (\Delta T)(\alpha)$$

Thus, the strain is assumed to vary linearly across the depth of the beam.

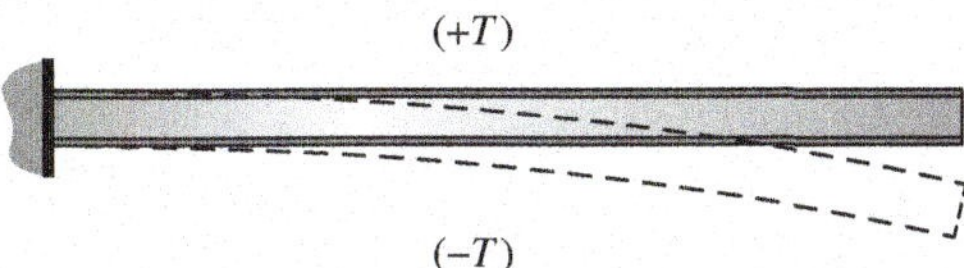

FIGURE 21.5 Beam experiences curvature when a temperature gradient is introduced.

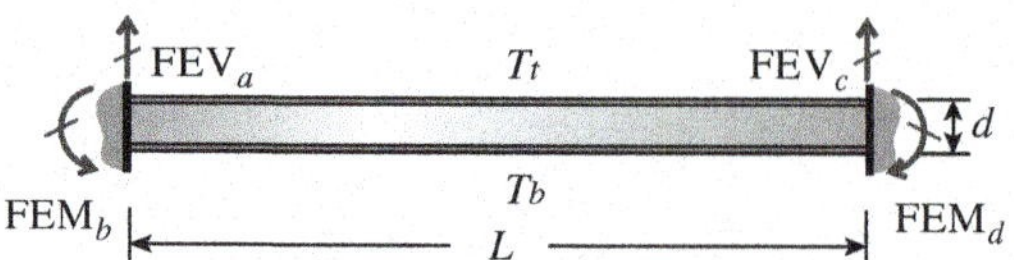

FIGURE 21.6 Beam FEFs due to temperature gradient.

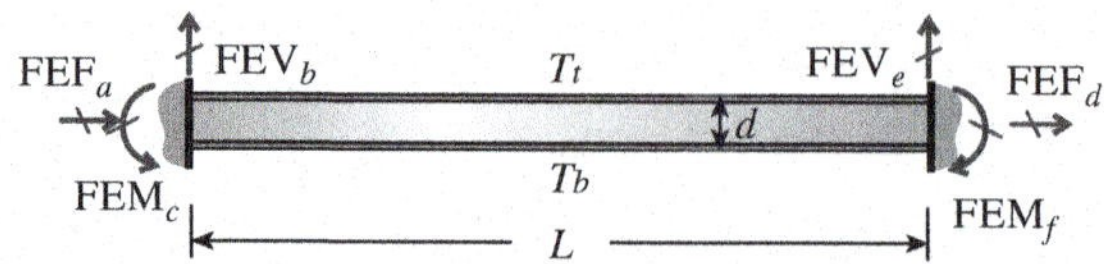

FIGURE 21.7 Frame FEFs due to temperature gradient.

NOTATION

- d = depth of the beam
- T_t = temperature at the top of the beam
- T_b = temperature at the bottom of the beam
- α = coefficient of thermal expansion
- L = length of beam
- I = moment of inertia
- E = modulus of elasticity

QUICK NOTE

The element FEFs shown in Figure 21.7 must be converted to the global coordinate system using the matrix method.

$$\left\{F_{\text{FEF}}\right\}^{(n)} = [T]_n \left\{Q_{\text{FEF}}\right\}^{(n)}$$

elsewhere.[1] Thus, the FEFs due to a temperature gradient are given in Figure 21.6 without derivation.

$$\begin{Bmatrix} \text{FEV}_a \\ \text{FEM}_b \\ \text{FEV}_c \\ \text{FEM}_d \end{Bmatrix} = \begin{Bmatrix} 0.0 \\ \frac{EI\alpha(T_b - T_t)}{d} \\ 0.0 \\ -\frac{EI\alpha(T_b - T_t)}{d} \end{Bmatrix}$$

If axial motion is also considered, as it is in frames, the analyst must include the terms for the axial DOFs.

$$\begin{Bmatrix} \text{FEF}_a \\ \text{FEV}_b \\ \text{FEM}_c \\ \text{FEF}_d \\ \text{FEV}_e \\ \text{FEM}_f \end{Bmatrix} = \begin{Bmatrix} AE\alpha\frac{T_b - T_t}{2} \\ 0.0 \\ \frac{EI\alpha(T_b - T_t)}{d} \\ -AE\alpha\frac{T_b - T_t}{2} \\ 0.0 \\ -\frac{EI\alpha(T_b - T_t)}{d} \end{Bmatrix}$$

EXAMPLE 21.3

The two-span steel beam is subjected to a temperature gradient as shown. Consider $\alpha = 6.5 \times 10^{-6}/°\text{F}$ and use the matrix approach to solve for the joint rotations, reactions, and member end forces.

110°F [1] 110°F [2]
d = 36 in.
70°F 70°F
80 ft 80 ft
E = 29,000 ksi I = 10,500 in^4

SOLUTION STRATEGY

The task is to find the FEFs due to the temperature gradient. The rest of the analysis is the same process demonstrated in the previous example.

[1] McGuire, W., Gallagher, R.H., and Ziemian, R.D. (2000) *Matrix Structural Analysis, 2nd Ed.*, John Wiley and Sons, Inc., New York, NY.

SOLUTION

Assemble the element and structure stiffness matrices based on member properties. Label the DOFs as shown.

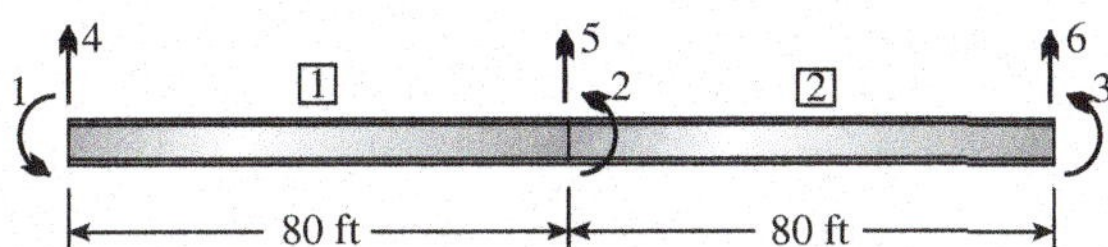

QUICK NOTE

The units used throughout this example are k, in., and °F.

$$K^{(1)} = \begin{array}{c} \\ 4 \\ 1 \\ 5 \\ 2 \end{array}\begin{array}{c} \begin{array}{cccc} 4 & 1 & 5 & 2 \end{array} \\ \begin{bmatrix} 5.9 & 2832 & -5.9 & 2832 \\ 2832 & 1{,}812{,}500 & -2832 & 906{,}250 \\ -5.9 & -2832 & 5.9 & -2832 \\ 2832 & 906{,}250 & -2832 & 1{,}812{,}500 \end{bmatrix} \end{array} \qquad K^{(2)} = \begin{array}{c} \\ 5 \\ 2 \\ 6 \\ 3 \end{array}\begin{array}{c} \begin{array}{cccc} 5 & 2 & 6 & 3 \end{array} \\ \begin{bmatrix} 5.9 & 2832 & -5.9 & 2832 \\ 2832 & 1{,}812{,}500 & -2832 & 906{,}250 \\ -5.9 & -2832 & 5.9 & -2832 \\ 2832 & 906{,}250 & -2832 & 1{,}812{,}500 \end{bmatrix} \end{array}$$

The following structure stiffness matrices now can be assembled.

$$K_{uu} = \begin{array}{c} \\ 1 \\ 2 \\ 3 \end{array}\begin{array}{c} \begin{array}{ccc} 1 & 2 & 3 \end{array} \\ \begin{bmatrix} 1{,}812{,}500 & 906{,}250 & 0.0 \\ 906{,}250 & 3{,}625{,}000 & 906{,}250 \\ 0.0 & 906{,}250 & 1{,}812{,}500 \end{bmatrix} \end{array} \qquad K_{ru} = \begin{array}{c} \\ 4 \\ 5 \\ 6 \end{array}\begin{array}{c} \begin{array}{ccc} 1 & 2 & 3 \end{array} \\ \begin{bmatrix} 2832 & 2832 & 0.0 \\ -2832 & 0.0 & 2832 \\ 0.0 & -2832 & -2832 \end{bmatrix} \end{array}$$

Assemble the FEF vectors for each member. In this problem, these vectors will be identical.

$$\frac{EI\alpha(T_b - T_t)}{d} = \frac{(29{,}000\text{ ksi})(15{,}000\text{ in}^4)(6.5 \times 10^{-6}/°\text{F})(70° - 110°)}{36\text{ in.}} = -3142\text{ k}\cdot\text{in.}$$

$$\left\{F_{\text{FEF}}\right\}^{(1)} = \left\{Q_{\text{FEF}}\right\}^{(1)} = \begin{array}{c} 4 \\ 1 \\ 5 \\ 2 \end{array}\begin{Bmatrix} 0.0 \\ -3142 \\ 0.0 \\ 3142 \end{Bmatrix}\begin{array}{l} \text{k} \\ \text{k}\cdot\text{in.} \\ \text{k} \\ \text{k}\cdot\text{in.} \end{array} \qquad \left\{F_{\text{FEF}}\right\}^{(2)} = \left\{Q_{\text{FEF}}\right\}^{(2)} = \begin{array}{c} 5 \\ 2 \\ 6 \\ 3 \end{array}\begin{Bmatrix} 0.0 \\ -3142 \\ 0.0 \\ 3142 \end{Bmatrix}\begin{array}{l} \text{k} \\ \text{k}\cdot\text{in.} \\ \text{k} \\ \text{k}\cdot\text{in.} \end{array}$$

Assembling this into a structure-level force vector produces:

$$\begin{Bmatrix} F_{\text{FEF}-u} \\ \hline F_{\text{FEF}-r} \end{Bmatrix} = \begin{array}{c} 1 \\ 2 \\ 3 \\ 4 \\ 5 \\ 6 \end{array}\begin{Bmatrix} -3142 \\ 0.0 \\ 3142 \\ \hline 0.0 \\ 0.0 \\ 0.0 \end{Bmatrix}\begin{array}{l} \text{k}\cdot\text{in.} \\ \text{k}\cdot\text{in.} \\ \text{k}\cdot\text{in.} \\ \text{k} \\ \text{k} \\ \text{k} \end{array}$$

QUICK NOTE

F_u for this beam is

$$\left\{F_u\right\} = \begin{array}{c} 1 \\ 2 \\ 3 \end{array}\begin{Bmatrix} 0.0 \\ 0.0 \\ 0.0 \end{Bmatrix}\begin{array}{l} \text{k}\cdot\text{in.} \\ \text{k}\cdot\text{in.} \\ \text{k}\cdot\text{in.} \end{array}$$

since no external loads are applied at the joints.

Joint rotations and computed reactions are easily calculated. Thus,

$$\left\{\Delta_u\right\} = \begin{Bmatrix} \theta_1 \\ \theta_2 \\ \theta_3 \end{Bmatrix} = \left[K_{uu}\right]^{-1}\left(\begin{Bmatrix} 0.0 \\ 0.0 \\ 0.0 \end{Bmatrix} - \begin{Bmatrix} -3142 \\ 0.0 \\ 3142 \end{Bmatrix}\right) = \underline{\underline{\begin{Bmatrix} 0.0017 \\ 0.0 \\ -0.0017 \end{Bmatrix}\begin{array}{l} \text{rad} \\ \text{rad} \\ \text{rad} \end{array}}}$$

$$\{F_r\} = \begin{Bmatrix} R_4 \\ R_5 \\ R_6 \end{Bmatrix} = [K_{ru}] \begin{Bmatrix} 0.0017 \\ 0.0 \\ -0.0017 \end{Bmatrix} + \begin{Bmatrix} 0.0 \\ 0.0 \\ 0.0 \end{Bmatrix} = \begin{Bmatrix} 4.9 \\ -9.8 \\ 4.9 \end{Bmatrix} \begin{matrix} \text{k} \\ \text{k} \\ \text{k} \end{matrix}$$

The member end forces are the last quantity to be computed, and this is done using the element stiffness matrices. So

$$\{Q\}^{(1)} = \begin{Bmatrix} Q_4 \\ M_1 \\ Q_5 \\ M_2 \end{Bmatrix} = [K]^{(1)} \begin{Bmatrix} 0.0 \\ 0.0017 \\ 0.0 \\ 0.0 \end{Bmatrix} + \begin{Bmatrix} 0.0 \\ -3142 \\ 0.0 \\ 3142 \end{Bmatrix} = \begin{Bmatrix} 4.9 \\ 0.0 \\ -4.9 \\ 4713 \end{Bmatrix} \begin{matrix} \text{k} \\ \text{k·in.} \\ \text{k} \\ \text{k·in.} \end{matrix}$$

$$\{Q\}^{(2)} = \begin{Bmatrix} Q_5 \\ M_2 \\ Q_6 \\ M_3 \end{Bmatrix} = [K]^{(2)} \begin{Bmatrix} 0.0 \\ 0.0 \\ 0.0 \\ -0.0017 \end{Bmatrix} + \begin{Bmatrix} 0.0 \\ -3142 \\ 0.0 \\ 3142 \end{Bmatrix} = \begin{Bmatrix} -4.9 \\ -4713 \\ 4.9 \\ 0.0 \end{Bmatrix} \begin{matrix} \text{k} \\ \text{k·in.} \\ \text{k} \\ \text{k·in.} \end{matrix}$$

21.4 Stiffness Formulation for Structures with Misfit Members

QUICK NOTE
A positive value of e represents a bar that is too long while a negative value represents a bar that is too short.

Many structures are fabricated with one or more members having incorrect lengths. If the structure is "forced" together during assembly, forces may develop in various portions of the structure—even when no external loads are applied. The procedure for determining these "locked-in" forces is essentially the same as has been described for members that experience temperature changes or support settlements. Once again, the only challenge is to come up with the correct FEFs to use in the analysis.

Let e equal the fabrication error in the length of a member. Then the axial force that develops in this member when it is forced into position will be

DID YOU KNOW?
A bar that is too long will be put into compression when it is forced into position. This is why the axial force equation uses $-e$.

$$N = \left(\frac{AE}{L}\right)(-e)$$

The FEFs can be readily assembled with this understanding.

EXAMPLE 21.4

QUICK NOTE
Temperature changes are not considered as part of this example, but if they were, the FEF vector would simply include the thermal effects and the misfit effects.

Consider the truss configuration presented in Example 21.2. If bar number 2 was fabricated to be 0.5 in. too long, use the matrix approach to find all joint displacements, reactions, and member forces. For all bars $E = 10,000$ ksi and $A = 2.25$ in^2. Use the DOF numbering given.

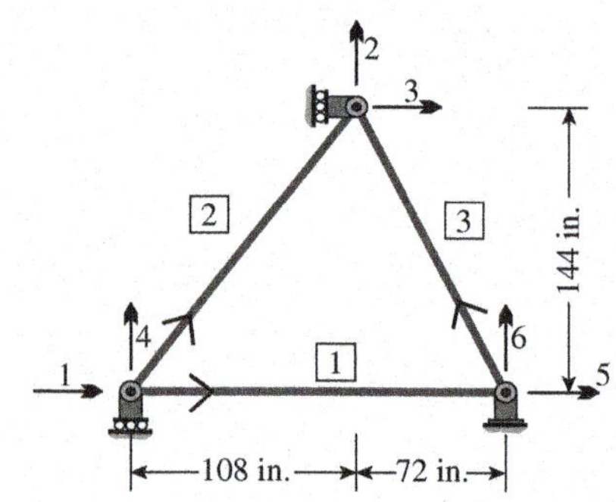

SOLUTION STRATEGY

Recognize that all stiffness and transformation matrices were already developed in Example 21.2. Therefore, the problem should start out by assembling the FEF vectors, and then proceed with the analysis as done previously.

FEF DUE TO MISFIT

If bar number 2 was fabricated to be too long by 0.5 in., the the fixed-end axial force that develops is

$$N_2 = \frac{(2.25\text{ in}^2)(10{,}000\text{ ksi})}{180\text{ in.}}(-0.5\text{ in.})$$

$$= -62.5\text{k}$$

SOLUTION

Recognizing that bar number 2 is in compression (see the side note), the FEF vector for bar 2 is as shown.

$$\left\{Q_{\text{FEF}}\right\}^{(2)} = \begin{matrix} a \\ b \\ c \\ d \end{matrix} \begin{Bmatrix} 62.5 \\ 0.0 \\ -62.5 \\ 0.0 \end{Bmatrix} \begin{matrix} \text{k} \\ \text{k} \\ \text{k} \\ \text{k} \end{matrix}$$

Now convert into global coordinates using $[T]_2$.

$$\left\{F_{\text{FEF}}\right\}^{(2)} = \left[T_2\right]^T \left\{Q_{\text{FEF}}\right\}^{(2)} = \begin{matrix} 1 \\ 4 \\ 3 \\ 2 \end{matrix} \begin{Bmatrix} 37.5 \\ 50.0 \\ -37.5 \\ -50.0 \end{Bmatrix} \begin{matrix} \text{k} \\ \text{k} \\ \text{k} \\ \text{k} \end{matrix}$$

The joint deflection now may be computed as

$$\Delta_u = \begin{Bmatrix} \Delta_1 \\ \Delta_2 \end{Bmatrix} = [K_{uu}]^{-1}\left(\{F_u\} - \{F_{\text{FEF}-u}\}\right)$$

$$\Delta_u = \begin{Bmatrix} \Delta_1 \\ \Delta_2 \end{Bmatrix} = [K_{uu}]^{-1}\left(\begin{Bmatrix} 0.0 \\ 0.0 \end{Bmatrix} - \begin{Bmatrix} 37.5 \\ -50.0 \end{Bmatrix}\right) = \begin{Bmatrix} -0.1445 \\ 0.2155 \end{Bmatrix} \begin{matrix} \text{in.} \\ \text{in.} \end{matrix}$$

QUICK NOTE

Because there are no external loads applied to the truss, $\{F_u\}$ is given as

$$\{F_u\} = \begin{Bmatrix} 0.0 \\ 0.0 \end{Bmatrix}$$

If there were also joint loads, this is where they would be added.

Reactions come next:

$$F_r = \begin{Bmatrix} R_3 \\ R_4 \\ R_5 \\ R_6 \end{Bmatrix} = [K_{ru}]\{\Delta_u\} + \left\{F_{\text{FEF}-r}\right\}$$

$$F_r = \begin{Bmatrix} R_3 \\ R_4 \\ R_5 \\ R_6 \end{Bmatrix} = [K_{ru}]\begin{Bmatrix} -0.1445 \\ 0.2155 \end{Bmatrix} + \begin{Bmatrix} -37.5 \\ 50.0 \\ 0.0 \\ 0.0 \end{Bmatrix} = \begin{Bmatrix} -30.11 \\ 24.09 \\ 30.11 \\ -24.09 \end{Bmatrix} \begin{matrix} \text{k} \\ \text{k} \\ \text{k} \\ \text{k} \end{matrix}$$

The last step is to compute the bar forces, being sure to note that, since bar 2 had the misfit, the final results are found through superposition. Thus,

RECALL

$$\{u\} = [T]\{\Delta\}$$

and

$$\{Q\} = [k]\{u\}$$

$$\{u\}^{(1)} = [T]_1 \begin{Bmatrix} -0.1445 \\ 0.0 \\ 0.0 \\ 0.0 \end{Bmatrix} = \begin{Bmatrix} -0.1445 \\ 0.0 \\ 0.0 \\ 0.0 \end{Bmatrix} \begin{matrix} \text{in.} \\ \text{in.} \\ \text{in.} \\ \text{in.} \end{matrix} \qquad \{u\}^{(2)} = [T]_2 \begin{Bmatrix} -0.1445 \\ 0.0 \\ 0.0 \\ 0.2155 \end{Bmatrix} = \begin{Bmatrix} -0.0867 \\ 0.1156 \\ 0.1724 \\ 0.1293 \end{Bmatrix} \begin{matrix} \text{in.} \\ \text{in.} \\ \text{in.} \\ \text{in.} \end{matrix}$$

$$\{u\}^{(3)} = [T]_3 \begin{Bmatrix} 0.0 \\ 0.0 \\ 0.0 \\ 0.2155 \end{Bmatrix} = \begin{Bmatrix} 0.0 \\ 0.0 \\ 0.1927 \\ -0.0964 \end{Bmatrix} \begin{matrix} \text{in.} \\ \text{in.} \\ \text{in.} \\ \text{in.} \end{matrix}$$

$$\{Q\}^{(1)} = [k]^{(1)} \begin{Bmatrix} -0.1445 \\ 0.0 \\ 0.0 \\ 0.0 \end{Bmatrix} = \begin{Bmatrix} -18.1 \\ 0.0 \\ 18.1 \\ 0.0 \end{Bmatrix} \begin{matrix} \text{k} \\ \text{k} \\ \text{k} \\ \text{k} \end{matrix} \qquad \{Q\}^{(3)} = [k]^{(3)} \begin{Bmatrix} 0.0 \\ 0.0 \\ 0.1927 \\ -0.0964 \end{Bmatrix} = \begin{Bmatrix} -26.9 \\ 0.0 \\ 26.9 \\ 0.0 \end{Bmatrix} \begin{matrix} \text{k} \\ \text{k} \\ \text{k} \\ \text{k} \end{matrix}$$

$$\{Q\}^{(2)} = [k]^{(2)} \begin{Bmatrix} -0.0867 \\ 0.1156 \\ 0.1724 \\ 0.1293 \end{Bmatrix} + \begin{Bmatrix} 62.5 \\ 0.0 \\ -62.5 \\ 0.00 \end{Bmatrix} = \begin{Bmatrix} 30.1 \\ 0.0 \\ -30.1 \\ 0.0 \end{Bmatrix} \begin{matrix} \text{k} \\ \text{k} \\ \text{k} \\ \text{k} \end{matrix}$$

21.5 Static Condensation

An important topic in the matrix analysis of structures is the procedure known as *static condensation*. This procedure permits an analyst to reduce the order of the set of stiffness equations to simplify the problem or to accomplish some other objective. In essence, it takes a structure that contains a certain number of DOFs and then reduces the number of DOFs used to describe that same structure.

QUICK NOTE
The next section will demonstrate a useful application of static condensation.

Consider the cantilever beam shown in Figure 21.8. It has a total of two unrestrained DOFs with the following stiffness matrix.

$$\left[K_{uu}\right] = \begin{matrix} & 1 & 2 \\ 1 & \frac{12EI}{L^3} & \frac{6EI}{L^2} \\ 2 & \frac{6EI}{L^2} & \frac{4EI}{L} \end{matrix}$$

QUICK NOTE
Admittedly, the true benefit of static condensation is not well recognized when using only a two-DOF structure. However, the procedure and its impact on results is easily seen on such a small structure.

If there is a load of $-P$ applied at DOF 1 and nothing at DOF 2, then the following joint displacements may be computed.

$$\begin{Bmatrix} \Delta_1 \\ \theta_2 \end{Bmatrix} = \left[K_{uu}\right]^{-1} \begin{Bmatrix} -P \\ 0 \end{Bmatrix} = \begin{Bmatrix} -\frac{PL^3}{3EI} \\ \frac{PL^2}{2EI} \end{Bmatrix}$$

In this case, the analyst may not have really been interested in knowing the rotation at DOF 2. This then begs the question, "Is it possible to represent this beam using only one unrestrained DOF instead of two?" The answer to this question is yes. We choose the DOF we want to suppress and write it in terms of the DOF we want to keep.

QUICK NOTE
DOFs we want to keep:

$p \equiv$ primary DOFs

DOFs we want to hide:

$s \equiv$ secondary DOFs

Static Condensation Procedure

When performing a condensation of the structural stiffness matrix, the analyst must identify the DOFs they want to keep. We will refer to these as the *primary* DOFs, which will be represented using a subscript of p. They will also need to identify the DOFs they want to suppress, which

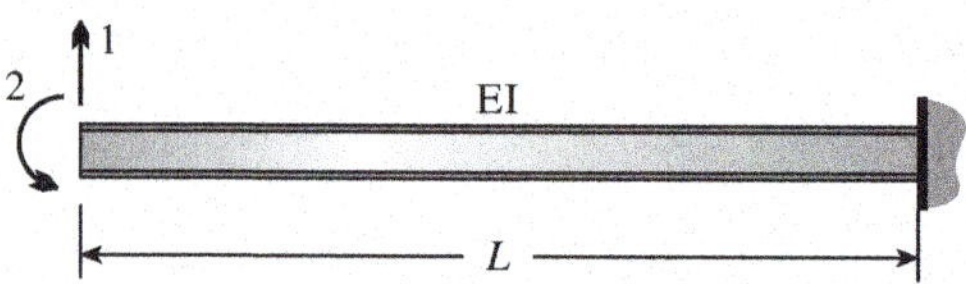

FIGURE 21.8 Cantilever beam with two unrestrained DOFs.

will be called *secondary* and denoted with a subscript of s. Thus, the $\left[K_{uu}\right]$ can conveniently be partitioned based on primary and secondary DOFs as

$$\left[K_{uu}\right] = \left[\begin{array}{c|c} K_{pp} & K_{ps} \\ \hline K_{sp} & K_{ss} \end{array}\right]$$

The full equilibrium equation of $[K_{uu}]\{\Delta_u\} = \{F_u\}$ then takes on the following form when we partition according to primary and secondary DOFs.

$$\left[\begin{array}{c|c} K_{pp} & K_{ps} \\ \hline K_{sp} & K_{ss} \end{array}\right] \left\{\begin{array}{c} \Delta_p \\ \hline \Delta_s \end{array}\right\} = \left\{\begin{array}{c} F_p \\ \hline F_s \end{array}\right\}$$

The goal here is to rewrite this equation so that $\{\Delta_s\}$ is written in terms of $\{\Delta_p\}$. Multiply out the lower row of the equilibrium equation and get

$$[K_{sp}]\{\Delta_p\} + [K_{ss}]\{\Delta_s\} = \{F_s\}$$

Now solve for the expression for $\{\Delta_s\}$, so

$$\{\Delta_s\} = [K_{ss}]^{-1}\left(\{F_s\} - [K_{sp}]\{\Delta_p\}\right)$$

QUICK NOTE

Don't get lost in the mathematics. If you need to review matrix multiplication, refer to Appendix C.

Take the original equilibrium equation and multiply out the top row this time. Thus,

$$[K_{pp}]\{\Delta_p\} + [K_{ps}]\{\Delta_s\} = \{F_p\}$$

Note that this expression is currently in terms of $\{\Delta_p\}$ and $\{\Delta_s\}$, but we only want it in terms of $\{\Delta_p\}$. Therefore, we substitute into this equation the previously derived expression for $\{\Delta_s\}$, giving

$$[K_{pp}]\{\Delta_p\} + [K_{ps}][K_{ss}]^{-1}\left(\{F_s\} - [K_{sp}]\{\Delta_p\}\right) = \{F_p\}$$

The next couple of steps will first distribute the $[K_{ps}][K_{ss}]^{-1}$ term and will then collect like terms.

Distribute

$$[K_{pp}]\{\Delta_p\} + [K_{ps}][K_{ss}]^{-1}\{F_s\} - [K_{ps}][K_{ss}]^{-1}[K_{sp}]\{\Delta_p\} = \{F_p\}$$

Collect Like Terms

$$\left([K_{pp}] - [K_{ps}][K_{ss}]^{-1}[K_{sp}]\right)\{\Delta_p\} = \{F_p\} - [K_{ps}][K_{ss}]^{-1}\{F_s\}$$

QUICK NOTE

Recognize that the force vectors are positioned on one side of the equation and the displacement vector $\{\Delta_p\}$ is on the other.

Make the following variable definitions. Let,

$$\begin{aligned} [\hat{K}_{pp}] &= [K_{pp}] - [K_{ps}][K_{ss}]^{-1}[K_{sp}] \\ \{\hat{F}_p\} &= \{F_p\} - [K_{ps}][K_{ss}]^{-1}\{F_s\} \end{aligned}$$

Therefore, the equation may be represented in the following simplified form.

$$[\hat{K}_{pp}]\{\Delta_p\} = \{\hat{F}_p\}$$

What this tells us is that we can find the displacements at our primary DOFs by using a condensed stiffness matrix $[\hat{K}_{pp}]$ and a condensed force vector $\{\hat{F}_p\}$.

EXAMPLE 21.5

For the cantilever beam shown in Figure 21.8, condense out the rotational DOF number 2, and then solve for the displacement at DOF 1 if a load of $-P$ is applied to 1.

SOLUTION STRATEGY

The $[K_{uu}]$ matrix is already given with Figure 21.8. Use the expressions previously derived for $[\hat{K}_{pp}]$ and $\{\hat{F}_p\}$. Use these to solve for the displacement at DOF 1.

SOLUTION

The $[K_{uu}]$ stiffness matrix is partitioned as follows because DOF 1 is a primary DOF and DOF 2 is a secondary DOF.

$$\left[K_{uu}\right] = \left[\begin{array}{c|c} K_{pp} & K_{ps} \\ \hline K_{sp} & K_{ss} \end{array}\right] = \begin{array}{c} \\ 1 \\ 2 \end{array}\begin{array}{c} \begin{array}{cc} 1 & \quad 2 \end{array} \\ \left[\begin{array}{c|c} \frac{12EI}{L^3} & \frac{6EI}{L^2} \\ \hline \frac{6EI}{L^2} & \frac{4EI}{L} \end{array}\right] \end{array}$$

DID YOU KNOW?
The force vector for the uncondensed structure is

$$\{F_U\} = \left\{\begin{array}{c} F_p \\ \hline F_s \end{array}\right\} = \left\{\begin{array}{c} -P \\ \hline 0.0 \end{array}\right\}$$

since there is no applied moment at DOF 2.

The condensed stiffness matrix is computed as

$$[\hat{K}_{pp}] = \frac{12EI}{L^3} - \left(\frac{6EI}{L^2}\right)\left(\frac{L}{4EI}\right)\left(\frac{6EI}{L^2}\right) = \frac{3EI}{L^3}$$

The condensed stiffness force vector is computed as

$$\{\hat{F}_p\} = -P - \left(\frac{6EI}{L^2}\right)\left(\frac{L}{4EI}\right)(0.0) = -P$$

The displacement at DOF 1 is now found with

$$\{\Delta_p\} = [\hat{K}_{pp}]^{-1}\{\hat{F}_p\}$$

$$\{\Delta_p\} = \left(\frac{L^3}{3EI}\right)(-P) = \underline{\underline{-\frac{PL^3}{3EI}}}$$

Notice that this is the same answer found before when the two-DOF formulation was used at the beginning of this section.

EXAMPLE 21.6

A simple portal frame is shown here and is modeled using four DOFs. Assume that the given 4×4 stiffness matrix and the associated force vector are as shown using units of *k* and *in*. Solve for the displacements at all DOFs. Then condense out the rotational DOFs (i.e., DOF 3 and 4) to obtain $[\hat{K}_{pp}]$ and $\{\hat{F}_p\}$. Use these condensed matrices to solve for the displacements at DOFs 1 and 2. Compare with the original solution.

$$[K_{uu}] = \begin{array}{c} \\ 1 \\ 2 \\ 3 \\ 4 \end{array}\begin{array}{c} \begin{array}{cccc} 1 & 2 & 3 & 4 \end{array} \\ \left[\begin{array}{cccc} 1762 & -1611 & 9063 & 0 \\ -1611 & 1762 & 0 & 9063 \\ 9063 & 0 & 1,208,333 & 241,667 \\ 0 & 9063 & 241,667 & 1,208,333 \end{array}\right] \end{array} \qquad \{F_u\} = \begin{array}{c} 1 \\ 2 \\ 3 \\ 4 \end{array}\left\{\begin{array}{c} 0.0 \\ 0.0 \\ 0.0 \\ 3600 \end{array}\right\}\begin{array}{l} \text{k} \\ \text{k} \\ \text{k·in.} \\ \text{k·in.} \end{array}$$

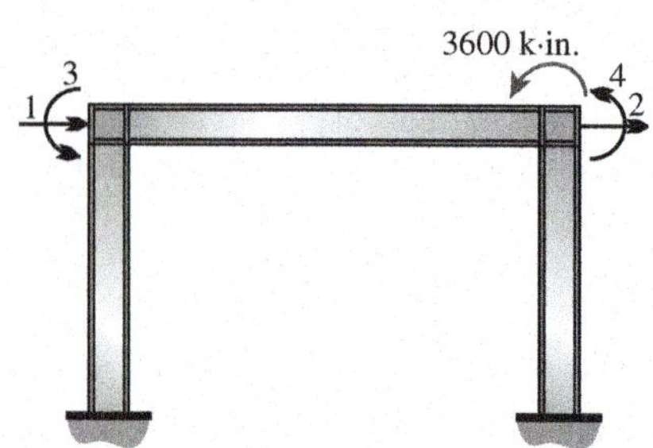

SOLUTION STRATEGY

Given the stiffness matrix and force vector, you can directly solve for the joint displacements using the usual approach. To perform condensation, partition the $[K_{uu}]$ matrix into the primary (1 and 2) and the secondary (3 and 4) DOFs. Follow the condensation procedure previously outlined.

SOLUTION

Find joint displacements for the four-DOF formulation.

$$\begin{Bmatrix} \Delta_1 \\ \theta_2 \\ \Delta_3 \\ \theta_4 \end{Bmatrix} = \begin{bmatrix} 1762 & -1611 & 9063 & 0 \\ -1611 & 1762 & 0 & 9063 \\ 9063 & 0 & 1,208,333 & 241,667 \\ 0 & 9063 & 241,667 & 1,208,333 \end{bmatrix}^{-1} \begin{Bmatrix} 0.0 \\ 0.0 \\ 0.0 \\ 3600 \end{Bmatrix} = \begin{Bmatrix} -0.114 \\ -0.124 \\ 0.00008 \\ 0.00390 \end{Bmatrix} \begin{matrix} \text{in.} \\ \text{in.} \\ \text{rad} \\ \text{rad} \end{matrix}$$

The $[K_{uu}]$ matrix can be partitioned according to the primary and secondary DOFS. This results in the following matrices.

$$[K_{pp}] = \begin{matrix} 1 \\ 2 \end{matrix} \overset{\begin{matrix} 1 & \quad 2 \end{matrix}}{\begin{bmatrix} 1762 & -1611 \\ -1611 & 1762 \end{bmatrix}} \qquad [K_{ps}] = \begin{matrix} 1 \\ 2 \end{matrix} \overset{\begin{matrix} 3 & \quad 4 \end{matrix}}{\begin{bmatrix} 9063 & 0 \\ 0 & 9063 \end{bmatrix}}$$

$$[K_{sp}] = \begin{matrix} 3 \\ 4 \end{matrix} \overset{\begin{matrix} 1 & \quad 2 \end{matrix}}{\begin{bmatrix} 9063 & 0 \\ 0 & 9063 \end{bmatrix}} \qquad [K_{ss}] = \begin{matrix} 3 \\ 4 \end{matrix} \overset{\begin{matrix} 3 & \quad 4 \end{matrix}}{\begin{bmatrix} 1,208,333 & 241,667 \\ 241,667 & 1,208,333 \end{bmatrix}}$$

The same partitioning must happen for the force vector as well.

$$\{F_p\} = \begin{matrix} 1 \\ 2 \end{matrix} \begin{Bmatrix} 0.0 \\ 0.0 \end{Bmatrix} \begin{matrix} \text{k} \\ \text{k} \end{matrix} \qquad \{F_s\} = \begin{matrix} 3 \\ 4 \end{matrix} \begin{Bmatrix} 0.0 \\ 3600.0 \end{Bmatrix} \begin{matrix} \text{k·in.} \\ \text{k·in.} \end{matrix}$$

The condensation of both the stiffness matrix and the force vector may now take place.

$$[\hat{K}_{pp}] = \begin{bmatrix} 1762 & -1611 \\ -1611 & 1762 \end{bmatrix} - \left(\begin{bmatrix} 9063 & 0 \\ 0 & 9063 \end{bmatrix} \begin{bmatrix} 1,208,333 & 241,667 \\ 241,667 & 1,208,333 \end{bmatrix}^{-1} \begin{bmatrix} 9063 & 0 \\ 0 & 9063 \end{bmatrix} \right)$$

$$= \begin{matrix} 1 \\ 2 \end{matrix} \overset{\begin{matrix} 1 & \quad 2 \end{matrix}}{\begin{bmatrix} 1691.2 & -1596.8 \\ -1596.8 & 1691.2 \end{bmatrix}}$$

$$\{\hat{F}_p\} = \begin{Bmatrix} 0.0 \\ 0.0 \end{Bmatrix} - \left(\begin{bmatrix} 9063 & 0 \\ 0 & 9063 \end{bmatrix} \begin{bmatrix} 1,208,333 & 241,667 \\ 241,667 & 1,208,333 \end{bmatrix}^{-1} \begin{Bmatrix} 0.0 \\ 3600 \end{Bmatrix} \right) = \begin{matrix} 1 \\ 2 \end{matrix} \begin{Bmatrix} 5.63 \\ -28.13 \end{Bmatrix} \begin{matrix} \text{k} \\ \text{k} \end{matrix}$$

The final step is to find the displacements at DOFs 1 and 2. Thus,

$$\begin{Bmatrix} \Delta_1 \\ \Delta_2 \end{Bmatrix} = \begin{bmatrix} 1691.2 & -1596.8 \\ -1596.8 & 1691.2 \end{bmatrix}^{-1} \begin{Bmatrix} 5.63 \\ -28.13 \end{Bmatrix} = \begin{Bmatrix} -0.114 \\ -0.124 \end{Bmatrix} \begin{matrix} \text{in.} \\ \text{in.} \end{matrix}$$

A quick comparison of the results from the four-DOF formulation and the two-DOF formulation shows agreement. One should note, however, that the condensation procedure did eliminate our direct knowledge of $\{\Delta_s\}$. If one desires this, the following equation would need to be used.

$$\{\Delta_s\} = [K_{ss}]^{-1}(\{F_s\} - [K_{sp}]\{\Delta_p\})$$

COMMON MISTAKE

Students often assume that if they want a reduced number of DOFs that they simply need to partition $[K_{uu}]$ and retain $[K_{pp}]$ and $\{F_p\}$ without using the condensation procedure. With this current example, one may quickly see that using only $[K_{pp}]$ and $\{F_p\}$ does not produce the correct values of $\{\Delta_p\}$.

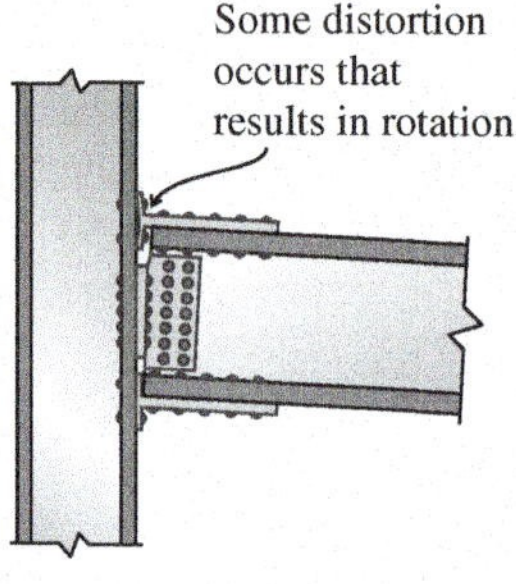

(a) Deformed moment connection

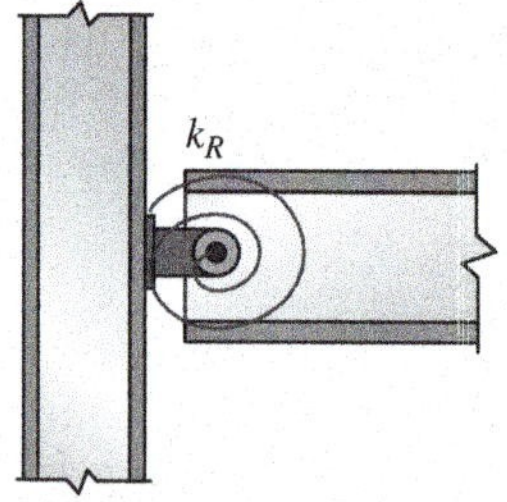

(b) Connection model able to capture deformation

FIGURE 21.9 Partially restrained connection.

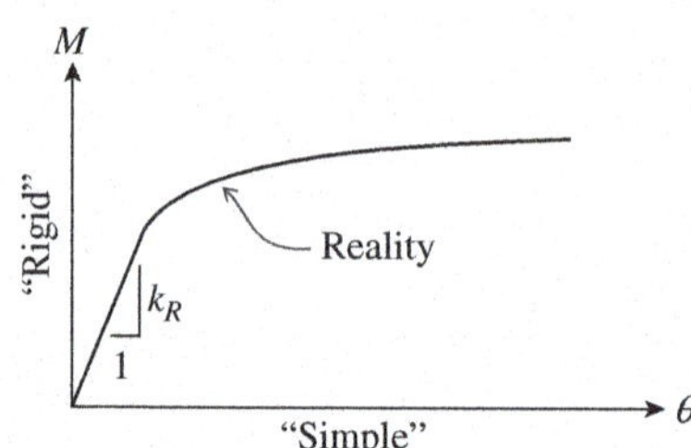

FIGURE 21.10 Moment-rotation behavior of connections.

21.6 Partially Restrained Connections

In analysis, it is common to assume that connections between members can be ideally modeled as pinned (i.e., "simple") or as moment resisting (i.e., "rigid"). For many analyses, these assumptions can give reasonable results. However, when one looks closer at these connections, they will recognize that some relative rotation can develop between the members due to distortions in the connections. Figure 21.9a illustrates this relative rotation.

Recognizing that these distortions occur when the connection is subjected to moment, we can more closely capture the real behavior by inserting a rotational spring as demonstrated in Figure 21.9b. The rotational spring is a convenient mathematical way to describe the *moment–rotation* $(M-\theta)$ relationship between the two members. Figure 21.10 illustrates the $M-\theta$ behavior of this spring.

DID YOU KNOW?
The idea of ideal versus non-ideal connections was first introduced in Section 5.3.

DID YOU KNOW?
k_R represents the stiffness of the rotational spring in the linear region of Figure 21.10. If k_R is set to 0.0, the connection behaves as a "simple" connection. If k_R is set to ∞, the connection becomes "rigid". Anywhere between these two values is known as a *partially restrained* (PR) connection.

Rotational Spring Stiffness

The stiffness matrix for a rotational spring is easily assembled. Recall that to derive the matrix, one would lock the DOFs (see Figure 21.11 for labeling). Then impose a unit rotation at each DOF independently and find the reactions. Doing so would produce the following stiffness matrix. Since the rotation θ in the local coordinates and the global coordinates is the same, then the local and global stiffness matrices are the same—$[k_{\text{spring}}] = [K_{\text{spring}}]$. No transformation matrix is needed.

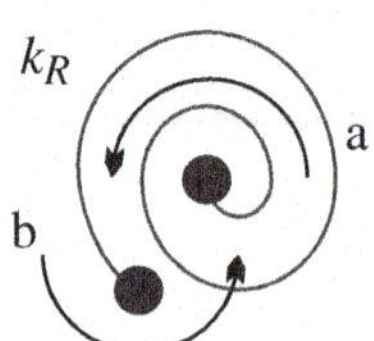

FIGURE 21.11 Rotational spring.

$$[k_{\text{spring}}] = [K_{\text{spring}}] = \begin{matrix} & a & b \\ a & k_R & -k_R \\ b & -k_R & k_R \end{matrix}$$

QUICK NOTE
The stiffnesses of the rotational springs shown in Figure 21.12 are labeled as k_b and k_e, where the subscripts represent the spring at the *beginning* and at the *end*, respectively.

Spring–Beam–Spring (SBS) Element

It is possible to accommodate the PR connections in a beam by incorporating a rotational spring at each end of the beam element, as shown in Figure 21.12. You will quickly note that this creates an assemblage that has six DOFs, whereas a typical beam element would only have four.

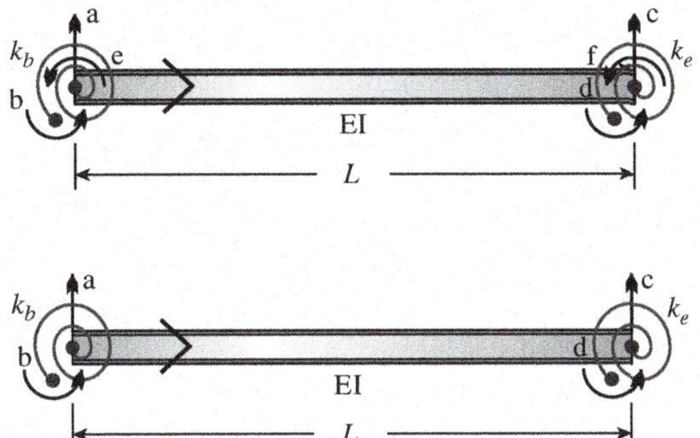

FIGURE 21.12 Beam with PR connections.

FIGURE 21.13 Beam with PR connections with only four DOFs.

The stiffness matrices for the individual elements are given as

$$
\left[k_{\text{beam}}\right] = \begin{array}{c} \\ a \\ e \\ c \\ f \end{array}
\begin{array}{c}
\begin{array}{cccc} \text{a} & \text{e} & \text{c} & \text{f} \end{array} \\
\begin{bmatrix}
\frac{12EI}{L^3} & \frac{6EI}{L^2} & -\frac{12EI}{L^3} & \frac{6EI}{L^2} \\
\frac{6EI}{L^2} & \frac{4EI}{L} & -\frac{6EI}{L^2} & \frac{2EI}{L} \\
-\frac{12EI}{L^3} & -\frac{6EI}{L^2} & \frac{12EI}{L^3} & -\frac{6EI}{L^2} \\
\frac{6EI}{L^2} & \frac{2EI}{L} & -\frac{6EI}{L^2} & \frac{4EI}{L}
\end{bmatrix}
\end{array}
$$

$$
\left[k_{\text{beginning}}\right] = \begin{array}{c} b \\ e \end{array}
\overset{\begin{array}{cc} \text{b} & \text{e} \end{array}}{\begin{bmatrix} k_b & -k_b \\ -k_b & k_b \end{bmatrix}}
\qquad
\left[k_{end}\right] = \begin{array}{c} d \\ f \end{array}
\overset{\begin{array}{cc} \text{d} & \text{f} \end{array}}{\begin{bmatrix} k_e & -k_e \\ -k_e & k_e \end{bmatrix}}
$$

QUICK NOTE

The DOF labeling may look a little strange to begin with, but this is done so that the final matrix is set up for condensation.

With the individual stiffness matrices in hand, we can now assemble the stiffness matrix for the SBS assembly. This will produce the following matrix.

$$
\left[k_{\text{SBS}}\right] = \begin{array}{c} \\ a \\ b \\ c \\ d \\ e \\ f \end{array}
\begin{array}{c}
\begin{array}{cccccc} \text{a} & \text{b} & \text{c} & \text{d} & \text{e} & \text{f} \end{array} \\
\left[\begin{array}{cccc|cc}
\frac{12EI}{L^3} & 0 & -\frac{12EI}{L^3} & 0 & \frac{6EI}{L^2} & \frac{6EI}{L^2} \\
0 & k_b & 0 & 0 & -k_b & 0 \\
-\frac{12EI}{L^3} & 0 & \frac{12EI}{L^3} & 0 & -\frac{6EI}{L^2} & -\frac{6EI}{L^2} \\
0 & 0 & 0 & k_e & 0 & -k_e \\
\hline
\frac{6EI}{L^2} & -k_b & -\frac{6EI}{L^2} & 0 & \frac{4EI}{L} + k_b & \frac{2EI}{L} \\
\frac{6EI}{L^2} & 0 & -\frac{6EI}{L^2} & -k_e & \frac{2EI}{L} & \frac{4EI}{L} + k_e
\end{array}\right]
\end{array}
$$

RECALL

When a matrix is partitioned for condensation, these assignments are made:

$$[k] = \left[\begin{array}{c|c} k_{pp} & k_{ps} \\ \hline k_{sp} & k_{ss} \end{array}\right]$$

Condensation then takes place with the operations:

$$[\hat{k}_{pp}] = [k_{pp}] - [k_{ps}][k_{ss}]^{-1}[k_{sp}]$$

The rotational DOFs e and f are *internal* or *secondary* DOFs. If we condense out these two DOFs, we are left with a 4×4 beam stiffness matrix that has PR connections already incorporated into it (see Figure 21.13). This condensation is facilitated using the partitioning, which is already shown in the $[k_{\text{SBS}}]$ matrix.

EXAMPLE 21.7

Develop the 4×4 stiffness matrix for an SBS element if the following properties are given: L, E, I, and $k_b = k_e = \frac{3EI}{L}$.

SOLUTION STRATEGY

Use the known properties to develop the 6×6 stiffness matrix. Then condense out DOFs e and f.

SOLUTION

Substituting in the given properties results in the following 6 × 6 matrix.

$$
\left[k_{\mathrm{SBS}}\right] = \begin{array}{c} \\ a \\ b \\ c \\ d \\ e \\ f \end{array}
\begin{array}{c}
\begin{array}{cccccc} a & b & c & d & e & f \end{array} \\
\left[\begin{array}{cccc|cc}
\frac{12EI}{L^3} & 0 & -\frac{12EI}{L^3} & 0 & \frac{6EI}{L^2} & \frac{6EI}{L^2} \\
0 & \frac{3EI}{L} & 0 & 0 & -\frac{3EI}{L} & 0 \\
-\frac{12EI}{L^3} & 0 & \frac{12EI}{L^3} & 0 & -\frac{6EI}{L^2} & -\frac{6EI}{L^2} \\
0 & 0 & 0 & \frac{3EI}{L} & 0 & -\frac{3EI}{L} \\
\hline
\frac{6EI}{L^2} & -\frac{3EI}{L} & -\frac{6EI}{L^2} & 0 & \frac{7EI}{L} & \frac{2EI}{L} \\
\frac{6EI}{L^2} & 0 & -\frac{6EI}{L^2} & -\frac{3EI}{L} & \frac{2EI}{L} & \frac{7EI}{L}
\end{array}\right]
\end{array}
$$

DID YOU KNOW?

A generic form of this 4 × 4 matrix may be developed in terms of k_b and k_e, so that it is very versatile. This is what commercial structural analysis packages like SAP2000 do. However, the generic form is not presented in this textbook because it is quite complex.

With the $[k_{\mathrm{SBS}}]$ matrix partitioned into $[k_{pp}]$, $[k_{ps}]$, $[k_{sp}]$, and $[k_{ss}]$, the condensed matrix can be computed as

$$
[\hat{k}_{\mathrm{SBS}}] = [k_{pp}] - [k_{ps}][k_{ss}]^{-1}[k_{sp}] = \begin{array}{c} \\ a \\ b \\ c \\ d \end{array}
\begin{array}{c}
\begin{array}{cccc} a & b & c & d \end{array} \\
\left[\begin{array}{cccc}
\frac{4EI}{L^3} & \frac{2EI}{L^2} & -\frac{4EI}{L^3} & \frac{2EI}{L^2} \\
\frac{2EI}{L^2} & \frac{8EI}{5L} & -\frac{2EI}{L^2} & \frac{2EI}{5L} \\
-\frac{4EI}{L^3} & -\frac{2EI}{L^2} & \frac{4EI}{L^3} & -\frac{2EI}{L^2} \\
\frac{2EI}{L^2} & \frac{2EI}{5L} & -\frac{2EI}{L^2} & \frac{8EI}{5L}
\end{array}\right]
\end{array}
$$

EXAMPLE 21.8

Develop the 4 × 4 stiffness matrix for an SBS element if the following properties are given: L, E, I, $k_b = \infty$, and $k_e = 0$.

SOLUTION STRATEGY

Use the known properties to develop the 6 × 6 stiffness matrix. Then condense out DOFs e and f.

QUICK NOTE

When the rotational stiffness is set to ∞, the connection becomes "rigid", and when the stiffness is set to 0, the connection becomes "simple"—in other words a pin.

SOLUTION

Substituting in the given properties results in the following 6 × 6 matrix.

$$
\left[k_{\mathrm{SBS}}\right] = \begin{array}{c} \\ a \\ b \\ c \\ d \\ e \\ f \end{array}
\begin{array}{c}
\begin{array}{cccccc} a & b & c & d & e & f \end{array} \\
\left[\begin{array}{cccc|cc}
\frac{12EI}{L^3} & 0 & -\frac{12EI}{L^3} & 0 & \frac{6EI}{L^2} & \frac{6EI}{L^2} \\
0 & \infty & 0 & 0 & -\infty & 0 \\
-\frac{12EI}{L^3} & 0 & \frac{12EI}{L^3} & 0 & -\frac{6EI}{L^2} & -\frac{6EI}{L^2} \\
0 & 0 & 0 & 0 & 0 & -0 \\
\hline
\frac{6EI}{L^2} & -\infty & -\frac{6EI}{L^2} & 0 & \infty & \frac{2EI}{L} \\
\frac{6EI}{L^2} & 0 & -\frac{6EI}{L^2} & -0 & \frac{2EI}{L} & \infty
\end{array}\right]
\end{array}
$$

With the $[k_{SBS}]$ matrix partitioned into $[k_{pp}]$, $[k_{ps}]$, $[k_{sp}]$, and $[k_{ss}]$, the condensed matrix can be computed as

$$[\hat{k}_{SBS}] = [k_{pp}] - [k_{ps}][k_{ss}]^{-1}[k_{sp}] = \begin{array}{c} a \\ b \\ c \\ d \end{array}\overset{\begin{array}{cccc} a & b & c & d \end{array}}{\begin{bmatrix} \frac{3EI}{L^3} & \frac{3EI}{L^2} & -\frac{3EI}{L^3} & 0 \\ \frac{3EI}{L^2} & \frac{3EI}{L} & -\frac{3EI}{L^2} & 0 \\ -\frac{3EI}{L^3} & -\frac{3EI}{L^2} & \frac{3EI}{L^3} & 0 \\ 0 & 0 & 0 & 0 \end{bmatrix}}$$

This produces a stiffness matrix for a beam with a rigid connection on the left and a pin connection on the right.

QUICK NOTE

Evaluating a matrix with ∞ in it can be a bit tricky. One way to handle this is to simply assign the following values $L = 1$, $EI = 1$, and k_b to be a very large number. You can then do the calculations and see what the asymptotic values are for each stiffness term.

EXAMPLE 21.9

The following beam with PR connections should be evaluated for joint displacements and reactions. Each rotational spring has the stiffness $k_R = 3\left(\frac{EI}{L}\right)_2$. Use the DOF labeling shown in the figure.

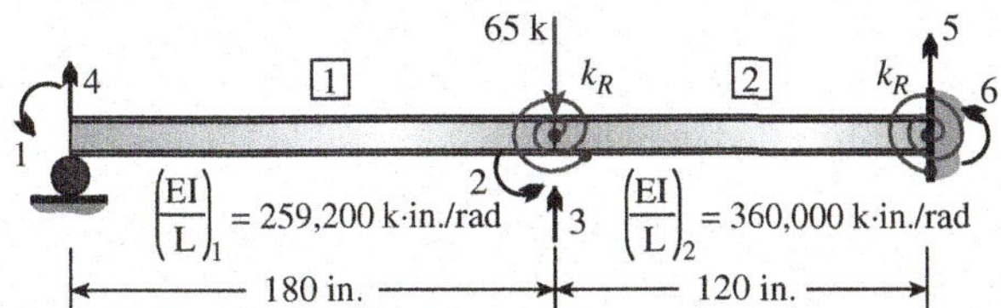

SOLUTION STRATEGY

Recognize that beam element 1 is a standard beam element, but that element 2 can use the SBS element that was developed in Example 21.7. Assemble the global stiffness matrix and execute the required matrix operations to solve for joint displacements and reactions.

SOLUTION

The element stiffness matrices are computed using the properties given. Element 1 was previously developed in Section 20.3 of this text. The matrix for element 2 was previously derived in Example 21.7.

$$K^{(1)} = \begin{array}{c} 4 \\ 1 \\ 3 \\ 2 \end{array}\overset{\begin{array}{cccc} 4 & 1 & 3 & 2 \end{array}}{\begin{bmatrix} 96 & 8640 & -96 & 8640 \\ 8640 & 1,036,800 & -8640 & 518,400 \\ -96 & -8640 & 96 & -8640 \\ 8640 & 518,400 & -8640 & 1,036,800 \end{bmatrix}} \quad K^{(2)} = \begin{array}{c} 3 \\ 2 \\ 5 \\ 6 \end{array}\overset{\begin{array}{cccc} 3 & 2 & 5 & 6 \end{array}}{\begin{bmatrix} 100 & 6000 & -100 & 6000 \\ 6000 & 576,000 & -6000 & 144,000 \\ -100 & -6000 & 100 & -6000 \\ 6000 & 144,000 & -6000 & 576,000 \end{bmatrix}}$$

The following structure level stiffness matrices now may be assembled.

$$K_{uu} = \begin{array}{c} 1 \\ 2 \\ 3 \end{array}\overset{\begin{array}{ccc} 1 & 2 & 3 \end{array}}{\begin{bmatrix} 1,036,800 & 518,400 & -8640 \\ 518,400 & 1,612,800 & -2640 \\ -8640 & -2640 & 196 \end{bmatrix}}$$

$$K_{ru} = \begin{array}{c} 4 \\ 5 \\ 6 \end{array}\overset{\begin{array}{ccc} 1 & 2 & 3 \end{array}}{\begin{bmatrix} 8640 & 8640 & -96 \\ 0 & -6000 & -100 \\ 0 & 144,000 & 6000 \end{bmatrix}}$$

QUICK NOTE

The force vector for this structure is

$$\{F_u\} = \begin{Bmatrix} 0.0 \\ 0.0 \\ -65.0 \end{Bmatrix} \begin{array}{l} \text{k·in.} \\ \text{k·in.} \\ \text{k} \end{array}$$

The requested response quantities are computed as

$$\begin{Bmatrix} \theta_1 \\ \theta_2 \\ \Delta_3 \end{Bmatrix} = [K_{uu}]^{-1}\{F_u\} = \begin{Bmatrix} -0.0048 \\ 0.0007 \\ -0.5332 \end{Bmatrix} \begin{matrix} \text{rad} \\ \text{rad} \\ \text{in.} \end{matrix}$$

$$\begin{Bmatrix} R_4 \\ R_5 \\ M_6 \end{Bmatrix} = [K_{ru}]\{\Delta_u\} = \begin{Bmatrix} 15.7 \\ 49.3 \\ -3104 \end{Bmatrix} \begin{matrix} \text{k} \\ \text{k} \\ \text{k·in.} \end{matrix}$$

QUICK NOTE

This beam was originally analyzed without PR connections in Example 20.2, which gave these results:

$$\begin{Bmatrix} \theta_1 \\ \theta_2 \\ \Delta_3 \end{Bmatrix} = \begin{Bmatrix} -0.0031 \\ 0.0017 \\ -0.2713 \end{Bmatrix} \begin{matrix} \text{rad} \\ \text{rad} \\ \text{in.} \end{matrix}$$

$$\begin{Bmatrix} R_4 \\ R_5 \\ M_6 \end{Bmatrix} = \begin{Bmatrix} 13.7 \\ 51.3 \\ -3678 \end{Bmatrix} \begin{matrix} \text{k} \\ \text{k} \\ \text{k·in.} \end{matrix}$$

A quick comparison shows that the forces are distributed differently, depending on the relative stiffness of the connections.

21.7 Releases

The assemblage of a structure-level beam or frame stiffness matrix, as performed in Chapter 20, inherently assumes that members are connected using "rigid" connections. Many structures are connected with members that don't exhibit this behavior, so we may model them with PR connections (see Section 21.6) or may insert internal releases.

Typical internal releases are listed here and are depicted in Figure 21.14.

Moment Release Allows unrestrained relative rotation between connected members about an axis perpendicular to the members.

Shear Release Allows unrestrained relative translation between connected members in a direction perpendicular to the axis of the members.

Axial Release Allows unrestrained relative translation between connected members in a direction parallel to the axis of the members.

Torsion Release Allows unrestrained relative rotation between connected members about an axis parallel to the members (not relevant for 2-D analysis).

QUICK NOTE

When two elements share the same DOF number, we are stating that those two elements must move exactly together at that location/direction.

QUICK NOTE

A *torsion release* is not explicitly discussed in this text, but the principles introduced for implementing the other releases also can be applied to this.

Implementation

The implementation of internal releases is straightforward. There are two direct ways for accomplishing this.

1. Build your model using SBS elements that were discussed in Section 21.6. Then when you want a release inserted, you would simply need to define the spring stiffness to be zero as was done in Example 21.8. If you don't want a release, you can set the stiffness to be ∞ (i.e., numerically large number).

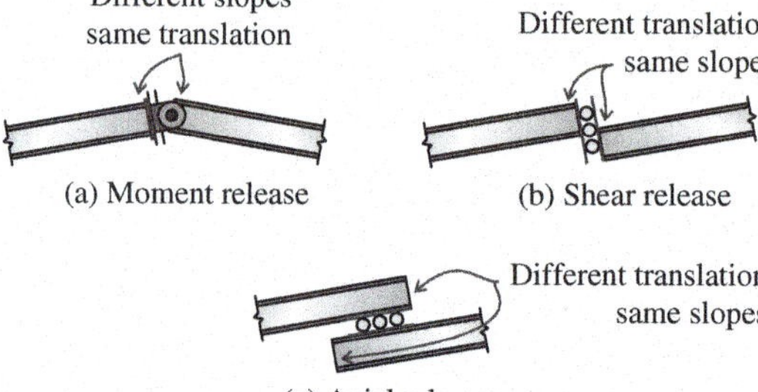

FIGURE 21.14 Description of allowed motions for typical internal release types.

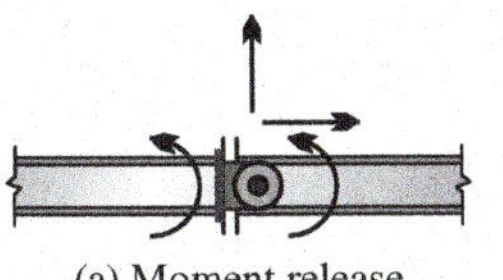
(a) Moment release

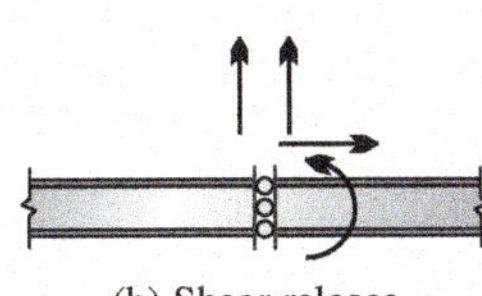
(b) Shear release

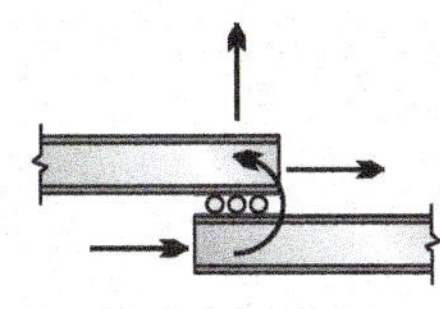
(c) Axial release

FIGURE 21.15 DOFs associated with different release types.

This approach requires rotational springs to implement moment releases and parallel and perpendicular translation springs to implement axial and shear releases, respectively. This is actually how commercial analysis software like SAP2000 approaches this problem.

2. Implement the release through strategic numbering of the DOFs associated with the release. This requires the addition of another DOF and will thus increase the overall dimensions of the stiffness matrix.
   Figure 21.15(a) shows the DOFs used at the joint of two members to achieve a moment release between the members. The DOFs for shear and axial releases are also demonstrated in Figure 21.15.

RECALL

A DOF describes a possible motion—whether rotation or translation. Thus, the double DOFs associated with each release is saying that the motion of interest (e.g., rotation for a moment release) can be different on either side of the release.

EXAMPLE 21.10

The following beam has an internal moment release between members 1 and 2. Using the DOF numbering shown, compute the joint displacements and the reactions for this beam.

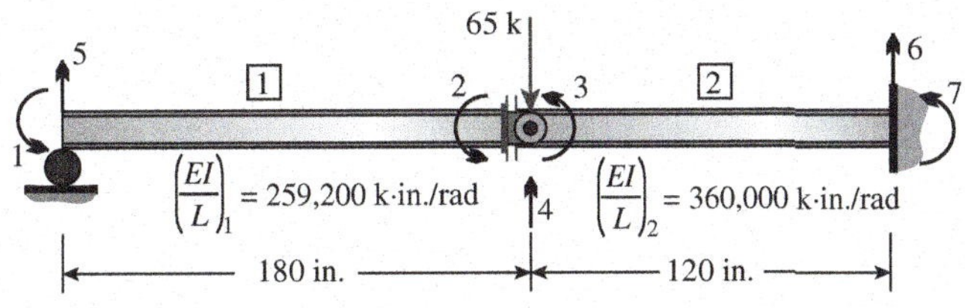

QUICK NOTE

The stiffness matrices for these beam elements were first developed in Section 20.3 of this textbook. Note that they are exactly the same as presented here—except for the way we numbered the DOFs.

SOLUTION STRATEGY

Assemble the element stiffness matrices as usual. Be sure that rotational DOF 2 is assigned to member 1 and rotational DOF 3 is assigned to member 2. All other analysis steps are carried out as usual.

SOLUTION

The element stiffness matrices are found using the standard beam-element stiffness matrix.

$$K^{(1)} = \begin{array}{c} \\ 5 \\ 1 \\ 4 \\ 2 \end{array} \begin{array}{c} \begin{array}{cccc} 5 & 1 & 4 & 2 \end{array} \\ \begin{bmatrix} 96 & 8640 & -96 & 8640 \\ 8640 & 1{,}036{,}800 & -8640 & 518{,}400 \\ -96 & -8640 & 96 & -8640 \\ 8640 & 518{,}400 & -8640 & 1{,}036{,}800 \end{bmatrix} \end{array} \qquad K^{(2)} = \begin{array}{c} \\ 4 \\ 3 \\ 6 \\ 7 \end{array} \begin{array}{c} \begin{array}{cccc} 4 & 3 & 6 & 7 \end{array} \\ \begin{bmatrix} 300 & 18{,}000 & -300 & -18{,}000 \\ 18{,}000 & 1{,}440{,}000 & 18{,}000 & 720{,}000 \\ -300 & 18{,}000 & 300 & -18{,}000 \\ -18{,}000 & 720{,}000 & -18{,}000 & 1{,}440{,}000 \end{bmatrix} \end{array}$$

Since we have an extra DOF located at the moment release, the entire structure has seven DOFs instead of the usual six.

$$K_{uu} = \begin{array}{c} \\ 1 \\ 2 \\ 3 \\ 4 \end{array} \begin{array}{c} \begin{array}{cccc} 1 & 2 & 3 & 4 \end{array} \\ \begin{bmatrix} 1{,}036{,}800 & 518{,}400 & 0 & -8640 \\ 518{,}400 & 1{,}036{,}800 & 0 & -8640 \\ 0 & 0 & 1{,}440{,}000 & 18{,}000 \\ -8640 & -8640 & 18{,}000 & 396 \end{bmatrix} \end{array}$$

QUICK NOTE

The force vector for this structure is

$$\{F_u\} = \begin{Bmatrix} 0.0 \\ 0.0 \\ 0.0 \\ -65.0 \end{Bmatrix} \begin{array}{l} \text{k·in.} \\ \text{k·in.} \\ \text{k} \\ \text{k} \end{array}$$

$$K_{ru} = \begin{matrix} & 1 & 2 & 3 & 4 \\ 5 \\ 6 \\ 7 \end{matrix}\begin{bmatrix} 8640 & 8640 & 0 & -96 \\ 0 & 0 & -18,000 & -300 \\ 0 & 0 & 720,000 & 18,000 \end{bmatrix}$$

POINT TO PONDER

Look at the response of member 1. Notice how the rotations at each end are equal and that the reaction at DOF 5 is 0.0 k. Does this seem unusual? Can you explain why there is no load going to member 1?

The requested quantities now may be computed.

$$\begin{Bmatrix} \theta_1 \\ \theta_2 \\ \theta_3 \\ \Delta_4 \end{Bmatrix} = [K_{uu}]^{-1}\{F_u\} = \begin{Bmatrix} -0.0048 \\ -0.0048 \\ 0.0108 \\ -0.87 \end{Bmatrix} \begin{matrix} \text{rad} \\ \text{rad} \\ \text{rad} \\ \text{in.} \end{matrix}$$

$$\begin{Bmatrix} R_5 \\ R_6 \\ M_7 \end{Bmatrix} = [K_{ru}]\{\Delta_u\} = \begin{Bmatrix} 0.0 \\ 65.0 \\ -7800 \end{Bmatrix} \begin{matrix} \text{k} \\ \text{k} \\ \text{k·in.} \end{matrix}$$

To help in understanding the results, the deflected shape of the beam was plotted using shape functions.

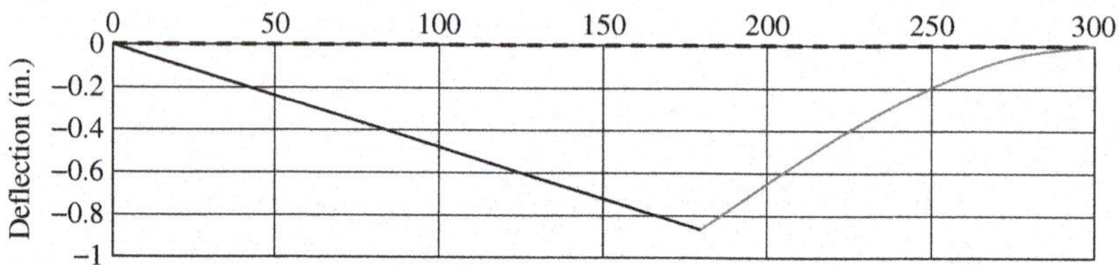

EXAMPLE 21.11

The following beam has an internal shear release between members 1 and 2. Using the DOF numbering shown, compute the joint displacements and the reactions for this beam.

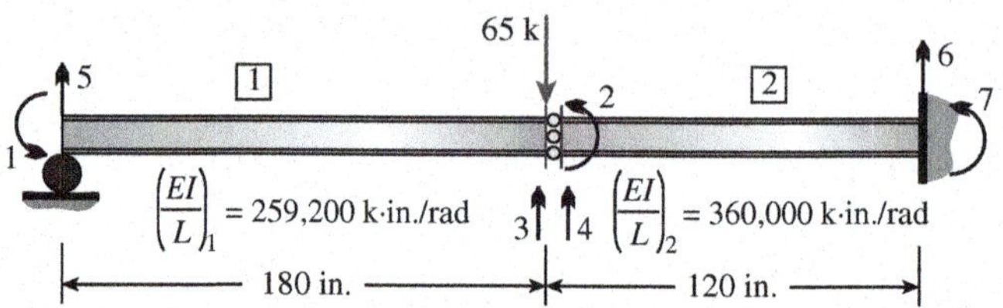

QUICK NOTE

The stiffness matrices for these beam elements were first developed in Section 20.3 of this textbook. Note that they are exactly the same as presented here—except for the way we numbered the DOFs.

SOLUTION STRATEGY

Assemble the element stiffness matrices as usual. Be sure that translational DOF 3 is assigned to member 1 and translational DOF 4 is assigned to member 2. All other analysis steps are carried out as usual.

SOLUTION

The element stiffness matrices are found using the standard beam-element stiffness matrix.

$$K^{(1)} = \begin{matrix} & 5 & 1 & 3 & 2 \\ 5 \\ 1 \\ 3 \\ 2 \end{matrix}\begin{bmatrix} 96 & 8640 & -96 & 8640 \\ 8640 & 1,036,800 & -8640 & 518,400 \\ -96 & -8640 & 96 & -8640 \\ 8640 & 518,400 & -8640 & 1,036,800 \end{bmatrix} \quad K^{(2)} = \begin{matrix} & 4 & 2 & 6 & 7 \\ 4 \\ 2 \\ 6 \\ 7 \end{matrix}\begin{bmatrix} 300 & 18,000 & -300 & -18,000 \\ 18,000 & 1,440,000 & 18,000 & 720,000 \\ -300 & 18,000 & 300 & -18,000 \\ -18,000 & 720,000 & -18,000 & 1,440,000 \end{bmatrix}$$

Since we have an extra DOF located at the shear release, the entire structure has seven DOFs instead of the usual six.

$$K_{uu} = \begin{array}{c} \\ 1 \\ 2 \\ 3 \\ 4 \end{array}\begin{array}{c} \begin{array}{cccc} 1 & 2 & 3 & 4 \end{array} \\ \begin{bmatrix} 1,036,800 & 518,400 & -8640 & 0 \\ 518,400 & 2,476,800 & -8640 & 18,000 \\ -8640 & -8640 & 96 & 0 \\ 0 & 18,000 & 0 & 300 \end{bmatrix} \end{array}$$

$$K_{ru} = \begin{array}{c} \\ 5 \\ 6 \\ 7 \end{array}\begin{array}{c} \begin{array}{cccc} 1 & 2 & 3 & 4 \end{array} \\ \begin{bmatrix} 8640 & 8640 & 0 & -96 \\ 0 & 0 & -18,000 & -300 \\ 0 & 0 & 720,000 & 18,000 \end{bmatrix} \end{array}$$

The requested quantities now may be computed.

$$\begin{Bmatrix} \theta_1 \\ \theta_2 \\ \Delta_3 \\ \Delta_4 \end{Bmatrix} = [K_{uu}]^{-1}\{F_u\} = \begin{Bmatrix} -0.0551 \\ -0.0325 \\ -8.56 \\ 1.95 \end{Bmatrix} \begin{array}{l} \text{rad} \\ \text{rad} \\ \text{in.} \\ \text{in.} \end{array}$$

$$\begin{Bmatrix} R_5 \\ R_6 \\ M_7 \end{Bmatrix} = [K_{ru}]\{\Delta_u\} = \begin{Bmatrix} 65.0 \\ 0.0 \\ 11700 \end{Bmatrix} \begin{array}{l} \text{k} \\ \text{k} \\ \text{k·in.} \end{array}$$

To help in understanding the results, the deflected shape of the beam was plotted using shape functions.

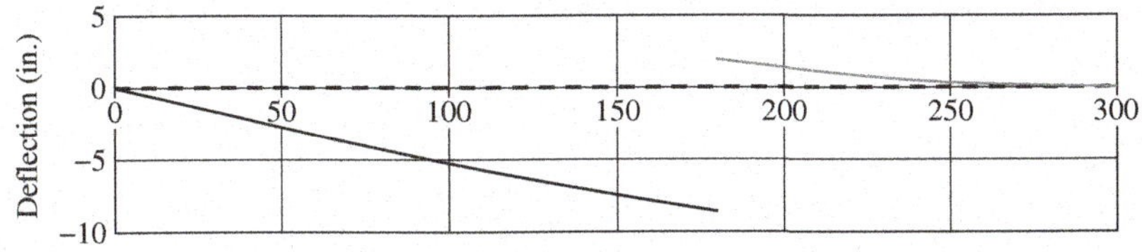

QUICK NOTE

The force vector for this structure is

$$\{F_u\} = \begin{Bmatrix} 0.0 \\ 0.0 \\ -65.0 \\ 0.0 \end{Bmatrix} \begin{array}{l} \text{k·in.} \\ \text{k·in.} \\ \text{k} \\ \text{k} \end{array}$$

This is because the point load is applied just to the left of the shear release at DOF 3.

POINT TO PONDER

Look at the displacements just to the left and right of the shear release—they are opposite in sign. Can you explain why the right side goes up when the applied load on the left is going down?

21.8 Inclined Supports

A particular fact that has been taken for granted throughout the matrix analysis of structures so far is that the restraints (i.e., supports) have been relatively easy to define. Typical translational DOFs for a structure are oriented in either a vertical or horizontal fashion, as seen in Figure 21.16. The direction of the restrained motion—for example the up-down direction in Figure 21.16—coincides nicely with the vertical DOF at that location.

What happens when the desired restraint direction doesn't match up with the default DOF orientation? Examine Figure 21.17(a), and ask yourself which of the translational DOFs you would restrain in order to get restraint in the desired direction (i.e., perpendicular to the roller surface). If you restrain only one of the translational DOFs, it will be free to move in the other

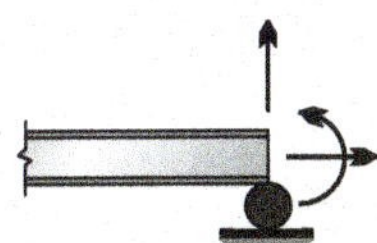

FIGURE 21.16 Typical DOFs at support location.

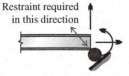

Restraint required in this direction

(a) DOFs and inclined support don't line up

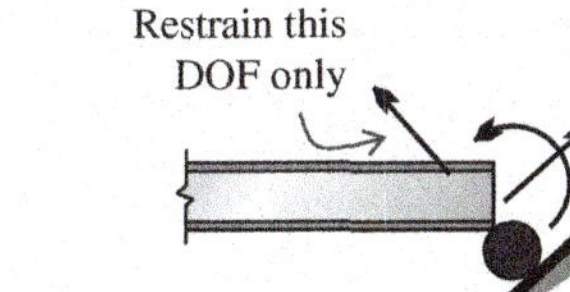

(b) DOFs and inclined support line up

FIGURE 21.17 Inclined support.

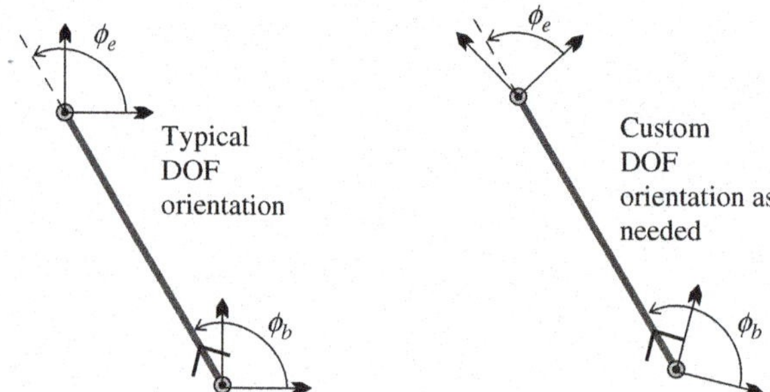

FIGURE 21.18 Identifying appropriate angles for the transformation matrix.

QUICK NOTE

ϕ_b represents the angle between the global DOF and the local DOF at the **beginning** joint of the member. ϕ_e represents the angle between the global and local DOFs at the **ending** joint of the member.

perpendicular direction. If you restrain both DOFs, you will be restraining all translational motion. Neither of these scenarios describes the inclined roller situation.

In Figure 21.17(b), the DOFs have been rotated so that they coincide with the orientation of the inclined roller support. With the DOFs in this new orientation, it becomes very clear which DOF needs to be restrained to get the desired behavior. The reorientation of the global DOFs is a simple solution to the issue, but we do need to explain the practical implementation in analysis.

Modified Transformation Matrix

QUICK NOTE

The angle ϕ_b is measured from the first DOF rotating counterclockwise to the actual member. The angle ϕ_e is measured from the first DOF rotating clockwise to the projection of the member that is depicted with the dashed line in Figure 21.18.

You will recall that the transformation matrix $[T]$ is a simple tool to translate DOFs from the local coordinate system to the global coordinate system. Indeed, the process was explicitly addressed in Sections 19.6 and 20.9 of this textbook. In both cases, the derivation of the transformation matrix assumed that the global DOFs were oriented in the positive X and Y directions. Thus, if the global coordinates at a particular location are positioned in an alternate orientation, we have to go back to the transformation matrix to make this adjustment. The basic transformation matrix for a truss is

$$[T]_{\text{truss}} = \begin{bmatrix} \cos\phi_b & \sin\phi_b & 0 & 0 \\ -\sin\phi_b & \cos\phi_b & 0 & 0 \\ 0 & 0 & \cos\phi_e & \sin\phi_e \\ 0 & 0 & -\sin\phi_e & \cos\phi_e \end{bmatrix}$$

COMMON CONFUSION

Students often think that, since the DOFs are not in the $X - Y$ direction, finding member end forces is going to be tricky. Actually, the process for finding these is the same as always.

$$\{u\} = [T]\{\Delta\} \qquad \{Q\} = [k]\{u\}$$

No special treatment is required.

The challenge, as usual, is to come up with the correct angles for ϕ_b and ϕ_e. Consider the truss bar shown in Figure 21.18. The beginning joint is indicated as usual with the large arrow head. Two scenarios are being presented in this figure. The first is the way we would measure angles if the two orthogonal DOFs at each end were oriented in the standard $X - Y$ directions. The second scenario depicts the situation where the DOFs are at different orientations.

Once the correct angles are identified, the transformation matrix is assembled, and all of the rest of the steps of the matrix method are carried out as usual.

EXAMPLE 21.12

MEMBER LENGTHS

$L_1 = 120$ in.

$L_2 = 60$ in.

$L_3 = 134.2$ in.

The truss assembly has an inclined roller at one location, as shown in the sketch. Use the matrix approach to compute the joint displacements, reactions, and member forces. Use the DOF labeling shown.

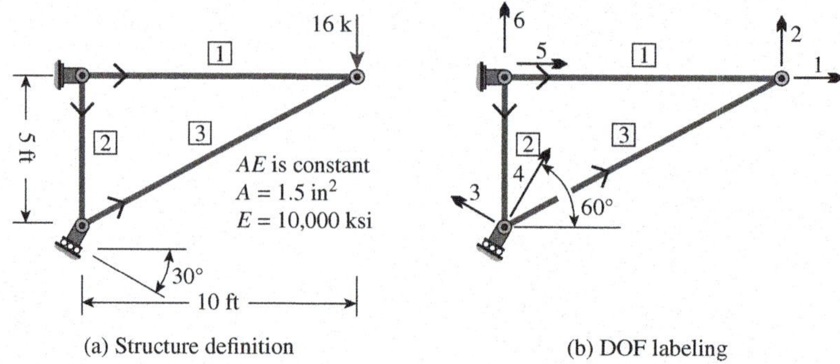

(a) Structure definition

(b) DOF labeling

SOLUTION STRATEGY

The process for analyzing this truss is the exact same as performed in Chapter 19. However, we need to adjust the element transformation matrix for each member connected to the inclined support. Make a sketch of members 2 and 3 to make sure correct angles are identified. Then proceed with analysis.

SOLUTION STRATEGY

As always, start with the element stiffness matrices in their local coordinate systems.

$$k^{(1)} = \begin{array}{c} a \\ b \\ c \\ d \end{array} \overset{\begin{array}{cccc} a & b & c & d \end{array}}{\begin{bmatrix} 125.0 & 0.0 & -125.0 & 0.0 \\ 0.0 & 0.0 & 0.0 & 0.0 \\ -125.0 & 0.0 & 125.0 & 0.0 \\ 0.0 & 0.0 & 0.0 & 0.0 \end{bmatrix}} \text{k/in.} \quad k^{(2)} = \begin{array}{c} a \\ b \\ c \\ d \end{array} \overset{\begin{array}{cccc} a & b & c & d \end{array}}{\begin{bmatrix} 250.0 & 0.0 & -250.0 & 0.0 \\ 0.0 & 0.0 & 0.0 & 0.0 \\ -250.0 & 0.0 & 250.0 & 0.0 \\ 0.0 & 0.0 & 0.0 & 0.0 \end{bmatrix}} \text{k/in.}$$

$$k^{(3)} = \begin{array}{c} a \\ b \\ c \\ d \end{array} \overset{\begin{array}{cccc} a & b & c & d \end{array}}{\begin{bmatrix} 111.8 & 0.0 & -111.8 & 0.0 \\ 0.0 & 0.0 & 0.0 & 0.0 \\ -111.8 & 0.0 & 111.8 & 0.0 \\ 0.0 & 0.0 & 0.0 & 0.0 \end{bmatrix}} \text{k/in.}$$

The transformation matrices for each element must be defined next. Member 1 does not coincide with any rotated DOFs. For this reason, its transformation matrix is developed the same way as before, resulting in

$$T_1 = \begin{array}{c} a \\ b \\ c \\ d \end{array} \overset{\begin{array}{cccc} 5 & 6 & 1 & 2 \end{array}}{\begin{bmatrix} 1.00 & 0 & 0 & 0 \\ 0 & 1.00 & 0 & 0 \\ 0 & 0 & 1.00 & 0 \\ 0 & 0 & 0 & 1.00 \end{bmatrix}}$$

QUICK NOTE

The orientation of member 3 is computed using basic trigonometry.

$$\phi = \arctan\left(\frac{5}{10}\right) = 26.6°$$

The transformation matrices for members 2 and 3 need to explicitly have their angles at each end defined. Refer to the sketches for relevant angles, and refer back to Figure 21.18 to identify how the beginning and ending angles are supposed to be defined.

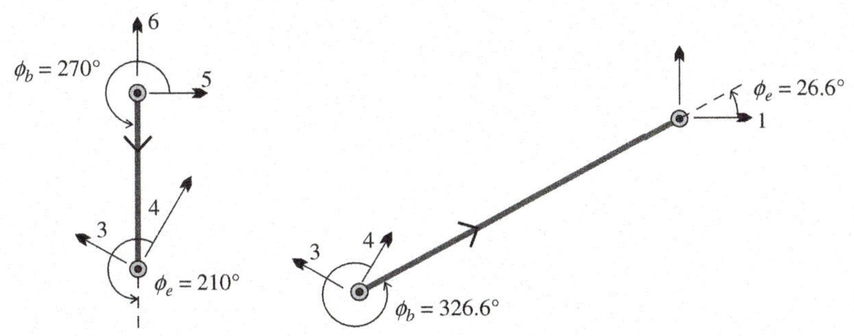

SOME TRIGONOMETRY CALCULATIONS

$\sin(270) = -1.000$
$\cos(270) = 0.000$
$\sin(210) = -0.500$
$\cos(210) = -0.866$
$\sin(326.6) = -0.550$
$\cos(326.6) = 0.835$
$\sin(26.6) = 0.448$
$\cos(26.6) = 0.894$

$$T_2 = \begin{array}{c} a \\ b \\ c \\ d \end{array} \overset{\begin{array}{cccc} 5 & 6 & 4 & 3 \end{array}}{\begin{bmatrix} 0 & -1.00 & 0 & 0 \\ 1.00 & 0 & 0 & 0 \\ 0 & 0 & -0.866 & -0.500 \\ 0 & 0 & 0.500 & -0.866 \end{bmatrix}} \quad T_3 = \begin{array}{c} a \\ b \\ c \\ d \end{array} \overset{\begin{array}{cccc} 4 & 3 & 1 & 2 \end{array}}{\begin{bmatrix} 0.835 & -0.550 & 0 & 0 \\ 0.550 & 0.835 & 0 & 0 \\ 0 & 0 & 0.894 & 0.447 \\ 0 & 0 & -0.447 & 0.894 \end{bmatrix}}$$

From here on forward, the analysis process is identical to that used in Chapters 19 and 20. The global-element stiffness matrices come next. Thus,

$$K^{(1)} = \begin{matrix} & 5 & 6 & 1 & 2 \\ 5 & 125.0 & 0.0 & -125.0 & 0.0 \\ 6 & 0.0 & 0.0 & 0.0 & 0.0 \\ 1 & -125.0 & 0.0 & 125.0 & 0.0 \\ 2 & 0.0 & 0.0 & 0.0 & 0.0 \end{matrix} \text{ k/in.} \qquad K^{(2)} = \begin{matrix} & 5 & 6 & 4 & 3 \\ 5 & 0.0 & 0.0 & 0.0 & 0.0 \\ 6 & 0.0 & 250.0 & -216.5 & -125.0 \\ 4 & 0.0 & -216.5 & 187.5 & 108.3 \\ 3 & 0.0 & -125.0 & 108.3 & 62.5 \end{matrix} \text{ k/in.}$$

$$K^{(3)} = \begin{matrix} & 4 & 3 & 1 & 2 \\ 4 & 77.9 & -51.4 & -83.5 & -41.7 \\ 3 & -51.4 & 33.9 & 55.0 & 27.5 \\ 1 & -83.5 & 55.0 & 89.4 & 44.7 \\ 2 & -41.7 & 27.5 & 44.7 & 22.4 \end{matrix} \text{ k/in.}$$

The relevant structure-level stiffness matrices and force vector are assembled next.

$$K_{uu} = \begin{matrix} & 1 & 2 & 3 \\ 1 & 214.4 & 44.7 & 55.0 \\ 2 & 44.7 & 22.4 & 27.5 \\ 3 & 55.0 & 27.5 & 96.4 \end{matrix} \text{ k/in.} \qquad K_{ru} = \begin{matrix} & 1 & 2 & 3 \\ 4 & -83.5 & -41.7 & 56.9 \\ 5 & -125.0 & 0.0 & 0.0 \\ 6 & 0.0 & 0.0 & -125.0 \end{matrix} \text{ k/in.}$$

$$\{F_u\} = \begin{matrix} 1 \\ 2 \\ 3 \end{matrix} \begin{Bmatrix} 0.0 \\ -16.0 \\ 0.0 \end{Bmatrix} \begin{matrix} \text{k} \\ \text{k} \\ \text{k} \end{matrix}$$

The matrix operations may now take place.

$$\begin{Bmatrix} \Delta_1 \\ \Delta_2 \\ \Delta_3 \end{Bmatrix} = [K_{uu}]^{-1}\{F_u\} = \begin{Bmatrix} 0.2560 \\ -1.6154 \\ 0.3151 \end{Bmatrix} \begin{matrix} \text{in.} \\ \text{in.} \\ \text{in.} \end{matrix}$$

$$\begin{Bmatrix} R_4 \\ R_5 \\ R_6 \end{Bmatrix} = [K_{ru}]\{\Delta_u\} = \begin{Bmatrix} 64.0 \\ -32.0 \\ -39.4 \end{Bmatrix} \begin{matrix} \text{k} \\ \text{k} \\ \text{k} \end{matrix}$$

Member end forces are computed by transforming global displacements into local displacements and then using the local-element stiffness matrices.

$$\{u\}^{(1)} = [T_1] \begin{Bmatrix} 0.000 \\ 0.000 \\ 0.256 \\ -1.615 \end{Bmatrix} = \begin{Bmatrix} 0.000 \\ 0.000 \\ 0.256 \\ -1.615 \end{Bmatrix} \begin{matrix} \text{in.} \\ \text{in.} \\ \text{in.} \\ \text{in.} \end{matrix} \qquad \{u\}^{(2)} = [T_2] \begin{Bmatrix} 0.000 \\ 0.000 \\ 0.000 \\ 0.315 \end{Bmatrix} = \begin{Bmatrix} 0.000 \\ 0.000 \\ -0.158 \\ -0.273 \end{Bmatrix} \begin{matrix} \text{in.} \\ \text{in.} \\ \text{in.} \\ \text{in.} \end{matrix}$$

$$\{u\}^{(3)} = [T_3] \begin{Bmatrix} 0.000 \\ 0.315 \\ 0.256 \\ -1.615 \end{Bmatrix} = \begin{Bmatrix} -0.173 \\ 0.263 \\ -0.493 \\ -1.559 \end{Bmatrix} \begin{matrix} \text{in.} \\ \text{in.} \\ \text{in.} \\ \text{in.} \end{matrix}$$

$$\{Q\}^{(1)} = [k]^{(1)} \begin{Bmatrix} 0.000 \\ 0.000 \\ 0.256 \\ -1.615 \end{Bmatrix} = \begin{Bmatrix} -32.0 \\ 0.0 \\ 32.0 \\ 0.0 \end{Bmatrix} \begin{matrix} \text{k} \\ \text{k} \\ \text{k} \\ \text{k} \end{matrix} \qquad \{Q\}^{(2)} = [k]^{(2)} \begin{Bmatrix} 0.000 \\ 0.000 \\ -0.158 \\ -0.273 \end{Bmatrix} = \begin{Bmatrix} 39.4 \\ 0.0 \\ -39.4 \\ 0.0 \end{Bmatrix} \begin{matrix} \text{k} \\ \text{k} \\ \text{k} \\ \text{k} \end{matrix}$$

$$\{Q\}^{(3)} = [k]^{(3)} \begin{Bmatrix} -0.173 \\ 0.263 \\ -0.493 \\ -1.559 \end{Bmatrix} = \begin{Bmatrix} 35.8 \\ 0.0 \\ -35.8 \\ 0.0 \end{Bmatrix} \begin{matrix} \text{k} \\ \text{k} \\ \text{k} \\ \text{k} \end{matrix}$$

21.9 SAP2000 Computer Applications

All of the special analysis procedures discussed in this chapter are readily accomplished within SAP2000. This section gives a brief explanation of how each analysis is to be defined.

Enforced Displacements

The first step is to apply a restraint at the location where you want to enforce the displacement. Referring to the case shown in Example 21.1, this would require a roller to be defined at DOF 3. Go to the following pull-down menu: *ASSIGN → JOINT LOADS → DISPLACEMENTS.* The menu in Figure 21.19 shows the resulting pop-up window.

QUICK NOTE

The displacements may be imposed in either global (*XYZ*) or local coordinates (123) of the joint. These automatically coincide—unless the local axes of the joint have been rotated, as was required in Section 21.8.

Imposing Thermal Loads

Thermal loads are applied to a member by first selecting the member and then by selecting the pull-down menu: *ASSIGN → FRAME LOADS → TEMPERATURE.* Figure 21.20 shows the pop-up window that results from this selection.

If a uniform temperature change is to be imposed on a member, such as in Example 21.2, the *TEMPERATURE* button should be selected and the temperature change (e.g., 80°F) should be entered into the *BY ELEMENT* entry box. This is illustrated in Figure 21.20(a).

If a temperature gradient needs to be applied, as in the case of Example 21.3, the temperature gradient button should be selected. Then in the *BY ELEMENT* entry box, the temperature gradient must be defined. This gradient is computed as the change in temperature divided by the depth of the member $\Delta T/d$. For Example 21.3, this temperature gradient is computed as 40°F/36 in. = 1.1111. Refer to Figure 21.20(b).

QUICK NOTE

Be sure to know what units are set for temperature (i.e., °F or °C. You need to also define the thermal coefficient of expansion α when you define the materials (see Figure 21.21).

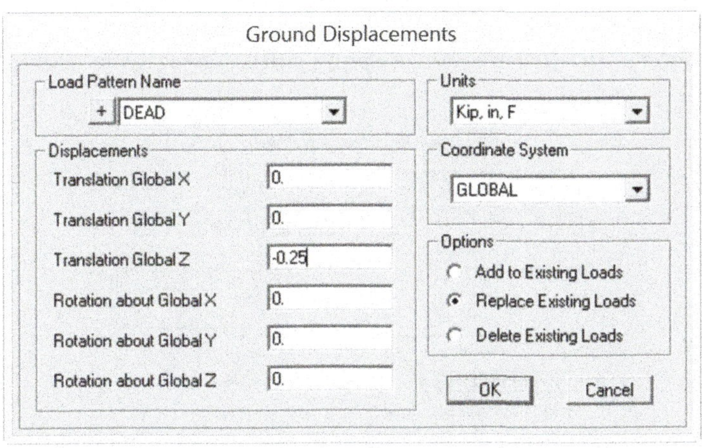

FIGURE 21.19 Pop-up window for imposing ground displacements.

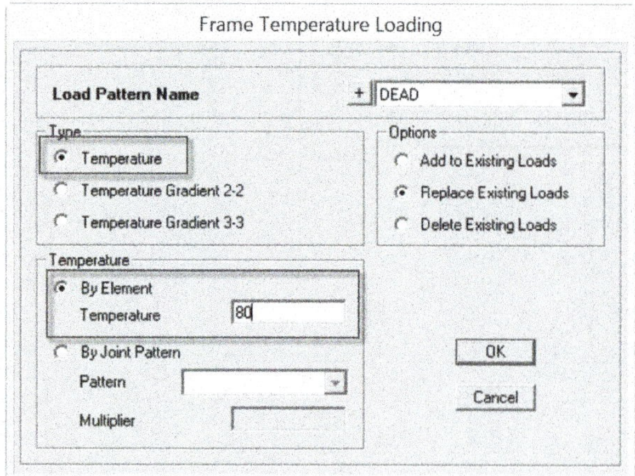

(a) Pop-up for Example 21.2

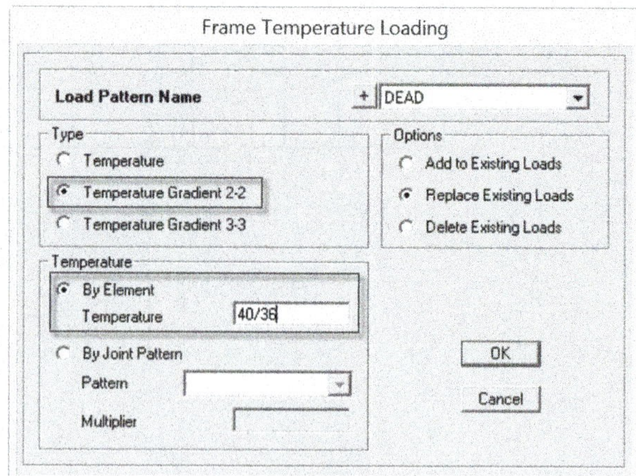

(b) Pop-up for Example 21.2

FIGURE 21.20 Pop-up for imposing thermal loads in SAP2000.

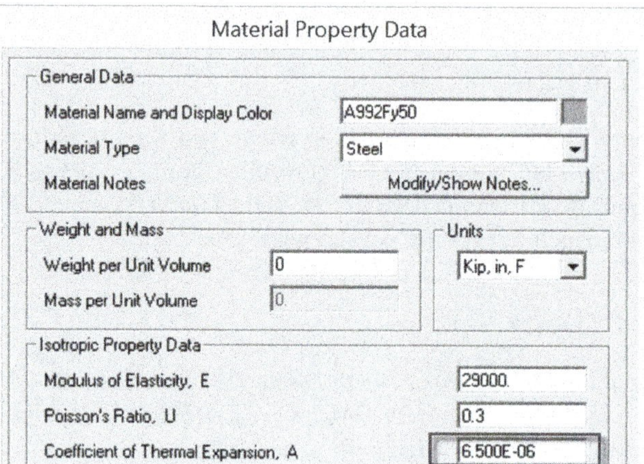

FIGURE 21.21 Pop-up window for defining coefficient of thermal expansion.

Misfit Members

Misfit members are defined in SAP2000 by first selecting the member of interest and then selecting the menu: *ASSSIGN* → *FRAME LOADS* → *DEFORMATION*. The only kind of deformation that can be applied is in the axial direction of the member (see Figure 21.22).

Partially Restrained Connections and Releases

The opportunity to define partially restrained connections or internal releases is actually done in SAP2000 using the same approach. In Section 21.7, we presented two options for including releases in a structure. The first was to set spring stiffnesses equal to zero, and the second

FIGURE 21.22 Pop-up window for defining member misfit.

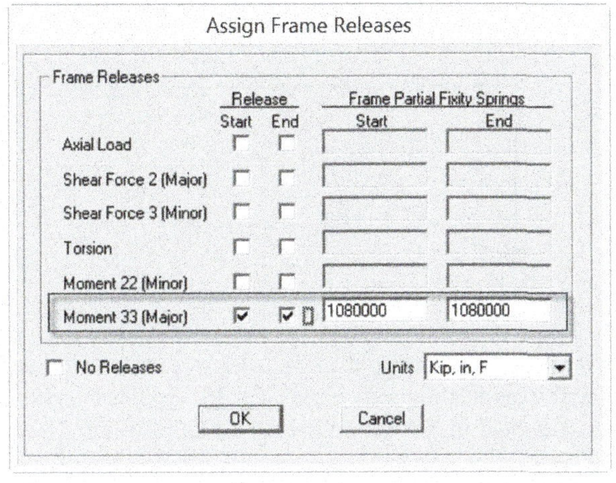

FIGURE 21.23 Pop-up window for defining PR connections or releases.

FIGURE 21.24 Rendering of releases and PR connections in SAP2000 (see Example 21.9).

DID YOU KNOW?
If you define a shear release in SAP2000, the rendering of the deflected shape will not be correct. Don't let this bother you. All of the internal and external forces—along with the joint displacements—will be correct.

approach was to add additional DOFs at the location of the release. SAP2000 actually takes the first approach for implementation.

First select the member in that you want to insert the PR connection or release. Select the following menu: *ASSIGN → FRAME → RELEASES/PARTIAL FIXITY*. This creates the pop-up window shown in Figure 21.23. This figure shows the input values used for Example 21.9. If you needed releases instead of PR connections, you simply insert a zero into the stiffness entry blocks. Figure 21.24 demonstrates the rendering of internal releases and PR connections.

DID YOU KNOW?
The starting and ending node of a member is defined in SAP2000 during the drawing phase of creating the model. When drawing the member, the beginning end will be the end you click first.

Inclined Supports

As was done in the matrix formulation for inclined supports in Section 21.8, the DOFs in SAP2000 need to be rotated to accommodate this inclination. While not strictly required, you may want to first turn on the view of the *local axes* for the joints. This will show the orientation of the joint axes which by default correspond with the global *XYZ* axes.

Select the joint of interest and then follow this menu: *ASSIGN → JOINT → LOCAL AXES*. The pop-up window shown in Figure 21.25 is then accessed. Define the angle of rotation—in degrees—for the local axes of the joint.

QUICK NOTE
Releases/PR connections only may be inserted at the ends of members. If you are needing to insert a release mid-member, you will first have to split the member before defining the releases.

QUICK NOTE
To view the local axes for a given joint, you may select the following menu: *DISPLAY → SHOW MISC ASSIGNS → JOINT*. Then click the box to view local axes. SAP2000 color codes the axes and has red, green, and blue corresponding to 1, 2, and 3, respectively.

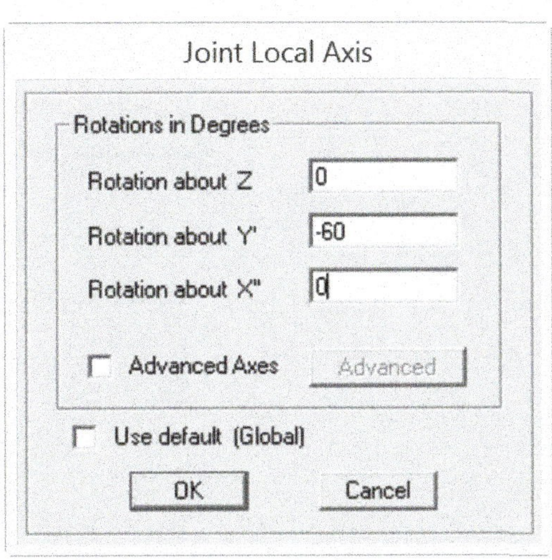

FIGURE 21.25 Pop-up window for defining the orientation of joint axes.

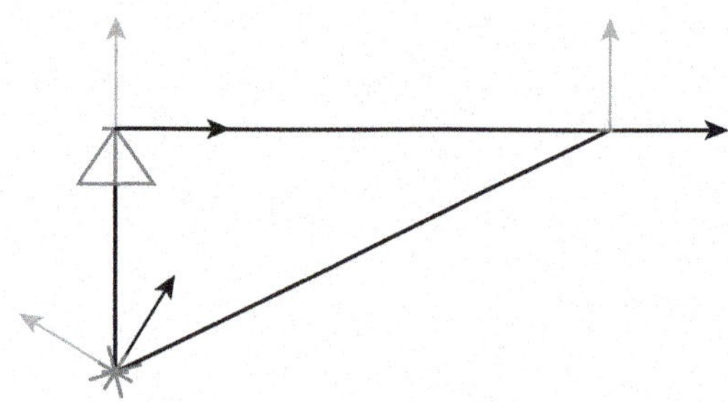

FIGURE 21.26 Rendering of joint local axes for the truss in Example 21.12 once joint rotation has been imposed.

To define the inclined support of Example 21.12, one would enter in an angle of −60°. The rendering of the joint local axes is shown in Figure 21.26. One now needs only to restrain the motion in the 1 direction (denoted in red within the SAP2000 user window).

21.10 Examples with Video Solution

VE21.1 The beam first introduced in Figure VE20.1 is now subjected to a temperature gradient in member 2. Figure VE21.1 shows this gradient and also the member depth. Assuming that $\alpha = 5.5 \times 10^{-6}/°F$, compute the joint displacements, reactions, and member end forces for the beam.

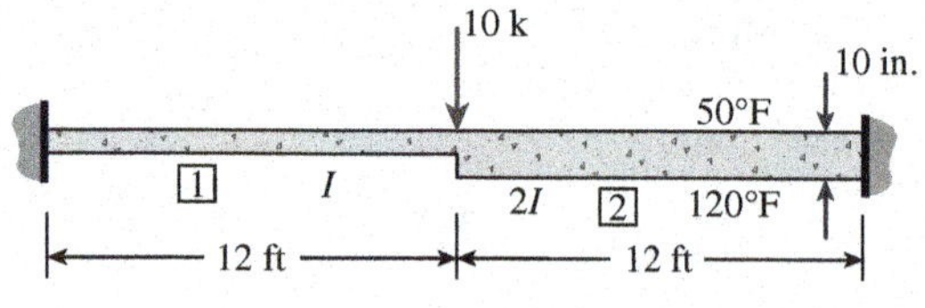

Figure VE21.1

VE21.2 Assume that the K_{uu} and F_u matrices for a certain truss are given as follows. First calculate the displacements Δ_u. Then condense out DOFs 3 and 4 and present the new stiffness and displacements at DOFs 1 and 2. How do these compare with the original displacements? Units are kN and mm.

$$K_{uu} = \begin{array}{c} \\ 1 \\ 2 \\ 3 \\ 4 \end{array} \begin{array}{c} \begin{array}{cccc} 1 & 2 & 3 & 4 \end{array} \\ \begin{bmatrix} 13.24 & -4.32 & -10.00 & 0 \\ -4.32 & 13.26 & 0 & 0 \\ -10.00 & 0 & 13.24 & 4.32 \\ 0 & 0 & 4.32 & 13.26 \end{bmatrix} \end{array}$$

$$F_u = \begin{array}{c} 1 \\ 2 \\ 3 \\ 4 \end{array} \begin{Bmatrix} 20 \\ 10 \\ 0 \\ -10 \end{Bmatrix} \text{kN}$$

VE21.3 Use the matrix approach to solve for the joint displacements, reactions, and member end forces for the beam with the internal shear release. By hand, sketch the shear and moment diagrams. *EI* is constant for both members.

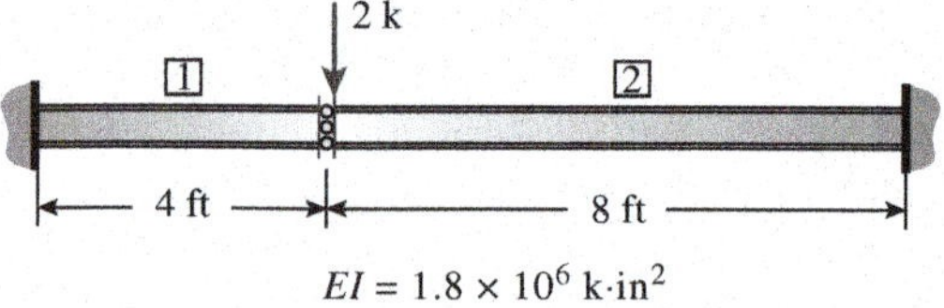

Figure VE21.3

VE21.4 Use the matrix approach to find the joint displacements for the truss shown in Figure VE21.4, which has an inclined roller.

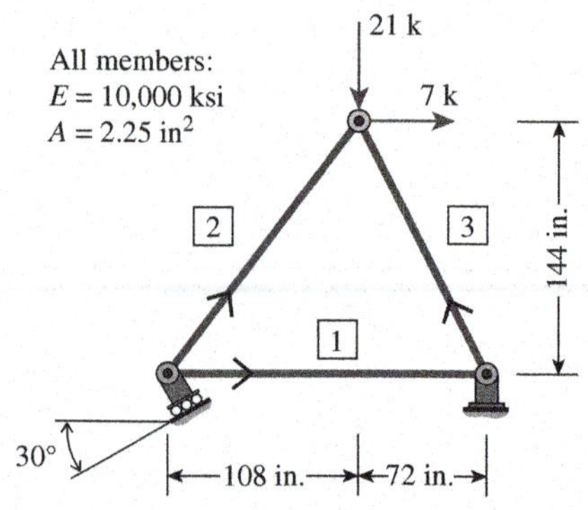

Figure VE21.4

21.11 Problems for Solution

Section 21.2

P21.1 A fixed-end beam has an intermediate support which is located 0.1 in. below the beam. Determine the nodal displacements and reactions that result if the beam is jacked into place. EI is constant = 1.0×10^6 k·in^2. (Ans: $\theta_B = -0.00313$ rad, $R_{By} = -73.24$ k)

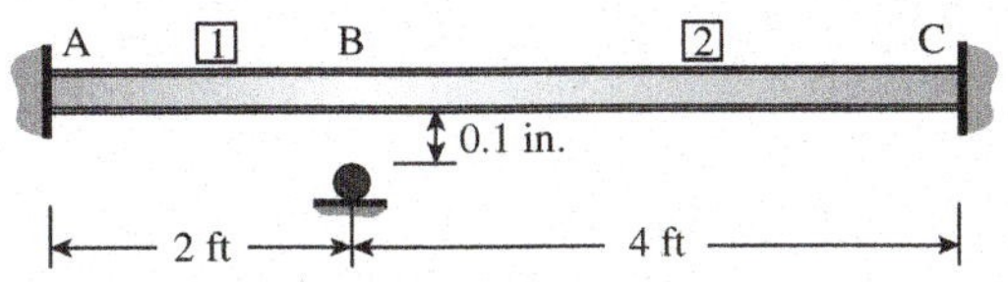

Problem 21.1

P21.2 The support located at the right end of a fixed-end beam has been built 0.005 rad out of position. Determine the nodal displacements and reactions that result if the beam is jacked into place. EI is constant = 2.6×10^6 k·in^2.

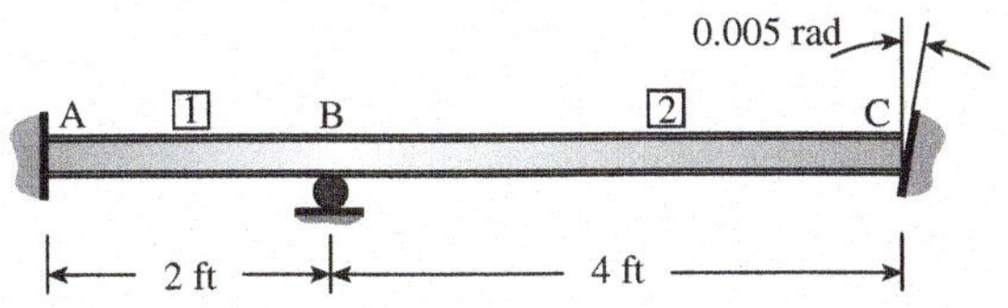

Problem 21.2

P21.3 The roller support located at joint 3 has been built 0.3 in. too low. Determine the nodal displacements and reactions that result if the truss is jacked into place. $E = 29,000$ ksi. For all bar areas except bars 4 and 5, $A = 1.5$ in^2. For bars 4 and 5, $A = 0.5$ in^2. (Ans: $\Delta_{1x} = 0.887$ in., $\Delta_{1y} = -0.576$ in.)

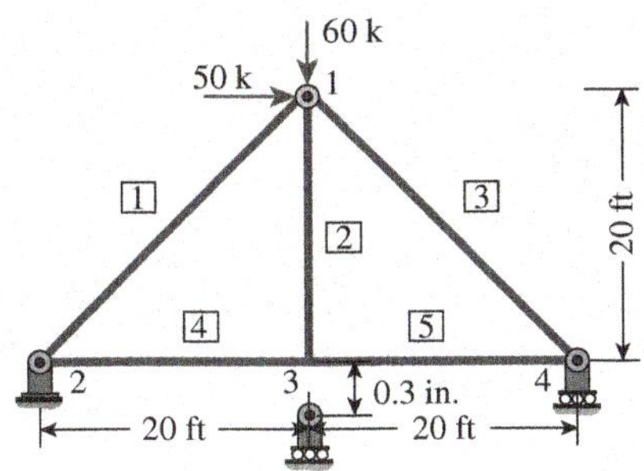

Problem 21.3

P21.4 The roller support located at the left end of a beam has been built 0.15 in. too low. Determine the nodal displacements and reactions that result if the beam is jacked into place. Compare these results with the case where the roller was not built incorrectly. $E = 29,000$ ksi and $I = 350$ in^4.

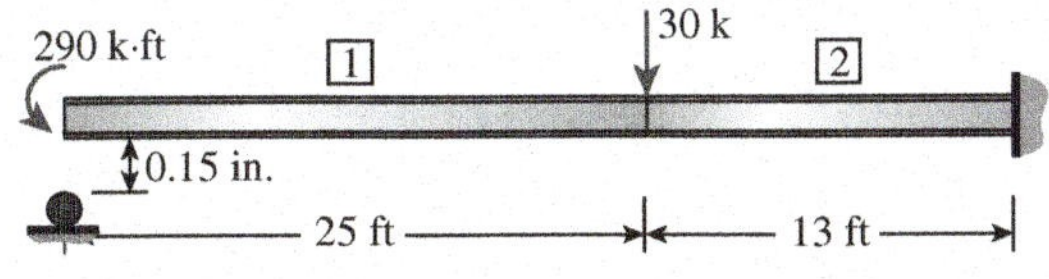

Problem 21.4

P21.5 The pin support located at the right end of a frame has been built 0.5 in. too far to the right. Determine the nodal displacements and reactions that result if the frame is jacked into place. (Ans: for the joint of members 1 and 2: $\Delta_x = 0.427$ in., $\Delta_y = -0.046$ in., $\theta = -0.00240$ rad)

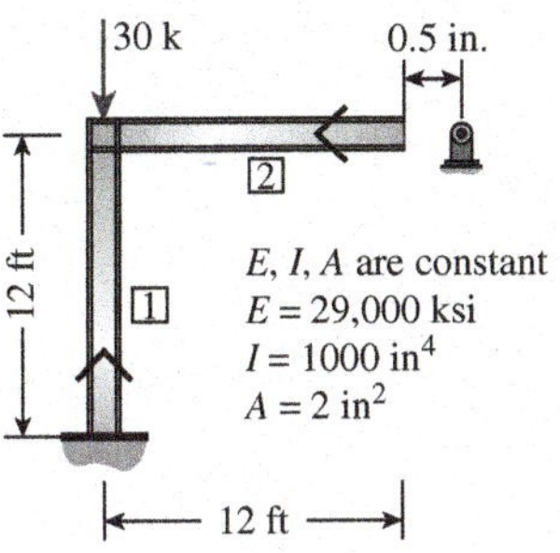

Problem 21.5

Section 21.3

P21.6 A fixed-end beam is exposed to a temperature gradient. The temperature at the top of the beam is 40°F and the temperature at the bottom is 65°F. Neglecting any axial effects, determine the nodal displacements and reactions that result from this gradient applied to both members. EI is constant = 1.0×10^6 k·in^2 and $\alpha = 6.5 \times 10^{-6}$/°F.

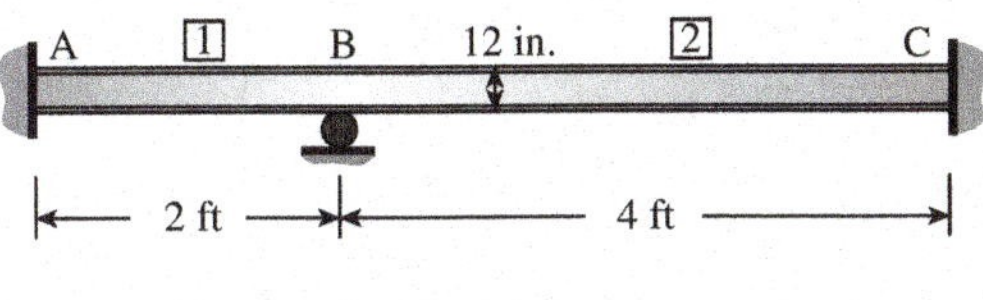

Problem 21.6

P21.7 Rework Problem 21.6 but have the temperature gradient only apply to member 1. (Ans: $R_{Ay} = 0.56$ in., $\theta_B = 0.00005$ rad)

P21.8 The temperature in bar 1 decreases by 30°F while the temperature in bar 3 increases by 20°F. The temperature in all other bars remain unchanged. Determine the nodal displacements and reactions that result from this temperature change in the truss members. $\alpha = 6.5 \times 10^{-6}$/°F and $E = 29,000$ ksi. All bar areas except bars 4 and 5, $A = 1.5$ in^2. For bars 4 and 5, $A = 0.5$ in^2.

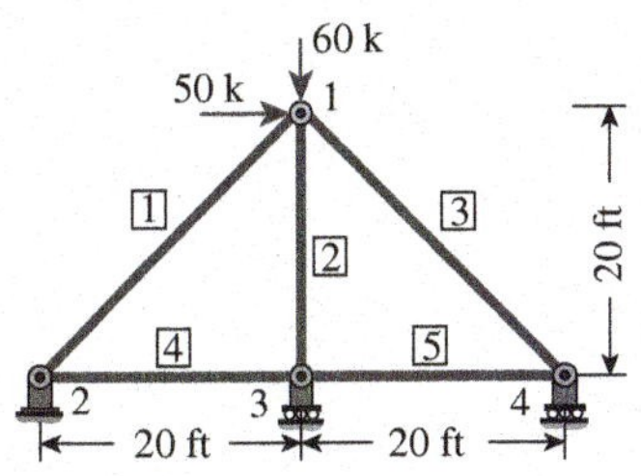

Problem 21.8

P21.9 The temperature in bar 2 increases by 60°F while the temperature in all other bars remain unchanged. Determine the nodal displacements and reactions that result from this temperature change in the truss member. $\alpha = 6.5 \times 10^{-6}/°F$ and $E = 29{,}000$ ksi. All bar areas except bars 4 and 5 have $A = 1.5$ in^2. For bars 4 and 5, $A = 0.5$ in^2. (Ans: $R_{2y} = -25.0$ k, $R_{4x} = -25.41$ k)

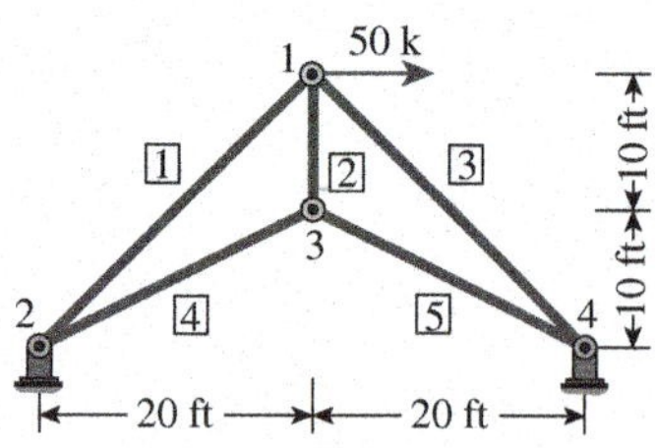

Problem 21.9

P21.10 Member 1 of the given frame experiences a uniform temperature increase (i.e., not a gradient) of 52°F while member 2 experiences no change at all. Determine the nodal displacements and reactions that result from this thermal change.

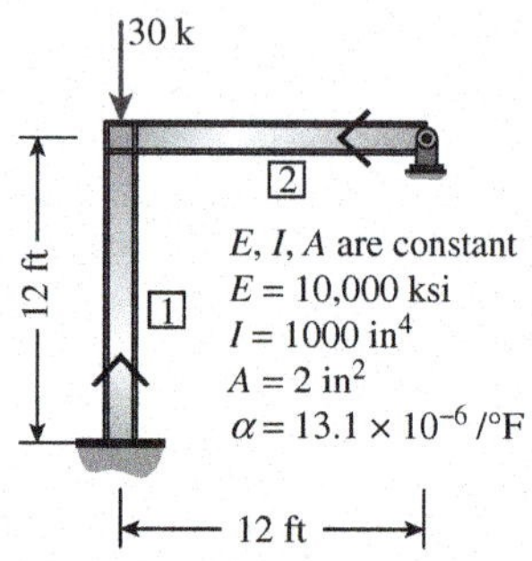

Problem 21.10

Section 21.4

P21.11 For the truss described in Problem 21.8 (without the thermal change), determine the nodal displacements and reactions that result if member 2 is made 0.5 in. too long (i.e., 240.5 in.). (Ans: $\Delta_{1x} = 0.581$ in., $\Delta_{3x} = 0.190$ in.)

P21.12 For the truss described in Problem 21.9 (without the thermal change), determine the nodal displacements and reactions that result if members 4 and 5 are each made 0.25 in. too short.

P21.13 For the frame described in Problem 21.10 (without the thermal change), determine the nodal displacements and reactions that result if both members are each made 0.35 in. too short. (Ans: for joint between members 1 and 2: $\Delta_x = 0.272$ in., $\Delta_y = -0.527$ in., $\theta = -0.00005$ rad)

Section 21.5

For Problems 21.14 through 21.16 the K_{uu} matrix and the F_u vector are given. First compute Δ_u using the given matrices. Next condense out the DOFs indicated, report $\hat{K}_{pp}$ and $\hat{F}_p$ then compute the nodal displacements using these reduced matrices.

P21.14 Condense out DOFs 4 through 7.

$$K_{uu} = \begin{array}{c} \\ 1\\2\\3\\4\\5\\6\\7 \end{array}\begin{array}{c} \begin{array}{ccccccc} 1 & 2 & 3 & 4 & 5 & 6 & 7 \end{array} \\ \begin{bmatrix} 27.4 & 0 & -13.7 & -22.9 & -13.7 & 22.9 & 0 \\ 0 & 76.2 & -22.9 & -38.1 & 22.9 & -38.1 & 0 \\ -13.7 & -22.9 & 77.8 & 45.7 & -50.3 & 0 & 0 \\ -22.9 & -38.1 & 45.7 & 76.2 & 0 & 0 & 0 \\ -13.7 & 22.9 & -50.3 & 0 & 100.2 & -33.3 & -13.7 \\ 22.9 & -38.1 & 0 & 0 & -33.3 & 83.1 & 22.9 \\ 0 & 0 & 0 & 0 & -13.7 & 22.9 & 38.9 \end{bmatrix} \end{array} \text{k/in.} \qquad F_u = \begin{Bmatrix} 20 \\ 0 \\ 40 \\ 0 \\ 0 \\ 0 \\ 0 \end{Bmatrix} \text{k}$$

P21.15 Condense out DOFs 4, 5, and 6.

$$K_{uu} = \begin{array}{c} \\ 1\\2\\3\\4\\5\\6 \end{array}\begin{array}{c} \begin{array}{cccccc} 1 & 2 & 3 & 4 & 5 & 6 \end{array} \\ \begin{bmatrix} 173.4 & 0 & 0 & 0 & -46.4 & -30.9 \\ 0 & 162.1 & 0 & -120.8 & 30.9 & -20.6 \\ 0 & 0 & 161.1 & 0 & -80.6 & 0 \\ 0 & -120.8 & 0 & 120.8 & 0 & 0 \\ -46.4 & 30.9 & -80.6 & 0 & 127.0 & 0 \\ -30.9 & -20.6 & 0 & 0 & 0 & 141.5 \end{bmatrix} \end{array} \text{k/in.} \qquad F_u = \begin{Bmatrix} -30 \\ -60 \\ 0 \\ -60 \\ 0 \\ 0 \end{Bmatrix} \text{k}$$

(Ans: $\Delta_1 = 0.170$ in., $\Delta_2 = -4.487$ in.)

P21.16 Condense out DOFs 3 and 4.

$$K_{uu} = \begin{array}{c} \\ 1 \\ 2 \\ 3 \\ 4 \end{array} \begin{array}{c} \begin{array}{cccc} 1 & 2 & 3 & 4 \end{array} \\ \begin{bmatrix} 128.2 & 0 & 0 & -64.1 \\ 0 & 309.4 & 0 & 64.1 \\ 0 & 0 & 120.8 & -60.4 \\ -64.1 & 64.1 & -60.4 & 124.5 \end{bmatrix} \end{array} \text{k/in.} \qquad F_u = \begin{Bmatrix} 50 \\ -60 \\ 0 \\ 0 \end{Bmatrix} \text{k}$$

P21.17 The beam assembly given has its DOFs labeled as shown. Assemble the global stiffness matrix for this assembly. Now condense out DOFs 5 and 6 and present the new stiffness matrix for this assemblage – $\hat{K}_{pp}$. (Ans: $\hat{K}_{11}$ = 17.26 k/in., $\hat{K}_{44}$ = 159,091(k·in.)/rad)

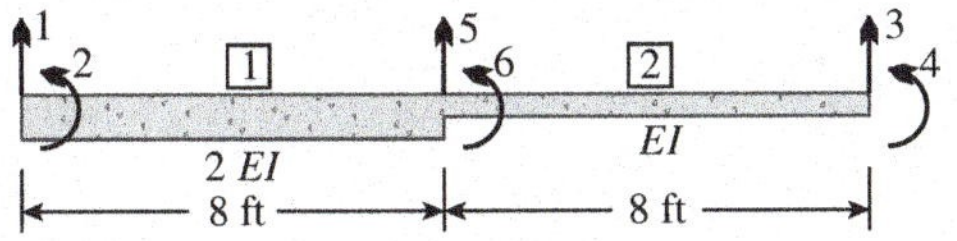

$EI = 7 \times 10^6$ k·in^2

Problem 21.17

P21.18 The beam assembly given has its DOFs labeled as shown. Assemble the global stiffness matrix for this assembly. Now condense out DOFs 5 through 8 and present the new stiffness matrix for this assemblage – $\hat{K}_{pp}$.

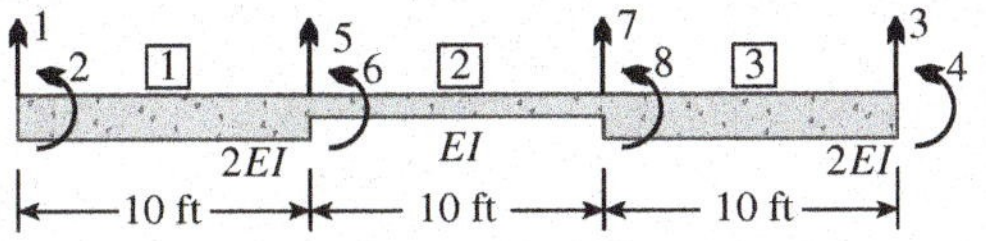

$EI = 8.4 \times 10^6$ k·in^2

Problem 21.18

P21.19 The frame assembly given has its DOFs labeled as shown. Assemble the global stiffness matrix for this assembly. Now condense out DOFs 7 through 9 and present the new stiffness matrix for this assemblage – $\hat{K}_{pp}$. (Ans: $\hat{K}_{11}$ = 20.27 k/in., $\hat{K}_{66}$ = 212,652 (k·in.)/rad)

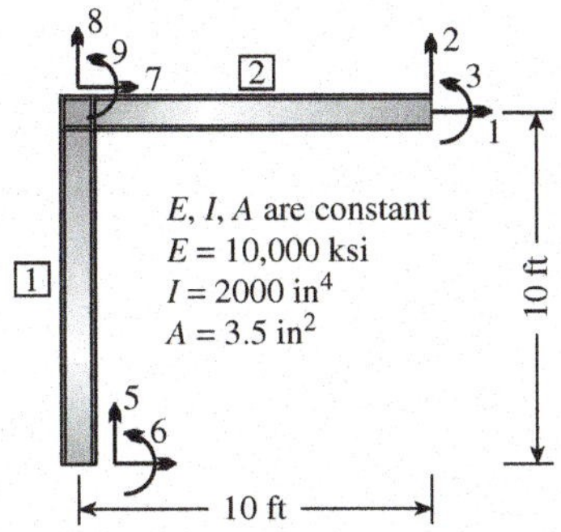

Problem 21.19

Section 21.6

For Problems 21.20 through 21.23 develop the 6-DOF spring-beam-spring element and then condense out DOFs *e* and *f* to get the equivalent 4-DOF stiffness matrix for the element.

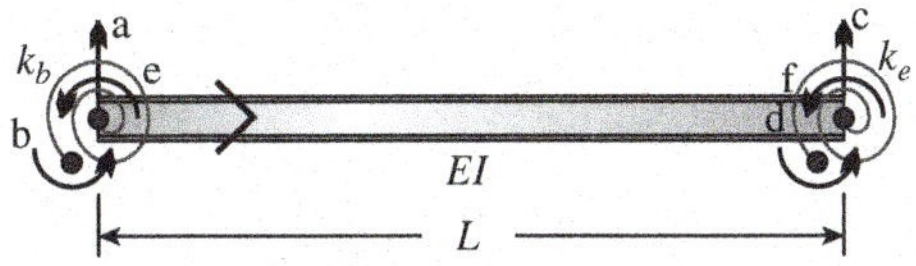

Problem 21.20 through 21.23

P21.20 Assume the following values. L = 10 ft, E = 29,000 ksi, I = 300 in^4, k_b = 0.0 (k·in.)/rad, and k_e = 0.0 (k·in.)/rad.

P21.21 Assume the following values. L = 16 ft, E = 10,000 ksi, I = 500 in^4, k_b = 300,000 (k·in.)/rad, and k_e = 150,000 (k·in.)/rad. (Ans: $\hat{K}_{aa}$ = 4.81 k/in, $\hat{K}_{bb}$ = 72,180 (k·in.)/rad)

P21.22 Assume the following values. L = 20 ft, E = 29,000 ksi, I = 1300 in^4, k_b = 0.0 (k·in.)/rad, and k_e = 1,900,000 (k·in.)/rad.

P21.23 Assume the following values. L = 12.5 ft, E = 12,000 ksi, I = 675 in^4, k_b = 40,000 (k·in.)/rad, and k_e = 40,000 (k·in.)/rad. (Ans: $\hat{K}_{aa}$ = 3.16 k/in., $\hat{K}_{bb}$ = 32,397 (k·in.)/rad)

P21.24 For the beam shown, assume the following values. E = 29,000 ksi, I = 350 in^4, and k_R = 150,000 (k·in.)/rad. Solve for all joint displacements and reactions due to the applied loads.

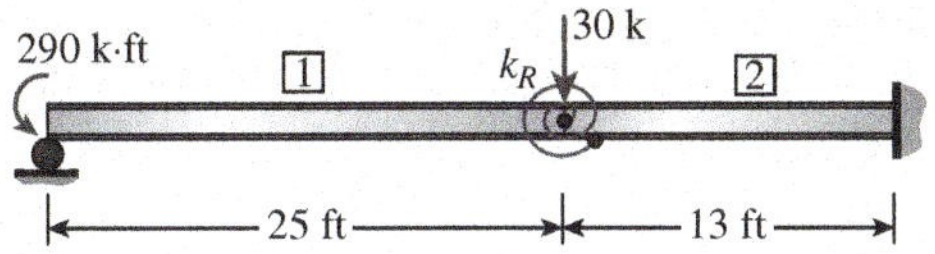

Problem 21.24

P21.25 For the beam shown, assume the following values. E = 5400 ksi, I = 5000 in^4, and k_R = 100,000 (k·in.)/rad. Solve for all joint displacements and reactions due to the applied loads. (Ans: Δ_{left} = –16.54 in., M_{left} = –18,126 k·in.)

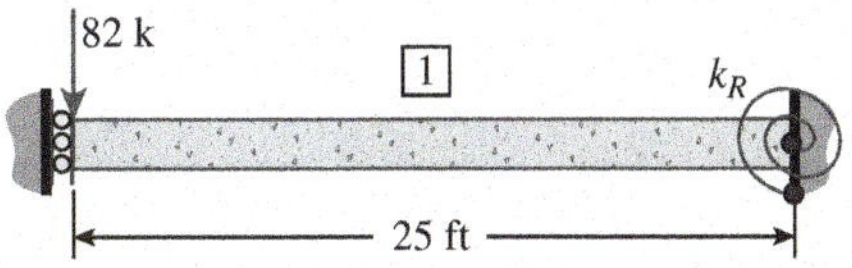

Problem 21.25

P21.26 For the beam shown, assume the following values. E = 29,000 ksi, I = 225 in^4, and k_R = 1,500,000 (k·in.)/rad. Solve for all joint displacements and reactions due to the applied loads.

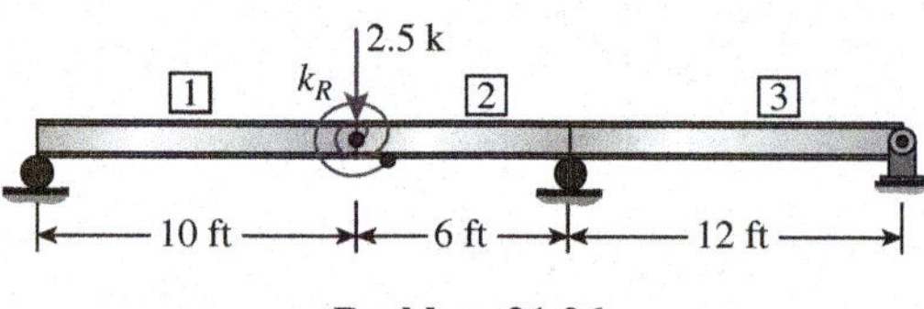

Problem 21.26

Section 21.7

P21.27 Rework Problem 21.24 using a spring stiffness k_R = 0.0 (k·in.)/rad. (Ans: $\theta_{\text{left}-1}$ = 0.0266 rad, R_{left} = 11.60 k)

P21.28 Rework Problem 21.25 using a spring stiffness k_R = 0.0 (k·in.)/rad.

P21.29 Rework Problem 21.26 using a spring stiffness k_R = 0.0 (k·in.)/rad. (Ans: $\theta_{\text{left}-1}$ = −0.0012 rad, R_{left} = 0.0 k)

For Problems 21.30 through 21.33, use the double-DOF approach to model the releases in the following beams. Solve for all joint displacements and reactions.

P21.30 For the beam shown, assume the following values. E = 29,000 ksi, I = 350 in^4.

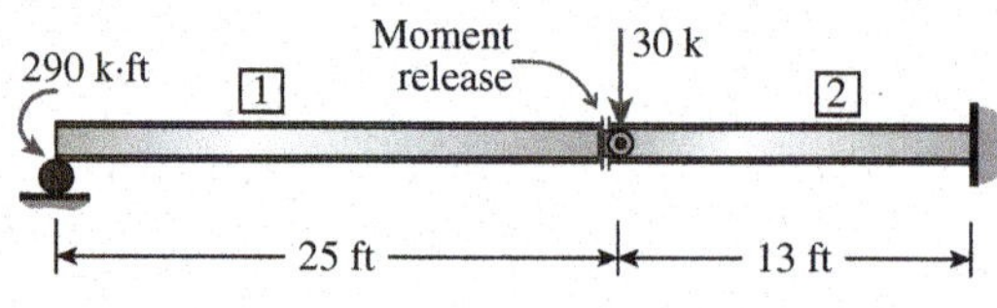

Problem 21.30

P21.31 For the beam shown, assume the following values. E = 29,000 ksi and I = 350 in^4. (Ans: $\theta_{\text{left}-1}$ = 0.192 rad, R_{left} = 0.0 k)

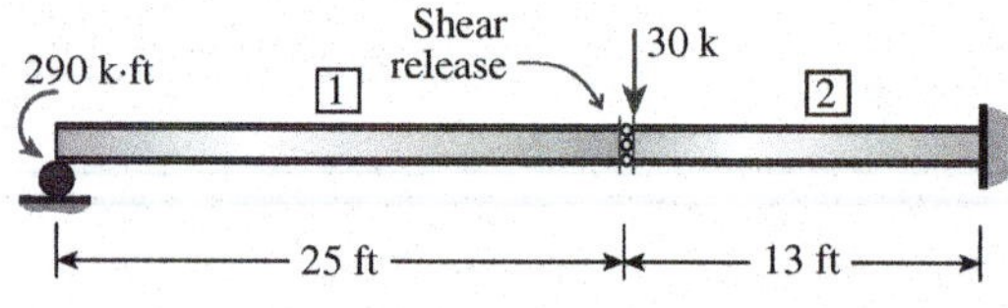

Problem 21.31

P21.32 For the beam shown, assume the following values. E = 29,000 ksi and I = 225 in^4.

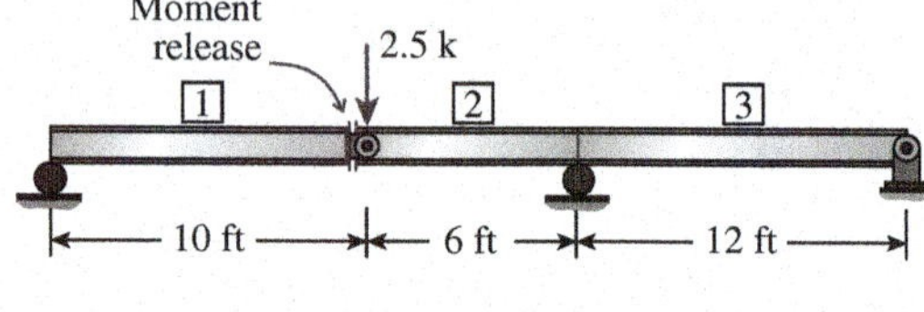

Problem 21.32

P21.33 For the beam shown, assume the following values. E = 29,000 ksi and I = 225 in^4. (Ans: $\theta_{\text{left}-1}$ = −0.0083 rad, R_{left} = 2.50 k)

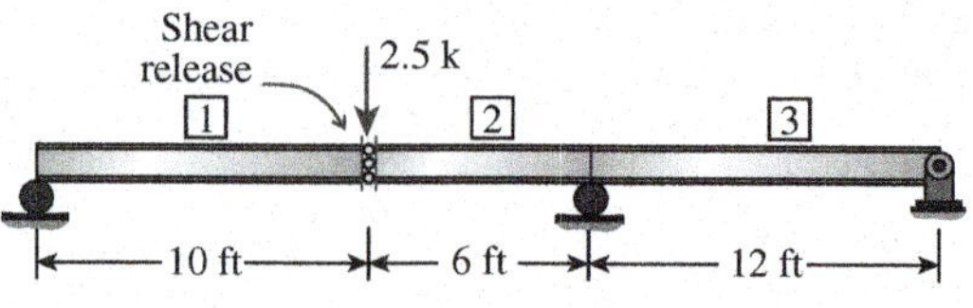

Problem 21.33

Section 21.8

P21.34 The roller at joint 4 is inclined at 20° for the given truss. Determine the nodal displacements and reactions that result. E = 29,000 ksi. All bar areas except bars 4 and 5 have A = 1.5 in^2. For bars 4 and 5, A = 0.5 in^2.

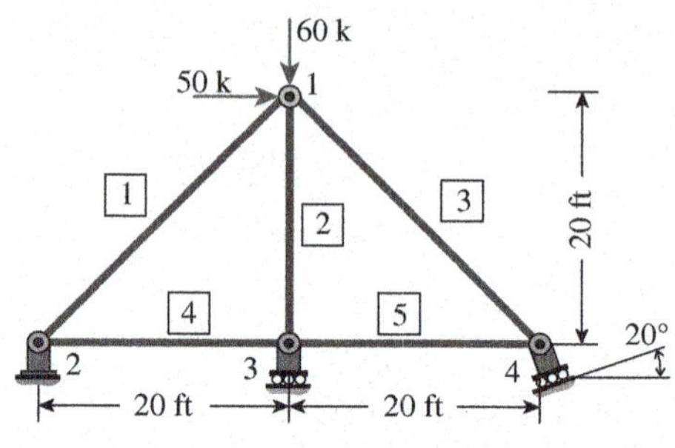

Problem 21.34

P21.35 The roller at joint 5 is inclined at 25° for the given truss. Determine the nodal displacements and reactions that result. E = 29,000 ksi. All bar areas A = 0.5 in^2. (Ans: Δ_{1x} = 3.078 in., R_{4y} = −33.32 k)

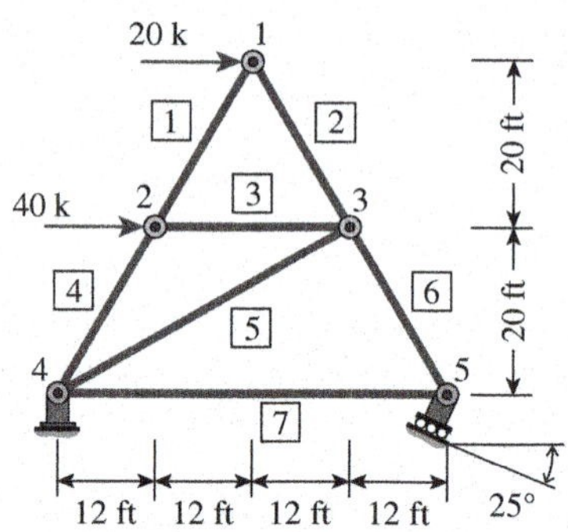

Problem 21.35

P21.36 The roller at the right end of the frame is inclined at 10°. Determine the nodal displacements and reactions that result.

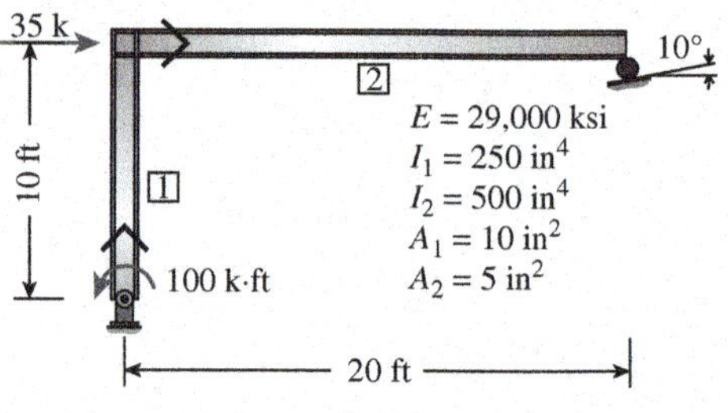

Problem 21.36

SAP2000

P21.37 Model the beam described in Problem 21.1. Find the reactions for following support misalignments at B: 0.05 in., 0.1 in., 0.15 in., and 0.20 in. Make an observation regarding the relationship between the reactions and the support misalignment.

P21.38 Model the beam described in Problem 21.6. Find the reactions and plot the deflected shape. Now apply the temperature gradient to both members. Provide the reactions and plot the deflected shape for this scenario. Make an observation regarding the change, if any, of these responses between the two scenarios.

P21.39 Model the truss described in Problem 21.8. Find the reactions and deflections of the joints. Now model the truss so that all bars experience a temperature increase of 30°F. Then model the truss without any thermal effects. Make an observation about the comparison of the reactions and the deflected shape for all three scenarios.

P21.40 Model the beam described in Problem 21.26. Plot the moment diagrams for the following four scenarios: k_R = 0.0 (k·in.)/rad, k_R = 150,000 (k·in.)/rad, k_R = 1,500,000 (k·in.)/rad, and k_R = 15,000,000 (k·in.)/rad. Compare the moment diagrams and make an observation regarding the influence of the spring stiffness on these diagrams.

P21.41 Model the beams described in Problems 21.30 and 21.31. Plot the shear and moment diagrams for each beam. Make an observation regarding the influence of these releases on the internal shear and moment in the beams.

P21.42 Model the truss described in Problem 21.34. Plot the deflected shape for the following five scenarios where the roller is inclined at: 0°, 22.5°, 45°, 67.5°, and 90°. Make an observation regarding the influence of the incline upon the deflected shape.

A

ASCE 7-16 Information

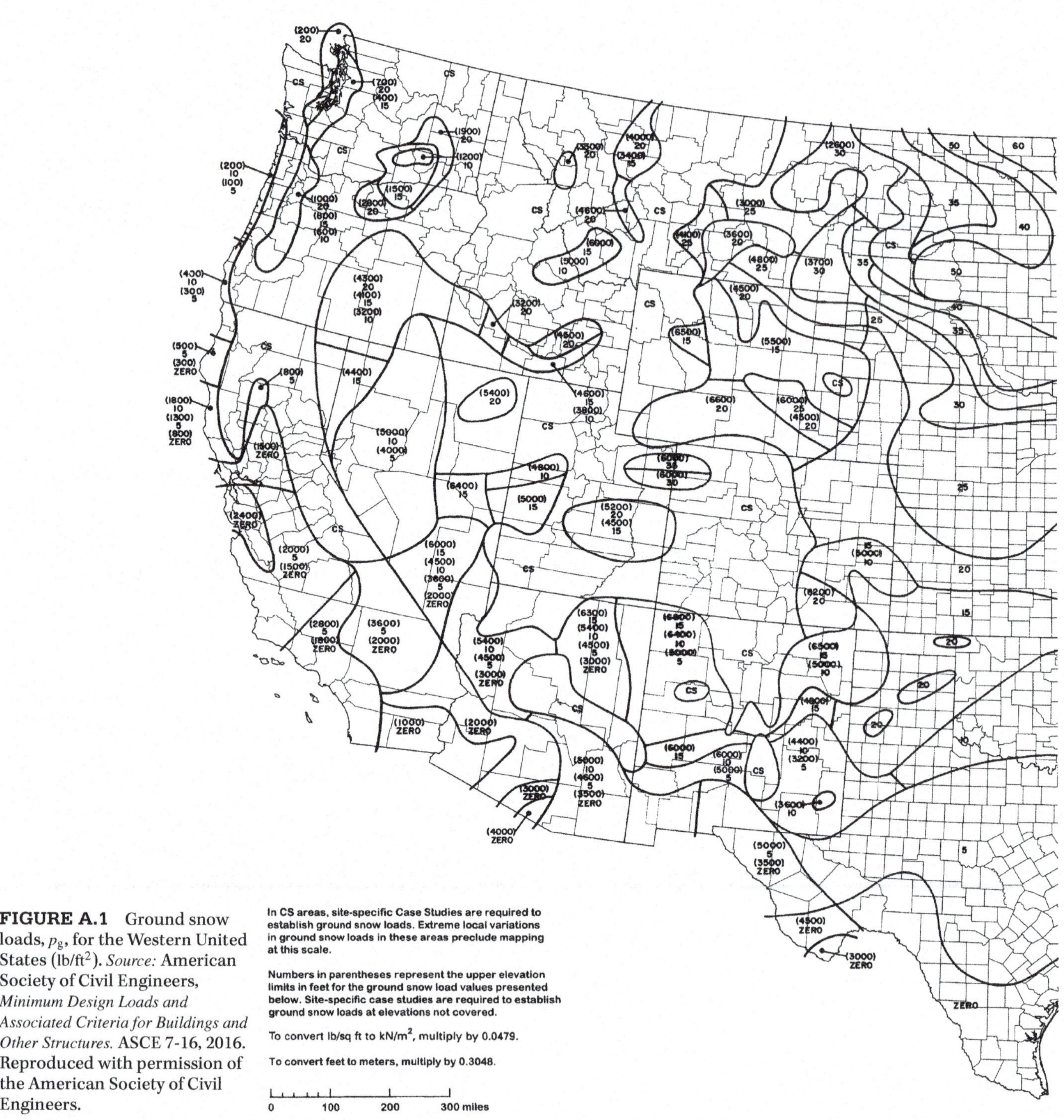

In CS areas, site-specific Case Studies are required to establish ground snow loads. Extreme local variations in ground snow loads in these areas preclude mapping at this scale.
Numbers in parentheses represent the upper elevation limits in feet for the ground snow load values presented below. Site-specific case studies are required to establish ground snow loads at elevations not covered.
To convert lb/sq ft to kN/m², multiply by 0.0479.
To convert feet to meters, multiply by 0.3048.
0
100
200
300 miles

FIGURE A.1 Ground snow loads, p_g, for the Western United States (lb/ft^2). *Source:* American Society of Civil Engineers, *Minimum Design Loads and Associated Criteria for Buildings and Other Structures.* ASCE 7-16, 2016. Reproduced with permission of the American Society of Civil Engineers.

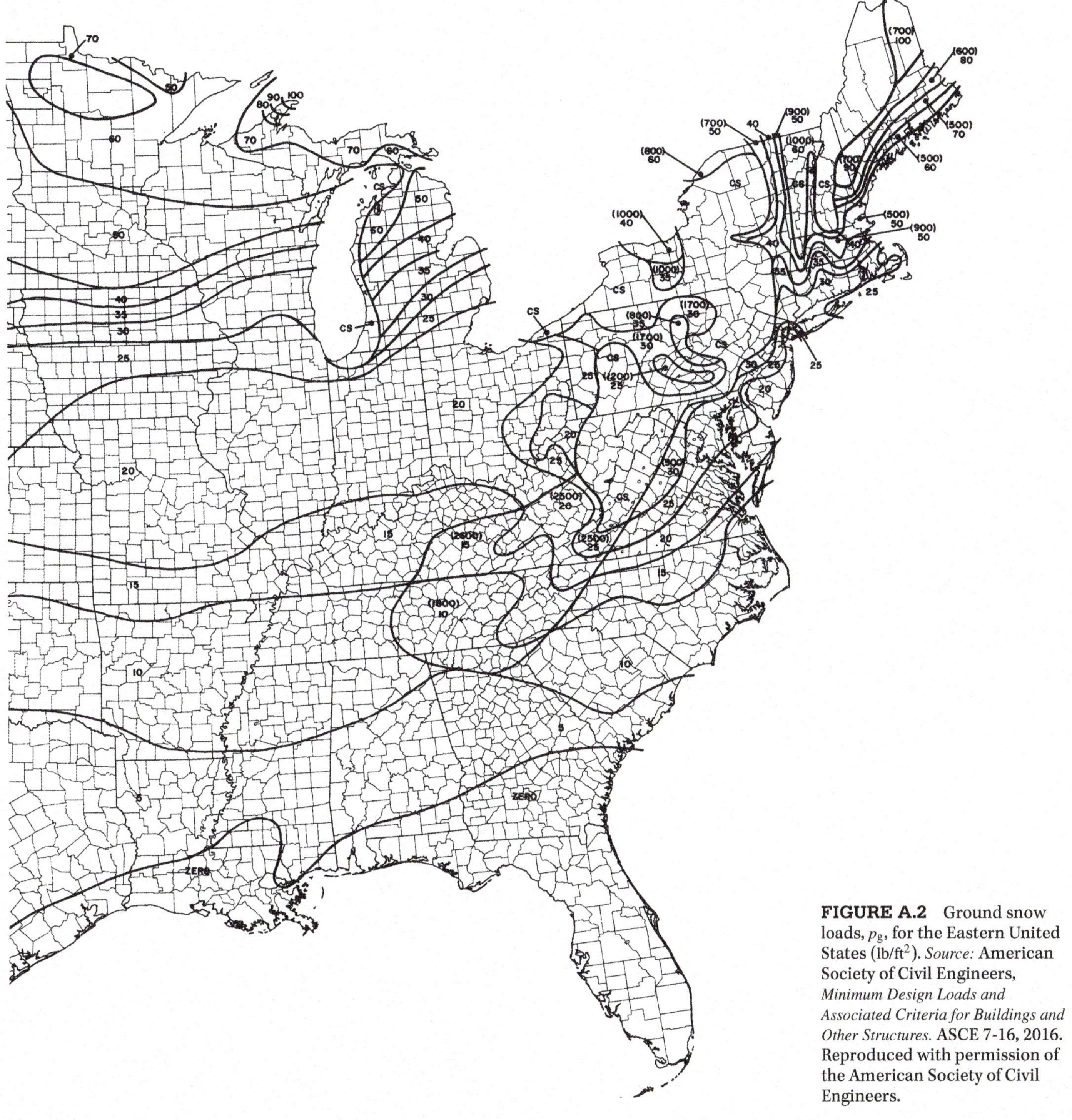

FIGURE A.2 Ground snow loads, p_g, for the Eastern United States (lb/ft^2). *Source:* American Society of Civil Engineers, *Minimum Design Loads and Associated Criteria for Buildings and Other Structures.* ASCE 7-16, 2016. Reproduced with permission of the American Society of Civil Engineers.

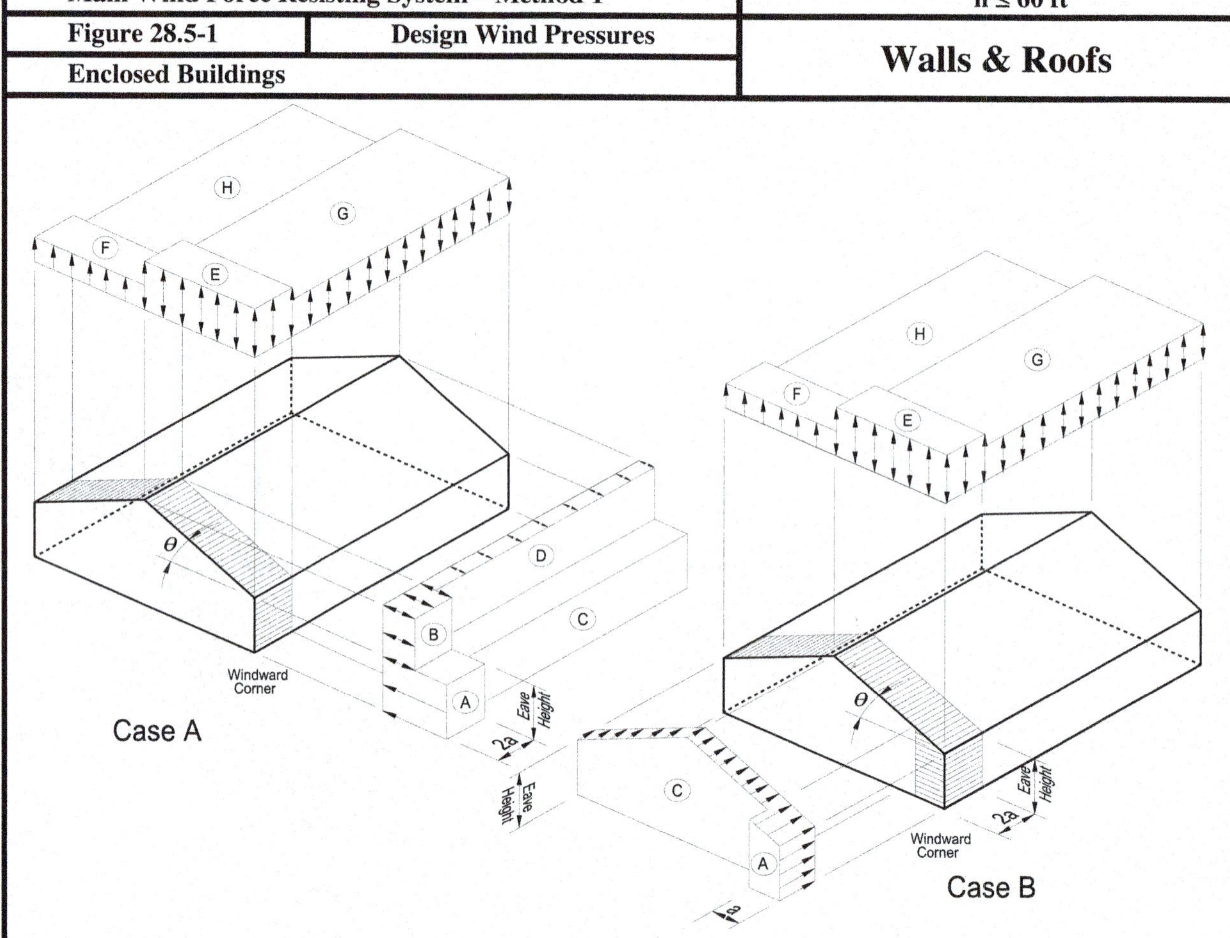

Notes:

1. Pressures shown are applied to the horizontal and vertical projections, for exposure B, at $h = 30$ ft (9.1 m). Adjust to other exposures and heights with adjustment factor λ.
2. The load patterns shown shall be applied to each corner of the building in turn as the reference corner.
3. For the design of the MWFRS use $\theta = 0°$.
4. Load cases 1 and 2 must be checked for $25° < \theta \leq 45°$. Load case 2 at 25° is provided only for interpolation between 25° and 30°.
5. Plus and minus signs signify pressures acting toward and away from the projected surfaces, respectively.
6. For roof slopes other than those shown, linear interpolation is permitted.
7. The total horizontal load shall not be less than that determined by assuming $p_S = 0$ in zones B & D.
8. Where zone E or G falls on a roof overhang on the windward side of the building, use E_{OH} and G_{OH} for the pressure on the horizontal projection of the overhang. Overhangs on the leeward and side edges shall have the basic zone pressure applied.
9. Notation:

   a: 10 percent of least horizontal dimension or 0.4h, whichever is smaller, but not less than either 4% of least horizontal dimension or 3 ft (0.9 m).

   Exception: For buildings with θ = 0 to 7° and a least horizontal dimension greater than 300 ft (90 m), dimension ‘a’ shall be limited to a maximum of 0.8 h.

   h: Mean roof height, in feet (meters), except that eave height shall be used for roof angles <10°.

   θ: Angle of plane of roof from horizontal, in degrees.

FIGURE A.3 Design wind pressures for the main wind force resisting systems. *Source:* American Society of Civil Engineers, *Minimum Design Loads and Associated Criteria for Buildings and Other Structures.* ASCE 7-16, 2016. Reproduced with permission of the American Society of Civil Engineers.

Main Wind Force Resisting System – Method 1		h ≤ 60 ft
Figure 28.5-1 (*Continued*)	Design Wind Pressures	Walls & Roofs
Enclosed Buildings		

Simplified Design Wind Pressure , p_{S30} (psf) *(Exposure B at h = 30 ft. with I = 1.0)*

Basic Wind Speed (mph)	Roof Angle (degrees)	Load Case	Zones									
			Horizontal Pressures				Vertical Pressures				Overhangs	
			A	B	C	D	E	F	G	H	E_{OH}	G_{OH}
90	0 to 5°	1	12.8	-6.7	8.5	-4.0	-15.4	-8.8	-10.7	-6.8	-21.6	-16.9
	10°	1	14.5	-6.0	9.6	-3.5	-15.4	-9.4	-10.7	-7.2	-21.6	-16.9
	15°	1	16.1	-5.4	10.7	-3.0	-15.4	-10.1	-10.7	-7.7	-21.6	-16.9
	20°	1	17.8	-4.7	11.9	-2.6	-15.4	-10.7	-10.7	-8.1	-21.6	-16.9
	25°	1	16.1	2.6	11.7	2.7	-7.2	-9.8	-5.2	-7.8	-13.3	-11.4
		2	-------	-------	-------	-------	-2.7	-5.3	-0.7	-3.4	-------	-------
	30 to 45	1	14.4	9.9	11.5	7.9	1.1	-8.8	0.4	-7.5	-5.1	-5.8
		2	14.4	9.9	11.5	7.9	5.6	-4.3	4.8	-3.1	-5.1	-5.8
100	0 to 5°	1	15.9	-8.2	10.5	-4.9	-19.1	-10.8	-13.3	-8.4	-26.7	-20.9
	10°	1	17.9	-7.4	11.9	-4.3	-19.1	-11.6	-13.3	-8.9	-26.7	-20.9
	15°	1	19.9	-6.6	13.3	-3.8	-19.1	-12.4	-13.3	-9.5	-26.7	-20.9
	20°	1	22.0	-5.8	14.6	-3.2	-19.1	-13.3	-13.3	-10.1	-26.7	-20.9
	25°	1	19.9	3.2	14.4	3.3	-8.8	-12.0	-6.4	-9.7	-16.5	-14.0
		2	-------	-------	-------	-------	-3.4	-6.6	-0.9	-4.2	-------	-------
	30 to 45	1	17.8	12.2	14.2	9.8	1.4	-10.8	0.5	-9.3	-6.3	-7.2
		2	17.8	12.2	14.2	9.8	6.9	-5.3	5.9	-3.8	-6.3	-7.2
110	0 to 5°	1	19.2	-10.0	12.7	-5.9	-23.1	-13.1	-16.0	-10.1	-32.3	-25.3
	10°	1	21.6	-9.0	14.4	-5.2	-23.1	-14.1	-16.0	-10.8	-32.3	-25.3
	15°	1	24.1	-8.0	16.0	-4.6	-23.1	-15.1	-16.0	-11.5	-32.3	-25.3
	20°	1	26.6	-7.0	17.7	-3.9	-23.1	-16.0	-16.0	-12.2	-32.3	-25.3
	25°	1	24.1	3.9	17.4	4.0	-10.7	-14.6	-7.7	-11.7	-19.9	-17.0
		2	-------	-------	-------	-------	-4.1	-7.9	-1.1	-5.1	-------	-------
	30 to 45	1	21.6	14.8	17.2	11.8	1.7	-13.1	0.6	-11.3	-7.6	-8.7
		2	21.6	14.8	17.2	11.8	8.3	-6.5	7.2	-4.6	-7.6	-8.7
115	0 to 5°	1	21.0	-10.9	13.9	-6.5	-25.2	-14.3	-17.5	-11.1	-35.3	-27.6
	10°	1	23.7	-9.8	15.7	-5.7	-25.2	-15.4	-17.5	-11.8	-35.3	-27.6
	15°	1	26.3	-8.7	17.5	-5.0	-25.2	-16.5	-17.5	-12.6	-35.3	-27.6
	20°	1	29.0	-7.7	19.4	-4.2	-25.2	-17.5	-17.5	-13.3	-35.3	-27.6
	25°	1	26.3	4.2	19.1	4.3	-11.7	-15.9	-8.5	-12.8	-21.8	-18.5
		2	-------	-------	-------	-------	-4.4	-8.7	-1.2	-5.5	-------	-------
	30 to 45	1	23.6	16.1	18.8	12.9	1.8	-14.3	0.6	-12.3	-8.3	-9.5
		2	23.6	16.1	18.8	12.9	9.1	-7.1	7.9	-5.0	-8.3	-9.5
120	0 to 5°	1	22.8	-11.9	15.1	-7.0	-27.4	-15.6	-19.1	-12.1	-38.4	-30.1
	10°	1	25.8	-10.7	17.1	-6.2	-27.4	-16.8	-19.1	-12.9	-38.4	-30.1
	15°	1	28.7	-9.5	19.1	-5.4	-27.4	-17.9	-19.1	-13.7	-38.4	-30.1
	20°	1	31.6	-8.3	21.1	-4.6	-27.4	-19.1	-19.1	-14.5	-38.4	-30.1
	25°	1	28.6	4.6	20.7	4.7	-12.7	-17.3	-9.2	-13.9	-23.7	-20.2
		2	-------	-------	-------	-------	-4.8	-9.4	-1.3	-6.0	-------	-------
	30 to 45	1	25.7	17.6	20.4	14.0	2.0	-15.6	0.7	-13.4	-9.0	-10.3
		2	25.7	17.6	20.4	14.0	9.9	-7.7	8.6	-5.5	-9.0	-10.3
140	0 to 5°	1	31.1	-16.1	20.6	-9.6	-37.3	-21.2	-26.0	-16.4	-52.3	-40.9
	10°	1	35.1	-14.5	23.3	-8.5	-37.3	-22.8	-26.0	-17.5	-52.3	-40.9
	15°	1	39.0	-12.9	26.0	-7.4	-37.3	-24.4	-26.0	-18.6	-52.3	-40.9
	20°	1	43.0	-11.4	28.7	-6.3	-37.3	-26.0	-26.0	-19.7	-52.3	-40.9
	25°	1	39.0	6.3	28.2	6.4	-17.3	-23.6	-12.5	-19.0	-32.3	-27.5
		2	-------	-------	-------	-------	-6.6	-12.8	-1.8	-8.2	-------	-------
	30 to 45	1	35.0	23.9	27.8	19.1	2.7	-21.2	0.9	-18.2	-12.3	-14.0
		2	35.0	23.9	27.8	19.1	13.4	-10.5	11.7	-7.5	-12.3	-14.0

Unit Conversions – 1.0 ft = 0.3048 m; 1.0 psf = 0.0479 kN/m²

FIGURE A.3 (*continued*)

Main Wind Force Resisting System – Method 1		$h \leq 60$ ft
Figure 28.5-1 (*Continued*)	Design Wind Pressures	**Walls & Roofs**
Enclosed Buildings		

Simplified Design Wind Pressure, p_{S30} (psf) *(Exposure B at h = 30 ft.)*

Basic Wind Speed (mph)	Roof Angle (degrees)	Load Case	Zones									
			Horizontal Pressures				Vertical Pressures				Overhangs	
			A	B	C	D	E	F	G	H	E_{OH}	G_{OH}
160	0 to 5°	1	40.6	-21.1	26.9	-12.5	-48.8	-27.7	-34.0	-21.5	-68.3	-53.5
	10°	1	45.8	-19.0	30.4	-11.1	-48.8	-29.8	-34.0	-22.9	-68.3	-53.5
	15°	1	51.0	-16.9	34.0	-9.6	-48.8	-31.9	-34.0	-24.3	-68.3	-53.5
	20°	1	56.2	-14.8	37.5	-8.2	-48.8	-34.0	-34.0	-25.8	-68.3	-53.5
	25°	1	50.9	8.2	36.9	8.4	-22.6	-30.8	-16.4	-24.8	-42.1	-35.9
		2	-------	-------	-------	-------	-8.6	-16.8	-2.3	-10.7	-------	-------
	30 to 45	1	45.7	31.2	36.3	25.0	3.5	-27.7	1.2	-23.8	-16.0	-18.3
		2	45.7	31.2	36.3	25.0	17.6	-13.7	15.2	-9.8	-16.0	-18.3
180	0 to 5°	1	51.4	-26.7	34.1	-15.8	-61.7	-35.1	-43.0	-27.2	-86.4	-67.7
	10°	1	58.0	-24.0	38.5	-14.0	-61.7	-37.7	-43.0	-29.0	-86.4	-67.7
	15°	1	64.5	-21.4	43.0	-12.2	-61.7	-40.3	-43.0	-30.8	-86.4	-67.7
	20°	1	71.1	-18.8	47.4	-10.4	-61.7	-43.0	-43.0	-32.6	-86.4	-67.7
	25°	1	64.5	10.4	46.7	10.6	-28.6	-39.0	-20.7	-31.4	-53.3	-45.4
		2	-------	-------	-------	-------	-10.9	-21.2	-3.0	-13.6	-------	-------
	30 to 45	1	57.8	39.5	45.9	31.6	4.4	-35.1	1.5	-30.1	-20.3	-23.2
		2	57.8	39.5	45.9	31.6	22.2	-17.3	19.3	-12.3	-20.3	-23.2
200	0 to 5°	1	63.4	-32.9	42.1	-19.5	-76.2	-43.3	-53.1	-33.5	-106.7	-83.5
	10°	1	71.5	-29.7	47.6	-17.3	-76.2	-46.5	-53.1	-35.8	-106.7	-83.5
	15°	1	79.7	-26.4	53.1	-15.0	-76.2	-49.8	-53.1	-38.0	-106.7	-83.5
	20°	1	87.8	-23.2	58.5	-12.8	-76.2	-53.1	-53.1	-40.2	-106.7	-83.5
	25°	1	79.6	12.8	57.6	13.1	-35.4	-48.2	-25.6	-38.7	-65.9	-56.1
		2	-------	-------	-------	-------	-13.4	-26.2	-3.7	-16.8	-------	-------
	30 to 45	1	71.3	48.8	56.7	39.0	5.5	-43.3	1.8	-37.2	-25.0	-28.7
		2	71.3	48.8	56.7	39.0	27.4	-21.3	23.8	-15.2	-25.0	-28.7

Adjustment Factor for Building Height and Exposure, λ

Mean roof height (ft)	Exposure		
	B	C	D
15	1.00	1.21	1.47
20	1.00	1.29	1.55
25	1.00	1.35	1.61
30	1.00	1.40	1.66
35	1.05	1.45	1.70
40	1.09	1.49	1.74
45	1.12	1.53	1.78
50	1.16	1.56	1.81
55	1.19	1.59	1.84
60	1.22	1.62	1.87

Unit Conversions – 1.0 ft = 0.3048 m; 1.0 psf = 0.0479 kN/m^2

FIGURE A.3 (*continued*)

Introduction to SAP2000

B

B.1 Introduction

The commercially available structural analysis software *SAP2000* is utilized extensively throughout this text. The main body of this book does contain some *how-to* information to use the software, but it is not enough information to generate a structural model from scratch. The producer of this software provides many resources, including tutorials, on how to use their product. You may go to their website for these resources:

`https://wiki.csiamerica.com/display/sap2000/Home`

Although these resources are very useful, a new user may feel overwhelmed with the extensive number of options and resources. For this purpose, the authors have put together a basic tutorial that will familiarize the reader with the modeling processes required to carry out the modeling presented in this text.

A tutorial video titled *A SAP2000 Tutorial* is available on the publisher's website. This is where the necessary explanations are provided. However, a copy of the presentation slides are provided within this text because experience has shown that having a ready reference to these slides during the learning stage is invaluable. A student version of SAP2000 may be downloaded from the textbook website at www.wiley.com/college/nielson.

B.2 Slides

Introduction to SAP2000

A commercially available structural analysis software

Nielson and McCormac

Disclaimer by Computers and Structures, Inc.

"Considerable time and expense have gone into the development and documentation of SAP2000. The program has been thoroughly tested and used. In using this program, however, the user accepts and understands that no warranty is expressed or implied by the developers or distributors on accuracy or the reliability of the program. The user must understand the assumptions of the program and must independently verify the results."

In other words, it becomes the structural engineer's responsibility to check the SAP2000 results via simple assumptions and hand calculations.

Computers and Structures, Inc.
Berkeley, California, USA

Nielson and McCormac

Basics

- Draw geometry
  - Nodes
  - Members
- Define member (section properties)
- Define material properties
- Define support conditions
- Define loads
- Define analysis
- Report results

Nielson and McCormac

SAP specifics

- Turn off self-weight
- Turn off shear deformations
- SAP2000 does carry units so be mindful of them throughout the model definition phase
- SAP2000 creates a lot of files so put your model in a folder before running and avoid using a network drive or jump drive.
- Once you run the model for the first time, you will need to unlock it before you can modify again

Nielson and McCormac

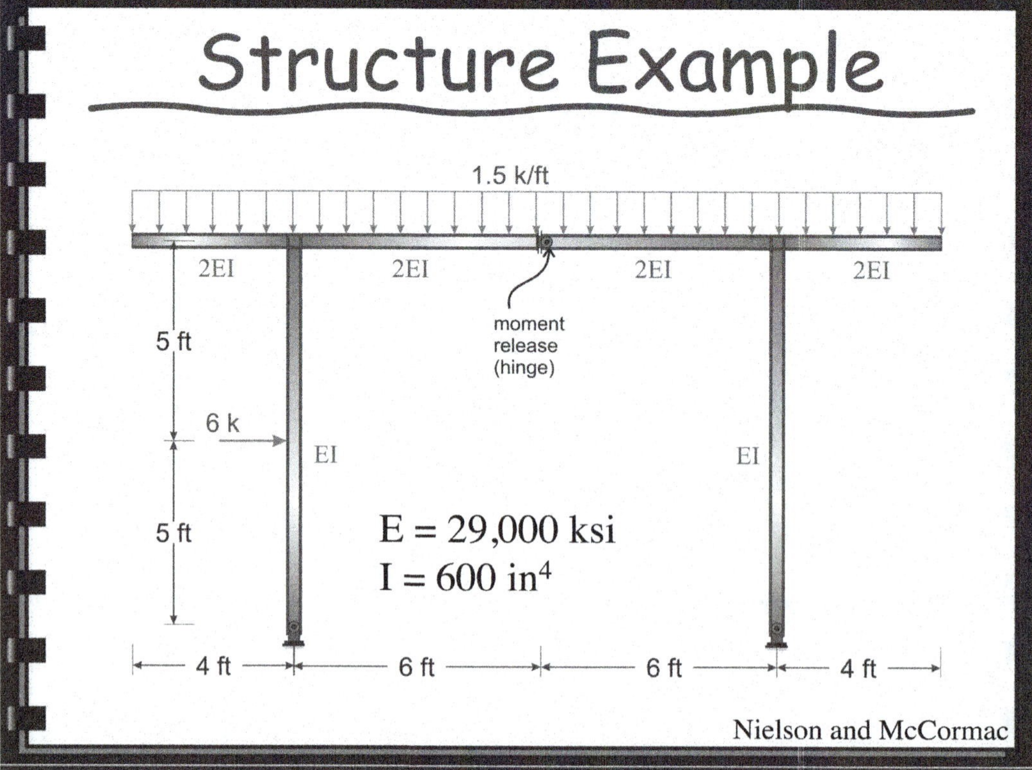

Define New Model

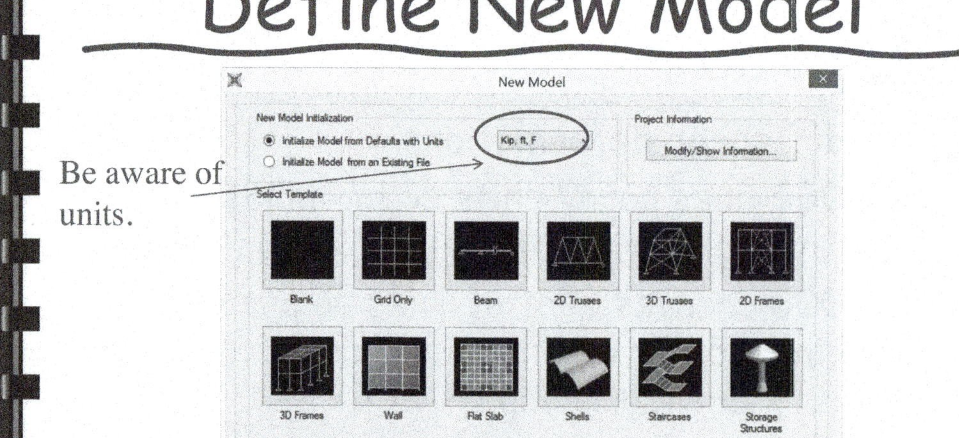

Choose: Grid

(you can use the wizard when appropriate)

Nielson and McCormac

Define Gridlines

•**ALWAYS** work in the X-Z plane. (Choosing another plane like X-Y will only confuse you at this stage.)

•Choose a grid spacing that will be helpful in model definition. You want gridlines to intersect anywhere you need a joint.

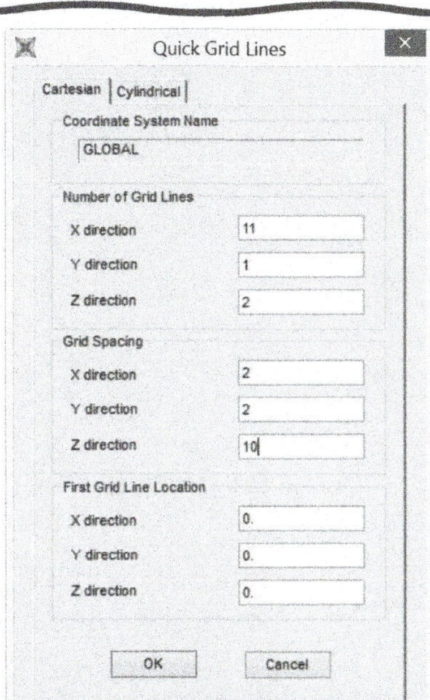

McCormac

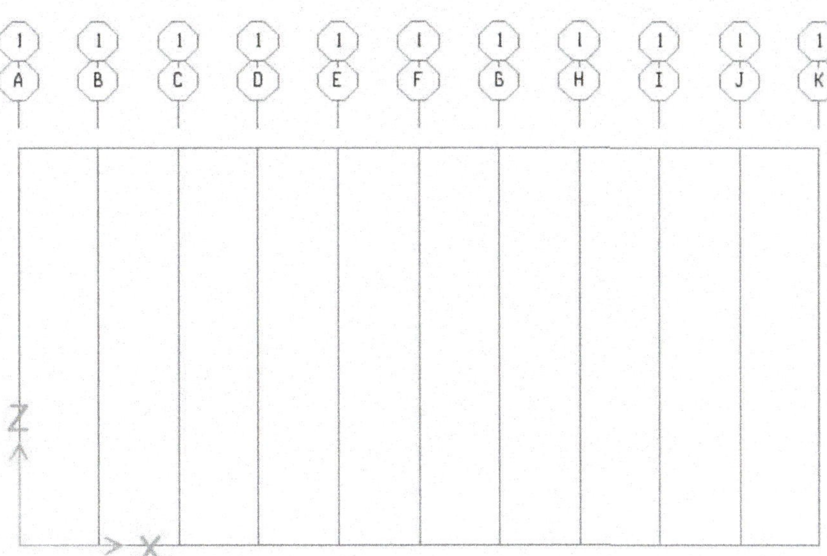

We now have grids spaced at 2 ft intervals in the x-direction and 10 ft interval in the z-direction. This gives a grid which is 20 ft wide by 10 ft tall.

Nielson and McCormac

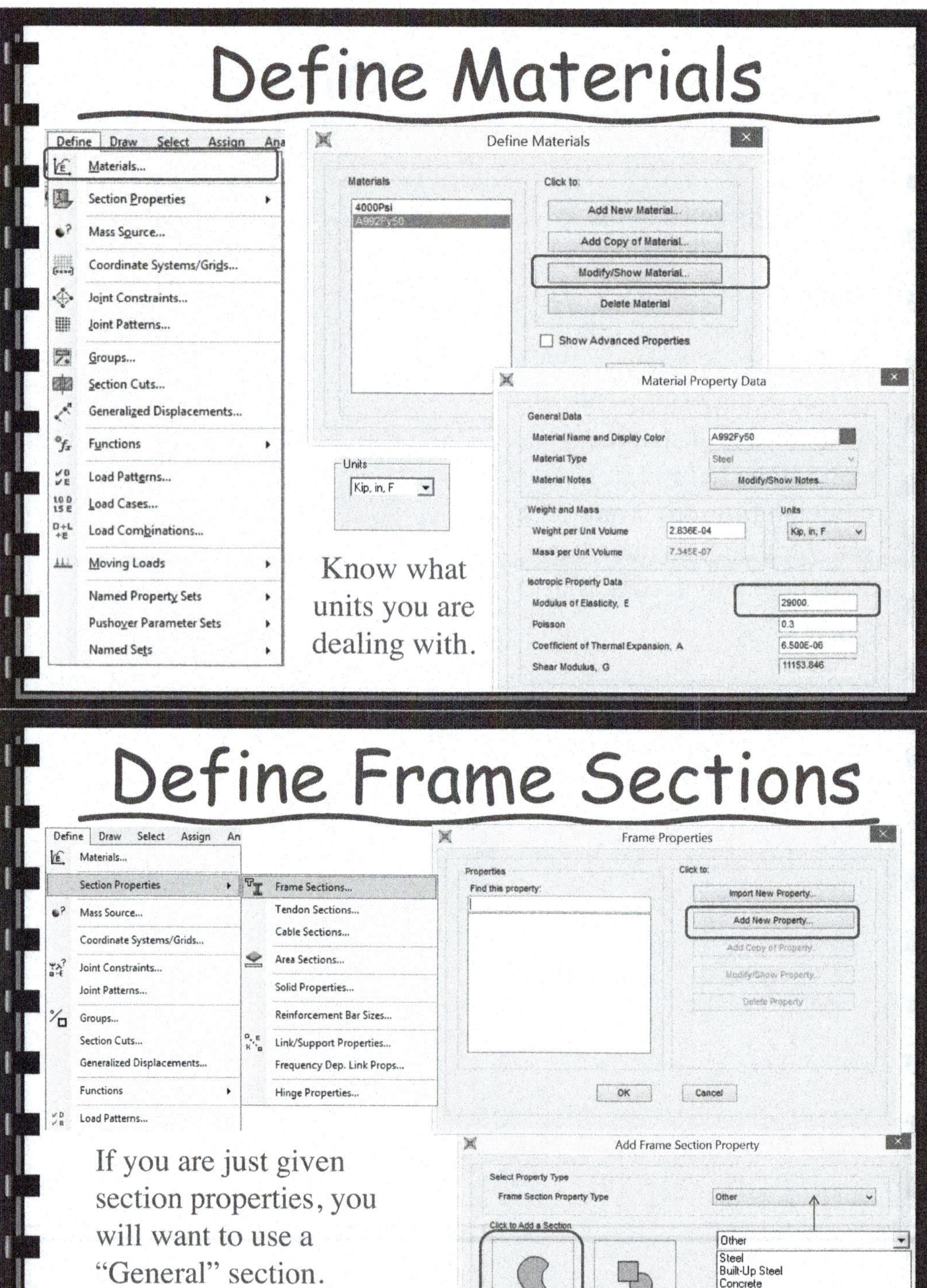
Define Materials
Define Draw Select Assign Ana
Materials...
Section Properties
Mass Source...
Coordinate Systems/Grids...
Joint Constraints...
Joint Patterns...
Groups...
Section Cuts...
Generalized Displacements...
Functions
Load Patterns...
Load Cases...
Load Combinations...
Moving Loads
Named Property Sets
Pushover Parameter Sets
Named Sets
Define Materials
Materials
4000Psi
A992Fy50
Click to:
Add New Material...
Add Copy of Material...
Modify/Show Material...
Delete Material
Show Advanced Properties
Units
Kip, in, F
Know what units you are dealing with.
Material Property Data
General Data
Material Name and Display Color
A992Fy50
Material Type
Steel
Material Notes
Modify/Show Notes...
Weight and Mass
Weight per Unit Volume
2.836E-04
Mass per Unit Volume
7.345E-07
Units
Kip, in, F
Isotropic Property Data
Modulus of Elasticity, E
29000.
Poisson
0.3
Coefficient of Thermal Expansion, A
6.500E-06
Shear Modulus, G
11153.846
Define Frame Sections
Define Draw Select Assign An
Materials...
Section Properties
Mass Source...
Coordinate Systems/Grids...
Joint Constraints...
Joint Patterns...
Groups...
Section Cuts...
Generalized Displacements...
Functions
Load Patterns...
Frame Sections...
Tendon Sections...
Cable Sections...
Area Sections...
Solid Properties...
Reinforcement Bar Sizes...
Link/Support Properties...
Frequency Dep. Link Props...
Hinge Properties...
Frame Properties
Properties
Find this property:
Click to:
Import New Property...
Add New Property...
Add Copy of Property...
Modify/Show Property...
Delete Property
OK
Cancel
If you are just given section properties, you will want to use a "General" section.
However, if you are given a defined steel section, select an import (AISC13.pro).
Add Frame Section Property
Select Property Type
Frame Section Property Type
Other
Click to Add a Section
General
Section Designer
Other
Steel
Built-Up Steel
Concrete
Aluminum
Cold Formed
Other
Nielson and McCormac

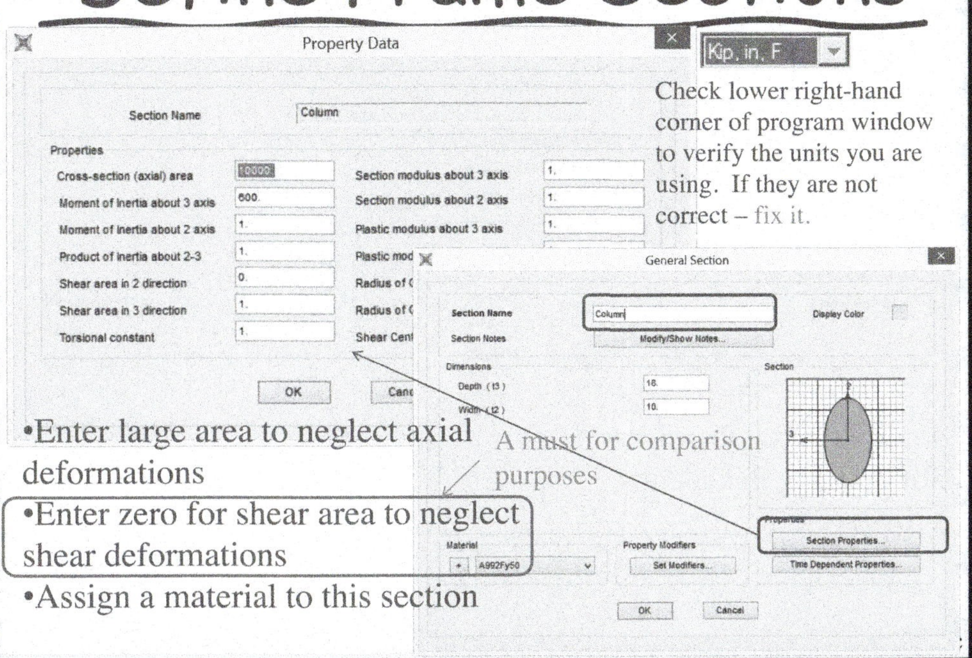
Define Frame Sections
Property Data
Section Name
Column
Properties
Cross-section (axial) area
Moment of Inertia about 3 axis
Moment of Inertia about 2 axis
Product of Inertia about 2-3
Shear area in 2 direction
Shear area in 3 direction
Torsional constant
Section modulus about 3 axis
Section modulus about 2 axis
Plastic modulus about 3 axis
OK
Kip, in, F
Check lower right-hand corner of program window to verify the units you are using. If they are not correct – fix it.
General Section
Section Name
Section Notes
Modify/Show Notes...
Display Color
Dimensions
Depth (t3)
Width (t2)
Section
Material
A992Fy50
Property Modifiers
Set Modifiers...
Section Properties...
Time Dependent Properties...
OK
Cancel
•Enter large area to neglect axial deformations
A must for comparison purposes
•Enter zero for shear area to neglect shear deformations
•Assign a material to this section

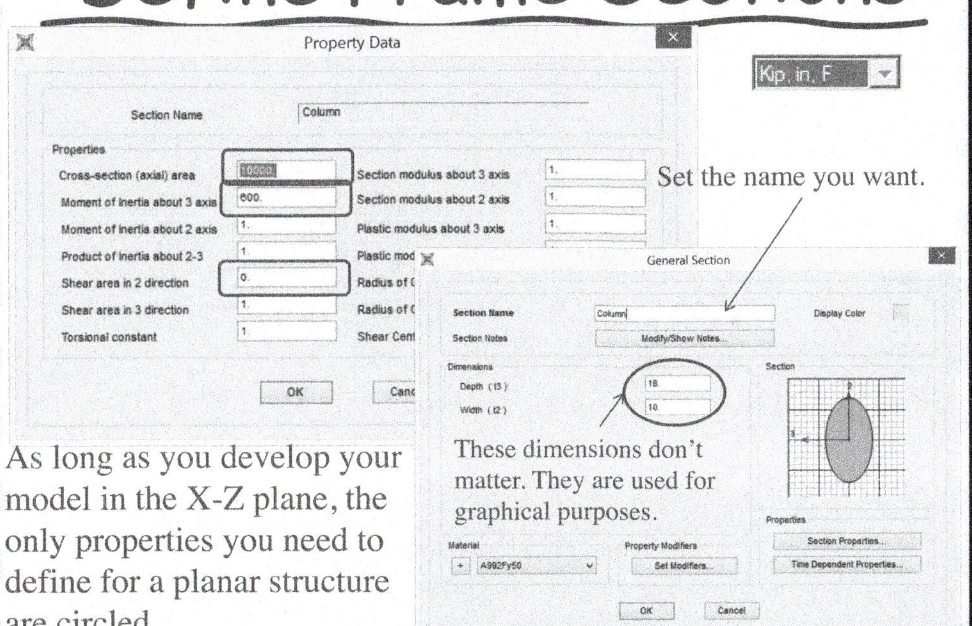
Define Frame Sections
Property Data
Section Name
Column
Properties
Cross-section (axial) area
Moment of Inertia about 3 axis
Moment of Inertia about 2 axis
Product of Inertia about 2-3
Shear area in 2 direction
Shear area in 3 direction
Torsional constant
Section modulus about 3 axis
Section modulus about 2 axis
Plastic modulus about 3 axis
OK
Kip, in, F
Set the name you want.
General Section
Section Name
Section Notes
Modify/Show Notes...
Display Color
Dimensions
Depth (t3)
Width (t2)
Section
These dimensions don't matter. They are used for graphical purposes.
Material
A992Fy50
Property Modifiers
Set Modifiers...
Properties
Section Properties...
Time Dependent Properties...
OK
Cancel
As long as you develop your model in the X-Z plane, the only properties you need to define for a planar structure are circled.

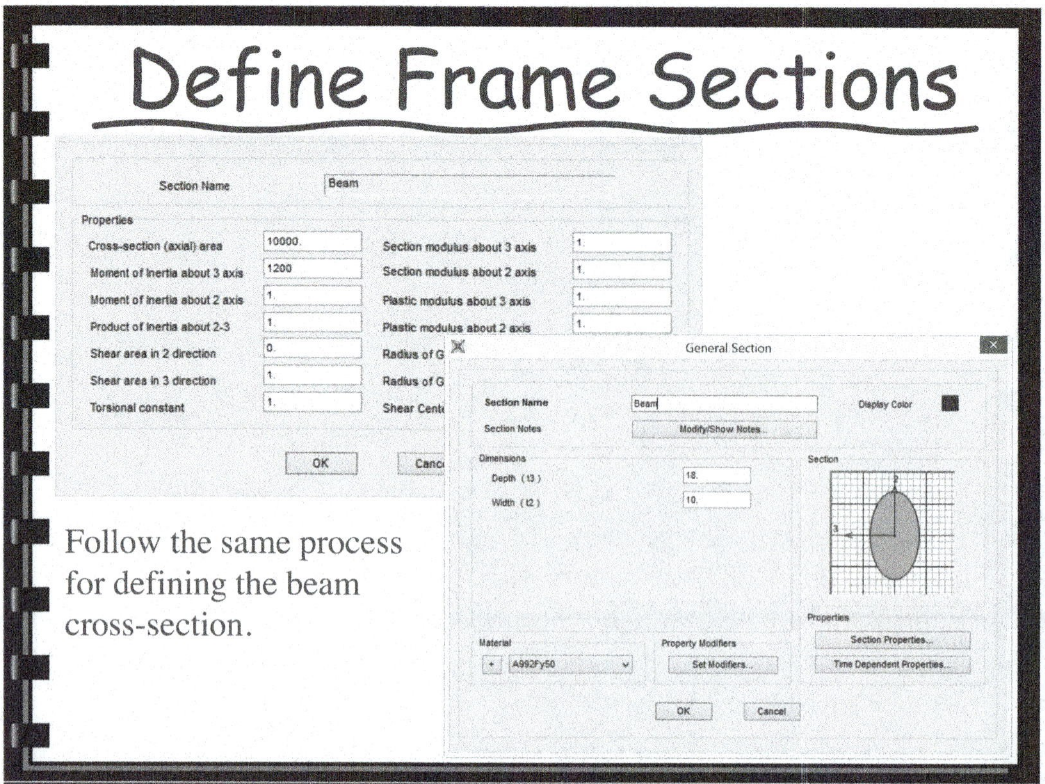

Define Frame Sections

Follow the same process for defining the beam cross-section.

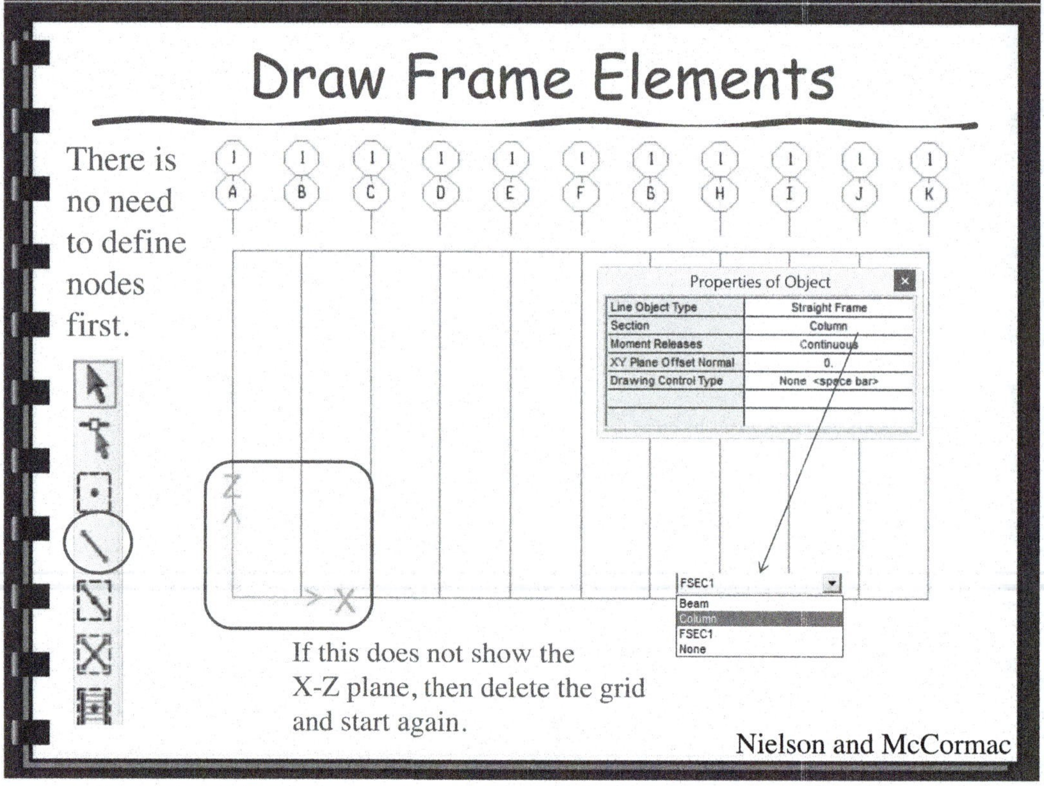

Draw Frame Elements

There is no need to define nodes first.

If this does not show the X-Z plane, then delete the grid and start again.

Nielson and McCormac

Draw Frame Elements

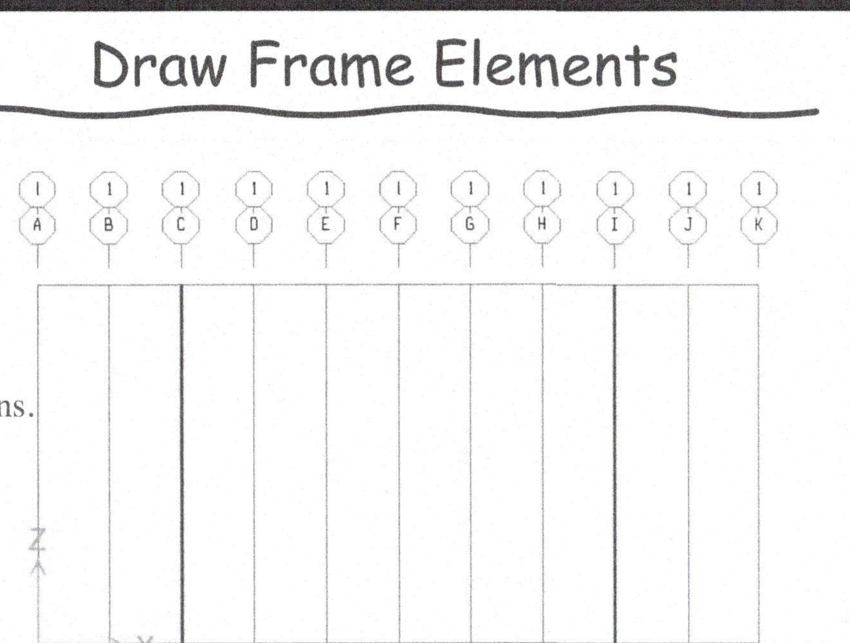

Draw the columns.

Nielson and McCormac

Draw Frame Elements

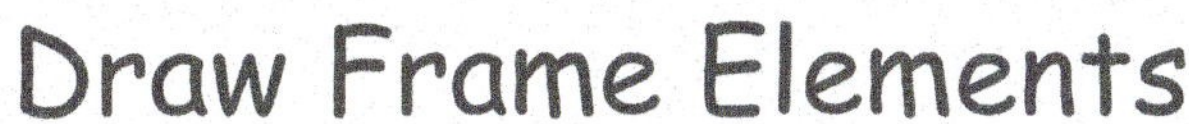

Draw the beams.
Four beam elements were drawn.
A-C
C-F
F-I
I-K

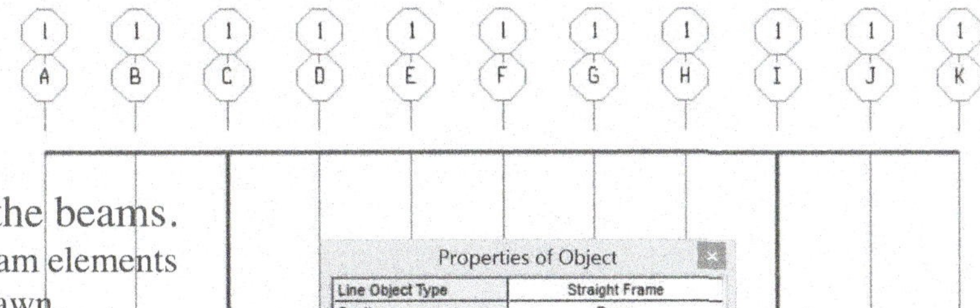

Properties of Object
Line Object Type | Straight Frame
Section | Beam
Moment Releases | Continuous
XY Plane Offset Normal | 0.
Drawing Control Type | None <space bar>

This lets the columns talk to the beams and also the insertion of a moment release (hinge) at F.

Common Mistake:
Students will often draw only one beam element from A – K. While this looks right, the columns and beams don't share the same nodes (joints).

Insert Moment Release

- Select frame element between C-F.

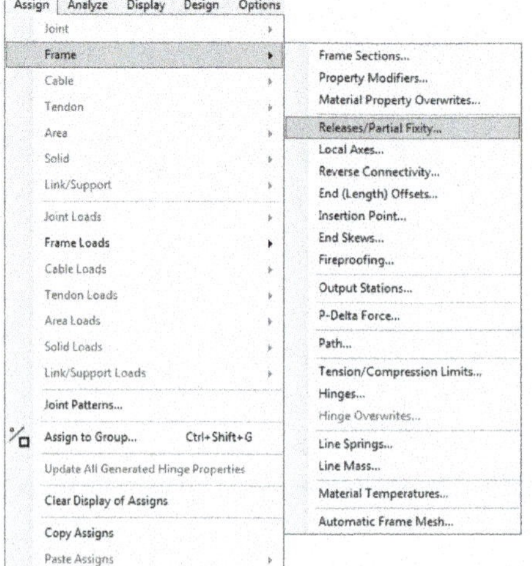

As long as you built your model in the X-Z plane then your releases should be in the 3-3 axis.

Place release at only the right end of the member.

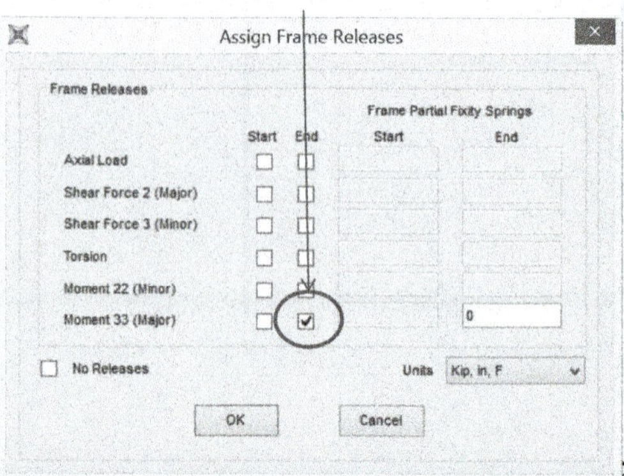

Insert Moment Release

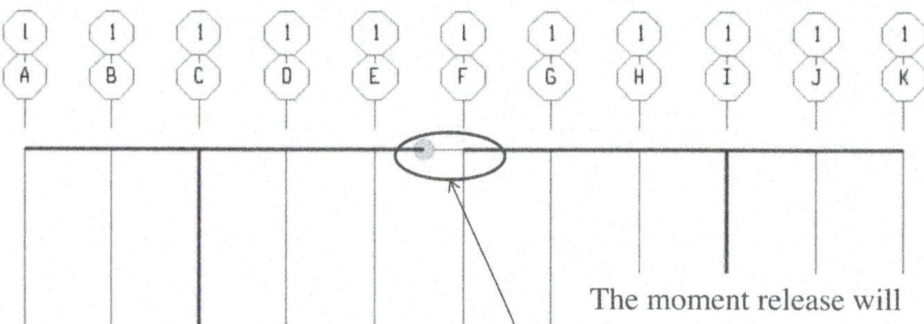

Quick Note:

The next action you take will make the release disappear off the screen. Don't be alarmed. The release is still there but just not being shown.

The moment release will show up where you have placed it. We can see from this that the release was placed in member C-F at the right end. You could have placed it in member F-I at the left end. Just don't place it in both members simultaneously.

Select the joints for which you want to add supports and then go to the following menu.

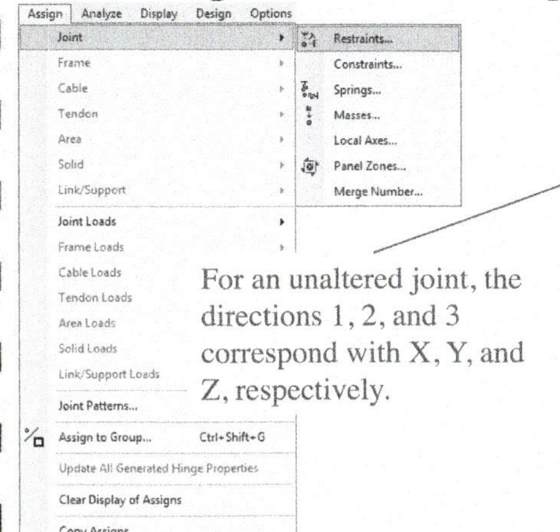

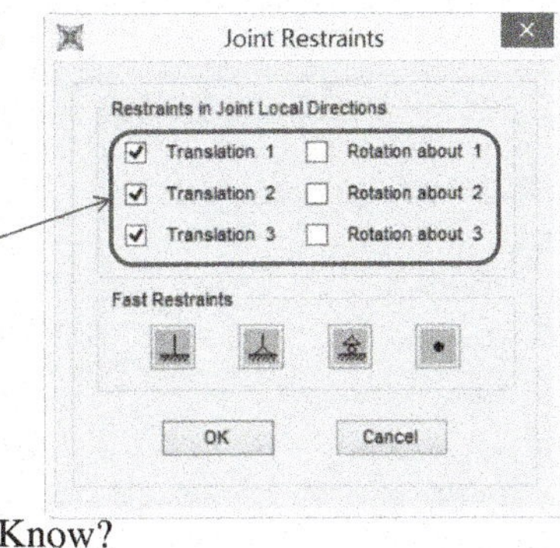

For an unaltered joint, the directions 1, 2, and 3 correspond with X, Y, and Z, respectively.

<u>Did You Know?</u>

A restraint describes a joint movement which is not allowed. For example, above we are not allowing translation of these joints in the X, Y, or Z direction. This is what a pin support would do.

Nielson and McCormac

Insert Supports

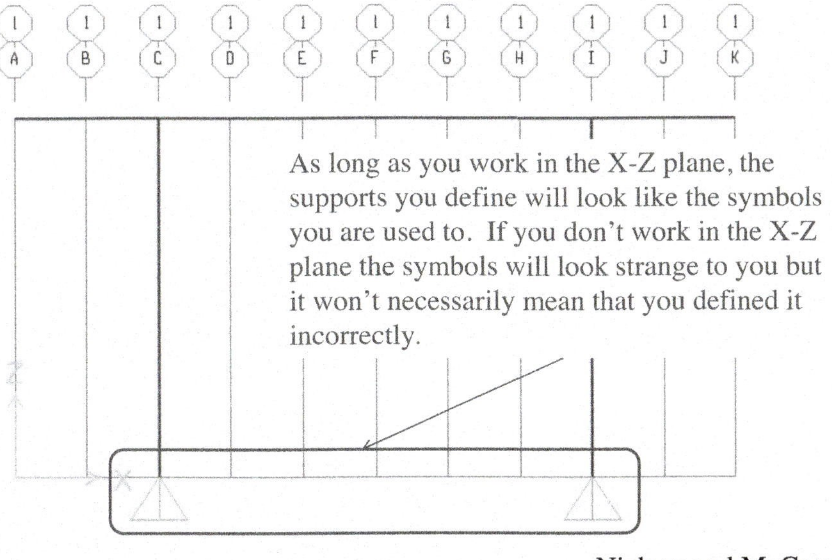

As long as you work in the X-Z plane, the supports you define will look like the symbols you are used to. If you don't work in the X-Z plane the symbols will look strange to you but it won't necessarily mean that you defined it incorrectly.

Nielson and McCormac

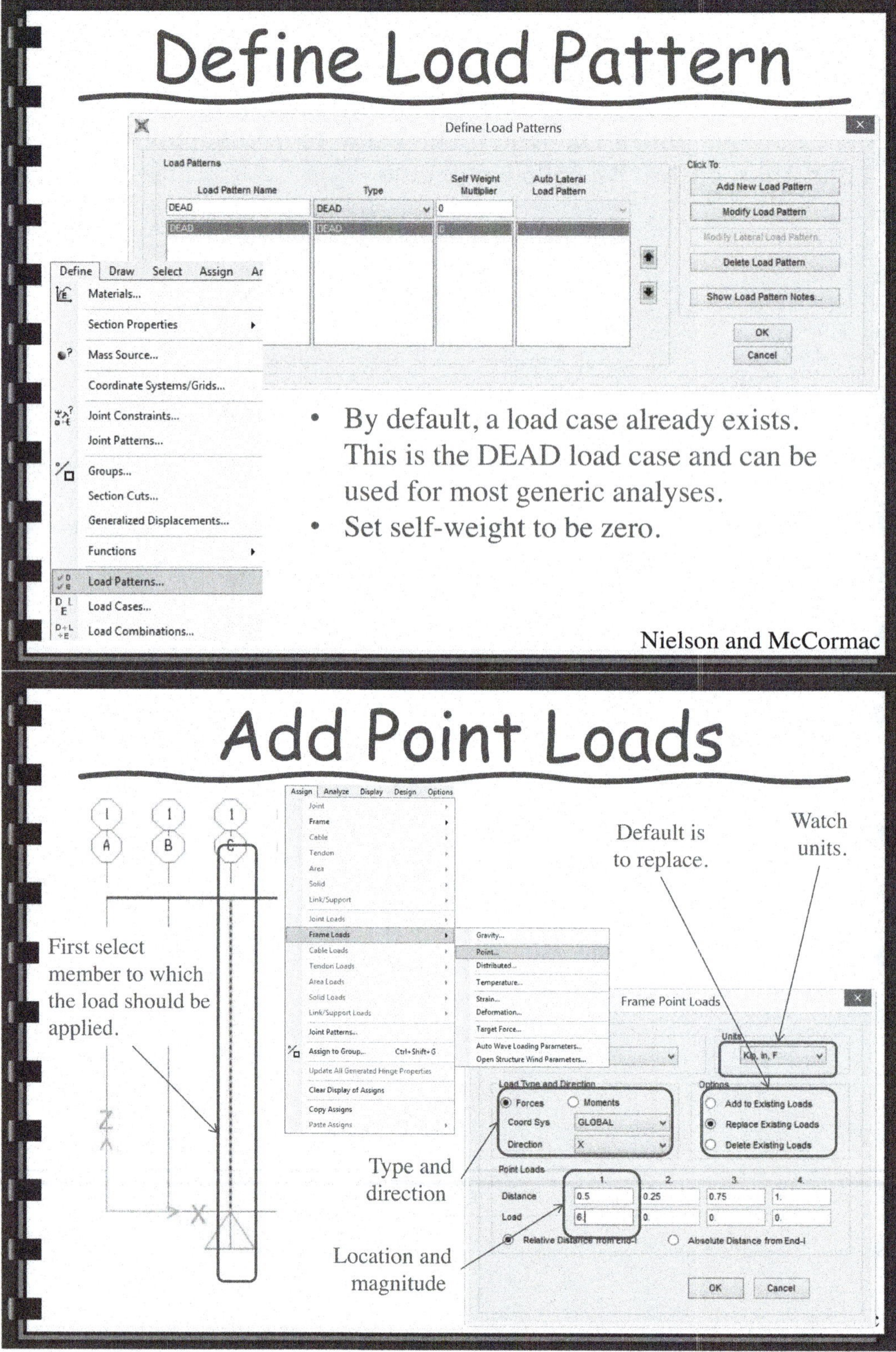
Define Load Pattern
Define Load Patterns
Load Patterns
Load Pattern Name
Type
Self Weight Multiplier
Auto Lateral Load Pattern
DEAD
DEAD
0
Click To:
Add New Load Pattern
Modify Load Pattern
Modify Lateral Load Pattern
Delete Load Pattern
Show Load Pattern Notes...
OK
Cancel
Define
Draw
Select
Assign
Materials...
Section Properties
Mass Source...
Coordinate Systems/Grids...
Joint Constraints...
Joint Patterns...
Groups...
Section Cuts...
Generalized Displacements...
Functions
Load Patterns...
Load Cases...
Load Combinations...
• By default, a load case already exists. This is the DEAD load case and can be used for most generic analyses.
• Set self-weight to be zero.
Nielson and McCormac
Add Point Loads
First select member to which the load should be applied.
Default is to replace.
Watch units.
Type and direction
Location and magnitude
Frame Point Loads
Units
Options
Add to Existing Loads
Replace Existing Loads
Delete Existing Loads
Forces
Moments
Coord Sys
GLOBAL
Direction
X
Point Loads
Distance
Load
Relative Distance from End-I
Absolute Distance from End-I
OK
Cancel

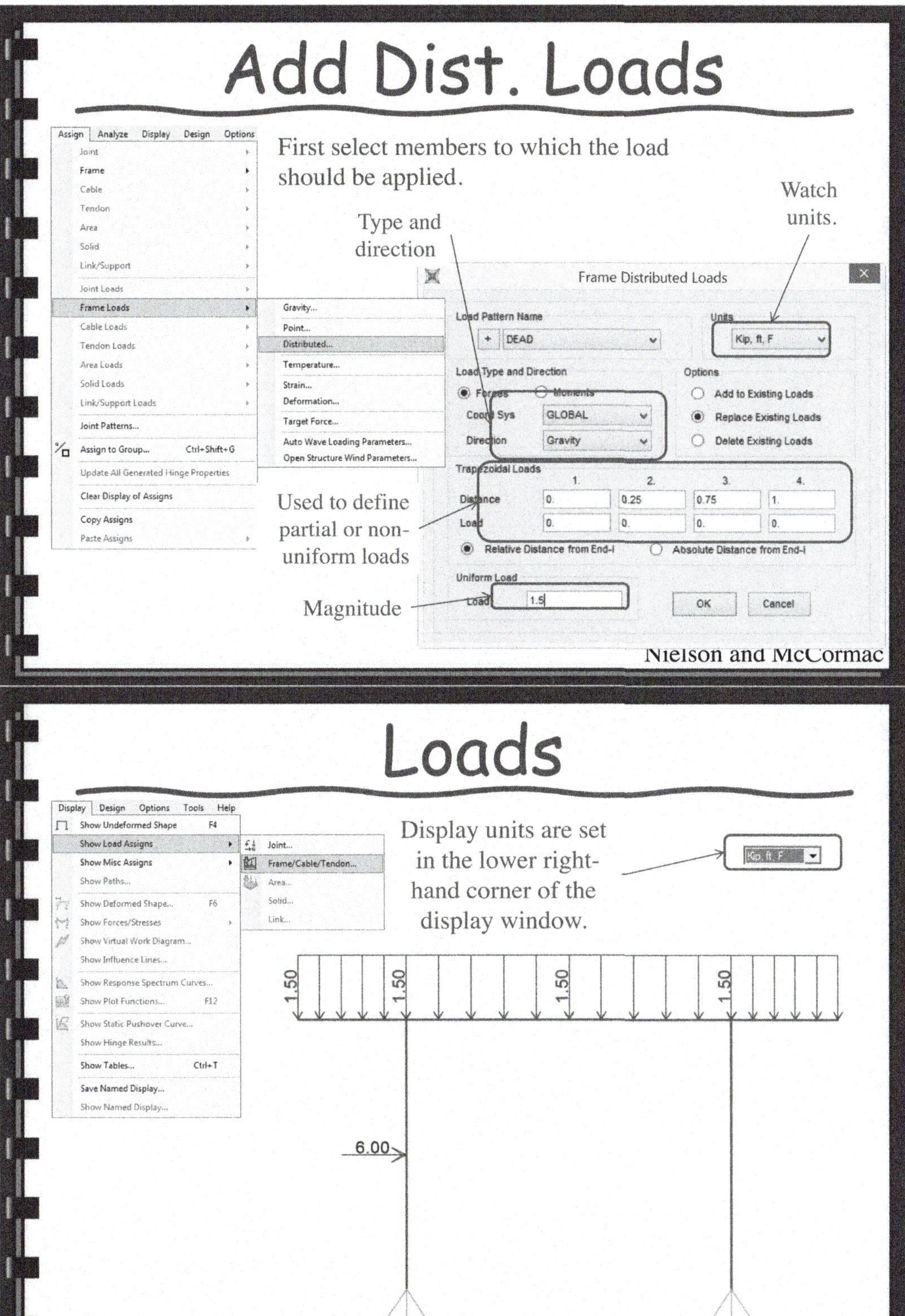
Add Dist. Loads
First select members to which the load should be applied.
Watch units.
Type and direction
Frame Distributed Loads
Used to define partial or non-uniform loads
Magnitude
Nielson and McCormac
Loads
Display units are set in the lower right-hand corner of the display window.
Kip, ft, F
1.50
6.00
Nielson and McCormac

Set Analysis

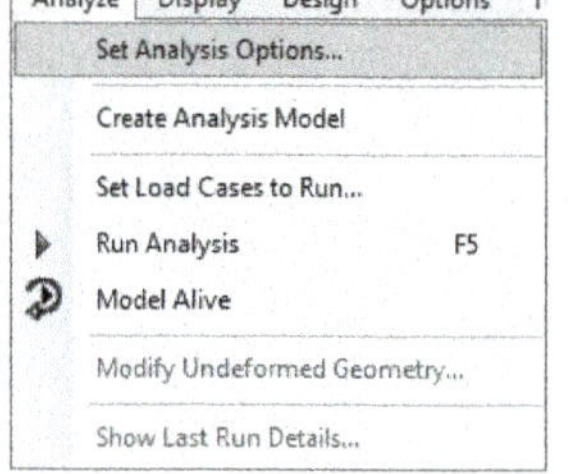

Here you tell it which motions you want analyzed. Another useful reason for using the X-Z plane to build your model.

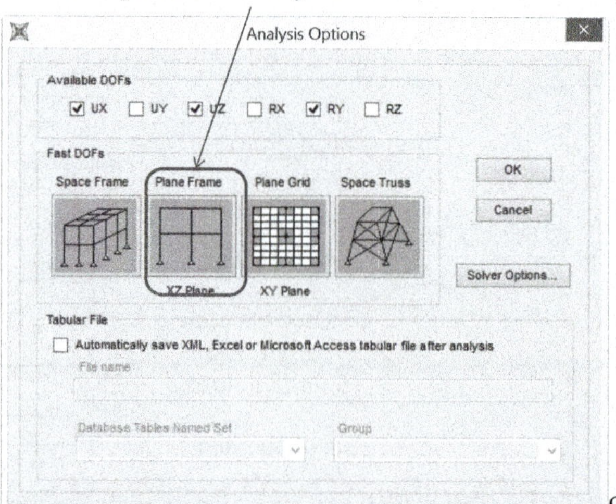

Run

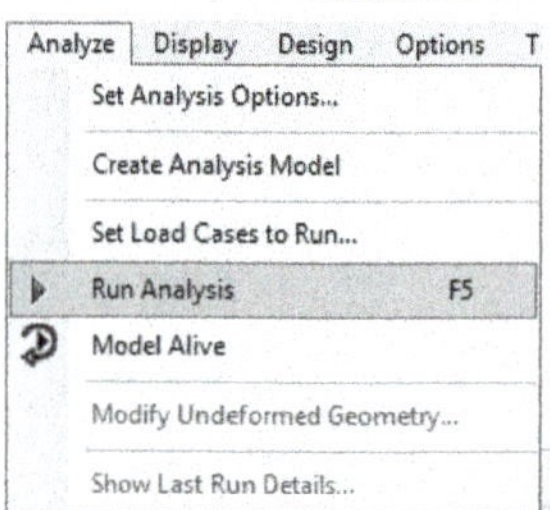

Two default cases appear in the dialog box - the static load case and a modal load case. The modal load case reports dynamic properties and is not needed for a static analysis. To cut down on the amount of output, I recommend you turn off this case (i.e., Do Not Run).

If you have not yet saved your model, it will prompt you to do so. Remember to save the model in a folder because it creates a lot of output files.

Case Name	Type	Status	Action
DEAD	Linear Static	Not Run	Run
MODAL	Modal	Not Run	Do Not Run

Click to:
Run/Do Not Run Case
Show Case...
Delete Results for Case
Run/Do Not Run All
Delete All Results
Show Load Case Tree...

Analysis Monitor Options
Always Show
Never Show
Show After 4 seconds

Model-Alive
Run Now
OK
Cancel

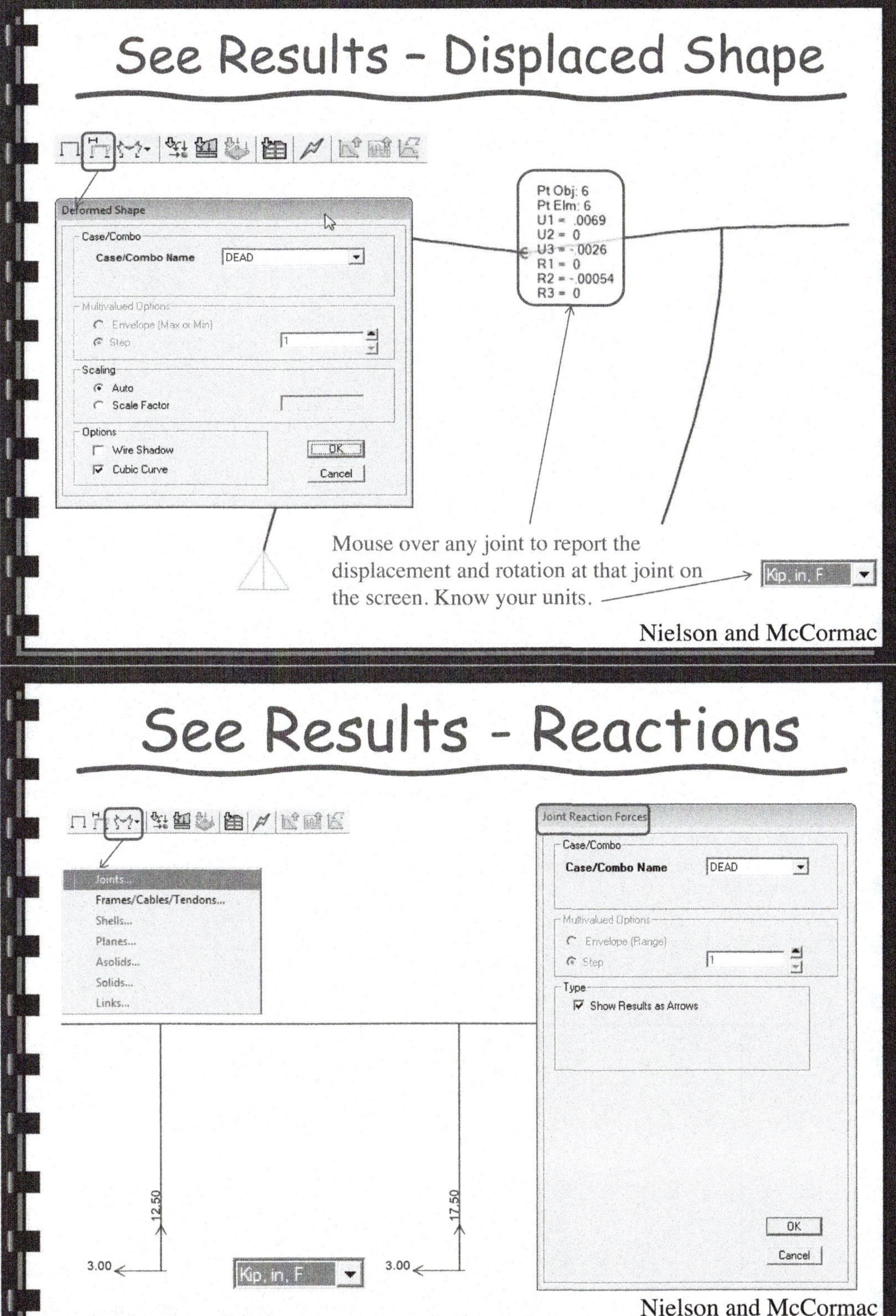
See Results - Displaced Shape
Deformed Shape
Case/Combo
Case/Combo Name
DEAD
Multivalued Options
Envelope (Max or Min)
Step
Scaling
Auto
Scale Factor
Options
Wire Shadow
Cubic Curve
OK
Cancel
Pt Obj: 6
Pt Elm: 6
U1 = .0069
U2 = 0
U3 = -.0026
R1 = 0
R2 = -.00054
R3 = 0
Mouse over any joint to report the displacement and rotation at that joint on the screen. Know your units.
Kip, in, F
Nielson and McCormac
See Results - Reactions
Joints...
Frames/Cables/Tendons...
Shells...
Planes...
Asolids...
Solids...
Links...
Joint Reaction Forces
Case/Combo
Case/Combo Name
DEAD
Multivalued Options
Envelope (Range)
Step
Type
Show Results as Arrows
OK
Cancel
12.50
17.50
3.00
3.00
Kip, in, F
Nielson and McCormac

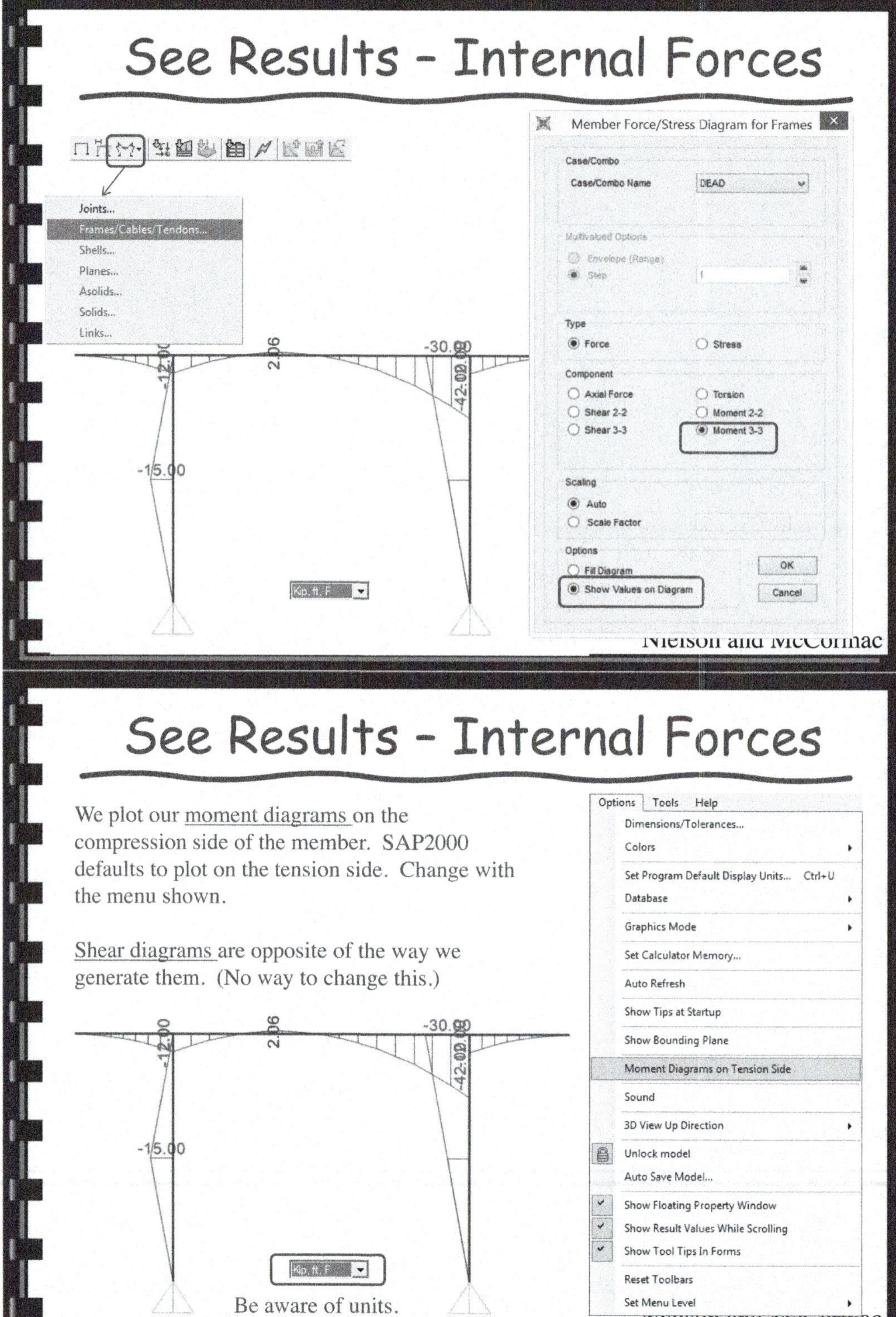
See Results - Internal Forces
Joints...
Frames/Cables/Tendons...
Shells...
Planes...
Asolids...
Solids...
Links...
Member Force/Stress Diagram for Frames
Case/Combo
Case/Combo Name
DEAD
Type
Force
Stress
Component
Axial Force
Shear 2-2
Shear 3-3
Torsion
Moment 2-2
Moment 3-3
Scaling
Auto
Scale Factor
Options
Fill Diagram
Show Values on Diagram
OK
Cancel
-12.00
2.06
-30.00
-42.00
-15.00
Kip, ft, F
Nelson and McCormac
See Results - Internal Forces
We plot our moment diagrams on the compression side of the member. SAP2000 defaults to plot on the tension side. Change with the menu shown.
Shear diagrams are opposite of the way we generate them. (No way to change this.)
Options
Tools
Help
Dimensions/Tolerances...
Colors
Set Program Default Display Units... Ctrl+U
Database
Graphics Mode
Set Calculator Memory...
Auto Refresh
Show Tips at Startup
Show Bounding Plane
Moment Diagrams on Tension Side
Sound
3D View Up Direction
Unlock model
Auto Save Model...
Show Floating Property Window
Show Result Values While Scrolling
Show Tool Tips In Forms
Reset Toolbars
Set Menu Level
-12.00
2.06
-30.00
-42.00
-15.00
Kip, ft, F
Be aware of units.
Nelson and McCormac

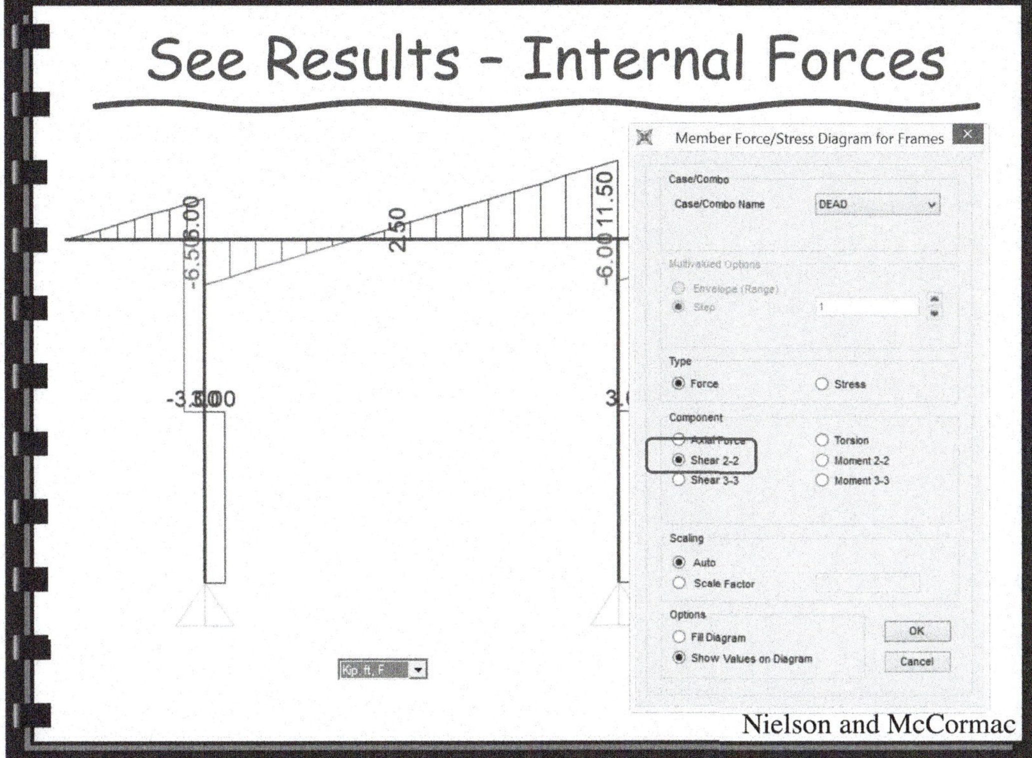
See Results – Internal Forces
Member Force/Stress Diagram for Frames
Case/Combo
Case/Combo Name
DEAD
Multivalued Options
Envelope (Range)
Step
1
Type
Force
Stress
Component
Axial Force
Shear 2-2
Shear 3-3
Torsion
Moment 2-2
Moment 3-3
Scaling
Auto
Scale Factor
Options
Fill Diagram
Show Values on Diagram
OK
Cancel
2.50
11.50
-6.00
Kip, ft, F
Nielson and McCormac

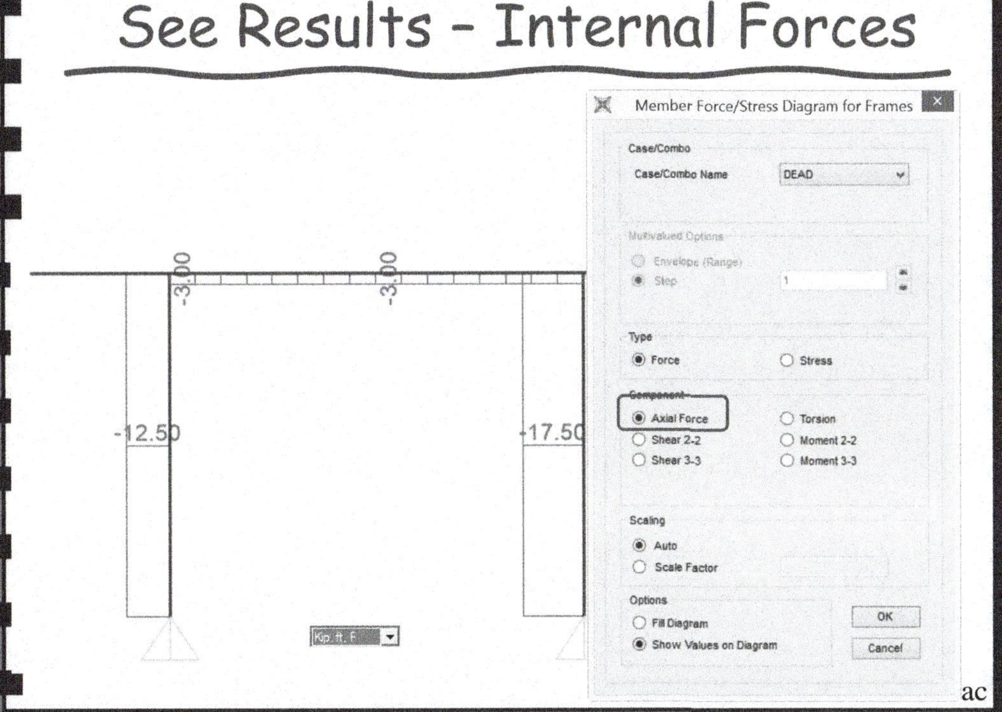
See Results – Internal Forces
Member Force/Stress Diagram for Frames
Case/Combo
Case/Combo Name
DEAD
Multivalued Options
Envelope (Range)
Step
1
Type
Force
Stress
Component
Axial Force
Shear 2-2
Shear 3-3
Torsion
Moment 2-2
Moment 3-3
Scaling
Auto
Scale Factor
Options
Fill Diagram
Show Values on Diagram
OK
Cancel
-3.00
-3.00
-12.50
-17.50
Kip, ft, F
ac

See Results – Internal Forces

The resolution of a diagram is affected by the number of stations defined. This can be changed by:

- Unlock the model
- Right click on the member (e.g., left column).
- Change station settings
- Rerun model.

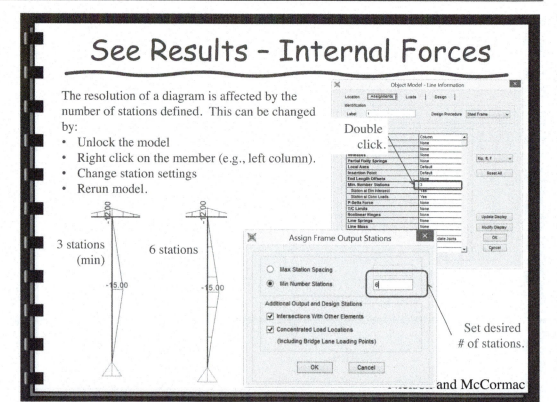

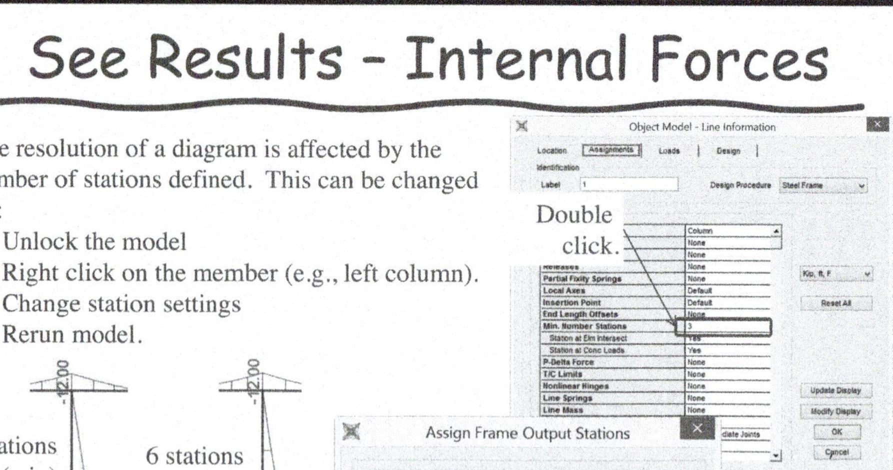

and McCormac

See Results – Internal Forces

You can see all the internal forces for single member if you just right-click on it. Example ➔ member C-F

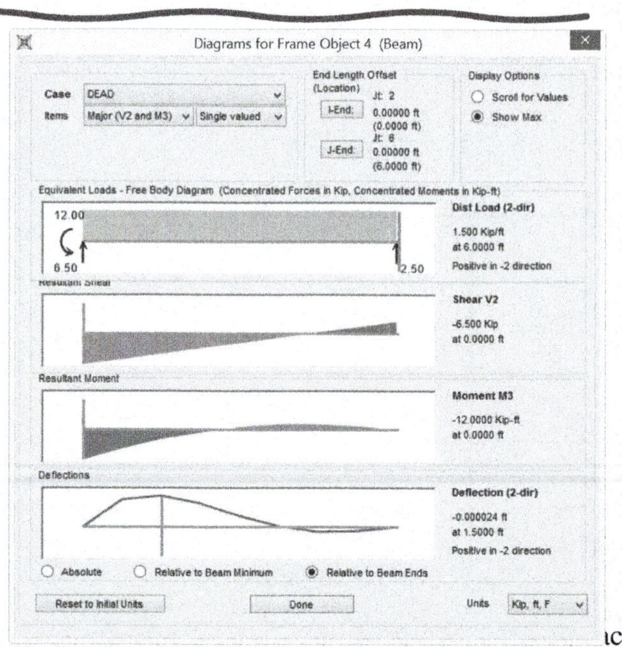

You will create tables that you can print.

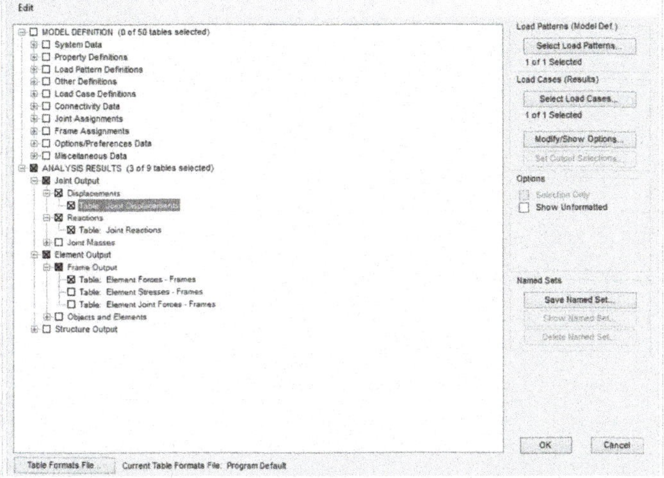

Only select the output results you want at this stage to minimize the data you get back.

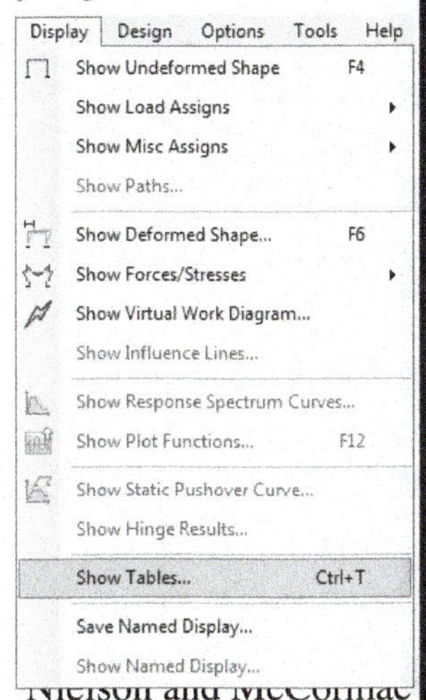

Nielson and McCormac

Print Results

You will create tables that you can print.

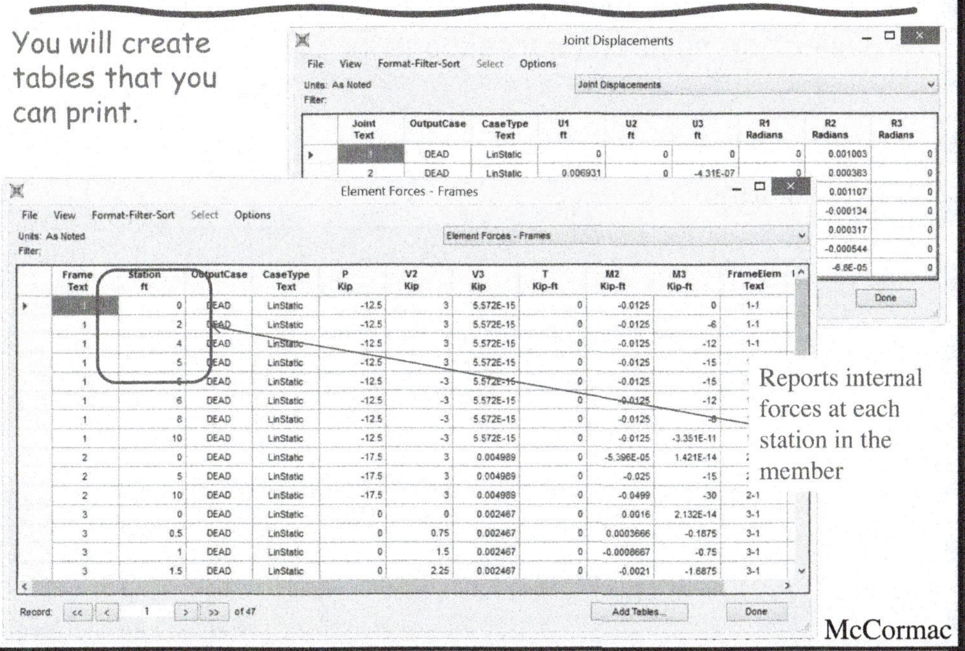

Joint Displacements

File View Format-Filter-Sort Select Options

Units: As Noted

Filter:

Joint Text	OutputCase	CaseType Text	U1 ft	U2 ft	U3 ft	R1 Radians	R2 Radians	R3 Radians
1	DEAD	LinStatic	0	0	0	0	0.001003	0
2	DEAD	LinStatic	0.006931	0	-4.31E-07	0	0.000383	0
							0.001107	0
							-0.000134	0
							0.000317	0
							-0.000544	0
							-6.8E-05	0

Done

Element Forces - Frames

File View Format-Filter-Sort Select Options

Units: As Noted

Filter:

Frame Text	Station ft	OutputCase	CaseType Text	P Kip	V2 Kip	V3 Kip	T Kip-ft	M2 Kip-ft	M3 Kip-ft	FrameElem Text
1	0	DEAD	LinStatic	-12.5	3	5.572E-15	0	-0.0125	0	1-1
1	2	DEAD	LinStatic	-12.5	3	5.572E-15	0	-0.0125	-6	1-1
1	4	DEAD	LinStatic	-12.5	3	5.572E-15	0	-0.0125	-12	1-1
1	5	DEAD	LinStatic	-12.5	3	5.572E-15	0	-0.0125	-15	
1	[illegible]	DEAD	LinStatic	-12.5	-3	5.572E-15	0	-0.0125	-15	
1	6	DEAD	LinStatic	-12.5	-3	5.572E-15	0	-0.0125	-12	
1	8	DEAD	LinStatic	-12.5	-3	5.572E-15	0	-0.0125	-6	
1	10	DEAD	LinStatic	-12.5	-3	5.572E-15	0	-0.0125	-3.351E-11	
2	0	DEAD	LinStatic	-17.5	3	0.004989	0	-5.396E-05	1.421E-14	
2	5	DEAD	LinStatic	-17.5	3	0.004989	0	-0.025	-15	
2	10	DEAD	LinStatic	-17.5	3	0.004989	0	-0.0499	-30	2-1
3	0	DEAD	LinStatic	0	0	0.002467	0	0.0016	2.132E-14	3-1
3	0.5	DEAD	LinStatic	0	0.75	0.002467	0	0.0003666	-0.1875	3-1
3	1	DEAD	LinStatic	0	1.5	0.002467	0	-0.0008667	-0.75	3-1
3	1.5	DEAD	LinStatic	0	2.25	0.002467	0	-0.0021	-1.6875	3-1

Record: << < 1 > >> of 47 Add Tables... Done

Reports internal forces at each station in the member

McCormac

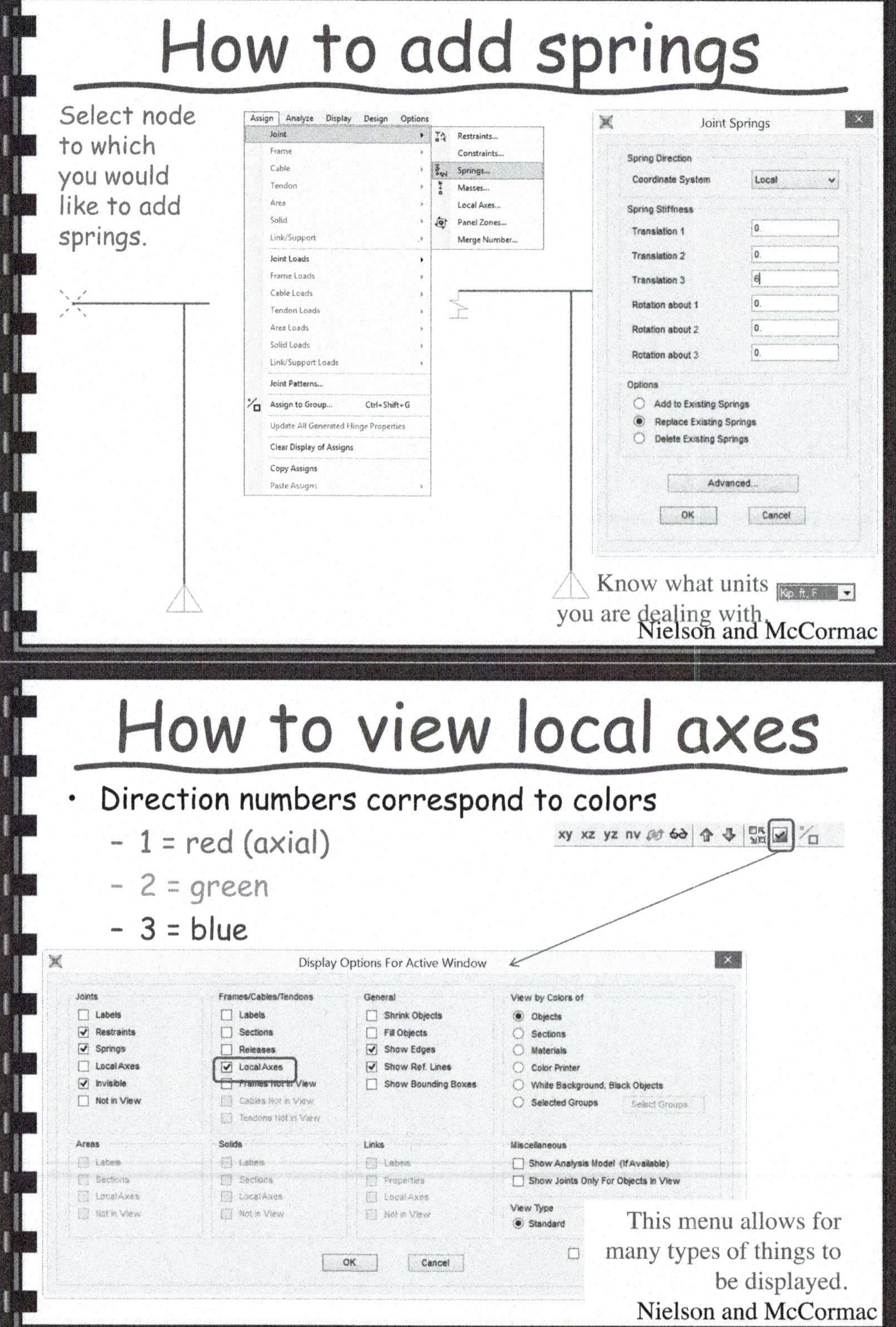
How to add springs
Select node to which you would like to add springs.
Assign Analyze Display Design Options
Joint
Frame
Cable
Tendon
Area
Solid
Link/Support
Joint Loads
Frame Loads
Cable Loads
Tendon Loads
Area Loads
Solid Loads
Link/Support Loads
Joint Patterns...
Assign to Group... Ctrl+Shift+G
Update All Generated Hinge Properties
Clear Display of Assigns
Copy Assigns
Paste Assigns
Restraints...
Constraints...
Springs...
Masses...
Local Axes...
Panel Zones...
Merge Number...
Joint Springs
Spring Direction
Coordinate System Local
Spring Stiffness
Translation 1 0.
Translation 2 0.
Translation 3 6
Rotation about 1 0.
Rotation about 2 0.
Rotation about 3 0.
Options
Add to Existing Springs
Replace Existing Springs
Delete Existing Springs
Advanced...
OK Cancel
Know what units you are dealing with.
Kip, ft, F
Nielson and McCormac
How to view local axes
• Direction numbers correspond to colors
– 1 = red (axial)
– 2 = green
– 3 = blue
xy xz yz nv
Display Options For Active Window
Joints
Labels
Restraints
Springs
Local Axes
Invisible
Not in View
Frames/Cables/Tendons
Labels
Sections
Releases
Local Axes
Frames Not in View
Cables Not in View
Tendons Not in View
General
Shrink Objects
Fill Objects
Show Edges
Show Ref. Lines
Show Bounding Boxes
View by Colors of
Objects
Sections
Materials
Color Printer
White Background, Black Objects
Selected Groups
Select Groups
Areas
Labels
Sections
Local Axes
Not in View
Solids
Labels
Sections
Local Axes
Not in View
Links
Labels
Properties
Local Axes
Not in View
Miscellaneous
Show Analysis Model (If Available)
Show Joints Only For Objects in View
View Type
Standard
OK Cancel
This menu allows for many types of things to be displayed.
Nielson and McCormac

How to view local axes

- Direction numbers correspond to colors
  - 1 = red (axial)
  - 2 = green
  - 3 = blue

Remember shear is in the 2-axis and moment is about the 3-axis.

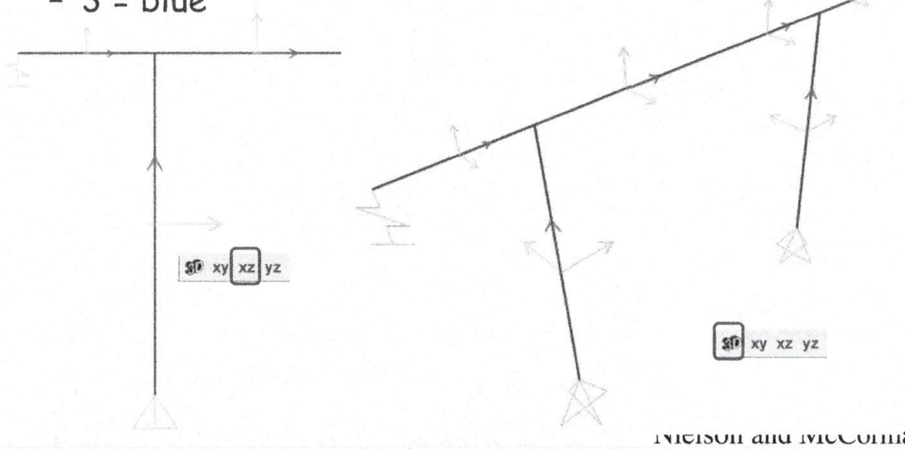

Modeling a Truss

When your draw the members, make sure you are using pin-ended members.

OR

Draw all members as normal and then add moment releases to all members (both ends).

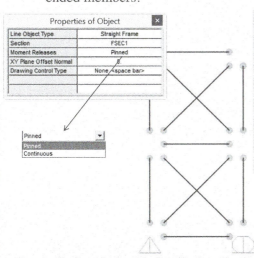

Properties of Object

Line Object Type	Straight Frame
Section	FSEC1
Moment Releases	Pinned
XY Plane Offset Normal	0.
Drawing Control Type	None <space bar>

Pinned
Pinned
Continuous

Assign Frame Releases

Frame Releases

	Start	End	Frame Partial Fixity Springs Start	End
Axial Load	☐	☐		
Shear Force 2 (Major)	☐	☐		
Shear Force 3 (Minor)	☐	☐		
Torsion	☐	☐		
Moment 22 (Minor)	☐	☐		
Moment 33 (Major)	☑	☑	0	0

☐ No Releases   Units Kip, ft, F

OK   Cancel

All other processes are the same as for beams and frames.

Nielson and McCormac

C Matrix Algebra

C.1 Introduction

An introduction to the basic rules of matrix algebra is presented here. The material is intended to provide introductory background of the subject. The focus of the material is on applications that may be of interest primarily to structural engineers.

C.2 Matrix Definitions and Properties

A *matrix* is defined as an ordered arrangement of numbers in rows and columns, as

$$[A] = \begin{bmatrix} a_{1,1} & a_{1,2} & \cdots & a_{1,n} \\ a_{2,1} & a_{2,2} & \cdots & a_{2,n} \\ \vdots & \vdots & \ddots & \vdots \\ a_{m,1} & a_{m,2} & \cdots & a_{m,n} \end{bmatrix} \tag{C.1}$$

A representative matrix $[A]$, such as shown in Equation C.1, consists of m rows and n columns of numbers enclosed in brackets. Matrix $[A]$ is said to be of order $m \times n$ and may be represented in short form as $[A]_{m \times n}$. The elements $a_{i,j}$ in the array are identified by two subscripts. The first subscript designates the row in which the element is located, and the second subscript designates the column. Thus, $a_{3,5}$ is the element located at the intersection of the third row and the fifth column of the matrix. Elements with repeated subscripts (for example, $a_{i,i}$) are located on the main diagonal of the matrix.

Although the number of rows and columns of a matrix may vary from problem to problem, two special cases deserve mention. When m is equal to 1, the matrix consists of only one row of elements and is called a row matrix or row vector. It is written as

$$\lfloor A \rfloor = \lfloor\, a_1 \quad a_2 \quad \cdots \quad a_n \,\rfloor \tag{C.2}$$

When $n = 1$, the matrix consists of only one *column* and is called a column matrix or column vector. It is written as

$$\{A\} = \begin{Bmatrix} a_1 \\ a_2 \\ \vdots \\ a_n \end{Bmatrix} \tag{C.3}$$

C.3 Special Matrix Types

Square Matrix

When the number of rows m and the number of columns n are equal, the matrix is said to be *square*. As an example, if $m = n = 3$, the square matrix $[A]$ may appear as

$$[A] = \begin{bmatrix} a_{11} & a_{12} & a_{13} \\ a_{21} & a_{22} & a_{23} \\ a_{31} & a_{32} & a_{33} \end{bmatrix} \tag{C.4}$$

Symmetric Matrix

A symmetric matrix is one in which the off-diagonal terms are reflected about the main diagonal. As such,

$$a_{ij} = a_{ji} \quad \text{for } i \neq j \tag{C.5}$$

When the elements of a square matrix obey this rule, the matrix is said to be *symmetric*, and the elements are arranged symmetrically about the main diagonal. An example of a symmetric 3×3 matrix is

$$[A] = \begin{bmatrix} 3 & -2 & 5 \\ -2 & 4 & 7 \\ 5 & 7 & 6 \end{bmatrix} \tag{C.6}$$

Symmetric matrices occur frequently in structural theory and play an important role in the matrix manipulations used to develop the theory.

Identity Matrix

When all of the elements on the main diagonal of a square matrix are equal to unity and all of the other elements are equal to zero, the matrix is called an *identity* matrix. Sometimes it is also called a *unit* matrix. Matrices of this type are identified with the symbol $[I]$. An example of a 3×3 identity matrix is

$$[I] = \begin{bmatrix} 1 & 0 & 0 \\ 0 & 1 & 0 \\ 0 & 0 & 1 \end{bmatrix} \tag{C.7}$$

Transposed Matrix

When the elements of a given matrix are reordered so that the columns of the original matrix become the corresponding rows of the new matrix, the new matrix is said to be the *transpose* of the original matrix. In this text, the transpose of matrix $[A]$ is given the symbol $[A]^T$. An example of a specific matrix and its transpose is

$$[A] = \begin{bmatrix} 1 & 3 & -2 \\ 5 & 4 & 7 \\ 8 & 2 & 6 \end{bmatrix} \qquad [A]^T = \begin{bmatrix} 1 & 5 & 8 \\ 3 & 4 & 2 \\ -2 & 7 & 6 \end{bmatrix} \tag{C.8}$$

The reader should note that if the matrix $[A]$ is a symmetric matrix, then

$$[A]^T = [A] \tag{C.9}$$

This property is used frequently in the development of structural theory using matrices. As special cases, the transpose of a row matrix becomes a column matrix and vice versa. Thus,

$$\lfloor A \rfloor^T = \{A\} \qquad \{A\}^T = \lfloor A \rfloor \tag{C.10}$$

C.4 Determinant of a Square Matrix

A determinant of a square matrix $[A]$ is given by the symbol $|A|$ and, in its expanded form, is written as

$$|A| = \begin{vmatrix} a_{11} & a_{12} & \cdots & a_{1n} \\ a_{21} & a_{22} & \cdots & a_{2n} \\ \vdots & \vdots & \ddots & \vdots \\ a_{m1} & a_{m2} & \cdots & a_{mn} \end{vmatrix} \tag{C.11}$$

This determinant is said to be of order m. Unlike matrix $[A]$, which has no single value, the determinant $|A|$ does have a single numerical value. The value of $|A|$ is easily found for a 2×2 array of numbers, for example,

$$\begin{vmatrix} 2 & 1 \\ 4 & 5 \end{vmatrix} = 2 \cdot 5 - 1 \cdot 4 = 6 \tag{C.12}$$

The value of $|A|$ in this example was determined by multiplying the numbers on the main diagonal and subtracting from this product the product of the numbers on the other diagonal. Unfortunately, this simple procedure does not work for determinants of an order greater than two.

A general procedure for finding the value of a determinant sometimes is called "expansion by minors." The first minor of the matrix $[A]$, corresponding to the element a_{ij}, is defined as the determinant of a reduced matrix obtained by eliminating the ith row and the jth column from matrix $[A]$. The minor is a specific number—like any other determinant.

As an illustration, several first minors are shown for a matrix $[A]$:

$$[A] = \begin{bmatrix} 1 & 2 & 3 \\ -2 & 3 & 4 \\ 1 & 5 & 2 \end{bmatrix}$$

$$\begin{aligned} \text{minor of } a_{11} &= \begin{vmatrix} 3 & 4 \\ 5 & 2 \end{vmatrix} = 3 \cdot 2 - 4 \cdot 5 = -14 \\ \text{minor of } a_{12} &= \begin{vmatrix} -2 & 4 \\ 1 & 2 \end{vmatrix} = -2 \cdot 2 - 4 \cdot 1 = -8 \\ \text{minor of } a_{23} &= \begin{vmatrix} 1 & 2 \\ 1 & 5 \end{vmatrix} = 1 \cdot 5 - 2 \cdot 1 = 3 \end{aligned} \tag{C.13}$$

When the proper sign is attached to a minor, the result is called a *cofactor* and is given the symbol A_{ij}. The sign of a minor is determined by multiplying the minor by $(-1)^{i+j}$. Several cofactors for the example matrix $[A]$ are

$$\begin{aligned} A_{11} &= (-1)^{(1+1)} \times \text{minor of } a_{11} = (1)(-14) = -14 \\ A_{12} &= (-1)^{(1+2)} \times \text{minor of } a_{12} = (-1)(-8) = 8 \\ A_{23} &= (-1)^{(2+3)} \times \text{minor of } a_{23} = (-1)(3) = -3 \end{aligned} \tag{C.14}$$

Now, to obtain the value of a general determinant, we can choose any arbitrary row i of matrix $[A]$ and expand according to the relation:

$$|A| = \sum_{j=1}^{m} a_{ij}A_{ij} \tag{C.15}$$

The value of a determinant can also be found by choosing an arbitrary column j and expanding according to the relation:

$$|A| = \sum_{i=1}^{m} a_{ij}A_{ij} \tag{C.16}$$

If the order of the original determinant is large, the procedure described does not appear to produce a simple solution. For example, a 15th-order determinant will still have first minors that are of the 14th order. However, the 14th-order minors can be reduced to 13th-order minors by the *expansion* process. This process can be repeated until the resulting minors are of the second order. These minors then can be evaluated readily using the procedure described in the initial illustration of this section.

Although the procedure for evaluating determinants may appear long and tedious, computer algorithms can be written that will perform the necessary algebraic operations. Other simplifying procedures are available that make use of special characteristics of determinants—discussion of which is beyond the scope of this text.

C.5 Adjoint Matrix

A special matrix exists called an *adjoint matrix* and is given the symbol adj$[A]$. To find the adjoint matrix corresponding to an original matrix $[A]$, first replace each element of $[A]$ with its cofactor; the adjoint matrix is then the transpose of this resultant matrix.

Symbolically, the adjoint matrix adj$[A]$ is written as

$$\text{adj}\,[A] = \begin{bmatrix} A_{11} & A_{12} & \cdots & A_{1m} \\ A_{21} & A_{22} & \cdots & A_{2m} \\ \vdots & \vdots & \ddots & \vdots \\ A_{m1} & A_{m2} & \cdots & A_{mm} \end{bmatrix} \tag{C.17}$$

A specific numerical example of a matrix [A] and its adjoint matrix is

$$[A] = \begin{bmatrix} 1 & 2 & 3 \\ 2 & 3 & 4 \\ 1 & 5 & 5 \end{bmatrix} \tag{C.18}$$

$$\begin{aligned} A_{11} &= -11 \quad & A_{12} &= -2 \quad & A_{13} &= 7 \\ A_{21} &= 9 & A_{22} &= 0 & A_{23} &= -3 \\ A_{31} &= -1 & A_{32} &= 2 & A_{33} &= -1 \end{aligned}$$

$$\text{adj}\,[A] = \begin{bmatrix} -11 & -2 & 7 \\ 9 & 0 & -3 \\ -1 & 2 & -1 \end{bmatrix}^T = \begin{bmatrix} -11 & 9 & -1 \\ -2 & 0 & 2 \\ 7 & -3 & -1 \end{bmatrix}$$

The adjoint matrix and the determinant both are used in the computation of the inverse of a matrix—a topic that is treated in a later section of this appendix.

C.6 Matrix Arithmetic

Equality of Matrices

Two matrices are equal only if the corresponding elements of the two matrices are equal. Thus, equality of matrices can exist only between matrices of equal dimensions.

Addition and Subtraction of Matrices

Two matrices may be added or subtracted only if they have the same dimensions. The addition of two matrices, $[A]$ and $[B]$, is performed as

$$[A] + [B] = [C] \tag{C.19}$$

where the elements of matrix $[C]$ are the sum of corresponding elements of $[A]$ and $[B]$, that is

$$c_{ij} = a_{ij} + b_{ij} \tag{C.20}$$

An example of the addition of two matrices is

$$\begin{bmatrix} 1 & 2 \\ 3 & 4 \end{bmatrix} + \begin{bmatrix} 5 & 7 \\ 6 & 8 \end{bmatrix} = \begin{bmatrix} 6 & 9 \\ 9 & 12 \end{bmatrix} \tag{C.21}$$

Subtraction of two matrices is performed similarly by subtracting corresponding elements.

Both the commutative and the associative laws hold for the addition and subtraction of matrices. Thus,

$$[A] + [B] = [B] + [A] \tag{C.22}$$

and

$$[A] + ([B] + [C]) = ([A] + [B]) + [C] \tag{C.23}$$

Scalar Multiplication of Matrices

To multiply a matrix by a scalar, α, each element of the matrix is multiplied by the scalar individually. Thus,

$$(\alpha)\,[A] = \begin{bmatrix} (\alpha)a_{11} & (\alpha)a_{12} & \cdots & (\alpha)a_{1n} \\ (\alpha)a_{21} & (\alpha)a_{22} & \cdots & (\alpha)a_{2n} \\ \vdots & \vdots & \ddots & \vdots \\ (\alpha)a_{m1} & (\alpha)a_{m2} & \cdots & (\alpha)a_{mn} \end{bmatrix} \tag{C.24}$$

Multiplication of Matrices

The product of two matrices exists only if the matrices are *conformable*. For the matrix product $[A][B]$, conformability means that the number of columns of $[A]$ equals the number of rows of $[B]$. The two matrices shown are conformable (in the order shown) and may be multiplied.

$$[A] = \begin{bmatrix} 1 & 3 \\ 2 & 5 \end{bmatrix} \quad [B] = \begin{bmatrix} 2 & 6 & 5 \\ 1 & 3 & -4 \end{bmatrix} \tag{C.25}$$

The number of columns in $[A]$ is equal to the number of rows in $[B]$. However, if the order of the matrix multiplication is reversed as $[B][A]$, the matrices are not conformable and the matrix product does not exist. Furthermore, even if the matrices $[A]$ and $[B]$ are square and thus conformable in the order $[A][B]$ and $[B][A]$, the two matrix products are normally not the same. In general,

$$[A][B] \neq [B][A] \tag{C.26}$$

The formal definition of a matrix product between matrices that are conformable is given as

$$[A]_{m\times k}[B]_{k\times n} = [C]_{m\times n} \tag{C.27}$$

where

$$c_{ij} = \sum_{q=1}^{k} a_{iq}b_{qj} \tag{C.28}$$

This is most easily remembered by noting that the inner dimensions, q, must be equal for both matrices, and the final dimensions of $[C]$ are the outer dimensions of the two matrices m and n.

Note that matrix $[A]$ is not of the same order as $[B]$, but the two matrices are conformable, since the number of columns of $[A]$ equals the number of rows of $[B]$. The order of the product matrix $[C]$ is $m \times n$.

A simple illustration of a matrix product is

$$\begin{bmatrix} 1 & 2 \\ -3 & 2 \end{bmatrix} \cdot \begin{bmatrix} 1 & 3 & 2 \\ 4 & 5 & 3 \end{bmatrix} = \begin{bmatrix} 9 & 13 & 8 \\ 5 & 1 & 0 \end{bmatrix} \tag{C.29}$$

In this matrix multiplication, the terms in the matrix $[C]$ are computed as

$$\begin{aligned} c_{11} &= (1)(1) + (2)(4) = 9 \\ c_{12} &= (1)(3) + (2)(5) = 13 \\ c_{13} &= (1)(2) + (2)(3) = 8 \\ c_{21} &= (-3)(1) + (2)(4) = 5 \\ c_{22} &= (-3)(3) + (2)(5) = 1 \\ c_{23} &= (-3)(2) + (2)(3) = 0 \end{aligned} \tag{C.30}$$

Although the order in which two matrices are multiplied may not be reversed, in general, without obtaining different results, both the associative and the distributive laws are valid for matrix products. Thus,

$$[A][B][C] = ([A][B])\,[C] = [A]\,([B][C]) \tag{C.31}$$

and

$$[A]\,([B] + [C]) = [A][B] + [A][C] \tag{C.32}$$

If a matrix $[A]$ is multiplied by the identity matrix $[I]$ (assuming that the matrices are conformable), the matrix $[A]$ remains unchanged. Thus,

$$\begin{bmatrix} 1 & 2 & 3 \\ -2 & 4 & 6 \\ 3 & 5 & 2 \end{bmatrix} \cdot \begin{bmatrix} 1 & 0 & 0 \\ 0 & 1 & 0 \\ 0 & 0 & 1 \end{bmatrix} = \begin{bmatrix} 1 & 2 & 3 \\ -2 & 4 & 6 \\ 3 & 5 & 2 \end{bmatrix} \tag{C.33}$$

In general,

$$\begin{aligned} [A]\,[I] &= [A] \\ [I]\,[A] &= [A] \end{aligned} \tag{C.34}$$

Transpose of a Product

If a matrix product is transposed, the result is the reverse product of transpose of the individual matrices. The result is shown symbolically as

$$([A]\,[B])^T = [B]^T\,[A]^T \tag{C.35}$$

The transpose of a triple matrix product may be found by using Equation C.35 and the associative law in several stages, as

$$\begin{aligned}([A]\,[B]\,[C])^T &= \{[A]\,([B]\,[C])\}^T \\ &= ([B]\,[C])^T\,[A]^T \\ &= [C]^T\,[B]^T\,[A]^T\end{aligned} \tag{C.36}$$

Note that in finding the transpose of a matrix product, the order of multiplication changes.

Matrix Inverse

Although matrix addition, subtraction, and multiplication have been defined in the preceding sections, no mention has been made of matrix division. In fact, division of matrices in the form $[A]/[B]$ does not exist. However, a matrix operation does exist that closely parallels algebraic division. This operation makes use of a *matrix inverse.*

The inverse of a square matrix $[A]$ is given the symbol $[A]^{-1}$. It is defined such that

$$[A]\cdot[A]^{-1} = [I] \tag{C.37}$$

Many techniques exist by which a matrix inverse may be determined. One formal technique is described by the relationship:

$$[A]^{-1} = \frac{\text{adj}[A]}{|A|} \tag{C.38}$$

As an example, consider the matrix $[A]$ given by Equation C.18. The adjoint matrix adj$[A]$ is also shown in Equation C.18. The determinant $|A|$ may be found by using the elements and first minors of the first column of $[A]$:

$$\begin{aligned}|A| &= a_{11}A11 + a_{21}A_{21} + a_{31}A_{31} \\ &= 1(-11) + 2(9) + 1(-1) \\ &= 6\end{aligned} \tag{C.39}$$

Thus,

$$[A]^{-1} = \frac{1}{6}\begin{bmatrix} -11 & 9 & -1 \\ -2 & 0 & 2 \\ 7 & -3 & -1 \end{bmatrix} \tag{C.40}$$

The correctness of the values given for the coefficients of $[A]^{-1}$ may be verified by forming the matrix product $[A][A]^{-1}$ and checking to see if the result is the identity matrix. For example,

$$[A]\,[A]^{-1} = \begin{bmatrix} 1 & 2 & 3 \\ 2 & 3 & 4 \\ 1 & 5 & 3 \end{bmatrix}\cdot\frac{1}{6}\cdot\begin{bmatrix} -11 & 9 & -1 \\ -2 & 0 & 2 \\ 7 & -3 & -1 \end{bmatrix} = \frac{1}{6}\begin{Bmatrix} 6 & 0 & 0 \\ 0 & 6 & 0 \\ 0 & 0 & 6 \end{Bmatrix} = [I] \tag{C.41}$$

A special situation involving the inverse of a matrix deserves attention. If the determinant of a matrix $[A]$ is equal to zero ($|A| = 0$), the division operation indicated by Equation C.38 cannot be performed. Under these circumstances, the inverse of matrix $[A]$ does not exist, and matrix $[A]$ is said to be *singular*. Singular matrices occur frequently in structural theory, and a reader should be aware of the meaning of this term. An example of a singular matrix is

$$[A] = \begin{bmatrix} 1 & 2 & 4 \\ -2 & 3 & 2 \\ 3 & 6 & 12 \end{bmatrix} \tag{C.42}$$

and

$$|A| = a_{11}A_{11} + a_{21}A_{21} + a_{31}A_{31} = 24 + 0 - 24 = 0 \tag{C.43}$$

The inverse of a matrix product may be found using rules that are very similar to those used in finding the transpose of a matrix product, as shown in Equation C.35. Specifically,

$$([A][B])^{-1} = [B]^{-1}[A]^{-1} \tag{C.44}$$

and

$$\begin{aligned} ([A][B][C])^{-1} &= ([B][C])^{-1}\,[A]^{-1} \\ &= [C]^{-1}[B]^{-1}[A]^{-1} \end{aligned} \tag{C.45}$$

Application of the Matrix Inverse

Consider a set of algebraic equations, each of which contains a number of unknown quantities, x_i. Namely,

$$\begin{aligned} a_{11}x_1 + a_{12}x_2 + a_{13}x_3 &= b_1 \\ a_{21}x_1 + a_{22}x_2 + a_{23}x_3 &= b_2 \\ a_{31}x_1 + a_{32}x_2 + a_{33}x_3 &= b_3 \end{aligned} \tag{C.46}$$

The set of algebraic equations may be cast in matrix form as

$$\begin{bmatrix} a_{11} & a_{12} & a_{13} \\ a_{21} & a_{22} & a_{23} \\ a_{31} & a_{32} & a_{33} \end{bmatrix} \begin{Bmatrix} x_1 \\ x_2 \\ x_3 \end{Bmatrix} = \begin{Bmatrix} b_1 \\ b_2 \\ b_3 \end{Bmatrix} \tag{C.47}$$

or symbolically as

$$[A]\{X\} = \{B\} \tag{C.48}$$

The solution for the unknown quantities x_1, x_2, and x_3 may be found by pre-multiplying both sides of Equation C.48 by $[A]^{-1}$. Namely,

$$\begin{aligned} [A]^{-1}\,[A]\,\{X\} &= [A]^{-1}\,\{B\} \\ [I]\,\{X\} &= [A]^{-1}\,\{B\} \\ \{X\} &= [A]^{-1}\,\{B\} \end{aligned} \tag{C.49}$$

Therefore, if $[A]^{-1}$ is known or can be computed, the values of x can be determined from a simple matrix product.

As a numerical example, consider the following algebraic equations. The coefficients in these equations are the same as those shown in matrix $[A]$ in Equation C.18. So

$$\begin{aligned} x_1 + 2x_2 + 3x_3 &= 100 \\ 2x_1 + 3x_2 + 4x_3 &= 140 \\ 1x_1 + 5x_2 + 3x_3 &= 110 \end{aligned} \tag{C.50}$$

Thus, the set of simultaneous equations written in matrix form is

$$\begin{bmatrix} 1 & 2 & 3 \\ 2 & 3 & 4 \\ 1 & 5 & 3 \end{bmatrix} \begin{Bmatrix} x_1 \\ x_2 \\ x_3 \end{Bmatrix} = \begin{Bmatrix} 100 \\ 140 \\ 110 \end{Bmatrix} \tag{C.51}$$

To solve for $\{X\}$, both sides are pre-multiplied by $[A]^{-1}$.

$$\begin{Bmatrix} x_1 \\ x_2 \\ x_3 \end{Bmatrix} = \begin{bmatrix} 1 & 2 & 3 \\ 2 & 3 & 4 \\ 1 & 5 & 3 \end{bmatrix}^{-1} \begin{Bmatrix} 100 \\ 140 \\ 110 \end{Bmatrix} \tag{C.52}$$

The inverse of $[A]$ has previously been given in Equation C.40. Therefore, Equation C.52 may be rewritten as:

$$\begin{Bmatrix} x_1 \\ x_2 \\ x_3 \end{Bmatrix} = \frac{1}{6} \begin{bmatrix} -11 & 9 & -1 \\ -2 & 0 & 2 \\ 7 & -3 & -1 \end{bmatrix} \begin{Bmatrix} 100 \\ 140 \\ 110 \end{Bmatrix} \tag{C.53}$$

The values of the unknowns are calculated as

$$\begin{aligned} x_1 &= \tfrac{1}{6}\,[(-11)(100) + (9)(140) + (-1)(110)] = 8.333 \\ x_2 &= \tfrac{1}{6}\,[(-2)(100) + (0)(140) + (2)(110)] = 3.333 \\ x_3 &= \tfrac{1}{6}\,[(7)(100) + (-3)(140) + (-1)(110)] = 28.333 \end{aligned} \tag{C.54}$$

Many other techniques exist for solving algebraic equations simultaneously. The use of the matrix inverse is a special technique that may be used at the option of the analyst.

Many computer programs and calculators today easily handle the requisite matrix math. Chapter 19 demonstrates how to use a spreadsheet to carry out these operations swiftly.

C.7 Matrix Partitioning

The manipulation of matrix equations is frequently made simpler by dividing the matrices into smaller matrices called *partitions*. Partitioning is indicated in this book by horizontal and vertical lines drawn between both the rows and the columns of the matrices. Illustrations of matrices that have been partitioned are given here.

$$[A] = \left[\begin{array}{cc|c} a_{11} & a_{12} & a_{13} \\ a_{21} & a_{22} & a_{23} \\ \hline a_{31} & a_{32} & a_{33} \end{array}\right] \tag{C.55}$$

$$\{A\} = \left\{\begin{array}{c} a_1 \\ a_2 \\ \hline a_3 \end{array}\right\} \tag{C.56}$$

Partitioning of a matrix equation is like this.

$$[A]\,\{X\} = \{B\}$$

$$\left[\begin{array}{cc|c} a_{11} & a_{12} & a_{13} \\ a_{21} & a_{22} & a_{23} \\ \hline a_{31} & a_{32} & a_{33} \end{array}\right] \left\{\begin{array}{c} x_1 \\ x_2 \\ \hline x_3 \end{array}\right\} = \left\{\begin{array}{c} b_1 \\ b_2 \\ \hline b_3 \end{array}\right\} \tag{C.57}$$

Note that the horizontal partition lines in Equation C.57 extend between the same two rows for each matrix in the equation. Furthermore, the column numbers that define the vertical partition line for the square matrix are the same as the row numbers that define the horizontal partition lines for the complete equation. Thus, if the horizontal partition lines run between the second and third rows, the vertical partition line runs between the second and third column of the square matrix. Be aware that the actual rows/columns where the partitioning needs to occur is dependent on the problem at hand. This illustration placed the lines between the second and third, but this is purely a demonstration and is not indicative of every problem. The use of partitioning in structural analysis is explained and implemented in Chapter 19.

As an illustration of the use of partitioning, consider the same matrix equation described in Equation C.51:

$$\left[\begin{array}{cc|c} 1 & 2 & 3 \\ 2 & 3 & 4 \\ \hline 1 & 5 & 3 \end{array}\right]\left\{\begin{array}{c} x_1 \\ x_2 \\ \hline x_3 \end{array}\right\} = \left\{\begin{array}{c} 100 \\ 140 \\ \hline 110 \end{array}\right\} \tag{C.58}$$

In Equation C.58, partition lines have been drawn between the second and third rows of each matrix and between the second and third columns of matrix $[A]$. When the matrix is partitioned, the partitions still are matrices and are manipulated as matrices. In this example, though, one of the partitions is a 1×1 matrix, which can be treated as a scalar. The original matrix equation may now be written as two matrix equations:

$$\begin{bmatrix} 1 & 2 \\ 2 & 3 \end{bmatrix}\begin{Bmatrix} x_1 \\ x_2 \end{Bmatrix} + \begin{Bmatrix} 3 \\ 4 \end{Bmatrix} x_3 = \begin{Bmatrix} 100 \\ 140 \end{Bmatrix} \tag{C.59}$$

and

$$\begin{bmatrix} 1 & 5 \end{bmatrix}\begin{Bmatrix} x_1 \\ x_2 \end{Bmatrix} + 3x_3 = 110 \tag{C.60}$$

Equation C.60 may be solved for x_3 in terms of x_1 and x_2 as

$$\begin{aligned} 3x_3 &= 110 - \begin{bmatrix} 1 & 5 \end{bmatrix}\begin{Bmatrix} x_1 \\ x_2 \end{Bmatrix} \\ x_3 &= \frac{110}{3} - \frac{1}{3}\begin{bmatrix} 1 & 5 \end{bmatrix}\begin{Bmatrix} x_1 \\ x_2 \end{Bmatrix} \end{aligned} \tag{C.61}$$

The expression of x_3 from Equation C.61 is substituted into Equation C.59, and the resultant equation is solved for x_1 and x_2:

$$\begin{bmatrix} 1 & 2 \\ 2 & 3 \end{bmatrix}\begin{Bmatrix} x_1 \\ x_2 \end{Bmatrix} + \begin{Bmatrix} 3 \\ 4 \end{Bmatrix}\left(\frac{110}{3} - \frac{1}{3}\begin{bmatrix} 1 & 5 \end{bmatrix}\begin{Bmatrix} x_1 \\ x_2 \end{Bmatrix}\right) = \begin{Bmatrix} 100 \\ 140 \end{Bmatrix} \tag{C.62}$$

Following some basic algebra, the following equation is formed. So

$$\begin{Bmatrix} x_1 \\ x_2 \end{Bmatrix} = \begin{bmatrix} 0 & -9 \\ 2 & -11 \end{bmatrix}^{-1}\begin{Bmatrix} -30 \\ -20 \end{Bmatrix} = \begin{Bmatrix} 8.333 \\ 3.333 \end{Bmatrix} \tag{C.63}$$

Note that the values of x_1 and x_2 are the same as those found in Equation C.54, as should be expected. These values are now substituted in Equation C.61 to obtain the value of x_3.

$$x_3 = \frac{110}{3} - \frac{1}{3}\begin{bmatrix} 1 & 5 \end{bmatrix}\begin{Bmatrix} 8.333 \\ 3.333 \end{Bmatrix} = 28.333 \tag{C.64}$$

One obvious advantage of partitioning is that the order of the matrices for which inverses must be found is reduced. However, this advantage is offset somewhat by the fact that additional algebraic manipulations are required when using partitioning schemes. Nonetheless, partitioning of matrix equations is used extensively in developing computer solutions for structural problems (see Chapter 19).

Reference Charts

D

D.1 Introduction

In the process of performing structural analysis, the work with both determinate and indeterminate structures can often be expedited and facilitated by taking advantage of information that has already been developed for basic structural components. Specifically, beam deflection charts and also beam fixed-end force charts can readily be used to solve problems using the force-based approach (Chapters 11 and 12), the moment distribution method (Chapters 13 and 14), and also the matrix-based stiffness method (Chapters 20 and 21). Figures D.1 and D.2 give beam deflections for typical cantilever beams and simply supported beams, respectively. Figure D.3 presents the fixed-end moments and fixed-end shears for beams which are either fixed-fixed or fixed-pinned. Finally, Figure D.4 presents useful geometric properties for some basic shapes. These properties can be useful when generating shear and moment diagrams in Chapter 6 and in calculating deflections in Chapter 9.

	Beam	Slope	Deflection	Elastic Curve
1	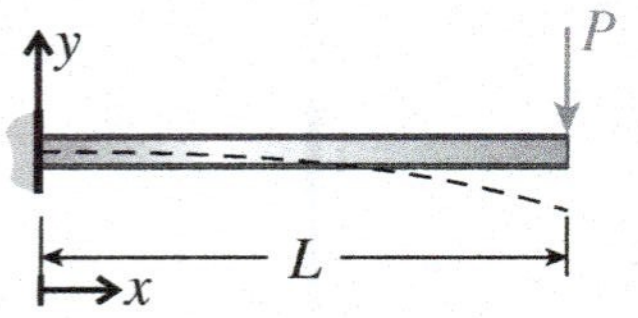	$\theta_{max} = -\dfrac{PL^2}{2EI}$ at $x = L$	$y_{max} = -\dfrac{PL^3}{3EI}$ at $x = L$	$y = -\dfrac{Px^2}{6EI}(3L - x)$ for $0 \le x \le L$
2	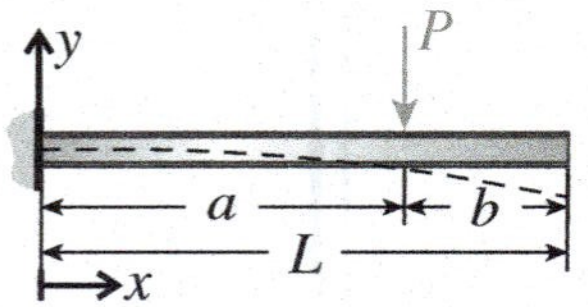	$\theta_{max} = -\dfrac{Pa^2}{2EI}$ at $a \le x \le L$	$y_{max} = -\dfrac{Pa^2}{6EI}(3L - a)$ at $x = L$	$y = \dfrac{P}{2EI}\left(\dfrac{x^3}{3} - ax^2\right)$ for $0 \le x \le a$ $y = \dfrac{Pa^2}{6EI}(-3x + a)$ for $a \le x \le L$
3	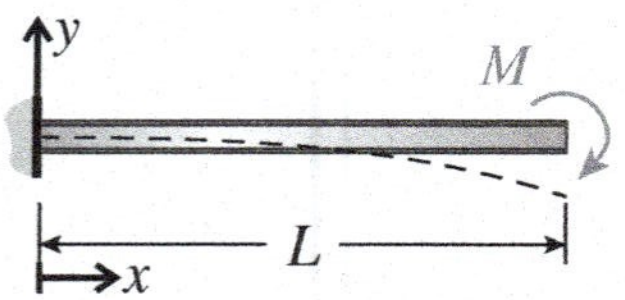	$\theta_{max} = -\dfrac{ML}{EI}$ at $x = L$	$y_{max} = -\dfrac{ML^2}{2EI}$ at $x = L$	$y = -\dfrac{Mx^2}{2EI}$ for $0 \le x \le L$
4	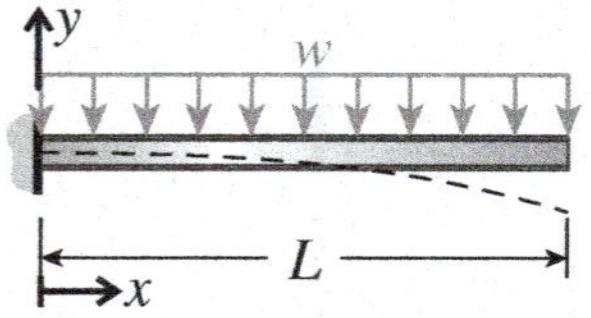	$\theta_{max} = -\dfrac{wL^3}{6EI}$ at $x = L$	$y_{max} = -\dfrac{wL^4}{8EI}$ at $x = L$	$y = -\dfrac{wx^2}{24EI}(6L^2 - 4Lx + x^2)$ for $0 \le x \le L$
5	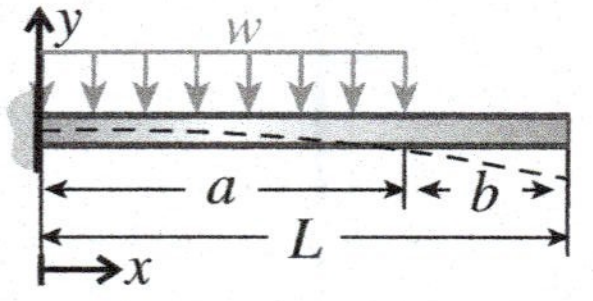	$\theta_{max} = -\dfrac{wa^3}{6EI}$ at $a \le x \le L$	$y_{max} = -\dfrac{wa^3}{24EI}(4L - a)$ at $x = L$	$y = -\dfrac{wx^2}{24EI}(6a^2 - 4ax + x^2)$ for $0 \le x \le a$ $y = -\dfrac{wa^3}{24EI}(4x - a)$ for $a \le x \le L$

FIGURE D.1 Deflections for cantilever beams.

	Beam	Slope	Deflection	Elastic Curve
6	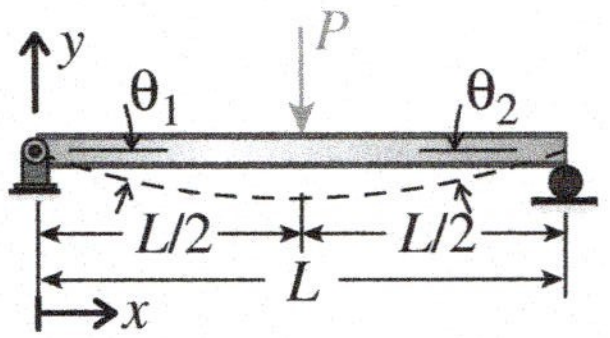	$\theta_1 = -q_2 = -\frac{PL^2}{16EI}$	$y_{max} = -\frac{PL^3}{48EI}$	$y = -\frac{Px}{48EI}(3L^2 - 4x^2)$ for $0 \le x \le \frac{L}{2}$ use symmetry for the region $\frac{L}{2} \le x \le L$
7	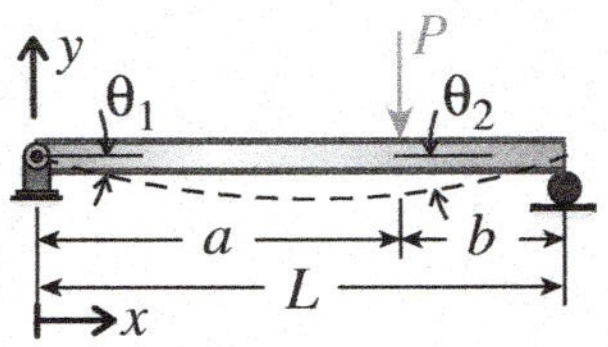	$\theta_1 = -\frac{Pb(L^2 - b^2)}{6LEI}$ $\theta_2 = \frac{Pa(L^2 - a^2)}{6LEI}$	point of max deflection changes location depending on the value of *a*	$y = \frac{-Pbx}{6LEI}(L^2 - b^2 - x^2)$ for $0 \le x \le a$ To find deflection in the region to the right of the load *P*, reverse the origin of *x* and the *a* and *b* values. Then use the equation given.
8	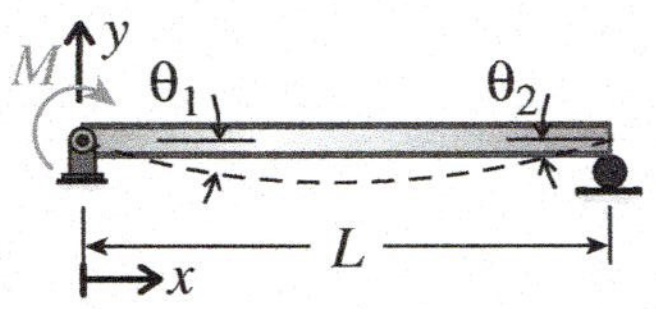	$\theta_1 = -\frac{ML}{3EI}$ $\theta_2 = \frac{ML}{6EI}$	$y_{max} = -\frac{ML^2}{9\sqrt{3}EI}$	$y = -\frac{Mx}{6LEI}(2L^2 - 3Lx + x^2)$ for $0 \le x \le L$
9	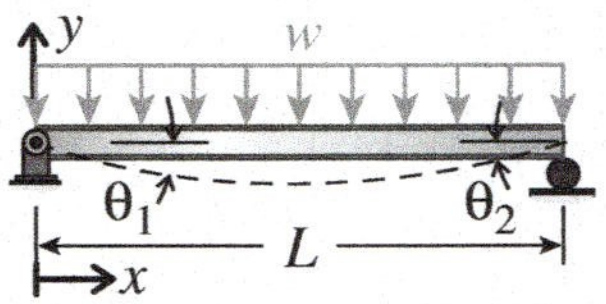	$\theta_1 = -\theta_2 = -\frac{wL^3}{24EI}$ at $x = L\left(1 - \frac{\sqrt{3}}{3}\right)$	$y_{max} = -\frac{5wL^4}{384EI}$ at $x = \frac{L}{2}$	$y = -\frac{wx}{24EI}(L^3 - 2Lx^2 + x^3)$ for $0 \le x \le L$
10	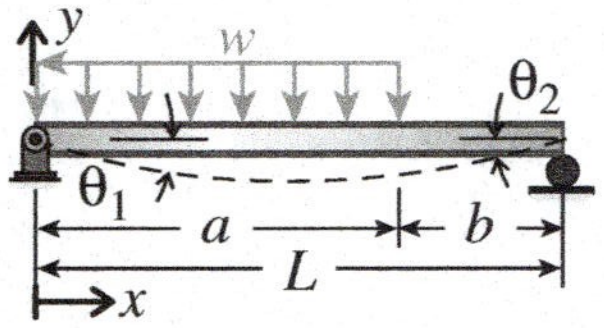	$\theta_1 = -\frac{wa^2}{24LEI}(2L - a)^2$ $\theta_2 = +\frac{wa^2}{24LEI}(2L^2 - a^2)$	point of max deflection changes location depending on the value of *a*	$y = -\frac{wx}{24EIL}\begin{pmatrix} Lx^3 - 4aLx^2 + 2a^2x^2 \\ +4a^2L^2 - 4a^3L + a^4 \end{pmatrix}$ for $0 \le x \le a$ $y = -\frac{wa^2}{24EIL}\begin{pmatrix} 2x^3 - 6Lx^2 + a^2x \\ +4L^2x - a^2L \end{pmatrix}$ for $a \le x \le L$

FIGURE D.2 Deflections for simply supported beams.

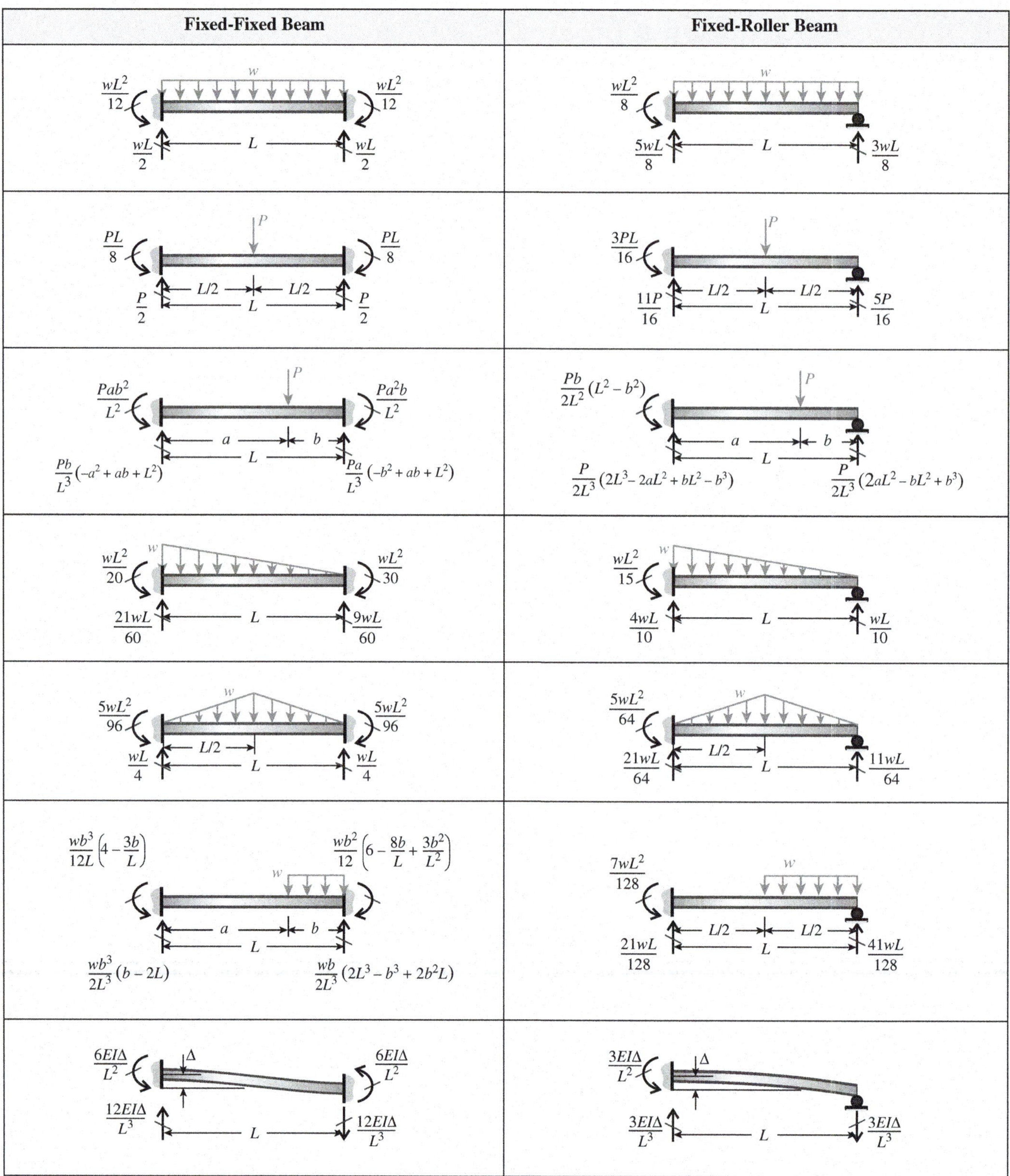

FIGURE D.3 End forces for fixed-ended beams.

<table>
<tr><th>Geometric Area</th><th>Area Properties</th></tr>
<tr><td>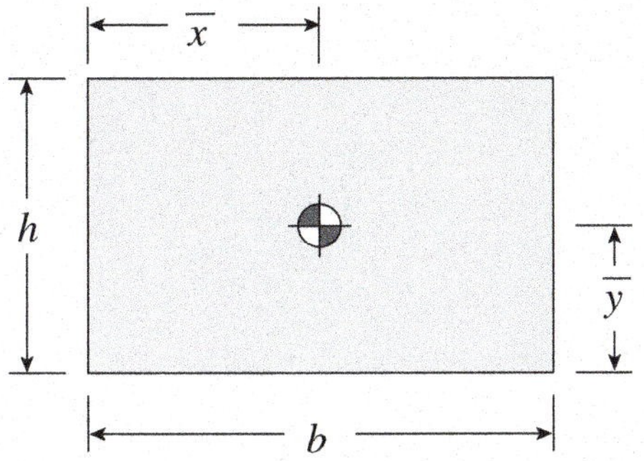
</td><td>$A = bh$
$\bar{x} = \frac{b}{2}$
$\bar{y} = \frac{h}{2}$</td></tr>
<tr><td>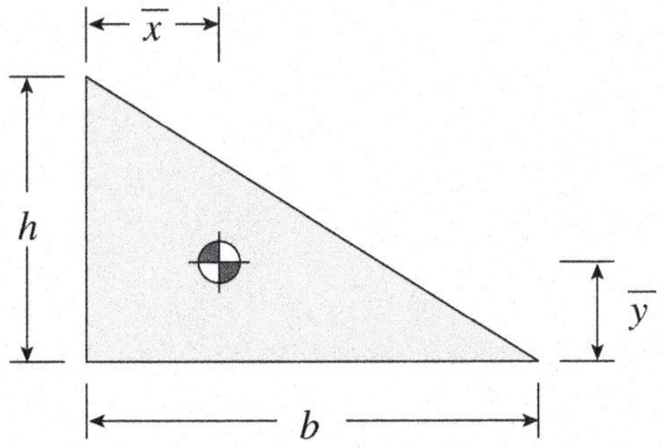
</td><td>$A = \frac{bh}{2}$
$\bar{x} = \frac{b}{3}$
$\bar{y} = \frac{h}{3}$</td></tr>
<tr><td>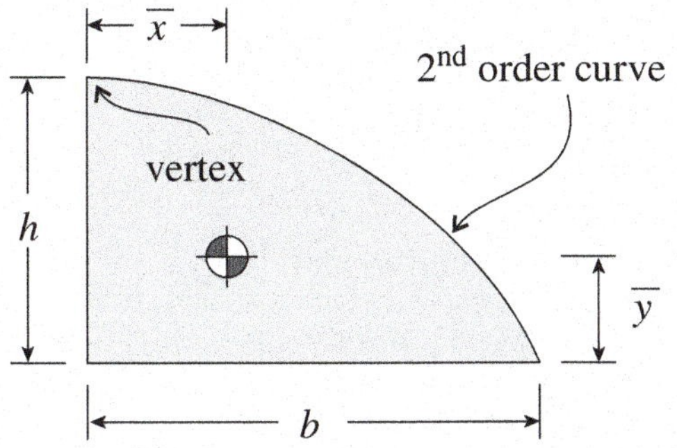
</td><td>$A = \frac{2bh}{3}$
$\bar{x} = \frac{3b}{8}$
$\bar{y} = \frac{2h}{5}$</td></tr>
<tr><td>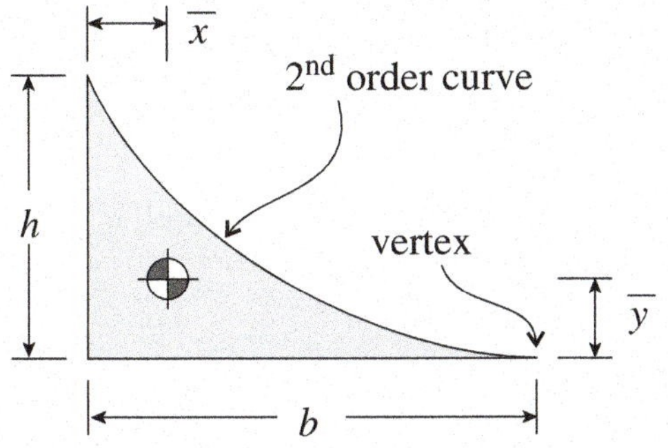
</td><td>$A = \frac{bh}{3}$
$\bar{x} = \frac{b}{4}$
$\bar{y} = \frac{3h}{10}$</td></tr>
</table>

FIGURE D.4 Useful properties for basic geometric shapes.

Index

F

N

O

P

S

CPSIA information can be obtained
at www.ICGtesting.com
Printed in the USA
BVOW04s0312151217
502350BV00006B/3/P